17th Conference on Applied Mathematics (APLIMAT 2018)

Bratislava, Slovak Republic
6-8 February 2018

Volume 1 of 2

Editors:

Dagmar Szarkova
Peter Letavaj
Daniela Richtarikova
Jana Gabkova

ISBN: 978-1-5108-6248-7

Printed from e-media with permission by:

Curran Associates, Inc.
57 Morehouse Lane
Red Hook, NY 12571

CURRAN ASSOCIATES INC. proceedings.com

Some format issues inherent in the e-media version may also appear in this print version.

Printed by Curran Associates, Inc. (2018)

Institute of Mathematics and Physics, Faculty of Mechanical Engineering
Slovak University of Technology in Bratislava
Namestie Slobody 17
812 31 Bratislava 1
Slovak Republic

Phone: +421 2 572 96 111
Fax: +421 2 529 25 749

www.stuba.sk

Additional copies of this publication are available from:

Curran Associates, Inc.
57 Morehouse Lane
Red Hook, NY 12571 USA
Phone: 845-758-0400
Fax: 845-758-2633
Email: curran@proceedings.com
Web: www.proceedings.com

SLOVAK UNIVERSITY OF TECHNOLOGY IN BRATISLAVA
FACULTY OF MECHANICAL ENGINEERING

17th CONFERENCE ON APPLIED MATHEMATICS APLIMAT 2018 PROCEEDINGS

February 6 - 8, 2018 Bratislava
Slovak Republic

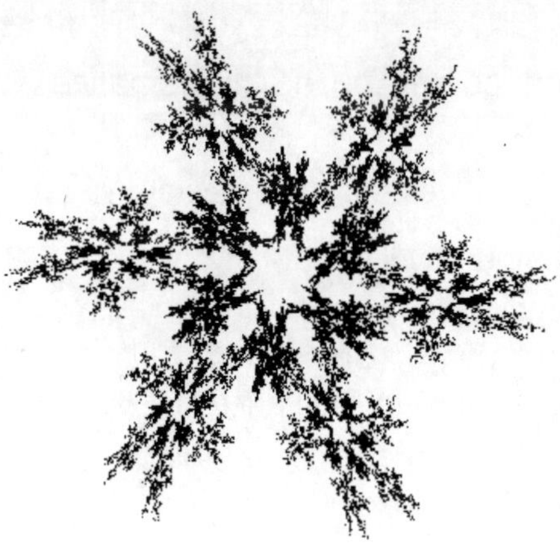

SPEKTRUM STU

STU
SLOVAK UNIVERSITY OF TECHNOLOGY IN BRATISLAVA

Slovenská spoločnosť SG pre Geometriu a Grafiku

Proceedings, 17th Conference on Applied Mathematics – APLIMAT 2018
First edition

Published by:
Slovak University of Technology in Bratislava in publishing house SPEKTRUM STU

Preface

This volume contains the papers presented at APLIMAT 2018: The 17th Conference on Applied Mathematics held on February 6 - 8, 2018 in Bratislava, Slovakia.

The purpose of the conference is to promote discussions and interactions between researchers and practitioners focused on cooperation in disciplinary, interdisciplinary and transdisciplinary issues, ideas, concepts, theories, methodologies and applications.

We are particularly interested in fostering the exchange of concepts, research ideas and other results which could contribute to the academic area and also benefit business and the industrial community.

Conference main topics

- Algebra and geometry and their applications
- Differential equations, dynamic systems and their applications
- Mathematics and art, section dedicated to the memory of Mauro Francaviglia
- Modeling and simulation in engineering and scientific computations
- New trends in university mathematics education
- Statistical methods in technical and economic sciences and in practise
- Network science, graphs and combinatorics
- Mathematics and physics in contemporary science and technique

Conference Chair

VELICHOVÁ Daniela	–	Slovak Republic

Conference Scientific Board

BAŠTINEC Jaromír	–	Czech Republic
CARKOVS Jevgeņijs	–	Latvia
FERREIRA Manuel A. M.	–	Portugal
JANIGA Ivan	–	Slovak Republic
LORENZI Marcella G.	–	Italy
NEHÉZ Martin	–	Slovak Republic
PĂUN Márius	–	Romania
TEDESCHINI LALLI Laura	–	Italy

Additional Reviewers

Baicoianu Alexandra
Bruni Fabio
De Martino Salvatore
Falcolini Corrado
Gabriele Stefano
Iannace Gino
Iltins Ilmars
Jančo Roland
Kozák Vladislav
Křivoňáková Naďa
Madaras Tomáš
Magrone Paola
Marin Marin
Matvejevs Andrejs
Mikeš Josef
Pavlenko Oksana
Rebelos Margareta
Salerno Ginevra
Sangeorzan Livia
Spadafora Giovanna
Vlase Sorin

CONTENTS

Slovak University of Technology in Bratislava
Faculty of Mechanical Engineering
17th Conference on Applied Mathematics **APLIMAT 2018**

Proceedings

CONTROL SMOOTHING SPLINES WITH INITIAL CONDITIONS

ASMUSS Svetlana (LV), BUDKINA Natalja (LV)

Abstract. This work is devoted to the problem of optimal control of a linear dynamic system with initial conditions. The main attention is paid to a controlled system reduced to the second order differential equation considered with the cost functional which controls the input function of this system to push a trajectory nearby a set of desired points. We show how the technique of smoothing splines can be adapted for construction of solutions of such problem.

Keywords: smoothing spline, control theory, dynamic system

Mathematics subject classification: Primary 65D07; Secondary 49J15, 65K10

1 Introduction

This work is devoted to the problem of determining control and state trajectories for a linear dynamic system over a period of time to minimize an objective function. To solve this problem in some special cases we use the results from the theory of splines. The obtained solutions are so called control theoretic smoothing splines whose characteristics depend on the dynamics of the control system.

Spline functions are well known and widely used for practical approximation of functions by the information of values of function at the given points. The smoothing spline is a smooth function s in a suitable function space that minimizes the objective functional with a weight parameter which controls the smoothing. We refer the reader to [4] and references therein for the properties of smoothing splines. From the control theory point of view, the smoothing spline model is closely related to the finite-horizon linear quadratic optimal control problem by treating derivatives of s as a control input. There are a lot of publications about relations between control theory and smoothing splines, which in this context called smoothing theoretic splines. It should be mentioned also, that in the most of the papers (see, e.g., [2],[3],[5]) about control splines the problems on splines are reduced to the problem of control theory. The aim of this paper is otherwise to reduce the problem of control theory to the problem of smoothing splines and to use for its analysis methods and results of the general theory of splines. It will be done for some special cases of the following control theory problem:

$$x' = Mx + \beta u, \;\; y = \gamma^{\top} x, \tag{1}$$

considered with additional initial condition $x(a) = \alpha$.

We consider system (1) as the curve $z = y(t)$ generator. Our aim is to find a control law $u \in L_2[a, b]$ which drives the scalar output trajectory close to a finite sequence of set points at fixed times

$$\{(t_i, z_i) : i = 1, 2, \ldots, n\}, \text{ where } a < t_1 < t_2 < \ldots < t_n < b.$$

and minimizes the objective functional with a positive weight ρ:

$$\int_a^b u^2(t)dt + \rho \sum_{i=1}^n (y(t_i) - z_i)^2 \rightarrow min.$$

The main attention in this paper is paid to the special case of problem (1):

$$M = \begin{pmatrix} 0 & 1 \\ -q & -p \end{pmatrix}, \quad \beta = \begin{pmatrix} 0 \\ 1 \end{pmatrix}, \quad x = \begin{pmatrix} x_1 \\ x_2 \end{pmatrix}, \quad \alpha = \begin{pmatrix} \alpha_1 \\ \alpha_2 \end{pmatrix}.$$

For this case the objective functional of the problem of optimal control could be rewritten as

$$\int_a^b (x_1''(t) + px_1'(t) + qx_1(t))^2 dt + \rho \sum_{i=1}^n (\gamma_1 x_1(t_i) + \gamma_2 x_1'(t_i) - z_i)^2 \longrightarrow \min_{x_1(a)=\alpha_1, x_1'(a)=\alpha_2}. \quad (2)$$

We also note that usually such problem has been considered without initial conditions.

2 Smoothing splines

Problem (2) corresponds to the following more general conditional minimization problem:

$$||Tg||^2 + \rho||Ag - z||^2 \longrightarrow \min_{Bg=\alpha}, \quad (3)$$

where linear operators $T : W_2^r[a, b] \rightarrow L_2[a, b]$, $A : W_2^r[a, b] \rightarrow I\!R^n$ and $B : W_2^r[a, b] \rightarrow I\!R^m$ are continuous, parameter $\rho > 0$ and vectors $z \in I\!R^n$ and $\alpha \in I\!R^m$ are given. We assume that $A(W_2^r[a, b]) = I\!R^n$, $B(W_2^r[a, b]) = I\!R^m$ and all functionals of A and B, i.e. A_i, B_j, $i = 1, \ldots, n$, $j = 1, \ldots, m$, are linearly independent and we recall the theorem (see, e.g., [1]) on the existence and characterization of solutions of problem (3).

Theorem 1 *Under the assumptions that $kerT + kerA$ is closed a solution of problem* (3) *exists. An element $s \in B^{-1}(\alpha)$ is a solution of this problem if and only if there exist vectors $\lambda \in I\!R^n$ and $\nu \in I\!R^m$ such that*

$$T^*Ts = A^*\lambda + B^*\nu \textit{ and } \lambda + \rho(As - z) = 0. \quad (4)$$

Under the additional assumption $kerT \cap kerB = \{0\}$ this theorem gives the uniqueness of solution.

The characterization theorem implies that a solution of problem (3) is a spline from the space

$$S(T, (A, B)) = \{s \in W_2^r[a, b] : \ < Ts, Tg >= 0 \text{ for all } g \in ker(A, B)\},$$

where (A, B) consists of all functionals of A and B. Here and in the sequel $<, >$ is the scalar product and $kerA$ is the kernel of operator A.

Let us denote by $P : W_2^r[a,b] \to kerT$ a projector to the kernel of T. By using P we rewrite the first equation of (4) as

$$< Ts, Tg >= \sum_{i=1}^{n} \lambda_i A_i(g - Pg) + \sum_{j=1}^{m} \nu_j B_j(g - Pg). \tag{5}$$

For applying (5) in this paper we consider $g - Pg$ written in an integral form which includes Tg:

$$g(t) = (Pg)(t) + \int_a^b (Tg)(\tau)K(t,\tau)d\tau. \tag{6}$$

Now we rewrite (5) in the form

$$\int_a^b (Tg)(\tau)(Ts)(\tau)d\tau = \int_a^b (Tg)(\tau)(\sum_{i=1}^{n} \lambda_i A_i(K(.,\tau)) + \sum_{j=1}^{m} \nu_j B_j(K(.,\tau)))d\tau$$

(here $A(K(.,\tau))$ means that $K(\cdot,\tau)$ is considered as the function of the first argument when the second argument is fixed as τ) which implies

$$(Ts)(\tau) = \sum_{i=1}^{n} \lambda_i A_i(K(.,\tau)) + \sum_{j=1}^{m} \nu_j B_j(K(.,\tau)) \text{ and}$$

$$s(t) = (Ps)(t) + \sum_{i=1}^{n} \lambda_i \int_a^b A_i(K(.,\tau))K(t,\tau)d\tau + \sum_{j=1}^{m} \nu_j \int_a^b B_j(K(.,\tau))K(t,\tau)d\tau.$$

Let us assume that $dim(kerT) = m$, $kerT \cap kerB = \{0\}$ and $g - Pg \in kerB$ for all g. Then the unique solution of problem (3) is

$$s(t) = (Ps)(t) + \sum_{i=1}^{n} \lambda_i \int_a^b A_i(K(.,\tau))K(t,\tau)d\tau \tag{7}$$

such that

$$A_i s + \frac{\lambda_i}{\rho} = z_i,\ i = 1,\ldots,n, \text{ and } B_j Ps = \alpha_j,\ j = 1,\ldots,m. \tag{8}$$

Let us note that vector $\nu \in I\!R^m$ (see Theorem 1) is such that

$$\sum_{i=1}^{n} \lambda_i A_i h + \sum_{j=1}^{m} \nu_j B_j h = 0 \text{ for all } h \in kerT. \tag{9}$$

3 Smoothing splines in control theory

We consider problem (2) as problem (3) with

$$Tg = g'' + pg' + qx_1 = u, \;\; A_i g = \gamma_1 g(t_i) + \gamma_2 g'(t_i), i = 1, ..., n, \;\; B_1 g = g(a), B_2 g = g'(a).$$

In this case operators T, A and B fulfill all assumptions mentioned in the previous section. In the sequel we obtain the view of solution of problem (2) and on this basis we describe special classes of control splines from $S(T, (A, B))$ depending on the roots r_1, r_2 of the equation $r^2 + pr + q = 0$.

- Class 1 (exponential splines with polynomial coefficients): $r_1 = r_2 \in I\!R \setminus \{0\}$.
- Class 2 (exponential splines): $r_1, r_2 \in I\!R, \; r_1 \neq r_2$.
- Class 3 (polynomial-exponential splines): $r_1, r_2 \in I\!R, \; r_1 \neq r_2, r_1 \neq 0, r_2 = 0$.
- Class 4 (polynomial splines): $r_1 = r_2 = 0$.
- Class 5 (trigonometric splines with polynomial coefficients): $r_{1,2} = \pm i\eta \neq 0$.
- Class 6 (trigonometric splines with exponential-polynomial coefficients): $r_{1,2} = \zeta \pm i\eta$.

The remaining part of this section is organized as follows: the subsections correspond to the classes mentioned above; we present a proof of the obtained result for the first class (Subsection 3.1) and give results for other classes (Subsections 3.2-3.6) without proofs due to the space limitation.

3.1 Exponential splines with polynomial coefficients

We consider the case $r_1 = r_2 \in I\!R \setminus \{0\}$, i.e., $q > 0, \; p^2 = 4q$ and $Tg = g'' - 2r_1 g' + r_1^2 g$. The kernel of operator T is $kerT = \{(C_1 + C_2 t)e^{r_1 t} \mid C_1, C_2 \in I\!R\}$. We use (6) with

$$(Pg)(t) = g(a)e^{r_1(t-a)} + (g'(a) - r_1 g(a))e^{r_1(t-a)}(t-a), \;\; K(t,\tau) = e^{r_1(t-\tau)}(t-\tau)_+$$

and obtain the following representation

$$s(t) = (\mu_1 + \mu_2(t-a))e^{r_1(t-a)} + \sum_{i=1}^{n} \lambda_i \int_a^b e^{r_1(t_i+t-2\tau)}((\gamma_1 + \gamma_2 r_1)(t_i - \tau)_+ + \gamma_2(t_i - \tau)_+^0)(t-\tau)_+ d\tau$$

for solution s of problem (2) by using (7). Applying (8) we obtain

$$\mu_1 = B_1 Ps = \alpha_1, \quad \mu_2 = B_2 Ps - r_1 B_1 Ps = \alpha_2 - r_1 \alpha_1, \tag{10}$$

$$s(t_i) + \frac{\lambda_i}{\rho} = z_i, i = 1, \ldots, n. \tag{11}$$

The expression of Ts gives the corresponding control function u :

$$u(t) = (Ts)(t) = \sum_{i=1}^{n} \lambda_i e^{r_1(t_i - t)}((\gamma_1 + \gamma_2 r_1)(t_i - t)_+ + \gamma_2(t_i - t)_+^0). \tag{12}$$

By using the equalities $(t_i-\tau)_+^k=(t_i-\tau)^k+(-1)^{k+1}(\tau-t_i)_+$ and

$$(\gamma_1+\gamma_2 r_1)\sum_{i=1}^{n}\lambda_i e^{r_1 t_i}+(\nu_1+\nu_2 r_1)e^{r_1 a}=0,$$

$$(\nu_1 a+\nu_2 r_1 a+\nu_2)e^{r_1 a}+\sum_{i=1}^{n}\lambda_i((\gamma_1+\gamma_2 r_1)t_i e^{r_1 t_i}+\gamma_2 e^{r_1 t_i})=0, \tag{13}$$

(two last equalities are written using (9) with $h(t)=e^{r_1 t}$ and $h(t)=te^{r_1 t}$) we rewrite s:

$$s(t)=(\mu_1+\mu_2(t-a))e^{r_1(t-a)}+\int_a^t e^{r_1(a+t-2\tau)}((\nu_1+\nu_2 r_1)(\tau-a)-\nu_2))(t-\tau)d\tau+$$

$$+\sum_{i=1}^{n}\lambda_i\int_a^t e^{r_1(t_i+t-2\tau)}((\gamma_1+\gamma_2 r_1)(\tau-t_i)_+-\gamma_2(\tau-t_i)_+^0)(t-\tau)d\tau.$$

The integration gives us the result:

$$s(t)=(\mu_1+\mu_2(t-a))e^{r_1(t-a)}+\frac{\nu_1}{4r_1^3}(r_1(t-a)(e^{r_1(a-t)}+e^{r_1(t-a)})+(e^{r_1(a-t)}-e^{r_1(t-a)}))+$$
$$+\frac{\nu_2(t-a)}{4r_1}(e^{r_1(a-t)}-e^{r_1(t-a)})+\sum_{i=1}^{n}\lambda_i(\frac{\gamma_1}{4r_1^3}(r_1(t-t_i)_+(e^{r_1(t_i-t)}+e^{r_1(t-t_i)})+$$
$$+(e^{r_1(t_i-t)}-e^{r_1(t-t_i)})(t-t_i)_+^0)+\frac{\gamma_2}{4r_1}(e^{r_1(t_i-t)}-e^{r_1(t-t_i)})(t-t_i)_+). \tag{14}$$

The following proposition is proved.

Proposition 1 *In the case* $r_1=r_2\in I\!R\setminus\{0\}$ *the unique solution of problem* (2) *is exponential spline* (14) *which polynomial coefficients fulfill conditions* (10), (11), (13). *The corresponding control function* u *is given by* (12).

3.2 Exponential splines

In the case $r_1,r_2\in I\!R,\ r_1\neq r_2, r_1\neq 0, r_2\neq 0$ we use $Tg=g''-(r_1+r_2)g'+r_1 r_2 g$. The kernel of this operator is $kerT=\{C_1e^{r_1 t}+C_2e^{r_2 t}\mid C_1,C_2\in I\!R\}$.

By analogy with the previous case we obtain the solution s of problem (2) in the following form:

$$s(t)=\frac{\alpha_2-\alpha_1 r_2}{r_1-r_2}e^{r_1(t-a)}-\frac{\alpha_2-\alpha_1 r_1}{r_1-r_2}e^{r_2(t-a)}+\frac{1}{2(r_1^2-r_2^2)}(\frac{(\nu_1-\nu_2 r_1)e^{r_1(t-a)}}{r_1}-\frac{(\nu_1-\nu_2 r_2)e^{r_2(t-a)}}{r_2}+$$
$$+\sum_{i=1}^{n}\lambda_i(\frac{\gamma_1}{r_1}e^{r_1|t-t_i|}+\gamma_2(e^{r_1(t_i-t)_+}-e^{r_1(t-t_i)_+})-\frac{\gamma_1}{r_2}e^{r_2|t-t_i|}-\gamma_2(e^{r_2(t_i-t)_+}-e^{r_2(t-t_i)_+}))), \tag{15}$$

where coefficients ν_1 and ν_2 are expressed using the following system:

$$(\gamma_1+\gamma_2 r_1)\sum_{i=1}^{n}\lambda_i e^{r_1 t_i}+(\nu_1+\nu_2 r_1)e^{r_1 a}=0,\quad(\gamma_1+\gamma_2 r_2)\sum_{i=1}^{n}\lambda_i e^{r_2 t_i}+(\nu_1+\nu_2 r_2)e^{r_2 a}=0. \tag{16}$$

Proposition 2 *In the case* $r_1, r_2 \in I\!R,\ r_1 \neq r_2, r_1 \neq 0, r_2 \neq 0$ *the unique solution of problem* (2) *is exponential spline* (15) *which coefficients fulfill conditions* (11), (16). *The corresponding control function* u *is given by*

$$u(t) = \sum_{i=1}^{n} \frac{\lambda_i (t_i - t)_+^0}{r_1 - r_2} (\gamma_1 ({}^{r_1(t_i-t)} - e^{r_2(t_i-t)}) + \gamma_2 (r_1 e^{r_1(t_i-t)} - r_2 e^{r_2(t_i-t)})). \tag{17}$$

3.3 Polynomial-exponential splines

Now we consider the case when one of the roots $r_1, r_2 \in I\!R$ is equal to zero. The following result can be obtained by minimal changes of the previous proof.

Proposition 3 *In the case* $r_1, r_2 \in I\!R,\ r_1 \neq r_2, r_1 \neq 0, r_2 = 0$ *the unique solution of problem* (2) *is polynomial-exponential spline*

$$s(t) = \tfrac{\alpha_2}{r_1} e^{r_1(t-a)} + \alpha_1 - \tfrac{\alpha_2}{r_1} + \tfrac{1}{r_1^3}((\nu_1 - \nu_2 r_1) e^{r_1(t-a)} - (\nu_1 + \nu_2 r_1) e^{r_1(a-t)} + 2 r_1 \nu_2 - 2 r_1 \nu_1 (t-a) +$$

$$+ \sum_{i=1}^{n} \lambda_i (t - t_i)_+^0 (-\gamma_1 r_1 (t - t_i) - \tfrac{\gamma_1 + \gamma_2 r_1}{2} e^{r_1(t_i - t)} + \tfrac{\gamma_1 - \gamma_2 r_1}{2} e^{r_1(t - t_i)} + \gamma_2 r_1))$$

which coefficients fulfill (11) *and the following system*

$$(\gamma_1 + \gamma_2 r_1) \sum_{i=1}^{n} \lambda_i e^{r_1 t_i} + (\nu_1 + \nu_2 r_1) e^{r_1 a} = 0, \quad \sum_{i=1}^{n} \lambda_i \gamma_1 + \nu_1 = 0,$$

The corresponding control function u *is given by* (17).

3.4 Cubic splines

In the case $r_1 = r_2 = 0$, i.e., $p = q = 0$, we use $Tg = g''$. This case corresponds to the classical smoothing problem in the theory of splines according to which a solution of (2) without additional conditions is a cubic spline. Taking into account the initial conditions and Theorem 1 we get the solution s of problem (2) and the corresponding control function u:

$$s(t) = \alpha_1 + \alpha_2 (t - a) + \tfrac{\nu_1}{6} (t - a)_+^3 - \tfrac{\nu_2}{2} (t - a)_+^2 + \sum_{i=1}^{n} \tfrac{\lambda_i}{6} (t - t_i)_+^3, \quad u(t) = s''(t),$$

with the coefficients which fulfill the conditions (11) and $\nu_1 + \sum_{i=1}^{n} \lambda_i = 0, \quad \nu_1 t_0 + \nu_2 + \sum_{i=1}^{n} \lambda_i t_i = 0.$

3.5 Trigonometric splines with polynomial coefficients

Now we consider the case when $r_{1,2} = \pm i\eta \neq 0$, i.e., $p = 0,\ q > 0,\ \eta = \sqrt{q}$. It means that we take $Tg = g'' + qg$. The kernel of this operator is $kerT = \{C_1 cos\, \eta t + C_2 sin\, \eta t \mid C_1, C_2 \in I\!R\}$.

Proposition 4 *In the case* $r_{1,2} = \pm \eta \neq 0$ *the unique solution of problem* (2) *is trigonometric spline*

$$s(t) = \alpha_1 cos\,(\eta(t-a)) + \frac{\alpha_2}{\eta} sin\,(\eta(t-a)) + \frac{\nu_1}{2\eta^3} (\eta(t-a) cos\,(\eta(t-a)) - sin\,(\eta(t-a))) +$$

$$+\frac{\nu_2}{2\eta}(t-a)sin\,(\eta(t-a))+\frac{1}{2}\sum_{i=1}^{n}\lambda_i(\frac{\gamma_1}{\eta^3}sin\,(\eta(t-t_i)_+)-\frac{(t-t_i)_+}{\eta^2}(\gamma_1 cos\,(\eta(t-t_i))+\gamma_2\eta sin\,(\eta(t-t_i)))$$

which polynomial coefficients fulfill (11) and

$$\sum_{i=1}^{n}\lambda_i(\gamma_1 sin\,(\eta t_i)+\gamma_2\eta cos\,(\eta t_i))+\nu_1 sin\,(\eta a)+\nu_2\eta cos\,(\eta a)=0,$$
$$\sum_{i=1}^{n}\lambda_i(\gamma_1 cos\,(\eta t_i)-\gamma_2\eta sin\,(\eta t_i))+\nu_1 cos\,(\eta a)-\nu_2\eta sin\,(\eta a)=0.$$

The corresponding control function u is given by

$$u(t)=\sum_{i=1}^{n}\lambda_i(\frac{\gamma_1}{\eta}sin\,\eta(t_i-t)_+ +\gamma_2(t_i-t)_+^0 cos\,\eta(t_i-t)).$$

3.6 Trigonometric splines with exponential-polynomial coefficients

In this section we consider the case $r_{1,2}=\zeta\pm i\eta$, i.e., $p^2-4q<0, \zeta\neq 0, \eta\neq 0$. We use operator $Tg=g''-2\zeta g'+(\zeta^2+\eta^2)g$ with the kernel $kerT=\{C_1e^{\zeta t}cos\,\eta t+C_2e^{\zeta t}sin\,\eta t \mid C_1, C_2\in I\!R\}$.

Proposition 5 *In the case $r_{1,2}=\zeta\pm i\eta, \zeta\neq 0, \eta\neq 0$ the unique solution of problem* (2) *is trigonometric spline*

$$s(t)=\alpha_1 e^{\zeta(t-a)}cos\,(\eta(t-a))+\frac{\alpha_2-\zeta\alpha_1}{\eta}e^{\zeta(t-a)}sin\,(\eta(t-a))+\sum_{i=1}^{n}\frac{\lambda_i(\tau-t_i)_+^0}{4\zeta\eta}((e^{\zeta(t_i-t)}-e^{\zeta(t-t_i)})\times$$
$$\times(\frac{\gamma_1\eta}{\zeta^2+\eta^2}cos\,\eta(t-t_i)+\gamma_2 sin\,\eta(t-t_i))+(e^{\zeta(t_i-t)}+e^{\zeta(t-t_i)})sin\,\eta(t-t_i)\frac{\gamma_1\zeta}{\zeta^2+\eta^2}+\frac{1}{4\eta}((e^{\zeta(a-t)}-$$
$$-e^{\zeta(t-a)})(\frac{\nu_1\eta}{\zeta^2+\eta^2}cos\,\eta(t-a)+\nu_2 sin\,\eta(t-a))+(e^{\zeta(a-t)}+e^{\zeta(t-a)})sin\,\eta(t-a)\frac{\zeta\nu_1}{\zeta^2+\eta^2}).$$

which exponential-polynomial coefficients fulfill (11) *and*

$$\sum_{i=1}^{n}\lambda_i e^{\zeta t_i}((\gamma_1+\gamma_2\zeta)sin\,\eta t_i+\gamma_2\eta cos\,\eta t_i)+e^{\zeta a}((\nu_1+\nu_2\zeta)sin\,\eta a+\nu_2\eta cos\,\eta a)=0,$$
$$\sum_{i=1}^{n}\lambda_i e^{\zeta t_i}((\gamma_1+\gamma_2\zeta)cos\,\eta t_i-\gamma_2\eta sin\,\eta t_i)+e^{\zeta a}((\nu_1+\nu_2\zeta)cos\,\eta a-\nu_2\eta sin\,\eta a)=0.$$

The corresponding control function u is given by

$$u(t)=\sum_{i=1}^{n}\frac{\lambda_i(t_i-t)_+^0}{\eta}e^{\zeta(t_i-t)}((\gamma_1+\gamma_2\zeta)sin\,\eta(t_i-t)+\gamma_2\eta cos\,\eta(t_i-t)).$$

4 Numerical example

We consider the problem (2) with $\gamma_1=1,\ \gamma_2=0,\ \alpha_1=1,\ \alpha_2=1,\ p=-3,\ q=2$, i.e., $r_1=1, r_2=2$, and desired points: $(0.1;1), (0.3;7), (0.4;2), (0.6;11), (0.7;9)$. We solve this problem on $[0:0.7]$ with three different values of ρ: $\rho_1=1000$, $\rho_2=7000$ and $\rho_3=30000$.

In this case the unique solution of (2) is the exponential spline (15) with coefficients from Tab.1. We note that $s(t_i)-z_i=\lambda_i/\rho, i=1,...,n$.

The graphs of s for ρ_1 (dash dot line), ρ_2 (dash long line) and ρ_2 (solid line) are shown in Figure 1 (a), the graphs of the corresponding control function u are given in Figure 1 (b).

	λ_1/ρ	λ_2/ρ	λ_3/ρ	λ_4/ρ	λ_5/ρ	ν_1	ν_2
ρ_1	-0.859	2.696	-3.203	2.244	-1.063	155.34	-16.64
ρ_2	-0.772	1.926	-2.215	1.359	-0.602	2064.43	-7.35
ρ_3	-0.408	0.911	-1.006	0.578	-0.251	5181.63	12.86

Tab. 1. Coefficients of the solution.

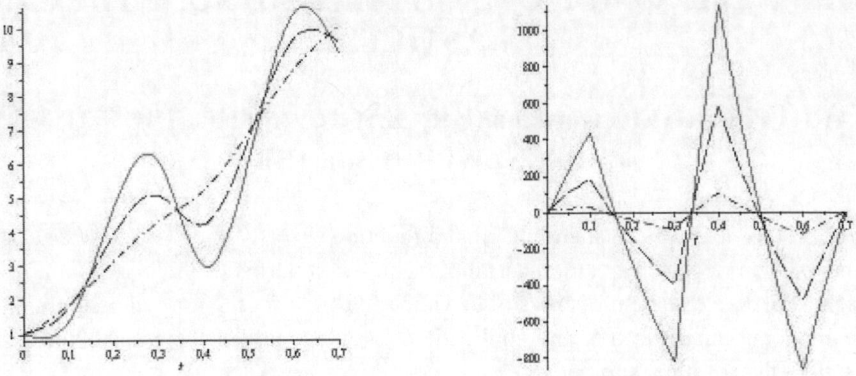

(a) (b)

Fig. 1. Graphs of solution.

References

[1] ASMUSS, S., BUDKINA, N.: On Some Generalization of Smoothing Problems. *Mathematical Modelling and Analysis*, V. 15, Nr. 3, 2015: 11-28.

[2] EGERSTEDT, M., MARTIN, C.: *Control Theoretic Splines*, Princeton University Press, 2010.

[3] SHEN, J., LEBAIR, J., MARTIN, C.: Splines and Linear Control Theory. *Automatica*, V. 53, 2015: 216-224.

[4] WAHBA, G.: *Spline Models for Observational Data*, in CBMS-NSF Regional Conference Series in Applied Mathematics, Philadelphia: SIAM, 1990.

[5] ZHANG, Z., TOMLINSON, T. M.: Shape Restricted Smoothing Splines via Constrained Optimal Control and Nonsmooth Newton's Methods. In *Acta Applicandae Mathematica*, V. 49, Nr. 1, 1997: 1–34.

Current address

Asmuss Svetlana, Dr. Math.
University of Latvia
Zellu street 25, Riga, LV-1002, Latvia
Institute of Mathematics and Computer Science of University of Latvia
Rainis blvd. 29, Riga, LV-1459, Latvia
E-mail: svetlana.asmuss@lu.lv

Budkina Natalja, Dr. Math.
Faculty of Computer Science and Information Technology
Riga Technical University
Daugavgrivas street 2, Riga, LV-1048, Latvia
E-mail: natalja.budkina@rtu,lv

Slovak University of Technology in Bratislava
Faculty of Mechanical Engineering
17th Conference on Applied Mathematics **APLIMAT 2018**

Proceedings

ON NORMAL ASSUMPTIONS ON DEMAND FUNCTION AND ITS ELASTICITY

BARIĆ PISAROVIĆ Gordana (HR), RAGUŽ Andrija (HR), VOJVODIĆ ROZENZWEIG Višnja (HR)

Abstract. In this note we consider the demand function $D = D(p)$, where p is price of a certain good and we introduce some natural assumptions on D in terms of the corresponding elasticity coefficient. We derive some properties of D which follow from the normal assumptions, and we provide economic interpretation and application. We also present some classes of of functions which satisfy the normal assumptions.

Keywords: Elasticity of function, demand function, mathematical economics, mathematical analysis

Mathematics subject classification: Primary 34E15, 34C25, 35M10; Secondary 49J45

1 Introduction

Economist often study how demand for a certain commodity reacts to a change of price. It can be measured by how many units the quantity demanded will change per unit of money increase in price. However, there is unsatisfactory aspect to this way of measuring the sensitivity of demand to price changes because one unit of money increase in the price may be considerable for some commodities (bread or milk, for instance) whereas it can be insignificant for other commodities (such as cars). This difficulty can be eliminated if relative changes are used instead. It can be asked by what percentage the quantity demanded changes when the price increases by 1 percent. The number obtained in this way is independent of the units in which both quantity and price are measured. This number is called the price elasticity of demand, measured at a given price. Assume now that the demand for a commodity can be described by the function $x = D(p)$, where p denotes the price. The change of price from p to $p + \Delta p$, yields the change of quantity of demand, the absolute change of x being $\Delta x = D(p + \Delta p) - D(p)$. Hence, the relative (or proportional) change of x is given by $\frac{\Delta x}{x} = \frac{D(p+\Delta p)-D(p)}{D(p)}$. The ratio of two aforementioned relative changes equals to

$$\frac{\frac{\Delta x}{x}}{\frac{\Delta p}{p}} = \frac{p}{D(p)} \frac{D(p + \Delta p) - D(p)}{\Delta p} .$$

If $\Delta p = \frac{p}{100}$, the parameter p increases by 1 percent, and we get

$$\frac{\frac{\Delta x}{x}}{\frac{\Delta p}{p}} = \frac{\Delta x}{x} \cdot 100 ,$$

whereby the right-hand side in the equation above represents the change of quantity of demand measured in terms of percentages of x. This quantity is called the average elasticity of x in the interval

$[p, p+\Delta p]$, and it depends both on the price change Δp and on the price p, but is unit free. Therefore it makes no difference whether the quantity of demand is measured in tons, kilograms or whatever units, or if the price is measured in dollars, euros, or any other currency. To define the actual elasticity of $D(p)$ with respect to p (denotes by $E_p D(p)$) in a manner that it does not depend on the amount of increase in p, we consider the limit as $\Delta p \longrightarrow 0$, getting

$$E_p D(p) := \frac{p}{D(p)} \cdot D'(p)$$

Since the increase of the price p for normal goods results in decrease in the demand $D(p)$, we expect $E_p D(p)$ to be negative for any admissible p.

The price elasticity of demand varies enormously from one commodity to another. The more substitutes there are for a commodity and the closer they are, the greater will be the price elasticity of demand (ignoring the negative sign). The reason is that consumers will be able to switch to the substitutes when the price of the commodity increases. For example, the price elasticity of demand for a particular brand of a product will probably be fairly high, especially if there are many other similar brands. By contrast the demand for a product in general should be less sensitive to a change in price. As well, the higher the proportion of consumers' income is spent on a certain commodity, the more they will have to reduce their consumption of it following a rice in price i.e. the more elastic will be the demand. Economists are interested in knowing the sensitivity of demand not just to a change in price but also to changes in other variables such as consumers' incomes (income elasticity of demand) or expenditure on a particular advertising campaign or other forms of promotion. Furthermore, they consider elasticities of supply, elasticities of revenue, cost elasticity and several other kinds of elasticity. It is therefore logical and certainly helpful to define elasticity for a general differentiable function $f = f(x)$.

- If $|E_x f(x)| > 1$, then f is elastic at x,
- If $|E_x f(x)| = 1$, then f is unit elastic at x,
- If $|E_x f(x)| < 1$, then f is inelastic at x,
- If $|E_x f(x)| = 0$, then f is perfectly inelastic at x,
- If $|E_x f(x)| = +\infty$, then f is perfectly elastic at x.

If the demand is price elastic, quantity demanded changes proportionately more than price, therefore a rise in price will lead to a reduction in consumers' expenditure on the commodity and hence to a reduction in the total producers' revenue. When the demand is price inelastic, price changes proportionately more than quantity demanded, so a rise in price will lead to an increase in total expenditure and revenue. If the price elasticity of demand is unity then the quantity demanded changes by the same proportion as the price. There are some rules for elasticities of sums, products, quotients and composite functions might be useful. It can easily be shown that if $f = f(x)$ and $g = g(x)$ are positive valued differentiable functions of x and A is a constant, then the following rules hold:

- $E_x A = 0$,
- $E_x(fg) = E_x(f) + E_x(g)$,

- $E_x(\frac{f}{g}) = E_x(f) - E_x(g)$,
- $E_x(f+g) = \frac{f}{f+g}E_x f + \frac{g}{f+g}E_x g$,
- $E_x(f-g) = \frac{f}{f-g}E_x f - \frac{g}{f-g}E_x g$,
- $E_x(f \circ g) = E_u f(u) \cdot E_x u$, where $u = g(x)$.

Also, we note that, if $y = f(x)$ admits an inverse function $x = f^{-1}(y)$, then it holds that $E_y f^{-1} = \frac{1}{E_x f}$. We also use the notation $E_{g,x}$ instead of $E_x f$, which is sometimes more convenient. Then we can re-state some of the following results as follows:

Proposition 1.1. If $x = x(p)$ (i.e., $p = p(x)$), it holds that:

(i) $E_{px,p} = 1 + E_{x,p}$,

(ii) $E_{px,x} = 1 + E_{p,x}$,

(iii) $E_{p,x}E_{x,p} = 1$.

2 Examples

Further, we provide six typical examples of demand functions (cf. [2]).

$$x = a - bp^c, a, b, c > 0 \tag{1}$$
$$x = a + bp^{-c}, a, b, c > 0 \tag{2}$$
$$x = \frac{b}{p+a}, a, b > 0 \tag{3}$$
$$x = \sqrt{a - bp}, a, b > 0 \tag{4}$$
$$x = ae^{-bp}, a, b > 0 \tag{5}$$
$$x = p^a e^{-b(p+c)}, a, b, c > 0 \tag{6}$$

The function defined by (1) is usually considered as a demand function for $c = 1$ (linear function), $c = 2$(quadratic function) and $c = \frac{1}{2}$ (square root). All of the functions (1)-(6) satisfy the standard assumptions on the demand functions, but, on top of that, also share the common feature (with the exception of (2)): for all admissible values of the price p determined by $\mathcal{D}(x) = \{p \geq 0 : x(p) \geq 0, x'(p) \leq 0\}$ it holds that the elasticity coefficient $E_{x,p}$ is a decreasing function of the price p, i.e., as the price p increases, the demand function monotonically makes then transition from the interval of non-elasticity to the interval of elasticity, or, for short, $\frac{dE_{x,p}}{dp} \leq 0$.

3 Demand Function and Revenue Function

Since the affine function is instructive and simple example, we describe the connection between in the case of the demand function (1). In such a case, we have

$$x = a - bp \quad \text{or} \quad p = \frac{a - x}{b} \quad \text{for } a, b > 0 \, .$$

The demand function is well-defined for $p \in [0, \frac{a}{b}]$. The elasticity coefficient is

$$E_{x,p} = -\frac{bp}{x} \, .$$

The inelasticity interval is $p \in [0, \frac{a}{2b}\rangle$, whereas the elasticity interval is $p \in \langle \frac{a}{2b}, \frac{a}{b}]$. Provided

$$p = \frac{a}{2b}, \quad x = \frac{a}{2}, \quad \text{it results} \quad E_{x,p} = -1 \,.$$

We have

$$\frac{dE_{x,p}}{dp} = -\frac{ab}{x^2} \leq 0 \,.$$

Then the function of (total) revenue reads $R(p) = px = ap - bp^2$, and it attains the maximal value

$$R^* = \frac{a^2}{4b} \quad \text{for} \quad p = \frac{a}{2b}, \quad x = \frac{a}{2}, \quad E_{x,p} = -1 \,.$$

On the other hand, we can express the total revenue in terms of quantity of demand, getting

$$R(x) = \frac{ax - x^2}{b} \,;$$

whereby the marginal revenue is

$$MR = R' = \frac{dR}{dx} = \frac{a - 2x}{b} \,.$$

If the marginal revenue equals zero, i.e., $R' = 0$, the total revenue admits stationary point $x^* = \frac{a}{2}$ and $p(x^*) = \frac{a}{2b}$. We have $R' > 0$ ($R' < 0$, resp.), i.e., the total revenue increases (decreases, resp.) for $0 \leq x \leq \frac{a}{2}$ ($\frac{a}{2} \leq x \leq a$., resp.). Thus, the maximum of the total revenue is attained at $x = x^*$. In terms of elasticity this means that demand is elastic for $0 \leq x \leq \frac{a}{2}$ and inelastic for $\frac{a}{2} \leq x \leq a$. Hence,

$$\frac{dE_{x,p}}{dx} = \frac{a}{x^2} \geq 0 \,.$$

This provides as additional assumption on the demand function. That is to say, normal assumptions on the demand function follow from natural assumptions on the total revenue function $R = R(x) = px$. Indeed, we have:

$$\frac{dR}{dx} = \frac{R}{x}(1 + E_{p,x}). \tag{7}$$

We distinguish three cases:

(i) If $E_{p,x} \in \langle -1, 0\rangle$ then (7) we get $\frac{dR}{dx} \geq 0$. The interval where total revenue increases with respect to demand x coincides with the interval of elasticity of the demand function since it holds that $E_{x,p}E_{p,x} = 1$. We conclude:

$$E_{p,x} \in \langle -1, 0\rangle \Rightarrow E_{x,p} < -1 \,.$$

(ii) If $E_{p,x} = -1$, then (by (7)) $\frac{dR}{dx} = 0$ for $x = x^*$ and the total revenue admits stationary point.

(iii) If $E_{p,x} < -1$ then $\frac{dR}{dx} \leq 0$. The interval where total revenue decreases with respect to demand x coincides with the interval of inelasticity of the demand function, i.e., $E_{x,p}E_{p,x} = 1$ yields

$$E_{p,x} < -1 \Rightarrow E_{x,p} > -1 \,.$$

We recall that natural assumption on the total revenue function provides that for a small quantity of demand x the total revenue increases, which means that the demand grows faster then the price declines to some eventual stationary pointx^*, where the total revenue attained its maximum. After the point x^*, total revenue decreases. In terms of elasticity this means that the demand function which depends on the price p is decreasing. The consideration above allows us to formulate normal assumptions on the demand function:

Normal Assumptions on Demand Function:

$$p \geq 0 \,, \tag{8}$$

$$x \geq 0 \,, \tag{9}$$

$$x' \leq 0 \,, \tag{10}$$

$$\frac{dE_{x,p}}{dp} \leq 0 \,. \tag{11}$$

4 Elasticity of Demand Function

Definition 4.1 The elasticity coefficient of the second order of $x = x(p)$ is denoted by $E^{(2)}_{x,p}$ and it is the elasticity coefficient of the function $E = E_{x,p}$, i.e.,

$$E^{(2)}_{x,p} := E_{E,p} = \frac{p}{E}\frac{dE}{dp}. \tag{12}$$

Definition 4.2 The elasticity coefficient of order $k+1, k \in \mathbb{N}$ is defined by $E^{(k)} = E^{(k)}_{x,p}$, i.e.,

$$E^{(k+1)}_{x,p} := E_{E^{(k)},p} = \frac{p}{E^{(k)}}\frac{dE^{(k)}}{dp}. \tag{13}$$

Now we can rewrite condition (11), $\frac{dE_{x,p}}{dp} \leq 0$, coupled with the condition $E_{x,p} \leq 0$ in terms of the elasticity coefficient of the second order, getting

$$E^{(2)}_{x,p} \geq 0. \tag{14}$$

Further, we can rewrite the properties of typical demand functions in terms of higher-order elasticity coefficients.

1. The demand function (1)

$$x = a - bp^c, a, b, c > 0$$

is defined for $p \in [0, (\frac{a}{b})^{\frac{1}{c}}]$. Then it holds that:

$$E = E_{x,p} = -bcp^c x^{-1} \leq 0 \,,$$
$$E^{(2)} = E_{E,p} = c - E \,,$$
$$E^{(3)} = E_{c-E,p} = -E \,,$$
$$E^{(4)} = E_{-E,p} = c - E \,.$$

Notice that we have $c > 0, -E \geq 0 \Rightarrow E^{(2)} = E_{E,p} = c - E \geq 0$, and so normal assumptions are fulfilled. If

$$p^* = (\frac{a}{b(c+1)})^{\frac{1}{c}}$$

we get $E_{x,p} = -1$, which is the price for which demand monotonically changes from inelasticity interval to the elasticity interval. On the other hand, total revenue increases from $p = 0$ do $p = p^*$, and at $p = p^*$ attains its maximal value, whereby for $p > p^*$ we have decrease of total revenue.

2. The demand function (3)

$$x = \frac{b}{p+a}, a, b > 0$$

is defined for $p \geq 0$.

$$\begin{aligned} E =& E_{x,p} = -b^{-1}px \leq 0 \, , \\ E^{(2)} =& E_{E,p} = 1 + E \, , \\ E^{(3)} =& E_{1+E,p} = E \, , \\ E^{(4)} =& E_{E,p} = 1 + E \, . \end{aligned}$$

This function is inelastic for any non-negative price p, i.e., $0 \geq E_{x,p} > -1$, so that $E^{(2)} = E_{E,p} = 1 + E > 0$.

3. The demand function (4)

$$x = \sqrt{a - bp}, a, b > 0,$$

is defined for $p \in [0, \frac{a}{b}]$.

$$\begin{aligned} E =& E_{x,p} = -0.5bpx^{-2} \leq 0 \, , \\ E^{(2)} =& E_{E,p} = 1 - 2E \, , \\ E^{(3)} =& E_{1-2E,p} = -2E \, , \\ E^{(4)} =& E_{E,p} = 1 - 2E \, . \end{aligned}$$

Since $-E \geq 0 \Rightarrow E^{(2)} = E_{E,p} = 1 - 2E \geq 0$, the normal assumptions are satisfied. For

$$p^* = \frac{2a}{3b}$$

we get $E_{x,p} = -1$, and so p^* is the price for which demand monotonically changes from inelasticity interval to the elasticity interval. On the other hand, total revenue increases from $p = 0$ do $p = p^*$, and at $p = p^*$ attains its maximal value, whereby for $p > p^*$ we have decrease of total revenue.

4. The demand function (5)

$$x = ae^{-bp}, a, b > 0,$$

is defined for $p \geq 0$. Computation gives

$$\begin{aligned} E =& E_{x,p} = -bp \leq 0 \, , \\ E^{(2)} =& E_{E,p} = 1 > 0 \, , \\ E^{(3)} =& E_{1,p} = 0 \, . \end{aligned}$$

Provided

$$p^* = \frac{1}{b} \, ,$$

we get $E_{x,p} = -1$ and so p^* is the price for which demand monotonically changes from inelasticity interval to the elasticity interval. On the other hand, total revenue increases from $p = 0$ do $p = p^*$, and at $p = p^*$ attains its maximal value, whereby for $p > p^*$ we have decrease of total revenue.

5. The demand function (6)

$$x = p^a e^{-b(p+c)}, a, b, c > 0$$

is defined for $p \geq 0$.
The elasticity coefficient is

$$E = E_{x,p} = a - bp$$

and so

$$E = E_{x,p} = a - bp \leq 0 \Rightarrow p \geq \frac{a}{b}.$$

Thus for $p \in [\frac{a}{b}, \infty\rangle$ we obtain

$$\begin{aligned} E^{(2)} =& E_{E,p} = -\frac{bp}{E} = 1 - \frac{a}{E} > 0 \,, \\ E^{(3)} =& E_{-\frac{bp}{E},p} = \frac{a}{E} \,, \\ E^{(4)} =& E_{\frac{a}{E},p} = -E_{E,p} = -E^{(2)} \,. \end{aligned}$$

For

$$p^* = \frac{a+1}{b}$$

we get $E_{x,p} = -1$, and so p^* is the price for which demand monotonically changes from inelasticity interval to the elasticity interval. On the other hand, total revenue increases from $p = \frac{a}{b}$ to $p = p^*$, and at $p = p^*$ attains its maximal value, whereby for $p > p^*$ we have decrease of total revenue.

Finally, we formulate natural requirements for the demand function with respect to its elasticity.

Normal assumptions on demand function in terms of elasticity

$$\begin{aligned} p \geq& 0 \,, \\ x \geq& 0 \,, \\ E_{x,p} \leq& 0 \,, \\ E^{(2)}_{x,p} \geq& 0 \,. \end{aligned}$$

Theorem 4.1 If $x = x(p), p, x \geq 0$, $x' \leq 0$, $x(0) > 0$ and

$$E^{(2)}_{x,p} = \alpha + \beta E, \alpha > 0, \beta \neq 0$$

then the function $x = x(p)$ satisfies normal assumptions on the demand function.
Proof. Starting from $E^{(2)}_{x,p} = \alpha + \beta E$, according to (12) we get

$$\frac{E'}{E(\alpha + \beta E)} = \frac{1}{p}$$

or, equivalently,

$$\frac{dE}{E(\alpha+\beta E)} = \frac{1}{p}dp.$$

The solution E of this ordinary differential equations is

$$\frac{E}{\alpha+\beta E} = cp^{\alpha}$$

which can be written in its explicit form as

$$E = \frac{\alpha c p^{\alpha}}{1-c\beta p^{\alpha}}\,. \tag{15}$$

Next, we have

$$\frac{x'}{x} = \frac{\alpha c p^{\alpha-1}}{1-c\beta p^{\alpha}}$$

whereby

$$x = b(1-c\beta p^{\alpha})^{-1/\beta}\,. \tag{16}$$

To begin with, since $x(0) > 0$, we get $b > 0$. Then it follows that $x \geq 0, b > 0 \Rightarrow 1 - c\beta p^{\alpha}$.

Case 1.

If it holds that $c\beta < 0$, then $1 - c\beta p^{\alpha} > 0$ for every $p \geq 0$.

From (15) and $x' \leq 0$ we deduce $c < 0$, and also from (15) we recover

$$E' = \frac{\alpha^2 c p^{\alpha-1}}{(1-c\beta p^{\alpha})^2}\,. \tag{17}$$

But then from $c < 0$ it results $E' \leq 0$.

Case 2.

If it holds that $c\beta > 0$, $\alpha > 0$ and $1 - c\beta p^{\alpha} \geq 0$ provide $0 \leq p \leq (c\beta)^{-\frac{1}{\alpha}}$, $x(0) = b > 0$.

Hence, $E \leq 0 \Rightarrow c < 0$. Now from $c\beta > 0$ and $c < 0$ we have $\beta < 0$. Therefore, for $\alpha > 0, \beta < 0$ we have$E^{(2)}_{x,p} = \alpha + \beta E \geq 0$.

Q.E.D.

Conclusion

One example of the class of demand functions which satisfy normal assumptions is:

$$x = b(1-c\beta p^{\alpha})^{-1/\beta}\,,$$

where

$$b > 0, c < 0, \alpha > 0, \beta \neq 0.$$

5 General setting

In a broader context, our goal is to examine and describe the structure of elementary functions of one variable according to the properties of their generalized elasticity coefficients. For a given open interval $D \subset (0, +\infty)$ we introduce the class of functions X_D, defined by

$$X_D := \{g : D \longrightarrow (0, +\infty) : g^{(n)}(x) > 0,\ n \in \mathbf{N} \cup \{0\}\,, x \in D\}\,.$$

For $f, g \in X_D$ we set $f \approx g$ iff there exists $\lambda \in \mathbf{R}\backslash\{0\}$ such that $g = \lambda f$, $[g] := \{f \in X_D : f \approx g\}$, $\overline{X}_D := X_D/\approx$. Then $E : \overline{X}_D \longrightarrow X_D$, $E[g](x) := E_{g,x}$ is well-defined injection.

Definition 5.1. We say that function g is

(i) φ_n-type recursive function if there exists $n_0 \in \mathbf{N}$ such that for every $n \in \mathbf{N}$, $n \geq n_0$, it holds that $E_g^{(n+1)} = \varphi_n(E_g^{(n)})$; if for some φ it holds that $\varphi_n = \varphi$ for every $n \geq n_0$, we say that g (where $g = g(\varphi)$) is φ-type recursive function,

(ii) is n_0-static, if φ_n-type recursive function g satisfies $\varphi_n(\xi) = \xi$ for every $n \geq n_0$.

Example 5.1. If we set $n_0 := 2$ and

$$\varphi_n(\xi) := \begin{cases} -\xi, & \text{if } n \geq 3 \text{ is odd} \\ c - \xi, & \text{if } n \text{ is even,} \end{cases}$$

$$\left(\psi_n(\xi) := \begin{cases} \xi, & \text{if } n \geq 3 \text{ is odd} \\ 1 + \xi, & \text{if } n \text{ is even,} \end{cases} \quad \text{resp.}\right)$$

then $g(x) := a - bx^c$ ($g(x) := b(x-a)^{-1}$, resp.) is φ_n-type (ψ_n-type, resp.) recursive function. Similarly, if we set

$$\tilde{\varphi}_n(\xi) := \begin{cases} -2\xi, & \text{if } n \geq 3 \text{ is odd} \\ 1 - 2\xi, & \text{if } n \text{ is even,} \end{cases}$$

$$\left(\tilde{\psi}_n(\xi) := \begin{cases} \frac{a}{\xi}, & \text{if } n \geq 3 \text{ is odd} \\ 1 - \frac{a}{\xi}, & \text{if } n \text{ is even,} \end{cases} \quad \text{resp.}\right)$$

then $g(x) := (a - bx)^{1/2}$ ($g(x) := x^a e^{-b(x+c)}$, resp.) is $\tilde{\varphi}_n$-type ($\tilde{\psi}_n$-type, resp.) recursive function. We note that, if g is fixed point of operator $E^{(1)}$ (i.e., if $g(x) = -\frac{1}{ln(Cx)}$, $C > 0$), then g is n_0-static for every $n_0 \in \mathbf{N}$. On the other hand, we do not know if there exists a function g which is φ-type recursive for any $n_0 \in \mathbf{N}$ provided $\varphi \neq id$, where $id(\xi) := \xi$ is identity function.

Definition 5.2. We say that function a set of functions A is $E^{(n)}$-stable for given $n \in \mathbf{N}$ if it holds that $E^{(n)}(A) \subset A$.

It is easy to see that the set of all rational (irrational, resp.) functions is $E^{(1)}$-stable and therefore $E^{(n)}$-stable for every $n \in \mathbf{N}$. We conjecture that there is no nontrivial set of functions which is $E^{(n)}$-stable for some $n \in \mathbf{N}$, $n > 1$, and which is not $E^{(1)}$-stable.

We note that homogeneity properties of the elasticity coefficients are inherited from homogeneity properties of the function itself.

Theorem 5.1. $x \mapsto g(x)$ is k-homogeneous function for some $k \in \mathbf{R}$ iff $x \mapsto E_{g,x}$ is 0-homogeneous function.

We have not been able to provide direct economic interpretation of higher order elasticity coefficient of general function g. Instead, we introduce some particular classes of functions for which such interpretation is possible. For instance, this is the case for functions g which are φ_n-type recursive with weak dependence of φ_n with respect to n (cf. Example 5.1).

References

[1] Allen, R. G. D.: *Mathematical Analysis for Economists*, MacMillan and co. Limited, London, 1938.

[2] Martić, Lj.: *Matematičke metode za ekonomske analize*, Volume 1, Edition 8, Volume 153 of Biblioteka Udžbenici, Narodne novine, Zagreb, 1990, ISBN 8633701561, 9788633701563.

Current address

Baric Pisarovic Gordana, BSc.
Zagreb School of Economics and Management
Jordanovac 110, 10000 Zagreb, Croatia
E-mail: gbaric@zsem.hr

Raguz Andrija, prof.
Zagreb School of Economics and Management
Jordanovac 110, 10000 Zagreb, Croatia
(Luxembourg School of Business, 19 Rue Eugene Ruppert L-2453, Luxembourg)
E-mail: araguz@zsem.hr

Vojvodic Rozenzweig Visnja, prof.
Zagreb School of Economics and Management
Jordanovac 110, 10000 Zagreb, Croatia
E-mail: vvojvodi@zsem.hr

Slovak University of Technology in Bratislava
Faculty of Mechanical Engineering
17th Conference on Applied Mathematics **APLIMAT 2018**

Proceedings

TEACHING PROCESS OF AN APPLIED STATISTICS COURSE FOR STUDENTS OF THE ANTROPOLOGY STUDY PROGRAMME AT MASARYK UNIVERSITY

BENDOVÁ Veronika (CZ), BUDÍKOVÁ Marie (CZ)

Abstract. The point of these proceedings is to describe the teaching process of an applied statistics course that is offered by the Institute of Mathematics and Statistics on the Masaryk University in Brno to students of the Anthropology study programme. Teaching statistics to students of non-mathematical study programmes is in general a difficult matter. Therefore, we aim to make it as easy for the students as possible. To fulfill this goal, we employ a collection of solved statistics exercises, worksheets, along with friendly and open mindset. We succeed in motivating students to use statistical methods in practice despite raised difficulty of the course.

Keywords: course of Applied Statistics, antropology, R software

Mathematics subject classification: Primary 97K80; Secondary 97Dxx

1 Introduction

An important task of the Institute of Mathematics and Statistics in Brno is to develop statistical skill set of students of non-mathematical study programmes, among which are students of the anthropology study programme. Graduates of said programme have to be able to understand the contents of scientific studies that are based on quantitative data obtained by measurement of real objects. It is also expected that they will be able to present results of these studies in a way that is both statistically correct and anthropologically comprehensible. Because of these requirements, two mandatory courses have been integrated into the curricula of anthropology study programmes: Applied statistics I has been integrated into the Bachelor's study programme, and Applied statistics II has been integrated into the follow-up Master's programme.

Lectures of courses Applied statistics I and II are open to all student of non-mathematical study programmes. Illustrative examples, discussed during the lectures, are based on topics from various fields of study, e.g., economics, psychology, and medicine. The STATISTICA software is used for computations of the examples. This software is also used during practical seminars designed for students of all study programmes with the exception of anthropology students.

A special seminar that caters to specific requirements of anthropology students has been created for Applied statistics I and II, since their dominant computation tool is the R statistical software [1]. During these seminars, students analyze one-dimensional and multidimensional data using the R software. Practice real world data, used in the seminars, have been meticulously selected in collaboration with the Department of Anthropology. Students learn on model examples how to get a basic overview of the dataset using methods of descriptive statistics, how to correctly formulate goals and hypotheses, how to choose the right statistical test, and how to interpret obtained results.

We consider knowledge of basic terminology and methods of mathematical statistics, as well as a degree of proficiency in practical skill in working with the R software, to be a key part of absolvent's profile of the modern anthropology study programme for the 21st century.

2 Syllabus of the course

The special seminar for students of the anthropology study programme, that encompasses usage of the R software, is currently taking place only for the Applied statistics I course. A similarly designed practical seminar for the follow-up course Applied statistics II is in preparation, and will be offered during the spring term of 2018. Therefore, we will only consider the former course in this text. Lectures of the Applied statistics I course, as well as the practical seminars, take place each week for 2 hours. As the compulsory literature is used the text [2] or [3]. Students should be able to comprehend advanced statistical literature [4] after completing this course.

The syllabus of Applied statistics I is as follows: 1. Exploratory data analysis, diagnostic plots. 2. Probability space, classic and conditional probability, independent events. 3. Random variable and description of their probability distribution, numerical characteristics, selected discrete and continuous distribution. 4. Basic terms of mathematical statistics – random selection, statistics derived from random selection, point and interval estimations of parameters. 5. Normality tests. 6. Parametric tests with one, two, or more independent random selections. 7. Non-parametric tests with one, two, or more independent random selections. 8. Dependence analysis of two random samples – dependence in contingency tables, Spearman's and Pearson's correlation coefficient.

3 Teaching methods

All study materials, i.e. texts of the lectures, notes for the seminars, practical examples, R-scripts written during the seminars, additional practice examples and files containing associated datasets, are available to the students in the Information system of the Masaryk University.

Each lecture starts with motivation, followed by explanation of theoretical terms, description of statistical methods, and presentation of several examples. Practical seminars take place in computer seminar rooms. The most important theoretical matter is revised during each seminar. Then, its application is shown on practical examples. The examples are granted to the students in the form of worksheets, with which they work. Solution to the examples is created by students in cooperation with the seminar tutor. First, the assigned example is decomposed on the whiteboard. The tutor asks questions that lead to the correct decomposition of the example. Students provide answers, and thus actively participate on the decomposition. The R software is then used for statistical computations, which are carried out by the tutor and visualized on the projector. Students type the R code, projected by the tutor, on their computers, which leads to the students getting familiar with the R software. Formulation of conclusions of the analysis, as well as interpretation of results, is done collectively by students under the supervision of the seminar tutor.

4 Course completion

The course concludes with an exam, which consists of a theoretical and a practical part. Students are allowed to use all study materials from lectures and practical seminars during the exam. In the theoretical part, students solve 8 examples for no more than 60 minutes. The point of the practical part is to present results of a dataset analysis that is carried out independently by each student. The presentation is evaluated on the basis of choosing the right statistical methods, the quality of contents and graphics of the presentation, and the performance of the speaker themselves.

The dataset for the presentation is obtained by the students in different ways. Some use the dataset from their bachelor's thesis, while others download freely available data from the internet. The remaining students ask their lecturer or seminar tutor. Presentation topics tend to vary. For example, these were the topics of some of the presentations from the fall term of 2016: 1. Processing of length and width dimensions of skulls and their comparison. 2. Circumference of hand and height of newborns. 3. Comparison of birth weight of first born and second born boys. 4. Determination of correlation dependence between length of ear and length of nose. 5. Incidence of dermatoglyphic patterns and ridge-count values.

5 Midterm homework

Another requirement for successful completion of the course is to solve a set of examples, assigned to the students as homework during the second half of the term. The point of this homework is to motivate students to use the R software in a self-reliant way and to think about an assigned examples in conditions similar to those that resemble conditions of scientific work, i.e., calm environment with the availability of all materials, as well as with the potential help of one's colleagues.

The homework contains 6 to 8 examples (the count is based on the difficulty of considered examples) that encompass all of the matter that was taught up to the point of assignment of the homework. Students successfully pass the homework if they get a score of at least $75\,\%$. The assignment is scored on the basis of correctness of the solution, graphical quality of plots, readability of the R code, and interpretation of results. Examples of the homework are assigned along with correct numerical results, as well as a view on required plots. Students can work on the assignment together. The ultimate point is to make the students think about the examples together and discuss them. The only practice that is forbidden is copying of the R-code. The time limit for handing in the solution of the homework is 14 days.

6 Sample of used examples

Since teaching of statistics supported by the R software bears a large challenge for students of non-mathematical study programmes (and thus for student of anthropology), we have decided to create a collection of examples with guides for solving each one using the R software. Usage of statistical methods, which are shown during each lecture, is always illustrated on an solved example that strongly emphasizes correct formulation of hypotheses on the basis of research goals, as well as valid interpretation of results received in numeric, tabular, or graphic manner. Students have sufficient number of examples for each statistical method available along with results, but without a solution guide.

Below are sample of three examples used in the course: An example from the collection, an example from the worksheet used in the practical seminars, and an example from the homework.

6.1 A solved example from the collection

We have english set of documented skelets, which includes osteometric data of clavicles (Parsons, 1916). The expected value ($\mu = 151.74$ mm) and the standard deviation ($\sigma = 11$ mm) of the right-side clavicle length of males were estimated from the data set. Suppose the dataset have normal distribution, compute the probability, that the right-side clavicle length of males is (a) equal to 150 mm; (b) less than 140 mm; (c) greater than 160 mm.

a) Because the data have normal distribution, the probability $\Pr(X = x) = 0$, and also

$$\Pr(X = x) = \Pr(X = 150) = \mathbf{0}.$$

b) We have two ways to compute the probability that right-side clavicle length of males is less than 140 mm. The first way is to use cumulative distribution function for the normal distribution $N(151.74, 11^2)$, the second is to use cumulative distribution function for the standardized normal distribution $N(0, 1)$.

$$\Pr(X < 140) = \Pr(X \leq 140) = \mathsf{pnorm}(140, 151.74, 11) = \mathbf{0.1429}$$

$$\begin{aligned}\Pr(X < 140) = \Pr(X \leq 140) &= \Pr\left(\frac{X-\mu}{\sigma} \leq \frac{140-\mu}{\sigma}\right) = \\ &= \Pr\left(\frac{X-\mu}{\sigma} \leq \frac{140-151.74}{11}\right) = \\ &= \Pr(U \leq -1.067) = \Phi(-1.067) = \\ &= \mathsf{pnorm}(-1.067, 1, 0) = \mathbf{0.1429}.\end{aligned}$$

c) We have two ways to compute the probability that right-side clavicle length of males is greater than 160 mm. The first is to use cumulative distribution function for the normal distribution $N(151.74, 11^2)$, second way is to use cumulative distribution function for the standardized normal distribution $N(0, 1)$.

$$\Pr(X > 160) = 1 - \Pr(X \leq 160) = 1 - \mathsf{pnorm}(160, 151.74, 11) = \mathbf{0.2264}$$

$$\begin{aligned}\Pr(X > 160) = 1 - \Pr(X \leq 160) = 1 - &\Pr\left(\frac{X-\mu}{\sigma} \leq \frac{160-\mu}{\sigma}\right) = \\ &= \Pr\left(\frac{X-\mu}{\sigma} \leq \frac{160-151.74}{11}\right) = \\ &= \Pr(U \leq 0.7509) = \Phi(0.7509) = \\ &= \mathsf{pnorm}(0.7509, 1, 0) = \mathbf{0.2264}.\end{aligned}$$

Interpretation of the results:

1. Because the data have normal distribution, the probability, that right-side clavicle length of males is equal to 150 mm, is 0 %.
2. The probability, that right-side clavicle length of males is less than 140 mm, is 14.29 %.
3. The probability, that right-side clavicle length of males is greater than 160 mm, is 22.64 %.

6.2 An example from the worksheet used in the practical seminars

We have origin dataset of the craniometric data of skull width (in mm) of 13 females and 7 males from ancient egyptian population (Schmidt 1888, archival materials, adjusted for this example). We can see the data in the table 1.

ID	1	2	3	4	5	6	7	8	9	10	11	12	13
females	133	134	132	141	135	135	136	137	137	136	139	118	120
males	132	132	133	128	149	132	137						

Tab. 1: Skull width of ancient egyptian females and males.

On the significance level $\alpha = 0.05$ test the null hypothesis, that skull width of ancient egyptian females is the same as the skull width of ancient egyptian males. Draw the boxplot (see the figure 1).

Normality test

```
## [1] 0.004327
## [1] 0.024478
```

The p-value of normality test for skull width of females ($p = 0.0043$) is less than significance level $\alpha = 0.05$, so H_0 is rejected. The p-value of normality test for skull width of males ($p = 0.0245$) is also less than significance level $\alpha = 0.05$, so H_0 is rejected. Both the datasets have not normal distribution.

Because the datasets have not normal distribution, we have to use ~~parametric~~ / non-parametric method to test the null hypothesis, that skull width of females is equal to the skull width of males.

H_0 : ... $x_{0.5} - y_{0.5} = 0$ ...

H_1 : ... $x_{0.5} - y_{0.5} \neq 0$ ...

```
## [1] "U1=36; U2=55"
## [1] "W: (-infty ; 20>"
## [1] "CI: (-11.9999 ; 5)"
## [1] "p-value: 0.4487"
```

1. Rejected region approach:
The minimum of test statistics U_1, U_2 is ... *36* ..., the sample size of females is n_1= ... *13* ..., the sample size of males is n_2 = ... *7* ..., rejected region W is ... $(-\infty; 20\rangle$ ... . Because the value U_1 ... $36 \notin W$ ..., the null hypothesis of medians equality ... *is not rejected* ... on significance level $\alpha =$ ... *0.05* ... .

2. Confidence interval approach:
The confidence interval is ... $(-8; 5)$ ... . Because ... $c = 0 \in CI$ ..., the null hypothesis of medians equality ... *is not rejected* ... on significance level $\alpha =$ ... *0.05* ... .

3. The p-value approach:
Because the p-value = *0.4487* is *greater* than $\alpha =$ 0.05 the null hypothesis H_0 of medians equality *is not rejected* on the significance level α.

Boxplot

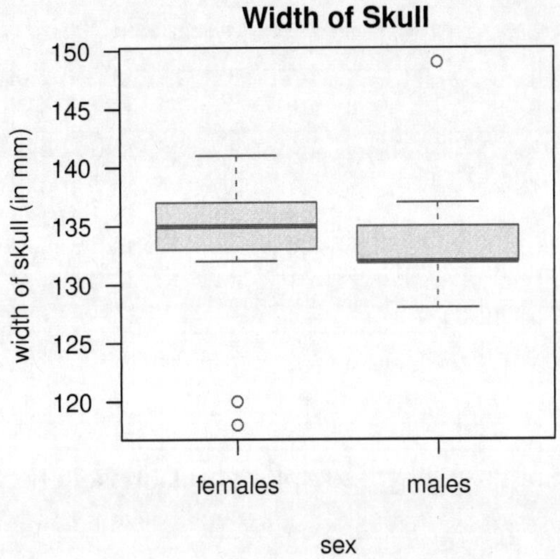

Fig. 1: Boxplot of skull width of ancient egyptian females and males.

Interpretation of the results:
There *is not* statistically significant difference between skull width of females and skull width of males. That means, the skull width of females *is the same* as skull width of males from ancient egyptian population.

6.3 An example from the homework

The dataset *102-nose-length-and-width.txt* includes data about nose length and nose width of the malaysian and peruvian population.

1. Compute the value of median, upper quartile and lower quartile of nose length of the peruvian population. (3 points)

2. Compute the value of arithmetic mean, standard deviation, skewness and kurtosis of nose length of the peruan population. (4 points)

3. The characteristics computed in part (1) and (2) insert to the table. (2 points)

4. Draw the histogram and boxplot of nose length of the peruvian population (see the Figure 2). (8 points)

5. Compute the appropriate type of correlation coefficient for association between nose length and nose width of the peruvian population. (2 points)

6. Draw a dotplot describing the relationship between nose length and nose width of the peruvian population. (4 points)

7. Be sure to sufficiently comment on all results and plots. (5 points)

```
##      median l.quartile h.quartile    mean st.dev skewness kurtosis
## 50%      51         48         53 50.4565 3.0267  -0.3134  -0.4115
```

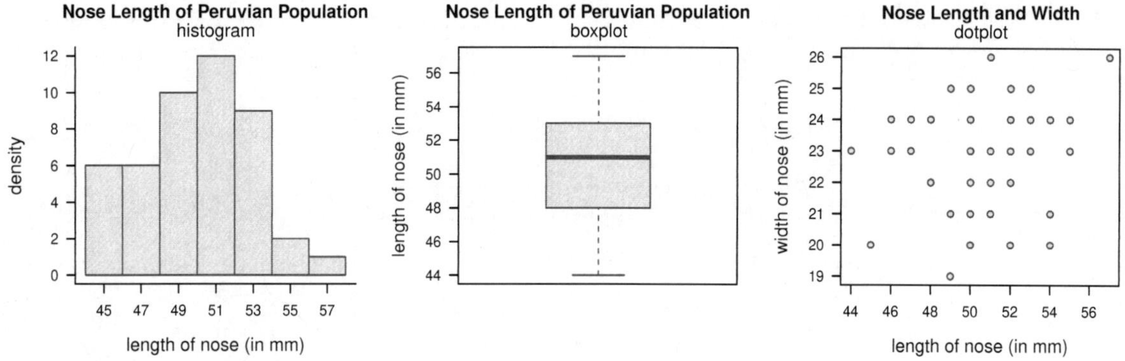

Fig. 2: Histogram and boxplot of nose length; dotplot of relationship between nose length and nose width of peruvian population.

```
## [1] "Pearson correlation coefficient: 0.1371"
```

7 Conclusion

Excellent anthropological research cannot do in this day and age without basic knowledge of statistics and data analysis. That is why the Applied statistics I course has been added to the list of mandatory courses of the anthropology Bachelor's study programme at the Masaryk University, while the anthropology Master's study programme was enriched by the mandatory Applied statistics II course. In these courses, students gradually get familiar with statistical methods suitable for analysis of anthropological data, along with the R statistical software. Learning statistics and the R software is, however, fraught with difficulty for students of non-mathematical study programmes. We therefore aim to ease the comprehension of statistical matter using a collection of solved problems and worksheets used in practical seminars. We also prefer friendly and open approach to every student. Our goal is to motivate students to use statistical methods and to accept the usefulness of statistics in their future careers. That is why students like to contact us after successfully completing the course to consult with us the statistical methods for analyzing datasets for their Bachelor's or Master's thesis.

References

[1] R CORE TEAM. R: *A Language and Environment for Statistical Computing. The R Project for Statistical Computing*, 2017 [cited 2017 Nov 28]. URL: https://www.R-project.org/

[2] BUDÍKOVÁ M., KRÁLOVÁ M., MAROŠ B.: *Průvodce základními statistickými metodami*, Praha, Grada, 2010, ISBN 978-80-210-7752-2, 272 s.

[3] BUDÍKOVÁ M., LERCH T., MIKOLÁŠ Š.: *Základní statistické metody*, Brno, Masarykova Univerzita, 2009, ISBN 978-80-210-3886-8, 170 s.

[4] KATINA S., KRÁLÍK M., HUPKOVÁ A.: *Aplikovaná štatistická inferencia I.*, Brno: MUNI Press, 2015, ISBN 978-80-210-7752-2, 306 s.

Current address

Bendová Veronika, Mgr.
Institute of Mathematics and Statistics
Faculty of Science
Masaryk University in Brno
Kotlářská 2, 611 37 Brno, Czech Republic
E-mail: bendova.veroonika@gmail.com

Budíková Marie, RNDr., Dr.
Institute of Mathematics and Statistics
Faculty of Science
Masaryk University in Brno
Kotlářská 2, 611 37 Brno, Czech Republic
E-mail: budikova@math.muni.cz

Slovak University of Technology in Bratislava
Faculty of Mechanical Engineering
17th Conference on Applied Mathematics **APLIMAT 2018**

Proceedings

DALÍ EXPERIENCE: SALVADOR DALÍ AND MATHS

BENVENUTI Silvia (IT)

Abstract. Dalí experience is the title of an exhibition held in winter 2016/17 at Palazzo Belloni, a prestigious historic building in Bologna. The exhibition collected about 200 works, from the "The Dalí Universe" Collection, one of the richest documentations of Dalí's artistic history, exhibited in an interactive and multimedia path, with the aim of inviting the visitor to an engaging and participatory experience. A good fortune for lovers of ... maths!

Keywords: golden section, topology, fourth dimension, catastrophe theory, art

Mathematics Subject Classification: Primary 00A66; Secondary 00A06, 00A69

1 Introduction

Maths?!
Maths, yes.
(And art, of course!)

As people generally ignore, but Aplimat lovers know very well, art has a lot to do with maths. Consider for instance Salvador Dalí, one of the most revolutionary artists of his time. Just take a look at his mustache in one of the photographs in Figure 1: don't you think our genius shows some interest in the subject considered by many the most arid of the universe?

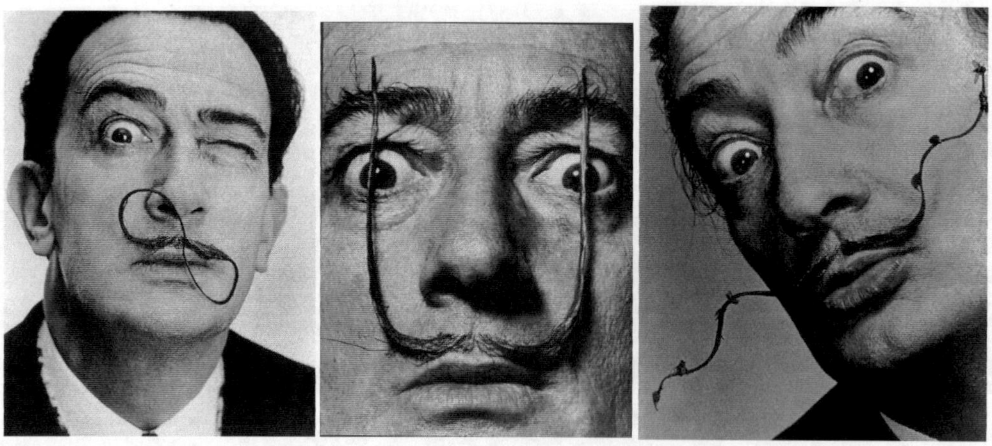

Fig. 1. Mathematical forms in the mustache of Salvador Dalí.

But Salvador Felipe Jacinto Dalí y Domènech (1904-89) goes a step further: analyzing his work is indeed a great way to support the thesis that mathematics can be, in addition to a good technical support for artists, also an exceptional creative stimulus. Starting from his statement, contained in his book *50 Secrets of Magic Craftsmanship*, according to which
«You must, especially as a young man, use geometry as a guide to symmetry in the composition of your works. I know that more or less romantic painters argue that these mathematical scaffolds kill the artist's inspiration, giving him too much to think and reflect. Do not hesitate for a moment to respond promptly that, on the contrary, it is just not to have to think and reflect on certain things that you use them».
Paraphrasing, Dalí suggests to the young artist a "recipe" for beauty: put close geometric constraints at the base of a picture, and then let your creativity flow, sure that the result will be aestetically harmonic and nice to see. This recipe is made explicit by the preparatory sketches of Dalí, one among all that of the *Atomic Leda* (figure 2). Comparing the picture with its sketch, we realize that in the picture nothing is left to chance: the right foot of Gala-Leda ends in one of the five points of the Pythagorean star described by the diagonals of a pentagon inscribed in the circumference whose diameter delimits the width of the picture; the circumference is tangent to the bottom of the stool on which Leda is sitting, and the woman's head is entirely contained in the center point of the star. «Some days I have the impression that I will die of an overdose of satisfaction» - he used to say. He must have thought that while signing that picture: how to contradict him?

Fig. 2. Atomic Leda (1949 - left) and its preparatory study (right).

2 Golden section

Leda is one of the examples of paintings based on the concept of *golden section*, which underlies the construction of the pentagon: in a regular pentagon, in fact, the relationship between each diagonal and the side is the same as this number, Φ, known in the centuries as *gold number*, *golden ratio*, *golden section*, *golden mean* (look here https://www.youtube.com/watch?v=kkGeOWYOFoA to learn more). All artists have been fascinated by this number, and Dalí is conformist in this respect: golden rectangles and related objects pervade his work even when it does not seem.
It is quite clear, for instance, that the golden rectangle governs the proportions of *The Last Supper* (figure 3): the table is positioned so as to divide the total height in two parts that are in aureus ratio; the disciples to the right and left of Christ divide the picture into two vertical

strips of amplitudes in the golden ratio one with the other. In the background we can see the dodecahedron, which with its twelve pentagonal faces is a living incarnation of Φ .

Fig. 3. Golden Section in The Last Supper (1955). The ratio between the blue and white segments, both vertical and horizontal, is the golden one.

Equally explicit is the presence of the number Φ in *Giant Flying Mocca Cup with an Inexplicable Five Metre Appendage* (1944 - figure 4), of which we particularly like the title, with reference to the appendage that only serves to adjust the dimensions of the picture.

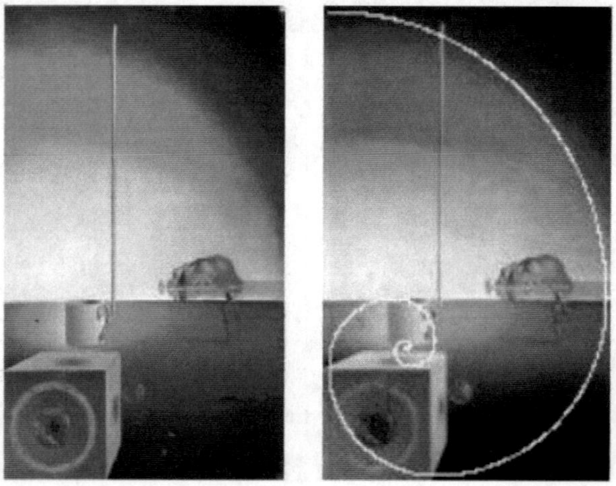

Fig. 4. Golden Spiral in *Giant Flying Mocca Cup with an Inexplicable Five Metres Appendage*. As shown in the figure on the right, the length of the pendant is given by the request that it ends at the intersection point between the vertical and the golden spiral centered in the cup and passing through the top left corner.

3 Topology

That of the golden section is a too popular maths for a poliedric artist such as Dalí, which in fact extends its interest in much more highbrow maths subjects.
In the Dalí Experience exposition in Bologna, for instance, one could appreciate the Dalínian soft watches, made famous by *The persistence of memory* (Fig. 6 - left). This is a representation of reality that we can define as a *topological* one, with reference to a relatively modern discipline, topology. This kind of geometry is a bit different from the one we study at school, where all objects have their own size, with rigid and crushproof contours. A topologist

thinks instead of objects as if they were made of pongo (or Didò, or plasticine, depending on his birth year): two objects are topologically equivalents if, thinking of them as made of plasticine, we can deform one into the other without cutting or gluing. A dish, for instance, is equivalent both to a bowl and a glass, while a cup with a handle is not the same as the previous objects, and a two-handled cup is not equivalent to either the first or the second. The cup with a handle, on the other hand, is equivalent to a lifebuoy, or to a donut (see https://commons.wikimedia.org/wiki/File:Mug_and_Torus_morph.gif)

Dalí's watches "trickle" as if they were made of plasticine, in a reality in which objects have changing contours, but their memory persists and rememebrs somehow the shape. Most of the works of Dalí's last period are topological even in the title: think about *Topological Contortion of a Female Figure* (1983, Figure 5 - left), and *Topological contortion of a female figure becoming a cello* (1983, fig. 5 - right)

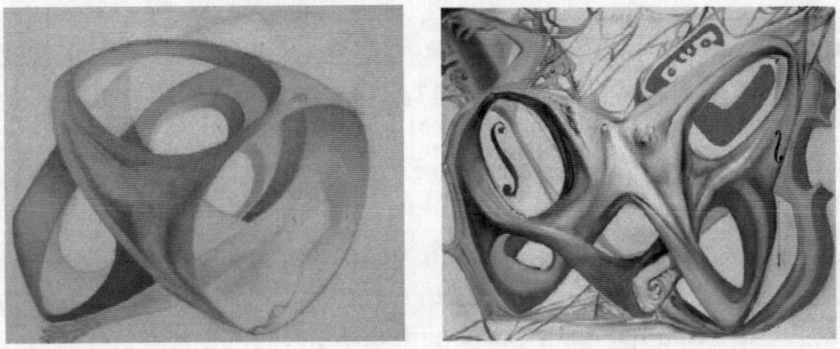

Fig. 5. *Topological Contortion of a Female Figure* (left) and *Topological contortion of a female figure becoming a cello* (right).

It would be embarrassing, though mathematically significant, to get into the details of the topological genus of the female figure represented in these two paintings, and perhaps some reader would seek help from Freud. That is why we self-censor and take a step forward, from topology to quantum mechanics. This will lead us to leave Freud for Heisenberg, and the soft clocks for their disintegration: as Dalí writes in his *Manifesto antimaterico*, «during the surrealist period I wanted to create the iconography of the inner world and the wonderful world conceived by my father Freud. Today, however, the outer world and the physical world have surpassed that of psychology. Today my father is Dr. Heisenberg. » As a consequence, the *persistence* of 1931 *disintegrates* in 1954, decomposing into blocks who summarize the author's recognition of the new frontiers of science (see Fig. 6 - right).

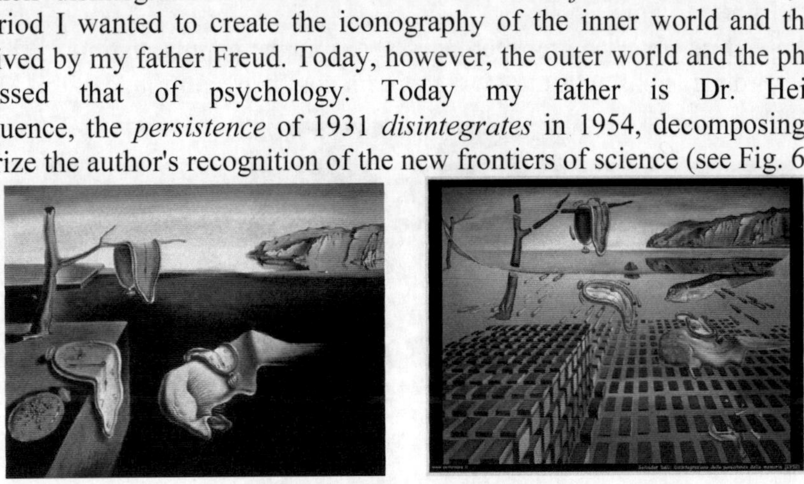

Fig. 6. *The Persistence of Memory* (l) vs *The Disintegration of The Persistence of Memory* (r).

Before leaving the concept of topology, if you have been so lucky to visit it, think about the Dalí experience exhibition and reflect on the modalities of the exhibition itself: we can define it in a certain sense "intrinsically topological" because it "trickles" on the city, in paths that pour out from Palazzo Belloni to inundate all Bologna, in a deliberately interactive experience that softens boundaries, deforming and contaminating. Something that definitely Dalí would have liked a lot.

4 Fourth dimension

But let's go further: if you talk about Dalí's relationship with mathematics, you can not help but mention his passion for the fourth dimension, embodied in the beautiful representation of the hypercube that serves as a cross to the Christ of *Corpus Hypercubus* (1954). Just as the cross represents the planar development of a cube, the figure to which the dalinian Christ is crucified represents the three-dimensional development of its analogue in dimension 4, which mathematicians call *hypercube*: the artist uses this concept to communicate the idea of the transcendence of Christ, who lives in an higher dimension than the ones the human beings are able to conceive.

Let's explain something more: imagine a square. If we "cut" the two sides converging to one of its vertices, we can "open" the square up to extend it along a straight line, which occupies 4 consecutive segments: this figure represents the 1-dimensional development of a square. It lives in the plan, in 2 dimensions. Now let's take a cube, for instance a cubic box of cardboard. If we cut the box over three consecutive sides, and then carefully cut it along other sides (which ones?!), we can, as before with the square, open the cube and spread it on a table: the cross-shaped figure we get represents the 2-dimensional development of the cube, which lives in the 3-dimensional space instead. Finally, let's take an hypercube: none of us can imagine it (at least we think, with artists is better not to be too categorical), because such an object lives in 4 dimensions, where our perception does not arrive. However, we can think of doing the same as before with the square and then with the cube, that is "cutting" the hypercube ... along what? .... Think about that, and you will conclude that it is along a set of squares that bound cubes, which in turn bound the hypercube. If you cut correctly, you can "open" your hypercube and "spread it" in the ordinary space, that is in the 3-dimensional space, where we all can see it. What we will see is actually the 3-dimensional development of the hypercube, which lives in a 4-dimensional space, but is so kind to visit Dalí in a smaller dimension so that he can draw it!

See more, play and get much more references about maths popularization of the concept of fourth and higher dimensions at http://www.loopmm.com/benvenuti/index.html [20].

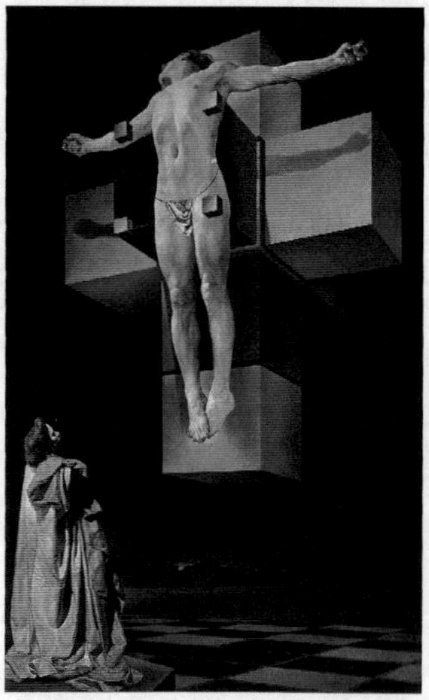

Fig. 7. Fourth Dimension in Dalí: Corpus hypercubus.

5 Chatastrophe theory

But let's go further: while visiting the Bologna exhibition, you will surely dream to rest a second on the object that looks so good in one of the central rooms (see fig. 8). Definitely, it is a sofa. But certainly not a usual one, with that color, vivid red, and with that peculiar shape.

Fig. 8. Mae West sofa (1935?).

As Salvador Dalí itself designed it, it cannot be a trivial object. In your living room it would be perfect, next to the window door, though no placement can match the one our man conceived in his Figueres home-museum, to draw a smile in the serious face of the actress Mae West. But the most surprising thing is that this object, with its comfortable sensuality, is not inspired (perhaps) by the fleshy lips of Gala or those of Mae West itself, but rather by the cryptic sensuality of ... a mathematical concept. Concept, besides, which is not at all a trivial one: that of "singularity", or "bifurcation", or "catastrophe". Dalí heard about that probably through the mathematician René Thom, his friend and Fields medal winner, author of the book *Structural Stability and Morphogenesis*, the Bible of Chatastrophs.
Let us consider a mathematical object, for example the curve in Figure 9a, described by an equation containing one or more parameters. By making them change gradually, the parameters can be used as "knobs" to adjust the position and shape of the curve: sometimes small variations cause small changes, as can be seen from the situation in fig. 9a to that of Fig. 9b; in other cases, minimal variations cause the transition between two qualitatively different situations, such as in the passage from the curve made by a single piece of figure 9b to the one composed of two pieces shown infigure 9d. We then say that a bifurcation, or a catastrophe, occurred. Specifically, the catastrophe curve is that of figure 9c, from which any minimal variation in the parameter leads to the two qualitatively different situations depicted in the adjacent figures. Thom, in his book, lists a finite number of "elementary catastrophes", with evocative names such as swallowtail, butterfly, fold, cusp, etc, which generate all the others, including the lips we met above.

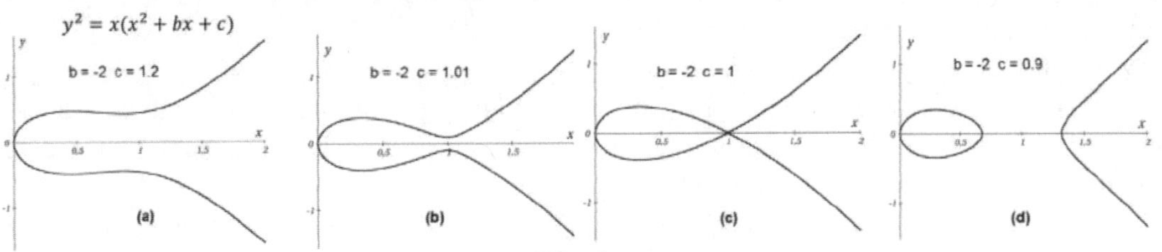

Fig. 9

Well: if Thom's book stays on Dalí's night table starting from 1984, while Mae West's sofa is designed in 1935, it's really difficult for the latter to be really inspired by the former - although, being Dalí a visionary man, nobody knows. Imagine, however, Dalí's satisfaction in seeing his sofa emerging from Thom's original pages as a «section of the continuous surface of two cusps contained in the image of a swallowtail catastrophe». It looks like a tongue-twister, in fact. But figure 10 explains it better.

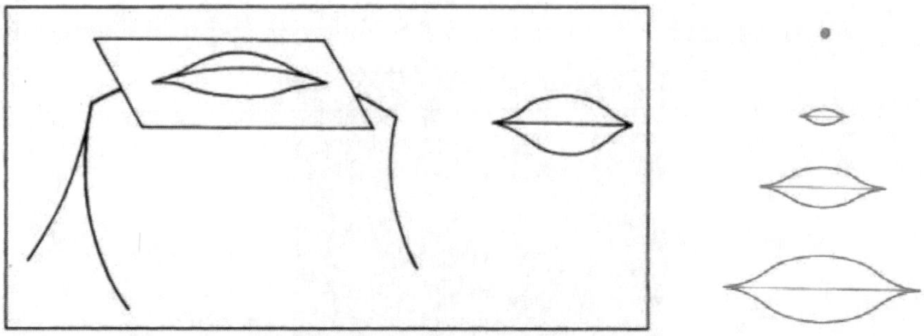

Fig. 10. Formation of the lips singularity (Tito Tonietti - left); same singularity in the book of Thom (right).

If that of the lips singularity is an *a posteriori* interpretation, there is no doubt that Dalí explicitly relies on Thom's theory in his last works: the catastrophe known as the swallowtail, for instance, is the main character of both his last picture , *Swallowtail and cello* (1983 - fig. 11, left) and of *Topological abduction of Europe. Tribute to René Thom* (1983 - fig. 11, right).

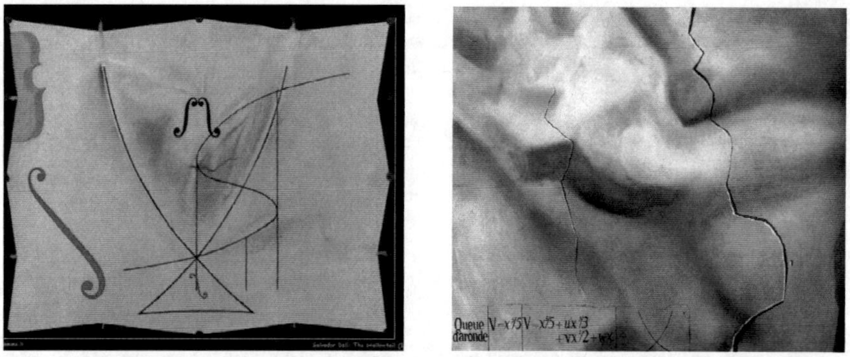

Fig. 11. *Swallowtail and cello* (left) and *Topological abduction of Europe.* (right).

After all, the artist defined the Catastrophe Theory as «the most beautiful aesthetic theory in the world». And it is not a matter of a senile dementia: as we are proving, Dalí has been deeply fascinated by maths throughout all his life, and his works deeply reflect its passion.

6 But there's a lot more...

But there is a lot more: Salvador Dalí not only understands the importance of mathematics, so much to use it deliberately in his paintings; moreover, he studied it a lot, thanks also to his friendship with important mathematicians like René Thom, Matila Ghyka, Thomas Banchoff, Martin Gardner. He was also irresistibly attracted by many other sciences: he personally knew Einstein, studied theory of relativity and quantum physics, his home-museum is full of representations of the model of the hydrogen atom; his library contained hundreds of science books, with its annotations on the margin to witness fervent reading; he was until his death subscribed to many scientific journals; he did not even miss one of the six major lectures, of important physicists, mathematicians, astrophysics and chemists, of the symposium *Cultura y ciencia: determinismo e liberdad*, celebrated in 1985 in his museum; one of his paintings, in tribute to Crick and Watson, is titled *Galacidalacidesoxiribonucleicacid* ... do you need other evidence of his fascination for ANY science?

One mention would be due to a thousand other mathematical themes, including anamorphisms, stereoscopy, optical illusions and holograms. But the range of math arguments touched by Dalí is so vast that it is unavoidable to operate a selection, democratically carried out according to the preferences of the writer.

Hence, let me conclude with a topic perhaps more neglected, but that you have noticed for sure: the ubiquity of the egg. The eggs in Dalí can be found everywhere: the wall of his house in Figueres (fig. 14) have monumental spherical and ovoidal decorations; eggs appear in many of his paintings and sculptures; egg-shaped are the faces of many of his statues. Dalí apparently considered the egg to be so fundamental to use them as a cabin for his *Ovocipede*, a peculiar means of locomotion that in his times was perhaps not properly appreciated, but constitutes today the joy of thousands of children (including the author's) and the despair of thousands of moms (including myself) when we meet it in the playgrounds (Figure 12). Both the dalinian and the modern ovocipede have a propulsion system based on the driver's movement: essentially the driver runs like a hamster making it roll, and then releasing sweaty and soaking wet, but extremely happy (at least until he is 8 years old ...).

Fig. 12. Dalinian (left) and modern (right) ovocipede.

For a topologist, an egg is completely indistinguishable from a sphere (and also from a cube, or a dodecahedron); moreover, an ostrich egg is topologically equivalent to one of quail, as well as any intermediate egg size. If, however, we forget the topology and consider the objects in accordance with their metric properties, then everything changes: the egg and the sphere are distinct objects, and the ostrich egg is metrically different from the hen one. What, however, combines eggs of different shapes is that all of them are obtained by turning a flat curve around its axis: if we want to get a sphere we will rotate a circle, if we need an ellipsoid we will use an ellipse, if we really want an egg ... well, it gets a bit more complicated. Go to the fridge and take any egg: you will notice that while the sections made with planes perpendicular to the axis are all circumferences of different sizes (which tells us that the egg it is a rotation surface), those obtained by cutting with planes containing the vertical axis are neither circumferences nor ellipses, but strange curves in which a half is more paunchy than the other. Obviously, a mathematician can not cook an egg while ignoring the equation of the curve that generates it by rotation: hence here come the ovals of Cassini, Descartes and

Kepler, as well as the egg of Granville, that of Hügelschäffer and that of Moss, to end up with the double egg.

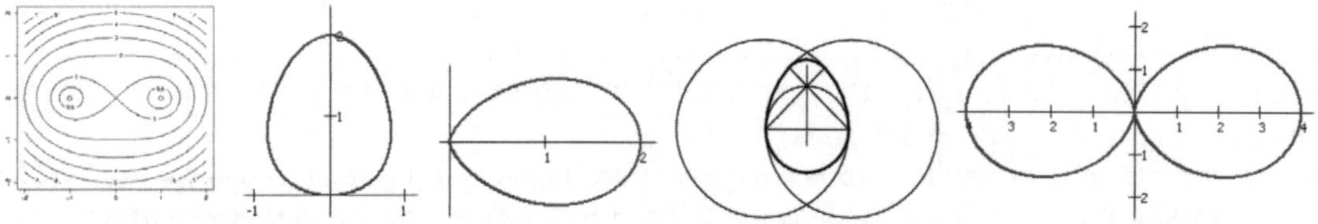

Fig. 13. Mathematical Eggs: Cassini, Granville, Keplero, Moss, Double Eggs.

As we said, the egg is ubiquitous in Dalí: he draws and sculpts it in all the possible shapes and sizes, and it comes spontaneously to wonder why. One possible explanation is that the egg is in fact also present in many classical paintings as a symbol of resurrection, as it is intrinsically linked to the birth of a new life: we find it for example in the famous *Pala di Montefeltro*, also called *Madonna of the Egg*, by Piero della Francesca, that Dalí loved so much to reproduce it in his peculiar way in the famous *Madonna of Port Lligat* (1950). It is well-known that Dalí had a deep fascination for classics, who he reinterpreted in unespected ways, for example by making a surrealist landscape on Michelangelo's Giuliano dei Medici's cheek, or strategically perforating his pictorial reproduction of the Pietà (see http://www.mostradalipisa.it).
«Begin by drawing and painting like the ancient masters», he used to say with regard to that. «After that you can do what you want: everyone will respect you».
A second reason for the persistence of the egg in Dalí's work is probably his "intrauterine egg theory": he was convinced to remember perfectly of his life in the maternal uterus, which he transforms into a kind of gourmet paradise where *huevos fritos* dominate.
In a more prosaic way we can also see in that ubiquity a manifestation of its Catalan spirit: in Barcelona and the surrounding area, in fact, the *ou* is eaten in all restaurants, from the greasy spoon to the starred restaurant, *ferrat* and accompanied by potatoes and *pernil de gla* in its working class' version, or "cooked at low temperature on potato parmentier and black butifarra foam" from 4 stars up. Besides, it is certain that Dalí wanted to pay homage to his land by immortalizing some of its culinary excellence: just think of the Empordà loaf with which he rakes the walls of his museum (fig. 14), that with the *ou ferrat* is absolutely delightful.

Fig. 14. Eggs and loaves in the walls of the Teatro Museo di Figueres.

Moreover, for Dalí, who at six years wanted to become a (female) chef, cooking is a fundamental pleasure, so much as to push him to write and illustrate even a recipes book, entiltled of course *Gala Recipes*, not precisely dietetic ones: «If you are a follower of these calorie weighters who transform the joys of a meal into a punishment, close the book», he writes in the preface. At this point it is impossible, even if blasphemous, to resist the temptation of demystifying our genius by imagining it in slippers in front of a half-finished canvas, while Gala with the apron mess about at the stove and, when the eggs are cooked, calls him: "Cariiiinyo, lunch is ready when you aaaaareeeee, run that it gets cooold!"

References

[1] BANCHOFF, T. , lezione magistrale, *La quarta dimensione e Salvador Dalí*, Festival della Matematica di Roma, 2008.

[2] BENVENUTI, S., *La sregolatezza regolata di Salvador Dalí.* In: AA. VV.. Rivista Mate. Matematica da zero a infinito, vol VI. p. 66-69, Milano, Centauria, 2016.

[3] BENVENUTI, S., *Catastroficamente creativo: Salvador Dalí e la matematica.* In: AA.VV. (a cura di): Claudio Mazzanti, Dalí Experience. p. 56-65, Bologna:con-fine ed., 2016

[4] BENVENUTI, S., *Genio e regolatezza: le passioni matematiche di Salvador Dalí*, http://maddmaths.simai.eu/divulgazione/salvador-Dalí/, 2017.

[5] BENVENUTI, S., *Catastrophically creative: Salvador Dalí and Maths*, http://mathematics-in-europe.eu/?p=966, 2017.

[6] BISCHI, G. I., *Biforcazioni,* Lettera Pristem 100, 2016.

[7] BLASCO, F., *Matemáticas en Dalí*, Matematicalia vol 7, n. 4, 2011.

[8] CEPEDA FUENTES, M., *Il surrealismo in cucina tra il pane e l'uovo*, Il Leone Verde, 2004.

[9] CHAPARRO, L., *La obsesión de Dalí por una musa llamada ciencia*, Sinc, 2013.

[10] DALÍ, S., *Fifty Secrets of Magic Crafts,* 1948.

[11] DE ROSE, L., LORENZI M.G., FRANCAVIGLIA M., *From the "cosmic egg" to the big bang: a short excursus on the origin of the universe, between history, mathematics and art,* Aplimat v. 4 n. 4 , 2011.

[12] GARDNER, M., *Anamorphic Art*, in *Time, Travel and Other Mathematical Bewilderments*, W.H. Freeman, 1988.

[13] IORFIDA, V., FRANCAVIGLIA, M., LORENZI, M.G., *Algebraic varieties in the art of the nineteenth century,* Aplimat v. 4 n. 4 , 2011.

[14] PÉREZ GÓMEZ, R., *¿Paranoia o topología trascendental? Salvador Dalí, 100 años*, Gaceta de la RSME, vol 7.3, 2004.

[15] RUIZ, C., *Salvador Dalí y la ciencia, más allá de una simple curiosidad*, Centro de Estudios Dalínianos. Fundación Gala-Salvador Dalí, Pasaje a la Ciencia pp. 4-13, 2010.

[16] SAMOILA, G.S., *The armonic geometry of art,* Aplimat v. 3 n. 1 , 2010.

[17] THOM, R., *Stabilità strutturale e morfogenesi*, Einaudi, 1980.

[18] TONIETTI, T., *Catastrofi. Il preludio alla complessità*, Bari, Dedalo, 2002.

[20] http://www.loopmm.com/benvenuti/index.html, animation by Gian Marco Todesco, texts by Silvia Benvenuti, graphics by Lavinia Cenci, Guardian Core.

Current address

Benvenuti Silvia, PhD.
Sezione di Matematica, Scuola di Scienze e Tecnologie
Università di Camerino
via Madonna delle Carceri, 62032, Camerino (MC), Italy
E-mail: silvia.benvenuti@unicam.it

Slovak University of Technology in Bratislava
Faculty of Mechanical Engineering
17th Conference on Applied Mathematics **APLIMAT 2018**

Proceedings

WAGE LEVELS BY REGIONS IN THE CZECH REPUBLIC AND PPP-ADJUSTED WAGES IN SELECTED EUROPEAN AND NON-EUROPEAN COUNTRIES

BÍLKOVÁ Diana (CZ)

Abstract. The present paper focuses on a comparison of wage levels of the Czech Republic regions. Similar wage-level clusters were created using the methods of cluster analysis. Three regions with the highest and lowest wage levels, respectively, were selected. The wage distribution model for the above regions was created to allow for the comparison of wage development over the past seven years, three-parameter lognormal curves representing the basis of the theoretical wage distribution.

Keywords: cluster analysis, farthest neighbour method, Euclidean distance metric, lognormal curves, maximum likelihood method

Mathematics Subject Classification: Primary 60E05, 62H12; 62H30; Secondary 91C20

1 Introduction

Income levels of the population have been constantly researched by economists in the developed countries mainly due to their connection with the living standards of the population. Knowledge of the wage distribution and its comparison from various socio-economic and time-spatial aspects is a precondition for the assessment of living standards, social security and equality in the division of material values produced by the society. Statistical analysis of the wage distribution also forms the basis for government social policy, taxation and budgetary decisions.

The primary aim of this paper is to create the clusters of Czech Republic regions and those of selected European countries based on wage level similarities. The farthest neighbour clustering method and the Euclidean distance metric were used for the creation of those clusters. The key objective lies in modelling the wage distribution in individual regions so that its development since 2009 can be traced. Three-parameter lognormal curves represent the basis of the model distribution with parameters estimated employing the maximum likelihood method. The accuracy of the models obtained is assessed applying Akaike and

Bayesian Information criteria. (Such models for particular European and some other countries could not have been constructed due to a different type of data.)

The comparison of wage levels in selected European and non-European countries is another principal aim of the study. The annual average wage in USD converted to purchasing power parity was used for this purpose since it represents real living expenses and price levels in different countries.

Alongside the above research agenda, the present paper examines the differences between male and female wage levels in the Czech Republic and other selected countries.

2 Database

Data for the present study are collected from the official website of the Czech Statistical Office (CSO) and Eurostat and OECD data files.

The CSO database contains the total wage distribution for the period 2009–2015 covering all employees in the Czech Republic broken down by regions. In individual regions, employees were divided into groups of men and women and the gender wage gap was examined. Annual data are related to gross monthly nominal wages in the respective years, the average (median) wage, for instance, representing average (median) gross monthly wage over the year.

There are also data in the form of the interval frequency distribution with uneven and extreme open intervals. Neither more detailed nor individual data have been currently available. Since only nominal wage data are provided by the CSO, the obtained average and median nominal wages had to be converted to average and median real wages using the CSO-reported inflation rate data.

Having calculated the differences in average and median wages (both nominal and real) between men and women, the trends in these differences were identified, the gross monthly nominal and real wage along with the male-female differences in average and median wage (nominal and real) being research variables. Data on average gross monthly (nominal) wages in the selected European countries are gathered from the Eurostat website.

Only the data on nominal wages having been available, inflation rates had to be used for the conversion to a real wage that reflects purchasing power allowing for a comparison of the wage development without inflation effects in the research period. The rate of inflation is derived from the consumer price index (CPI), based on the Laspeyres price index. The real wage was calculated using the real wage index, the nominal wage index being divided by the CPI (living cost index). The data were processed utilizing SAS and Statgraphics statistical programme packages and Microsoft Excel spreadsheets.

The late 1990s marked the culmination of Czech economy transformation to the market economy. Enterprises were being privatized, industries restructured and prices liberalized. These processes necessitated a change in the methodology of statistical analysis, the combination of exhaustive and sample surveys becoming more appropriate since the number of small businesses was growing considerably. A high rate of inflation was recorded, namely 8.5–10.7 per cent between 1996 and 1998. In 2003, on the other hand, the inflation rate fell as low as 0.1 per cent, rising to 6.3 per cent in 2008, when the recession triggered a sharp slowdown in real wage growth. Currently, the rate of inflation remains at a very low level (0.3 and 0.7 per cent in 2015 and 2016, respectively).

The research data include wages and salaries paid to employees for work performed in the private (business) and public (state budget, non-business) sectors, respectively. In terms of the data presented on the CSO website, “wages” cover remuneration for work done in both the sectors.

Eurostat and OECD databases comprise the wage data for the period 2000–2015 containing all employees in selected European and non-European countries. However, the nature of the data is somewhat different from that of the CSO, and moreover, total wage distributions are not available. After the conversion into purchasing power parity, US dollar wages were used for comparison of wage levels in countries within and outside of Europe, the adjusted wages taking into account different living costs and price levels.

An impact of the 2008 financial crisis and subsequent economic downturn was clearly noticeable in the Czech Republic. Having recorded a decline of 4.8 per cent in 2009, the economy revived slightly over the following two years, the GDP growth reaching maximum of 2.3 per cent. However, since firms did not have enough time to recover and boost their investments, the Czech economy declined further by 0.8 and 0.7 per cent in 2012 and 2013, respectively. Despite the 2010 warnings of independent analysts, a fall into a protracted recession was not prevented. (This double recession is considered even more severe than the 1997–1998 slump. Having fallen by less than one percentage point of GDP, the Czech economy reported a GDP growth of 1.4 per cent in 1999.) In 2014 and 2015, however, we witnessed a GDP growth of 2 and 4.3 per cent, respectively. For due reasons, the present study spans just the seven-year period 2009–2015.

3 Theory and Methods

3.1 Cluster Analysis

Cluster analysis was used to divide the Czech regions into relatively homogeneous groups according to their respective gross monthly wage levels. Multivariate data analysis, which is often done to process economic data (see, e.g. Lukáš Malec [8]), may include other approaches to statistical data analysis, namely that of canonical correlation. (Particular aspects of cluster analysis are dealt with Elena Makhalova, and Iva Pecáková [7] or Hana Řezanková, and Tomáš Löster [9].)

Multidimensional observations can be used when classifying a set of objects into several relatively homogeneous clusters. We have a data matrix $\boldsymbol{X}$ of n X p type, where n is the number of objects and p is the number of variables. Assuming various decompositions $S^{(k)}$ of the set of n objects into k clusters, we look for the most appropriate decompositions. The aim is to find the objects within certain clusters that are as similar as possible to those from other clusters. Only decompositions with disjunctive clusters and tasks with a specified number of classes are conceded.

3.1.1 Criteria for Assessing the Quality of Decomposition

The general task is to assess to what extent the cluster analysis aim has been achieved in a given situation, while applying a specific algorithm. Several criteria – decomposition functions – are proposed for this purpose. The most frequently used ones exhibit the following characteristics. They are the matrices of internal cluster variance

$$\boldsymbol{E} = \sum_{h=1}^{k}\sum_{i=1}^{n_h}(\boldsymbol{x}_{hi} - \overline{\boldsymbol{x}}_h)\cdot(\boldsymbol{x}_{hi} - \overline{\boldsymbol{x}}_h)' \qquad (1)$$

and between-cluster variance

$$\boldsymbol{B} = \sum_{h=1}^{k} n_h\cdot(\overline{\boldsymbol{x}}_h - \overline{\boldsymbol{x}}\,)\cdot(\overline{\boldsymbol{x}}_h - \overline{\boldsymbol{x}}\,)', \qquad (2)$$

whose sum is the matrix of total variation

$$\boldsymbol{T} = \sum_{h=1}^{k}\sum_{i=1}^{n_h}(\boldsymbol{x}_{hi} - \overline{\boldsymbol{x}}\,)\cdot(\boldsymbol{x}_{hi} - \overline{\boldsymbol{x}}\,)'. \qquad (3)$$

There are vectors of the observations for the ith object and h^{th} cluster $\boldsymbol{x}_{hi}$, the averages for the hth cluster $\overline{\boldsymbol{x}}_h$ and those for the total set $\overline{\boldsymbol{x}}$. There are p^{th}-membered vectors, $\boldsymbol{E}$, $\boldsymbol{B}$ and $\boldsymbol{T}$ being symmetric square matrices of the pth order. The principal aim, consisting in the creation of mutually distant compact clusters, is fulfilled by reaching the minimum of the total sum of the deviation squares of all values of corresponding cluster averages

$$C_1 = \mathrm{st}\,\boldsymbol{E} = \sum_{h=1}^{k}\sum_{i=1}^{n_h}\sum_{j=1}^{p}(x_{hij} - \overline{x}_{hj})^2, \qquad (4)$$

i.e. the Ward criterion. Since the st $\boldsymbol{T}$ is the same for all decompositions, the minimization of the st $\boldsymbol{E}$ means the same as that of the st $\boldsymbol{B}$. In order to become independent on the used units of measurement (or, more generally, the invariance to the linear transformations), it is recommended to minimize the determinant of the matrix of the internal cluster variance

$$C_2 = |\boldsymbol{E}|$$

or to maximize the trace criterion

$$C_3 = \mathrm{st}\,(\boldsymbol{BE}^{-1}) \text{ or else } C_4 = \mathrm{st}\,(\boldsymbol{BT}^{-1}).$$

The criteria mentioned above are employed not only retrospectively to assess the decomposition quality accomplished, changes in criterion values also guiding the creation of clusters. Since the criteria ultimately reach the limits (C_1 and C_2 the minimum, C_3 and C_4 the maximum) at $k = n$, it is necessary to find the extreme of the purpose function that properly includes the loss following from the growth in the number of clusters. The Ward criterion, for instance, is proposed to move towards the minimization of the quantity

$$Z_1 = C_1 + z \cdot k \ , \tag{5}$$

where constant z represents the loss resulting from an increase in the number of clusters by one.

3.1.2 Distance and Similarity of Objects

Having selected the variables characterizing the properties of the clustered objects and found their values, we decided on the method of the evaluation of distance or similarity of objects, the calculation of appropriate measures for all pairs of objects often being the initial stage of clustering algorithm implementation. The symmetric square matrix of n X n type has zeros or ones on the diagonal, depending on whether it is the matrix of distance $\boldsymbol{D}$ measures or that of similarity $\boldsymbol{A}$ measures, respectively.

Let us now focus on measuring the distance of the objects described by quantitative variables. The Hemming distance can be used when individual variables are roughly on the same level or at least expressed in the same units of measurement

$$D_{\mathrm{H}}(x_i, x_{i'}) = \sum_{j=1}^{p} \left| x_{ij} - x_{i'j} \right| . \tag{6}$$

The Euclidean distance can be applied in the same case

$$D_{\mathrm{E}}(x_i, x_{i'}) = \sqrt{\sum_{j=1}^{p} (x_{ij} - x_{i'j})^2} \tag{7}$$

as well as the Chebyshev distance

$$D_{\mathrm{C}}(x_i, x_{i'}) = \max_j \left| x_{ij} - x_{i'j} \right| . \tag{8}$$

All the above mentioned measurements have some common drawbacks – the dependence on the used measuring units that sometimes hinders the meaningful acquisition of any sum for different variables and the fact that if the variables are considered in sum with the same weights, the strongly correlated variables have a disproportionately large effect on the outcome. The starting point is the transformation of variables. The adverse effect of the measuring units can be removed by dividing all the values by the balancing factor, which can be presented with the corresponding average $\bar{x}_j$, standard deviation s_j or the range after deletion of extremes

$$\max_i x_{ij} - \min_i x_{ij} .$$

Particular variables can be also assigned more weight – having decided subjectively or on the basis of relevant information – their values then appearing in the formulas for the calculation of distance.

Other measurements of distance and similarity of objects for numerical, ordinal, nominal and alternative variables are described in the professional literature. When dealing with variables of a different type, the Lance-Williams distance is recommended

$$D_{LW}(x_i, x_{i'}) = \frac{\sum_{j=1}^{p} |x_{ij} - x_{i'j}|}{\sum_{j=1}^{p} (x_{ij} + x_{i'j})}. \tag{9}$$

3.1.3 Algorithm for the Creation of Hierarchical Sequence of Decompositions

The creation of a hierarchical sequence of decompositions belongs to the most widely used techniques applied in the cluster analysis, occurring sequentially in the following steps:

1) $\boldsymbol{D}$ matrix calculation of appropriate measurements of distances;
2) the start of the decomposition process $\boldsymbol{S}^{(n)}$ from n clusters, each of them containing one object;
3) the assessment of the symmetric matrix $\boldsymbol{D}$ (a lower or upper triangle), finding two clusters (the hth and h'^{th} ones) whose distance $D_{hh'}$ is minimal;
4) the connection of the hth and h'^{th} clusters into a new g^{th} cluster, the replacement of the hth and h'^{th} row and column in the matrix $\boldsymbol{D}$ with those of the new cluster, the order of the matrix being reduced by one;
5) renumbering of the order of the cycle $l = 1, 2, \ldots, n - 1$, the identification of the connected objects h, h' and the level of the connection $d_l = D_{hh'}$;
6) returning to step (3) if the creation of decompositions has not been completed by connecting all objects into a single cluster $S^{(1)}$.

A divisive hierarchical procedure, contrary to the agglomerative hierarchical one, is less-used, starting from a single cluster $S^{(1)}$, splitting one of the clusters into two in each step and obtaining $S^{(n)}$ at the end of the process. The results of hierarchical cluster procedures can be effectively displayed in the form of a graphical tree dendrogram.

Given the choice of variables $x_1, x_2, \ldots, x_p$ and the matrix of distances $\boldsymbol{D}$, the results of applying the described algorithm vary according to the way the distance between clusters is evaluated.

Nearest Neighbour Method

Within the nearest neighbour method, both clusters, whose connection is considered, are represented by objects that are the closest to each other. The $D_{hh'}$ distance between the hth and h'^{th} clusters therefore represents the minimum of all $q = n_h n_{h'}$ distances between their objects, the procedure of the third phase of the above algorithm thus being specified. In the fourth step, the hth and h'^{th} rows and columns in the distance matrix are replaced with the new g^{th} cluster's row and column of distances. In the lth cycle, total $n - l - 1$ distances determined by

$$D_{gg'} = \min (D_{g'h}, D_{g'h'})$$ (10)

can be written.

If the way of evaluation of the proximity or similarity of clusters is given, which also determines the conversion of the distance matrix in each cycle, the above algorithm allows for the creation of a hierarchical sequence of decompositions and construction of the dendrogram.

When using this method, even considerably distant objects can get together in the same cluster if a large number of other objects create a kind of bridge between them. This typical chaining of objects is considered as a drawback, especially if there is a reason for the clusters to acquire the usual elliptical shape with a compact core. This method, however, possesses many positive features that outweigh the above disadvantage.

Farthest Neighbour Method

The method of the farthest neighbour is based on the opposite principle. The criterion for the connection of clusters is the maximum of q possible between-cluster distances of objects. When editing the matrix of distances, we proceed according to

$$D_{gg'} = \max (D_{g'h}, D_{g'h'}) .$$ (11)

An adverse chain effect does not occur in this case. On the contrary, there is a tendency towards the formation of compact clusters, not extraordinarily large, though.

Average Linkage Method (Sokal-Sneath Method)

As a criterion for the connection of clusters, this method applies an average of the q possible between-cluster distances of objects. When recalculating the distance matrix, we use

$$D_{gg'} = \frac{n_h \cdot D_{g'h} + n_{h'} \cdot D_{g'h'}}{n_h + n_{h'}} .$$ (12)

The method often leads to similar results as the farthest neighbour one.

Centroid method (Gower method)

Unlike the above methods, this one is not based on summarizing the information on between-cluster distances of objects, the criterion being the Euclidean distance of centroids

$$D_E(\bar{x}_h, \bar{x}_{h'}) = \sum_{j=1}^{p} (\bar{x}_{hj} - \bar{x}_{h'j})^2 .$$ (13)

The recalculation of the distance matrix is done as follows

$$D_{gg'} = \frac{1}{n_h + n_{h'}} \cdot \left(n_h \cdot D_{g'h} + n_{h'} \cdot D_{g'h'} - \frac{n_h \cdot n_{h'}}{n_h + n_{h'}} \cdot D_{hh'} \right) . \tag{14}$$

Ward Method

The method uses a functional of the decomposition quality C_1 in formula (4). The criterion for the cluster connection is an increment to the total intra-group sum of the squares of observation deviations from the cluster average, thus

$$\Delta C_1 = \sum_{i=1}^{g} \sum_{j=1}^{p} (x_{gij} - \bar{x}_{gj})^2 - \sum_{i=1}^{h} \sum_{j=1}^{p} (x_{hij} - \bar{x}_{hj})^2 - \sum_{i=1}^{h'} \sum_{j=1}^{p} (x_{h'ij} - \bar{x}_{h'j})^2 . \tag{15}$$

The increment is expressed as a sum of squares in an emerging cluster which is reduced by the sums of squares in both vanishing clusters. Using arithmetic modifications, the expression can be simplified into the form

$$\Delta C_1 = \frac{n_h \cdot n_{h'}}{n_h + n_{h'}} \cdot \sum_{j=1}^{p} (\bar{x}_{hj} - \bar{x}_{h'j})^2 . \tag{16}$$

This equation is a product of the Euclidean distance between the centroids of clusters considered for the connection and a coefficient depending on the cluster size. The value of this coefficient grows with an increasing size of clusters, and for fixed $n_h + n_{h'}$ it represents the maximum in the case of same-size $(n_h = n_{h'})$ clusters. Since we create the connections to ensure the minimization of the criterion ΔC_1, the Ward method tends to eliminate small clusters, i.e. to form those of roughly the same size, which is often a desirable property. Starting from the matrix of Euclidean distances between objects in the process of its modification, we can use the formula

$$D_{gg'} = \frac{1}{n_h + n_{h'} + n_{g'}} \cdot [(n_h + n_{g'}) \cdot D_{hg'} + (n_{h'} + n_{g'}) \cdot D_{h'g'} - n_{g'} \cdot D_{hh'}] . \tag{17}$$

3.2 Lognormal Distribution

Theoretical aspects of the lognormal distribution have been amply dealt with in the statistical literature in terms of the course, characteristics and construction of a lognormal curve and methods of point parameter estimation; see, e.g., Norman L. Johnson, Samuel Kotz, and Narayanaswamy Balakrishnan [4] or Christian Kleiber, and Samuel Kotz [6]. The importance of the lognormal distribution as a sample distribution model is beyond all discussion. It has many practical applications in various fields, e.g. economics, sociology, technology or astronomy. The lognormal model allows for capturing differentiating features, such as random-to-systematic variance changes, sequential effects of interdependent factors or trends towards geometrically sequential development.

In the field of economics, wages and incomes of the population are among the many phenomena that the lognormal model enables to interpret. When choosing a curve for modelling frequency distribution, it is necessary to meet the following requirements. The curve is supposed to

- reflect the given shape of the frequency distribution, being fully compliant with the relevant distribution modelled according to its basic characteristics, i.e. location, variability, skewness and kurtosis;
- have a relatively simple shape so that it can be easily manipulated, depending on a small number of parameters estimated using a suitable method of point parameter estimation;
- show interpretable parameters allowing for the prediction of their values without using the methods of statistical time series analysis, especially in those cases when sufficiently long time series are not available.

Every option is always a compromise between the above requirements. The parameter functions of lognormal curves allow for an easy interpretation. In the case of a three-parameter lognormal curve, the parameter θ represents the minimum of the curve (the beginning of distribution, theoretical minimum), the expression $\exp(\mu)$ denote the distance of the median wage (income) from this theoretical minimum, parameters μ and σ^2 representing the expected value and variance of logarithms of wage (income) distances from the theoretical minimum θ.

An old notion that in the area of economy the logarithms of the distances of variables from the theoretical minimum θ are normally distributed stems from the fact that the effects of a large number of various stimuli, resulting in the value of a given quantity, are proportional to that quantity at the corresponding time.

A strong concordance of the model with the global wage or income distribution does not mean, however, that a lognormal distribution will suffice for all circumstances or extremely homogeneous subgroups of employees or households categorized in minute detail according to the selected demographic and socio-economic indicators. If the latter is not the case, it is possible to model the wage or income distribution accurately enough with the use of a lognormal curve, parameters of the lognormal distribution being appropriately estimated from the sample. Alternatively, a shift of the curve can be made with either a subjectively determined wage/income minimum or another (shift) parameter, which is estimated from the sample. This solution led to positive results in the construction of wage and income models for a nationwide scale and large relatively homogeneous groups roughly categorized by some demographic and socio-economic indicators. The lognormal model, on the other hand, is not suitable for subsets formed from minutely classified employees or households. However, this is not the case of the present study that focuses on the nationwide wage distribution in the Czech and Slovak Republics.

3.2.1 Three-Parameter Lognormal Distribution

A continuous random variable *X* has a three-parameter lognormal distribution with parameters μ, σ^2 and θ, where $-\infty < \mu < \infty$, $\sigma^2 > 0$, $-\infty < \theta < \infty$, if its probability density function has the form

$$f(x;\mu,\sigma^2,\theta) = \frac{1}{\sigma\cdot(x-\theta)\cdot\sqrt{2\pi}}\cdot\exp\left[-\frac{[\ln(x-\theta)-\mu]^2}{2\sigma^2}\right], \quad x>\theta. \tag{18}$$
$$= 0, \quad \text{else.}$$

The three-parameter lognormal distribution with parameters μ, σ^2 and θ is marked LN(μ, σ^2, θ), where parameter θ is the beginning of the distribution (theoretical minimum). The probability density function of the three-parameter lognormal distribution is asymmetric, positively skewed. Figures 1 and 2 display the graphs of the probability density function of the three-parameter lognormal distribution depending on the values of parameters of this theoretical probability distribution.

The probability density function of the three-parameter lognormal distribution is sometime introduced in the form

$$f(x;\gamma,\delta,\theta) = \frac{\delta}{(x-\theta)\cdot\sqrt{2\pi}}\cdot\exp\left\{-\frac{1}{2}[\gamma+\delta\cdot\ln(x-\theta)]^2\right\}, \quad x>\theta, \tag{19}$$
$$= 0, \quad \text{else,}$$

where it is valid between the expressions of probability density functions (18) and (19) that

$$\mu = -\frac{\gamma}{\delta} \quad \text{and} \quad \sigma = \frac{1}{\delta}.$$

The distribution function of the three-parameter lognormal distribution has a form

$$F(x) = \Phi\left[\frac{\ln(x-\theta)-\mu}{\sigma}\right], \quad x>\theta. \tag{20}$$

If the random variable X has a three-parameter lognormal distribution LN(μ, σ^2, θ), then the random variable

$$Y = \ln(X-\theta) \tag{21}$$

has a normal distribution N(μ, σ^2) and the random variable

$$U = \frac{\ln(X-\theta)-\mu}{\sigma} = \gamma + \delta\cdot\ln(X-\theta) \tag{22}$$

has a standardized normal distribution N(0; 1). Parameter μ is then the expected value of a random variable (21), parameter σ^2 being its variance. Parameter θ is the beginning of the distribution, i.e. theoretical minimum of the random variable X.

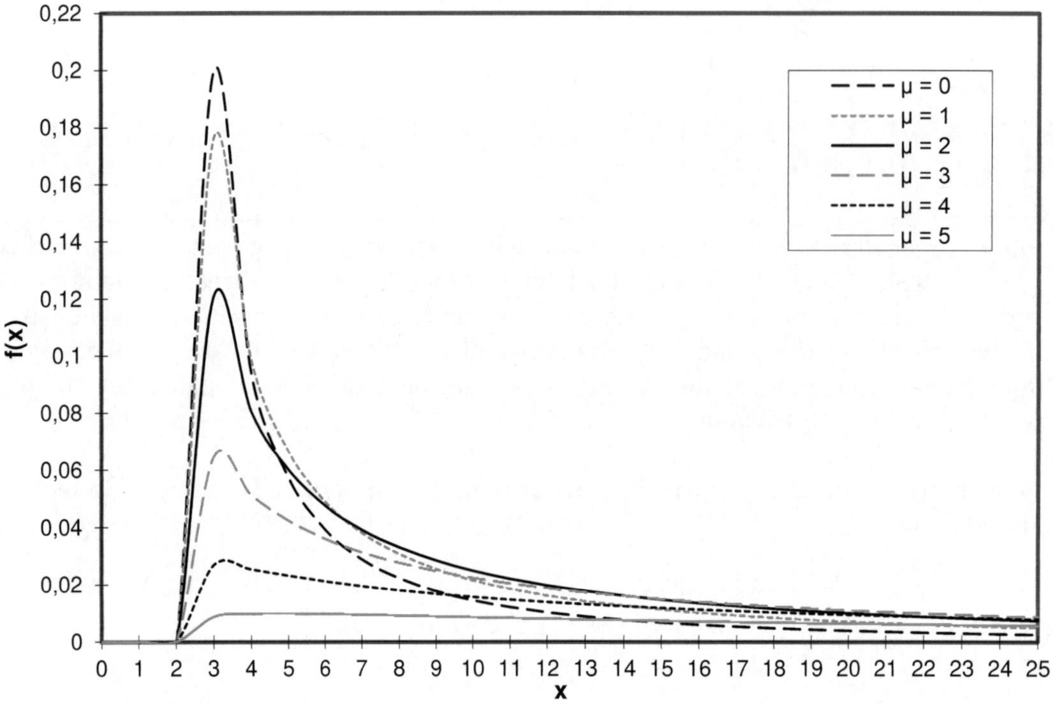

Source: Own research

Fig. 1. Probability density function of lognormal distribution for parameter values $\sigma = 2$ ($\sigma^2 = 4$); $\theta = 2$.

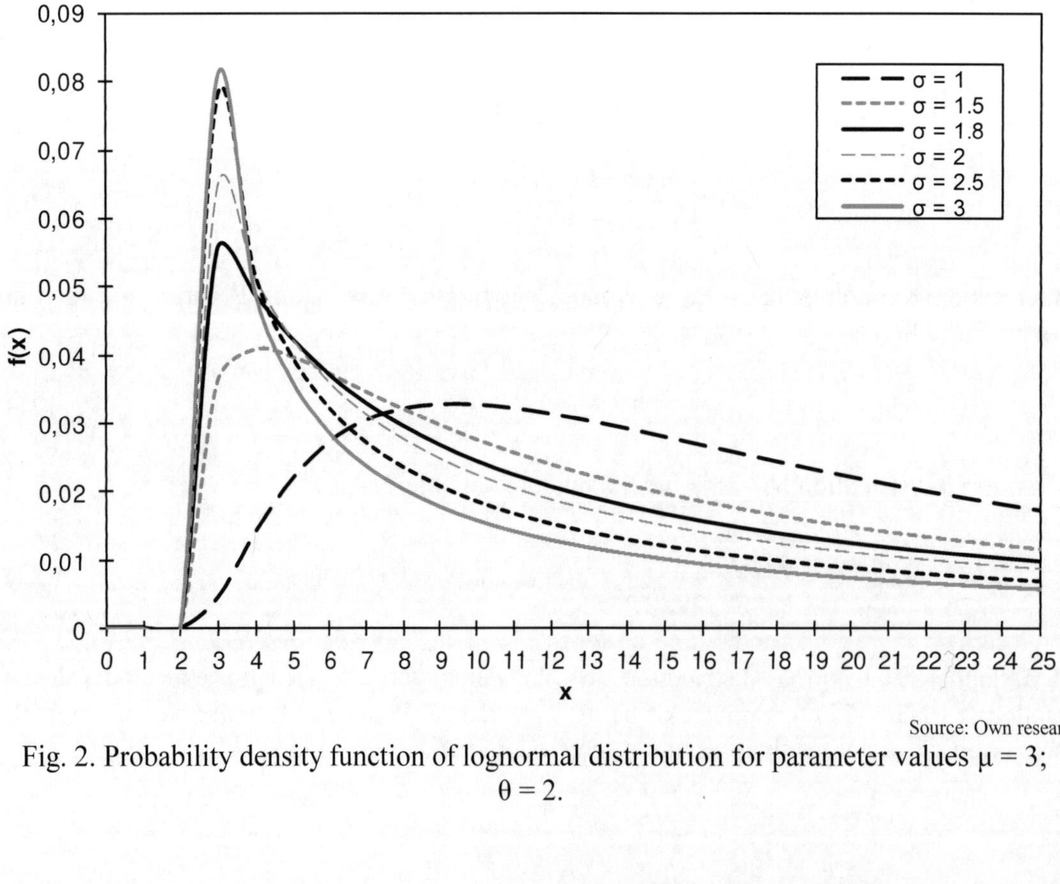

Source: Own research

Fig. 2. Probability density function of lognormal distribution for parameter values $\mu = 3$; $\theta = 2$.

For $\omega = \exp(\sigma^2)$, the rth common and central moments of the three-parameter lognormal distribution have the forms

$$\mu_r^{\backslash} = E(X^r) = \theta + \exp\left(r \cdot \mu + \frac{r^2 \sigma^2}{2}\right), \tag{23}$$

$$\mu_r = E[(X - \mu_1^{\backslash})^r] = \omega^{r/2} \cdot \left[\sum_{j=0}^{r} (-1)^j \cdot \binom{r}{j} \cdot \omega^{(r-j)\cdot(r-j-1)/2}\right] \cdot \exp(r \cdot \mu), \tag{24}$$

specifically,

$$\mu_3 = \omega^{3/2} \cdot (\omega - 1)^2 \cdot (\omega + 2) \cdot \exp(3 \cdot \mu), \tag{25}$$

$$\mu_4 = \omega^2 \cdot (\omega - 1)^2 \cdot (\omega^4 + 2\omega^3 + 3\omega^2 - 3) \cdot \exp(4 \cdot \mu). \tag{26}$$

We obtain the expressions for the expected value and the variance of the random variable X with a three-parameter lognormal distribution from the expressions (23) and (24)

$$E(X) = \theta + \exp\left(\mu + \frac{\sigma^2}{2}\right), \tag{27}$$

$$D(X) = \exp(2\mu + \sigma^2) \cdot [\exp(\sigma^2) - 1] = \exp(2\mu) \cdot \omega \cdot (\omega - 1). \tag{28}$$

The formula for median

$$Median(X) = \theta + \exp(\mu) \tag{29}$$

is derived from the relationship for the 100-percent-quantile of the distribution

$$x_P = \theta + \exp(\mu + \sigma \cdot u_P). \tag{30}$$

The three-parameter lognormal distribution is unimodal, having a single mode

$$Mode(X) = \theta + \exp(\mu - \sigma^2) = \theta + \frac{\exp(\mu)}{\omega}. \tag{31}$$

The relationship between the expected value, median and mode follows from the expressions (27), (29) and (31)

$$E(X) > Median(X) > Mode(X), \tag{32}$$

which is typical for a positively skewed frequency distribution.

The coefficient of variation of the three-parameter lognormal distribution is a function of all three parameters

$$V(X) = \frac{\exp\left(\mu + \frac{\sigma^2}{2}\right)\sqrt{\exp(\sigma^2) - 1}}{\theta + \exp\left(\mu + \frac{\sigma^2}{2}\right)} = \frac{\exp\left(\mu + \frac{\sigma^2}{2}\right)\sqrt{\omega - 1}}{\theta + \exp\left(\mu + \frac{\sigma^2}{2}\right)}. \tag{33}$$

The Gini coefficient of differentiation also depends on the values of all three parameters μ, σ^2 and θ of the distribution

$$G = \frac{\exp\left(\mu + \frac{\sigma^2}{2}\right) \cdot \operatorname{erf}\left(\frac{\sigma}{2}\right)}{\theta + \exp\left(\mu + \frac{\sigma^2}{2}\right)}. \tag{34}$$

Moment measurements of skewness and kurtosis depend on a single parameter σ^2

$$\beta_1 = \sqrt{\exp(\sigma^2) - 1} \cdot [\exp(\sigma^2) + 2] = \sqrt{\omega - 1} \cdot (\omega + 2)\,, \tag{35}$$

$$\beta_2 = [\exp(4\sigma^2) + 2\exp(3\sigma^2) + 3\exp(2\sigma^2) - 3] = (\omega^4 + 2\omega^3 + 3\omega^2 - 3). \tag{36}$$

3.2.2 Maximum Likelihood Method

The substance of the maximum likelihood method of point parameter estimation is explained in detail in statistical literature, e.g., in Norman L. Johnson, Samuel Kotz, and Narayanaswamy Balakrishnan [5].

Let a random sample of size n be taken from the three-parameter lognormal distribution with probability density function (18). The likelihood function then has the form

$$L(\boldsymbol{x}; \mu, \sigma^2, \theta) = \prod_{i=1}^{n} f(x_i; \mu, \sigma^2, \theta) =$$

$$= \frac{1}{(\sigma^2)^{n/2} \cdot (2\pi)^{n/2} \cdot \prod_{i=1}^{n} (x_i - \theta)} \cdot \exp\left\{ \sum_{i=1}^{n} - \frac{[\ln(x_i - \theta) - \mu]^2}{2\sigma^2} \right\}. \tag{37}$$

We determine the logarithm of the likelihood function

$$\ln L(\boldsymbol{x}; \mu, \sigma^2, \theta) =$$

$$= \sum_{i=1}^{n} -\frac{[\ln(x_i - \theta) - \mu]^2}{2\sigma^2} - \frac{n}{2}\cdot\ln\sigma^2 - \frac{n}{2}\cdot\ln(2\pi) - \sum_{i=1}^{n}\ln(x_i - \theta). \tag{38}$$

Then we set the first partial derivatives of the likelihood function logarithm equal to zero

$$\frac{\partial \ln L(\boldsymbol{x}; \mu, \sigma^2, \theta)}{\partial \mu} = \frac{\sum_{i=1}^{n}[\ln(x_i - \theta) - \mu]}{\sigma^2} = 0,$$

$$\frac{\partial \ln L(\boldsymbol{x}; \mu, \sigma^2, \theta)}{\partial \sigma^2} = \frac{\sum_{i=1}^{n}[\ln(x_i - \theta) - \mu]^2}{2\sigma^4} - \frac{n}{2\sigma^2} = 0.$$

After adjustments, the maximum likelihood estimates of parameters μ and σ^2 for the given parameter θ are obtained as follows

$$\hat{\mu}(\theta) = \frac{\sum_{i=1}^{n}\ln(x_i - \theta)}{n}, \tag{39}$$

$$\hat{\sigma}^2(\theta) = \frac{\sum_{i=1}^{n}[\ln(x_i - \theta) - \hat{\mu}(\theta)]^2}{n}. \tag{40}$$

If the value of the parameter θ is known, we get the maximum likelihood estimates of the remaining two parameters of the three-parameter lognormal distribution using the expressions (39) and (40). However, if the value of the parameter θ is unknown, the situation is more complicated. It can be proved that if the parameter θ is close to $\min\{X_1, X_2, ..., X_n\}$, then the maximum likelihood approaches infinity. Moreover, the maximum likelihood method is often combined with the Cohen method, the smallest sample value being set equal to the 100-$(n+1)^{-1}$-percent quantile

$$x_{\min}^{V} = \hat{\theta} + \exp(\hat{\mu} + \hat{\sigma} \cdot u_{(n+1)^{-1}}). \tag{41}$$

Equation (41) is then combined with equations (39) and (40).

3.2.3 Akaike and Bayesian Information Criteria

Let us consider L as the maximum value of the likelihood function for an assumed model of data, k and n denoting the number of parameters estimated and the sample size, respectively. The Akaike information criterion (AIC) has the form

$$AIC = 2k - 2\ln L \tag{42}$$

and the Bayesian information criterion (BIC) is defined as

$$BIC = k \ln n - 2 \ln L. \tag{43}$$

The model with minimal AIC or BIC values is preferred over other alternatives, AIC and BIC criteria also including a penalty which is an increasing function of the number of estimated parameters.

3.3 Other Methods Used

The wage distribution can be also recorded by simple descriptive measurements; see, e.g., Gerald M. Barber [1] or Mario F. Triola [10]. As for the gender wage gap development and forecasting, the trend analysis of time series was performed, interpolation criteria being applied to choose an appropriate trend function; for details on economic time series analysis, see Peter J. Brockwell, and Richard A. Davis [2] or Paul S. B. Cowperwait, and Andrew V. Metcalfe [3].

4 Results and Discussion

4.1 Situation in the Czech Republic

Figures 3 and 4 provide information on the geographical location of each region of the Czech Republic (their official names are presented in Table 1) and the respective level of the gross monthly wage. The figures clearly show a substantially higher wage level in the region of the capital Prague. A relatively high level of wages in Central Bohemian and Pilsen regions is noticeable, low levels, on the other hand, being reported in Karlovy Vary, Zlin and Olomouc regions.

The region of the capital Prague is not food and energy self-sufficient. Its macroeconomic statistics, however, considerably exceed those of other regions, wages earned in Prague resembling those received in more developed countries. One of the reasons is the concentration of industries with higher labour productivity such as finance and informatics. Corporate policies also play an important role, since Prague-based firms often produce values outside the capital but divide the profits at a place where they are headquartered. As expected, the three regions with the highest wage levels are identical to those with the lowest unemployment rates in the same order. However, the order of the regions at the bottom wage levels is not the same as that of the regions with top rates of unemployment, neither the Moravian-Silesian Region nor Usti Region belonging to the three lowest wage level areas; for details, see Table 2.

Figures 5–8 provide an overview of the results of regional cluster analysis according to the wage level employing the method of the farthest neighbour and Euclidean distance metric.

The first cluster always contains only one element – the Capital Prague Region – in the case of both the average and median monthly wage (i.e. three- and five-cluster analysis, respectively), due to markedly higher wage levels in this respective region.

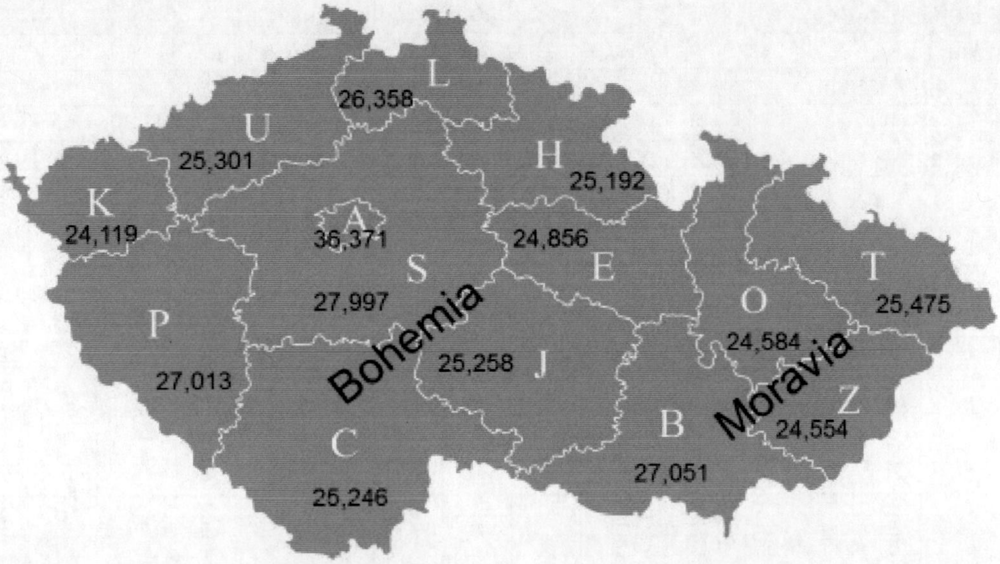

Source: www.czso.cz; own customization

Fig. 3. Average gross monthly wages (in CZK) in respective regions of the Czech Republic in 2015.

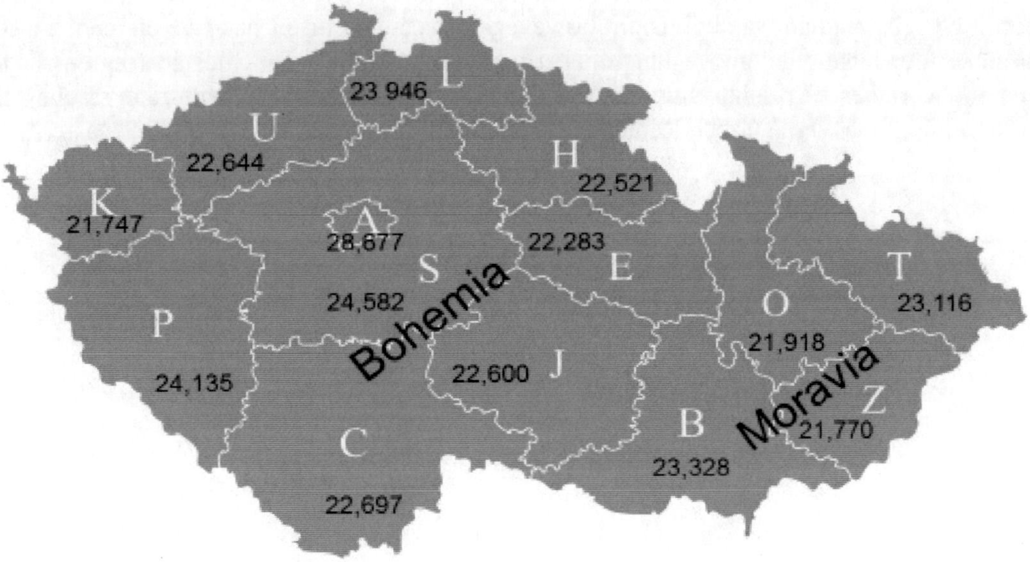

Source: www.czso.cz; own customization

Fig. 4. Median gross monthly wages (in CZK) in respective regions of the Czech Republic in 2015.

Region	Code	Region	Code
Capital Prague Region	A	Hradec Kralove Region	H
Central Bohemian Region	S	Pardubice Region	E
South Bohemian Region	C	Vysocina Region	J
Pilsen Region	P	South Moravian Region	B
Karlovy Vary Region	K	Olomouc Region	O
Usti Region	U	Zlin Region	Z
Liberec Region	L	Moravian-Silesian Region	T

Source: www.mdcr.cz

Tab. 1. Official names[1)] of regions of the Czech Republic[2)].

	Region													
	Central Prague Region	Central Bohemian Region	South Bohemian Region	Pilsen Region	Karlovy Vary Region	Usti Region	Liberec Region	Hradec Kralove Region	Pardubice Region	Vysocina Region	South Moravian Region	Olomouc Region	Zlin Region	Moravian-Silesian Region
Unemploy-ment rate	2.8	3.5	4.0	3.8	6.7	7.6	5.5	5.6	4.6	4.7	5.0	5.9	4.7	8.1

Source: www.czso.cz

Tab. 2. Average unemployment rates (in %) in regions of the Czech Republic in 2015.

Within the division of the regions into three clusters by the average wage, the second cluster includes four elements – Central Bohemian, Pilsen, Liberec and South-Moravian regions. According to the median wage division, however, the second cluster has five elements; along with those four mentioned above, this cluster contains the Moravian-Silesian Region, which seems to be rather surprising since this region's general unemployment rate reaches the highest value of the whole Czech Republic. The remaining regions form the third clusters.

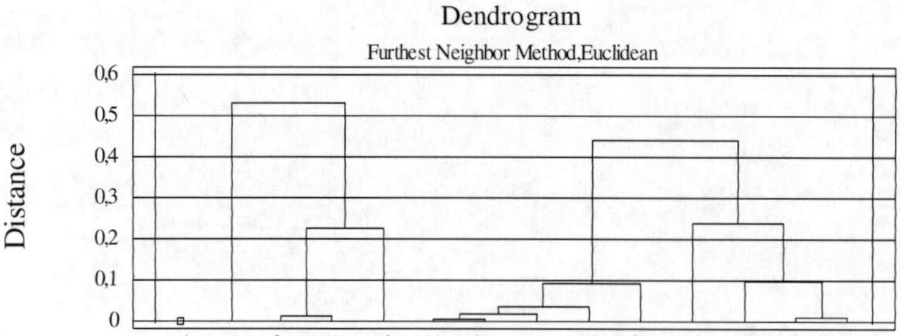

Source: Own research

Fig. 5. Cluster analysis using three clusters, farthest neighbour method and Euclidean distance metric; 2015 average wage.

[1)] The names of most regions match those of their respective capitals.

[2)] Different backgrounds distinguish the Bohemian regions (grey) from Moravian ones (white).

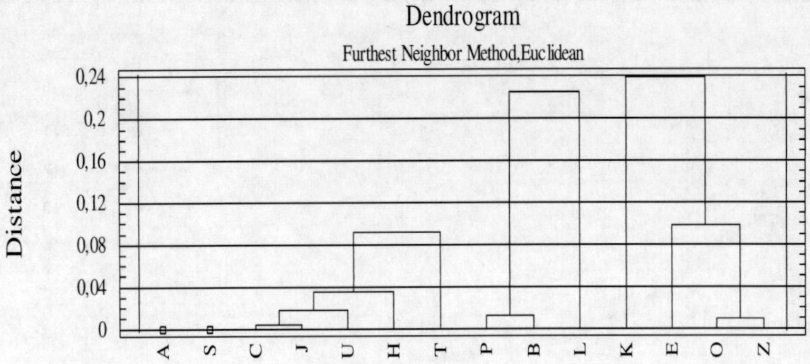

Source: Own research

Fig. 6. Cluster analysis using five clusters, farthest neighbour method and Euclidean distance metric; 2015 average wage.

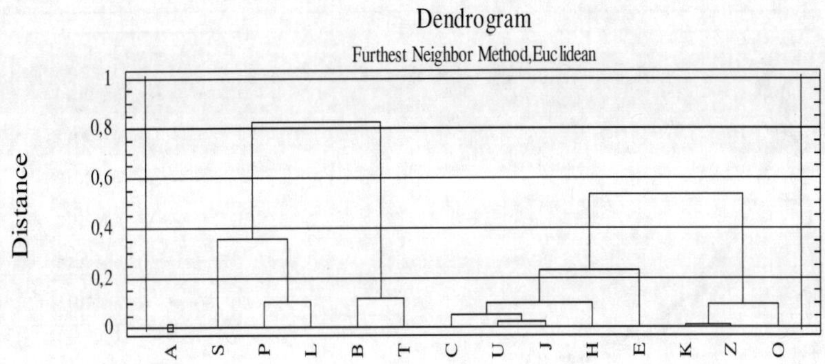

Source: Own research

Fig. 7. Cluster analysis using three clusters, farthest neighbour method and Euclidean distance metric; 2015 median wage.

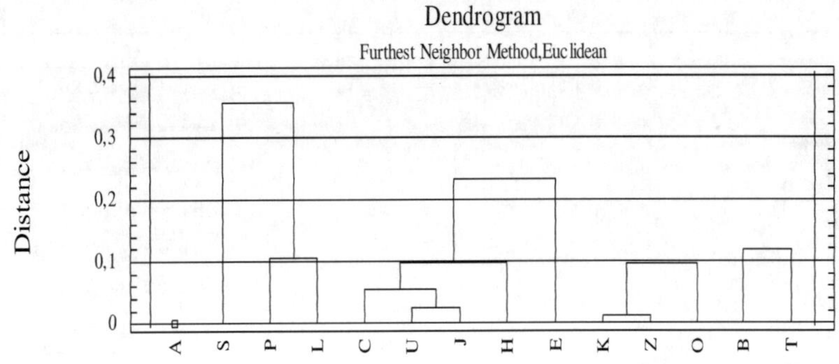

Source: Own research

Fig. 8. Cluster analysis using five clusters, farthest neighbour method and Euclidean distance metric; 2015 median wage.

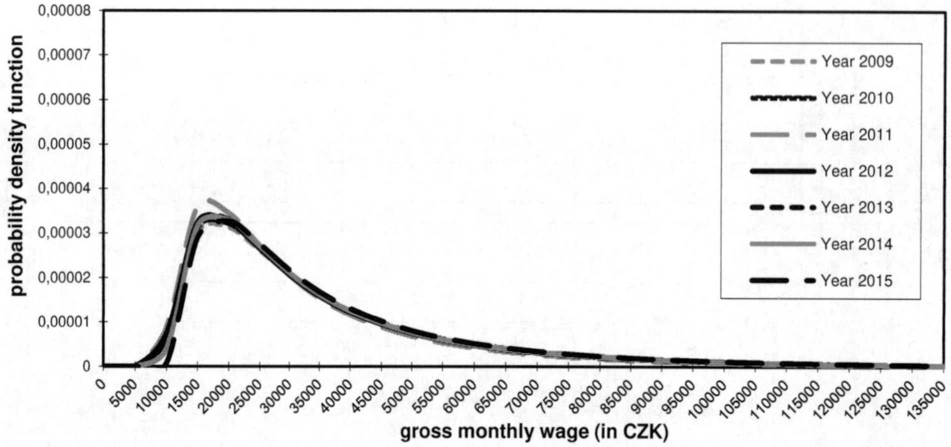

Source: Own research

Fig. 9. Development of model wage distributions – Capital Prague Region.

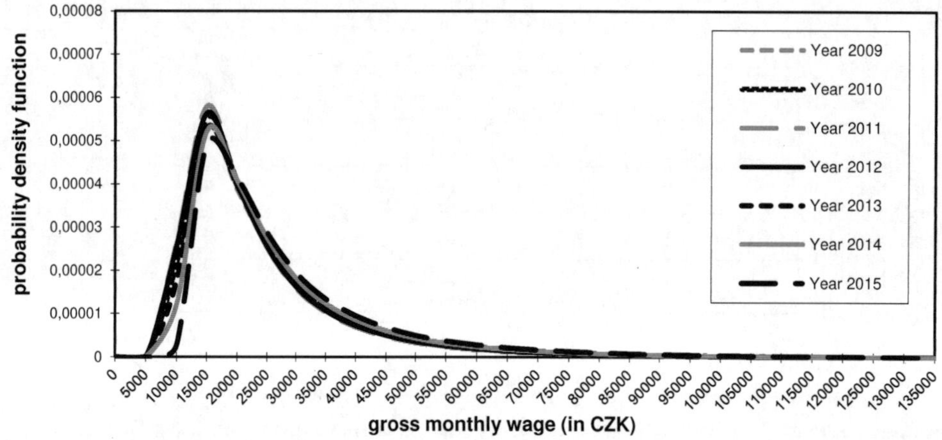

Source: Own research

Fig. 10. Development of model wage distributions – Central Bohemian Region.

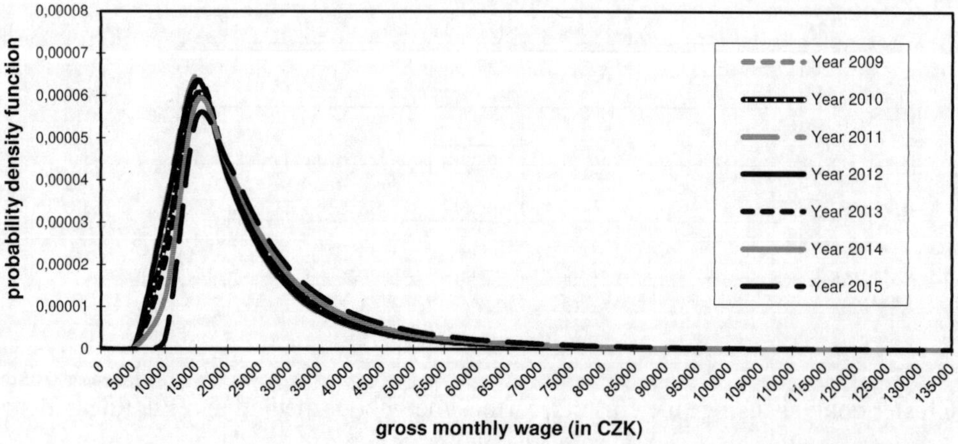

Source: Own research

Fig. 11. Development of model wage distributions – Pilsen Region.

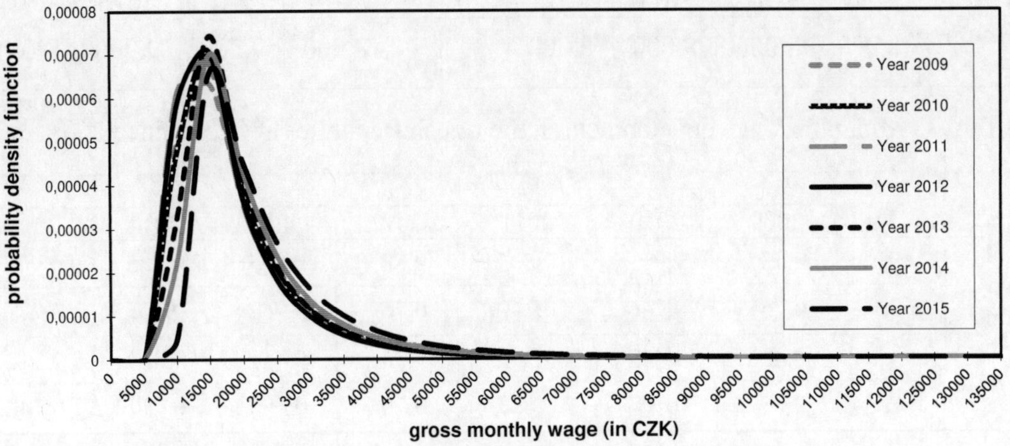

Source: Own research

Fig. 12. Development of model wage distributions – Karlovy Vary Region.

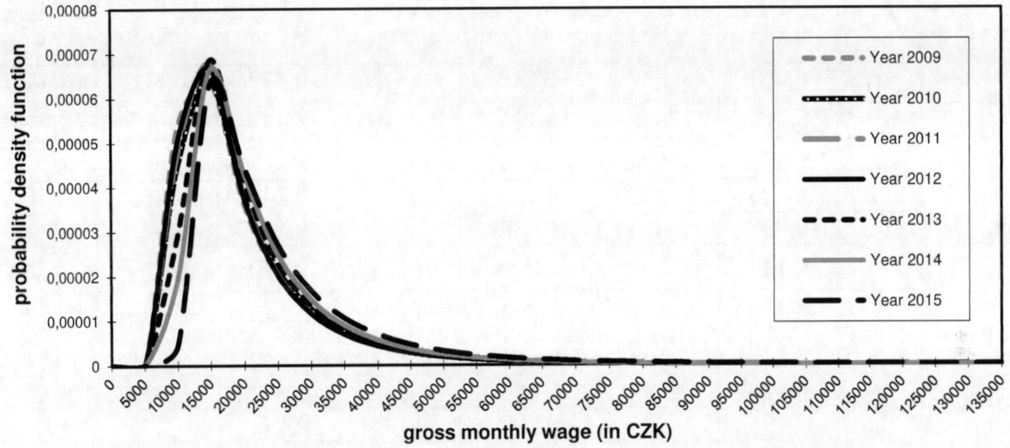

Source: Own research

Fig. 13. Development of model wage distributions – Zlin Region.

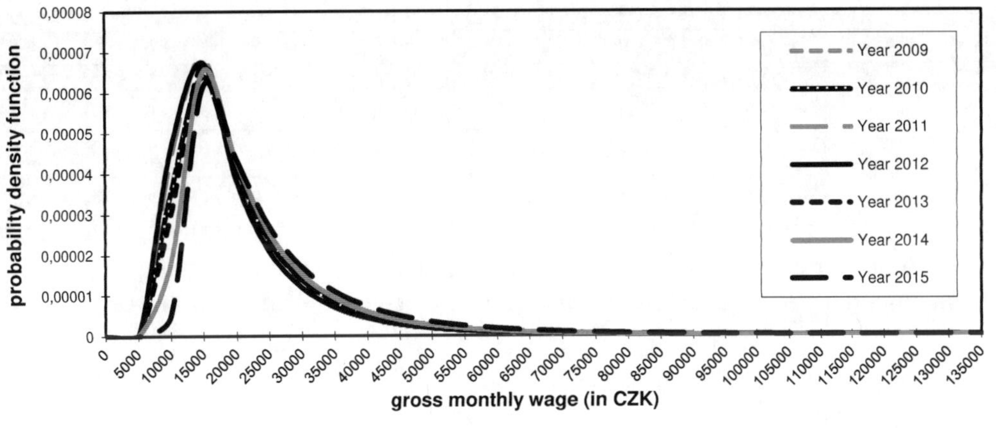

Source: Own research

Fig. 14. Development of model wage distributions – Olomouc Region.

Year	2009	2010	2011	2012	2013[3]	2014	2015	2016	2017
Minimum wage	8,000	8,000	8,000	8,000	8,000[4] 8,500[5]	8,500	9,200	9,900	11,000

Source: www.mpsv.cz

Tab. 3. Minimum wage development in the Czech Republic (in CZK) since 2009.

		Year						
Region	Est.	2009	2010	2011	2012	2013	2014	2015
Capital Prague Region	μ	9,937	9,911	9,803	9,885	9,875	9,885	9,907
	σ^2	0,765	0,746	0,740	0,761	0,752	0,765	0,777
Central Bohemian Region	μ	9,426	9,443	9,342	9,366	9,412	9,467	9,512
	σ^2	0,732	0,707	0,683	0,702	0,699	0,711	0,742
South Bohemian Region	μ	9,112	9,156	9,124	9,092	9,176	9,250	9,312
	σ^2	0,701	0,688	0,633	0,633	0,627	0,632	0,668
Pilsen Region	μ	9,298	9,316	9,202	9,226	9,275	9,371	9,426
	σ^2	0,653	0,634	0,639	0,649	0,630	0,639	0,682
Karlovy Vary Region	μ	9,113	9,054	8,978	8,963	9,060	9,149	9,239
	σ^2	0,737	0,627	0,686	0,635	0,577	0,616	0,670
Usti Region	μ	9,264	9,280	9,130	9,169	9,216	9,266	9,337
	σ^2	0,711	0,674	0,671	0,664	0,637	0,654	0,707
Liberec Region	μ	9,326	9,292	9,146	9,182	9,247	9,308	9,392
	σ^2	0,950	0,656	0,636	0,625	0,618	0,626	0,689
Hradec Kralove Region	μ	9,144	9,201	9,087	9,139	9,190	9,259	9,304
	σ^2	0,640	0,671	0,623	0,625	0,620	0,636	0,661
Pardubice Region	μ	9,225	9,168	9,105	9,118	9,160	9,231	9,306
	σ^2	0,920	0,697	0,642	0,660	0,642	0,650	0,704
Vysocina Region	μ	9,195	9,204	9,093	9,133	9,193	9,255	9,322
	σ^2	0,754	0,691	0,616	0,640	0,605	0,627	0,685
South Moravian Region	μ	9,358	9,394	9,268	9,311	9,361	9,404	9,460
	σ^2	1,028	1,025	1,028	1,027	1,018	1,014	1,009
Olomouc Region	μ	9,204	9,203	9,088	9,086	9,164	9,239	9,296
	σ^2	0,661	0,659	0,632	0,657	0,651	0,633	0,719
Zlin Region	μ	9,077	9,139	9,075	9,065	9,149	9,215	9,277
	σ^2	0,739	0,704	0,656	0,680	0,626	0,633	0,686
Moravian-Silesian Region	μ	9,196	9,251	9,220	9,233	9,262	9,290	9,345
	σ^2	0,681	0,664	0,665	0,662	0,656	0,654	0,702

Source: Own research

Tab. 4. Parameter estimates of three-parameter lognormal distribution using maximum likelihood method (parameter θ equalling respective annual minimum wage).

[3] In 2013, the beginning of lognormal curves was determined proportionally, i.e

$$8\ 000 + 7 * \frac{8\ 500 - 8\ 000}{12} \doteq 8\ 292 .$$

[4] From 1st January 2013 to 31st July 2013.

[5] From 1st August 2013 to 31st December 2013.

Region	Crit.	Year						
		2009	2010	2011	2012	2013	2014	2015
Capital Prague Region	*AIC*	1,715,847	1,605,510	1,514,319	1,456,900	1,362,167	1,291,967	1,219,753
	BIC	1,715,870	1,605,533	1,514,342	1,456,923	1,362,191	1,291,990	1,219,776
Central Bohemian Region	*AIC*	544 ,441	536,317	528,511	543,060	546,067	556,645	577,484
	BIC	544,462	536,338	528,533	543,082	546,089	556,666	577,505
South Bohemian Region	*AIC*	297,991	296,670	282,364	284,622	285,256	289,311	303,276
	BIC	298,011	296,691	282,384	284,643	285,277	289,331	303,296
Pilsen Region	*AIC*	269,898	267,230	272,194	278,447	275,599	281,651	298,363
	BIC	269,919	267,250	272,215	278,468	275,619	281,671	298,383
Karlovy Vary Region	*AIC*	127,711	115,055	123,387	117,626	110,162	116,655	124,583
	BIC	127,730	115,074	123,406	117,644	110,180	116,674	124,602
Usti Region	*AIC*	348,582	336,708	336,174	334,146	324,956	331,734	350,898
	BIC	348,603	336,728	336,195	334,167	324,977	331,755	350,919
Liberec Region	*AIC*	231,143	185,448	183,454	183,282	183,845	187,689	203,032
	BIC	231,162	185,468	183,474	183,302	183,864	187,709	203,052
Hradec Kralove Region	*AIC*	240,158	250,588	239,702	242,212	242,731	249,249	258,341
	BIC	240,178	250,609	239,722	242,233	242,751	249,270	258,361
Pardubice Region	*AIC*	275,403	235,724	226,401	234,568	233,737	239,552	257,057
	BIC	275,423	235,744	226,421	234,588	233,757	239,572	257,077
Vysocina Region	*AIC*	238,726	228,425	213,311	222,088	215,872	224,336	241,838
	BIC	238,746	228,445	213,331	222,108	215,892	224,356	241,858
South Moravian Region	*AIC*	786,921	788,918	793,299	796,116	795,440	796,875	797,641
	BIC	786,943	788,940	793,321	796,138	795,462	796,897	797,663
Olomouc Region	*AIC*	255,008	261,376	260,425	274,584	279,522	280,684	313,825
	BIC	255,028	261,396	260,445	274,604	279,543	280,705	313,846
Zlin Region	*AIC*	289,572	282,403	271,177	279,996	265,977	270,105	288,029
	BIC	289,593	282,423	271,197	280,017	265,997	270,125	288,050
Moravian-Silesian Region	*AIC*	581,796	574,637	578,439	579,575	578,706	580,905	613,021
	BIC	581,817	574,659	578,461	579,597	578,728	580,927	613,042

Source: Own research

Tab. 5. Akaike and Bayesian information criteria values.

Within the division of the regions into five clusters by the average wage, the second cluster contains only one element, namely the Central Bohemian Region. However, according to the median wage, the second cluster consists of three elements – Central Bohemian, Pilsen and Liberec regions. The third cluster always comprises five elements – South Bohemian, Usti, Hradec Kralove, Vysocina and Moravian-Silesian regions by the average wage and South Bohemian, Usti, Hradec Kralove, Pardubice and Vysocina regions, respectively, according to the median wage. The fourth cluster has only three elements in both cases – Pilsen, Liberec and South Moravian regions according to the average wage, and Karlovy Vary, Olomouc and Zlin regions by the median wage, the latter being those with the lowest wage levels. The fifth clusters formed by the average and median wage contain the four and two remaining regions, respectively; for details, see Figures 5–8).

Theoretical models for the wage distribution of each region from 2009 onwards have been constructed. They are based on the use of the probability density function of three-parameter lognormal curves, whose parameters were estimated using the maximum likelihood method. The beginning of these curves is represented by the value of the minimum wage in respective years; see Table 3.

The accuracy of the models obtained was compared applying the Akaike and Bayesian information criteria, both of which take a number of the corresponding wage model parameters into account.

Theoretical wage models using three-parameter lognormal curves and the maximum likelihood method of parameter estimation were created. Table 4 indicates the values of parameters estimated, Table 5 presenting the values of the Akaike and Bayesian information criteria, which enable to assess the estimation accuracy. Using Figures 9–14, the theoretical wage models capture the three regions with the highest (Prague, Central Bohemian and Pilsen regions) and the other three with the lowest wage levels (Karlovy Vary, Zlin and Olomouc regions).

The above figures allow for comparison of the wage distribution development of the regions with the highest and lowest wages over the last seven years. As observed in the figures, distributions with a high wage level are characterized by higher variability than those with a low wage level. Moreover, distributions with a low level of wages are more skewed and have higher kurtosis than those with high wage levels. A probability model, usually representing a simple approximation of a rather complex empirical distribution and the knowledge of the development trend of its parameters allow for the estimation of the whole wage distribution for future research purposes. As we can see from Table 5, wage models for the capital Prague region show the lowest accuracy, while Karlovy Vary region with its lowest wage level indicates the best accuracy of wage models, the number of model parameters being implicated in both information criteria (AIC and BIC). The relationship between the model accuracy and the wage level in the respective region is obvious. It holds that low model accuracy corresponds to a high wage level and vice versa.

Figures 15 and 16 display the time series of the gender wage gap between men's and women's average and median (nominal) wages in Prague region and the Central Bohemian region over the period 2009–2015. An adaptive method of modelling the trend component was applied to construct the trend model using both parametric and non-parametric approach. An appropriate (i.e. quadratic) trend function was chosen employing interpolation criteria in the former and moving averages (of the three respective lengths) in the latter case. Figures 17–28 show an analogous development of the gender wage gap and its models for the other regions.

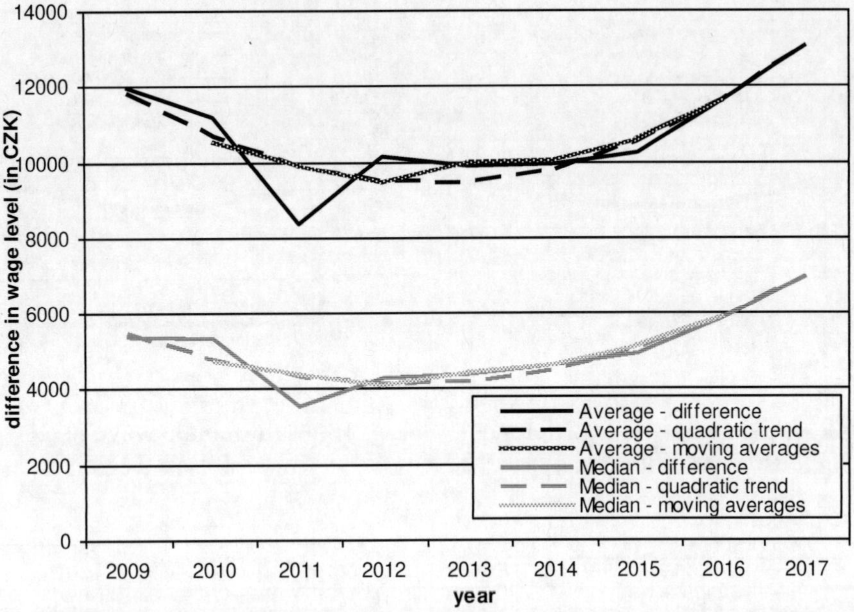

Source: Own research

Fig. 15. Male-female wage level differences: Capital Prague Region.

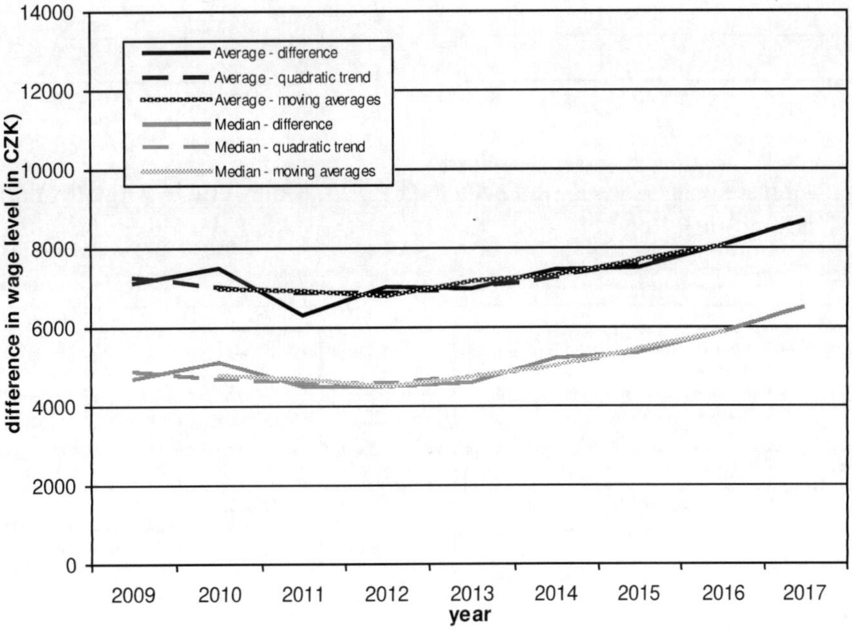

Source: Own research

Fig. 16. Male-female wage level differences: Central Bohemian Region.

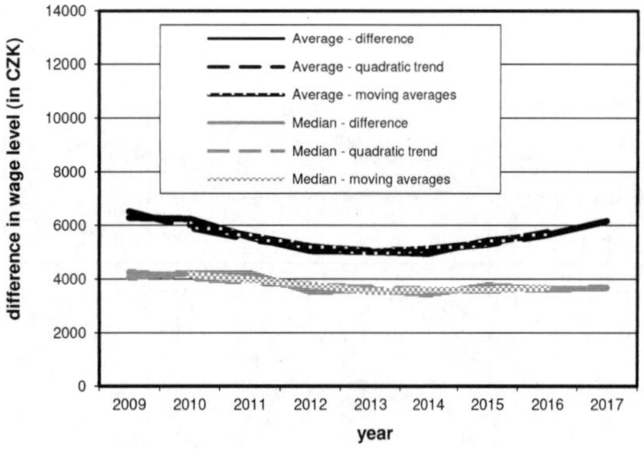

Source: Own research

Fig. 17. Male-female wage level differences: South Bohemian Region.

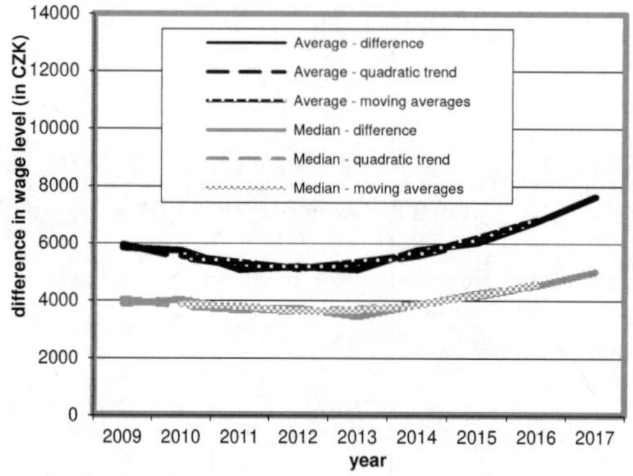

Source: Own research

Fig. 18. Male-female wage level differences: Pilsen Region.

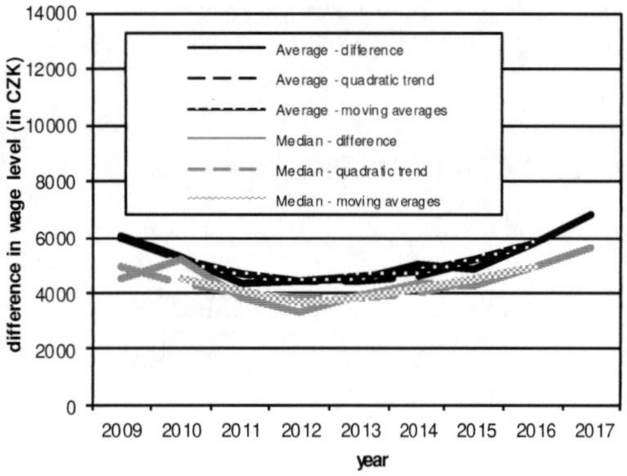

Source: Own research

Fig. 19. Male-female wage level differences: Karlovy Vary Region.

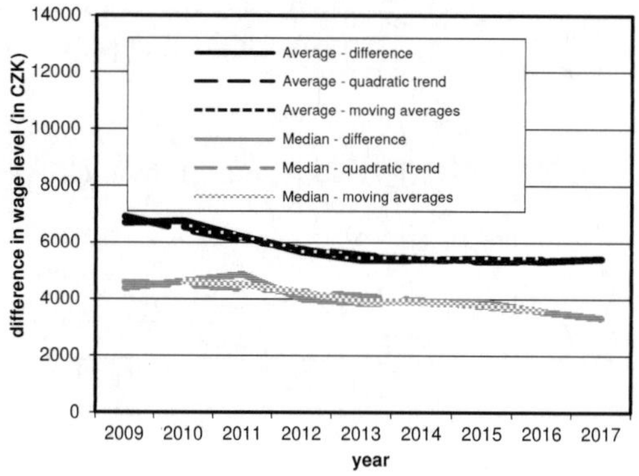

Source: Own research

Fig. 20. Male-female wage level differences: Usti Region.

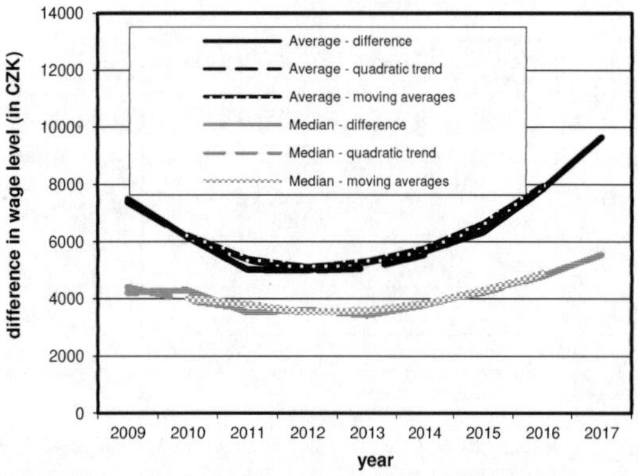

Source: Own research

Fig. 21. Male-female wage level differences: Liberec Region.

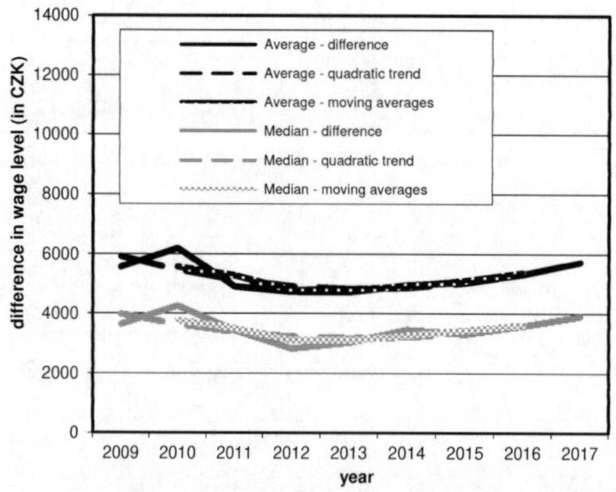

Source: Own research

Fig. 22. Male-female wage level differences: Hradec Kralove Region.

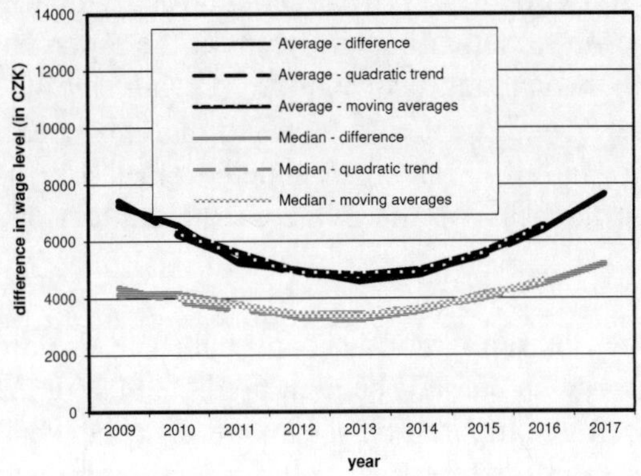

Source: Own research

Fig. 23. Male-female wage level differences: Pardubice Region.

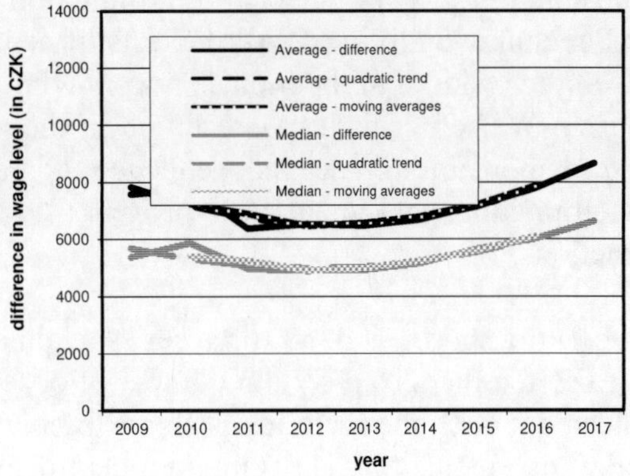

Source: Own research

Fig. 24. Male-female wage level differences: Vysocina Region.

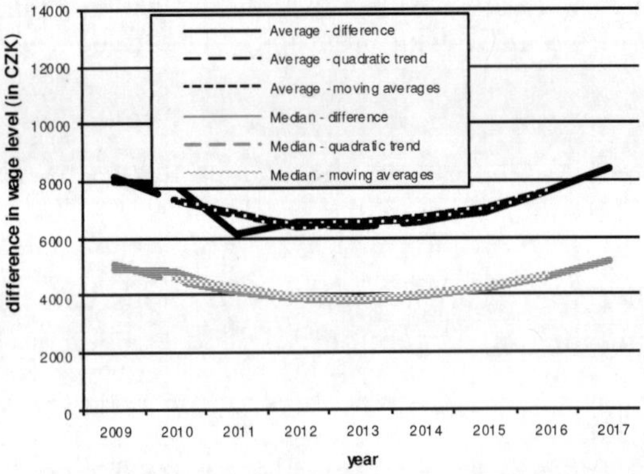

Source: Own research

Fig. 25. Male-female wage level differences: South Moravian Region.

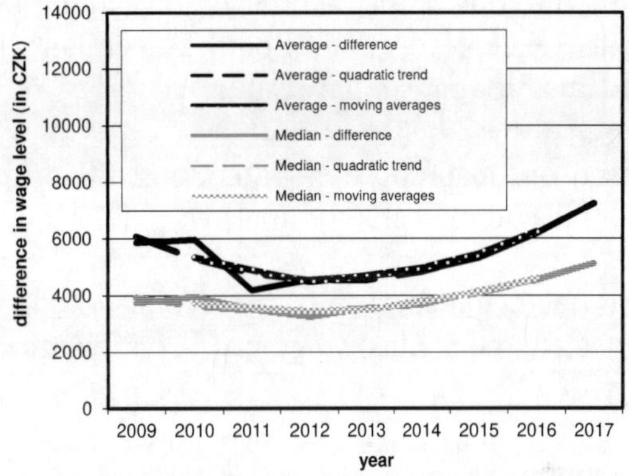

Source: Own research

Fig. 26. Male-female wage level differences: Olomouc Region.

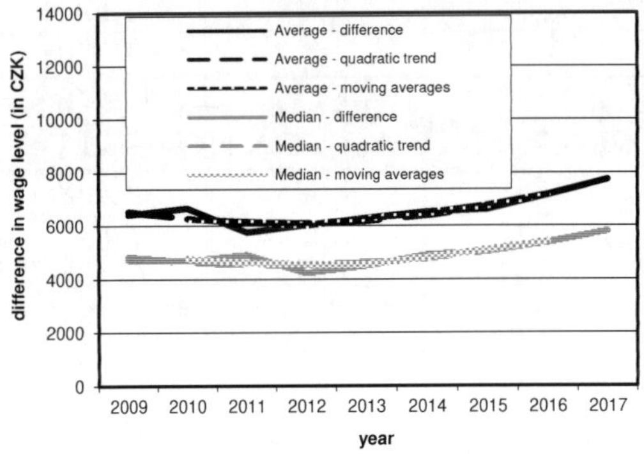

Source: Own research

Fig. 27. Male-female wage level differences: Zlin Region.

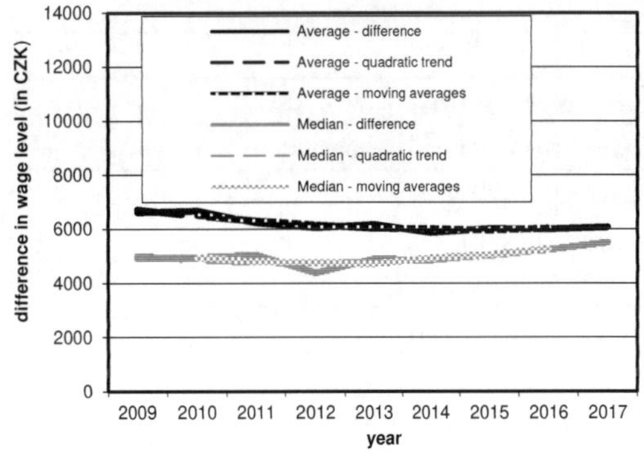

Source: Own research

Fig. 28. Male-female wage level differences: Moravian-Silesian Region

The suitability of the selected trend function was subsequently verified using an individual t-test and an overall F-test. All the tests turned out to be statistically significant at the 5 percent level, the residues of the models not showing any significant autocorrelation. All values of the Durbin-Watson statistics were found in the interval (1.4, 2.6), being equal to the two values and demonstrating the independence of random failures. The calculated values of the determination coefficient also proved the appropriateness of the selected quadratic trend functions.

Due to the shortness of the time series available, their extrapolation was not made (i.e. we did not shorten them by a certain number of observations which could be then predicted using the selected model, the accuracy of the extrapolation being compared by means of, e.g., the Theil index of inequality). The same applies to gender wage differences in the real average and median wage.

It is apparent from all figures 15–28 that in the vast majority of regions, the gender wage gap was narrowing at the beginning of the research period, presumably due to the global economic downturn. The recession having ceased, the gender wage inequality started rising again (except for Usti, South Bohemian and Moravian-Silesian regions). The real wage analysis brought similar results.

Based on quadratic trend functions, 2016 and 2017 predictions of the gender inequality in average and median (nominal) wages for individual regions were made; see Table 6.

This table indicates that a widening of the gender wage gap in almost all regions can be expected, the median wage gap in Usti Region being the only exception.

	Average		Median	
Region	Year 2016	Year 2017	Year 2016	Year 2017
Capital Prague Region	11,607	13,074	5,866	6,931
Central Bohemian Region	8,071	8,683	5,850	6,460
South Bohemian Region	5,649	6,162	3,623	3,684
Pilsen Region	6,765	7,650	4,535	5,014
Karlovy Vary Region	5,812	6,774	4,913	5,627
Usti Region	5,329	5,415	3,549	3,328
Liberec Region	7,851	9,652	4,795	5,546
Hradec Kralove Region	5,312	5,726	3,580	3,913
Pardubice Region	6,384	7,643	4,505	5,188
Vysocina Region	7,796	8,672	6,019	6,631
South Moravian Region	7,572	8,443	4,630	5,204
Olomouc Region	6,180	7,256	4,542	5,131
Zlin Region	7,172	7,740	5,372	5,794
Moravian-Silesian Region	5,986	6,058	5,232	5,503

Source: Own research

Tab. 6. 2016 and 2017 predictions of male-female wage level differences (in CZK).

4.2 Situation in the World

Figures 29–31 illustrate the above statements using the results of cluster analysis for 31 selected European countries. These states were divided into three, five and seven clusters according to their average gross monthly (nominal) wage in 2015 using method of the farthest neighbour and Euclidean distance metric.

Figure 29 shows the segmentation of these countries into three clusters. The first cluster consists of countries with medium wage levels in Europe. There are the following fourteen states: Austria, Belgium, Finland, France, Germany, Iceland, Ireland, Italy, Luxembourg, Netherlands, Spain, Sweden, Switzerland and United Kingdom. The second cluster consists of states with rather lower wage levels, to where the countries of the former socialist bloc (Bulgaria, Croatia, Czech Republic, Estonia, Hungary, Latvia, Lithuania, Poland, Romania, Slovakia and Slovenia) belong, inter alia (Cyprus, Greece, Malta and Portugal). Two countries with the highest wage levels make up the third cluster. There are Denmark and Norway.

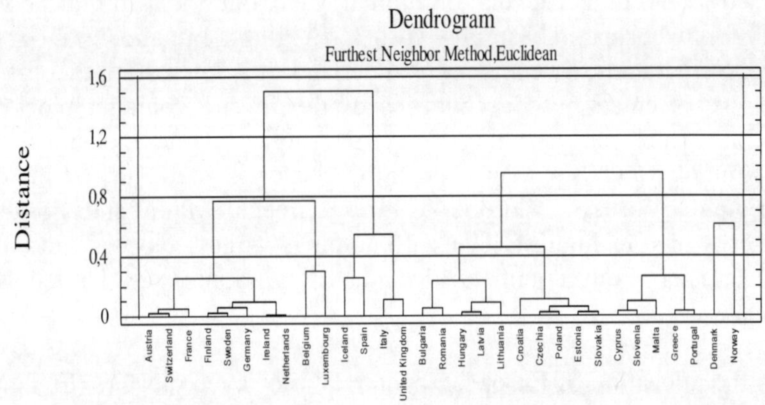

Source: Own research

Fig. 29. Cluster analysis using three clusters, method of the farthest neighbour and Euclidean distance metric, nominal average wage in 2015.

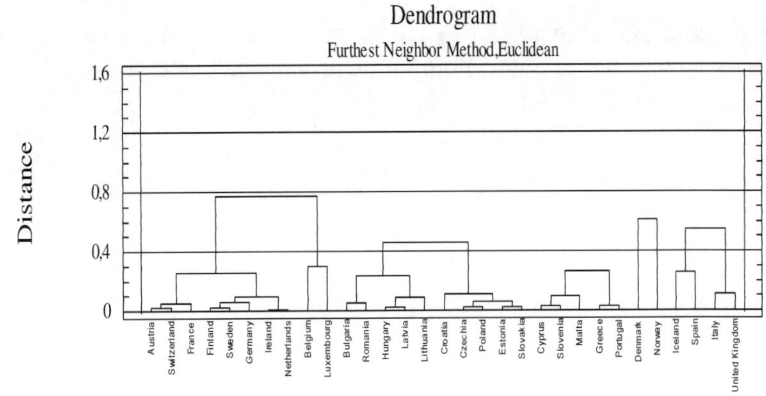

Source: Own research

Fig. 30. Cluster analysis using five clusters, method of the farthest neighbour and Euclidean distance metric, nominal average wage in 2015.

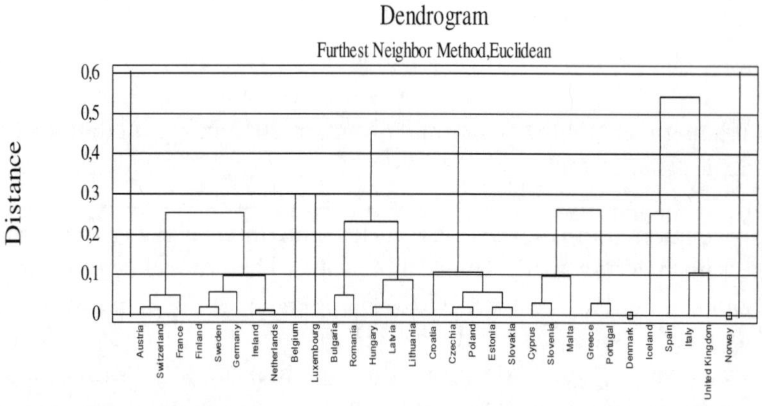

Source: Own research

Fig. 31. Cluster analysis using seven clusters, method of the farthest neighbour and Euclidean distance metric, nominal everage wage in 2015.

Figure 30 represents the classification of these states into five clusters. The first cluster is formatted by states with rather higher levels of nominal wage, but not with the countries with the highest wage levels in Europe. These are Austria, Belgium, Finland, France, Germany Ireland, Luxembourg, Netherlands, Sweden and Switzerland. It is interesting that the whole of the second cluster is represented by the countries of the former socialist block with the exception of Slovenia: Bulgaria, Croatia, Czech Republic, Estonia, Hungary, Latvia, Lithuania, Poland, Romania and Slovakia. The third cluster is made up of rather poorer countries of the original "Western" Europe (Cyprus, Greece, Malta and Portugal) plus Slovenia. Two states with the highest levels of nominal wages, Norway and Denmark, represent the fourth cluster. Four countries; Iceland, Italy, Spain and United Kingdom, represent the fifth cluster.

Figure 31 represents the allocation of European countries into seven clusters. Eight advanced countries of Western and Northern Europe constitute the first cluster. There are Austria, Finland, France, Germany, Ireland, Netherlands, Sweden and Switzerland. Only two states with very high level of nominal wages; Belgium and Luxembourg, form the second cluster. European countries of the former socialist bloc, with the exception of Slovenia, represent the entire third cluster. There are Bulgaria, Croatia, Czech Republic, Estonia, Hungary, Latvia, Lithuania, Poland, Romania and Slovakia. Cluster number four is constituted by somewhat less developed countries of the Western and Southern Europe, such as Cyprus, Greece, Malta, Portugal and Slovenia. The fifth and the seventh clusters have only one country with the highest level of nominal wages. Denmark constitutes the fifth cluster and Norway the seventh cluster. The remaining four countries, such as Iceland, Italy, Spain and United Kingdom form the sixth cluster.

Figures 32 and 33 allow for some comparison of material aspects of the standard of living in 36 selected countries. Figure 32 presents annual average gross and minimum wages in 2015 after the conversion into purchasing power parity (PPP) considering the different living costs and price levels in each country. Figure 33 shows the ratio of the minimum wage to the average gross wage and that of the old-age pension to net income before retirement in 2015. (Zero minimum wages and their ratios in respective figures mean that this institute has not been enacted in the given countries.) The above figures indicate an economic lag in terms of living standards in the Eastern European countries representing the former socialist bloc compared to the developed Western European countries.

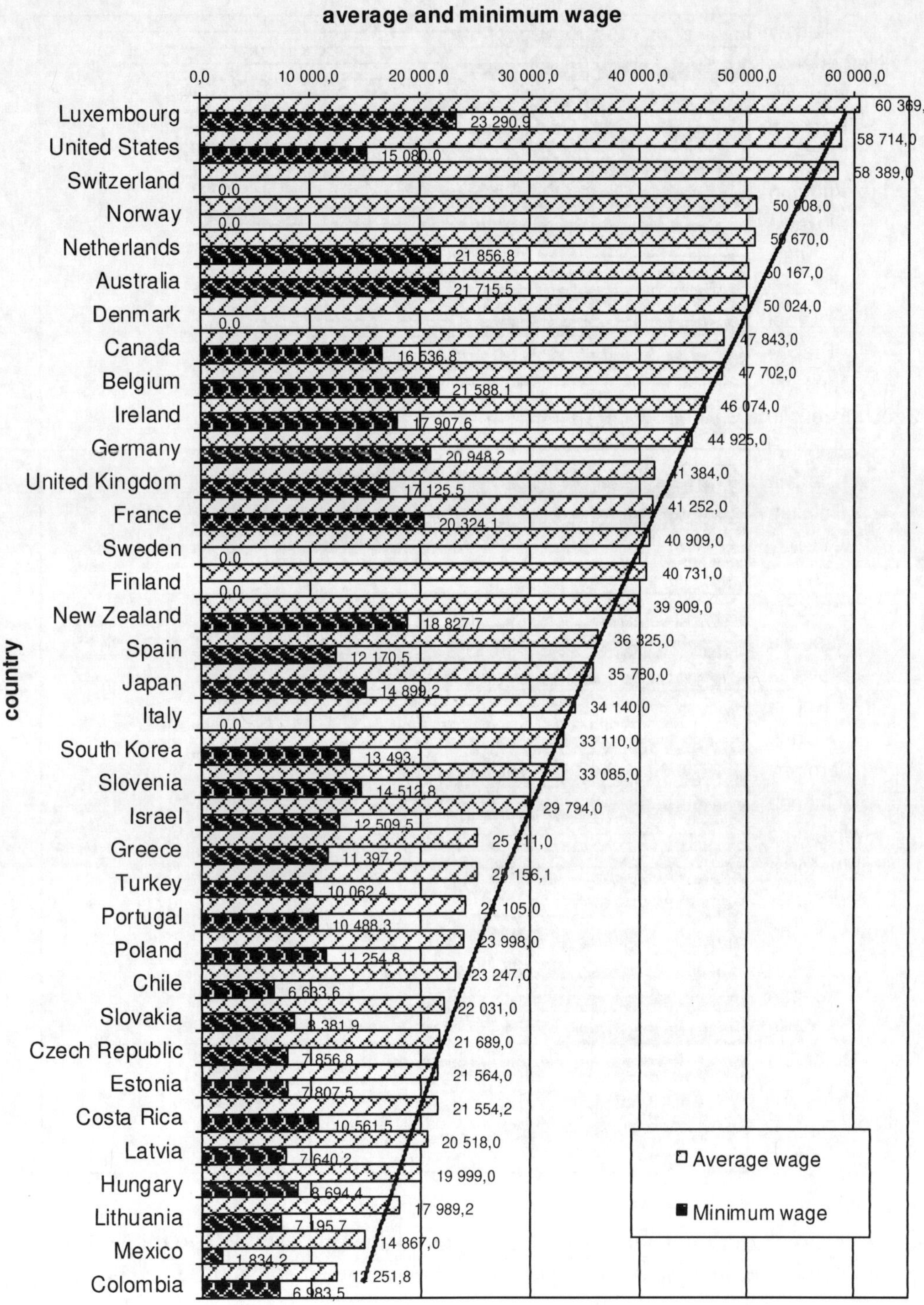

Source: stats.oecd.org; own customization

Fig. 32. Annual average gross and minimum wages – PPP-adjusted in USD of constant 2015 prices – in selected countries in 2015.

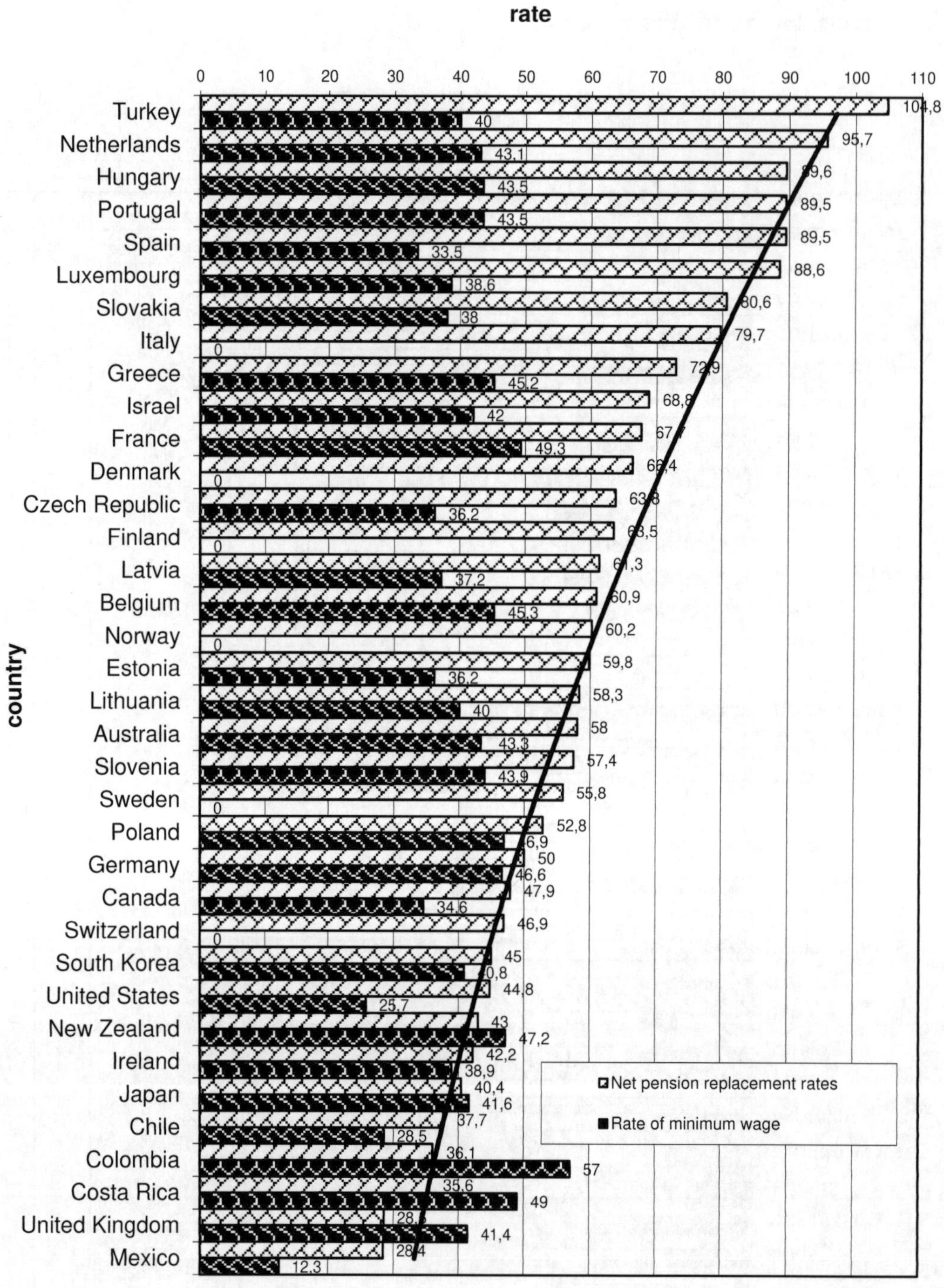

Source: stats.oecd.org; own customization

Fig. 33. Ratio of minimum wage to average gross wage (in %) and net old-age pension divided by net income before retirement (in %) in selected countries in 2015.

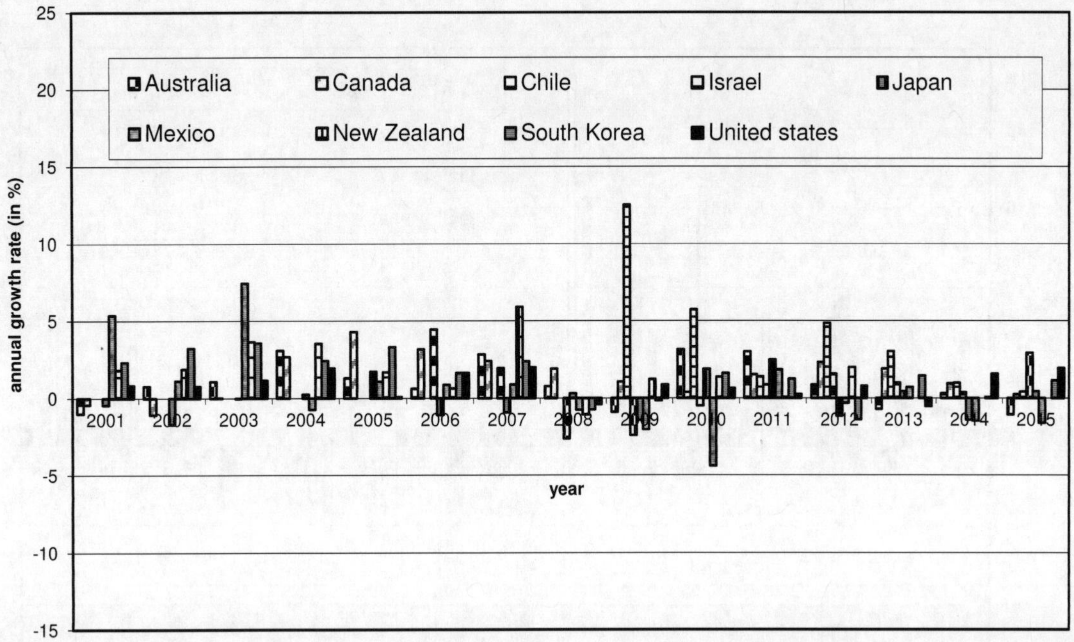

Source: Own research

Fig. 34. Annual growth rates (in %) of average gross wages – PPP-adjusted in USD of constant 2015 prices – in selected non-European countries from 2001 to 2015.

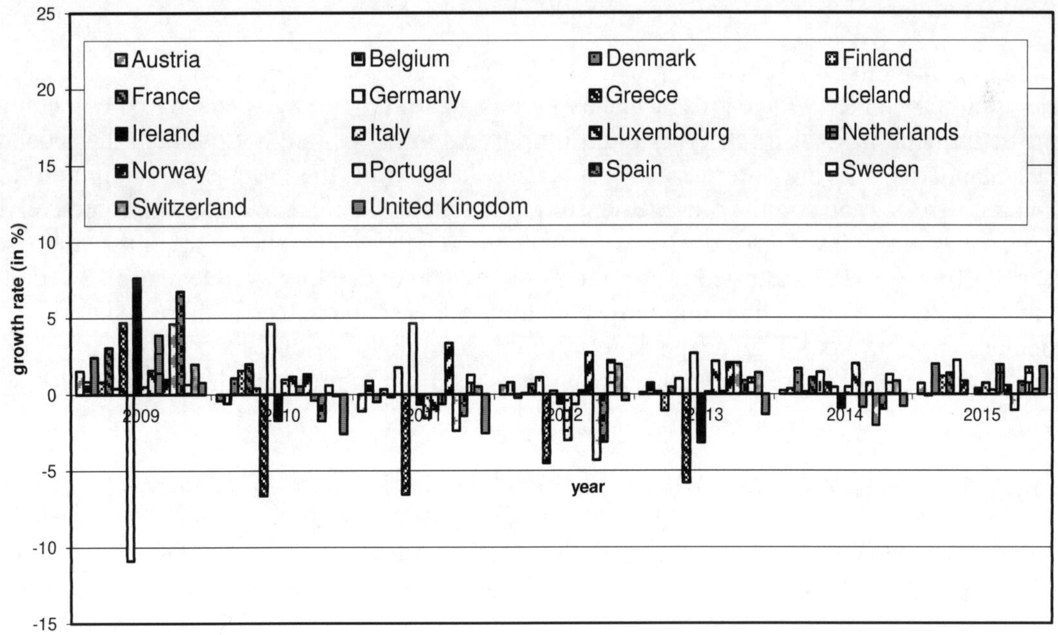

Source: Own research

Fig. 35. Annual growth rates (in %) of average gross wages – PPP-adjusted in USD of constant 2015 prices – in selected Western European countries from 2009 to 2015.

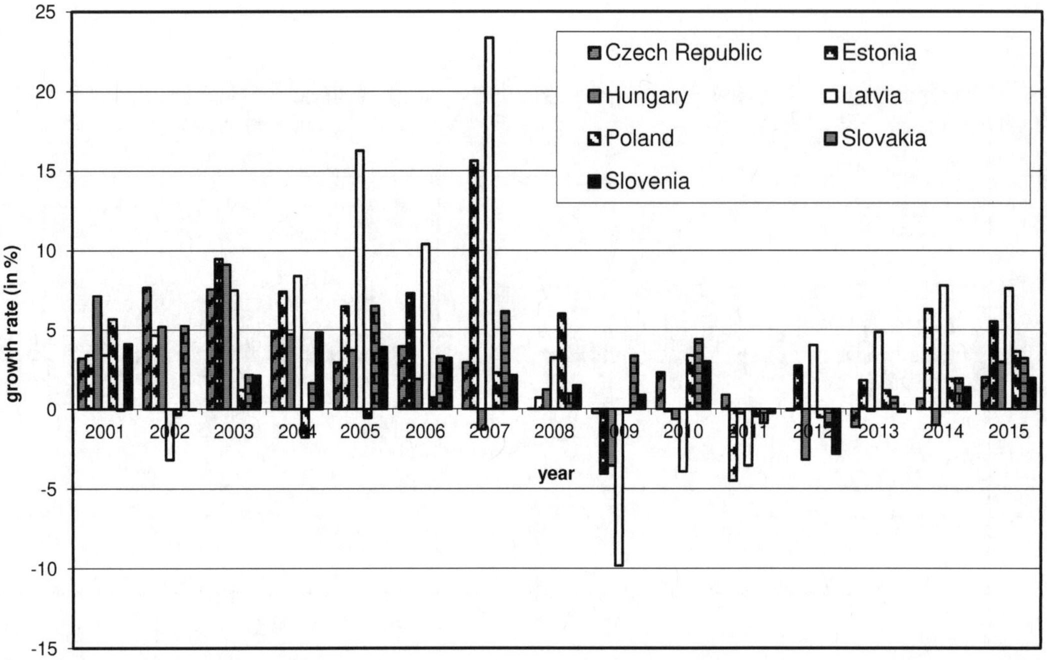

Source: Own research

Fig. 36. Annual growth rates (in %) of average gross wages – PPP-adjusted in USD of constant 2015 prices – in selected Eastern European countries (former socialist bloc) from 2001 to 2015.

It is obvious from Figure 32 that employees in Luxembourg earn the highest wages, the ratio of the minimum to the average gross wage as well as that of the old-age pension to net income before retirement being also relatively high. Employees in the United States reach the second highest standards of living in terms of the wage level, however, the ratios of the minimum to the average gross wage and the old-age pension to net income before retirement, respectively, are relatively low. Switzerland, Norway and the Netherlands show the next highest purchasing power parity-adjusted wage rates. In Switzerland, Norway, Denmark, Sweden, Finland and Italy, however, no minimum wage threshold is enacted, the first three countries in particular reporting very high wage levels. Employees in less developed Latin American states, on the other hand, receive rather poor wages, the minimum wage threshold being also low.

Figures 34–36 present the annual growth rates[6] in three groups of countries from the beginning of the 21st century. They provide available data on wage level developments after the conversion into purchasing power parity in selected countries outside of Europe (Fig. 34), the developed Western European countries (Fig. 35) and the former socialist countries (Fig. 36).

---

6) If y_t and y_{t-1} are the values of time series in t and t–1 points of time, respectively, the growth rate is defined

$$100 \cdot \frac{y_t}{y_{t-1}} - 100 \text{ (in \%)}.$$

Its 2000 calculation requiring the value of the time series for 1999. That is why the values of the growth rate for 2000 are not presented here.

The 2008–2015 data available indicate a rather rapid wage growth in Chile, the average annual growth rate of PPP-adjusted wages being about 4 per cent. The same growth was recorded in Estonia and Latvia, the latter country reaching an average of almost 5 percent annual growth throughout the period. With the onset of the economic recession, the wage levels decreased significantly in some European countries (see Fig. 36), the yearly decline in average wage in Latvia, for example, being almost 10 per cent in 2009.

Figure 37 provides information on the gender wage gap in chosen European countries over the period 2009–2015. It is apparent from the figure that the gap does not change in time within the individual countries. The difference between male and female wages is the largest in Estonia (almost 30 per cent in 2012 and 2013), the Czech Republic, Austria and Germany, the smallest gap being in Slovenia, where women even earned almost one per cent more on average than men in 2009.

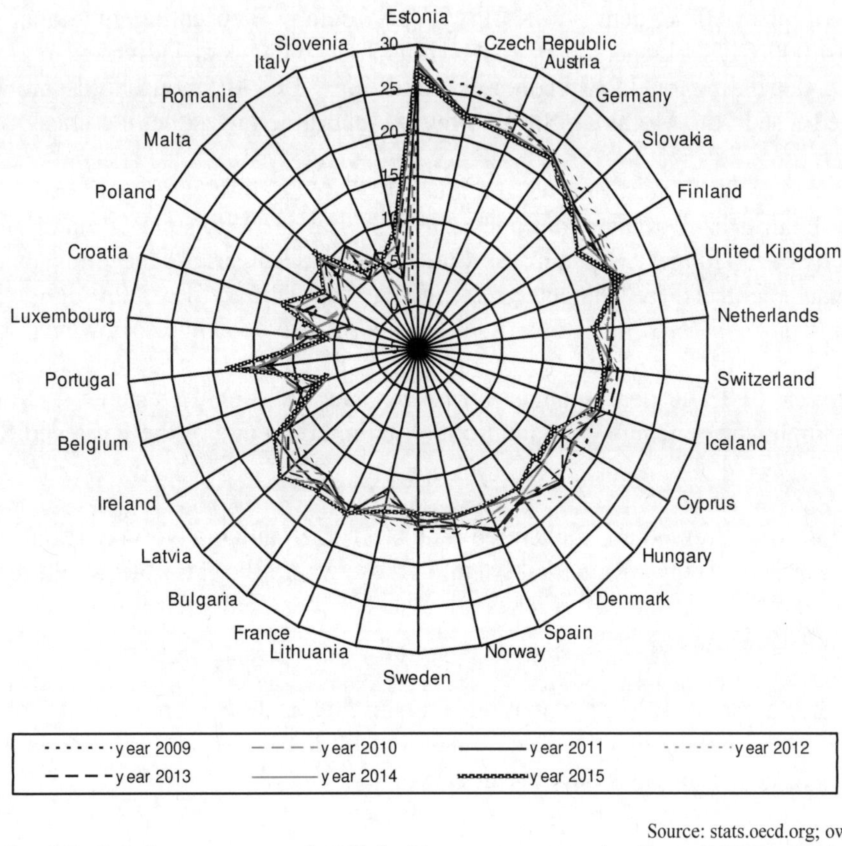

Source: stats.oecd.org; own customization

Fig. 37. Gender wage gap (in %) in European countries from 2009 to 2015.

5 Conclusion

Currently, the wage level is increasing in all regions of the Czech Republic, the wages of both men and women growing. Having slowed down between 2010 and 2013, the annual wage growth has accelerated in the whole of the Czech Republic since then, the average wage increasing by more than 2,300 CZK in most regions. The slowest growth was recorded in

Moravian-Silesian and Usti regions (by 1,859 and 2,125 CZK, respectively), the fastest in the Central Bohemian region (by 3,260 CZK). The lowest wage level has long been reported in Karlovy Vary region – the average wage was a third lower than in the capital of Prague in 2015, the difference between the two regions gradually diminishing, the greatest one having been in 2011.

The highest and lowest wages are reported in Prague and Karlovy Vary regions, respectively, the average gross monthly wage amounting to 36,371 CZK in the former, compared to only 24,119 CZK in the latter region in 2015. Residents of Central Bohemian, Pilsen and South Moravian regions receive relatively high wages, averaging 27,997, 27,013 and 27,051 CZK, respectively, in the same year. High-income regions, however, are also characterized by relatively wide gender wage gaps.

It follows from the results of the 2015 Living Conditions Survey that net household income per capita increased by 4,400 CZK on average in 2014 (latest data available), compared to the previous year. The annual fall in income occurred only in Vysocina region, the sharpest increase – more than 10 per cent – in Karlovy Vary region, representing an annual growth of more than 14,000 CZK per person in cash. Household income in individual regions varies considerably, depending on local economic conditions. Best-off households are in Prague region, while those in the Moravian-Silesian region report the lowest net income.

Looking at the map of Eastern Europe, the area of relatively low wages stretches from north to south. It begins in Estonia, continues through Latvia, Lithuania, Poland, the Czech Republic, Slovakia, Hungary, Croatia, Serbia, Romania, Bulgaria, Macedonia and ends in Turkey. These countries are characterized by a cheap labour force, through which they apparently try to lure foreign investors, much cheaper Asian economies, however, effectively competing with them. Labour costs are higher in all countries west of us, reaching the maximum (over 40 EUR per hour) in Denmark and two non EU states – Norway and Switzerland, high wages being symptomatic for Luxembourg and in the Paris and Stockholm areas.

The populations of Luxembourg, the United States, Switzerland and Norway reach the highest material standards of living. (Remarkably, in Turkey, where the ratio of the old-age pension to net income before retirement reaches almost 105 per cent, pensioners receive higher amounts than wage earners.)

Low labour costs themselves do not reduce unemployment. (In Macedonia, for instance, unemployment rates have long exceeded 30 per cent.) It is obvious that the competitiveness is affected by average labour costs only marginally.

Acknowledgement

This paper was subsidized by the funds of institutional support of a long-term conceptual advancement of science and research number IP400040 at the Faculty of Informatics and Statistics, University of Economics, Prague, Czech Republic.

References

[1] BARBER, G. M. *Elementary Statistics for Geographers.* New York: The Guilford Press, 1988.

[2] BROCKWELL, P. J., DAVIS, R. A. *Introduction to Time Series and Forecasting*. New York: Springer, 2002.

[3] COWPERWAIT, P. S. B., METCALFE, A. V. *Introductory Time Series with R*. New York: Springer, 2009.

[4] JOHNSON, N. L., KOTZ, S., BALAKRISHNAN, N. *Continuous Univariate Distributions 1*. New York: John Wiley & Sons, 1994.

[5] JOHNSON, N. L., KOTZ, S., BALAKRISHNAN, N. *Continuous Univariate Distributions 2*. New York: John Wiley & Sons, 1995.

[6] KLEIBER, C., KOTZ, S. *Statistical Size Distributions in Economics and Actuarial Sciences*. New York: John Wiley & Sons, 2003.

[7] MAKHALOVA, E., PECÁKOVÁ, I. The Fuzzy Clustering Problems and Possible Solutions. In *Proceedings of the 9th International Days of Statistics and Economics – MSED*, Prague, Czech Republic, 2015, ISBN 978-80-87990-06-3, pp. 1052–1061.

[8] MALEC, L. Some Remarks on the Functional Relation Between Canonical Correlation Analysis and Partial Least Squares. *Journal of Statistical Computation and Simulation*, 2016, 86(12): pp. 2379–2391.

[9] ŘEZANKOVÁ, H., LÖSTER, T. Shluková analýza domácností charakterizovaných kategoriálními ukazateli (Cluster Analysis of Households Characterized by Categorical Indicators). *E+M Ekonomie a Management*, 2013, 16(3): pp. 139–147.

[10] TRIOLA, M. F. *Elementary Statistics*. Redwood City: The Benjamin/ Cummings Publishing Company, 1989.

Current address

Bílková Diana, doc. Ing., Dr.
Department of Statistics and Probability, Faculty of Informatics and Statistics
University of Economics, Prague
Sq. W. Churchill 1938/4, 130 67 Prague, Czech Republic
E-mail: diana.bilkova@vse.cz

Slovak University of Technology in Bratislava
Faculty of Mechanical Engineering
17th Conference on Applied Mathematics **APLIMAT 2018**

Proceedings

BASIC PRINCIPLES OF ORTHOGONAL AXONOMETRY WITH THE USE OF GEOGEBRA

BÍMOVÁ Daniela (CZ), PIRKLOVÁ Petra (CZ)

Abstract. The contribution introduces the use of dynamic applets created in GeoGebra in teaching the orthogonal axonometry for students of the third year of a Bachelor's degree Mathematics with a focus on Education. The dynamic applets are used with a big advantage in teaching for students with a lower level of spatial imagination, or for part-time students. The presented dynamic applets show chosen basic principles of the orthogonal axonometry.

Keywords: teaching, orthogonal axonometry, GeoGebra, dynamic applets

Mathematics Subject Classification: Primary 97B50, 68U05; Secondary 51N05

1 Introduction

Besides all, the students of the third year of a Bachelor's degree Mathematics with a focus on Education, Faculty of Science, Humanities and Education, Technical University of Liberec study the course Geometry 2. There are taught four themes – axiomatic principles of the spatial geometry, stereometry, Monge projection, and orthogonal axonometry - in this course. Unfortunately, at least half of the students of every year has got troubles with studying this course. A lower level of the spatial imagination and insufficient experience of solving spatial problems from the high schools are the main reasons for their troubles in studying the course Geometry 2. As a result, the students are lagging behind in creating images of spatial situations and problems, and consequently, they are mostly not able to solve the spatial problems correctly.

It is possible to show some appropriate spatial situations using models during the lectures and lessons. On the contrary, there are some spatial situations that are not easy to model them. E.g. the constructions in orthogonal axonometry are more difficult for modelling. For that reason, we started to create the dynamic applets in GeoGebra which bring the given spatial situations closer.

2 The dynamic applets created in GeoGebra

GeoGebra is the freeware mathematical dynamic software. Using various tools of GeoGebra, the users are able to create many different and useful dynamic applets. The applets can be dynamic either due to the main tool of the program – the slider or due to the free objects (points, lengths of line segments, radii of circles, ...) constructed "freely" in GeoGebra.

The slider also allows the connection between the texts inserted in one 2D window, the objects drawn in the second 2D window, and objects constructed in the 3D window. Consequently, it is possible to show either basic principles, or constructive problems drawn in the 2D window and also in the 3D window at the same moment. The users can move or rotate with the spatial images in the 3D window by pressing the right button of the mouse and by dragging the mouse at once. That´s why the spatial images can be seen from different angles of view. As a result, the users can set the position of the spatial image under such an angle of view that fits best to them and in which they are able to see and understand the spatial situation best.

The green buttons titled "*Parallel view onto the spatial situation*" and used in the below-introduced applets are set by the authors under such an angle of view that fits best to them. The settings of the buttons are not able to be changed by the users.

2.1 Sharing of the dynamic applets created in GeoGebra

If the dynamic applets are ready, the task is how to use them in the teaching process. The teacher is able to use the GeoGebra source files of the dynamic applets during the lectures or lessons. On the other hand, there is a question if the students can work with the applets themselves during the lessons or at home. The answer is positive. There are at least two possibilities how the students gain the created dynamic applets. The teacher sends (e.g. via email) the GeoGebra source files of the dynamic applets to the appropriate group of the students or the teacher exports the dynamic applets into his GeoGebra account on the GeoGebra web page [1], see Fig. 1.

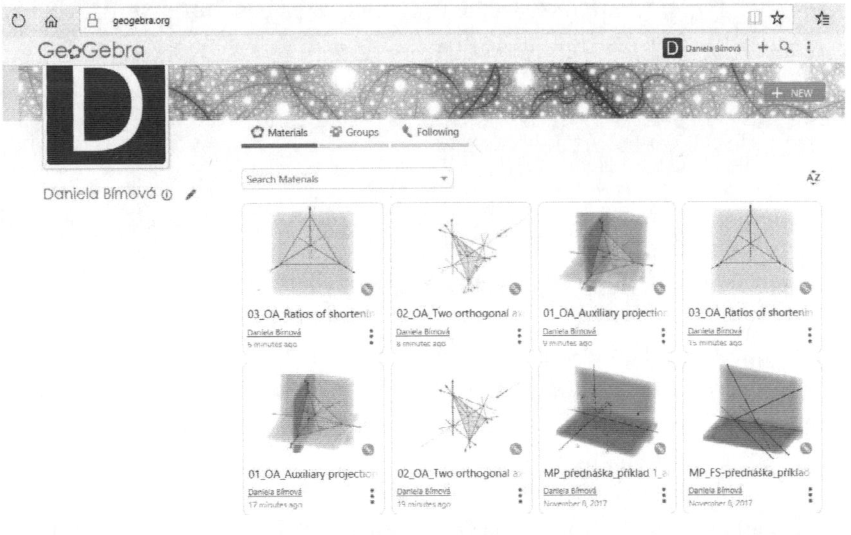

Fig. 1. Overview of the dynamic applets inserted on the GeoGebra web page of the author.

2.1.1 Kinds of sharing the dynamic applets on the GeoGebra web page

There are three different possibilities of sharing the dynamic applets created in GeoGebra on the GeoGebra web page:

- public visibility – everyone can find and view the dynamic applet;
- shared with link – only users who have a link can view the dynamic applet;
- private – only the author of the applet can find and view it.

Before setting the sharing of the dynamic applets as the "public visibility", it is better to share them with a link. In this case, the teacher has got the opportunity to send the links only to chosen groups of students and after that to test the functions and usefulness of the applets together with the students in the teaching process. If all the things work perfectly in the applets, the teacher can set the sharing of the applets as "public visibility".

2.2 Work with the dynamic applets created in GeoGebra

If the lessons of the course Geometry 2 take place in a PC laboratory, every student can find the appropriate applet through the usage of the link typed in the web browser on the computer. If the lessons take place in the classroom, the tablets can be used. Only the internet connection is needed. On the contrary, the internet connection is widespread on all the university campus of TUL. The students could also use their own electronic types of equipment.

GeoGebra was generated for various kinds of electronic equipment such as for PCs, tablets, or for mobile phones. It was also generated for different systems such as i.e. for Windows, Mac, Linux, or Android. For that reason, the dynamic applets can be used by the students in a lot of different electronic equipment. Let´s say, it can be used everywhere where there is the internet connection and it can be used by all the users that have the link. It means, every student should have the opportunity to find out the dynamic applets if he has obtained the links.

The work with the dynamic applets created in GeoGebra is quite intuitive. *Checkboxes, buttons,* or *sliders* help students how to follow the process of construction or solution in the dynamic applets. There are introduced three dynamic applets created in GeoGebra concerning the basic principles of orthogonal axonometry in the next section.

3 Dynamic applets of some basic principles of orthogonal axonometry in GeoGebra

3.1 Basic elements of the orthogonal axonometry

Orthogonal axonometry is a drawing method in which there is orthogonally drawn onto axonometric projection plane and onto three auxiliary projection planes – the top-view, the front-view, and the side-view auxiliary projection planes. It is true that all three auxiliary projection planes are orthogonal to each other. We can coincide the top-view projection plane with the coordinate plane (xy), further the front-view projection plane with the coordinate plane (yz), and finally the side-view projection plane with the coordinate plane (xz). Next, it must be true that the axonometric projection plane is concurrent with all three auxiliary planes. It intersects the axes *x*, *y*, and *z* in a row in points *X*, *Y*, and *Z*. The triangle *XYZ* is called an axonometric triangle. The orthogonal projections x_a, y_a, z_a of the coordinate axes *x*, *y*, *z* are the heights of the triangle *XYZ*. The point of intersection O_a of the heights of the triangle *XYZ* is the orthogonal projection of the origin of the Cartesian coordinate system {*O*, *x*, *y*, *z*} onto the axonometric plane. Above-mentioned, the axonometric triangle *XYZ* lies in the axonometric projection plane.

There are shown all the mentioned basic elements of the orthogonal axonometry in Fig. 2. Fig. 2 displays the dynamic applet available on the GeoGebra website using the link [2].

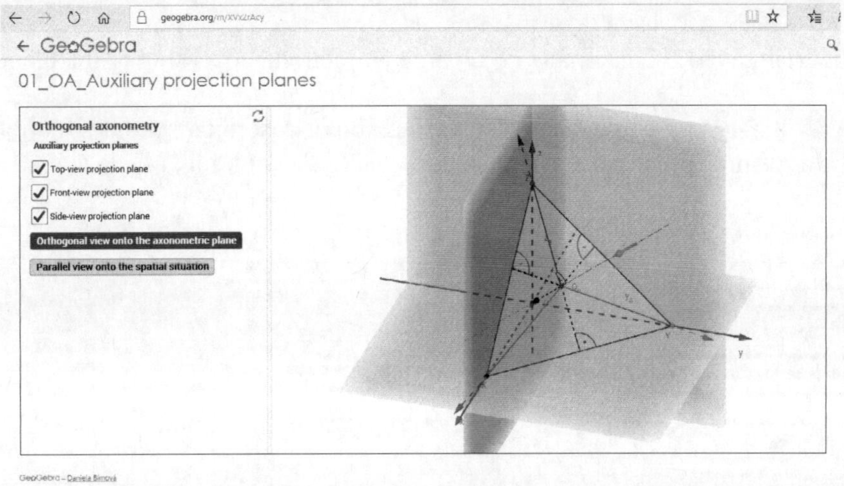

Fig. 2. Figure of the dynamic applet containing basic elements of the orthogonal axonometry.

The dynamic applet drawn in Fig. 2 consists of two windows. There is the 2D window on the left side and the 3D window on the right side of the Fig. 2. Besides of the titles, three checkboxes, that show or hide the particular auxiliary planes in the 3D window, and two buttons, pressing them, the direction of the point of view onto the spatial image is changed, are in the 2D window. The set "*Parallel view onto the spatial situation*" by the author is shown in the 3D window in Fig. 2. The preview of the dynamic applet in "*Orthogonal view onto the axonometric plane*" in a version for anaglyphic glasses is displayed in Fig. 3. The applet in version for anaglyphic glasses can be found on GeoGebra website by the link [3].

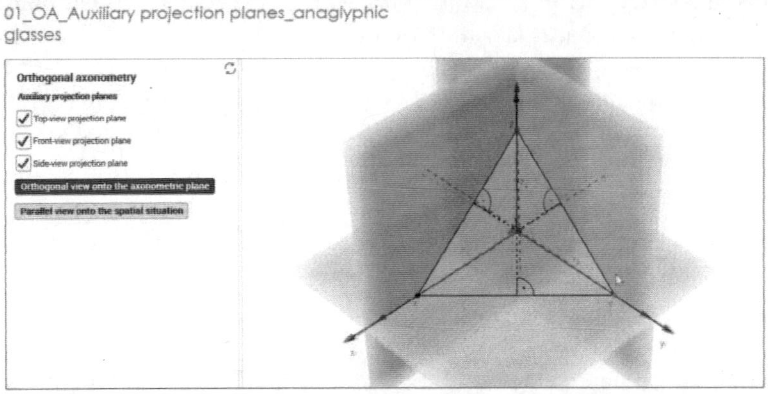

Fig. 3. The preview of the dynamic applet in the version for anaglyphic glasses.

3.2 Two orthogonal axonometries

In general, the axonometric projection plane coincides with the drawing plane. If there is given an acute-angle triangle *XYZ* in the drawing plane, two orthogonal axonometries exist for which the given triangle *XYZ* is the axonometric triangle. Further, if there are given three straight-lines passing through one point in such a way that every of the given straight-lines passes through the obtuse angle of the rest two straight-lines, two orthogonal axonometries exist for which three given straight-lines are the straight-lines of the axonometric axial cross.

It means that there are two different Cartesian coordinate systems $\{O, x, y, z\}$ and $\{O', x', y', z'\}$ which axes have their axonometric projections $x_a \equiv x'_a$, $y_a \equiv y'_a$, $z_a \equiv z'_a$ in the heights of the triangle *XYZ*. The points *O, O'* are symmetrical according to the plane *XYZ*.

The preview of the dynamic applet displaying the above described spatial situation is shown in Fig. 4. The dynamic applet itself can be opened using the web link [4].

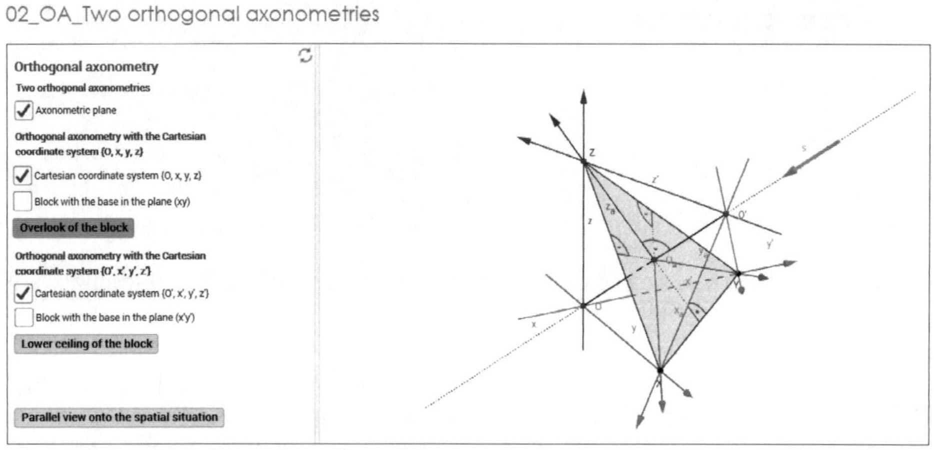

Fig. 4. The preview of the dynamic applet showing two Cartesian coordinate systems corresponding to one axonometric triangle *XYZ*.

We can imagine the axonometric projection plane as the vertical plane, after that one of the points *O, O'* is situated in front of the axonometric plane in the space. It is the point *O'* in our case. In the same half-space, there is also the observer. Follows, the second point is set behind the axonometric plane in the space. It is the point *O*.

Let's imagine that there is situated a block in each of two given Cartesian coordinate systems. If we think of the Cartesian coordinate system $\{O, x, y, z\}$ and if we look at the spatial situation orthogonally in the direction to the axonometric projection plane, we see the overlook of the given block. See Fig. 5. On the contrary, if we think of the Cartesian coordinate system $\{O', x', y', z'\}$ and if we look at the spatial situation orthogonally in the direction to the axonometric projection plane, we see the lower ceiling of the given block. In Fig. 6 there is displayed the lower ceiling of such a block.

Both the Cartesian coordinate systems together with the blocks are incorporated into one dynamic applet available using the above-mentioned link [4].

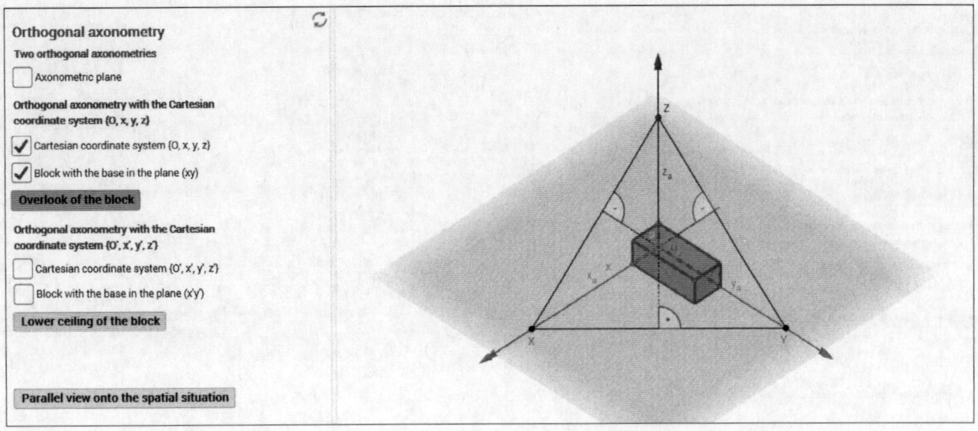

Fig. 5. The overlook of the block with one of its bases lying in the plane (*xy*)

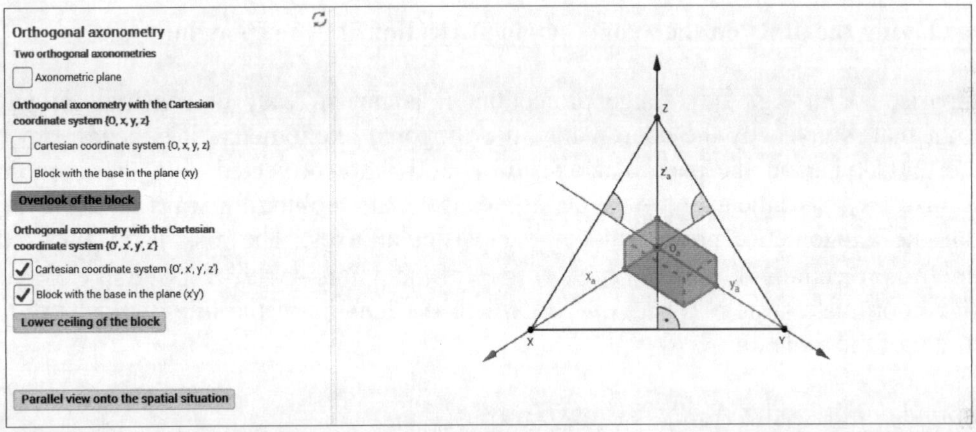

Fig. 6. The lower ceiling of the block with one of its bases lying in the plane (*x'y'*)

There are used five checkboxes in the applet. The corresponding objects of the spatial situation are interconnected with the appropriate checkbox. Activating the particular checkbox, the interconnected objects appear on the right side of the applet in the 3D window. The checkbox is activated if the check mark is in it. See e.g. Fig. 6. Further, there are two buttons - "*Overlook of the block*" and "*Lower ceiling of the block*" – that enable to switch between the overlook and the lower ceiling views of the block. The third button of the applet – "*Parallel view onto the spatial situation*" is programmed in such a way to all the drawn objects rotate into the set position in which the spatial situation is better seen according to the author.

The dynamic applet in the version for anaglyphic glasses was also generated for displaying two orthogonal axonometries with one corresponding axonometric triangle *XYZ*. It can be found out using the web link [5]. Fig. 7 shows the preview of "*Parallel view onto the spatial situation*" of overlooking of the block in the version for anaglyphic glasses.

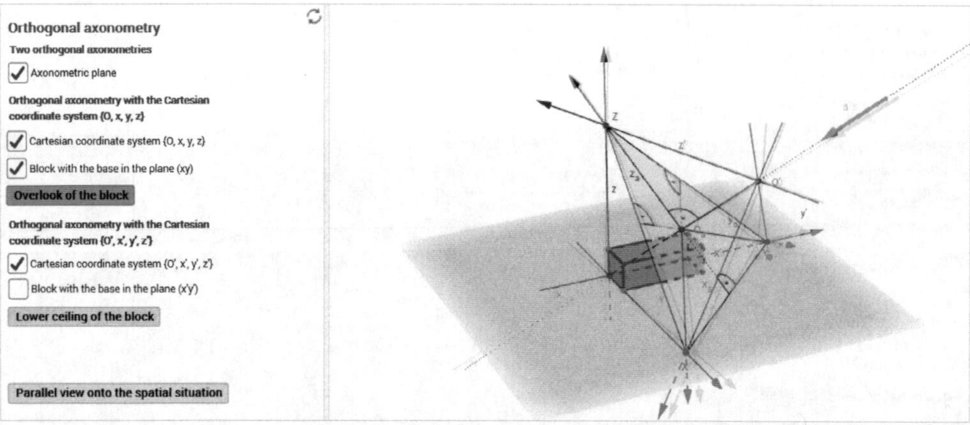

Fig. 7. Parallel view onto the spatial situation of overlooking of the block in the version for anaglyphic glasses.

3.3 Displaying the units on the axonometric projections of the coordinate axes

In comparison with e.g. the Monge projection, it is not so easy to construct projections of a point that is given by the coordinates in orthogonal axonometry. It is needful to know how the units lying on the particular axes x, y, and z are projected onto the axonometric projections x_a, y_a, z_a of the axes x, y, and z. It means we are looking for ways of displaying the units on the axonometric projections of the particular axes. The task is to find out the axonometric projections $A_a \in x_a$, $B_a \in y_a$, $C_a \in z_a$ of the points $A \in x$, $B \in y$, and $C \in z$ of the Cartesian coordinate system $\{O, x, y, z\}$ in which the lengths of the line segments OA, OB, and OC are of the same unit.

We can find out the points A_a, B_a, C_a in two ways.

Firstly, the axonometric projections $A_a \in x_a$, $B_a \in y_a$, $C_a \in z_a$ of the points $A \in x$, $B \in y$, and $C \in z$ can be constructed by rotation of the auxiliary planes round the lines of intersection between auxiliary projection planes and axonometric projection plane into the axonometric projection plane. We put the unit in its true length on the rotated axes x_0, y_0 in the axonometric projection plane. As a result, we receive the points A_0, B_0 on the axes x_0, y_0. It is true that $|A_0O_0| = |B_0O_0| = j$. Using the planar axial affinity, we construct the axonometric projections A_a, B_a of the points A, B on the axonometric projections x_a, y_a of the axes x and y. For better imagination, the reader can see Fig. 10.

Secondly, the axonometric projections $A_a \in x_a$, $B_a \in y_a$, $C_a \in z_a$ of the points $A \in x$, $B \in y$, and $C \in z$ can be drawn by folding of the plane passing through the corresponding axis and its axonometric projection. In Fig. 11 there is shown the step by step process of folding of the plane (zz_a) into the axonometric projection plane. The unit j is put on the folded axis (z) in its true length in the axonometric projection plane. Consequently, we receive the point (C) on the axis (z) for which it is true that $|(O)(C)| = j$. Using the planar axial affinity, we draw the axonometric projection C_a of the point C onto the axonometric projection z_a of the axis z.
Both the described ways of constructions are inserted into one dynamic applet created in GeoGebra. There are three parts in this dynamic applet. On the left side, there is one of the 2D

windows. The user can find two checkboxes in this 2D window after opening the applet. Each checkbox corresponds to one way of the above-described constructions. Next, three buttons for changing the angle of view onto the spatial situation are at the bottom of the left 2D window. The second 2D window is situated in the middle of the applet. There are drawn constructions in orthogonal axonometry in this middle 2D window. On the right side, there is the 3D window in which 3D images of the axonometric constructions are displayed. See Fig. 8.

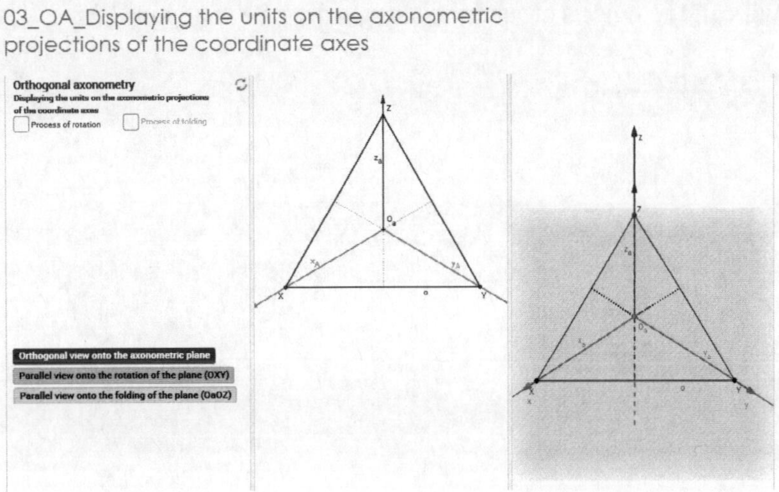

Fig. 8. Preview of the dynamic applet created in GeoGebra showing two ways of the constructions of displaying the units on the axonometric projections of the coordinate axes after its opening.

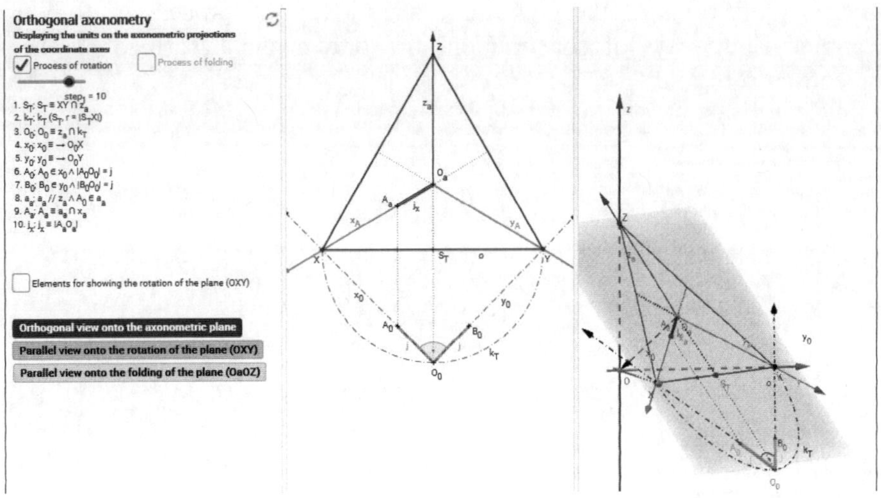

Fig. 9. The tenth step of the process of the construction using the way of rotation.

In this GeoGebra applet, there is used the abovementioned connection between 2D, 3D constructions and symbolic construction entries. The objects in 2D, 3D constructions and the particular steps of the symbolic construction entries are connected together with the usage of the dynamic tool of GeoGebra – with the slider Moving the slider, the particular steps of the symbolic construction entries appear in the left 2D window and the corresponding 2D and 3D objects appear in the middle 2D window and in the right 3D window in one moment. The tenth step of the process of construction of the axonometric projections A_a, B_a of the points A, B on the axonometric projections x_a, y_a of the axes x, y is shown in Fig. 9..

Activating the checkbox "*Elements for showing the rotation of the plane* (*OXY*)", the objects for showing the process of rotation in 3D construction appear. Moving the point O_0, the plane (*OXY*) rotates with all the objects around the straight-line *XY*. See Fig. 10.

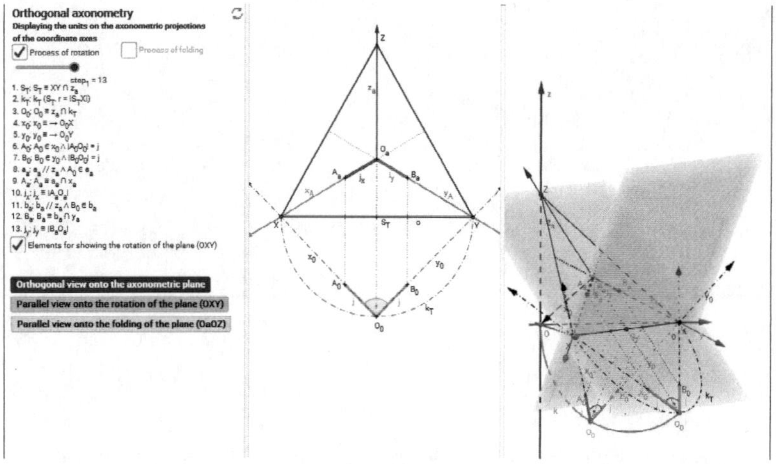

Fig. 10. Rotation of the plane (*OXY*) around the axis *XY* in 3D construction.

The similar process of solution is created also for the second way of displaying the units on the axonometric projections of the coordinate axes - for the process of folding - in the GeoGebra applet. Both ways of constructions are drawn and described by the symbolic construction entries in Fig. 11.

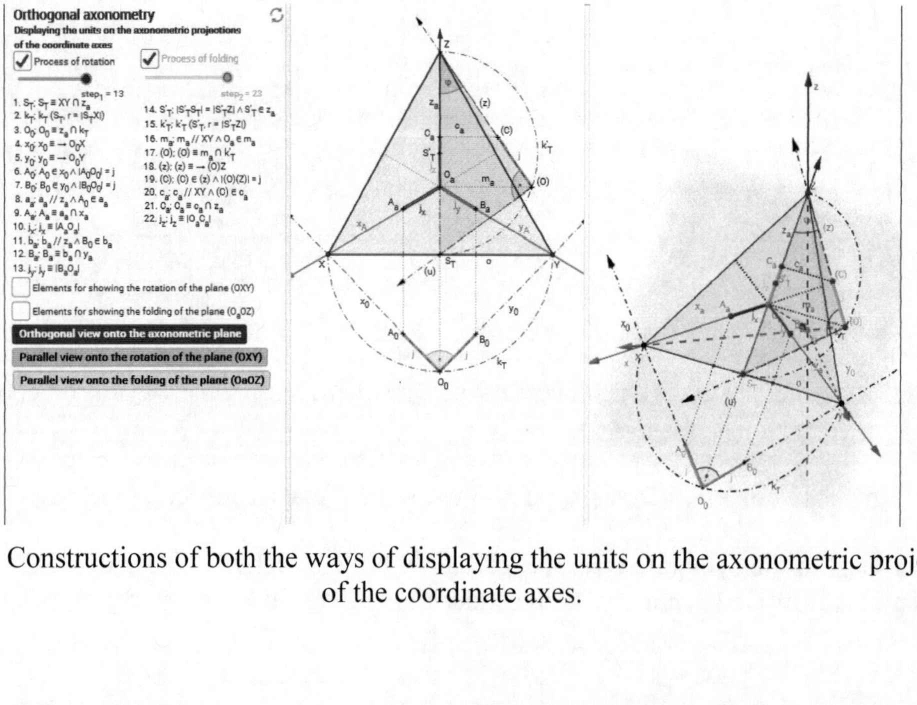

Fig. 11. Constructions of both the ways of displaying the units on the axonometric projections of the coordinate axes.

The possibility of showing only the construction of folding the plane (zz_a) also exists. The checkbox „*Process of rotation*" must be deactivated in this case.

The above-mentioned applet can be opened via the web link [6]. The same applet generated in the version for anaglyphic glasses is possible to found on the web link [7].

4 Conclusion

In summary, we are able to say on the basis of our experiences that the best way of teaching the various kinds of projection methods of the descriptive geometry (e.g. orthogonal axonometry) is the combination of classical interpretation of the principles and problems of the projection method using a marker and a whiteboard, of electronic presentation containing a text and static images of spatial situations and solved problems, and of using the e-learning material. Dynamic applets created in GeoGebra can be used with a big advantage as the e-learning material that helps students to see better the spatial situations and that shows students the step by step processes of solving the spatial problems. The dynamic applets are very useful especially for part-time students of a Bachelor's degree Mathematics with a focus on Education during their study at home.

Acknowledgement

The paper was supported by OP VVV grant of the registration number CZ.02.3.68/0.0/0.0/16_011/0000664 entitled "Společenství praxe pro podporu vzdělávání – SPPPV".

References

[1] URL, http://www.geogebra.org/.
[2] URL, https://www.geogebra.org/m/XVx2rAcy.
[3] URL, https://www.geogebra.org/m/faJCWHEc.
[4] URL, https://www.geogebra.org/m/YFRRJqHH.
[5] URL, https://www.geogebra.org/m/rPn9WVHw.
[6] URL, https://www.geogebra.org/m/Zda2jvV4.
[7] URL, https://www.geogebra.org/m/ABNxwqQy.
[8] URBAN, A., *Deskriptivní geometrie 1*. SNTL, Praha 1965, 263 p.

Current address

Bímová Daniela, Mgr. Ph.D.
Department of Mathematics and Didactics of Mathematics; Faculty of Science Humanities and Education
Technical University of Liberec
Studentská 2, 461 17 Liberec, Czech Republic
E-mail: daniela.bimova@tul.cz

Pirklová Petra, Mgr. Ph.D.
Department of Mathematics and Didactics of Mathematics; Faculty of Science
Humanities and Education
Technical University of Liberec
Studentská 2, 461 17 Liberec, Czech Republic
E-mail: petra.pirklova@tul.cz

Slovak University of Technology in Bratislava
Faculty of Mechanical Engineering
17th Conference on Applied Mathematics **APLIMAT 2018**

Proceedings

PROPERTIES OF REDUCED MEET IDEALS IN LOWER BCK–SEMILATTICES

BORDBAR Hashem (IR), NOVÁK Michal (CZ), CRISTEA Irina (SI), JUN Young Bae (KR)

Abstract. In our paper we include some new results on lower BCK–semilattices, which are a special class of BCK-algebras. We concentrate on a specific c lass o f i deals o f lower BCK–semilattices. We make use of some recent results on weak closure operations on ideals of lower BCK–semilattices or relative annihilators in lower BCK–semilattices obtained by Jun et al.

Keywords: BCK–algebra, ideal, lower BCK–semilattice, relative annihilator

Mathematics subject classification: Primary 06F35, 06A12

1 Introduction

In our paper we include some new results in the area of BCK*–algebras*. This concept was introduced in 1966 by Iséki [16] and brought to a shape by Iséki and Tanaka in papers such as [17]. BCK–algebras are a special case of BCI–algebras, named so because their axioms work with certain combinators used in combinatory logic. For a collection of results on BCI– / BCK–algebras see [15, 20]; for the underlying theoretical concepts see e.g. papers by Bundler, Meyer or Hindley [6, 7, 14] and book [12] written by Curry and Feys.

Definition 1. *An algebra* $(X; *, 0)$ *of type* $(2, 0)$ *is called a* BCI–algebra *if it, for all* $x, y, z \in X$, *satisfies the following conditions:*

(i) $((x * y) * (x * z)) * (z * y) = 0$,

(ii) $(x * (x * y)) * y = 0$,

(iii) $x * x = 0$,

(iv) *simultaneous validity of* $x * y = 0$ *and* $y * x = 0$ *implies that* $x = y$.

In BCI–algebras we can define, for all $x, y \in X$, a relation "$\leq$", called a *BCI–ordering*, by setting

$$x \leq y \text{ whenever } x * y = 0. \tag{1}$$

It is easy to show that "$\leq$" is a partial ordering on X.

The nature of BCI–algebras is such that the operation "$*$" can often *intentionally* be neither associative nor commutative in the usual sense of the word. In the special case of *associative BCI–algebras*, i.e. BCI–algebras $(X; *, 0)$ such that, for all $x, y, z \in X$, there is $(x * y) * z = x * (y * z)$, we have that $x * y = y * x$ and $0 * x = x$ for all $x, y \in X$ (all these statements are in fact equivalent).

Definition 2. *A BCI–algebra is called a* BCK–algebra *if, for all $x \in X$, there is $0 * x = 0$. A BCK–algebra is called* commutative *if, for all $x, y \in X$ there is $x * (x * y) = y * (y * x)$. A BCK–algebra is called a* lower BCK–semilattice *if $(X, \leq)$, where "$\leq$" is a BCI–ordering, is a lower semilattice.*

Thus, lower BCK–semilattices are a special – yet distinct and important – class of lower semilattices. *Bounded BCI/BCK*–algebras are such BCI/BCK–algebras that have the greatest element, which is usually denoted by 1. In a bounded commutative BCK–algebra the least upper bound of an arbitrary pair of elements $x, y \in X$ satisfies

$$x \vee y = 1 * ((1 * x) \wedge (1 * y)),$$

which means that, in this case, $(X, \leq)$ is a *distributive lattice*. Notice that bounded commutative BCK–algebras are in fact *MV–algebras* introduced by Chang [11].

In our paper we focus on *lower BCK–semilattices* and their *ideals*. The concept of an *ideal* is one of the corner stones of the theory of BCI– / BCK–algebras (see [20]).

Definition 3. *A subset A of a BCK/BCI-algebra X is called an* ideal *of X if, for all $x, y \in X$, there is*

$$0 \in A, \tag{2}$$

$$x * y \in A, y \in A \Rightarrow\ x \in A. \tag{3}$$

Note that – obviously – every ideal A of a BCK/BCI–algebra X satisfies, for all $x, y \in X$, the following implication:

$$x \leq y,\ y \in A\ \Rightarrow\ x \in A. \tag{4}$$

The following terminology is standard for numerous algebraic concepts.

Definition 4. *For any subset A of X, the* ideal generated by A *is defined to be the intersection of all ideals of X containing A. We denote it $\langle A \rangle$. If A is finite, we say that $\langle A \rangle$ is a* finitely generated ideal *of X.*

In what follows we by $\mathcal{I}(X)$ and $\mathcal{I}_f(X)$ mean the set of all ideals of X and the set of all finitely generated ideals of X, respectively. By $(X, \wedge, \leq)$ we mean, unless specified otherwise, a *lower BCK–semilattice*.

2 Preliminaries

In Section 3 we make use of some results on *weak closure operations* on ideals of BCK–algebras or *relative annihilators* in lower BCK–semilattices obtained by Jun et al. [3, 4, 5]. Notice that closure operations or (relative) annihilators are studied for various types of combinatory logic algebras – see e.g. Chajda and Rachůnek [10]. For some results concerning lattices within the theory of BCK–algebras see e.g. papers by Chajda and Länger such as [8, 9].

For the proofs of results included in this section see [4, 5].

Definition 5. *For any nonempty subsets A and B of $(X, \wedge, \leq)$, we denote*

$$A \wedge B := \{a \wedge b \mid a \in A, b \in B\}, \tag{5}$$

which is called the meet set *of X generated by A and B. If $A = \{a\}$, then $\{a\} \wedge B$ is denoted by $a \wedge B$. Also, if $B = \{b\}$, then $A \wedge \{b\}$ is denoted by $A \wedge b$.*

The following example shows that the set $A \wedge B$ need not always be an ideal of X if A, B are *arbitrary* subsets of X.

Example 1. *Consider the lower BCK-semilattice $X = \{0, 1, 2, 3, 4\}$ with the following Cayley table.*

$*$	0	1	2	3	4
0	0	0	0	0	0
1	1	0	1	0	1
2	2	2	0	2	0
3	3	1	3	0	3
4	4	4	4	4	0

X has six ideals: $A_0 = \{0\}$, $A_1 = \{0, 1, 3\}$, $A_2 = \{0, 2\}$, $A_3 = \{0, 1, 2, 3\}$, $A_4 = \{0, 2, 4\}$ and $A_5 = X$. For $A = \{2, 3\}$ and $B = \{1, 4\}$, we have

$$A \wedge B = \{a \wedge b \mid a \in A,\ b \in B\} = \{0, 1, 2\},$$

which is not an ideal of X.

Therefore, we provide conditions for the meet set $A \wedge B$ to be an ideal.

Theorem 1. *If A and B are ideals of X, then so is their meet set $A \wedge B$.*

Proposition 1. *If A, B and C are ideals of X, then*

(i) $A \wedge 0 = 0$.

(ii) $A \wedge B = A \cap B$.

(iii) $A \wedge B) \wedge C = A \wedge (B \wedge C) = \{a \wedge b \wedge c \mid a \in A, b \in B, c \in C\}$.

Obviously, $A \wedge B = B \wedge A$ for an arbitrary pair of nonempty subsets A and B of X. For any nonempty subsets A, B and C of X, we have

$$A \subseteq B,\ A \subseteq C \Rightarrow A \subseteq B \wedge C. \tag{6}$$

The following example shows that there exist subsets A, B and C of X such that $A \subseteq B$ and $A \subseteq C$, but $B \wedge C \nsubseteq A$.

Example 2. *Consider the lower BCK-semilattice $X = \{0, 1, 2, 3, 4\}$ with the following Cayley table (different from the one in Example 1).*

*	0	1	2	3	4
0	0	0	0	0	0
1	1	0	1	0	1
2	2	2	0	2	0
3	3	3	3	0	3
4	4	4	4	4	0

For subsets $A = \{0, 1\}$, $B = \{0, 1, 2, 3\}$ and $C = \{0, 1, 2, 4\}$ of X, we have

$$B \wedge C = \{b \wedge c \mid b \in B,\ c \in C\} = \{0, 1, 2\} \nsubseteq \{0, 1\} = A.$$

Definition 6. [4] *For any nonempty subsets A and B of X, we define a set*

$$(A :_\wedge B) := \{x \in X \mid x \wedge B \subseteq A\} \tag{7}$$

which is called the relative annihilator *of B with respect to A.*

Given a lower BCK-semilattice X, note that if $A = \{0\}$, then

$$\begin{aligned}(\{0\} :_\wedge B) &= \{x \in X \mid x \wedge B \subseteq \{0\}\} \\ &= \{x \in X \mid x \wedge b = 0,\ \forall b \in B\} \\ &= B^*\end{aligned} \tag{8}$$

which is the annihilator of B (see Huang [15]). Hence the concept of the *relative annihilator of B with respect to A* is a generalization of the concept of the *annihilator of B*.

Proposition 2. [4] *For any nonempty subsets A and B of X, we have*

(i) If A is an ideal of X, then $A \subseteq (A :_\wedge B)$.

(ii) If $B_1 \subseteq B_2$ in X, then $(A :_\wedge B_2) \subseteq (A :_\wedge B_1)$.

Lemma 1. [4] *If A and B are ideals of X, then the relative annihilator $(A :_\wedge B)$ of B with respect to A is an ideal of X.*

For any nonempty subsets A and B of X, let us now consider the set

$$F := \{x \in X \mid x \wedge B \subseteq A \wedge B\} \tag{9}$$

where $x \wedge B$ and $A \wedge B$ are meet sets. The following example shows that such a set need not be an ideal.

Example 3. *Consider the lower* BCK*-semilattice* $X = \{0, 1, 2, 3, 4\}$ *with the following Cayley table (different from the one in Example 2).*

$*$	0	1	2	3	4
0	0	0	0	0	0
1	1	0	1	0	0
2	2	2	0	2	0
3	3	1	3	0	1
4	4	4	4	4	0

We have five ideals of X*:* $A_0 = \{0\}$, $A_1 = \{0, 1, 3\}$, $A_2 = \{0, 2\}$, $A_3 = \{0, 1, 2, 3\}$, *and* $A_4 = X$. *For subsets* $A = \{2, 3\}$ *and* $B = \{4\}$ *of* X, *we have*

$$\{x \in X \mid x \wedge B \subseteq A \wedge B\} = \{1, 2, 3\},$$

which is not an ideal of X.

3 Some new results on ideals of lower BCK–semilattices

First of all, we provide conditions for the set $F = \{x \in X \mid x \wedge B \subseteq A \wedge B\}$ to be an ideal.

Theorem 2. *If* A *and* B *are ideals of a lower* BCK*-semilattice* X, *then the set*

$$F = \{x \in X \mid x \wedge B \subseteq A \wedge B\}$$

is an ideal of X.

Proof. Since A and B are ideals of X, using Theorem 1 shows that $A \wedge B$ is an ideal. So $0 \in A \wedge B$. Also, $0 \wedge B = 0 \in A \wedge B$. Hence, $0 \in F$.

Let $x, y \in X$ be such that $x * y \in F$ and $y \in F$. Then $(x * y) \wedge B \subseteq A \wedge B$ and $y \wedge B \subseteq A \wedge B$, that is,

$$(x * y) \wedge b \in A \wedge B \tag{10}$$

and

$$y \wedge b \in A \wedge B \tag{11}$$

for all $b \in B$. Since $b \in B$, we have $\langle b \rangle \subseteq B$. It follows that $y \wedge \langle b \rangle \subseteq y \wedge B$ and $(x * y) \wedge \langle b \rangle \subseteq (x * y) \wedge B$. Thus, $y \wedge \langle b \rangle \subseteq A \wedge B$ and $(x * y) \wedge \langle b \rangle \subseteq A \wedge B$. It means that,

$$y \in (A \wedge B :_{\wedge} \langle b \rangle) \tag{12}$$

and

$$x * y \in (A \wedge B :_{\wedge} \langle b \rangle) \tag{13}$$

Using Lemma 1, we conclude that $(A \wedge B :_{\wedge} \langle b \rangle)$ is an ideal of X. So (12), (13) and (3) show that

$$x \in (A \wedge B :_{\wedge} \langle b \rangle)$$

Hence, or all $b \in B$, $x \wedge b \in x \wedge \langle b \rangle \subseteq A \wedge B$. Therefore,

$$x \wedge B \subseteq A \wedge B,$$

which means that $x \in F$. □

In the following text we are going to investigate properties of such ideals. The name, which we choose for them, stresses the fact that the set A is reduced to a one element set.

Definition 7. *For two ideals A and B of X, the ideal*

$$\{x \in X \mid x \wedge B \subseteq A \wedge B\}$$

*is called A-*reduced meet ideal based on *B and denoted by $\frac{A\wedge B}{B}$.*

The following example shows that the converse of Theorem 2 is not true in general, which means that there exist subsets A and B of X which are not ideals, but $\frac{A\wedge B}{B}$ is an ideal of X.

Example 4. *Consider the lower BCK-semilattice $X = \{0, 1, 2, 3, 4\}$ with the following Cayley table (different from the ones in Example 2 or Example 3).*

$*$	0	1	2	3	4
0	0	0	0	0	0
1	1	0	1	0	1
2	2	2	0	2	0
3	3	1	3	0	3
4	4	4	2	4	0

We have four ideals of X: $A_0 = \{0\}$, $A_1 = \{0, 1, 3\}$, $A_2 = \{0, 2, 4\}$ and $A_3 = X$. For subsets $A = \{0, 1\}$ and $B = \{1, 2\}$ of X which are not ideals, we have

$$\frac{A \wedge B}{B} = \{0, 1, 3\},$$

which is the ideal A_1 of X.

Proposition 3. *For any ideals A and B of X, we have*

(i) $A \subseteq \frac{A\wedge B}{B}$ *and* $B \subseteq \frac{A\wedge \frac{A\wedge B}{B}}{\frac{A\wedge B}{B}}$.

(ii) $\frac{A\wedge X}{X} = A$ *and* $\frac{A\wedge A}{A} = X$.

(iii) $\frac{A\wedge B}{B} = X \Leftrightarrow B \subseteq A$.

Proof. (i) Suppose an arbitrary $x \in A$. Then $x \wedge B \subseteq A \wedge B$ for all $B \in \mathcal{I}(X)$. Thus $x \in \frac{A\wedge B}{B}$, and so $A \subseteq \frac{A\wedge B}{B}$.

Let $x \in B$ be an arbitrary element. For any arbitrary element $y \in \frac{A\wedge B}{B}$, we have $y \wedge B \subseteq A \wedge B$. Since $x \in B$, it follows that $y \wedge x \in A \wedge B$. Thus,

$$x \wedge \frac{A \wedge B}{B} \subseteq A \wedge B.$$

Besides,

$$A \wedge B \subseteq A = A \wedge \frac{A \wedge B}{B}.$$

Therefore, $x \wedge \frac{A\wedge B}{B} \subseteq A \wedge \frac{A\wedge B}{B}$, implying that

$$x \in \frac{A \wedge \frac{A\wedge B}{B}}{\frac{A\wedge B}{B}}.$$

(ii) Let $x \in A$. Then $x\wedge X \subseteq A\wedge X$, and so $x \in \frac{A\wedge X}{X}$. Conversely, let $x \in \frac{A\wedge X}{X}$. Then $x\wedge X \subseteq A\wedge X$. Since $A \wedge X = A$ and $x \wedge x = x$, we have $x \in A$. Therefore,

$$\frac{A \wedge X}{X} = A.$$

Clearly,

$$\frac{A \wedge A}{A} = \{x \in X \mid x \wedge A \subseteq A \wedge A\} = X.$$

(iii) Suppose that $\frac{A\wedge B}{B} = X$. Let b be an arbitrary element of B. Since $B \subseteq X$, clearly we have $b \wedge B \subseteq A \wedge B$, and so

$$b = b \wedge b \in A \wedge B.$$

Thus $B \subseteq A \wedge B$. Also, we always have $A \wedge B \subseteq B$. Therefore, $A \wedge B = B$ which means that $B \subseteq A$.

Conversely, suppose that $B \subseteq A$. Let $x \in X$ and $b \in B$ be an arbitrary element. Then $x \wedge b \leq b$, and thus by using (4), $x \wedge b \in B \subseteq A$, i.e. $x \wedge b \in A \wedge B$ and $x \in \frac{A\wedge B}{B}$. Thus $X \subseteq \frac{A\wedge B}{B}$, and so $X = \frac{A\wedge B}{B}$. □

The following example shows that the converse of Proposition 3 (i), is not true in general, meaning that there exist ideals A and B of X such that $\frac{A\wedge B}{B} \not\subseteq A$ and $\frac{A\wedge \frac{A\wedge B}{B}}{\frac{A\wedge B}{B}} \not\subseteq B$.

Example 5. *Consider the lower BCK-semilattice $X = \{0, 1, 2, 3, 4\}$ with the following Cayley table (again different from all the above examples).*

$*$	0	1	2	3	4
0	0	0	0	0	0
1	1	0	0	0	0
2	2	2	0	2	0
3	3	3	3	0	0
4	4	4	4	4	0

We have six ideals of X: $A_0 = \{0\}$, $A_1 = \{0, 1\}$, $A_2 = \{0, 1, 2\}$, $A_3 = \{0, 1, 3\}$, $A_4 = \{0, 1, 2, 3\}$ and $A_5 = X$. For ideals $A = A_2 = \{0, 1, 2\}$ and $B = A_1 = \{0, 1\}$ of X we have

$$\frac{A \wedge B}{B} = A_5,$$

and clearly,

$$\frac{A \wedge B}{B} \not\subseteq A.$$

Also,

$$A \wedge \frac{A \wedge B}{B} = A \wedge A_5 = A.$$

Thus,

$$\{x \in X \mid x \wedge \frac{A \wedge B}{B} \subseteq A \wedge \frac{A \wedge B}{B}\} = \{x \in X \mid x \wedge A_5 \subseteq A\} = \{0, 1, 2\},$$

and hence,

$$\frac{A \wedge \frac{A \wedge B}{B}}{\frac{A \wedge B}{B}} \not\subseteq B.$$

Proposition 4. *For any ideals A and B of X such that $A \subseteq B$, we have*

$$\frac{A \wedge B}{B} \cap B = A.$$

Proof. Let $x \in A$. Then clearly $x \in B$ and $x \wedge B \subseteq A \wedge B$. Thus

$$x \in \frac{A \wedge B}{B} \cap B.$$

Now suppose that $x \in \frac{A \wedge B}{B} \cap B$. Then $x \in B$ and $x \in \frac{A \wedge B}{B}$. So $x \wedge B \subseteq A \wedge B$. Since $A \subseteq B$ and $x \in B$, we have $A \wedge B = A$ and

$$x = x \wedge x \in A \wedge B = A.$$

Therefore, $x \in A$ and

$$\frac{A \wedge B}{B} \cap B = A.$$

□

Proposition 5. *Let B be an ideal of X. For any family $\{A_\lambda \mid \lambda \in \Lambda\}$ of ideals, we have*

$$\left(\frac{\bigcap_{\lambda \in \Lambda} A_\lambda \wedge B}{B}\right) = \bigcap_{\lambda \in \Lambda} \left(\frac{A_\lambda \wedge B}{B}\right). \tag{14}$$

Proof. Let $x \in \left(\frac{\bigcap_{\lambda \in \Lambda} A_\lambda \wedge B}{B}\right)$. Then $x \wedge B \subseteq \bigcap_{\lambda \in \Lambda} A_\lambda \wedge B$. So for an arbitrary element $y \in B$, $x \wedge y \in \bigcap_{\lambda \in \Lambda} A_\lambda \wedge B$. Hence there exist $a \in \bigcap_{\lambda \in \Lambda} A_\lambda$ and $b \in B$ such that $x \wedge y = a \wedge b$. Thus, for every $\lambda \in \Lambda$, $a \wedge b \in A_\lambda \wedge B$ and so $x \wedge y \in A_\lambda \wedge B$. Since y is an arbitrary element of B, we have $x \wedge B \subseteq A_\lambda \wedge B$. Hence, for every $\lambda \in \Lambda$,

$$x \in \left(\frac{A_\lambda \wedge B}{B}\right)$$

which means that

$$x \in \bigcap_{\lambda \in \Lambda} \left(\frac{A_\lambda \wedge B}{B}\right).$$

Therefore,

$$\left(\frac{\bigcap_{\lambda \in \Lambda} A_\lambda \wedge B}{B}\right) \subseteq \bigcap_{\lambda \in \Lambda} \left(\frac{A_\lambda \wedge B}{B}\right).$$

Now suppose that $x \in \bigcap_{\lambda\in\Lambda} \left(\frac{A_\lambda \wedge B}{B}\right)$. Then for every $\lambda \in \Lambda$, $x \in \frac{A_\lambda \wedge B}{B}$ and so $x \wedge B \subseteq A_\lambda \wedge B$. For an arbitrary element $y \in B$ and every $\lambda \in \Lambda$, $x \wedge y \in A_\lambda \wedge B$. So for $\lambda_1 \in \Lambda$, there exist $a_{\lambda_1} \in A_{\lambda_1}$ and $b_1 \in B$ such that $x \wedge y = a_{\lambda_1} \wedge b_1$. Similarly, for $\lambda_i \in \Lambda$ $(2 \leq i \leq n)$, there exist $a_{\lambda_i} \in A_{\lambda_i}$ and $b_i \in B$ such that $x \wedge y = a_{\lambda_i} \wedge b_i$. Now we have

$$\begin{aligned} x \wedge y =&(x \wedge y) \wedge (x \wedge y) \wedge \ldots \wedge (x \wedge y) \\ &= (a_{\lambda_1} \wedge b_1) \wedge (a_{\lambda_2} \wedge b_2) \wedge \ldots \wedge (a_{\lambda_n} \wedge b_n) \\ &= (a_{\lambda_1} \wedge a_{\lambda_2} \wedge \ldots \wedge a_{\lambda_n}) \wedge (b_1 \wedge b_2 \ldots \wedge b_n) \end{aligned}$$

Clearly, $a_{\lambda_1} \wedge a_{\lambda_2} \wedge \ldots \wedge a_{\lambda_n} \in \bigwedge_{\lambda\in\Lambda} A_\lambda$ and $b_1 \wedge b_2 \ldots \wedge b_n \in B$. Using Proposition 1, we conclude that

$$a_{\lambda_1} \wedge a_{\lambda_2} \wedge \ldots \wedge a_{\lambda_n} \in \bigcap_{\lambda\in\Lambda} A_\lambda.$$

Thus, $x \wedge y \in \bigcap_{\lambda\in\Lambda} A_\lambda \wedge B$. Since y is an arbitrary element of X, we have

$$x \in \left(\frac{\bigcap_{\lambda\in\Lambda} A_\lambda \wedge B}{B}\right).$$

Therefore,

$$\left(\frac{\bigcap_{\lambda\in\Lambda} A_\lambda \wedge B}{B}\right) = \bigcap_{\lambda\in\Lambda} \left(\frac{A_\lambda \wedge B}{B}\right).$$

□

Theorem 3. *For any ideals A and B of X, we have*

$$A \wedge B \subseteq \frac{A \wedge B}{B} \subseteq (A :_\wedge B). \tag{15}$$

Moreover, if $A \subseteq B$, then

$$\frac{A \wedge B}{B} = (A :_\wedge B). \tag{16}$$

Proof. Let $x \in A \wedge B$ be an arbitrary element. Then there exist $a \in A$ and $b \in B$ such that $x = a \wedge b$. For an arbitrary element $b' \in B$ we have

$$x \wedge b' = (a \wedge b) \wedge b' = a \wedge (b \wedge b').$$

Since $b \wedge b' \leq b$ and $b \in B$, using (4) we conclude that $b \wedge b' \in B$. Thus, $a \wedge (b \wedge b') \in A \wedge B$. Therefore, $A \wedge B \subseteq \frac{A \wedge B}{B}$.

Now let $x \in \frac{A \wedge B}{B}$. Then for every element $b \in B$, $x \wedge b \in A \wedge B$. Since $A \wedge B \leq A$, we have $x \wedge b \in A$. Hence $x \wedge B \subseteq A$ and $x \in (A :_\wedge B)$. Therefore, $\frac{A \wedge B}{B} \subseteq (A :_\wedge B)$.

If $A \subseteq B$, then $A \wedge B = A$ and equation (16) is clear. □

Theorem 4. *For ideals A, B_1 and B_2 of X, if $B_1 \subseteq B_2$, then*

$$\frac{A \wedge (A :_\wedge B_1)}{(A :_\wedge B_1)} \subseteq \frac{A \wedge (A :_\wedge B_2)}{(A :_\wedge B_2)}. \tag{17}$$

Proof. Suppose that $B_1 \subseteq B_2$ and $x \in \frac{A \wedge (A :_\wedge B_1)}{(A :_\wedge B_1)}$. Then

$$x \wedge (A :_\wedge B_1) \subseteq A \wedge (A :_\wedge B_1).$$

Using Proposition 2, we have

$$A \wedge (A :_\wedge B_1) = A = A \wedge (A :_\wedge B_2),$$
$$(A :_\wedge B_2) \subseteq (A :_\wedge B_1).$$

Thereby,

$$x \wedge (A :_\wedge B_2) \subseteq x \wedge (A :_\wedge B_1) \subseteq A \wedge (A :_\wedge B_1) = A = A \wedge (A :_\wedge B_2).$$

Therefore, $x \wedge (A :_\wedge B_2) \subseteq A \wedge (A :_\wedge B_2)$ and thus

$$x \in \frac{A \wedge (A :_\wedge B_2)}{(A :_\wedge B_2)}.$$

□

The following example shows that the converse of inclusion (17) is not true in general, which means that there exist ideals A, B_1 and B_2 of X such that $B_1 \subseteq B_2$, but

$$\frac{A \wedge (A :_\wedge B_2)}{(A :_\wedge B_2)} \not\subseteq \frac{A \wedge (A :_\wedge B_1)}{(A :_\wedge B_1)}$$

Example 6. *Consider the lower BCK-semilattice $X = \{0, 1, 2, 3, 4\}$ with the following Cayley table.*

$*$	0	1	2	3	4
0	0	0	0	0	0
1	1	0	1	1	1
2	2	2	0	2	2
3	3	3	3	0	3
4	4	4	4	4	0

For ideals $A = \{0, 1\}$, $B_1 = \{0, 3\}$ and $B_2 = \{0, 2, 3\}$ of X, we have

$$(A :_\wedge B_1) = \{x \in X \mid x \wedge B_1 \subseteq A\} = \{0, 1, 2, 4\},$$

and

$$(A :_\wedge B_2) = \{x \in X \mid x \wedge B_2 \subseteq A\} = \{0, 1, 4\}.$$

Thus,

$$\frac{A \wedge (A :_\wedge B_1)}{(A :_\wedge B_1)} = \{x \in X \mid x \wedge (A :_\wedge B_1) \subseteq A \wedge (A :_\wedge B_1)\} = \{0, 1\},$$

and

$$\frac{A \wedge (A :_\wedge B_2)}{(A :_\wedge B_2)} = \{x \in X \mid x \wedge (A :_\wedge B_2) \subseteq A \wedge (A :_\wedge B_2)\} = \{0, 1, 2, 3\}.$$

Clearly,

$$\frac{A \wedge (A :_\wedge B_1)}{(A :_\wedge B_1)} \subseteq \frac{A \wedge (A :_\wedge B_2)}{(A :_\wedge B_2)}.$$

But

$$\frac{A \wedge (A :_\wedge B_2)}{(A :_\wedge B_2)} \not\subseteq \frac{A \wedge (A :_\wedge B_1)}{(A :_\wedge B_1)}.$$

Proposition 6. *For ideals A, B_1 and B_2 of X, we have*

$$\frac{A \wedge (A :_\wedge B_1)}{(A :_\wedge B_1)} \cap \frac{A \wedge (A :_\wedge B_2)}{(A :_\wedge B_2)} \subseteq \frac{A \wedge (A :_\wedge \langle B_1 \cup B_2 \rangle)}{(A :_\wedge \langle B_1 \cup B_2 \rangle)}. \tag{18}$$

Proof. Since $B_1 \subseteq \langle B_1 \cup B_2 \rangle$, using Theorem 4, we conclude that

$$\frac{A \wedge (A :_\wedge B_1)}{(A :_\wedge B_1)} \subseteq \frac{A \wedge (A :_\wedge \langle B_1 \cup B_2 \rangle)}{(A :_\wedge \langle B_1 \cup B_2 \rangle)}.$$

Similarly, we have

$$\frac{A \wedge (A :_\wedge B_2)}{(A :_\wedge B_2)} \subseteq \frac{A \wedge (A :_\wedge \langle B_1 \cup B_2 \rangle)}{(A :_\wedge \langle B_1 \cup B_2 \rangle)}.$$

Therefore,

$$\frac{A \wedge (A :_\wedge B_1)}{(A :_\wedge B_1)} \cap \frac{A \wedge (A :_\wedge B_2)}{(A :_\wedge B_2)} \subseteq \frac{A \wedge (A :_\wedge \langle B_1 \cup B_2 \rangle)}{(A :_\wedge \langle B_1 \cup B_2 \rangle)}.$$

□

4 Future work

The roots of BCI–, BCK–, MV– and other types of algebras lie in combinatory logic and information sciences. Notice that recently Jun and Song [18] and Flaut [13] linked BCK–algebras to block codes used in channel encoding in earlier mobile communication systems. Matrices of block codes are studied in Saeid et al. [21]. Since, in a special case, we obtain lattices (or semilattices), whenever multivalued aspects are employed we can make use of results of the hyperstructure theory and concepts such as e.g. EL–hyperstructures (see e.g. Křehlík and Novák [19] where sets of matrices and lattices are studied) or even classical results of Varlet [22] who provided the link between distributive lattices and hyperstructure theory. In our future work we shall concentrate on finding links between our results and the above mentioned concepts, or those connected with closure operations on BCK–algebras [1, 2].

Acknowledgement

Michal Novák was supported by the project FEKT-S-17-4225 of Brno University of Technology. Irina Cristea acknowledges the financial support from the Slovenian Research Agency (research core funding No. P1 – 0285).

References

[1] BORDBAR, H., ZAHEDI, M. M.: A finite type of closure operations on BCK-algebra, *Appl. Math. Inf. Sci. Lett.*, **4**(2) (2016), pp. 1–9.

[2] BORDBAR, H., ZAHEDI, M. M.: Semi-prime closure operations on BCK-algebra, *Commun. Korean Math. Soc.*, **30**(5) (2015), pp. 385–402.

[3] BORDBAR, H., ZAHEDI, M. M., AHN, S. S., JUN, Y. B.: Weak closure operations on ideals of BCK-algebras, *J. Comput. Anal. Appl.* (in press).

[4] BORDBAR, H., ZAHEDI, M. M., JUN, Y. B.: Relative annihilators in lower BCK-semilattices, *Math. Sci. Lett.*, **6**(2), (2017), pp. 1–7.

[5] BORDBAR, H., ZAHEDI, M. M., AHN, S. S., JUN, Y. B.: Semiring structures based on meet and plus ideals in lower BCK-semilattices, *J. Comput. Anal. Appl.*, (2017). (in press).

[6] BUNDLER, M. W.: Corrections to some results for **BCK** logics and algebras, *Logique et Analyse*, **31**(121/122) (1988), pp. 115–122.

[7] BUNDLER, M. W., MEYER, R. K.: A result for combinators, **BCK** logics and **BCK**–algebras, *Logique et Analyse*, **109** (1985), pp. 33–40.

[8] CHAJDA, I.: A structure of BCI–algebras, *Int. J. Theor. Phys.*, **53**(10) (2014), pp. 3391–3396.

[9] CHAJDA, I., LÄNGER, H.: Weak Lattices, *Ital. J. Pure Appl. Math.*, **30** (2013), pp. 125–140.

[10] CHAJDA, I., RACHŮNEK: J: Annihilators in normal autometrized algebras, *Czech. Math. J.*, **51**(1) (2001), pp. 111–120.

[11] CHANG, C. C.: Algebraic analysis of many valued logics, *Trans. Amer. Math. Soc.*, **88** (1958), pp. 467-490.

[12] CURRY, H. B., FEYS, R.: Combinatory Logic, North-Holland, Amsterdam, 1958.

[13] FLAUT, C.: BCK–algebras arising from block codes, *J. Intell. Fuzzy Syst.*, **28**(4) (2015), pp. 1829–1833.

[14] HINDLEY, R.: The principal type-scheme of an object in combinatory logic, *Trans. Amer. Math. Soc.*, **146** (1969), pp. 29–60.

[15] HUANG, Y.: BCI-algebra, Beijing: Science Press, 2006.

[16] ISÉKI, K: An algebra related with a propositional calculus, *Proc. Japan. Acad.*, **42** (1966), pp. 26–29.

[17] ISÉKI, K: An introduction to the theory of BCK–algebras, *Math. Japon.*, **23** (1978), pp. 1–26.

[18] JUN, Y. B., SONG, S. Z.: Codes based on BCK–algebras, *Inform. Sci.*, **181** (2015), pp. 5102–5109.

[19] KŘEHLÍK, Š., NOVÁK, M: From lattices to H_v–matrices, *An. Şt. Univ. Ovidius Constanţa*, **24**(3) (2016), pp. 209–222.

[20] MENG, J., JUN, Y. B., BCK-algebras, Seoul: Kyung Moon Sa Co. 1994.

[21] SAEID, A. E. et al.: Some connections between BCK–algebras and n–ary block codes, *Soft Comput.* (2017) (online ready).

[22] VARLET, J. V.: Remarks on distributive lattices, *Bull. de l'Acad. Polonnaise des Sciences, Serie des Sciences Math., Astr. et Phys.*, **23** (1975), pp. 1143–1147.

[23] VASSILEV, J. C.: Structure on the Set of Closure Operations of a Commutative Ring, *J. Algebr.*, **321**(10) (2009), pp. 2737-2753.

Current address

Bordbar Hashem, Ph.D.
Faculty of Mathematics, Statistics and Computer Science
Shahid Beheshti University
Tehran, Iran
E-mail: bordbar.amirh@gmail.com

Novák Michal, RNDr. Ph.D.
Faculty of Electrical Engineering and Communication
Brno University of Technology
Technická 8, 616 00 Brno, Czech Republic
E-mail: novakm@feec.vutbr.cz

Cristea Irina, prof. dr.
Center for Information Technologies and Applied Mathematics
University of Nova Gorica
Vipavska 13, SI-5000 Nova Gorica, Slovenia
E-mail: irinacri@yahoo.co.uk

Jun Young Bae, Prof. Emeritus
Department of Mathematics Education (and RINS)
Gyeongsang National University
Jinju 52828, Korea
E-mail: skywine@gmail.com

Slovak University of Technology in Bratislava
Faculty of Mechanical Engineering
17th Conference on Applied Mathematics **APLIMAT 2018**

Proceedings

A MAGIC FORMULA OF NATURE

BRANDI Primo (IT), SALVADORI Anna (IT)

Abstract. The *superformula of nature*, introduced by Dutch botanist Johan Gielis to describe a wide range of shapes ([3]), is examined from the educational point of view. According to our experience, it reveals an optimal chance to propose important knowledge also in university courses of "weak" mathematics. Students become acquainted with graphic representation of functions , polar coordinates and the concept of mathematical model; moreover, they train themselves for a conscious use of parameters. It also offers the opportunity of a critical use of technology.

Keywords: elementary mathematical models, teaching innovation

Mathematics Subject Classification: Primary 97D40; Secondary 97U50, 97U70

1 Introduction

For centuries, scientists have been trying to express the forms of nature in mathematical terms; the description of the forms is one of the major problems of biology, especially when dealing with complicated shapes such as leaves or shells.
In 2003 the Dutch botanist Johan Gielis proposed a formula that can describe a wide range of natural shapes ([3]). Gielis' *superformula* represents a tool for analysis and comparison, moreover it can be used in reconstruction and recognition programs.
Despite its aspect

$$\rho = \rho(\varphi) = R(\varphi)\left(\left|\frac{1}{a}\cos\left(\frac{m}{4}\varphi\right)\right|^{p_2} + \left|\frac{1}{b}\sin\left(\frac{m}{4}\varphi\right)\right|^{p_3}\right)^{-\frac{1}{p_1}}$$

the equation reveals rather simple: it is a modified version of the equation for a circle, depending of seven parameters.
Changing one of the parameters the proportions of the shapes change, going from a perfect circle to an ellipse. By changing another one, you change the axes of symmetry, going from a circle to a triangle, a square, a pentagon, and so on. By changing both the proportions and the symmetry, shapes are produced with any number of smooth or irregular sides.
You can also produce non-biological forms, such as snowflakes or crystals.
Of course also a three-dimensional version of the formula hold, with many interesting applications.

In our contribution we wish to focalize a different aspect: the potentiality of the formula for educational purposes. Students, fascinated by the wonderful shapes, let themselves be led into the mathematical world where they acquire a wide range of knowledge without difficulty and become aware of the important unifying properties of the discipline.
For didactical purpose we have chosen to present superformula step by step as subsequent changes to the equation of a circumference. In this way, introducing one parameter at a time, we have the opportunity to show clearly the role played by each parameter.
More precisely the paper is organized as follows.

Section 2. From circumference to super-circumferences (parameter p).

Section 3. From super-circumferences to super-ellipses (parameters a and b).

Section 4. From Cartesian to polar coordinates (parameters r and φ).

Section 5. Three different exponential parameters p_1 p_2 p_3 take the place of p.

Section 6. The argument φ is revised as phase parameter $\frac{m}{4}\varphi$.

Section 8. Scale parameter r is replaced by a function $R(\varphi)$.

Sections 7 and 9 are devoted to applications of superformula to natural shapes. See also Section 3.3, 8.1 and 8.2 for other examples.

The graphics in the present paper were implemented by the Authors with Mathematica ver.7.

2 The *squared* circle

For a long time the circle and the square have been considered as "opposed" figures, basic models with completely different peculiarities. At the beginning of the nineteenth century French mathematician Gabriel Lamé (Tours 1795 – Paris 1870) revolutionized this view by proving that circles and squares could be represented by a single equation.
As it is well known, the equation of the circumference with center in the origin and radius r is the following

$$x^2 + y^2 = r^2 \tag{1}$$

Lamé introduced the following generalization of the circumference equation

$$|x|^p + |y|^p = r^p \qquad \text{with} \qquad p > 0 \tag{1'}$$

which today is known as **Lamé circumference**. The introduction of parameter p allows to extend widely the shapes that can be represented by the equation, as we will show in the next section.

2.1 The role of parameter *p*

Of course, for $p = 2$ we get usual circumference equation, but as the parameter $p \in R^+$ varies, other interesting figures are described, as the following images show

Lamé super circumferences

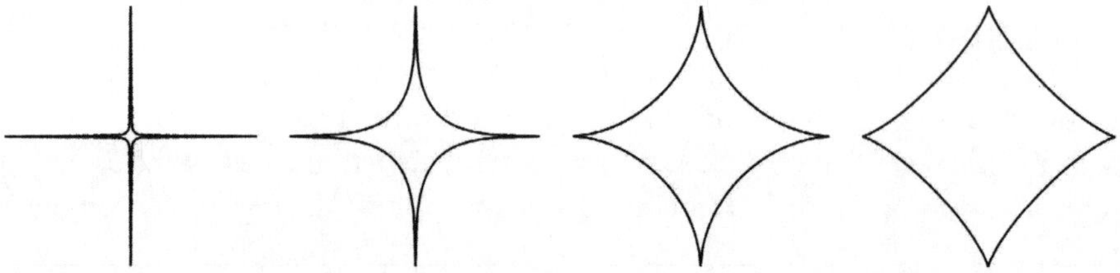

Fig. 1. p = 0.2 Fig. 2. p = 0.4 Fig. 3. p = 0.6 Fig. 4. p = 0.8

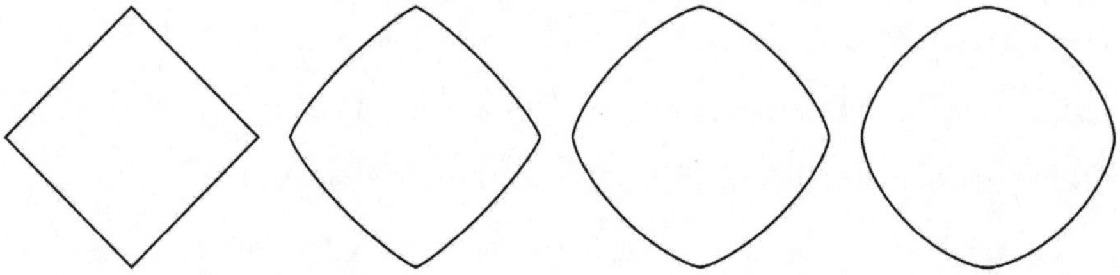

Fig. 5. p = 1 Fig. 6. p = 1.2 Fig. 7. p = 1.6 Fig. 8. p = 2

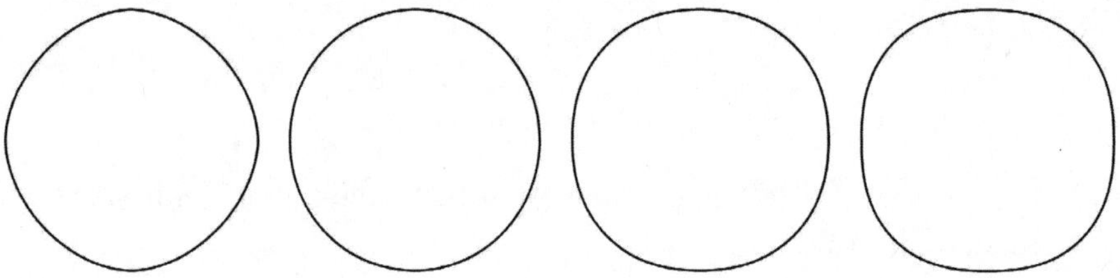

Fig. 9. p = 2.4 Fig. 10. p = 3 Fig. 11. p = 6 Fig. 12. p = 10

For very small values of the parameter $(p \cong 0)$ C_p is constituted by two orthogonal segments of length 2 which intersect in the origin; as the values of p increases, we get "asteroid" figures which gradually evolve first to the rhombus C_1, then to the circumference C_2 and finally, for great values of the parameter C_p they are getting closer to the square with the sides parallel to the axes $[-1,1]^2$.

3 Super-ellipses

Ellipse can be considered as a generalization of the circumference

$$(x/a)^2 + (y/b)^2 = 1 \tag{2}$$

where $a > 0$ and $b > 0$ are the length of the axes. If $a = b$ it reduces to a circumference. Thus Lamé proposed also the equations of super-ellipses

$$|x/a|^p + |y/b|^p = 1 \quad p > 0 \tag{2'}$$

which depends on three parameters a, b, p.
The contribution of each parameter can be easily deduced from the previous remark.

3.1 Super-ellipses in the real world

For more than one century, Lamé equation did not appear to be of great interest until it was re-discovered by Dutch architect Piet Hein and has founded a major role in architecture and design. We wish to recall just a meaningful example.

In 1959, the city of Stockholm was facing a building design problem whose brilliant solution would have provided an innovative idea from which many designers have drawn inspiration. In Sergel's Torg two large north-south and east-west traffic arteries intersected, creating a vast rectangular space (about 200 yards long) that was intended for commercial use. The designers faced a problem that was more difficult than expected: the shape of the building.

A rectangular construction would optimize the available space for the shopping but the corners would interfere with the surrounding traffic fluidity. An ellipsoidal design would have sacrificed space for business, it would not be harmonized with the rectangular surface of the available space and would not have solved the traffic difficulties of cars satisfactorily.

Hein choose a shape that could satisfied all the requests: the super-ellipse with $a/b=6/5$ and $p=5/2$ (see Fig. 13 and 14).

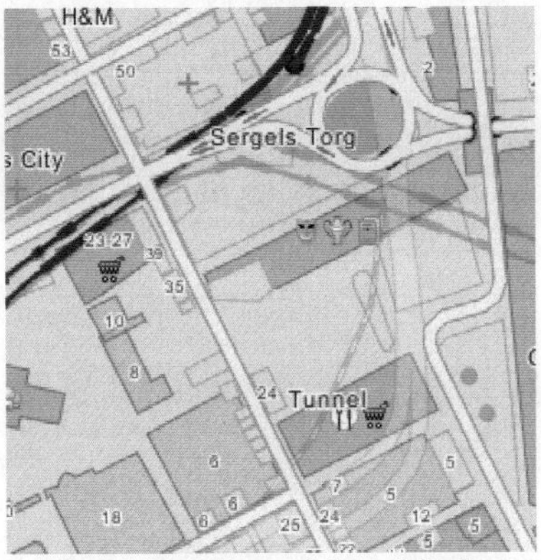

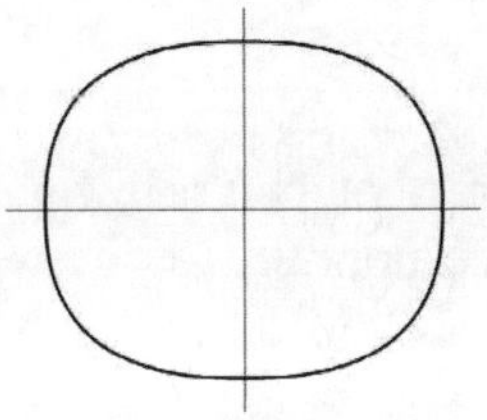

$a/b=6/5$
$p=5/2$

Fig. 13. Map of Stockholm. Fig. 14. super-ellipse. Fig. 15. Sergel's Torg.

After the brilliant solution proposed by Hein, super-ellipses often met the favor of designers and architects in Canada, France, Japan, the United States and Mexico.

Piet Hein's collaboration with Bruno Mathsson, a famous Swedish designer, produced several super-elliptical artifacts: glasses, plates, desk lamps (very popular in the sixties), chairs and beds. (see http://www.piethein.dk/)

An interesting link between super-ellipses and nature are the sections of bamboo plants. Some species of bamboo cane have *square* sections at the bottom (Chimono bambusa quadrangularis), more precisely the sections have the form of super-circumferences. These square bamboos are more resistant than circular ones and are widely used for buildings in the east countries.

Fig. 16 Chimono bambusa quadrangularis [2] .

4 The first step toward super-formula

In 2003 ([2],[3]) Dutch biologist Johan Gielis gave new life to Lamé equation by introducing some modifications that produced a very ductile formula. As we will show in the following, Gielis equation is able to generate a multiplicity of forms of plants and living beings.

4.1 From Cartesian to Polar coordinates.

If we pass to polar coordinates $\begin{cases} x = \rho\cos\varphi \\ y = \rho\sin\varphi \end{cases}$ super-ellipses equation becomes

$$\rho = \rho(\varphi) = r\left(\left|\frac{1}{a}\cos\varphi\right|^{p} + \left|\frac{1}{b}\sin\varphi\right|^{p}\right)^{-\frac{1}{p}} \tag{3}$$

This formula adds two more parameters to super ellipses. Let us show the contribution of each parameter on the shape of the corresponding figure.

4.2 The role of parameter *r*

Of course r is a *scale* parameters that controls figure dimension. The following image shows a family of super-ellipses corresponding to different values of the parameter r (from 1 to 5).

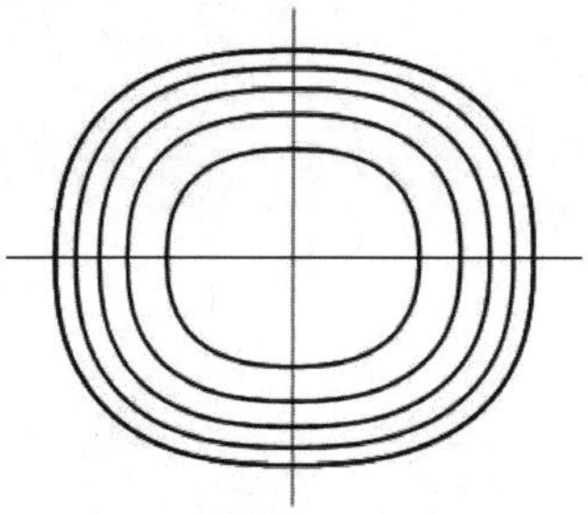

Fig. 17. Super-ellipses.

4.3 The role of parameter φ

We can consider the particular case $a = b = r = 1$
Observe that function $\rho = \rho(\varphi)$ $\varphi \in [0, 2\pi]$ represent the length of the vector ray corresponding to angle φ.
For this reason, the local minima and maxima of ρ play a fundamental for the figure shape. Since that function ρ is the reciprocal of function

(3') $$f_p(\varphi) = \left(|\cos\varphi|^{p} + |\sin\varphi|^{p}\right)^{\frac{1}{p}} \quad \varphi \in [0, 2\pi].$$

the local maximum and minimum points of ρ correspond to the local minimum and maximum points of f ; in fact we have

$$\rho(\varphi_0) \le \rho(\varphi) \quad \forall\varphi \in I(\varphi_0) \quad \Leftrightarrow \quad f_p(\varphi_0) \ge f_p(\varphi) \quad \forall\varphi \in I(\varphi_0).$$

4.4 Again on the role of parameter *p*

As the parameter p varies, $\rho_p(\varphi) = \left(|\cos\varphi|^p + |\sin\varphi|^p\right)^{-\frac{1}{p}}$ assume different shapes, corresponding to the different values of the maxima e minima of function

$$f_p(\varphi) = \left(|\cos\varphi|^p + |\sin\varphi|^p\right)^{\frac{1}{p}} \quad \varphi \in [0, 2\pi].$$

For $p = 1$ (Manhattan metric) it can easily proved that function

$$f_1(\varphi) = |\cos\varphi| + |\sin\varphi| \quad \varphi \in [0, 2\pi]$$

admits

$$\min f_1 = 1 \qquad \max f_1 = \sqrt{2}$$

and presents: 4 minimum points $\varphi_j = j\frac{\pi}{2}$; 4 maximum points $\varphi_j = (2j+1)\frac{\pi}{4}$ $j = 0,1,2,3$.

The following images shows as the shape of f_1 is transmitted to that of ρ_1 .

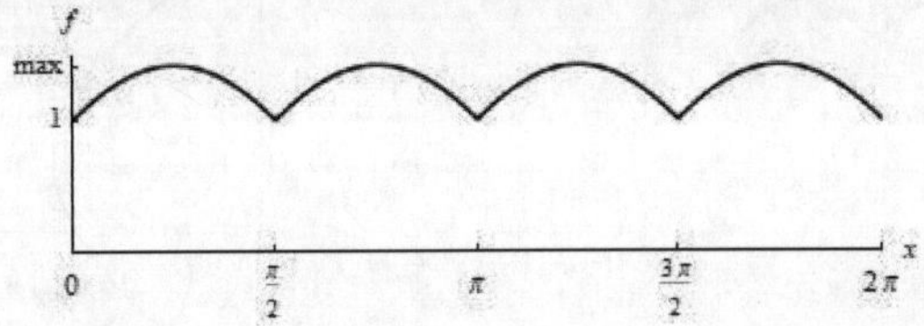

Fig. 18. Function f_1 $\max f_1 = \sqrt{2}$.

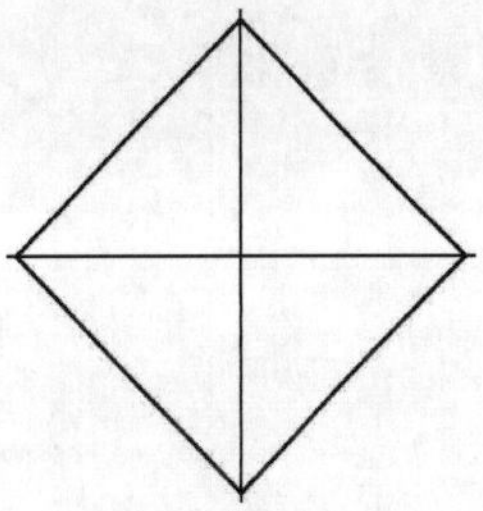

Fig. 18'. Function ρ_1 .

It is easy to prove that functions $f_p(\varphi)$ still admits 4 minimum points and 4 maximum points for every value of parameter p .

For $0 < p < 2$ they are the same of f_1, while for $p > 2$ they are inverted

minimum points: $\varphi_j = (2j+1)\frac{\pi}{4}$ maximum points $\varphi_j = j\frac{\pi}{2}$ $j = 0,1,2,3$.

Let us summarize the results in the following table.

parameter	min f_p	max f_p	
$0 < p < 1$	1	It progressively decreases as p increases	
$p = 1$	1	$\sqrt{2}$	(Manhattan metric)
$1 < p < 2$	1	It progressively decreases as p increases	
$p = 2$	1	1 (constant function)	(Euclidean metric)
$p > 2$	It progressively decreases as p increases	1	

The following images illustrate the results of the table.

The role of parameter p

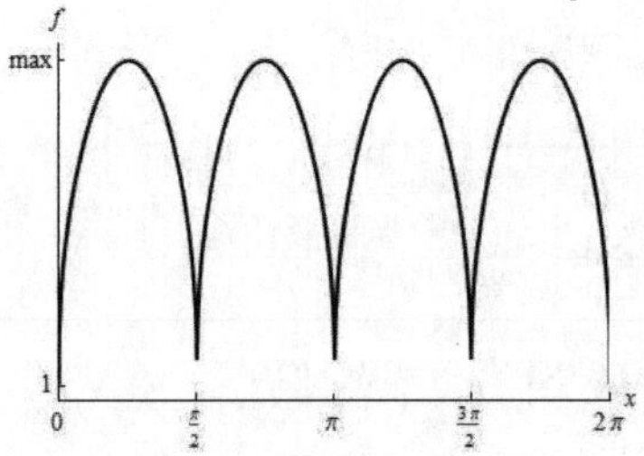

Fig. 19. $p = 0.1$ $\max f \cong 22.62$ $\min f = 1$

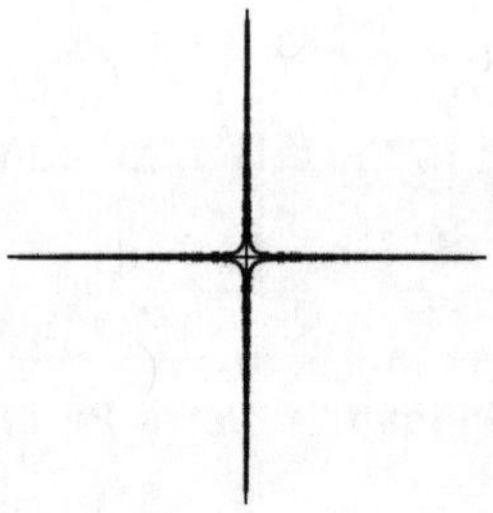

Fig. 19'.

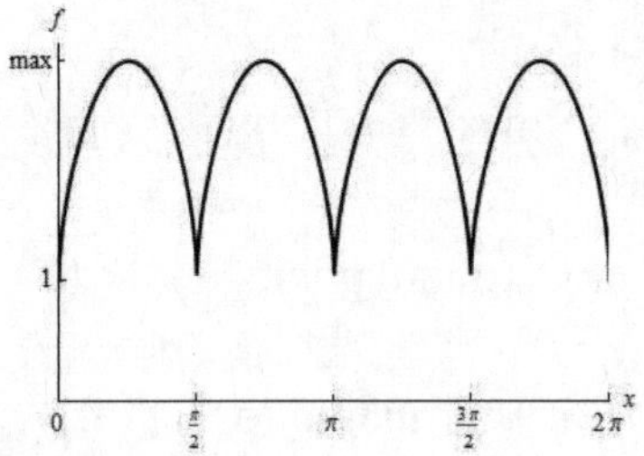

Fig. 20. $p = 0.5$ $\max f \cong 2.82$ $\min f = 1$

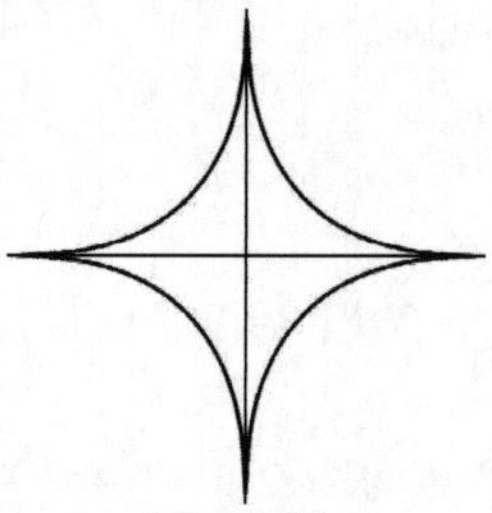

Fig. 20'.

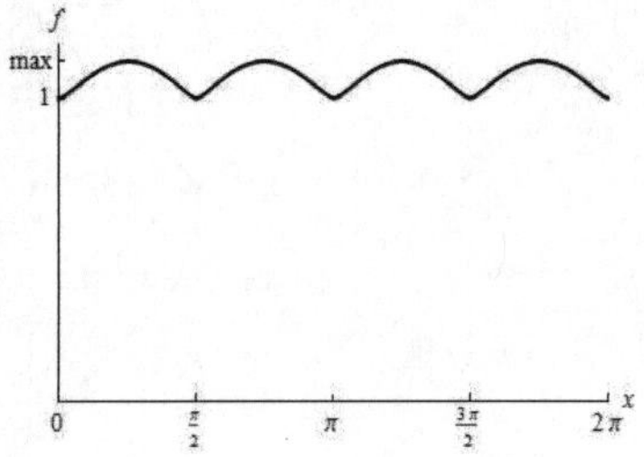

Fig. 21. $p = 1.5$ $\max f \cong 1.12$ $\min f = 1$

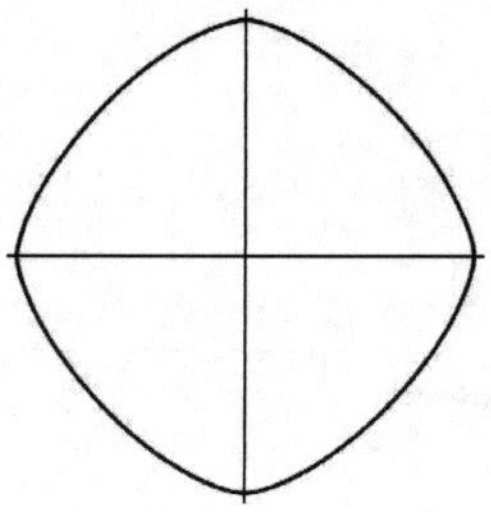

Fig. 21'.

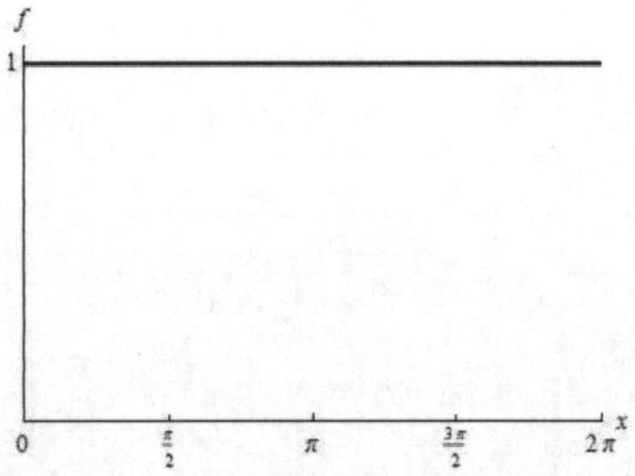

Fig. 22. $p = 2$ $\max f = \min f = 1$

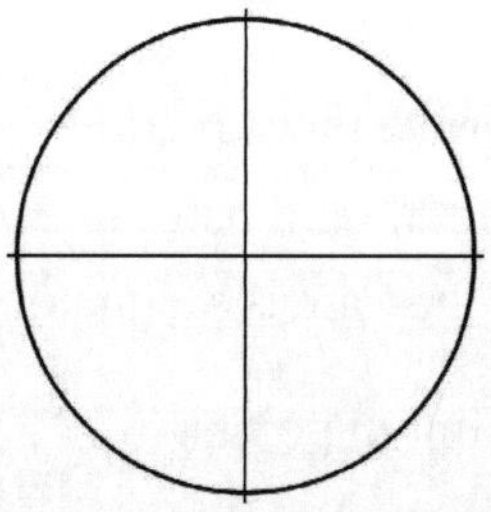

Fig. 22'.

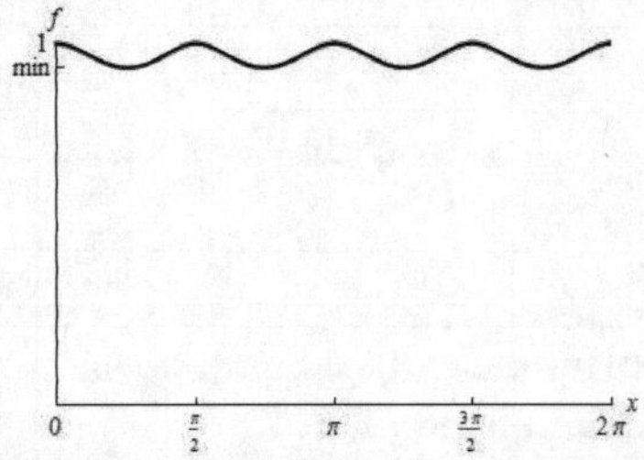

Fig. 23. $p = 2.5$ $\max f = 1$ $\min f \cong 0.93$

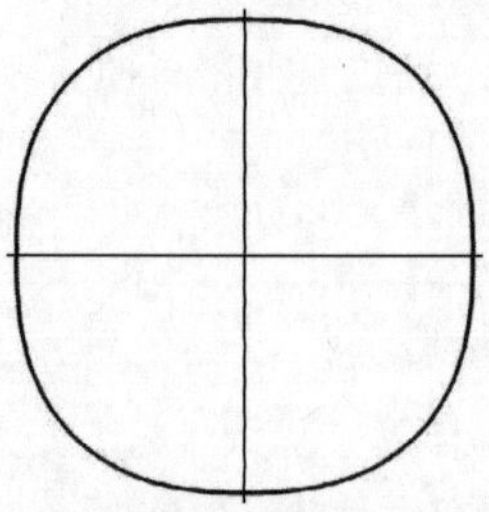

Fig. 23'.

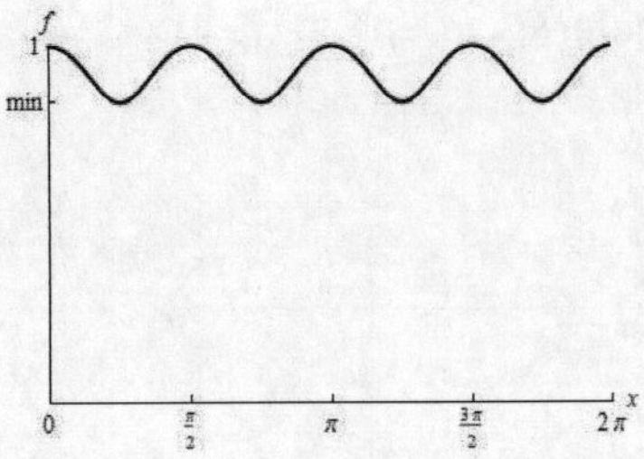

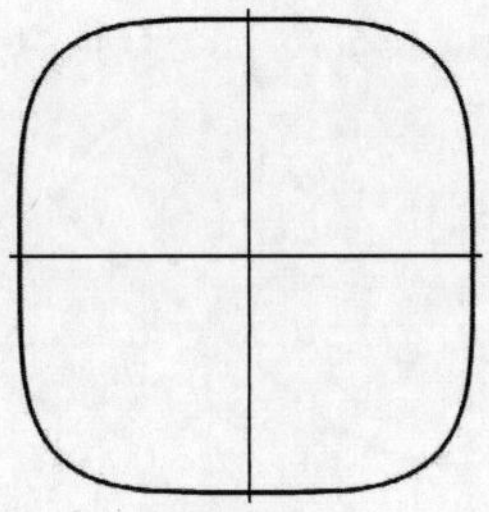

Fig. 24. $p = 4$ $\max f = 1$ $\min f \cong 0.84$

Fig. 24'.

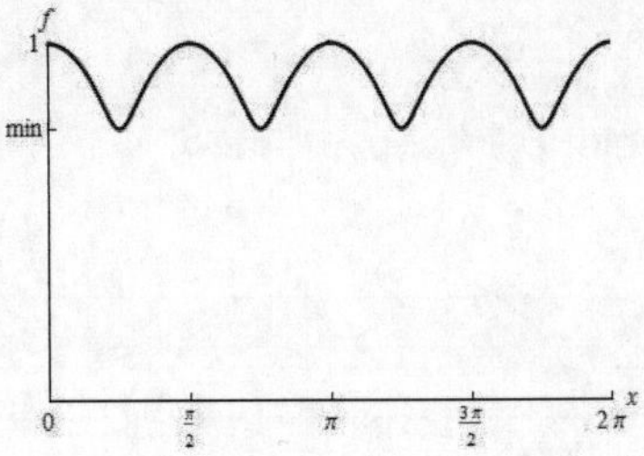

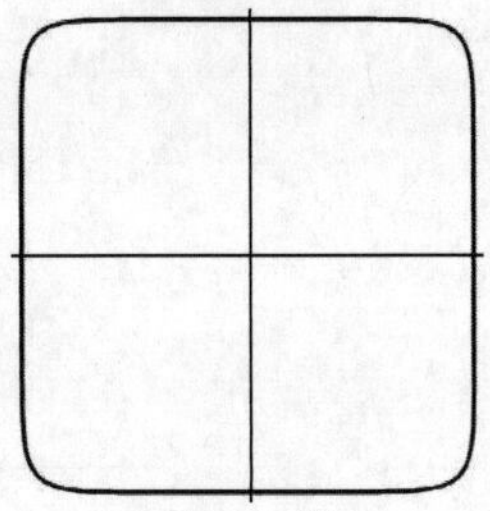

Fig. 25. $p = 10$ $\max f = 1$ $\min f \cong 0.75$

Fig. 25'.

5 A second first step toward super-formula

A further extension of super-ellipses equation proposed by Gielis consists on adopting three different values for the exponents p_i $i = 1, 2, 3$

$$\rho = r\left(\left|\frac{1}{a}\cos\varphi\right|^{p_2} + \left|\frac{1}{b}\sin\varphi\right|^{p_3}\right)^{-\frac{1}{p_1}} \tag{4}$$

Let us show the strategic power of this innovation.

5.1 The role of parameters p_1, p_2, p_3

We can consider again the particular case $a = b = r = 1$.
First of all, note that parameter p_2 acts on the horizontal component, while parameter p_3 on the vertical component.
More precisely, each parameter produces the effect of a non-linear transformation, respectively

$$\begin{cases} x' = x^{p_2} \\ y' = y \end{cases} \qquad \begin{cases} x' = x \\ y' = y^{p_3} \end{cases}$$

Let us assume $p_1 = 1$ and show the role of parameters p_2 and p_3 by means of some images.

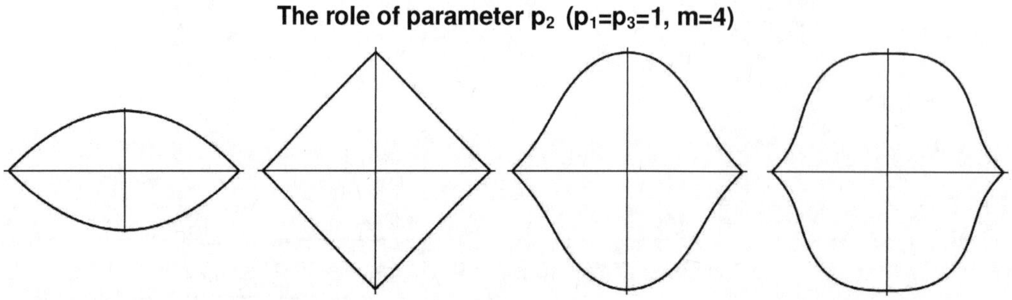

Fig. 26. $p_2 = 0$ Fig. 27. $p_2 = 1$ Fig. 28. $p_2 = 2$ Fig. 29. $p_2 = 3$

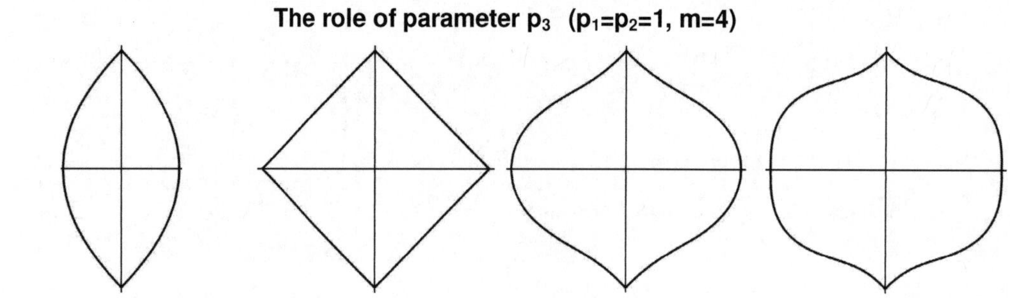

Fig. 30. $p_3 = 0$ Fig. 31. $p_3 = 1$ Fig. 32. $p_3 = 2$ Fig. 33. $p_3 = 3$

Note that parameter p_1 produces an effect of multidirectional deformation.
Let us see some examples, assuming $p_2 = p_3 = 1$.

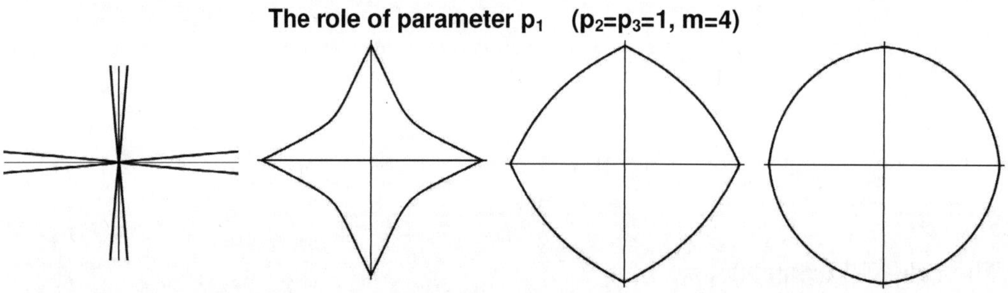

Fig. 34. $p_1 = 0.01$ Fig. 35. $p_1 = 0.5$ Fig. 36. $p_1 = 2$ Fig. 37. $p_1 = 10$

6 A third step: the introduction of phase parameter

Gielis observes that all the figures above have the disadvantage of a limited symmetry, as it happen with super-ellipses. Precisely they have a rotational symmetry of order four, while in nature there are organisms with hexagonal or pentagonal symmetry.
For this reason he adds a *phase parameter* k :

$$\rho(\varphi) = r\left(\left|\frac{1}{a}\cos(k\varphi)\right|^{p_2} + \left|\frac{1}{b}\sin(k\varphi)\right|^{p_3}\right)^{-\frac{1}{p_1}} \qquad (5)$$

To show the role of phase parameter, we can consider the particular case $a = b = r = 1$ $p_1 = p_2 = p_3 = 1$

$$\rho_k(\varphi) = \left(|\cos(k\varphi)| + |\sin(k\varphi)|\right)^{-1}$$

Note that function

$$f_k(\varphi) = |\cos(k\varphi)| + |\sin(k\varphi)| \quad \varphi \in [0, 2\pi] \tag{5'}$$

admits the same maxima and minima of functions f_1 in Section 4.4 (see Fig. 18), but the number of maximum and minimum points is now 4k. They are respectively

$$\text{minimum points: } \varphi_j = \frac{j}{k}\frac{\pi}{2} \quad j = 0, 1, \ldots, 4k-1$$

$$\text{maximum points } \varphi_j = \frac{2j+1}{k}\frac{\pi}{4} \quad j = 0, 1, \ldots, 4k-1.$$

The role of parameter k

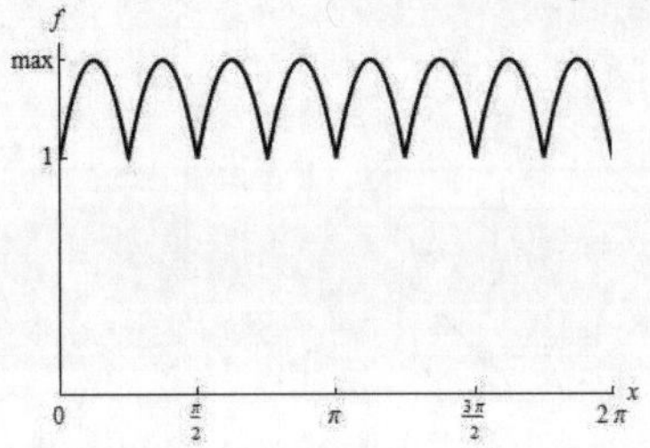

Fig. 38. $k = 2 \quad \max f = \sqrt{2} \quad \min f = 1$

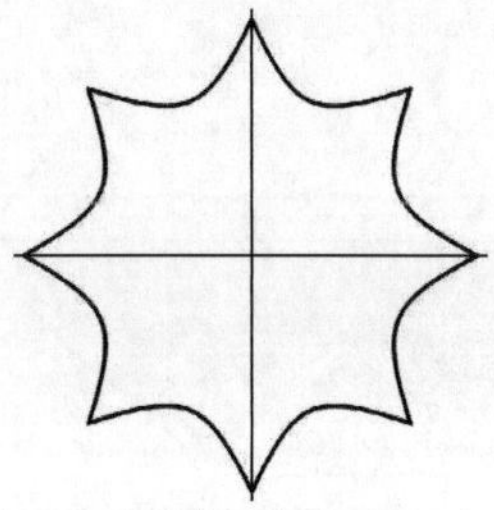

Fig. 38'.

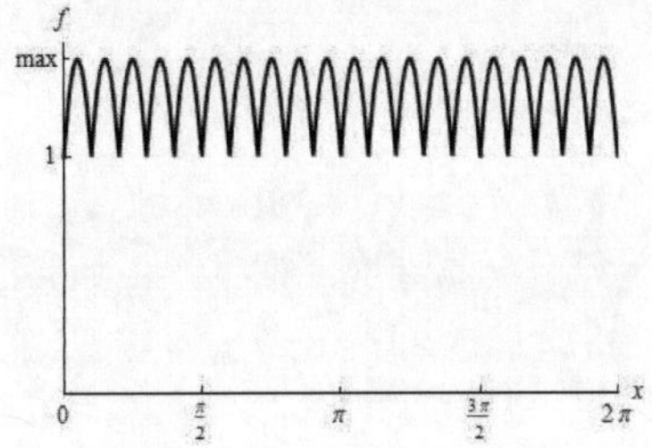

Fig. 39. $k = 5 \quad \max f = \sqrt{2} \quad \min f = 1$

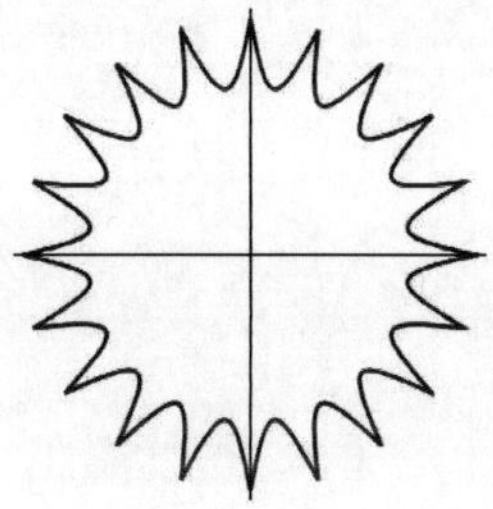

Fig. 39'.

For this reason Gielis proposed to chose as phase parameter the number

$$k = \frac{m}{4}.$$

Thus the equation becomes

$$\rho = \rho(\varphi) = r\left(\left|\frac{1}{a}\cos\left(\frac{m}{4}\varphi\right)\right|^{p_2} + \left|\frac{1}{b}\sin\left(\frac{m}{4}\varphi\right)\right|^{p_3}\right)^{-\frac{1}{p_1}} \tag{6}$$

6.1 The role of parameter *m*

Let us discuss the fundamental role played by phase parameter m, assuming $a = b = r = 1$, $p_1 = p_2 = p_3 = 1$.
We start form the case of integer values for parameter *m* (see Fig. 40-48)

***m* integer**

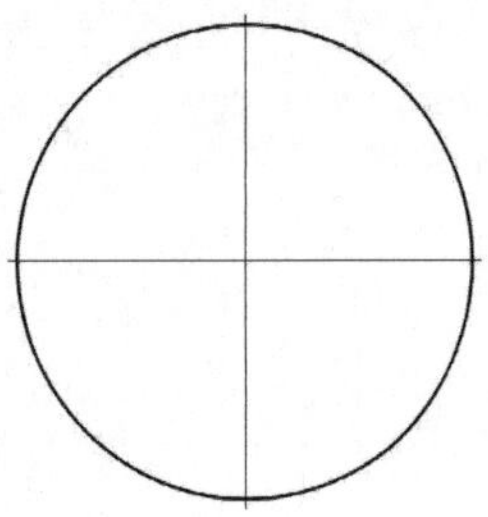

Fig. 40. $m = 0$

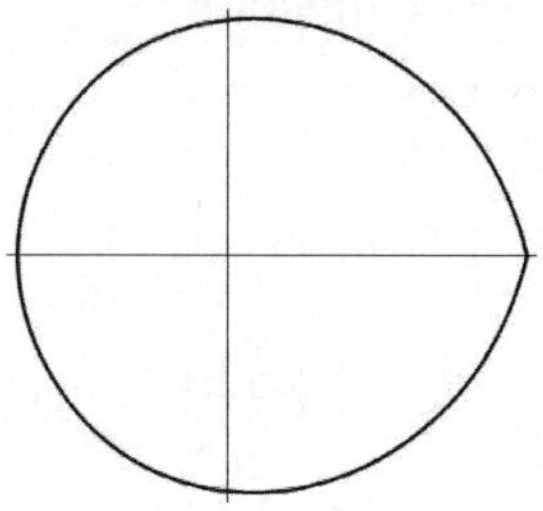

Fig. 41. $m = 1$

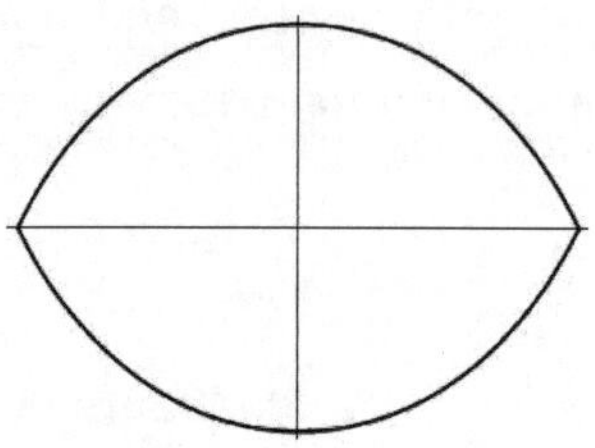

Fig. 42. $m = 2$

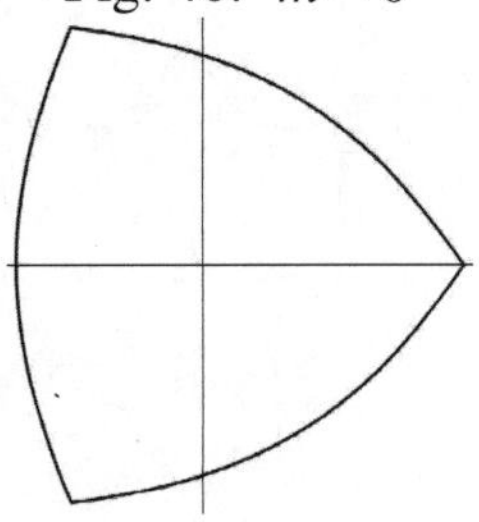

Fig. 43. $m = 3$

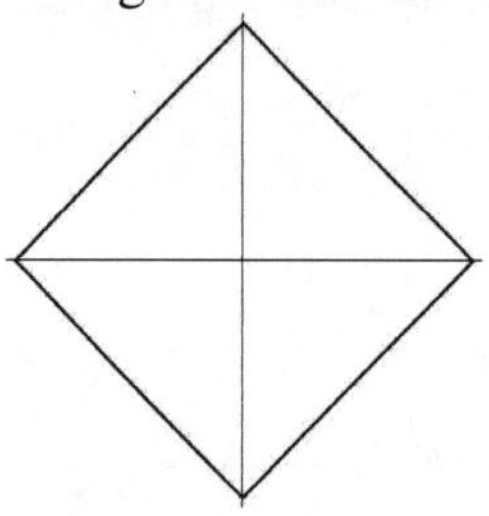

Fig. 44. $m = 4$

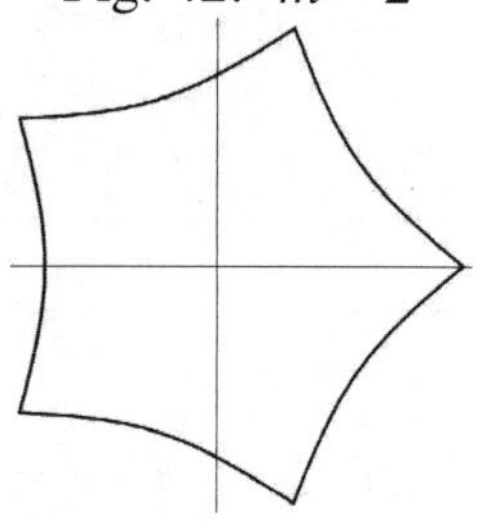

Fig. 45. $m = 5$

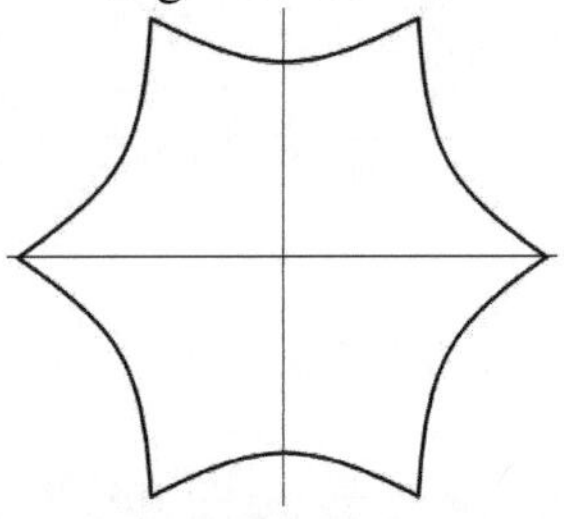

Fig. 46. $m = 6$

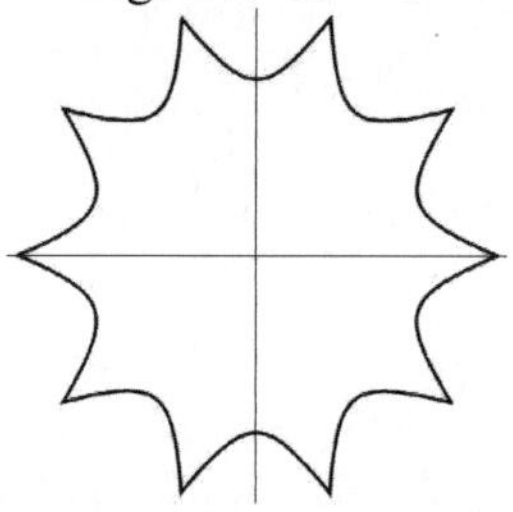

Fig. 47. $m = 10$

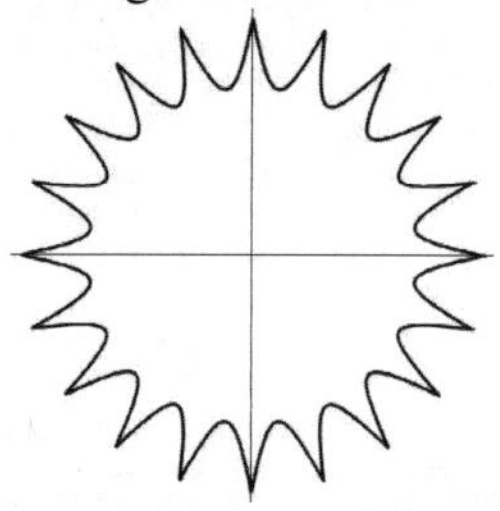

Fig. 48. $m = 20$

For Lamè super-ellipses the plane is ideally divided into four parts (the four quadrants) and all the figures have the same four fixed points on the circle.
Now the plane can be divided into *m* sectors. As we can see from the previous images, the phase parameter allows the sides to bend in and out like a fan. Moreover *m* determines the number of fixed points on the unit circle and the space between them. In other words m represents the number of rotational symmetries.
Assume now that $m = \mu/\upsilon$ is a rational number. The figure will close after υ rotations, while μ determines the number of angles. In particular, if *m* is integer, after a rotation the figure closes and in the subsequent rotations the figure remains unchanged (see Fig. 40-54).

m rational

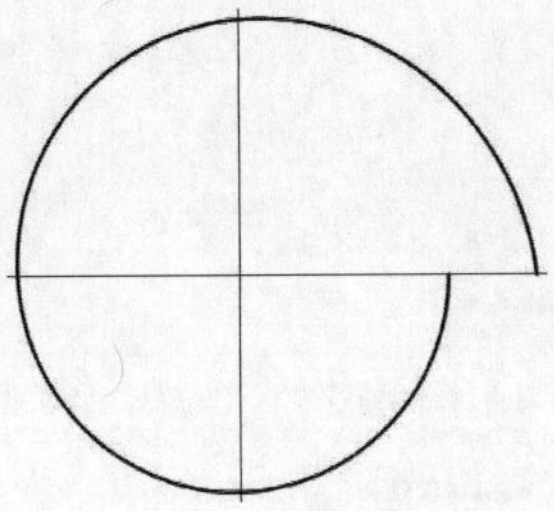

Fig. 49. $m = 1/2$
$0 \le \varphi \le 2\pi$

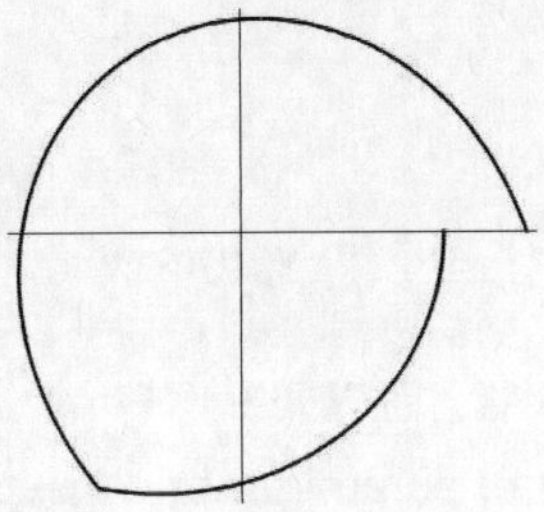

Fig. 50. $m = 3/2$
$0 \le \varphi \le 2\pi$

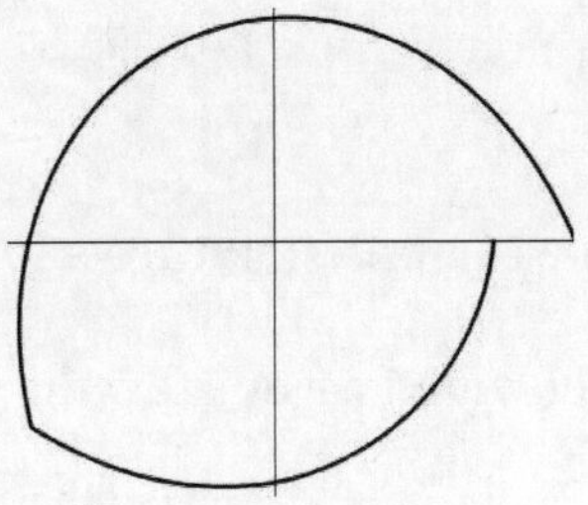

Fig. 51. $m = 5/3$
$0 \le \varphi \le 2\pi$

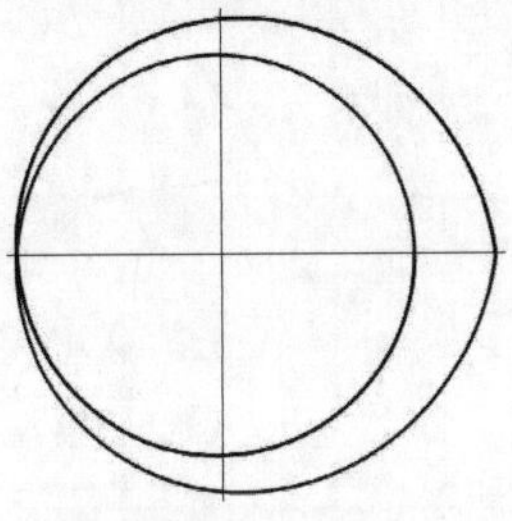

Fig. 52. $m = 1/2$
$0 \le \varphi \le 2 \cdot 2\pi$

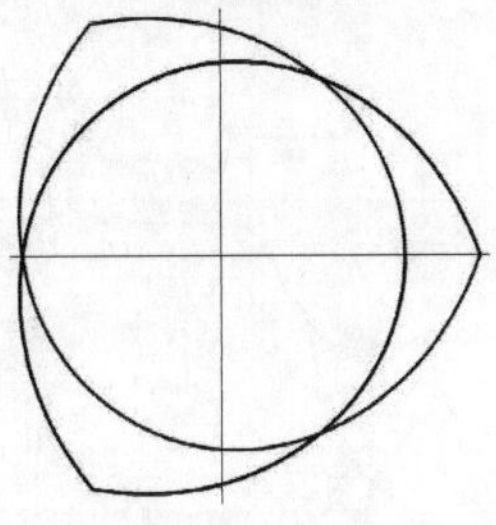

Fig. 53. $m = 3/2$
$0 \le \varphi \le 2 \cdot 2\pi$

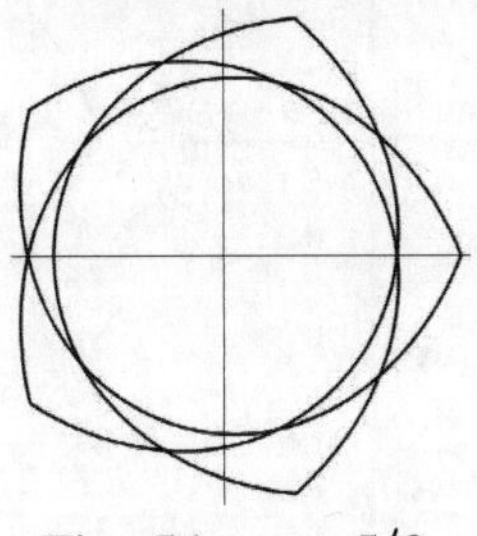

Fig. 54. $m = 5/3$
$0 \le \varphi \le 2 \cdot 3\pi$

While if m is not rational, the figure generated after a single rotation does not close, as the following examples show (see Fig. 55-57).

m not rational

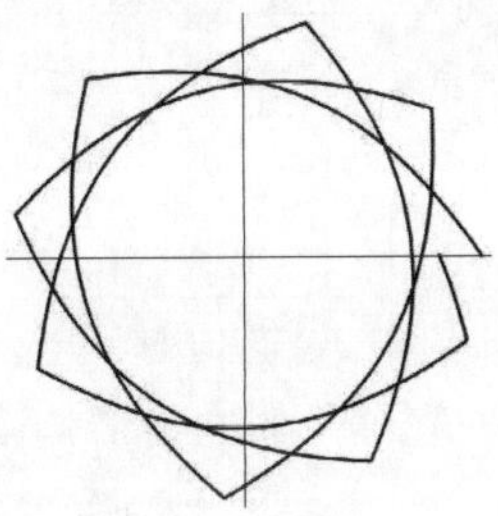

Fig. 55. $m = e$
$0 \le \varphi \le 6\pi$

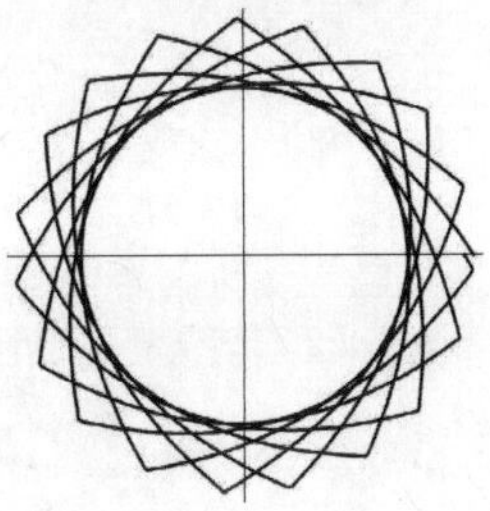

Fig. 56. $m = e$
$0 \le \varphi \le 14\pi$

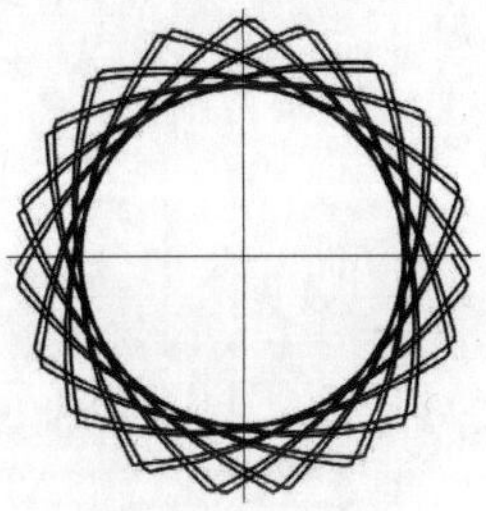

Fig. 57. $m = e$
$0 \le \varphi \le 28\pi$

Of course the number of possible shapes increase greatly if we assume different values for the exponent p.

In the particular case $m = 1$, we have a circle for $p = 2$, but for every $p \ne 2$ the figures present an angular points (see Fig. 58, 59). Precisely, for $0 < p < 2$ the angular points is on the right, while for $p > 2$ it is on the left (rotated by π).

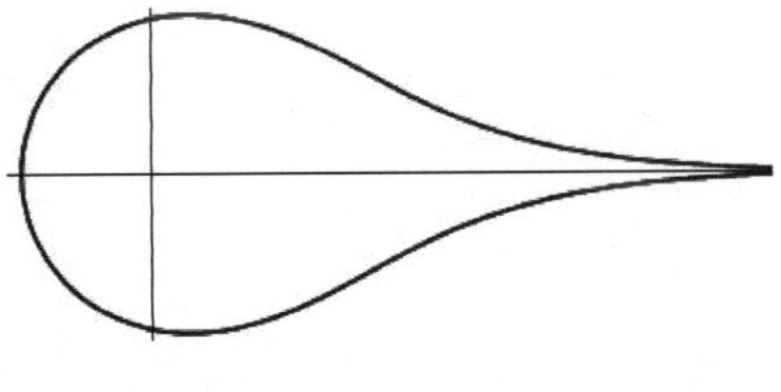

Fig. 58. $m = 1 \quad p = 0.3$

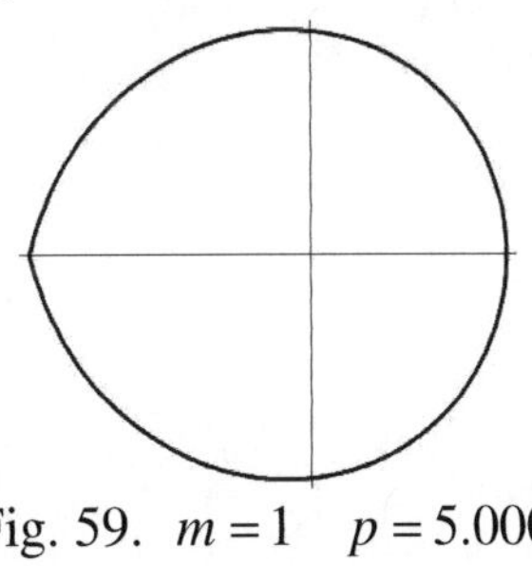

Fig. 59. $m = 1 \quad p = 5.000$

In the particular case $m = 2$ figures similar to the previous case are obtained, but with two angular points instead of one.

7 Some examples: nature's forms and polygons

The following images show the great potential offered by the variability of the parameters

The role of parameter p_1 (p_2=p_3=1, m=4)

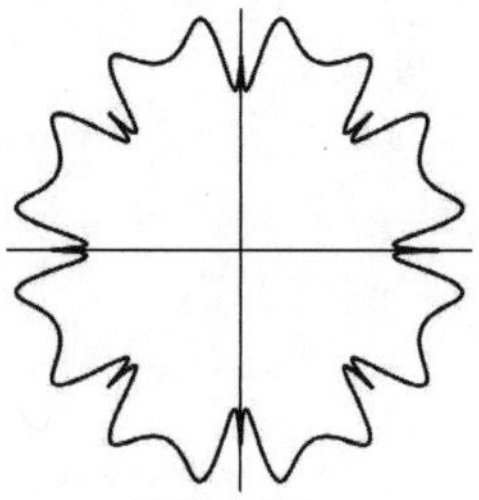

Fig. 60.
$m = 16\ p_1 = 1$
$p_2 = 10\ p_3 = 0.3$

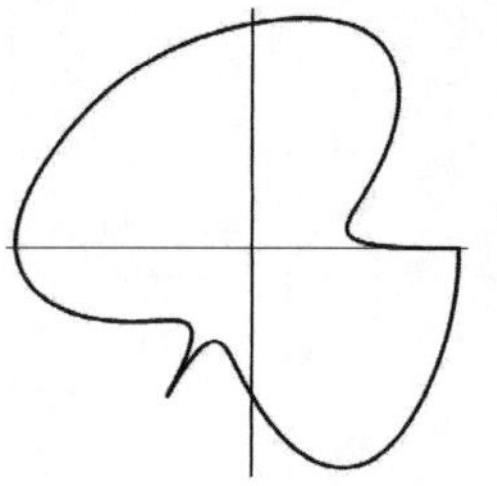

Fig. 61.
$m = 3\ p_1 = 0.5$
$p_2 = 10\ p_3 = 0.3$

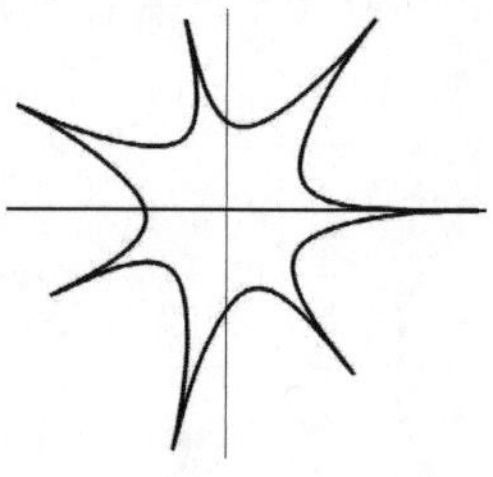

Fig. 62.
$m = 7\ p_1 = 0.5$
$p_2 = 0.5\ p_3 = 0.3$

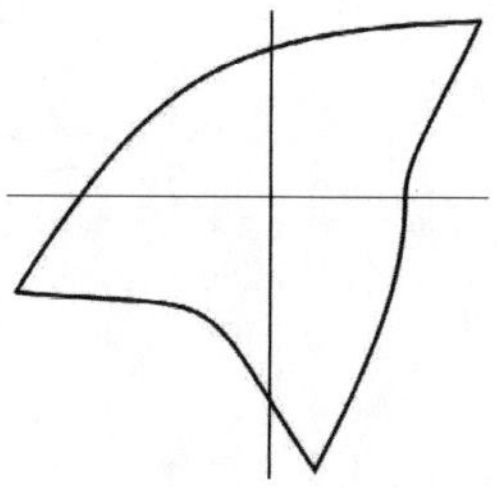

Fig. 63.
$m = 3\ p_1 = 10000$
$p_2 = p_3 = 2018$

As we have already observed, by means of formula (6) we can describe the circles, ellipses and rectangles. Note that we can also represent all the regular polygons together with their concave and convex versions (i.e. sub-polygons and super-polygons), as the following images show.

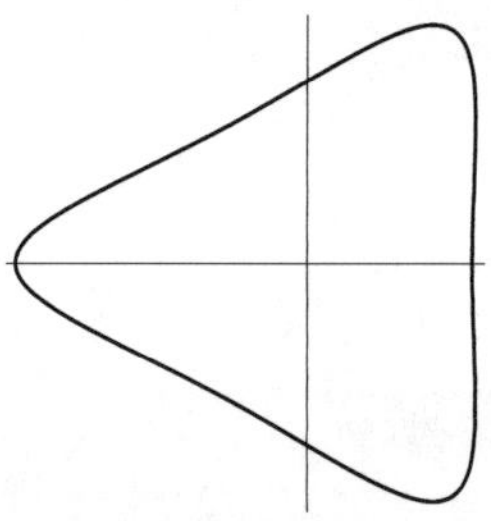

Fig. 64.
$m = 3\ \ p_1 = 5$
$p_2 = p_3 = 10$

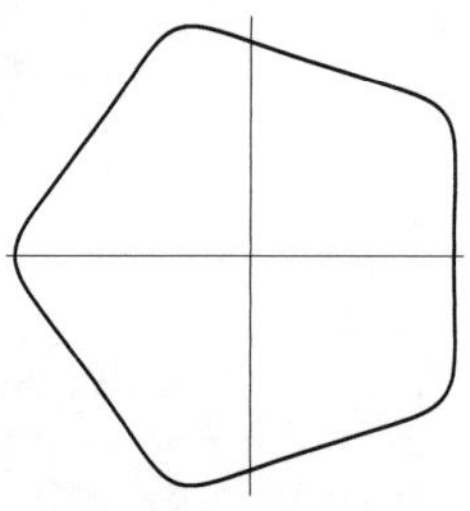

Fig. 65.
$m = 5\ \ p_1 = 6.5$
$p_2 = p_3 = 4.65$

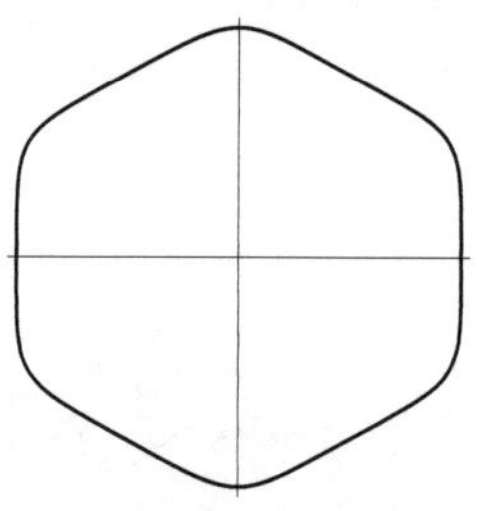

Fig. 66.
$m = 6\ \ p_1 = 10.65$
$p_2 = p_3 = 4.65$

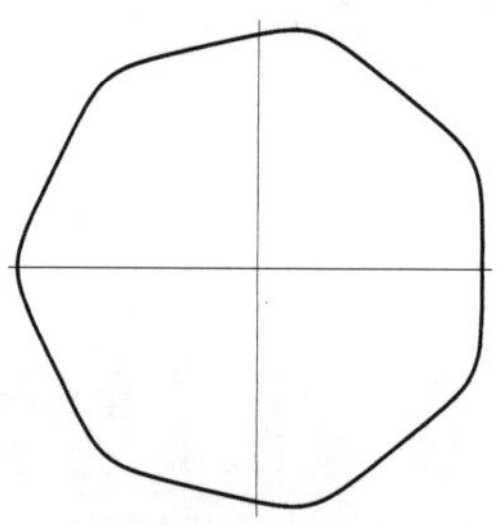

Fig. 67.
$m = 7\ p_1 = 15$
$p_2 = p_3 = 4.65$

8 Superformula

The final Gielis' generalization of equation (6) consists on replacing parameter r with a function of the variable φ

$$\rho = \rho(\varphi) = R(\varphi)\left(\left|\frac{1}{a}\cos\left(\frac{m}{4}\varphi\right)\right|^{p_2} + \left|\frac{1}{b}\sin\left(\frac{m}{4}\varphi\right)\right|^{p_3}\right)^{-\frac{1}{p_1}} \tag{8}$$

Of course, this innovation improves the performance . Let us discuss two remarkable particular cases.

8.1 Spirals

Let us consider the particular case of the function

$$R(\varphi) = \varphi$$

which is constant in each period, but whose argument increases with increasing periods. Inserted in the superformula it leads to the representation of spirals, as the following examples show.

Fig. 68.
$m = 6\ p_1 = p_2 = p_3 = 100$
$0 \le \varphi \le 12\pi$

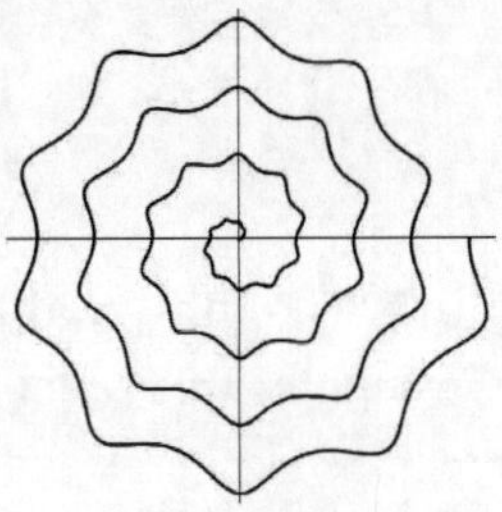

Fig. 69.
$m = 10 \quad p_2 = p_3 = 5$
$p_1 = 8 \quad 0 \le \varphi \le 8\pi$

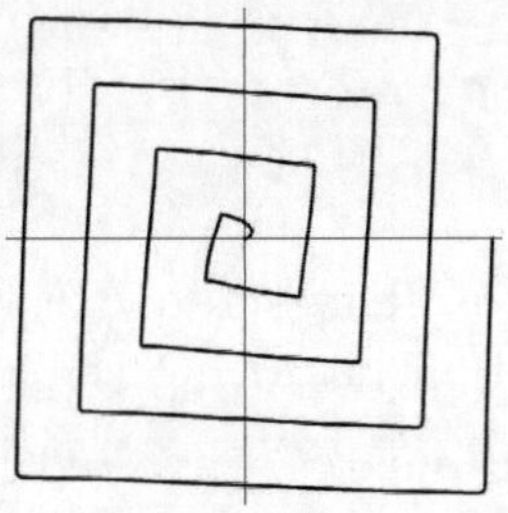

Fig. 70.
$m = 4\ p_1 = p_2 = p_3 = 100$
$0 \le \varphi \le 8\pi$

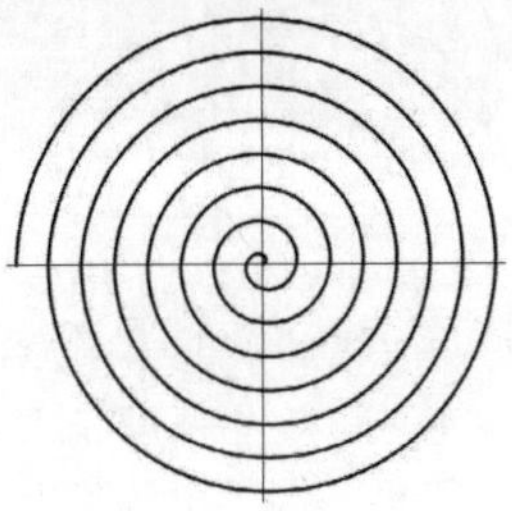

Fig. 71.
$m = 6 \quad p_2 = p_3 = 1$
$p_1 = 100 \quad 0 \le \varphi \le 15\pi$

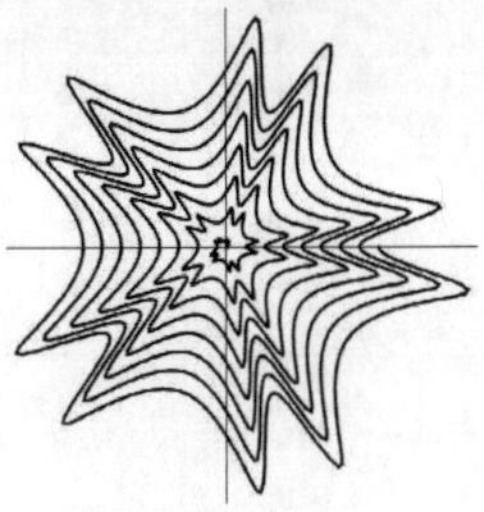

Fig. 72.
$m = 10 \quad p_2 = 50\ p_3 = 5$
$p_1 = 8 \quad 0 \le \varphi \le 16\pi$

Fig. 73.
$m = 10\ p_2 = 0{,}5\ p_3 = 2$
$p_1 = 1 \quad 0 \le \varphi \le 16\pi$

8.2 Flowers

Function $$R(\varphi) = \left|\cos\frac{m}{2}\varphi\right|$$

admits $\max R(\varphi) = 1$, $\min R(\varphi) = 0$ and presents m maximum points and m minimum ones. Inserted in the superformula it allows to represent flower petals, as the following examples show.

Fig. 74. $m = 5$
$p_1 = p_2 = p_3 = 1$

Fig. 75. $m = 10$
$p_1 = p_2 = p_3 = 1$

Fig. 76. $m = 16$
$p_1 = 1\ p_2 = p_3 = 5$

Fig. 77. $m = 5$
$p_3 = p_1 = 1 \quad p_2 = 10$

Fig. 78. $m = 8$
$p_2 = 5 \; p_3 = 0.3 \; p_1 = 1$

Fig. 79. $m = 12$
$p_2 = 0.1 \; p_3 = 5 \; p_1 = 1$

The previous figures show that a different values of the parameters p_2 and p_3 produces asymmetries between the petals. Parameter p_1 adjusts the roundness, as the following sequences of images shows (where $m = 5 \; p_2 = p_3 = 1$).

Fig. 80. $p_1 = 0.3$

Fig. 81. $p_1 = 3$

Fig. 82. $p_1 = 30$

9 A selected list of examles

Let's now propose some examples, taken from ([1]), that testify superformula potentiality on a faithful reproduction of various forms of nature.

$a = b = 10 \quad m = 5$
$p_1 = p_3 = 2 \quad p_2 = 1$
$0 \le \varphi \le 2 \cdot 2\pi$

Fig. 83. Calys and sepals of roses.

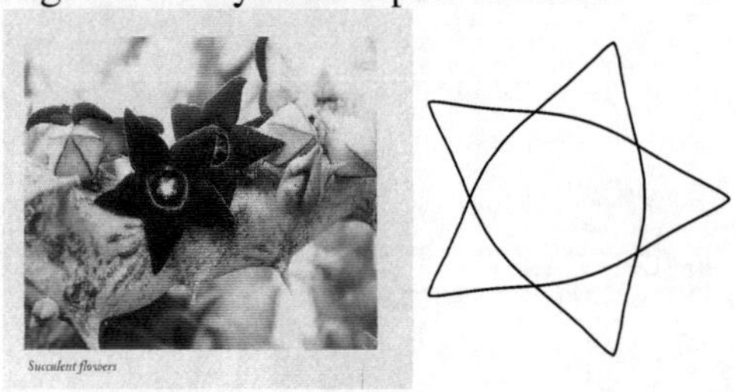

$a = b = 10 \quad m = 5$
$p_1 = 3.6 \; p_3 = 2.5 \; p_2 = 1.4$
$0 \le \varphi \le 2\pi$

Fig. 84. Succulent flowers.

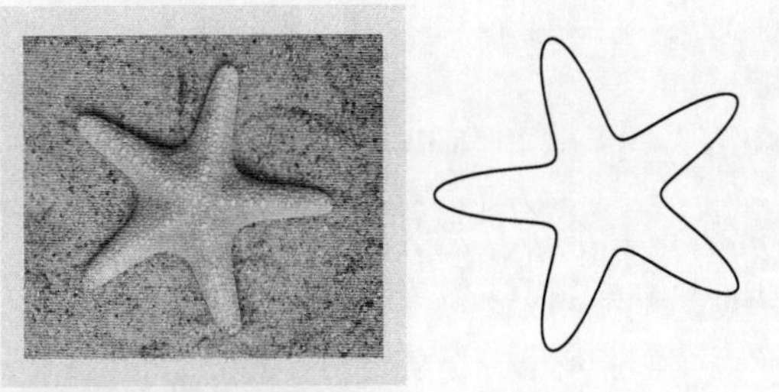

$a = b = 10 \quad m = 5$

$p_2 = p_3 = 5 \quad p_1 = 1$

$0 \le \varphi \le 2 \cdot 2\pi$

Fig. 85. Starfish.

$a = b = 10 \quad m = 5$

$p_1 = 1 \; p_3 = 5.9 \; p_2 = 4.6$

$0 \le \varphi \le 2 \cdot 2\pi$

Fig. 86. Flower.

$a = b = 1 \quad m = 10$

$p_2 = p_3 = 5 \quad p_1 = 8$

$0 \le \varphi \le 14 \cdot 2\pi$

$R(\varphi) = \varphi^{2.55}$

Fig. 87. Pleuroploca trapezium.

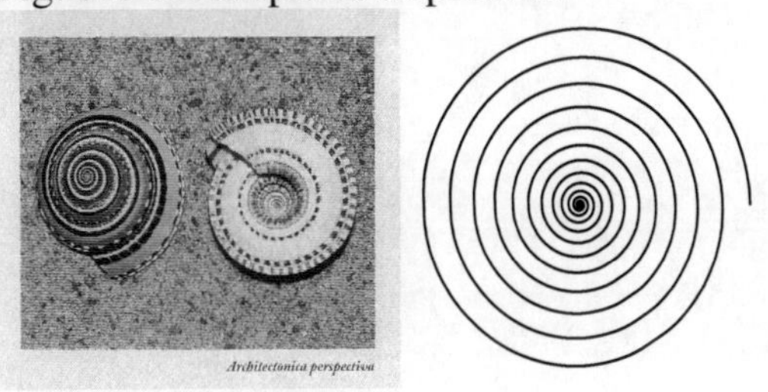

$a = b = 1 \quad m = 6$

$p_2 = 0 \; p_3 = p_1 = 100$

$0 \le \varphi \le 14 \cdot 2\pi$

$R(\varphi) = \varphi^{2.4}$

Fig. 88. Architectonica perspectiva.

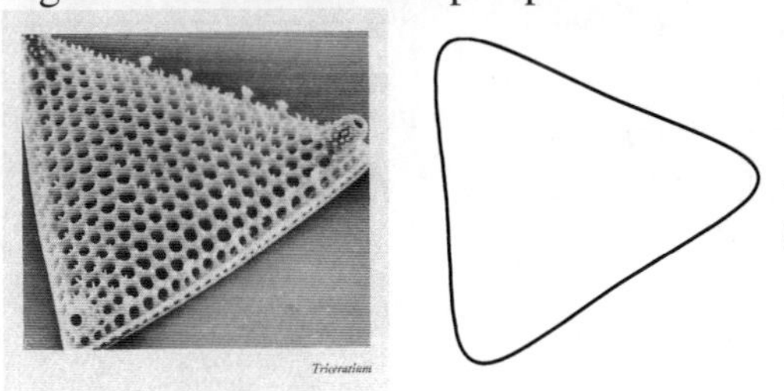

$a = b = 10 \quad m = 3$

$p_1 = 5 \; p_3 = p_2 = 10$

$(11/36) \cdot 2\pi \le \varphi \le (11/36) \cdot 2\pi + 2\pi$

Fig. 89. Diatomee.

References

[1] BRACCI, P., *La superformula della natura*, Tesi di laurea in Matematica, Università di . Perugia (2013/14)
[2] GIELS, J., A *generic geometric transformation that unifies a wide range of natural and abstract shapes*, American Journal of Botany, 90, 333-338, 2003
[3] GIELS, J., *Inventing the circle: the geometry of nature*, Geniaal, 2003
[4] URL, http://www.genicap.com/

Current address

Brandi Primo
Department of Mathematics and Informatics
University of Perugia
via Vanvitelli 1, 06123 Perugia, Italy
E-mail: primo.brandi@unipg.it

Salvadori Anna
Department of Mathematics and Informatics
University of Perugia
via Vanvitelli 1, 06123 Perugia, Italy
E-mail: anna.salvadori@unipg.it

Slovak University of Technology in Bratislava
Faculty of Mechanical Engineering
17th Conference on Applied Mathematics **APLIMAT 2018**

Proceedings

THE ABILITY OF PRIMARY SCHOOL PUPILS TO SOLVE COMBINATORIAL TASKS

BŘEHOVSKÝ Jiří (CZ), PŘÍHONSKÁ Jana (CZ)

Abstract. The paper describes the results of a survey which investigated the success rate of primary school pupils in solving basic types of combinatorial tasks that can be used at elementary school and the methods used to solve them. A part of the paper also includes the information on the result of a questionnaire survey focused on primary school teachers in terms of their approach to including and solving combinatorial tasks in teaching mathematics.

Keywords: ability to solve problems, combinatorial problems, primary education

Mathematics Subject Classification: Primary 97-02

1 Introduction

Combinatorics is one of the oldest branches of discrete mathematics, dating back to the 16th century when games of chance played a key role in society life (Abramovich & Pieper, 1996). Combinatorics is a significant component of the mathematics curriculum, comprising a rich structure of powerful principles that underline several other areas such as counting, computation, and probability (Borovcnik, Peard 1996). Solving of combinatorial problems helps pupils to develop the use of heuristic strategies for solving problems. Applying these unconventional ways of solving problems also improves the communication among pupils. (Doulík & all 2016) Despite its importance in the mathematics curriculum, combinatorics continues to remain neglected, particularly at the elementary school level (English 2005).

The paper is one of the outputs of the scientific project SGS 21162 "Combinatorics at Elementary School - Activating Activities for the Development of Logical and Combinatorial Reasoning of Students". The research that is being carried out as part of the project's solution is focused on generating activating activities that will help develop pupils logical and combinatorial thinking. As it turns out, solving combinatorial tasks seems rather difficult for many pupils. Therefore, one of the objectives of the project is to verify these activities in the teaching of mathematics at primary school level. Another aim of this project is to use the results to improve the educational process of future elementary school teachers at the Faculty of Science, Humanities and Education at the Technical University of Liberec. We have been monitoring not only the pupils success in solving combinatorial tasks but also the selected solving strategies.

In order to gain a more comprehensive overview of the whole situation, the survey also included a questionnaire survey focused on primary school teachers. Its goal was to map the approach of elementary school teachers to including and solving combinatorial problems in teaching mathematics.

2 Theoretical background

Combinatorics is an essential component of discrete mathematics and, as such, it has an important role to play in school mathematics. In 1970, Kapur presented the following reasons, which could still be valid, to justify the teaching of elementary Combinatorics at school:

- Since it does not depend on calculus, it has suitable problems for different grades; usually very challenging problems can be discussed with pupils, so that they discover the need for more mathematics to be created;
- It can be used to train pupils in enumeration, making conjectures, generalization and systematic thinking; it can help the development of many concepts, such as equivalence and order relations, function, sample, etc.
- Many applications in different fields can be presented.

Therefore, our research focuses on the development of pupils´ combinatorial reasoning at primary school level. School combinatorics is an essential part of the mathematical culture of education. Many combinatorial problems can be formulated very easily, however, their solution is often very difficult. The importance of children's combinatorial reasoning in analyzing sample space has been evident in several studies (e.g. Benson & Jones 1999; Johnson, Jones, Thornton, Langrall, & Rous 1998; Nisbet et al., 2000; Zimmermann & Jones, 2002). The results of these studies show the difficulties faced by pupils in solving problems that require pupils to employ combinatorial reasoning.

In order to focus on developing pupils competencies in the area of solving combinatorial tasks, it was first necessary to map the current state which included not only the pupils ability to solve combinatorial tasks but also finding out what strategies they use most often to solve them. Various studies (e.g. Inhelder & Piaget 1975), which used "colored counter" tasks, dealt with the verification of pupils ability to solve combinatorial problems.

The colored counters task was part of an investigation into children's development of the idea of chance. In another task, children were presented with sets of counters: each set was a different color, and the children were asked to create as many different pairs of counters as possible.

The pupils performance in experiment suggested that pupils generate combinations only in an empirical manner by randomly associating two elements at a time (i.e., there is a lack of systematic method).

In one such study (English, 1991), 50 children aged between 4.5 years and 9.8 years were individually administered a series of 7 novel tasks that involved the dressing of cardboard toy bears (placed on stands) in all possible different outfits. The results of the study revealed a series of solution paths used by the children in solving the set of problem tasks. These paths ranged from random item selection through to a systematic pattern in item choice, reflecting increasing sophistication in solution procedure.

A combinatorial task/problem is a problem that leads towards the creation of various configurations and diagrams (Fuchs 2000). We can create configurations and diagrams by e.g. selecting elements from a predetermined finite set. In that case, it is important to distinguish whether the order of the selections matters or not. However, it does not necessarily have to be a selection of elements we are talking about. A combinatorial task might also involve a reorganisation of the set, change of its properties, etc. Such combinatorial problems also facilitate the development of enumeration processes, as well as conjectures, generalisations, and systematic thinking (English 2005).

3 The questionnaire survey

The questionnaire survey focused on primary school teachers in terms of their approach to including and solving combinatorial tasks in teaching mathematics. A total of 52 teachers, teaching at primary school level, participated in the survey. Their teaching experience ranged from 2 to 35 years. 7 elementary schools in the Liberec region joined the survey. The questionnaire was anonymous. It comprised a total of 6 questions, including 3 with closed and 3 with open answers (Table 1).

Sex:						
Taught Year:						
Teaching Practice Length:						
1	**What mathematics textbooks (publishing house) do you use in teaching - mark with a cross**					
	Prodos	Alter	SPN	Fortuna	Studio 1+1	Other
	Specify other used textbooks:					
2	**What types of task do you assign most often - select the appropriate option, mark with a cross**					
	Counting (arithmetic)	Word problem	Geometric	Other		

	Specify other assigned tasks:					
3	**Do you think that textbooks include enough combinatorial tasks? Select the appropriate option, mark with a cross.**					
	Yes		**No**		**Unable to say**	
4	**Specify what types of combinatorial tasks you solve most often with your pupils:**					
5	**What manner (method) do you employ to solve combinatorial tasks?**					
6	**How often do you solve combinatorial tasks with your pupils?**					

Tab. 1. Questionnaire for teachers.

When assessing the data, we primarily focused on the following two questions:

a) "How often do you solve combinatorial tasks with pupils?"
From the available data it is clear that the frequency of combinatorial tasks was split up into four groups: daily, once a week, seldom and never. Graph 1 shows the percentage of individual groups.

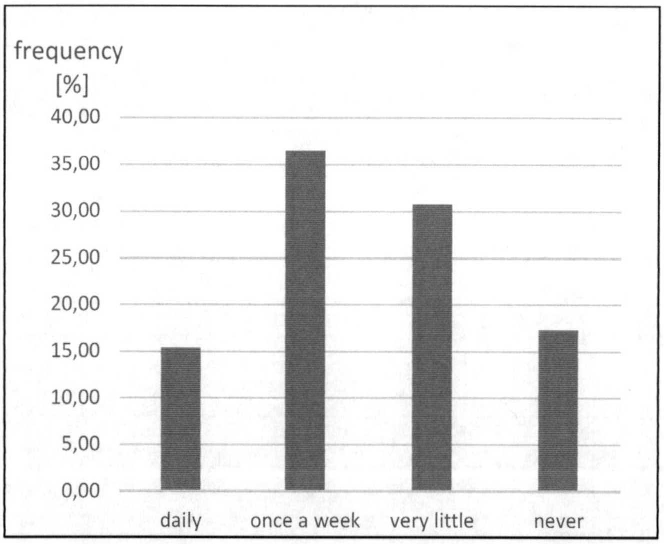

Graph 1. Percentage frequency of use of combinatorial tasks.

It is clear from Graph 1 that 48% of the participating teachers seldom or never assign combinatorial tasks to pupils. This is partly due to the kind of textbooks, used by teachers. The analysis of textbooks (Břehovský, Příhonská 2017) shows that the average percentage representation of combinatorial tasks in the textbooks is 6.3%. The average total number of all tasks in a textbook is 216, meaning the average number of combinatorial tasks in each textbook is 13. The questionnaire analysis shows a certain correlation between the combinatorial assignment frequency and the textbooks used.

b) "What manner (method) do you use to solve combinatorial tasks?"

It is clear from the questionnaire evaluation that teachers use three methods to solve combinatorial tasks: list of elements, consideration and graphical representation, and trial and error strategy. Graph 2 shows the percentage frequency of the individual solutions used. It is evident that in more than 70% of cases, teachers use the *consideration and graphical representation* strategy to solve combinatorial tasks.

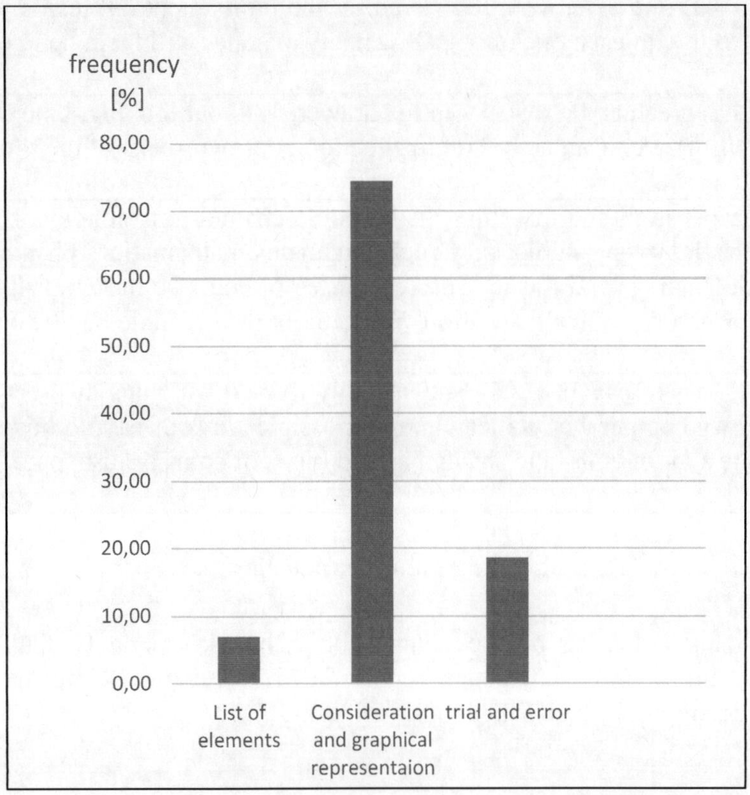

Graph 2. Percentage frequency of combinatorial tasks solution strategy.

The available data did not enable a clear identification of the type of combinatorial tasks that teachers solve with the pupils most often. The survey results suggest that the selection of tasks depends primarily on the type of textbook used. The analysis of textbooks (Břehovský, Příhonská 2017) shows that diversity of combinatorial tasks types in mathematics textbooks for the primary school level is not large.

3 Testing pupils

The test was primarily focused on determining the success rates of primary school pupils in solving basic types of combinatorial tasks. Another output of the test was the information on the manner of solving these tasks. We wanted to get an idea of what strategies pupils use to solve combinatorial tasks. Altogether, 72 pupils participated in testing, of which 48 girls and 24 boys. Their age ranged from 10 to 11 years. Testing took place at six different elementary schools. Twelve pupils were selected from each school by a stratified selection. The pupils with specific learning disabilities were not selected and the sample was homogeneous in terms of pupils skills in mathematics. The test contained four open-answer tasks. The complete assignment of all four tasks is shown in Table 2. The tasks used were selected on the basis of a series of pretests to reduce undesirable effects that could adversely affect test results. Tasks were contextually different and more strategies could be used to solve them.

Task No.	Assignment
1.	A dodgeball tournament is held at the school. Five teams from five classes (3 A, 3 B, 4 A, 4 B and 5 A) were registered. In the tournament, all teams will play each other (each with each one once). How many matches will there be in total?
2.	Jindra has green and red socks in his drawer. Without looking, he took three socks out of the drawer. Can he be sure that he took out two socks of the same color?
3.	The teacher is ill and therefore the schedule changes on Monday. The pupils will have the following subjects: Czech language, mathematics, physical education, English language and music. The second lesson must be the English language and the fifth lesson physical education. What might the schedule be like on Monday?
4.	Maruška has to pay 11 CZK. She has only five-crown coins, two-crown coins and one-crown coins in her wallet. Find all possible ways of paying this amount when there are a lot of coins and she can use all kinds of coins or just some.

Tab. 2. Tasks assignment in the test.

The success of the solution has been evaluated for each task separately. Only the complete solution with all the options stated was considered to be correct. Graph 3 illustrates the percentage success of pupils in solving individual tasks.

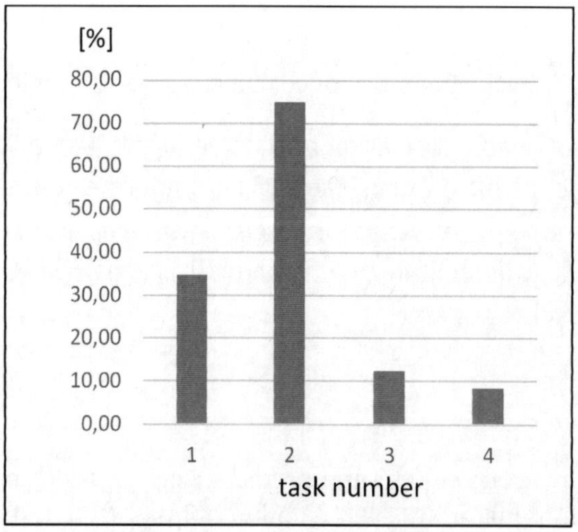

Graph 3. Percentage success of pupils in solving individual tasks.

Task number 2 had the highest solution success rate. It was correctly solved by 77% of pupils. In many cases, the pupils correctly substantiated their claims, largely using a picture. Two-thirds of pupils who responded correctly drew all variants of the solution. Task number 1 was solved correctly by 34% of pupils. The pupils who used a spreadsheet or graph depiction of the break-down of all matches were more successful in solving the task. For the remaining two tasks, the success rate was 12.5% for the third task and 8% for the fourth task. It was the fourth task where the pupils achieved the smallest success rate. The pupils were unable to find all the options for solving the task. Most often, 4 or 7 options were stated as the correct answer. In both cases, this answer was stated by 18% of pupils. Only six pupils found all 11 options and they all used systematic search to do so.

The evaluation of available data shows that the pupils chose the following solution strategies to solve combinatorial tasks: using the table to find all solutions (Tasks 1 and 3); using the color picture and visualization of the situation (Tasks 1 and 2); using systematic recording when searching for all options (Tasks 3 and 4). In each case, the node graph (Task 1) and symbolic recording (Task 2) were successfully used. Figures 1 through to 4 show specific pupils solutions using different strategies.

Task 1. A dodgeball tournament is held at the school. Five teams from five classes (3 A, 3 B, 4 A, 4 B and 5 A) were registered. In the tournament, all teams will play each other (each with each one once). How many matches will there be in total?

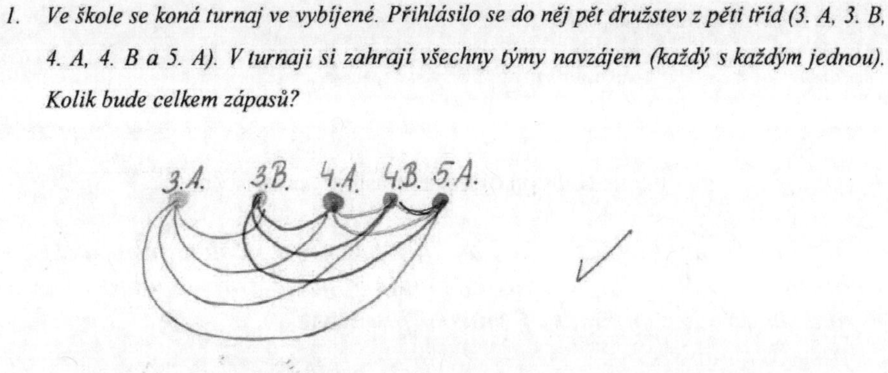

Fig. 1. Pupils solution of Task 1, use of visualization and node graph.

Task 2. Jindra has green and red socks in his drawer. Without looking, he took three socks out of the drawer. Can he be sure that he took out two socks of the same color?

Yes
either he took out
   a) 2 red socks and 1 green sock, or
   b) 2 green socks and one red sock

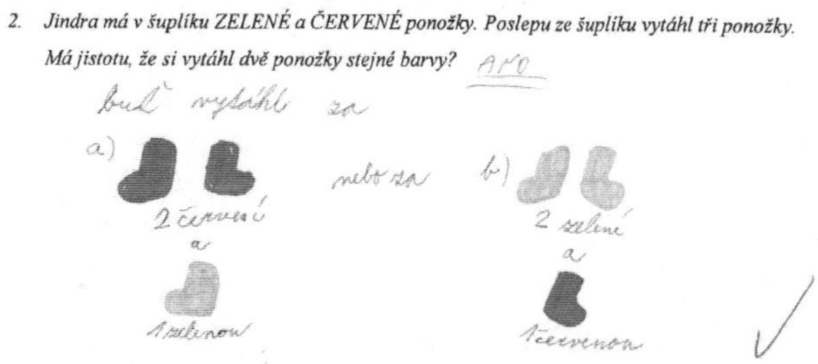

Fig. 2. Pupils solution of Task 2, use of visualization.

Task 3. The teacher is ill and therefore the schedule changes on Monday. The pupils will have the following subjects: Czech language, mathematics, physical education, English language and music. The second lesson must be the English language and the fifth lesson physical education. What might the schedule be like on Monday?

The answer was: “The Monday schedule might look like this.”

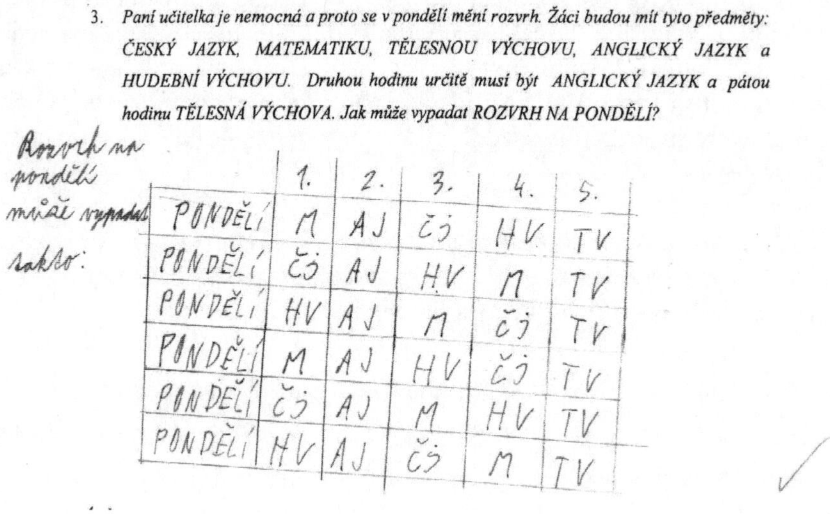

Fig. 3. Pupils solution of Task 3, use of spreadsheet.

Task 4. Maruška has to pay 11 CZK. She has only five-crown coins, two-crown coins and one-crown coins in her wallet. Find all possible ways of paying this amount when there are a lot of coins and she can use all kinds of coins or just some.

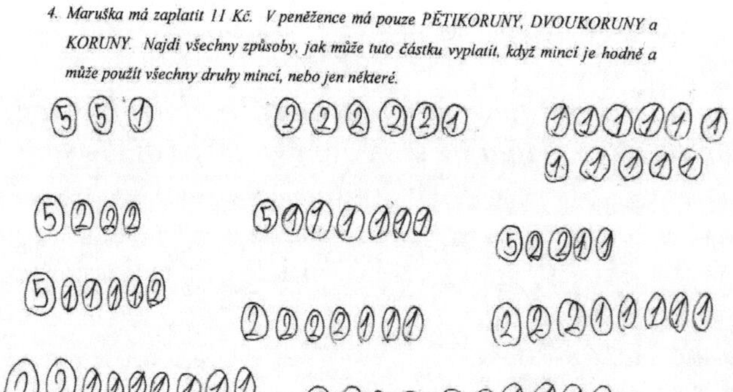

Fig. 4. Pupils solution of Task 4, use of systematic writing.

Conclusion

In our survey we looked into the results of a questionnaire survey focused on primary school teachers and their approach to including and solving combinatorial tasks in teaching mathematics. The available data clearly show that 48% of the participating teachers seldom or never assign combinatorial tasks to pupils. The analysis of the questionnaire also points to a certain correlation between the frequency of combinatorial tasks assignment and the textbooks used by teachers during tuition. It is clear from the questionnaire evaluation that teachers use three methods to solve combinatorial tasks with pupils: list of elements, consideration and graphical representation, and trial and error strategies. Consideration and graphical representation are employed even in over 70% of solutions.

The second part of the survey was the determination of success rate of primary school pupils in solving the basic types of combinatorial tasks and their method of solving. For this purpose, a test was created that included four typological and contextually different tasks. Each task was evaluated separately, both in terms of success and in terms of the use of solution strategies. Solely the complete solution, stating all the options, was considered to be correct. The highest success rate was achieved by pupils in the second task, which was 75%. They did quite well in solving the first task with 35% success rate. For the remaining tasks, success rate was small, only 12.5% and 8%. Pupils successfully used different strategies to solve the tasks: spreadsheet, systematic listing of all options, visualization using color pictures or node graph. The most effective strategy to find solutions is to use systematic recording when pupils create possible groups of elements through a logical process (for example, in Task 4 they determine the first correct possibility to use eleven one-crown coins and then gradually interchange coins so that the sum still equals 11). Similar findings are also reported by English in his study (1991).

The results of the surveys allow us to focus more in the following research on the creation of those activities that develop pupils logical and combinatorial thinking and that bring greatest difficulty in solution to pupils. We will also pay attention to the use of a larger variety of solving strategies. Last but not least, we would like to use these activities to motivate teachers to include the combinatorial tasks more often in teaching.

The results of the research will be farther used to improve the educational process of future elementary school teachers at the Faculty of Science, Humanities and Education at the Technical University of Liberec. They will be particularly used to promote constructive approaches in teaching mathematics. The teachers will be also offered specific and proven problem tasks which they can use in their practice.

Acknowledgement:

The research was supported by SGS project 2017 at FP TUL in Liberec.

References

[1] ABRAMOVICH, S., PIEPER, A.: Fostering recursive thinking in combinatorics through the use of manipulatives and computer technology. In The Mathematics Educator. 1996. Volume 7. Issue 1. pp 4-12.

[2] BENSON, C. T., JONES, G. A.: Assessing students' thinking in modelling probability contexts. In The Mathematics Educator. 1999. Volume 4. Issue 2. pp 1-21.

[3] BOROVCNIK, M., PEARD, R.: Propability. In Internacional Hnadbook in Mathematics Education. 1996. Dordrecht, Netherlands. Part 1, pp 239-288. ISBN 9789401071550.

[4] BŘEHOVSKÝ, J., PŘÍHONSKÁ, J.: Combinatorial problems of mathematics for elementary school. Source: *Aplimat 2017*. Pages 206 – 214. Bratislava 2017, ISBN 987-80-227-4650-2. Conference: 16th Conference of Applied Mathematics 2017.

[5] DOULÍK, P., EISENMANN, P., PŘIBYL, J., ŠKODA, J.: Unconventional Ways of Solving Problems in Mathematics Classes. Source: *The New Educational Review*, 2016, Vol. 43, No. 1, 53 – 67. ISSN 1732-6729.

[6] ENGLISH, L. D. Young children's combinatoric strategies. *Educational Studies in Mathematics.* 1991. Volume 22, pp 451-474.

[7] ENGLISH, L. D.: Combinatorics and the Developement of Childrens Combinatorial Reasoning. In Mathematics Education Library, *Exploring Probability in School*. 2005. Volume 40. pp 121-142. ISSN 0924-4921.

[8] FUCHS, E.: Diskrétní matematika a Teorie množin pro učitele. 2000. [CD ROM]. Brno: Masarykova univerzita.

[9] INHELDER, B., & PIAGET, J.: *The growth of logical thinking: From childhood to adolescence.* 1958. (A. Parsons & S.Milgram, Trans.). London: Routledge and Kegan Paul.

[10] JOHNSON, T. M., JONES, G. A., THORNTON, C. A., LANGRALL, C. W., ROUS, A.: Students' thinking and writing in the context of probability. In Written Communication, 1998. Volume 15. Issue 2. pp 203-229.

[11] KAPUR, J. N.: Combinatorial Analysis and School Mathematics. In Educational Studies in Mathematics 3. 1970. pp 111-127.

[12] NISBET, S., JONES, G. A., LANGRALL, C. W., & THORNTON, C. A.: A dicey strategy to get your M & Ms. In Australian Primary Mathematics Classroom. 2000. Volume 5. Issue 3. pp 19-22.

[13] ZIMMERMANN, G. M., & JONES, G. A.: Probability simulation: What meaning does it have for high school students? In Canadian Journal of Science, Mathematics and Technology Education. 2002. Volume 2. Issue 2. pp 221-236.

Current address

Břehovský Jiří, Mgr, Ph.D.
Faculty of Sciences, Humanities and Education, Technical University of Liberec, Studentská 1402/2, 461 17 Liberec 1, Czech Republic
Tel. +420 48 535 2923, e-mail: jiri.brehovsky@tul.cz

Příhonská Jana, Doc, RNDr., Ph.D.
Faculty of Sciences, Humanities and Education, Technical University of Liberec, Studentská 1402/2, 461 17 Liberec 1, Czech Republic
Tel. +420 48 535 2870, e-mail: jana.prihonska@tul.cz

Slovak University of Technology in Bratislava
Faculty of Mechanical Engineering
17th Conference on Applied Mathematics **APLIMAT 2018**

Proceedings

TALL BUILDINGS IN MILANO. PROPORTIONS AND MORPHOLOGY BETWEEN ANALOGICAL AND DIGITAL DESIGN PROCEDURE. SOME CONSIDERATIONS ABOUT TWO CASE HISTORIES OF CONTEMPORARY CONSTRUCTION SITES: FROM THE XX LEGACY TO THE XXI NEW CENTURY URBAN LANDSCAPE

BRUNETTI Federico Alberto (IT)

Abstract. The design of tall architectures building assumed for modern architects a particular compositive challenges: not only following the new technical potentiality derived from innovative materials for construction (i.e. iron, steel, reinforced concrete etc.), but also evocating different essential symbols through monolithic shapes. Some recent building sites in Milan offer the opportunity to focus the attention on some evidences in the evolution of design techniques toward the composition of these buildings, in comparison between the XX century analogue orthogonal drafting machine and the actual complex digital 3D (/4D) tools. These two drawing techniques could be considered as the output devices of different algorithms approaching the same gravity challenge to which any architecture design must afford. The verticality of the building structures allows the overlapping of the horizontal floors following the static sciences: the construction and available materials technique could obtain different solutions essentially in reason of calculation process and aesthetical approaches defined by design concept and drawing procedures.

Keywords: mathematics and visual arts, mathematics modelling, application of mathematics, music, language, architecture computer science applications, mathematics and architecture

Mathematics Subject Classification: Primary: 00A66, 97B70; Secondary: 97M80, 00A67, 00A05, 01A05, 97U99, 51N05, 01A05

1 Milan skyline

The Milanese urban landscape has profoundly changed in these last decades. Considering the most significant episodes of post-war architectural construction, and in particular the most characteristic building typology of modernity - particularly skyscrapers or tall buildings[1] - we can see how, between the generation of twentieth century buildings and those of the last decade, is recognizable a substantial diversity. [1]

Fig. 1. Milan skyline with modern and historical building (in first evidence the Torre Velasca and Duomo) and a sequence of contemporary new skyscrapers (from left to right: Unicredit tower; Solaria tower; Regione Lombardia tower; Diamante tower; Torre Galfa; torre piazza Repubblica e Grattacielo Pirelli) (photography of architecture by the author).

Fig. 2. a, b, c, d. Large Milano Town model (skyline view), and a sequence of morphological models of urban tall building in Milan after the post-2nd war urban development: actually exibited at Urban Centre of di Milano in piazza Scala, previously realized for the Biennale di Venezia (editor Arch. Cino Zucchi) (photography of architecture by the author, with the courtesy of: Urban Center).

2 Analogical composition and digital morphogenesis

It is not only a question of style or construction procedures or building materials that have radically changed by a few decades of difference, although many techniques have highly evolved and refined. I will try to demonstrate and recognize how this evident design modification, above all is dued to the innovation of the design and calculation methods of the architectural project [2]. In fact the first generations of skyscrapers were designed and calculated according to analogical procedures, with the mathematical support of structuralists and experts of static load, with three-dimensional physical models in scale for the static testing of resistances exceeding the pure weight of the building itself (like the wind, vibrations or other stressing factors) [3].
The analogic composition - as a stabilized and harmonious equilibrium between the parts- have been progressively approached, matched and replaced by the digital elaboration of morphologies - produced according to the multiple parameters of algorithms - selecting the choice as the best solution. A symbolic representation of the former analogic algorithm and desktop, in its procedure and devices (see: Fig 3 and quoted in §3, fig 5 and 6.). The former hypothesis evaluation procedure (heuristic sketches and formal intuition), typical of the analogical project, has progressively moved to digital designed parameters that are analytically evaluated and processed, up to the massive production of results from which the designer "selects" the one in which it recognizes as the most suitable for his concept.

Fig. 3 a, b. The mechanical joint with a protractor's goniometer knob, allows the draftsman to maintain the orthogonal Cartesian coordinates on the whole plane of the drawing template, possibly varying the coherent inclinations. While the left hand of the draftsman guides the device to convey the accuracy of the analog algorithm to the field of the draw on which to operate, the right hand will perform the calibrated traces by the two right-angle millimeter scales.

An intermediate generation of projects (years 1990 - 2000) has acquired the calculation procedures according to the first generation of digital softwares, which have effectively allowed to accelerate the phases of design and calculation, being able to interlace directly the drawn shapes and the numerical values as parameters that could be examined by the designer into the same computational process.
First of all, this allowed to verify immediately - correct, modify and validate- the various phases of the original project - bidimensionally in plan and sections - and following in a digitalized process that, setting the drawing as templates of 3D matrix virtual spatialized data, make able to modify the three-dimensional view -being a virtual model- in a visually effective way, essentially due to an automated generated geometric perspective, on the monitor display or by a printer.

Fig. 4 a, b, c, d. Some examples of educational projects of small architectures composed and rendered according to a first choice of material surfaces, simulated in perspective view and chiaroscuro the environment in which they are inserted. Softwares: Sketchup and Kubity for VR with QRC interactive database (didactic design exercises at the Brera Art School, Laboratory of Design: Prof. Federico Brunetti, with the collaboration of Arch. Marco Valentino A.S. : 2016-2017).

In this first phase, the digitalization of the project was soon appreciated both for the dematerialization of the design process that transformed drawings on paper into digital elaborated files obtaining their effective transmissibility and sharing. Besides also for the easy visual illumination rendering and surface texture simulations that progressively made the design presentation simulated as useful photorealistic images substitute for the evaluation of the project.
Subsequently - over the course of a few decades and with considerable speed - the digital relationship between the different design skills (architects – static engineers - plant engineering, energy engineering, aerodynamic etc.) were able to integrate with the relative complexity of specific approach and features, thanks to new softwares that were able to align different kind of drawing files into complex layered systems. These new generation of softwares -often originally conceived into high-tech laboratories in function of more advanced technologies such as aerospace- become a shareable project tool.
Also for this reason, today we can see futuristic and eterogeneous projects and realizations, rather detached from the usual image and usual morphology of the urban fabric: each building demonstrates not only the design concept of the architect and the team that conceived it, but also the computational innovations that made them possible drawing and calculation of shapes. We face with the apparent paradox of highly technological design solutions that assume assonances of biological or natural morphologies, since managed by digital algorithm able to propose dynamic solutions to an univocal original problem.
It should be noted that such process of integrated evolved design, is finalized to compare the competences (and the various specific softwares) in order to forecast accurately not only the dynamic structural stability, but also the thermodynamic behavior of the building as a whole environmental impact (defined “Multiphysics”), its potential and planned and effective maintenance procedures (BIM), and the visual and potentially ecosystemic relations of the building with in its environment. It should also be remembered that the digital technologies of VR / AR (stereo interactive virtual or augmented reality) have a remarkable application precisely for the perceptual, interactive and immersive simulation into virtual space and environmental qualities for the foreshadowed fruition of design

solutions and decisions [4]. Related to this same devices, some new paradigm of visual experience of the space has been also introduced by digital spherical cameras –and the sw needed for the image matching- that allow to perceive and broadcast a full 360° view of the environment, even in remote without the presence of the shooting operator [5].

3 Analogic drawing desktop to Digital drawing desktop

Merely as a symbolic representation of the changing occurred from *Analogic drawing desktop* to *Digital drawing desktop*, are following presented two allegoric moke up of scenes representing the methodologies in these two era, so near to us, but deeply evolved.
In the first are in evidenced the typical books of the modern architect education and debate, referring the proportional modularization into the building composition (Fig: 5). The modern architectural movement sustained the actuality and legitimation of its methodology as possible application concerning the standardization format for industrial production of objects and architectures. Avoiding to recognize any explicit validity about traditional architecture proportioning methods (inspired by to archetipical principles of *utilitas, firmitas venistas* – i.e. functional efficiency, structural stability, symbolic beauty-) Le Corbusier in his famous treatise - manifesto *"Le Modulor"* recognizes the golden sections as a compositional methodology conceived of universal value for modern architecture and rediscovered the theoretical and practical significance of proportions. This approach relegated to archaeological researches the simple recognition of their traditional value in historical monuments [6]. In fact the modern design drawing practice, avoiding to be based on the imitation of recurrent stylistic models – considered as endless arguments of academic discussions- rediscovered the power of analogical algorithm, as golden section or even graphic computation, to organize efficient composition of plans, sections and volumes. The structural calibration of reinforced concrete, or steel, constructions were definitely delegated to static engineering competence, and the architectural composite research believed to be free to be concerned on harmonic and socially useful research [7]. But this creative and productive approach has also some strong ideological and self - referenced features, in fact any effective debate for the unification of modern Arts around golden proportion in mathematics had some explicit failure [8]. For that generation of architects the icon of the New York skyline was surely a strong challenge in comparison with Italian post war architects possibilities, but the wise use of well balanced harmonies, by means of the analogic algorithm of orthogonal drawing, aided by the proportional computation by *slide rul*e and the use of *compass*, were able to realize emblematic solution, still admired after more than half a century. Some *other typical analogic devices*, as a typewriter machine (a real digital device !), a photographic telemeter camera and a stereoscopic slides viewer, were other usual tools of that ingenius age [9] [10] [11].
On the ***digital drawing desktop*** environment some new procedure has been taking place (Fig: 6). The shape of objects can be obtained as the result of optimized data of efficiency from the forecast of dynamic phenomena. The horizontal plans and the vertical development of a building can be modified progressively obtaining the best possible results matching static resilience with symbolic evolutive design, environmental energy consumption. The ermeneutical inspiration sources are more rarely books, but huge databases of virtual models catalogues reach the designer desktop everywhere a good web connection can arrive. The support of the drawing, since now as essential documentation material, in no more a sheet of paper, but a cooled digital *stocking memory*, where the virtual models of tridimensional building are computed, stored and renderizable in any interactive visualization. The pencil or the compass tracks are potentially substituted by the hand *gestures of the designer* by a mouse or a drawing tablet, and the drawing attitude of the designer is traced by a scanner pen. The sharpness of the drawing is substituted by the precision of the pixels, pure positional numbers, and being comparable to a *laser light ray* able to scan or measure any close or remote detail. The proportionate experience of the sight on the environment is *augmented*

by full field *spherical cameras* and *stereoscopic smartglasses devices*, whose overlapping vision can be activated by *QRC* targets [12]. The computation performances of digital algorithm can expand our knowledge toward infinitesimal and complex shapes preview, may be potentially closer than ever before, very similar to the naturalistic biological evolving configurations –as fractals- where matter, energy time and space are inestricably intewined.

4 A XX th century emblematic modern architecture: The Galfa tower

The first case study here presented is a building realized after the second World War economic development, when tall buildings and towers represented the first new skyline of the directional business center near the Central Railways station. The Torre Galfa (1956-1959) is an emblematic example of modern building -currently undergoing a "revitalization" by the actual Real Estate property Unipol Insurance company, following the project of studio BG&K associati - in the area near the the famous Pirelli skyscraper designed by Gio Ponti. In this work Arch. Melchiorre Bega (1898-1976) designed it as an elegant example of a particularly proportioned and refined composition in its prismatic essentiality in the context of a lively debate, and other coeval examples of the Milanese modern architectural movement,. Originally designed and calculated in the era of the analogical project (with Papini (1924-71), P.A. Rognoni 1921), ing. Luigi Antonietti (1924-2010), Arturo Danusso (1880-1968)), the building is articulated by a hypogeum plate with a wide structural grid, extended on the entire available urban block, from which it rises with a relation to the lower floors of a body made by low and long-range transversal factory of services. The construction of the tower - initially planned for a 29-story hypothesis high 99.44 m, results in the final version 102.37 m. and 30 floors. The essential and almost hermetic prismatic shape has a plant of 37.50 x 15.75 m., compose the floor with 21 modules in depth and 50 length by a square grid of 75 cm. from the side. The supporting structure of the floors is not resolved in pillars, but with vertical septa (75 cm thick at the base of the building) that support it with a variable pitch between 9.70 and 3.90m. From the 24th, these sects diverge into two thinner pillars. Just as the plant and the section show a decidedly accurate modular composition and numerology (of which in the sketches that we present only some initial interpretative hypotheses), the original façade was designed with a planar articulation by a series of aluminum modules with windows and vertical / horizontal tilting windows, designed with an exemplary rhythmic sequence both in the single module and in the overall surface area. The morphological characteristic of the building in its prismatic essentiality lies both in the planimetric proportions on the facades, and in the whole volume, expresses an unexceptionable order. The compositional rules of this way of planning are rooted in the classicism of golden sections, which now we can only deduce some interpretative hypotheses, while the studies on the geometric articulation of proportions are normally not published and implicit in the final executive project. The analogical algorithms were not less precise in the graphic composition of golden proportions, basing their mathematics on the use of the compass line and circle to define the rhythm of the partitions, although they remained in field of mutual relations and three-dimensional orthogonality, indeed finding a special form of modern abstract beauty and visual harmony in this complex simplicity.

Fig. 5. Analogic drawing desktop, symbolic representation (picture of the author) [13].

Fig. 6. Digital drawing desktop allegory, symbolic representation
(picture of the author) [14].

The current state of revitalization of this building has allowed us to observe and document it both before the works, followed after a long period of decommissioning [15], and during the phases of its skeletal transparency, when the previous facades, currently technically obsolete, were replaced by new ones cells in glass doors and windows that - together with other technological, typological and functional upgrades- bring to actual thermo-acoustic insulation standards requirements._In this building we can observe the continuity of some fundamental concepts. The ancient "*canon*" of the Greek classicism in which *téchne (i.e.:* - art / technique) were merged (in heritage of thousand-year tradition of the golden proportion analogic algorithm that connect in the drawing activity the human intelligence with the re-creation in the project (nature - morphology / geometry - beauty), and the modern relationship built on the rationality between order and form, derived from LeCorbusier's Modulor where efficiency and anthropocentrism tried to establish a new form of equilibrium (ergonomic standard - reproduction / production) [16].

Fig. 7 a, b, c, d. Galfa Tower, 2015-2017: phases of revitalizing of the building, works in progress of refurbishing; façade details, models (Urban Center and study maquette).

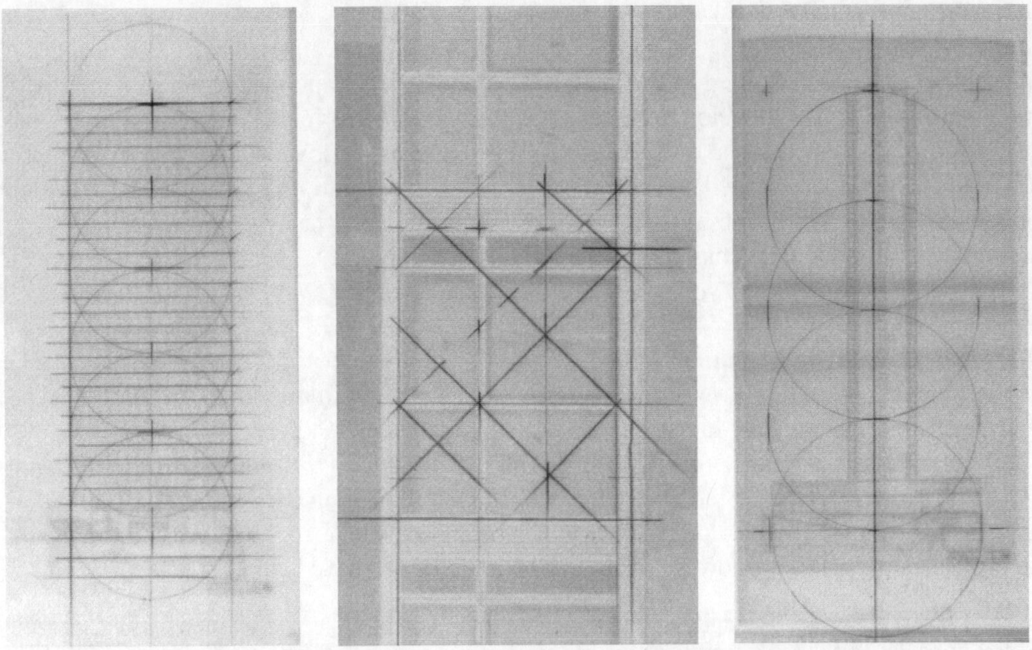

Fig. 8 a, b, c. Galfa Tower: Authors' heuristic hand drawn proportional sketches, with hypothetical first interpretation, deduced from edited original drawing of GALFA tower. (some inevitable approximations are certainly intrinsic in these sketches, due also to the fact that they are not based on original copies of the project drawing, but simply derived from reproductions edited in recent publications; other further studies are in progress) .

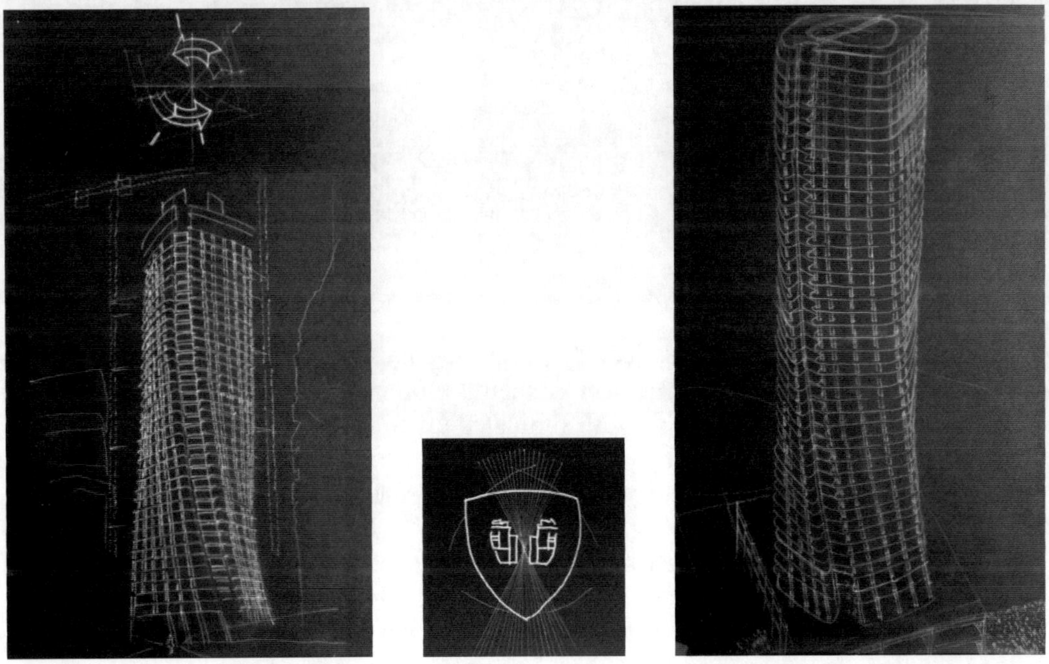

Fig. 9 a, b, c. Hadid Tower: Authors' heuristic proportional didactical sketches, hand drawn on blackboard, following Zaha Hadid studio presentation, as seen at Generali Hadid Tower during the press event book presentation on 17th.10.2017.

5 Hadid Generali Tower

The second case study belong to the last generation of skyscrapers designed for the town. The site where this urban development take place was the former Fiera di Milano location, previously here located since quite a century ago, and now moved to a more infrastrucurated areas in external part of Milan **[17]**. The project, contextual the dismission of the former setting towards the proposal for the new urban design, had been the occasion for an international architectural challenge where the most important "Archistar" were requested to work, in collaboration with young Italian architectural ateliers and urban national developer, to propose their solution.

This tower is a part of a complex setting defined by the winner for the development challenge: the Citylife group. The tower is settled in the visual landmark long distance perspective fulcrum, where the actual urban development introduces a three towers system, visually strictly related to the near connection to the highway approach to the town, that assumes the role of new symbols in the iconic architectural constellation in Milan metropolitan area. The first two skyscrapers: Zaha Hadid tower and Arata Isozaki towers have been already finished, the third Daniel Libeskind tower is going to be in starting phase [18] [19] [20].

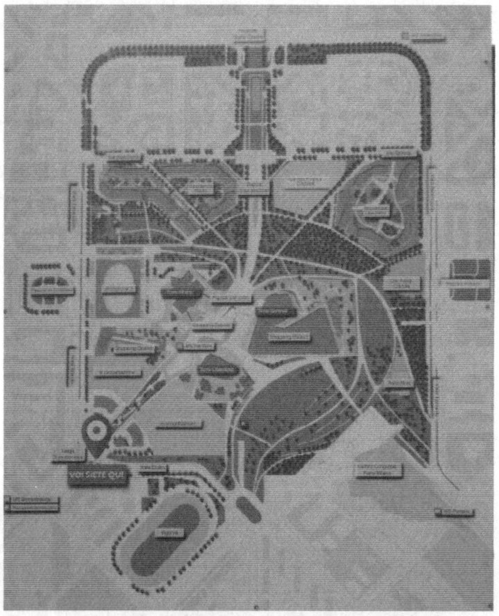

Fig. 10 (a). (left): The original discrict map: a square area settled with diagonals organizes the subdivision of the space. A XIX century isometric military exercitations field becomed and ideal exibitions platform location, with simulated urban roads partition and large covered spaces designed to host the Fair events. This "Fair machine" worked efficently here since 1926, being already based in Milan a former great background: the Universal Exposition in 1906.

Fig. 10 (b). (Right): the actual urban development plan, as presented in an orientation panel oudoor the public park entrance, where the three towers, and block of residences at their neighbors, have been built.

The Citylife towers urban development is located in a strong infrastructural, surface and underground, lines area. The social impact in this part of the town, when all the functions will be operative, will be quite important. Some block of residential buildings, and a large green area and a market district have been previously realized.

The skyscrape here in evidence has been conceived and realized by the Zaha Hadid studio and it is characterized by a clear and impressive feature in the vertical torsion that appear in its vertical

façade and that needed important structural research in the design of pillar systems, disposed in a proportionated inclination, following the general shape.
This master design concept concerned the structural engineer to elaborate a specific algorithm to generate a complex and dynamic virtual model to forecast and balance any consequence of this choice. The double symmetric torsion of this vertical grid of pillars , also sustained by an inner structural core tower where all vertical direct communications are concentrated, results so particularly strong and elegant even if morphologically unique in the town landscape. The digital procedure here adopted allowed to realize, and verify by laser scanner survey and drone inspections, all the phases of construction.
Finally the façade glass surface has been realized following the general mathematic curve of each windows position. Hadid architect studio, REDESCO engineering and CMB construction company, during the inauguration of the building, defined this design algorithm as a 5D procedure, since includes the 3D basic dimensions, but even the timing and building sequence aspects, and all related documentation attaches and BIM model and information for maintenance.

Fig. 11 a, b, c, d. Hadid Tower at Citylife, Milano 2017: final view in urban skyline and façade detail. Some figures of this skyscraper: Height: Architectural: 177.4 m / 582 ft; Floors Above Ground 44; 3.200 people workplaces; Parking Spaces 300; Owner Generali Group; DesignZaha Hadid Architects; Structural Engineer Redesco Progetti srl, MEP Engineer , Max Fordham LLP; Manens-Tifs; Façade:Arup; LEED

Fig. 12. Hadid Tower at Citylife, Milano 2017- 2014: final view in urban skyline, and former phase of the construction of
(photography of architecture by Federico Brunetti)

6.1 First Conclusion. Some methodological relations about proportions and morphology between analogical and digital design procedure.

The verticality of the building structures allows the overlapping of the horizontal floors following the static sciences: the construction and available materials technique could obtain different solutions essentially in reason of calculation process and make possible options in aesthetical approach by innovative design concept and drawing procedures.

As I could examine through the selected case study, the elaboration of the design procedure of an architectonic artefact has deeply modified in the last half century. The example of some tall building construction, in reason of its extreme challenge toward gravity forces, exemplify how the change in the approach and innovation of different ***drawing algorithm***, from analogic to digital, driving to diverging result. The same project procedure has deeply modified each of its steps: from the technology of ***survey*** in the environment for the capture of measures, data, images, that now allows a full analytic experience of the site [21], to the changes in the ***design concept*** that, thanks to new impredicatable power in computational process that make possible to dematerialize ***drawing procedures*** and invent unpredictable ***aestetical*** approaches, also by a broader source of information as ***ermeneutics*** of the project.
The result of design can be immediately verified not only on flat technical drawing, certainly useful for the building site executive realization, but directly in comparison into virtual modelling. This fact makes possible any check about interdisciplinary technique interference before and during the construction phases and to preview for maintenance and energetic behaviour of the artefact in the environment. The model itself, for all the defined feature that are located in its virtual structure, can be considered as a shared benchmark for any control in execution phases.

design **concept**
computational **process**
↓
Modelling (analogic – virtual – BIM)
case study - observation **survey** → **drawing algorithm** → architectural **project** - executive **realization**
effective control test – function & maintenance

↑
Drawing **procedures**
aestetical approaches.
ermeneutics

As many other masterful scholars already put in evidence [22], it is quite clear how much the digital innovation can optimized the design procedure, but it is also evident that the actual challenge and important opportunity to develop new paradigms in proportion and composition of architecture, that can be more easily finalized to the human requirements into the complex balance with the natural world. The conformative effects in progress for the new mathematics of drawing algorithm [23] and the new responsibility for designer creativity and social effect of architecture is in evident development.

6.2 Open conclusion. The " 3 worlds"; town and the value of interpersonal communication.

The experience of practice and theory about design drawing algorihtm is deeply related to the problems of our age, since every sketches that is drawn can potentially become a sign for a future project.
It becomes more evident what the great philosopher of Science and Communication Karl Popper pointed out some decades ago: the World 3 (theorical concepts or virtual enviromment) can clearly effect retroactive consequence on the World 2 (sensorial subjective experience), and cause following human consequence to the care of World 1 (raw material reality: i.e.: Nature) [24].
Far away to be considered these as mere theorethical considerations like here to quote the conclusion remarked at the 2017 International Tall Buildings Conference held in Milano, by Steve Watts, CTBUH Treasurer Steve Watts, Partner, Alinea Consulting. He closed the conference and shared his thoughts on the modern cyber-psychologist: *"...People concerned about smart technologies and the intensive use of internet and social media, who could, as a result, be destroying social, physical relationships. Eighty-three percent of young people in London succumb to loneliness, and that is one reason why developers there tend to invest heavily in common areas rather than private ones. There is a vision of encouraging social intercourse. The event closed with a final point of emphasis: skyscraper development is related to technology, but also to other aspects that create additional profit avenues. Despite this, even the biggest projects can fail: not because of lack of technology, but because of failures in interpersonal communication."* [25]. We must consider and appreciate, in our specific case of study, how much the sorrounding environment of the town has recieved, during the years, a good impact of these tall bulding. In particular for Citylife three towers development, where actually a residential buildings complex around the towers, stylish commercial center and a very large green area sorround the skyscrapers. Anyway an accurate planning of well combined mix of urban functions must be organized with this kind of tall building, in relation to the very strong social impact that they can introduce into the metropolitan style of life. The widespread success of skyscrapers in urban new development, overall in new developing countries, must be followed through an adequate consideration of their symbolic power, and with the deep meaning of humans as citizens. Technology innovations and antropological philosophy can here find new, or old, matter of dialogue [26] [27]

Reference

[1] COPPA, A., TENCONI, L.: *Grattanuvole: un secolo di grattacieli a Milano*, Santarcangelo di Romagna, 2015.
CRIPPA, M. A., ZANZOTTERA, F.: *Milano si alza: torri, campanili e grattacieli in città*, Milano, 2004.
BRUNETTI, F.: *BBPR; La torre Velasca*. Introduzione di Lodovico Belgioioso. Ed. Alinea, 2000, Firenze.; Presentazione: Torre Velasca, 23 marzo 2000, con L. Belgiojoso, A. Rossari, G. Denti.

[2] EISELE, J., KLOFT, E.: *High-Rise manual: typology and design, construction and technology,* Basel: Boston: Berlin, 2002

[3] BOLOGNESI, C.; CATTANI, G.: *Grattacieli: architetture per il XXI. secolo*: Lucca, Fondazione Ragghianti, catalogo della mostra Lucca: Fondazione Ragghianti studi sull'arte, 2005.

[4] BRUNETTI, F. A.: *Experience of Three-Dimensional Vision in the Era of Digital Interaction. New devices opportunity and challenges for applied mathematics and visual design.* in: APLIMAT 2014. 13th Conference on Applied Mathematics. pp.44-68 in: Proceedings Publishing House STU. Slovak University of Technology Bratislava, 2014,
BRUNETTI, F. A.: *Through the screen. Concepts and typologies of digital vision experience.* In: 15th Conference APLIMAT 2016, Proceedings, 2-4 February 2016- p.115- 125. Editors: Ľudovit Balko,

Dagmar Szarkova; Daniela Richtarikova, Vydavatel'stvo STU Institute of Mathematics and Physics Faculty of Mechanical Engineering, Slovak University of Technology in Bratislava Bratislava SLOVAKIA, 2016;
BRUNETTI, F. A.: *Into the frame. The legacy of perspective in the technological design of visual digital devices*. in: Proceedings of: ICGG2016 17th International Conference on Geometry and Graphics ©2016 ISGG 4–8 august 2016, BEIJING; article No. 149.

[5] BRUNETTI, F. A.: *Image as algorithm; visions of science in the digital era.* In *Scientific and Technological culture 14th International Art and Technology Meeting (#14.ART): Art and Human Enhancement,* University of Aveiro, Portugal, October 7-11, 2015.joint organization between Universities of: Aveiro (PT), Brasília (BR) and Goiás (BR); Procedings editor: Paulo Bernardino Bastos; 2015 University of Aveiro;
BRUNETTI,-F.: *Vedere, comporre, restituire. L'architettura inquadrata,* pp.51-73. - *Genius Loci – Geo Tag. Esperienza del luogo e immagine come informazione.* pp.91-103. In : BRUNETTI-F. (a cura e portfolio fotografico) *Asnago e Vender. L'edificio di via Rossini a Milano. Un palinsesto del moderno.* Milano, 2014.

[6] BAIRATI, C.: *La simmetria dinamica: scienza ed arte nell'architettura classica.* Milano 1952.

[7] BRUNETTI, F. A.: *Topology and proportion in the drawing of the a modern architectural space design: the Parco Sempione Library in Milano (Ico Parisi, Silvio Longhi; X Triennale di Milano 1954). Geometry and Arts in an environmental experience, between radial symmetry as covering structure and spiral as interior growth dynamic symbol. An analogical didactic re-composition and tracing of a building place* Proceedings, pp.:08-18. In: *Aplimat 2013. 12th Conference on Applied Mathematics*; February 5-7, 2013, Bratislava, Slovak Republic Institute of Mathematics and Physics Faculty of Mechanical Engineering, Slovak University of Technology in Bratislava .

[8] IRACE, F., CIMOLI, C.: *De Divina Proportione. Triennale 1951*, Milano 2007

[9] Elvia Redaelli *Un architetto con la Leica. L'archivio fotografico personale di Gian Luigi Banfi. Lo strumento notazionale e l'atlante visivo di un protagonista dei B.B.P.R.* (Tesi di Laurea , Prof. Giuliana Ricci, co-Rel.: Federico Brunetti)

[10] BRUNETTI, F.: *Brevi note, ed alcuni quartetti visivi (pp. 180-188),* In: *Lodovico Belgiojoso Architetto 1909-2004. La ricerca di un'Italia "altra". Atti convegno studi nell'anniversario della nascita Politecnico di Milano, Facoltà di Architettura e Società,* 2009; A cura di G. Bertelli e M. Ghilotti. Ginevra-Milano 2013.

[11] BRUNETTI,-F.: *La fotografia di Ico Parisi nella metafora del taccuino, della cornice e dello specchio*; pp.97-111, in: D'Amia G. e Tenconi L., *Ico Parisi, architettura fotografia e design. L'immagine come progetto.* Presentazione: Gualdoni F.; Introduzione: Cavadini L., Como 2012

[12] BRUNETTI, F. A.: *Augmented Realities, p.: 83-100. In:* Marco Hammerling, Luigi Cocchiarella, *Informed Architecture. Computational Strategies in Architectural Design* . Atti del convegno e mostra didattica internazionale: Exhibition *Computation is the new Black. Updating the Making of Architecture, organized by Prof. Marco Hemmerling* at the School of Architecture Urban Planning Construction Engineering (AUIC) of the Politecnico di Milano in January 2016.)
BRUNETTI, F.: *Landscape digital memory. Urban transition in the contemporary landscape and digital iconographical survey: photographic archives of post-industrial areas in Milano by GPS integrated database Files.* (pp.1-8. in .pdf, DVD.) In*: Paisaje Cultural-Paysage Culturel-Cultural Landascape. EURAU'08*. Universidad Politecnica de Madrid, 2007; 4th «European Symposium on Research in Architecture and Urban Design EURAU'08»

[13] GRECO, L., MORNATI, S.: *La torre Galfa di Melchiorre Bega: architettura e costruzione* , Roma, 2012
CARRESE, V., FARABOLA, F.: Milano 1945 – 1955. Comune di Milano, Milano, 1955
VALENTI, F.: rel. Tiziana Poli; correl. Vittorio Bega *Architetture verticali: storia, evoluzione e contesto italiano: restauro di Torre Galfa. a Milano*: tesi di laurea / Milano, Politecnico, 2007/08
YANG, SHUANG; TANG, JIANAN: *The skin for Torre Galfa. A proposal for the sustainable renovation of a high rise building curtain wall.* Tesi di Laurea DEL CURTO, D. Politecnico di Milano 2014
[14] JODIDIO, P.: *Hadid. Complete Works 1979–today,* Basel, 2013.
VISINI, A, MOLINARI, L.: *Complex structures, free thinking : theory and projects by Redesco Structural Engineering; Milano, 2017.*
[15] http://www.skyscrapercenter.com/building/torre-galfa/4962
http://global.ctbuh.org/resources/papers/download/245-confronting-the-question-of-demolition-or-renovation.pdf
[16] LE CORBUSIER: *Modulor 2 (la parole est aux usagers)* 1955: *suite de Le Modulor* 1948 Architecture d'aujourd'hui , Boulogne, 1959.
BRUNETTI, F.: *Apprendre a voir. L'itinerario di Le Corbusier a Villa Adriana, ripercorso attraverso gli schizzi dei carnets comparati con le vedute attuali* . In: *disegno della citta: opera aperta nel tempo.* Contributo in: *Atti del Convegno internazionale AED*, 28/29/30 giugno 2002, San Gimignano, Itali: raccolta dei contributi'. S. Giminiano, 28/29/30 giugno 2002, vol. Tomo 1, Materia e Geometria, p. 529-535, Firenze 2003.
[17] https://www.fondazionefieramilano.it/it/fiere-e-congressi/fiera-milano.html
https://www.fondazionefieramilano.it/en/historical-archives/history.html
[18] http://www.libeskindtower.it/#s1
[19] ANDREINI, L.: *Arata Isozaki,* Milano, 2008
[20] AA. VV. *Libeskind tower* , Milano 2015
[21] BRUNETTI, F.: *Spazio, tempo ed architettura. Esperienze di luoghi, forme e costruzioni* attraverso *le percezioni dinamiche della restituzione visiva multimediale: dall'architettura in video al cantiere nell'era digitale.* In: *Studiare con l'architettura, PRATELLI A. , GAGLIARDI*
M. L.(a cura) Atti convegno: AAA Arti Artefatti Architetture, Video, Architettura, Città. Udine, 2005. Università degli Studi, Corso di Studi in Architettura, Consorzio Universitario del Friuli. *Colloqui di Architettura* 2004/2005 Udine 2009. *pp. 55-74.*
[22] UNALI, M.: *Atlante dell'abitare virtuale: Il Disegno della Città Virtuale, fra Ricerca e Didattica*, Roma 2014;
SACCHI, L. ; *Disegnare idee immagini* n° 38/2009: Rivista semestrale del Dipartimento di Rappresentazione e Rilievo dell'Università degli studi di Roma "La Sapienza"
[23] GIELIS, J. A.: *Generic geometric transformation that unifies a wide range of natural* and *abstract shapes. American Journal of Botany*, 90, 333 - 338, (2003).
[24] POPPER, K., JOHN, C.: Eccles, *L'io e il suo cervello. 3*: *Dialoghi aperti tra Popper ed Eccles,* Roma: A. Armando, 1981
[25] http://www.ctbuh.org/TallBuildings/CTBUHCompetitions/tabid/3068/language/en-US/Default.aspx
[26] COCCHIARELLA, L.: *Back to the Babel Tower, Warning and Archetype;* section: Mathematics and art. 16th *Conference* on Applied Mathematics; pp. 387- 392, APLIMAT 2017 Proceedings. Ed.: Dagmar Szarkova; Daniela Richtarikova, STU Institute of Mathematics and Physics Faculty of Mechanical Engineering, Slovak University of Technology in Bratislava Bratislava, SLOVAKIA.
[27] PETROSINO, S., GARLASCHELLI, E.: *Lo stare degli uomini. Sul senso dell'abitare e il suo dramma,* Milano 2012. **PETROSINO S.**, *Capovolgimenti. La casa non è una tana, l'economia non è il* business, Milano 2008. **PETROSINO S.**, ***Babele***. *Architettura, filosofia e linguaggio di un delirio,* Genova 2003
PETROSINO, S.: *The Need for an Anthropology of Wealth,* in AA.VV., *The crisis conundrum. How To Reconcile Economy and Society*, Palgrave Macmillan, New York 2017, pp. 207-217

Acknowledgements

The authors gratefully acknowledge the Scientific Grant Agency: Design Dept., Politecnico di Milano. *-This paper follows the friendly gratitude to Mauro Francaviglia that take care from the beginning, with accurate and eminent supervision and encouragement, to my interest in science communication and visual design of scientific iconography. With his rare and enthusiastic vision of dialogue between science and arts, he has been a rare mentor for me. Now, in an impossible direct colloque, we simply try to look to his original example, exploring the challenges of the present.*

Current Adress:

Brunetti Federico Alberto, Dott. Arch. PhD.
Politecnico di Milano Dept. of Design – School of Design;
Via Durando 38/a, 20133 Milano, Italy
Visiting professor: Elementi di Percezione Visiva - Laboratorio di elementi visivi del progetto
E-mail: federico.brunetti@polimi.it
www.federico.brunetti.it; www.sciencecommunicationdesign.eu

Slovak University of Technology in Bratislava
Faculty of Mechanical Engineering
17th Conference on Applied Mathematics **APLIMAT 2018**

Proceedings

THE IDEAS' MACHINE.

THE BIG SCIENCE IN SEARCH FOR YOUNG GENERATION CONCEPTS ABOUT IMAGINATION IN PHYSICS. THE EUROPEAN CERN – CMS CHALLENGE: "ARTS & SCIENCE ACROSS ITALY "

BRUNETTI Federico Alberto (IT)

Abstract. The debate about the possible interactions between Science disciplines and Artistic research has recently increased, not only as mere communication method to disseminate complex knowledge into widespread language, but with the aim to generate a matching between these two apparently opposite procedure that realize innovative representation of the world. Some of the most important international Institutions organized conferences, challenges and workshops to improve this bilateral understanding. The CERN (Geneve) promoted a digital Art residential programme named Art@CERN that gathered experts from every country to suggest and propose their experience. But even at high school educational level, several local experiences followed this master example.

The *CERN – CMS challenge "Arts & Science across Italy* " is part of the European CREATIONS project of Horizon 2020 and is organized by the Italian INFN (National Institute of Nuclear Physics) and the CMS Experiment of CERN in Geneva. The project is aimed at high school students of the third and fourth year of Milan, Florence, Padua / Venice and Naples, lasting two years (2017-2018) and has the aims to bring high school students closer to the world of Science and Research and in particular to the sub- nuclear. The first iconographical and modelling results of the national competition and best result selection has been exhibited in the Museo della Scienza e Tecnologia Leonardo da Vinci in Milan, and further selections are ongoing in other main town of Italy. A first survey of the different tendencies in interpretation of Science research proposed by young students is proposed in this manuscript.

Keywords: mathematics and visual arts, music, language, architecture, computer science applications, mathematics in vocational training and career education, computer science education

Mathematics Subject Classification: Primary 00A66; Secondary 97M80, 97M20, 00A05, 01A05; 97U99 51N05; 01A05

1 Big Science and knowledge legacy

The contemporary scientific research has developed its own investigation into fields and sizes of Nature that are by far to our own world and our imagination capabilities. In this research, three different contexts are called to interact in a highly articulated interplay: the abstract features of theoretical speculation, the avant-garde technical solutions adopted in the experiments carried out to test their hypotheses, and finally a huge amount of data to be examined as a result, to extract coherent or conflicting meaning respect to the theory.

The digital methodologies - themselves the result of this intrinsic capability to bond scientific discoveries with technological innovations - have powerfully extended and exponentially amplified observation capabilities and the data processing of information with more and more sensitive tools, and algorithms of an increasingly complexity. These contexts and methodologies have also made it possible and necessary to structure a sort of "large community" - called "Big Science" - where researchers, tools and data are shared to keep the ongoing investigations under uninterrupted multilateral matching. Indeed, this new scenario has made it possible to achieve new cognitive levels previously unimaginable, namely to investigate problems, case studies and environments completely inaccessible to human perception, both in terms of space (micro – macro ; ie: particles - Cosmos) that of time (ie the remote conditions of the origin of the universe, the stratifications of history, the infinitesimal partitions of time)

A "biological-social" condition of this new “augmented condition” undoubtedly concerns the temporal and inter-generational extension of current -and foreseeably future- research projects in the transmission of knowledge, that must not only retain their own acquired knowledge, but also knowing exactly handing down and communicating problems that have not yet been resolved. We can consider how, in the past, the techniques of writing and the libraries were up to the modernity, institutionally endowed with this task, even in different forms and in different societies

2 Naming, images, metaphors

Another not negligible aspect, both in the present and in the origins of modern science, concerns the visualization of data and, in broader terms, the use of images in scientific thought. [1]

In fact, both the graphical-visual representation, or the illustration, of the observable phenomena - either directly or by any instrument - (i.e. from the telescope to the microscope in the analogue era, and today to the digital technologies and through the sensing, amplification and visualization computation of the detected data) allow the use of images, even in broader terms - such as metaphorical or literary expressions - has found an important use in presenting conditions that science has met or hypothesized, but which do not yet have the most accurate description. "Black holes", "antimatter" or "dark matter" are now well-known literary images examples of "unknown lands" to which, in the meantime, we define our field of relative "incompetence" [2]. As we can guess these metaphoric images concern ambivalently both outreach packaged labels for communication destined to the general public debate around science, but also define as conjectures useful to speculate in first theoretical discussion, on which the disciplinary debate remains wide open, until better hypothesis. In this context, so briefly outlined, I would like to refer here to a cultural event of particular value that both educational and intergenerational, both in the iconic and metaphorical research in the in the arduous task of "giving names to things". An extensive methodological debate explores the original ambivalence between purely quantitative interpretations of the data of experiments and theories, and the corresponding search for qualitative representations (drawing, images, metaphors,) for the understanding of the invisible phenomena of science ("vision")

3 Relation Art - Science and Mathematics - Art

It is worthwhile briefly to dwell on a clarification - which alone would imply a much broader dissertation - in relation to the relationship between Art and Science (as the title of the competition chosen as a case study) and Mathematics - Art (as defined in the topic of this specific session of research in APLIMAT2018). The analogy and the differences between these two pairs of relationships may seem subtle, or merge into mere resemblance, but may underlie an interesting heterogeneity. On the one hand the binomial *Art - Science* can refer however to the objective datum of Nature and to the attempts of human intelligence engaged in the observation - direct or through the instrumental data - of the phenomena in the real world. On the other hand, the contiguity between *Mathematics - Art* deals with two conceptual universes, both originally abstract or intuitive, initially deriving (and only sometimes) the starting point from the real world, but moving forward and sublimating its elaborations towards a conceptual vision or a perceptive dimension.
In both cases, it seems evident that the disciplinary structure of the *logical-formal reasoning* remain essential at the basis of the theoretical or interpretative speculation, as for the artistic side it is fundamental the open the research to *intuitive, symbolic and creative solutions*.
One can easily understand in an experiential and subjective way - as is increasingly demonstrated by the researches of neuroscience - how the two *logical* and *creative* universes of human thought belong to two hemispheres - disciplinary and anatomical - specific and different, contiguous and synergistic. Both constitute the banks of knowledge of the world and of oneself at work: in the act of *observing, recognizing, naming and inventing* [3].

4 "Arts & Science across Italy". Context and format of the competition

The competition we are discussing as a case study is part of a broad research path with valid methodological and institutional references that legitimize its experimentation on a large scale: "*Art & Science across Italy is part of the European CREATIONS project of Horizon 2020 and is organized by the National Institute of Nuclear Physics and the CERN Experiment of CERN in Geneva. The project is aimed at high school students of the third and fourth year of Milan, Florence, Padua / Venice and Naples, and lasts for two years (2017-2018). The project aims to bring high school students closer to the world of Science and Research and in particular to the sub- nuclear. The project was conceived by Pierluigi Paolucci (INFN Napoli), Angelos Alexopoulos (CERN), Vincenzo Napolano (INFN Communication Manager), Michael Hoch (Vienna), who co-ordinate it with the heads of the four locations: D. Menasce (INFN Milano), S. Paoletti (INFN Florence), E. Vannuccini (INFN Florence), M. Fedi (INFN Florence) and M. Michelotto (INFN Padova).*" [4]

In fact, since some years at CERN has triggered a broad discussion of mutual awareness and interest in quality view of knowledge, so not limited to the visual arts, but just the chance to

try to understand and represent through every expression of the features arts invisible to reality that is the subject of scientific research. The Geneva-based research group ARTS@CERN [5] has developed a series of collaborations inviting professional artists, on proposals of their projects through an international call, thus inviting a period of research and residence in contact with scientists and their work. The characteristics of these residential research art projects are among the most varied and have this intent: ***"Arts at CERN** is the leading art and science programme, promoting the dialogue between artists and particle physics. It fosters the creation of new expert knowledge in the arts by extending artists' practice in connection with fundamental research."* [1] In fact both the characteristics of conceptual innovation and the digital modalities of the executive techniques have been a recurrent motive of the various experiences, all very interesting, until now realized. Evidently this kind of residential collaboration and interaction is very selective and highly experimental.
The competition *"Arts & Science Across Italy",* born as a specific initiative promoted by the CMS experiment, has chosen a different path of a broad base relationship and supported by a staff of experienced referents in some Italian cities in which INFN is present with researchers and some experimental research centers. Supported by an international team of experts in scientific communication, who organized presentation seminars, traveling exhibitions, visits to experiments, talks with scientists in schools. Just to define the activated concern we can summarize the "project numbers**":** 38 high schools with 108 classes and more than 3010 students from the cities of Milan, Venice, Padua, Florence and Naples. 140 hours in two years of school-work alternation. To date: 100 seminars, 40 visits to laboratories and city museums. 28 days of the exhibition *"The colors of the Higgs boson"* and 140 artistic compositions. There are 150 local winners, 30 national winners, 24 places for the master at CERN. The session that I followed personally closer was the one in Milan. In fact, the tour of the competition began from this city, with the support of some Milanese high schools (Artistic, Classical and scientific Lyceums) institutions such as Milan Bicocca University and Leonardo da Vinci Museum of Science and Technology.
The competition in question has been credited as an experience of participation in the kind of scientific work thanks to a specific agreement with INFN with MIUR (the Italian State Minister for Education University and Research) and local Schools and Institution , and therefore these activities have resulted as an integral part of the course for the classes of students who have participated in this title. Naturally, each interested student was authorized to participate, having previously identified their own reference teacher. This initiative has received particular momentum thanks also to the concomitance of a normative innovation of the teaching called *" School-Work Alternation",* thanks to which the students are engaged to carry out an important part of their activities in the final three years (200 hours in three final years), carrying out collaborations in approaching professional methodologies with and from "external bodies" (i.e.: agency, authority, body, bureau, corporation, organization etc.).
In Italy, following the passionate and concerned activity of some pioneers of Physics Communication such as prof. Franco Fabbri and, since 2008, by the former INFN Press Office leader Romeo Bassoli (1954-2013), and nowadays by the full staff that here was formed, many activity have been developed: as the *Assimmetrie* monographic magazine, the

---

[1] Monica Bello, Head of Ars at Cern , followed by the **Cultural Advisory Board:** Andrea Bellini, Director of Centre d'Art Contemporain de Genève (CAC).; Frédérick Bordry, Director of Accelerators and Technology CERN.; Bilge Demirköz, Associate Professor at Middle East Technical University (METU), Ankara.; Ariane Koek, founder and former Head of Arts at CERN.; Laurent Le Bon, President of Picasso Museum in Paris.; Charlotte Lindberg Warakaulle, Director for International Relations CERN.

science mass media press release related with ANSA, the AR multimedia interactive installation research group leaded by Vincenzo Napolano. [6]

5 *"Arts & Science Across Italy"*. A methodological commentary on the lines of creative project

The Milanese experience began with a conference by Dario Menasce (INFN Milano Bicocca) at the Liceo Artistico di Brera with all the groups of students involved. The photographic exhibition of the INFN volume *"The rings of knowledge"* [7] was the background of this first communication and at the opening seminar day in which all the participants of the various Milan high schools were invited to the University of Milano Bicocca and to the survey of local experiments at Faculty of Physics. From this moment on, other similar presentation meetings have continued in the other high schools that have begun the researches and in-depth enquiries into the contents to be understood and visualized by the groups of students.
The results were scheduled to be able to converge in April 2017 in a selection and exhibition of all the participants at the Museum's large cloisters, where a bilingual conference was held between local authorities and CERN scientists. Here participated, as well as staff of the competition, the Rector of the University of Milano Bicocca prof. Cristina Messa, the Councilor for Education Dott.ssa Anna Scavuzzo (who mentioned her basic training in physics).
It is worth mentioning some interpretative trends that emerged from the works presented.
Thanks to the worldwide scientific communication in regard to some emblematic images - selected from the scientific iconography in the reconstruction of the data of the collisions significant for the candidate collisions - the three-dimensional diagrams for the search for the Higgs Boson have become, from enigmas for experts, as a sort of logos for the contemporaneity. Despite this widespread communication, so effective and for the general public, the attempt to imagine and represent the extreme conditions of the initial state of space-time re-generated in the experiments of the LHC has profoundly affected and influenced the imagination of the candidates works. In particular, as can be consulted in the web archive edited by D. Menasce, they can be found examples of different resulting "iconographic" solutions and trends, both in forms and in concepts and titles. Certainly some constructive aspects related to the phenomenal characteristics of the experiment machines have hit the imagination of young artists; moreover, the recurring representation of the central perspective has found a particular correspondence with the dynamic accentuation in the figures of the accelerator or of the detectors. But the most interesting works, both of the winners but also of several other participants, focus their attention on the innovative and further temporal space qualities opened by the search for new particles. The three dimensional and complex interpenetration of force fields and punctual elements has profoundly influenced young students towards new models open to the surrounding space to which they powerfully allude. These works that construct complex and asymmetrical, or essential and bipolar forms in the model, which seek to trigger into the observer a reflection on the usual mental models of space. Some works, by the apparent custom of box objects, suggest the impression of enigmatic situations related to the *external-internal dichotomy*, or the other equally crucial opposition between *visible and invisible*. Some authors allude to the *virtual quality of the images of science*, using both the suggestions of specular materials, and some video projection techniques or digital video. The *emblematic elements of the human figure* is sometimes included as a symbol of the observer becoming as a part of the nature that

investigates . Several compositions finally discuss the *fragmentation* as an argument of the work, deconstructing recognizable wellknown simple forms in small disintegrated universes. Finally, some authors defined their concepts in the attempt to construct *devices* that - through mechanisms or visual dynamics in digital elaboration even with limited technical means, could simulate the complexity of the unstable and probabilistic relationships of the quantum dimensions of the subnuclear worlds. Exceeding of stereotypes and models; Perspective as a subjective space-time device; Box as an enigma; Diagrams of fragmentation; Models and plasticity of the world of particles: these are synthetically the lines of research developed by these first brave young authors.

The experience of the *"Arts & Science Across Italy"* competition is so demonstrating a particular intellectual vitality of the young Italian generations regarding the themes of the conceptual visualization of the invisible. Invisible conceived not as a metaphysical category, but as a fertile environment in which the mental imagination of these not directly perceptible dimensions of reality can find -in the open mind of the researcher - images able to match the extreme and fundamental qualities of Nature. This method of imaginative involvement is evidently matching the interest of young people, and their willing tutors, far beyond the usual disciplinary interpretations. The definition suggested in the title of this paper: "Machine of Ideas" (derived from the physical definition of "Machine" as transformer of one form of energy into another) indicate precisely the process of intuitive and conceptual processing here so happily activated.

The question now perhaps concerns the reverse path: that is how and when the world of Big - Science can take advantage of these new intellectual resources so laboriously triggered.

6 Training and education in the millennials digital era

An emphasis must be made on the fundamental support of teachers and tutors concerned at every level of this training. In this epochal transition, after having gradually accompanied digital techniques between languages and intercommunication methods, they are also involved in the commitment to support the scenarios of innovations and discoveries that, thanks to these technologies, are making themselves available for research. But it is no longer only metaphorical to achieve some "dimensional scales" of space-time in which to "zoom in" new morphologies of Nature (in micro or macro), but having to challenge yourself with these mode of being of reality that *transcend our sensorial perception and empiric logic*, in order to introduce other categories, which were often struggle to define the language of commonly treated areas, otherwise relatively confined to scholars.

If our generation of adults - researchers, teachers, parents - has crossed the century-old threshold between analogue and digital. The new generation of *"millennials"* is already *"digital native"* and looks at the concepts and shapes of thinking of the XX th century as a classicism to be inspired, in the same way that they observes, Design object - perceived once as originally innovative – considering them today already as "vintage". Their perception of the continuous metamorphic innovation of digital technologies, and of the scientific exploration of Nature - in the era of the Higgs -, unfolds with aptitudes of a very open mental openness. At the same time they are surrounded to a hypertrophic quantity of information that risks to complicate every possible decision. But in such a complex scenario their ability to connect past and future, images and words - far beyond the humanistic / scientific schemas of scholastic disciplines - appears as though they were merging multiple data synaptically from information platform, leaving us amazed and admired. The precious and unrepeatable task of the adults that have the educational and training role in accompanying the growth of these generations is at the fulcrum of a really challenging and fascinating intergenerational challenge.

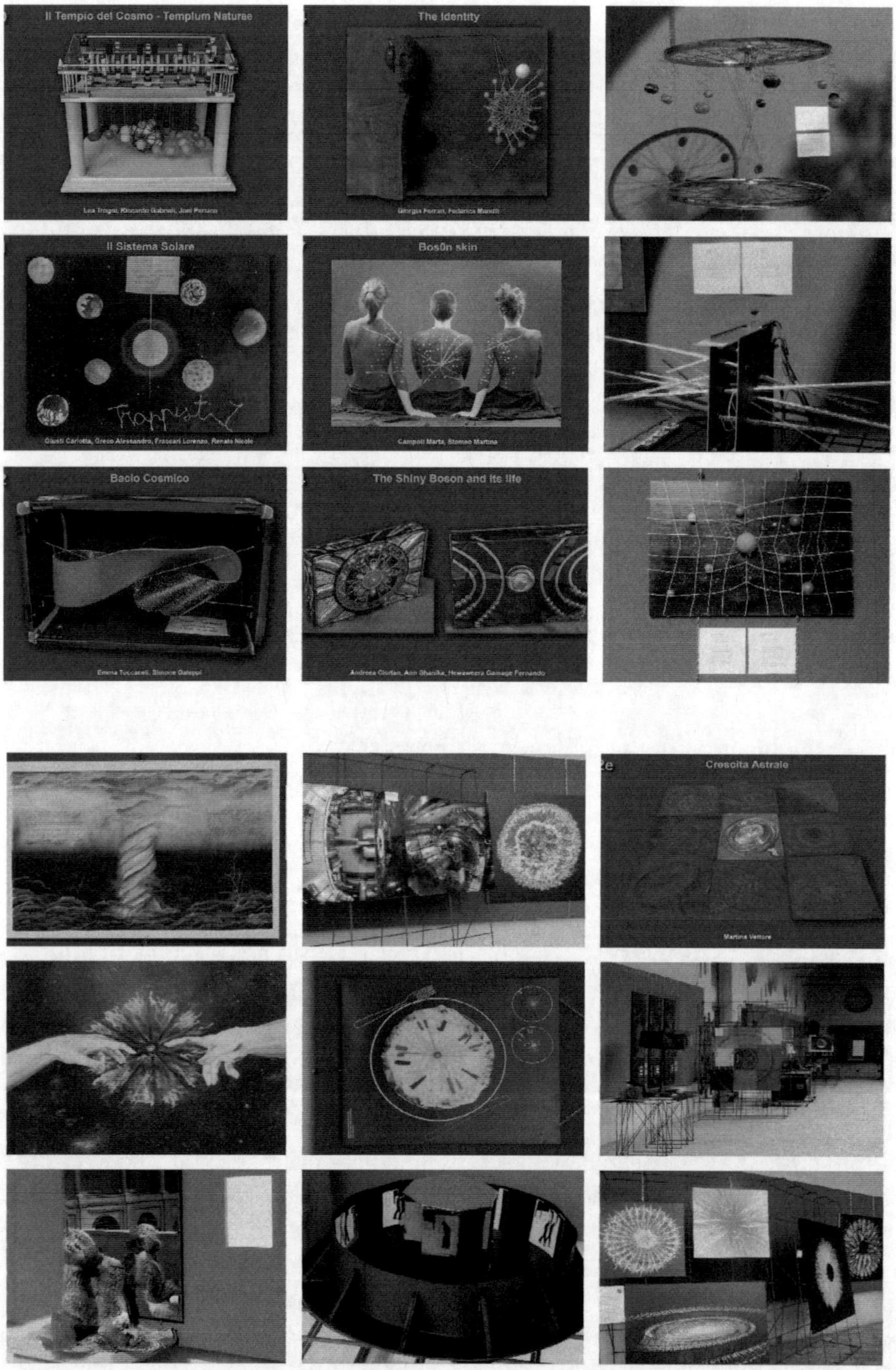

Fig. 1. Milano, 2017. Synthetic framing of works exhibited at Museo delle Scienza e della Tecnologia Leonardo da Vinci. _Some of these pictures have been also taken by the author. A complete catalogue of works , titles and authors is available in the website created by Dario Menasce, (INFN Milano-Bicocca) Slide-show of the works of art exposed in Milano at the Museo Nazionale della Scienza e della Tecnologia Leonardo da Vinci in Milano. [8]

Fig. 2. Milano, Università degli Studi Bicocca: seminar and exhibition, INFN laboratory survey; Museo delle Scienza e della Tecnologia Leonardo da Vinci : prizegiving. Pictures and spherical views by the author.

Fig. 3. Posters and "icons" of the Milan events; Science Museum, Liceo di Brera, Higgs Boson's pizza. [9]

Their "soft-skills" are readily more up-to-date than the adults around them, and we hope that in a forthcoming hypothetical society 4.0 they will be able to re-invent regenerative and responsible professions that their parents might lose in the processes automation, in a relationship that is already defined as "bottom-up": i.e. where young people can be able to train adults using new technologies. Probably the most honest knowledge that we can share with them is to match the scenarios of the unpredictable, or exploring the multiple shapes of the unknown that is always in front of us. The "sense of beauty" - as a splendid film-documentary [10] - that has chosen the life of scientists at CERN as a case study - is also the ultimate meaning that moves the creativity of every research: that of the exact sciences as those of artistic research. The motto "already and not yet" can therefore represent both the interminable work of the sciences in the exploration of Nature - at all scales of its being - and the one of the Arts - in all its expressions - in giving shape to the invisible of the human thought.

7 Conclusions. Science & Arts: cross - science methodology experiences.

The human kind of our century is opened towards knowledge challenges the are extremely innovative and impredictible. The intergenerational legacy in the self-styled developed world is very hard to assume in all its responsible consequences, and the questions opened by the science research are so wide that only a very concerned communication and educational work can give the chance to be able to follow, share and implement these results. Nature -with its still unknown qualities- patiently wait for our enquiry, but science and art seem to get close to the evidence of the reality through by different paths. A new learning approach, in respect of any specific disciplines, seems to emerge from this experience of didactic challenge –Italian based but essentially international -, gathering high level scientist, high school concerned students, interdisciplinary tasks teachers. The results of this extraordinary involving experiment between Science & Arts go beyond any effective participants work, and open a new evident possibility for the development in approaching to specific disciplines by cross - science methodology experiences.

Acknowledgements

The authors gratefully acknowledge the Scientific Grant Agency: Design Dept., Politecnico di Milano.
-My gratitude is first for the team leader Pierlugi Paolucci and all staff, for the kind opportunity that they allow me to present their work in this first report, even during the public developing of the challenge. A special thanks is addressed to Dario Menasce, who lead the Milan events of the challenge and offered his expert and kind overview to this manuscript.
-This paper follows the friendly gratitude to Mauro Francaviglia that take care from the beginning, with accurate and eminent supervision and encouragement, to my interest in science communication and visual design of scientific iconography. With his rare and enthusiastic vision of dialogue between science and arts, he has been a rare mentor for me. Now, in an impossible direct colloque, we simply try to look to his original example, exploring the challenges of the present.

References

[1] BRUNETTI, F., (Politecnico di Milano - Scuola del Design), *L'immaginario dell'invisibile. Iconografie della Fisica contemporanea tra macrocosmo e microcosmo: un caso di studio per il Design della comunicazione scientifica. Pp. 250-254* In: *ComunicareFisica2012 Museo Regionale di Scienze Naturali, Torino,*. Frascati Physics Series - Italian Collection Collana: Scienza Aperta Vol. IV (2014) ComunicareFisica2012; Atti 4° Convegno "Comunicare Fisica e altre Scienze", Torino 2012. ISBN 978-88-86409780.
BRUNETTI, F., *La visione scientifica e le immagini della ricerca,* in: Alberto Peruzzi (a cura di), *Pianeta Galileo 2011. Arti visive e nuove frontiere della cosmologia*, Atti del convegno, Accademia delle Arti e del Disegno, Firenze 3 novembre 2011, Consiglio Regionale della Toscana, Firenze, 2012, pp. 323-346. ISBN 978-88-89365-10-6.
BRUNETTI, F., *Iconography of science representations ad visual concepts in the digital era; first outlines. pp.119-123.; Physics research as a project process. Aesthetics in science, representation and technology, design concepts from new Physics. A visual report from LHC - CERN -* pp.151-154_In: a cura Luigi Cocchiarella. *The Visual Language of Technique between Science and Art - Heritage and Expectations in Research and Teaching -The Visual Language of Technique,* Vol.1 - History and Epistemology. Proceedings, Politecnico di Milano 150° International conference: 6 Maggio 2013, Scuola di Architettura e Società del Politecnico di Milano ISBN 978-3-319-05350-9. Ed, Spinger 2014, ISBN 978-3-319-05340-0 ISBN 978-3- 319-05341-7 (eBook); DOI 10.1007/978-3-319-05341-7 Library of Congress Control Number: 2014949295. In: *The visual language of technique, v.1 History and epistemology, The visual language of technique,* by Luigi Cochiarella, editor; ISBN 978-3-319-053257
BRUNETTI, F., *La visione scientifica e le immagini della ricerca;* in A. Peruzzi (a cura di), Pianeta Galileo 2011, Consiglio regionale della Toscana, Firenze 2012 (pp.323-346). Conference Proceeding: Firenze 3 novembre 2011. *"Arti visive e nuove frontiere della cosmologia",* Accademia delle Arti e del Disegno. Firenze 2012, ISBN 978-88-89365-10-6.

[2] BRUNETTI, F., *"Disegno, Immagini e metafore nella comunicazione scientifica" in: CommunicareFisica.07.* 2007.10.1-6 Trieste *Conferenza-Workshop nazionale sulle tematiche e sulle metodologie della comunicazione della fisica e delle altre scienze. Linguaggi e strumenti della comunicazione scientifica verso il pubblico. Sponsor: I:N.F.N. Istituto Nazionale di Fisica Nucleare; Università di Trieste; ICTC International Centre for theoretical Physics, Sincrotrone Trieste S.C. p.a.; Università degli studi di Udine.* in*: Proceedings,* FRASCATI PHYSICS SERIES Italian Collection – Scienza Aperta Volume II Atti 2° Convegno Comunicare Fisica e altre Scienze *a cura di* Francesco Longo e Erica Novacco, pp.63-64)
www.ts.infn.it/eventi/ComunicareFisica/presentazioni/brunetti.pdf

[3] http://www.artescienzaeconoscenza.it/

[4] https://web.infn.it/artandscience/index.php/en/

[5] http://arts.cern/

[6] http://home.infn.it/it/comunicazione; http://home.infn.it/it/editoria ; https://www.asimmetrie.it/ ; http://home.infn.it/it/mostre/installaioni

[7] BRUNETTI, F., *Gli anelli del sapere. INFN x LHC. Il contributo italiano alla più grande ricerca sulla fisica delle particelle al CERN di Ginevra. The rings of knowledge. INFN x LHC The italian contribution to the world's largest particle physics research project at CERN, Geneve* (Istituto Nazionale di Fisica Nucleare/ CERN / Dept. Indaco Politecnico

di Milano / ed. Abitare Segesta RCS Mediagroup) Milano 2009; ISBN 978-88-8611-693-thematic essays in the volume:_BRUNETTI F. *LHC between science and architecture*, pp.12-13. *Physics as design. Beauty, representation, vision.* pp.28-37 *Drawing, images, metaphores in scientific communication, pp.42-43.*

[8] *https://dariomenasce.wixsite.com/artandscience; http://www.museoscienza.org/news/dettaglio.asp?idnotizia=1070 https://web.infn.it/artandscience/index.php/en/*

[9] *https://web.infn.it/artandscience/index.php/en/gruppi-di-lavoro/87-news/122-primo-incontro-con-gli-studenti-a-milano http://www.liceoartisticodibrera.com/wp-content/uploads/2017/02/Gli-anelli-del-sapere.pdf http://home.infn.it/it/comunicazione/news/1914-arriva-al-cern-la-pizza-del-bosone-di-higgs*

[10] *Il senso della bellezza,* writtend and directed by: Valerio Jalongo, Photography : Alessandro Pesci, Leandro Monti ; post-production: Massimo Fiocchi , Original Musics: Maria Bonzanigo, Carlo Crivelli – by Orchestra della Svizzera Italiana directed by Kevin Griffiths, Sound: Christophe Giovannoni, Masaki Hatsui, Producers : Tiziana Soudani, Silvana Bezzola Rigolini, M. Letizia Mancini,: Camilo Cienfuegos per La Frontera Video & Film; Tina Boillat; Amka Films, RSI Radiotelevisione Svizzera SRG SSR, Ameuropa International with Rai Cinema; Distribution: Officine UBU

References – event and report Web sources:
https://web.infn.it/artandscience/index.php/en/
http://www.na.infn.it/it/7-notizie/390-presentazione-del-progetto-art-science-across-italy
Milano, Museo della Scienza e dalle Tecnologia Leonardo da Vinci . 12.04 – 1.05 2017.
https://web.infn.it/artandscience/index.php/en/milano-aprile-2017
https://web.infn.it/artandscience/index.php/en/milano-aprile-2018
https://dariomenasce.wixsite.com/artandscience
Venezia. Palazzo del Casinò di Venezia Lido dal 4 -12. 07 2017:
https://www.facebook.com/artandscienceacrossitaly/videos/1365665633555127/
Napoli 18 ottobre 2017:
https://www.gazzettadinapoli.it/istruzione/universita/art-science-across-italy-lunedi-20-alla-federico-ii/
https://web.infn.it/artandscience/index.php/en/87-news/168-giornata-formativa-di-napoli-materiale
https://agenda.infn.it/contributionListDisplay.py?confId=14132
https://agenda.infn.it/conferenceTimeTable.py?confId=14132#20171018
(http://home.infn.it/it/mostre/archivio-mostre/1329-il-dono-della-massa).
Padova Università 23 novembre 2017:
https://www.facebook.com/artandscienceacrossitaly/?ref=bookmarks
http://home.infn.it/it/news-3/news-uffcom/2660-art-science-across-italy-i-vincitori-della-tappa-padovana

Current Address:

Brunetti Federico Alberto, Dott. Arch. PhD.
Politecnico di Milano Dept. of Design – School of Design
Visiting professor: Elementi di Percezione Visiva - Laboratorio di elementi visivi del progetto
Via Durando 38/a, 20133 Milano, Italy
E-mail: federico.brunetti@polimi.it
www.federico.brunetti.it; www.sciencecommunicationdesign.eu

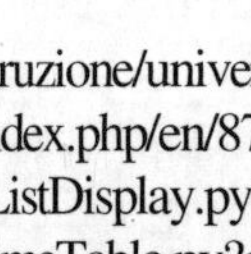

HYDROLOGICAL AND METEOROLOGICAL DATA AND MODIFICATIONS OF STATISTICAL DISTRIBUTIONS - HEURISTIC

BUDÍK Ladislav (CZ)

Abstract. The presented paper is dedicated to new approaches of fitting theoretical curves (transformed survival function) to daily average discharges, daily average air temperatures and air humidity and daily precipitation, using generalized log-normal distribution using 5 parameters and modified normal distribution.

Keywords: LN5 distribution, probability optimization method, discharge, daily air temperature, air humidity, precipitation

Mathematics Subject Classification: 62P12

1 Introduction

It is necessary in hydrology to use methods of mathematical statistics and numerical mathematics. They can effectively help even if data does not strictly satisfy all conditions, which are necessary for using such methods. One problem being solved using such mathematical methods is fitting theoretical curves with cumulative frequency curves from measured data.

Cumulative frequency function is an inverse function to the survival function.

By fitting theoretical curve one can obtain several estimated parameters for a given profile. It is then possible to easily analyze data and deduce values that cannot be directly measured – extrapolate beyond measured data (occurrence of rare phenomena such as floods). Deduced values can be used for deriving cumulative frequency curves in unobserved profiles using geographical and physical conditions in profiles (for example using regression equations for parameters of distribution).

2.1 Theoretical frequency curves and methods of parameters optimization on theoretical frequency curves of discharges

It is now necessary to describe the transformation of normal distribution into the so-called LN5 distribution. It is a generalized LN3 transform. LN5 distribution is a logarithmic-exponential transformation of normal distribution. In LN5 transformation is in contrast to

LN3 transformation used similarity of curves using parameter a. This transformation is given by the formula $Y = a \cdot e^{sign(\sigma X + \mu)|\sigma X + \mu|^b} + y_0$, where a, b and y_0 are parameters, σ is variance of the normal distribution, μ is mean shift of this distribution and X is standardized normal distribution [1], [2]. If one sets the parameters a and b to be equal to 1, then one obtains the LN3 distribution.

Various different metrics that optimize quality of fit are used for fitting theoretical curves. The quality of the different metrics can be compared with the help of simulations or their application using large data sets. In this way one can assess the quality of results using different metrics.

The only metric used in hydrology is the least square method derived for symmetrical distribution or weighted least square metrics used for LN2, LN3 or Pearson distribution. In climatology, it is used for other distributions, e. g. the Weibull distribution.

There is one problem with the weighted least square method. When values close to zero are used, one divides differences between measured values and theoretical values by value close to zero and then the smallest values (though they have usually lower accuracy) have greater weight than the other values. In an ideal case, the weight given to the highest and smallest values would decrease slightly. All the known metrics try to optimize the sum (usually to minimum) of metric using some relation difference between values of theoretical curve and measured one in the same probabilities for being equal or exceeding their values. This method is used for optimization of any function with measured values that characterize it.

An experiment has been done for creating a method not based on minimization of some function of differences between measured and theoretical data but which uses minimization of sum of differences between probability of measured frequency and the same theoretical value. Such a method could work generally with different frequency curves. It could be universally applied and allow deriving theoretical function quality. Such method could be usable even for survival functions because it is nearly the same. It would be possible to obtain more parameters of theoretical distribution, which could subsequently be analysed further.

When using method of probability optimization one does not minimize expression $\sum_i |F(p_i) - y_i|^d$, where $F(p_i)$ is a theoretical frequency curve, p_i is a probability for being equal or exceeding their values, y_i are aligned to measured values and d is an exponent (usually second power).In case of even numbers one does not need absolute value.

Using the method of probability optimization the parameters of frequency curve by heuristic iteration process can be obtained. It is necessary to choose convenient starting parameters that largely depend on user's experience.

Description of probability optimization method: suppose we have $r \geq 1$ groups of the same values y_i in a sample of n measured values. They have different empiric probabilities for being equal or exceeding their values. Such situation is caused mostly by inaccuracy of measurements. In an ideal case same values occur with zero probability. In i group we have n_i equal values y_i. Their empirical probabilities for being equal or exceeding probability values are $p_{i1}, \ldots, p_{in_i}$. Theoretical probability for equal or exceeding value y_i is marked p_i. Then $F(p_i) = y_i$. We choose convenient starting parameters for LN5 distribution and by iteration process find parameter estimation of such distribution so that the expression $\sum_{i=1}^{r}\sum_{j=1}^{n}\left|p_{ij}-p_i\right|^d$ is minimal.

We have a lot of experience with working with hydrological data and we know that it is necessary to norm such data to average of 1. This is necessary to do if one wants to compare parameters of different curves. This is necessary to do in order to have the possibility to obtain relationships of the physical-geographical conditions of catchments. The value of normed average 1 follows from transforming of average of normal distribution, which is 0. After an exponential transformation it is $e^0 = 1$.

2.2 Characteristics of probability optimization method

Above described method of parameters estimation seems to possess several interesting characteristics, but they are not accurately mathematically proven.

a) When using simulated data it was shown that the parameters estimation is very close to the real parameters of distribution regardless of symmetry or asymmetry of distribution.
b) The results seem to better match the course of densities similarly to the method of maximal likelyhood. However, the probability optimization method is known to work even with additive and multiplicative coefficients in expressions of distribution (it is important even in LN3 and especially LN5 distribution).
c) Method of maximal reliability has tendencies to fail under parameter estimation for data sets where groups of same values exist. Probability optimization method solves this problem very well.
d) Except for one case, in all so far tested discharge data sets (about 100 data sets) the estimation of parameters was independent of the size of d power of the difference between theoretical and empirical frequency probabilities in the range of 0.1 to 4. When using the commonly used methods for estimation based on differences between measured and theoretical values it depends on power used.
e) A simulation experiment was performed using nine different parameters. Then relative differences between simulated and theoretical values of extremely high and low values were compared. The data range in each simulation of LN5 distribution was 824 values. Frequency probability for extremely high values was chosen to be 1.214×10^{-7} and for extremely low values $1 - 1.214 \times 10^{-7}$. Parameter estimations of the frequency curve were obtained by both the method of optimization of probability and the weighted least square method. Results are shown in Table 1.

Method	Average relative difference for extremely low discharges	Average relative difference for extremely high discharges
Optimization probability	0.253550	0.149132
Weighted least square	0.295576	0.172721

Tab. 1. Average relative differences - simulated and theoretical values.

As can be seen, the method of probability optimization provides better results in both cases.

f) Further analyzed were removed 10 % and then 20 % values at both tails and their effects. Frequency probabilities of the other values were retained. Data represented flows on the river Balinka in Baliny in the years 1997-2016. Estimations made by the relative least square method and method of probability optimization was depicted in Tab. 3. First row of the table gives estimations for all values, in the 2nd row 10 % values are omitted at the start and end of frequency curve, in the 3rd row 20 % is omitted as well. Last column gives extrapolated value of frequency probability 2.7397×10^{-5}. Course of frequency curves is shown on Fig. 2 and 4.

Probability optimization method						
	Est. a	Est. b	Est. μ	Est. σ	Est. y_0	Extrapolation [$m^3.s^{-1}$]
All data	2.1952	0.9123	-1.2300	1.0003	-0.0541	20,78
20 % omitted	2.0364	0.9606	-1.2337	1.0334	0.0012	23.79
40 % omitted	2.0475	0.9968	-1.2737	1.0410	0.0260	25.76
Weighted least square method						
	Est. a	Est. b	Est. μ	Est. σ	Est. y_0	Extrapolation [$m^3.s^{-1}$]
All data	11.7674	1.1309	-2.5912	0.8008	-0.0135	15.81
20 % omitted	14.4738	0.6889	-5.1105	2.1082	-0.0667	100.25
40 % omitted	14.2035	0.6825	-5.1623	2.2018	-0.0592	111.63

Est. = estimation.

Tab. 2. Parameters LN5 distribution estimations obtained by two methods under releasing of values on the tails.

Using the table one can see that at this point on the river the least square method gives the least stable estimates for the parameters of theoretical frequency curve compared to the relative least square method. Same goes for the extrapolated values.

Similar tests were done for the other six profiles – with one exception the relative least square method has always been the least stable one.

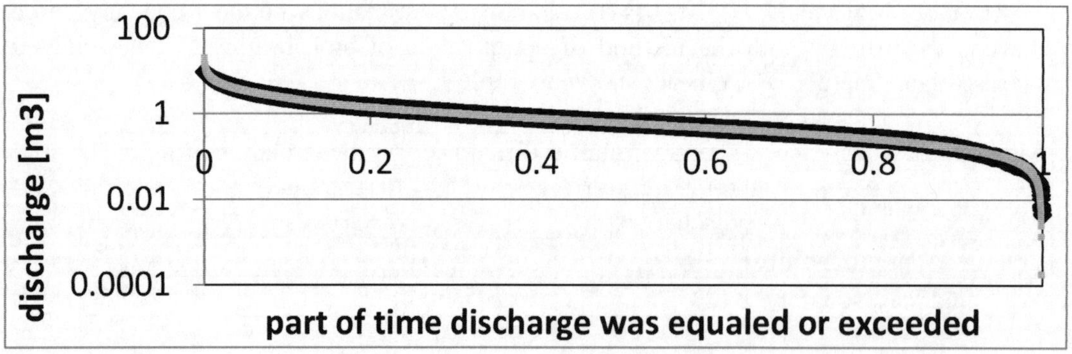

Fig. 1. Fitting theoretical distribution LN5 on daily discharges of river Balinka in Baliny between 1997-2016, probability optimization method.

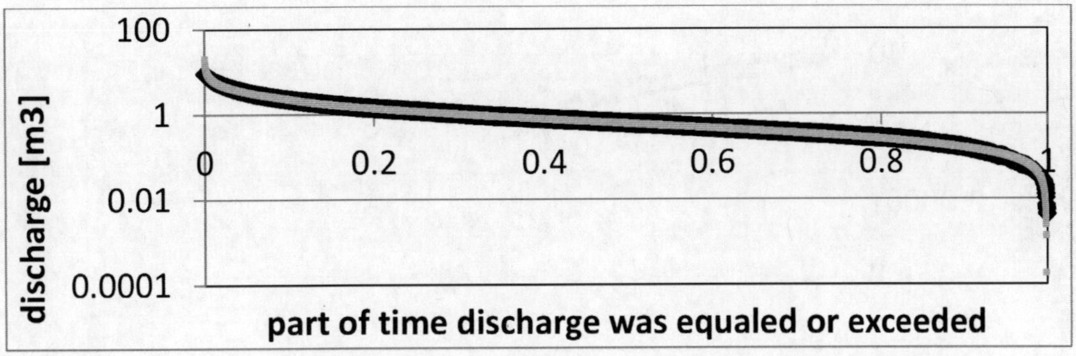

Fig. 2. Example of fit when omitting 20 % data at the tails, weighted least square method/

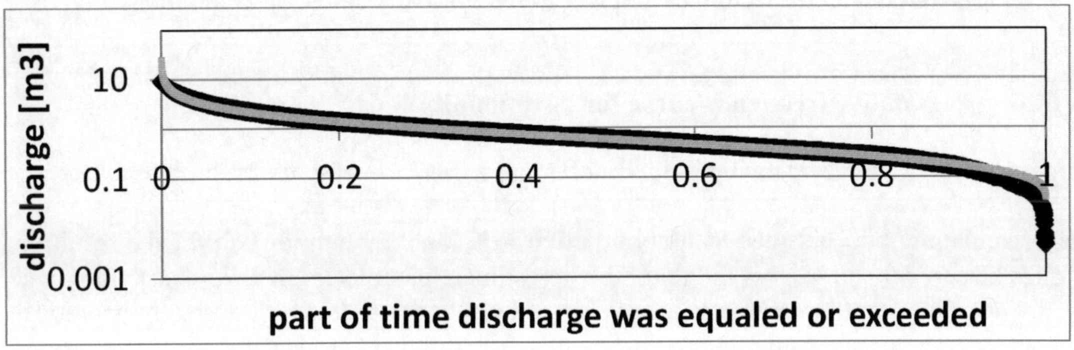

Fig. 3. Example of fit when omitting 20 % data at the tails, probability optimization method.

2.3 Fitting cumulative frequency curve for daily precipitation

To pass fit theoretical cumulative frequency curve through daily precipitation is relatively easy. Using the described methods the fit can be made including dry days. Days with precipitation are fitted by part of LN5 where values of transformed normal distribution are greater than zero. Negative values, which after transformation give very negligible values near zero, represent dry part of the curve. LN5 distribution seems to be two - parametric distribution for precipitation. Only parameters a and μ vary. The other three seem to be, at least for the region of Czech Republic, constant.

Using all five parameters gives insignificantly better result, but the question is whether the difference is only result of errors in measurements.

The values of "constant" parameters are $b = 0.18708$, $\sigma = 277769.0352$, $y_0 = 0$.

Fig. 4 shows the cumulative frequency curve of the daily precipitation at Babice nad Svitavou station between 1997 – 2016. Data are in logarithmic scale. Parameter estimation a = 0.00008899 and μ = -81874.33072 were obtained by the probability optimization method.

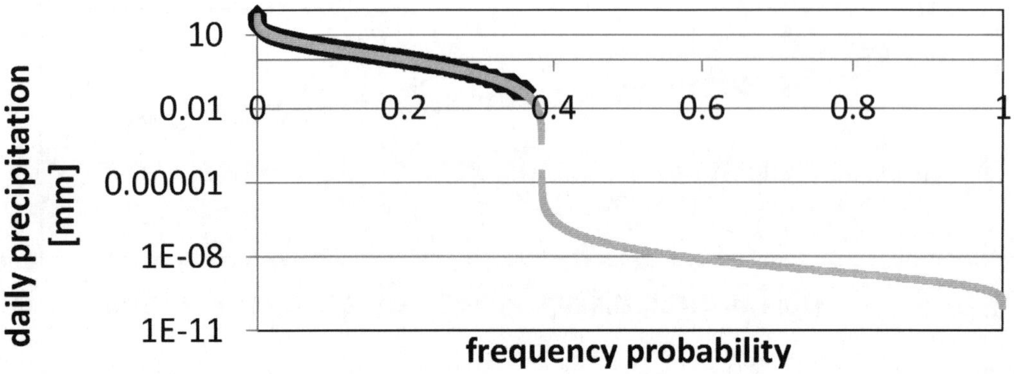

Fig. 4. Example of theoretical frequency curve of daily sum of precipitations.

2.4 Fitting cumulative frequency curve for air humidity

In order to fit air humidity data in % it was necessary to perform easy data adjustment.

When calculating ratio between saturation deficit in % and air humidity in %, the resulting ratio lies between 0 and ∞. It is then possible to estimate parameters of LN5 distribution and then divide values on theoretical frequency curve by saturation deficit and obtain theoretical frequency curve for air humidity data. Resulting theoretical curve can only have values between 0 % and 100 %. An example of such curve is depicted on Fig. 5. Data comes from the station in Pec pod Sněžkou. Parameters estimation for ratio between saturation deficit and air humidity are: $a = 0.001376$, $b = 3.0635$, $\mu = 6.7442$, $\sigma = 2.60119$, $y_0 = 0.01505$.

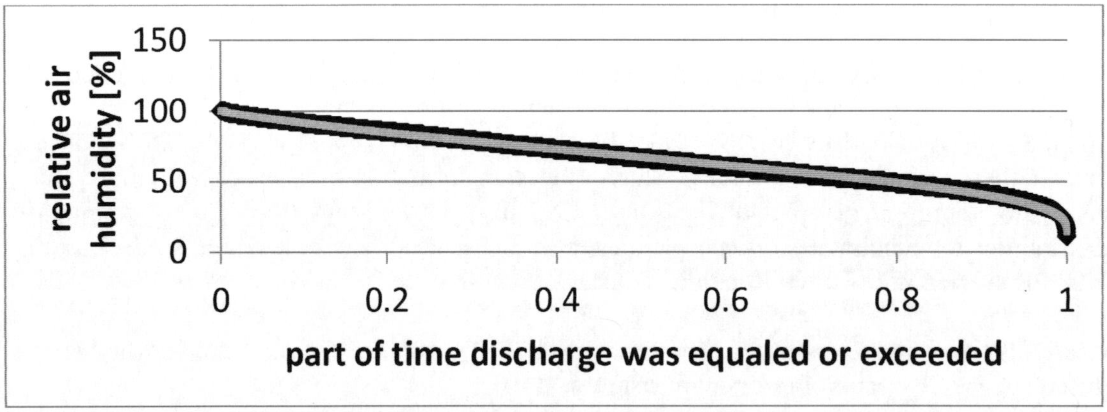

Fig. 5. Course of air humidity theoretical frequency curve.

2.5 Fitting theoretical frequency curve for daily air temperatures

Air temperatures measured in °C are usually fitted by normal distribution though they are a little asymmetric. They have extended end towards the lower values. It is possible to make them symmetrical using the Stephan – Boltzmann law: warm energy emitted from an absolutely black body with absolute temperature T_1 to the space with temperature T_2 is proportional to the difference of fourth powers of absolute temperatures of the body and the space (in Kelvin). For easier work with data the result, it is convenient to divide it by 100 000 000. Frequency curve of such data is symmetric (Fig. 6). The particular values are from station in the city Jihlava. If it was to be fitted with a similar quality as described above it is necessary to use power transformation of normal distribution with the power marked as *b* (it is assumed that $b \neq 1$).

$$Y = sign[(X+1)\sigma + \mu] \cdot a \cdot |(X+1)\sigma + \mu|^{b} + y_0 - 1 \text{ for } X \geq 0$$

$$Y = sign[(X-1)\sigma + \mu] \cdot a \cdot |(X-1)\sigma + \mu|^{b} + y_0 + 1 \text{ for } X < 0$$

Subtraction or addition of 1 ensures that positive and negative part of theoretical frequency curve will be continuous. After parameter estimation of transformed data it is possible to make reverse transformation from values proportional to the emitted energy – Fig. 6 to the temperature in °C – Fig. 7. It turned out that similarly to daily precipitation it is not necessary to use all parameters but apparently at least in the Czech Rep., only one parameter is necessary, and it is the parameter *a*. Using all five parameters has the same consequences as precipitation.

Values of "constant" parameters are approximately $b = 0.33967$, $\mu = 0$, $\sigma = 0.0001$, $y_0 = 1$.

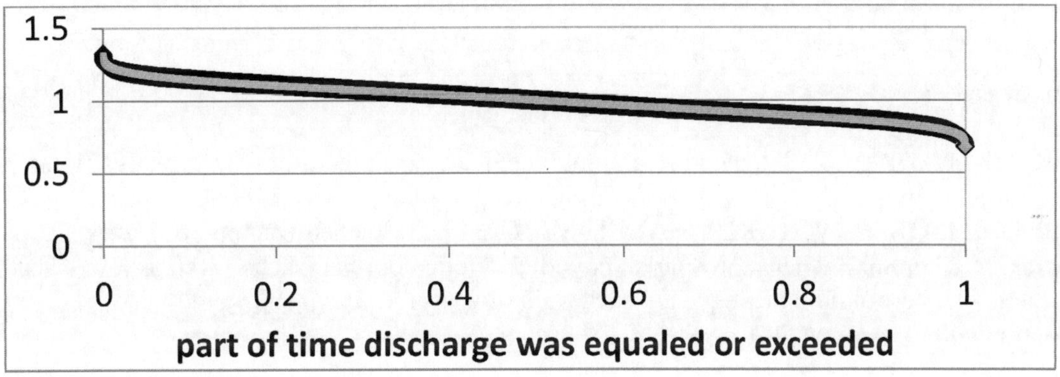

Fig. 6. Theoretical curve of exceeding values relative to emitted energy $a = 10.45342$.

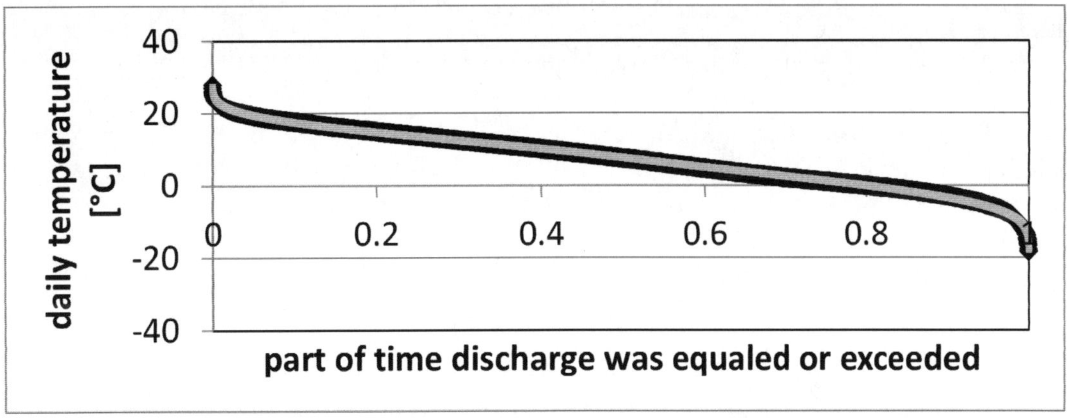

Fig. 7. Theoretical curve of exceeding temperature in °C.

3 Conclusion

It can be said that theoretical frequency curves LN5 and power transformation of the normal distribution are more accurate than the theoretical frequency curves currently used in hydrology and climatology. If probability optimization method is used, estimations of parameters are fairly stable against omitting data at the tails of theoretical frequency curve from optimization process. Their role in theoretical frequency curves must be preserved. This also means that probabilities of the measured values of the frequency curves must be preserved as well. Further advantage is that potential extrapolations of the theoretical frequency curve do not reach nonsense values even when using probabilities in the order of the age of the Earth. This is a consequence of the parameter b being < 1. If this way extrapolated data are accurate remains unknown. In addition, the Earth's climate is constantly changing. Above stated conclusion would only be correct if current climate was the same in the past 5.109 years. In the meantime, it is possible to say that these results have physical basis. Except for air temperatures where this is clearly given by type of transformation of input data (physical basis of the modification of the standard normal distribution remains questionable), this is not completely clear. More questions remains than there are answers.

Acknowledgement

I would like to thank my wife Marie for help with the mathematical part of text and revision of the whole text, thank Gejza Wimmer for a math consultation and Jáchym Brzezina for revision text translated to English language.

References

[1] BUDÍK, LADISLAV, Skládání křivek překročení časových řad průtoků (Superposition of cumulative frequency curves of flows) Bratislava, Aplimat 2015, 134-141

[2] BUDÍK, LADISLAV, KUKLA, PAVEL, ŠERCL, PETR, Odhady parametrů křivek překročení průtoků v nepozorovaných povodích a jejich optimalizace – výsledky, (Parametres estimation of frequency curves in unmeasured catchments and their optimization), Aplimat 2013

Current address

Budík Ladislav
Czech Hydrometeorological Institute, Praha-Komořany
Branch Brno
Kroftova 43, 616 67 Brno, Czech Republic
E-mail: budik@chmi.cz

Slovak University of Technology in Bratislava
Faculty of Mechanical Engineering
17th Conference on Applied Mathematics **APLIMAT 2018**

Proceedings

VIRTUAL RECOMPOSITION AND ANALYSIS OF A SURVEYED MODEL OF THE WOODEN CRUCIFIX OF ST. PETER'S BASILICA

CANCIANI Marco (IT), FALCOLINI Corrado (IT), PASTOR ALTABA Maria (IT)

Abstract. The wooden Crucifix of St. Peter's Basilica in Rome has been recently restored by a group coordinated by L. D'Alessandro and G. Capriotti and under the supervision of P. Zander for the Fabbrica di San Pietro. The work has involved a detailed survey with range based and image based methods for a three-dimensional model useful for virtual recomposition, measures, analysis and documentation. The study includes geometrical models, parametric curves and numerical algorithms.

Keywords: 3D survey, virtual recomposition, point clouds, object restoration, model analysis

Mathematics Subject Classification: Primary 65K05, 53A04; Secondary 97N50

1 Introduction

The wooden Crucifix of Saint Peter's Basilica in Rome, a marvellous artwork ascribed to the school of Pietro Cavallini in the 14th century, has been recently brought back to its original splendour by a team of restorers coordinated by Lorenza D'Alessandro and Giorgio Capriotti under the supervision of Pietro Zander for the Fabbrica di San Pietro [1]. After a temporary placement, at the end of the 2016 Jubilee of Mercy (see Fig. 1), it is now possible to admire it in the Chapel of the Blessed Sacrament (see Fig. 3a).

The restoration has involved, among other disciplines, also a detailed survey, with range based and image based methods, to construct a three-dimensional model useful for virtual recomposition, measures, mathematical analysis and documentation.

The main results of the survey procedures have been a very detailed virtual model of the Crucifix, useful for special analysis and monitoring aimed at preservation or documentation, and some specific support to restorer's study and work (as in previous collaboration using similar techniques, see [2, 3]). In particular, the visualization of different historical restoration phases (elaborated with photographic textures applied on a triangulated mesh), the virtual re-

Fig. 1. The restored Crucifix on the main altar near a column of St. Peter's Baldachin.

position of the wooden statue on the new made Cross (with data elaboration on volumes, forces, barycenter and equilibrium conditions), the left arm joint description and the wood breakings analysis to speculate on trunk diameter and drying time evolution.

2 The model

The 3D model (see Fig.2. for the detail of the head) is the result of several point cloud

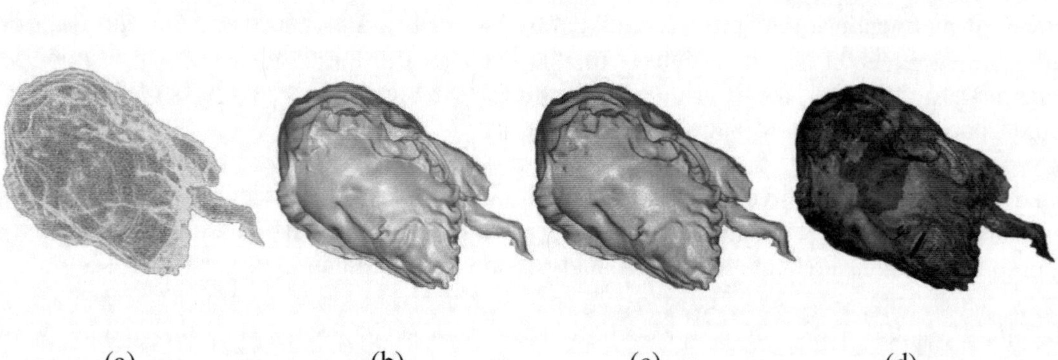

(a) (b) (c) (d)

Fig. 2. Detail models of the head. (a) The connected point cloud. (b) The surface shape. (c) The triangulated surface. (d) The photographic texture mapped on the mesh.

surveys, made during the restoration process, using laser scanner and photogrammetric procedures: a laser scanner instrument returns millions of points coordinates of the object in a given reference frame whereas several pictures from different viewpoints capture its shape (with a similar point cloud) and its true colours.

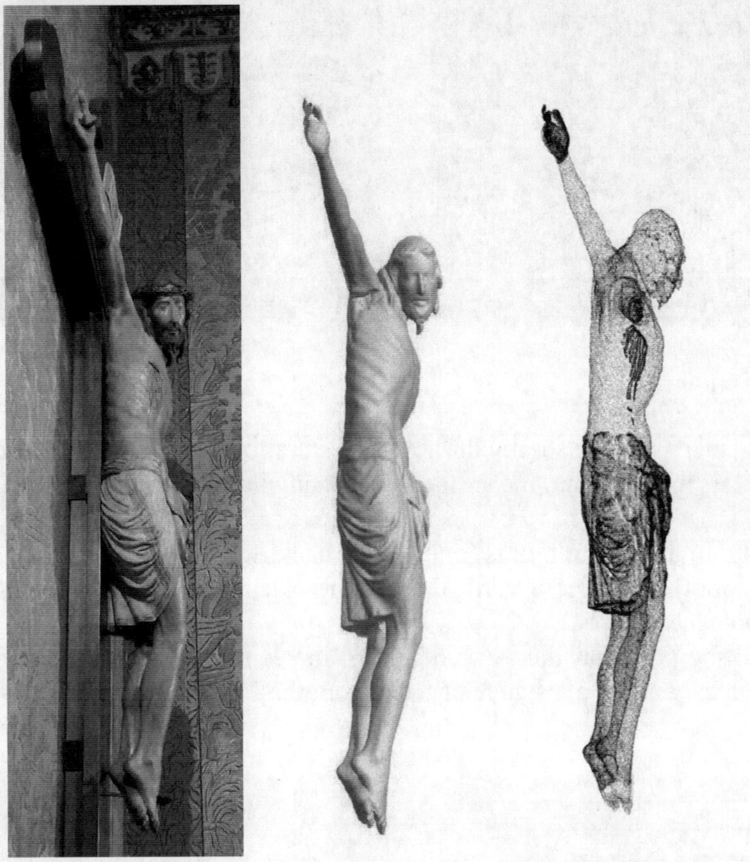

(a) (b) (c)

Fig. 3. (a) The restored Crucifix placed in its actual position. (b) A smooth 3D model determined as a triangulated surface of a surveyed point cloud (c). Note the greater point density of the hand, the drapery and the blood leak.

Out of different surveys, and by an alignment of the corresponding point clouds, a single 3D model of 230.000 points has been worked out as a base for orienteering, scaling and aligning specific parts with much greater point density and accuracy: with more than 800 pictures it has been possible to produce 50 different working models and 12 final models for specific parts with a total of 10 millions points per model.

3 Data analysis

Among other elaborations, we mention here: 1) the analysis of the virtual left arm joint (see Fig.4b) together with the corresponding shoulder hole to gain some information on their relative positions and frame; 2) horizontal and vertical planar section curves [4] to check the position of the body barycentre (see Fig. 5) and a volume estimate out of the surveyed point cloud (see Fig. 6); 3) principal wood breakings analysis to speculate on weight, trunk diameter and drying time evolution.

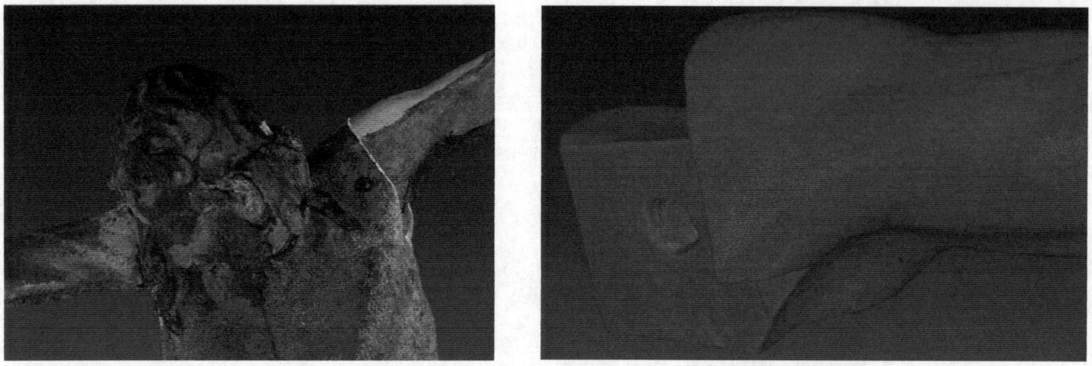

(a) (b)

Fig. 4. (a) Photogrammetric survey during the restoration process: image of the 3D model with true colors mapped onto the surface. (b) Triangulated surface of the left arm joint.

The position of two anchoring points (one on the back, with a metal support, taken as a pole and another on foot) identifies a vertical axis: the calculated barycentre is sloping 7° with respect to the pole (see Fig. 5a).

The back anchoring point has the same height of the centre of mass of the upper part of the body, on the vertical axis, at a distance of 8 cm from the Cross surface: this datum gives a

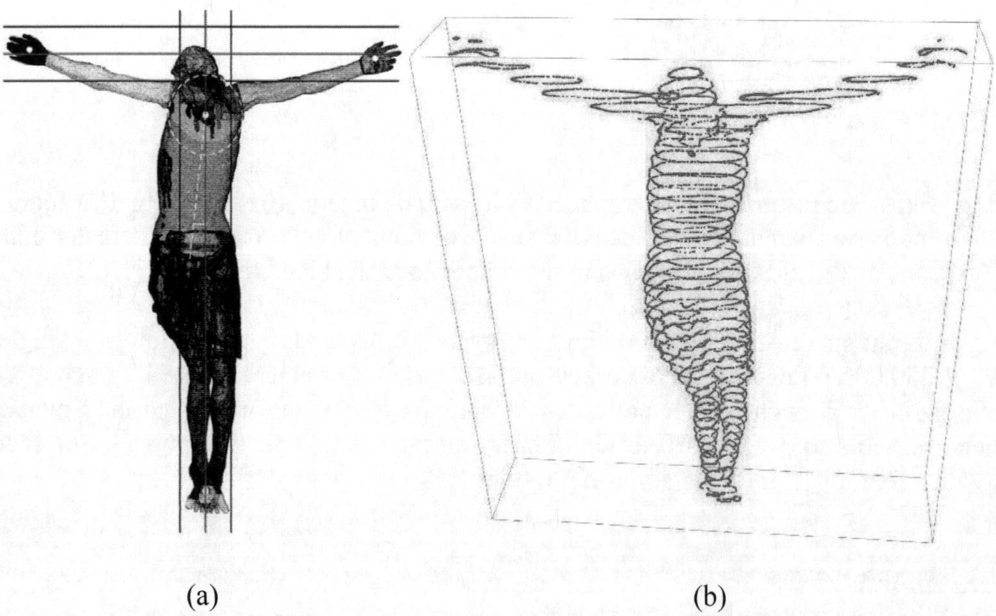

(a) (b)

Fig. 5. (a) Position on the Cross: the barycentre (red dot) is slightly sloping with respect to the vertical axis through the anchoring points (white dots). (b) Horizontal section curves.

measured explanation of the position of the security support.

Given the dimensions of the Crucifix (height 214 cm, wide 197 cm), the estimated volume of the wooden sculpture has been calculated from cylindrical slices of horizontal section curves (see Fig. 6): their number has been increased for high accuracy to give a volume of 137 dm^3. The surface area has been evaluated directly on the triangulated model in 288 dm^2.

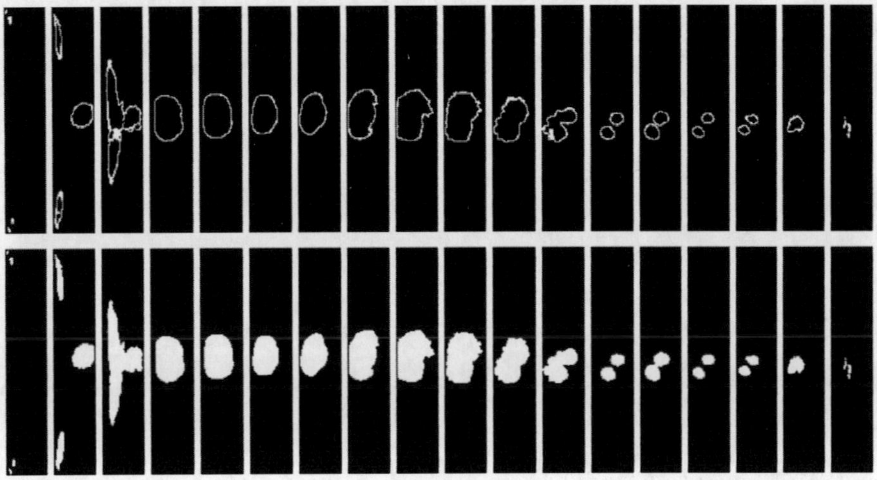

Fig. 6. Horizontal sections and the corresponding cylindrical slices to get a volume estimate out of the point cloud model.

The weight evolution curve (see Fig. 7a) is a rational interpolation of the known data of the new made Cross (of the same walnut wood of the Crucifix, dryed for 33 years) and data elaborated from the virtual model.

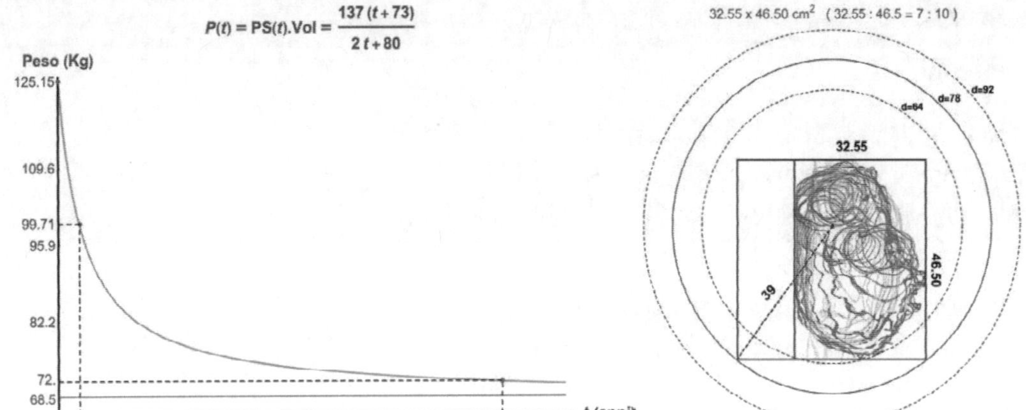

(a) (b)

Fig. 7. (a) Drying time evolution of the wooden Crucifix. The known data are the three red dots. (b) Walnut trunk diameter hypothesis given the central position of the marrow and the red sections of Fig.5b.

With the simplifying hypothesis of a rational function of type f(t) = (a t+b)/(c t+d) one can fix the parameters using the calculated wooden sculpture volume (137 dm^3) and an estimate on its actual weight (72 kg), the volume and weight of the new Cross (elaborated from its surveyed shape: 82.44 dm^3 and 60 kg) together with its measured drying time and the standard intervals for specific weight of walnut green (0.91-0.92) and dryed (0.44-0.68) wood.
The variation inside the two intervals depends on wood density and therefore is increasing going from center to the trunk surface (see Fig. 7b).

The simplified model of the weight evolution as a function of time (in years) can be given as the simple expression f(t) = 137 (t+73)/(2t+80).

References

[1] ZANDER P., (Ed.) The wooden Crucifix of Saint Peter's Basilica. *Volume supported and promoted through the Order of the Knights of Columbus.* Copyright © 2017 *Fabbrica di San Pietro in Vaticano.* ISBN 978-0-9837297-4-7

[2] CANCIANI M., CAPRIOTTI G., D'ALESSANDRO L., FALCOLINI C., Algoritmi rinascimentali: il progetto di image modeling in 3D per il restauro della scultura in terracotta del Sant'Andrea di Stiffe (AQ), *Lo Stato dell'Arte – Atti del XIV Congresso Nazionale IGIIC – Accademia Di Belle Arti di L'Aquila, 2016*

[3] CANCIANI M., CAPRIOTTI G., D'ALESSANDRO L., FALCOLINI C., SACCONE M., The recomposition of fragmented objects: the case study of St. Andrea statue at Stiffe, L'Aquila, *Le vie dei Mercanti*, XIII Forum Internazionale di Studi, Capri, 2015

[4] FALCOLINI C., Parametric Planar Sections of a Point Cloud with applications in Cultural Heritage. In *Proceedings of 16th Conference on Applied Mathematics Aplimat 2017*, 510-517, 2017

Current address

Canciani Marco, Associate Professor
Dipartimento di Architettura, Università Roma
Tre Via Madonna dei Monti, 40 - 00184 Roma (Italy)
E-mail: marco.canciani@uniroma3.it

Falcolini Corrado, Associate Professor
Dipartimento di Architettura, Università Roma
Tre Via Madonna dei Monti, 40 - 00184 Roma (Italy)
E-mail: falco@mat.uniroma3.it

Pastor Altaba Maria, Phd student
Dipartimento di Architettura, Università Roma
Tre Via Madonna dei Monti, 40 - 00184 Roma (Italy)
E-mail: mpaltaba@gmail.com

Slovak University of Technology in Bratislava
Faculty of Mechanical Engineering
17th Conference on Applied Mathematics **APLIMAT 2018**

Proceedings

ON SUBJECTED TO RAPID RANDOM EXTRACTIONS TWO-SEX POPULATION GROWTH

CARKOVS Jevgeņijs (LV), MATVEJEVS Andrejs (LV)

Abstract: The paper analyzes a model for isolated population with sexual reproduction under assumption that the extractions are random and going at random time moments. Applying the approximative procedures of stochastic analysis we construct an ordinary differential equation for population dynamics in the mean and a linear stochastic differential equation for deviations on the mean trajectories. This approximative model permits to analyze a population growth as the Gaussian process with mean and variance given by ordinary differential equations.

Keywords: two-sex population; random extractions; stochastic approximation.

Mathematics Subject Classification: Primary 92B05,60H30; Secondary 60G57, 37H10.

1 Introduction

The dynamics of the two-sex populations has been discussed by many authors as the different forms of difference or differential equations (see, for example, [1 - 12] and references there). As has been mentioned by authors of the above papers and books the mathematical models may be used for qualitative and quantitative analysis of ecosystem, but also to manage the problems of agriculture, forest, animal husbandry and fishery ecosystem. The model construction is based on analysis of the increments $x(t_{k+1})-x(t_k)$ and $y(t_{k+1})-y(t_k)$ for male and female densities $\{x(t_k), k\in \mathbf{N}\}$ and $\{y(t_k), k\in \mathbf{N}\}$, where $\{0 < t_1 < t_2 < ... < T\}$ are real numbers. Our paper analyze the popular in mathematical biology the *Liu* two-sex population model [9] with jump type random extractions at the below defined (1.1) time moments. We assume that the dynamics of population is given by the system of equations for the male and female densities $\{x_\varepsilon(t), y_\varepsilon(t), t\geq 0\}$. The corresponding to our model dynamical system may be described as follows:

- at time moment τ_k the phase coordinates have jumps to point

$$x_\varepsilon(\tau_k) = x_\varepsilon(\tau_k -) - \varepsilon h(\xi_k) x_\varepsilon(\tau_k -),\ y_\varepsilon(\tau_k) = y_\varepsilon(\tau_k -) - \varepsilon g(\xi_k) y_\varepsilon(\tau_k -), \tag{1.1}$$

where $x(\tau_1-):=\lim_{t\uparrow\tau_1} x(t), y(\tau_1-)=\lim_{t\uparrow\tau_1} y(t)$;

- further, at any time interval $(\tau_{k-1},\tau_k), \mathrm{k}\in\mathbf{N}$ the population dynamics is given by differential equation

$$\frac{dx_\varepsilon(t)}{dt}=\alpha_1 y_\varepsilon(t)-\beta_1 x_\varepsilon(t)-K^{-1}(x_\varepsilon(t)+y_\varepsilon(t))x_\varepsilon(t), \frac{dy_\varepsilon(t)}{dt}=(\alpha_2-\beta_2)y_\varepsilon(t)-K^{-1}(x_\varepsilon(t)+y_\varepsilon(t))y_\varepsilon(t), \tag{1.2}$$

where α_1 and α_2 are per capita birth rate for males and females, β_1 and β_2 are per capita death rate for males and females, K - is carrying capacity of population, $\{\xi_k, k\in\mathbf{N}\}$ are uniform R(0,1) distributed random variables, $\{h(\xi), g(\xi), \xi\in[0,1]\}$ are bounded nonnegative functions,

$$\begin{gathered}\mathbf{E}\{h(\xi_k)\}\equiv\int_0^1 h(\xi)d\xi=\gamma_1, \mathbf{E}\{g(\xi_k)\}\equiv\int_0^1 g(\xi)d\xi=\gamma_2, \mathbf{E}\{h^2(\xi_k)\}\equiv\int_0^1 h^2(\xi)d\xi=\sigma_1^2, \\ \mathbf{E}\{g^2(\xi_k)\}\equiv\int_0^1 g^2(\xi)d\xi=\sigma_2^2, \mathbf{E}\{h(\xi_k)g(\xi_k)\}\equiv\int_0^1 h(\xi)g(\xi)d\xi=\sigma_{12}^2\end{gathered} \tag{1.3}$$

and $\{\tau_k-\tau_{k-1}, k\in\mathbf{N}\}$ are independent on $\{\xi_k, k\in\mathbf{N}\}$ independent identically exponentially distributed with parameter ε^{-1}. In the next section applying stochastic averaging procedure [15] we proof that $\{x_\varepsilon(t), y_\varepsilon(t)\}$ are such continuously dependent on parameter ε functions , that there exists

$$p\lim_{\varepsilon\to 0} x_\varepsilon(t)=\bar{x}(t), p\lim_{\varepsilon\to 0} y_\varepsilon(t)=\bar{y}(t) \tag{1.4}$$

and derive the ordinary differential equations for $\bar{x}(t), \bar{y}(t)$. In the third section we proof that for any *T>0* the distributions of the random processes $\left\{\frac{x_\varepsilon(t)-\bar{x}(t)}{\sqrt{\varepsilon}}, t\in[0,T]\right\}$ and $\left\{\frac{y_\varepsilon(t)-\bar{y}(t)}{\sqrt{\varepsilon}}, t\in[0,T]\right\}$ may be approximated by distributions of Gaussian processes, which satisfy the stochastic differential equations. The impulsive differential equation (1.1)-(1.2) defines two dimensional homogeneous Markov process. This process may be uniquely defined by the infinitesimal generator [13], which one can find as the limit

$$(L(\varepsilon)v)(x,y):=\lim_{t\downarrow 0}\frac{1}{t}\ \mathbf{E}\{v(x_\varepsilon(t), y_\varepsilon(t))/x_\varepsilon(0)=x, y_\varepsilon(0)=y\}-v(x,y)$$

for an arbitrary sufficiently smooth bounded function $v(x,y)$. Using the asymptotic equalities

$$\mathbf{P}(\xi(t)\le x/\xi(0)=\eta,t<\Delta_1)=\mathbf{P}(\eta\le x)\mathbf{P}(t<\Delta_1)=\mathbf{P}(\eta\le x)e^{-\varepsilon^{-1}t}=\mathbf{P}(\eta\le x)+o(t),$$
$$\mathbf{P}(\xi(t)\le x/\xi(0)=\eta,t\ge\Delta_1)=\mathbf{P}(\xi_1\le x)\ 1-e^{-\varepsilon^{-1}t}\ +o(t)=\varepsilon^{-1}t\mathbf{P}(\xi_1\le x)+o(t)$$

we can find the above infinitesimal operator as follows:

$$(L(\varepsilon)v)(x,y)=v'_x(x,y)\left[\alpha_1 y-\beta_1 x-K^{-1}(x+y)x\right]+v'_y(x,y)\left[(\alpha_2-\beta_2)y-K^{-1}(x+y)y\right]+$$
$$+\varepsilon^{-1}\left[\int_0^1 \mathbf{E}\{v(x-\varepsilon h(\xi)x,y-\varepsilon g(\xi)x)\}d\xi-v(x,y)\right] \quad (1.5)$$

2 The first approximation

As it has been proved in [14,15] if at any point $\{x,y\}\in\mathbf{R}^2$ for any sufficiently smooth function $v(x,y)$ there exists limit $\lim_{\varepsilon\to 0}(L(\varepsilon)v)(x,y):=Lv(x,y)$ and L is the first order linear differential operator $L=a(x,y)\frac{\partial}{\partial x}+b(x,y)\frac{\partial}{\partial y}$ then we can approximate accurate within $o(\varepsilon)$ the expectations with the solutions of the smooth dynamical system with the vector flow $\{a(x,y),b(x,y)\}$. Corresponding to (1.5) limit operator for our Markov dynamical system (1.1)-(1.2) has a form

$$(Lv)(x,y)==\left[\alpha_1 y-(\beta_1+\gamma_1)x-K^{-1}(x+y)x\right]\frac{\partial}{\partial x}v(x,y)+\left[(\alpha_2-\beta_2-\gamma_2)y-K^{-1}(x+y)y\right]\frac{\partial}{\partial y}v(x,y) \quad (2.1)$$

Therefore, the above mentioned approximative dynamical system has a form:

$$\begin{cases}\dfrac{d\overline{x}(t)}{dt}=\alpha_1\overline{y}(t)-(\beta_1+\gamma_1)\overline{x}(t)-K^{-1}(\overline{x}(t)+\overline{y}(t))\overline{x}(t),\\[2mm] \dfrac{d\overline{y}(t)}{dt}=(\alpha_2-\beta_2-\gamma_2)\overline{y}(t)-K^{-1}(\overline{x}(t)+\overline{y}(t))\overline{y}(t)\end{cases} \quad (2.2)$$

If $\{\overline{x}(t),t\ge 0\}$ and $\{\overline{y}(t),t\ge 0\}$ are the solutions of the equation (2.2) with initial conditions $\overline{x}(0)=\mathbf{E}\{x_0(0)\},\overline{y}(0)=\mathbf{E}\{y_0(0)\}$ then $\lim_{\varepsilon\to 0}\sup_{0\le t\le T}\ |\mathbf{E}\{x_\varepsilon(t)\}-\overline{x}(t)|+|\mathbf{E}\{y_\varepsilon(t)\}-\overline{y}(t)|\ =0$ for any T>0. The equation (2.2) has two equilibrium points: $A_1=\{0,0\}$ and $A_2=\{\overline{x},\overline{y}\}$, where

$$\overline{x}=\frac{K\alpha_1(\alpha_2-\beta_2-\gamma_2)}{\alpha_1+\beta_1+\gamma_1+\alpha_2-\beta_2-\gamma_2},\qquad \overline{y}=\frac{K(\alpha_2-\beta_2-\gamma_2)(\beta_1+\gamma_1+\alpha_2-\beta_2-\gamma_2)}{\alpha_1+\beta_1+\gamma_1+\alpha_2-\beta_2-\gamma_2} \quad (2.3)$$

The numbers (2.3) are positive if and only if the birth rate for females is sufficiently large: $\alpha_2>\beta_2+\gamma_2$. Not so difficult to ensure that under the above assumption the point $A_1=\{0,0\}$ is unstable knot and point A_2 is stable knot. The Fig. 1 contains two graphics: the solid lines are the solutions of equation (2.2) and dot lines are sample trajectories for solution of random impulsive differential equation (1.1) - (1.2) for parameters

$$\varepsilon = 0.01, K = 10,\ \alpha_1 = 0.7, \beta_1 = 0.4, \gamma_1 = 0.2, \alpha_2 = 1, \beta_2 = 0.4, \gamma_2 = 0.2 . \qquad (2.4)$$

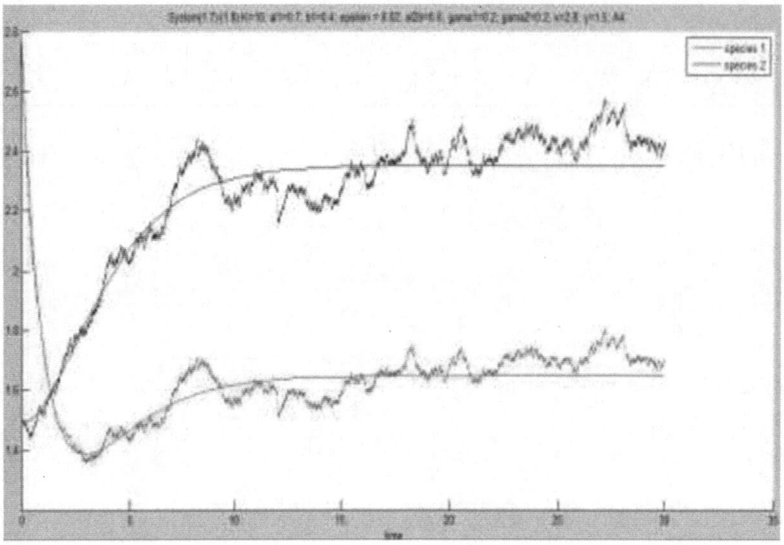

Fig. 1. The solution of equation (2.2) and the sample trajectories for solutions of the Markov dynamical system (1.1)-(1.2) for parameters (2.4).

As we can see there are sufficiently large deviations on population size averaged trajectory. To estimate these deviation, we need next step of the stochastic approximation procedure.

3 The normalized deviations on the first approximation

The second step for asymptotic analysis of the Markov impulsive differential equations (1.1)-(1.2) is diffusion approximation of the *normalized deviations* [14,15]:

$$z_\varepsilon(t) := \frac{x_\varepsilon(t) - \overline{x}(t)}{\sqrt{\varepsilon}}, u_\varepsilon(t) := \frac{y_\varepsilon(t) - \overline{y}(t)}{\sqrt{\varepsilon}} \qquad (3.1)$$

The impulsive differential dynamical system for these processes has form ordinary differential equations:

$$\begin{aligned} \frac{dz_\varepsilon(t)}{dt} &= \frac{1}{\sqrt{\varepsilon}}[\gamma_1 \overline{x}(t)] + \{\alpha_1 u_\varepsilon(t) - \beta_1 z_\varepsilon(t) - K^{-1}[(\overline{y}(t) + 2\overline{x}(t)) z_\varepsilon(t) + u_\varepsilon(t)\overline{x}(t)]\} - \\ &- \sqrt{\varepsilon} K^{-1}\{z_\varepsilon(t) + u_\varepsilon(t) z_\varepsilon(t)\} \\ \frac{du_\varepsilon(t)}{dt} &= \frac{1}{\sqrt{\varepsilon}}[\gamma_2 \overline{y}(t)] + \{(\alpha_2 - \beta_2) u_\varepsilon(t) - K^{-1}[(\overline{x}(t) + 2\overline{y}(t)) u_\varepsilon(t) + z_\varepsilon(t)\overline{y}(t)] - \\ &- \sqrt{\varepsilon} K^{-1}(z_\varepsilon(t) + u_\varepsilon(t)) u_\varepsilon(t) \end{aligned} \qquad (3.2)$$

for $t \in (\tau_{k-1}, \tau_k), k \in \mathbf{N}$, and equations for jumps:

$$
\begin{aligned}
z_\varepsilon(\tau_k) &= z_\varepsilon(\tau_k -) - \sqrt{\varepsilon} h(\xi_k)\overline{x}(\tau_k) - \varepsilon h(\xi_k) z_\varepsilon(\tau_k -), \\
u_\varepsilon(\tau_k) &= u_\varepsilon(\tau_k -) - \sqrt{\varepsilon} g(\xi_k)\overline{y}(\tau_k) - \varepsilon g(\xi_k) u_\varepsilon(\tau_k -)
\end{aligned}
\tag{3.3}
$$

for $k \in \mathbf{N}$. The system of equations (2.2)-(3.2)-(3.3) defines on the probability space $(\Omega, \mathfrak{F}, \mathfrak{F}^t, \mathbf{P}, t \geq 0)$ four dimensional homogeneous Markov process $\{\overline{x}(t), \overline{y}(t), z_\varepsilon(t), u_\varepsilon(t), t \geq 0\}$. The weak infinitesimal operator for this process is defined by formula:

$$
\begin{aligned}
&(\mathcal{L}(\varepsilon)v)(x,y,z,u) = \lim_{t \downarrow 0} \frac{1}{t}\Big[\mathbf{E}_{xy}^{zu}\{v(\overline{x}(t), \overline{y}(t), z_\varepsilon(t), u_\varepsilon(t))\} - v(x,y,z,u)\Big] = \\
&= [\alpha_1 y - (\beta_1 + \gamma_1)x - K^{-1}(x+y)x]\frac{\partial}{\partial x}v(x,y,z,u) + [(\alpha_2 - \beta_2 - \gamma_2)y - K^{-1}(x+y)y]\frac{\partial}{\partial y}v(x,y,z,u) + \\
&+ \{\alpha_1 u - \beta_1 z - K^{-1}[(2x+y)z + ux]\}\frac{\partial}{\partial z}v(x,y,z,u) + \{(\alpha_2 - \beta_2)u - K^{-1}[(x+2y)u + zy]\}v\frac{\partial}{\partial u}(x,y,z,u) + \\
&+ \frac{1}{\sqrt{\varepsilon}}[\gamma_1 x]\frac{\partial}{\partial z}v(x,y,z,u) + \frac{1}{\sqrt{\varepsilon}}[\gamma_2 y]\frac{\partial}{\partial u}v(x,y,z,u) + \\
&+ \varepsilon^{-1}\int_0^1 v(x, y, z - \sqrt{\varepsilon}h(\xi)x - \varepsilon h(\xi)z, u - \sqrt{\varepsilon}g(\xi)y - \varepsilon g(\xi)u)\, d\xi
\end{aligned}
\tag{3.4}
$$

where $\mathbf{E}_{xy}^{zu}\{\bullet\} := \mathbf{E}\{\bullet / \overline{x}(0) = x, \overline{y}(0) = y, z_\varepsilon(0) = z, u_\varepsilon(0) = u\}$ and $v(x,y,z,u)$ sufficiently smooth bounded function. To apply stochastic approximation procedure [12] we have to calculate the limit in the formula (3.4) as $\varepsilon \to 0$. Under assumption that $v(x,y,z,u)$ is sufficiently smooth bounded function we can pass to a limit as $\varepsilon \to 0$:

$$
\begin{aligned}
&(\hat{\mathcal{L}}v)(x,y,z,u) := \lim_{\varepsilon \to 0}(\mathcal{L}(\varepsilon)v)(x,y,z,u) = [\alpha_1 y - (\beta_1 + \gamma_1)x - K^{-1}(x+y)x - \gamma_1 z]\frac{\partial}{\partial x}v(x,y,z,u) + \\
&+ [(\alpha_2 - \beta_2 - \gamma_2)y - K^{-1}(x+y)y - \gamma_2 u]\frac{\partial}{\partial y}v(x,y,z,u) + xy\sigma_{12}^2\frac{\partial^2}{\partial z \partial u}v(x,y,z,u) + \\
&+ \{\alpha_1 u - \beta_1 z - K^{-1}[(2x+y)z + ux]\}\frac{\partial}{\partial z}v(x,y,z,u) + \frac{1}{2}\sigma_1^2 x^2\frac{\partial^2}{\partial z^2}v(x,y,z,u) + \\
&+ \{(\alpha_2 - \beta_2)u - K^{-1}[(x+2y)u + zy]\}\frac{\partial}{\partial u}v(x,y,z,u) + \frac{1}{2}\sigma_2^2 y^2\frac{\partial^2}{\partial u^2}v(x,y,z,u)
\end{aligned}
\tag{3.5}
$$

The operator $\hat{\mathcal{L}}$ may be interpret [13] as the weak infinitesimal operator of the diffusion Markov process $\{z(t), u(t), t \geq 0\}$ given by the linear nonhomogeneous Ito stochastic differential equations:

$$\begin{cases} dz(t) = [-K^{-1}(2\overline{x}(t)+\overline{y}(t))-\beta_1]z(t)dt+[\alpha_1-K^{-1}\overline{x}(t)]u(t)dt+ \\ \quad +\overline{\sigma}_1(t)dw_1(t)+\overline{\sigma}_{12}(t)dw_2(t), \\ du(t) = -K^{-1}\overline{y}(t)z(t)dt+[(\alpha_2-\beta_2)-K^{-1}(\overline{x}(t)+2\overline{y}(t))]u(t)dt+ \\ \quad +\overline{\sigma}_2(t)dw_1(t)+\overline{\sigma}_{12}(t)dw_2(t) \end{cases} \tag{3.6}$$

with initial condition $z(0)=0, u(0)=0$, where $\{\overline{x}(t),\overline{y}(t),t\geq 0\}$ are solution of the equation (2.2) with initial conditions $\overline{x}(0)=x_\varepsilon(0), \overline{y}(0)=y_\varepsilon(0)$, $w_1(t),w_2(t)$ are the standard independent Wiener processes, and coefficients $\overline{\sigma}_1(t),\overline{\sigma}_2(t),\overline{\sigma}_{12}(t)$ are the nonnegative solutions of the matrix equations

$$\begin{pmatrix} \overline{\sigma}_1(t) & \overline{\sigma}_{12}(t) \\ \overline{\sigma}_{12}(t) & \overline{\sigma}_2(t) \end{pmatrix}^2 = \begin{pmatrix} \sigma_1^2\overline{x}^2(t) & \sigma_{12}^2\overline{x}(t)\overline{y}(t) \\ \sigma_{12}^2\overline{x}(t)\overline{y}(t) & \sigma_2^2\overline{y}^2(t) \end{pmatrix}.$$

The covariance matrix $Q(t)=\begin{pmatrix} q_{11}(t) & q_{12}(t) \\ q_{12}(t) & q_{22}(t) \end{pmatrix}$ for the solution of equation (3.6) with zero initial condition satisfies the differential equation

$$\frac{d}{dt}Q(t) = AQ(t)+Q(t)A^T+\Sigma^2 \tag{3.7}$$

with initial condition $Q(0)=0$. Not so difficult to ensure that under assumption $\alpha_2>\beta_2+\gamma_2$ there exists $\lim_{t\to\infty}Q(t)=\hat{Q}$, where matrix $\hat{Q}=\begin{pmatrix} \hat{q}_{11} & \hat{q}_{12} \\ \hat{q}_{12} & \hat{q}_{22} \end{pmatrix}$ satisfies an algebraic equation

$$A\hat{Q}+\hat{Q}A^T=-\Sigma^2.$$

As it has been proved in [14,15] on any finite dimensional interval [0,T] the finite dimensional distributions of the processes $\{x_\varepsilon(t),y_\varepsilon(t),0\leq t\leq T\}$ may be approximated by the Gaussian distributions of the processes $\{\overline{x}(t)+z(t)\sqrt{\varepsilon},\overline{y}(t)+u(t)\sqrt{\varepsilon},0\leq t\leq T\}$. As it has been mentioned the solution $\{\overline{x}(t),\overline{y}(t)\}$ of the average equation converges to the equilibrium point $\{\overline{x},\overline{y}\}$ given by equation (2.3). Therefore, with the course of time the population concentrates in a neighborhood of equilibrium in the mean (2.3) and the equations (3.6) have more simple form:

$$d\vec{z}(t)=A\vec{z}(t)dt+\Sigma d\vec{w}(t), \tag{3.8}$$

where

$$\vec{z}(t) = \begin{pmatrix} z(t) \\ u(t) \end{pmatrix}, \vec{w}(t) = \begin{pmatrix} w_1(t) \\ w_2(t) \end{pmatrix}, \Sigma = \begin{pmatrix} \bar{\sigma}_1 & \bar{\sigma}_{12} \\ \bar{\sigma}_{12} & \bar{\sigma}_2 \end{pmatrix},$$

$$A = \begin{pmatrix} -K^{-1}(2\bar{x}+\bar{y})-\beta_1 & \alpha_1 - K^{-1}\bar{x} \\ -K^{-1}\bar{y} & \alpha_2 - \beta_2 - K^{-1}(\bar{x}+2\bar{y}) \end{pmatrix}$$

and coefficients $\bar{\sigma}_1, \bar{\sigma}_2, \bar{\sigma}_{12}$ are defined as the positive solution of the algebraic equation:

$$\bar{\sigma}_1^2 + \bar{\sigma}_{12}^2 = \sigma_1^2 \bar{x}^2,\ \bar{\sigma}_1 + \bar{\sigma}_2\ \ \bar{\sigma}_{12} = \sigma_{12}^2 \overline{xy}, \bar{\sigma}_2^2 + \bar{\sigma}_{12}^2 = \sigma_2^2 \bar{y}^2 .$$

The solution of the equation (3.8) with initial condition $\vec{z}(t_0) = \vec{z}_0$ is the two dimensional Gaussian vector process given by formula:

$$\vec{z}(t, t_0, z_0) = \exp\ (t - t_0)A\ \vec{z}_0 + \int_{t_0}^{t} \exp\ (t-s)A\ \Sigma d\vec{w}(s)\ . \tag{3.9}$$

Using the solutions (3.9) with initial values $t_0 = 0, \vec{z}_0 = \begin{pmatrix} 0 \\ 0 \end{pmatrix}$ we can approximate the solution $\vec{X}_\varepsilon(t) = \begin{pmatrix} x_\varepsilon(t) \\ y_\varepsilon(t) \end{pmatrix}$ of the impulsive differential equation (1.1) - (1.2) as follows:

$$\vec{X}_\varepsilon(t) \approx \vec{\bar{X}}(t) + \sqrt{\varepsilon} \int_0^t \exp\ (t-s)A\ \Sigma d\vec{w}(s), \tag{3.10}$$

where $\vec{\bar{X}}(t)$ is the vector-solution of the equation (2.2) with initial condition $\vec{\bar{X}}(0) = \vec{X}_\varepsilon(0)$.

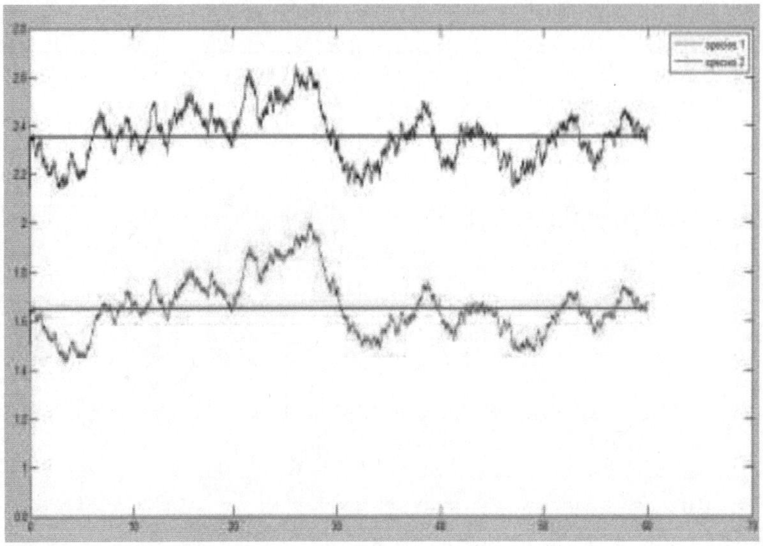

Fig. 2. The solution of equation (2.2) with initial condition (2.3)
and the sample trajectories for the Gaussian approximation.

As has been mentioned in the second section, if $\alpha_2 > \beta_2 + \gamma_2$ then the eigenvalues of the matrix A are negative. Therefore the stochastic integral in (3.9) converges as $t_0 \to -\infty$ with probability one and [14] there exists satisfying to (3.8) stationary stochastic process

$$\lim_{t_0 \to -\infty} \vec{z}(t,t_0,z_0) = \vec{Z}_0(t) = \begin{pmatrix} \hat{z}(t) \\ \hat{u}(t) \end{pmatrix} = \int_{-\infty}^{t} \exp\ (t-s)A\ \Sigma d\vec{w}(s) \tag{3.11}$$

with zero mean and the symmetric constant covariance matrix $\hat{Q}$.

4 Conclusions

Remember that by definition the process $\vec{Z}_0(t)$ is the solution of the stochastic differential equation (3.8) with initial conditions $\vec{z}_0(0) = \int_{-\infty}^{0} \exp\ -sA\ \Sigma d\vec{w}(s)$. Therefore $\vec{V}(t) = \vec{Z}_0(t) - \vec{z}(t,t_0,z_0)$ is deterministic vector-function that satisfies a homogeneous ordinary differential equation in $\mathbf{R}^2$ and may be given as the matrix exponent $\vec{V}(t) = \hat{Z}(0)\exp\{At\}$, where A is a matrix with negative eigenvalues. This means that under assumption $\alpha_2 > \beta_2 + \gamma_2$ the population size stabilizes at the $\sqrt{\varepsilon}$ – neighbourhood of the defined by formula (2.3) equilibrium point $\vec{x} = \begin{pmatrix} \overline{x} \\ \overline{y} \end{pmatrix}$ and for sufficiently large $t > 0$ the random variable $\{x_\varepsilon(t), y_\varepsilon(t)\}$ may be analyzed as two dimensional normal distributed random variable with mean $\vec{x}$ and covariance matrix $\varepsilon\hat{Q}$.

References

[1] ELLNER, S.P., GUCKENHEIMER, J., *Dynamic Models in Biology.* Princeton University Press, Princeton NJ, 2006.

[2] KENDALLD, G., *Stochastic processes and population growth.* Journal of the Royal Statistical Society. Series B(Methodological) 11, 1949, 230-264.

[3] HADELER, K.P., R.WALDSTËATTER, R. and WËORZ-BUSEKROS, A., *Models for pair formation in bisexual populations.* Journal of Mathematical Biology 26, 1988, 635-649.

[4] GARIBALDI, E., SOBOTKA, M., *A nonsmooth two-sex population model.* Mathematical Biosciences. 253, 2014, 1-10.

[5] A.M. de ROOS and PERSSON, L, *Population and Community Ecology of Ontogenetic Development.* Monographs in Population Biology. 51. Princeton University Press. Princeton. 2013.

[6] FISHER, R.A., *The Genetical Theory of Natural Selection.* 2nd ed. New York: Dover Publications,1958.

[7] ENNETT, J.H., (ed.) *The Genetical Theory of Natural Selection.* Variorum Edn. Oxford: Oxford University Press, 1999.

[8] CASTILLO-CHAVEZ, C. and HUANG, W., *The logistic equation revisited: the two-sex case.* Mathematical Biosciences. 128, 1995, 299-316.

[9] HANWU LIU. *Population dynamics of different sex with different birth and death rate.* Chinese journal of Ecology. 22, 2003, 63-65.

[10] BAZYKIN, A.D., *Nonlinear Dynamics of Interacting Populations.* World Scientific. Singapore. 1998.

[11] MURRAY, J.D., *Mathematical Biology I: An Introduction.* vol. I. Springer-Verlag. 3rd edition. 2003.

[12] YUEJIAN JIE, YUAN YUAN, *Model Stability Analysis of Marine Ecosystem. International journal of biology.* Vol. 1, No. 2. 2009. 22 - 27.

[13] DYNKIN, YE.B., *Theory of Markov Processes.* Prentice-Hall. *1961.*

[14] SKOROKHOD, A.V., *Asymptotic methods in the theory of stochastic differential equations.* (Transl. Math. Monographs). 78. American Mathematical Society. Providence. RI. 1989.

[15] TSARKOV, YE. (J.CARKOVS), *Asymptotic methods for stability analysis Markov impulse dynamical systems.* Nonlinear dynamics and system theory 9 (2002). no. 2(1). 103–115.

[16] SAMOILENKO, A.M. and PERESTYUK, N.A., *Impulsive Differential Equations.* Vol.14 of World Scientific Series on Nonlinear Science. Series A: Monographs and Treatises. World Scientific. Singapore. 1995.

Current address

Carkovs Jevgeņijs, Professor

Department of Probability Theory and Mathematical Statistics
Faculty of Computer Science and Information Technology, Riga Technical University
1 Kaļķu Street, Riga, Latvia, LV-1658
E-mail: carkovs@latnet.lv

Matvejevs Andrejs, Dr.sc.ing., professor
Department of Probability Theory and Mathematical Statistics
Faculty of Computer Science and Information Technology, Riga Technical University
1 Kaļķu Street, Riga, Latvia LV-1658,
E-mail: Andrejs.Matvejevs@rtu.lv

Slovak University of Technology in Bratislava
Faculty of Mechanical Engineering
17th Conference on Applied Mathematics **APLIMAT 2018**

Proceedings

ON INTERSPECIFIC COMPETITION OF LOGISTIC POPULATIONS WITH THE ASSUMPTION OF RAPID RANDOM CONTACTS

CARKOVS Jevgeņijs (LV), PAVLENKO Oksana (LV)

Abstract: We analyze a logistic type Markov impulsive differential model for interspecific competition of two populations under assumption that they come into contact at random time moments only but between the contacts they grow independently. Assuming the intervals be sufficiently small we apply the stochastic averaging procedure and construct an ordinary 2-dimensional differential equation for population dynamics in the mean and a linear 2-dimensional stochastic differential equation for deviations on the mean trajectories.

Keywords: interspecific competition; random contacts; stochastic approximation

Mathematics Subject Classification: Primary 92B05,60H30; Secondary 60G57.

1 Introduction

The famous mathematical model for competition of two populations for the first time appears in the papers [1,2]. Using the notations $x(t)$ for the mass density of the first species and $y(t)$ for the mass density of the second species at time moment t the mentioned authors assume that the dynamics of populations may be studied as the system of ordinary differential equations:

$$\frac{dx}{dt} = r_1 x(K_1 - x - \alpha y)K_1^{-1}, \frac{dy}{dt} = r_2 y(K_2 - y - \beta x)K_2^{-1} \tag{1}$$

where $r_1 > 0$ and $r_2 > 0$ are the birth rates in absence of intraspecific and interspecific competition, $K_1 > 0$ and $K_2 > 0$ are the asymptotic masses of the first and of the second species when they grow separately, and the coefficients $\alpha > 0, \beta > 0$ show the influences of one species to another, that defines decreasing opportunity of growth. Later this model was analyzed in detail by many authors (see, for example, [3,4,5,6,7] and references there). The phase portrait of the dynamical system (1) has a very simple structure. There are only four stationary states

$$A_1:\bar{x}_1=0,\bar{y}_1=0;\ A_2:\bar{x}_2=0,\bar{y}_2=K_2;\ A_3:\bar{x}_3=K_1,\bar{y}_3=0;A_4:x_4=\frac{K_1-\alpha K_2}{1-\alpha\beta},y_4=\frac{K_2-\beta K_1}{1-\beta\alpha} \quad (2)$$

These points define the following possibilities for the longtime behavior of the populations: A_1 - both populations become extinct; A_2 - the *x*-population converges to a certain value, bet *y*-population disappears; A_3 - the *y*-population converges to a certain value, bet *x*-population disappears; A_4 - both population co-exist. As it has been shown in [8] the defined by the equations (1) dynamics of populations for any initial values $x(0)>0, y(0)>0$ may be described in the following way:

1. the point A_1 is unstable, that is, for a long time exists at least one of the above populations;
2. if $r_1K_1>\alpha K_2, r_2K_2<\beta K_1$ then the point A_2 is asymptotically stable, that is, $\lim_{t\to\infty}x(t)=K_1, \lim_{t\to\infty}y(t)=0$;
3. if $r_1K_1<\alpha K_2, r_2K_2>\beta K_1$ then the point A_3 is asymptotically stable, that is, $\lim_{t\to\infty}x(t)=0, \lim_{t\to\infty}y(t)=K_2$;
4. if $r_1K_1>\alpha K_2, r_2K_2>\beta K_1$ then the point A_4 is asymptotically stable, that is, $\lim_{t\to\infty}x(t)=\frac{K_1-\alpha K_2}{1-\alpha\beta}, \lim_{t\to\infty}y(t)=\frac{K_2-\beta K_1}{1-\beta\alpha}$;
5. if $r_1K_1<\alpha K_2, r_2K_2<\beta K_1$ then subject to initial conditions $\lim_{t\to\infty}x(t)=K_1, \lim_{t\to\infty}y(t)=0$ or $\lim_{t\to\infty}x(t)=0, \lim_{t\to\infty}y(t)=K_2$.

Naturally, in reality the above analyzed model (1) gives just a simplified view on the population dynamics. The deterministic approach to analysis of ecosystems for the first time was questioned by Feller [4]. Analyzing the logistic model for the real logistic type population dynamics he found that population trajectories could equally well be fitted by scaled normal distribution curves with time-dependent argument. Therefore, later the deterministic model (1) was converted by many authors (see, for example, [4] and references there) into the Ito stochastic differential equations. In our paper we also analyze the stochastic modification of the two-species competition model, using as basis the following assumptions. Note that the model (1) supposes a time-continuous contact of populations. This means that within a very small time interval $[t,t+\Delta)$ the impact of population *y* on population *x* may be given by term $\alpha r_1K_1^{-1}x(t)y(t)\Delta$ and the impact of population *x* on population *y* by term $\beta r_2K_2^{-1}x(t)y(t)\Delta$. Our model supposes that the populations come in contact at the random time moments $\tau_1<\tau_2<\cdots<\tau_n<\cdots$ and time-intervals $\tau_n-\tau_{n-1}$ are independent exponentially distributed random variables, $\mathbf{E}\{\tau_n-\tau_{n-1}\}=\varepsilon$, where ε is small positive parameter. The impacts of the populations against each other at the time moment τ_n also are sufficiently small and random: the impact of the population *y* on population *x* is given by term $\varepsilon a(\xi_n)r_1K_1^{-1}x(\tau_n-)y(\tau_n-)$ and the impact of the population *x* on population *y* by term $\varepsilon b(\xi_n)r_2K_2^{-1}x(\tau_n-)y(\tau_n-)\varepsilon$, where $\{\xi_n, n\in\mathbf{N}\}$ is the independent on $\{\tau_n, n\in\mathbf{N}\}$ sequence of identically Uniform(0,1)-distributed random variables. Specify that here and later we use a notation $f(t-)$ for the left limit, that is, $f(t-) := \lim_{\tau \uparrow t} f(\tau)$. The formal definition of the above-mentioned model will be proposed in the next section.

2 The model

As it has been mentioned in the previous section our paper deals with interspecific competition model for populations $\{x_\varepsilon(t), t \geq 0\} \subset \mathbf{R}$ and $\{y_\varepsilon(t), t \geq 0\} \subset \mathbf{R}$ under assumption that the populations come in contacts only at random time moments given by the increasing sequence $\mathrm{T} = \{\tau_k, k \in \mathbf{N}\}$. Index $\varepsilon \in (0, \varepsilon_0)$ is a small positive parameter, which we will use for asymptotic analysis of population dynamics. The sequence T consists of the switching time moments of the Markov process $\{\xi(t), t \geq 0\}$, which is defined on the probability space $(\Omega, \mathfrak{F}, \mathbf{P})$ by the infinitesimal operator [8]

$$(Qv)(\xi) = \varepsilon^{-1} \int_0^1 (v(z) - v)(\xi)) dz \tag{3}$$

where $\{v(z), z \in [0,1]\}$ is an arbitrary real continuous function. We assume that in absence of contacts the populations evolve independently in compliance with the logistic law:

$$\frac{dx_\varepsilon(t)}{dt} = r_1 x_\varepsilon(t)(K_1 - x_\varepsilon(t))K_1^{-1}, \frac{dy_\varepsilon(t)}{dt} = r_2 y_\varepsilon(t)(K_2 - y_\varepsilon(t))K_2^{-1} \tag{4}$$

and at moments of contacts they have the jumps given by the equations:

$$x_\varepsilon(t) - x_\varepsilon(t-) = -\varepsilon K_1^{-1} r_1 a(\xi(t-)) x_\varepsilon(t-) y_\varepsilon(t-), y_\varepsilon(t) - y_\varepsilon(t-) = -\varepsilon K_2^{-1} r_2 b(\xi(t-)) x_\varepsilon(t-) y_\varepsilon(t-) \tag{5}$$

where $a(z), b(z)$ are real continuous functions on segment [0,1]. Let $\{\mathcal{F}^t, t \geq 0\}$ be the minimal filtration [8] for the Markov process $\{\xi(t), t \geq 0\}$. By definition the two-dimensional random process $\{x_\varepsilon(t), y_\varepsilon(t), t \geq 0\}$ is $\mathcal{F}^t$-adopted and possesses the Markov property [8]. Besides, the dynamical characteristics given by equations (4)-(5) are homogeneous in time. Therefore, the process $\{x_\varepsilon(t), y_\varepsilon(t), t \geq 0\}$ may be uniquely defined by the infinitesimal generator [8], which one can find calculated the limit

$$(L(\varepsilon)v)(x, y) := \lim_{t \downarrow 0} \frac{1}{t} [\mathbf{E}\{v(x_\varepsilon(t), y_\varepsilon(t)) / x_\varepsilon(0) = x, y_\varepsilon(0) = y\} - v(x, y)] \tag{6}$$

for an arbitrary sufficiently smooth bounded function $v(x, y)$. Remind that by definition [8] of the Poisson process with infinitesimal generator (3) the time-intervals $\{\Delta_k = \tau_k - \tau_{k-1}, k \in \mathbf{N}, \tau_0 = 0\}$ are independent, do not dependent on the sequence $\{\xi(\tau_k), k \in \mathbf{N}\}$ and have an exponential distribution with parameter ε^{-1}. Not so difficult to calculate the limit (6):

$$(L(\varepsilon)v)(x,y) = v'_x(x,y)\left[rx(K_1 - x)\right]K_1^{-1} + v'_y(x,y)\left[ry(K_2 - y)\right]K_2^{-1}y +$$
$$+\varepsilon^{-1}\left[\int_0^1 \mathbf{E}\{v(x - \varepsilon r_1 a(\xi)xy, y - \varepsilon r_2 b(\xi)xy)\}d\xi - v(x,y)\right] \tag{7}$$

3 Stochastic asymptotic analysis of population dynamics

The assumption that the populations have rapid contacts implies the smallness of time intervals Δ_k that have the exponential distribution with parameter ε. This parameter we will use for approximation of trajectories $\{x_\varepsilon(t), y_\varepsilon(t), t \ge 0\}$ applying the stochastic approximation procedure proposed in [5,6,7]. The first step of this procedure is a construction of the ordinary differential equations for the mean values $\bar{x}(t) := \mathbf{E}\{x_\varepsilon(t)\}$ and $\bar{y}(t) := \mathbf{E}\{y_\varepsilon(t)\}$ of populations species. For that we should pass to limit in (7) rushing ε to zero: $\lim_{\varepsilon\to 0}(L(\varepsilon)v)(x,y) = v'_x(x,y)xr_1\left[(1 - K_1^{-1}x) + \alpha y\right] + v'_y(x,y)yr_2\left[y(1 - K_2^{-1}y) + \beta x\right]$

where $\alpha = \mathbf{E}\{a(\xi)\} = \int_0^1 a(\xi)d\xi, \beta = \mathbf{E}\{b(\xi)\} = \int_0^1 b(\xi)d\xi$. This means that the mean values of the populations species satisfy the same equations as (1):

$$\frac{d\bar{x}}{dt} = r_1\bar{x}(K_1 - \bar{x} - \alpha\bar{y})K_1^{-1}, \frac{d\bar{y}}{dt} = r_2\bar{y}(K_2 - \bar{y} - \beta\bar{x})K_2^{-1} \tag{8}$$

Not so difficult to ensure that A_1, A_2 and A_3 are equilibrium points for the dynamical system (4)-(5) and the system (8). As it has been proved in [6] for sufficiently small $\varepsilon > 0$ the dynamical systems (4)-(5) and (8) have the same local stability properties of the above equilibriums. The sample trajectories of the solutions of (4)-(5) jointly with solutions of equations (8) (magnitudes of parameters $r_1 = 1,\ r_2 = 1,\ K_1 = 40,\ \varepsilon = 0.05, K_2 = 50$) are shown below on the Fig. 1. As we can see the solutions of the system (4)-(5) preserve behavior pattern of the corresponding solutions of the equation (8). The equilibrium point A_5 of the system (8) is not an equilibrium point for the dynamical system (4)-(5). If $r_1K_1 > \alpha K_2, r_2K_2 > \beta K_1$, then as we can see at the Fig.1, the sample trajectories of the dynamical system (4)-(5) have relatively large random deviations $\{x_\varepsilon(t) - \bar{x}(t), y_\varepsilon(t) - \bar{y}(t)\}$ from the corresponding trajectories of the average equations (8). To estimate the above mentioned deviations we need to do the second step of the diffusion approximation procedure.

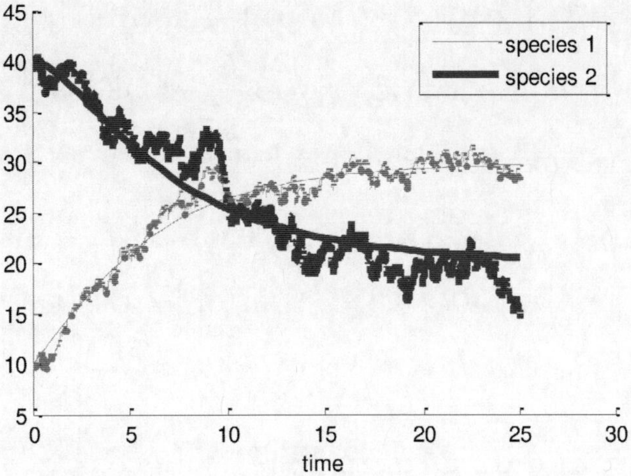

Fig. 1. $r_1K_1 < \alpha K_2, r_2K_2 < \beta K_1, x(0) = 10, y(0) = 40, \alpha = 0.5, \beta = 1$

As it has been shown in [6] for this aim we have to introduce the two-dimensional random processes of *the normalized deviations:*

$$z_\varepsilon(t) := \frac{x_\varepsilon(t) - \overline{x}(t)}{\sqrt{\varepsilon}}, u_\varepsilon(t) := \frac{y_\varepsilon(t) - \overline{y}(t)}{\sqrt{\varepsilon}} \tag{9}$$

which by definition satisfies differential equations

$$\begin{cases} \dfrac{dz_\varepsilon(t)}{dt} = \dfrac{r_1K_1^{-1}}{\sqrt{\varepsilon}}\left[\alpha\overline{x}(t)\overline{y}(t) + \sqrt{\varepsilon}z_\varepsilon(t)(K_1 - 2\overline{x}(t)) - \varepsilon\left(z_\varepsilon(t)\right)^2\right], \\ \dfrac{dy_\varepsilon(t)}{dt} = \dfrac{r_2K_2^{-1}}{\sqrt{\varepsilon}}\left[\beta\overline{x}(t)\overline{y}(t) + \sqrt{\varepsilon}u_\varepsilon(t)(K_2 - 2\overline{y}(t)) - \varepsilon\left(u_\varepsilon(t)\right)^2\right] \end{cases} \tag{10}$$

for $t \notin \mathrm{T}$ and have the jumps

$$\begin{cases} z_\varepsilon(t) - z_\varepsilon(t-) = -\sqrt{\varepsilon}K_1^{-1}r_1a(\xi(t-))\left[\overline{x}(t)\overline{y}(t) + \sqrt{\varepsilon}\left(z_\varepsilon(t-))\overline{y}(t) + \overline{x}(t)u_\varepsilon(t-)\right) + \varepsilon z_\varepsilon(t-)u_\varepsilon(t-)\right], \\ u_\varepsilon(t) - u_\varepsilon(t-) = -\sqrt{\varepsilon}K_2^{-1}r_2b(\xi(t-))\left[\overline{x}(t)\overline{y}(t) + \sqrt{\varepsilon}(u_\varepsilon(t-))\overline{x}(t) + z_\varepsilon(t-)\overline{y}(t)) + \varepsilon z_\varepsilon(t-)u_\varepsilon(t-)\right] \end{cases} \tag{11}$$

at time moments $t \in \mathrm{T}$. $t \in \mathrm{T}$. Like in previous section we can prove that defined by equations (8)-(10)-(11) four-dimensional process $\{\overline{x}(t), \overline{y}(t), z_\varepsilon(t), u_\varepsilon(t), t \geq 0\}$ on the probability space $(\Omega, \digamma, \digamma^t, \mathbf{P}, t \geq 0)$ is homogeneous and possess the Markov property. To avoid cumbersome formulas, we will use the following notation $\mathbf{E}_{xy}^{zu}\{\bullet\} = \mathbf{E}\{\bullet / \overline{x}(0) = x, \overline{y}(0) = y, z_\varepsilon(0) = z, u_\varepsilon(0) = u\}$
for conditional expectation. The weak infinitesimal operator of this Markov process $\{\overline{x}(t), \overline{y}(t), z_\varepsilon(t), u_\varepsilon(t), t \geq 0\}$ should be calculated by formula:

$$(\mathcal{L}(\varepsilon)v)(x,y,z,u)=\lim_{t\downarrow 0}\frac{1}{t}\left[\mathbf{E}_{xy}^{zu}\{v(\bar{x}(t),\bar{y}(t),z_\varepsilon(t),u_\varepsilon(t))\}-v(x,y,z,u)\right]=$$
$$=v'_x(x,y,z,u)r_1x(K_1-x-\alpha y)K_1^{-1}+v'_y(x,y,z,u)r_2y(K_2-y-\beta x)K_2^{-1}+$$
$$+\frac{1}{\sqrt{\varepsilon}}[v'_z(x,y,z,u)r_1K_1^{-1}\alpha xy+\beta xyv'_u(x,y,z,u)r_2K_2^{-1}\beta xy]+$$
$$+v'_z(x,y,z,u)r_1K_1^{-1}z(K_1-2x)+v'_u(x,y,z,u)r_2K_2^{-1}u(K_2-2y)-$$
$$-\sqrt{\varepsilon}[v'_z(x,y,z,u)r_1K_1^{-1}z^2+v'_u(x,y,z,u)r_2K_2^{-1}u^2]+$$
$$+\varepsilon^{-1}\mathbf{E}\left\{v\left(x,y,z-\sqrt{\varepsilon}K_1^{-1}r_1a(\xi)\left[xy+\sqrt{\varepsilon}\left(zy+xu\right)+\varepsilon zu\right],u-\sqrt{\varepsilon}K_2^{-1}r_2b(\xi)\left[xy+\sqrt{\varepsilon}(ux+zy)+\varepsilon zu\right]\right)\right\}-$$
$$-\varepsilon^{-1}v(x,y,z,u)$$

According to stochastic approximation procedure [5] we should pass to limit in the previous formula for $\varepsilon\to 0$, assumed that $v(x,y,z,u)$ is sufficiently smooth function:

$$\begin{aligned}(\hat{\mathcal{L}}v)(x,y,z,u)&:=\lim_{\varepsilon\to 0}(\mathcal{L}(\varepsilon)v)(x,y,z,u)=\\&=v'_x(x,y,z,u)r_1x(K_1-x-\alpha y)K_1^{-1}+v'_y(x,y,z,u)r_2y(K_2-y-\beta x)K_2^{-1}+\\&+v'_z(x,y,z,u)r_1K_1^{-1}[z(K_1-2x)-\alpha\left(zy+xu\right)]+v'_u(x,y,z,u)r_2K_2^{-1}[u(K_2-2y)-\beta\left(zy+xu\right)]+\\&+\frac{1}{2}(xy)^2\left[K_1^{-2}r_1^2\sigma_1^2v''_{zz}(x,y,z,u)+K_2^{-2}r_2^2\sigma_2^2v''_{uu}(x,y,z,u)+2K_1^{-1}K_2^{-1}r_1r_2\sigma_{12}v''_{uz}(x,y,z,u)\right]\end{aligned}\tag{12}$$

where $\sigma_1^2=\mathbf{E}\{a^2(\xi)\},\sigma_2^2=\mathbf{E}\{b^2(\xi)\},\sigma_{12}=\mathbf{E}\{a(\xi)b(\xi)\}$. The defined by this formula operator $\hat{\mathcal{L}}$ may be interpreted [5,7] as the weak infinitesimal operator of the Markov process $\{\bar{x}(t),\bar{y}(t),\hat{z}(t),\hat{u}(t),t\geq 0\}$. This infinitesimal operator may be used to define on the probability space $(\Omega,\mathfrak{F},\mathbf{P})$ the systems of Ito stochastic linear inhomogeneous differential equations with dependent on $\{\bar{x}(t),\bar{y}(t)\}$ drift and diffusion:

$$\begin{cases}d\hat{z}=r_1K_1^{-1}[\hat{z}(K_1-2\bar{x}(t))-\alpha\left(\hat{z}\bar{y}(t)+\bar{x}(t)\hat{u}\right)]dt+\frac{1}{\sqrt{2}}\bar{x}(t)\bar{y}(t)\left[K_1^{-1}r_1\sigma_1dw_1(t)+\sqrt{2K_1^{-1}K_2^{-1}r_1r_2\sigma_{12}}dw_2(t)\right],\\d\hat{u}=r_2K_2^{-1}[\hat{u}(K_2-2\bar{y}(t))-\beta\left(\hat{z}\bar{y}(t)+\bar{x}(t)\hat{u}\right)]dt+\frac{1}{\sqrt{2}}\bar{x}(t)\bar{y}(t)\left[K_2^{-1}r_2\sigma_2dw_1(t)+\sqrt{2K_1^{-1}K_2^{-1}r_1r_2\sigma_{12}}dw_2(t)\right]\end{cases}\tag{13}$$

where $w_1(t)$ and $w_2(t)$ are standard independent Wiener processes and $\{\bar{x}(t),\bar{y}(t),t\geq 0\}$ is the solution of the equation system (8) with initial condition $\bar{x}(0)=x_\varepsilon(0)$, $\bar{y}(0)=y_\varepsilon(0)$ and $\{\hat{z}(t),\hat{u}(t),t\geq 0\}$ be the solution of the system (13). The two-dimensional process $\{X_\varepsilon(t):=\bar{x}(t)+\hat{z}(t)\sqrt{\varepsilon},\mathrm{Y}_\varepsilon(t)=:\bar{y}(t)+\hat{u}(t)\sqrt{\varepsilon},t\geq 0\}$ where $\{\hat{z}(t),\hat{u}(t)\}$ is the solution of the equations (13) with the trivial initial conditions $\hat{z}(0)=\hat{u}(0)=0$ is called *the Gaussian approximation* of the process $\{x_\varepsilon(t),y_\varepsilon(t),t\geq 0\}$.

Therefore $\mathrm{E}\{\hat{z}(t)\}\equiv 0$ and $\mathrm{E}\{\hat{u}(t)\}\equiv 0$. Not so difficult to ensure that if $\lim_{t\to\infty}\bar{x}(t)\bar{y}(t)=0$ (the cases 1,2,3 and 5, section 1, the second page of this article) then $\lim_{t\to\infty}\mathrm{E}\{\hat{z}^2(t)\}=0$ and $\lim_{t\to\infty}\mathrm{E}\{\hat{u}(t)\}=0$. This means if $\lim_{t\to\infty}\bar{x}(t)\bar{y}(t)=0$ then the Gaussian approximation process has the same behavior as the corresponding solutions of the initial impulsive differential equations (4)-(5). But if $r_1K_1 > \alpha K_2, r_2K_2 > \beta K_1$ then $\lim_{t\to\infty} \bar{x}(t) = \frac{K_1 - \alpha K_2}{1-\alpha\beta} := \bar{\bar{x}}$ and $\lim_{t\to\infty} \bar{y}(t) = \frac{K_2 - \beta K_1}{1-\beta\alpha} := \bar{\bar{y}}$, that is, the drift and diffusion coefficients of the equation (12) stabilize. The corresponding to this limit stochastic differential equation for the vector $\vec{z}(t) = \begin{pmatrix} \hat{z}(t) \\ \hat{u}(t) \end{pmatrix}$ may be given as follows:

$$d\vec{z}(t) = A\vec{z}(t)dt + \Sigma d\vec{w}(t) \tag{14}$$

where $\vec{z}(t) = \begin{pmatrix} \hat{z}(t) \\ \hat{u}(t) \end{pmatrix}, \vec{w}(t) = \begin{pmatrix} w_1(t) \\ w_2(t) \end{pmatrix}$, $\Sigma = \frac{1}{\sqrt{2}}\overline{\overline{xy}}\begin{pmatrix} K_1^{-1}r_1\sigma_1 & \sqrt{2K_1^{-1}K_2^{-1}r_1r_2\sigma_{12}} \\ K_2^{-1}r_2\sigma_2 & \sqrt{2K_1^{-1}K_2^{-1}r_1r_2\sigma_{12}} \end{pmatrix}$,

$$A = \begin{pmatrix} r_1K_1^{-1}(K_1 - 2\bar{\bar{x}} - \alpha\bar{\bar{y}}) & -r_1K_1^{-1}\alpha\bar{\bar{x}} \\ -\beta r_2K_2^{-1}\bar{\bar{y}} & r_2K_2^{-1}(K_2 - 2\bar{\bar{y}} - \beta\bar{\bar{x}}) \end{pmatrix} \tag{15}$$

Applying the Cauchy formula [5] we can find a vector solution of the equation (14) with initial condition $\vec{Z}(t_0) = \vec{Z}_0$ as follows:

$$\vec{Z}(t) = \exp\{(t-t_0)A\}\vec{Z}_0 + \int_{t_0}^{t} \exp\{(t-s)A\}\Sigma d\vec{w}(s) \tag{16}$$

Recall that we should apply the formula (16) with $t_0 = 0$ and $\vec{Z}(0) = \vec{0}$. The sample trajectories for the coordinates $\{X_\varepsilon(t), Y_\varepsilon(t)\}$ of corresponding to magnitudes of parameters $r_1 = 1$, $r_2 = 1$, $K_1 = 40$, ε = 0.05, $K_2 = 50, \alpha = 0.5$, $\beta = 1$, $x(0)$ = 10, $y(0)$ = 40 Gaussian approximation are shown below on the Fig. 2.

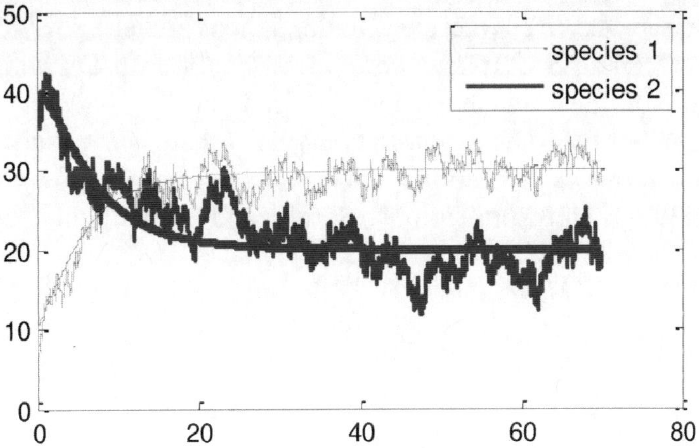

Fig.2. The sample trajectories of the Gaussian approximation.

Under assumptions $r_1K_1 > \alpha K_2, r_2K_2 > \beta K_1$ the eigenvalues of the matrix A are negative. Therefore, with probability one there exists a limit [5]:

$$\lim_{t_0 \to -\infty} \vec{Z}(t) = \tilde{Z}(t) := \begin{pmatrix} \tilde{z}(t) \\ \tilde{u}(t) \end{pmatrix} = \int_{-\infty}^{t} \exp\{(t-s)A\} \Sigma d\vec{w}(s) \tag{17}$$

Not so difficult to ensure that this integral defines the two-dimensional Gaussian process with zero mean and constant covariance matrix $\hat{Q} = \int_0^{\infty} \exp\{tA\} \Sigma^2 dt$. This means that for sufficiently large *t* >*0* the probability distribution function for the coordinates of process (17) may be approximated as a two-dimensional Gaussian distribution
$P(X_\varepsilon(t) \le X, Y_\varepsilon(t) \le Y) \approx P(\bar{x}(t) + \sqrt{\varepsilon}\xi \le X, \bar{y}(t) + \sqrt{\varepsilon}\eta \le Y)$ where random vector $\{\xi, \eta\}$ has the zero mean and covariance matrix $\hat{Q}$ defined by equation $A\hat{Q} + \hat{Q}A^T = -\Sigma^2$.

4 Conclusion

Applying the asymptotic methods of the contemporary stochastic analysis [5,6,7] we have constructed the average differential equations for the described in Section 2 Markov impulsive differential model for interspecific competition of two populations in a form (1) and the stochastic differential equations for the Gaussian deviations of the random trajectories on the corresponding solutions of the average equations.

References

[1] LOTKA, A. J. *Elements of Physical Biology.* Baltimore, Williams and Wilkins, 1925

[2] VOLTERRA, V. *Lefons sur la thiorie mathematique de la lutte pour la vie.* Paris, 1931.

[3] BAZYKIN, A.D. *Nonlinear Dynamics of Interacting Populations*. World Scientific, Singapore, 1998.

[4] FELLER, W. *On the Logistic Law of Growth and its Empirical Verifications in Biology.* Acta Biotheor. 5, 1939, pp. 51-66.

[5] SKOROKHOD, A.V. *Asymptotic methods in the theory of stochastic differential equations* (Transl. Math. Monographs **78**). American Mathematical Society, Providence, RI, 1989

[6] TSARKOV, Ye. (CARKOVS, J.) *Asymptotic methods for stability analysis Markov impulse dynamical systems.* Nonlinear dynamics and system theory 9 (2002), no. 2(1), pp. 103–115.

[7] TSARKOV, Ye. (CARKOVS, J.) *Asymptotic methods for stability analysis of Markov impulse dynamical systems*. Advances of Stability Theory of the End of XXth Century. Stability and Control: Theory, Methods and Applications. Gordon and Breach Science Publishers, London, **13** (2000).- pp. 251-264.

[8] DYNKIN, Ye. B. *Theory of Markov Processes.* Prentice-Hall, 1961.

Current address

Carkovs Jevgeņijs, Professor
Dep. of Prob. Theory and Math. Statistics
Faculty of Computer Science and Information Technology, Riga Technical University
1 Kaļķu Str., LV-1658, Latvia
E-mail: carkovs@latnet.lv

Pavlenko Oksana, Dr.math.
Dep. of Prob. Theory and Math. Statistics
Faculty of Computer Science and Information Technology, Riga Technical University
1 Kaļķu Str., LV-1658, Latvia
E-mail: Oksana.Pavlenko@rtu.lv

Slovak University of Technology in Bratislava
Faculty of Mechanical Engineering
17th Conference on Applied Mathematics **APLIMAT 2018**

Proceedings

STOCHASTIC APPROXIMATION PROCEDURE FOR ISOLATED POPULATION IN FAST OSCILLATED RANDOM ENVIRONMENT

CARKOVS Jevgeņijs (LV), ŠADURSKIS Kārlis (LV), GOLDŠTEINE Jolanta (LV)

Abstract. The paper deals with a logistic type population mathematical model given in a form of Markov impulsive differential equation controlled by a fast oscillating ergodic Poisson process. Applying the stochastic approximation procedure, we construct and analyse ordinary differential equations for the mean value of population size and stochastic Ito differential equation for random deviations on the mean values.

Key words: stochastic modelling, logistic equations, stochastic approximation

Mathematics Subject Classification: Primary 92B05,60H30,08A72; Secondary 60G57, 37H10

1 Introduction

In classical mathematical models of population dynamics, as a rule, an assumption is made that external conditions are constant. Therefore, the logistic model [1] is the most popular, for example, for describing the dynamics of an isolated population. It assumes that for any time moment $t \geq 0$ for a sufficiently small time interval Δ the increment $\Delta x(t) = x(t+\Delta) - x(t)$ of the biomass $x(t)$ of the population can be represented in the form of an asymptotic equality:

$$\Delta x(t) = rx(t)(1 - K^{-1}x(t))\Delta - mx(t)\Delta + o(\Delta)\,, \tag{1.1}$$

where K is the capacity of the habitat, $r(1-K^{-1}x)$ is fertility, m is the death rate. In this case, we can divide both sides of (1.1) by Δ, approach to the limit as $\Delta \to 0$ and use the ordinary differential equation

$$\frac{dx}{dt} = rx(1 - K^{-1}x) - mx \tag{1.2}$$

as a mathematical model.

The main advantage of this model is the ability to solve the equation explicitly and to make sure that the behaviour of the population depends only on the sign of the difference $r-m$. If this difference is negative, then the population is bound to become extinct, if it is positive, then the population size increases and monotonically approaches the value $Kr^{-1}(r-m)$. Unfortunately, this forecast does not always come true: in the real population environment there are random irregular perturbations, and the fertility $r(1-K^{-1}x)$ and mortality mx in the formula (1.1) actually are regression lines. Therefore, one cannot rely on the conclusions made only based on the analysis of the parameters r and m of these regression lines. Under the influence of random perturbations the population may leave the neighbourhood of a stable equilibrium $Kr^{-1}(r-m)$ and may enter the neighbourhood of zero and, in fact, disappear. This leads to the necessity for specifying increments in (1.1) in the form of the time-dependent random functions dependent on the phase coordinate x and the time interval Δ . Moreover, it is desirable to randomize the model retaining the predictability property of mathematical model of population dynamics. Therefore, papers have been published (see, for example, [3- 8] and references there) in which increments of a logistic type population are given in the form

$$\Delta x(t) = rx(t)(1-K^{-1}x(t))\Delta - mx(t)\Delta + x(t)B(x(t))(w(t+\Delta)-w(t)) \tag{1.3}$$

where $B(x)=b+b_1x$, and $w(t)$ is the standard Brownian motion process, that is, a process with independent normally $N(0,\Delta)$ distributed increments $w(t+\Delta)-w(t)$. The formula for increments (1.3) allows us to propose a stochastic differential equation as a randomized logistic model

$$dx(t) = rx(t)(1-K^{-1}x(t))dt - mx(t)dt + x(t)B(x(t))dw(t) \tag{1.4}$$

The proposed mathematical model has the predictability property because the equation (1.4) determines the Markov process [5]. Using the apparatus of modern stochastic analysis, one can draw conclusions about the qualitative behaviour of the trajectories of solutions (1.4) (the existence and asymptotic stability of the stochastic equilibrium, the tendency of approaching zero, the time of staying in a given interval, etc.). A shortcoming of the model (1.4) is the inability to simulate an abrupt change in the population size because any trajectory of the solution of equation (1.4) with the probability of one is continuous. This feature precludes the use this model (1.4) for populations subject to possible losses at random time moments (selection such as fishing or hunting, predator attacks, etc.). This model (1.4) similarly cannot be used for small populations, since the trajectory $x(t)$ must become discontinuous at the moment of death of even a single individual. We analyse a model with discontinuous trajectories whose dynamic characteristics depend on a piecewise constant ergodic Poisson process $y(t)$ with values in some set $U \subset \mathbf{R}^m$, invariant probability measure $\pi(du)$ and switching time moments $\mathbf{T}=\{\tau_k, k\in \mathbf{N}\}$. We propose that the initial size of the population at the time moment $t=0$ is equal to $x(0)=x_0$. At each interval $[\tau_{k-1},\tau_k), k\in N$, the aggregate biomass $\{x(t), t\geq 0\}$ of individuals of the population varies logistically

$$\frac{dx(t)}{dt} = rx(t)(1-K^{-1}x(t)) - mx(t) \tag{1.5}$$

where ε is a small positive parameter, and at each moment $t \in \mathbf{T}$ the mass of the population goes from the point $x(t-) = \lim_{s \downarrow 0} x(t-s)$ to the point

$$x(t) = x(t-) + \varepsilon x(t-) b(y(\varepsilon^{-1}t-), x(t-)) . \tag{1.6}$$

We briefly describe the stochastic approximation procedure proposed in [12] for the dynamical system with Markov switching, and we apply these results to analyse the dynamics of the logistic model given by equations of the form (1.5) - (1.6) for sufficiently small ε. We have already used this method in [11] to study a simpler impulsive Markov dynamical system given by a stochastic equation of the form

$$dx(t) = rx(t)(1 - K^{-1}x(t))dt - \varepsilon m \int_0^1 x(t)c(u)\mu(\varepsilon^{-1}, dt, du)$$

where $\mu(\varepsilon^{-1}, dt, du)$ is Poisson measure with parameter $\varepsilon^{-1} dtdu$.

2 Stochastic approximation procedure for Markov impulsive dynamical system

Let $(\Omega, \mathrm{F}, \mathrm{P})$ be a probability space, let $\mathrm{Y} := \{y_k, k \in \mathbf{N}\}$ be the ergodic Markov chain on the measurable space $(U, \Sigma_U), U \subset \mathbf{R}^m$, with transition probability $p(y, du)$ and invariant measure $\pi(du)$, let $\mu(\lambda, dt, du)$ be the Poisson type measure [5] on the measurable space $(U \times \mathbf{R}, \Sigma_{U\times\mathbf{R}})$ with parameter $\lambda p(y, du)dt$. The stochastic differential equation

$$dy = \int_U (u - y)\mu(\lambda, dt, du) \tag{2.1}$$

defines [5] the ergodic piecewise constant Markov process $\{y(t), t \geq 0\}$ on (U, Σ_U) with the weak infinitesimal operator

$$Qv(y) = \lambda \left[\int_U v(u)p(y, du) - v(y) \right] \tag{2.2}$$

and the invariant measure $\pi(dy)$. This means that there exists a sequence of i.i.d. random variables $\mathrm{D} := \{\delta_k, k \in \mathbf{N}\}$ such that $y(t) \equiv y(\tau_{k-1})$ for any $k \in \mathbf{N}, t \in [\tau_{k-1}, \tau_k)$, where $\tau_k = \sum_{n=1}^{k} \delta_n, \tau_0 = 0$, and $y(\tau_k) = y_k$. This sequence is independent on Markov chain Y and distributed exponentially with parameter λ. Using the above notations and definitions we define impulsive Markov dynamical system with phase coordinate $\{x_\varepsilon(t), t \geq 0\}$ in $\mathbf{R}$ as follows:

$$\tau_{k-1} < t < \tau_k, k \in N: \ \frac{dx_\varepsilon(t)}{dt} = F(x_\varepsilon(t)), \tag{2.3}$$

$$t \in \mathrm{T}: \ x_\varepsilon(t) = x_\varepsilon(t-) + \varepsilon G(x_\varepsilon(t-), y_\varepsilon(t-)) \tag{2.4}$$

where $y_{\varepsilon}(t) = y(\varepsilon^{-1}t)$, and$\varepsilon \in (0, \varepsilon_0)$ is a small parameter, which we will use for approximative analysis of dynamical system (2.1) - (2.2). The weak infinitesimal operator [9] for the Markov process $\{y_{\varepsilon}(t), x_{\varepsilon}(t), t \geq 0\}$ may be presented in a following form:

$$(\varepsilon)v(x,y) = F(x)\frac{\partial}{\partial x}v(x,y) + \varepsilon^{-1}Qv(x,y) + \\ + \varepsilon^{-1}\lambda\left[v(x+\varepsilon G(x,y),y) - v(x,y)\right] \tag{2.5}$$

The first approximative dynamical system for (2.3) - (2.4) is an ordinary differential equation

$$\frac{d}{dt}\overline{x}(t) = f(\overline{x}(t)) \tag{2.6}$$

where $f(x) := \int_U f_0(x,y)\pi(dy), f_0(x,y) = F(x) + \lambda G(x,y)$. By definition [9] of the weak infinitesimal operator (2.2) the equation

$$QV\left(x,y\right) = f(x) - f_0(x,y) \tag{2.7}$$

has the solution $V\left(x,y\right) = \Pi f_0(x,y)$ that satisfies an identity

$$\int_U \Pi f_0(x,y)\pi(dy) \equiv 0 \tag{2.8}$$

Now we can take a limit

$$\lim_{\varepsilon\to 0} (\varepsilon)\left[v(x) + \varepsilon\Pi f_0(x,y)\frac{\partial}{\partial x}v(x)\right] = \left\{\lim_{\varepsilon\to 0} F(x,y)\left[\frac{\partial}{\partial x}v(x) + \varepsilon\Pi f_0(x,y)\frac{\partial}{\partial x}v(x)\right] + \right.$$

$$+\lim_{\varepsilon\to 0}\varepsilon^{-1}Q\left[v(x) + \varepsilon\Pi f_0(x,y)\frac{\partial}{\partial x}v(x)\right] + \lim_{\varepsilon\to 0}\varepsilon^{-1}\lambda\left[v(x+\varepsilon G(x,y)) - v(x)\right] +$$

$$\left. +\lambda\lim_{\varepsilon\to 0}\left[\Pi f_0(x+\varepsilon G(x,y)) - \Pi f_0(x,y)\right]\frac{\partial}{\partial x}v(x)\right\} =$$

$$= F(x)\frac{\partial}{\partial x}v(x) + \lambda G(x,y)\frac{\partial}{\partial x}v(x) + Q\Pi f_0(x,y)\frac{\partial}{\partial x}v(x) = f(x)\frac{\partial}{\partial x}v(x)$$

which provides

$$\lim_{\varepsilon\to 0}\{v(x) + \varepsilon\Pi f_0(x,y), \quad (\varepsilon)[v(x) + \varepsilon\Pi f_0(x,y)]\} = \left\{v(x), f(x)\frac{d}{dx}v(x)\right\} \tag{2.9}$$

for any sufficiently smooth function $v(x)$. Therefore [10]

$$\lim_{\varepsilon\to 0}\sup_{|z|<r} \mathrm{P}\left(\sup_{0\leq t\leq T}|x_{\varepsilon}(t) - \overline{x}(t)| > C\right) = 0$$

for any $C>0$, $r>0$, $T>0$, $x_{\varepsilon}(0) = \overline{x}(0) = x$ and $y(0) = y$. Besides, if is there exists such an isolated point $\hat{x}$ that $f_0(\hat{x}, y) \equiv 0$ and $f'(\hat{x}) < 0$ then this point is locally asymptotically stable equilibrium of the initial Markov dynamical system (2.3)-(2.4) for any realization of the given by equation (2.1) Markov process.

The second step of the stochastic approximation procedure is the analysis of the normal deviations

$$X_\varepsilon(t)=\frac{x_\varepsilon(t)-\overline{x}(t)}{\sqrt{\varepsilon}} \tag{2.10}$$

After substitution $x_\varepsilon(t)=\overline{x}(t)+\sqrt{\varepsilon}X_\varepsilon(t)$ in equation (2.3) - (2.4)

$$\tau_{k-1}<t<\tau_k, k\in N: \frac{d}{dt}X_\varepsilon(t)=\frac{F(\overline{x}(t)+\sqrt{\varepsilon}X_\varepsilon(t))-f(\overline{x}(t))}{\sqrt{\varepsilon}} \tag{2.11}$$

$$t\in \mathrm{T}: X_\varepsilon(t)=X_\varepsilon(t-)+\frac{1}{\sqrt{\varepsilon}}G(\overline{x}(t)+\sqrt{\varepsilon}X_\varepsilon(t-), y_\varepsilon(t-)) \tag{2.12}$$

The equations (2.1) - (2.6) - (2.11) - (2.12) define three-dimensional Markov process on $\mathbf{R}^2\times U$ with the weak infinitesimal operator

$$(\varepsilon)v(\overline{x},x,y)=f(\overline{x})\frac{\partial}{\partial\overline{x}}v(\overline{x},x,y)+\frac{F(\overline{x}(t)+\sqrt{\varepsilon}X_\varepsilon(t))-f(\overline{x}(t))}{\sqrt{\varepsilon}}\frac{\partial}{\partial x}v(\overline{x},x,y)+$$
$$+\varepsilon^{-1}Qv(\overline{x},x,y)+\varepsilon^{-1}\lambda\left[v_\varepsilon(\overline{x},x+\sqrt{\varepsilon}G(\overline{x}(t)+\sqrt{\varepsilon}X_\varepsilon(t-),y_\varepsilon(t-)),y)-v(\overline{x},x,y)\right] \tag{2.13}$$

To derive a formula for the weak infinitesimal operator of the approximative Markov process for normalized deviations we will apply the same as before. Let us consider $\{(\varepsilon)v_\varepsilon, v_\varepsilon\}$, where

$$v_\varepsilon(\overline{x},x,y)=v(\overline{x},x)+\sqrt{\varepsilon}v_1(\overline{x},x,y)+\varepsilon v_2(\overline{x},x,y) \tag{2.14}$$

$v\in\mathbf{C}^2(\mathbf{R}^2)$ is an arbitrary positive function,

$$v_1(\overline{x},x,y)=\Pi f_0(x,y)\frac{\partial}{\partial x}v(\overline{x},x) \tag{2.15}$$

$$v_2(\overline{x},x,y)=x\Pi f_0(x,y)\frac{\partial^2}{\partial x\partial\overline{x}}v(\overline{x},x)+\lambda\frac{\Pi G^2(\overline{x},y)}{2}\frac{\partial^2}{\partial x^2}v(\overline{x},x) \tag{2.16}$$

After substitution (2.14) in (2.13) and decomposition accurate within $o\left(\sqrt{\varepsilon}\right)$ we can write the formula

$$(\ (\varepsilon)v_\varepsilon(\overline{x},x,y)\ =f(\overline{x})\frac{\partial}{\partial\overline{x}}\ v(\overline{x},x)+x\frac{\partial}{\partial\overline{x}}\ [F(\overline{x},y)+\lambda G(\overline{x},y)]\frac{\partial}{\partial x}\ v(\overline{x},x)+$$
$$+\lambda\frac{G^2(\overline{x},y)}{2}\frac{\partial^2}{\partial x^2}v(\overline{x},x)+Qv_2(\overline{x},x,y)+$$
$$+\frac{1}{\sqrt{\varepsilon}}\left\{[F(\overline{x},y)+\lambda G(\overline{x},y)-f(\overline{x})]\frac{\partial}{\partial x}v(\overline{x},x)+Qv_1(\overline{x},x,y)\ \right\}+o\left(\sqrt{\varepsilon}\right)$$

Now subject to the formulae (2.15) - (2.16) we can to ensure that

$$\lim_{\varepsilon\to 0}G(\varepsilon)v_\varepsilon(\overline{x},x,y)=\hat{L}v(\overline{x},x)$$

where

$$(\ \hat{}\ v(\overline{x},x)=f(\overline{x})\frac{\partial}{\partial\overline{x}}v(\overline{x},x)+x\frac{\partial}{\partial\overline{x}}f(\overline{x})\frac{\partial}{\partial x}v(\overline{x},x)+\lambda\frac{\sigma^2(\overline{x})}{2}\frac{\partial^2}{\partial x^2}v(\overline{x},x) \tag{2.17}$$

and $\sigma^2(\bar{x}) = \int_U G^2(\bar{x}, y)\pi(dy)$. Operator $\hat{}$ defines [5] a two-dimensional dynamical system composed of the ordinary differential equation (2.6) and the stochastic Ito differential equation

$$dX(t) = X(t) f'(\bar{x}(t))dt + \sigma(\bar{x}(t))\sqrt{\lambda}dw(t) \tag{2.18}$$

Thus for a sufficiently small $\varepsilon > 0$ for any initial conditions $x_\varepsilon(0) = x, y_\varepsilon(0) = y$, the finite dimensional distributions of the process $\{x_\varepsilon(t), t \geq 0\}$ given by equations (2.3)-(2.4) may be approximated [10,14] by the corresponding finite dimensional distributions of the process

$$\hat{x}_\varepsilon(\mathrm{t}) = \bar{x}(t) + \sqrt{\varepsilon}X(t) \tag{2.19}$$

where $\{\bar{x}(t), t \geq 0\}$ is the solution of equation (2.6) with the initial condition $\bar{x}(0) = x$ and where $\{X(t), t \geq 0\}$ is the solution of equation (2.18) with the initial condition $X(0) = 0$.

3 Stochastic approximation for logistic impulsive Markov model

Let us study a logistic type model with impulse type random perturbations given by the equations (1.5) - (1.6) and Poisson process with infinitesimal operator (2.1). This stochastic dynamical system may be defined as two-dimensional system of stochastic Poisson type differential equations

$$\begin{aligned} dx_\varepsilon(t) &= \left\lfloor r x_\varepsilon(t)(1 - K^{-1}x_\varepsilon(t)) - m x_\varepsilon(t) \right\rfloor dt + \\ &+ \varepsilon \int_U x_\varepsilon(t)\left[\alpha(y(t-)) + \beta(y(t-))x_\varepsilon(t-)\right]\mu(\varepsilon^{-1}\lambda, dt, du) \end{aligned} \tag{3.1}$$

$$dy = \int_U (u - y)\mu(\varepsilon^{-1}\lambda, dt, du) \tag{3.2}$$

where $\lambda \in \mathbf{R}_+, r > 0$, $m \geq 0, \varepsilon \in (0, \varepsilon_0)$ is a small positive parameter, $\mu(\varepsilon^{-1}\lambda, dt, du)$ is a Poisson measure on $U \subset \mathbf{R}_+ \times \mathbf{R}^m$ with parameter $\varepsilon^{-1}\lambda dt \pi(du)$ $\pi(du)$ is a probabilistic measure on U, $\{\alpha(y), \beta(y)\}$ are bounded measurable functions on U,

$$\int_U \alpha(y)\pi(dy) = 0, \int_U \beta(y)\pi(dy) = 0, a = \int_U \alpha^2(y)\pi(dy), b = \int_U \beta^2(y)\pi(dy), \int_U \alpha(y)\beta(y)\pi(dy) = 0$$

Making use of stochastic averaging procedure we can derive an equation in the form of the classical logistic differential equation (1.2) for the mean of the population aggregate biomass:

$$\frac{d\bar{x}}{dt} = r\bar{x}(1 - K^{-1}\bar{x}) - m\bar{x} \tag{3.3}$$

For $\bar{r} = \bar{m}$ the solution of (3.3) is hyperbola

$$\bar{x}(t) = \frac{Kx(0)}{K + r\bar{x}(0)t} \tag{3.4}$$

and for $r \neq m$ it may be presented as follows:

$$\bar{x}(t) = \frac{x}{1 + xrK^{-1}(r - m)^{-1}\left(\exp\{(r - m)t\} - 1\right)}\exp\{(r - m)t\} \tag{3.5}$$

If $r < m$ then the function (3.5) monotonically decreases to zero. If $r \geq m$ then it increases to $\hat{x} = \dfrac{r-m}{rK^{-1}}$. Therefore [12] if the mean value of the birth rate is sufficiently lower than the mean value of the death rate, the population will almost surely go extinct, but for $r - m = O(\varepsilon)$ we may not apply an averaging procedure for asymptotic analysis of the equations (2.3) - (2.4). Applying formula (2.18) we can approximate the finite dimensional distribution functions of the normalized deviations $X_{\varepsilon}(t) = \dfrac{x_{\varepsilon}(t) - \overline{x}(t)}{\sqrt{\varepsilon}}$ by the corresponding distribution functions of the solution of the stochastic differential equation

$$dX(t) = (r - m - 2rK^{-1}\overline{x}(t))X(t)dt + \sigma(\overline{x}(t))dw(t) \tag{3.6}$$

with the initial condition $X(0) = 0$. The above solution is the Gaussian process [5] with zero mean and variance that satisfies the equation

$$\frac{d}{dt}\mathrm{E}\{X^2(t)\} = 2(r - m - 2rK^{-1}\overline{x}(t))\mathrm{E}\{X^2(t)\} + \overline{x}^4(t)(a + b\overline{x}^2(t))^2\lambda \tag{3.7}$$

where $\sigma(x) = x^2(a + bx^2)\sqrt{\lambda}$. If $r > m$ then $\lim_{t\to\infty}\overline{x}(t) = \dfrac{r-m}{rK^{-1}} := \tilde{x}$ and for asymptotic analysis of the Markov process defined by the equation (3.6) we can simplify this equation as follows:

$$d\tilde{X}(t) = (m - r)\tilde{X}(t)dt + \sigma(\tilde{x})dw(t) \tag{3.8}$$

The stationary diffusion process $\hat{X}(t) = \sigma(\tilde{x})\int_{-\infty}^{t}\exp\{(m-r)(t-s)\}\,dw(s)$ satisfies the equation (3.8), has zero mean, the ergodic distribution $N\left(0, \dfrac{\sigma^2(\tilde{x})}{2(r-m)}\right)$, and has the covariance function $K(t,\tau)=\mathrm{E}\{\hat{X}(t)\hat{X}(\tau)\} = \dfrac{\sigma^2(\tilde{x})}{2(r-m)}\exp\{2(m-r)(t-\tau)\}$. The difference of any other solution $\tilde{X}(t)$ of (3.8) on $\hat{X}(t)$ exponentially tends to zero.

Conclusion

Stochastic approximation procedure provides a useful and informative way to explore a logistic type population mathematical model given in a form of Markov impulsive differential equation with discontinuities in a form of a fast oscillating ergodic Poisson process. This method allows to analyse the ordinary differential equation describing the mean value of population size and stochastic Ito differential equation for random deviations. Probabilistic characteristics of the resulting processes are found in explicit form.

References

[1] BAZYKIN, A.D., *Nonlinear dynamics of the interacting populations*. Moscow, IKI, 2003 (rus.).

[2] SVIREZHEV, Yu.M., LOGOPHED, D.J., *Stability of biological community*. Moscow, Nauka, 1978, (rus.)

[3] KHODABIN, M., KIAEE, N., Stochastic dynamical logistic population growth model. *Journal of Mathematical Sciences: Advances and Applications*, vol. 11, Nr. 1, 2011, pp. 11-29.

[4] TUCKWELL, H., KKOZIOL, J., Logistic population growth under random dispersal, *Bulletin of Mathematical Biology* Vol. 49, No. 4, pp. 495-506, 1987.

[5] GIKHMAN, I.I. , SKOROKHOD, A.V., *Stochastic Differential Equations*, Springer-Verlag, 1972.

[6] MAY, R., *Stability and Complexity in Model Ecosystems*. Princeton, NJ: Princeton University Press, 1973.

[7] LUDWIG, D. 1975. Persistence of Dynamical Systems under Random Perturbations. *SIAM Rev*. 17, 605-640.

[8] MURRAY, J., *Mathematical Biology I: An Introduction*, volume I. Springer-Verlag, 3rd edition, 2003.

[9] DYNKIN, Ye.B. *Theory of Markov Processes*. Prentice-Hall. 1961.

[10] SKOROKHOD, A.V. *Asymptotic methods in the theory of stochastic differential equations* (Transl. Math. Monographs 78), American Mathematical Society, Providence, RI, 1989.

[11] CARKOVS, J., ŠADURSKIS, K. Dynamics of a population subject to impulse type random loss. In *Proceedings of the Latvian Academy of sciences*. Section B, Vol. 71 (2017), No. 4 (709), pp. 298–302.

[12] TSARKOV, Ye.F., (CARKOVS, J.), Asymptotic methods for stability analysis Markov impulse dynamical systems, *Nonlinear dynamics and system theory*, 9 (2002), No. 2(1), 103–115.

Current address

Carkovs Jevgeņijs, prof, Dr.hab.math.
Department of Probability Theory and Mathematical Statistics
Riga Technical University
Kaļķu iela 1, LV-1658, Riga, Latvia
E-mail: carkovs@livas.lv

Šadurskis Kārlis, prof., Dr.math.
Department of Probability Theory and Mathematical Statistics
Riga Technical University
Kaļķu iela 1, LV-1658, Riga, Latvia
E-mail: karlis.sadurskis@gmail.com

Goldšteine Jolanta, doc, Dr.math.
Department of Probability Theory and Mathematical Statistics
Riga Technical University
Kaļķu iela 1, LV-1658, Riga, Latvia
E-mail: Jolanta.goldsteine@lezards.lv

Slovak University of Technology in Bratislava
Faculty of Mechanical Engineering
17th Conference on Applied Mathematics **APLIMAT 2018**

Proceedings

MATHEMATICAL MODELS ON MILK, COFFEE AND BEER

CEPPITELLI Rita (IT), MARCONI Ombretta (IT)

Abstract. The simple differential equation $y' = ky, k \in \mathbb{R}$ is an important mathematical model. It is well known in Population Dynamics as the *Malthusian Equation* and it is familiar to food technologists as *Bigelow's First Law*, which describes the microbial mortality at constant lethal temperatures. In the paper we focus on the process of thermal sterilization of milk. The Malthusian Equation is also a mathematical model for Newton's law of cooling or heating. The rate of heat loss of a cooling body is proportional to the difference in temperatures between the body and its surroundings. In the paper we analyze the data of an experimental study on the cooling of coffee conduced by "Mathematics & Real Life" Project. The measurements of the temperature were taken using a centesimal temperature sensor connected to a computer for automatic data capture.
Another simple differential equation is Darcy's equation, a fundamental mathematical model in the filtration t heory. In the paper we are interested in mathematical models derived from Darcy's law, which are applied to the cross flow m icrofiltration at co nstant pr essure of beer. These models are then tested using experimental data provided by the Italian Brewing Research Centre of the University of Perugia.

Keywords: Malthusian Equation, Darcy's Equation, Newton's law of cooling, First Bigelow model, cross flow microfiltration

Mathematics subject classification: Primary 97M10; Secondary 97M50, 34A34

1 Introduction

The *Malthusian Equation* $y' = ky, k \in \mathbb{R}$ and *Darcy's law* $Q = \frac{PA}{R\mu}$ are important mathematical models in Food Technologies.
Microbial mortality at constant lethal temperatures is traditionally considered a process following a first-order constant rate kinetics reaction (*First Bigelow model*). According to this model the organism's survival can be described by the Malthusian Equation where $y(t)$ is the microbial population survived after t seconds of thermal process and $k < 0$ is the constant rate of microbial destruction relating to homogeneous microbial population. During heat treatment several food compounds undergo changes. In the specific case of thermal sterilization of milk an usual side-effect is the production of *furosine*, a typical *Maillard reaction* that induces a yellow-brown colour and smell of cooked milk.

The Italian law sets the limit at 2.9 mg/ls for the furosine in the milk. Luckily, the formation of furosine follows a linear growth production. According to the Collision Theory, the constant rate k of each reaction increases with temperature. The results of our models prove that the optimization of sterilization effect, i.e the sanitization of the product with minimizing the degradation of milk quality, is achieved by increasing the temperature and reducing time, in agreement with the acronyms found in products on the market: UHT (Ultra High Temperature) and HTST (High Temperature Short Time).
The Malthusian equation is also a mathematical model for Newton's law of cooling or heating, which is applied in the processes of pasteurization and sterilization to evaluate the temperature during the heating of the product. Newton's law can be described by the equation $T' = k(T - T_a), k < 0$, where T_a is the temperature of the surrounding environment. The constant rate depends on physical and environmental factors. In the paper we calculated the value of the constant rate using the data from an experimental test on cooling of a cup of coffee. The measurements of the temperature were taken using a centesimal temperature sensor connected to a computer for automatic data capture.
Darcy's equation is a fundamental mathematical model in the filtration theory. In the cross flow microfiltration, Darcy's equation $Q = \frac{PA}{R\mu}$ describes the flow rate Q as a function of resistance R and membrane area A where P is the trans-membrane pressure and μ is the viscosity of the solution, which are constant values. In the paper we studied some of the prominent mathematical models derived from Darcy's law that are proposed in the literature. Precisely we focused on the classical four-constant pressure Hermia's models (basic models) [7] and their feasible combinations (combined models). These models are typically used to describe fouling. We validated the models with experimental trials of the cross flow microfiltration of beer monitoring the filtrate volume and the flow rate by a pilot plant equipped with a ceramic membrane. The data were provided by the Italian Brewing Research Centre of the University of Perugia. None of the basic models has been able to explain fouling. This indicates that multiple mechanisms participate in the fouling process and suggests the development of the combined models. The combination of two basic models, the standard pore blocking model and the complete pore blocking, yields the better fit in the first phase of the process. This result can be explained considering the beer composition that consists of different dimensions particles. In the second phase of the process, the CPB+SPB model does not fit the experimental data because it does not take in account the back-flush washing performed during the experimental trials. Regarding the flow rates, the results obtained by the basic models are more encouraging.
Our research project is to continue these studies, improve the combined mathematical models and evaluate other experimental tests, even considering filtration of different fluids or other types of membranes.

2 Malthusian equation

The equation

$$y' = ky, \quad k \in \mathbb{R}$$

whose solution is

$$y(t) = y(0)e^{kt}, \quad y(0) \text{ initial data at } t = 0$$

is an important dynamic model. It is known as the Malthusian equation by the name of the English economist Thomas Malthus (1766 - 1834). What was commonly accepted in Malthus's analysis was the idea that in the absence of external constraints (food or environmental constraints, pollution, social-economic constraints, economic constraints) populations tend by nature to unlimited growth, directly proportional to the number of the population, with a constant growth rate. In many cases the equation appears in the discrete formulation, as a difference equation.

The finite difference method is a technique used to obtain numerical solutions to Differential Equations, i.e. solutions that are known only at a finite number of points. In general, increasing the number of points increases the accuracy of the numerical solution.
The solution of the discrete formulation of Malthusian equation $y_{n+1} - y_n = ky_n$ is the geometric sequence $y_n = y(0)(1+k)^n$, with mesh $\triangle t = 1$.
If the mesh is reduced $\triangle t = 1/r, r > 0$ the geometric sequence becomes $y_n = y(0)(1+k/r)^{rn}$ from which it is deduced that the corresponding continuum of the geometric sequence is the exponential law as shown in the following Figure 1:

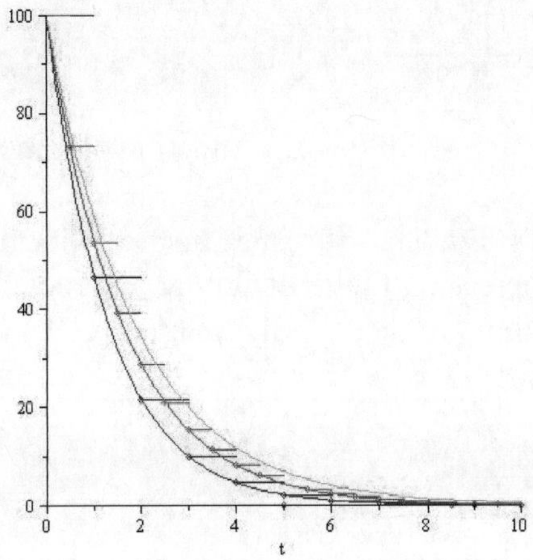

Fig. 1.

- $y_n = y(0)(1-k)^n, \quad k = 0.536, \quad y(0) = 100, \quad \triangle t = 1$.
- $y_n = y(0)(1-k/2)^{2n}, \quad k = 0.536, \quad y(0) = 100, \quad \triangle t = 0.5$.
- $y(t) = y(0)e^{-kt}, \quad k = 0.536, \quad y(0) = 100$.

If k_F is the constant rate of difference equation, then the constant rate k of related differential equation is $k = \ln(k_F + 1) \simeq k_F$ $[k_F = e^k - 1 \simeq k]$ for values of the constant rate close to zero. So if the differential equation $y'(t) = 0.02y(t), \quad t$ years, describes the growth of a population, with a good approximation we can say that the population grows by 2% a year.

2.1 First Bigelow model

In food technologies the equation $y' = -ky, \quad k > 0$ is knows as *First Bigelow model* and describes the microbial mortality at constant lethal temperatures [9]. $y(t) = y(0)e^{-kt}$ is the microbial population survived after t seconds of thermal process and k is the constant rate of microbial destruction in relation to a homogeneous microbial population. In the specific case of thermal sterilization of milk the constant rate (at 120 ^0C) is $1.10 \cdot 10^{-2}$ (s^{-1}), and the commercial sterility is estimated at 10^{-4} (spore/l). According to the Collision Theory, the constant rate k of each reaction depends on the temperature. Raising the temperature, the rate of reaction increases [6].
The temperature dependence of the kinetic parameters is adequately described by the *Arrhenius model*. In Table 1 we can read the values of the constant rates when the temperature of the treatment is varied [10]. The same table also shows the data on the formation of furosine, a negative side effect that occurs during the heat treatment of the milk. The formation of furosine is a typical *Maillard reaction* that follows a linear growth and leads to the production of compounds that gives the yellow-brown colour and smell of cooked milk ([1, 12]).

temperature	k (s^{-1}) First Bigelow model	k (mg/(l s)) kinetics of furosin production	seconds of treatment for commercial sterility	mg/l of furosine produced
120 °C	0.011	0.126	1249.03	157.37
125 °C	0.041	0.172	332.91	57.412
130 °C	0.141	0.234	91.19	21.359
135 °C	0.415	0.315	25.78	8.141
140 °C	0.841	0.422	7.51	3.176
141 °C	0.904	0.447	5.89	1.628

Tab.1. Constant rates values as a function of temperature.

The Italian law sets to 2.9 mg/l the limit for furosine in the milk, hence the optimization of the sterilization effect is achieved by increasing the temperature and reducing time. These results are in agreement with the acronyms found in products on the market: UHT (Ultra High Temperature) and HTST (High Temperature Short Time) [3].

2.2 Newton's law of cooling

The Malthusian equation is a mathematical model for Newton's law of cooling or heating. The rate of cooling or heating of a body is proportional to the difference in temperatures between the body and its surroundings. Newton's law can be described by the equation $T' = k(T - T_a), k < 0$, whose solutions are $T(t) = Ce^{kt} + T_a$ (for cooling) and $T(t) = T_a - Ce^{kt}$ (for heating), where T_a is the temperature of the surrounding environment.
The discrete formulation of Newton's law of cooling is $T_{n+1} - T_n = k(T_n - T_a)$.
Put $S_n = T_n - T_a$ the thermal jump, we find the discrete formulation of Malthusian equation $S_{n+1} - S_n = kS_n$.
Clearly, the rate constant depends on many factors, both physical and environmental. We calculated the value of the rate constant using the data from an experimental test on cooling of a cup of coffee conduced by "Mathematics & Real Life" Project, July 2006.
In July, a cup of coffee at the initial temperature of $70 - 75\ ^0C$ was placed in an environment at the temperature of 29 ^0C and by using a centesimal temperature sensor connected to a computer for automatic data capture, the temperature was measured as shown in Table 2.

Room temperature	29 °C
Duration of each test	45 minutes
Number of readings per test	4500
Estimation of reading error	0.01 °C
Initial coffee temperature	70 -75 °C
Final measured temperature	34 -38 °C

Tab.2. Experimental trials on the cooling of a cup of coffee.

The graph in Fig.2 shows the temperatures measured every twelve seconds starting from $61.36\ ^0C$ i.e. the temperature measured 2.45 minutes after the beginning of the experiment.

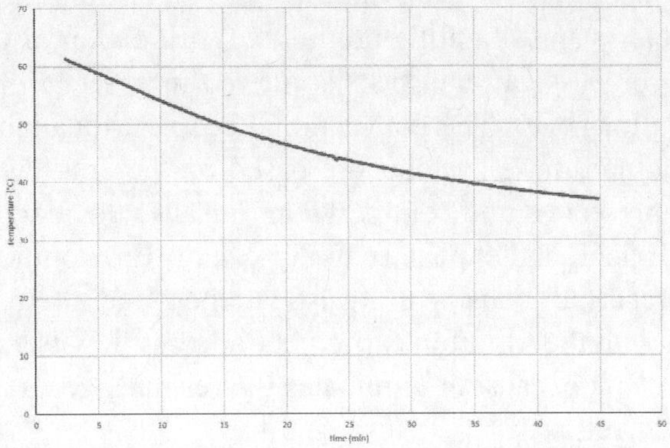

Fig. 2. Experimental temperature data.

By using the semilogarithmic coordinates in Fig.3, we can estimate the value of the relative cooling rate: $k = -0.0338\ (\mathrm{min}^{-1})$

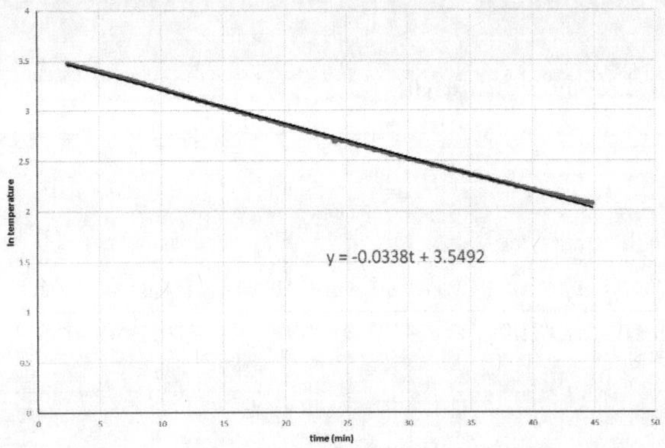

Fig. 3.

The solution $T(t) = 29 + (61.36 - 29)e^{-0.0338t}\ (^0C)$ of the mathematical model $T' = -0.0338(T - 29)$ is a good approximation of the experimental data (see Fig.4).

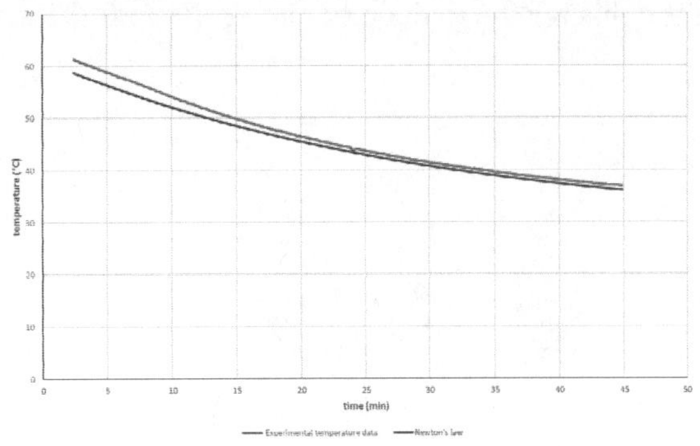

Fig. 4. Validation of the model.

3 Darcy Equation

Darcy's law is a fundamental mathematical model in the cake filtration theory. In this paper we will be focused in the filtration of beer. The brewing industry has an ancient tradition and is a dynamic sector open to modern technology and scientific progress. Clarification of the beer is an important operation during the brewing process. Rough beer is filtered in order to eliminate yeast cells and colloidal particles responsible for haze. The conventional beer clarification process employs filter press or pressure vessel filters and these systems are based on the employ of diatomaceous earth. This type of process, though, presents significant problems because the diatomaceous earth is a non-renewable material which is difficult and expensive to dispose of. Environmental pressure on the use of diatomaceous earth has forced the industry to investigate new and alternative technologies. The membrane separation process with the filtration carried out in cross flow is a promising alternative to the traditional clarification process because it eliminates the residues generated by the conventional treatment and the need for filter aids.

An important research topic related to the cross flow microfiltration processes is the study of the flux decline due to membrane fouling and many different models have been proposed in the literature [2, 4, 5, 7, 8, 11]. It may be useful to classify fouling as in-depth pore fouling, pore plugging and cake formation. In this context, prominent models are the classical four - constant pressure Hermia's models, that are typically used to describe fouling : standard blocking model, intermediate blocking model, cake filtration model, and complete blocking model [7]. These models derive from Darcy's law.

In the crossflow microfiltration Darcy's equation

$$Q = \frac{PA}{R\mu}$$

describes the flow rate Q (m^3/s) as a function of resistance R (m^{-1}) and membrane area A (m^2) where P (N/m^2) is the trans-membrane pressure and μ (Ns/m^2) is the viscosity of the solution, which are constant values.

In the first two models, Complete and Intermediate pore blocking, the study of the fouling depends mainly on the reduction of the surface filtration in time. Denoted by Q_0, A_0, R_0 the initial data, we can assume $\frac{Q}{Q_0} = \frac{A}{A_0}$. In the other two models the increasing of the resistance assumes greater importance and they are based on the assumption $\frac{Q}{Q_0} = \frac{R_0}{R}$.

Complete pore blocking model (CPB model) is based on the premise that the particles are larger than the pore size of the membrane and this results in the particles sealing off the membrane and preventing the flow.

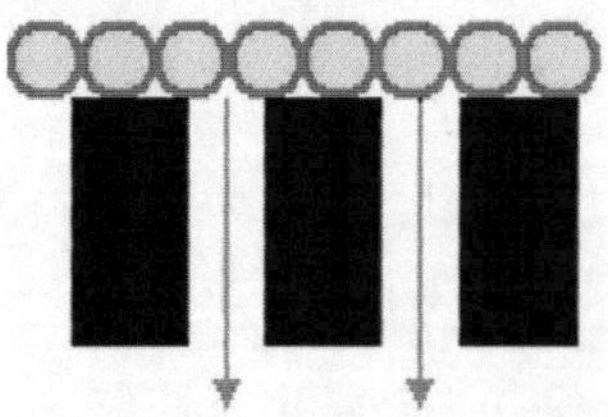

Fig. 5. Complete pore blocking model.

The available membrane area decreases with respect to permeate volume according the equation $A = A_0(1 - \frac{k_b}{Q_0}V)$ that combined with Darcy's law gives $Q = \frac{dV}{dt} = Q_0 - k_b V$ from which it is derived

$$V(t) = \frac{Q_0(1 - e^{-k_b t})}{k_b} \quad \text{and} \quad Q(t) = Q_0 e^{-k_b t}$$

where $k_b = \frac{P\sigma}{R\mu}$ (s^{-1}) and σ (m^{-1}) is the blocked area per unit filtrate volume.

Intermediate pore blocking model (IPB model) assumes that a portion of the particles seal off pores and the rest accumulate on top of other deposited particles.

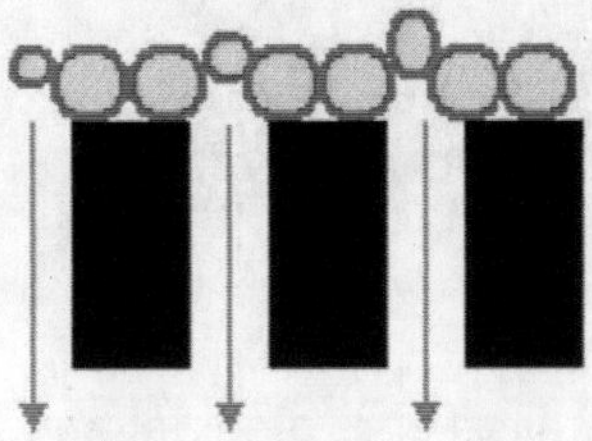

Fig. 6. Intermediate pore blocking model.

This model uses a successive approximation method to membrane area reduction and to permeate volume. The reduction of the available membrane with respect to the permeate volume can be expressed as a function of volume by $A = A_0 e^{-k_i V}$. By Darcy's law we obtain $Q = \frac{dV}{dt} = Q_0 e^{-k_i V}$ from which

$$V(t) = \frac{\ln(1 + k_i Q_0 t)}{k_i} \quad \text{and} \quad Q(t) = \frac{Q_0}{1 + k_i Q_0 t}$$

where $k_i = \frac{\sigma}{A_0}$ (m^{-3}).

Cake filtration model (CF model) occurs when particles accumulate on the surface of a membrane in a permeable cake of increasing thickness that adds resistance to flow.

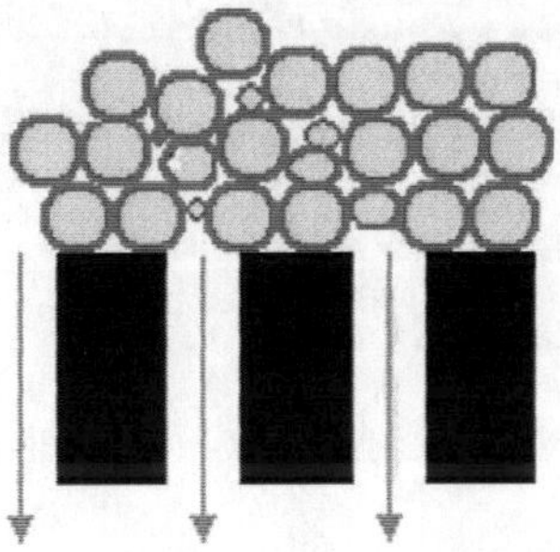

Fig. 7. Cake filtration model.

The total resistance R is the sum of the membrane resistance and the cake resistance; it can be expressed by the function $R = R_0(1 + k_c Q_0 V)$ that together with Darcy's law gives

$Q = \frac{dV}{dt} = \frac{Q_0}{1+k_c Q_0 V}$. As a result, the permeate volume and the flow rate as a function of the time is

$$V(t) = \frac{\sqrt{1+2k_c Q_0^2 t}-1}{k_c Q_0} \quad \text{and} \quad Q(t) = \frac{Q_0}{\sqrt{1+2k_c Q_0^2 t}}$$

where $k_c = \frac{S}{AR_0 Q_0}$ (sm^{-6}), S (m^{-2}) depends on the cake specific resistance and on filtrate density.

Standard pore blocking model (SPB model) assumes that particles accumulate inside membranes on the walls of straight cylindrical pores.

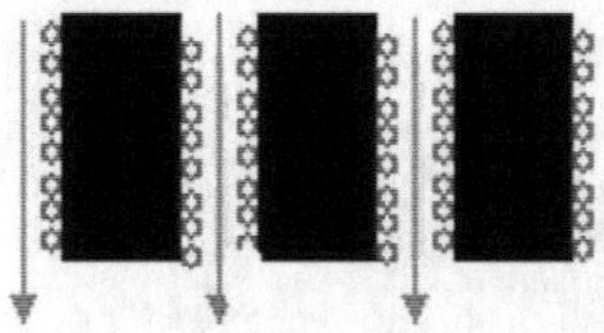

Fig. 8. Standard pore blocking model.

The rays of the pores decrease due to the solid matter that accumulates on the pore walls, so the growth of resistance as a function of the permeate volume can be expressed by $R = R_0(1 - \frac{k_s V}{2})^{-2}$ and, from Darcy's law, $Q = \frac{dV}{dt} = Q_0(1 - \frac{k_s V}{2})^2$. Through the integral we have

$$V(t) = (\frac{k_s}{2} + \frac{1}{Q_0 t})^{-1} \quad \text{and} \quad Q(t) = \frac{Q_0}{(1+\frac{k_s Q_0 t}{2})^2}$$

where $k_s = \frac{2C}{LA}$ (m^{-3}) and C is the volume of solid particles retained per unit filtrate volume, L (m) is the membrane thickness.

Combining the previous basic models, other models can be developed which describe the fouling taking into account both membrane area reduction and increasing of resistance [2, 4]. Darcy's law, in these models, describes the flow rate as a function of the resistance and the membrane area that vary with permeate volume, under constant pressure conditions

$$\frac{Q}{Q_0} = \frac{R_0 A}{R A_0}.$$

In the various combinations, the permeate volume as function of the time is described by

$V(t) = \frac{Q_0}{k_b}(1 - e^{\frac{-k_b\sqrt{1+2k_c Q_0^2 t-1}}{k_c Q_0^2}})$ (CPB and CF models)

$V(t) = \frac{1}{k_i} ln(1 + \frac{k_i}{k_c Q_0}(\sqrt{1+2k_c Q_0^2 t}-1))$ (IPB and CF models)

$V(t) = \frac{Q_0}{k_b}(1 - e^{\frac{-2k_b t}{2+k_s Q_0 t}})$ (CPB and SPB models)

$V(t) = \frac{1}{k_i}\ln(1 + \frac{2k_i Q_0 t}{2+k_s Q_0 t}$ (IPB and SPB models).

In the present paper, basic models as well as combined models have been used to explain experimental observations.

3.1 Experimental trials

The technological trials of cross flow microfiltration of beer were conducted using a pilot plant equipped with a ceramic membrane. The membrane presented the following characteristics (Tab.3):

Material	Aluminium Oxide ceramic fiber
Structure	symmetric
Porosity	$0.80\mu m$
Filtration surface	$0.8m^2$
Transmembrane pressure	5.5 bar
Number of capillaries	290
Internal diameter of capillary	3.0 mm

Tab.3. Constructive characteristics of the ceramic membrane.

Two technological replications of the cross flow microfiltration of beer were conducted monitoring the filtrate volume (m^3) (Fig. 9) and the volumetric flow rate (m^3/s) (Fig.10) every five minutes for a total filtration cycle of 180 minutes and performing back-flush washing each three minutes.

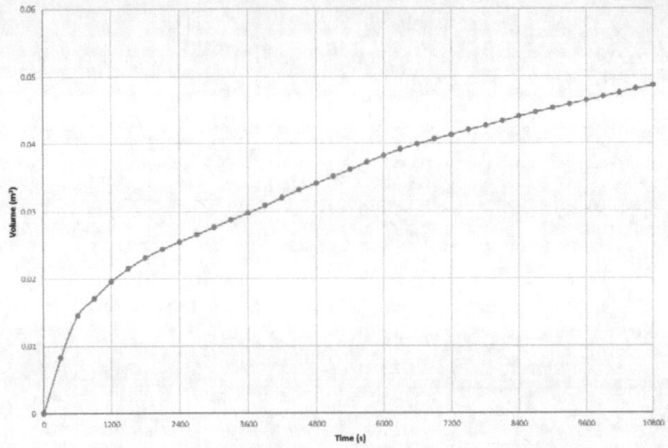

Fig. 9. Experimental data of the cumulate volume.

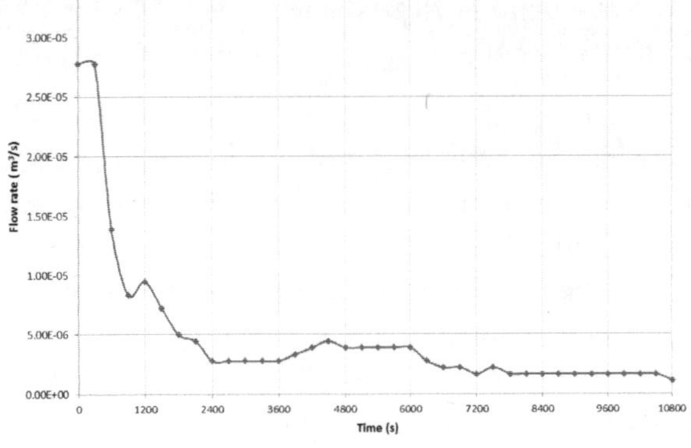

Fig. 10. Experimental data of the volumetric flow rate.

The experimental data were used for the estimate of the parameters k_b, k_i, k_c, k_s related to the previous described models. The values of these parameters depend on the type of membrane and the composition of beer (see Tab.4).

$k_b(s^{-1})$	$4.84 \cdot 10^{-4}$
$k_i(m^{-3})$	58.56
$k_c(sm^{-6})$	$11.01 \cdot 10^{6}$
$k_s(m^{-3})$	30.17

Tab.4. Empirical parameters associated to the models.

The results related to the filtrate volume described by the basic models are not satisfactory (Fig. 11). None of the basic models is able to explain fouling. The intermediate pore blocking model provided the best fit for the data in the first thirty minutes. However, significant deviations are observed in the following minutes. This indicates that multiple mechanisms participate in the fouling process and suggests the development of the combined models.

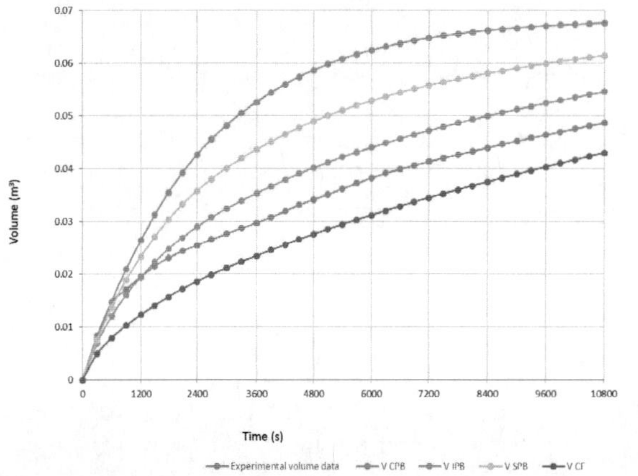

Fig. 11. Experimental cumulate volume and estimated volume by basic models.

As can be seen in Fig. 12, the combination of the standard pore blocking model and the complete pore blocking model yields the better fit in the first ninety minutes of the process.

This result can be explained considering the beer composition. Beer is a complex matrix containing different dimension particles as yeast cells, proteins, no starch polysaccharides as beta-glucan and arabinoxylan, and other compounds that derived from raw materials. The larger particles are subjected to superficial retention on the membrane while the smaller particles cause the occlusion of the membrane pores. The combined model CPB+SPB simultaneously takes into account the reduction of the filter membrane surface and the increase in resistance and describes in a more appropriate manner the experimental studied process. In the second phase of the process, the model CPB+SPB does not fit the experimental data because it does not take in account the back-flush washing performed during the experimental trials.

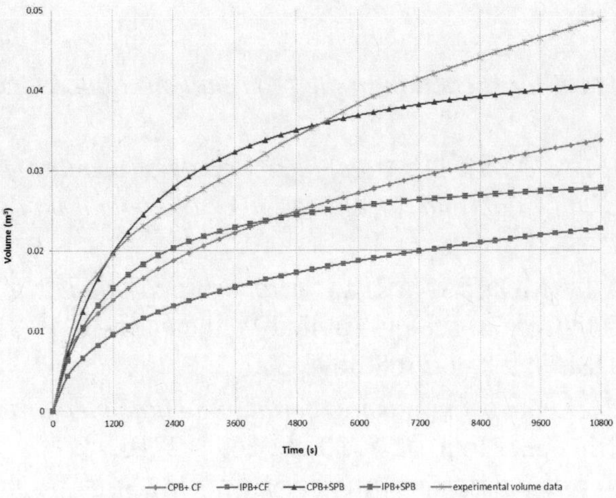

Fig. 12. Experimental cumulate volume and estimated volume by combined models.

Regarding the flow rates, the results obtained by the basic models are more encouraging. In particular, except for the CPB model, the other functions describe the experimental data with a good approximation (see Fig. 13).

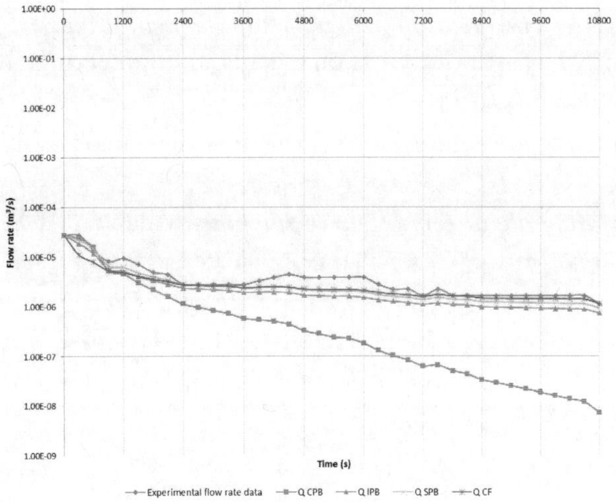

Fig. 13. Experimental flow rate data and estimated flow rate by basic models.

In the brewing industry the cross flow filtration technique is a valid alternative to traditional filtration processes. The application of this technology is still under study and development and further research and experimentation are needed to optimise the production process and obtain the best performance. The formulation of mathematical models that interpret these processes is a valid aspect of research in this area.

Our research project is to continue these studies, improve the combined mathematical models and evaluate other experimental tests, even considering filtration of different fluids or other types of membranes.

References

[1] AMES J. M.: *Application of Maillard reaction in the food Industry*, Food Chemistry, **62** (1998), 431–439.

[2] BOLTON G., KURIYEL R., LACASSE D.: *Combined models of membrane fouling: development and application to microfiltration and ultrafiltration of biological fluids*, Journal of Membrane Science, **277** (2006), 75-84.

[3] CEPPITELLI R., LA TONA L., MONTANARI L.: *Mathematical models for food sanitization process*, APLIMAT 2017 - 16th Conference on Applied Mathematics 2017, Bratislava, Slovakia Proceedings 2017, 334–342, ISBN: 978-1-5108-3698-3.

[4] CIMINI A., MORESI M.: *Beer clarification using ceramic tubular membranes*, Food and Bioprocess Technology: An International Journal, **7** (2014), 2694–2710.

[5] CIMINI A., MARCONI O., PERRETTI G., MORESI M.: *Novel Procedure for Lager Beer Clarification and Stabilization using Sequential Enzymatic, Centrifugal, Regenerable PVPP and Crossflow Microfiltration Processing*, Food and Bioprocess Technology: An International Journal, **7** (11) (2014), 3156-3165.

[6] CUNHA L.M., OLIVEIRA F.A.R.: *Optimal experimental design for estimating the kinetic parameters of processes described by first-order Arrhenius model under linearly increasing temperature profiles*, J. Food Engineering, **46** (2000), 53–60.

[7] HERMIA J.: *Constant pressure blocking filtration laws-application to power law non-newtonian fluids*, Transaction of the Institution of Chemical Engineers, **60** (1982),183–187.

[8] DE OLIVEIRA R.C., MENDES E.S., INNOCENTI T.D., DE BARROS S.T.D., ALICIEO T.V.R., PEREIRA N.C.: *Beer clarification with polysulfone membrane and study of fouling mechanism*, ENPROMER, 2nd Mercosur Congress on Chemical Engineering, 4th Mercosur Congress on Process Systems Engineering, 2011.

[9] PELEG M.: *Mathematical interpretation of dose-reponse curves*, Bull. Math. Biology, **59** (1997), 747–761.

[10] PERI C., ZANONI B.: *Manuale di Tecnologie Alimentari* Milano, (1998).

[11] SAMPATH M., SHUKLA A., RATHORE A.S.: *Modeling of Filtration Processes—Microfiltration and Depth Filtration for Harvest of a Therapeutic Protein Expressed in Pichia pastoris at Constant Pressure*, Bioengineering, **1** (2014), 260-277.

[12] VAN BOECKEL M.A.J.S.: *Formation of flavour coumpounds in the Maillard reaction*, Biothecnology Advances, **24** (2006), 230–233.

Current address

Ceppitelli Rita, Associate Professor
Department of Mathematics and Informatics
University of Perugia,
Via L. Vanvitelli 1, 06123 Perugia, Italia
E-mail: rita.ceppitelli@unipg.it

Marconi Ombretta, Assistant Professor
Department of Agricultural, Food an Environomental Sciences
University of Perugia
Via San Costanzo, 06126 Perugia, Italia
E-mail: ombretta.marconi@unipg.it

Slovak University of Technology in Bratislava
Faculty of Mechanical Engineering
17th Conference on Applied Mathematics **APLIMAT 2018**

Proceedings

GOOD KNOWLEDGE OF BASIC MATHEMATICS – A SUCCESSFUL PREREQUISITE TO STUDY IN RIGA TECHNICAL UNIVERSITY

CERNAJEVA Sarmite (LV), VOLODKO Inta (LV),
ILTINA Marija (LV), ILTINS Ilmars (LV)

Abstract. The issue viewed in this paper, which affects a multitude of technical universities: an insufficient level of basic mathematical knowledge among students. The 2016/17 centralized State examination, as well as, basic mathematical knowledge examinations done by Riga Technical University have been analyzed. The results show that a large part of students do not have the required level of knowledge in basic mathematics to successfully study calculus. Measures that RTU is using to avert these issues are mentioned.

Keywords: higher education, attitude, mathematics, teaching approach, educational process

Mathematics Subject Classification: 97D50, 97C70, 97D60

1 Introduction

Engineers are highly valued and demanded in the labor market, which means the quality of education of young engineers has a crucial role in the future of the country. Evidence shows that it is much easier to study if the students already have a high level of knowledge of exact sciences. Sadly, evaluating the knowledge of basic mathematics among students accepted into the university, it shows that many of them are not knowledgeable enough to successfully study calculus. For years now, we have been looking for ways to help these students stay in university and successfully finish their studies.

Working as lecturers in four faculties of Riga Technical University (RTU): Faculty of Computer Science and Information Technology, Faculty of Power and Electrical Engineering, Faculty of Civil Engineering, Faculty of Mechanical Engineering, Transport and Aeronautics, the application of mathematical knowledge among specialists is very important in our area of work. We look for answers to the questions: what it is, what should be taught, and how it should be done in order for this obtained knowledge to be useful in professional work.

2 Centralized State examination in mathematics

The obligatory state centralized examination (CE) in mathematics was introduced in the academic year of 2008/2009. This exam is also done by vocational school pupils, and it is made in such a way that many different schools can use it.

The exam consists of three parts. The first part, "Knowledge and basic skills," contains 15 questions with multiple choice answers and 10 questions with short answers. In total, there are 25 questions, each worth one point. The second part, "Application of knowledge in standard situations," contains 10 expanded problems (worth 40 points in total). The third part "Application of knowledge in non-standard situations" consists of three expanded problems (worth 15 points in total).

How are exams graded?

For example: Mathematics – overall score 85% (Knowledge and basic skills 96%, application of knowledge in standard situations 80%, application of knowledge in non-standard situations 80%). 24 out of 25 possible points have been earned in the 1st part (knowledge and basic skills), which is 96%. 32 out of 40 points have been earned in the 2nd part (application of knowledge in standard situations), which is 80%. 12 out of 15 points have been earned in the 3rd part (application of knowledge in non-standard situations), which is 80%. And then the overall score is calculated (24+32+12): (maximum of 80 possible points), or 85%.

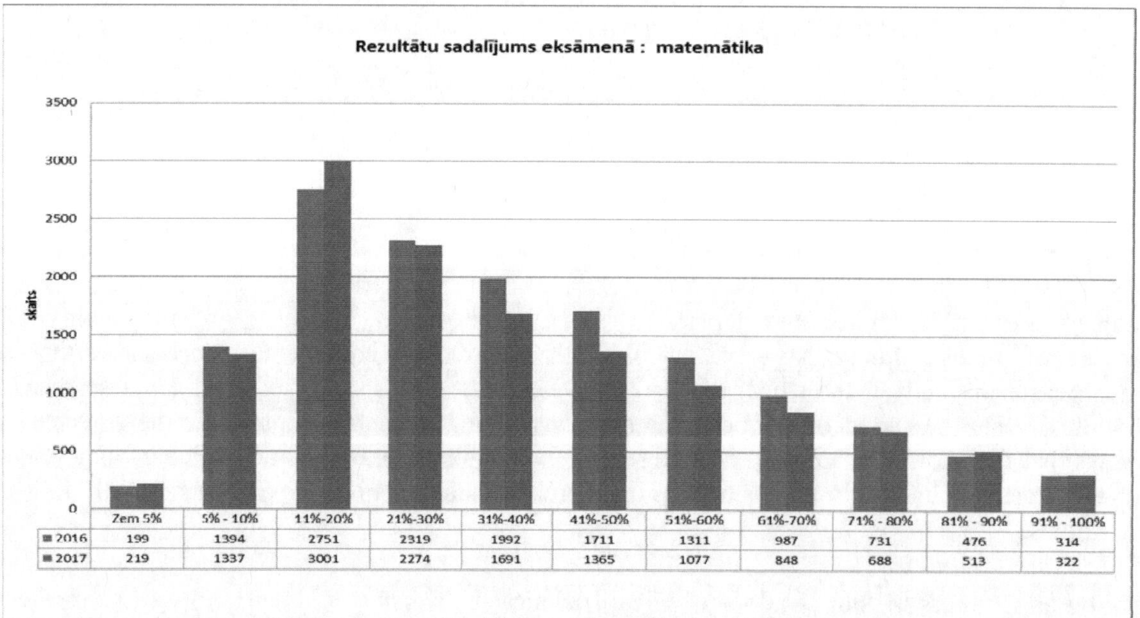

Fig. 1. The distribution of scores in the state centralized examination in mathematics in the academic year of 2016/17.

A little over 13 300 students did these centralized state exams in the academic year of 2016/17. According to the information given by National Centre of Education of the Republic of Latvia, the average overall score in these exams for students in the 12th grade was 34.9%, which is 1.3% less than the last academic year. The distribution of scores in the centralized state examination in mathematics in the academic year of 2016/17 is shown in Figure 1.

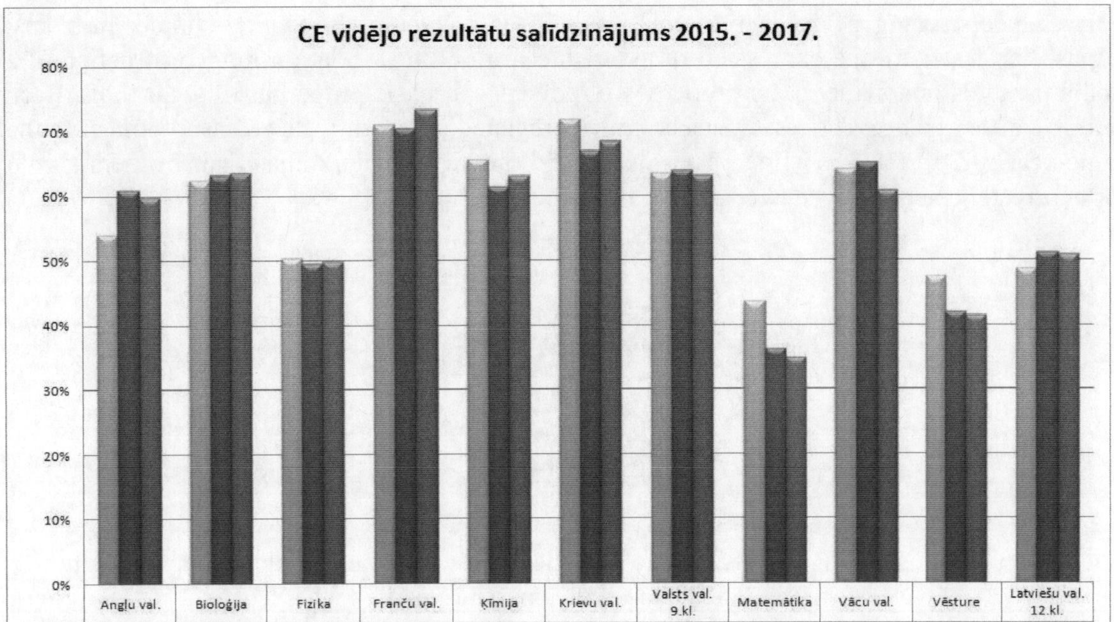

	Anglu val.	Bioloģija	Fizika	Franču val.	Ķīmija	Krievu val.	Valsts val. 9.kl.	Matemātika	Vācu val.	Vēsture	Latviešu val. 12.kl.
2015	54.1%	62.5%	50.3%	71.1%	65.6%	71.8%	63.4%	43.6%	64.2%	47.5%	48.7%
2016	61.0%	63.3%	49.6%	70.6%	61.5%	67.0%	63.9%	36.2%	65.2%	41.9%	51.3%
2017	59.7%	63.7%	49.9%	73.3%	63.1%	68.6%	63.1%	34.9%	60.9%	41.5%	50.9%

Fig. 2. A comparison of state centralized examination scores in the years 2015-2017.

In the second graph, the comparison of centralized state examinations in different subjects is shown in the last three years. The 1st column shows the results in English, 2nd – biology, 3rd – physics, 4th – French, 5th – chemistry, 6th – Russian, 7th – Latvian (9th grade), 8th – mathematics, 9th – German, 10th – history, 11th – Latvian (12th grade).

It is clearly visible that the worst scores are in the mathematics examination, and are only getting worse each year. Students' results over many years show that simple mathematical skills (dealing with algorithms), which every teacher tries to teach and develop, are still not at a good enough level. The same can be said about students' results in problems where conceptional knowledge is required. The inability to deal with logical tasks among students is frightening. However, it isn't anything new, as previous analysis show that students have no issues learning given materials, but encounter problems when asked to use them. Students' failure to successfully complete logical tasks shows that a change in **methodology of teaching** is necessary.

3 Basic mathematical examination done by RTU

If a student has had difficulties with basic mathematics in high school or exams, they are given the chance to attend additional lessons in RTU. The offer of these additional lessons for students with weaker understanding of basic mathematics has been in use for over five years now. Though it was offered as more of an offer for those who wished, starting with the 2016/17 academic year, it's been marked as obligatory for all students who fail a specific test at the beginning of the year.

As experience shows, a lot of the students accepted into the university do not have required level of knowledge in basic mathematics to successfully finish their studies in calculus and other technical subjects. During the first mathematics lesson of studies, students are given tests to determine how good their understanding of basic mathematics is. This test consists of five simple problems: operations with fractions, the expression of a variable from a linear relationship, construction of a line and a parabola in a plane, basic properties of powers and basic properties of logarithms. Each problem is worth 2 points. The test is successfully completed if student gets at least 4 points. In the academic year of 2017/18 1553 first year students did this test. Sadly, almost half of them (38%) couldn't correctly complete even two of these five problems.

Fig. 3. A pie chart showing the percentage of students that could and could not successfully complete the test with a passing mark.

Only 6% of students solved all the problems correctly and received the maximums 10 points, however 7% got 0 points, as in, they couldn't even begin to solve these problems correctly. The most common mark is 2 points, which means that the student completed one problem successfully, or did two problems halfway. The Figure 4 shows the percentages of marks.

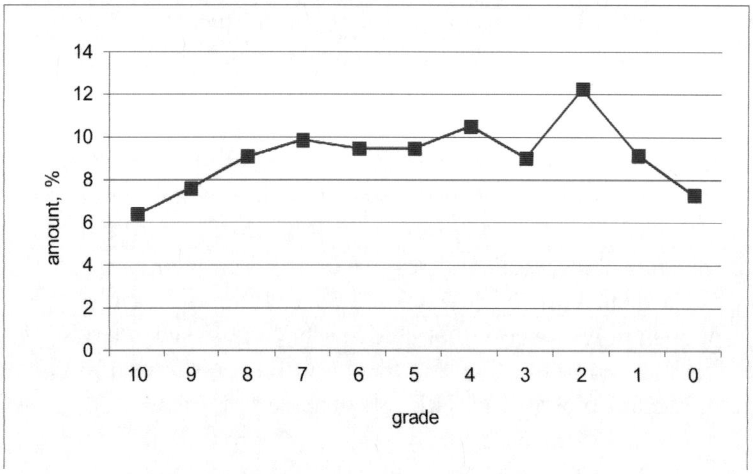

Fig. 4. Percentages of marks among students.

Comparing these results to the last five years (which are shown in Figure 5), we can see that they change every year, getting better one year, but worsening the next one.

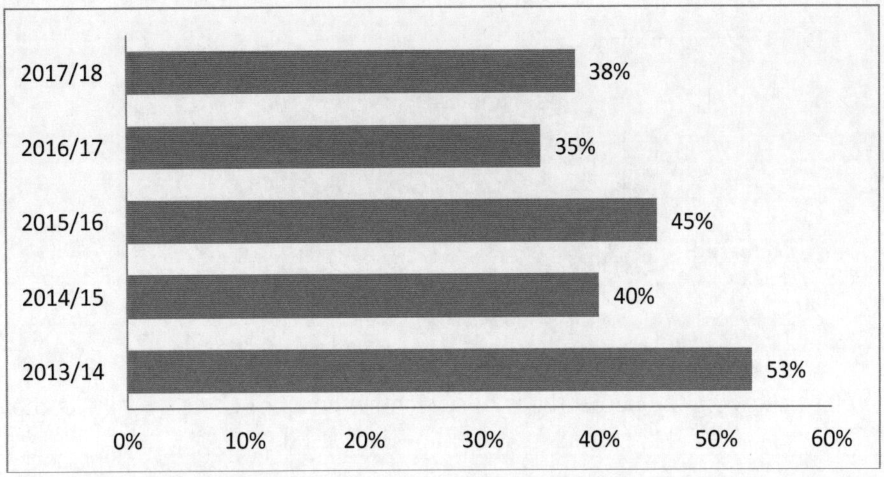

Fig. 5. A comparison of unsuccessful students in the last 5 years.

We made a comparison between the CE results that students receive in school and the tests students did in one of RTU's faculties (Faculty of Computer Science and Information Technology). We chose the faculty where results were the best. The comparison is seen in Figure 6. The horizontal axis shows CE results, but the vertical one shows the RTU test results. Considering that the tests done in RTU are much easier than the centralized state examination, the results should be better. Drawing a diagonal over the graph, we can see that there are more points under it than above it, which means that most students had worse marks than the CE results.

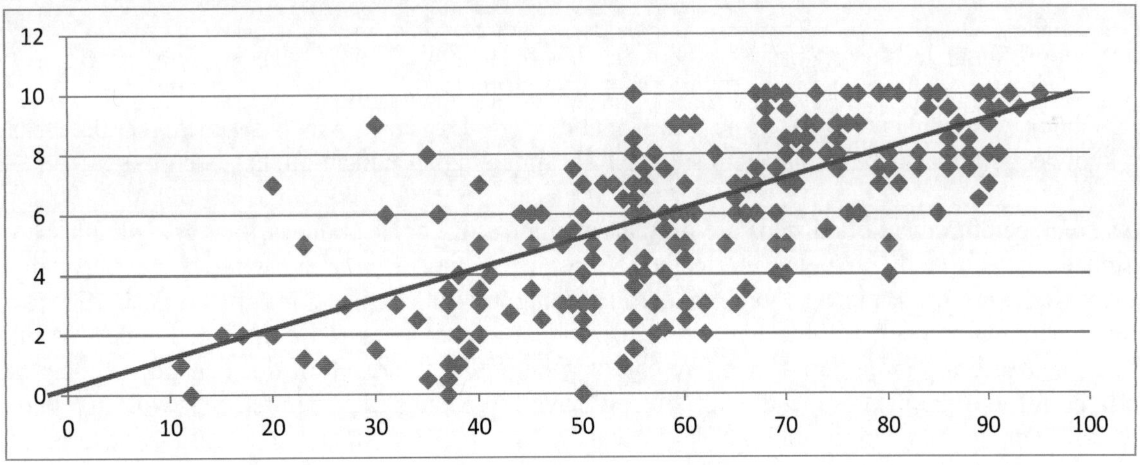

Fig. 6. A comparison between students' CE results and the basic mathematics test results.

Roughly comparing 100% on the CE to 10 points in the test done by RTU, 90% - 9 points etc., we can see that the marks of 47% of students get worse. From this we can conclude that the knowledge of students is unstable, and, instead of preparing for studies, students instead manage to forget what they've learned. This leads to the aforementioned conclusion, that it is necessary to change the methodology of teaching.

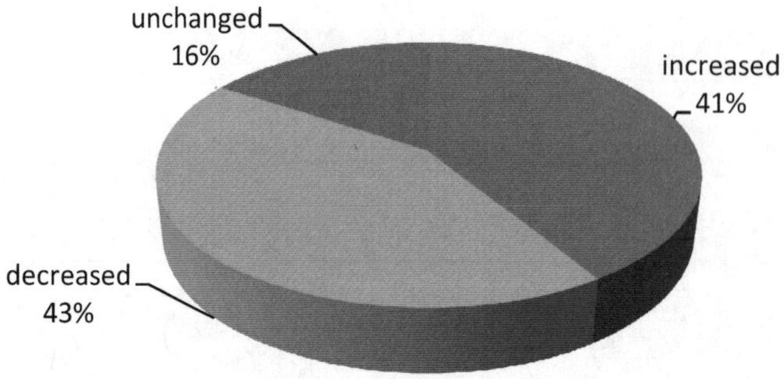

Fig. 7. A pie chart showing the percentages of students whose RTU test results, compared to CE results increased or decreased.

4 Search for the solution

The lecturers of RTU Department of Engineering Mathematics have taken a series of steps to help those students that do not have enough knowledge of basic mathematics.

1) Two lessons a week are dedicated to studying basic mathematics for first year students. It is obligatory for those who failed the basic mathematics test at the beginning of the year, but optional for all other students.
2) A video lecture course that is freely available for all on the YOUTUBE website has been created.
3) A support course covering basic mathematics has been created on the website mooc.rtu.lv.
4) There are materials available on the RTU website ORTUS for students to use.
5) A book "Fundamentals of Basic Mathematics" (in Latvian), which teaches the necessary knowledge of basic mathematics to successfully finish higher education, has been issued.

However, the aforementioned steps are only an attempt to fix consequences. The goal should be to eliminate the causes by changing the methodology of teaching. Beginning with the next academic year, a gradual transition into a thorough learning content will take place in Latvia. In accordance to it, the teaching approach will be changed to ensure that pupils develop the skills and understanding of science necessary in modern times. We can only hope that the result of this change will be what we hope for and the next students will not only have the basic skills required, but will also know how to adapt to new circumstances, develop the better parts of their personalities and create a specific system of values.

5 Conclusions

1) The state centralized examination should be divided into two levels. One would be for those who wish to further study in university, the second – for everyone else. The exam should be made much more serious for the first group. However, if a person decides to study in university after graduation, it should be possible to take the exam at a later time.
2) Almost half of the students accepted into RTU do not have the necessary level of knowledge in basic mathematics to successfully finish their studies. Not to be forced to leave university during the first session, they should be given extra help with learning basic mathematics.

3) Concerning suggestions in bettering the mathematics programs, it is recommended to integrate mathematics applications (MathCad and MatLab) into the study process and teach students the practical uses of mathematics.

References

[1] http://visc.gov.lv/vispizglitiba/eksameni/statistika/2017/
[2] https://www.skola2030.lv/
[3] http://www.iespejutilts.lv/

Current address

Cernajeva Sarmite, lecturer
Department of Engineering Mathematics, Institute of Applied Mathematics, Faculty of Computer Science and Information Technology, Riga Technical University
2 Daugavgrivas street, Riga, LV-1007, Latvia
Tel. number: (+371) 67089528, e-mail: sarmite.cernajeva@rtu.lv

Volodko Inta, professor
Department of Engineering Mathematics, Institute of Applied Mathematics, Faculty of Computer Science and Information Technology, Riga Technical University
2 Daugavgrivas street, Riga, LV-1007, Latvia
Tel. number: (+371) 67089528, e-mail: inta.volodko@rtu.lv

Iltina Marija, associate professor
Department of Engineering Mathematics, Institute of Applied Mathematics, Faculty of Computer Science and Information Technology, Riga Technical University
2 Daugavgrivas street, Riga, LV-1007, Latvia
Tel. number: (+371) 67089528, e-mail: marija.iltina@rtu.lv

Iltins Ilmars, professor
Department of Engineering Mathematics, Institute of Applied Mathematics, Faculty of Computer Science and Information Technology, Riga Technical University
2 Daugavgrivas street, Riga, LV-1007, Latvia
Tel. number: (+371) 67089528, e-mail: ilmars.iltins@rtu.lv

Slovak University of Technology in Bratislava
Faculty of Mechanical Engineering
17th Conference on Applied Mathematics **APLIMAT 2018**

Proceedings

CONFORMAL MAPPINGS OF RIEMANNIAN MANIFOLDS PRESERVING THE GENERALIZED EINSTEIN TENSOR

CHEREVKO Yevhen (UA), BEREZOVSKI Vladimir (UA), CHEPURNA Olena (UA)

Abstract. We study conformal mappings preserving the generalized Einstein tensor. We have derived corresponding partial differential equations and their integrability conditions. In addition to the generalized Einstein tensor we got other invariants of the mappings. Also we have proved that orientable compact manifolds equipped by positive definite metric, do not admit conformal mappings preserving the generalized Einstein tensor.

Keywords: Affine connection, Riemannian manifolds, conformal mappings, generalized Einstein tensor

Mathematics subject classification: Primary 53B20; Secondary 53C21, 53B50

1 Introduction

Diffeomorphisms preserving certain geometric objects are being given much attention of many researchers in the differential geometry realm.

In particular, conformal mappings which preserved the Einstein tensor

$$E_{ij} = R_{ij} - \frac{Rg_{ij}}{n}$$

studied in [1]. Preserving the stress-energy tensor

$$S_{ij} = R_{ij} - \frac{Rg_{ij}}{2}$$

by conformal mappings was explored in [4], [2]. It's worth for noting that in many classical issues e. g. [7, p. 359], just the latter is referred to as the Einstein tensor. Let us refer to

$$\mathfrak{E}_{ij} \stackrel{\text{def}}{=} R_{ij} - \kappa R g_{ij}. \tag{1}$$

as **the generalized Einstein tensor**. Here κ is a constant.

2 Conformal mappings of Riemannian manifolds

Let (M^n, g) and $(\overline{M}^n, \overline{g})$ be n-dimensional Riemannian manifolds with metric tensors g_{ij} and $\overline{g}_{ij}$ respectively. Both metrics are defined in a common coordinate system (x^i).

Definition *The correspondence between (M^n, g) and $(\overline{M}^n, \overline{g})$ is conformal, if the fundamental tensors g_{ij} and $\overline{g}_{ij}$ of two manifolds M^n and $\overline{M}^n$ are in the relation*

$$\overline{g}_{ij}(x) = e^{2\varphi(x)} g_{ij}(x), \tag{2}$$

where $\varphi(x)$ is a function of the x's.

Connections Γ^k_{ij} and $\overline{\Gamma}^k_{ij}$ compatible with the metrics g_{ij} and $\overline{g}_{ij}$ respectively must satisfy the equations

$$\overline{\Gamma}^k_{ij} = \Gamma^k_{ij} + \delta^k_i \varphi_j + \delta^k_j \varphi_i - \varphi^k g_{ij}, \tag{3}$$

where $\varphi_i = \frac{\partial \varphi}{\partial x^i}$. Also we have the equations [3], [6]

$$\overline{R}^h_{ijk} = R^h_{ijk} + \delta^h_k \varphi_{ij} - \delta^h_j \varphi_{ik} + g^{hl}(\varphi_{lk} g_{ij} - \varphi_{lj} g_{ik}) + (\delta^h_k g_{ij} - \delta^h_j g_{ik})\Delta_1 \varphi, \tag{4}$$

$$\overline{R}_{ij} = R_{ij} + (n-2)\varphi_{ij} + (\Delta_2 \varphi + (n-2)\Delta_1 \varphi) g_{ij}, \tag{5}$$

$$\overline{R} = e^{-2\varphi}(R + 2(n-1)\Delta_2 \varphi + (n-1)(n-2)\Delta_1 \varphi). \tag{6}$$

Here $\varphi_i = \partial_i \varphi$, $\Delta_1 \varphi = \varphi_i \varphi_j g^{ij}$, $\varphi_{ij} = \nabla_j \varphi_i - \varphi_i \varphi_j$. Also R^h_{ijk} and $\overline{R}^h_{ijk}$ are the Riemann tensors of the manifolds M^n and $\overline{M}^n$ correspondingly. We denote as $R_{ij} = R^\alpha_{ij\alpha}$ and $\overline{R}_{ij} = \overline{R}^\alpha_{ij\alpha}$ their Ricci tensors. Finally, $R = R_{ij} g^{ij}$ and $\overline{R} = \overline{R}_{ij} \overline{g}^{ij}$ are their scalar curvatures.

3 Conformal mappings preserving generalized Einstein tensor

It follows from (5) and (6) that the function φ must satisfy the system [3, p. 114]

$$\begin{aligned} \nabla_j \varphi_i = \varphi_i \varphi_j - \frac{1}{2} g_{ij} \Delta_1 \varphi + \frac{1}{n-2}\left(\overline{R}_{ij} - \frac{\overline{R}\overline{g}_{ij}}{2(n-1)}\right) \\ - \frac{1}{n-2}\left(R_{ij} - \frac{R g_{ij}}{2(n-1)}\right). \end{aligned} \tag{7}$$

It follows from (1) that the deformation of the generalized Einstein tensor can be written as

$$\overline{\mathfrak{E}}_{ij} - \mathfrak{E}_{ij} = \overline{R}_{ij} - \kappa \overline{R}\overline{g}_{ij} - R_{ij} + \kappa R g_{ij}. \tag{8}$$

Taking account of the preservation requirement, i. e. $\overline{\mathfrak{E}}_{ij} = \mathfrak{E}_{ij}$, from (8) we get

$$\overline{R}_{ij} - R_{ij} = \kappa \overline{R}\overline{g}_{ij} - \kappa R g_{ij}. \tag{9}$$

Since (9) holds we can rewrite (7) as

$$\nabla_j \varphi_i = \varphi_i \varphi_j - \frac{1}{2} g_{ij} \Delta_1 \varphi + \lambda \overline{R}\overline{g}_{ij} - \lambda R g_{ij}, \tag{10}$$

where $\lambda = \frac{2\kappa(n-1)-1}{2(n-1)(n-2)}$. Differentiating (10) covariantly with respect to x^k and the connection Γ of the manifold (M^n, g) we get

$$\begin{aligned}\nabla_k\nabla_j\varphi_i =&\nabla_k\varphi_i\varphi_j + \varphi_i\nabla_k\varphi_j - \varphi_m g^{ml}\nabla_k\varphi_l g_{ij} + \lambda\partial_k\overline{R}\overline{g}_{ij} + 2\lambda\varphi_k\overline{R}\overline{g}_{ij} - \lambda\partial_k R g_{ij} = \\ =&\left(\varphi_i\varphi_k - \frac{1}{2}g_{ik}\Delta_1\varphi + \lambda\overline{R}\overline{g}_{ik} - \lambda R g_{ik}\right)\varphi_j + \varphi_i\nabla_k\varphi_j - \\ &- \varphi_m g^{ml}\left(\varphi_l\varphi_k - \frac{1}{2}g_{lk}\Delta_1\varphi + \lambda\overline{R}\overline{g}_{lk} - \lambda R g_{lk}\right)g_{ij} + \\ &+ \lambda\partial_k\overline{R}\overline{g}_{ij} + 2\lambda\varphi_k\overline{R}\overline{g}_{ij} - \lambda\partial_k R g_{ij} = \varphi_i\varphi_k\varphi_j - \frac{1}{2}g_{ik}\Delta_1\varphi\varphi_j + \\ &+ \lambda\varphi_j\overline{R}\overline{g}_{ik} - \lambda\varphi_j R g_{ik} + \varphi_i\nabla_k\varphi_j - \frac{1}{2}g_{ij}\Delta_1\varphi\varphi_k + \lambda\varphi_k\overline{R}\overline{g}_{ij} + \\ &+ \lambda\varphi_k R g_{ij} + \lambda\partial_k\overline{R}\overline{g}_{ij} - \lambda\partial_k R g_{ij}\end{aligned} \tag{11}$$

Alternating (11) in j and k and using the Ricci identity, we obtain

$$\varphi_\alpha R^\alpha_{ijk} = 2\lambda(\frac{1}{2}\partial_k\overline{R}\overline{g}_{ij} - \frac{1}{2}\partial_j\overline{R}\overline{g}_{ik} - \frac{1}{2}\partial_k R g_{ij} + \frac{1}{2}\partial_j R g_{ik} + R(\varphi_k g_{ij} - \varphi_j g_{ik})). \tag{12}$$

The condition (12) can be rewritten as

$$\varphi_\alpha R^\alpha_{ijk} - 2\varphi_\alpha\lambda R(\delta^\alpha_k g_{ij} - \delta^\alpha_j g_{ik}) = \lambda(\partial_k\overline{R}\overline{g}_{ij} - \partial_j\overline{R}\overline{g}_{ik} - \partial_k R g_{ij} + \partial_j R g_{ik}),$$

or

$$\varphi_\alpha Z^\alpha_{ijk} = \lambda(\partial_k\overline{R}\overline{g}_{ij} - \partial_j\overline{R}\overline{g}_{ik} - \partial_k R g_{ij} + \partial_j R g_{ik}), \tag{13}$$

where

$$Z^h_{ijk} \stackrel{\text{def}}{=} R^h_{ijk} - \frac{2\kappa(n-1)-1}{(n-1)(n-2)}R(\delta^h_k g_{ij} - \delta^h_j g_{ik}). \tag{14}$$

We can express φ_{ij} using (10)

$$\varphi_{ij} = -\frac{1}{2}g_{ij}\Delta_1\varphi + \lambda\overline{R}\overline{g}_{ij} - \lambda R g_{ij}.$$

If the expressions for φ_{ij} are substituted in (4), we find that

$$\overline{Z}^h_{ijk} = Z^h_{ijk},$$

where Z^h_{ijk} is defined in (14). Hence we obtain

Lemma *If manifolds (M^n, g) and $(\overline{M}^n, \overline{g})$, $(n > 3)$ are in the conformal correspondence and the mapping preserves the tensor $\mathfrak{E}_{ij} = R_{ij} - \kappa R g_{ij}$, $(\kappa = const)$, then the condition*

$$\varphi_\alpha Z^\alpha_{ijk} = \frac{2\kappa(n-1)-1}{2(n-1)(n-2)}(\partial_k\overline{R}\overline{g}_{ij} - \partial_j\overline{R}\overline{g}_{ik} - \partial_k R g_{ij} + \partial_j R g_{ik})$$

holds. Also, the tensor

$$Z^h_{ijk} = R^h_{ijk} - \frac{2\kappa(n-1)-1}{(n-1)(n-2)}R(\delta^h_k g_{ij} - \delta^h_j g_{ik})$$

is also preserved by the mapping.

Now we prove that for preserving of the generalized Einstein tensor more strong conditions must be satisfied. Let us contract the tensor Z^h_{ijk} for h and k.

$$Z^{\alpha}_{ij\alpha} = R^{\alpha}_{ij\alpha} - \frac{2\kappa(n-1)-1}{(n-1)(n-2)} R(\delta^{\alpha}_{\alpha} g_{ij} - \delta^{\alpha}_{j} g_{i\alpha}) = R_{ij} - \frac{2\kappa(n-1)-1}{(n-2)} R g_{ij}.$$

Obviously, the tensor $Z^{\alpha}_{ij\alpha} = R_{ij} - \frac{2\kappa(n-1)-1}{(n-2)} R g_{ij}$ formed by contraction also must be preserved by the mappings. Hence we obtain that if $\kappa \neq \frac{1}{n}$ the product Rg_{ij} also must be preserved, as a linear combination (constant coefficients) of the preserved tensors.

$$\mathfrak{E}_{ij} - Z^{\alpha}_{ij\alpha} = R_{ij} - \kappa R g_{ij} - R_{ij} + \frac{2\kappa(n-1)-1}{(n-2)} R g_{ij} = \frac{1-\kappa n}{n-2} R g_{ij}.$$

Similarly, it can be shown that the conformal mapping also must preserve the Ricci tensor R_{ij}. Hence the system (7) can be written in the form

$$\nabla_j \varphi_i = \varphi_i \varphi_j - \frac{1}{2} g_{ij} \Delta_1 \varphi. \tag{15}$$

Let us differentiate (15) covariantly with respect to x^k and the connection Γ of the manifold (M^n, g). We get

$$\begin{aligned}
\nabla_k \nabla_j \varphi_i =& \nabla_k \varphi_i \varphi_j + \varphi_i \nabla_k \varphi_j - \varphi_m g^{ml} \nabla_k \varphi_l g_{ij} = \\
=& \left(\varphi_i \varphi_k - \frac{1}{2} g_{ik} \Delta_1 \varphi \right) \varphi_j + \varphi_i \nabla_k \varphi_j - \varphi_m g^{ml} \left(\varphi_l \varphi_k - \frac{1}{2} g_{lk} \Delta_1 \varphi \right) g_{ij} = \\
=& \varphi_i \varphi_k \varphi_j - \frac{1}{2} g_{ik} \Delta_1 \varphi \varphi_j + \varphi_i \nabla_k \varphi_j - \frac{1}{2} g_{ij} \Delta_1 \varphi \varphi_k.
\end{aligned} \tag{16}$$

Alternating (16) in j and k and using the Ricci identity, we obtain

$$\varphi_\alpha R^{\alpha}_{ijk} = 0. \tag{17}$$

We can express φ_{ij} using (15)

$$\varphi_{ij} = \nabla_j \varphi_i - \varphi_i \varphi_j = -\frac{1}{2} g_{ij} \Delta_1 \varphi,$$

and substitute in (4). Collecting the terms we have

$$\overline{R}^h_{ijk} = R^h_{ijk},$$

hence

Theorem 3.1 *If manifolds* (M^n, g) *and* $(\overline{M}^n, \overline{g})$, $(n > 3)$ *are in the conformal correspondence and the mapping preserves the tensor* $\mathfrak{E}_{ij} = R_{ij} - \kappa R g_{ij}$, *and* $\kappa \neq \frac{1}{n}$, *then the function* φ *generating the mapping, must satisfy the system of PDE's*

$$\nabla_j \varphi_i = \varphi_i \varphi_j - \frac{1}{2} g_{ij} \Delta_1 \varphi,$$

whose conditions of integrability are

$$\varphi_\alpha R^\alpha_{ijk} = 0.$$

Also, the Riemann tensor R^h_{ijk}, the Ricci tensor R_{ij}, and the product Rg_{ij} are preserved by the mapping.

It's worth for noting that some of presented results have been obtained in [4], in particular, the equations (15) and (17).

Let us differentiate (10) covariantly with respect to x^k

$$\varphi_{\alpha,l} R^\alpha_{ijk} + \varphi_\alpha R^\alpha_{ijk,l} = 0.$$

Because of (15) and (17) it follows

$$-\frac{1}{2}\Delta_1\varphi R_{lijk} + \varphi_\alpha R^\alpha_{ijk,l} = 0. \tag{18}$$

If an explored manifold (M_n, g) is locally symmetric, then

$$R^h_{ijk,l} = 0.$$

Hence it follows from (18) that

$$\frac{1}{2}\Delta_1\varphi R_{lijk} = 0. \tag{19}$$

If a manifold (M_n, g) is recurrent, i. e. the covariant derivative of the Riemann tensor respect to x^l satisfies

$$R^h_{ijk,l} = \rho_l R^h_{ijk},$$

then because of (17), it follows from (18) that (19) holds.

Now, transvecting (17) and (18) with g^{ij} we get for Einstein maniflds

$$\frac{1}{2}\Delta_1\varphi R_{lk} = 0. \tag{20}$$

Thus, following (19) and (20), we have

Theorem 3.2 *Recurrent(in particular, symmetric) manifolds (M^n, g) $(n > 3)$, equipped by positive definite metric with the Riemann tensor which is not equal to zero, do not admit conformal mappings preserving the generalized Einstein tensor $\mathfrak{E}_{ij} = R_{ij} - \kappa R g_{ij}$, $(\kappa \neq \frac{1}{n})$. Also, Einstein manifolds equipped by positive definite metric, do not admit conformal mappings preserving the generalized Einstein tensor if they are not Ricci flat.*

Also for compact manifolds we have

Theorem 3.3 *Orientable compact manifolds (M^n, g) $(n > 2)$ equipped by positive definite metric, do not admit conformal mappings preserving the generalized Einstein tensor $\mathfrak{E}_{ij} = R_{ij} - \kappa R g_{ij}$, $(\kappa \neq \frac{1}{n})$.*

proof. Let a compact orientable manifold (M^n, g) $(n > 2)$ admits conformal mappings preserving the generalized Einstein tensor $\mathfrak{E}_{ij} = R_{ij} - \kappa R g_{ij}$, $(\kappa \neq \frac{1}{n})$. Then, the generating function φ must satisfy the equations (15):

$$\nabla_j \varphi_i = \varphi_i \varphi_j - \frac{1}{2} g_{ij} \Delta_1 \varphi.$$

Transvecting this with g^{ij}, we find

$$\nabla_i \varphi^i = -\frac{n-2}{2}\Delta_1 \varphi.$$

On the other hand, according to the Theorem of Green [9, p. 21]

$$\int\limits_{M^n} \nabla_i \varphi^i d\tau = 0,$$

where $d\tau$ is the volume element

$$d\tau = \sqrt{g} d\xi^1 \wedge d\xi^2 \wedge ... \wedge d\xi^n.$$

In this case we obtain

$$\frac{n-2}{2}\int\limits_{M^n} \Delta_1 \varphi d\tau = 0,$$

that is impossible for nontrivial conformal mappings of a manifold equipped by positive definite metric g. The theorem is proved.

4 Local structure of Riemann manifolds admitting conformal mapping preserving the generalized Einstein tensor

Let us multiply the both sides of (15) by $-e^{-\varphi}$ [5]. We get

$$-e^{-\varphi}(\nabla_j \varphi_i - \varphi_i \varphi_j) = e^{-\varphi}\frac{1}{2} g_{ij} \Delta_1 \varphi.$$

Then, putting $u = e^{-\varphi}$ from (15) we obtain

$$\nabla_j u_i = \frac{\Delta_1 u}{2u} g_{ij}. \tag{21}$$

Differentiating the multiplier $\frac{\Delta_1 u}{2u}$ we see that because of (21) any partial derivative of the multiplier is equal identically to zero

$$\frac{\partial}{\partial x^i}\left(\frac{\Delta_1 u}{2u}\right) = 0.$$

Thus, if a manifold admits nontrivial conformal mappings which preserve the generalized Einstein tensor $\mathfrak{E}_{ij} = R_{ij} - \kappa R g_{ij}$, $(\kappa \neq \frac{1}{n})$, then there exists a function u, that

$$\nabla_j u_i = c g_{ij}, \qquad c = const. \tag{22}$$

Suppose conversely that on a certain manifold (M^n, g) there exists a function satisfying (22). Multiplying the both sides of (22) by the vector u^i, we get

$$u^i \nabla_j u_i = c u_j,$$

or

$$\frac{1}{2}\frac{\partial}{\partial x^i}(\Delta_1 u) = c\frac{\partial u}{\partial x^i}.$$

It's obvious that

$$\frac{1}{2}\Delta_1 u = cu + c_2,$$

where c_2 is an arbitrary constant. We can express c using the equation and substitute the expression in (22)

$$\nabla_j u_i = \frac{\Delta_1 u}{2\big(u + \frac{c_2}{c}\big)} g_{ij}. \tag{23}$$

Since the constant c_2 is arbitrary, we can replace $u + \frac{c_2}{c}$ in (23) by $\tilde{u}$, satisfying the condition $\tilde{u} > 0$. Thus,

$$\nabla_j \tilde{u}_i = \frac{\Delta_1 \tilde{u}}{2\tilde{u}} g_{ij}. \tag{24}$$

Supposing that $\tilde{u} = e^{-\varphi}$, we derive that (15) holds. We suppose that $c > 0$, since the scalar field u can always be determined so that the conditions $c > 0$ and $u > 0$ hold. For example we can replace $u \longrightarrow u + C$, or $u \longrightarrow -u + C$, where C is a positive new constant. Obviously, the conditions of integrability of (22) are

$$u_t R^t_{ijk} = 0.$$

Thus we have

Theorem 4.1 *In order that a manifold (M^n, g) admit conformal mappings preserving the generalized Einstein tensor $\mathfrak{E}_{ij} = R_{ij} - \kappa R g_{ij}$, ($\kappa \neq \frac{1}{n}$) it is necessary and sufficient that there exist a scalar field u such that*

$$\nabla_j \nabla_i u = c g_{ij}, \qquad c = const \neq 0.$$

The problem of existence of such field was explored by P. A. Shirokov [8, p. 287]. Taking account of the Shirokov's results and using the theorem 4.1 , we have obtain

Theorem 4.2 *In order that a manifold (M^n, g) which is not Eclidean, admit conformal mappings preserving the generalized Einstein tensor $\mathfrak{E}_{ij} = R_{ij} - \kappa R g_{ij}$, ($\kappa \neq \frac{1}{n}$) it is necessary and sufficient that (M^n, g) must be irreducible and the metric g has the form*

$$ds^2 = (dx^1)^2 + (x^1)^2 h_{ts} dx^t dx^s, \qquad t, s = \overline{2, n}, \tag{25}$$

where coefficients h_{ts} do not depend on x^1. In that case only solution of the system (15) is the function generating the mapping

$$\varphi = ln \frac{1}{C_1 (x^1)^2 + C_2}, \tag{26}$$

where C_1 and C_2 are positive arbitrary constants.

If the metric g has the form (25), then the metric is said to define an Euclidean family of concentric spheres. As an example [8, p. 291] we can consider 3-dimensional space which contains such family. The metric g has the form

$$ds^2 = (dx^1)^2 + (x^1)^2 \big(g_{22}(dx^2)^2 + g_{33}(dx^3)^2\big),$$

and the function generating the conformal mapping preserving the generalized Einstein tensor is defined by the equation (26).

References

[1] BAI, Z.: *On the Riemannian space with some independent concircular transformations*, Acta Math. Sinica, vol. 14, no. 1, (1964), p. 6274.

[2] CHEREVKO, Y. V.: *Conformal transformations preserving the stress-energy tensor*, Proc. Inter. Geom. Center, vol. 5, no. 3-4, (2012), pp. 51–58.

[3] EISENHART, L. P.: *Riemannian geometry*, Princeton Landmarks in Mathematics, Princeton University Press, Princeton, NJ, 1997, ISBN 0-691-02353-0, x+306 pp., eighth printing, Princeton Paperbacks.

[4] KIOSAK, V. A.: *Conformal mappings preserving the stress-energy tensor*, Izvestia PSPU, vol. 26, (2011), pp. 98–104.

[5] MIKEŠ, I., GAVRILCHENKO, M. L., GLADYSHEVA, E. I.: *Conformal mappings onto Einstein spaces*, Vestnik Moskov. Univ. Ser. I Mat. Mekh., , no. 3, (1994), pp. 13–17, 97, ISSN 0579-9368.

[6] MIKEŠ, J., STEPANOVA, E., VANŽUROÁ, A., et al.: *Differential geometry of special mappings*, Palacký University Olomouc, Faculty of Science, Olomouc, 2015, ISBN 978-80-244-4671-4, 568 pp.

[7] PETROV, A. Z.: *New methods in the general theory of relativity*, Izdat. "Nauka", Moscow, 1966, 496 pp.

[8] SHIROKOV, P.: *Selected investigations on geometry*, 1966.

[9] YANO, K.: *Differential geometry on complex and almost complex spaces*, International Series of Monographs in Pure and Applied Mathematics, Vol. 49, A Pergamon Press Book. The Macmillan Co., New York, 1965, xii+326 pp.

Current address

Cherevko Yevhen
Dept. of Economic Cybernetics
Odessa National Economic University
Odessa 65082, Ukraine
e-mail: cherevko@usa.com

Berezovski Vladimir, Doc. CSc.
Dept. of Mathematics,
Uman National University of Horticulture
Institutskaya 1, Uman, Ukraine
e-mail: berez.volod@gmail.com

Chepurna Olena, Doc. CSc. Department
of Mathematics
Odessa State Economic University
Odessa 65082, Ukraine
e-mail: chepurna67@gmail.com

Slovak University of Technology in Bratislava
Faculty of Mechanical Engineering
17th Conference on Applied Mathematics **APLIMAT 2018**

Proceedings

GROUPS AND HYPERGROUPS OF ARTIFICIAL NEURONS

CHVALINA Jan (CZ), SMETANA Bedřich (CZ)

Abstract. When we study structure of the most used artificial neural network - multilayer perceptron and functionality of artificial neuron, there is possibility using several ways to describe function and neural network properties on the basis of known algebraic structures, vector spaces and graphs theory or properties of relations. Using certain analogy with relations between descriptions of differential equations certain quality there is developed access to new view point on these subjects. In this paper some concepts of description and modelling systems of neurons are investigated.

Keywords: Neural network, transposition hypergroups, linear ordinary differential operators, groups of neurons

Mathematics subject classification: Primary 92 B 20; Secondary 20 N 20, 43 A 62, 47 E 05

1 Introduction

The beginning of the establishment of neural networks is considered to work Warren McCulloch and Walter Pitts of 1943, which created a very simple mathematical model of a neuron, which is the basic cell of the nervous system [3, 13, 14, 25]. The numerical values of the parameter in this model were predominantly bipolar, i.e. from the set $\{-1, 0, 1\}$. They showed that the simplest types of neural networks can in principle compute any arithmetic or logic function.

A neuron called also as artificial or formal neuron is the basic stone of the mathematical model of any neural network. Its design and functionalites are derived from observation of a biological neuron that is basic building block of biological neural networks (systems) which includes the brain, spinal cord and peripheral ganglia. In case of artificial neuron the information comes into the body of an artificial neuron via inputs that are weighted (i.e. each input can be individualy multiplied with a weight). The body of an artificial neuron then sums the weighted inputs, bias and "processes" the sum with a transfer function. At the end an artificial neuron passes the processed information via outputs.

Neuron activity can be described mathematically: Capturing signal and transmission in neurons, there is created potential P:

$$P = w_1 * x_1 + w_2 * x_2 + ... + x_n * w_n n$$

If the potential is sufficiently large, the neuron transmits a signal y:

$$y = 1, if\ P > w_0,\ otherwise\ y = 0.$$

The condition that $P > w_0$ can be overridden by activating function $f(P)$. The entire activity of neurons can then enroll in one mathematical relationship where w_0 is a negative number that represents the threshold that shall overcome potential. Formally, the transfer function can have a zero threshold and a neural boundary with the negative sign being understood as the weight, the so-called bias $w_0 = {}^-\theta$ of another formal input $x_0 = 1$ with a constant unit value. The value of the internal potential y_in,where

$$\mathbf{y\_in} = \sum_{\mathbf{i=1}}^{\mathbf{n}} \mathbf{w_i x_i}.$$

after reaching the value $\mathbf{b(w0 = b - bias)}$ it invokes the output state **y** of the neuron **Y**, axon pulse. Increasing the output values $y = y\_in$ when the potential value of **b** is given by the activating (transfer) function f. In general, a single-layer neural network is not capable of solving all tasks. Therefore the most commonly used type is a multilayer feedforward network with a backpropagation learning method. From paper focus is interesting the linear transfer function. The output of a linear transfer function is equal to its input:

$$a = n,$$

as illustrated in Figure.

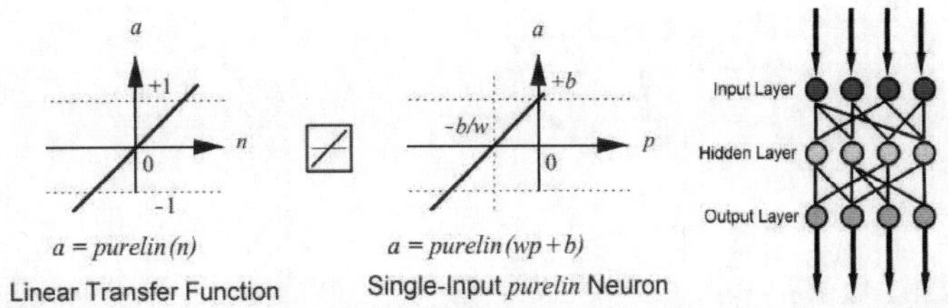

Fig. 1. Feedforward neural network and linear transfer function.

Neurons with this transfer function are used in the ADALINE networks. The output neuron expression can be written in matrix form:

$$\mathbf{y} = \mathbf{W} * \mathbf{x} + \mathbf{b},$$

where the matrix for the single neuron case has only one row. Output and input product of artificial neurons can be the same way interpreted as vectors of input or output linear vector spaces. Now the neuron output of multilayer neural network can be written according terms of matrix form description as:

$$\mathbf{y^n} = \mathbf{f^n}(\mathbf{W^n f^{n-1}}(\mathbf{W^{n-1} f^{n-2}} ... (\mathbf{W^1 x} + \mathbf{b^1}) + \mathbf{b^2}) ... + \mathbf{b^n})$$

for n-layer neural network. For further development is necessary describe basic terms from usage linear ordinary differential operators.

Artificial neural networks can be viewed as a weghted directed graphs in which artificial neurons are nodes and directed edges with weight are connections between neuron outputs and neuron inputs.

Recall that in the framework of Artificial neural networks perceptrons is a network of simple neurons called percetrons. The basic concept of a single perceptron was introduced by Rosenblatt in the year 1958. The perceptron computes a single output from multiple real-valued inputs by forming a linear combination according to its input weights and then possibly putting the output through some nonlinear activation function. As usually mathematically this can be written as:

$$y = \varphi\left(\sum_{i=1}^{n} w_i x_i + b\right) = \varphi(\vec{w}^T \vec{x} + b),$$

where $\vec{w} = (w_1, \ldots, w_n)$ denotes the vector of weights, $\vec{x} = (x_1, \ldots, x_n)$ is the vector of inputs, b is the bias and φ is the activation function.

2 Groups and hypergroups of linear differential operators

Let us mention from the history of sixties and seventies from the past century when Otakar Borůvka and his collaborators and successors begun with the investigation of differential equations using the algebraic and geometrical approach.

The substanding representative of the mentioned school František Neuman wrote in his paper [15]: "Algebraic, topological and geometrical tools together with the methods of the theory of dynamical systems and functional equations make possible to deal with problems concerning global properties of solutions by contrast to the previous local investigations and isolated results." Influence of mentioned ideas is a certain motivating factor of our investigations.

So, we consider linear ordinary differential operators of the form:

$$L_n = \sum_{k=0}^{n} p_k(x) D^k,$$

where $D_k = \frac{d^k}{dx^k}$, $p_k(x)$ is a continuous function on some open interval $J \subset \mathbb{R}$, $k = 0, 1, \ldots, n-1$, $p_n(x) \equiv 1$, i.e. $L_n(y) = 0$ which is a linear homogenous ordinary differential equation of the form:

$$y^(n)(x) + \sum_{k=0}^{n-1} p_k(x) y^{(k)}(x) = 0.$$

By an ordered group we mean (as usually) a triad $(G, \cdot, \leq)$, where $(G, \cdot)$ is a group and $\leq$ is a reflexive, symmetrical and transitive binary relation on the set G such that for any triad $x, y, z \in G$ with the property $x \leq y$ also $x \cdot z \leq y \cdot z$, $z \cdot x \leq z \cdot y$ is satisfied. Further, $[a)_{\leq} = \{x \in G; a \leq x\}$ is the principal end generated by $a \in G$. To any element $a \in G$ there is assigned a pair of mappings $\lambda_a : G \to G, \rho_a : G \to G$, which are called a left translation, a right translation, respectively, determined by the element $a \in G$, i.e. $\lambda_a(x) = a \cdot x$, $\rho_a(x) = x \cdot a$.(Of course, in the case of a commutative group $(G, \cdot)$ we have $\lambda_a \equiv \rho_a$). Notice, that a group with an ordering $(G, \cdot, \leq)$ is an ordered group if and only if all its left and right translations $\lambda_a, \rho_a, a \in G$ are order-preserving mappings, i.e. isotone selfmaps of the ordered set $(G, \leq)$.

The following lemma which is crutial for further constructions is proved in [5, 7] (the Czech version is proved in [6], pp. 146, 147).

Lemma 1. *Let* $(G, \cdot, \leq)$ *be an ordered group. Define a hyperoperation* $* : G \times G \to \mathcal{P}(G)^*$ *by*

$$a * b = [a, b)_{\leq}(= \{x \in G; a \cdot b \leq x\})$$

for all pairs of elements $a, b \in G$. Then $(G, *)$ is a hypergroup which is commutative if and only if the group $(G, \cdot)$ is commutative.

Application of the above lemma and many new results are obtained in papers of Michal Novák. See at least titles [20, 21, 22].

For present following results we use similar notation published in [7]. So there $\mathbb{R}$ stands for the set of all reals, $J \subset \mathbb{R}$ is an open interval (bounded or unbounded) of real numbers, $\mathbb{C}^k(J)$ is the ring (with respect to usual addition and multiplication of functions) of all real functions with continuous derivatives up to the order $k \geq 0$ including. We write $\mathbb{C}(J)$ instead of $\mathbb{C}^0(J)$. For a positive integer $n \geq 2$ we denote by $\mathbb{A}_n$ the set of all linear homogeneous differential equations of the *n*-th order with continuous real coefficients defined on *J*, i.e.

$$y^{(n)} + p_{n-1}(x)y^{(n-1)} + \cdots + p_0(x)y = 0,$$

(cf. [5, 7, 14, 15, 19]), where $p_k \in \mathbb{C}(J)$, $k = 0, 1, \cdots, n-1$, $p_0(x) > 0$ for any $x \in J$ (this is not essential restriction). Denote $L(p_0, \cdots, p_{n-1}) : \mathbb{C}^n(J) \to \mathbb{C}^n(J)$ the above defined linear operator defined by

$$L(p_0, \cdots, p_{n-1})(y) = y^{(n)} + p_{n-1}(x)y^{(n-1)} + \cdots + p_0(x)y$$

and put

$$\mathbb{LA}_n(J) = \{L(p_0, \cdots, p_{n-1}); p_k \in \mathbb{C}(J), p_0 > 0\}.$$

Further $\mathbb{N}_0(n) = \{0, 1, \cdots, n-1\}$ and δ_{ij} stands for the Kronecker δ, $\overline{\delta_{ij}} = 1 - \delta_{ij}$. For any $m \in \mathbb{N}_0(n)$ we denote by $\mathbb{LA}_n(J)_m$ the set of all linear differential operators of the $n-$th order $L_0(p_0, \cdots, p_{n-1}) : \mathbb{C}^n(J) \to \mathbb{C}(J)$, where $p_k \in \mathbb{C}(J)$ for any $k \in \mathbb{N}_0(n)$, $p_m \in \mathbb{C}_+(J)$, (i.e. $p_m(x) > 0$ for each $x \in J$). Using the vector notation $\vec{p}(x) = (p_0(x), \cdots, p_{n-1}(x))$, $x \in J$ we can write $L_n(\vec{p}_0)y = y^{(n)} + (\vec{p}(x), (y, y', \cdots, y^{(n-1)}))$, (i.e. a scalar product).

We define a binary operation "$\circ_m$" and a binary relation "$\leq_m$ on the set $\mathbb{LA}_n(J)_m$ in this way:

For arbitrary pair $L(\vec{p})$, $L(\vec{q}) \in \mathbb{LA}_n(J)_m$, $\vec{p} = (p_0, \cdots, p_{n-1})$, $\vec{q} = (q_0, \cdots, q_{n-1})$ we put $L(\vec{p}) \circ_m L(\vec{q}) = L(\vec{u})$, $\vec{u} = (u_0, \cdots, u_{n-1})$, where

$$u_k(x) = p_m(x)q_k(x) + (1 - \delta_{km})p_k(x),\ x \in J$$

and $L(\vec{p}) \leq L(\vec{q})$ whenewer $p_k(x) \leq q_k(x)$, $k \in \mathbb{N}_0(n)$, $p_m(x) = q_m(x)$, $x \in J$. Evidently, $(\mathbb{LA}_n(J)_m, \leq_m)$ is an ordered set.

In paper [7] there is presented the sketch of the proof of the following lemma:

Lemma 2. *The triad* $(\mathbb{LA}_n(J)_m, \circ_m, \leq_m)$ *is an ordered (noncommutative) group.*

In what follows we will construct a group and hypergroup of artificial neurons using the above mentioned approach.

3 Groups and hypergroups of artificial neurons

As it is mentioned in the dissertation [3] neurons are the atoms of neural computation. Out of those simple computational units all neural networks are build up. The output computed by a neuron can be expressed using two functions $y = g(f(w, x))$. The details of computation consist in several steps: In a first step the input to the neuron, $x := \{x_i\}$, is associated with the weights of the neuron, $w := \{w_i\}$, by involving the so-called propagation function f. This can be thought as computing the activation potential from the pre-synaptic activities. Then from that result the so-called activation function

g computes the output of the neuron. The weights, which mimic synaptic strenght, constitute the adjustable internal parameters of the neuron. The process of adapting the weights is named learning.

From the biological point of view it is advisible to use an integrative propagation function. And therefore convenient choice would be to use the weighted sum of the input $f(w,x) = \sum_i w_i x_i$, that is the activation potential equals to the scalar product of input and weights. In fact, the most popular propagation function since the dawn of neural computation, however it is ofen used in the slightly different form:

$$f(w,x) = \sum_i w_i x_i + \Theta, \qquad (*)$$

The special weight Θ is called bias. Applying $\Theta(x) = 1$ for $x > 0$ and $\Theta(x) = 0$ for $x < 0$ as the above activation function yields the famous perceptron of Rosenblatt. In that case the function Θ works as a threshold.

Besides $(*)$ there are, of course, many other possible propagation functions. If $(*)$ is supplemented with the identity as activation function and real-valued domains are given a real linear neuron $\mathsf{y} = \sum_i w_i x_i + \Theta$ is obtained. This real linear neuron can be seen as an example of a Clifford neuron. Denoting by $\mathfrak{C}_{p,q,r}$ the unique universal Clifford algebra corresponding to a standard quadratic space $(\mathbb{R}^{p+q+r}, Q)$, $Q(x) = x^2$ and by $\otimes_{p,q,r}$ the geometric product of the algebra $\mathfrak{C}_{p,q,r}$ (cf. [3] ch. 2, part 2.1) we can say that a Clifford Neuron(CN) computes the following function from $(\mathfrak{C}_{p,q,r})^n$ to $\mathfrak{C}_{p,q,r}$:

$$\mathsf{y} = \sum_{i=1}^{n} w_i \otimes_{p,q,r} x_i + \Theta.$$

([3], Definition 3.1).

It is to be noted that deep neural networks contain multiple non-linear hidden layers and this makes them very expressive models that can learn very complicated relationships between their inputs and outputs.

To estimate networks parameters Hinton et all [25] proposed the wake-sleep algorithm. As models serve $q_i = \sigma\left(\sum_j s_i \Phi_{ij} + \Phi_{0j}\right)$ for the position "wake"
and $p_j = \sigma\left(\sum_k s_k \Theta_{kj} + \Theta_{0j}\right)$ for the position "sleep". Similar probability functions occure within the model description of the socalled Restricted Boltzmann Machines - [25].

The formal description of the architecture of the general Neural Abstractions Pyramid contains formally similar function mentioned above, given by the mathematical model of an artificial neuron. The basic processing element consists of the P_{kl} projection units and a single output unit. The activity $a^t_{ijkl} \in \mathbb{R}$ of the feature cell at position (i,j) for feature array k in layer l at time t is computed as follows:

$$a^t_{ijkl} = \psi_{kl}\left(\sum_{p=1}^{P_{kl}} v^p_{kl} b^{l_p}_{ijkl} + v^0_{kl}\right).$$

The output unit computes a weighted sum of the projection potentials $b^{tp}_{ijkl} \in \mathbb{R}$ with the weighting factors described by $v^p_{kl} \in \mathbb{R}$. A bias value of v^0_{kl} is also added to the sum before it is passed through the output transfer function ψ_{kl}.

The computation of the individual projection potentials is described by:

$$b_{ijkl}^{tp} = \varphi_{kl}\left(\sum_{q=1}^{Q_{kl}^{p}} w_{kl}^{pq} a_{i\acute{}j\acute{}k\acute{}l\acute{}}^{t\acute{}} + w_{kl}^{p0}\right).$$

The processing element that computes a feature cells consists of P_{kl} projection units and output unit that produces the activity a_{ijkl}^{t}. The output unit computes the weighted sum of the potentials b_{ijkl}^{tp} of the individual projections and passes this sum through a transfer function ψ_{kl}. Each projection unit computes the weighted sum of activities $a_{i\acute{}j\acute{}k\acute{}l\acute{}}^{l\acute{}}$ with the weighting factors described by $w_{kl}^{pq} \in \mathbb{R}$. The number of contributions to a projection p is Q_{kl}^{p}. In addition, a bias value of w_{kl}^{p0} is added before the sum is passed trough the projection transfer function ϕ_{kl}^{p}.

So, recall the well-known mathematical description of a formal neuron:

Let $F : \mathbb{R} \to \mathbb{R}$ be a general non-linear (or piece-wise linear) transfer function. Then the action of a neuron can be expressed as this model:

$$y(k) = F\left(\sum_{i=1}^{m} w_i(k)x_i(k) + b\right),$$

where $x_i(k)$ is input value in discrete time k where i goes from 0 to m, $w_i(k)$ is weight value in discrete time where i goes from 0 to m, b is bias, $y_i(k)$ is output value in discrete time k.

Notice that in some very special cases the transfer function F can be also linear. Transfer function defines the properties of artificial neuron and can be any mathematical function. Usually it is chosen on the basis of problem that artificial neuron(artificial neural network) needs to solve and in most cases it is taken(as mentioned above) from the following set of functions: step function, linear function and non-linear(sigmoid) function.

In what follows we will consider a certain generalization of classical artificial neurons mentioned above consisting in such a way that inputs x_i and weight w_i will be functions of an argument t belonging into a linearly ordered (tempus) set T with the least element 0. As the index set we use the set $\mathbb{C}(J)$ of all continuous functions defined on an open interval $J \subset \mathbb{R}$. So, denote by W the set of all non-negative functions $w : T \to \mathbb{R}$ forming a subsemiring of the ring of all real functions of one real variable $x : \mathbb{R} \to \mathbb{R}$. Denote by $Ne(\vec{w}_r) = Ne(w_{r1}, \ldots, w_{rn})$ for $r \in \mathbb{C}(J)$, $n \in \mathbb{N}$ the mapping

$$y_r(t) = \sum_{k=1}^{n} w_{r,k}(t)x_{r,k}(t) + b_r$$

which will be called the artificial neuron with the bias $b_r \in \mathbb{R}$. By $\mathbb{AN}(T)$ we denote the collection of all such artificial neurons.

Neurons are usually denoted by capital letters X, Y or X_i, Y_i, nevertheless we use also notion $Ne(\vec{w})$, where $\vec{w} = (w_1, \ldots, w_n)$ is the vector of weights.

We suppose - for the sake of simplicity - that transfer functions (activation functions) φ, $\sigma(or\ f)$ are the same for all neurons from the collection $\mathbb{AN}(T)$ or the role of this function plays the identity function $f(y) = y$.

Now, similarly as in the case of the collection of linear differential operators above, we will construct a group and hypergroup of artificial neurons.

Denote by δ_{ij} the so called Kronecker delta, $i, j \in \mathbb{N}$, i.e. $\delta_{ii} = \delta_{jj} = 1$ and $\delta_{ij} = 0$, whenever $i \neq j$.

Suppose $Ne(\vec{w}_r)$, $Ne(\vec{w}_s) \in \mathbb{AN}(T)$, $r, s \in \mathbb{C}(J)$, $\vec{w}_r = (w_{r1}, \ldots, w_{r,n})$, $\vec{w}_s = (w_{s1}, \ldots, w_{s,n})$, $n \in \mathbb{N}$. Let $m \in \mathbb{N}$, $1 \leq m \leq n$ be a such an integer that $w_{r,m} > 0$. We define

$$Ne(\vec{w}_r) \cdot_m Ne(\vec{w}_s) = Ne(\vec{w}_u),$$

where

$$\vec{w}_u = (w_{u,1}, \ldots, w_{u,n}) = (w_{u,1}(t), \ldots, w_{u,n}(t)),$$

$$\vec{w}_{u,k}(t) = w_{r,m}(t) w_{s,k}(t) + (1 - \delta_{m,k}) w_{r,k}(t), t \in T$$

and, of course, the neuron $Ne(\vec{w}_u)$ is defined as the mapping $y_u(t) = \sum\limits_{k=1}^{n} w_k(t) x_k(t) + b_u$, $t \in T$, $b_u = b_r b_s$. Further for a pair $Ne(\vec{w}_r)$, $Ne(\vec{w}_s)$ of neurons from $\mathbb{AN}(T)$ we put $Ne(\vec{w}_r) \leq_m Ne(\vec{w}_s)$, $w_r = (w_{r,1}(t), \ldots, w_{r,n}(t))$, $w_s = (w_{s,1}(t), \ldots, w_{s,n}(t))$ if $w_{r,k}(t) \leq w_{s,k}(t)$, $k \in \mathbb{N}$, $k \neq m$ and $w_{r,m}(t) = w_{s,m}(t)$, $t \in T$ and with the same bias. Evidently $(\mathbb{AN}(T), \leq_m)$ is an ordered set. A relationship (compatibility) of the binary operation "$\cdot_m$" and the ordering $\leq_m$ on $\mathbb{AN}(T)$ is given by this assertion analogical to the above one.

Lemma 3. *The triad* $(\mathbb{AN}(T), \cdot_m, \leq_m)$ *(algebraic structure with an ordering) is a non-commutative ordered group.*

Sketch of the proof:

1. The operation "$\cdot_m$" is evedently asociative.

2. Let $\vec{w}(t) = (w_1(t), \ldots, w_n(t))$, where $w_k(t) = \delta_{km}$ for any $t \in T$. Then the neuron $Ne(\vec{w})$ determined by $y(t) = \sum\limits_{k=1}^{n} w_k(t) x_k(t) + 1$ is the neutral element of the semigroup $(\mathbb{AN}(T), \cdot_m)$, i.e. the unit with respect to the binary operation "$\cdot_m$".

3. Inverse elements: Define

$$\bar{w}_k(t) = (w_m(t))^{-1}(w_k(t) + 1)\delta_{km} - w_k(t),$$

   $t \in T$ and any $k = 1, 2, \ldots, n$. Then the inverse element to the neuron $Ne(\vec{w})$, with $y(t) = \sum\limits_{k=1}^{n} w_k(t) x_k(t) + \mathfrak{b}$ within the semigroup $(\mathbb{AN}(T), \cdot_m)$ is the neuron $Ne(\vec{\bar{w}})$, determined by $\bar{y}(t) = \sum\limits_{k=1}^{n} \bar{w}_k(t) \bar{x}_k(t) + \mathfrak{b}^{-1}$.

4. Compatibility of the ordering relation $\leq_m$ with the binary operation "$\cdot_m$" (the substitution property):

   Suppose $Ne(\vec{w}_r)$, $Ne(\vec{w}_s)$, $Ne(\vec{w}_u) \in \mathbb{AN}(T)$ are neurons such that $Ne(\vec{w}_r) \leq_m Ne(\vec{w}_s)$, i.e. $w_{r,m}(t) \equiv w_{s,m}(t)$, $w_{r,k}(t) \leq w_{s,k}(t)$ for any index $k \in \{1, 2, \ldots, n\}$,
   $k \neq m$, $t \in T$. Denote $Ne(\vec{w}_a) = Ne(\vec{w}_r) \cdot_m Ne(\vec{w}_u)$, $Ne(\vec{w}_b) = Ne(\vec{w}_s) \cdot_m Ne(\vec{w}_u)$, where $\vec{w}_a(t) = (w_{a,1}(t), \ldots, w_{a,n}(t))$, $\vec{w}_b(t) = (w_{b,1}(t), \ldots, w_{b,n}(t))$, $t \in T$.

   For any index $k \in \{1, 2, \ldots, n\}$ we have

$$w_{a,k(t)} = w_{r,m}(t) w_{u,k}(t) + (1 - \delta_{km}) w_{r,k}(t),$$

$$w_{b,k}(t) = w_{s,m}(t)w_{u,k}(t) + (1 - \delta_{km})w_{s,k}(t),\ t \in T,$$

Since $w_{r,m}(t) \equiv w_{s,m}(t),\ w_{r,k}(t) \le w_{s,k}(t),\ k \ne m,\ t \in T$, for $k = m$ there holds:

$$w_{a,m}(t) = w_{r,m}(t)w_{u,k}(t) = w_{s,m}(t)w_{u,m}(t),\ t \in T$$

and for $k \ne m$ we have

$$w_{a,k}(t) = w_{r,m}(t)w_{u,k}(t) + w_{r,k}(t) \le w_{s,m}(t)w_{u,k}(t) + w_{s,k}(t) = w_{b,k}(t),\ t \in T,$$

thus

$$Ne(\vec{w}_r) \cdot_m Ne(\vec{w}_u) \le_m Ne(\vec{w}_s) \cdot_m Ne(\vec{w}_u).$$

Similarly we obtain that

$$Ne(\vec{w}_u) \cdot_m Ne(\vec{w}_r) \le_m Ne(\vec{w}_u) \cdot_m Ne(\vec{w}_s).$$

Denoting

$$\mathbb{AN}_1(T)_m = \{Ne(\vec{w}); \vec{w} = (w_1, \dots, w_n),\ w_k \in \mathbb{C}(T),\ k = 1, \dots, n,\ w_m(t) \equiv 1\},$$

we get the following assertion:

Proposition 1. *Let $T = \langle 0, t_0) \subset \mathbb{R},\ t_0 \in \mathbb{R} \cup \{\infty\}$. Then for any positive integer $n \in \mathbb{N},\ n \ge 2$ and for any integer m such that $1 \le m \le n$ the semigroup $(\mathbb{AN}_1(T)_m, \cdot_m)$ is an invariant subgroup of the group$(\mathbb{AN}(T)_m, \cdot_m)$.*

1. Suppose $Ne(\vec{w}_r),\ Ne(\vec{w}_s) \in (\mathbb{AN}_1(T)_m$ are arbitrary neurons. Then

$$\vec{w}_r = (w_{r,1}(t), \dots, w_{r,m}(t), \dots, w_{r,n}(t)),$$

$$\vec{w}_s = (w_{s,1}(t), \dots, w_{s,m}(t), \dots, w_{s,n}(t)),$$

where $w_{r,m}(t) \equiv 1,\ w_{s,m}(t) \equiv 1$. Evidently the formal neuron $Ne(\vec{w})$, where $\vec{w}(t) = (w_1(t), \dots, w_n(t))$ with $w_k(t) = \delta_{km},\ t \in T$, acting by

$$y(t) = \sum_{k=1}^{n} w_k(t)x_k(t) + 1,$$

which is, of course, neutral element of the group $(\mathbb{AN}(T)_m, \cdot_m)$ also belongs to $\mathbb{AN}_1(T)_m$. Denoting $Ne(\vec{w}_v) = Ne(\vec{w}_r).\ Ne(w_s),\ \vec{w}_v = (w_{v,1}, \dots, w_{v,n})$ we have

$$w_{v,m}(t) = w_{r,m}(t)(w_{s,m}(t))^{-1} = 1$$

for any $t \in T$, thus $Ne(\vec{w}_v) \in \mathbb{AN}_1(T)_m$, consequently we obtain that $(\mathbb{AN}_1(T)_m, \cdot_m)$ is a subgroup of the group $(\mathbb{AN}(T)_m, \cdot_m)$.

2. Now suppose $Ne(\vec{w}_r) \in \mathbb{AN}(T)_m,\ Ne(\vec{w}_s) \in \mathbb{AN}_1(T)_m$, where $\vec{w}_r(t),\ \vec{w}_s(t)$ are vectors of function as above. Denoting

$$Ne(\vec{w}_u) = Ne^{-1}(\vec{w}_r) \cdot_m Ne(\vec{w}_s) \cdot_m Ne(\vec{w}_r),$$

where $\vec{w}_u(t) = \vec{w}(t) = (w_{u,1}(t), \dots, w_{u,n}(t)),\ t \in T$, then $\vec{w}_{u,m}(t) = (\vec{w}_{r,m}(t))^{-1} \cdot_m \vec{w}_{s,m}(t) \cdot_m \vec{w}_{r,m}(t) = \vec{w}_{s,m}(t) = 1$ for any $t \in T$, thus $Ne(\vec{w}_u \in \mathbb{AN}_1(T)_m$, which means that

$$Ne^{-1}(\vec{w}_s) \cdot_m \mathbb{AN}_1(T)_m \cdot_m Ne(\vec{w}_s) \subset \mathbb{AN}_1(T)_m,$$

therefore the group $(\mathbb{AN}_1(T)_m, \cdot_m)$ is an invariant subgroup of the group $(\mathbb{AN}(T)_m, \cdot_m)$.

If $m,\, n \in \mathbb{N},\, 1 \leq m \leq n-1$, then a certain relationship between groups $(\mathbb{AN}_r(T)_m,\, \cdot_m)$, $(\mathbb{LA}(T)_{m+1},\, \circ_{m+1})$ is contained in the following proposition:

Proposition 2. *Let $t_0 \in \mathbb{R},\, t_0 > 0,\, T = \langle 0,\, t_0) \subset \mathbb{R}$ and $m,\, ,n \in \mathbb{N}$ are integers such that $1 \leq m \leq n-1$. Define a mapping $F : \mathbb{AN}_n(T)_m \to \mathbb{LA}_n(T)_{m+1}$ by this rule: For an arbitrary neuron $Ne(\vec{w}_r \in \mathbb{AN}_n(T)_m$, where $\vec{w}_r = (w_{r,1}(t), \ldots,\, w_{r,n}(t)) \in [\mathbb{C}(T)]^n$ we put $F(Ne(\vec{w}_r)) = L(w_{r,1}, \ldots,\, w_{r,n}) \in \mathbb{LA}_n(T)_{m+1}$ with the action :*

$$L(w_{r,1}, \ldots,\, w_{r,n})y(t) = \frac{d^n y(t)}{dt^n} + \sum_{k=1}^{n} w_{r,k}(t) \frac{d^{k-1}(t)}{dt^{k-1}},\; y \in \mathbb{C}^n(T).$$

Then the mapping $F : \mathbb{AN}_n(T)_m \to \mathbb{LA}_n(T)_{m+1}$ is a homomorphism of the group $(\mathbb{AN}_n(T)_m,\, \cdot_m)$ into the group $(\mathbb{LA}_n(T)_{m+1},\, \circ_{m+1})$.

Consider $Ne(\vec{w}_r),\, Ne(\vec{w}_s) \in \mathbb{AN}_n(T)_m$ and denote $F(Ne(\vec{w}_r)) = L(w_{r,1}, \ldots,\, w_{r,n})$, $F(Ne(\vec{w}_s = L(w_{s,1}, \ldots,\, w_{s,n})$. Denote $Ne(\vec{w}_u) = Ne(\vec{w}_r) \cdot_m Ne(\vec{w}_s)$. There holds

$$F(Ne(\vec{w}_r) \cdot_m Ne(\vec{w}_s)) = F(Ne(\vec{w}_u)) = L(w_{u,1}, \ldots,\, w_{u,n}),$$

where

$$L(w_{u,1}, \ldots,\, w_{u,n})y(t) = y^{(n)}(t) + \sum_{k=1}^{n} w_{u,k}(t) y^{(k-1)}(t).$$

Here $w_{u,k}(t) = w_{r,m+1}(t) w_{s,k}(t) + w_{r,k}(t),\, k \neq m$, and $w_{u,m+1}(t) = w_{r,m+1}(t) w_{s,m+1}(t)$. Then $L(w_{u,1}, \ldots,\, w_{u,n}) = L(w_{r,1}, \ldots,\, w_{r,n}) \cdot_m L(w_{s,1}, \ldots,\, w_{s,n}) = F(Ne(\vec{w}_r)) \cdot_m F(Ne(\vec{w}_s))$. The neutral element $Ne(\vec{w}) \in \mathbb{AN}_n(T)_m$ is also mapped onto the neutral element of the group $(\mathbb{L}_n\mathbb{A}(T)_{m+1},\, \cdot_{m+1})$, thus the mapping $F : (\mathbb{AN}_n(T)_m,\, \cdot_m) \to (\mathbb{L}_n\mathbb{A}(T)_{m+1},\, \circ_{m+1})$ is a group homomorphism.

Now, using the construction described in the above Lemma we obtain the final transpozition hypergroup (called also non-commutative join space). Denote by $\mathbb{P}(\mathbb{AN}(T)_m)^*$ the power set of $\mathbb{AN}(T)_m$ consisting of all nonempty subsets of the last set and define a binary hyperoperation

$$*_m : \mathbb{AN}(T)_m \times \mathbb{AN}(T)_m \to \mathbb{P}(\mathbb{AN}(T)_m)^*$$

by the rule

$$Ne(\vec{w}_r) *_m Ne(\vec{w}_s) = \{Ne(\vec{w}_u);\, Ne(\vec{w}_r) \cdot_m Ne(\vec{w}_s) \leq_m Ne(\vec{w}_u)\}$$

for all pairs $Ne(\vec{w}_r),\, Ne(\vec{w}_s) \in \mathbb{AN}(T)_m$. More in detail if $\vec{w}(u) = (w_{u,1}, \ldots,\, w_{u,n}),\, \vec{w}(r) = (w_{r,1}, \ldots,\, w_{r,n}),\, \vec{w}(s) = (w_{s,1}, \ldots,\, w_{s,n})$, then $w_{r,m}(t) w_{s,m}(t) = w_{u,m}(t),\, w_{r,m}(t) w_{s,k}(t) + w_{r,k}(t) \leq w_{u,k}(t)$, if $k \neq m,\, t \in T$. Then we have that $(\mathbb{AN}(T)_m,\, *_m)$ is a non-commutative hypergroup. The above defined invariant (termed also normal) subgroup $(\mathbb{AN}_1(T)_m,\, \cdot_m)$ of the group $(\mathbb{AN}(T)_m,\, \cdot_m)$ is the carried set of a subhypergroup of the hypergroup $(\mathbb{AN}(T)_m,\, *_m)$ and it has certain significant properties.

Using certain generalization of methods from [7] we obtain after investigation of constructed structures this result:

Let $T = \langle 0,\, t_0) \subset \mathbb{R},\, t_0 \in \mathbb{R} \cup \{\infty\}$. Then for any positive integer $n \in \mathbb{N},\, n \geq 2$ and for any integer m such that $1 \leq m \leq n$ the hypergroup $(\mathbb{AN}(T)_m,\, *_m)$,where

$$\mathbb{AN}(T)_m = \{Ne(\vec{w}_r);\, \vec{w}_r = (w_{r,1}(t), \ldots,\, w_{r,n}(t)) \in [\mathbb{C}(T)]^n,\, w_{r,m}(t) > 0,\, t \in T\},$$

is a transposition hypergroup (i.e. a non-commutative join space) such that $(\mathbb{AN}(T)_m,\, *_m)$ is its subhypergroup, which is

- invertible (i.e. $Ne(\vec{w}_r)/Ne(\vec{w}_s) \cap \mathbb{AN}_1(T)_m \neq \emptyset$ implies $Ne(\vec{w}_s)/Ne(\vec{w}_r) \cap \mathbb{AN}_1(T)_m \neq \emptyset$ and $Ne(\vec{w}_r)\ Ne(\vec{w}_s) \cap \mathbb{AN}_1(T)_m \neq \emptyset$ implies $Ne(\vec{w}_s)\ Ne(\vec{w}_r) \cap \mathbb{AN}_1(T)_m \neq \emptyset$ for all pairs of neurons $Ne(\vec{w}_r),\ Ne(\vec{w}_s) \in \mathbb{AN}_1(T)_m$,

- closed (i.e. $Ne(\vec{w}_r)/Ne(\vec{w}_s) \subset \mathbb{AN}_1(T)_m$, $Ne(\vec{w}_r) \setminus Ne(\vec{w}_s) \subset \mathbb{AN}_1(T)_m$ for all pairs $Ne(\vec{w}_r), /, Ne(\vec{w}_s) \in \mathbb{AN}_1(T)_m$,

- reflexive (i.e. $Ne(\vec{w}_r)\ \mathbb{AN}_1(T)_m = \mathbb{AN}_1(T)_m/Ne(\vec{w}_r)$ for any neuron $Ne(\vec{w}_r) \in \mathbb{AN}(T)_m$ and

- normal (i.e.$Ne(\vec{w}_r) * \mathbb{AN}_1(T)_m = \mathbb{AN}_1(T)_m * Ne(\vec{w}_r)$ for any neuron $Ne(\vec{w}_r) \in \mathbb{AN}(T)_m$.

Remark A certain generalization of the formal (artificial) neuron can be obtained from expression of linear differential operator of the n-th order. Recall the expression of formal neuron with inner potential $y_{-in} = \sum\limits_{k=1}^{n} w_k(t)x_k(t)$, where $\vec{x}(t) = (x_1(t), \dots, x_n(t))$ is the vector of inputs, $\vec{w}(t) = (w_1(t), \dots, w_n(t))$ is the vector of weights. Using the bias b of the considered neuron and the transfer function σ we can expressed the output as $y(t) = \sigma\left(\sum\limits_{k=1}^{n} w_k(t)x_k(t) + b\right)$.

Now consider a tribal function $u : J \to \mathbb{R}$, where $J \subseteq \mathbb{R}$ is an open interval; input are derived from the function $u \in \mathbb{C}^n(J)$ as it follows: Inputs $x_1(t) = u(t), x_2 = \frac{du(t)}{dt}, \dots, x_n(t) = \frac{d^{n-1}(t)}{dt^{n-1}}$, $n \in \mathbb{N}$. Further the bias $b = b_0\frac{d^n u(t)}{dt^n}$. As weights we use the continuous function $w_k : J \to \mathbb{R}$, $k = 1, \dots, n-1$.

Then formula

$$y(t) = \sigma\left(\sum_{k=1}^{n} w_k(t)\frac{d^{k-1}u(t)}{dt^{k-1}} + b_0\frac{d^n u(t)}{dt^n}\right)$$

is a description of the action of the neuron Dn which will be called a formal(artificial) differential neuron. This approach allows to use solution spaces of corresponding linear differential equations.

4 Conclusion

Neural nets and neural computation form wide topics with interesting history and with many applications in science and number of technical utilizations. In references there is presented a certain, but very restricted choice of publications [2, 3, 12-14, 25-30]. The new chapter in the history of neural computation is attributed with the names Rumelhart and McClelland, knowned as members of PDP group, co-authors of algorithm backpropagation. Around the time neural networks were widely recognized as leading directly forwards real artificial inteligence.Our considerations are based on algebraic approach using classical structures in which are investigated in the present time. These investigations allow further development.

Acknowledgement The second author was supported by a Brno University of Technology grant No. FEKT-S-17-4225.

References

[1] BERÁNEK, J., CHVALINA, J.: *Noncommutative join hypergroups of affine transformations of ordered fields*, In: Univ. South Bohemia, České Budějovice, Depth. Math. Report Series Vol.10 (2002), pp. 15-22.

[2] BEHNKE, S.: *Hierarchical Neural Networks for Image Interpretation*, Lecture Notes in Computer Science, vol. 2766, Springer, Heidelberg (2003), pp. 245.

[3] BUCHHOLZ, S.: *A theory of neural computation with Clifford-algebras*, Technical Report Number 0504, Christian-Albrechts-Universität zu Kiel, Institut für Informatik und Praktische Mathematik (2005), pp.1-135.

[4] CHVALINA, J.: *Commutative hypergroups in the sense of Marty and ordered sets*, Proceedings of the Summer School an General Algebra and Ordered Sets, Olomouc (Czech Republic),1994, pp.19-30.

[5] CHVALINA, J.: *Join spaces of linear ordinary differential operators of the second order*, Colloquium on Differential and Difference Equations, CDDE 2002, Proceedings (M. Bartušek, O. Došlý, ed.) Folia Fac. Sci. Nat. Univ. Masarykianae Brunensis, Mathematica 13, MU Brno 2003, p. 77-86. Masaryk Univ. Brno 1995, 305 pp. (in Czech).

[6] CHVALINA, J.: *Functional Graphs, Quasi-ordered Sets and Commutative Hypergroups*,Masaryk Univ. Brno 1995, pp. 205 (in Czech).

[7] CHVALINA, J., CHVALINOVÁ, L.: *Modelling of join spaces by n- th order linear ordinary differential operators.* In: Fourth International Conference APLIMAT 2005. Bratislava: Slovak University of Technology, Bratislava, 2005. pp. 279-284. ISBN: 80-969264-2- X.

[8] CHVALINA, J., HOŠKOVÁ, Š.: *Modelling of join spaces with compatible proximities by first-order linear partial differential operators.* Italian Journal of Pure and Applied Math. 2004, pp. 177-190.

[9] CORSINI, P.: *Prolegomena of Hypergroup Theory.* Aviani Editore Tricesimo 1993.

[10] CORSINI, P., LEOREANU, V.: *Applications of Hyperstructure Theory.* Kluwer Academic Publishers, Dordrecht/Boston/London 2003.

[11] GOODMANN, G., S., HOFFMANN-JORGENSEN, J.: *Remarks on the calculus of set-valued mappings*, Preprint Series 1968/69 No. 10, Mat. Institut Aarhus Univ., pp. 3.

[12] HAGAN, M., DEMUTH, H., BEALE, M.: *Neural Network Design*, Boston, MA: PWS Publishing, 1996.

[13] KRENKER, A., BEŠTER, J., KOS, A.: *Introduction to the artificial neural networks.* K. Suzuki (Ed.), Artificial neural networks – Methodological advances and biomedical applications, In: Tech, Rijeka (2011), pp. 3-18.

[14] McCULLOCH, W., PITTS, W.: *A logical calculus of the ideas immanent in nervous activity.* Bulletin of Mathematical Biophysics, vol. 5, 1943, pp. 115-133.

[15] NEUMAN, F.: *From local to global investigations of linear differential equations of the n-th order.* Jahrbuch Uberlicke Mathematik 1984, pp. 55-80.

[16] NEUMAN, F.: *Global Properties of Linear Ordinary Differential Equations* Academia Praha - Kluwer Academic Publishers Dordrecht/Boston/London 1991, pp. 320.

[17] NEUMAN, F.: *Global theory of ordinary linear homogenerous differential equations in the real domain.* Aequationes Math. 33 (1987), pp. 123-149.

[18] NEUMAN, F.: *Global theory of ordinary linear homogeneous differential equations in the real domain II.* Aequationes Math. 34 (1987), pp. 1-22.

[19] NEUMAN, F.: *Ordinary linear differential equations - a survey of the global theory.* Equadiff 6, Proc. Internat. Conf. Differential Equations and their Appl., Brno 1985, pp. 59-70.

[20] NOVÁK, M.: *n-ary hyperstructures constructed from binary quasi- ordered semigroups.* Analele Stiintifice Ale Universitatii Ovidius Constanta, Seria Matematica, 2014, roč. 2014 (22), č. 3, pp. 147-168. ISSN: 1224 - 1784.

[21] NOVÁK, M.: *Some basic properties of EL- hyperstructures.* European Journal of Combinatorics, 2013, roč. 2013 (34), č. 2, pp. 446-459. ISSN: 0195 - 6698.

[22] NOVÁK, M.: *On EL - semihypergroups.* European Journal of Combinatorics, 2015, roč. 44 (2015), č. B, Feb. 2015, pp. 274-286. ISSN: 0195 - 6698.

[23] RAČKOVÁ, P.: *Hypergroups of symmetric matrices*, 10th International Congress of Algebraic Hyperstructures and Applications, Proceedings of AHA 2008, University of Defence, Brno (2009), pp. 267-272.

[24] ROSENBERG, I., G.: *Hypergroups and join spaces determined by relations*, Ital. J. Pure Appl. Math 4 (1998), pp. 93 - 101.

[25] SRIVASTAVA, N., HINTON, G., KRIZHEVSKY, A., SUTSKEVER, I., SALAKHUTDINOV, R.: *Dropout: a simple way to prevent neural networks from overfitting.* J. Machine Learning Res. 15, 2014, pp. 1929–1958.

[26] STANĚK, D.: *Rough Sets on State Spaces of Automata.* In Studen EEICT, Proceedings of the 22nd conference. Brno: Vysoké učení technické v Brně, FEKT, 2016. pp. 753-757. ISBN: 978-80-214-5350- 0.

[27] TUČKOVÁ, J.: *Comparison of two Approaches in the Fundamental Frequency Control by Neural Nets.* Proceedings of the 6th Czech-German Workshop in Speech Processing, Praha, September 1996, pp.37.

[28] TUČKOVÁ, J., BOREŠ, P.: *The Neural Network Approach in Fundamental Frequency Control.* Speech Processing: Forum Phoneticum (Ed.: Wodarz,H.-W.), Frankfurt am Main, Hector Verlag. 1997, pp.143-154, ISSN 0341-3144, ISBN 3-930220-10-5.

[29] TUČKOVÁ, J., ŠEBESTA, V.: *Data Mining Approach for Prosody Modelling by ANN in Text-to-Speech Synthesis.* Proc. of the IASTED Inernational Conference on "Artificial Intelligence and Applications - AIA 2001", Ed. M. H. Hamza, ACTA Press Anaheim-Calgary-Zurich, ISBN: 0-88986-301-6, ISSN: 1482-7913, pp. 164-166, September 2001, Marbella, Spain.

[30] VOLNÁ, E.: *Neuronové sítě 1. 2.* vyd. Ostrava: Ostravská univerzita, 2008, pp. 86.

[31] WARNER, M., W.: *Semigroup, group quotient and homogeneous automata*, Inform. Control 47 (1980), pp. 59-66.

Current address

Chvalina Jan, Prof., RNDr., DrSc.
Brno University of Technology, Faculty of Electrical Engineeering and Communication, Department of Mathematics, Technická 8, 616 00 Brno, Czech Republic
Tel. number, e-mail: +420 541146058, chvalina@feec.vutbr.cz

Smetana Bedřich, RNDr.
Brno Univerzity of Defence, Faculty of Military Leadership, Department of Econometrics, Kounicova 65, 662 10 Brno, Czech Republic
Tel. number, e-mail: +420 973443429, bedrich.smetana@unob.cz

Slovak University of Technology in Bratislava
Faculty of Mechanical Engineering
17th Conference on Applied Mathematics **APLIMAT 2018**

Proceedings

AN INNOVATIVE EDUCATIONAL PATH ON MATHEMATICS AND FILM CRITICS FOR THE FLIPPED CLASSROOM

CIVERIATI Elisa (IT), TORRE Matteo (IT)

Abstract. The paper talk about an educational path that link film critics and mathematical beauty developed in a 1st class of Italian High School. The two film used to introduce the mathematical concepts are π and *Vertigo*. Some innovative aspects are contained in this educational path and the teaching method is the flipped classroom.

Keywords: scientific film, art, π, spiral

Mathematics Subject Classification: Primary 01A20, 97A40, Secondary 97A80

1 Introduction

But what ther's in a million? A lot of zeros. A nothing, a circle with a hole in the middle. This abutment, comes from the film "Sabrina" of Billy Wilder, bears witness to the way in which the film world evoked mathematics until many years ago. Yet it had greatly helped the cinema, in his birth and growth. Various mathematicians, moreover, it had loved: Gödel there went frequently, loving in particular the animated film of Walt Disney and seeing three times "Snow White". Ludwig Wittgenstein, that was not exactly a mathematician but in which philosophical thought mathematics has always been well present, adored the western and the Hollywood musicals; Norbert Wiener was so passionate about cinema to write in 1952 a letter to Alfred Hitchcock to persuade them to make a film on its Mexican adventure; Renato Caccioppoli was for many years president of the cinema club of Naples [1].

The cinema instead, until not so long ago, has ignored the aid received in its technological evolve from the mathematics and the admiration paid by many mathematicians its expressive forms. For some years, however, the situation appears to be changed radically. The interest in mathematics rose considerably. *The mathematics has become fashionable subject of theatrical performances, topic of successful film* [2]. Why this sudden interest of cinema toward the most misunderstood, and defamed discipline of human knowledge? As always happens in the explanation of a historical and social phenomenon, it is difficult to find a single cause. Recently it has been found for example that among the public of the frequenters of exhibitions and festivals as well as between that of readers of books, the interest in mathematics has exploded with force, so that the initiatives related to the World Year of mathematics, proclaimed in 2000 by UNESCO, have often become a mass phenomenon.

Moreover the role of mathematics in contemporary society has grown immensely and the cinema, mirror more or less faithful (as all the mirrors) culture of his age, it was finally noticed that the mathematicians are directly and usefully dealing with matters of biology, economy, information technology, safety, ethics, ... and many other fields of knowledge and human making, essential to the social and economic life of the world started to become globalized.

And the didactics of mathematics is prudent of the cinema? If you exclude Emmer and few other exceptions in the Italian panorama is rare to find examples of integration of cinema in the didactics of mathematics if not in only a few cases (often interesting) wherein theater performances or short films are used for the dissemination of mathematical concepts.

In this paper we will discuss an innovative and didactic proposal, experienced in the school year 2016-2017 in a first class of Liceo Scientifico "B. Cavalieri" in Verbania, composed by 28 pupils, in the context of the educational project "Adotta science and art in your classroom" [3]. The didactic proposal foresaw the introduction and the anticipation of mathematical concepts considered complex but important for ministerial guidelines, through the use of film which functioned as "lessons" according to the spirit of the flipped classroom teaching model. The teaching methodology of the flipped classroom was chosen for the numerous possibilities that it offers to all students (in particular those LD) of approach to new topics according to the times and manner determined by the student himself, as well as to be able to capture the interest of students through means of communication that they use on a daily basis. The selected films will assume, in this experimentation, the role of "real task", a category of item being sought by the flipped didactics, not easy to define and difficult to place within the Ministerial guidelines.

2 The Flipped Classroom

Before describing the merits of the didactic proposal in question, we grant you a short detour on the meaning of Flipped Classroom. It consists in reversing the traditional didactic moments, allowing students to follow the explanations at home and performing the exercises at school, to promote active learning of the student. This choice has substantially two reasons, one social and one didactic.

In the two moments envisaged by the method Flipped Classroom both teacher and student interpret new and different roles than the current ones.The teacher has no longer necessarily the role of dispenser of knowledge, but rather should be able to direct the students on the correct road that leads them to the conquest of such knowledge. The student, on the other hand, is no longer the passive auditor of a lecture, but he/she becomes responsible for his/her own learning.

The theoretical bases of this teaching model are based on doctrines of John Dewey and Maria Montessori, beyond that in modern pedagogy that prefers the appearance workshop of learning. Dewey speaks of *active* school, that school can put the child in contact with the difficulties of the outside world, encouraging and guiding it in the strategies and actions that he is caused to perform. His school is also known as *progressive*, because the child lives and develops for degrees with the succession of experiences that lead to an education ever deeper. Dewey argued that *Education is not a preparation for life, education is life itself*. The Flipped Classroom can therefore be considered a revisitation in a modern key of Dewey's active school.

The Montessori method is mainly based on the idea that the learner must be free: free to experiment spontaneously, free to express his innate creativity, free to cultivate his own authentic interests and even free to move to reach the mastery of himself and of his own body. Also in this case the education is focused on the needs of the learner and the educator plays, just as in the flipped Classroom, a role of guidance and support.

3 The didactic path

As previously anticipated, we used the Seventh Art to speak about mathematical concepts that are present in the didactic program of the first year of high school and to introduce, in an absolutely original, the concept of beauty from the mathematical point of view. The vision of the films was assigned to the students as home work; the critical comment and analysis of mathematical aspects from this emerged were treated at school. In the following we will present a quick analysis of the two films, "π" and "Vertigo", we chose for our educational path and the motivations that led us to select them.

The first film, π (as is already apparent from the title), speaks of mathematics and science in a strict way, there are no hidden references to uncover and was chosen to introduce the concept of irrational number, both for the fascination that the number π, together with other irrational, has always aroused mathematical in evoking the concept of beauty.

In fact, exist a class of Numbers that since the Antiquity fascinate human's mind, leading them to the research of symmetries and rules even where these do not exist. Numbers represent the known and the rational for excellence. Nevertheless, from Irrational Numbers of Pythagoreans to the Golden Section, many Mathematics "mysteries" have been investigated. Several of these ideas and cultural traditions have a significant influence even on our Society, so that it is interesting to wonder which are their origins and which processes do bring their diffusion.

We have chosen to propose the integral vision of the film, but it could be appropriate to submit a shortened version, operating perhaps cuts to lighten the vision of some steps of the film and to adapt the usability at the level of the class.

The second film, "Vertigo", is decidedly more "difficult" and its analysis was possible only thanks to the skills of teachers to synthesize, in a limited time, the stylistic and cinema elements that distinguish it as a film by scientific flap. The Hitchcock's film in fact, for the complex internal architecture subtended to the whole frame (not particularly complicated to follow), lends itself well to be used as a practical "laboratory" to begin the guys at the rudiments of the filmic analysis, introduced timidly at school by enthusiasts teachers, but that is slowly taking more and more a structured shape. The Hitchcock's masterpiece is not precisely a film on mathematics, but it approaches to scientific sphere in the moment when it transforms the golden spiral in a whole narration element, making it emblem and symbol. This bond, evocative and fascinating, was useful for students to approach with a more critical and attentive look to the numerous interdisciplinary links that exist between science and art, as well as to recognize that knowledge is not made to closed pouches, but it is necessary to seek those interconnections that make the knowledge complete and dynamic.

"12:45 Utter again my theories.
1 The nature speaks through mathematics.
2 All that surrounds us can be represented and understood through the numbers.
3 By plotting the graph of any numeric system it follows a diagram, then
everywhere in nature exist schemas".

This is the mantra that Max Choen, the protagonist of the film π, insistently repeats to spectators. Independent US movie with a very low budget and directed by Darren Aronofsky, π is presented in 1998 at Sundance Film Festival, the prestigious exhibition dedicated to the independent cinema, obtaining a good success with the critics and the public.

The stylistic figure of the film seems to be just very daring and characterizing: shot in 16 mm and with a black and white triturated that accentuates the contrasts, π catapult immediately the spectator inside its oneiric and alienating dimension and through a frenzied and anxiogenous race makes it sink in the mental universe of the protagonist.

Why π is in fact a "not history", is a journey inside and outside the mind brilliant and frustrated mathematician Max Choen, that, as a modern Icarus, moves too close to the sun - both physically, when he tells and relive the experience of him the child, when literally blinded for having wanted to look directly and too long to sunlight, both metaphorically, when he thinks he has found the essence of God translated in numerical sequence - and precipitates dramatically.

Icarus loses its wings of wax, Max loses the capacity to reduce everything to a schema to a "pure reason": can no longer dominate his mind, then decides to destroy it.
Despite the film departs from the canons of cinematography more traditional, you can find some topos of film genre.

Since the first sequences in fact we understand without a shadow of doubt that the mathematician Max Choen is a creature particular, above the lines, alienated and a little foolish (he suffers from very strong headaches that cause off stabbing pains and hallucinations); he has no significant contacts with other human beings and find security only in a reality that he can understand and control, which may thus be translated numerically.
Genius and madness in short, as in the best tradition.

But in this film this aspect is brought to exasperation: point of view is confusing, the spectator is accompanied with a swirling movement into and out of the hallucinations of Max, so increasingly excited and confused, until (filmic) reality and (oneiric) vision interweave, making the spectator a chaotic movement inside and outside the mind of Max, movement stressed masterfully by a soundtrack hammer and visionary, consisting mainly of electronic sounds. The paranoia and the subjectivity of the protagonist are underlined by the stylistic choice of film direction shooting always over the shoulders of Max, excluding the point of view of any other character.

Obsessed with the study of π and from research of mathematical models in nature, Max seeks at all costs to find a system able to predict any event that will happen.
His Holy Grail is a number of 216 digits that could open the door to the mysteries of the Universe, a mystical number that may lead to the essence of God as the refers the Rabbi Choen (his namesake), or that could drag to collapse the global financial system.

Stops the brain Max. Entrusted to instinct, use intuition, gently says his mentor Sol, mathematician of Russian origins, also a scholar of the sequence numbers of π. Max is a Sol which is not given for winning, are black and white, just like the tiles of the table of go, game of Chinese origin playing which the two friends compete.

Life is not only made of mathematics - says Sol to his former pupil -. *I have spent more than forty years in search of diagrams decisive and I have not found anything.*

In this duality of views which do not seem to be able to meet, the game of Go becomes a symbol of that universe on which collide with the different points of view of two mathematicians: The goban (the Go game table) is a veritable micro-word; first simple and tidy, as the white stones and black are positioned arrives chaos. The possible combinations of the game are unimaginable (though not unlimited) and unpredictable, so as to be able to say that no game has ever been played twice.

For Sol a universe uncontrollable then to Max a reality that you can manipulate, win, predict.
The more the game progresses, more moves become predictable, he says in support of his theory and of course not only refers to the fact impressed on the game board. *Life is not only made of mathematics* ... or maybe yes?

"11:15, restate my assumptions
1. Mathematics is the language of nature
2. Everything around us can be represented and understood through numbers
3. If you graph these numbers, patterns emerge. Therefore: There are patterns everywhere in nature".

The golden spiral, Leonardo and the absolute perfection of the pythagorean *tetraktys*, the Fibonacci numbers are essential topics that are strokes in the plot of the movie (Fig. 1); And in the relationships between mathematics and the cinema, precisely these issues are of particular importance and, as has already been on several occasions written [4], have a close relationship with nature, architecture and art. Without going into too much depth in fact, they are subjects easily explained and therefore possess the great advantage of being able to be shown with effectiveness to the cinema.

And for the non-mathematical undoubtedly constitute an element of considerable charm, aspect that, united to the emotional power that the filmic image is able to elicit, instils in the spectator the desire to learn more, encouraging them to discover fascinating of perfection hidden in the world to which we belong.

Fig. 1. Max draws a spiral with the sequence of 216 numbers of Fibonacci's Serie (frame by π movie).

In the film the trace of sublime beauty and perfection hidden in the world is represented by the shape of the spiral, a real obsession for the protagonist and in which he encounters continuously and inevitably. The shell found on the seashore, the milk that comes into contact with the surface of the coffee and draws sinuous elements, the ordered structure of an organism analyzed under a microscope that contains within itself this perfect shape, the wobbling motion of the leaves falling from the trees: the camera insists on these elements in a manner that always clearest, almost to say (to the protagonist, but also to all of us) that perfection is inherent in the nature, is beauty, is harmony, create itself and lasts forever and cannot therefore be reduced to mere rational model.

In the table of go, in the farewell of Sol at its most brilliant pupil, the tiles are arranged to form a spiral, almost to say that even in the chaos there is perfection, but is ineffable, unreachable. Only now Max includes, but it is too late. It has moved too close to the sun and not remains other than to dissolve his wings.

The quest for the perfect beauty, the desire to recreate a reality that cannot exist and the interest for the theme of the spiral unites and approaching the film π to another film light years away for epoch, writing, style, kind and budget. And fame.
It is "Vertigo", film of 1958 directed by Alfred Hitchcock (Fig. 2), now considered one of the masterpieces of master of thrill and one of the most important and significant films in the history of cinema.

Fig. 2. “Vertigo” movie poster.

As will be known to most of us, “Vertigo” tells the story of Scottie Ferguson, a former policeman who suffers from acrophobia, who agrees to tracking the sensual and mysterious Madeleine, the wife of one of his old friend and he considered mad; the woman in fact, believing herself the reincarnation of a suicide ancestor, is thrown by a bell tower, for then reappear under other semblances. In truth - and the policeman will discover this at its own expense - everything was a deception of the friend to get rid, unpunished, the rich wife.

Exactly as happens to great literary works, much has been written about this film and much more you could write; the risk to go beyond the things that are of interest to us is very high, then we shall limit ourselves to a careful and schematic analysis, comprehensive compared to what concerns us, but not exhaustive with respect to the completeness of the cinematographic work in itself. We will limit ourselves to emphasize how this show reminders in some way the mathematical language for essentially three aspects: the iconicity, the centrality of the theme of the spiral and the scientific rigor with which each single part was connected to the other. The iconicity is given by some peculiar choices that have helped make the work a milestone in the history of cinema.

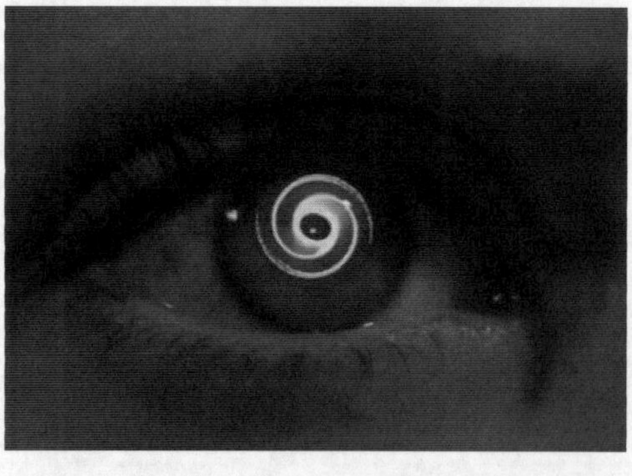
Fig. 3. the Judy’s eye in the maintitles.

One of these choices is the willingness to brighten the story in San Francisco, that with its straight and boundless roads, uphill and then downhill, makes it perfectly the idea of a movement that seems to make progress, but in reality only rotates on itself, just as the movement of the spiral, of which we will talk a little later. Iconic is the choice to make the female protagonist, Madeleine/Judy, the woman who lived twice, to quote the title with which the film was distributed in Italy, the archetype of the perfect beauty. Always remaining in this furrow, iconic is also the hairstyle of Madeleine, hair gathered with a chignon that refers to the shape of the spiral, where the gaze in subjective of protagonist lingers long.

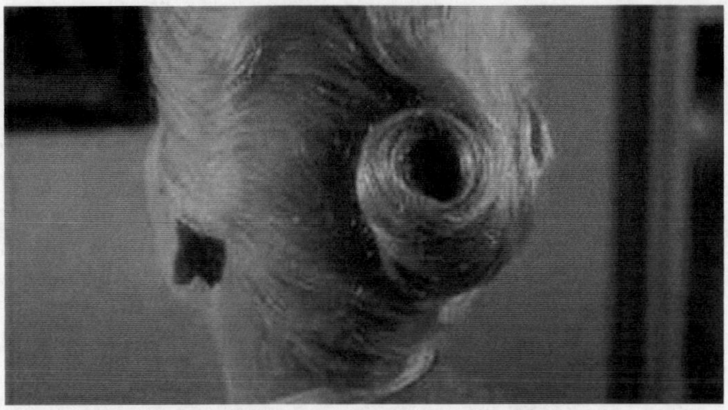

Fig. 4. Judy/Madeleine spiral chignon.

Iconic are opening credits, that never as in this case, become an integral part of the narration and indeed provide the key to access the deep meaning of the film. Comparable to the overture of an opera, represent a succession of spiral figures which were confused between them and entrain the spectator toward that dizziness of which the film is permeated [5].

The spirals and shapes that are interwoven and rotate to the infinity, which describe a movement which can be repeated, but not completed: It is no coincidence that the spiral movement is introduced and arrested by the same shot (on the first floor of a female eye) almost as if wanting to melt together the beginning and the end.

Fig. 5. The belfry staircases where Madeleine falls, taken with a particular and innovative (at the time) technique thanks to which is simulated the movement of vertiginous fall, that emembers the continuous movement of the spiral of the maintitles (frame by "Vertigo").

Throughout the film Scottie goes in search of someone who is not there, someone who pretends to be someone else. Just as Max Choen sought perfection in numbers, the protagonist of "Vertigo" search for perfection in feminine form of a woman but has never existed, but that he needs to recreate and pretend to be able to possess, though accepting to believe in a lie. Judy has almost completed the metamorphosis that will lead to be again Madeleine, but to the perfection of woman there is something missing: not enough clothes, hair color... Scottie needs to rediscover that perfection which has lost and asks to Judy a last act of submission; And Judy tidies up her hair in spiral chignon (Fig. 4) that distinguishes the hair of the woman loved by the protagonist, which now returns in all its evocative power.

Just as in the film previously analyzed, even here the unreachable perfection is plastically represented by a spiral. But even here the sublime beauty, or better, the perception of beauty, leads to ruin. The supreme act is an ascent that turns into ruinous fall, everything ends as is started, nothing more can be explained and all that remains is to persist in the oblivion of conscience. Just as for the explanation of complicated mathematical theorems it is essential to cut each step to have a clear vision and rigorous of everything, to understand deeply this film must disassemble and then rebuild it: only by joining together the various parts you can grasp the magnificent all of the film; for this reason we speak of scientific rigor, which is achieved with the willingness of the director to create a secret inner harmony that leads to the understanding of the theorem proposed by Hitchcock. Only a careful and trained spectator is able to grasp.

Everything becomes a circle, but the circle does not dose. Thus Erich Rohmer – the great French filmmaker - has defined "Vertigo" and perhaps really there is no most iconic way to describe what is the most mysterious, abstract and painful film of the ingenious "master of thrill".

4 Analysis of the educational impact and future perspectives

Make use of a movie, even not expressly mathematician, to introduce abstract and complex concepts is not new, but original was the choice of these two films. In the specific literature there is no trace of educational paths that integrate "Vertigo" with the introduction of analytic functions and geometry: this audacious attempt liked students involved in this experimentation, who have taken full advantage of the potential offered by the flipped methodology to deepen, also individually, the theme of beauty from the mathematical, artistic and cinema point of view, bringing in class numerous deep questions that have kindled a debate rich of content.

But we do not want to stop here. Our intention is to seek in the panorama of cinema films do not strictly mathematical, but that offer the possibility to be interpreted in a scientific key. Our idea is to be able to disseminate mathematical aspects complex even without resorting to biographical film on mathematicians, which too often put the emphasis on the eccentricity of the character and too little on scientific disclosure. Always bearing in mind that mathematics is synonymous with beauty. It is not by chance that the reason for which it is considered necessary to adopt in the architecture and in the art the fee pythagorean mathematician was born from the need to get as close as possible to the ideal of beauty, an objective which has always engaged the man who is preparing to theorising the things of the world.

Proportion, symmetry, harmony, order and beauty are the qualities to which man is naturally and that positively impact the psyche with a sense of wellness, balance and security.

The Swiss architect Max Bill (1908-1994) wrote: "*When we ask for what reason man tries to establish precise reports, rules and the exact measurements, we approach directly to the main problem […]: because man seeks the harmony? Because it goes in search of constructions based on exact measurements? Why organizes space in commensurate measures? The answer is very simple: man seeks to protect themselves against the unknown, against the uncertain*".

We want to try to transmit the same certainties of which he speaks to our students.

References

[1] EMMER M., *Visibili armonie*, Bollati Boringhieri, Torino, 2006

[2] BOTTAZZINI U., *Il flauto di Hilbert*, UTET, Torino 2003.

[3] BOCCARDI B., CAVICCHI V., FABBRI F. L., GIURGOLA G., LORENZI M.G., PAROLINI G., SARTORI A., SOLARI A., TORRE M., Adotta Scienza e Arte nella tua classe, progetto didattico-divulgativo per l'anno scolastico 2012-2013, *XCVIII Congresso Nazionale SIF*, Napoli 17-21 Settembre 2012.

[4] AMADEI D., FRANCAVIGLIA M., LORENZI M.G., TORRE M., The art of being irrational: $\sqrt{2}$ and φ in mathematics and architecture from Vitruvius to Leon Battista Alberti, in *Proceedings of 12th Conference on Applied Mathematics* APLIMAT 2013, Bratislava, Feb. 2013.

[5] TETI M., La vertigine in una spirale. La centralità della figura spiraliforme in "La donna che visse due volte", *Annali Online di Lettere*, Vol. 1, pp. 141-157, Ferrara, 2010.

[6] URL, http://www.miktex.org.

[7] URL, http://www.tug.org/texlive.

[8] URL, http://www.linkiesta.it/it/article/2015/11/21/vertigo-di-hitchcock-e-lossessione-per-la-donna-che-visse-due-volte

[9] URL, http://www.pellicolascaduta.it/wordpress/?p=834

[10] URL, http://www.900letterario.it/cinema/temi-stile-vertigo-hitchcock/

[11] URL, http://it.federicopistono.org/filmreview/recensione-pi-greco-il-teorema-del-delirio-1998

[12] EMMER, M., *Dionysus ex machina*, IV (2013), pp. 447-472

[13] VIGANO', D., *La camera oscura: il cinema tra memoria e immaginario*, vol. 1, 2002

[14] SESTI, M., *In quel film c'è un segreto: raccontare al cinema, raccontare il cinema*, Feltrinelli, Milano 2006.

Current address

Civeriati Elisa, high school teacher
I.C. "Tortona B"
Viale Kennedy 12, 15057 Tortona (AL), Italy
E-mail: elisa.cive@virgilio.it

Torre Matteo, high school teacher, free researcher
Liceo Scientifico Statale "L. B. Alberti"
Via Pontecurone 17, 15142 Valenza (AL), Italy
E-mail: matteo.torre1984@gmail.com

Slovak University of Technology in Bratislava
Faculty of Mechanical Engineering
17th Conference on Applied Mathematics **APLIMAT 2018**

Proceedings

ON THE IDENTITY OF SIGHT(S) IN SOME WESTERN AND EASTERN PROJECTIVE REPRESENTATIONS

COCCHIARELLA Luigi (IT)

Abstract. In the book *Vie et mort de l'image,* literally *life and death of the image* [1], the French journalist and writer Régis Debray proposed a challenging overview on iconographies and their native cultural contexts. Which calls into question the identity *behind* the image [2]. In this paper we will focus on some aspects of the adoption of axonometric and perspective representations in the Western and in the (Far) Eastern world, in search for clues or evidence revealing *identities behind.*

Keywords: geometry and graphics, symbolic forms, perspective, axonometrics, eastern and western representation, iconography, image

Mathematics Subject Classification: Primary 00A05, 00A66, 01A05, 51N05; Secondary 97U99

1 Introduction

As a part of language, graphic representation is related to the conditions under which it appears, for instance, a specific place at a specific time. Its graphic style reflects this connection. In our discussion we will consider an image as a *medium*, or a kind of bridge, between two subjects (or groups of subjects): the author(s) who makes it, and the spectator(s) who admires it. They both, together with the image, are immersed in a communication context. In order to be understood, pictorial codes embedded *inside the image* have to match the cognitive codes of the spectator *in front of it* (Fig. 1). The fact is that, since author and spectator normally live in different geographic and historical situations, the communication context is not stable, rising a barrier between those *behind the image* (authors) and those *in front of the image* (spectators). By the way, barriers are also frequent when author and spectator belong to the same community. Arthur Edwards [3] has pointed out the psychological and aesthetic recurrences among genetics and cognition, as well as their links with the collective unconscious, including visual and spatial archetypes generated and accepted in the various places and times, which form the main focus of this work. These recurrences seem the only stable point in the communication process, therefore we will try to take advantage of them, starting from the representational forms and styles here considered.

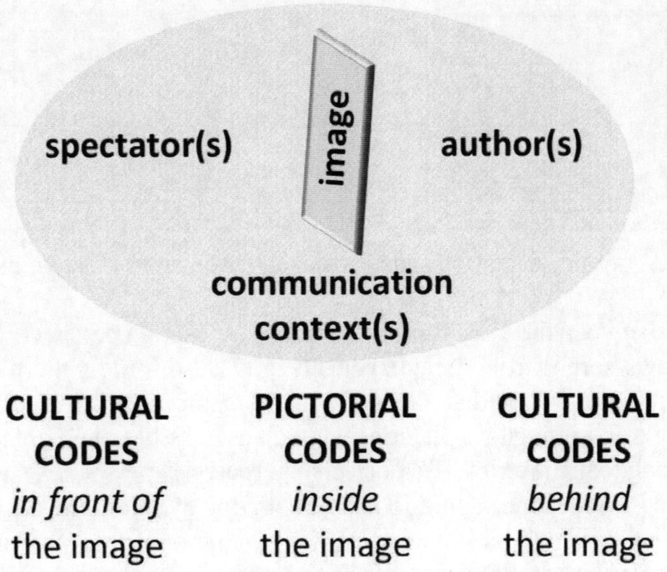

Fig. 1. The approach (Drawn by author).

2 Patterns

We all know how perspective and axonometric patterns look (Fig. 2). From former times they have been stable graphic references, intuitively used in various regions of the world, far before Projective and Descriptive Geometry were developed. In this sense, they are visual archetypes for human beings. Despite the drastic reduction of the three-dimensional space into bi-dimensional diagrams, the objects, spaces, and scenes represented are easily recognizable.

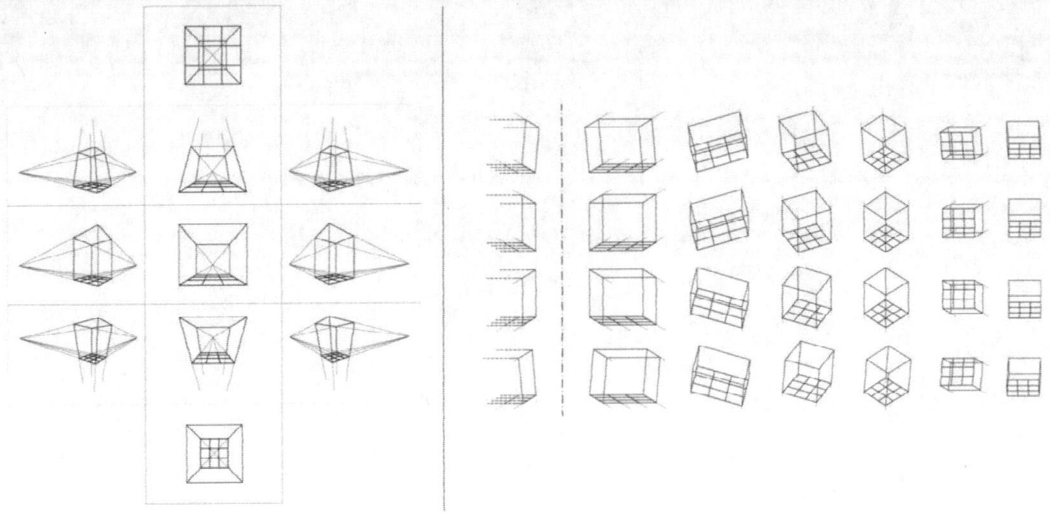

Fig. 2. Some variations in the patterns involved (Drawn by author).

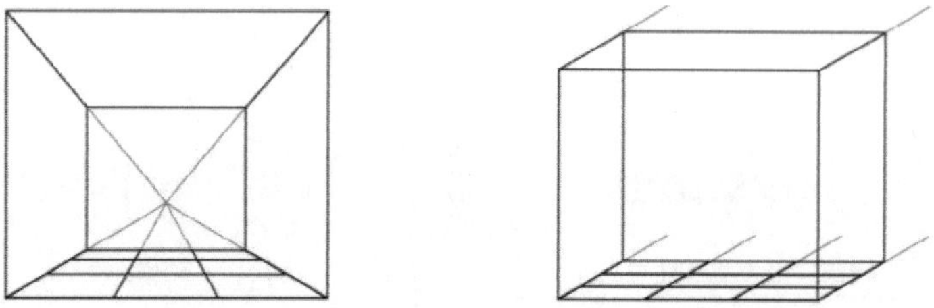

Fig. 3. The graphic sets among the most frequently used (Drawn by author).

A close comparison between the most familiar graphic sets used in perspective and axonometric representations, shows some strong peculiarities (Fig. 3). Both the patterns produce efficient images, but with different distortions. Generated by a conic projection, perspective shows a polarized graphic system aiming at replacing, as much as possible, the human visual pattern, at the expenses of 'euclidean' metrics. To correctly perceive a perspective image and the full spatial illusion it can provide, one should look at it with one eye from the designated point only. Any other viewpoint offers approximated views. Generated by a cylindrical projection, axonometrics preserves the parallelism of the lines, keeping angular distortions constant, at the expenses of similarity to visual perception. Strictly speaking, an axonometric image should be perceived from the infinity, where the projective centre is located. In praxis, due the parallelism of the projector rays, any position of the viewpoint is aligned with the centre of projection, therefore any viewpoint is supposed to be correct.

3 Symbolic forms

Fig. 4. Narrating by images: two paradigmatic cases.
Image above: Paolo Uccello (1397-1475), *Miracle of the Desecrated Host* (perspective, about 1468, detail, squared painting; source:web). Image below: Nishikawa Sukenobu (1671-1751), *Scenes in the four seasons* (axonometrics, detail, hand scroll painting; source: [5]).

However, the implications of these differences involve not only Geometry. They seem to deal with two distinguishable visions and approaches to the world. Although in the long term history both the forms have been used either in Western or in Eastern areas, perspective has been a real obsession in Italy and Europe, as well as axonometrics in the Eastern countries. Starting from Euclid, who developed Optics as a new math's branch inspired by the geometry of vision, and thanks to Filippo Brunelleschi, who formalized its projective principle, in Western Art perspective has been more than a mere representational form, it has worked as a metaphorical way of seeing the world, putting emphasis on the role of a designated human point of view in the process of knowledge. Spatial properties are forced according to the coercive graphic distortion appearing in the image, which represent a visual shot, instantaneous and ephemeral, mathematically calculated. For centuries people have been used to perceiving visual art with perspective eyes, reason for the well-known opinion by Erwin Panofsky [4], who defined perspective as a *symbolic form*. Moreover, the use of mathematics (objective system of properties) to correctly represent the visual perception (that is a subjective field) suggested Panofsky to consider perspective as *an objectification of the subjectivity*. Which is consistent with the Renaissance idea of mankind as the centre of universe, very well symbolized by Leonardo Da Vinci in the *homo ad circulum* diagram. Based on a similar reasoning, considering the wide predilection of axonometric representations in the Eastern countries, we should consider axonometrics as the *symbolic form* of the Eastern Art, reflecting a different approach. Axonometrics does not need a designated viewpoint, and due to more regular distortions occurring in the image, it suggests a visual experience more similar to the one that people can have in the real world. Perception can start from any point and proceed in any direction, changing distance of sight at any moment according to the positions of the spectator. Eastern paintings indeed, are normally painted as hand scroll paintings. Perceptual time is called into question, since the objective world represented remains independent of any prefixed optical distortion, allowing any subjective experience. For this reason, opposite to perspective, the use of axonometrics in Eastern Art seems to show the aim at *a subjectification of the objectivity*. The above mentioned arguments should be clarified by the comparison between the two artworks appearing in Figure 4. Both the paintings tell a story, but while Paolo Uccello proceeds by separate shots, hard to connect graphically, Nishikawa Sukenobu shows a continuous graphic environment, allowing a video camera effect. As proof of the fact that these representational strategies can be regarded as symbolic form, let us look out of the canvases at the real world. Just to refer to very well known paradigmatic examples, we can compare the Italian and Japanese gardens appearing below in the Figure 5.

Fig. 5. Symbolic forms beyond the canvas (source: web).
Image on the left: *Gamberaia Garden*, Florence (Italian garden, XVIII and XX Centuries).
Image on the right: *Taizo-in Temple Garden*, Kyoto (Japanese garden, XX Century).

4 Projective insights in retrospect

Fig. 6. Shadowing or (apparently) not shadowing. Image on the left: Caravaggio (1571-1610), *Saint Matthew's vocation*, 1599-1600 (squared painting; source: web). Image on the right: *Feast at the Kiyomizu Temple*, about 1615 (squared painting; source: [5]).

The story of perspective and axonometrics between Western and Eastern cultures has been sometimes controversial, sometimes characterized by cross fertilization [5]. However, the two cultures keep their profound specificities. An intriguing point is the relation between symbolic forms and religious aspects, since symbols provide iconographical support to the sacred, as clearly stated by Mircea Eliade [6]. Of course we will not enter this huge topic, but the two coeval images in the Figure 6, both realized at the beginning of the XVII century, could help us. On the left we see a masterpiece painted by Michelangelo Merisi, that is, Caravaggio. The vocation of Matthew is there masterfully symbolized by an apotheosis of light and shadows, where light means the presence of God touching and converting Matthew. The scene is perspective, showing the specific moment and the specific place of the event, from a designated viewpoint. The image on the right shows a feast in the Kiyomizu Temple, and people along a dizzy timber structure, emphasized by the axonometric pattern adopted. As usual in the Eastern paintings [7], no shadows are represented in the scene, which seems to conflict with the aim at making reliable images of the world. Many hypotheses have been proposed about this apparent lack of an important natural element. Among those, in an interesting volume on the Eastern representations, Agostino De Rosa [8] suggests that in these images the light source coincides with the centre of projection, therefore shadows exist but are hidden to the spectator, since they are formed beyond the visible objects depicted. As far as the hypothesis is convincing from a projective point of view, the proposed sacred meaning is simply astonishing. If we suppose deity be at infinity (human beings could never reach that position), therefore god himself is the projective source producing the images. Consequently, these images without shades can be seen as divine emanations. Nothing can be more different from Western perspective. There, man is in the centre, here god is in the centre.

5 Conclusions

Karl Kraus [9] stated that any painting shows its author, that is, it reveals who is *behind* the canvas. We also saw that it shows a specific cultural identity and aesthetic sensitivity [10], that is, who and what is *around* the author. At the end of this discussion, the thesis proposed by Regis Debray is clearly confirmed, since looking at history and geography we recognize some specific *visual ages* or, if we prefer, some *ages of sight*. On his opinion, indeed, the Western perspective generated a *graphosphere*, where image is considered a unique result of the work of an author, inspired by the *cult of beauty*. In the Eastern Art, instead, a *logosphere* arises, where image aims at representing the essence of things regardless of the specific author, and following a pre existing style, as in a ritual inspired by the *cult of saint*. These ideas are clear if we compare two representative icons of the two worlds: the mentioned *homo ad circulum* diagram by Leonardo Da Vinci, consistent with an anthropocentric sight, and the *Taijitu* diagram, consistent with a naturalistic sight. That is, two *identities of sight* behind the image.

References

[1] DEBRAY, R., *Vie et mort de l'image: Une histoire du regard en Occident*. Paris: Gallimard, 1992, ISBN 2070328724.

[2] COCCHIARELLA, L., Behind the Image: Codes and Suggestions Between Eastern and Western Representation. In *Proceedings of the 14th International Conference on Geometry and Graphics*, Kyoto, 2010, ISBN 978-4-9900967-1-7, CD-ROM, pp. 1-9.

[3] EDWARDS, A.M., *Images of Eden. An Inquiry into Psychology of Aesthetics*. Welwyn: Skylark, 2003, ISBN 0953611604.

[4] PANOFSKY, E., Die Perspektive als "symbolische Form". In *Worträge der Bibliothek Warburg 1924-1925.* Leipzig-Berlin, 1927, pp. 99-167.

[5] CALZA, G.C., MENEGAZZO, R. (Editors), *Giappone: potere e splendore. 1568/1868.* Milano: Federico Motta Editore, 2010, ISBN 978-88-7179-626-0.

[6] ELIADE, M., *Traité d'histoire des religions.* Paris: Payot, 1948.

[7] ANDERSON, W., *The pictorial arts of Japan*. London: L. Lowe, Marston, Searle, and Rivington, 1886.

[8] DE ROSA, A., *L'infinito svelato allo sguardo. Forme della rappresentazione estremo-orientale*. Milano: Città Studi Edizioni, 1998, ISBN 88-251-7226-5.

[9] KRAUS, K., "Nachts". In *Beim Wort genommen*. München, 1955.

[10] SARTWELL, C., *Six names of beauty.* New York: Routledge, 2004, ISBN 0415965586.

Current address

Cocchiarella Luigi, Architect, PhD., Associate Professor
Department of Architecture and Urban Studies (DASTU), School of Architecture Urban Planning Construction Engineering (AUIC)
Politecnico di Milano, Campus Leonardo
Via Bonardi 9, Bld. 14 "La Nave", 20133 Milano, Italy
E-mail: luigi.cocchiarella@polimi.it

Slovak University of Technology in Bratislava
Faculty of Mechanical Engineering
17th Conference on Applied Mathematics **APLIMAT 2018**

Proceedings

DO IMPORTANT NETWORK ACTORS FORM IMPORTANT TIES?

CORONIČOVÁ HURAJOVÁ Jana (SK), MADARAS Tomáš (SK)

Abstract. In this contribution, we analyze the location of vertices and edges with respect to maximum betweenness centrality within a graph. We present a construction of graphs where the distance of the set of vertices and the set of edges with the maximum betweenness (the betweenness separation) can be arbitrarily high, together with sufficient conditions for graphs in which this distance is zero. We also discuss the betweenness separation in real-world networks and several models of pseudo-random graphs which use to model complex networks.

Keywords: betweenness, edge betweenness, prism of graph, random graph, Barabási-Albert model

Mathematics subject classification: Primary 05C82; Secondary 05C12, 91D30

1 Introduction

Throughout this paper, we consider connected simple graphs (without loops or multiple edges). In a graph $G = (V, E)$, the distance $d_G(u, v)$ of two vertices $u, v \in V$ is the length of the shortest $u - v$-path in G. For two sets $X, Y \subseteq V$, we define $d_G(X, Y) = \min_{u \in X, v \in Y} d_G(u, v)$. The diameter diam$(G)$ of G is the largest distance of vertices of G. Other notation and terminology used here is taken from [12].

When analyzing a social network from the point of view of information spreading, one is interested in determining both the actors through which the most of information flows (because then they get more informed and have more possibilities to control this flow) as well as the links most frequently used in mutual actor communication (because its breaking can badly influence the access of actors to information). As suitable measure of amount of information flow through actors and links, one may consider betweenness centrality indices, which are defined in the following way (see [3] and [7]): for a vertex x and an edge yz of a graph $G = (V, E)$, the vertex betweenness of x is

$$B_G(x) = \sum_{\substack{u,v \in V \\ u \neq x \neq v}} \frac{\sigma^G_{u,v}(x)}{\sigma^G_{u,v}}$$

where $\sigma^G_{u,v}$ is the number of shortest paths of G between two distinct vertices u, v and $\sigma^G_{u,v}(x)$ is the number of those shortest $u-v$-paths in G having x as an internal vertex (the subscript is omitted if G is clear from context). The edge betweenness of yz is

$$B_G(yz) = \sum_{u,v \in V} \frac{\sigma^G_{u,v}(yz)}{\sigma^G_{u,v}}$$

where $\sigma^G_{u,v}(yz)$ is the number of those shortest $u-v$-paths of G that pass through yz (here u and v may coincide with y, z). These invariants are often applied in social network analysis and, also, their graph-theoretical properties are recently investigated, see [4], [5], [10] and the survey chapter [6].

Within a network, an actor with the highest betweenness appears on many shortest paths, and thus one might expect that, of the links that this actor forms, some shall have high edge betweenness too (or, less formally, important actors should form also important links). Formally, we quantify the "closeness" of vertices and edges with the maximum vertex / edge betweenness in a graph G by *betweenness separation* $\mathfrak{s}(G)$ which is equal to the distance $d_G(\mathcal{B}_v(G), \mathcal{B}_e(G))$ where $\mathcal{B}_v(G)$ is the set of all vertices of G with the maximum vertex betweenness, and $\mathcal{B}_e(G)$ is the set of endvertices of all edges of G with the maximum edge betweenness. While we show that, in general, the value of $\mathfrak{s}(G)$ may be arbitrarily large, we present the results of computations on large collections of graphs randomly generated according to particular graph distributions which suggest that, for real-world networks, the above paradigm is often true or, at least, the vertices and edges of maximum betweenness are not too distant.

2 Theoretical results

In this section, we present several results on the values of $\mathfrak{s}$ for graphs in general.

Example 1: Consider the graph K_n^{--} ($n \geq 5$) obtained from the complete graph K_n on the vertex set $\{v_1, \ldots, v_n\}$ by deleting its two nonadjacent edges, say v_1v_2 and v_3v_4. Observe that only two pairs of vertices – namely, $\{v_1, v_2\}$ and $\{v_3, v_4\}$ – add a positive contribution (equal to $\frac{1}{n-2}$) to the betweenness of a vertex of K_n^{--}; thus, the betweenness of vertices of K_n^{--} is $\frac{2}{n-2}$ or $\frac{1}{n-2}$, and there are $n-4$ vertices of maximum betweenness (the ones with the maximum degree). Similarly, for an edge v_iv_j of K_n^{--}, the only positive contributions to its edge betweenness come from the pair $\{v_i, v_j\}$ (the value 1) and, possibly, from the pairs $\{v_1, v_2\}$ and $\{v_3, v_4\}$ (each with the value $\frac{1}{n-2}$). We then conclude that the maximum betweenness of edges of K_n^{--} is $1 + \frac{1}{n-2} + \frac{1}{n-2} = \frac{n}{n-2}$ and is attained for the four edges with endvertices from $\{v_1, v_2, v_3, v_4\}$. This example also shows that the betweenness separation can be nonzero even in very dense graphs. ■

Next, we present the construction of graphs which have arbitrarily large betweenness separation.

Example 2: Given positive integers $n, d \geq 2$, let $B_{n,d}$ be the graph obtained from the n-vertex path $x_1x_2\ldots x_n$ by connecting the vertex x_n to d new vertices $y_1, \ldots, y_d$. By routine calculations, we compute the betweenness of vertices and edges of $B_{n,d}$:

$B(x_1) = B(y_i) = 0$ for each $i = 1, \ldots, d$,
$B(x_i) = (i-1)(n+d-i)$ for each $i = 2, \ldots, n-1$,
$B(x_n) = \binom{d}{2} + d(n-1)$,
$B(x_ny_i) = n+d-1$ for each $i = 1, \ldots, d$,
$B(x_ix_{i+1}) = i(n+d-i)$ for each $i = 1, \ldots, n-1$

Let k be a positive integer. Choose now n, d such that $n - d > 2(k+1)$ and $d > \frac{n}{2}$. Then the maximum edge betweenness is attained for the edge $x_i x_{i+1}$ with $i = \frac{n+d}{2}$ if $n+d$ is even, or for the pair of edges $x_{i-1}x_i, x_i x_{i+1}$ with $i = \frac{n+d-1}{2}$ otherwise. Furthermore, for $i = 2, \ldots, n-1$, the maximum of $B(x_i)$ is equal to $\left(\frac{n+d-1}{2}\right)^2$ being attained for $i = \frac{n+d+1}{2}$; the condition $d > \frac{n}{2}$ now implies that, for $i = 2, \ldots, n-1$, $B(x_n) - \max_{i=2,\ldots,n-1} B(x_i) \geq \binom{d}{2} + d(n-1) - \left(\frac{d+n+1}{2}\right)^2 = \frac{d^2}{4} + \frac{dn}{2} - d - \frac{n^2}{4} + \frac{n}{2} - \frac{1}{4} > \frac{n^2}{16} - \frac{1}{4} \geq 0$ since $n \geq 2$. Thus x_n has the maximum vertex betweenness among all vertices of $B_{n,d}$. The distance of the vertex x_n and the maximum betweenness edge is equal to $n - \left(\frac{n+d}{2} + 1\right) = \frac{n-d-2}{2} > k$. ■

On the other hand, the betweenness separation of many graphs is equal to zero. This is clearly true for vertex-transitive or edge-transitive graphs, and for nontransitive graphs of many specific families. For example, examining (by help of Wolfram Mathematica) the grids (that is, the Cartesian products $P_m \Box P_n$ of two paths on m and n vertices), one finds that the set $\mathcal{B}_v(P_m \Box P_n)$ consists either of the single vertex, or pair of adjacent vertices, or else four vertices forming 4-cycle (depending on whether m, n are both odd, or exactly one of them is odd, or both are even), which corresponds to Cartesian product of graphs induced by $\mathcal{B}_v(P_m)$ and $\mathcal{B}_v(P_n)$ (the central vertex or the pair of central vertices in path). Similarly, one finds that $\mathcal{B}_e(P_m \Box P_n)$ consists either of four edges with common endvertex (when $m = n$ is odd), or four edges forming 4-cycle (when $m = n$ is even), or two nonadjacent edges (when $m \neq n$, both being even), or two adjacent edges (when $m \neq n$, both being odd), or else a single edge. In each case, one obtains $\mathfrak{G}(P_m \Box P_n) = 0$. An analogous behaviour may be observed also in higher–dimensional grids.

The above observations for grids suggest the following sufficient condition for zero betweenness separation:

Lemma 1 *Let G be a connected k-regular graph such that the subgraph induced on $\mathcal{B}_e(G)$ contains k edges with a common vertex. Then $\mathfrak{G}(G) = 0$.*

Proof: We use the fact (see [6]) that, for any vertex $x \in V(G)$,

$$B(x) = \frac{1}{2}\left(\sum_{xy \in E(G)} B(xy) + 1 - |V(G)|\right).$$

From this, it follows that $\max_{x \in V(G)} \sum_{xy \in E(G)} B(xy)$ is attained for a vertex z of $\mathcal{B}_e(G)$ having k neighbours in $\mathcal{B}_e(G)$; thus z has the maximum betweenness among vertices of G and is incident with edges of maximum edge betweenness. □

In this lemma, the number of k maximum betweenness edge with a common vertex cannot be decreased – the cubic graph G on Figure 1 has $\mathcal{B}_v(G) = \{8, 10\}$ and $\mathcal{B}_e(G) = \{\{3,7\}, \{7,12\}\}$.

The comparison of sets of graph elements with maximum betweenness in graphs and their Cartesian products might also suggest that, for any connected graphs G and H, $\mathcal{B}_v(G \Box H) = \mathcal{B}_v(G) \times \mathcal{B}_v(H)$ and $\mathcal{B}_e(G \Box H) \subseteq \mathcal{B}_e(G) \times \mathcal{B}_e(H)$ (the symbol $\times$ here refers to Cartesian product of sets). However, these relations are not true in general: in the graph on Figure 2, $\mathcal{B}_v(G) = \{1\}$ and $\mathcal{B}_e(G) = \{\{1,5\}\}$ while $\mathcal{B}_v(G \Box K_2) = \{2, 9\}$ and $\mathcal{B}_e(G \Box K_2) = \{\{2,9\}\}$ (the vertex 9 being the counterpart of the

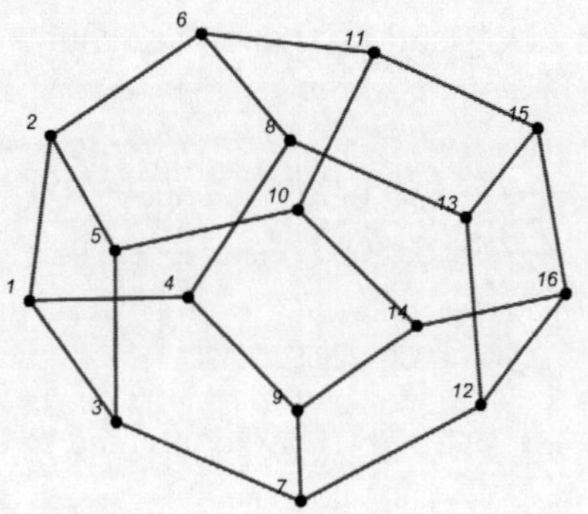

Fig. 1. A cubic graph G with $\mathfrak{d}\!\mathfrak{z}(G) = 0$ with $\mathcal{B}_e(G)$ consisting of two adjacent edges

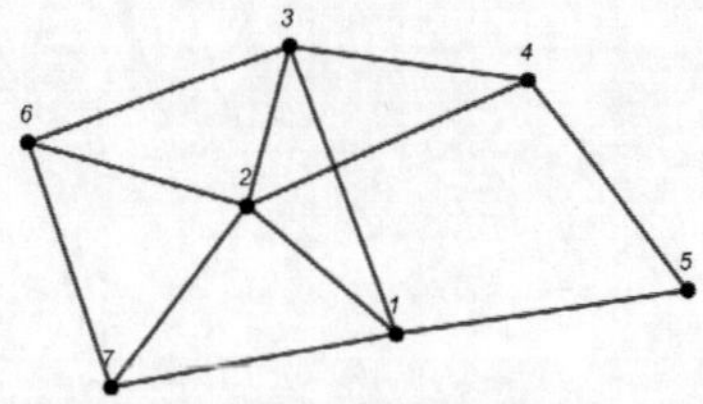

Fig. 2. A graph G whose $\mathcal{B}_v(G)$ and $\mathcal{B}_e(G)$ does not extend to $\mathcal{B}_v(G\Box K_2)$ resp. $\mathcal{B}_e(G\Box K_2)$

vertex 2 in the second copy of G in $G\Box K_2$). Using the Wolfram Mathematica computer algebra system, we have verified that this is the unique smallest graph with such property.

These relations are generally not true even for trees, as seen from the example on Figure 3: here $\mathcal{B}_v(T) = \{6\}$ while $\mathcal{B}_v(T\Box K_2) = \{1, 16\}$ (the vertex 16 being the counterpart of the vertex 2 in the second copy of T in $T\Box K_2$). We have also verified that this is the smallest tree with such property, besides other three trees on the same number of vertices.

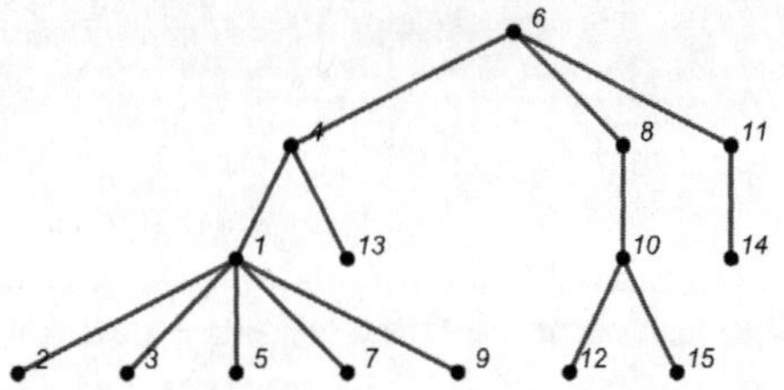

Fig. 3. A tree T whose $\mathcal{B}_v(T)$ does not extend to $\mathcal{B}_v(T\Box K_2)$

More precise information on values of vertex betweenness in graph prisms is provided by

Theorem 2 *Let G be a connected graph and $x \in V(G\Box K_2)$. Then*

$$B_{G\Box K_2}(x) = 2B_G(x) + \sum_{u,v\in V(G)\setminus\{x\}} \left(\frac{1}{d_G(u,v)+1}\right)\frac{\sigma^G_{u,v}(x)}{\sigma^G_{u,v}} + \sum_{u\in V(G)\setminus\{x\}} \frac{1}{d_G(u,x)+1}.$$

Proof: Set $H = G\Box K_2$ with $V(H) = V(G_1) \cup V(G_2)$ where G_1, G_2 are layered copies of G in H. Without loss of generality, let $x = x_1 \in V(G_1)$, $x_2 \in V(G_2)$ be the counterpart of x in G_2 (in general, to distinguish the vertices of G_1 and G_2 which are counterparts of each other, we will use subscripts 1 and 2). Then

$$B_H(x) = \sum_{u,v\in V(H)\backslash\{x\}} \frac{\sigma^H_{u,v}(x)}{\sigma^H_{u,v}} = \sum_{u,v\in V(G_1)} \frac{\sigma^H_{u,v}(x)}{\sigma^H_{u,v}} + \sum_{u,v\in V(G_2)} \frac{\sigma^H_{u,v}(x)}{\sigma^H_{u,v}} + \sum_{u\in V(G_1),v\in V(G_2)} \frac{\sigma^H_{u,v}(x)}{\sigma^H_{u,v}} =$$

$$A + B + C.$$

If $u, v \in V(G_1)$ then all shortest $u - v$-paths in H are contained in G_1, thus

$$A = \sum_{u,v\in V(G_1)} \frac{\sigma^H_{u,v}(x)}{\sigma^H_{u,v}} = \sum_{u,v\in V(G_1)} \frac{\sigma^{G_1}_{u,v}(x)}{\sigma^{G_1}_{u,v}} = B_G(x)$$

and (because $x \in V(G_1)$)

$$B = \sum_{u,v\in V(G_2)} \frac{\sigma^H_{u,v}(x)}{\sigma^H_{u,v}} = 0.$$

Now write

$$C = \sum_{u\in V(G_1),v\in V(G_2)} \frac{\sigma^H_{u,v}(x)}{\sigma^H_{u,v}} = \sum_{u\in V(G_1)\backslash\{x\},v\in V(G_2)\backslash\{x\}} \frac{\sigma^H_{u,v}(x)}{\sigma^H_{u,v}} + \sum_{u\in V(G_1)\backslash\{x\}} \frac{\sigma^H_{u,x_2}(x)}{\sigma^H_{u,x_2}}$$

If $\sigma^G_{u,v}(x) = 0$ for some $u, v \in V(G)$, then $\sigma^H_{u_1,v_2}(x) = 0$. Let $\sigma^G_{u,v}(x) > 0$ and $d_G(u, v) = d = i + j$ where $i = d_G(u, x)$ and $j = d_G(x, v)$. Then $\sigma^H_{u_1,v_2} = (d + 1)\sigma^G_{u,v}$ (because any shortest $u - v$-path of G may be considered as a "projection" of certain $d + 1$ shortest $u_1 - v_2$-paths of H) and, similarly, $\sigma^H_{u_2,v_1} = (d + 1)\sigma^G_{u,v}, \sigma^H_{u_1,v_2}(x) = (j + 1)\sigma^G_{u,v}(x), \sigma^H_{u_2,v_1}(x) = \sigma^G_{u,v}(x)$. Hence

$$\sum_{u\in V(G_1)\backslash\{x\},v\in V(G_2)\backslash\{x\}} \frac{\sigma^H_{u,v}(x)}{\sigma^H_{u,v}} = \sum_{u,v\in V(G)\backslash\{x\}} \left(\frac{j+1}{d+1} \cdot \frac{\sigma^G_{u,v}(x)}{\sigma^G_{u,v}} + \frac{i+1}{d+1} \cdot \frac{\sigma^G_{u,v}(x)}{\sigma^G_{u,v}} \right) =$$

$$\sum_{u,v\in V(G)\backslash\{x\}} \frac{d+2}{d+1} \cdot \frac{\sigma^G_{u,v}(x)}{\sigma^G_{u,v}} = B_G(x) + \sum_{u,v\in V(G)\backslash\{x\}} \frac{1}{d+1} \cdot \frac{\sigma^G_{u,v}(x)}{\sigma^G_{u,v}}$$

Moreover, for each $u_1 \in V(G_1)$, exactly one shortest $u_1 - x_2$-path passes through x, so

$$\sum_{u\in V(G_1)\backslash\{x\}} \frac{\sigma^H_{u,x_2}(x)}{\sigma^H_{u,x_2}} = \sum_{u\in V(G_1)\backslash\{x\}} \frac{1}{d(u,x)+1}.$$

Summing the obtained expression for A, B, C then yields the result. □

Note that, in the above expression for $B_{G\Box K_2}(x)$, the middle and the last sum are very similar to length-scaled betweenness centrality (introduced in [2], see also [6]) and the reciprocal centrality (see [9]) – indeed, their denominators differ just by the additive factor 1. Thus, the mutual relation of $\mathcal{B}_v(G\Box K_2)$ and $\mathcal{B}_v(G)$ is influenced by the mutual location of the central vertices with respect to betweenness, length-scaled betweennes and the reciprocal centrality (where one can expect that there exist graphs in which the sets of those central vertices are mutually disjoint).

Network	ʤ	diameter	order	size
ZacharyKarateClub	0	5	34	78
DolphinSocialNetwork	0	8	62	159
DavisSouthernWomen	1	4	32	89
EurovisionVotes	1	∞	46	467
JazzMusicians	1	6	198	2742
LesMiserables	0	14	77	254
September11Terrorists	0	5	37	85
TaggedTestImages	3	8	71	188
USPoliticsBooks	2	7	105	441
WordAdjacencies	0	5	112	425
WorldCup1988	0	∞	35	118
AmericanCollegeFootball	2	4	115	613

Tab. 1. Betweenness separation of selected real-world networks (the individual networks are obtained, in Wolfram Mathematica, by evaluating the command ExampleData[{"NetworkGraph","<network_name>"}] where <network_name> is an entry from the first column)

3 Experimental results

It is interesting that, for many real-world networks, the betweenness separation is zero or very small compared to network characteristics. This claim is supported by exhibiting the selected graphs from the collection of example graphs provided by Wolfram Research data server (their full list can be obtained, in environments supporting Wolfram Language (like Wolfram Mathematica or Wolfram Programming Lab), by evaluating the procedure `ExampleData ["NetworkGraph"]`; in total, there are 228 networks, and we studied those ones having not many vertices). The obtained results are summarized in Table 1.

In addition, we tested large collections of graphs generated for various graph distributions, some of them being related to real-world networks. Particularly, three graph distributions were considered:

- Uniform graph distribution with parameters n, m – corresponds to the set of all graphs on n vertices and m edges.
- Bernoulli graph distribution with parameters n, p – the graphs are constructed from n independent vertices by selecting each edge independently with probability p.
- Barabási-Albert graph distribution with parameters n, k – the graphs are constructed by consecutive adding new vertices of degree k (as an initial graph, one can take, for example, a $(k + 1)$-cycle) until reaching n vertices in total in such a way that the k edges of the newly added vertex are attached to existing vertices at random, but with the probability proportional to the number of edges that the existing nodes already have.

For each of these graph distributions, we generated, using pseudo-random generating procedures provided by Wolfram Mathematica, lists of 10000 graphs on 30 and 60 (and, for the Barabási-Albert distribution, on 120) vertices. The numbers of vertices were chosen to reach a compromise between the performance of betweenness separation testing algorithm and the need to work with large number of not-so-trivial graphs (while sparse graphs were relatively easy to handle, for random graphs with high edge density, the algorithm has performed slowly). The lists were generated in straightforward way by evaluating the commands `RandomGraph[{m,n},10000]` for pseudo-random graphs on n vertices and m edges, and

```
Table[RandomGraph[BernoulliGraphDistribution[n, p]],10000]
```

or

```
Table[RandomGraph[BarabasiAlbertGraphDistribution[n,k]],10000]
```

for random n-vertex graphs with edge probability p or n-vertex graphs resulted from Barabási-Albert process, respectively. For uniform graph distribution, we have also chosen different values of m (50, 100, 200, 300, 400 and $\binom{30}{2} - 2$ for 30-vertex graphs, and twice much edges for 60-vertex graphs) and, for Bernoulli graph distribution, four edge probabilities p ($0.1, 0.25, 0.5$ and 0.75). The Barabási-Albert graph distribution was considered with parameter $k = 2, 3, 5, 10$.

The histograms for distribution of $\mathfrak{z}(G)$ in uniform distribution-generated graphs are shown on Figure 4. One may observe that $\mathfrak{z}$ is small in general, and tends to be mostly 0 for sparse graphs. But, when a graph gets more edges, its betweenness separation increases, reaching eventually a peak for certain edge density after which its values decrease again; if a graph G is just two edges apart from the complete one, then $\mathfrak{z}(G) \in \{0, 1\}$ with majority of graphs having the value 1. This is consistent with the fact that the fraction of labelled n-vertex graphs with $\binom{n}{2} - 2$ edges is equal to $\frac{\frac{1}{2}\binom{n}{2}\binom{n-2}{2}}{\binom{\binom{n}{2}}{2}} \to 1$ for $n \to +\infty$.

Figure 5 contains the results for randomly generated graphs with different edge probability. The simulations suggest that the edge probability at least $\frac{1}{2}$ yields the betweenness separation equal to 1 for majority of random graphs regardless on the number of vertices; for low probabilities, however, the frequency of individual values for $\mathfrak{z}(G)$ seems to depend also on the order of G. Nevertheless, the overall values again tend to be small.

For pseudo-random graphs generated from Barabási-Albert graph distribution (see Figure 6), one can also observe the trend for small values of $\mathfrak{z}$, with majority of graphs having zero value; nevertheless, with the increasing degree of vertices being added, significant portion of graphs posesses nonzero values.

4 Concluding remarks

The above results show that to answer the question in this paper's title is not easy if we choose the betweenness as the measure of importance of actors and ties. The situation perhaps might be less complicated under another centrality measure, however, one has to consider the vertex centrality which also extends, in some generic way, to the edges. So far, we are aware only of one other such centrality, namely the current-flow betweenness, see [1]. It would be interesting to consider an analogue of vertex closeness $C(x)$ (defined as the reciprocal of the sum of distances from selected vertex x to every other vertex) for edges (could be defined, for example, as the sum of closenesses of edge endvertices, or as the sum of all distances from the set of ednvertices of given edge to sets of endvertices of all other edges) and to define, in similar way as at the beginning of Section 2, the closeness separation between the graph elements of the maximum closeness in connected graphs. Note that, in contrast with Theorem 2, it is easy to show that $C_{G\square K_2}(x) = \frac{1}{2\frac{1}{C_G(x)} + |V(G)|}$ and

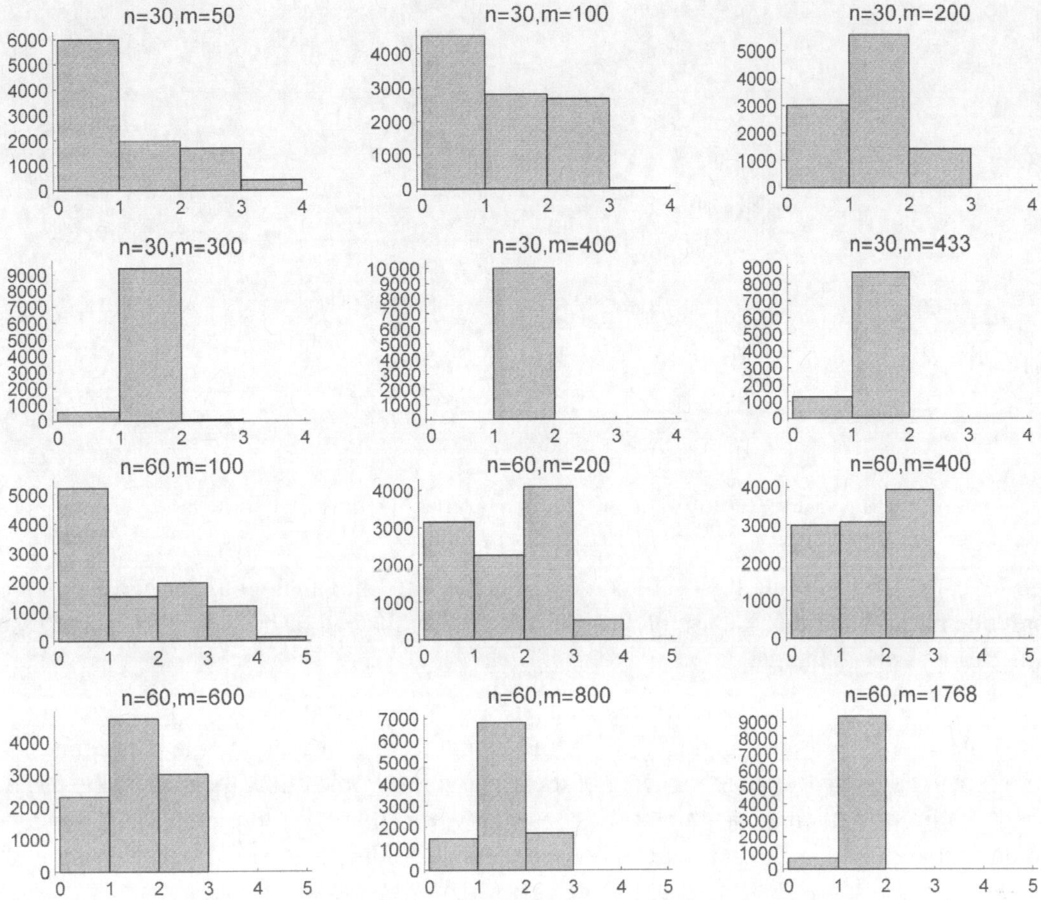

Fig. 4. Histograms for uniformly generated graphs

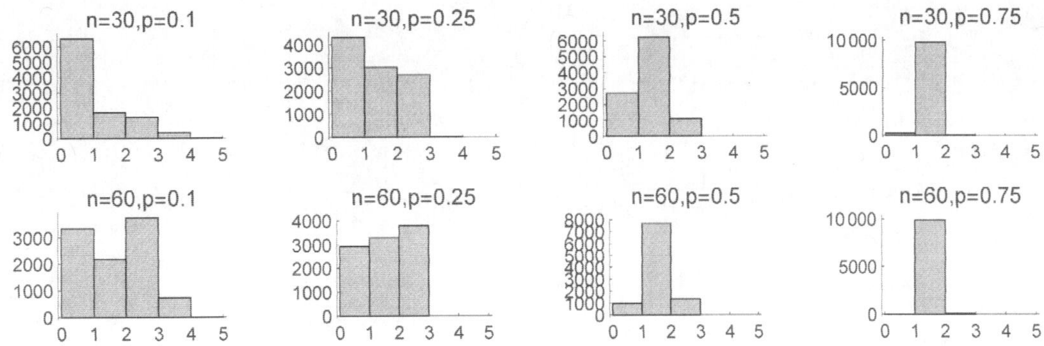

Fig. 5. Histograms for random graphs

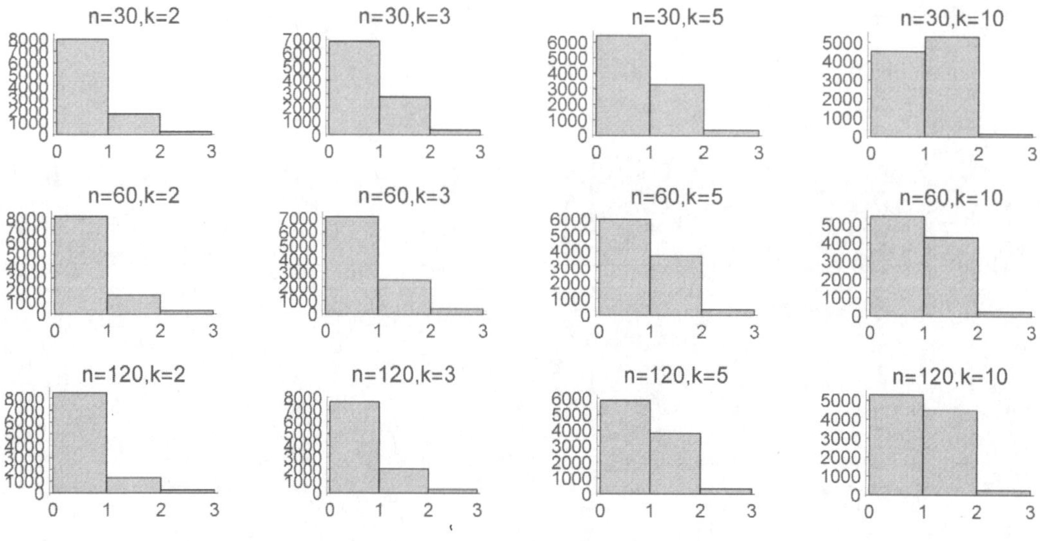

Fig. 6. Histograms for graphs of Barabási-Albert distribution

thus, the set of vertices of highest closeness in G naturally lifts into analogous set in the prism of G; therefore, one might hope that the use of closeness centrality instead of betweenness would result in problems being easier to handle.

As the betweenness centrality is often used in analyses of biological networks like protein-protein interaction networks (see for example [8]), it would be interesting to examine the betweenness separation of them. The relative sparsity of some of these networks would allow the reasonable use of our developed Wolfram Mathematica framework; note, however, that for more complicated examples, one should probably use, instead of built-in betweenness algorithms, a more efficient code.

Acknowledgement. This paper was developed with support of operating program Research and development for the project: "Univerzitný vedecký park Technicom pre inovačné aplikácie s podporou znalostných technológií – II. fáza" (University Science Park Technicom for innovative applications with support of knowledge technologies – II. phase), code ITMS: 313011D232, co-financed from European funds.

References

[1] BRANDES, U., FLEISCHER, D.: *Centrality Measures Based on Current Flow*, Proceedings of the 22nd Annual Symposium on Theoretical Aspects of Computer Science (STACS), volume 3404 (2005) 533–544

[2] EVERETT, M.G., BORGATTI, S.P.: *A graph-theoretic perspective on centrality*, Social Networks 28 (2006) 466–484

[3] FREEMAN, L.: *A set of measures of centrality based on betweenness*, Sociometry 40 (1977) 35–41

[4] GAGO, S., HURAJOVÁ, J., MADARAS, T.: *Notes on the betweenness centrality of a graph*, Mathematica Slovaca 62 No. 1 (2012) 1–12

[5] GAGO, S., CORONIČOVÁ HURAJOVÁ, J., MADARAS, T.: *On betweenness-uniform graphs*, Czechoslovak Math. Journal 63 (138) (2013) 629–642

[6] GAGO, S., CORONIČOVÁ HURAJOVÁ, J., MADARAS, T.: *Betweenness centrality in graphs*, in: Quantitative graph theory: Mathematical Foundations and Applications (M. Dehmer and F. Emmert-Streib, Eds.), CRC Press, 2014

[7] GIRVAN, M., NEWMAN, M.E.J.: *Community structure in social and biological networks*, Proc. Natl. Acad. Sci. USA 99 (2002) 7821–7826

[8] JOY, M.P., BROCK, A., INGBER, D.E., HUANG, S.: *High-betweenness proteins in the yeast protein interaction network*, Journal of Biomedicine and Biotechnology 2005:2 (2005) 96–103

[9] KNOR, M., MADARAS, T.: *On farness- and reciprocally-selfcentric antisymmetric graphs*, Congressus Numerantium 171 (2004) 173–178

[10] KLISARA, J., CORONIČOVÁ HURAJOVÁ, J., MADARAS, T., ŠKREKOVSKI, R.: *Extremal graphs with respect to vertex betweenness for certain graph families*, Filomat 30:11 (2016) 3123–3130

[11] WATTS, D. J., STROGATZ, S. H.: *Collective dynamics of 'small-world' networks*, Nature 393 (6684) (1988) 440–442

[12] WEST, D.: *Introduction to Graph Theory*, 2nd ed., Prentice Hall, 2001

Current address

Coroničová Hurajová Jana RNDr., PhD.
Faculty of business economics with seat in Košice, University of Economics
Tajovského 2, 04001 Košice
Slovak Republic
E-mail: jana.coronicova.hurajova@euke.sk

Madaras Tomáš, doc. RNDr., PhD.
Institute of Mathematics, Faculty of Science, P.J. Šafárik University in Košice
Jesenná 5, 04001 Košice
Slovak Republic
E-mail: tomas.madaras@upjs.sk

Slovak University of Technology in Bratislava
Faculty of Mechanical Engineering
17th Conference on Applied Mathematics **APLIMAT 2018**

Proceedings

NOTES TO THE PROBLEMS WITH FIGURES AT SCHOOL MATHEMATICS

CSACHOVÁ Lucia (SK), JUREČKOVÁ Mária (SK)

Abstract. The aim of this paper is to analyse the achieved level of the special mathematical competence at Slovak lower secondary schools in recent years. The attention will be given to the analysis of the pupil knowledge in understanding and correct solution of mathematical tasks which we have named *problems with figures.* The identification of problematic thematic units of mathematics at schools is good feedback for the preparation of the future mathematics teachers at the universities.

Keywords: statistical analysis, teaching mathematics, evaluation

Mathematics Subject Classification: 97B10, 97C70, 97D60.

1 Introduction

Currently, people are constantly inundated with new information of all kinds. On the road to knowledge, the development of the pupils' ability to understand and correctly process information from the text implies therefore an increasing importance of the role of schools in this process. Moreover, mathematical education at schools creates a space for understanding the text, which contains charts, diagrams and other forms of images.

In our contribution, we focused on the level of graduates of our secondary schools in the area of understanding and the subsequent correct solution of mathematical tasks of this type, which we have named *problems with figures*. [1]

The analysis was made on the basis of nationwide testing T9 from the last six years. The nationwide testing takes a special place in evaluating of the learning outcomes. Testing T9 is aimed at 15-year-old pupils before leaving the lower secondary school. From 2013 the tested content units included in mathematical part of T9 have been following (in accordance with [7]):

- Numbers, variables and arithmetic operations with numbers,
- Relations, functions, tables, charts,

- Geometry and measurement,
- Combinatorics, probability, statistics, logic, reasoning, evidence.

In the paper we present some of our findings in this area: a) what is the trend in success in solving problems of this type, b) what the problems with figures can "say" about the mathematical education. Finally, we propose some recommendations connected with these problems for higher education of future mathematics teachers.

The low overall success rate in this testing is largely due to poor results of the content unit Geometry and measurement. In more detailed qualitative analysis of T9 tasks, we had focused on geometry tasks, but we were also interested in something else. The text passages in the assignment of some tasks are completed with "figures". In these tasks, the part of the input data or data in options in closed test items are graphically visualized by means of the figures (scheme, graph, table, picture, map). Such group of tasks is at the centre of our interest in the following part of this paper.

2 Problems with figures

At first sight, the problems with figures can be understood as the problems from the content unit Geometry and measurement. But not all "geometric" problems in T9 are with figures and some problems with figures are not geometric and come from other content units (see Fig. 1).

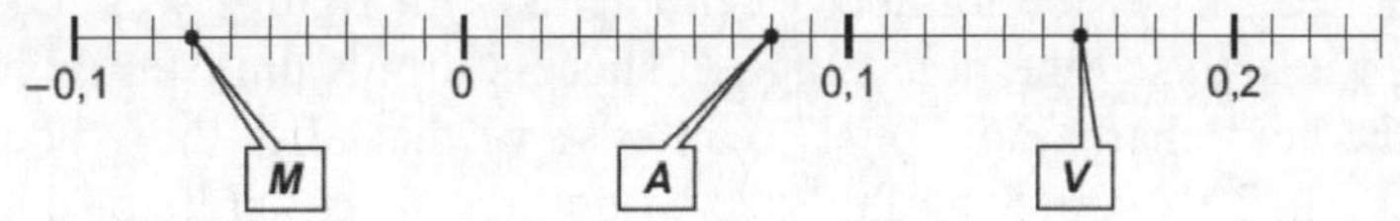

Fig. 1. The figure from the test item 02/2017.
(Numbers M, A, V are shown on the number axis. Calculate $M + A + V$.)

The numbers of the problems with figures in each year of the period 2012–2017 are in the Tab. 1.

Year	2012	2013	2014	2015	2016	2017
Number of problems with figures	8	9	5	5	7	7

Tab. 1. The numbers of the problems with figures in years 2012–2017.
(Total number of the tasks is 20 in each year.)

The most of the problems with figures in T9 focus on the following areas and aims:
- *angles* – calculation of the measure of the marked angles in planar objects,
- *planar objects* – determination of the area or the circuit of a given planar object using formulas or a square grid,

- *spatial objects* – determination of the surface area or the volume of a given spatial object using formulas or determination of the surface area of a composite 3D object composed of identical cubes,
- *number axis* – determination of the position of a given number on the number axis,
- *part and whole* – determination of the part of the whole as a fraction or a percentage and vice-versa,
- *charts* – reading the numerical data from the chart and determination of some numerical characteristics (e. g. average) or truth values of given statements.

3 Quantitative statistical analysis

The statistical analysis we made with the aim to compare the level of the mathematical competence in the last six years in Slovakia. The source of the data was the anonymized database provided by NICEM. It contains information concerning the results of the tests of 242 830 pupils of the ninth grade of lower secondary schools across Slovakia from years 2012–2017. For our analysis, we used a point-based evaluation of all tasks in the tests in the selected period as well as the information on the sex of pupils and the founders of the schools that they attended (state, private, church). Besides the overall evaluation, we special paid attention to the results achieved in the group of problems with figures. The statistical analysis was carried out in the statistical program IBM SPSS and all charts and tables in this section are the outputs of this program.

As it was mentioned above, the number of problems with figures was not the same in each year. Therefore, we analysed the percentage of success in this unit. Initial information can be obtained from the next chart (Fig. 2). We can observe that after the gradual increase of the selected parameter in 2012–2015, a rapid decrease in the success rate of pupils came in 2016. In 2017 the success rate reached the level of 2012. In the next part we selected one sample from the problems with figures from T9 in 2016 with a small successfulness and tried to explain what could cause their unsuccessfulness (apart from reduction of mathematical skills of pupils).

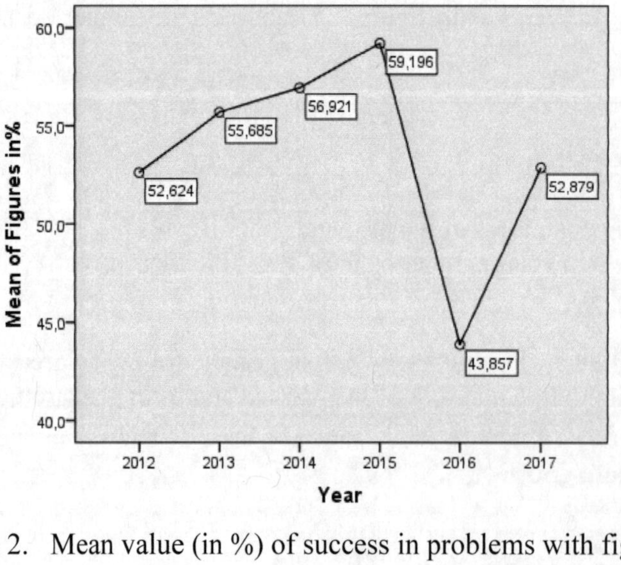

Fig. 2. Mean value (in %) of success in problems with figures.

From the following chart (Fig. 3) we can compare the total success in T9 and the success in the problems with figures. The chart shows that only in years 2014 and 2015 the pupils' success of the problems with figure was higher than their overall success. It looked like a good sign that the figure in the assignment of the task is very important, it helps pupils to process better the data in the assignment, or the illustrative image in the assignment can help them solve the task. However, we can observe, that in the last two years, there has been a significant decline in success in examples with figures.

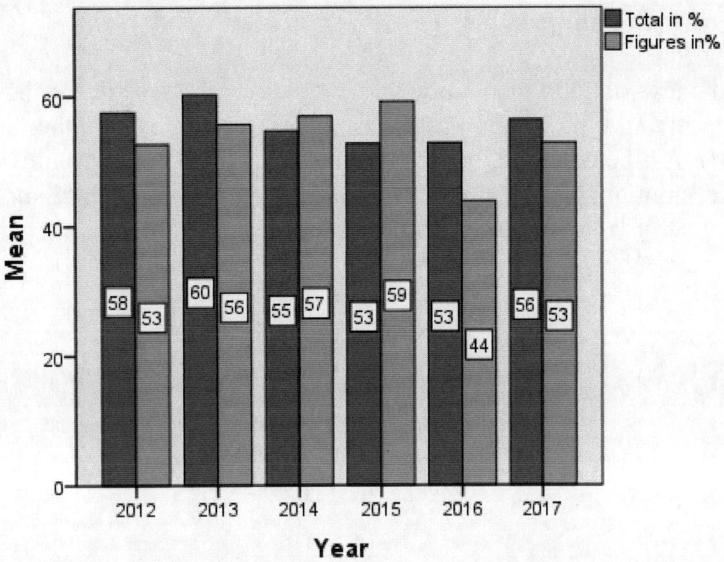

Fig. 3. Mean value (in %) of results in T9 and problems with figures.

In the case of the total success in T9 we observed that the distribution of values is for boys and girls alike. However, for problems with figures, it is a clear significant difference of the distribution of gained points (Tab. 2). Another more detailed analysis has shown that even one of the reference years girls did not exceed the average success of boys.

Hypothesis Test Summary

	Null Hypothesis	Test	Sig.	Decision
1	The distribution of total is the same across categories of sex.	Independent-Samples Mann-Whitney U Test	,311	Retain the null hypothesis.
2	The distribution of figures is the same across categories of sex.	Independent-Samples Mann-Whitney U Test	,000	Reject the null hypothesis.

Asymptotic significances are displayed. The significance level is ,05.

Tab. 2. Mann-Whitney U test for the distribution of total T9 and of problems with figures across sex.

Subsequently we compared the average successfulness in the problems with figures for each category of the founder. Considering the test results, we can conclude that especially in the last two reporting year, private schools showed significant success in solving the problems with the figures in comparison with state and church schools. We can also say that public schools in any particular year showed the lowest mean value of achieved points in the problems with figures.

4 Selected problems with figures in T9

For a more detailed view of individual task and typology of tasks, it can be noticed that recurring tasks (or recurring type of task) in T9 are characterized by a higher success of the solution (see [2], [4]). As the example, the test item 5/2017 from content unit Geometry and measurement can be mentioned: *The ray o shown in the figure is the axis of the angle β. Calculate the magnitude of the angle δ in degrees.* (Fig. 4)

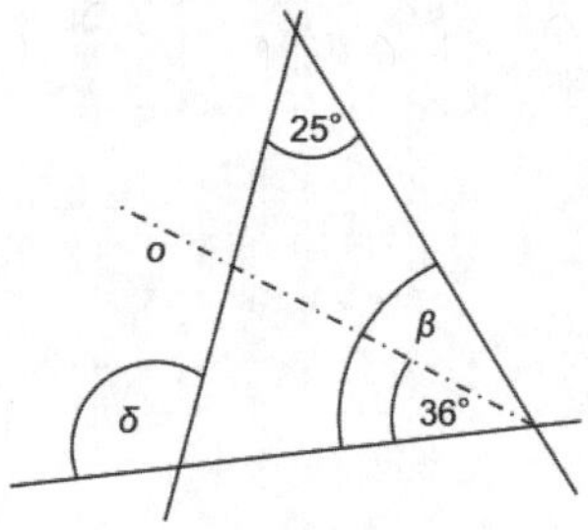

Fig. 4. The triangle (test item 5/2017).

Solution of this task is based on the knowledge of the sum of the internal angles' values in the triangle and the sum of the supplementary angles, and its successfulness was 52,2%. Similar tasks occurred in 2012, 2013, 2014, and 2015. This type of task is included to the typical geometric tasks in school mathematics, it is also used in textbooks for a long time, and the pupils can solve it because it is trained for many times.

On the other side, there are types of tasks which have not usually occurred in T9 (and also in the school mathematics) and pupils often fail in their solutions. The following problem is the example of the non-occurring type of test item with a very low successfulness: *The net of a cuboid is shown in the figure; two dimensions are 11 cm and 4,5 cm. Calculate the third dimension of this cuboid in centimetres.* (Fig. 5)

The success rate of this test item's solution was the second lowest one in 2017 – 29,1%. Its solution is based on right "reading" of the given figure (the net of the cuboid), and one of the possibilities how to solve this problem can be strategy "trial and error". It requires neither a formula nor a special algorithm, only the spatial imagination.[1]

[1] The similar task was occurred in external part of final examination at the secondary school in 2016 (18/2016), but the success rate was higher – 73,3%.

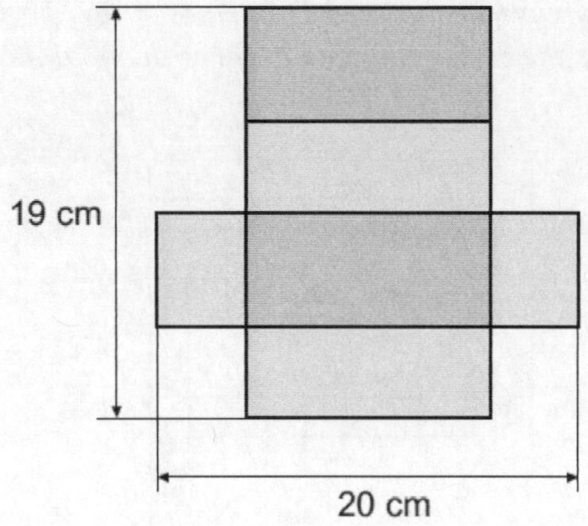

Fig. 5. The net of the given cuboid (test item 09/2017).

During the lessons of the subject Didactics of mathematics, our students solved some types of tasks from T9. The students of mathematics teaching had no problems with the previous problem. But it was problematic for the students of primary teaching. Only 23 % students solved it right (e. g. see Fig. 6), 18 % of them did not solve it at all (the answer of one student: "I cannot be oriented in the figure, and therefore I do not know which third dimension (what a part) should be counted."). The rest of students did not solve it right, e. g. in 18 % cases there is 4.75 cm written as the result (19 cm : 4 = 4.75 cm).

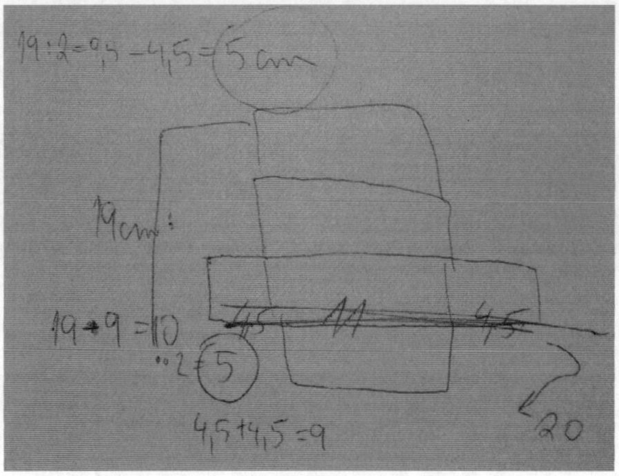

Fig. 6. The right solution of the task 09/2017.

Tasks focused on calculating the area of a planar object drawn in a square grid belong to a recurring standard task type. They occurred in 2013 (two times) and 2015, and the last time in 2016. The success of this type was acceptable, which, we assume, was the result of drilling. In 2016, however, there was a sharp decline, when only 22 % of pupils solved such a problem successfully and up to 23 % of all pupils did not solve it at all. We would like to show the assignment of this task:

2016/17 The area of each square in the grid is 25 mm² (Fig. 7). Calculate the area of the triangle DEF in cm². Express the result rounded to three decimal places.

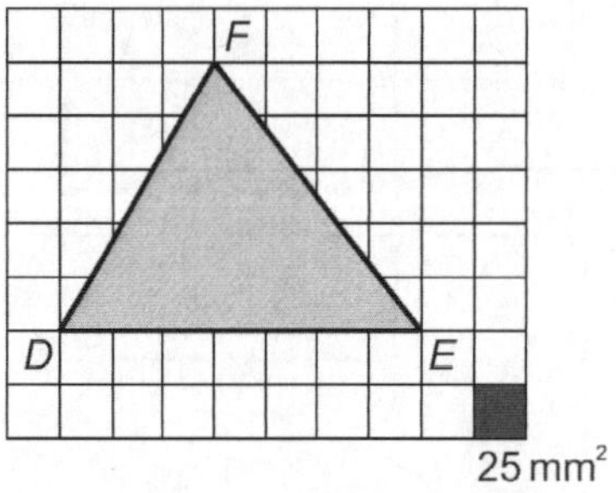

Fig. 7. Figure from the test item 2016/07.

If we make a comparison with the task of the same type (see Fig. 8), we can see that the difference in the input data lies in the dimensions of the square unit in the grid, respectively in their area, and in units in which the area is required.

In our opinion, the low success rate of this test item was due to the following reasons:

- a pupil correctly determines the length of the square side in millimetres, he/she correctly calculates the area of the triangle using the formula. But at the end a pupil lets the value of the area in square millimetres or makes a mistake in converting units of the length;
- a pupil mistakenly uses the given information – the value "25 mm²" is understood as the "25 mm", it means as the length of the square side, and farther calculates with this value.

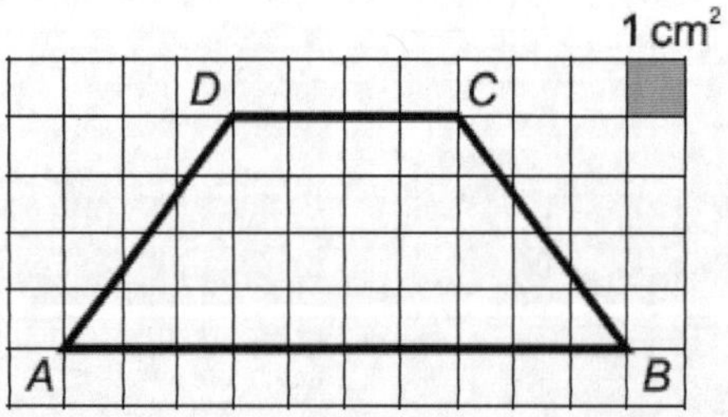

Fig. 8. The figure from the test item 2015/20.

5 The critical areas of school mathematics

The main purpose of Slovak nationwide testing T9 is to:

- compare the performance of individual pupils and schools in tests,
- obtain an image of the performance of pupils at the exit of the second level school (ISCED 2),

- monitor the level of the readiness of pupils for further study (ISCED 3),
- provide schools, decision makers and the general professional public with feedback on the level of knowledge and skills of pupils in the subjects tested, which will assist in improvement of teaching (see [8], [9], [10], [11], [12], [13]) .

The results of this testing can additionally offer another option except of the above-mentioned ones. Obtained statistical data can be more quantitative analysed in the individual testing content units or in the selected sets of test tasks. On the selected tasks it can be done deeper qualitative analyse. Finally, on the base of the analysis it can be identified strong and weak parts of school mathematics, especially the weak ones. The concept *critical areas of school mathematics* was introduced for such issue in [3] and named the areas in which students often and repeatedly fail. In this publication the school geometry has been identified as the critical area in Czech Republic and the research was based on the results of international research testing TIMSS 2007 and interviews with the teachers. A similar situation has also found in Slovakia, where the low level of results in the content unit Geometry and measurement in T9 proves the low level of competences of pupils in this field.

It is important for the teachers of all levels of mathematics education to know about the critical areas of school mathematics. In these critical areas, it would be appropriate and necessary to change teachers' approach to education. This requires cooperation with teachers from practice.

6 Conclusion

Reports to the Slovak nationwide testing of T9 as well as various studies for further testing show that the geometry is a critical area of school mathematics. Teachers often identify school geometry with geometrical constructions (see [3], [5], [6]) or well-trained procedures for solving calculations concerning the properties of planar or spatial objects. But the ability to perceive the geometry of the world is lost. In school geometry it is important to introduce modelling, discovering and real-world geometry problems. At this time, the next important area concerning school geometry is the visual representations of data, because we are surrounded by "fast" data, we need to tell as much information as possible. Images, diagrams or charts can be used for this purpose.

The role of mathematical drilling and training algorithms in a given type of task is important, but not the most important in school mathematics. The main aim of school mathematics is not to achieve good test results, but to develop mathematical thinking, not to calculate as many tasks as possible, but to understand the mathematical context in the world around us and to be able to solve different problems. The results of the test should only reflect the level of the pupil's achievement and the effectiveness of the cooperation between teacher and pupil.

In the preparation of future mathematics teacher, we suggest to:

- demonstrate the problems with the figures as specific tasks with different types of data (information),
- focus on using images in solving different types of tasks,
- devote to the creation new assignments of the problems with figures,

- create a sequence of tasks supporting elimination of the negative impact of the drill.

If we want to prepare high-quality mathematics teachers at universities, it is necessary to have information from primary and secondary schools about the issues that are problematic and properly warn future teachers of mathematics to these topics.

Acknowledgement

The authors would like to thank the National Institute for Certified Educational Measurements in Bratislava (NICEM) for providing the database of the results of T9 tests.

References

[1] GUNČAGA, J., CSACHOVÁ, L., JUREČKOVÁ, M., Educational Research on the Mathematical Competence. In Patterson, K. (Ed.) *Focus on Mathematics Education Research*, NOVA science publishers, 2017, pp. 31-62. ISBN 978-1-53611-826-1.

[2] PICHANIČOVÁ, I., POLGÁRYOVÁ E. et al., *Zbierka úloh pre vzdelávací stupeň ISCED 2: Matematika, Slovenský jazyk a literatúra, Slovenský jazyk a slovenská literatúra, Maďarský jazyk a literatúra.* Bratislava: NÚCEM, 2013, ISBN 978-80-89638-04-8.

[3] RENDL, M., VONDROVÁ, N., Kritická místa v matematice u českých žáků na základě výsledků šetření TIMSS 2007. In *Pedagogická orientace,* 2014, roč. 24, č. 1, ISSN 1211-4669, s. 22–57.

[4] REPOVSKÝ, M., KUBIŠ, T. et al. *Zbierka uvoľnených úloh z testovania matematickej a čitateľskej gramotnosti pre 2. stupeň základných škôl a 1. -4. ročník osemročných gymnázií.* Bratislava: NÚCEM, 2012, ISBN 978-80-970261-8-9.

[5] VANÍČEK, J., Role technologií pro rozvoj matematické gramotnosti. In A. Hošpesová a kol., *Matematická gramotnost a vyučování matematice.* České Budějovice: Jihočeská univerzita, 2011, ISBN 978-80-7394-259-5, pp. 151-184.

[6] VONDROVÁ, N., RENDL, M. a kol., *Kritická místa matematiky základní školy v řešení žáků.* Praha: Karolinum, 2016, ISBN 9788024632346.

[7] *Rámcový učebný plán pre základné školy s vyučovacím jazykom slovenským.* In: http://www.statpedu.sk/sites/default/files/dokumenty/statny-vzdelavaci-program/rup2.pdf

[8] *Testovanie 9-2012 Priebeh, výsledky analýzy.* In: http://www.nucem.sk/documents//26/testovanie_9_2012/Sprava_T9_2012_v2.3.pdf

[9] *Testovanie 9-2013 Priebeh, výsledky analýzy.* In: http://www.nucem.sk/documents//26/testovanie_9_2013/Sprava_T9_2013_v2.3.pdf

[10] *Testovanie T9-2014 – priebeh, výsledky a analýzy.* In: http://www.nucem.sk/documents//26/testovanie_9_2014/Sprava_T9_FINAL_2.pdf

[11] *Testovanie T9-2015 – priebeh, výsledky a analýzy.* In: http://www.nucem.sk/documents//26/testovanie_9_2015/Sprava-T-9-FINAL_2015.pdf

[12] *Testovanie T9-2016 – priebeh, výsledky a analýzy.* In: http://www.nucem.sk/documents//26/testovanie_9_2016/SPRAVA_T_9-2016_final.pdf

[13] *Testovanie T9-2017 – priebeh, výsledky a analýzy.* In: http://www.nucem.sk/documents//26/testovanie_9_2017/Sprava_T9-2017_09_final.pdf

Current address

Csachová Lucia, RNDr., PhD.
Faculty of Education, Catholic University in Ružomberok,
Hrabovská 1, 034 01 Ružomberok, Slovak Republic
E-mail: lucia.csachova@ku.sk

Jurečková Mária, doc. RNDr., CSc.
Faculty of Education, Catholic University in Ružomberok,
Hrabovská 1, 034 01 Ružomberok, Slovak Republic
E-mail: maria.jureckova@ku.sk

Slovak University of Technology in Bratislava
Faculty of Mechanical Engineering
17th Conference on Applied Mathematics **APLIMAT 2018**

Proceedings

NETWORK FLOWS AS A TOOL FOR SOLVING LOCATION-SCHEDULING PROBLEM TO OPTIMIZE CHARGING INFRASTRUCTURE FOR ELECTRIC VEHICLES

CZIMMERMANN Peter (SK), BUZNA Ľuboš (SK), KOHÁNI Michal (SK)

Abstract. We propose an MIP model to design a private charging infrastructure for a fleet of electric vehicles operating in large urban areas. We derive an IP problem and we show that it is equivalent to a network flow model. We illustrate that the flows in networks can be used in the study of properties of location-scheduling problems related to the design of charging infrastructure.

Keywords: electric vehicles, charging infrastructure, mixed integer programming, flows in networks

Mathematics subject classification: Primary 90B06; Secondary 94C15, 68R10

We propose mixed integer programming (MIP) and integer programming (IP) models to design a private charging infrastructure for a fleet of electric vehicles operating in large urban areas. Examples of such a fleet include taxicabs and small vans used in city logistics or with shared vehicles. The fleet is composed of vehicles equipped with an internal combustion engine, but the operator is wishing to replace them with fully electric vehicles in the near future. Hence it is required to design a private network of charging stations that will be specifically adjusted to the operation of the fleet. We use GPS traces to characterize actual travel patterns of individual vehicles. We formulate location-scheduling optimization models that determine the maximum number of vehicles that can be recharged without affecting their routes and parking behaviour (MIP model) or minimize the number of charging stations under the condition that all vehicles can be recharged (IP model). The MIP model assumes that all vehicles possess complete information about all other vehicles. To study the role of available information, we evaluate the resulting designs considering uncoordinated charging when vehicle drivers know only the actual occupation of charging points at the time when they are choosing a charging station. Tests show a gap between optimization and uncoordinated charging. We suggest a network flow model that allows us to study the gap and properties of the location-scheduling problem. We also show that the network flow model can be formulated as the IP model.

1 Introduction

Advances in battery technologies and decreasing prices of electric vehicles (EVs) stimulate more interest in converting large fleets of vehicles into electric. Benefits could be considerable due to high utilization of such vehicles. Recently, mathematical optimization has been often proposed to support the design of charging infrastructure. The special class of models was developed to cover trajectories of vehicles [1, 6]. This approach is applicable in the design of the charging infrastructure along motorways to cover long distance trips. The extension of this approach that considers multiple paths connecting origins and destinations of trips was proposed in [8].

We assume that the operation of the fleet should remain unaffected, i.e. for the recharging of batteries only parking events are used. Here we propose an extension of the approach [4, 5], where optimization can select vehicles that are successfully converted to electric, we consider charging points with different charging speeds and evaluate the role of available information when choosing a charging point.

We propose a mixed integer programming (MIP) model of the problem (Section 2). To investigate the role of available information for the MIP model, we introduce a procedure for uncoordinate charging of vehicles (Section 3). The results of numerical experiments can be found in Section 4. It follows from the experiments that the MIP model represents minimal requirements for charging infrastructure. We suggested an integer programming (IP) model (Section 5) and equivalent flow model (Section 6) to analyse further behaviour of the proposed problem.

2 Mathematical model

The fleet of vehicles is represented by the set C and each vehicle is equipped with a battery of capacity β (for simplicity, we suppose that all vehicles have the same capacity that determines the maximum driving range). The set of candidate locations to locate charging stations is denoted by I. We discretise the time by dividing it into the set of non-overlapping intervals T of equal size. For each vehicle $c \in C$, we extract from GPS data an ordered sequence of parking events R_c and N_{cr} denotes the list of time intervals $t \in T$ that have an overlap with the parking event $r \in R_c$. To simplify notation, for each vehicle we add to R_c the fictional introductory parking event 0 and fictional terminal parking event r_c. The set of charging speeds is denoted by S (here we consider $|S| = 3$). The price to build a charging point of type $s \in S$ is p_s and γ denotes large positive constant. From GPS data we extract driving distances and we estimate the battery capacity consumption when vehicle $c \in C$ drives towards stop $r \in R_c$ as $u_{rc} \geq 0$. $a_{ct} \in \langle 0, 1\rangle$ is the fraction of the time interval $t \in T$ when vehicle $c \in C$ is parking. To simplify the description of the model, we define $B_{itc} \in \{0, 1\}$, where $B_{itc} = 1$ if vehicle $c \in C$ is parking closer than the distance ρ_{max} from location $i \in I$ during the time interval $t \in T$, and $B_{itc} = 0$ otherwise.

Decisions are described by the following variables:

- $w_{is} \in Z^+$ represents the number of charging points of charging speed $s \in S$ allocated to station $i \in I$,
- $z_c \in \{0, 1\}$, when vehicle $c \in C$ exceeds driving range and thus cannot be successfully converted to electric $z_c = 1$, otherwise $z_c = 0$,
- $x_{cts} \in \{0, 1\}$, if vehicle $c \in C$ is being charged with speed $s \in S$ during the time interval $t \in T$, then $x_{cts} = 1$, and otherwise $x_{cts} = 0$,

- $q_{cts} \geq 0$ represents the part of interval $t \in T$ when vehicle $c \in C$ is being charged with speed $s \in S$,
- $d_{cr} \geq 0$ corresponds to the state of charge of the vehicle $c \in C$ at the beginning of the parking event $r \in R_c$.

We consider the objective function minimizing the number of vehicles that exceed driving range, i.e.:

$$\text{Minimize} \sum_{c \in C} z_c \tag{1}$$

$$\text{subject to} \sum_{c \in C} B_{itc} x_{cts} \leq w_{is} \qquad \text{for } i \in I, t \in T, s \in S \tag{2}$$

$$d_{c0} \leq 0.5\beta + z_c \gamma \qquad \text{for } c \in C \tag{3}$$

$$d_{c,r_c} + z_c \gamma \geq 0.5\beta \qquad \text{for } c \in C \tag{4}$$

$$d_{cr} + \sum_{s \in S, t \in N_{cr}} q_{cts} s \leq \beta \qquad \text{for } c \in C, r \in R_c - \{0\} \tag{5}$$

$$d_{cr} \leq d_{c,r-1} - u_{cr} + \sum_{s \in S, t \in N_{c,r-1}} q_{cts} s + z_c \gamma \qquad \text{for } c \in C, r \in R_c - \{0\} \tag{6}$$

$$\sum_{s \in S} x_{cts} \leq 1 \qquad \text{for } c \in C, t \in T, s \in S \tag{7}$$

$$q_{cts} \leq x_{cts} a_{ct} \qquad \text{for } c \in C, t \in T, s \in S \tag{8}$$

$$\sum_{i \in I, s \in S} w_{is} p_s \leq G \tag{9}$$

Constraints (2) make sure that at each station and each time interval we cannot use more charging points of a given type than available. Constraints (3) (constraints (4)) ensure that initial (final) state of charge of each vehicle is not more (less) than 50% of the battery capacity. Constraints (5) make sure that the battery capacity cannot be exceeded. Constraints (6) ensure the contiguity in charging and discharging of vehicles. Constraints (7) make sure that a vehicle cannot be charged simultaneously at more than one speed. Constraints (8) ensure that a vehicle is being charged within the time interval $t \in T$ only when vehicle is parking and $x_{cts} = 1$. Finally, constraint (9) keeps the costs to set up the charging infrastructure to be less or equal than the budget limit G.

3 Evaluating the role of available information

In previous section, we supposed that complete information about requests of customers has been available. But reality is often different, hence, to evaluate designs of charging stations found by the model, we consider approach, which works with uncoordinated charging strategy where the users know only the occupancy of charging stations when they chose a charging station, but have no information on when the charging stations are freed. The set of charging stations found by model (1)-(9) is denoted by $\overline{I}$. K_i is the set of charging points placed at station $i \in \overline{I}$ (we suppose that $0 \notin K_i$ and we set $k_{max} = 0$). $R = \cup_{c \in C} R_c$ is the set of possible charging events, P is the set of movements of vehicle, and $E = R \cup P$ is the set of all events. The charging speed of point $k \in K_i$ is denoted as s_k. If $e \in P$, then $l(e)$ is the driving distance.Thus, we assume that vehicles are in one of the two modes, either parking or driving. Vehicle associated with event $e \in E$ is denoted by $c(e)$. The start time and the end time of the event $e \in E$ are denoted by $b(e)$ and $z(e)$, respectively. Level of the battery of vehicle $c \in C$ is d_c. Value ρ_{max} is the maximum acceptable proximity of a vehicle from a station to make the charging still possible, and T_k is the end time of the most recent charging event at point $k \in K_i$, for $i \in \overline{I}$.

3.1 Evaluation procedure

We evaluate the performance of charging infrastructure by running the following algorithm:
Step 1: (Initialization)
For $c \in C$ set $d_c = \beta/2$. Order events from E ascendingly with respect to $b(e)$. Set $T_k = 0$ for $k \in K_i$ and $i \in \overline{I}$.
Step 2: (Event list processing)
For each $e \in E$ do:

If $e \in R$, then order the set $\overline{I}$ descendingly with respect to the sum of speeds s_k over charging points $k \in K_i$ that are free in time $b(e)$.
For $i \in \overline{I}$ do:

If the distance of vehicle $c(e)$ at time $b(e)$ from station i is less than ρ_{max}, then process the charging event following an uncoordinated charging strategy.

If $e \in P$, then set $d_{c(e)} = d_{c(e)} - l(e)$.
Step 3: (Evaluation)
Evaluate feasibility of all vehicles for $c \in C$. If $d_c \geq 0$ all the time during the run of the algorithm and $d_c \geq \beta/2$ at the time when the algorithm is terminated, then $c \in C$ is feasible.
Output of the procedure are decisions about feasibility of each vehicle.

Strategy of uncoordinated charging

We suppose that drivers have only information about the occupancy of charging points and unplug their vehicles at the time of departure for the next trip.

Find $k_{max} \in argmax(s_k)$[1], where $k \in K_i$ and $T_k \leq b(e)$.
If $k_{max} \neq 0$:

Set $d_{c(e)} = \min\{\beta, d_{c(e)} + (z(e) - b(e))s_{k_{max}}\}$. Set $T_{k_{max}} = z(e)$, $k_{max} = 0$ and continue with step 2, and process the next event.

4 Numerical experiments

In the case study, we consider the fleet C of 500 taxicabs operating in the area of great Stockholm, in Sweden, whose positions were recorded from 01/May/2014 until 14/May/2014. Each vehicle reported on average every 90 seconds its id, GPS position, time-stamp and information whether it is hired or not.[2] When solving optimization model we split the time period covered by the data into the set T of 1344 time intervals, each of length 15 min. We used GPS data to deduce two sets of candidate locations (with $|I| = 5$ and $|I| = 20$), the sets R_c, N_{cr} and P using the procedure described in [5], while setting parameters $V_{max} = 0.1$ m/s and $M = 150$. To test the proposed approach, we performed numerical experiments with the following values of parameters: the driving range of all vehicles was set to $\beta = 300$ km and vehicles can be charged only if they park closer than $\rho_{max} = 500$

[1] If $argmax$ returns a non-empty set, then k_{max} can be an arbitrary element from this set.
[2] Data have been provided by M. Cebecauer from KTH Royal Institute of Technology in Stockholm.

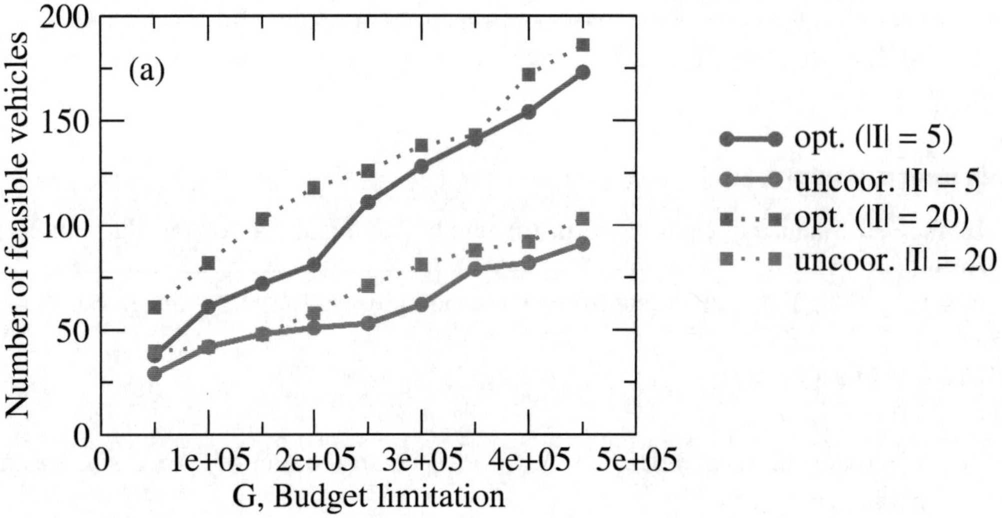

Fig. 1. The number of feasible vehicles as a function of budget limit G obtained by solving the optimization model and applying the evalution procedure using uncoordinated charging strategy, each for $|I| = 5$ and $|I| = 20$. The values achieved for the largest value of G remain further unchanged when G is increased.

metres from the charging station. We considered three types of charging points characterized by the set of charging speeds $S = \{5.3, 21.3, 74.6\}$ km/hour (we call them slow, medium and fast) and the corresponding set of installation costs $p = \{500, 2\,500, 25\,000\}$ USD.

For $|I| = 5$ ($|I| = 20$) and unlimited budget G, we obtained 8 (2) slow, 15 (5) medium and 16 (21) fast charging points. In Figure 1, we show the dependency between the number of feasible vehicles obtained from mathematical model and obtained by running the evaluation procedure of uncoordinated charging as a function of the budget limit G. Gap between results obtained by the optimization model and unccordinated charging strategy gives an idea about the role of available information. The number of feasible vehicles that can be turned into electric is systematically significantly higher for the optimization model. The optimization model assumes availability of perfect information about other vehicles, when drivers choose a charging stations. While in the case of uncoordinated charging strategy, information about other vehicles is limited to the occupancy of charging stations only.

5 Discrete model

Our results indicate that this approach can be used to estimate the minimal requirements to set up the charging infrastructure. When we want to better explain the gap between optimization and uncoordinate charging procedure, we need to understand the behaviour of the studied problem. This is the reason, why we simplify and suggest an IP model, which has very similar structure, but it is posed differently. We do not look for the maximum number of vehicles that can be served by the limited charging infrastructure, but for a given number of vehicles we search for a minimal infrastructure that can serve all vehicles. We work with the following variables: $y_i \in \{0, 1\}$, where $y_i = 1$ when a charging station is placed at the location $i \in I$ and $y_i = 0$ otherwise; $x_{ct} \in \{0, 1\}$, where $x_{ct} = 1$ when vehicle $c \in C$ is charged during the time interval $t \in T$ and $x_{ct} = 0$ otherwise; and $d_{cr} \in Z_0^+$ is the distance that the vehicle $c \in C$ is able to drive at the beginning of the parking event $r \in R_c$. Constant β is the capacity of battery, m is the maximum number of charging points. We estimate the

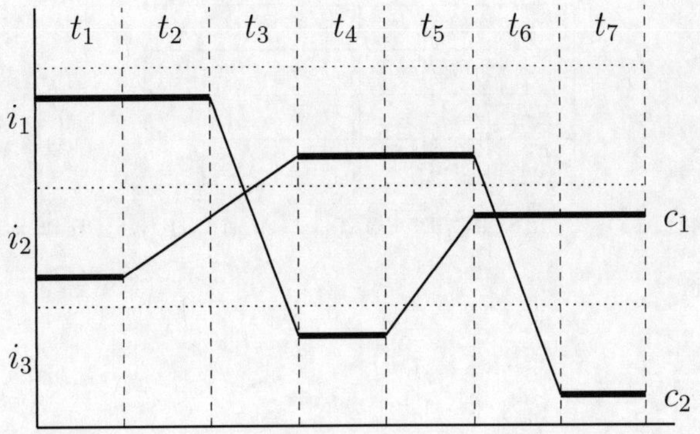

Fig. 2. Diagram illustraiting parking events that represent charging opportunities (thick horizontal lines) and transfer of vehicles between charging stations (thin diagonal lines).

battery capacity consumption when vehicle $c \in C$ drives towards stop $r \in R_c$ as $u_{rc} \in Z^+$.

$$\text{Minimize} \sum_{i \in I} y_i \tag{10}$$

$$\text{subject to} \sum_{c \in C} B_{itc} x_{ct} \leq m y_i \qquad \text{for } i \in I, t \in T \tag{11}$$

$$d_{cr} + \sum_{t \in N_{c,r}} x_{ct} \leq \beta \qquad \text{for } c \in C, r \in R_c \cup \{r_c\} \tag{12}$$

$$d_{cr} \leq d_{c,r-1} - u_{cr} + \sum_{t \in N_{c,r-1}} x_{ct} \qquad \text{for } c \in C, r \in R_c \cup \{r_c\} \tag{13}$$

In the objective function (10), we minimize the number of charging stations. Constraints (11) ensure that we cannot use more charging points at each location and in each time interval than available. Constraints (12) ensure that battery capacity is never exceeded and constraints (13) ensure the contiguity of charging and discharging of batteries.

6 Flow model

In [2], we introduced a flow model approximating this problem. In this section we amend the model to be equivalent with IP model. We construct a network G. The set of vertices of G is the union of the sets:
$V_1 = \{v_{it} : i \in C, t \in T\}$, where v_{it} represents the vehicle i in time t,
$V_2 = \{u_{it} : i \in I, t \in T\} \cup \{w_{it} : i \in I, t \in T\}$, where u_{it} and w_{it} represent the candidate location i in time t,
$V_3 = \{s, z, w\}$, where s is the source, z is the sink and w an additional vertex of the network.

The set of edges of G is the union of the sets:
$E_1 = \{(s, v_{i,1}) : i \in C\}$,
$E_2 = \{(s, u_{i,1}) : i \in I\}$,
$E_3 = \{(v_{i,t-1}, v_{i,t}) : i \in C, t \in T - \{1\}\} \cup \{(v_{i,n}, z) : i \in C, n = \max(T)\}$,
$E_4 = \{(u_{i,t-1}, u_{i,t}) : i \in I, t \in T - \{1\}\} \cup \{(u_{i,n}, z) : i \in I, n = \max(T)\}$,
$E_5 = \{(u_{i,t}, w_{i,t}) : i \in I, t \in T\} \cup \{(w_{i,t}, u_{i,t+1}) : i \in I, t \in T - \{n\}\} \cup \{(w_{i,n}, z) : i \in I\}$,

E_1	$\langle \beta, \beta \rangle$	E_5	$\langle 0, m \rangle$
E_2	$\langle 0, \|T\| \rangle$	E_6	$\langle 0, 1 \rangle$
E_3	$\langle 0, \beta \rangle$	E_7	$\langle 1, 1 \rangle$
E_4	$\langle 0, \|T\| \rangle$	E_8	$\langle 0, \infty)$

Tab. 1. Lower and upper bounds for the flows on edges.

$E_6 = \{(w_{i,t}, v_{j,t}) : i \in I, t \in T, j \in C, B_{i,t,j} = 1\}$,
$E_7 = \{(v_{j,t}, w) : j \in C, t \in T, \sum_{i \in I}\} B_{i,t,j} = 0$,
$E_8 = \{(w, z)\}$.
Every edge has assigned to it a lower and upper bound for the flow, which are represented by the interval $\langle l, u \rangle$. These bounds are given in Table 1. It is possible to show that the optimal solution of IP model is equivalent to the feasible flow in G with the minimum number of non-zero flows on the edges of the set E_2. We will call this problem: a sparse feasible flow. If the flow on the edge $(s, u_{i,1}) \in E_2$ is non-zero, then we place a charging point in location i. Problems of finding a feasible flow in the network with lower bounds on edges and its minimisation are solvable in polynomial time [3]. We suppose that the problem to find a sparse feasible flow in network is NP-hard problem. Hence we suggest heuristic method to solve it:
Description of the algorithm. We construct a network G' (it is subnetwork of G):
Its vertex set is $V_1' \cup V_2' \cup V_3'$, where
$V_1' = \{v_{i,t} : i \in C, t \in T\} \cup \{z\}$,
$V_2' = \{u_{i,t} : i \in I, t \in T\} \cup \{w_{i,t} : i \in I, t \in T\}$.
The edge set is the union of the sets:
$E_1' = \{(v_{i,t-1}, v_{i,t}) : i \in C, t \in T - \{1\}\} \cup \{(v_{i,n}, z) : i \in C\}$,
$E_2' = \{(u_{i,t-1}, u_{i,t}) : i \in I, t \in T - \{1\}\}$,
$E_3' = \{(u_{i,t}, w_{i,t}) : i \in I, t \in T\} \cup \{(w_{i,t}, u_{i,t+1}) : i \in I, t \in T - \{n\}\} \cup \{(w_{i,n}, z) : i \in I\}$,
$E_4' = \{(w_{i,t}, v_{j,t}) : i \in I, t \in T, j \in C, B_{i,t,j} = 1\}$,
$E_5' = \{(u_{i,n}, z) : i \in I, n = \max(T)\}$.
Lower and upper bounds of the edges are taken from G.

Algorithm.
Input is the network G with feasible flow x and subnetwork G'.
For each pair of vertices $u_{i,2}, u_{j,2} \in V_2$ such that $0 < x(s, u_{i,2}) \leq x(s, u_{j,2}) < |T|$ **do**:

while there is an augmenting path $P(j, i)$ with reserve r from $u_{j,2}$ to $u_{i,2}$ in G' **do**:

add the edges $(s, u_{j,2})$ and $(s, u_{i,2})$ to $P(j, i)$ to form a (non-oriented) cycle
$C = (s, u_{j,2}, \ldots, u_{i,2}, s)$,
change the flow x in C as follows:
if (u, v) is the forward edge in C, **then** $x(u, v) = x(u, v) + r$,
if (u, v) is the reverse edge in C, **then** $x(u, v) = x(u, v) - r$,

process another pair of vertices.

Output is a new feasible flow.

A polynomial algorithm for finding an augmenting path in network is presented in [7].
This approach does not guarantee that given solution is optimal. We aim to test the algorithm and

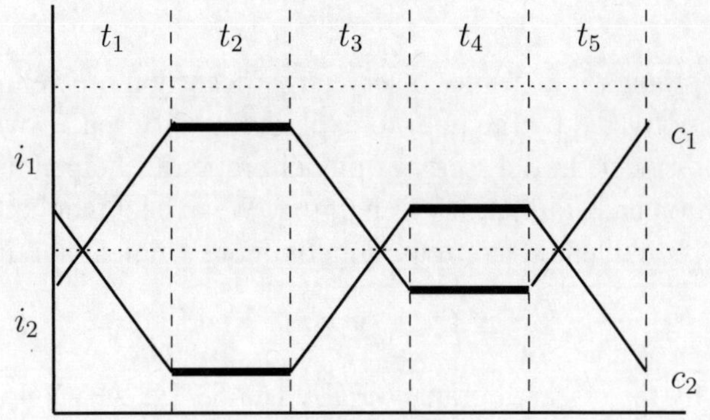

Fig. 3. Diagram from Example 1.

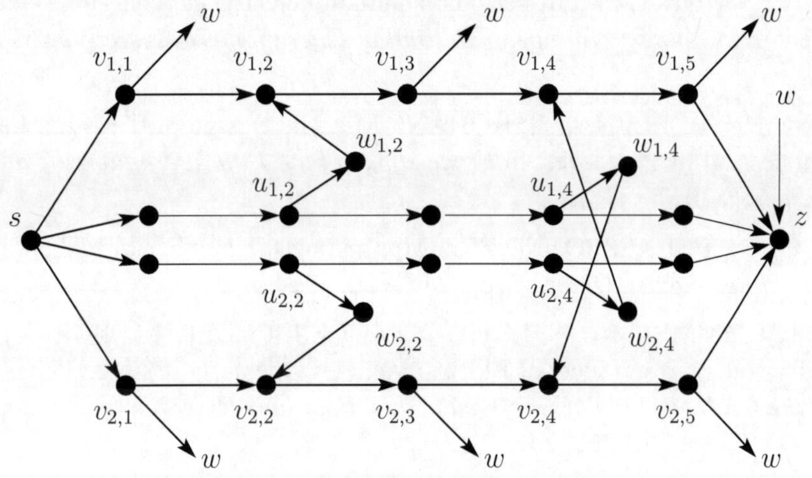

Fig. 4. Network G from Example 1.

study its properties in our future works.

Example 1.
To illustrate, how the graph G can be constructed, we consider situation with two vehicles, two candidate locations and five time intervals as can be seen in Figures 3 and 4. Existence of a path $u_{k,j} \rightarrow w_{k,j} \rightarrow v_{i,j}$ means that the vehicle i is parking in location k at the time interval j.

What is the cause of the gap in the number of feasible vehicles obtained from the optimization problem and simulation procedure?

We suppose that the main reason is in the available information. When we use an optimization algorithm, we work with the whole network G. However, in the simulation procedure, at time $t \in T$, we only have information about edges which start in vertices $u_{i,t}$. These constraints do not allow us to find an optimal solution by the evaluation procedure. The open questions are: How do we design the charging infrastructure especially in cases when the limited level of information is given? How can we spread the available information (from O-D matrices, probability models) to obtain a lower number of unfeasible vehicles?

7 Conclusions

Our results indicate that optimization is able to serve significantly more vehicles than uncoordinated charging strategy. Being motivated by the goal to explain this differences we introduced simpler IP model and associated flow model. Initial study of model properties helped us to gain intuition about the gap between optimization and simulation of charging. We expect that better understanding of the behaviour of these models could lead to computationally more efficient solving methods.

Acknowledgements

This work was supported by the research grants VEGA 1/0463/16, APVV-15-0179 and it was facilitated by the FP 7 project ERAdiate [621386].

References

[1] CHUNG, S., H., KWON, C.: Multi-period planning for electric car charging station locations: A case of Korean Expressways. *European Journal of Operational Research, 242(2)*, pp. 677–687, 2015.

[2] CZIMMERMANN, P., BUZNA, Ľ., KOHÁNI, M.: The Design of Charging Infrastructure for Electric Vehicles and Its Properties. In *Proceedings of the 15th International Symposium on Operational Research*, 2017.

[3] DANTZIG, G., B., FULKERSON, D., R.: Minimizing the number of tankers to meet a fixed schedule. *Naval Res. Logist. Quart., 1*, pp. 217–222, 1954.

[4] KOHÁNI, M., CZIMMERMANN, P., VÁŇA, M., CEBECAUER, M., BUZNA, Ľ.: Designing Charging Infrastructure for a Fleet of Electric Vehicles Operating in Large Urban Areas. In *Proceedings of the 6th International Conference on Operations Research and Enterprise Systems - Volume 1:ICORES*, pp. 360–368, 2017.

[5] KOHÁNI, M., CZIMMERMANN, P., VÁŇA, M., CEBECAUER, M., BUZNA, Ľ.: Location-scheduling optimization problem to design private charging infrastructure for electric vehicles. In *Operations Research and Enterprise Systems, ICORES 2017, Revised Selected Papers.* Communications in Computer and Information Science, Springer. (Accepted for publication.)

[6] MIRHASSANI, S., A., EBRAZI, R.: A Flexible Reformulation of the Refueling Station Location Problem. *Transportation Science, 47(4)*, pp.617–628, 2013.

[7] NEMHAUSER, G., L., WOLSEY, L., A.: Integer and Combinatorial Optimization. *Wiley-Interscience series in discrete mathematics and optimization.* ISBN 0-471-82819-X, pp. 65, 1988.

[8] YILDIZ, B., ARSLAN, O., KARASAN, O., E.: A branch and price approach for routing and refueling station location model. *European Journal of Operational Research, 248(3)*, pp.815–826, 2016.

Current address

Czimmermann Peter (corresponding author)
Department of Mathematical Methods and Operation Research
Faculty of Management Science and Informatics, University of Žilina
Univerzitná 8215/1, 010 26 Žilina, Slovak Republic
E-mail: Peter.Czimmermann@fri.uniza.sk

Buzna Ľuboš
Department of Mathematical Methods and Operation Research
Faculty of Management Science and Informatics, University of Žilina
Univerzitná 8215/1, 010 26 Žilina, Slovak Republic
E-mail: Lubos.Buzna@fri.uniza.sk

Koháni Michal
Department of Mathematical Methods and Operation Research
Faculty of Management Science and Informatics, University of Žilina
Univerzitná 8215/1, 010 26 Žilina, Slovak Republic
E-mail: Michal.Kohani@fri.uniza.sk

Slovak University of Technology in Bratislava
Faculty of Mechanical Engineering
17th Conference on Applied Mathematics **APLIMAT 2018**

Proceedings

NEW TRENDS IN TEACHING FUNCTIONS AT UNIVERSITY LEVEL

DILLINGEROVÁ Monika (SK), VANKÚŠ Peter (SK)

Abstract. In our paper we focus on one of the key concepts of Mathematical Analysis – the functions. We present results of our survey, realized on the sample of 91 students. During this survey we analysed their solutions of selected tasks and identified some major misconceptions the students have in the area of the functions. In the paper we then propose some new trends those can help to avoid these misconceptions and improve students' knowledge.

Keywords: functions, mathematical analysis, misconceptions, teaching methods, ICT

Mathematics Subject Classification: Primary 97D40, 97C70; Secondary 97I99

1 Introduction

Authors of the paper are teaching Seminars on Mathematical Analysis at the Faculty of Mathematics, Physics and Informatics. In this paper we describe qualitative study of some mistakes the students makes in this topic and we propose some methods those can be helpful in overcoming these problems.

Functions are important concept that is essential for many other concepts in Analysis, so the good understanding of them is needed. Our paper surveys knowledge of our students in the area of functions and tries to point to the most common misconceptions they have. Afterwards, we propose various strategies and new trends of teaching. These should avoid described mistakes.

2 Theoretical framework

2.1 Analysis of students' misconceptions

Analysis of students' misconceptions is the process of identifying and reviewing students' errors to determine whether an error pattern exists – that means whether several students are making the same type of error. If a pattern does exist, the teacher can identify students' misconceptions or skill deficits and subsequently design and implement instruction to address them.

Research on error analysis is not new: Researchers around the world have been conducting studies on this topic for decades. Error analysis has been shown to be an effective method for identifying patterns of mathematical errors for any student, who is struggling in mathematics, as is stated in [1].

In the research process we have to analyze students' errors, attempt to understand them, explain what they consist in, and find what causes them. Depending on the conclusions of such an analysis, we should select corrective strategies and teaching methods in order to overcome these misconceptions.

Typically, student mathematical errors fall into three categories: factual, procedural, and conceptual. Each of these errors is related either to a student's lack of knowledge or a misunderstanding [2].

Factual errors occur when students lack factual information (e.g., definitions, formulas, vocabulary, digit identification, place value identification, etc.).
Procedural errors are associated with procedural knowledge and conceptual errors are associated with conceptual knowledge. Procedural knowledge refers to mastery of computational skills and knowledge of procedures for identifying mathematical components, algorithms, and definitions. Conceptual knowledge refers to knowledge of the underlying structure of mathematics – the relationships and interconnections of ideas that explain and give meaning to mathematical procedures [3], [4].

Not every error is the result of a lack of knowledge or skill. Sometimes, a student will make a mistake simply because he was fatigued or distracted (i.e., careless errors). Because conceptual and procedural knowledge often overlap, it is sometimes difficult to distinguish conceptual errors from procedural errors [5].

2.2 Analysis a priori and analysis a posteriori

Analysis a priori and analysis a posteriori are notions commonly used in the theory of didactic situations by Brousseau [6]. They are essential part of analysis of students' errors therefore we use them in our research.

Objective of analysis a priori of a mathematical task is to predict as accurately as possible the students' reactions and attitudes (obstacles, misconceptions and mistakes, correction of and further work with these mistakes), possible solving strategies (correct and incorrect), knowledge prerequisite for the use of the different solving strategies and also the teacher's reactions. E.g. it is the concept of an examiner work while making distractors to multiple-choice questions. In a posteriori analysis, a priori analysis is compared with experience from students' solutions of given task [7]. In our research we focus mostly on analysis of students' errors therefore this is the focus of both our analysis a priori and a posteriori of the tasks those were given to students.

3 Empirical results

3.1 Research design and analysis apriory

Our research was experiment with the sample of 91 students in the first semester of study at the Faculty of Mathematics, Physics and Informatics, Comenius University Bratislava. In the survey we did analysis of students' solutions of four tasks addressing the functions. Those tasks were part of a test from the subject Mathematical Analysis 1. All students wrote the test at the same time at 25.10. 2017, they all had the same variant of the test. We analyzed their results with the focus on the errors that students did and we summarized them based on the theoretical framework mentioned in previous chapter. This summarization was our further basis for recommendations of teaching strategies to overcome these errors and to strengthen students' knowledge of functions.

As the first stage of our qualitative analysis we did *analysis a priory* of the tasks.

Task 1: Composition of which elementary functions gives the function $y = \sqrt[3]{\dfrac{1}{\cos(x+2)}}$?

Use variables notions that indicate the order in which are functions composed.

Analysis apriory of the Task 1: We were expecting mostly good solutions; similar problems were part of seminars. Most common error we expected was incorrect order of functions during composition.

Task 2: State the function which graph we get from the graph of the function $y = \ln(2x-1)$ by:

a) Its shift by 3 units to the right.

b) Its shrink two-times (that means point $[x, y]$ we substitute with the point $\left[\dfrac{x}{2}, y\right]$).

We have standard position of the axes; we recommend testing your solutions for example by the positions of intersections with x axis.

Analysis apriory of the Task 2: This task is one of common tasks, which are part of mathematics already at upper secondary school. Therefore, we were expecting minimal mistakes mostly composed by the wrong direction of function shifting (left instead of right and stretching instead of shrinking).

Task 3: Sketch the graphs of these functions (each as one picture): $f : y = x^3$, $g : y = 0{,}3^x$, $F : y = \tan x$, $G : y = \log_{\frac{1}{2}} x$.

Analysis apriory of the Task 3: This task we perceived as the easiest one. Students just needed to use knowledge they already have from the upper secondary school and we used them also during our seminars. So we were expecting minimal errors.

Task 4: Sketch the graph of the function: $f(x) = \begin{cases} x-2, & x < 0 \\ 2x+1, & x \geq 0 \end{cases}$

Find its range and inverse function f^{-1}.

Analysis apriory of the Task 4: The task is from our point of view not difficult. We were expecting minor problems with domain of the inverse function, but otherwise we believed students will have no problems to solve this task.

3.2 Analysis a posteriori

In this paragraph we present actual students' errors in selected tasks. We compare them with our analysis apriory and indicate main problems.

Task 1: 33 students solved the task correctly. 13 students did not solve the task at all. Most common errors of students are stated in table 1.

Name of error	Number of students	Type of the error
Some functions they used were not elementary	26	Factual error
They do not fully understand process of function composition	7	Conceptual error
They did not write the order in which functions are composed	5	Procedural error
They used wrong order of functions in composition	4	Conceptual error, Procedural error
They used bad way how to write the composition	3	Factual error

Tab. 1. Errors in the task 1.

Task 2: Part a) of the task was solved correctly by 16 students; 6 students did not solve it at all. Part b) was solved correctly by 20 students; 9 students did not even start to solve it. Most common errors of students are stated in table 2.

Name of error	Number of students	Type of the error
Part a): Students did remember they 'need to subtract 3 to move the graph to the right', but did that in the wrong way	51	Conceptual error
Part a): Students add 3 in different ways	9	Factual error, conceptual error
Part b): Students did remember they 'need to multiply by 2 to shrink the graph two-times', but did that in the wrong way	27	Conceptual error
Part b): Students divide by 2 in different ways	35	Factual error, conceptual error

Tab. 2. Errors in the task 2.

Task 3: 38 students graphed all 4 functions correctly. Table 3 gives an overview of the mistakes.

Name of error	Number of students	Type of the error
Bad shape of graph $y = 0{,}3^x$	26	Factual error
Bad shape of graph $y = \log_{\frac{1}{2}} x$	24	Factual error
Bad shape of graph $y = x^3$	19	Factual error
Bad shape of graph $y = \tan x$	5	Factual error

Tab. 3. Errors in the task 3.

Task 4: 24 students solved this task correctly. The mistakes of the students are listed in the table 4. In mistakes in graph of original function we commonly indentified that students did not change slope of linear function between $x - 2$ and $2x + 1$.

Name of error	Number of students	Type of the error
Bad domain of inverse function	36	Conceptual error
Bad range of original function	34	Conceptual error
Domain of the inverse function is missing	26	Conceptual error
Mistake in inverse function formula	22	Conceptual error, Procedural error
Mistake in graph of original function	17	Conceptual error, Procedural error

Tab. 4. Errors in the task 4.

3.3 Research results and discussion

In this chapter we will sum up results of our survey and discuss them.

Comparison between analysis a priori and analysis a posteriori in all tasks showed that students were doing much more errors that we have expected.

In tasks one the most common error was factual error that students did not know which functions are elementary ones. We also noticed conceptual error – some students did not understand process of function composition at all.

In task two the most common error was conceptual one. For example students remembered mechanically that they need 'subtract 3 to move the graph to the right' but they did that in wrong way: either just subtract three: $y=\ln(2x-1)\rightarrow y=\ln(2x-1-3)$; or subtracting 3 from the whole function: $y=\ln(2x-1)\rightarrow y=\ln(2x-1)-3$.

Task 3 showed that students have factual problems even with most common graphs of elementary functions. This problem was the most visible when these functions were not the most common ones – in the case of logarithmic function and exponential function with the base lower than one.

In task 4 most students' errors were in finding domain of the inverse function, or not writing it altogether. The students have no conceptual connection that domain of function is necessary part of its definition. Students also showed some formal procedural errors in finding inverse function: some of them mechanically changed x with y but did not express inverse function formula. Many students also did mistake in the range of original function and its graph. In mistakes in graph of original function we commonly indentified conceptual error that students did not change slope of linear function between $x-2$ and $2x+1$.

These most common mistakes show us some deep conceptual, procedural and even factual errors in the area of functions. These errors can make obstacles in understanding other concepts in Mathematical Analysis. Therefore, in the next chapter of our article we will propose some trends in teaching to overcome these misconceptions and help students to get rid of them.

4 New trends of teaching to overcome students' errors

In this chapter we propose some strategies and problems those have potential to address issues found out during our survey. We will list them according the misconceptions they are connected with.

4.1 Shape of graphs

These strategies are connected with errors in changing the shape of graphs (task 2 and partially task 4 of our survey).

Students meet with drawing of functions graphs during their secondary education. But, the tasks they solve are more focused on making graphs from the function formulas. In our test we wanted the reverse operation. We knew that graph is shifted and shrank and we needed to find the function formula. In our survey we found out, that students did not have connection between shifting the graph by 3 units to the right and function formula $f(x-3)$.

To make this connection we propose to start with this introductory task:
1A) Sketch the graph of the function $f(x)=5x$. Create the graph of the function $g(x)=f(x-3)$.

Important is also the connection to the formula of function $g(x)$, which means the ability to create formulas of new functions. Therefore, we need during the process of teaching to integrate tasks in which we have a formula and we have to sketch the graph. Nevertheless, we need also tasks when we have graph and formula of the function and we need to find formula of function with a changed shape. In this point it is important to use ICT technologies, because with just pen and paper we can manage only a few tasks. To make enough tasks in order to achieve deeper understanding of students we can sketch the graphs by computer, using e.g. freeware software Geogebra.

Thus, we can solve number of tasks with the problem:
1B) We have function $f(x)$. Sketch its graph and compare it with the graph of the function $g(x) = f(\text{some changes of argument})$.

Then, after solving several tasks of this type, we can move to the new tasks, again using ICT, in order to project the graphs to illustrate the solutions for the students. For example:
1C) In the figure 1 are graphs of $f(x)$ and $g(x)$. *a)* Write the formula of $g(x)$ using formula of $f(x)$, that means in format $g(x) = f(\text{some changes of argument})$. *b)* Using the points of its graph find the formula of $f(x)$. *c)* Using the points of its graph find the formula of $g(x)$. Compare your results with results from *a)* and *b)*.

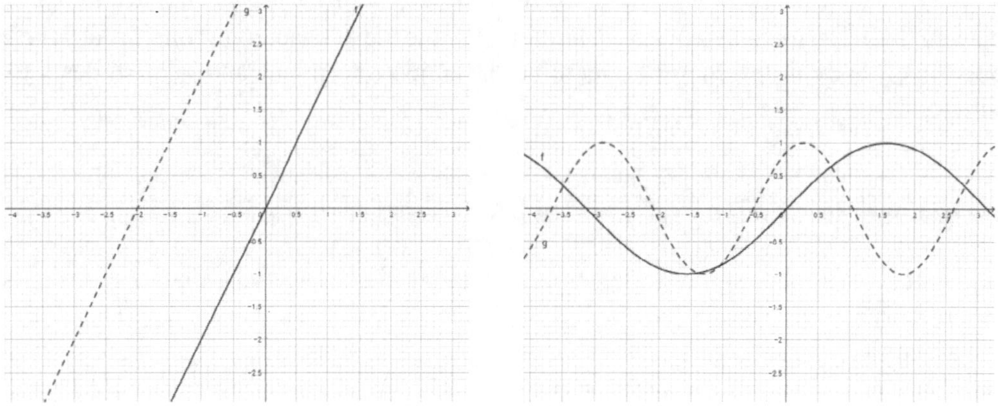

Fig. 1

During these tasks students find out importance of each aspect of function formula and graph. The need to be precise in these aspects, as e.g. the slope, is then apparent. This was a problem in task 4 of our survey.

4.2 Elementary functions

In this chapter we propose strategies to overcome errors connected with elementary functions and their graphs (task 1 and task 3 of our survey).

As we have seen in our survey, students do not understand notion elementary function. They meet this term first time at the university. It is not a part of any definition they need to know, it just begins to appear. Such usage of this notion without its clarification is considered propaedeutic in the theory of mathematics education. Our opinion is that this propaedeutic should come at secondary school and the creation of knowledge should come at the university level. This is factual error, so we need to fill this gap during our seminars. We need to point out that elementary function is the simplest form of function that we can call the class of the

functions. For example for the class of quadratic functions we have elementary function $f(x) = x^2$, which is the simplest quadratic function.
The tasks for students should look like:

2A) Which classes of functions do you know?

2B) What is the simplest form of square root function?

2C) Find out, which of the following functions are elementary:

$$f(x) = \frac{1}{x},\ g(x) = (x+1)^3,\ h(x) = \frac{1}{\cos x},\ i(x) = \sin x,\ j(x) = \log_3 x,\ k(x) = \ln x$$

Each of presented functions should be represented also by the sketch of its graph and verification of this graph using graphing software. This could be done during the presentation of the problems, if the teacher is skilled enough. Otherwise, the teacher should have prepared these graphs before the lesson in order to avoid unnecessary delays. The graphs could be then projected or created in the same time as we sketch them on the blackboard.

In order to get rid of the errors in the task 3, connected with the graphs of exponential and logarithmic function with argument lower than 1, the students should meet with these functions more often. Here can help just careful selection of easy tasks leading to sketching of such graphs during the seminars.

Another problem for students was the composition of functions. We have partially discussed this already in previous chapter. Now is the time to focus students' attention on this. We can add in the task 2C instruction: If some of the functions is not elementary one, write the elementary functions whose composition creates this function. E.g. for function $g(x) = (x+1)^3$ we should get functions $y_1 = x^3$, $y_2 = (x+1)$. How to composite them can be again find out with the help of computer. We will create graphs of $g(x), y_1, y_2, y_1 \circ y_2, y_2 \circ y_1$ and we will compare compositions of functions with the graph of $g(x)$. The teacher should have all these graphs prepared, see figure 2.

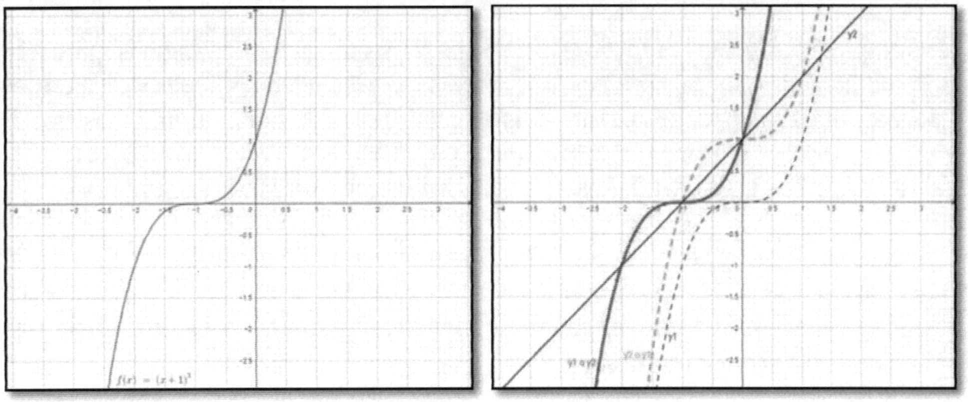

Fig. 2

In the same way we solve also function $h(x)$ from the task 2C. Then, we continue with more complex functions. The process of the composition of the functions should be always checked by the graphs! If we compose three functions, we should have prepared graphs of possible compositions and actual correct function.

For example task:

2D) Find out elementary functions those give as composition the function $y = \dfrac{1}{\sin(x - \frac{\pi}{4})}$ and write, how you need to compose them.

In this task we should get the functions $y_1 = x - \frac{\pi}{4}$, $y_2 = \sin x$, $y_3 = \frac{1}{x}$. Students will probably state as correct ones compositions $y_1 \circ y_2 \circ y_3$ and $y_3 \circ y_2 \circ y_1$. Teacher needs the graphs of these two compositions and the graph of the given function. This verification of the result is for students very important. Such graphical representations will help avoid students' misconceptions discussed in this chapter.

4.3 Domain of function and inverse function

In this chapter we will deal with errors in the task 4 of our survey. These are connected with the domain of the functions and also formal problems in the creation of inverse function.

Domain of the function appears to be not useful notion in the beginning of the secondary school mathematics education. Students meet at first with linear and quadratic functions. Importance of the domain is apparent just with logarithmic and goniometric functions. Controversially, in the whole document ISCED 3A (Educational program for upper secondary school education) [8] we cannot find the notion domain of the function. In the textbook for second year of upper secondary education [9] in the chapter about the graph of quadratic functions (page 50) is nothing about the domain or the range of these functions. In this way domain does not become automatically part of the function for the students.

However, despite these facts, the domain of the function and its range is part of upper secondary mathematics education. These concepts are included in the external tests from mathematics during the upper secondary school leaving exams. For example these tasks were directly focused on the domain of the function: year 2016, test 5178, task 24; year 2015, test 1203, task 21; year 2013, test 8103, task 27; year 2011, test 3306, task 25; etc. [10]

As explained, the domain and range of the function are not perceived by the students as integral part of the function. We should accept this situation at our seminar and focused the tasks on them – read them from the graphs of functions and also look for domain of each function we deal with from its formula.

Formal problems in the creation of the inverse function can be avoided by the application tasks from physics which do not use variables x and y.

Typically, these tasks can look like:
3A) Write the formula for the distance (in meters) depending on the time (in seconds) when object performs uniform motion at the speed 5 ms^{-1}. Sketch the graph of this relationship. Then, write the relationship of the time depending on the distance in the same motion and graph it. What did you noticed comparing these two graphs?

Similarly, we can do the same with the formula of the free fall motion, when the function is not just a linear one ($h = \frac{1}{2}gt^2$). We can find a lot of such problems in the physics. In this way students will see necessity to express variable from the formula and will be not fixed just on variables x and y. The task for the teacher is to point out that these problems are connected with the inverse functions.

5 Conclusion

In our paper we presented our survey, realized on the sample of 91 students in the first semester of university study at Faculty of Mathematics, Physics and Informatics, Comenius University in Bratislava. This survey was focused on functions as one of very important concepts in the university level mathematics. We analysed students' solutions of selected mathematical tasks and identified some major misconceptions connected with the functions. In order to avoid these misconceptions we then presented some trends of the teaching. These trends include proposals of the problems, instructions for using of ICT in the education of this topic and also recommendations for teachers useful to improve students' knowledge.

References

[1] BROWN J., SKOW K., IRIS CENTER, *Mathematics: Identifying and addressing student errors.* 2016, URL, http://iris.peabody.vanderbilt.edu/case_studies/ics_matherr.pdf.

[2] FISHER, D., FREY, N., Making time for feedback. *Feedback for Learning*, 70(1), p. 42–46.

[3] EISENHART, M., BORKO, H., UNDERHILL., R., BROWN, C., JONES, D., AGARD, P., Conceptual knowledge fall through the cracks: Complexities of learning to teach mathematics for understanding. *Journal for Research in Mathematics Education*, 24(1), 1993, p. 8–40, ISSN 0021-8251.

[4] MORU, E.K., QHOBELA, M., POKA, W., NCHEJANE, J., Teacher knowledge of error analysis in differential calculus. *Pythagoras*, 35(2), 2014, ISSN 2223-7895.

[5] RICCOMINI, P. J., *Identifying and using error patterns to inform instruction for students struggling in mathematics*. Webinar series, 2014, URL, http://www.ohioregion14.org/perspectives/?p=1005

[6] BROUSSEAU, G., *Theory of Didactical situations in mathematics 1970-1990.* [Edited and translated M. Cooper, N. Balacheff, R. Sutherland and V. Warfield.], Dordrecht: Kluwer Academic Publishers, 1997.

[7] MORAOVÁ, HANA; NOVOTNÁ, J., Lesson planning in CLIL in the Light of Brousseau's Theory of Didactical Situations in Mathematics. *Sborník konference Kompetence v cizích jazycích jako důležitá součást profilu absolventa vysoké školy IV.*, Brno: Univerzita obrany, 2014, p. 203 - 210.

[8] URL, http://www.statpedu.sk/files/articles/dokumenty/statny-vzdelavaci-program/matematika_isced3a.pdf.

[9] KUBÁČEK, Z., *Matematika pre druhý ročník gymnázií. Prvá časť*. Bratislava: Orbis Pictus Istropolitana, 2009.

[10] URL, https://www.matura.sk/maturitne-testy/matematika.

Current address

Dillingerová Monika, RNDr., PhD.
Faculty of Mathematics, Physics and Informatics, Comenius University in Bratislava
Mlynská dolina, 842 48 Bratislava, Slovak Republic
E-mail: monika.dillingerova@fmph.uniba.sk

Vankúš Peter, PaedDr., PhD.
Faculty of Mathematics, Physics and Informatics, Comenius University in Bratislava
Mlynská dolina, 842 48 Bratislava, Slovak Republic
E-mail: peter.vankus@fmph.uniba.sk

Slovak University of Technology in Bratislava
Faculty of Mechanical Engineering
17th Conference on Applied Mathematics **APLIMAT 2018**

Proceedings

OUR EXPERIENCE WITH THE INVOLVEMENT OF STUDENTS IN THE CREATION OF STUDY MATERIALS

DLOUHA Dagmar (CZ), HAMRIKOVA Radka (CZ)

Abstract. In past year we joined a project of the statutory city of Ostrava, which is called The Program focused mainly on supporting education and talent-management in the field of technical and natural science in the statutory city of Ostrava for the year 2017 and 1st trimester of 2018. Within this project we could develop co-operation with our talented students. Currently we have eight students from full-time bachelor, continuing masters and doctoral study.

Keywords: students, cooperation, mathematics, video

Mathematics subject classification: Primary 65D17; Secondary 53A04, 14Q05

1 Introduction

The first activity in which students are participaing is the co-operation on the creating of teaching material from maths and decriptive geometry for their subjects of study.

The second activity is recording commented videos of exercises from maths for students of continuing study of Faculty of Mining and Geology.

2 Motivational exercises

The common question of students from bachelor study is - what is math good for? So we engaged students of higher years of bachelor and master study in preparation of excercises, where the themes from bachelors maths and descriptive geometry can be found. Students of three different fields created for their study subject five excercises from professional subjects of theirs study. The parts of descriptive geometry and maths, which are taught within the bachelors or masters study, are included in those excercises. We want to use these excercises on lectures and practices to answer the neverending question of our students which is "Why do we have to learn this?"

Students had to choose such a exercises, which are being taught in professional subjects and the knowledges of descriptive geometry and high-school maths are necessary for the solution. The information in which year and in which subject is this specific problematic being taught is given in every exercise.

2.1 The example of motivational excercise

Application of rectangular axonometry to the central part of the Rozna mining area.

In the 2nd year of the subject Descriptive Geometry you will deal with plotting in rectangular axonometry. Axonometric rendering of the mine is used for spatial display of underground spaces, eg for emergency plans, mine ventilation maps, etc.

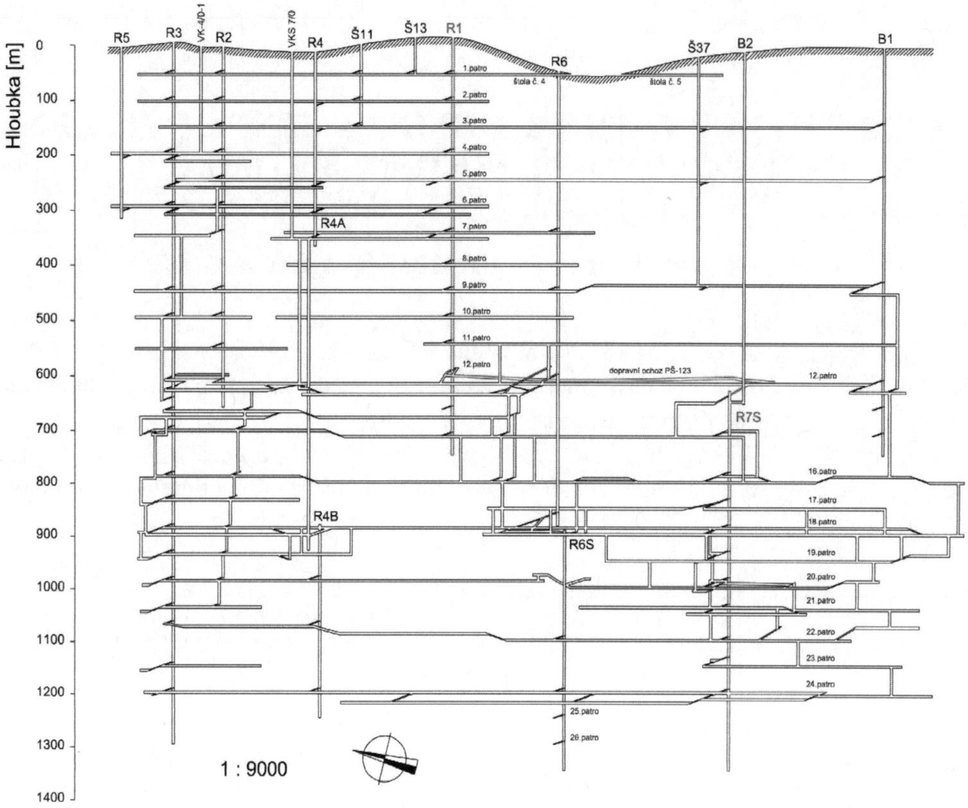

Fig. 1. Scheme of mine

2.2 Use of rectangular axonometry in practice

We can show the use of rectangular axonometry in practice on simplified task of transport in mine: The journey of miner. For the correct plotting you will need to know: scale conversion, rectangular axonometry - isometric staging triangle, display points and lines.

2.2.1 Assignment

The miner works at the Rozna I mine in Dolni Rozinka and needs to get to the mining. [5] The journey takes about half an hour and is as follows:

- on the surface boarding the cage of R1 pit and going to the 12th floor,
- on the 12th floor he is hauled on the rail transport, which takes him to the rim of the hidden pit R7S

- on the 12th floor of the R7S pit, he goes back to the cage and goes to the 24th floor
- on the 24th floor there will be a trekking run

2.2.2 Task

Draw the path of the miner in rectangular axonometry, respectively. in isometric (equilateral isometric triangle) according to the following coordinates (in meters):

Pit R1	$A = [0; 0; 0]$	$B = [0; 0; -648]$
Pit R7S	$C = [2400; 65; -494]$	$D = [2400; 65; -1154]$
Corridor on the 12th floor	$E = [0; 0; -548]$	$F = [2400; 65; -534]$
Corridor on the 24th floor	$G = [2400; 65; -1104]$	$H = [3815; 55; -1040]$

Tab. 1. Coordinates

2.2.3 Task

Draw a isometric diagram of a mining route at a scale of $1 : 20,000$. Do not consider the width of mine workings. Scale conversion $1 : 20000$. To do so, you divide the given coordinate by the scale and convert it to the extractor unit (in our case, per cm). Example of Pit Counting R1:

$$B_z = \frac{-648\ m}{20000} = -0,0324\ m = -3,24\ cm \tag{1}$$

Pit R1	$A = [0; 0; 0]$	$B = [0; 0; -3.24]$
Pit R7S	$C = [12; 3.25; -2.47]$	$D = [12; 3.25; -5.77]$
Corridor on the 12th floor	$E = [0; 0; -2.74]$	$F = [12; 3.25; -2.67]$
Corridor on the 24th floor	$G = [12; 3.25; -11.04]$	$H = [19.08; 2.75; -5.20]$

Tab. 2. Conversion of individual point coordinates

2.3 Sketch of the feature

Before we go into the resulting feature, we will create a sketch in which we verify that the displayed situation fits to the selected paper size. We will not develop the situation in general axonometry but in isometry. Its advantage is easy to apply co-ordinates and we can go along without an axonometric triangle. We choose the location of the origin of the coordinate system, roughly specifying points. If we do not like the layout, we can adjust it by moving the origin of the coordinate system. Once the situation is well represented, we can go into the resulting feature. See Fig. 2.

2.4 Feature

First, bring the isometric triangle into the upper left corner, see the sketch of the feature. Then convert the recalculated coordinates x and y of all points, according to the respective triangle axes, into the ground plan, ie points with index 1 (B1, C1, ...). Once you get the intersection with the surface, apply the z coordinate to get the depth of the point. You get the lines by linking the relevant points. See Fig. 3.

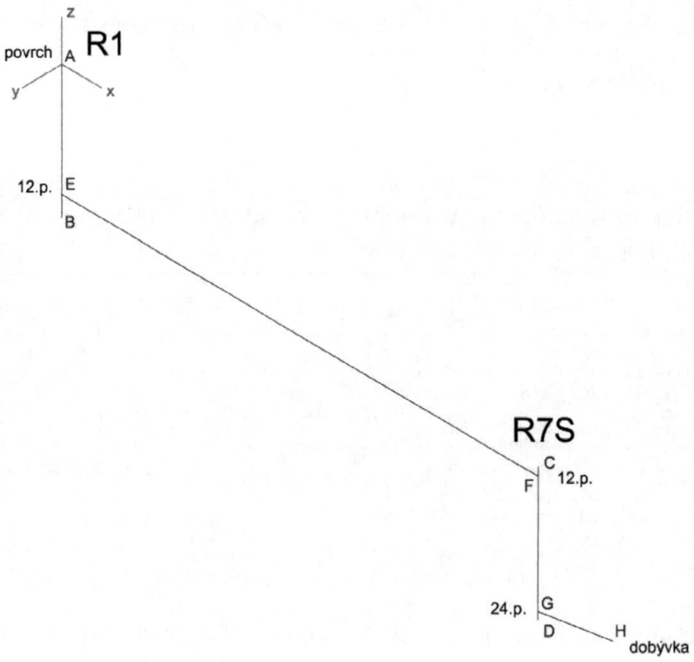

Fig. 2. Sketch

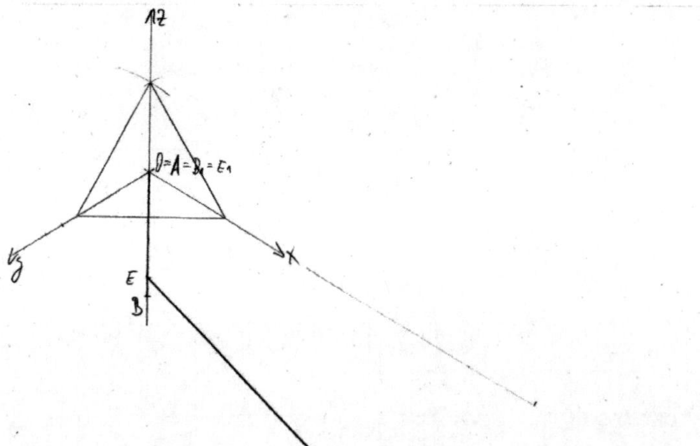

Fig. 3. Feature

3 Videos for students

Students of master's programs come to the VSB - Technical University from various specializations and the level of their knowledge of mathematics is therefore diametrically different. In this article, we will focus on a variety of students from present and combined following bachelors studies of Faculty of Mining and Geology. We run the Engineering Mathematics and Selected chapters in mathematics for this faculty. The lessons take place in Ostrava and on the detached workplace in Most. Both courses require a good knowledge of a Bachelor's Mathematics I. and II. Not every student has the required knowledge to follow these courses. The combined form of study is divided into 18 hours of classes, which are running in two or three blocks. We have students with different levels of mathematic attending these classes, as they absolved the Bachelor's maths in different schools, where the level was lower than on VSB-TU Ostrava. Students can use the scripts or the web address called "studijni opory" - learning support (www.studopory.vsb.cz). These pages are not only for teaching maths. They were created in 2006 - 2008 in the Learning Support project with preponderant distant elements for subjects of the theoretical basis of the study, also video materials on the web of the department and the collection of unsolved exercises for training. If the student approaches a difficulty, he has the possibility of personal consultation with his tutor. If the personal meeting cannot be arranged, we can consult the student over the phone or Skype.

But because most of our students come from distant places and they have to commute to school or study on our detached workplace in Most, personal consultations are almost impossible for them. In these cases, we can help by leading the consultation via Skype or Phone. And because students understand the explanation from another student better than professional definition from teachers, we approached our excellent students and offered them the possibility to take a part in this activity.

3.1 Commented videos

As a base for video-consultations we used the collection of unsolved exercises for training. From every thematic field, we've chosen one exercise and recorded a video tutorial of solving the problem with the spoken comment. Our students are already used to this form of selfstudy support, as there is already a bank of exercises from the Bachelors Mathematics I, II and Descriptive geometry. For video recording we use interactive board, program for operating the board, which is called the Device manager, then the Camtasia studio for cut and sounding and a graphic calculator GeoGebra for functions and 3D mathematic, which is free to download. This form of study is very popular among our students. We've created a questionary and it was answered by half of our students approx. From their answers, it was clear, that our videos are very important part of their preparation for an exam. They appreciate their accessibility on the internet and our willingness to record another videos. From our experiences, we already know, that when solving a certain exercise, students play the commented video of a similar problem in the first place and they try to understand each step of the solution. We record our videos such as the exercise is not only solved correctly but also economically. Too long explaining of each step could discourage the more talented students and would cut down the independent activity of another ones. If any student shall need further and closer explaining of any steps, he can send us a question. For communication with students, we use the school e-mail or comment underneath the videos on YouTube.

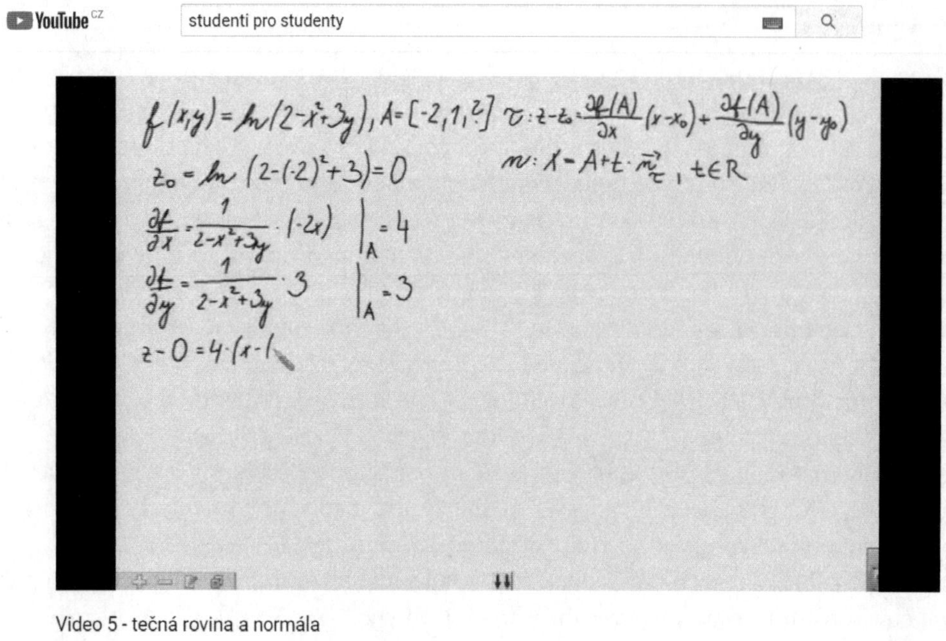

Fig. 4. Students for students - youtube channel

3.2 Youtube channel

At first, we planned to put those materials on school web, where they also were for two semesters and students were using them actively. After we started to do video-consultations we realized that the space on our web is highly insufficient, so after considering another options we decided to move our materials on YouTube. We don't have many views so far, but we believe that students will watch our materials here too, because the rules of videoconsultations are explained in the presentation. All students, who have the certain subject written in their study plan have received an email with a new link to all materials.

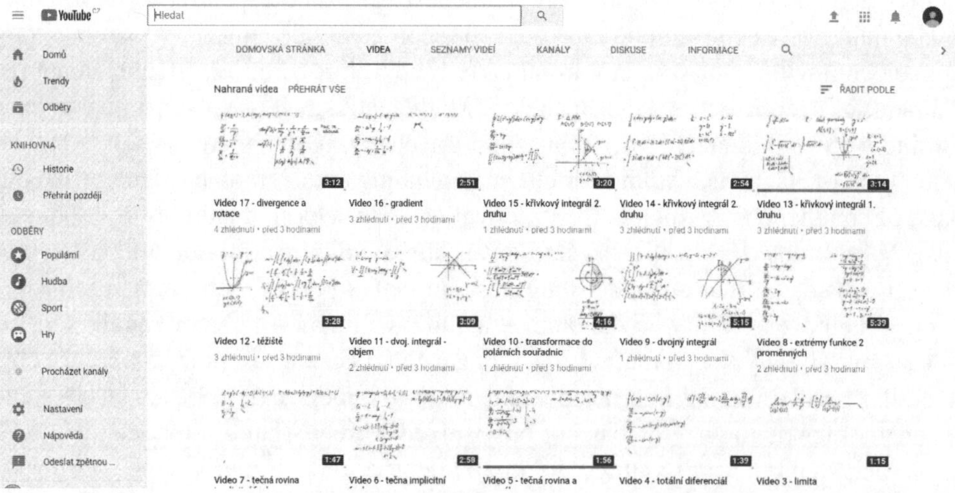

Fig. 5. Students for students - all videos

3.3 Videoconsultations

Now we will be showing how the usual communication between students and tutors looks like. One of the situation is extremes of a function of two variables (local, dependent and global). In the time allocation of 1 hour are students introduced to the theory, they will have the example of exercise calculated and will receive the link to the web address with materials to self-study. Unfortunately, the majority of students is only re-writing the information from the board and are thinking about the meaning back at home, where there is no one to explain what they didnt understand. Many students also cannot attend all classes due to high workloads. Therefore they often find difficulties when solving the exercises from the collection of unsolved exercises as a preparation for the exam, and they are unable to solve them on their own. The collection has 17 themes. We recorded a video of one example of the theme and the students are to solve another exercises by themselves.

4 Conclusion

It is a great contribution to school for the excellent students who work with us to stay at school as postgraduate students, and they can be expected to remain at school as well as scientists even though they are from different regions of the Czech Republic and the Slovak Republic. We want them to be helpful and prepare them for their pedagogical or scientific career by preparing them for the work of the educator, for their publishing activities and ensuring their participation in professional conferences.

References

[1] URL: http://www.studopory.vsb.cz/.

[2] URL: https://www.youtube.com/channel/UCK_YTyx_ZpwaJXYJ0qniEVg/videos .

[3] POLACEK, J.: *Zaklady deskriptivni a konstruktivni geometrie*, Ostrava: VSB-Technicka univerzita, 1996. ISBN 80-7078-388-5.

[4] DOLEZAL, J., POLACEK, J.: *Pravouhla axonometrie - sbirka resenych uloh*, Ostrava: VSB-Technicka univerzita, 2013. ISBN 979-80-248-2989-0.

[5] VOKURKA, M.: *Koncepce odvodnovani dolu Rozna I v Dolni Rozince Obecna charakteristika loziska Rozna.*, Ostrava: Vysoka skola banska, 2016.

[6] HAMRIKOVA, R., DLOUHA, D.: *Youtube in teaching geometry.* In: Systemy wspomagania w inzenyrii produkvji. Gliwice: P.A.NOVA S.A., 2017, s. 252-256. ISBN 978-83-65265-14-2.

[7] HAMRIKOVA, R., DLOUHA, D.: *Video tutorials for Master's program.* In: . PRAHA: CENTER HIGHER EDUCATION STUDIES. Conference: 12th International Conference on Towards Open Education and Information Society (DisCo) Location: Prague, CZECH REPUBLIC Date: JUN 26-27, 2017

[8] HAMRIKOVA, R. *Interaktivni tabule pri tvorbe e-learningovzch kurzu.* In Moderni matematicki metody v inzenyrstvi : sbornik z 22. seminare : [3.-5. cervna 2013 v Beskydech v Horni Lomne u Jablunkova]. Ostrava : VSB - Technicka univerzita Ostrava, 2013, s. 32-35.

[9] DLOUHA, D., HAMRIKOVA, R., DOLEZAL, J. *Videopomucky pro DG s vyuzitim GeoGebry.* In Uziti pocitacu ve vyuce matematiky : sbornik prispevku 7. konference : 5. - 7. listopadu 2015, Ceske Budejovice. Ceske Budejovice : Jihoceska univerzita, Pedagogicka fakulta, 2015, s. 73-78.

Current address

Dlouha Dagmar, Mgr., Ph.D.
Institute of Mathematics and Descriptive geometry
VSB - TU Ostrava
17. listopadu 15, Ostrava 70833
E-mail: dagmar.dlouha@vsb.cz

Hamrikova Radka, Mgr., Ph.D.
Institute of Mathematics and Descriptive geometry
VSB - TU Ostrava
17. listopadu 15, Ostrava 70833
E-mail: radka.hamrikova@vsb.cz

Slovak University of Technology in Bratislava
Faculty of Mechanical Engineering
17th Conference on Applied Mathematics **APLIMAT 2018**

Proceedings

MODELLING OF MATRIX CONVERTER SYSTEM WITH FICTITIOUS INTERLINK FUNCTION USING INSTANTANEOUS STATE CALCULATION

DOBRUCKÝ Branislav (SK), CHERNOYAROV Oleg V. (RU), MARČOKOVÁ Mariana (SK), ŠTEFANEC Pavol (SK)

Abstract. The paper deals with the analysis of matrix converter system using discrete instantaneous state calculation method. Because of impulse character of matrix converter, the space-state equations are transformed and discretized into discrete form. A virtual model of the matrix converter system uses fictitious interlink, and due to non-linear loads, the exciting fictitious functions are used for the creation of the discrete model. A new modified method of solution makes possible to determine the state of the system instantaneously. Results of theoretical analysis are confirmed by numerical computer simulation.

Keywords: matrix converter, modeling and simulation, circuit analysis, state-space equation, linear discrete control

Mathematics Subject Classification: Primary 15A05; Secondary 65R10.

1 Introduction - basic description of matrix converter system

Matrix converters are an important part of the field of conversion and conditioning of electrical energy [1]. One of them is matrix converter system (MCS) which converts the three-phase system to five-phase one ([3x5]). Such a system can be considered as a direct or indirect system [2-3]. The basic topology of [3x5] matrix converter system is depicted in fig. 1.

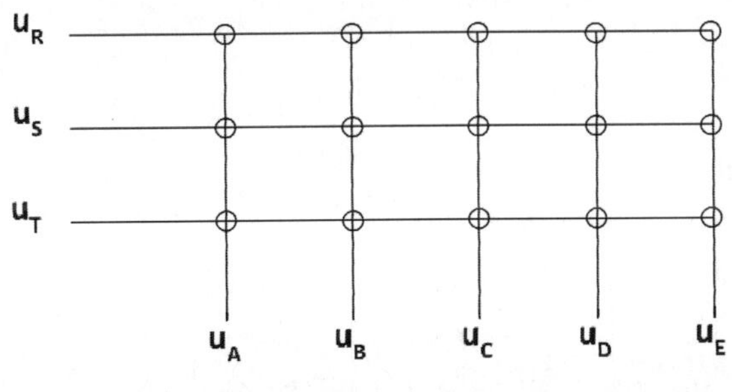

Fig. 1. Matrix topology of [3x5] matrix converter system (every single node comprises a bidirectional electronic switch).

Corresponding equivalent electrical scheme of [3x5] MCS is depicted in fig. 2.

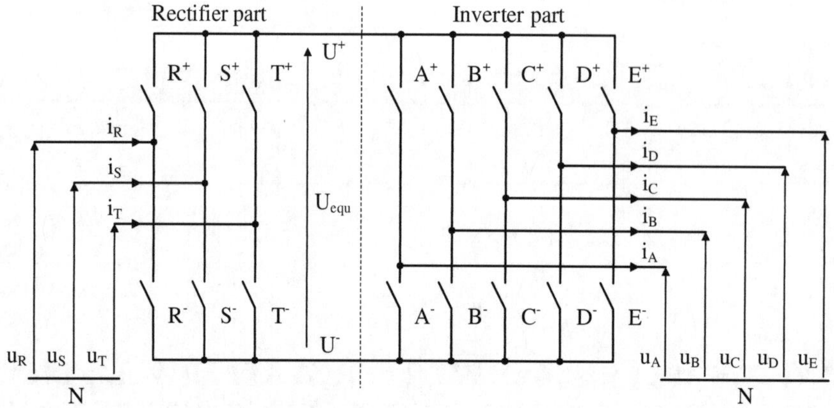

Fig. 2. Schematic model of indirect [3x5] MCS with fictitious interlink with voltage U_{equ}.

The input voltages of the system can be defined as time variable functions

$$u_R = U_R \sin(\omega_0 t),$$

$$u_S = U_S \sin\left(\omega_0 t - \frac{2\pi}{3}\right),$$

$$u_T = U_T \sin\left(\omega_0 t + \frac{2\pi}{3}\right),$$

where U_R, U_S, U_T mean maximal values of the functions and ω_0 is input angle frequency.

Our goal is to calculate output currents i_A, i_B, i_C, i_D, i_E under linear or non-linear passive and active loads.

2 Fictitious interlink and output voltages of matrix converter

The equivalent fictitious interlink voltage function $u_{equ}(t)$ can be calculated as difference of two potentials U^+ and U^- (see fig. 2)

$$u_{equ}(t) = U^+(t) - U^-(t).$$

Difference of U^+ and U^- potentials depends on the maximal and minimal values of input phase voltages u_R, u_S, u_T

$$\begin{pmatrix} U^+(t) \\ U^-(t) \end{pmatrix} = \begin{pmatrix} R^+ & S^+ & T^+ \\ R^- & S^- & T^- \end{pmatrix} \begin{pmatrix} u_R(t) \\ u_S(t) \\ u_T(t) \end{pmatrix},$$

where $R^+ \ldots T^-$ are logical variables for the maximal and minimal input phase voltages.

Then output voltages u_A, u_B, u_C, u_D, u_E of the system from fig. 2 will be [4]

$$\begin{pmatrix} u_A(t) \\ u_B(t) \\ u_C(t) \\ u_D(t) \\ u_E(t) \end{pmatrix} = \begin{pmatrix} u_{m1}(t), & 1 - u_{m1}(t) \\ u_{m2}(t), & 1 - u_{m2}(t) \\ u_{m3}(t), & 1 - u_{m3}(t) \\ u_{m4}(t), & 1 - u_{m4}(t) \\ u_{m5}(t), & 1 - u_{m5}(t) \end{pmatrix} \begin{pmatrix} U^+(t) \\ U^-(t) \end{pmatrix},$$

where modulation voltages

$$u_{mk}(t) = r\cos(\varphi)\sin\left[\omega_0 t - \frac{2\pi}{5}(k-1)\right] + \frac{1}{2}$$

for $k = 1,\ldots, 5$, $r = \frac{\cos(\varphi-\pi/3)}{\cos(\varphi)}$ and $\varphi = \omega t|_{\mathrm{mod}(\frac{\pi}{3})} - \frac{\pi}{6}$ as it is explained in detail in [4].

3 Modelling of the MCS system with linear passive and active loads

Example 1, considering linear passive resistive-inductive (RL) load

Voltage equation can be expressed using Clarke transform as complex time function in vector form for three-phase system [5-6]

$$\boldsymbol{u}_3(t) = \frac{2}{3}[u_A(t)\boldsymbol{a}^0 + u_B(t)\boldsymbol{a}^1 + u_C(t)\boldsymbol{a}^2]$$

where $\boldsymbol{a}$ is the unit vector $\boldsymbol{a} = \exp\left(j\frac{2\pi}{3}\right) = e^{j\frac{2\pi}{3}}$ and $\frac{2}{3}$ is transform constant for three-phase system.

Thus, in our case, for five-phase system

$$\boldsymbol{u}_5(t) = \frac{2}{5}[u_A(t)\boldsymbol{a}^0 + u_B(t)\boldsymbol{a}^1 + u_C(t)\boldsymbol{a}^2 + u_D(t)\boldsymbol{a}^3 + u_E(t)\boldsymbol{a}^4],$$

where $\boldsymbol{a}$ is the unit vector $\boldsymbol{a} = \exp\left(j\frac{2\pi}{5}\right) = e^{j\frac{2\pi}{5}}$ and $\frac{2}{5}$ is transform constant for five-phase system.
As complex time function the voltage $\boldsymbol{u}$ is presented as follows

$$\boldsymbol{u}(t) = \mathrm{Re}\{\boldsymbol{u}(t)\} + j\mathrm{Im}\{\boldsymbol{u}(t)\} = u_\alpha(t) + ju_\beta(t),$$

and it can be decomposed after Clarke transform into three scalar functions

$$u_{5,\alpha}(t) = \frac{2}{5}\left[u_A(t) + u_B(t)\cos\frac{2\pi}{5} + u_C(t)\cos\frac{2\pi}{5} + u_D(t)\cos\frac{2\pi}{5} + u_E(t)\cos\frac{2\pi}{5}\right]$$

$$u_{5,\beta}(t) = \frac{2}{5}\left[u_B(t)\sin\frac{2\pi}{5} + u_C(t)\sin\frac{2\pi}{5} + u_D(t)\sin\frac{2\pi}{5} + u_E(t)\sin\frac{2\pi}{5}\right]$$

$$u_{5,0}(t) = \frac{1}{5}\left[u_B(t)\sin\frac{2\pi}{5} + u_C(t)\sin\frac{2\pi}{5} + u_D(t)\sin\frac{2\pi}{5} + u_E(t)\sin\frac{2\pi}{5}\right] = 0$$

for symmetrical system.

Then, considering the same *RL* load of *α,β* phases

$$\frac{\mathrm{d}}{\mathrm{d}t}\begin{pmatrix} i_\alpha(t) \\ i_\beta(t) \end{pmatrix} = \boldsymbol{A}\begin{pmatrix} i_\alpha(t) \\ i_\beta(t) \end{pmatrix} + \boldsymbol{B}\begin{pmatrix} u_{5,\alpha}(t) \\ u_{5,\beta}(t) \end{pmatrix}, \quad (1)$$

where $\boldsymbol{A}, \boldsymbol{B}$ are stationary matrices of the system under *RL* load

$$\boldsymbol{A} = \begin{pmatrix} -\frac{R}{L}, & 0 \\ 0, & -\frac{R}{L} \end{pmatrix}; \quad \boldsymbol{B} = \begin{pmatrix} -\frac{1}{L}, & 0 \\ 0, & -\frac{1}{L} \end{pmatrix}.$$

By time discretization of (1) with the step Δ creating discrete dynamical model [7]

$$\begin{pmatrix} i_\alpha \\ i_\beta \end{pmatrix}_{k+1} = \boldsymbol{F}_\Delta\begin{pmatrix} i_\alpha \\ i_\beta \end{pmatrix}_k + \boldsymbol{G}_\Delta\begin{pmatrix} u_{5,\alpha} \\ u_{5,\beta} \end{pmatrix}_k$$

for $k \in< 0, \infty)$ where k is integer.

It is possible to get the matrices $\boldsymbol{F}_\Delta$ and $\boldsymbol{G}_\Delta$ using some numerical method, e.g.

$$\boldsymbol{F}_\Delta = (\mathbf{1} - \Delta \boldsymbol{A})\,; \quad \boldsymbol{G}_\Delta = \Delta \boldsymbol{B}$$

where the integration step Δ due to possible numerical instability must be sufficiently small, in any case, shorter than the lowest time constant $\frac{L}{R}$ of the MCS system, and also as a step of the sequence of voltage pulses $\frac{T}{5}$, [7-8].

Using modified method for numerical solution [10-12], it is possible to get the result at any time discrete instant k

$$\begin{pmatrix} i_\alpha \\ i_\beta \end{pmatrix}_k = \boldsymbol{F_\Delta}^k \begin{pmatrix} i_\alpha \\ i_\beta \end{pmatrix}_0 + \boldsymbol{G}_\Delta \sum_{l=0}^{k-1} \boldsymbol{F_\Delta}^l \begin{pmatrix} u_{5,\alpha} \\ u_{5,\beta} \end{pmatrix}_{k-(l+1)}$$

for $l \in< k-1, \infty)$ where l is integer.

Real current of the A-phase i_A is equal i_α. Other phase currents i_B, i_C, i_D, i_E can be obtained using inverse Clarke transform $i_{\alpha,\beta,0} \leftrightarrow i_A, i_B, i_C, i_D, i_E$. In steady-state, the other phase currents are shifted against previous one by electric angle $\frac{2\pi}{5}$. Time waveforms of phase voltage and current are shown in fig. 3.

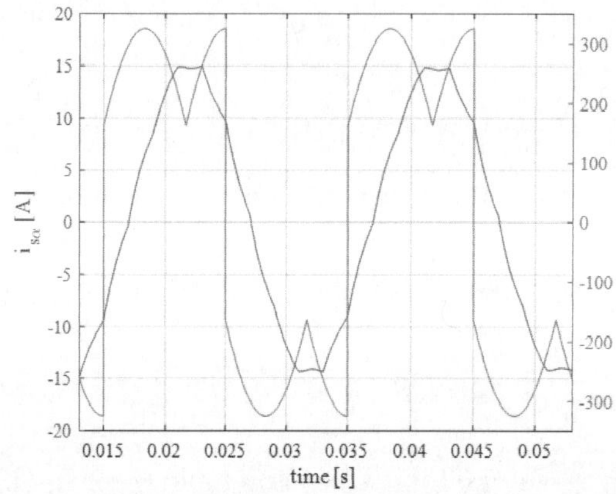

Fig. 3. Phase voltage and current time waveforms of [3x5] matrix converter system at steady-state operation.

Note, that by using this method matrices $\boldsymbol{A}$ and $\boldsymbol{B}$ should be stationary, and time-independent ones. It practically means that elements of these matrices should be constants.

Example 2, considering nonlinear active motoric load - five-phase induction motor

Now, we concentrate our attention to the problems of stator quantities of the motor calculation. It can be expressed using Clarke transform as in the previous example, i.e.

$$\boldsymbol{u}_s(t) = \frac{2}{5}\left[u_A(t) + u_B(t)e^{j\frac{2\pi}{5}} + u_C(t)e^{j\frac{4\pi}{5}} + u_D(t)e^{j\frac{6\pi}{5}} + u_E(t)e^{j\frac{8\pi}{5}}\right],$$

where the voltage is expressed in a component form as

$$\boldsymbol{u}_s(t) = u_{s\alpha}(t) + ju_{s\beta}(t).$$

Then the vectors of stator and rotor currents are possible to be obtained from a dynamic model of the motor [9]

$$\frac{\mathrm{d}}{\mathrm{d}t}\begin{pmatrix} i_{s,\alpha}(t) \\ i_{s,\beta}(t) \\ i_{r,\alpha}(t) \\ i_{r,\beta}(t) \end{pmatrix} = \boldsymbol{A}_{non}\begin{pmatrix} i_{s,\alpha}(t) \\ i_{s,\beta}(t) \\ i_{r,\alpha}(t) \\ i_{r,\beta}(t) \end{pmatrix} + \boldsymbol{B}_{non}\begin{pmatrix} u_{5,\alpha}(t) \\ u_{5,\beta}(t) \\ 0 \\ 0 \end{pmatrix} \tag{2}$$

and torque equation

$$\frac{\mathrm{d}}{\mathrm{d}t}\omega_m = \frac{T_{elmg}(t) - T_{load}}{J_m}. \tag{3}$$

Matrices $\boldsymbol{A}_{non}$, $\boldsymbol{B}_{non}$ are in this case non-stationary, they comprise a state-variable ω (see appendix), and $T_{elmg}(t)$ is electro-magnetic torque, T_{load} is load torque, J_m is moment of inertia. System of differential equations (2)-(3) is nonlinear one due to nonlinear functions such as multiplications of state-variables. So, the method mentioned above can not be used.

Linearization of the system equations (2)-(3) can be done using exciting fictitious functions: non-stationary elements of the matrices $\boldsymbol{A}_{non}$, $\boldsymbol{B}_{non}$ are withdrawn from them, and fictisious exciting functions are created. Then

$$\frac{\mathrm{d}}{\mathrm{d}t}\begin{pmatrix} i_{s,\alpha}(t) \\ i_{s,\beta}(t) \\ i_{r,\alpha}(t) \\ i_{r,\beta}(t) \end{pmatrix} = \boldsymbol{A}_{lin}\begin{pmatrix} i_{s,\alpha}(t) \\ i_{s,\beta}(t) \\ i_{r,\alpha}(t) \\ i_{r,\beta}(t) \end{pmatrix} + \boldsymbol{B}_{lin}\begin{pmatrix} u_{5,\alpha}(t) \\ u_{5,\beta}(t) \\ f(\omega, i) \\ f(\omega, i) \end{pmatrix}.$$

Now, we can generate new, linearized system of differential equations, and after time discretization the numerical solution can be used:

$$\begin{pmatrix} i_{s,\alpha}(t) \\ i_{s,\beta}(t) \\ i_{r,\alpha}(t) \\ i_{r,\beta}(t) \end{pmatrix}_{k+1} = \boldsymbol{F}_{lin}\begin{pmatrix} i_{s,\alpha}(t) \\ i_{s,\beta}(t) \\ i_{r,\alpha}(t) \\ i_{r,\beta}(t) \end{pmatrix}_{k} + \boldsymbol{G}_{lin}\begin{pmatrix} u_{5,\alpha}(t) \\ u_{5,\beta}(t) \\ f(\omega, i) \\ f(\omega, i) \end{pmatrix}_{k}.$$

Note, that linearized system with $\boldsymbol{A}_{lin}$, $\boldsymbol{B}_{lin}$ should be checked for stability and controllability before numerical computations. Time waveforms of phase voltage and current are shown in fig. 4.

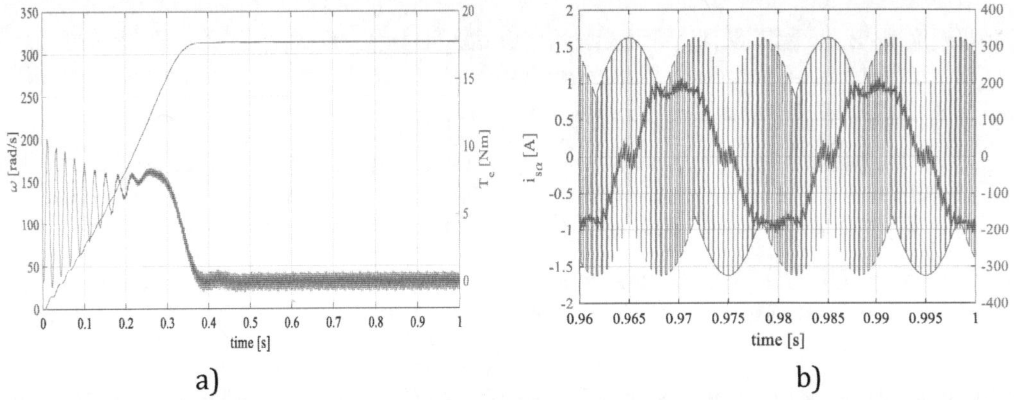

Fig. 4. Start-up of the motor which is fed by [3x5] matrix converter system for speed and torque (a), detailed waveforms of phase voltage and current at steady-state (b) gained by computer simulation.

7 Conclusion

The method introduced in the paper presents good mean for matrix converter system solution under periodic non-harmonic impulse supply. The method makes possible to obtain the instantaneous state of the system if the system of matrices are stationary ones. Otherwise, one can use approximated approach. So, a new modified method of solution makes possible to determine the state of the system instantaneously at any discrete time instant. Worked-out results have confirmed that theoretical assumptions and verifying computer simulations are in good agreement.

Appendix

$$\boldsymbol{A}_{non} = \begin{pmatrix} -\frac{R_r}{L_r} & \omega_r & M\frac{R_r}{L_r} & 0 \\ -\omega_r & -\frac{R_r}{L_r} & 0 & M\frac{R_r}{L_r} \\ \frac{MR_r}{\sigma L_s L_r^2} & \frac{M}{\sigma L_s L_r}(\omega_s - \omega_r) & -\frac{R_s}{\sigma L_s} & \omega_s \\ \frac{M}{\sigma L_s L_r}(\omega_s - \omega_r) & \frac{MR_r}{\sigma L_s L_r^2} & -\omega_s & -\frac{R_s}{\sigma L_s} \end{pmatrix};$$

$$\boldsymbol{B}_{non} = \begin{pmatrix} 1 & 0 & 0 & 0 \\ 0 & 1 & 0 & 0 \\ -\frac{M}{\sigma L_s L_r} & 0 & \frac{1}{\sigma L_s} & 0 \\ 0 & -\frac{M}{\sigma L_s L_r} & 0 & \frac{1}{\sigma L_s} \end{pmatrix},$$

where elements of matrices are derived from motor parameters [9], except for the angular velocity ω_s or ω_r, respectively, which is one of the state variables.

Acknowledgement

The reported study was supported by Slovak Grant Agency VEGA through the grant No. 1/0928/15 and by Russian Science Foundation through the Grant No. 14-49-00079P.

References

[1] DOBRUCKÝ, B., ALTUS, J., ŠPÁNIK, P. Synergic influence of Power Electronics, Electric Traction, and Electro-Energetics. *Communications-Scientific Letters of University of Žilina*, Vol. 3, no. 2-3, 2001, pp. 16-28, ISSN 1335-4205.

[2] CHLEBIŠ, P., ŠIMONÍK, P., KABAŠTA, M. The Comparison of Direct and Indirect Matrix Converters. In *Proceedings of PIERS´10 International Conference*, Cambridge: The Electromagnetics Academy, 2010, pp. 310-313, ISBN 978-1-934142-14-1.

[3] JUSSILA, M., TUUSA, H. Comparison of Direct and Indirect Matrix Converters in Induction Motor Drive. In *Proceedings of IEEE International Conference IECON´06*, Paris, France, June 2006, pp. 1621-1626, ISBN 1-4244-0136-4.

[4] REZAOUI, M. M., KOUZOU, A., MAHMOUDI, M.O., NEZLI, L. Comparison Performances between Two Matrix Converters [3x3] and [3x5] Supplying a Multi-Phases Induction Machine. *Journal of Electrical Engineering*, Published Online URL: http://www.jeee.ro.

[5] DOBRUCKÝ, B., SOLÍK, I. Contribution to state analysis of power semiconductor systems. *Journal of Electrical Engineering*, Vol. XXXXIII, no. 3, 1992, pp. 65-71, ISSN 1335-3632.

[6] DOBRUCKÝ, B., MARČOKOVÁ, M., KABAŠTA, M. Using Orthogonal Transform for Solution of Matrix Converter Power Circumstances in Mathematica ® Environment. In *Proceedings of the 7th International Conference on Applied Mathematics APLIMAT 2008*, Bratislava, 2008, pp. 735-739, ISBN 978-80-89313-03-7.

[7] MANN, H. *Using computers for electrotechnics designs* (in Czech). SNTL, Prague, 1984, ISBN 04-539-84.

[8] GÖLDNER, K. *Mathematical Basics of System Analysis Vol. 2* (in German: Mathematische Grundlagen der Systemanalyse). VEB Fachbuchverlag, Leipzig, 1982, ISBN 114-210-96-82.

[9] DOBRUCKÝ, B., KAŠČÁK, S., PRAŽENICA, M., PAVLÁSEK, P. Direct AC-AC Propulsion System using Matrix Converter and 5Φ traction motor for HEV Vehicle. *Communications-Scientific Letters of the University of Žilina*, Vol. 20, no. 1, 2018, , ISSN 1335-4205 – to appear.

[10] DOBRUCKÝ, B., ŠTEFANEC, P., BEŇOVÁ M., CHERNOYAROV O.V., POKORNÝ, M. Analysis of Higher Order System with Impulse Exciting Functions in Z-Domain. *Circuits and Systems*, 2016, no. 7, pp. P1-P11, ISSN 2153-1285.

[11] STREJC, V. *State-space theory of linear discrete control* (in Czech). Academia, Prague, 1978, ISBN 509-21-857.

[12] DOBRUCKÝ, B., MARČOKOVÁ, M., ŠTEFANEC, P. Analytic Solution of the Multi-Resonant Electrical Circuit under Impulse Supply. In *Proceedings of the 16th International Conference on Applied Mathematics APLIMAT 2017*, Bratislava, 2017, pp. 467-472, ISBN 978-80-227-4650-2.

Current addresses

Dobrucký Branislav, professor
University of Žilina, 01026 Žilina, Slovakia
Tel. number: +421415131602, e-mail: branislav.dobrucky@fel.uniza.sk
and
National Research University - Moscow Power Engineering Institute, 111 250 Moscow, Russia

Chernoyarov Oleg Vyacheslavovich, associate professor
National Research University - Moscow Power Energetic Institute, 111 250 Moscow, Russia
Tel. number: +79154008373, e-mail: o.v.ch@mail.ru

Marčoková Mariana, associate professor
University of Žilina, 01026 Žilina, Slovakia
Tel. number: +421914254980, e-mail: mariana.marcokova@gmail.com
and
National Research University - Moscow Power Energetic Institute, 111 250 Moscow, Russia

Štefanec Pavol, software engineer
Siemens s.r.o, Lamačská cesta 3/A, 841 04 Bratislava, Slovakia
Tel. number: +421949612886, e-mail: stefanec.pavol@gmail.com

Slovak University of Technology in Bratislava
Faculty of Mechanical Engineering
17th Conference on Applied Mathematics **APLIMAT 2018**

Proceedings

THE ROLE OF RTU STUDENTS' SURVEY IN ENSURING THE QUALITY OF MATHEMATICS STUDIES

DZENITE Ilona (LV), CERNAJEVA Sarmite (LV), MATVEJEVS Aleksandrs (LV)

Abstract. In the current process of change around the world, questions concerning the improvement of the efficiency and quality of the education system are very appropriate. The quality of higher education needs to be developed and improved for the benefit of students and other beneficiaries of higher education [1]. The authors' research is dedicated to finding ways of improving studies in Mathematics at Riga Technical University on the basis of surveyed students' opinions and suggestions.

Keywords: higher education, study process, students' survey, mathematical subjects

Mathematics Subject Classification: 97B40, 97C30, 97D50, 97D60.

1 Introduction

In a changing world, society's thoughts and understanding about education are transforming, and people's needs for education are changing. Therefore, everywhere in the world, questions about improving the education system, about improving the quality of education, are very appropriate and important. A special place in the education system is given to higher education because its task is to develop creative thinking personalities. In response to the rapid changes, the higher education institutions in Latvia are improving the content of study programs and the quality of studies by ensuring the modernization of the education system and significantly improving the basis for acquiring modern knowledge and qualifications. This means that changes are also taking place in the higher education system.

In the 2011 EU Council conclusions on modernizing higher education, the member states were invited "to promote the systematic development of effective strategies, to provide access to education for under-represented groups and to step up efforts to reduce early school leaving rates, by improving the quality, usefulness and attractiveness of training courses, in particular, regarding student-centred learning, and in the support, advice and consultations provided to students after joining a higher education institution" [3]. Changes in education are needed for a new quality in teaching, which is formed in the development of society and manifests itself as an ongoing process [2].

2 The origins of the research

Taking into account that the educational needs of students nowadays are changing and the nature of knowledge has changed, becoming more universally accessible, we wanted to analyze our own learning process, and ask ourselves what it is that we want to provide? We thought about various forms of learning, such as lectures, exercises, consultations, structured knowledge, practical knowledge, seminars, discussions, effective learning approaches, discussion management and innovative thinking strategies. Although each student takes these steps differently, each student has an important place in the learning process, and our job is to offer optimal support to everyone. The learning path is precisely determined. The information is selected by the lecturer, prepared and offered. The ability to orient oneself in the information environment, in fact, is one of the most important skills that must be learned today. We are focused particularly on the students who are seriously searching for their paths in education and their future profession. We try to restructure our teaching time in order to effectively use the components that make up our learning process.

Since the 2007/2008 academic year Riga Technical University (RTU) has used the ORTUS portal, a unified e-learning system, which is based on the MOODLE (Modular Object-Oriented Dynamic Learning Environment) e-learning program. In Latin the word "ortus" means "a new beginning". The RTU ORTUS portal (www.ortus.lv) facilitates the life of every RTU student who wants to make his/her teaching process faster, modern and more efficient. This ORTUS portal is available anywhere and at any time via internet access. In the ORTUS portal, the Study Department of RTU, in cooperation with the Information Technology Service, has implemented the Student Questionnaire System, which conducts regular, once per semester, student surveys on the quality of the content of acquired studies and on the quality of the teaching staff's work. Through the RTU ORTUS portal, every student receives a questionnaire for each semester's acquired study subject. The results of the questionnaire for each semester can be found in the appendixes and are fully accessible to each particular academic staff member (regarding their study subject), to the head of the structural unit of the teaching staff and - upon request - to the study program directors of the academic staff involved in the implementation of their study program and to representatives of the Students' Parliament. Survey results are carefully analyzed, evaluated, accumulated and compared with the results of the previous period in order that appropriate decisions are made to improve the quality of studies.

3 How this research began at the Department of Engineering Mathematics

Every semester surveys are mandatory and every RTU student has to answer all survey questions before the beginning the next semester. The existence of the "deadline" and the obligation to go through the whole survey, in our opinion, significantly influences the honesty of answers given by badly-organized students doing the survey at last moment, who may not think about the questions. The existence of such the survey, which aims to improve Mathematical studies through knowing what students really think, and to understand what can help them learn Mathematics better, prompted us, several Professors of Mathematics of the Department of Engineering Mathematics, to create our own survey with questions which are specific to Mathematical subjects.

We have a particular innovation this year. Due to the initiative of the Students' Parliament, starting from this 2017/2018 academic year, all RTU Professors of Mathematics are obliged to grade the course for a student not only upon the student's final exam results, but also upon their work in the semester. The following formula for calculating the semester mark is applied from the current academic year. The formula gives 50% of the mark for the final exam, taken during the exam period; 40% of the mark for tests and assignments, taken during the semester; 10% of the mark for homework solved at home. Such an approach encourages students to learn steadily and deeply, so that the results rely on a student's knowledge, not on luck during the exam.

The aim of the students 'survey was to find out the students' view of the study program they are studying, as well as the progress of studies at the higher education institution. In total, 317 students of the Faculty of Civil Engineering, Faculty of Mechanical Engineering, Transport and Aeronautics, Faculty of Power and Electrical Engineering, were surveyed. The survey was given for the 1st year students of the basic Mathematics course, and for the 2nd year students of the course "Supplementary Mathematics", both for Latvian and foreign groups of students. The survey sample for the 2nd year students is presented below.

For students of the study subject „**Supplementary Mathematics**"

SURVEY

"Students' opinion and suggestions for improving lessons on Mathematics"

1) Did YOU like Mathematics at school?

___ YES; ___ NO.

2) Has YOUR **ATTITUDE** to Mathematics changed over time?

___ YES, (at present, I ____);

___ NO.

3) Have YOU **FAILED EXAMS** for the 1st year subject „Mathematics I, II"?

___ YES, I have 1 failed exam for the _______semester;

___ YES, I have failed exams for BOTH semesters;

___ NO, I do not have any failed exams. Besides,

My mark for the 1st semester is ______, and I passed the exam on the _______ attempt;

My mark for the 2nd semester is ______, and I passed the exam on the _______ attempt.

4) In YOUR opinion, are **HOME WORK ASSIGNMENTS** needed?

___ YES, since it helps students to master the material and to be better prepared for the tests;

___ YES, to increase the semester final grade, since it contributes 0% - 10% to the semester's final mark on mathematical subjects in RTU;

___ YES, but MANDATORY home work is needed (this means that without having a positive evaluation of work made at home, students are not allowed to come and take the exam);

___ NO, it isn't needed, if it does not influence the final grade of the semester.

5) In YOUR opinion, are classroom written **CONTROL TESTS** on Mathematics needed?

___ YES, since it helps students to master the material and to be better prepared for the exams;

___ YES, to increase the semester final grade, since it contributes 40% - 50% to the semester's final mark on mathematical subjects in RTU;

___ YES, but MANDATORY control tests are needed (this means that without having a positive evaluation of all classroom tests, students are not allowed to come and take the exam);

___ NO, it isn't needed, if it does not influence the final grade of the semester.

6) In YOUR opinion, does the division of the semester final exam into **MIDTERM EXAMS** facilitate passing exams on mathematical subjects?

___ YES, (since ____);

___ NO, (since ____);

___ NOT important, it can be the choice of the lecturer.

7) In YOUR opinion, which **LECTURE** style would be better?

___ A lecturer DICTATES rapidly all the theory, so that it is the personal responsibility of a student to write it down or not. The lecturer solves all task examples on the blackboard;

___ All theory is given in SLIDES projected onto the screen so that a lecturer just explains it, but students are to write the notes over their personal printed copy of the slides. The lecturer solves all task examples on the blackboard;

___ All theory and a part of solved examples are given in slides so that a lecturer just explains it. And only a PART of solving examples is solved by the lecturer on the blackboard;

___ A lecturer writes the lecture key points and formulas, and solves all task examples on a screen which projects it onto a big wall screen;

___ YOUR suggestion: ____.

8) What is YOUR attitude to the **INVITING** students **to** the **BLACKBOARD** to solve tasks during tutorials on Mathematics?

___ FOR such a method, since it increases the level of understanding of mathematics;

___ AGAINST such a method in mathematical subjects, since ______________________________ ;

___ is BETTER, but only if the lecturer solves all task examples on the blackboard so that many solved examples can be considered during every tutorial;

___ NOT important, it can be the choice of the lecturer.

9) In YOUR opinion, should the **NUMBER** of **LECTURES** and **TUTORIALS** on Mathematics be increased?

___ YES, it definitely must be increased, since the number of solved tasks on each topic is not enough;

___ YES, it would be good;

___ NO, the present number of lessons is enough.

10) In YOUR opinion, is a **5-min BREAK** during the mathematical lessons needed?

___ YES, it definitely is needed in the middle of the LECTURE;

___ YES, it definitely is needed in the middle of the TUTORIAL;

___ NO, not needed, since in this case the lesson will be finished 5 min earlier;

___ NOT important, it can be the choice of the lecturer.

11) Do YOU have any other **SUGGESTIONS**?

___ YES, these are ____;

___ NO, I don't have.

THANK YOU FOR THE RESPONSE! YOUR OPINION IS VERY IMPORTANT!

Fig. 1. Survey questions for the 2nd year students.

4 Analysis and interpretation of the results

What is the optimal way to provide students with each of the necessary components of the learning process during their studies?

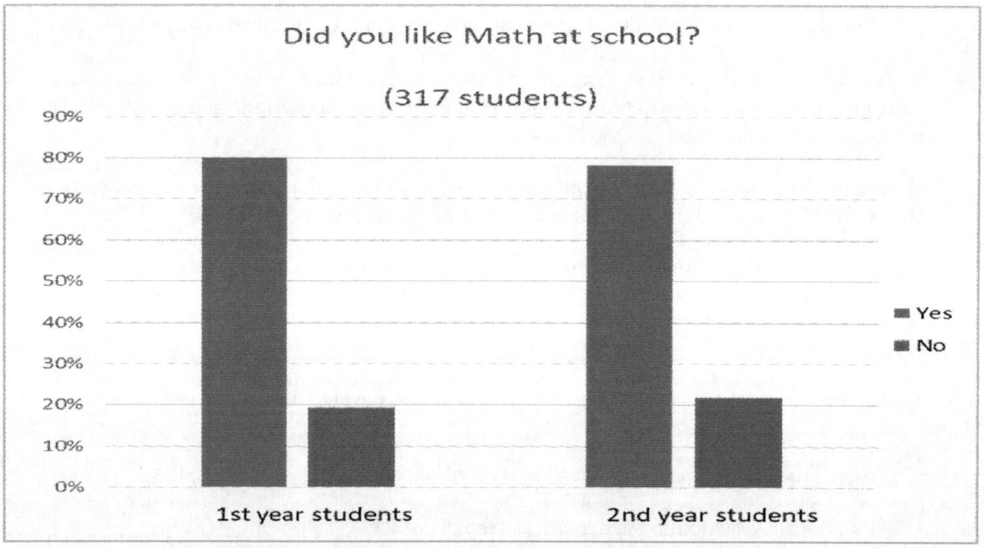

Fig. 2. Survey question: "Did you like Math at school?".

As expected the majority of students of the 1st year of studies in technical specialties of RTU liked school mathematics! The volume of work in higher mathematics in the first year probably encourages some the 2nd year students to reconsider their appreciation for school mathematics. The number of 2nd year students who liked mathematics in school is less than the number of similar students of the first year (see Fig. 2).

The point of view of the 1st year students about their success in school mathematics as a whole, corresponds to their assessment in mathematics in their secondary school diploma. Some reassessment of abilities is observed for students who have a "good" grade in their high school diploma, which is expected (it is possible that the school underestimated them) (see Fig. 3).

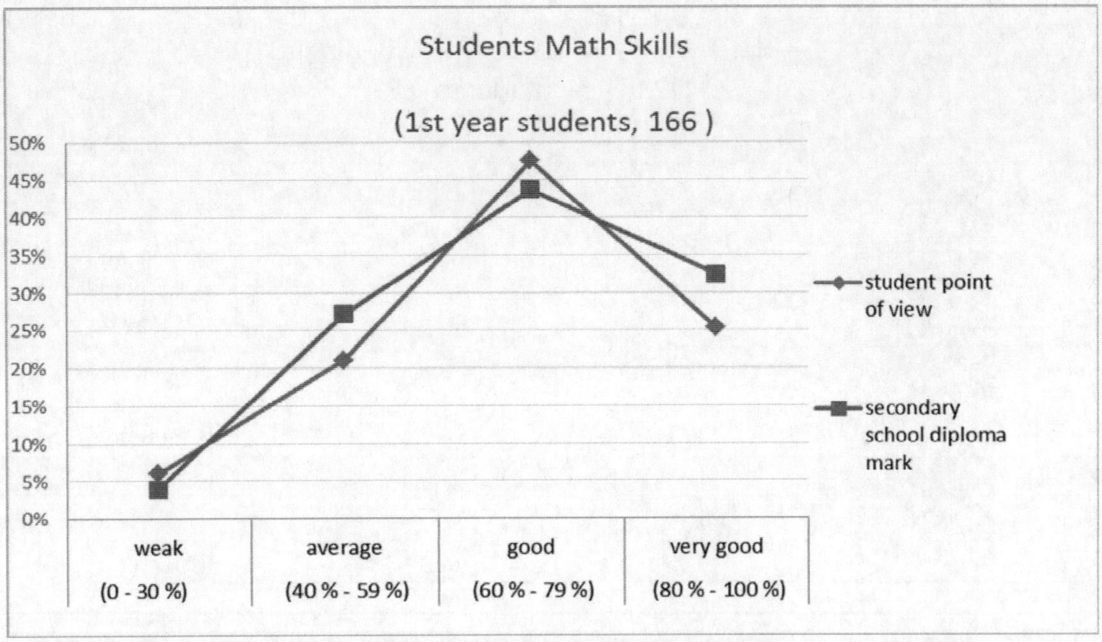

Fig. 3. Students Math Skills.

The majority of students, from both of the 1st and 2nd year, are sure that homework and classroom tests in higher mathematics contribute to the better preparation of students for examinations in higher mathematics at university (see Fig. 4 and Fig. 5).

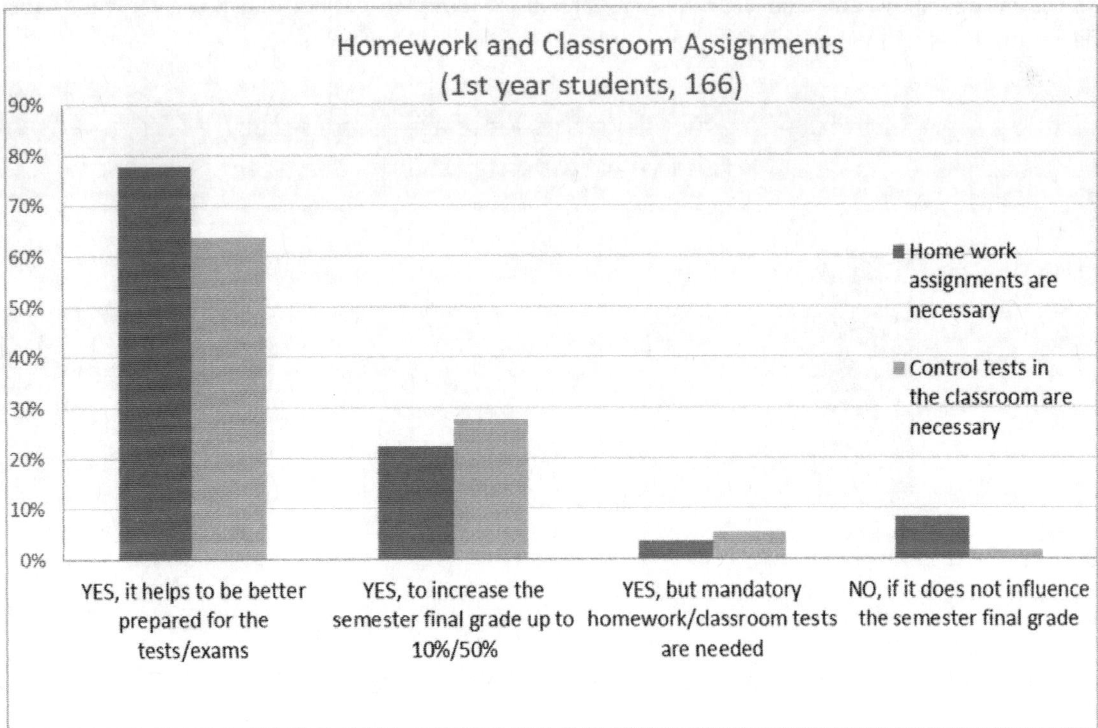

Fig. 4. Necessity of homework and classroom tests (1st year students).

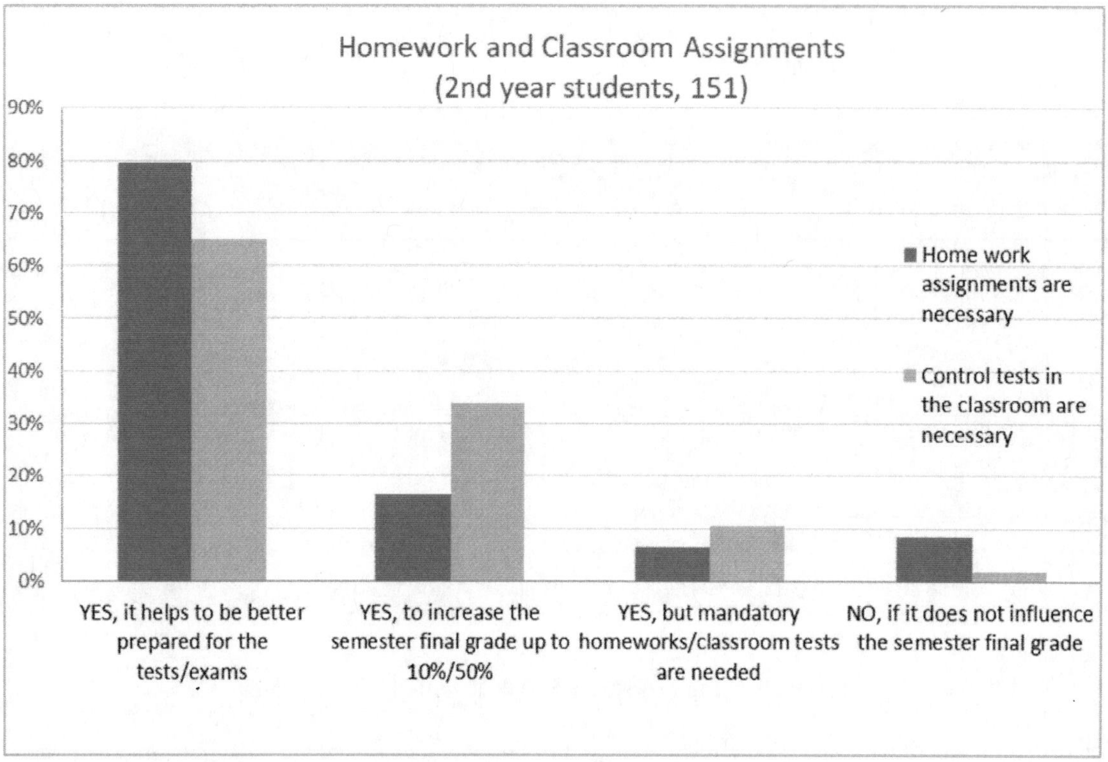

Fig. 5. Necessity of homework and classroom tests (2nd year students).

As the graphs show, the 2nd year students are more likely to expect that positive grades for classroom tests will increase their final grades at the end of the semester.

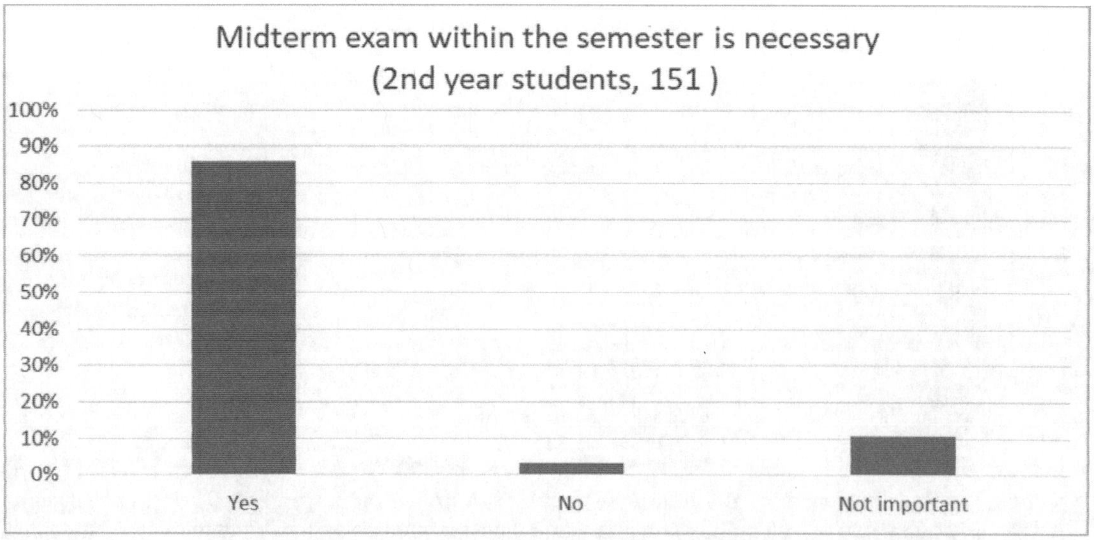

Fig. 6. Necessity of a midterm exam within the semester.

Students agree that it is easier for them to study the material for half a semester and pass a midterm exam in the middle of the semester.

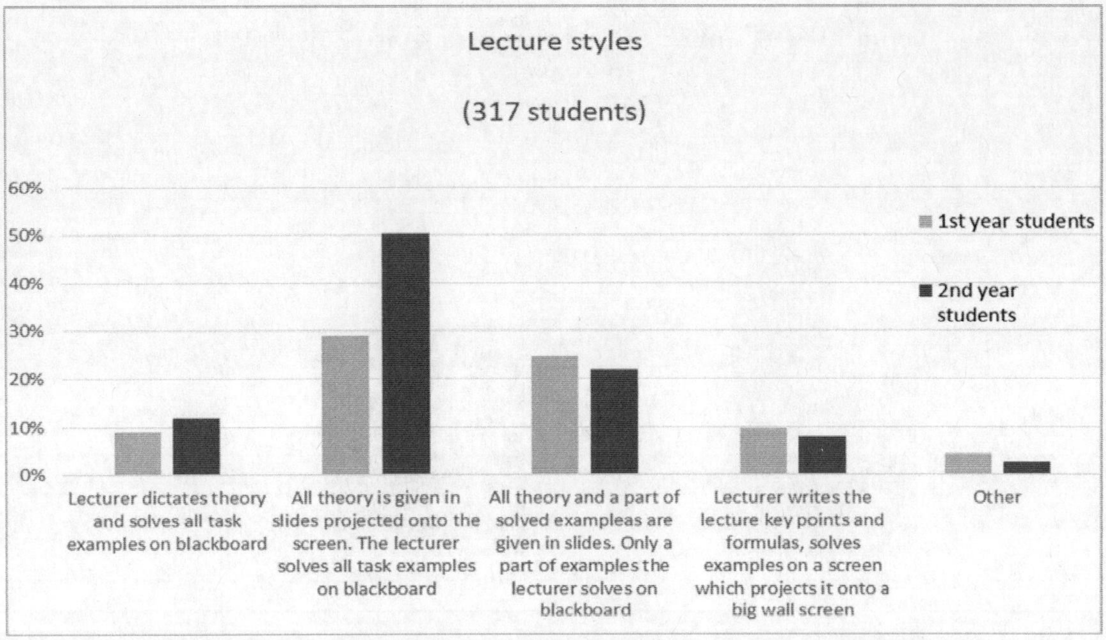

Fig. 7. Lecture preferred styles.

In general, students are satisfied with the study of lectures from slides and solved examples from the blackboard. The 2nd year students have already found internships and workplaces, so they try to use the classroom time for maximum acquisition of mathematics.

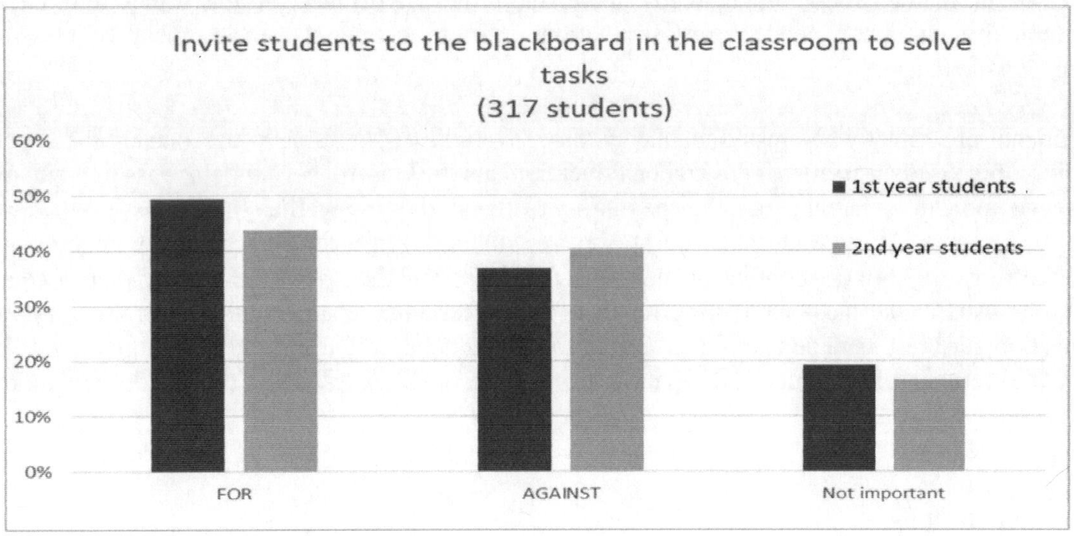

Fig. 8. Invitation of students to the blackboard.

Unfortunately, not all students like it when they are called to the blackboard to solve problems during tutorials. This is even more noticeable to the 2nd year students, as these graphs show (see Fig. 8). Although the 2nd year students already understand the importance of independent work, the mathematical problems are already very complicated and need time-consuming work, which is better done at the desk.

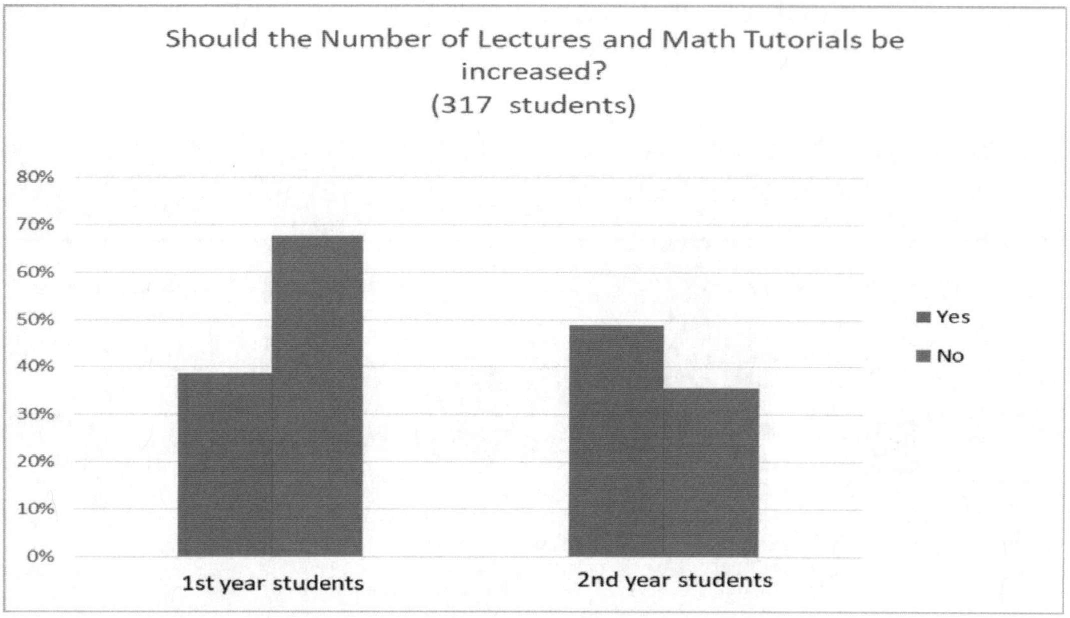

Fig. 9. Should the number of lectures and Math tutorials be increased?

According to the program of the course of higher mathematics in RTU, the number of planned classes is much more in the 1st year than in the 2nd. Obviously, for this reason, the number of students who are against the increase in the number of lectures is much larger in the 1st year than in the 2nd. More than a half of the 2nd year students want to increase the number of classes in mathematics: they already understand the importance of the topics and their complexity, and they need more explanations, but, unfortunately, the number of classes remains unchanged!

Students are directly interested in the quality of studies [4]. Surveys are one of the most commonly used instruments in higher education institutions to get students' assessment of various indicators that characterize the quality of the studies, as well as the professional work of the lecturer. At the end of the questionnaire, students can write their own comments and suggestions on both the content of the subject, teaching methods and the teacher themselves. This is usually the most interesting result of the questionnaire. Of course, the reliability of questionnaire results depends on the number of students participating in the questionnaire [5]. As the number of students participating in the questionnaire is high, the results are quite reliable.

5 Conclusions

1. Education cannot be provided with a "One size fits all" mentality.

2. Active learning methods contribute to the acquisition of knowledge-based learning, which makes this knowledge more significant and lasting.

3. Nowadays, the best option for the implementation of the study process is to work continuously on its improvement and development, according to the situation.

4. As a result of the student-centred learning process, the involvement of students in the formation of the learning process is increased, and the student's responsibility grows.

5. The still unanswered question is: what do students think about these questionnaires? Are they filled in only by dissatisfied students or, vice versa, by those who are delighted with the teachers? Do students find it useful to fill out these questionnaires?

References

[1] BABBLE, E., *Practice of Social Research (11th Edition).* Cengage Learning, 2006, 243.p.

[2] CELMA, D., *Skolas direktora loma pārmaiņu vadībā.* Rakstukrājums: Pedagoģija – teorija un prakse II, Liepāja, Liepa, Latvia, 2003, (in Latvian).

[3] Council of the European Union (20.12.2011.). Council Resolution on a renewed European adult education program (2011/C 372/01).

[4] EISENBERGER, K., KRAMER, J. W., *Möglichkeiten und Grenzen der Lehrevalution an einerHochschule.* In: Kramer, J.W. (Hrsg.) Hochschulen im Spannungsfeld zwischen Lehre und Forschung. München und Mering: Rainer Kampp Verlag, 2005, pp. 39-74.

[5] STUNDA-ZUJEVA, A., Priekšmetu anketēšanas rezultātu ietekmējošie faktori. Atgriezeniskās saites analīze 21.gadsimtā. *RTU Metodiskā konference*, 2016.gada 22. martā, Rīga, Latvija, (in Latvian).

Current address

Dzenite Ilona, Dr.math, Assistant Professor
Department of Engineering Mathematics, Institute of Applied Mathematics
Faculty of Computer Science and Information Technology, Riga Technical University
2 Daugavgrivas street, Riga, LV-1007, Latvia
e-mail: Ilona.Dzenite@rtu.lv

Cernajeva Sarmite, Mg.paed., Lecturer
Department of Engineering Mathematics, Institute of Applied Mathematics
Faculty of Computer Science and Information Technology, Riga Technical University
2 Daugavgrivas street, Riga, LV-1007, Latvia
e-mail: Sarmite.Cernajeva@rtu.lv

Matvejevs Aleksandrs, Dr.math, Assistant Professor
Department of Engineering Mathematics, Institute of Applied Mathematics
Faculty of Computer Science and Information Technology, Riga Technical University
2 Daugavgrivas street, Riga, LV-1007, Latvia
e-mail: Aleksandrs.Matvejevs@rtu.lv

Slovak University of Technology in Bratislava
Faculty of Mechanical Engineering
17th Conference on Applied Mathematics **APLIMAT 2018**
Proceedings

B3/S23 DESCENDING A STAIRCASE NO. 2

FEIJS Loe (NL)

Abstract. This artwork features gliders in cellular automatons playing Conway's game of life. Each playing field is one basic tile from the fashion pattern known as Pied de poule or Houndstooth. The pattern's basic tile has a contour consisting of straight line segments and staircases. In the context of the tessellation typology, it must be considered a hexagon. In our work of art, the basic tile is invisibly glued in a manner which is dictated by the tessellation type, thus implementing a Klein bottle topology, as an alternative to the more classical torus. The Klein bottle appears invisible but it comes to life because of the glider. The work is a tribute to the famous painting Nude Descending a Staircase No. 2, created by Marcel by Duchamp in 1912. Several other hints to Duchamp's work are embedded in the artwork. The author has a personal passion for Pied de poule, which is a rich source of aesthetic qualities in fashion and at the same time a playing field for various types mathematical recreations.

Keywords: topology, Klein bottle, Pied de poule, Marcel Duchamp, game of life

Mathematics Subject Classification: Primary: 00A66; Secondary:54E99, 05B45, 37B15.

1 Introduction

In this artwork, we propose a synthesis of Conway's game of life, topology, and art history, in particular, elements from Marcel Duchamp's work. The Pied de poule is an element added by the author and it serves to integrate the other elements of the synthesis. In this section, we describe various aspects of these ingredients and in the next section we describe the first version of our artwork, which is based on a torus topology. In the section after that, we describe B3/S23 Descending a Staircase No. 2, which is a set of Klein bottles.

The work fits in the category of "Generative art", embracing what the late Professor Mauro Francaviglia, to whom this session on Mathematics and Art is dedicated, and Marcella Giulia Lorenzi [1] wrote in 2007: (quote) *The defining trait of Generative Art is rather that the artist establishes a system, which can generate a number of possible forms rather than one single finished form. The role of the artist is to construct, initiate or merely select the frame of procedures for the generation of possible expressions; the constructions of art-making systems substitute the making of static forms while in other cases the systems rely on input from human actors or information feeds* (end quote). In 2017 we presented already a generative art-based work [2] in which Pied de poules came to life inside a cellular automaton. In a certain way, the present proposal works the other way around: the cellular automatons are alive inside Pied de poules.

1.1 Conway's Game of Life

A well-known form of entertainment for programmers and mathematicians is the development of various versions of Conway's game of Life and studying the behavior of Life forms such as gliders, oscillators, and still-lives. Usually, the playing field's cell structure is considered infinite or has a torus topology. Gardner described Life in 1970 [3].

Each cell has eight neighbors, which is known as a Moore neighborhood. The rule of Life is well-known (the B3/S23 rule): a new cell is Born on a blank square if and only if it is surrounded by exactly three alive cells; a cell Survives if and only if it is surrounded by two or three alive cells; if a cell is surrounded by less than two or more than three cells it dies (by loneliness or overpopulation, respectively). All kinds of patterns "live" in this artificial world, one of the simplest examples being the glider found by Richard K. Guy in 1970. In the artwork proposed (Sections 2 and 3), we use such gliders. Each glider has five alive cells and operates on a moving area of 18 cells: the live cells and their relevant neighborhood cells. See Fig. 1. The author has met John Conway in 2015 and is still fascinated by his mathematical creativity. Conway did not only invent the game of Life, but also the surreal numbers and much more [4].

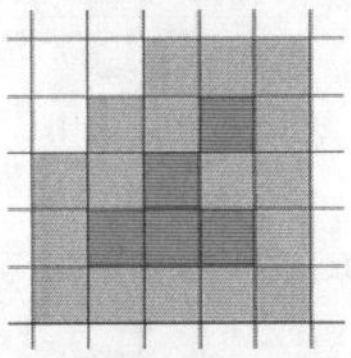

Fig. 1. Glider and its neighbourhood.

1.2 Topology

We assume the reader is somewhat familiar with general topology and knows constructions such as the torus and the Klein bottle. The torus has important applications in computer games, for example the Pacman game has the torus topology as its playing field. What topology and Conway's game of Life have in common is the notion of "neighborhood". Life has a discrete neighbourhood (on the grid). Topology is based on open sets, defining a topological space as a set of points with their neighborhoods for each point, satisfying certain axioms. As the notion of neighborhood works locally, there is considerable freedom in choosing the long-distance form-giving of a playing field. The idea of running Life on special topologies is not new (for example on www.youtube.com/watch?v=lxIeaotWIks, Life on a torus is presented). The practical advantage of the torus topology is that there is no need for special Life rules coping with "what to do at the edge?"

1.3 Pied de Poule

The essential mathematics of Pied de poule was described by Feijs in 2012 [5]. There is not a single pattern, but a family: one for each N = 1, 2, 3, etc., as shown in Fig. 2. The pattern arises naturally in weaving when a so-called twill binding is used (two over/two under, or three over/three under, etcetera).

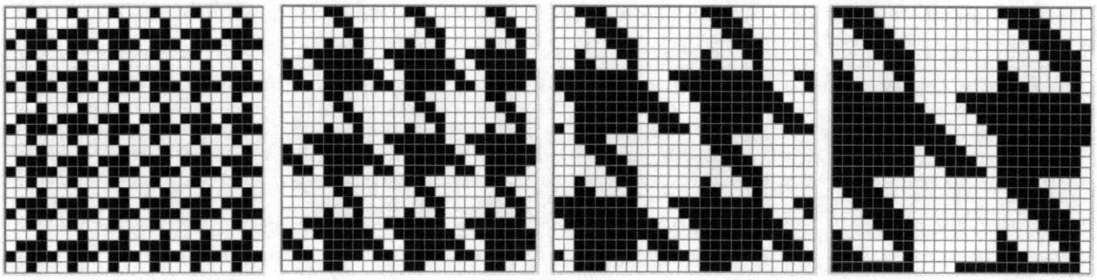

Fig. 2. Pied de poule patterns for $N = 1$, $N = 2$, $N = 3$ and $N = 4$.

In contemporary fashion, Pied de poule is still very much alive. It is an important cultural and mathematical phenomenon.

The Pied de poule patterns can be considered as tessellations, and in the 2012 Bridges paper, the typology according to Heesch and Kienzle's tessellation theory [6] is clarified. Classical Pied de poule has Heesch and Kienzle type TTTTTT, three pairs of Translated edges[1]. We show the tessellation network, omitting the black and white colour filling, in Fig. 3. For the purposes of the present paper it is important to note that the contour of each basic tile consists of six edges and six vertices. Topologically it is a hexagon.

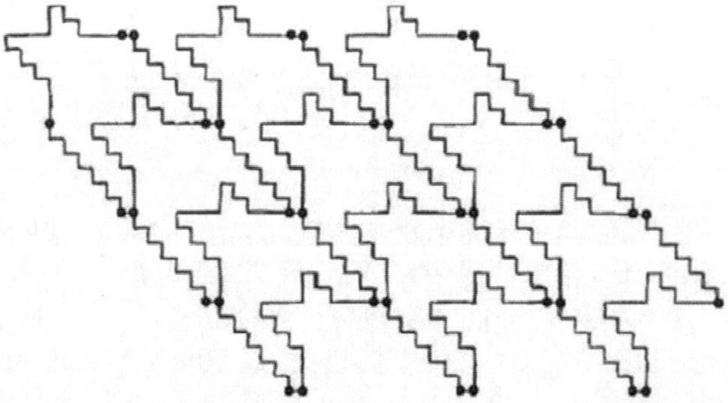

Fig. 3. Pied de poule ($N = 3$) viewed as a tessellation.

1.5 Marcel Duchamp

Marcel Duchamp (1887 - 1968) is famous for his proposition of "ready-mades", artworks which are in fact everyday objects but become a work of art, just by the artist's choice. We

[1] Heesch and Kienzle give the following recipe for this type: (quote) Verschiebe die Willkürliche Linie *AB* nach *DC*. Verschiebe die willkürliche Linie *AF* (*F* beliebig) nach *EC* (Translationsvektor *AE*). Verschiebe eine dritte willkürliche Linie *BE* nach *FD* (Translationsvektor *BF*). (end quote)

shall come back to a few of the ready-mades later. Here we like to mention the famous artwork Nude Descending a Staircase, No. 2 (French: Nu descendant un escalier n° 2) he painted in 1912. It was refused from an art show in Paris and later caused a lot of sensation in New York. Fig. 4 is a reproduction (the original is 147 cm × 89.2 cm).

Fig. 4. Nude Descending a Staircase, No. 2 (1912).

2 The artwork (initial version)

The artwork is a projected two-dimensional program output. It shows one basic Pied de poule tile, whose edges are invisibly glued in a torus topology and it shows a glider who lives in the rectilinear grid inside the tile. The dead cells are white, the alive cells are in various shades of skin color-like pink. The gluing of the edges is precisely reflecting the tessellation type of classical Pied de poule.

The tile is projected onto a manikin, as shown in Fig. 5. She wears an accessory in Pied de poule motive (a western bow tie). In this version, there is just one basic tile (but in Section 3 we propose a more sophisticated version where the manikin is “dressed” by a multi-tile set of Life universes).

Fig. 5. B3/S23 Descending a Staircase No. 1 (1917).

For the paper, the dynamic behavior is shown here as a sequence of six snapshots in Fig. 6. The wiggling behavior of the glider appears precisely like walking down a staircase. This staircase is formed by the leftmost tail of the Pied de poule tile.

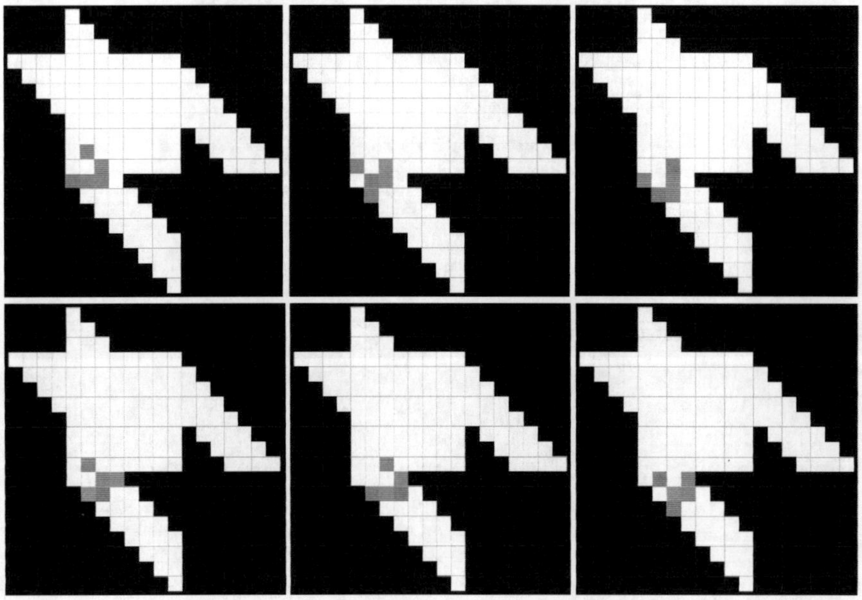

Fig. 6. Pink glider walking down the staircase.

The title of my work refers to the famous work of Marcel Duchamp, which has been re-interpreted in many forms and many media. The glider is the nude, although it is not anthropomorphic, its color suggests nudity and the glider walks down the staircase. It is a strong dynamic image. Note that pixelation is well-known technique for hiding nudity (see Fig. 7).

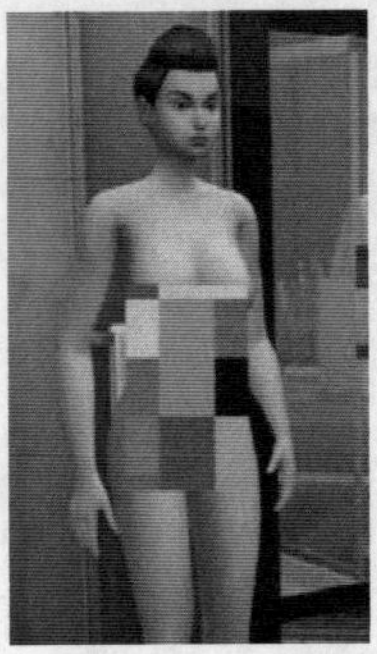

Fig.7. Pixelation hiding a Sim's nudity.

Although it appears obvious that the glider can walk the stairs, this is actually not trivial, as the neighborhood folds over the edges of the basic tile. Yet it goes well because of the glued topology which has no edges.

One would guess that the gluing of the edges is the same as gluing a rectangular playing field into a torus, as in Pacman. But we must be careful because the tessellation type tells us that the tile is not a quadrilateral but resembles a hexagon of type $abca^{-1}b^{-1}c^{-1}$. The general solution for gluing the edges of a hexagon topologically into a torus is shown in Fig. 8. The picture is adapted from www.math.cornell.edu/~mec/Winter2009/Victor/part1.htm. The technique is called an "Indian burn" in math.stackexchange.com/questions/53454.

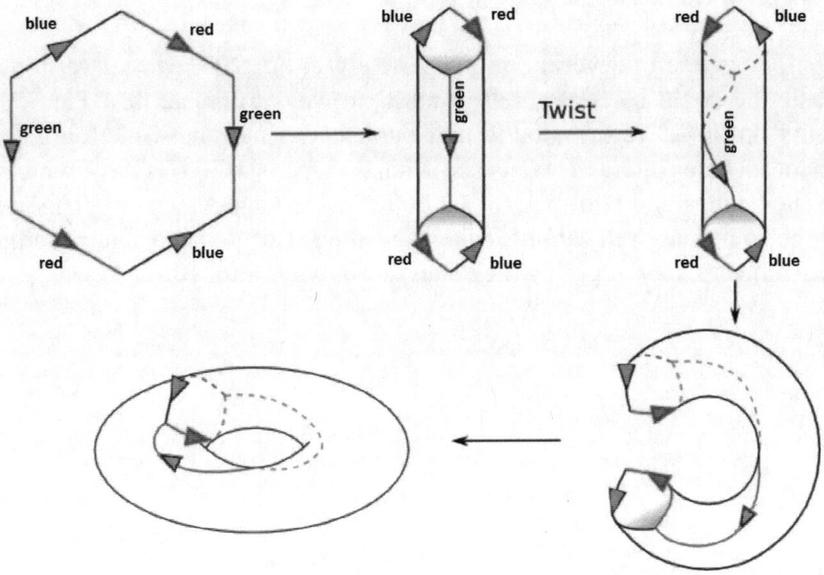

Fig. 8. Gluing the edges of a hexagon into a torus.

For the Pied de poule hexagon, only a small twist would suffice (the sides of the hexagon are not equally long). The hexagon of the pied de poule tile is abstractly shown in Fig. 8 (upper left). The edges to be glued in Fig. 8 are shown in the same color (blue to blue, red to red and green to green, no reversals). For the Life simulation, the connections are made in software, rewiring the grid of cells. In a computer program it works as follows:

```
cells[4][11].west ← cells[11][3];
cells[4][11].south ← cells[12][4];
cells[5][12].west ← cells[12][4];
cells[5][12].south ← cells[13][5];
etc.
```

The following figure (Fig. 9) shows the detailed edges to be glued as indicated by the colors: blue goes to blue, red to red, green to green, without any reversals.

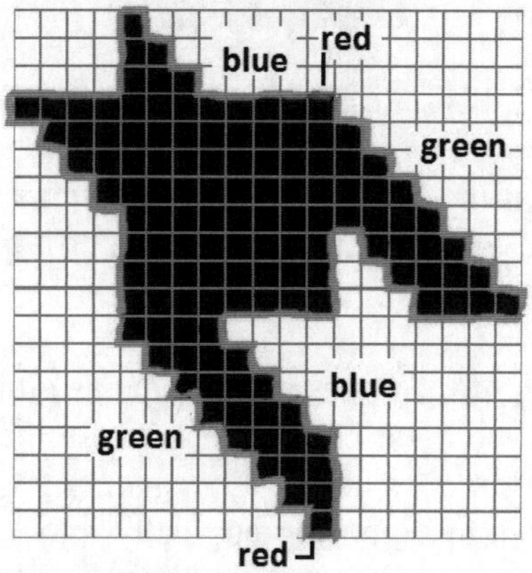

Fig. 9. Corresponding sides of the Pied the poule tile ($N = 4$).

3 Second version of the artwork

The torus topology is one of several options. Already in our 2012 work, we noticed that the very same basic tile could be configured in multiple ways to create new Pied de poule-like fashion patterns. In 2012 we thus found and published a new network topology. Here we propose yet another new pattern. The Heesch-Kienzle type is $TG_1G_2TG_2G_1$, which means one pair of Translated edges and two pairs of Glide reflected edges[2], see [6]. This new network topology can be folded as well, gluing edges and then it becomes a Klein bottle! The new pattern, which defines a way of gluing the edges, is shown in Fig. 10.

[2] Heesch and Kienzle give the following recipe for this type, in German: (quote) Verschiebe die Willkürliche Linie *AB* nach *DC*. Gleitspiegele die weitere willkürliche Linie *AF* (*F* beliebig) nach *EB*. Gleitspiegele eine dritte willkurliche Linie *FD* nach *CE* (Gleitspiegelachsen H_1I_1, H_2I_2 senkrecht *AD* in gleichen Abständen von *B* und *F*, bzw. von *D* und *E*) (end quote)

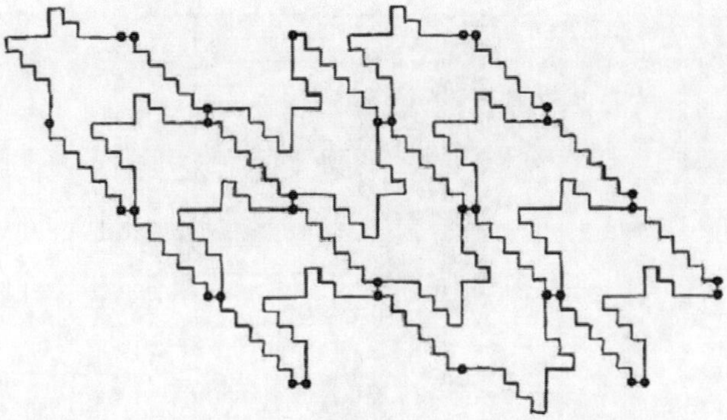

Fig. 10. Klein bottle-topology "Pied de poule" ($N = 3$).

Filling the tiles alternatingly with black and white, a new "Pied de poule" appears, as shown in Fig. 11 (for N = 4). Superficially it looks the same as the innovative pattern of 2012 [5], but it has another network type.

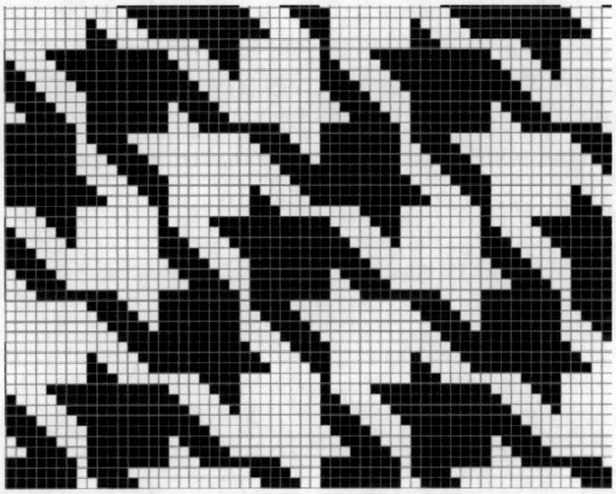

Fig. 11. New "Pied de poule" based on a Klein bottle-topology network ($N = 4$).

Some of the basic tiles are mirrored now. If we consider them as puzzle pieces, they would be upside down. But if we fold everything to a single Klein bottle, these upside-down puzzle pieces correspond to the backside of the surface (just like on a Möbius strip, there are continuous walks where one passes along the other side of the visible surface). In the program output, the active backside ("verso") cells are shown dark green (instead of pink).

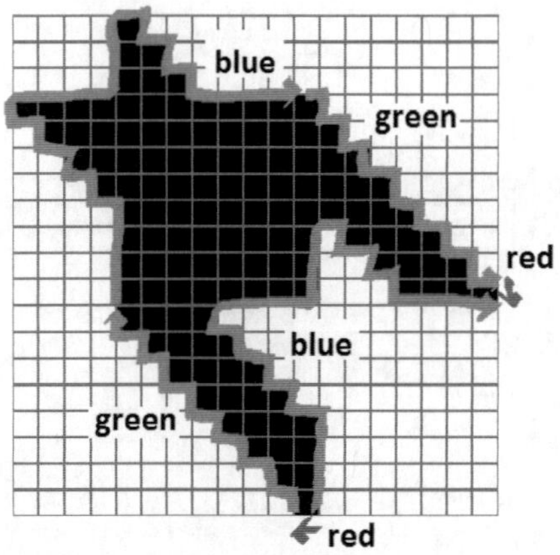

Fig. 12. Matching edges for Klein bottle-topology "Pied de poule" ($N = 4$). The blue edges have the same direction, whereas the red and green edge pairs have to be glued reversely.

It is well-known how to fold a reverse-edges square or rectangle into a Klein bottle. The construction is given in Fig. 13 (the figure is adapted from quibb.blogspot.nl/2011/04/). But here we are dealing with a hexagon again: the tessellation type dictates that the tile is essentially a hexagon of type $abca^{-1}cb$. Referring to Fig.12, a is blue (the asymmetric two-horned head), b is green (the staircase flanks) and c is red (the one-step end of the tails). We cannot simply take each pair of green and red edges (b and c) together to behave as one edge, as bc differs from cb. But we can apply the Indian burn method demonstrated in Section 2 (Fig. 8) once more, and then the very same Klein bottle construction applies again.

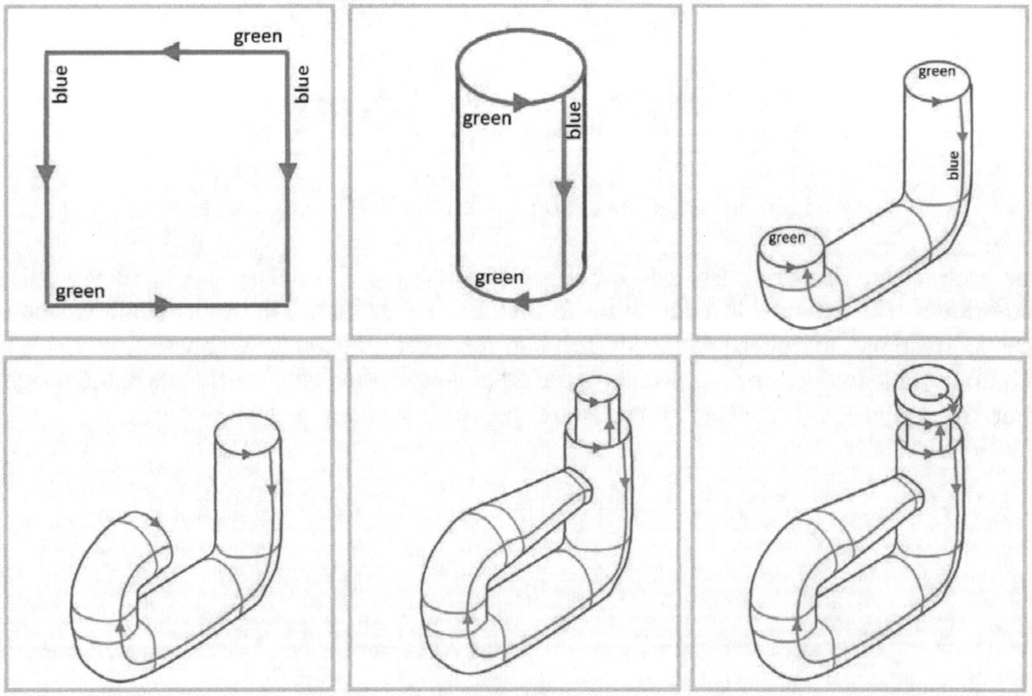

Fig. 13. Gluing a square into a Klein bottle.

Turning to the installation: instead of one basic tile, we project ten of them. The upper tile, larger than the others, shows the glider (the nude) descending the staircase. Although the topology is a Klein bottle now, not a torus, the glider walks along the staircase as before.

Then how can we "see" the Klein bottle? That is what the other basic tiles do. Each tile is a Life universe. They are not connected, but all run as an isolated Life-on-a-Klein bottle. The initial position of the gliders on the 2nd till 10th tile are different. There is also a beehive (the "eye" still-life on the verso side of I the second tile) and a "blinker" oscillator (in the third tile, verso). Instead of the South-east going nude, most of these gliders are North-east or North-west going. The latter gliders show a variety of pathways along the Klein bottle and some of them are seen to travel both the front side and the verso side of the surface. The variety of gliders on the ten Klein bottles visually confirm that the basic tiles are Klein bottles. We see the pathways of the gliders, sometimes jumping when they pass the outer zigzags and changing from pink to dark green. To make it explicit that the tiles are not connected to their neighbors, the Life execution threads run asynchronously.

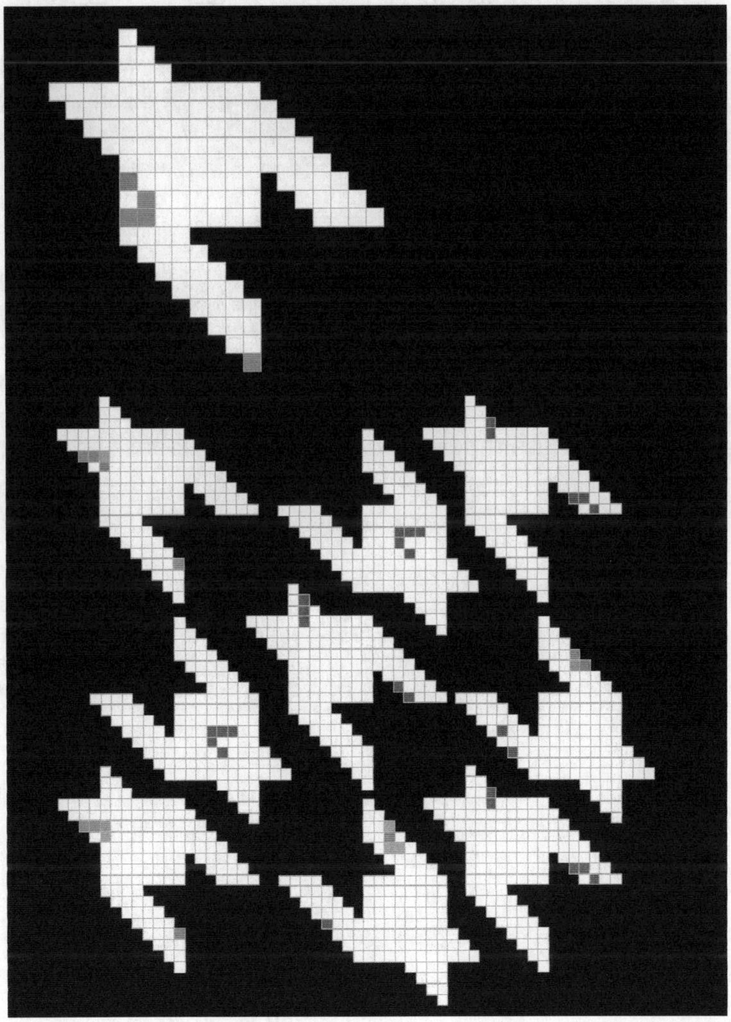

Fig. 14. Snapshot for the installation: B3/S23 Descending a Staircase No. 2 (2018). It is to be projected again, somewhat similar to No. 1.

The work has several other links to various works of Duchamp:

- The bottle rack, a ready-made produced by Duchamp in 1914. In a sense, our installation is a bottle rack, as it holds ten (Klein) bottles.
- The Fountain, in fact, a porcelain urinal, a ready-made by Duchamp in 1917. It does not take great imagination to see a connection between the Fountain and the Klein bottle. Duchamp was certainly aware of the Klein bottle. As Giunti [7] puts it: (quote) *Let's now consider the famous Fountain (1917) [ ]. It seems to me that it can be seen as a transversal section of a kleinian bottle. The neck for the connection with the water pipe (in the foreground of the historic photo) would be the equivalent of the kleinian bottle neck--plunging in its own belly (here the sectioned part) would reconnect with the holes of the draining (corresponding to the introverted bottom of the kleinian bottle)* (end quote).
- The Bride Stripped Bare by Her Bachelors, Even (La mariée mise à nu par ses célibataires, même), most often called The Large Glass (Le Grand Verre), an artwork produced by Marcel Duchamp from 1915 to 1923. In our work, there is one nude (the bride) and nine bachelors (the nine other Klein bottles, whose gliders are mostly moving upward rather than downwards). As Hopkins [8] discusses, it remains unclear whether the nude in Nude Descending a Staircase N° 2 was intended male or female, but here we imagine the glider of the upper tile to be a female.

We conclude with a few implementation notes. The Life engine is tailor-made for this application using object-oriented programming in Processing (essentially Java). There are classes for cells and for Pied de poules. Each cell *c* has fields *c*.north, *c*.east, *c*.south and *c*.west. On a torus, the neighborhood of cell *c* consists of *c*.north, *c*.north.east, *c*.east, *c*.south, etc. (8 cells). On the Klein bottle, some special care is needed because once an edge is crossed, the roles of north and east are swapped and similarly for the roles of south and west. The program consists of about 750 lines of code. Efficiency is not a problem (using an HP ZBook laptop).

Acknowledgements

The author is grateful for the inspiration and support by Marina Toeters of by-wire.net and Troy Nachtigall of TU/e.

References

[1] LORENZI, MARCELLA GUILIA and FRANCAVIGLIA, MAURO. Painting with light: generative artworks or "setting in motion" the fourth dimension. In *Proceedings of the 10th Generative Art Conference*, GA2007, Milano (pp. 12-14).

[2] FEIJS, LOE and TOETERS, MARINA. A Cellular Automaton for Pied-de-poule (Houndstooth). *Proceedings of Bridges 2017: Mathematics, Art, Music, Architecture, Education, Culture* (pp 403--406). Tessellations Publishing, Phoenix, Arizona.

[3] GARDNER, MARTIN (October 1970). "Mathematical Games – The fantastic combinations of John Conway's new solitaire game "life"". *Scientific American*. 223: 120–123.

[4] ROBERTS, SIOBHAN. Genius At Play: *The Curious Mind of John Horton Conway*. Bloomsbury USA (2015).

[5] FEIJS, LOE (2012). Geometry and computation of Houndstooth (Pied-de-poule). In Robert Bosch, Reza Sarhangi & Douglas McKenna (Eds.), *Proceedings of the Bridges Towson Conference 2012: Mathematics, Music, Art, Architecture, Culture*.

[6] HEESCH, HEINRICH and KIENZLE, OTTO (1963). *Flächenschluss; System der Formen lückenlos aneinanderschliessender Flachteile*. Springer Verlag, Berlin.

[7] GIUNTI, ROBERTO. *R.rO.S.E.Sel.A.Vy* by Roberto Giunti, Section 4b. Bottle in Art in Bottle (p. 13), www.toutfait.com/issues/volume2/issue_4/articles/giunti/giunti13.html

[8] HOPKINS, DAVID. *Marcel Duchamp and Max Ernst: The Bride Shared*. Clarendon Press (1998).

Current address

Feijs Loe
Department of Industrial Design, LG 1.78
Technische Universiteit Eindhoven Postbus *513*
5600 MB Eindhoven, Netherlands

Feijs Loe, Prof.dr.ir.
Department of Industrial Design, LG 1.78
Technische Universiteit Eindhoven, Netherlands
Tel. +31 6 2222079, e-mail: l.m.g.feijs@tue.nl

Slovak University of Technology in Bratislava
Faculty of Mechanical Engineering
17th Conference on Applied Mathematics **APLIMAT 2018**

Proceedings

CONSTRUCTIONS OF A SQUARE AND A REGULAR PENTAGON ONLY WITH A COMPASS

FERDIÁNOVÁ Věra (CZ), BRANNY Michael (CZ)

Abstract. In 1797, Lorenzo Mascheroni proved that a compass used for constructions has the equal power as using a ruler altogether with the compass. Therefore, it is possible to construct all ruler-and-compass constructions only with the compass itself. The aim of this article is not to prove the theorem but to introduce interesting constructions of a square and a regular pentagon.

Keywords: Mascheroni construction, square, regular pentagon

Mathematics subject classification: Primary 51N05; Secondary 01A50, 97G80

1 Introduction

Constructional and descriptive geometry are integral parts of university studies of the future Maths teachers, engineers, constructors and architects. Students must be able to interconnect and apply geometric knowledge and use their imagination. [6] The ruler-and-compass constructions (sometimes called Euclidean constructions) and their slight modifications are taught as a part of constructional geometry. However, there are other methods equal to Euclidean constructions, using different means. One of them are Mascheroni constructions which allow only the use of a compass. Lorenzo Mascheroni was an important Italian mathematician and even Napoleon himself was impressed with Mascheroni's work at that time. The most important part of his work is the so-called Mohr-Mascheroni Theorem, stating that every geometric construction carried out by a compass and a ruler can be done without a ruler. An indication of the theorem can be found in Euclidia Danubis, a book by Georg Mohr. Another proof based on a circle inversion, was created by Adler [1] or Kostovskii [3].

2 Georg Mohr and Lorenzo Mascheroni

Georg Mohr (Jorgen Mohr in Danish) was a mathematician, born in Copenhagen in 1640. Because there is a lack of information about his life, it is assumed he left Denmark to the Netherlands to study one of their universities as they were of high prestige at that time. His book Euclides Danicus, published in 1672, described the main idea of Mascheroni's theory in Danish, thus there was a belief it is only a translation of Euclid's work Elements. It was only in 1927 when professor Johan Hjemslevov discovered the importance of the work thanks to one his students who had found a copy of Euclides Danicus in a secondhand bookshop. [5] The second book of Mohr's, Compendium Euclides Curiosi, was published in 1673 and it is based on Elements again. Interestingly, a letter dated on May 12th, 1676 from Leibnitz to H. Oldenburg, a secretary of the Royal Society in London is still in existence, stating: "Georgius Mohr Danus, in geometria et analysi versatissimus". Leibnitz spend some time in Paris and it is possible he and Mohr met or Leibnitz read Mohr's work. [2] In 1681, Mohr travelled back to Denmark and applies for a job at Danish Royal Court. Although the king offered him supervision of building royal navy, he declined it. Mohr got married in 1687 and moved to the Netherlands supposedly due to a dispute with the king. During the last years of his life Mohr accepted Tschirnhaus's offer of job and moved with his family to Kieslingwald in Germany. [5] Nearly 200 years later, Lorenzo Mascheroni (1750-1800), Italian priest, poet and mathematician, was born in Bergamo. At his 17, he was ordained a priest and continued in his studies of rhetorics, which he began to teach in Bergamo 5 years later. Philosophy was his another field of interest - it included basics of logics and physics at that time. In 1786, he starts to work as a professor of algebra and geometry at university in Pavia. In the same year, he published a book called Nuove ricerche sull' equilibrio delle volte (1786) and later other important works, e.g. Adnotationes ad calculium integralem Euleri (1792), Problemi per gli agrimensori (1793) a Geometria del Compasso (1797). The latter work is the main one in our perspective as Mascheroni depicts Euclidean constructions which can be constructed only with compasses. [7]

Fig. 1. Lorenzo Mascheroni: La Geometria del Compasso [4]

La Geometria del Compasso is a complete work written in the Old Italian language by Mascheroni. The book is divided into 12 main chapters, each dealing with issues of the given topic. The structure of chapters is methodically arranged: a task, a description of a construction, a proof of construction, and an alternative solution (if there is one). The final constructions are to be found in the appendix of the book. Mascheroni used not only circles, but also arcs to achieve better arrangement of schemes. Each chapter moves from the easiest constructions to the most difficult ones. For instance, in a chapter, which focuses on division and lengthening of line segments, Mascheroni starts with a division of a line segment into 2 parts of the same length, then continues with a division into 4, 8 parts etc. Then he shows a division into an odd amount of parts and ends with construction of division of a line segment into n parts.

3 A square of a side a and an unknown diagonal u

Problem Let $AB,\ |AB| = a$ be a side of a square. Construct the square $ABCD$.

The problem can be solved in more ways. First of all, we must focus on a construction of two isosceles triangles ABC and ABD. To do that, we have to geometrically construct the side AC, which fuctions as the diagonal u of the square $ABCD$. If a side of the square is a then $|AC| = a\sqrt{2}$. Thus our goal is a construction of a line segment, i.e. a diameter of a circle of the length $a\sqrt{2}$.

Construction protocol

0. Points A, B; $|AB| = a$; circle k; $k(B, r = a)$
1. Point X; $X \in k \wedge |AX| = 2a$
   a) Circle $l_1; l_1(A, r = a)$
   b) Point $X_1; X_1 \in k \cap l_1$
   c) Circle $l_2; l_2(X_1, r = a)$
   d) Point $X_2; X_2 \in k \cap l_2$
   e) Circle $l_3; l_3(X_2, r = a)$
   f) Point $X; X \in k \cap l_3$
2. Circle $m_1; m_1(X, r = |XX_1|)$
3. Circle $m_2; m_2(A, r = |AX_2|$
4. Point $Y; Y \in m_1 \cap m_2$
5. Circle $n_1; n_1(B; r = |BY|)$
6. Point $D; D \in l_1 \cap n_1$
7. Circle $n_2; n_2(D; r = a)$
8. Point $C; C \in k \cap n_2$
9. Square $ABCD$

The line segment $a\sqrt{2}$ has occured because of the right-angled triangle ABY with side AB of length a, whereas BY is the value we have been seeking. However, the side AY of length $a\sqrt{3}$ is the essential one. But how the length $a\sqrt{3}$ was constructed? The circle $k(B, r = a)$ of radius a and center B is the so-called Thales' circle, i.e. any triangle ABQ, $Q \in k$, is right-angled. Let's focus on AXX_2, although we could discuss AXX_1 as well. The length of AX is $2a$ as AX represents doubled AB in the construction. XX_2 has the length a because the point X is the intersection of two

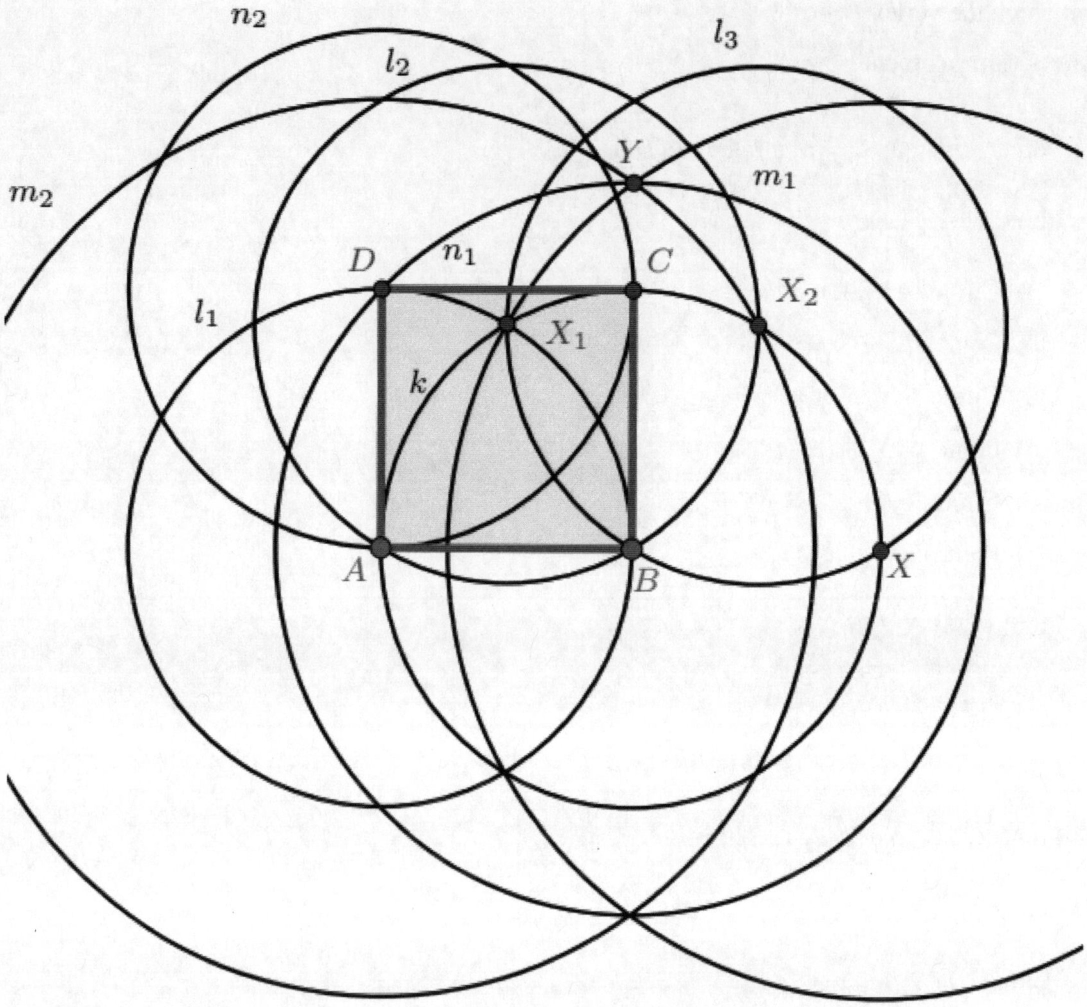

Fig. 2. The square $ABCD$ (A square of a side a and an unknown diagonal u)

circle, one of which is $l_3(X_2, r = a)$. The last side of the triangle is AX_2, whose side length can be calculated with Pythagorean theorem (with the right angle at X_2):

$$|AX|^2 = |XX_2|^2 + |AX_2|^2$$
$$(2a)^2 = a^2 + |AX_2|^2$$
$$3a^2 = |AX_2|^2$$
$$a\sqrt{3} = |AX_2|$$

$|AX_2|$, and $|XX_1|$ as well, figures as the radius of the circles which created the intersection Y so that $|AX_2| = |XX_1| = |AY|$.

The second version of construction is based on the construction of a perpendicular to AB in the point A, on which the vertex D lies.

Construction protocol

0. Points A, B; $|AB| = a$; circle k; $k(B, r = a)$
1. Point Y; $AY \perp AB$
    a) Circle l_1; $l_1(A, r = a)$
    b) Point S; $S \in k \cap l_1$
    c) Circle l_2; $l_2(S, r = a)$
    d) Point X; $X \in l_1 \cap l_2$
    e) Circle l_3; $l_3(X, r = a)$
    f) Point Y; $Y \in l_2 \cap l_3$
2. Point W; $W \in l_1 \cap l_3$
3. Circle m_1; $m_1(W, r = |SW|)$
4. Circle m_2; $m_2(B, r = |BX|)$
5. Point Z; $Z \in m_1 \cap m_2$
6. Circle n_1; $n_1(W, r = |AZ|)$
7. Point D; $D \in l_1 \cap n_1$
8. Circle n_2; $n_2(C, r = a)$
9. Point C; $C \in k \cap n_2$
10. Square $ABCD$

This construction constists of several simpler Mascheroni constructions. First, the perpendicular AY is constructed, using the properties of Thales' circle l_2. Doubling the segment BS, the point Y was created and BY functions as the hypotenuse of the triangle ABY. Therefore there must be a right angle at A and $AY \perp AB$. Points X, S have occured during the construction. Both points lie on the circle k, forming the arch SX. As the radius of the circe l_1 equals a and the triangle ASX is isosceles with sides of enght a, the centre of the arch SX must be the vertex D of the square. Here we divide the arch using Mascheroni construction. The triangle ASX is transformed into two parallelograms $ABSX$ and $ASWX$. Advantageously, the construction continues just as the preceding one. If we doubled AB into AX and constructed the point Y in order to get BY as the diagonal $a\sqrt{2}$, now we double AW into BW and search the point Z that $AZ = a\sqrt{2}$. Both version of solving the problem eventually merge into one although each of them uses different Mascheroni constructions.

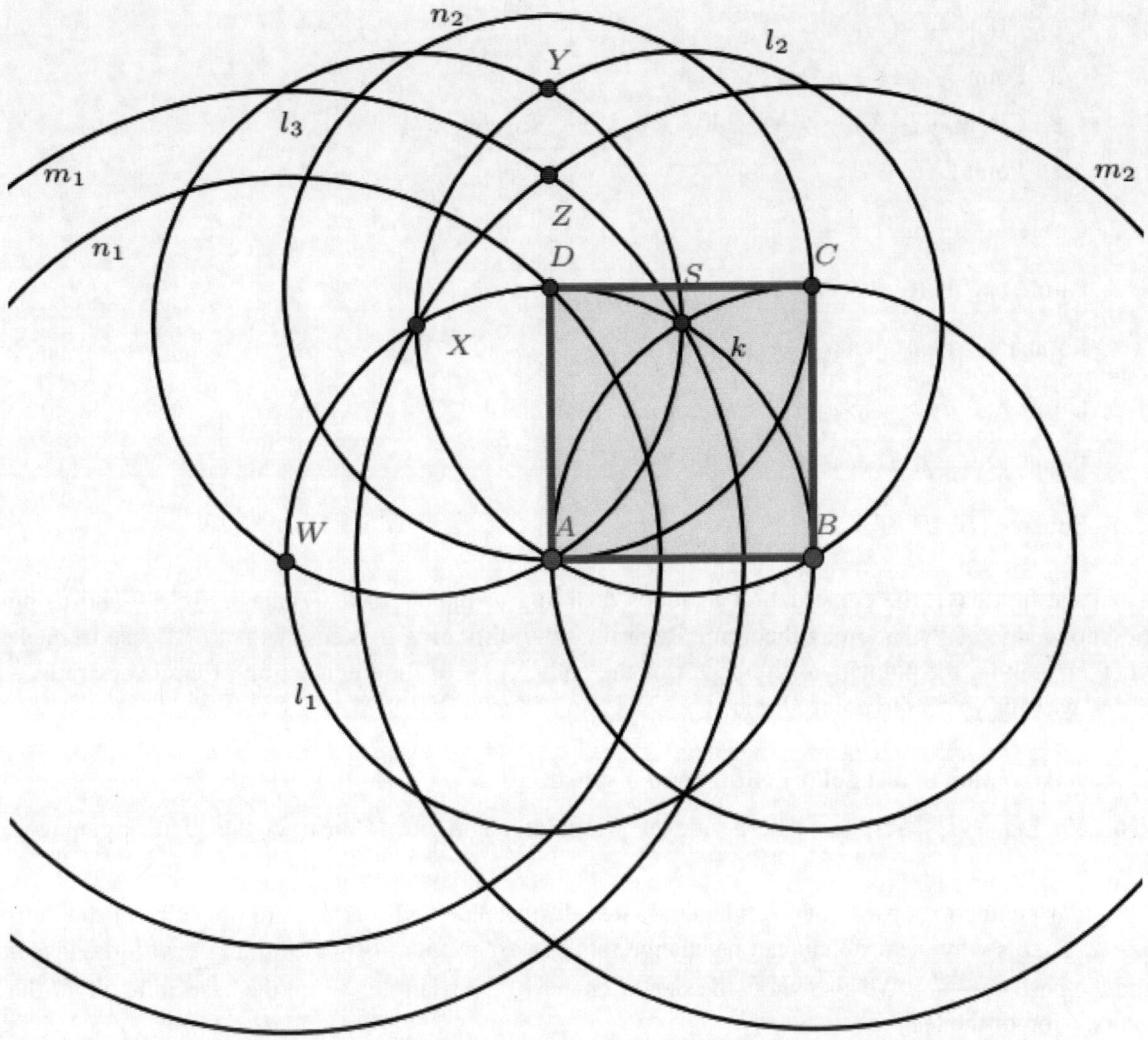

Fig. 3. The square $ABCD$ (A square of a side a and an unknown diagonal u)

4 A square of a diagonal u and an unknown side a

Problem Let AC, $|AC| = u$ be a diagonal of a square. Construct the square $ABCD$ with the diagonal AC.

Since we know the length $u = a\sqrt{2}$, we can change the problem to a construction of a square whose circumscribed circle has the radius $\frac{u}{2} = \frac{a\sqrt{2}}{2}$.

Construction protocol

0. Point S; circle k; $k(S, r = \frac{u}{2})$; bod A; $A \in k$
1. Point C; $C \in k \wedge |AC| = 2|AS|$
   a) Circle l_1; $l_1(A, r = \frac{u}{2})$

b) Point $X_1; X_1 \in k \cap l_1$

c) Circle $l_2; l_2(X_1, r = \frac{u}{2})$

d) Point $X_2; X_2 \in k \cap l_2$

e) Circle $l_3; l_3(X_2, r = \frac{u}{2})$

f) Point $C; C \in k \cap l_3$

2. Circle $m_1; m_1(A, r = |AX_2|)$

3. Circle $m_2; m_2(C, r = |CX_1|)$

4. Point $Y; Y \in m_1 \cap m_2$

5. Circle $n; n(A, r = |SY|)$

6. Points B, D; $B, D \in n \cap k$

7. Square $ABCD$

As in the first case, the construction includes doubling a line segment, Thales' circle k and using the knowledge of Pythagorean theorem. Regarding the difficulty, it is easier than first case because whether seeking the point B as the center of the arch X_1X_2 or the intersection of circles of radii $\frac{u}{2}$ and a, we always get the point D, too.

5 A construction of a regular pentagon of a side a.

Problem Let AB, $|AB| = a$ be a side of a regular pentagon. Construct the regular pentagon $ABCDE$.

A regular pentagon consists of five identical isosceles triangles whose legs are the sides of the pentagon. That is why the problem can be change into construction an isosceles triangle with the legs a and the base $a(\frac{1+\sqrt{5}}{2})$. We also know the size of angles of the triangle: $36°$ at the base and $108°$ at the intersection of the legs.

Construction protocol

0. Points A, B; $|AB| = a$; circle c; $k(A, r = a)$

1. Circle d; $d(B, r = a)$

2. Point $C_1; C_1 \in c \cap d$

3. Point $D_1, I, K; loopsteps 1, 2, 3$

4. Circle $p; p(B, r = |BD_1| = a\sqrt{3})$

5. Circle $q; q(I, r = a\sqrt{3})$

6. Point $J; J \in p \cap q$

7. Circle $r; r(D_1, r = |AJ|)$

8. Circle $t; t(K, r = |AJ|$

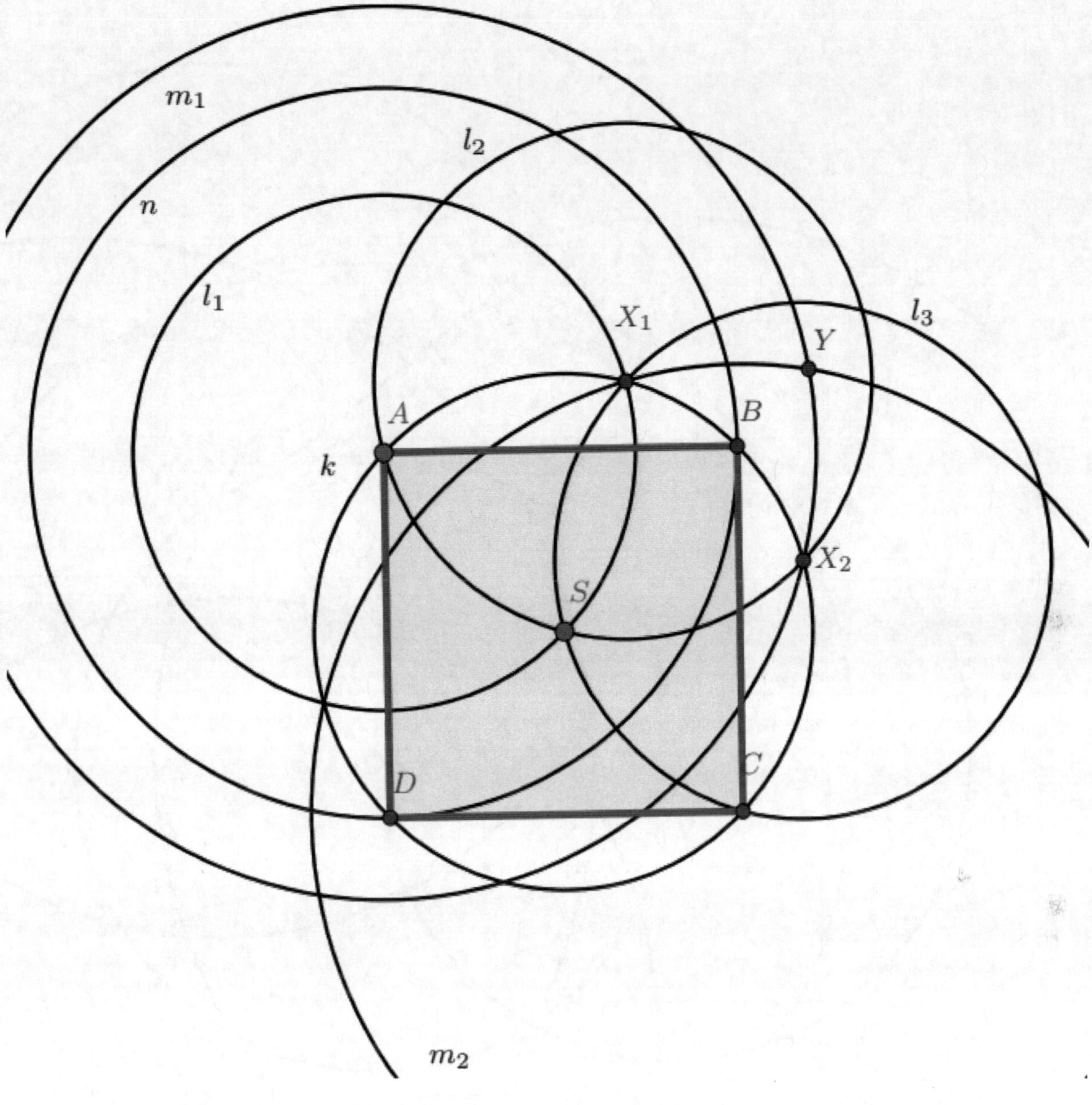

Fig. 4. The square $ABCD$ (A square of a diagonal u and an unknown side a)

9. Point $L; L \in r \cap t$
10. Circle $c_1; c_1(A; r = |IL|)$
11. Point $C; C \in d \cap c_1$
12. Pentagon $ABCDE$

During the construction, we use Mascheroni constructions of doubling a segment and division of an arch, Pythagorean theorem and the law of cosines. The aim is to construct a segment of length $a(\frac{1+\sqrt{5}}{2})$. We use the triangle BIJ in which $|BI| = |IJ| = |BD_1| = a\sqrt{3}$. Therefore the segment

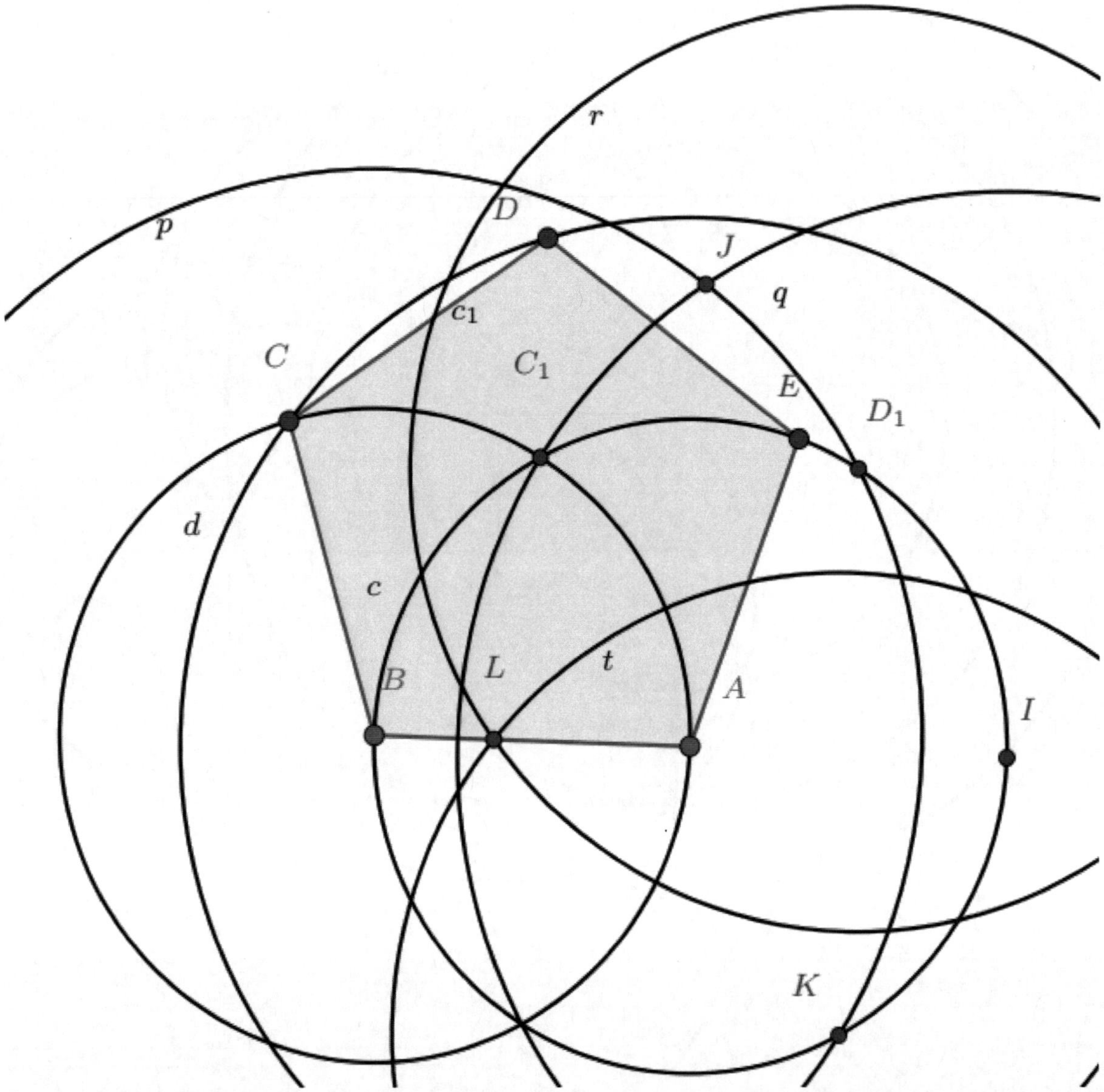

Fig. 5. The pentagon $ABCDE$

JA has the length $a\sqrt{2}$. Because $|D_1I| = |IK| = a$ a $|D_1K| = a\sqrt{3}$ for the triangle D_1IK, the length of IM[1] equals $\frac{1}{2}a$. As $|D_1L| = |KD_1| = \frac{\sqrt{3}}{2}a$ in the created triangle D_1LK, the length of LM equals $a\frac{\sqrt{5}}{2}$. Therefore $|IM|+|LM| = a(\frac{1+\sqrt{5}}{2})$ which is the segment we have been looking for.

[1] The point M is a center point of segment D_1K and is in the segment LI. We don't build the point M, because we use it only for a proof.

6 Conclusion

Mascheroni constructions were very popular in the past as seafarers or astronomers used them. Even today these constructions can appeal to us and can serve as motivational problems in constructional geometry. In the article, a few interesting construction of a square and a regular pentagon were introduced using GeoGebra. The constructions were slightly modified and may differ from the original ones in the book Geometria Del Compasso by Mascheroni.

7 Acknowledgement

The paper was supported by grant from SGS01/PřF/2018-2019: Numbers, Geometry and Physics.

References

[1] ADLER, A.: *Theorie der geometrischen Konstruktionen.* De Gruyter, 1906.

[2] GERHARDT, K.: *Leibniz in London.* Oxford University Press, 1917.

[3] KOSTOVSKII, A.N.: *Geometrical Constructions Using Compasses Only*, Blaisdell Publication Company, 1961.

[4] MASCHERONI, L.: *Geometria del Compasso.* Pávia, 1797.

[5] Mohr George: The Galileo Project [online]. URL: galileo.rice.edu/Catalog/NewFiles/mohr.html.

[6] POMP, M., VÁCLAVÍKOVÁ, Z.: Faktory ovlivňující geometrickou představivost žáků ZŠ. In: *Acta mathematica 13.* Nitra: Fakulta prírodných vied Univerzita Konštantína filozofa, 2010. Fakulta prírodných vied Univerzita Konštantína filozofa, 2010. s. 197-202. ISBN 978-80-8094-781-1.

[7] VOJTĚCH, J.: Theorie Geometrických konstrukcí. *Časopis pro pěstování matematiky a fyziky.* Vol. 31, 1902.

Current address

Ferdiánová Věra, Mgr., Ph.D.
Department of Mathematics
Faculty of Science
University of Ostrava
30.dubna 22, Ostrava, Czech Republic
E-mail: vera.ferdianova@osu.cz

Slovak University of Technology in Bratislava
Faculty of Mechanical Engineering
17th Conference on Applied Mathematics **APLIMAT 2018**

Proceedings

FINANCIALLY DEPENDENT PENSIONS FUNDS MAINTENANCE APPROACH THROUGH BROWNIAN MOTION PROCESSES

FERREIRA Manuel Alberto M. (PT)

Abstract. The situation of some pensions funds that are not appropriately auto financed and are thoroughly maintained with an outside financing effort is considered in this paper. To represent the unrestricted reserves value process of this kind of funds, a time homogeneous diffusion stochastic process is proposed. Then it is projected a financial tool that regenerates the diffusion at some level with positive value. So, the financing effort may be modeled as a renewal-reward process. The relevant costs are studied when the unrestricted reserves value process behaves as a generalized Brownian motion process.

Keywords: pensions fund, diffusion process, Brownian motion process, first passage time, perpetuity, renewal equation

Mathematics Subject Classification: 60J70

1 Introduction

The sustainability of a pensions fund study is performed in [4, 6 and 11], in the scope of queuing theory. Two infinite servers' queues are considered: one with the contributors to the fund, which service time is the time during which they pay to the fund; the other with the pensioners which service time is the time during which they receive the pension. In both queues there is no distinction between the customer and its server in queuing terms technical sense.

The funds portrayed herein are the so-called distributive pension's funds. State pension's funds are generally of this kind. Their income is the contributions of the employees that are distributed by pensioners in the form of retirement pensions. The most important result of these studies is that, in order for the fund to be in balance, the average pension should be equal - or lesser than - to the average contribution. However, due to the demographic imbalance that is happening in contemporary societies, with the number of taxpayers consecutively decreasing while the number of pensioners increases inexorably, it has meant that if the funds were to be autonomously balanced the contributions of the workers would have to assume insupportable values.

What usually is happening is the injection of capital into these funds through transfers from the State Budget, whenever necessary. Therefore in a disorganized way and, in general, in unforeseen situations that may coincide with large financial difficulties. This study purpose is to try to make these situations more predictable, both in relation to the time of occurrence and the amount needed, so that the protection of these funds occurs so smoothly as possible.

Along this paper, that problem is approached bearing in mind the protection cost present value expectation for a non-autonomous pension's fund. Two problems are considered in this context:
-The above mentioned expectation when the protection effort is perpetual,
-The protection effort for a finite time period.
It is admitted that the unrestricted fund reserves behavior may be modeled as a time homogeneous diffusion process and use then a regeneration scheme of the diffusion to include the effect of an external financing effort.

A similar work may be seen in [15], where has been considered a Brownian motion process conditioned by a particular reflection scheme. Less constrained, but in different conditions, exact solutions were then obtained for both problems.

Part of this work was presented at the Fifth International Congress on Insurance: Mathematics & Economics, see [12]. Other works on this subject are [3, 7, 9 and 10].

2 Diffusion Processes and Pensions Funds

Be $X(t), t \geq 0$ the reserves value process of a pensions fund given by an initial reserve amount $a, a > 0$ added to the difference between the total amount of contributions received up to time *t* and the total amount of pensions paid up to time *t*. It is assumed that $X(t)$ is a time homogeneous diffusion process, with $X(0) = a$, defined by drift and diffusion coefficients:

$$\lim_{h\to 0}\frac{1}{h}E[X(t+h)-X(t)|X(t)=x]=\mu(x),$$

$$\lim_{h\to 0}\frac{1}{h}E\left[\left(X(t+h)-X(t)\right)^2|X(t)=x\right]=\sigma^2(x).$$

Call S_a the first passage time of $X(t)$ by 0, coming from *a*. The funds to be considered in this work are non-autonomous funds. So

$$E[S_a] < \infty, \text{for any } a > 0 \qquad (2.1),$$

that is: funds where the pensions paid consume in finite expected time any initial positive reserve and the contributions received, so that other financing resources are needed in order that the fund survives.

The condition (2.1) may be fulfilled for a specific diffusion process using criteria based on the drift and diffusion coefficients. Begin stating that $P(S_a < \infty) = 1$ if the diffusion scale function is

$$q(x) = \int_{x_0}^{x} e^{-\int_{x_0}^{z}\frac{2\mu(y)}{\sigma^2(y)}dy}\,dz,$$

where x_0 is a diffusion state space fixed arbitrary point, fulfilling $q(\infty) = \infty$. Then the condition (2.1) is equivalent to $p(\infty) < \infty$, where

$$p(x)=\int_{x_0}^{x}\frac{2}{\sigma^2(z)}e^{\int_{x_0}^{z}\frac{2\mu(y)}{\sigma^2(y)}dy}dz,$$

is the diffusion speed function. It is thought that, whenever the exhaustion of the reserves happens, an external source places instantaneously an amount $\theta, \theta > 0$ of money in the fund so that it may keep on performing its function.

The reserves value process conditioned by this financing scheme is represented by the modification $\check{X}(t)$ of $X(t)$ that restarts at the level θ whenever it hits 0. Note that since $X(t)$ was defined as a time homogeneous diffusion, $\check{X}(t)$ is a regenerative process. Call $T_1, T_2, T_3, \ldots$ the sequence of random variables where T_n denotes the nth $\check{X}(t)$ passage time by 0. It is obvious that the sequence of time intervals between these hitting times $D_1 = T_1, D_2 = T_2 - T_1, D_3 = T_3 - T_2, \ldots$ is a sequence of independent random variables where D_1 has the same probability distribution as S_a and $D_2, D_3, \ldots$ the same probability distribution as S_θ.

3 First passage times Laplace transforms

Call $f_a(s)$ the probability density function of $S_a(D_1)$. The corresponding probability distribution function is denoted by $F_a(s)$. The Laplace transform of S_a is

$$\varphi_a(\lambda)=E\left[e^{-\lambda S_a}\right]=\int_0^{\infty}e^{-\lambda s}f_a(s)ds, \lambda>0.$$

Consequently, the density, distribution and transform of S_θ $(D_2, D_3, \ldots)$ will be denoted by $f_\theta(s), F_\theta(s)$ and $\varphi_\theta(\lambda)$, respectively.

The transform $\varphi_a(\lambda)$ satisfies the second order differential equation

$$\frac{1}{2}\sigma^2(a)u_\lambda''(a)+\mu(a)u_\lambda'(a)=\lambda u_\lambda(a), u_\lambda(a)=\varphi_a(\lambda), u_\lambda(0)=1, u_\lambda(\infty)=0 \quad (3.1).$$

For details see [7].

4 Perpetual Maintenance Cost Present Value

Consider the perpetual maintenance cost present value of the pension's fund that is given by the random variable

$$V(r,a,\theta)=\sum_{n=1}^{\infty}\theta e^{-rT_n}, r>0,$$

where r represents the appropriate discount rate. Note that $V(r,a,\theta)$ is a random perpetuity: what matters is its expected value which is easy to get using Laplace transforms. Since the T_n Laplace transform is $E\left[e^{-\lambda T_n}\right]=\varphi_a(\lambda)\varphi_\theta^{n-1}(\lambda)$,

$$v_r(a,\theta)=E[V(r,a,\theta)]=\frac{\theta\varphi_a(r)}{1-\varphi_\theta(r)} \quad (4.1).$$

It is relevant to note[1] that

[1] See again [7] for details.

$$\lim_{\theta \to 0} v_r(a,\theta) = \frac{u_r(a)}{-u_r'(0)} \qquad (4.2).$$

5 Brownian Motion Process

Consider that the diffusion process $X(t)$ underlying the reserves value behavior of the pensions fund is a generalized Brownian motion process, with drift $\mu(x) = \mu, \mu < 0$ and diffusion coefficient $\sigma^2(x) = \sigma^2, \sigma > 0$. Observe that the setting satisfies the conditions that were assumed initially. Namely $\mu < 0$ point toward condition (2.1). Everything else remaining as previously stated, it will be proceeded to present the consequences of this particularization. In general it will be added a symbol $(*)$ to the notation used before because it is intended to use these specific results later.

To get the first passage time S_a Laplace transform it must be solved, remember (3.1),

$\frac{1}{2}\sigma^2(a)u_\lambda^{*''}(a) + \mu(a)u_\lambda^{*'}(a) = \lambda u_\lambda^*(a), u_\lambda^*(a) = \varphi_a(\lambda), u_\lambda^*(0)=1, u_\lambda^*(\infty) = 0$.

This is a homogeneous second order differential equation with constant coefficients, which general solution is given by

$$u_\lambda^*(a) = \beta_1 e^{\alpha_1 a} + \beta_2 e^{\alpha_2 a}, with\ \alpha_1, \alpha_2 = \frac{-\mu \pm \sqrt{\mu^2 + 2\lambda\sigma^2}}{\sigma^2}.$$

Condition $u_\lambda^*(\infty) = 0$ implies $\beta_1 = 0$ and $u_\lambda^*(0)=1$ implies $\beta_2=1$ so that the particular solution is achieved:

$$u_\lambda^*(a) = e^{-K_\lambda a} \left(= \varphi_a^*(\lambda)\right), K_\lambda = \frac{\mu + \sqrt{\mu^2 + 2\lambda\sigma^2}}{\sigma^2} \qquad (5.1).$$

In this case, the perpetual maintenance cost present value of the pensions fund is given by, following (4.1) and using (5.1),

$$v_r^*(a,\theta) = \frac{\theta e^{-K_r a}}{1 - e^{-K_r \theta}} \qquad (5.2).$$

Note that $v_r^*(a,\theta)$ is a decreasing function of *a* and an increasing function of θ. In particular

$$\lim_{\theta \to 0} v_r^*(a,\theta) = \frac{e^{-K_r a}}{K_r} \qquad (5.3).$$

To reach an expression for the finite time period maintenance cost present value, start by the computation of $k_r^*(\theta)$, finding a positive *k* satisfying $e^{-K_r\theta}e^{-K_{-\lambda}\theta} = 1$ or $K_r + K_{-\lambda} = 0$. This identity is verified for the value of *k*

$$k_r^*(\theta) = \frac{\mu^2 - \left(-2\mu - \sqrt{\mu^2 + 2r\sigma^2}\right)^2}{2\sigma^2}, \text{if } \mu < -\sqrt{\frac{2r\sigma^2}{3}} \qquad (5.4).$$

Note that the solution is independent of θ in these circumstances. An easy solution, independent of *a* and θ , for $c_r^*(a,\theta)$ was also obtained:

$$c_r^*(a,\theta) = \frac{2\sigma^2\left(-2\mu - \sqrt{\mu^2 + 2r\sigma^2}\right)}{\mu^2 - \left(-2\mu - \sqrt{\mu^2 + 2r\sigma^2}\right)^2} \qquad (5.5).$$

So it is observable that the asymptotic approximation for this particularization reduces to $w_r^*(t;a,\theta) \approx v_r^*(a,\theta) - \pi_r(t)$, where the function $\pi_r(t)$ is, considering (5.4) and (5.5),

$$\pi_r(t) = \frac{2\sigma^2\left(-2\mu - \sqrt{\mu^2 + 2r\sigma^2}\right)}{\mu^2 - \left(-2\mu - \sqrt{\mu^2 + 2r\sigma^2}\right)^2} e^{-\frac{\mu^2 - \left(-2\mu - \sqrt{\mu^2+2r\sigma^2}\right)^2}{2\sigma^2}t}, \text{if } \mu < -\sqrt{\frac{2r\sigma^2}{3}} \qquad (5.6).$$

6 Representation of the Assets and Liability Behaviour

Now an application of the results obtained earlier to an asset-liability management scheme of a pensions fund will be presented. Assume that the assets value process of a pensions fund may be represented by the geometric Brownian motion process

$$A(t) = be^{a+(\rho+\mu)t+\sigma B(t)} \text{ with } \mu < 0 \text{ and } ab\rho + \mu\sigma > 0,$$

where $B(t)$ is a standard Brownian motion process. Suppose also that the liabilities value process of the fund performs as the deterministic process $L(t) = be^{\rho t}$. Under these assumptions, consider now the stochastic process $Y(t)$ obtained by the elementary transformation of $A(t)$

$$Y(t) = ln\frac{A(t)}{L(t)} = a + \mu t + \sigma B(t).$$

This is a generalized Brownian motion process exactly as the one studied before, starting at a and with drift μ and diffusion coefficient σ^2. Note also that the firs passage time of the assets process $A(t)$ by the mobile barrier T_n, the liabilities process, is the first passage time of $Y(t)$ by 0-with finite expected time under the condition, stated before, $\mu < 0$.

Consider also the pensions fund management scheme that raises the assets value by some positive constant θ_n when the assets value falls equal to the liabilities process by the nth time. This corresponds to consider the modification $\bar{A}(t)$ of the process $A(t)$ that restarts at times T_n when $A(t)$ hits the barrier $L(t)$ by the nth time at the level $L(T_n) + \theta_n$. For purposes of later computations it is a convenient choice the management policy where

$$\theta_n = L(T_n)\left(e^{\theta} - 1\right), \text{for some } \theta > 0 \quad (6.1).$$

The corresponding modification $\tilde{Y}(t)$ of $Y(t)$ will behave as a generalized Brownian motion process that restarts at the level $ln\frac{L(T_n)+\theta_n}{L(T_n)} = \theta$ when it hits 0 (at timesT_n).

Proceeding this way, it is reproduced via $\tilde{Y}(t)$ the situation observed before when the Brownian motion example was treated. In particular the Laplace transform in (5.1) is still valid.

Similarly to former proceedings, the results for the present case will be distinguished with the symbol $(\#)$. It is considered the pensions fund perpetual maintenance cost present value, as a consequence of the proposed asset-liability management scheme, given by the random variable:

$$V^{\#}(r,a,\theta) = \sum_{n=1}^{\infty} \theta_n e^{-rT_n} = \sum_{n=1}^{\infty} b\left(e^{\theta}-1\right)e^{-(r-\rho)T_n}, r>\rho$$

where r represents the appropriate discount interest rate. To obtain the above expression it was only made use of the $L(t)$ definition and (6.1). It is possible to express the expected value of the above random variable with the help of (5.2) as

$$v_r^{\#}(a,\theta) = \frac{b\left(e^{\theta}-1\right)}{\theta} v_{r-\rho}^{*}(a,\theta) = \frac{b\left(e^{\theta}-1\right)e^{-K_{r-\rho}a}}{1-e^{-K_{r-\rho}\theta}} \qquad (6.2).$$

As $\theta \rightarrow 0$

$$\lim_{\theta\rightarrow 0} v_r^{\#}(a,\theta) = \frac{be^{-K_{r-\rho}a}}{K_{r-\rho}} \quad (6.3).$$

In a similar way, the maintenance cost up to time t in the above mentioned management scheme, is the stochastic process

$$W^{\#}(t;r,a,\theta) = \sum_{n=1}^{N(t)} b\left(e^{\theta}-1\right)e^{-(r-\rho)T_n}, \; W^{\#}(t;r,a,\theta) = 0 \text{ if } N(t) = 0,$$

with expected value function

$$w_r^{\#}(t;a,\theta) = \frac{b\left(e^{\theta}-1\right)}{\theta} w_{r-\rho}^{*}(t;a,\theta) \qquad (6.4).$$

7 Conclusions

The whole work established depends on the possibility of solving equation (3.1) to obtain the first passage times Laplace transforms. Unfortunately, the solutions are known only for rare cases. An obvious case for which the solution of the equation is available is the one of the Brownian motion diffusion process. The main results concerning this particularization are formulae (5.2) and (5.6). Certain transformations of the Brownian motion process that allowed use of the available Laplace transform may be explored as it was done in section 6. Formulae (6.2) and (6.4) are this application most relevant results.

References

[1] ANDRADE, M., FERREIRA, M. A. M., FILIPE, J. A. (2012), "Representation of reserves through a Brownian motion model", *SPMCS 2012, JPCS-Journal of Physics: Conference Series* 394 (2012) 012036, IOP Publishing. http://dx.doi.org/10.1088/1742-6596/394/1/012036

[2] ANDRADE, M., FERREIRA, M. A. M., FILIPE, J. A., COELHO, M. (2012), "The study of a pensions fund equilibrium through an infinite servers nodes network". *International Journal of Academic Research Part A*, 4(3), 205-208.

[3] FERREIRA, M. A. M. (2012),"Non-autonomous pensions funds maintenance costs study through a diffusion process." *International Journal of Academic Research Part A*, 4(6), 51-56.**DOI**: 10.7813/2075-4124.2012/4-6/A.7.

[4] FERREIRA, M. A. M. (2016)," Results and applications in statistical queuing theory".*15th Conference on Applied Mathematics 2016, APLIMAT 2016*; Bratislava; Slovakia; 362-375.

[5] FERREIRA, M. A. M. (2017), "First order differential equations induced by the infinite servers queue with Poisson arrivals transient behavior probability distribution parameters as time functions". *16th Conference on Applied Mathematics, APLIMAT 2017*; Bratislava; Slovakia; 535-544

[6] FERREIRA, M. A. M., ANDRADE, M., FILIPE, J. A. (2012), "Studying pensions funds through an infinite servers nodes network: a theoretical problem", *SPMCS2012, JPCS-Journal of Physics: Conference Series* 394 (2012) 012035, IOP Publishing. http://dx.doi.org/10.1088/1742-6596/394/1/012035

[7] FERREIRA, M. A. M., FILIPE, J. A. (2013), "Costs study through a diffusion process of pensions funds held with an outside financing effort". International Mathematical Forum, **8** (28), 1359 -1368. http://dx.doi.org/10.12988/imf.2013.36113

[8] FERREIRA, M. A. M., FILIPE, J. A. (2017), "Preferences relations and consumer theory". *16th Conference on Applied Mathematics, APLIMAT 2017*; Bratislava; Slovakia; 545-552.

[9] FERREIRA, M. A. M. (2017), "Searching for answers to the maintenance problem of insufficiently financed, financially dependent pension funds through stochastic diffusion processes". In Webb, Alexandra (Ed.), *Pensions: Global Issues, Perspectives and Challenges,* 978-1-53612-467-5, Nova Science Publishers, Hauppauge, New York, USA; 113-126.

[10] FIGUEIRA, J. (2003), *Aplicação dos processos de difusão e da teoria do renovamento num estudo de reservas aleatórias*, Ph.D. Thesis presented at ISCTE-IUL, Lisboa, Portugal.

[11] FIGUEIRA, J., FERREIRA, M. A. M. (1999), "Representation of a pensions fund by a stochastic network with two nodes: an exercise". *Portuguese Review of Financial Markets*, 2(1), 75-81.

[12] FIGUEIRA, J., FERREIRA, M. A. M. (2001), "Maintenance cost of non-autonomous pensions funds: an evaluation via a regeneration scheme of diffusion processes". *Fifth International Congress in Insurance: Mathematics & Economics, Proceedings,* Pennsylvania State University, USA.

[13] FIGUEIRA, J., FERREIRA, M. A.M.(2006),"Resources dynamics from the point of view of the regenerated Ornstein-Uhlenbeck model". *5th International Conference APLIMAT 2006;* Bratislava; Slovakia; 433-436.

[14] FILIPE, J. A., FERREIRA, M. A. M., ANDRADE, M. (2012), "Reserves represented by random walks", *SPMCS 2012, JPCS-Journal of Physics: Conference Series* 394 (2012) 012034, IOP Publishing. http://dx.doi.org/10.1088/1742-6596/394/1/012034.

[15] GERBER, H. U., PRFUMI, G. (1998), "Stop-loss a tempo continuo e protezionedinamica di un fondo d'investimento". *Revista di Matematica per le Scienze Economiche e Sociale*, 2, 125-146.

Current address

Ferreira Manuel Alberto M., Professor Catedrático
INSTITUTO UNIVERSITÁRIO DE LISBOA (ISCTE-IUL) BRU – IUL, ISTAR-IUL
Av. das Forças Armadas, 1649-026 Lisboa, Portugal
Tel.: + 351 21 790 37 03, FAX: + 351 21 790 39 41
E-mail: manuel.ferreira@iscte.pt

Slovak University of Technology in Bratislava
Faculty of Mechanical Engineering
17th Conference on Applied Mathematics **APLIMAT 2018**

Proceedings

STUDY OF A COLLECTION OF SERVICE TIME DISTRIBUTIONS AND IMPACT IN THE RESPECTIVE M|G|∞ SYSTEM BUSY PERIOD PARAMETERS

FERREIRA Manuel Alberto M. (PT), FILIPE José António (PT)

Abstract. The problems arising when computing the moments of a particular service time distributions collection, for which the M|G|∞ queue system busy period becomes very easy to study, are presented and it is shown how to overcome them. Some results, precisely about the moment's computation of random variables with distribution functions given by this collection are given. The busy period "peakedness" and "modified peakedness" for the M|G|∞ queue in the case of those service time distributions are also computed.

Keywords: service time, collection, distributions, moments, M|G|∞ queue, busy period

Mathematics Subject Classification: 60K35

1 Introduction

If, in the M|G|∞ queue system, the service time length is a random variable with a distribution function belonging to the collection

$$G(t)=1-\frac{\left(1-e^{-\rho}\right)\left(\lambda+\frac{\lambda p+\beta}{1-p}\right)}{\lambda e^{-\rho}\left(e^{\left(\lambda+\frac{\lambda p+\beta}{1-p}\right)t}-1\right)+\lambda},t\geq 0,-\lambda\leq\beta\leq\frac{\lambda\left(1-pe^{\rho}\right)}{e^{\rho}-1},0\leq p<1 \qquad (1.1),$$

the busy period length probability distribution is exponential with an atom at the origin, see Ferreira (2005) and (Ferreira and Andrade, 2009). But although it is so easy to study the busy period in this situation it is very difficult to compute the service time moments.
Some results, precisely about the moment's computation of random variables with distribution functions given by this collection are given.
In the end are presented formulae that give the "peakedness" and the "modified peakedness" to the busy period of the M|G|∞ system for those service time distributions, see Ferreira (2004, 2013,2013a).
This work is built on a part of the presented in Ferreira (2007) which is so corrected, generalized and updated.

2 Moments Calculation Problems

Be $G(t), t \geq 0$ a distribution function and $g(t) = \frac{dG(t)}{dt}$. The differential equation $(1-p)\frac{g(t)}{1-G(t)} - \lambda p - \lambda(1-p)G(t) = \beta$, where $\lambda > 0$ and $-\lambda \leq \beta \leq \frac{\lambda(1-pe^{\rho})}{e^{\rho}-1}$, $0 \leq p < 1$ ($\rho = \lambda\alpha$, being α the mean of $G(t)$) has (1.1) as solution (see Ferreira (2005)). If, in (1.1), $G_i(t)$ is the solution associated to ρ_i, $i = 1,2,3,4$ it is easy to see that

$$\frac{G_4(t)-G_2(t)}{G_4(t)-G_1(t)} \cdot \frac{G_3(t)-G_1(t)}{G_3(t)-G_2(t)} = \frac{e^{-\rho_4}-e^{-\rho_2}}{e^{-\rho_4}-e^{-\rho_1}} \cdot \frac{e^{-\rho_3}-e^{-\rho_1}}{e^{-\rho_3}-e^{-\rho_2}} \tag{2.1}$$

as it had to happen since the differential equation dealt with is a Riccati equation. And computing,

$$\int_0^{\infty}[1-G(t)]dt = \int_0^{\infty} \frac{\left(1-e^{-\rho}\right)\left(\lambda + \frac{\lambda p + \beta}{1-p}\right)}{\lambda e^{-\rho}\left(e^{\left(\lambda + \frac{\lambda p+\beta}{1-p}\right)t} - 1\right) + \lambda} dt =$$

$$= \frac{\left(1-e^{-\rho}\right)\left(\lambda + \frac{\lambda p + \beta}{1-p}\right)}{\lambda} \int_0^{\infty} \frac{1}{e^{-\rho}\left(e^{\left(\lambda + \frac{\lambda p+\beta}{1-p}\right)t} - 1\right) + 1} dt =$$

$$= \frac{\left(1-e^{-\rho}\right)\left(\lambda + \frac{\lambda p + \beta}{1-p}\right)}{\lambda} \int_0^{\infty} \frac{e^{-\left(\lambda + \frac{\lambda p+\beta}{1-p}\right)t}}{e^{-\rho} - e^{-\rho}e^{-\left(\lambda + \frac{\lambda p+\beta}{1-p}\right)t} + e^{-\left(\lambda + \frac{\lambda p+\beta}{1-p}\right)t}} dt =$$

$$= \frac{\left(1-e^{-\rho}\right)\left(\lambda+\dfrac{\lambda p+\beta}{1-p}\right)}{\lambda}\int_0^{\infty}\frac{e^{-\left(\lambda+\frac{\lambda p+\beta}{1-p}\right)t}}{e^{-\rho}+\left(1-e^{-\rho}\right)e^{-\left(\lambda+\frac{\lambda p+\beta}{1-p}\right)t}}dt =$$

$$= \frac{\left(1-e^{-\rho}\right)\left(\lambda+\dfrac{\lambda p+\beta}{1-p}\right)}{\lambda}\cdot\frac{-1}{\left(1-e^{-\rho}\right)\left(\lambda+\dfrac{\lambda p+\beta}{1-p}\right)}\cdot\left[\log\left(e^{-\rho}+\left(1-e^{-\rho}\right)e^{-\left(\lambda+\frac{\lambda p+\beta}{1-p}\right)t}\right)\right]_0^{\infty} =$$

$$= -\frac{1}{\lambda}\left(\log e^{-\rho}-\log 1\right)=\frac{-\rho}{-\lambda}=\alpha$$

as it had to be because are considered positive random variable. The density associated to $G(t)$ given by (1.1) is

$$g(t)=\frac{\left(1-e^{-\rho}\right)e^{-\rho}\left(\lambda+\dfrac{\lambda p+\beta}{1-p}\right)^2 e^{-\left(\lambda+\frac{\lambda p+\beta}{1-p}\right)t}}{\lambda\left[e^{-\rho}+\left(1-e^{-\rho}\right)e^{-\left(\lambda+\frac{\lambda p+\beta}{1-p}\right)t}\right]^2}, t>0, -\lambda\le\beta\le\frac{\lambda\left(1-pe^{-\rho}\right)}{e^{\rho}-1}, 0\le p<1 \quad (2.2).$$

So,

$$\int_0^{\infty}t^n g(t)dt=\frac{\left(1-e^{-\rho}\right)e^{-\rho}\left(\lambda+\dfrac{\lambda p+\beta}{1-p}\right)^2}{\lambda}\cdot\int_0^{\infty}t^n\frac{e^{-\left(\lambda+\frac{\lambda p+\beta}{1-p}\right)t}}{\left[e^{-\rho}+\left(1-e^{-\rho}\right)e^{-\left(\lambda+\frac{\lambda p+\beta}{1-p}\right)t}\right]^2}dt\cdot$$

But, $\displaystyle\int_0^{\infty}t^n\frac{e^{-\left(\lambda+\frac{\lambda p+\beta}{1-p}\right)t}}{\left[e^{-\rho}+\left(1-e^{-\rho}\right)e^{-\left(\lambda+\frac{\lambda p+\beta}{1-p}\right)t}\right]^2}dt\ge\int_0^{\infty}t^n e^{-\left(\lambda+\frac{\lambda p+\beta}{1-p}\right)t}dt=$

$$=\frac{1}{\lambda+\dfrac{\lambda p+\beta}{1-p}}\frac{n!}{\left(\lambda+\dfrac{\lambda p+\beta}{1-p}\right)^n}, \beta\ne-\lambda.$$ And,

$$\int_0^\infty t^n \frac{e^{-\left(\lambda+\frac{\lambda p+\beta}{1-p}\right)t}}{\left[e^{-\rho}+\left(1-e^{-\rho}\right)e^{-\left(\lambda+\frac{\lambda p+\beta}{1-p}\right)t}\right]^2}dt \le e^{2\rho}\int_0^\infty t^n e^{-\left(\lambda+\frac{\lambda p+\beta}{1-p}\right)t}dt =$$

$$= \frac{e^{2\rho}}{\lambda+\frac{\lambda p+\beta}{1-p}} \frac{n!}{\left(\lambda+\frac{\lambda p+\beta}{1-p}\right)^n}, \beta \ne -\lambda .$$

So, calling T the random variable corresponding to $G(t)$:

$$\frac{\left(1-e^{-\rho}\right)e^{-\rho}}{\lambda} \frac{n!}{\left(\lambda+\frac{\lambda p+\beta}{1-p}\right)^{n-1}} \le \mathrm{E}\left[\mathrm{T}^n\right] \le \frac{e^{\rho}-1}{\lambda} \frac{n!}{\left(\lambda+\frac{\lambda p+\beta}{1-p}\right)^{n-1}},$$

$$,-\lambda < \beta \le \frac{\lambda\left(1-pe^{-\rho}\right)}{e^{\rho}-1}, 0 \le p < 1, n = 1,2,... \qquad (2.3).$$

Notes:

- The expression (2.3), giving bounds for $\mathrm{E}\left[\mathrm{T}^n\right]$, guarantees its existence,
- For $n=1$ the expression (2.3) is useless since $\mathrm{E}[\mathrm{T}]=\alpha$. Note, curiously, that here the upper bound is $\frac{e^{\rho}-1}{\lambda}$, the M|G|∞ system busy period mean value, evidently disparate,
- For $n=2$, subtracting to both bounds α^2, it is possible get from expression (2.3) bounds for $VAR[\mathrm{T}]$,
- For $\beta=-\lambda, \mathrm{E}\left[\mathrm{T}^n\right]=0, n=1,2,...,$ evidently.

See, however, that (1.1) can be written like:

$$G(t) = \frac{1+\frac{\frac{\lambda p+\beta}{1-p}}{\lambda}\left(1-e^{\rho}\right)e^{-\left(\lambda+\frac{\lambda p+\beta}{1-p}\right)t}}{1-\left(1-e^{\rho}\right)e^{-\left(\lambda+\frac{\lambda p+\beta}{1-p}\right)t}}, t \ge 0, -\lambda \le \beta \le \frac{\lambda\left(1-pe^{\rho}\right)}{e^{\rho}-1}, 0 \le p < 1 \qquad (2.4)$$

and, for $\rho < \log 2$,

$$G(t)=\left(1+\frac{\frac{\lambda p+\beta}{1-p}}{\lambda}\left(1-e^{\rho}\right)e^{-\left(\lambda+\frac{\lambda p+\beta}{1-p}\right)t}\right)\cdot\sum_{k=0}^{\infty}\left(1-e^{\rho}\right)^{k}e^{-k\left(\lambda+\frac{\lambda p+\beta}{1-p}\right)t},$$

$$,t\geq 0,-\lambda\leq\beta\leq\frac{\lambda\left(1-pe^{\rho}\right)}{e^{\rho}-1},0\leq p<1 \qquad (2.5).$$

After (2.5) it is easy to derive the T Laplace Transform for $\rho<\log 2$. And, so,

- For $\rho<\log 2$

$$\mathrm{E}\left[\mathrm{T}^{n}\right]=-\left(1+\frac{\frac{\lambda p+\beta}{1-p}}{\lambda}\right)n!\sum_{k=1}^{\infty}\frac{\left(1-e^{\rho}\right)^{k}}{k\left(\lambda+\frac{\lambda p+\beta}{1-p}\right)^{n}},-\lambda<\beta\leq\frac{\lambda\left(1-pe^{\rho}\right)}{e^{\rho}-1},0\leq p<1,$$
$$,n=1,2,\ldots \qquad (2.6).$$

Notes:

$$-\ \mathrm{E}[\mathrm{T}]=-\left(1+\frac{\frac{\lambda p+\beta}{1-p}}{\lambda}\right)\sum_{k=1}^{\infty}\frac{\left(1-e^{\rho}\right)^{k}}{k\left(\lambda+\frac{\lambda p+\beta}{1-p}\right)}=\frac{1}{\lambda}\sum_{k=1}^{\infty}(-1)^{k+1}\frac{\left(1-e^{\rho}\right)}{k}=$$

$$=\frac{1}{\lambda}\log e^{\rho}=\frac{\rho}{\lambda}=\alpha.$$

- For $n\geq 2$ only a finite number of parcels can be considered in the infinite sum. Calling M this number, to get an error lesser than ε it must be fulfilled simultaneously

a) $\mathrm{M}>\frac{1}{\lambda+\frac{\lambda p+\beta}{1-p}}-1,$

b) $\mathrm{M}>\log_{\left(e^{\rho}-1\right)}\frac{\varepsilon e^{\rho}\lambda}{n!\left(\lambda+\frac{\lambda p+\beta}{1-p}\right)}-1.$

So it is evident now that this distributions collection moment's computation is a complex task. This was already true for the study of Ferreira (1998) where the results presented are a particular situation of these ones for $p = 0$.To consider the approximation

$$\mathrm{E}_m^n = \sum_{k=1}^{\infty} \left(\frac{k}{m}\right)^n \left[G\left(\frac{k}{m}\right) - G\left(\frac{k-1}{m}\right) \right], -\lambda < \beta \leq \frac{\lambda(1 - pe^{\rho})}{e^{\rho} - 1}, 0 \leq p < 1, n = 1,2,... \quad (2.7)$$

may be helpful since $\lim_{m \to \infty} \mathrm{E}_m^n = \mathrm{E}[\mathrm{T}^n], n = 1,2,...$ (Ferreira and Andrade, 2012c) that allow the moments numerical computation .

3 The "Peakedness" and the "Modified Peakedness" for the M|G|∞ Queue Busy Period

The M|G|∞ queue busy period "peakedness" is the Laplace transform of its length at $1/\alpha$, Ferreira (2013,2013a) . It is a parameter that characterizes the busy period distribution length and contains information about all its moments. For the collection of service distributions (1.1) the "peakedness", named pi , is

$$pi = \frac{e^{-\rho}(\lambda + \beta)(\rho + 1) - \lambda p - \beta}{\lambda\left(e^{-\rho}(\rho + \alpha\beta) + 1 - p\right)}, -\lambda \leq \beta \leq \frac{\lambda(1 - pe^{\rho})}{e^{\rho} - 1}, 0 \leq p < 1 \quad (3.1).$$

In Ferreira (2013,2013a) is also introduced another measure, the "modified peakedness" got after the "peakedness" taking out the terms that are permanent for the busy period in different service distributions and putting over the common part. Calling it qi:

$$qi = pi \frac{\rho}{e^{\rho} - \rho - 1} + 1$$

and so, for the distributions given by collection (1.1):

$$qi = \frac{e^{-\rho}(\lambda + \beta)(\rho + 1) - \lambda p + \beta}{\lambda\left(e^{-\rho}(\rho + \alpha\beta) + 1 - p\right)} \frac{\rho}{e^{\rho} - \rho - 1} + 1, -\lambda \leq \beta \leq \frac{\lambda(1 - pe^{\rho})}{e^{\rho} - 1}, \ 0 \leq p < 1 \quad (3.2).$$

References

[1] ANDRADE, M. (2010). A note on foundations of probability. *Journal of Mathematics and Technology*, 1(1), 96-98.

[2] CARRILLO, M. J. (1991). Extensions of Palm's theorem: a review. *Management Science*, 37(6), 739-744.

[3] FERREIRA, M. A. M., RAMALHOTO, M. F. (1994). Estudo dos parâmetros básicos do período de ocupação da fila de espera $M|G|\infty$. *A Estatística e o Futuro e o Futuro da Estatística*. Actas do I Congresso Anual da S.P.E., Edições Salamandra, Lisboa, 47-60.

[4] FERREIRA, M. A. M. (1998). Momentos de Variáveis Aleatórias com Função de Distribuição dadas pela Colecção $G(t)=1-\frac{(1-e^{-\rho})(\lambda+\beta)}{\lambda e^{-\rho}\left(e^{(\lambda+\beta)}-1\right)+\lambda}, t \geq 0, \quad -\lambda \leq \beta \leq \frac{\lambda}{e^{\rho}-1}$. Revista Portuguesa de Gestão, II. ISCTE, Lisboa, 67-69.

[5] FERREIRA, M. A. M. (1998a). Application of Riccati equation to the busy period study of the $M|G|\infty$ system. *Statistical Review*, 1st Quadrimester, INE, 23-28.

[6] FERREIRA, M. A. M. (1998b). Computational simulation of infinite servers systems. *Statistical Review*, 3rd Quadrimester, INE, 23-28.

[7] FERREIRA, M. A. M. (2001). $M|G|\infty$ queue heavy-traffic situation busy period length distribution (power and Pareto service distributions). *Statistical Review*, 1st Quadrimester, INE, 27-36.

[8] FERREIRA, M. A. M. (2002). The exponentiality of the $M|G|\infty$ queue busy period. *Actas das XII Jornadas Luso-Espanholas de Gestão Científica*, Volume VIII-Economia da Empresa e Matemática Aplicada. UBI, Covilhã, Portugal, 267-272.

[9] FERREIRA, M. A. M. (2004), *M|G|∞ Queue Busy Cycle Renewal Function for Some Particular Service Time Distributions. Proceedings of Quantitative Methods in Economics (Multiple Criteria Decision Making XII)*. Virt, Slovakia, 42-47.

[10] FERREIRA, M. A. M. (2005). Differential equations important in the $M|G|\infty$ queue system transient behavior and busy period study. *Proceedings of 4th International Conference APLIMAT 2005*, Bratislava, Slovakia, 119-132.

[11] FERREIRA, M. A. M. (2005a), A Equação de Riccati no Estudo da Ocupação de um Sistema M|G|∞ (Uma Generalização). *Actas do I Congresso de Estatística e Investigação Operacional da Galiza e Norte de Portugal/VII Congreso Galego de Estatística e Investigación de Operacións*. Guimarães, Portugal.

[12] FERREIRA, M. A. M. (2007). M|G|∞ system parameters for a particular collection of service time distributions. *Proceedings of 6th International Conference APLIMAT 2007*, Bratislava, Slovakia, 131-137.

[13] FERREIRA, M. A. M., ANDRADE, M., FILIPE, J. A. (2009). Networks of queues with infinite servers in each node applied to the management of a two echelons repair system. *China-USA Business Review*, 8(8), 39-45 and 62.

[14] FERREIRA, M. A. M., ANDRADE, M. (2009). The ties between the $M|G|\infty$ queue system transient behavior and the busy period. *International Journal of Academic Research*, 1(1), 84-92.

[15] FERREIRA, M. A. M., ANDRADE, M. (2010). Looking to a $M|G|\infty$ system occupation through a Riccati equation. *Journal of Mathematics and Technology*, 1 (2), 58-62.

[16] FERREIRA, M. A. M., ANDRADE, M. (2010a). $M|G|\infty$ system transient behavior with time origin at the beginning of a busy period mean and variance. *APLIMAT-Journal of Applied Mathematics*, 3(3), 213-221.

[17] FERREIRA, M. A. M., ANDRADE, M. (2011).Fundaments of theory of queues. *International Journal of Academic Research*, 3(1), part II, 427-429.

[18] FERREIRA, M. A. M., ANDRADE, M. (2012). Busy period and busy cycle distributions and parameters for a particular $M|G|\infty$ queue system. *American Journal of Mathematics and Statistics*, 2(2), 10-15. http://article.sapub.org/10.5923.j.ajms.20120202.03.html

[19] FERREIRA, M. A. M., ANDRADE, M. (2012a). Queue networks models with more general arrival rates. *International Journal of Academic Research*, 4(1), part A, 5-11.

[20] FERREIRA, M. A. M., ANDRADE, M. (2012b). Transient behavior of the $M|G|$m and $M|G|\infty$ system. *International Journal of Academic Research*, 4(3), part A, 24-33.

[21] FERREIRA, M. A. M., ANDRADE, M. (2012c). A methodological note on the study of queuing networks. *Journal of Mathematics and Technology*, 3(1), 4-6.

[22] FERREIRA, M. A. M., ANDRADE, M., FILIPE, J. A (2012). The age or excess of the M/G/∞ queue busy cycle mean value. *Computer and Information Science*, 5(5), 93-97. http://dx.doi.org/10.5539/cis.v5n5p93

[23] FERREIRA, M. A. M. (2013). A $M|G|\infty$ queue busy period distribution characterizing parameter. *Computer and Information Science*, 6 (1), 83-88. http://dx.doi.org/10.5539/cis.v6n1p83

[24] FERREIRA, M. A. M. (2013a). The modified peakedness as a M/G/∞ busy cycle distribution characterizing parameter. *International Journal of Academic Research*, 5(2), part A, 5-8.

[25] FERREIRA, M. A. M. (2016). Results and applications in statistical queuing theory. *APLIMAT 2016 - 15th Conference on Applied Mathematics. Proceedings*. Bratislava, Slovakia, 362-375.

[26] FERREIRA, M. A. M. (2017). A particular collection of service time distributions parameters study and impact in some M|G|∞ system busy period and busy cycle parameters. *Acta Scientiae et Intellectus,* 3(1), 31-39.

[27] FERREIRA, M. A. M., FILIPE, J. A. (2017). In the search for the infinite servers queue with Poisson arrivals busy period distribution exponential behaviour, *Int. J. Business and Systems Research*, 11(4), 453–467.

[28] FIGUEIRA, J., FERREIRA. M. A. M. (1999). Representation of a pensions fund by a stochastic network with two nodes: an exercise. *Portuguese Revue of Financial Markets*, 1(3).

[29] HERSHEY, J. C., WEISS, E. N., MORRIS, A. C. (1981). A stochastic service network model with application to hospital facilities. *Operations Research*, 29(1), 1-22, 1981.

[30] KENDALL, M. G., STUART, A. (1979). *The advanced theory of statistics. Distributions theory*. London, Charles Griffin and Co., Ltd. 4th Edition.

[31] KLEINROCK, L. (1985). *Queueing systems*. Vol. I and Vol. II. Wiley- New York.

[32] RAMALHOTO, M. F., FERREIRA, M. A. M. (1996). Some further properties of the busy period of an M|G|∞ queue. *Central European Journal of Operations Research and Economics*, 4(4), 251-278.

[33] ROSS, S. (1983). *Stochastic Processes*. Wiley, New York.

[34] STADJE, W. (1985). The busy period of the queueing system M|G|∞. *Journal of Applied Probability*, 22, 697-704.

[35] SYSKI, R. (1960). *Introduction to congestion theory in telephone systems*. Oliver and Boyd, London.

[36] SYSKI, R. (1986). *Introduction to congestion theory in telephone systems*. North Holland, Amsterdam

[37] TAKÁCS, L. (1962). *An introduction to queueing theory*. Oxford University Press, New York.

[38] WHITT, W. (1984). On approximations for queues, I: extremal distributions. *A T &T Bell Laboratories Technical Journal*, 63(1), 115-138.

Current address

Ferreira Manuel Alberto M., Professor Catedrático
INSTITUTO UNIVERSITÁRIO DE LISBOA (ISCTE-IUL)
BRU – IUL, ISTAR-IUL
Av. das Forças Armadas, 1649-026 Lisboa, Portugal
Tel.: + 351 21 790 37 03. Fax: + 351 21 790 39 41
E-mail: manuel.ferreira@iscte.pt

Filipe José António, Professor Auxiliar
INSTITUTO UNIVERSITÁRIO DE LISBOA (ISCTE-IUL)
BRU – IUL, ISTAR-IUL
Av. das Forças Armadas, 1649-026 Lisboa, Portugal
Tel.: + 351 21 790 34 08, Fax: + 351 21 790 39 41
E-mail: jose.filipe@iscte.pt

Slovak University of Technology in Bratislava
Faculty of Mechanical Engineering
17th Conference on Applied Mathematics **APLIMAT 2018**

Proceedings

THE ALEXANDER MOSAIC AND THE HOUSE OF THE FAUN (POMPEII VI 12, 1-8) GEOMETRY PROPORTIONS AND ART OF COMPOSITION

FERRO Luisa (IT)

Abstract. This text analyzes the celebrated Alexander Mosaic, found in the House of the Faun in Pompeii starting from the premise that it may hide the traces of a compositional process based on rules that built its origin, the Ground Zero of any work of art as such. A painstaking process of calculation and precision based on regulators, even surprising symmetries, and aesthetic precision in the placement of figures that projects the represented scene in a well-defined, memorable, self-sufficient and "representational" form. Finally the text analyses the relationship between mosaic and the composition process of the House.

Keywords: geometry and art, composition, Alexander Mosaic Pompeii

Mathematic Subject Classification: Primary 01A20

1 The *Gran Mosaico*

On October 24, 1831, a surprising discovery was made in the so-called *Domus* of the Faun. The mosaic found at the center of a room with two red columns entered through a Nilotic-themed threshold was clearly "of a quite eminent value ... And its style and execution are finer than any other similar work from the ancient times, including the five other works discovered in the same house and described in the press. 9 pl. wide and 19 ½ long, this large mosaic made of tiny *tesserae* of colorful marbles depicts a battle between Greeks and Barbarians". [9]
While other mosaics had actually been found in the house – apparently belonging to the same figurative scheme [29] and made of tiny *tesserae* – the Alexander Mosaic was immediately recognized for its exceptional value [18] [19] [24]: "... in the combination of so many fine and valuable elements within a very large composition where every group, figure, and line resulted from the most accurate research and endless precautions, the highest value seems to be that no trace of a study, nor the clue of a pre-ordained intention seem to emerge".
(fig. 1)
The art historian, the militant critic or simply the scholar who looks at the battle offers a different interpretation; and perhaps, the painter had a meaning in mind that none of them can see, or he just conceived it as a riddle. The game of coincidences seems to increase the myth.

As of today, there are still many hypotheses on the table, all of them deserving the benefit of the doubt. There is still no consensus about which battle the scene represents. Neither is there consensus about who might have designed the mosaic. Some argue it is the copy of a painting, and therefore are more interested in discovering the name of its author [7] [29].
These questions consequently lead to a third question, itself unanswered yet, about the mosaic's actual place of origin.
The Hellenistic origin of the mosaic seems quite clear due to the absence of recognizably Roman elements in it. There may be a relation with a painting made during Alexander's life [7] [29].
That the figure depicted is indeed Alexander seems equally out of question, as proved by the other mosaics found in the House of the Faun (itself Hellenistic in terms of type) [17] [28] that narrate a precise history, from the hall through the room with two columns where the Mosaic itself was placed [21] [23] [29].
The mosaics in the House of the Faun (except for the one left in situ) seem to result from the artisanal skill of mosaicists based in Alexandria in Egypt [3] [12] [23] [29]. This is not surprising given the close cultural relationship, now clearly established, between Greek Hellenism, Egyptian Hellenism and Pompeii. [13]
Therefore, if the mosaic indeed came from Alexandria, it could only belong to the royal circle due to its value and because the artists who worked at the Museum of the Palace were exclusively employed by the King and therefore belonged to his "stable" [11] [25].

Fig. 1
(Courtesy MANN Archaeological Museum of Naples)

1.1 The Art of composition and its rules. Within the symphony of the Alexander Mosaic

This could be any one of the battles fought by Alexander against Darius during his quest to submit the Persian Empire and thus reach "the ends of the world". This (now) iconic image of Alexander summarizes memorable facts and makes us glimpse, for a moment, into the void of those obliterated actions, the facts that do not exist anymore, the lost events nobody will ever see again. Yet, in the absence of certain proof, what could be the right key to approach this work with a less superficial admiration (other than a generic appreciation of its "beauty"), or a

critical point of view that, unburdened by the complexity of a historical/critical assessment, frames it in a more contemporary dimension that is closer to our understanding?
Which methodological meditations could arise from this artwork?
Perhaps the definition of "copy" of a lost painting, often used for this mosaic, is not appropriate because it inherently (although involuntarily) lessens its value. There was probably a cartoon or a sketch, or the theme was inspired by a then celebrated painting. Nevertheless, the reproduction of a work from a technique to another is neither elementary nor automatic.
Therefore, the mosaic should be viewed as a unique artwork in its own right, particularly because it was executed with the technique of a painting and, given the highly skillful use of tiny tesserae; it was indeed defined as a painting. Since for Pliny (Naturalis Historia books XXXIII-XXXVII) *opus vermiculatum* mosaics were indeed paintings, the resulting ambiguity may have led to the interpretation of a painting as a mosaic and the other way around.
An artwork results from a process during which the composition develops towards its completion in a calculated and precise way necessarily based on hidden regulators, possibly surprising symmetries and an aesthetical precision in the placement of figures. Thus, the vision of the battle is crystallized (fixed) in a frame in a well-defined, memorable, self-sufficient, "representational" form.
The artwork might have followed the rules of *skiagraphia*, a technique mentioned by Pliny and Quintilian due to which painting rises above mere drawing (Moreno 2000). All this would be complemented by Aristotle's studies on the perception of colors.
In the Alexander Mosaic, the cold hues (blue and green) are missing. As we know from Pliny, classic painting only relied on *tetrachromy*, or the four earth tones – white, black, red, and yellow including all their varying hues – obtained from the natural soil.
The subtleties and high refinement encapsulated in the battle moment should hide a less apparent plane made of measures and a clearly defined grid. Art is supposed to produce *the precision and rigid consequence of a mathematical problem*: axes, circles, right angles are the principles of geometry and the effects our eye measures and recognizes – without these elements, there would be nothing but randomness, anomaly, discretion. The Ancients relied on this necessary tool. But very few artists revealed how they composed their works, and rather scattered around the clues of some mysterious game.
The outer frame, probably applied at a later stage with larger *tesserae*, except for the flowers that match those found in Abuqir/Canopus and here in the frame ideally fix the work to its backing [5] should be considered as external to the painting's space.
The horizontal band executed in one color (perhaps originally bearing a phrase like the one of the Thmuis mosaic found in the Nile Delta) should be recognized, instead, as part of the painting's space. Often forgotten and erased from reproductions, this band actually plays a key role in the painting's geometrical proportions. For this reason, it should be considered as a silent place, an empty but necessary space.
When viewing the work, our look immediately goes to the fragment that depicts Alexander and his horse because the figure has become a universal icon and the fragment is the survivor of a lost mosaic surface. The horse is not Bucephalus.
Alexander only rode his beloved but quite old black horse at the end of the battle.
The profile follows clear geometrical rules: eyes, chin, and ear create an equilateral triangle one side of which is adjacent to the vertical line I will illustrate thereafter.
If you consider a square built on the painting's shorter side (not including the silent band), Alexander's face is placed above the two diagonals, precisely on the vertical line on which the Macedonian king's profile is built.

The horizontal line that corresponds to Alexander's gaze is slightly above that of all the human figures in the background, all aligned on the same horizontal line.
This square encloses the true fate of the battle. With his spear, Alexander pierces one of the enemy generals whose face corresponds to the quarter of circle built on the corner of the square.

If, starting from the opposite short side, one builds a square that has the same size as the previous one and mirrors it, the diagonal of this square defines the position of the face and the inclination of the body of the standing enemy general (Darius) and the chariot's horses.
The arm of the charioteer who turns around is parallel to the quarter of circle pointed on the second square's bottom corner.
The line that unites the gazes of Alexander and Darius forms the hypotenuse of a right-angled triangle the shorter side of which falls on the quarter of circle on the right-hand square.
The spears in the foreground are parallel to the diagonal.
The vertical band formed between the two specular and symmetrical squares has the same size as the monochrome one at the bottom of the painting. This band is equally silent, a sort of free area where nothing happens.
The axis of symmetry is hidden in the silent band. **(Fig. 2)**
The axis is perhaps the earliest manifestation of man. The axis provides order.
Once this silent and empty zone is recognized, that is where the gaze lingers:
this is the painting's point of balance.
The human figure found in this point is the only one looking right at the viewer.
While seeming to turn his back, the figure suddenly turns around and directly engages the viewer. If you look at him carefully, you will notice that he is tangent to rather than precisely on the axis of symmetry. With a shrewd theatrical device, he seems to glance at Alexander. Actually, he shows us the exact position of the axis of symmetry.
This character reveals the uttermost secret – the center of the composition.
The order underlying the entire work.
Who is he? The artist? Look at his hat. Rather than a helmet like the ones the other Greeks wear (it has no metallic glint), it looks like a fabric or leather cap.
He is an intruder in the scene who looks at us to reveal a secret. Oddly, nobody ever noticed him before.

Look at the only two figures who stand above Alexander's gaze, the most important being the Persian (Darius) who stands on the chariot, apparently in retreat – this is nothing but a device to establish a symmetry with the dead tree (an element that was interpreted in a number of ways: [26]. The tree's branches perfectly mirror the standing general's arm and that of his charioteer. Being the losing party now retreating, they are likewise a dead tree, the symbol of a dying empire.
If you carefully observe the plane geometrical figures thus obtained, you will find countless other correspondences with the figures in the painting. The fallen figure who looks at his reflection in his shield lies within a right-angled triangle obtained from the diagonal. Famously, this figure was put in the wrong place when the mosaic was reconstructed in the House of the Faun.
Keep on studying the geometric space of the mosaic.
The horizontal line that runs through the gaze of the central figure (the hypothetical author) cuts the length of mosaic's bottom side (including the horizontal monochrome band) according to the Euclidian principle of division of a segment into *extreme and mean ratio*.

Otherwise said, the golden section. **(Fig. 3-4)**
Therefore, the above mentioned short side of the mosaic is bisected by the horizontal line generated by the gaze of the central figure according to the concept of the *proportional medium* (Euclid, *Elementi*, libro VI), and the ratio between the unequal lengths obtained with the total measure is exactly the irrational number 1.6180 – the golden ratio, or Phidias constant.
The mosaic's longer side is three times the longer segment of the proportional medium obtained on the short side. In other words, the entire surface comprises three golden rectangles (as we would call them now), the longer side of which coincides with the mosaic's short side.
The Alexander Mosaic reveals a remarkable skill in the use of geometry and space as mandated by the geometrical theories of Euclid, the "Composer of Elements" (a contemporary of Alexander's) who was called by Ptolemy I Soter to work in Alexandria at the time when the great Library and the Museum were established [11].
The axis of vertical symmetry and the horizontal secant line obtained through the rule of *proportional medium* (the point of intersection of which is indicated by the unknown character) are the matrix of the painting's geometries. The distance between the parallel lines that define the main symmetrical squares provides the module that divides the space occupied by the battle scene. The rotation of Alexander's spear also finds an explanation within this modular grid. **(Fig. 5)**
But that is not all.
The enemy army only seems to be advancing. This deceiving impression results from the point of view. Look at the figures in the background aligned along the central horizontal line, the Greek helmets in the third row are beyond the symmetry axis and superposed to the Persian helmets. The structure of the painting is designed both frontally and in plan, with the placement of the armies and Alexander who breaks the ranks and charges on.
The center of symmetry results from a three-dimensional imagination that transcends the plane figures. The drawing is imagined in the three spatial dimensions (we might say in 3D). And, according to the composition's measures, the figure that reveals the enigma plays a central role in the other spatial dimensions as well.
It is clear, by now, that this is a unique work in its own right rather than the replica of an existing artwork. Those who discovered it recognized this immediately. Its author was so convinced he was involved in an extraordinary achievement that he unexpectedly introduced his own portrait in the mosaic.
As a painting, it would not even be placed on the floor, just like many of the artworks that are called paintings would not be necessarily paintings at all. Indeed, at that time, the word *téchne* had a meaning that comprehensively embraced art, skill or even manual skill in a way that never fully separated what deserves to be called art from what fails to earn such qualification [36].
Now, let's go back to the figure at the center of the composition: who is this man?
Having established measurements and geometries, one can only delve into the realm of probabilities and statistics. As mentioned before, he could be someone who wanted to leave a sign of his own presence in a work that he recognized as special. The fact is that, except for the descriptions of lost works, no work found so far is comparable to the mosaic.

Fig. 2-3-4-5
(© LuisaFerro)

If the Mosaic indeed comes from Alexandria and is related to a painting from Alexander's time – 330-310 (Cohen 1997; Moreno 2000) – then some clues could emerge from the names of the artists who worked at the court of Ptolemy I Soter at the time. [1] [11] [14] [16] [25]: aside from Euclid there was the painter Antiphilus of Naucratis. Philoxenus of Eretria, a painter of the Theban-Attic school was also in Alexandria [2]. Some scholars believe that certain elements in the mosaic would intrinsically point to Apelles as its author [19].
Finally, Helena also worked in Alexandria: Helena, the daughter of Timon the Egyptian, is perhaps a poetic invention [5]. Or not.

2 The Alexander Mosaic and the House of the Faun

The House of the Faun, as we see it today, is the result of two main constructive moments both dating back to the 2nd century BC [8] [15]. The first plan dates back to 180 BC, the second to the last quarter of a century, when the house was completely renovated, enlarged, reworked in typology, in the whole of the geometric relationships, in the studied sequences. This is the moment when the precious cargo with the Alexandrian mosaics comes into play.
The Architect who carried out (in Ancient times) the restoration has found a house with a Tuscan *atrium* (connected to a second *atrium* also Tuscan but not open to the street). The Tuscan *atrium* has golden proportions and a *tablinum* as usual (Vitruvio, *De Arhitectura*) is arranged along the longitudinal axis. The first peristyle was already part of the house. Here the new intervention is limited to the transformation of the entrance hall and the arrangement of new stuccoes and new decorations and of course the insertion of the well-known mosaics mentioned above.
The part of the building containing the *tablinium* (to which the new three-dimensional figure floor is inserted) perhaps had already the depth visible now. The new project also connects the second *atrium* to the street, turning it into a Tetrastyle *atrium*. The latter remains an *atrium* related to the servant side of the house and, through a long corridor, is connected to the new great peristyle. A perpendicoular body is added that closes the first peristyle identical in thickness to the *tablinium*. Here the *exedra distila* which houses the mosaic of Alexander facing the first peristyle and the *tablinum* and a detached room with the lion's mosaic facing the new great peristyle. The latter closes with a built wall containing *lararii* and a small temple. At the center of the peristyle have been found statues probably referring to the rites that were celebrated in the small temple. Among other things, the columns in front of the temple are spaced the same size, probably to allow the view of the center of the garden of the great peristyle.
The architect built two itineraries in the house each of them related to the respective *atrium* on which relates specific perspective point of view. The first itinerary is a real museum tour and the second that seems to be linked to the service functions of the house is also the direct connection to the second peristyle. **(Fig. 6-7)**
The museum sequence describes a story related to the Egypt of the Ptolemaic court: scenes are followed by Dionysian connotations and typical settings of the Nilotic landscape (Alexandria and Canopus); erotic scenes, Dionysian scenes. So starting from the Tuscan atrium (a theatrical scene with false double height, also found in large Ptolemaic ships to amplify the space) the visitor is guided through a story.
In the central *tablinum* there is a geometric mosaic with three-dimensional figures, composed with rare and precious stones of east origin. On the visual perspective axis the exedra distila with the red columns (royal color). The culmination of the route is the distila hall (esedra)

with the Alexander mosaic that opens towards the peristyle. And the exedra is the crucial point of the museum itinerary. The two red columns are the arrival point of the visual itinerary of the whole house. Inside the Great Mosaic. The two rooms next to the exedra open towards the second peristyle, a place of sacred representation. The body that contains the exedra appears to have been on two floors. A covered porch was arranged on the three rooms and overlooked the large peristyle.

2.1 The Iconic light of Geometric Relationships

At this point, however, it is permissible to ask what was the process of project construction, the work of composition that underlies the figure of the house with its paths.
A modular grid has been identified, very similar to that used for large Hellenistic buildings [7] [8] [15]. The house is divisible into geometrically coherent parts and divided into two by a middle line tangent to the south face of the part of building containing the distyle room.
The two *atriums*, with their respective rooms form two identical squares. The two transverse bodies (depth equal) respectively included with the two relative peristyles form a square and a rectangle [7]. It is also evident an elementary reference grid that is the sub-multiple of the previously described geometrical ensemble, very similar to the modular references of the great Hellenistic buildings.
These important considerations are necessary prerequisites, but we need to clarify other aspects of the project, for example the relationship between the transverse building blocks and the module, the geometric role of the distiye room (which after all contains the most important artwork) and the layout of the Alexander mosaic. Finally, the relationship between mosaic and distyla room, between distyla room and the House.
Starting from the previous considerations, the following relationships can be noted.
- The square geometrical figures quoted have a modular dimension with a sequence of square numbers (Pythagoras): 2x2 + 2x2 (Atriums); 3x3 (first Peristyle); 4x4 + 2x2 (second Peristilio)
- The west side of the Tuscan *atrium*, first peristyle and second peristyle are aligned and coincide with the module as previously identified.
- The dystile room is built with golden proportions.
- By projecting the size of the *tablinium* in the Tuscanic *atrium* it is possible to verify the alignment with the west wall of the distyle room. The distance of the west wall from the module is the same as the linear projection of the east wall of the tablinium with the east wall of the distyle room. The east wall of the hall is also aligned with the east side of the Tuscan *atrium*. **(Fig. 8)**
- The large mosaic (the original) is not placed centered on the room but adjacent in its longer side above the module (note well that the position of the mosaic is not that given to the modern copy now in the house, but the one designed by Fausto and Felice Niccolini at the time of discovery). This alignment forms the side of an isosceles triangle whose vertex coincides with the intersection of the diagonals of the rectangle formed by this alignment and the three sides of the peristyle. This triangle is subdivided into two triangles (almost sacred triangles with proportions identified by Pythagoras with 3, 4, 5 units). The intersection between the apex and the diagonals of the Peristyle is the central point referring to the small temple on the north side Peristyle. **(Fig. 9)**
Constructing a circle of radius corresponding to the side of the isosceles triangle are obtainedthe position of the left column of the Exedra, the position of the fountain of the first

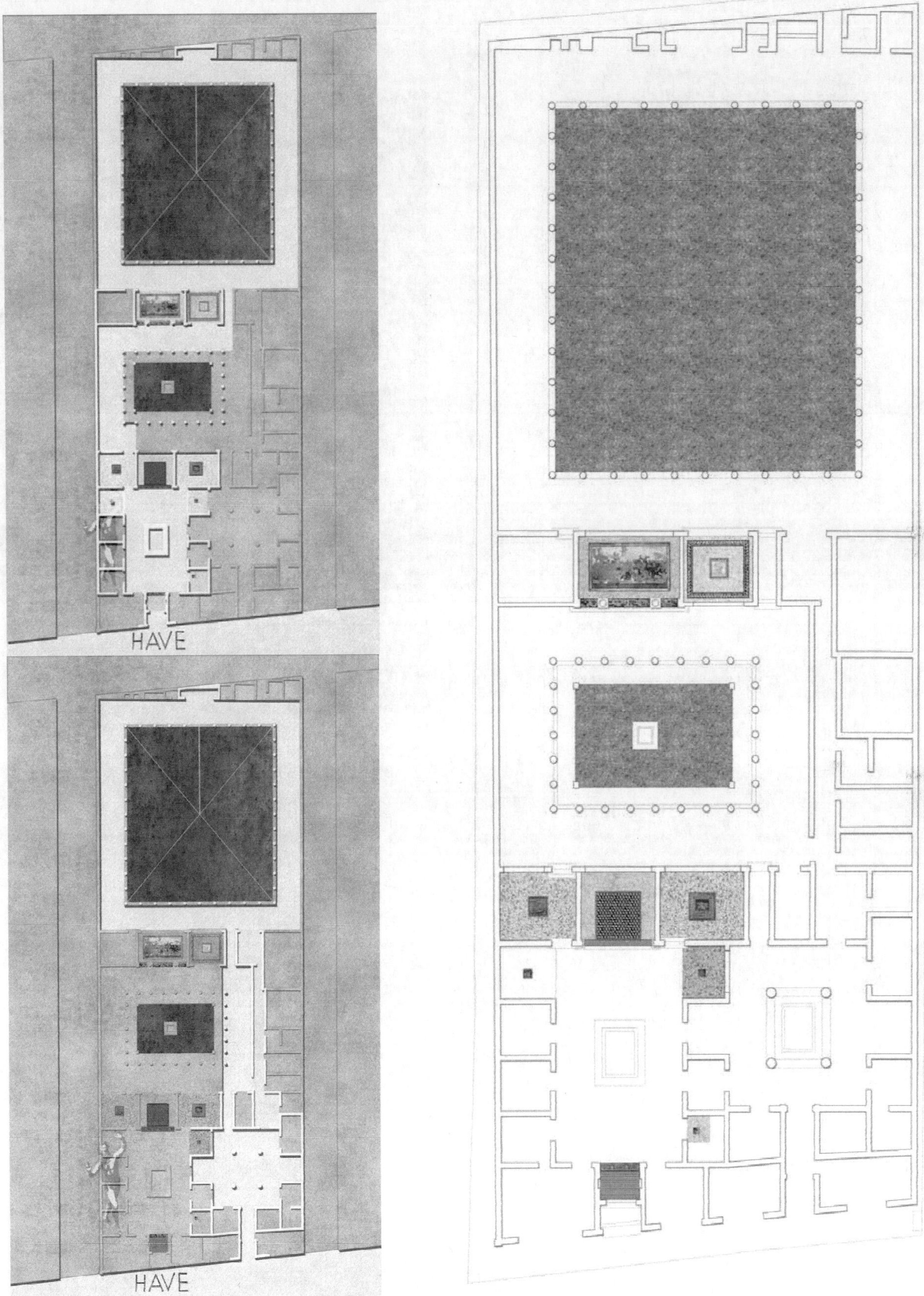

Fig. 6-7
(© LuisaFerro)

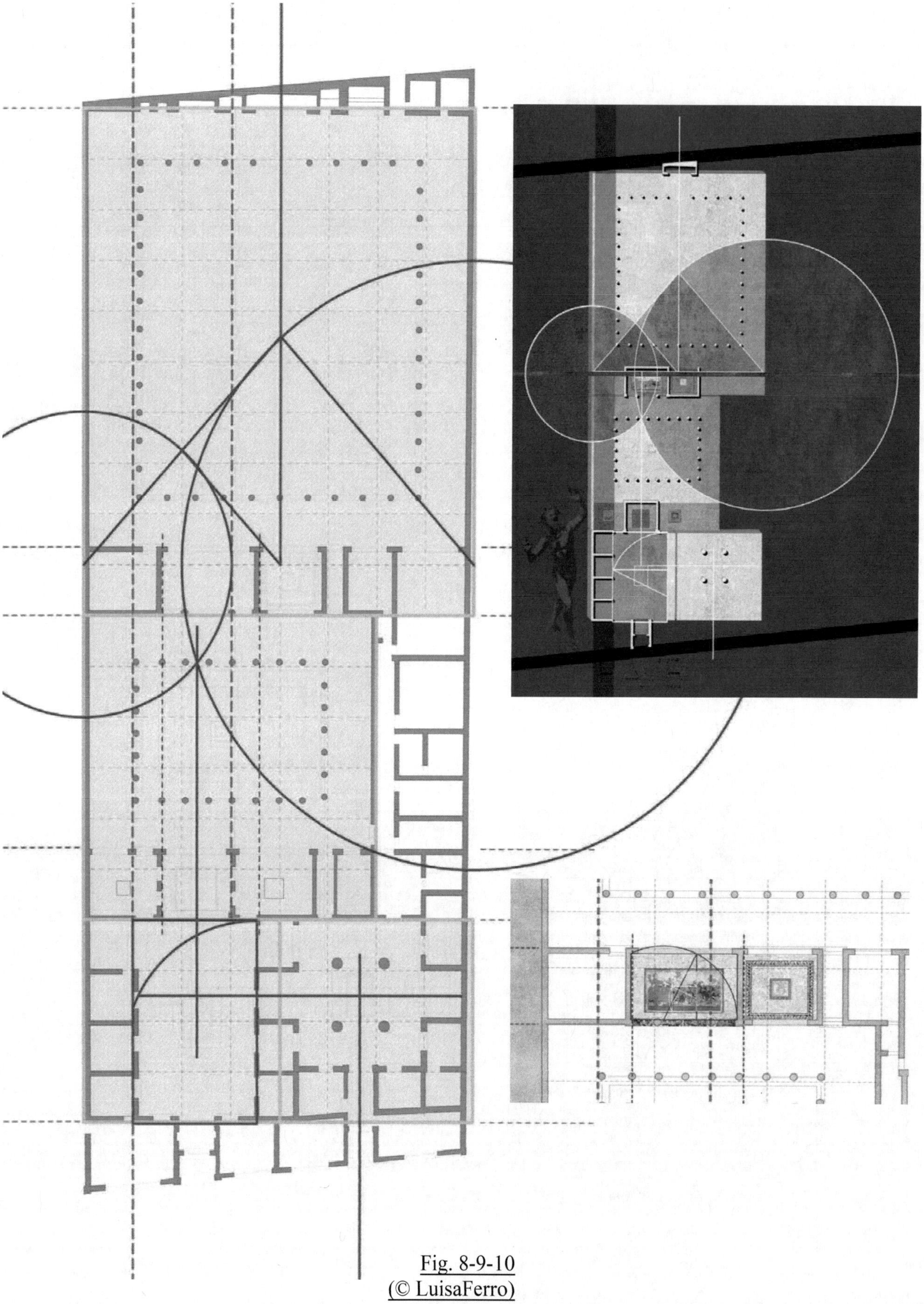

Fig. 8-9-10
(© LuisaFerro)

peristyle (which wrongly seemed arranged without geometric reasons) and the southeast corner of the first peristyle. The radius finally identifies the position of the corner column of the south side of the peristyle, which before seemed to be the only side not coincident with the modular grid.
A smaller circle is constructed with the radius given by the side of a second isosceles triangle with a base from the median of the first triangle to the extreme west of the mosaic alignment. The intersection with the major circle coincides with the axis of the tuscan *atrium*.
- The second column of the distyle hall corresponds to the modular grid.
All this demonstrate the geometric centrality of the *exedra* (distyle room) containing the Alexaender mosaic, a sort of casket that not only contains a very precious artwork, but is the geometric secret of the whole house, its ideal construction. A hinge between the existing (modified) part of the house, the new buildings, the new great Peristyle.
Going back to the construction of the golden proportion of the distyle hall, considring the mosaic in its original position, the average proportion of the room coincides with the golden rectangle of the mosaic on which Dario is depicted, while the axis of the tuscan atrium (around which all the mosaics of the cycle are arranged) coincides major side of the golden rectangle on which Alexander is depicted. **(Fig. 10)**

Acknowledgement

This work is part of a remarkably wider research (entitled *Archeology and Architecture*) that covers the case studies of Athens and Alexandria among many others. Over the last few years, the research has addressed the case study of Pompeii with the cooperation of Professor Massimo Osanna, Director of the Superintendency of Pompeii and Professor Giulio Magli, Director of the Department of Mathematics, Polytechnic of Milan, for their steady help. Finally, E. Ciapparelli and M. Saldarini, R. Acerbis, A. Colombo, M. Felici who helped me with the drawings. All pictures and measure drawings are by the author.

References

[1] ARBORIO MELLA A. Federico 1976: *L'Egitto dei faraoni. Storia, civiltà, cultura* (Milan, Mursia).
[2] BIANCHI BANDINELLI Ranuccio 1980: La pittura antica (Milan, Ghibli).
[3] BRECCIA, A. Evaristo (Naples) 1937-40: *Egitto greco e romano* (Pisa, Nistri-Lischi Editori).
[4] BRECCIA, A. Evaristo 1932: *Le Musée Gréco-Roman 1925-1931*. Municipalité d'Alexandrie (Bergamo, Istituto Italiano d'Arti grafiche).
[5] COHEN, Ada 1997: The Alexander Mosaic. Stories Victory and Defeat, (Cambridge University Press).
[6] DE CARO, Stefano 2001: I mosaici, la casa del Fauno (Naples, Soprintendenza Archeologica di Napoli e Caserta, Electa).
[7] DWYER Eugene 2001: The Unified Plan of the House of the Faun. *Journal of the Society of Architecural Historians* 60 (3).
[8] FABER, A., HOFFMANN, A., Die Casa del Fauno in Pompeji (VI 12) 1, Wiesbaden: Reichert Verlag, 2009 ISBN 9-86888-02

[9] FERROa Luisa MAGLI Giulio 2012: The Astronomical Orientation of the Urban Plan of Alexandria. *Oxford Journal of Archaeology* 31 (4).
[10] FIORELLI, Giuseppe (ed) 1860-64: Pompeianarum Antiquitatum Historia vol. I (Naples).
[11] FRASER P.M. 1972: Ptolemaic Alexandria, vol I (Oxford, Clarendon Press).
[12] GALLO Paolo 2016: se l'Egitto dei Romani è la costa Alessandrina. In *Il Nilo a Pompei. Visioni d'Egitto nel mondo romano* (Modena, Franco Panini Editore).
[13] GRECO Christian OSANNA Massimo, 2016: L'Egitto nella letteratura Greca. In *Il Nilo a Pompei. Visioni d'Egitto nel mondo romano* (Modena, Franco Panini Editore).
[14] GREEN Peter 1924: Alexander to Actium. The Historical Evolution of the Hellenistic Age (Los Angeles, University of California Press 1990).
[15] HOFFMANN, A., Die Casa del Fauno, in Basileia, Berlin, 1996
[16] Magi, Filippo (Treccani 1929) 1958: Filone. *Enciclopedia dell'arte antica classica e orientale* vol. I (Rome Istituto dell'Enciclopedia Italiana).
[17] MAIURI Amedeo 1951: La casa a Pompei. Corso Universitario di Antichità Pompeiane dell'anno 1950-51 (Naples, Raffele Pironti Editore).
[18] Monnier Marc 1864: Pompéi et les Pompéiens, in "Le tour du monde", I semestre.
[19] Moreno, Paolo 2000: Apelle. La battaglia di Alessandro (Milan Skira).
[20] NICCOLINI, Antonio 1852: Quadro in musaico scoperto in Pompei a di 24 ottobre 1831 (Naples, Stamperia Reale).
[21] NICCOLINI, Fausto e Felice 1854-96: Le case ed i monumenti di Pompei disegnati e descritti (Naples).
[22] PAPPALARDO Umberto, CIARDIELLO Rosaria 2010: Mosaici greci e romani. Tappeti di pietra in età ellenistico-romana (Verona, Arsenale Editrice).
[23] PESANDO Fabrizio 1996: Autocelebrazione aristocratica e propaganda politica in ambiente privato: la casa del Fauno a Pompei. *Cahier du Centre Gustave Glotz* 7.
[24] Quatremère de QUINCY, Antoine Chrystosome 1824-38: *Dissertation sur la mosaïque dite d'Alexandr à Arbelles*. In Mazois C.F., *Les Ruines de Pompéi* (4 voll, Paris).
[25] RUSSO Lucio, 2008 (terza edizione): La rivoluzione dimenticata. Il pensiero scientifico greco e la scienza moderna (Milan, Feltrinelli).
[26] TUCCINARDI Margherita 2015: L'albero secco nel mosaico pompeiano di Alessandro Magno. *Engramma. La tradizione classica nella memoria occidentale* 136.
[27] Valdecio Diunilgo 1786: Le donne illustri che nel mondo fiorirono viventi nel nostro secolo esposte in rima da Diunilgo Valdecio Pastor Arcade (Turin, Presso Francesco Prato libraio avanti la chiesa di S. Teresa).
[28] ZANKER Paul 1993: Pompei, Italian translation by Zambrini A. (Turin, Eunaudi)
[29] ZEVI, Fausto 1998: I mosaici della casa del Fauno a Pompei (Naples, Luciano Pedicini Archivio dell'Arte).

Current address

Ferro Luisa, Architect, Associate Professor in Architectural and Urban Composition
Department of Architecture, Built Environment and Construction Engineering
Politecnico di Milano
Piazza Leonardo da Vinci, 32, Milan 20132, Italy
E-mail: luisa.ferro@polimi.it

Slovak University of Technology in Bratislava
Faculty of Mechanical Engineering
17th Conference on Applied Mathematics **APLIMAT 2018**

Proceedings

EVALUATION OF THE QUALITATIVE PARAMETERS OF HUMUS AND SELECTED HEAVY METALS CONTENT

FESZTEROVÁ Melánia (SK), HUDEC Michal (SK), PORUBCOVÁ Lýdia (SK), NOGA Henryk (PL)

Abstract. Heavy metals get into the soil primarily through human and environmental activity. The effects of heavy metal pollution on land are processes that are long-lasting. To assess the relationship between chemical factors and heavy metal contents in the soil we used Pearson correlation analysis, which was tested at $p < 0.05$ and $p < 0.01$. The content of Pb and Cd heavy metals were compared using a Student t-test to calculate the differences between the averages to a significance level of $\alpha = 0.01$.

Keywords: Pearson correlation analysis, Student t-test, heavy metals, lead, cadmium, humic substances, Kremnické and Štiavnické mountains

Mathematics Subject Classification: Primary 62F; Secondary 92E

1 Introduction

Inorganic impurities are primarily introduced into the soil through human activity or the activity of the natural environment. According to Apea and Ephraim (2012) and Perelomova (2011), the distribution of trace elements in the soil is related to the sorption process. The impact of heavy metal soil pollution is long-term due to the relatively strong adsorption of many metals by humus and the impact of anthropogenic environmental factors and rock structure. Volcanic rocks are responsible for the presence of metals in soil and water (Aiuppa et al., 2000; Kelepertsis et al., 2001). According to Delmelle and Stix (2000) and Durand et al. (2004), it is volcanic activity that is responsible for the release of metals such as lead (Pb), aluminium (Al), magnesium (Mg), copper (Cu) and zinc (Zn), among others.

In the case of contaminated soils, it is very important that heavy metals are not absorbed by plants, not only on agricultural land but also on meadows and pastures, due to the fact that heavy metals can then enter the food chain. Humic substances can help improve soil properties, increase production productivity and improve nutrient intake (Tobiášová, 2017). Organic soil inputs such as mature compost containing a high proportion of the aggregate organic matter may help reduce the bioavailability of heavy metals in the soil by adsorbing them and forming stable complexes with humic substances (Takamatsu & Yoshida, 1978). These procedures and processes allow the restoration of vegetation on contaminated sites (Tordoff et al., 2000).

It is important to examine the mechanisms of sorption and the distribution of heavy metals Pb and Cd in the soil between the fractions of humic substances. Burlakovs et al. (2013) claim that humic substances have a good ability to reduce the content of bioavailable heavy metals, and the ability to form complexes with metal ions depends on the type of soil, the type of metal, and the concentration of humus in the soil.

The aim of this paper is to describe the characteristics of the qualitative parameters of humus and the selected heavy metal content (Pb, Cd) in the soil. To assess the relationship between chemical factors and heavy metal contents in the soil we used Pearson correlation analysis, which was tested at $p < 0.05$ and $p < 0.01$ levels. The content of lead and cadmium heavy metals extracted using HNO_3 ($c_{HNO3} = 2$ mol.dm^{-3}) and aqua regia was compared using a Student t-test to calculate the differences between the averages to a significance level of $\alpha = 0.01$.

2 Material and Methods

The geological composition of the monitored area is a set of volcanic rocks represented mainly by andesites, rhyolites, tuffs and volcanic breccia. The geological footwall is composed from the lava flows of pyroxenic and pyroxene-amphibolic andesites.

The analyzed soil samples were collected in the form of open borrow pits in order to determine the profile distribution of total organic carbon, in the autumn of 2011, 2012, 2013 and in the spring of 2012, 2013 and 2014. There was selected area 5 x 5 m on each location and it was realized (conducted) soil sampling in 6 different sampling sites. The opened borrow pits are used for soil for determination of soil characteristics, such as soil profile description, their depth, the stratigraphy (the arrangement of genetic horizons) and morphological characteristics. The study was conducted at Kremnické and Štiavnické vrchy Mountains (Fig. 1.). Kremnické and Štiavnické vrchy Mountains are volcanic mountain ranges located in central Slovakia. Soil samples were collected from 3 main soil types (Eutric Andosol, Eutric Cambisol, Eutric Planosol) of the selected horizons.

Soil was air-dried at room temperature and sieved (< 2 mm). We used to analysis selected parameters in soil samples and a sample was taken using standard procedures.
Total organic carbon (TOC) was measured using the Tjurin method modified by Nikitina according to Orlov and Grišina (1981) and from that we calculated the humus content (Humus = TOC • 1.724).
Soil reaction was determined in distilled water as active soil reaction (pH_{H2O}) and in a solution of 1 mol.dm^{-3} KCl as exchange soil reaction (pH_{KCl}). The ratio of soil and solution was 1 : 2.5 (van Reeuwijk, 2002).
The content of humic substances (HS), as well as the ratio humic acid : fulvic acid (HA : FA) were determined by group composition of humic substances using the Belčiková-Kononová method (Kononová & Belčiková, 1962). The humification degree (DH) of humified

substances was calculated from the relation DH=CHA /TOC.100 [%] (Grišina, 1986). Humic substances were extracted into solution $Na_4P_2O_7$ (c = 0.1 mol.dm^{-3}) and adjusted to pH = 13 with NaOH (c = 1.0 mol.dm^{-3}) and the samples were left for infusion for 24 hours. UV/VIS spectra were measured by Varian Cary 50 spectrophotometer in the range from 300 nm to 700 nm.

Quantitative determination of heavy metals (lead, cadmium) was performed by ET-AAS technique on atomic absorption spectrometer Spectra AA–200 (Varian, Mulgrave Virginia, Australia) equipped with deuterium background correction with GTA-100 module.

Potentially bioavailable forms of lead (Pb) and cadmium (Cd) were extracted with the solution 2 mol.dm^{-3} nitric acid (HNO_3) in a 1:10 (soil/ nitric acid).

Total concentration of heavy metals was determined using aqua regia (aqua regia = HCl : HNO_3 in a 3 : 1) extraction method. 3 g of soil sample had been digested for 2 h. at 180 °C (ISO 11466:1995).

The measured data was statistically processed using the software STATISTICA.

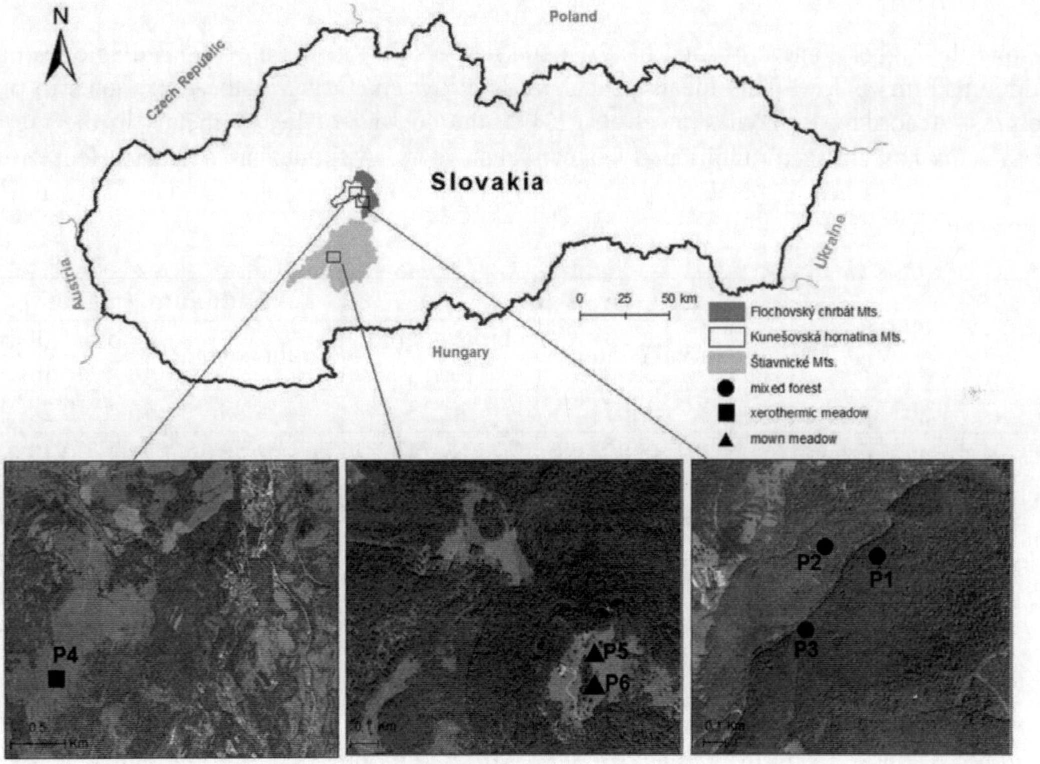

Fig. 1. Localisation of sampling sites.

3 Results and Discussion

Biologically available forms of lead and cadmium in the extract of 2 mol.dm^{-3} HNO_3 are assessed in to the decision of the Ministry of Agriculture No. 531/1994-540 “Limit values of hazardous inorganic substances in soil”. The limit with regard to lead content is 30 mg.kg^{-1} while for cadmium it is 0.3 mg.kg^{-1} of dry matter. The total lead content in the soil samples ranged from 30.1 mg kg^{-1} to 65.3 mg kg^{-1}, while the cadmium content ranged from 1.4 mgkg^{-1} to 5.3 mg kg^{-1} (Tab. 1.). This means that the lead and cadmium content was above the limit values in the soil samples.

In the case of the total lead content in the aqua regia extract, the limit values were not exceeded. The limit value for lead in the a extract is, according to the Act of the National Council of the Slovak Republic No. 220/2004 Coll. on the Protection and Use of Agricultural Land and on the Amendment of Act No. 245/2003 Coll. on Integrated Prevention and Control of Environmental Pollution and on amendments to certain acts for sand-clay and clay soil, 70 mg.kg^{-1}. Also the limit value of cadmium in the aqua regia effluent for sand-clay and clay soils - 0.7 mg.kg^{-1} - was not exceeded (Tab. 1.).

The content of potentially bioavailable forms of lead in the soil samples ranged from 10.8 mg.kg^{-1} to 38.1 mg.kg^{-1}, while the cadmium content ranged from 0.05 mg.kg^{-1} to 0.27 mg.kg^{-1} (Tab. 1.). The content of potentially bioavailable forms of lead and cadmium in the soil samples are low, and they do not exceed the limit values.

Despite the higher content of total heavy metal in the soil, the share of bioavailable lead and cadmium forms is low. This means that there is a low entry risk of these elements in plant biomass. According to Tomaškin et al. (2013), the content of heavy metals in the rays of ecosystems has changed significantly over recent years. Altitude has a great effect on the concentrations of heavy metals in the soil as their content increases with increasing altitude (Tomaškin et al., 2013).

Profile	**Soil types**	**Lead [mg.kg^{-1}]**		**Cadmium [mg.kg^{-1}]**	
		total content	bioavailable forms	total content	bioavailable forms
P1	Eutric Andosols	50.07	29.23	4.83	0.24
P2	Eutric Cambisols	45.23	16.23	2.78	0.12
P3	Eutric Planosols	45.85	14.25	1.95	0.09
P5	Eutric Cambisols	51.53	22.83	4.80	0.16
P6	Eutric Cambisols	50.60	19.07	3.13	0.19

Tab. 1. Average heavy metals content in soil.

The average content of cadmium in the Slovak soil is below 1.0 mg.kg^{-1}. Compared to lead, cadmium belongs among the elements that are strongly mobile in acidic soil, whereas lead is not very mobile (Árvay et al., 2012).

	Pb_{HNO3}	**Cd_{HNO3}**	**Pb_{AR}**	**Cd_{AR}**
pH_{H2O}	-0.443	-0.450	-0.437	-0.374
pH_{KCl}	0.309	0.082	0.364	0.287

$p < 0.05$ pH_{H2O} – active soil reaction, pH_{KCl} – exchange soil reaction, Pb_{HNO3} – lead in leachate 2 mol.dm^{-3} HNO_3, Cd_{HNO3} – cadmium in leachate 2 mol.dm^{-3} HNO_3, Pb_{AR} – lead in the aqua regia extract, Cd_{AR} – cadmium in the aqua regia extract

Tab. 2. Correlation between heavy metal content and soil reaction.

This assertion is consistent with the results of the correlation analysis. There was a negative correlation that was not statistically significant (Tab. 2.) between the values of the active soil reaction and lead and cadmium content in the aqua regia extract and in the HNO_3 solution ($c_{HNO3} = 2$ mol.dm^{-3}). The negative influence of the acidic soil reaction on the increased mobility of heavy metals was confirmed by Styk (2000).

We have found a positive correlation between the chemical factors of humic substances and the heavy metal content in soil (Tab. 3.) using Pearson correlation. The bond formed between heavy metals and organic matter in the soil tends to be relatively strong, with its strength decreasing in the order of $Cr^{3+} > Pb^{2+} > Cu^{2+} > Ni^{2+} > Cd^{2+} > Zn^{2+} > Mn^{2+}$ (Kováčik, 2000). According to Tomáš et al. (2001), the mobility and immobility of metals in soils influences the soil reaction and the make up of humic substances. The total amount of organic carbon is positively correlated with the total cadmium content ($r = 0.493$, $p < 0.05$) and with bioavailable lead forms ($r = 0.708$, $p < 0.01$) and with cadmium ($r = 0.734$, $p < 0.01$). (Tab. 3.) The content of humus substances was positively correlated with the bioavailable forms of lead ($r = 0.709$, $p < 0.01$) and with cadmium ($r = 0.692$, $p < 0.01$). We found a positive correlation between the content of humic acids and the bioavailable form of cadmium ($r = 0.499$, $p < 0.05$), and between the content of fulvic acids and bioavailable forms of lead ($r = 0.743$, $p < 0.01$) and cadmium ($r = 0.700$, $p < 0.01$) (Tab. 3). According to Tobiášová (1996), the stability of metal complexes with humic substances decreases in the order $Cd^{2+} > Zn^{2+} > Ni^{2+} > Cu^{2+} > Pb^{2+}$. The stability of the complexes depends strongly on pH values, with more stable complexes occurring in weakly acidic and neutral areas (Mestek & Volka, 1993).

Parameters	pH_{H2O}	pH_{KCl}	**Pb** Total content	**Cd** Total content	**Pb** Bioavailability forms	**Cd** Bioavailability forms
TOC	..-0.757**	.-0.521*	.0.344	...0.493*	0.708**	0.734**
HS	..-0.795**	.-0.568*	.0.299	.0.448	0.709**	0.692**
HA	-0.507*	.-0.496*	.0.299	.0.214	0.352	0.499*
FA	..-0.813**	.-0.559*	.0.290	.0.471	0.743**	0.700**
Q_{HL}	..-0.804**	.-0.585*	.0.343	.0.429	0.734**	0.664**
HA:FA	0.140	.0.153	.0.075	-0.443	-0.448	-0.228
DH	0.087	-0.090	-0.057	...-0.774**	-0.635**	-0.449

* $p < 0.05$ ** $p < 0.01$ pH_{H2O} – active soil reaction, pH_{KCl} – exchange soil reaction, Pb – lead, Cd – cadmium, TOC – total organic carbon, HS – humic substances, HA – humic acids, FA – fulvic acids, Q_{HL} – color quotient of humic substances, DH – humification degree

Tab. 3. Correlation between carbon parameters, soil reaction and heavy metal content.

Figure 2. shows the impact of the exchange soil reaction on the concentration of lead and cadmium heavy metals. Dependency is represented by point graphs translated using polynomial with regression equation. From the data obtained, it can be deduced that lead shows higher phytotoxicity at low values of the exchange soil reaction ($pH_{KCl} = 4.1 - 5.3$). The results found correspond with the results of Árvay et al. (2012), indicating an interchange interval from 4.4 to 5.6. In the case of cadmium, its phytotoxicity is fairly equal at all exchange soil response values. For all elements, we recorded positive correlations between the comparison parameters, which were not statistically significant (Tab. 2.). This does not correspond to the results of Huynh et al. (2008) and Pehlivan et al. (2009). However, a negative correlation was found in the case of an active soil reaction. However, this was also not statistically significant (Tab. 2.).

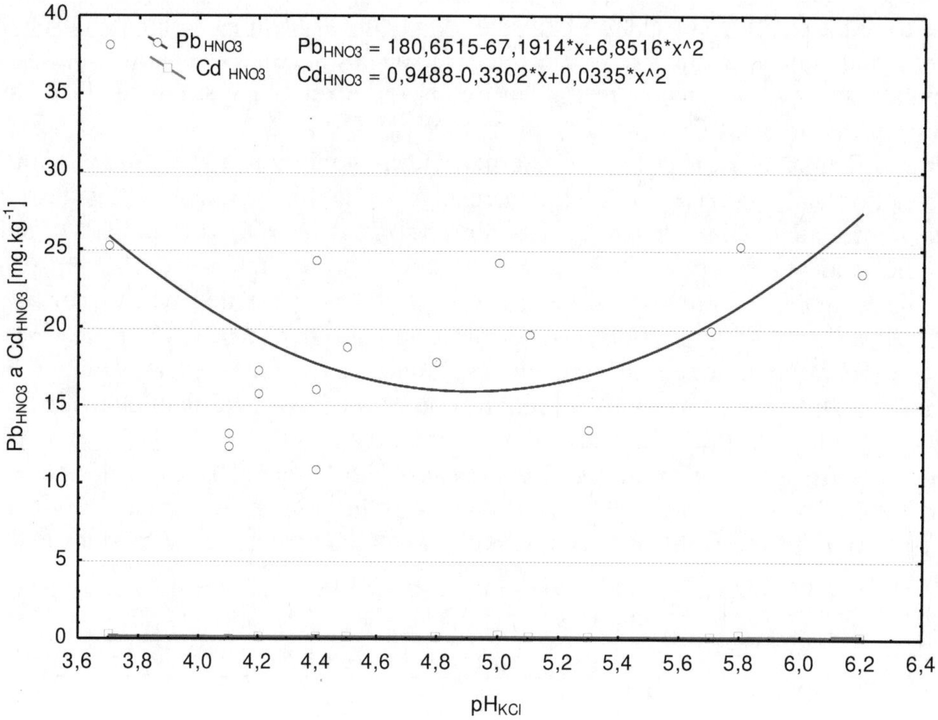

Fig. 2. Concentration of mobile forms Pb, Cd in the soil of the monitored sites depending on the exchange soil reaction values.

A correlation relationship was also found between the bioavailable forms of heavy metals and the color quotient of humic substances. (Tab. 3.)

The heavy metal contents in the extracted fractions are shown in Tab. 4. By statistical testing, significant differences were found between the cadmium and lead contents extracted with HNO_3 ($c_{HNO3} = 2$ mol.dm^{-3}) and that extracted with aqua regia, with Cd ($t = -4.516$, $p = 0.001$) and Pb ($t = -10.086$, $p = 0.000$).

The results also show the high sorption capacity of heavy metals to humic acids. The heavy metal distribution in the humic acid fraction (HA) was in the order of Pb > Cd, and in the fulvic acid fraction (FA) in the order Cd > Pb, with no statistically significant difference between the lead contents in HA and FA. The complexes Pb and Cu with humic acids are more stable than the Cd and Zn complexes (Waller & Pickering, 1993; Pinheiro et al., 1994; Gao et al., 1999). Higher Cd content bound to FA shows that this fraction is affected by the level of pollution (the Cd content to HA was lower). This might indicate a not very different sorption mechanism of Pb to FA or the presence of specific sorption sites on FA.

Fractions of heavy metals	Parameters	Pb	Cd
Aqua regia (total content heavy metals)	Average	48.49	3.46
	SD	9.69	1.25
2 mol.dm^{-3} HNO_3, extract (content of bioavailability forms of heavy metals)	Average	20.13	0.16
	SD	6.25	0.06
NaOH/$Na_4P_2O_7$extract	Average	2.56	0.02
	SD	3.46	0.01
Humic acids	Average	0.88	0.01
	SD	1.03	0.02
Fulvic acids	Average	0.56	0.07
	SD	0.33	0.05

Tab. 4. Average values and standard deviation (SD) of the extracted fractions of heavy metals in soil (mg.kg^{-1}).

The results according to Adekunle et al. (2007) show that the concentration of the metals bound to humic acids was in the order Pb > Cd, which corresponds to our results. Angehrn-Bettinazzi et al. (1989) report the high ability of lead to form complexes with insoluble humus substances, whereas cadmium and zinc form complexes with low molecular weight organic molecules.

The cadmium content for HA and FA was very low in most of the analyzed soil samples. This may cause an error in interpreting the results. Árvay et al. (2012) report the ability of humus to bind and favour Cd rather than Pb, which could affect its accumulation and the binding strength of humus substances. In her study, Barančiková (2009) found significant differences in the sorption of Cd to humic acids for different soil types and pH values. The lead content in the FA fraction correlated positively with the fulvic acid content in the soil ($r = 0.582$, $p < 0.05$) and correlated negatively with the values of the active soil reaction ($r = -0.599$, $p < 0.05$) (Tab. 5.). These results are confirmed by Adriano (2001), who states that lead forms stable compounds with soil organic matter at low pH values. Low pH values increase the lead sorption on both fractions of humus substances. The strength of the metals that are complex-bound to humic acids is influenced by soil pH values and ionic strength (Ladonin & Margolin, 1997; Adriano, 2001; Evangel & Marsi 2001).

	Pb	Cd	Pb_{HA}	Cd_{HA}	Pb_{FA}	Cd_{FA}
	NaOH/$Na_4P_2O_7$ extract		Humic acids		Fulvic acids	
TOC	0.657*	0.715*	0.728*	0.065	0.534*	0.571*
HA	0.695*	0.724*	0.730*	-0.088	0.269	0.643*
FA	0.637*	0.725*	0.731*	0.066	0.582*	0.564*
pH_{H2O}	-0.260	-0.263	-0.220	0.081	-0.599*	-0.170
pH_{KCl}	-0.126	-0.048	0.297	0.508	0.571*	0.168

Tab. 5. Correlation between the heavy metal contents bound in HA and FA and the chemical properties of the soil.

Soil contamination involving heavy metals is long-term due to the relatively strong adsorption of many metals to humus, and also to the influence of anthropogenic factors of the environment and the structure of rocks. The TOC content was positively correlated with all forms of heavy metals with the exception of cadmium in the humic acid fraction ($r = 0.065$; $p < 0.05$). This was because the cadmium content in humic acids was very low. This could increase the relative error and the interpretation of the results (Tab. 5.).

The contents of both heavy lead and cadmium metals in the FA fraction are closely correlated with the total metal content in the extract of aqua regia, Pb ($r = 0.662$, $p < 0.05$) and Cd ($r = 0.644$; $p < 0.05$). The content of Pb and Cd in the aqua regia extract was positively correlated with the content of metals found in humic substances, Pb ($r = 0.638$; $p < 0.05$) and Cd ($r = 0.717$, $p < 0.05$). There was no significant correlation at $p < 0.05$ between the heavy metal content in the aqua regia and Pb ($r = 0.474$) and Cd ($r = 0.305$) content in humic acids. A correlation was also found between the lead content extracted with HNO_3 ($c_{HNO3} = 2$ mol.dm^{-3}) and the content of the lead extracted with aqua regia ($r = 0.985$) and tetrasodium pyrophosphate ($Na_4P_2O_7$) ($r = 0.674$) at the $p < 0.05$ level. A positive correlation was also found between the lead content in the aqua regia and in the tetrasodium pyrophosphate ($r = 0.638$; $p < 0.05$). The cadmium content found in the tetrasodium pyrophosphate did not correspond to the cadmium content in a solution of 2 mol.dm^{-3} HNO_3 ($r = 0.126$) and in aqua regia ($r = 0.305$) for $p < 0.05$. A positive correlation was found only between the cadmium extracted with HNO_3 ($c_{HNO3} = 2$ mol.dm^{-3}) and aqua regia ($r = 0.907$; $p < 0.05$).

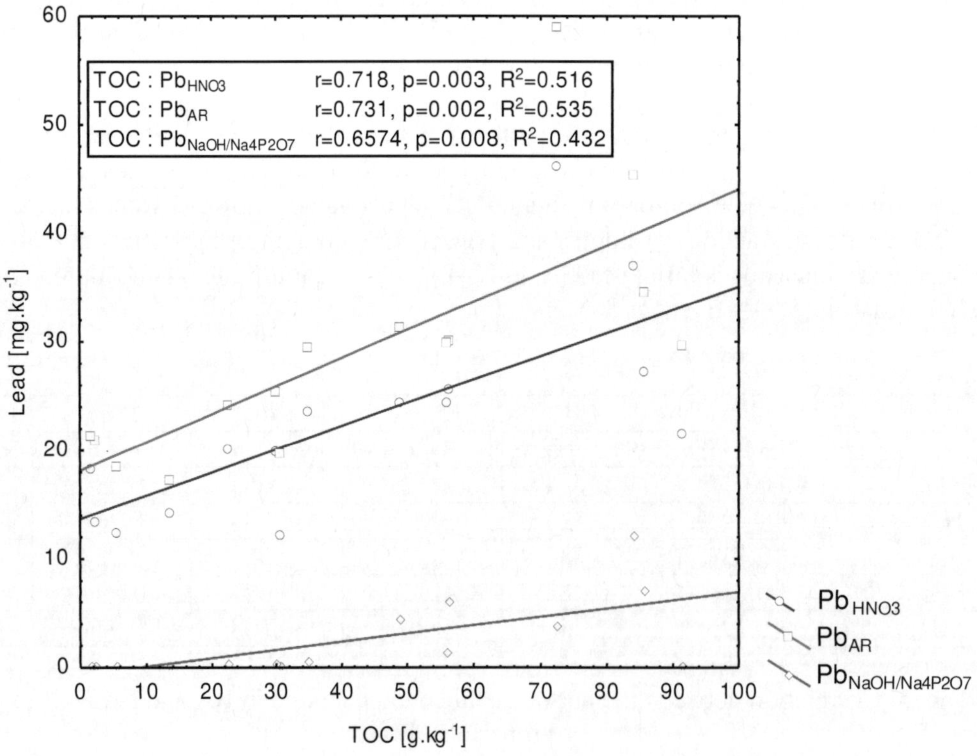

Fig. 3. Correlation between TOC and lead content in different extracts.

The contents of selected heavy metals (Pb, Cd) found in leachate 2 mol.dm^{-3} HNO_3 and in leachate aqua regia and tetrasodium pyrophosphate leachate (NaOH / $Na_4P_2O_7$) also had a positive effect on the total organic carbon content of the soil (Fig. 3. and Fig. 4.). Several studies show that organic matter can be used as a biosorbent due to its high sorption capacity, and the formation of insoluble complexes which prevent the intake of heavy metals by plants (Wang et al., 2007; Bose & Bhattecharyya, 2008).

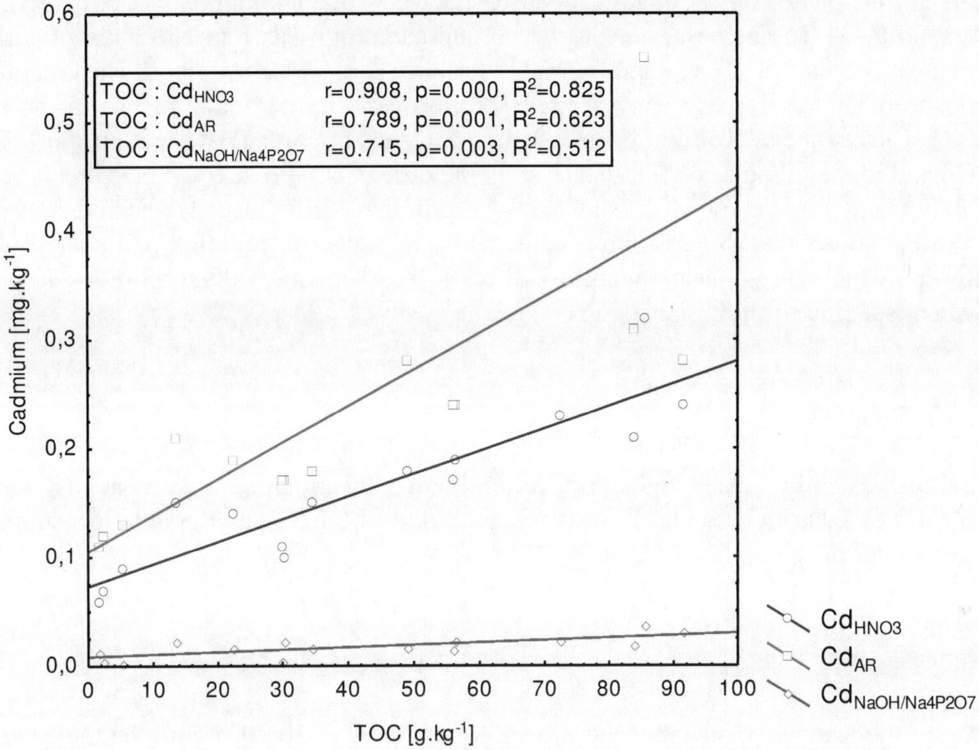

Fig. 4. Correlation between TOC and cadmium content in different extracts.

The strong bond between soil organic matter and heavy metals in soil is described by Deka and Sarma (2012). According to McBride et al. (1999), the higher the amount of dissolved organic matter in the soil the greater the increase in the mobility of metals in the soil / soil solution. Ashworth and Alloway (2004) claim that soluble organic matter reduces sorption but increases the mobility of heavy metals in the soil. According to Sherene (2010), soil organic matter is important for the absorption of heavy metals into the soil. This agrees with the results we have achieved. The increase in solubility of Pb in soil relates to the dissolution of humic acids (HA) as components of organic matter. This means that when the organic material solution is specified, the distribution of metals at a higher pH is dominant This is because, at a high pH values, the TOC in the soil increases due to the presence of dissolved humic acids (Sherene, 2010).

4. Conclusion

High content of lead were found in the studied soils, presumably due to human activities or the activities of the natural environment. Humic substances can contribute to improving the adverse soil properties. The ability of humic substances to form stable complexes with soil inorganic and organic compounds plays a key role in environmental pollution and also in migration and transformation of pollutants. By statistic tests significant differences of Cd ($t = -4.516$, $p = 0.001$) and Pb ($t = -10.086$, $p = 0.000$) were found between the cadmium and lead content extracted with 2 $mol.dm^{-3}$ HNO_3 and extracted from the aqua regia. TOC has a positive effect on all forms of heavy metals extracted with aqua regia 2 $mol.dm^{-3}$ HNO_3 and $NaOH$ / $Na_4P_2O_7$. The highest content of lead and cadmium was found in andosol soil with the highest content of TOC, and humic substances, as well as the lowest degree of humification. The results also show a high sorption capacity of heavy metals to the humic acids. The heavy metal distribution in the humic acids fraction (HA) was in the order of Pb > Cd and in the fulvic acids fraction (FA) in the order Cd > Pb, taking between contents of lead in HA and FA we found no statistically significant difference. Higher content of Cd bound to FA shows that this fraction is affected by the level of pollution (Cd content to HA was lower), which may indicate a slightly different mechanism of Pb sorption on FA or the presence of specific sorption sites on FA.

Acknowledgement

The authors gratefully acknowledge the Cultural and Educational Grant Agency (KEGA) of the Ministry of Education, Science, Research and Sport of the Slovak Republic for supporting this work under the Grant No. 044UKF-4/2017.

References

[1] Act of the National Council of the Slovak Republic No. 220/2004 Coll. on the Protection and Use of Agricultural Land and on the Amendment of the Act No. 245/2003 Coll. on Integrated Prevention and Control of Environmental Pollution and on amendments to certain acts.

[2] ADEKUNLE, I. M., AROWOLO, T. A., NDAHI, N. P., BELLO, B., OWOLABI, D. A., Chemical characteristics of humic acids in relation to lead, copper and cadmium levels in contaminated soils from south west Nigeria. In *Annals of Environmental Science*, vol. 1, 2007: p. 23 - 34.

[3] ADRIANO, D. C., *Trace elements in terrestrial environments*. Biogeochemistry, bioavailability, and risks of metals. 2nd ed. New York: Springer-Verlag, 2001.

[4] AIUPPA, A., ALLARD, P., D'ALESSANDRO, W., MICHEL, A., PARELLO, F., TREUIL, M., VALENZA, M., Mobility and fluxes of major, minor and trace metals during basalt weathering and groundwater transport at Mt. Etna volcano (Sicily). In *Geochim. Cosmochim. Acta*, vol. 64, 2000: p. 1827-1841.

[5] ANGEHRN-BETTINAZZI, C., THONI, L., HERTZ, J., An attempt to evaluate some factors affecting the heavy metal accumulation in a forest stand. In *Int. J. Environ. Anal. Chem.*, vol. 35, 1989: p. 69 - 79.

[6] APEA, O. B., EPHRAIM, J. H., Effect of humic acid on the kinetics and mechanism of

copper adsorption in soil-solution system. In *Journal of Applied Sciences in Environmental Sanitation*, vol. 7, 2012: p. 137-146.

[7] ÁRVAY, J., BAJČAN, D., TOMÁŠ, J., *Kvalita zložiek životného prostredia v alúviu rieky Štiavnica*: vedecká monografia. 1. vyd. Nitra: SPU, 2012, ISBN 978-80-552-0817-6, 221 s.

[8] ASHWORTH, D. J., ALLOWAY, B. J. 2004. Soil mobility of sewage sludge–derived dissolved organic matter, copper, nickel and zinc. In *Environ. Pollut.*, vol. 127, no. 1, p. 137-144.

[9] BARANČÍKOVÁ, G., Proposal of evaluation of soil organic matter sorption capacity. In *Humic Substances in Ecosystems 8*. Nitra: SPU; Bratislava : VÚPOP, 2009, pp. 11 - 16. ISBN 978-80-89128-60-0.

[10] BOSE, S., BHATTACHARYYA, A. K. Heavy metals accumulation in wheat plant grown in soil amended with industrial sludge. In *Chemosphere,* vol. 70, 2008: p. 12.

[11] BURLAKOVS, J., KLAVINŠ, M., OSINSKA, L., PURMALIS, O., The Impact of Humic Substances as Remediation Agents to the Speciation Forms of Metals in Soil. In APCBEE Procedia, vol. 5, 2013: p. 192-196.

[12] Decision of the Ministry of Agriculture No. 531/1994-540 "Limit values of hazardous inorganic substances in soil".

[13] DEKA, J., SARMA, H. P., Heavy metal contamination in soil in an industrial zone and its relation with some soil properties. In *Archives of Applied Science Research*, vol. 4, no. 2. 2012: p. 831-836.

[14] DELMELLE, P., STIX, J., Volcanic gases. In: *Encyclopedia of Volcanoes*, 2000. pp. 803-816, (Sigurdsson, H., Houghton, B. F., McNutt, S. R., Rymer, H., and Stix, J., Eds), San Diego, Academic Press.

[15] DURAND, M., FLORKOWSKI, C., GEORGE, P., WALMSLEY, T., WEINSTEIN, P., COLE, J., Elevated trace element output in urine following acute volcanic gas exposure. In *J. Volcanol. Geotherm.*, vol. 134, 2004: p. 139 - 148.

[16] EVANGELOU, V.P., MARSI, M., Composition and metal ion complexation behavior of humic fractions derived from corn tissue. In *Plant Soil*, vol. 229, 2001: p. 13 -24.

[17] GAO, K.Z., PEARCE, J., JONES, J., TAYLOR, C., Interaction between peat, humic acid and aqueous metal ions. In *Environ. Geochem. Helth*, vol. 21, 1999: p. 13 - 26.

[18] HUYNH, T. T., LAIDLAW, W. S., SINGH, B., GREGORY, D., BAKER, A. J. M., Effect of phytoextraction on heavy metal concentration and pH of pore-water of biosolids determined using an in situ sampling technique. In Environmetal pollution, vol. 156, Issue 3, 2008: p. 874-882.

[19] ISO 11466, Soil quality – Extraction of trace elements soluble in aqua regia. International Organization for Standardization, Geneva, Switzerland, 1995: 6 p.

[20] KELEPERTSIS, A., ALEXAKIS, D., KITA, I., Environmental geochemistry of soils and waters of Susaki area, Korinthos, Greece. In *Environ. Geochem. Health,* vol. 23, 2001: p. 117-135.

[21] KOVÁČIK, J., *Rizikové faktory potravového reťazca človeka*. Nitra: SPU Nitra, 2000, 143 s. ISBN 80-7137-796-1.

[22] KONONOVÁ, M. M., BELČIKOVÁ, N. P, Uskorennyje metody opredelenija sostava gumusa minerálnych počv. In *Počvovedenije*, 10, 1962: p. 75- 87.

[23] LADONIN, D.V., MARGOLINA, S. E., Interaction between humic acids and heavy metals. In *Eurasian Soil Sci.*, vol. 30, 1997: p. 710 - 715.

[24] McBRIDE, M. B., RICHARDS, B. K., STEENHUIS, T., SPIERS, G., Long-term leaching of trace elements in a heavily sludge-amended silty clay loam soil. In *Soil Sci.*, vol. 164, no. 9. 1999: p. 613-623.
[25] MESTEK, O., VOLKA, K., Interakce těžkých kovů s přírodními složkami. In *Chemické listy*, roč. 87, č. 11, 1993: s. 795-801.
[26] ORLOV, D. S., GRIŠINA, L. A., Praktikum po chemiji gumusa. Moskva : Izdateľstvo Moskovskovo univesiteta, 1981: 272 p.
[27] GRIŠINA, L. A., *Gumusoobrazovanije i gumusnoe sostojanije počv*. Moskva, Izd. MU, 1986, 242 p.
[28] PEHLIVAN, E., ALTUN, T., PARLAYICI, S., Utilization of barley straws as biosorbents for Cu^{2+} and Pb^{2+} ions. In *Journal of Hazardous Materials*, vol. 164, 2009: p. 982-986.
[29] PERELOMOV, L. U., PINSKIY, D. L., VIOLANTE, A., Effect of organic acids on the adsorption of copper, lead and zinc by goethite. In *Eurasian Soil Science*, vol. 44, 2011: p. 22 - 28.
[30] PINHEIRO, J. P., MOTA, A. M., GONCALVES, M. L. S., Complexation study of humic acids with cadmium(II) and lead(II). In *Anal. Chim. Acta*, vol. 284, 1994: p. 525 - 537.
[31] SHERENE, T., Mobility and transport of heavy metals in polluted soil environment. In *Biological Forum - An International Journal*, vol. 2, no. 2. 2010: p. 112-121.
[32] STYK, J., Problém ťažkých kovov (kadmium, olovo, meď, zinok) v pôdach Štiavnických vrchov a ich príjem trávnymi porastmi. In *Pedodisertation*, Bratislava : VÚPOP. 2001, 136 s.
[33] TAKAMATSU, T., YOSHIDA, T., Determination of stability constants of metal-humic acid complexes by potentiometric titration and ion-selective electrodes. In *Soil Sciences*, vol. 125, 1978: p. 377 - 386.
[34] TOBIÁŠOVÁ, E., *Komplexy humínových kyselín a fulvokyselín s ťažkými kovmi*: DP, Nitra: SPU Nitra, 1996, 86 s.
[35] TOBIÁŠOVÁ, E., *Organický podiel pôdy Percento pe život,* Nitra: SPU Nitra, 2017, 87 s.
[36] TOMÁŠ, J., HRONEC, O., TÓTH, T., ANDREJOVSKÝ, P., ADAMIŠIN, P., LAZOR, P., DAŇOVÁ, M., MUŠECINOVÁ, A., *Poškodzovanie pôd a rastlín ľudskými činnosťami* : vedecká monografia. Nitra: SPU, 2007, 110 s. ISBN 978-80-8069-902-4.
[37] TOMÁŠ, J., TÓTH, J., LAZOR, P., TÓTH, T., VOLLMANOVÁ, A., Heavy metals content and distribution in soil in relation to soil hygiene. In *Poľnohospodárstvo (Agriculture),* vol. 47, no. 1. 2001:p. 11-26.
[38] TOMAŠKIN, J., TOMAŠKINOVÁ, J., KMEŤOVÁ, J., DRIMAL, M., The Concentration of Heavy Metals in Grassland Ecosystems of th Central Slovakia National Parks. In *Carpathian Journal of Earth and Environmental Sciences*, vol. 8, no. 4, 2013: p. 35 - 40.
[39] TORDOFF, G. M., BAKER, A. J. M., WILLIS, A. J., Current approaches to the revegetation and reclamation of metalliferous mine wastes. In *Chemosphere*. vol. 41, 2000: p. 219 - 228.
[40] VAN REEUWIJK, L. P., *Producer for Soil Analysis*. International Soil reference and Information Centre, Wageningen. 2002.
[41] WALLER, P A., PICKERING, W. F., The effect of pH on the lability of lead and cadmium sorbed on humic acid particles. In *Chem. Spec. Bioavailab.*, vol. 5, 1993: p. 11 - 22.

[42] WANG, X., YANG, K., TAO, S., WING, B., Sorption of aromatic organic contaminants by biopolymers: effect of pH, copper (II) complexation and cellulose coating. In *Environmental Scientific Technology*, vol. 41, no. 1, 2007: p. 185-191.

Current address

Feszterová Melánia, doc. Ing., PhD.
Faculty of Natural Sciences
Constantine the Philosopher University
Tr. A. Hlinku 1, 949 74 Nitra, Slovak Republic
Tel. number: 00 421 903 456 414, e-mail: mfeszterova@gmail.com

Hudec Michal, PaedDr., PhD.
Piaristické gymnázium sv. Jozefa Kalazanského
Piaristická 6, 949 01 Nitra, Slovak Republic
Tel. number: 00 421 907 615 135

Porubcová Lýdia, Ing.
Faculty of Natural Sciences
Constantine the Philosopher University
Tr. A. Hlinku 1, 949 74 Nitra, Slovak Republic
Tel. number: 00 421 903 456 414

Noga Henryk, dr. hab. professor UP
Faculty of Mathematics, Physics and Technical Science
Pedagogical University of Cracow
Podchorążych 2, 31-015 Cracow, Poland
Tel. number: 00 48 604 895 789

Slovak University of Technology in Bratislava
Faculty of Mechanical Engineering
17th Conference on Applied Mathematics **APLIMAT 2018**

Proceedings

CRYSTAL, MOLECULAR AND ELECTRONIC PROPERTIES OF (*E*)-1-((5´-BROMO-[2,2´-BITHIOPHEN]-5-YL)METHYLENE)-2-(PENTAFLUOROPHENYL)HYDRAZINE

FRONC Marek (SK), SIVÝ Július (SK), BORTŇÁK Dušan (SK), VÉGH Daniel (SK), KRÁLIK Marián (SK)

Abstract. The pentafluorophenyl group is an important constituent not only in biological chemistry, but also in analytical, material and polymer chemistry, catalysis, dynamic combinatorial chemistry, and reaction development. We reported here the structures of new Schiff base 1-((5´-bromo-[2,2´-bithiophen]-5-yl)methylene)-2-(pentafluorophenyl) hydrazine (I), $C_{15}H_6BrF_5N_2S_2$, which crystallizes as racemate in the space group P-1. The C— S, C— F and the N—H bonds are presented. The dihedral angle between the thiophene rings is 3.33 (0.19)° and the side ring (C1 to C6) is 9.29 (18) and 10.73 (18)°, respectively. The crystal structure is slightly stabilized by the intramolecular N—H···F and intermolecular N—H···F, C—H···F hydrogen bonds as H-atoms donor, link the molecules into dimers along the *b* axis. Calculations of the NBO analyses (DFT/B3LYP/cc-pVDZ, single point geometry) were performed.

Keywords: crystal structure, DFT/B3LYP/cc-pVDZ, hydrazine, *R* factor, single crystal, T = 298 K, X-ray diffraction, *wR* factor

Mathematics Subject Classification: 74E15; 37N20

1 Comments

The pentafluorophenyl group is an important constituent not only in biological chemistry, but also in analytical, material and polymer chemistry, catalysis, dynamic combinatorial chemistry, and reaction development. Despite this widespread interest, as part of our studies on the substituent effects on the structures countaining pentafluorophenyl group and 5´-bromo-[2,2´-bithiophen]-5-yl)methylene fragment we investigated the hitherto not prepared Schiff base 1-((5´-bromo-[2,2´-bithiophen]-5-yl)methylene)-2-(pentafluorophenyl)hydrazine (**I**) is known to exhibit a wide variety of activities. We report here the crystal, molecular structure and electronic properties of the title compound (Fig. 1).

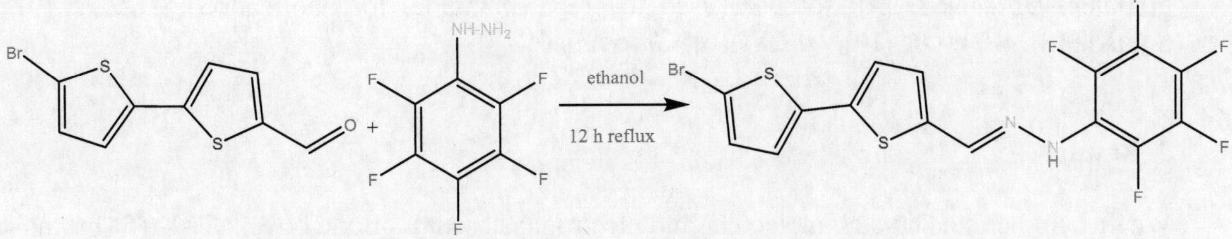

Fig. 1. Synthesis of (*E*)-1-((5′-bromo-[2,2′-bithiophen]-5-yl)methylene)-2-(pentafluorophenyl)hydrazine.

2 Experimental

The reaction of pentafluorophenylhydrazine with 5′-bromo-(2,2′-bithiophene)-5-carbaldehyde in ethanol gave the desired Schiff base 1-((5′-bromo-[2,2′-bithiophen]-5-yl)methylene)-2-(pentafluorophenyl)hydrazine with high yield and purity were characterized by elemental analyses, spectroscopic methods and X-ray diffraction analyses [1, 2, 3]. According to the reaction scheme (Fig. 1), the corresponding aldehyde (0.01 mol) was dissolved in 60 ml ethanol which contained solution of 0.011 mol pentafluorophenylhydrazine. The reaction mixture was refluxed for 12 h with the addition of 2 drops of concentrated HCl. The solvent was evaporated and the obtained crude product was purified by column chromatography to give the pure product in 68% yield. M.p. 149-151 ^{0}C.
^{1}H NMR (300 MHz, dmso) δ 10.37 (s, 1H), 8.22 (s, 1H), 7.20 (m, 4H).
^{13}C NMR (75 MHz, dmso) δ 139.11, 138.06, 137.62(m) 137.21(m), 136.71, 135.48, 133.71(m), 131.67, 129.16, 125.02, 124.73, 120.87 (m,), 110.67.
^{19}F NMR (282 MHz, dmso) δ -155.69 (2F, d, J = 24.0 Hz), -164.35 (2F, td, J = 21.6, 2.7 Hz), -169.87 (1F, tt, J = 23.3, 5.7 Hz).

2.1 Refinement

Hydrogen atoms were placed in calculated positions with C—H distances of 0.93 Å and constrained to ride on their parent atoms. The amide H atom was visible in a difference map and refined with the N—H distance restrained to 0.86 (2) Å. The U_{iso}(H) values were set at 1.2U_{eq}. Friedel pairs were merged. Refinement on F^2, least-squares matrix: full, $R[F^2 > 2\sigma(F^2)]$ = 0.039, $wR(F^2)$ = 0.090, S = 1.03, 3921 reflections, 226 parameters, 0 restraint, primary atom site location: structure-invariant direct methods, secondary atom site location: difference Fourier map, Hydrogen site location: inferred from neighbouring sites, H atoms treated by a mixture of independent and constrained refinement, $\Delta\rho_{max}$ = 0.28 eÅ^{-3}, $\Delta\rho_{min}$ = −0.32 eÅ^{-3}, Triclinic, P-1, a = 6.2562 (3), b = 9.0987 (8), c = 14.7830 (12) (Å), α = 102.555 (8), β = 93.116 (6), γ = 101.968 (7) (°).

2.1.1 Computing details

Data collection: *CrysAlis CCD* (Oxford Diffraction, 2009); cell refinement: *CrysAlis CCD*; data reduction: *CrysAlis RED* (Oxford Diffraction, 2009); program(s) used to solve structure: *SHELXS* [9], *Sir2014* ([5]; program(s) used to refine structure: *SHELXL* [9]; molecular graphics: *DIAMOND* [4]; software used to prepare material for publication: *SHELXL* [9], ShelXle [8], *PLATON* [10], *OLEX2* [6], *Gaussian 09* [7].

3 Results

We present here the crystal, molecular and electronic structure of one novel (*E*)-1-((5′-bromo-[2,2′-bithiophen]-5-yl)methylene)-2-(pentafluorophenyl)hydrazine (**I**). Compound (**I**) crystallizes in triclinic system, centrosymmetric space group *P-1*. The dihedral angle between the thiophene rings is 3.33 (0.19)° and the side ring (C1 to C6, pentafluorophenyl) is 9.29 (18) and 10.73 (18)°, respectively. Crystallographic data for the structure reported in this paper will be available from the Cambridge Crystallographic Data Centre.The hydrogen atom of the NH is involved in intramolecular hydrogen bonding and intermolecular hydrogen bonding . The crystal structure is stabilized by the intramolecular N2—H2···F1 bond and two intermolecular N2—H2···F1^{i}, F1···H2—N2^{i} hydrogen interactions as H-atoms donor, link the molecules into dimers along the *b* axis, we also see intermolecular C13—H13···F3^{ii} and C13—H13···F4^{ii} hydrogen interactions (Figs. 2, 3 and Tab. 1). Table 3 shows us the distance between ring centroids (Å) of Cg(1) (S1-C8-C11-C10-C9, Cg(2) (S2-C12-C13-C14-C15, Cg(3) (C1-C2-C3-C4-C5-C6). The bond lengths of the hydrazine group N1=C7 is 1.274 (4) Å somewhat shorter than typical bond [1]. This may be due to the fact that C7H7 is not participate in intra- and intermolecular hydrogen bonds. Atom N2 is sp^2-hybridized, as evidenced by the sum of the valence angles around it (359.99°). These data are consistent with conjugation of the lone-pair electrons on N2, similar to what is observed for amides. The electronic structure of the molecule of the title compound (**I**) was calculated by the *Ab Initio* method [7], DFT/B3LYP/cc-pVDZ basis set was involved into calculation. Calculation of the electronic structure of a compound provides several indices which characterize the distribution of electron density in the molecule and the multiplicity of atomic bonds. The selected net charges give a Tab. 2 of the distribution of electron density in the molecule and the values of the Wiberg bond indices enable one to estimate the multiplicity of individual atomic bonds. The net charge distribution in the molecule indicates that the positive charges are localized at the atoms N2 (0.049), S1 and S2 (0.159 and 0.183), whereas the negative net charges are located on the N1 (–0.100), F1 (–0.163), F2 (–0.143), F3 (–0.146), F4 (–0.143) and F5 (–0.135) . This charge distribution and the spatial arrangement (geometry) of the molecule govern its activity and are important for the overall stabilization of the crystal structure. It follows from the Wiberg index value for the bond N1—C7 ($\boldsymbol{I}_w$ = 1.695) is not a pure double bond but indicate the character of conjugated bond. The other bonds of the molecule have the character of single bonds (Tab. 2). That π-electrons are delocalized in the region of the C1 to C6 atoms and atoms S1, S2, Br1. The results of these calculations are in good agreement with the experimental values of the bond lengths found by the X-ray structure analyses. It was found by evaluating experimental interactions of the structure (**I**) in the solid phase in this work that the intra- and intermolecular interactions are nevertheless important in the parts of the molecule (**I**), (Tab. 1, Tab. 2).

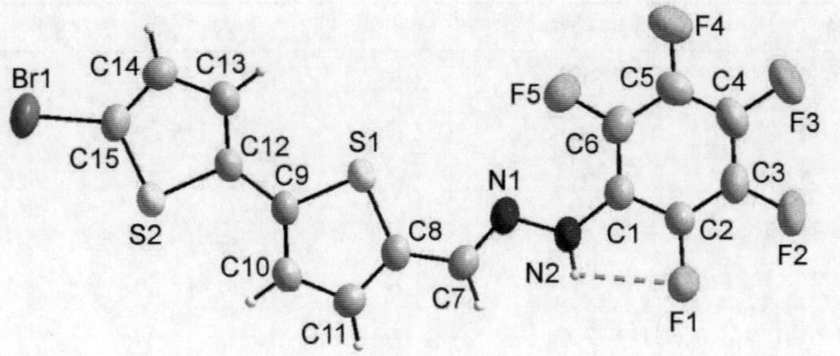

Fig. 2. Molecular structure of the title compound (**I**) with the atomic numbering scheme. Displacement ellipsoids are drawn at the 50 % probability level. H atoms are represented as small spheres of arbitrary radii.

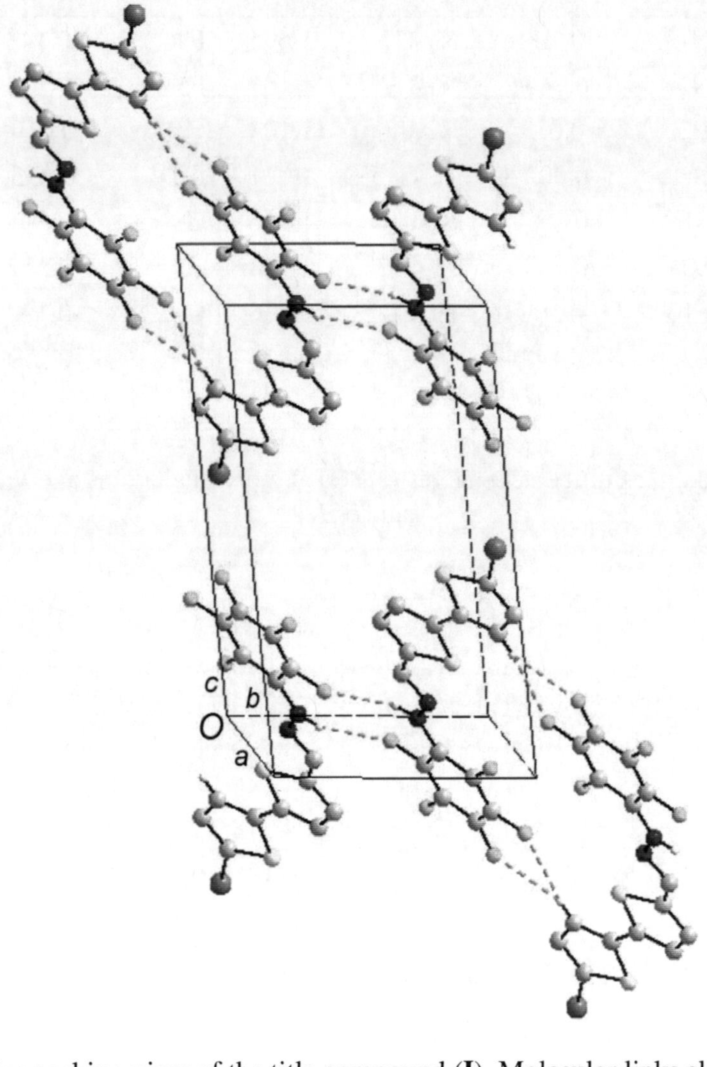

Fig. 3. Molecular packing view of the title compound (**I**). Molecular links along *c*-axis are generated by N—H · · · F, F · · · H—N hydrogen bonds which are shown by dashed lines. The rest of H atoms have been omitted for clarity.

D—H···*A*	D—H	H···*A*	*D*···*A*	*D*—H···*A*
N2—H2···F1	0.86	2.28	2.645(3)	106
N2—H2···F1^{i}	0.86	2.32	3.119(2)	155
F1···H2—N2^{i}	0.86	2.32	3.119(2)	155
C13—H13···F3^{ii}	0.93	2.59	3.246(3)	128
C13—H13···F4^{ii}	0.93	2.87	3.746(3)	173

Symmetry code: (**i**) 2-x, 1-y, 2-z; (**ii**) -x, -y, 2-z

Tab. 1. Hydrogen-bond geometry (Å, °) for (**I**).

Atoms	Charge, *q*	Bond	$\boldsymbol{I}_w$	Atoms	Charge, *q*
N1	-0.100	N1-C7	1.695	F4	-0.143
N2	0.049	N1-N2	1.107	F5	-0.135
H2	-0.019	N2-C1	1.103	C1	0.019
Br1	0.002	Br1-C15	1.061	C7	0.255
S1	0.159	S1-C9	1.166	C8	-0.205
S2	0.183	S2-C12	1.148	C9	-0.112
F1	-0.163	F1-C2	0.876	C12	-0.122
F2	-0.143	F3-C4	0.896	C15	-0.279
F3	-0.146	F5-C6	0.902		

Tab. 2. Selected values of the net charges (NBO) at individual atoms and Wiberg bonding indices $\boldsymbol{I}_w$.

Cg(1) - Cg(2)^{iii}	3.714(2)
Cg(1) - Cg(3)^{iv}	4.046(2)
Cg(2) - Cg(1)^{v}	3.714(2)
Cg(2) - Cg(2)^{vi}	3.975(2)
Cg(3) - Cg(1)^{iv}	4.046(2)
Cg(3) - Cg(3)^{vii}	4.314(2)

Symmetry code: (**iii**) 1+x, y, z; (**iv**) 1-x, 1-y, 2-z; (**v**) -1+x, y, z; (**vi**) -1-x, -y, 1-z; (**vii**) 1-x, -y, 2-z

Tab. 3. Distance between ring centroids (Å).

Acknowledgement

This work was financially supported by from the EU structural funds as part of the Security Research Centre of Excellency ITMS 26240120034, grant from the Research Center for

Industrial Synthesis of Drugs, ITMS 26240220061, supported by the Research & Development Operational Programme funded by the ERDF. This work was also supported by the Grant Agency of the Slovak Republic (*KEGA*, Grant No. 035STU-4/2017 and *VEGA* Project 1/0871/16 and Research and Development Agency of the Slovak Republic under the contracts No APVV-15-0079). We are grateful to the HPC centre at the Slovak University of Technology in Bratislava, which is a part of the Slovak Infrastructure of High Performance Computing (SIVVP project, ITMS code 26230120002, funded by the European region development funds, ERDF), for the computational time and the resources. This article was created with the support of the Ministry of Education, Science, Research and Sport of the Slovak Republic within the Research and Development Operational Programme for the project "University Science Park of STU Bratislava", ITMS 26240220084, co-funded by the European Regional Development Fund.

References

[1] LUKEŠ, V., MICHALÍK, M., POLIAK, P., CAGARDOVÁ, D., VÉGH, D., BORTŇÁK, D., FRONC, M., KOŽÍŠEK, J., *Theoretical and experimental study of model oligothiophenes containing 1-methylene-2-(perfluorophenyl)hydrazine terminal unit.* In *Synthetic Metals* 219, 2016, pp. 83–92.

[2] NOSHIRANZADEH, N., BIKAS, R., ŚLEPOKURA, K., SHAABANI, M., LIS, T., *Synthesis and characterization of cobalt complexes with pentafluorophenylhydrazine: Nucleophilic attack of phenolic oxygen to pentafluorophenyl ring during condensation of two Schiff base ligands.* In *Journal of Fluorine Chemistry* 160, 2014, pp. 34–40.

[3] JIANLI HUA, J., ZHANG, W., LI, Z., QIN, J., SHEN, Y., ZHANG, Y., LUY, Z., *Arylaldehydes-pentafluorophenyl Hydrazones as Second-order Nonlinear Optical Chromophores: A Novel Approach for Remarkably Defeating the Nonlinearity-transparency Trade-off.* In *Chemistry Letters*, 2002, 2, pp. 232–233.

[4] BRANDENBURG, K., 1999, *Diamond.* Version 2.1c. Crystal Impact GbR Bonn, Germany.

[5] BURLA, M. C., CALIANDRO, R., CARROZZINI, B., CASCARANO, G. L., GIACOVAZZO, C., MALLAMO, M., MAZZONE, A., POLIDORI, G., 2016, *Sir2014*, ver. 14.10. In *J. Appl. Cryst.* (2015). 48, pp. 306–309.

[6] DOLOMANOV, O. V., BOURHIS, L. J., GILDEA, R. J., HOWARD, J. A. K., PUSCHMANN, H., 2009, *OLEX2.* In *Journal of Applied Crystallography* **42**, pp. 339-341.

[7] FRISCH, M. J., TRUCKS, G. W., SCHLEGEL, H. B., SCUSERIA, G. E., ROBB, M. A., CHEESEMAN, J. R., SCALMANI, G., BARONE, V., MENNUCCI, B., PETERSSON, G.A., NAKATSUJI, H., CARICATO, M., LI, X., HRATCHIAN, H.P., IZMAYLOV, A. F., BLOINO, J., ZHENG, G., SONNENBERG, J. L., HADA, M., EHARA, M., TOYOTA, K., FUKUDA, R., HASEGAWA, J., ISHIDA, M., NAKAJIMA, T., HONDA, Y., KITAO, O., NAKAI, H., VREVEN, T., MONTGOMERY, JR. J. A., PERALTA, J. E., OGLIARO, F., BEARPARK, M., HEYD, J. J., BROTHERS, E., KUDIN, K. N., STAROVEROV, V. N., KEITH, T., KOBAYASHI, R., NORMAND, J., RAGHAVACHARI, K., RENDELL, A., BURANT, J.C., IYENGAR, S. S., TOMASI, J., COSSI, M., REGA, N., MILLAM, J. M., KLENE, M., KNOX, J. E., CROSS, J. B., BAKKEN, V., ADAMO, C., JARAMILLO, J., GOMPERTS, R., STRATMANN, R. E., YAZYEV, O., AUSTIN, A. J., CAMMI, R., POMELLI, C., OCHTERSKI, J. W., MARTIN, R. L., MOROKUMA, K., ZAKRZEWSKI, V. G., VOTH, G. A., SALVADOR, P., DANNENBERG, J. J., DAPPRICH, S., DANIELS, A. D., FARKAS, O.,

FORESMAN, J. B., ORTIZ, J. V., CIOSLOWSKI, J., FOX, D. J., 2013, *Gaussian09*. Gaussian. INC., Wallingford CT.
[8] HÜBSCHLE, Ch. B., SHELDRICK, G. M., DITTRICH, B., 2011, In *J. App. Cryst.*, **44**, pp. 1281–1284.
[9] SHELDRICK, G. M. 2015, In *Acta Cryst. C***71**, pp. 3-8.
[10] SPEK, A. L. 2009, In *Acta Cryst.* D**65**, pp. 148–155.

Current address

Fronc Marek, Ing., PhD.
Institute of Physical Chemistry and Chemical Physics
Faculty of Chemical and Food Technology, Slovak University of Technology
Radlinského 9, SK-812 37 Bratislava, Slovak Republic
E-mail: marek.fronc@stuba.sk

Sivý Július, doc. RNDr., PhD.
Institute of Mathematics and Physics,
Faculty of Mechanical Engineering, Slovak University of Technology
Nám. slobody 17, SK-812 31 Bratislava, Slovak Republic
E-mail: julius.sivy@stuba.sk

Bortňák Dušan, Ing., PhD.
Institute of Organic Chemistry, Catalysis and Petrochemy
Faculty of Chemical and Food Technology, Slovak University of Technology
Radlinského 9, SK-812 37 Bratislava, Slovak Republic
E-mail: dusan.bortnak@stuba.sk

Végh Daniel, Ing., DSc.
Institute of Organic Chemistry, Catalysis and Petrochemy
Faculty of Chemical and Food Technology, Slovak University of Technology
Radlinského 9, SK-812 37 Bratislava, Slovak Republic
E-mail: daniel.vegh@stuba.sk

Králik Marián, doc. Ing., PhD.
Institute of Manufacturing Systems, Environmental Technology and Quality Management
Faculty of Mechanical Engineering, Slovak University of Technology
Nám. slobody 17, SK-812 31 Bratislava, Slovak Republic
E-mail: marian.kralik@stuba.sk

Slovak University of Technology in Bratislava
Faculty of Mechanical Engineering
17th Conference on Applied Mathematics **APLIMAT 2018**

Proceedings

REALISING SOME TYPES WHILE OMITTING OTHERS

GLIVICKÁ Jana (CZ)

Abstract. We prove a theorem that combines omitting and realizing of types in one structure. Given two collections of types, we formulate a condition ensuring existence of a model realizing types from the first collection while omitting types from the second.

Keywords: saturated models, omitting types, realizing types

Mathematics subject classification: Primary 03C50; Secondary 06F50

1 Preliminaries

Let us first recall some important notions and fix notation that will be used throughout the paper.

1.1 Basic notation

By κ we denote infinite cardinal numbers; ω, resp. ω_1, denote the smallest infinite, resp. the smallest uncountable, cardinal number.

In the next, unless explicitly stated otherwise, $\mathcal{L}$ denotes a countable language, $\mathcal{M}$ a structure for $\mathcal{L}$, M denotes the universe of $\mathcal{M}$, and T a theory in $\mathcal{L}$.

If $A \subseteq M$, then $\mathcal{M}_A$ denotes the expansion of $\mathcal{M}$ by adding names for elements from A as new constants.

$\mathrm{Th}(\mathcal{M})$ denotes the set of all $\mathcal{L}$-formulas true in $\mathcal{M}$. $\mathrm{Th}(\mathcal{M}_M)$ is called the elementary diagram of $\mathcal{M}$ and is denoted by $\Delta_e(\mathcal{M})$.

This paper was processed with contribution of long term institutional support of research activities by Faculty of Informatics and Statistics, University of Economics, Prague. The study was supported by the Charles University, project GA UK No. 270815. The work was supported by the grant SVV-2017-260456.

1.2 Types in theories, types in structures

By a type in T we mean a set $p(x)$ of $\mathcal{L}$-formulas in the free variable x, such that $p(x)$ is consistent with T (i.e. for any $\varphi_1(x), \dots, \varphi_n(x) \in p(x)$, the theory $T \cup \{\exists x(\varphi_1(x) \wedge \dots \wedge \varphi_n(x))\}$ is a consistent). We say that $p(x)$ is isolated if there exists an $\mathcal{L}$-formula $\psi(x)$ consistent with T and such that $\psi(x) \rightarrow \varphi(x)$ is provable in T for any $\varphi(x) \in p(x)$.

By a type over A in $\mathcal{M}$ we mean a type in the theory $\mathrm{Th}(\mathcal{M}_A)$.

1.3 Realizing and omitting

We say that $p(x)$ is realized in $\mathcal{M}$ by an element $m \in M$ if $\varphi(m)$ holds true in $\mathcal{M}$ for every $\varphi(x) \in p(x)$. If $p(x)$ is not realized by any $m \in M$, we say that $p(x)$ is omitted in $\mathcal{M}$.

A structure $\mathcal{M}$ is said to be saturated if for any $A \subseteq M$, such that $|A| < |M|$, every type in $\mathcal{M}$ over A is realized in $\mathcal{M}$ by some $m \in M$.

2 Motivation

Given a theory T, one might ask how its models differ from one another with respect to the types realized or omitted in them. The next two theorems are well-known results that describe two opposite approaches to this question. They can be found in [1] as Theorem 4.2.4 and Corollary 4.3.14.

In a complete theory, all the isolated types are always realized. Apart from that, there is no other restriction on what types can be omitted, as expressed in the Omitting Types Theorem.

Theorem 1 (Omitting Types Theorem). *Let $\mathcal{L}$ be a countable language, T a consistent theory in the language $\mathcal{L}$. Let Ω be a countable collection of nonisolated types in T. Then there exists a countable model $\mathcal{M}$ of T omitting all the types from Ω.*

On the other hand, we can find a model realizing all the types we could wish for, under some assumptions.

Theorem 2. *Let $\mathcal{L}$ be a countable language, T a complete theory in $\mathcal{L}$ with infinite models. Suppose that $2^\kappa = \kappa^+$. Then there exists a saturated model of T of size κ^+.*

In our main theorem, we assume that the continuum hypothesis (CH) holds true, i. e. $2^\omega = \omega_1$. Therefore, in our case, the previous theorem ensures the existence of saturated structures of cardinality ω_1.

Of course there may be, in general, other models "in between" those two cases, i. e. models with some nonisolated types realized and others omitted. We concern ourselves with a situation when we are given two collections of types—collection Ω of types to be omitted and collection Σ of types to be realized. Our theorem is therefore filling in the gap between the theorem on existence of saturated models and the Omitting Types Theorem.

The motivation for our theorem comes not only from a theoretical interest, but from the study of real closed fields and their substructures, especially the so called integer parts, as presented in [2]. In particular, it is possible to study how automorphisms of real closed fields act on the substructures in question. A natural connection between automorphisms of a structure and saturation of its substructures is given by the following proposition.

Proposition 3. *If $\mathcal{A}$ is a structure and a its automorphism, then any saturated substructure of $\mathcal{A}$ is mapped by a onto a saturated substructure of $\mathcal{A}$.*

In our specific case, we were led to look for substructures of real closed fields that are saturated in the language containing only the order relation $\leq$, but not saturated when the whole language of arithmetic is considered. We believe that this is an example of the most natural application of our theorem—the collections Σ, of types to be realized, and Ω, of types to be omitted, are distinguished by the language; Σ contains types only in certain sublanguage, while the types from Ω can be in the full language in question.

3 Main result

Before we proceed to a formulation of our main theorem, we need one more notion. As we have indicated above, our aim is to find a structure $\mathcal{M}$ realizing all the types from the collection Σ. We would like to allow parameters from M in those types. For this reason, we allow countably many parametric variables v_i to appear in the formulas.

Definition 4. *A countably parametrized set of $\mathcal{L}$-formulas in the variable x is a set of $\mathcal{L}$-formulas whose all free variables belong to $\{x\} \cup \{v_i; i \in \omega\}$.*

In the next, we denote such countably parametrized sets of formulas by $s(x, \bar{v})$. Once we have obtained some structure $\mathcal{M}$ and are given a sequence of its elements $\bar{m} \in \mathcal{M}^\omega$, it is possible to "substitute" $\bar{m}$ for $\bar{v}$ and hence understand $s(x, \bar{m})$ as a set of formulas with parameters from M. Such an $s(x, \bar{m})$ may, or may not, become a type in $\mathcal{M}$, depending on the choice of $\bar{m}$.

If Σ is a collection of such countably parametrized sets of formulas and $e : \omega_1 \to \Sigma$ is its enumeration, we write $s_\alpha(x, \bar{v})$ for $e(\alpha)$.

Theorem 5. *(CH) Let $\mathcal{L}$ be a countable language, T a consistent theory. Let Ω be a countable collection of nonisolated types in T and let Σ be a collection of countably parametrized sets of $\mathcal{L}$-formulas. Furthermore, let us fix an enumeration of Σ.*

Suppose that for any $\beta < \omega_1$ and any countable model $\mathcal{M}$ of T omitting all types from Ω, it holds that no $p(x) \in \Omega$ is isolated in the theory

$$\bigcup\{s_\alpha(c_{\alpha,\bar{m}}, \bar{m}); \alpha \leq \beta, \bar{m} \in M^\omega, s_\alpha(x, \bar{m}) \text{ is a type of } \mathcal{M}\} \cup \Delta_e(\mathcal{M}),$$

where $\{c_{\alpha,\bar{m}}; \alpha \leq \beta, \bar{m} \in M^\omega\}$ is a set of fresh constants.

Then there exists a model $\mathcal{M}$ of T omitting every $p(x) \in \Omega$ and realizing every type in $\mathcal{M}$ of the form $s(x, \bar{m})$, where $s(x, \bar{v}) \in \Sigma$ and $\bar{m} \in M^\omega$.

Proof. First, we fix a set U of cardinality ω_1 and an enumeration $\{\bar{u}_\gamma\}_{\gamma \in \omega_1}$ of U^ω. The universe of the model $\mathcal{M}$ we look for will be a subset of U.

We construct an elementary chain $\mathcal{M}_0 \preccurlyeq \mathcal{M}_1 \preccurlyeq \ldots \preccurlyeq \mathcal{M}_\beta \preccurlyeq \ldots$ of length ω_1 satisfying the following conditions:

1. For any $\beta < \omega_1$, $\mathcal{M}_\beta$ is a countable model of T omitting all types from the collection Ω, $M_\beta \subseteq U$.

2. Let $\beta < \omega_1$, and let $\alpha, \gamma < \beta$. If there is $\delta < \beta$ such that $\bar{u}_\gamma \in (M_\delta)^\omega$ and $s_\alpha(x, \bar{u}_\gamma)$ is a type of $\mathcal{M}_\delta$, then it is realized in $\mathcal{M}_\beta$.

By the Omitting Types Theorem, there exists a countable model $\mathcal{M}_0$ of T omitting all types from Ω. Moreover, it can be chosen so that $M_0 \subseteq U$. Such an $\mathcal{M}_0$ clearly satisfies the conditions above.

Suppose λ is a limit ordinal and we have already constructed the elementary chain $\mathcal{M}_0 \preccurlyeq \mathcal{M}_1 \preccurlyeq \ldots \preccurlyeq \mathcal{M}_\beta \preccurlyeq \ldots$ of length λ satisfying all the above conditions up to λ. If we extend this elementary chain by adding $\mathcal{M}_\lambda := \bigcup_{\beta<\lambda} \mathcal{M}_\beta$ as its last element, it can be easily checked that such an extension satisfies the above condition up to and including λ.

Suppose we have already constructed the chain up to $\mathcal{M}_\beta$; we show how to obtain $\mathcal{M}_{\beta+1}$. Consider a theory

$$\bigcup\{s_\alpha(c_{\alpha,\bar{u}_\gamma}, \bar{u}_\gamma); \alpha, \gamma \leq \beta, \bar{u}_\gamma \in (M_\beta)^\omega, s_\alpha(x, \bar{u}_\gamma) \text{ is a type of } \mathcal{M}_\beta\} \cup \Delta_e(\mathcal{M}_\beta),$$

where $\{c_{\alpha,\bar{u}_\gamma}; \alpha, \gamma \leq \beta\}$ is a set of fresh constants.

It is a consistent theory in a countable language and, by the assumption of the theorem, no $p(x) \in \Omega$ is isolated in it. Hence, using again the Omitting Types Theorem, it has a countable model $\mathcal{M}_{\beta+1}$ omitting all types from Ω; moreover such that $M_{\beta+1} \subseteq U$. It is obvious that $\mathcal{M}_{\beta+1}$ satisfies condition 1. Condition 2 is satisfied as well, since the theory makes sure that the type $s_\alpha(x, \bar{u}_\gamma)$ is satisfied by the newly added constant $c_{\alpha,\bar{u}_\gamma}$.

Finally, let us define $\mathcal{M} := \bigcup_{\beta<\omega_1} \mathcal{M}_\beta$. It is a union of an elementary chain of models of T and therefore also a model of T. It ommits all types from Ω as all $\mathcal{M}_\beta$ do so. Let $\bar{m} \in M^\omega$, $\alpha < \omega_1$ and suppose $s_\alpha(x, \bar{m})$ is a type in $\mathcal{M}$. Then, by regularity of ω_1, there is some $\beta < \omega_1$ such that $\bar{m} \in (M_\beta)^\omega$. Put $\mu := max(\alpha, \beta)$; since $\mathcal{M}_\mu \preccurlyeq \mathcal{M}$, it holds that $s_\alpha(x, \bar{m})$ is also a type in $\mathcal{M}_\mu$. By condition 2, it is realized in $\mathcal{M}_{\mu+1}$ and therefore also in $\mathcal{M}$.

□

4 Application

We present one application of our main theorem as an illustration of its use. In this section, $\mathbb{N}$ denotes the set of all natural numbers and $\mathbb{P}$ denotes the set of all prime numbers.

Let Pr be the additive theory of the structure of natural numbers and $p(x)$ a type expressing that x is a nonzero element divisible by all primes; i. e. $\mathrm{Pr} = \mathrm{Th}(\langle \mathbb{N}, 0, 1, +, \leq \rangle)$ and $p(x)$ is the set $\{x \neq 0\} \cup \{p|x; p \in \mathbb{P}\}$.[1] We show how to obtain a model of Pr saturated in the language $\{0, 1, \leq\}$ omitting $p(x)$.

Lemma 6. *Let $\mathcal{M}$ be model of* Pr *and let $s(x)$ be a type in $\mathcal{M}$ in the language $\{0, 1, \leq\}$ omitted in $\mathcal{M}$. Let us fix a function $\tau : \mathbb{P} \to \mathbb{N}$, such that $\tau(p) \in \{0, 1, \ldots, p-1\}$. Then $s'(x) = s(x) \cup \{x \bmod p = \tau(p); p \in \mathbb{P}\}$ is also a type in $\mathcal{M}$.*

Proof. Let us fix $\mathcal{M}'$, a model of $\Delta_e(\mathcal{M})$ realizing $s(x)$ by some some element $m \in M'$ (such a model exists by Theorem 2). Then the $\mathbb{Z}$-component of m, i.e. the set $Z = \{m' \in M'; |m - m'| \in \mathbb{N}\}$, is disjoint from M (else m would be an element of $\mathcal{M}$). Therefore all the parameters from $s(x)$ lie outside of Z. By considering all the order-automorphisms of $\mathcal{M}'$ shifting the elements of Z while fixing all the other, in particular all parameters from $s(x)$, it is easy to see that any element of Z realizes $s(x)$. By the Chinese remainder theorem, every finite part of $s'(x)$ is realized by some element of Z. Therefore $s'(x)$ is a type. □

The lemma above tells us that we can control moduli of elements realizing types in the language $\{0, 1, \leq\}$. This allows us to avoid the type $p(x)$ from being realized, as stated in the next lemma.

[1]The formula $p|x$ can be expressed in the additive language as $(\exists y)(\underbrace{y + y + \cdots + y}_{p\text{-times}} = x)$.

Lemma 7. *Let $\mathcal{M}$ be a countable model of* Pr *omitting $p(x)$. Let Σ be a countable family of types in the language $\{0, 1, \leq\}$ in $\mathcal{M}$. Then there exists a model $\mathcal{M}'$ of $\Delta_e(\mathcal{M}) \cup \{s(c_s); s \in \Sigma\}$, where $\{c_s, s \in \Sigma\}$ is a set of fresh constants, omitting $p(x)$.*

If T is the theory Pr, $\Omega = \{p(x)\}$, and Σ is the set of all countably parametrized sets of formulas in the language $\{0, 1, \leq\}$, then, by Lemma 7, the assumption of Theorem 5 is satisfied. Therefore, as an application, we get the following result.

Theorem 8. *(CH) There exists a model $\mathcal{M}$ of* Pr *saturated in the language $\{0, 1, \leq\}$ and omitting the type $\{x \neq 0\} \cup \{p|x; p \in \mathbb{P}\}$.*

5 Further research

There are two basic areas that offer opportunity for further research.

First, one could try to get rid of the assumption of the continuum hypothesis and find a natural reformulation of our theorem with no additional set theoretic assumptions.

Second, one could try to come up with more algebraic applications of our theorem, apart from the application in the study of real closed fields and their integer parts we have already mentioned.

References

[1] MARKER, D.: *Model Theory: An Introduction*, Springer, 2002, ISBN 0-387-98760-6

[2] GLIVICKÁ, J., GLIVICKÝ, P.: Shepherdson's Theorems for Fragments of Open Induction. In *Proceedings, 16th Conference on Applied Mathematics Aplimat 2017*, Bratislava, 2017, pp. 583-590.

Current address

Glivická Jana, Mgr.
Department of Mathematics
Faculty of Informatics and Statistics
University of Economics, Prague
Ekonomická 957, 148 00 Praha 4, Czech Republic
E-mail: jana.glivicka@vse.cz

Slovak University of Technology in Bratislava
Faculty of Mechanical Engineering
17th Conference on Applied Mathematics **APLIMAT 2018**

Proceedings

A NOTE ON THE PROBLEM OF PRISONERS AND HATS

GLIVICKÝ Petr (CZ)

Abstract. We study the famous mathematical puzzle of prisoners and hats. We introduce a framework in which various variants of the problem can be formalized. We examine three particular versions of the problem (each one in fact a class of problems) and completely characterize them as to (non)existence of winning strategies.

Keywords: hat problems, prisoners and hats, game theory

Mathematics subject classification: Primary 91A05; Secondary 91A43, 00A08.

1 Introduction

Problems that can be collectively denoted as "puzzles of prisoners and hats" have been studied since 1960's (see e.g. [6]). The common idea of all such problems can be informally stated as follows (we give a more precise formulation below): A group of prisoners is asked to guess colors of hats they are wearing. In general, of course, they are not allowed to look at their own hats, but may be allowed to see some of the hats that others are wearing and hear the guesses of some of their colleagues. The goal of the prisoners usually is to maximize the number of correct guesses. No communication among the prisoners is allowed since the moment the hats are placed on their heads. However they have an opportunity to devise a common strategy in advance.

This may seem as just a recreational mathematical puzzle. Nevertheless certain variants of the problem have been studied and applied in contexts as distinct as coding theory [2], set theory [5] or theory of auctions [1].

A recent and rather informal introduction to the topic (focused on the infinite case) is [7]. The finite case is studied in more detail in [3]. An extensive treatise on the subject is [8].

In this paper we describe a framework in which the various versions of the problem of prisoners and hats can be formalized in a unified way (section 1.2). We use this framework to classify some of the important cases with regard to existence of a winning strategy for the prisoners. In

This paper was processed with contribution of long term institutional support of research activities by Faculty of Informatics and Statistics, University of Economics, Prague.

particular we will be interested in the "hear nothing, see all" case (section 2.1) and two cases where prisoners can be thought of as standing in a line — the "hear nothing, see forward" case (section 3.1) and the "hear backward, see forward" case (section 3.2). This, of course, is just the tip of the iceberg. A complete classification of prisoners and hats problems is an extensive but interesting project and an inspiration for further research.

1.1 Preliminaries and notation

1.1.1 Basic notation

${}^{x}y$ denotes the set $\{f; f : x \to y\}$ of all functions from x to y. id_X denotes the identity function on X. A natural number n and a function f are understood in the usual set-theoretical sense, i.e. $n = \{0, \dots, n-1\}$ and $f = \{(x, y); f(x) = y\}$.

1.1.2 Directed graphs

In a directed graph $G = \langle V, E\rangle$ a direction of an edge $(v', v) \in E$ is understood as from v' to v. A directed path in G is also called ascending path, while a descending path in G is a directed path in the graph $\langle V, E^{-1}\rangle$.

It is easy to see that if G has no infinite descending paths, then there is a well-ordering $\prec \supseteq E$ on V.

1.2 The prisoners and hats problem

The problem we want to investigate is based on the following story:

1.2.1 A story: Mathematicians and an evil hatter

A nonempty set M of mathematicians is captured by an evil hatter who announces that the next day they will play a hat guessing game with their freedom (if they win) or death (if they lose) at stake. The hatter will place differently colored hats, with colors coming from a nonempty set C, onto mathematicians heads and the unlucky prisoners will be forced to guess what color they will be wearing. The malevolent captor reveals some details about the rules in advance:

The game starts with the hatter choosing an assignment $a : M \to C$ from certain subset $A \subseteq {}^{M}C$ of allowed assignments of hat colors to the mathematicians. He then places hats accordingly onto mathematicians heads and decides that a mathematician m will be allowed to see the hat his colleague m' is wearing if and only if $(m', m) \in S$ for certain binary relation $S \subseteq M^2$.

The hatter then follows some well-ordering $\prec$ that extends the canonical partial ordering of certain directed graph $\langle T, H\rangle$ with no infinite descending paths whose vertices are labeled by mathematicians via some $\mathrm{m} : T \to M$ (not necessarily one-to-one nor onto) and at time $t \in T$ he asks the mathematician $\mathrm{m}(t)$ for her fateful guess (a color from C) which he then records as $g(t)$. Before $\mathrm{m}(t)$ is asked at time t, she will be allowed to hear the record of all guesses $g(t')$ such that $(t', t) \in H$. After her guess, $\mathrm{m}(t)$'s memory is erased so that she doesn't remember what she heard nor what her own guess was. (It is easy to see that from the point of view of the mathematicians, the choice of $\prec \supseteq H$ is irrelevant.)

After the whole collection $g : T \to C$ of guesses is obtained, the hatter compares it with the actual assignment a using an evaluation function $e : A \times {}^{T}C \to 2$ and he sets the mathematicians free if $e(a, g) = 1$ or executes them if $e(a, g) = 0$.

The hatter tells the mathematicians what C, A, S, $\langle T, H\rangle$, m and e he is going to use and leaves them for the night. They now have some time to find an optimal strategy that maximizes their chances of surviving.

1.2.2 The formalization of the problem

An instance (of the prisoners and hats problem) is an tuple $\mathcal{I} = (M, C, A, S, T, H, \mathrm{m}, e)$ where

- M and C are nonempty sets,
- $\emptyset \neq A \subseteq {}^M C$,
- $S \subseteq M^2$,
- $\langle T, H\rangle$ is a directed graph with no infinite descending paths,
- $\mathrm{m} : T \to M$,
- $e : A \times {}^T C \to 2$.

1.2.3 Strategies

There is a game $\mathcal{G}_{\mathcal{I}}$ of two players (the hatter and the group of mathematicians) with imperfect information that corresponds naturally (according to the story) to the instance $\mathcal{I}$.

We are not going to define $\mathcal{G}_{\mathcal{I}}$ here in detail. We just state, in a form of a definition, what a strategy for mathematicians in this game is:

A strategy (for $\mathcal{I}$) is any function $\sigma : X \to C$ where X is the set of all triples (t, α, γ) where $t \in T$, $\alpha = \alpha' \restriction S^{-1}[\mathrm{m}(t)]$ for some $\alpha' \in A$, $\gamma : H^{-1}[t] \to C$. Any strategy σ together with a choice of $a \in A$ uniquely determine the course of the game (i.e. the collection of all guesses $g = g_{a,\sigma} : T \to C$) and thus the result $e(a, g)$:

Lemma 1 *For any strategy σ and $a \in A$ there is the unique $g = g_{a,\sigma} : T \to C$ such that for every $t \in T$*

$$g(t) = \sigma(t, a \restriction S^{-1}[\mathrm{m}(t)], g \restriction H^{-1}[t]).$$

Proof. As $\langle T, H\rangle$ does not contain infinite descending paths, there is a well-ordering $\prec \supseteq H$ on T. Let $(t_\alpha)_{\alpha\in\delta}$ for some ordinal δ be an enumeration of T such that $t_\alpha \prec t_\beta$ if and only if $\alpha < \beta$. We construct $g : T \to C$ by recursion:

$$g(t_\alpha) = \sigma(t_\alpha, a \restriction S^{-1}[\mathrm{m}(t_\alpha)], g \restriction H^{-1}[t_\alpha]),$$

which is correct since $H^{-1}[t_\alpha] \subseteq \prec^{-1} [t_\alpha] \subseteq \{t_\beta; \beta < \alpha\}$.

The uniqueness of g can be easily proven by induction. □

A strategy σ for $\mathcal{I}$ is winning if $e(a, g_{a,\sigma}) = 1$ for every $a \in A$.

Strategies for "disjoint" instances can be combined to give a strategy for their "union": Let $I \neq \emptyset$ and for $i \in I$ let σ_i be a strategy for an instance $\mathcal{I}_i = (M_i, C_i, A_i, S_i, T_i, H_i, \mathrm{m}_i, e_i)$. Assume that the sets T_i, for $i \in I$, are pairwise disjoint and let

- $M = \bigcup_{i\in I} M_i$,

- $C = \bigcup_{i \in I} C_i$,
- $A \subseteq \{a \in {}^M C; (\forall i \in I)(a \restriction M_i \in A_i)\}$,
- $S \supseteq \bigcup_{i \in I} S_i$,
- $T = \bigcup_{i \in I} T_i$,
- $H \supseteq \bigcup_{i \in I} H_i$,
- $\mathrm{m} = \bigcup_{i \in I} \mathrm{m}_i$,
- $e : A \times {}^T C \to 2$.

Then there is a strategy σ for the instance $\mathcal{I} = (M, C, A, S, T, H, \mathrm{m}, e)$ given by

$$\sigma : (t, \alpha, \gamma) \mapsto \sigma_i(t, \alpha \restriction S_i^{-1}[\mathrm{m}_i(t)], \gamma \restriction H_i^{-1}[t])$$

where $i \in I$ is the unique index such that $t \in T_i$. We call such σ the ($\mathcal{I}$-)combination of strategies σ_i, $i \in I$, and denote it by $\bigsqcup_{i \in I}^{\mathcal{I}} \sigma_i$ or just $\bigsqcup_{i \in I} \sigma_i$.

2 A set-like case

In this paper, we are going to deal only with cases when every mathematician is asked for her guess exactly once, i.e. when $\mathrm{m} : T \to M$ is a bijection. Then we can simplify our notation by assuming $T = M$, $\mathrm{m} = id_T$ and by omitting T and m from the instances, writing them just as $\mathcal{I} = (M, C, A, S, H, e)$. Also we will always have $A = {}^M C$, which will allow for another simplification in notation later on.

The following evaluation functions will be used throughout the rest of this paper: For a cardinal number κ we define evaluation functions $e_{\geq \kappa}$ (at least κ correct guesses) and $e^{<\kappa}$ (less than κ incorrect guesses) by

$$e_{\geq \kappa}(a, g) = 1 \Leftrightarrow |\{m \in M; a(m) = g(m)\}| \geq \kappa,$$
$$e^{<\kappa}(a, g) = 1 \Leftrightarrow |\{m \in M; a(m) \neq g(m)\}| < \kappa.$$

The case, we are going to deal with in this section, requires no structure to be defined on the set M (we call such cases "set-like"). This distinguishes this case from the cases defined in the following section, where we will be assuming a fixed well-ordering on M ("well-ordered" cases).

2.1 The "hear nothing, see all" case

Let us denote by $\mathtt{hNsA}(M, C, e)$ the instance $\mathcal{I} = (M, C, A, S, H, e)$ with $A = {}^M C$, $S = M^2 - id_M$ and $H = \emptyset$ (that is every mathematician sees all hats except her own but hears no previous guesses).

This is a basic case, many subcases of which have been thoroughly analyzed elsewhere. For the convenience of the reader, we present here a compact selfcontained overview that gives a complete (at least from our point of view) characterization of existence of winning strategies.

2.1.1 Case: M finite

If both M and C are finite, then there is a strategy that guarantees at least $\lfloor |M|/|C| \rfloor$ correct guesses, but no strategy is able to guarantee more. This was probably first proven in [10] for just two colors and in [4] in full generality (see also [8, Theorem 2.3.1]):

Proposition 2 *Let M, C be finite, $n \in \mathbb{N}$. Then there is a winning strategy σ for the instance $\mathtt{hNsA}(M, C, e_{\geq n})$ if and only if $n \leq |M|/|C|$.*

Proof. "$\Leftarrow$": It is enough to find a winning strategy for $\mathtt{hNsA}(M', C', e_{\geq 1})$ where $|M'| = |C'|$. Indeed, M can be divided into n disjoint sets $M_0, \ldots, M_{n-1}$ of size $|C|$ and a possible leftover (as $n \leq |M|/|C|$). The combination $\sigma = \bigsqcup_{i<n} \sigma_i$ of winning strategies σ_i for $\mathtt{hNsA}(M_i, C, e_{\geq 1})$ is then a winning strategy for $\mathtt{hNsA}(M, C, e_{\geq n})$. (There is at least one correct guess on each of n disjoint sets M_i.)

Let us therefore assume that $|M| = |C|$ and $n = 1$. Without loss of generality, we can have $M = C \in \mathbb{N}$. Define, for $m \in M$ and $\alpha : M - \{m\} \to C$, the value of σ as

$$\sigma(m, \alpha, \emptyset) = m - \sum_{m' \neq m} \alpha(m'),$$

where the right side is computed modulo M. Then, for $a \in {}^M C$, let $m_a = \sum_{m' \in M} a(m')$ (modulo M). We get

$$g_{a,\sigma}(m_a) = \sigma(m_a, a \restriction (M - \{m_a\}), \emptyset) = m_a - \sum_{m' \neq m_a} a(m') = a(m_a),$$

thus guaranteeing $e(a, g) = 1$.

"$\Rightarrow$": Suppose that there is a winning strategy σ for $n > |M|/|C|$. Then

$$|\{(a, m); a \in A, m \in M, a(m) = g_{a,\sigma}(m)\}| \geq n|A| > \frac{|M||A|}{|C|},$$

hence for some $m \in M$ we get

$$|\{a \in A; a(m) = g_{a,\sigma}(m)\}| > |A|/|C|$$

and consequently

$$|\{a \in A; a \supseteq a', a(m) = g_{a,\sigma}(m)\}| > 1$$

for some $a' : M - \{m\} \to C$. But that means that for some $a_0 \neq a_1$ from A such that $a_i \supseteq a'$ for $i < 2$ we have $a_0(m) = g_{a_0,\sigma}(m) = \sigma(m, a', \emptyset) = g_{a_1,\sigma}(m) = a_1(m)$ and thus $a_0 = a_1$ — a contradiction. □

For infinite C, no strategy can guarantee even one correct guess:

Proposition 3 *Let M be finite, C be infinite. Then there is no winning strategy for the instance $\mathtt{hNsA}(M, C, e_{\geq 1})$.*

Proof. Suppose that a winning strategy σ exists. Let $C' \subseteq C$ such that $|C'| = |M| + 1$. Clearly, existence of σ implies existence of a winning strategy for $\mathtt{hNsA}(M, C', e_{\geq 1})$, which is impossible by the previous proposition. □

2.1.2 Case: M infinite

If $|C| = 1$, then obviously every strategy guarantees all guesses to be correct. For infinite M and $|C| > 1$ there is a strategy that guarantees at most finitely many incorrect guesses, but no fixed finite number of incorrect guesses can be guaranteed. The implication "$\Leftarrow$" in the following proposition is known as the Gabay-O'Connor theorem (see e.g., [8, Theorem 3.2.2]):

Proposition 4 *Let M be infinite, $|C| > 1$ and κ a cardinal. There is a winning strategy for the instance* $\mathtt{hNsA}(M, C, e^{<\kappa})$ *if and only if $\kappa \geq \omega$.*

Proof. "$\Leftarrow$": Clearly, it is enough to find a winning strategy for $\mathtt{hNsA}(M, C, e^{<\omega})$. Denote $\sim$ the equivalence on A such that $a \sim a'$ if and only if $a(m) \neq a'(m)$ for at most finitely many $m \in M$. Let $s : A/{\sim} \to A$ be a selector (i.e. $s([a]_\sim) \sim a$ for all $a \in A$). Then a winning strategy can be defined as

$$\sigma(m, \alpha, \emptyset) = s([a]_\sim)(m),$$

for $m \in M$ and $\alpha : M - \{m\} \to C$, where $a \in A$ is arbitrary such that $\alpha \subseteq a$ (the definition is correct as any $a, a' \supseteq \alpha$ may have different value only at m and thus $a \sim a'$). Indeed, σ is winning, as for all but finitely many $m \in M$ we get $g_{a,\sigma}(m) = \sigma(m, a \restriction (M - \{m\}), \emptyset) = s([a]_\sim)(m) = a(m)$.

"$\Rightarrow$": Suppose that there is a winning strategy σ for $\mathtt{hNsA}(M, C, e^{<n})$ for some $n \in \mathbb{N}$. Choose $C' = \{c_0, c_1\} \subseteq C$ with $c_0 \neq c_1$ and $M' \subseteq M$ such that $|M'| = 2n$. Define a strategy σ_0 for $\mathtt{hNsA}(M', C', e^{<n})$ by $\sigma_0(m, \alpha, \emptyset) = \sigma(m, \alpha \cup ((M - M') \times \{c_0\}), \emptyset)$ for $m \in M'$ and $\alpha : M' - \{m\} \to C'$. Since σ is winning, σ_0 is as well. That means that for each $a \in {}^{M'}C'$, we have $|\{m \in M'; a(m) = g_{a,\sigma_0}(m)\}| > |M'| - n = n$ and thus σ_0 is winning also for $\mathtt{hNsA}(M', C', e_{\geq n+1})$. But by Proposition 2 there is no winning strategy for $\mathtt{hNsA}(M', C', e_{\geq k})$ with $k > |M'|/|C'| = n$ — a contradiction. □

3 Well-ordered cases

We now move to cases where the set M is thought of as well-ordered and S and H are defined through this ordering.

Let $\mathcal{M} = \langle M, \prec \rangle$ be a nonempty well-ordering. By $\succ$ we mean $\prec^{-1}$. We define two subcases of the problem of prisoners and hats – the "hear nothing, see forward" case and the "hear backward, see-forward" case:

- $\mathtt{hNsF}(\mathcal{M}, C, e) = (M, C, A, S, H, e)$ where $A = {}^M C$, $S = \succ$ and $H = \emptyset$,
- $\mathtt{hBsF}(\mathcal{M}, C, e) = (M, C, A, S, H, e)$ where $A = {}^M C$, $S = \succ$ and $H = \prec$.

In less formal words, the mathematicians are standing in a (well-ordered) line facing the (possibly infinite) tail of the line. In $\mathtt{hNsF}(\mathcal{M}, C, e)$ every mathematician sees all hats in front of her but hears nothing, while in $\mathtt{hBsF}(\mathcal{M}, C, e)$ she sees all hats in front of her and hears all guesses of her colleagues behind her.

Further on we assume (without loss of generality) that M is a nonzero ordinal number and $\prec$ is the usual ordering $<$ of ordinals, i.e., $\mathcal{M} = \langle \beta, < \rangle$ for an ordinal $\beta > 0$.

3.1 The "hear nothing, see forward" case

3.1.1 Case: M finite

For finite M and $|C| > 1$ no strategy guarantees even one correct guess:

Proposition 5 *Let $|C| > 1$ and $\mathcal{M} = \langle n, < \rangle$ for $0 < n \in \mathbb{N}$. Then there is no winning strategy for the instance* $\texttt{hNsF}(\mathcal{M}, C, e_{\geq 1})$.

Proof. Assume, for contradiction, that σ is a winning strategy for $\texttt{hNsF}(\mathcal{M}, C, e_{\geq 1})$. We construct an assignment $a \in A$ such that $g_{a,\sigma}(m) \neq a(m)$ for all $m \in M$. This can be done easily by reverse induction on n: In the $(n-i)$th step we set $a(i) \neq \sigma(i, a \restriction \{i+1, \ldots, n-1\}, \emptyset) = g_{a,\sigma}(i)$, which is possible thanks to $|C| > 1$. □

3.1.2 Case: M infinite

If $\mathcal{M} = \langle \alpha, < \rangle$ for α infinite, then the situation is similar to the "hear nothing, see all" case — at most finitely many incorrect guesses can be guaranteed, but there is no finite upper bound on their number. The idea of the proof of the implication "$\Leftarrow$" in the following proposition comes from [8, Theorem 4.2.1], however our setting is different:

Proposition 6 *Let $|C| > 1$, $\mathcal{M} = \langle \beta, < \rangle$ for an ordinal $\beta \geq \omega$, and κ be a cardinal. There is a winning strategy for the instance* $\texttt{hNsF}(\mathcal{M}, C, e^{<\kappa})$ *if and only if $\kappa \geq \omega$.*

Proof. Analogous to the proof of Proposition 4.

"$\Leftarrow$": Clearly, it is enough to find a winning strategy σ for $\texttt{hNsF}(\mathcal{M}, C, e^{<\omega})$: For $m \in M$ we define the equivalence $\sim_m$ on $A = {}^M C$ by $a \sim_m a'$ if and only if $a(m') = a'(m')$ for all $m' > m$ (that is for all m' that m can see). We denote by s the selector on $\mathcal{P}(A) - \{\emptyset\}$ that for $\emptyset \neq X \subseteq A$ selects $s(X)$ as the $<^A$-least element of X, where $<^A$ is a fixed well-ordering of A. The strategy σ can then be defined as

$$\sigma(m, \alpha, \emptyset) = s([a]_{\sim_m})(m),$$

for $m \in M$ and $\alpha : \{m' \in M; m < m'\} \to C$, where $a \in A$ is arbitrary such that $\alpha \subseteq a$ (this is correct as any two such a's are $\sim_m$-equivalent).

We prove that σ is winning for $\texttt{hNsF}(\mathcal{M}, C, e^{<\omega})$: For contradiction, let there be $a \in A$ and $M' \subseteq M$ infinite such that for all $m \in M'$

$$a(m) \neq g_{a,\sigma}(m) = \sigma(m, a \restriction \{m' \in M; m < m'\}, \emptyset) = s([a]_{\sim_m})(m). \tag{1}$$

Let us choose an increasing sequence $m_0 < m_1 < \cdots$ of length ω in M'. We show that then $s([a]_{\sim_{m_0}}) >^A s([a]_{\sim_{m_1}}) >^A \cdots$ is a strictly $<^A$-decreasing sequence in A, which is in contradiction with $<^A$ being a well-ordering: From $m_i < m_{i+1}$ it follows $(\forall a' \in A)(a \sim_{m_i} a' \to a \sim_{m_{i+1}} a')$, i.e., $[a]_{\sim_{m_i}} \subseteq [a]_{\sim_{m_{i+1}}}$ and thus $s([a]_{\sim_{m_i}}) \geq^A s([a]_{\sim_{m_{i+1}}})$ by the definition of s. But the equality is not possible since that would mean $g_{a,\sigma}(m_{i+1}) = s([a]_{\sim_{m_{i+1}}})(m_{i+1}) = s([a]_{\sim_{m_i}})(m_{i+1}) = a(m_{i+1})$, contradicting (1).

"$\Rightarrow$": Assume for contradiction that there is a winning strategy σ for $\texttt{hNsF}(\mathcal{M}, C, e^{<n})$ for some $0 < n \in \mathbb{N}$. Then we can take $\mathcal{M}' = \langle n, < \rangle \subseteq \mathcal{M}$ and define a strategy σ_0 for

$\mathtt{hNsF}(\mathcal{M}', C, e^{<n})$ by $\sigma_0(m, \alpha, \emptyset) = \sigma(m, \alpha \cup ((M - M') \times \{c_0\}), \emptyset)$ for $m \in M'$ and $\alpha : \{m' \in M'; m < m'\} \to C$, where $c_0 \in C$ is fixed. From our assumption on σ it follows that σ_0 is winning for $\mathtt{hNsF}(\mathcal{M}', C, e^{<n})$, and thus (because $|M'| = n$) also for $\mathtt{hNsF}(\mathcal{M}', C, e_{\geq 1})$. This contradicts Proposition 5. □

3.2 The "hear backward, see forward" case

The approach we are going to present is an algebraic one. All the theorems that we use and that go beyond an elementary course of algebra can be found in [11]. The author owes the credit for the idea to Jan Šaroch [9].

Before stating the crucial Lemma 7, let us recall some notions concerning Abelian groups. For an (additively written) Abelian group G and an ordinal δ, we denote by G^δ the product of δ copies of G, that is the group whose underlying set is ${}^\delta G = \{f; f : \delta \to G\}$ and the group operation is defined pointwise. By $G^{(\delta)}$ we denote the direct sum of δ copies of G, that is the subgroup of G^δ whose underlying set is $\{f; f : \delta \to G \text{ and } f(\alpha) = 0 \text{ for all but finitely many } \alpha \in \delta\}$. There is the natural sum homomorphism $\Sigma : G^{(\delta)} \to G$ defined by $\Sigma(f) = f(\alpha_0) + \cdots + f(\alpha_{n-1})$ where $\alpha_0, \ldots, \alpha_{n-1}$ is an enumeration of all $\alpha \in \delta$ for which $f(\alpha) \neq 0$ and the sum is computed in G.

Lemma 7 *For any ordinal $\delta > 0$ and cardinal $\mu > 0$ there is an Abelian group G of size μ such that the sum homomorphism $\Sigma : G^{(\delta)} \to G$ can be extended to a homomorphism $\Sigma' : G^\delta \to G$.*

Proof. For $\mu \geq \omega$, let G be a divisible Abelian group of size μ, e.g. $G = \mathbb{Q}^{(\mu)}$. Then G is injective in the category of Abelian groups and thus Σ factors through the inclusion $id_{G^{(\delta)}} : G^{(\delta)} \to G^\delta$, yielding the required Σ'.

For $1 < \mu < \omega$, we take $G = \mathbb{Z}/\mu\mathbb{Z}$ (the additive group of integers modulo μ). Then G is purely injective (because finite). The inclusion $id_{G^{(\delta)}} : G^{(\delta)} \to G^\delta$ is pure (even an elementary embedding, see [11, Corollary 1.8]) and so Σ factors through it, yielding Σ' as before. □

Now we are ready to show that for the "hear backward, see forward" case there is always a strategy that guarantees at most one incorrect guess:

Proposition 8 *Let $|C| > 1$, $\mathcal{M} = \langle M, \prec \rangle$ be arbitrary, and κ be a cardinal. There is a winning strategy for $\mathtt{hBsF}(\mathcal{M}, C, e^{<\kappa})$ if and only if $\kappa \geq 2$.*

Proof. "⇒": Suppose for contradiction that there is a winning strategy σ for $\mathtt{hBsF}(\mathcal{M}, C, e^{<1})$, that is that σ guarantees all guesses correct. Then for the $\prec$-least element $m \in M$ we get $a(m) = g_{a,\sigma}(m) = \sigma(m, \alpha, \emptyset) = g_{a',\sigma}(m) = a'(m)$ for any $\alpha : M - \{m\} \to C$ and $\alpha \subseteq a, a' : M \to C$. This is of course a contradiction as we may take a, a' such that $a(m) \neq a'(m)$.

"⇐": It is enough to find a winning strategy for $\mathtt{hBsF}(\mathcal{M}, C, e^{<2})$. The case $|M| = 1$ is trivial. Further on we assume $|M| > 1$. Without loss of generality suppose that $M = \delta \cup \{-1\}$, for an ordinal $\delta > 0$, $\prec$ is the usual ordering of $\delta \cup \{-1\}$ and $C = \mu > 0$ a cardinal. By Lemma 7 there are an Abelian group G of size μ and a homomorphism $\Sigma' : G^\delta \to G$ extending the sum homomorphism $\Sigma : G^{(\delta)} \to G$. Again, without loss of generality, we may assume that the underlying set of G is $\mu = C$.

We define σ as follows:

$\sigma(-1, \alpha, \emptyset) = \Sigma'(\alpha)$ for $\alpha \in G^\delta = {}^M C$, and $\sigma(\beta, \alpha, \gamma) = \gamma(-1) - \Sigma'(\alpha \cup \{(\beta, 0)\} \cup \gamma \restriction \beta)$, for $\beta \in \delta$, $\alpha : \{\beta'; \beta < \beta' \in \delta\} \to C$, and $\gamma : \beta \cup \{-1\} \to C$, where the subtraction is computed in G.

We prove that

$$g_{a,\sigma}(\beta) = a(\beta) \tag{2}$$

for all $\beta \in \delta$ and $a \in A$ by induction on β: First, $g_{a,\sigma}(-1) = \sigma(-1, a \restriction \delta, \emptyset) = \Sigma'(a \restriction \delta)$ (which may differ from $a(-1)$). For the inductive step, let us assume (2) for all $\beta' < \beta$. Then

$$\begin{aligned} g_{a,\sigma}(\beta) &= \sigma(\beta, a \restriction \{\beta'; \beta < \beta' \in \delta\}, g_{a,\sigma} \restriction (\beta \cup \{-1\})) = \\ &= g_{a,\sigma}(-1) - \Sigma'(a_\beta) = \\ &= \Sigma'(a \restriction \delta) - \Sigma'(a_\beta) = \\ &= \Sigma'(a \restriction \delta - a_\beta) = \\ &= \Sigma(a \restriction \delta - a_\beta) = \\ &= a(\beta), \end{aligned}$$

where $a_\beta = a \restriction (\delta - \{\beta\}) \cup \{(\beta, 0)\}$, the first equality is just definition of $g_{a,\sigma}$, the second one holds due to the inductive assumption, the third one holds by the "-1st step", and the last three follow from the properties of Σ and Σ' and the definition of a_β. □

References

[1] AGGARWAL, G., FIAT, A., GOLDBERG, A. V., HARTLINE, J. D., IMMORLICA, N., SUDAN, M., Derandomization of auctions, in *Proceedings of the 37th Annual ACM Symposium on Theory of Computing, Baltimore, MD, USA, May 22-24, 2005*, 2005, pp. 619–625.

[2] BUSHI, J., *Optimal strategies for hat games*, master thesis, Portland State University, 2012.

[3] BUTLER, S., HAJIAGHAYI, M. T., KLEINBERG, R. D., LEIGHTON, T., Hat guessing games, *SIAM J. Discrete Math.*, 22(2), 2008:592–605.

[4] FEIGE, U., You can leave your hat on (if you guess the color), Technical report, The Weizmann Institute of Science, 2004.

[5] GALVIN, F., PRIKRY, K., Infinitary Jonsson algebras and partition relations, *Algebra Universalis*, 6(1), 1976:367–376.

[6] GARDNER, M., *The 2nd Scientific American Book of Mathematical Puzzles & Diversions*, Simon and Schuster, New York, 1961.

[7] HARDIN, C. S., TAYLOR, A. D., An introduction to infinite hat problems, *The Mathematical Intelligencer*, 30(4), 2008:20–25.

[8] HARDIN, C. S., TAYLOR, A. D., *The Mathematics of Coordinated Inference. A Study of Generalized Hat Problems*, Developments in Mathematics, Springer International Publishing, 2013.

[9] ŠAROCH, J., Úloha "Piráti s klobouky" – Algebraický přístup, *unpublished note*, 2010, in Czech.

[10] WINKLER, P., Games people don't play, in WOLFE, D., RODGERS, T., editors, *Puzzlers' Tribute: A Feast for the Mind*, A K Peters, Natick, 2001, pp. 301–313.

[11] ZIEGLER, M., Model theory of modules, *Annals of Pure and Applied Logic*, 26(2), 1984:149 – 213, special issue.

Current address

Glivický Petr, RNDr., Ph.D.
Department of Mathematics
Faculty of Informatics and Statistics
University of Economics, Prague
Ekonomická 957, 148 00 Praha 4, Czech Republic
E-mail: petrglivicky@gmail.com

Slovak University of Technology in Bratislava
Faculty of Mechanical Engineering
17th Conference on Applied Mathematics **APLIMAT 2018**

Proceedings

STOCHASTIC ANALYSIS OF POPULATION DYNAMICS SUBJECT TO CHANGES AT THE RANDOM TIME MOMENTS

GOLDŠTEINE Jolanta (LV), POLA Aija (LV), ŠADURSKIS Kārlis (LV)

Abstract. The paper presents a mathematical model of a predator-prey population dynamics given in a form of Markov dynamic impulse-type system, allowing discontinuities of trajectories at random time moments. The stochastic approximation procedure allows to analyse the qualitative behaviour of trajectories of the ecosystem.

Key words: stochastic modelling, logistic equations, stochastic approximation

Mathematics Subject Classification: Primary 92B05,60H30,08A72; Secondary 60G57, 37H10

1 Introduction

The method of qualitative analysis of predator-prey population behaviour by means of differential equations was first proposed at the beginning of the last century in [1,2]. These models assume continuous contact of predator population, with phase coordinate $\{y(t), t \geq 0\}$, with the prey population, with a phase coordinate $\{x(t), t \geq 0\}$. Subsequently, numerous papers on the analysis of a more general system of equations find a stable coexistence of populations (see the review in [3]). The most popular and quite common system of differential equations describing the dynamics of populations of predator-prey ecosystem is Kolmogorov model [4]:

$$\frac{dx}{dt} = \alpha(x)x - V(x)y, \qquad \frac{dy}{dt} = M(x)y, \tag{1.1}$$

with different types of right part of the equation. The simplest prey-predator model, the Lotka-Volterra model, has a form (1.1) with $a(x) \equiv \alpha, V(x) = \beta x$ and $M(x) = -\delta + \gamma\beta x$, and was analysed in detail by authors [1,2]. All the parameters in the equation (1.1) are positive, therefore the unique nontrivial equilibrium state of the system is the centre type fixed point $\{y = \alpha / \beta; x = \delta / \gamma\beta\}$ and all phase trajectories have the form of closed orbits. The equation (1.1) allows to explain the often observable fluctuations in predator-prey populations. However, this model is not robust i.e., small perturbations in the right-hand side of the equation (1.1) can significantly change the phase portrait. Later, a more realistic model with

$a(x) = r(1 - K^{-1}x), V(x) = \beta x$ and $M(x) = -\delta + \gamma\beta x$ was analysed in many papers (see the review in [3]). This model allows us to detect the asymptotically stable equilibrium and possibility of damped oscillations of predator and prey population size in case of the sufficiently high birth rate of the prey population. There are also the more complicated structurally stable prey-predator mathematical models of type (1.1) which can explain persistent oscillations in a form of a sable limit cycle. Most popular of this type model is proposed in paper [5] as a system of differential equations (1.1) with

$$a(x) = r(1 - K^{-1}x), V(x) = mx(A + x)^{-1} \quad \text{and} \quad M(x) = -\delta + \gamma mx(A + x)^{-1}$$

Despite the plausible qualitative description of the dynamics of populations using different variants of equations (1.1), it can be argued that this equation describes the behaviour of predator-prey ecosystem only as an average. In contrast, real populations are subject to random perturbations, which do not allow population size to remain constant. This makes it necessary to consider possible random perturbations in analysis. For such cases, our paper offers a population dynamic modelling method that employs Markov dynamic impulse-type systems approach, allowing for discontinuity of trajectories at random time moments.

2 Stochastic approximation procedure

Let us formally describe the proposed predator-prey model. We assume that the dynamics of populations are not only determined by the system of equations (1.1), but also depend on the stationary Markov Poisson piecewise constant random process $\{\xi(t), t \geq 0\}$ defined by its infinitesimal operator [6]:

$$(Q_\varepsilon v)(z) = \frac{1}{\varepsilon}\int_0^1 (v(u) - v(z))du \qquad (2.1)$$

where ε is the small positive parameter. It means that if at the time $t_0 \geq 0$ process $\xi(t)$ is equal to $\xi(t_0) = z$, then in the time interval $t \in [t_0, t_0 + \tau)$, where τ has an exponential distribution with the parameter ε^{-1}, it maintains its value until the moment $t_0 + \tau$ when the process jumps to the value $\xi(t_0 + \tau) = \eta$, where η is independent on τ uniformly on the segment [0, 1] distributed random variable. The random variables τ and η do not depend on history of the process $\{\xi(s), s \leq t\}$. Consequently, for small $\varepsilon > 0$ dynamical characteristics of the predator-prey system change very frequently. These changes happen only at the discrete random time moments $t \in \{\tau_k, k \in N\}$. The length of intervals $\tau_k - \tau_{k-1}$ have the exponential distribution with the parameter $\lambda = \varepsilon^{-1}$. Our model for dynamics of predator-prey populations is given as

- the differential equations for time intervals $\tau_{k-1} \leq t < \tau_k$;

$$\frac{dx(t)}{dt} = a(x(t))x(t), \quad \frac{dy(t)}{dt} = [-\delta + M(x(t))]y(t); \qquad (2.2)$$

- the conditions of jumps for $t \in \{\tau_k, k \in N\}, k \in \mathbf{N}$:

$$x(t) = x(t-) - \varepsilon y(t) G(\xi(t-), x(t-)) \tag{2.3}$$

where, $G(\xi, x)$ and $M(x)$ are nonnegative and

$$\mathbf{E}\{G(\xi(t), x)\} \equiv \int_0^1 G(z, x)dz = V(x), \mathbf{E}\{| G(\xi(t), x)|^2\} \equiv D(x).$$

Probabilistic properties of phase coordinates $\{x(t+s), y(t+s), s \geq 0\}$ are independent of the history of these processes before the time t. Therefore, a vector stochastic process $\{x(t), y(t), t > 0\}$ possesses the Markov property and its probability characteristics are completely determined by the infinitesimal operator [6]:

$$\left(L(\varepsilon)v\right)(x, y) := \lim_{s \downarrow 0} \frac{1}{s}[\mathrm{E}\{v(x(t+s), y(t+s)) / v(x(t) = x, y(t) = y\} - v(x, y)] \tag{2.4}$$

where $v(x, y)$ is a bounded and continuously differentiable function. The previous formula shows the dependence of the infinitesimal operator on the small parameter ε that defines the intensity of contacts between predator and prey. If $\tau \geq s$, there is no contact between predator and prey species in the time interval $[t, t+s)$. Therefore, under condition $x(t) = x, y(t) = y$ the increments on this interval are as follows: $x(t+s) - x(t) = sxa(x) + o(s)$, $y(t+s) - y(t) = s[-\delta + M(x)]y + o(s)$. If $\tau < s$, then in the time interval $(t, t+s)$ one contact between predator and prey individuals will take place and $x(t+s) - x(t) = sxa(x) - \varepsilon y G(\xi(t-), x) + o(s)$After substitution of these increments into formula (2.4) we can define the weak infinitesimal operator for the Markov process defined by equations (2.2) - (2.3) as:

$$\left(L(\varepsilon)v\right)(x, y) = xa(x)\frac{\partial}{\partial x}v(x, y) + \varepsilon^{-1}E\{v(x - \varepsilon y G(\xi(t), x), y) - v(x, y)\} +$$

$$+[-\delta + M(x)\}]y\frac{\partial}{\partial y}v(x, y)$$

It can be shown that there exists a limit $\lim_{\varepsilon \to 0}\left(L(\varepsilon)v\right)(x, y) = \bar{L}v(x, y)$, where

$$\bar{L}v(x, y) = \left[xa(x) - yV(x)\right]\frac{\partial}{\partial x}v(x, y) + [-\delta + M(x)]y\frac{\partial}{\partial y}v(x, y) \tag{2.5}$$

The system of differential equations

$$\frac{d\bar{x}}{dt} = \bar{x}a(\bar{x}) - yV(\bar{x}), \frac{d\bar{y}}{dt} = [-\delta + M(\bar{x})]\bar{y} \tag{2.6}$$

has been proven [7] to be an averaged system for the initial Markov dynamical system and the deviations of the solutions of the Markov dynamical system (2.2) - (2.3) from the corresponding solutions of the equation (2.6) have an order of $\sqrt{\varepsilon}$. Therefore, we can present the solution of the equations (2.2) - (2.3) in the form of a sum

$$x(t) = \bar{x}(t) + \sqrt{\varepsilon}X_\varepsilon(t), \quad y(t) = \bar{y}(t) + \sqrt{\varepsilon}Y_\varepsilon(t) \tag{2.7}$$

and can proceed to the construction of a system of equations for two-dimensional Markov nonhomogeneous process $\{X_\varepsilon(t), Y_\varepsilon(t)\}$. In order to calculate infinitesimal operator for this process we have: $\left(\Lambda(\varepsilon)v\right)(X, Y) = \left(\Lambda(t)v\right)(X, Y) + O(\sqrt{\varepsilon})$, where

$$(\Lambda(t)v)(X,Y) :=$$
$$= v'_X(X,Y)\{[a'(\bar{x}(t))\bar{x}(t)X + [a(\bar{x}(t))X - V(\bar{x}(t))]Y] - \bar{y}(t)V'(\bar{x}(t))\}X\} + \tag{2.8}$$
$$+ v'_Y(X,Y)\{XM'(\bar{x}(t))\}\bar{y}(t) + [-\delta + M(\bar{x}(t))\}Y]\} + \frac{1}{2}v''_{XX}(X,Y)(\bar{y}(t))^2 D(\bar{x}(t))\},$$

and $v(X,Y)$ is an arbitrary twice continuous differentiable function with bounded derivative. As proven in [7], for a sufficiently small the probabilistic characteristics of the process $\{X_\varepsilon(t), Y_\varepsilon(t)\}$ do not differ much from the corresponding characteristics of the Markov process $\{X(t), Y(t)\}$ defined by generator (2.8) on any finite time interval with the length of the order of $1/\sqrt{\varepsilon}$. By definition [6] the formula (2.8) corresponds to a two-dimensional diffusion Markov vector-process $\vec{X}(t) := \begin{pmatrix} X(t) \\ Y(t) \end{pmatrix}$, defined by the equation:

$$d\vec{X}(t) = A(\bar{x}(t), \bar{y}(t))\vec{X}(t)dt + \vec{b}(\bar{x}(t), \bar{y}(t))dw(t) \tag{2.9}$$

where

$$A(\bar{x}(t), \bar{y}(t)) = \begin{pmatrix} a'(\bar{x}(t))\bar{x}(t) + a(\bar{x}(t)) - \bar{y}(t)V'(\bar{x}(t)) & -V(\bar{x}(t)) \\ M'(\bar{x}(t))\bar{y}(t) & -\delta + M(\bar{x}(t)) \end{pmatrix}, \tag{2.10}$$

$$\vec{b}(\bar{x}(t), \bar{y}(t)) = \begin{pmatrix} \bar{y}(t)\sqrt{D(\bar{x}(t))} \\ 0 \end{pmatrix} \tag{2.11}$$

and $w(t)$ is a standard Wiener process. Using the definition [9] of the stochastic Ito differential equations we can derive the ordinary differential equation

$$\frac{d}{dt}Q(t) = A(\bar{x}(t), \bar{y}(t))\vec{X}(t) + \vec{X}(t)A^T(\bar{x}(t), \bar{y}(t)) + \vec{b}(\bar{x}(t), \bar{y}(t))\vec{b}^T(\bar{x}(t), \bar{y}(t)) \tag{2.12}$$

for covariance matrix of the solution of the equation (2.6).

3 Impulse type logistic equation

Suppose that the dynamics of the prey population $x(t)$ in absence of contacts with predators may be described by the logistic growth equation

$$\frac{dx(t)}{dt} = rx(t)(1 - K^{-1}x(t)). \tag{3.1}$$

This model assumes that in the absence of a predator the prey population has a potential carrying capacity $K > 0$ and develops according to the logistic law with an intrinsic growth rate $r > 0$. Predator and prey individuals interact at random, and these time moments of interaction $\mathrm{T} := \{\tau_k, k \in N\}$ are discontinuities for the piece wise constant Poisson process with the infinitesimal operator (1.4). If the predator-prey contact occurs at time t and predator population size at that moment is $y(t)$, then the prey's population size drops by

$$x(t) - x(t-) = -\varepsilon y(t) g(\xi(t-))x(t-)) \tag{3.2}$$

where $g(\xi(t)) > 0$, $\mathrm{E}\{g(\xi(t))\} = \beta > 0$, $\mathrm{E}\{g^2(\xi(t))\} = b^2$. The dynamics of the predator population at any time continuously changes according to

$$\frac{dy(t)}{dt} = [-\delta + \gamma\beta x(t)]\, y(t) \qquad (3.4)$$

where $\gamma > 0$. Later we will use the results of the previous section with $a(x) = rx(1 - K^{-1}x), M(x) = \gamma\beta x, G(\xi, x) = g(\xi)x$. The averaged dynamics of the predator-prey system may be given in the form of ordinary differential equations

$$\frac{d\overline{x}}{dt} = [r(1 - K^{-1}\overline{x}) - \beta\overline{y}]\overline{x}, \quad \frac{dy}{dt} = (-\delta + \gamma\beta\overline{x})\,\overline{y} \qquad (3.5)$$

This dynamical system has 3 fixed points [8]:

- the saddle point $A_1 = (0,0)$;
- the point $A_2 = (K,0)$ that is asymptotically stable for $\delta > K\gamma\beta$ and a saddle for $\delta < K\gamma\beta$
- the point $A_3 = \left(\frac{\delta}{\gamma\beta}, \frac{r(\gamma\beta - K^{-1}\delta)}{\gamma\beta^2}\right)$ that has positive coordinates and for $\delta < K\gamma\beta$ is asymptotically stable.

Following these conclusions about the averaged system, we can analyse normalized deviations from the averaged trajectory

$$\{X_\varepsilon(t+s), Y_\varepsilon(t+s)\} = \{\varepsilon^{-1/2}(x(t+s) - \overline{x}(t+s)),\ \varepsilon^{-1/2}(y(t+s) - \overline{y}(t+s))\} \qquad (3.6)$$

using an approximation by the equation (2.9) with

$$A(x,y)) = \begin{pmatrix} r(1-K^{-1}x)x - \beta y & -\beta x \\ \gamma\beta y & -\delta + \gamma\beta x \end{pmatrix}, \qquad \vec{b}(x,y) = \begin{pmatrix} bxy \\ 0 \end{pmatrix} \qquad (3.7)$$

If $\delta > K\gamma\beta$, then [8] $\lim_{t\to\infty} \overline{y}(t) = 0, \ \lim_{t\to\infty} \overline{x}(t) = K$,

$$\lim_{\to\infty t} A(\overline{x}(t), \overline{y}(t)) = \begin{pmatrix} -rK & -\beta K \\ 0 & \delta + \gamma\beta K \end{pmatrix} := H, \quad \lim_{\to\infty t} \vec{b}(\overline{x}(t), \overline{y}(t)) = \begin{pmatrix} 0 \\ 0 \end{pmatrix}$$

and the eigenvalues of matrix H are negative. This means that the covariance matrix Q(t) defined by the equation (2.19) tends to zero and random oscillations at the neighbourhood of the equilibrium A_2 decay.

If $\delta < K\gamma\beta$ then [8] any solution of equation (3.5) tends to the equilibrium point A_3 and

$$\lim_{t\to\infty} A(\overline{x}(t), \overline{y}(t)) = \hat{A} = \begin{pmatrix} -\frac{r\delta^2}{K\gamma^2\beta^2} & -\frac{\delta}{\gamma} \\ \frac{r(K\gamma\beta - \delta)}{K\beta} & 0 \end{pmatrix}, \ \lim_{t\to\infty} \vec{b}(\overline{x}(t), \overline{y}(t)) = \hat{b} = \begin{pmatrix} \frac{r(\gamma\beta - K^{-1}\delta)b}{\gamma^2\beta^3} \\ 0 \end{pmatrix}$$

Under assumption $\delta < K\gamma\beta$ the eigenvalues of matrix $\hat{A}$ are negative. This means that at a sufficiently small neighbourhood of the stable equilibrium point A_3 the process $\vec{X}(t)$, as defined by the equation (2.9), under zero initial condition is a diffusion process [8] with a covariance matrix $Q(t)$, as defined by equation (2.12). After further computations we may conclude that in the case of $\delta < K\gamma\beta$ the populations of the prey and the predator are stabilizing near the fixed point A3 and probabilistic characteristics of deviations from this equilibrium may be approximately considered as independent normally distributed random variables with variances correspondingly

$$\mathbf{D}(x_\varepsilon(t)) \approx \varepsilon \frac{b\delta r^2(\gamma\beta - K^{-1}\delta)(2\gamma\beta + K^{-1}\delta)}{2\gamma^3\beta^4}, \quad \mathbf{D}(y_\varepsilon(t)) \approx \varepsilon \frac{b\delta r^2(\gamma\beta - K^{-1}\delta)^2(2\gamma\beta + K^{-1}\delta)}{2\gamma^3\beta^5}$$

4 The random impulsive Holling type model

One of the most popular models of the dynamics of interacting predator and prey populations, such as those found within invertebrate and similar domains of life, in mathematical biology is the system of ordinary differential equations proposed in [5] and in detail analysed in [10]:

$$\frac{dx}{dt} = rx\left(1 - K^{-1}x\right) - \frac{mx}{A+x}y, \quad \frac{dy}{dt} = \left(-\delta + \gamma\frac{mx}{A+x}\right)y \tag{4.1}$$

where phase variables x and y denote the density of prey and predator populations respectively. In this model it is assumed that in the absence of a predator the prey population has a potential carrying capacity K and develops according to the logistic law with an intrinsic growth rate r. In the absence of a prey the predator population exponentially decreases to zero with the intrinsic growth rate δ. The mutual influence of changes in the densities of the prey and predator in model (4.1) is considered by the trophic function $mxy(A+x)^{-1}$, where the positive parameter m is the consumption rate or prey per unit of time by the predator. The positive parameter A reflects the saturation of the amount of prey consumed and, in addition, depends on the rate of reaction of the predator, i.e., the time between attacks on the prey. The parameter γ in formula (4.1) is a conversion factor that determines the effect of the prey consumed on the growth rate of the population of the predator. The model (4.1) is popular because under certain assumptions about positiveness of parameters r, K, m, A, δ and γ it is structurally stable [2] with a unique asymptotically stable periodic trajectory. In this section we propose an impulsive Markov type model (2.2) - (2.3) with

$$a(x) = r\left(1 - K^{-1}x\right), M(x) = \gamma\frac{mx}{A+x}, G(\xi, x) = \frac{q(\xi)x}{A+x} \tag{4.2}$$

where $q(\xi(t)) \geq 0, \mathrm{E}\{q(\xi(t))\} \equiv m > 0, \quad \mathrm{E}\{q^2(\xi(t))\} \equiv \sigma^2$, that makes it possible to take into account the stochastic character of the trophic function considered in the Holling type II model (4.1). Substituting

$$a(x) = r\left(1 - K^{-1}x\right), \int_0^1 G(z,x)dz = V(x) = \frac{mx}{A+x}, M(x) = \frac{\gamma mx}{A+x} \tag{4.3}$$

into the equation (2.8), the average population dynamics $\{\overline{x}(t), \overline{y}(t)\}$ may be analysed by that equation (4.1), and [7] for any positive number T and M there exists such positive numbers ε_0 that

$$\sup_{t<T\varepsilon^{-1}} \mathbf{E}\{|x_\varepsilon(t) - \overline{x}(t)| + |y_\varepsilon(t) - \overline{y}(t)|\} \leq M\varepsilon \tag{4.4}$$

for all $\varepsilon < \varepsilon_0$. As shown in [10,11], the system of equations (4.1) with the parameter values $m=5$, $A=18$, $\gamma=2$, $\delta=4$ has a unique asymptotically stable periodic trajectory, which corresponds to initial values $\overline{x}(0)=20, \overline{y}(0)=3$.

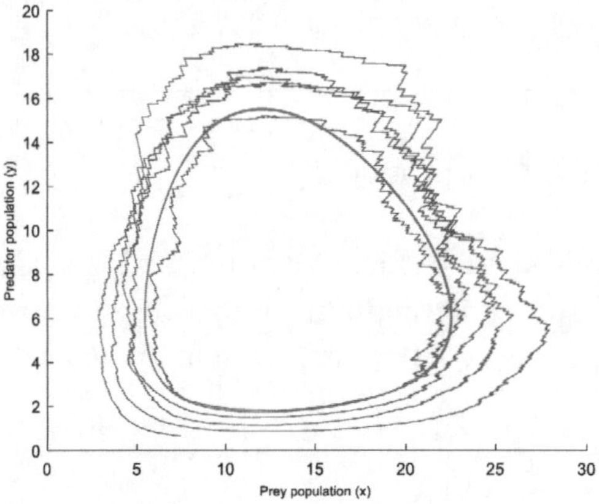

Fig. 4.1. Phase trajectory of Markov process (2.2) - (2.3) - (4.1) and the limit cycle.

In Figure 4.1 the trajectory shows the solution of the system of equations (2.2) - (2.3) - (4.1) for $\varepsilon = 0.01$ corresponding to the initial conditions above, and defined by equation (4.1) limit cycle. We may approximate the normalized deviations from the averaged trajectories

$$X_\varepsilon(t) = \frac{x_\varepsilon(t) - \overline{x}(t)}{\sqrt{\varepsilon}}, Y_\varepsilon(t) = \frac{y_\varepsilon(t) - \overline{y}(t)}{\sqrt{\varepsilon}} \tag{4.5}$$

by the vector solution $\vec{X}(t) := \begin{pmatrix} X(t) \\ Y(t) \end{pmatrix}$ of the equation (2.9)

$$d\vec{X}(t) = A(\overline{x}(t), \overline{y}(t))\vec{X}(t)dt + \vec{b}(\overline{x}(t), \overline{y}(t))dw(t) \tag{4.6}$$

Where

$$A(\overline{x}, \overline{y}) = \begin{pmatrix} \left(1 - \frac{2}{K}\overline{x}\right)r - \frac{A\overline{y}m}{(A+\overline{x}(t))^2} & -\frac{\overline{x}m}{A+\overline{x}} \\ \gamma\overline{y}\frac{mA}{(A+\overline{x})^2} & -\delta + \gamma\frac{\overline{x}m}{A+\overline{x}} \end{pmatrix}, \tag{4.7}$$

$$\vec{b}(\overline{x}, \overline{y}) = \sigma\begin{pmatrix} \sqrt{\frac{\overline{yx}}{2(A+\overline{x})}} \\ 0 \end{pmatrix} \tag{4.8}$$

and there are zero initial conditions. The covariance matrix $Q(t)$ for the process $\vec{X}(t)$ is the matrix solution of a system of ordinary differential equations

$$\frac{d}{dt}Q(t) = A(\overline{x}(t), \overline{y}(t))Q(t) + Q(t)A^T(\overline{x}(t), \overline{y}(t)) + \vec{b}(\overline{x}, \overline{y})\vec{b}^T(\overline{x}, \overline{y})$$

Now we can approximate the finite dimensional distributions of the process $\{x_\varepsilon(t), y_\varepsilon(t)\}$ by the corresponding distributions of the process $\{\tilde{x}_\varepsilon(t) = \overline{x}(t) + \sqrt{\varepsilon}X(t), \tilde{y}_\varepsilon(t) = \overline{y}(t) + \sqrt{\varepsilon}\ \ Y(t)$

However, the variances of the normalized deviations increase very rapidly. For example, $E\{|X(20)|^2\} = 10000$ but the aforementioned cycle ordinate does not exceed the value 25.

References

[1] LOTKA, A.J., *Elements of physical biology*, Williams and Wilkins, Baltimore, 1925.

[2] VOLTERRA, V., Variazioni e fluttuazioni del numero d'individui in specie animali conviventi, *Mem. R. Com. Tolassogr. Ital.* 131, 1927, p.1–142.

[3] FREEDMAN, H.I., *Deterministic Mathematical Models in Population Ecology*, New York, Marcel Dekker, 1980.

[4] KOLMOGOROV, A.N., Sulla teoria di Volterra della lotta per l`esistenza, In *Giornale Istituto Ital. Attuari,* 7, p.74-80, 1936.

[5] HOLLING, C.S., The components of predation as revealed by a study of small-mammal predation of the European pine sawfly, In *Can.Entomol.* 91, 1959, p. 293–320.

[6] DYNKIN, E.B., *Theory of Markov Processes.* Prentice-Hall, 1961.

[7] TSARKOV, Ye.F (CARKOVS, J.), Asymptotic methods for stability analysis of Markov impulse dynamical systems. In *Advances of Stability Theory of the End of XXth Century. Stability and Control: Theory, Methods and Applications*. Gordon and Breach Science Publishers, London, 13, 2000, p. 251-264.

[8] BAZYKIN, A.D., *Nonlinear Dynamics of Interacting Populations*, World Scientific, Singapore, 1998.

[9] SKOROKHOD, A.V., *Asymptotic methods in the theory of stochastic differential equations.* (Transl. Math. Monographs 78), American Mathematical Society, Providence, RI, 1989.

[10] GASULL, A., KOOIJ, R.E., TORREGROSA, J., Limit cycles in the Holling-Tanner model, *Publications Mathematiques,* Vol. 41, 1997, p. 149-167.

[11] CULLEN, M.R, The limit cycle in predator-prey models. http://www.cengage.com/math/book_content/0495108243_zill/projects_archive/debv4e/AModelingApplication6.pdf

Current address

Goldšteine Jolanta, doc, Dr.math.
Department of Probability Theory and Mathematical Statistics
Riga Technical University, Kaļķu iela 1, LV-1658, Riga, Latvia
E-mail: jolanta.goldsteine@lezards.lv

Pola Aija, lect.
Department of Probability Theory and Mathematical Statistics
Riga Technical University, Kaļķu iela 1, LV-1658, Riga, Latvia
E-mail: aija.pola@rtu.lv

Šadurskis Kārlis, prof., Dr.math.
Department of Probability Theory and Mathematical Statistics
Riga Technical University, Kaļķu iela 1, LV-1658, Riga, Latvia
E-mail: karlis.sadurskis@gmail.com

Slovak University of Technology in Bratislava
Faculty of Mechanical Engineering
17th Conference on Applied Mathematics **APLIMAT 2018**

Proceedings

CREATING AN INTERACTION WITH CELLULAR AUTOMATA FOR SCIENCE AND TECHNOLOGY MUSEUMS

HAUDE Lucas (NL)

Abstract. The development of technology creates a need for more students following science and technology curricula. Enthusiasm for these subjects is being raised for example by modern science and technology museums, and within schools increased attention is given to subjects like programming. This document describes the design process of an interaction with cellular automata. For interaction design, this means dealing with large sets of settings and a difficulty in offering a clear mental model to the user.

Keywords: cellular automata, interaction design, education

Mathematics Subject Classification: Primary 37B15

1 Introduction

The development of new technology brings changes to the way people live in multiple ways; current advances in robotics and artificial intelligence cause fears of ever more jobs being lost to a machine or an algorithm that performs better at lower cost. When Jacquard invented his groundbreaking automated loom weavers feared being put out of work and fiercely protested the new technology, not shunning violence and death to protect their interests. Technology and its monetary interests won; some weavers were indeed put out of work while others adopted the invention and became primordial programmers who punched holes in the cards that store instructions for the looms. This particular method of data storage and processing later inspired Babbage's analytical computers, forming the foundation for modern computing[1].
Around the advent of computers, John von Neumann tried to devise a machine that could reproduce itself; his approach contributed to what became known as cellular automata. Variants of cellular automata exist in different dimensions, but they share the attribute of living on a type of discrete grid. A square grid is most often used, but the principle can be used with other grid types such as hexagonal or triangular grids.

The state of each cell on the grid is updated according to the state of its neighbours: a ruleset determines the new state. Wolfram among others performed extensive research into elementary cellular automata which are one-dimensional in operation and whose output can be mapped in two dimensions; they work on a single line and leave a trace in a field. Three adjacent cells are used to determine the state of a cell on the next row. With the minimum two states per cell, this gives $2^{2^3} = 256$ different patterns, a particular instance of which called rule 30 can be used for random number generation[2]. Increasing the neighbourhood size to 5, the number of different outcomes explodes: $2^{2^5} = 4294967296$

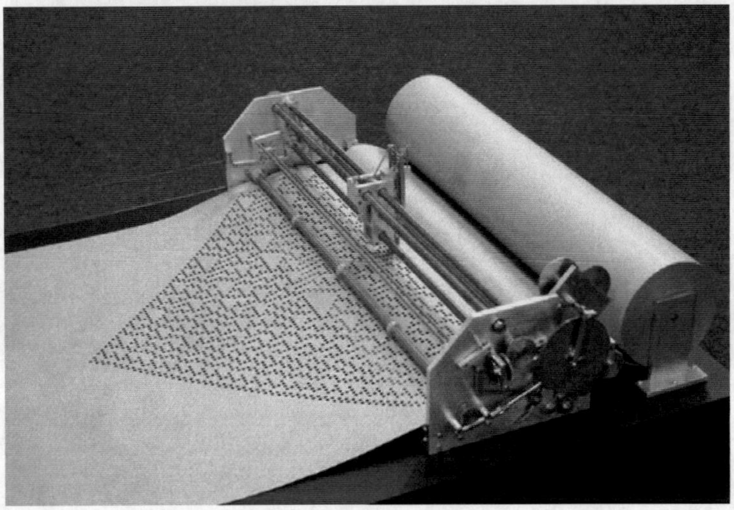

Fig. 1. Installation by K. Myskja that punches holes in a roll of paper according to rule 30.

Cellular automata are applied in science and design, for example to simulate chemical reactions, electrical circuits and traffic, or for generation of patterns[3, 4, 5]. An art display that mechanically recreates the aforementioned rule 30 shows the beauty of complex patterns from simple individual steps, as seen in figure 1.

These fascinating properties give an opportunity to spark interest in people unfamiliar with the principle of cellular automata. The graphical nature of the output lends itself to an interaction where the user influences the outcome; a type of generative design.

The following sections will describe the design process of an interaction with cellular automata within the context of a science and technology museum.

2.1 User interface with cellular automata

A simple starting point was chosen to create a first interaction with cellular automata, namely giving users direct control over the eight rules that govern an elementary cellular automaton. A board with eight switches controls a Processing sketch that iterates the ruleset. Each rule is displayed on screen together with its setting, while the other half shows the resulting pattern when the rule is iterated over a number of rows. Changing a rule updates the pattern to reflect

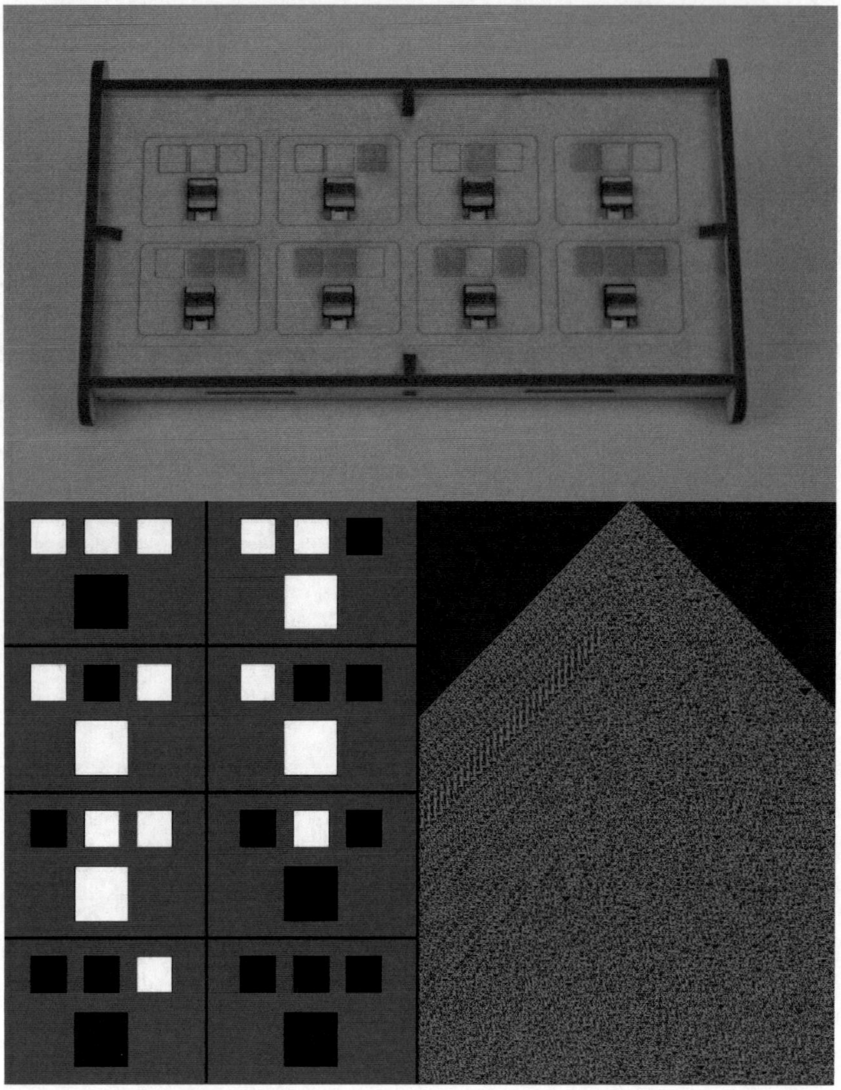

Fig. 2, 3. Switchboard with individual settings for the eight rules of the elementary automaton and the graphical interface showing the setting (rule 30 again) of each rule with the accompanying pattern.

the new settings. Figures 2 and 3 show the switchboard and the graphical interface respectively.

A small demonstration of this interaction was done within a lecture given at the faculty of Industrial Design at the Technical University Eindhoven. The general impression was that the cellular patterns are interesting to play with, but the overall GUI was somewhat confusing especially to people unfamiliar with the concept of cellular automata (CA).

The subsequent iteration should further clarify the principle of CA in a number ways: instead of instantly showing the cellular pattern iterated over hundreds of lines, a step-by-step approach should give more insight in how the patterns emerge.

2.2 Physical embodiment

Embodiment of interactions has been discussed and researched by amongst others Dourish [6]. The embodiment of an interaction determines the user's experience, and designers partly have influence over the embodiment. A substantial amount of an interaction's embodiment is determined by the user themselves: location, time, mood, social setting, and other factors are outside the sphere of the designer's influence.
Different frameworks for guiding the design of the remainder exist. This project undertakes the creation of a tangible interaction with cellular automata.; to make the abstract cellular automata more engaging, the embodiment needs to be brought from the digital to the physical world for both the CA input and output.

Through the aforementioned input device a form of interaction with cellular automata was created. Elements of the interaction could be classified as tangible: the eight rule inputs directly manipulate their digital counterparts. This connection was however not obvious because on the screen, the entire CA pattern was updated and displayed at once, making it difficult to see the individual rules and their characteristics in operation. The Processing sketch used to display the output could be altered to slow down the process and show the individual cell and line iterations, but that would not take the system output out of the digital realm. In order to bring the embodiment of the interaction further into the physical realm, the output would have to follow suit.

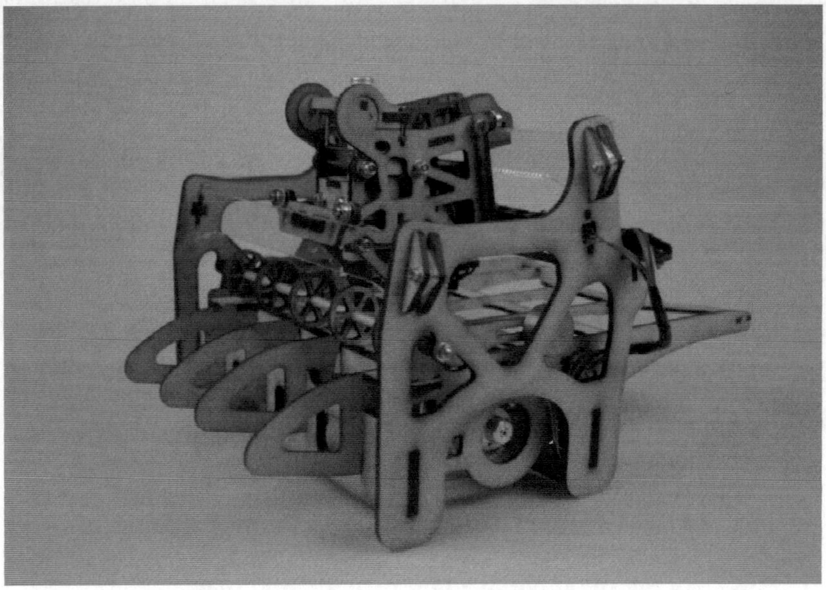

Fig. 4. Plotter designed and programmed to give a physical output to cellular patterns.

A decision was made to design a device with a mix of electronic and mechanical parts capable of creating a physical representation of CA patterns. The choice for this iteration was to create a plotter dedicated to printing CA patterns while enabling user interaction during the process;

figure 4 shows the plotter. A generic marker is operated by a solenoid while two stepper motors position the paper and pen across the two other axes.

2.3 User feedback & feedforward

In addition to the switchboard, serving as a device giving the user direct control over the rules, more interaction modalities were added. In the same vein as the switchboard, an accompanying panel - see fig. 5 - was designed with the following interactables:

- Indication lights of the current CA neighbourhood configuration
- Two-position switch to alter between 3-area and 5-area automata
- Dial to control the process speed.

The indication lights allow additional visual feedback on the process. They enable the user to see how the different rule configurations are applied individually, and follow the contribution each rule makes to the emerging pattern.

Switching between 3-area and 5-area cellular automata is intended to entice the user to try out the different outcome in operation and resulting patterns between the two modes. In the this iteration, a random ruleset is generated each time the mode is switched to the 5-area CA. The reasoning behind this decision is that the interaction with eight switches is already challenging to understand. The 5-area CA uses 32 rules - having that many switches would likely be overwhelming, and interaction would become somewhat haphazard experimentation rather than a concerted effort. Later iterations can deal with the larger sized automata.

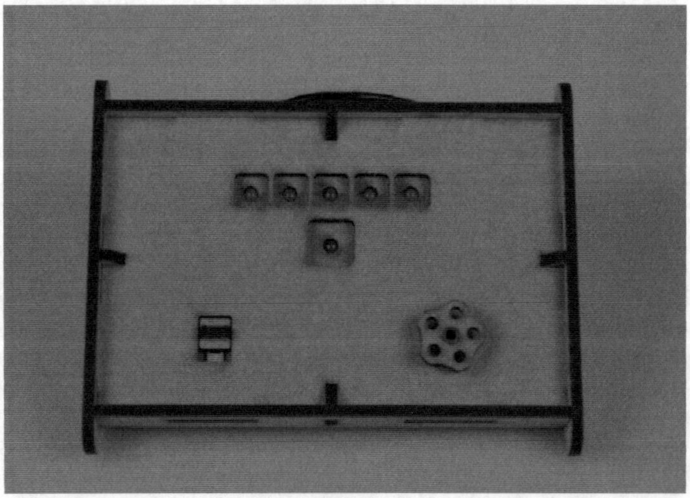

Fig. 5. User in/output in addition to the rule state switches.

A time component is often a part of interaction frameworks [7]. In tangible products, unity of location, direction, modality, and time are described as contributing to the coupling between user input and output. A time component can also influence the experience in a different way: *"Yet art forms with a strong temporal component (...) show that aesthetic experiences are built (...) on how the expression develops in a performance over time."* [8]

The user is given control over the process speed as the operation of cellular automata have a temporal component. While each individual cell state is determined, the user can slow down the process to gain a better understanding of each rule and the interaction between rules. To see how the rule configuration builds up over many iterations, the process can be accelerated to see the bigger picture.

2.4 Improving overall interaction quality

The first version of the plotter proved to work, but was not reliable enough. A redesign was made for an exhibition during the Maker Faire Eindhoven. The second iteration of the plotter allowed for more precise control over the size of the dots placed on the paper, faster operation due to improved stiffness of the construction and a better presentation in general. A visual comparison of both versions can be found in figures 6 and 7.

Fig. 6, 7. The first and second iteration of the cellular plotter.

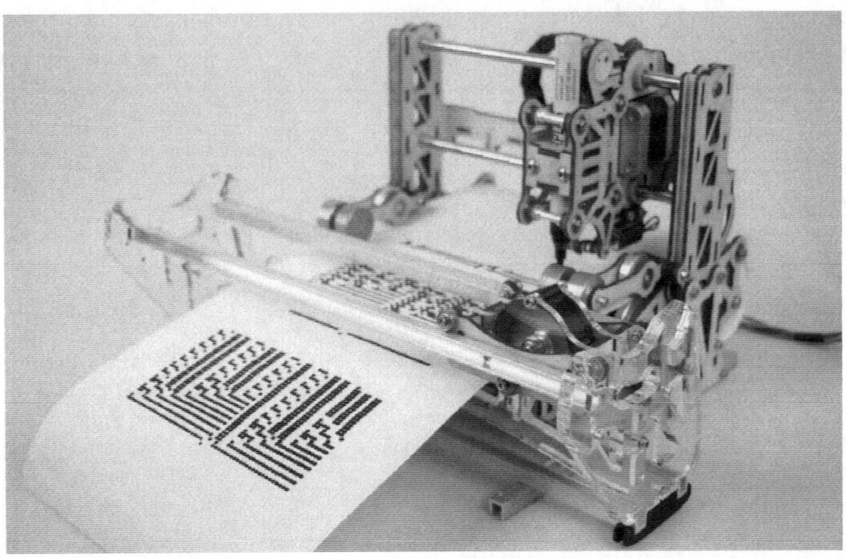

During the exhibition of the second iteration at the Maker Faire Eindhoven, it became clear that the project had potential and at the same time needed improvement, especially on the interaction side. Visitors were generally interested in the machine as well as the patterns, showing this combination of mathematics and technology has the ability to interest or even captivate both young and old visitors. It also showed the working of cellular automata was not necessarily clarified by the interaction, only what a their output patterns are like.
The most important remark made by users was their expectation of an immediate impact when changing the automaton rules. Another important observation was that the additional feedback and controls did not necessarily improve the user's understanding of how cellular automata operate. On the contrary, the flashing lights were seen as a nice distraction, but not contributing to their understanding of cellular automata without additional explanation. Switching between the 3 and 5 area automata also proved confusing as it takes some time for a new pattern to be take shape; there was too much of a disjunction between providing input and seeing output.

2.5 User interface for a large number of settings

Switching each individual rule as done with the elementary automaton does not translate well to automata with increased neighbourhood sizes; the switchboard with eight elements already proved to contain unpredictable or surprising results.
Taking into account the context of a museum, the 256 patterns offered by the elementary cellular automata are simply not numerous enough to provide sufficient variation for the amount of guests. Increasing the neighbourhood size to the next logical size of 5 suddenly solves this problem by offering over four billion different patterns with the comparatively low, four-fold increase in the number of rules.

From the observations of users interacting with the switchboard for the elementary CA, it didn't seem to be feasible to extrapolate this concept to the larger CA. A board containing 32 switches would most likely devolve to an interaction without meaning; research into interaction design within this context recommends taking a different approach.
Instead, an interaction was created using fewer controls. As an exploration a prototype was designed that contains three potentiometers and two switches. The analog to digital converter in Arduino boards is an 8-bit device, meaning it has a resolution of 1024 values. Combining the read-outs of three such potentiometers and two switches gives the exact amount of patterns that can be generated by a cellular automaton of neighbourhood size 5 while using five distinct settings.

$$2^{32} = 1024 * 1024 * 1024 * 4$$

A preview of the currently selected pattern is shown on an LCD, allowing users to scroll through the nearly endless variations; this immediate feedback is included to increase engagement through responsiveness.

Fig. 8. First iteration of an interaction with size 5 cellular automata.

This iteration of the project was exhibited during the Dutch Design Week; two modifications were made to the plotter. Instead of individual sheets of paper, the machine is fed by a roll of paper, making the machine more suitable for unattended expositions. Sections of paper can be cut off; this addition was made to give visitors the opportunity to take a souvenir of their experience.

3.1 Interaction design for museums

A non-exhaustive list of recommendations for interaction design within the context of a museum is given below. As the field of interaction design deals with sometimes subjective or hard to measure parameters, there is some disagreement on the best practices across different researches. Interacting with an artefact in a museum can be described as a sequence of steps, each with their particular recommendations. The final interaction design for this project is based on the previous iterations and these sources, arranged into sequences.

3.1.1 Trigger curiosity

One of the primary goals of an exhibit is attracting visitors. A sense of curiosity increases the likelihood a visitor engages with the exhibit [9]. If an exhibit does not invite visitors to further investigate, it can't be successful within the context of a science museum.

What constitutes a visually attractive exhibition differs across cultures and individuals; in order to make the object's attraction universal one can try to find a common ground, or one can make the decision to tailor their design to a specific user group. For example Hornecker and Stifter [10] recognise that older age groups have an interest in objects that evoke a nostalgic feeling, whereas the younger generation is more interested by an opportunity to play a game. Museums that promote technology among children will often have parents accompanying their children, so it makes sense in this context to create a design that appeals to both age groups.

One of Djajadiningrat's guidelines for interaction design[11] states "Don't hide, don't represent: show.", which is particularly applicable within a technology museum. Functional parts should be celebrated; the design of the plotter tries to adhere to this principle by being as transparent as possible about its working. This can be interpreted as an agreement with Simondon [12] who proposed the term techno-aesthetics for the beauty in forms or structures that are designed primarily from an engineer's perspective. Lattice bridges for example derive their structure from formulas describing loads and stresses instead of aesthetic considerations, but as it turns out this method of design has its own aesthetic merits. This design methodology is arguably more universal in its appeal; it also fits well within the theme of creating an interaction with a mathematical principle.

3.1.2 Engage in interaction

After the initial attraction visitors should be invited or persuaded to interact with the device. A classical approach would be to make use of clear affordances as popularised by Norman [13], something Djajadiningrat contradicts by proposing the affordances to be replaced by 'irresistables'. This step is strongly related to the first and sets apart art made purely to be thought-provoking and artefacts that offer an (educational) interaction. Djajadiningrat further argues for beauty in interaction and not in appearance. One could argue that beauty in appearance contributes in part to the 'irresistible' factor - however it's clear the interaction should be attractive in its own right to keep visitors engaged. The first few moments should provide incentive to continue or to capture the attention by offering a rich interaction instead of pressing buttons. Ease of use should take a backseat to enjoyment of the experience.

Giving immediate or direct feedback to the user is paramount; already during the Maker Faire exhibition this was noted as users expected an immediately visible effect when changing the automaton rules, and this was confirmed during the exhibition at the Dutch Design Week. Initial versions of the software only updated the LCD after a line was completely printed (for archaic reasons) and this turned away a substantial part of visitors who got the impression their input did not have an effect. Users were much more likely to try out different settings once their input was immediately processed and shown on the LCD.

3.1.3 Create experience

Once a visitor's interest is piqued, a longer interaction can be encouraged in different ways. Offering openings for their own creativity, following a story or narrative, and sharing ideas keep visitors engaged. Supporting information can be employed, but research recommends avoiding labels and large amounts of text as much as possible and instead design more around objects and artefacts. Enabling social interaction is also recommended as most museum visitors arrive in groups; visitors that are eager to learn more can achieve this goal by collaboration. Accommodating groups should not exclude visitors who interact by themselves however.

3.1.4 Closure

At some point the interaction will end, either because the user loses interest or the interaction scenario or narrative comes to a close. In the latter case users might choose to restart the interaction, perhaps with the intention to achieve a different outcome.
Interaction designers have the opportunity to define an end point or keep an open end, leaving it up to the user to decide on a moment to move on.
Within this design, a choice for an open end makes sense considering the context of free discovery. Had the design been made for a workshop or elective, it would be better to define a narrative or scenario and design the interaction with a clearly defined learning goal in mind.

After leaving, the visitor's experience becomes a memory that makes up part of the museum visit. Souvenirs can be offered in the gift shop or at individual exhibits. The latter works well when visitors can personalise the output of the exhibit - in this case, visitors are encouraged to take a piece of the paper they personalised.

4.1 Final interface

The final iteration of the user interface with the cellular pattern plotter combines experiences with the previous iterations and available literature. The design encompasses cellular automata with neighbourhood 5, and makes use of objects rather than switches or dials. The objects can be manipulated and represent the different combinations of cells that form the input of each iteration of a cellular automaton. Collaboration and social interaction is enabled, although an increase in scale would emphasise this aspect.

Each combination of states, a maximum of 32 in total, relates to a rule that can be switched 'on' or 'off' through an action inspired on the rich interaction paradigm [14]. The resulting pattern is updated on an LCD; when the user is satisfied with their setting of the rules, they can send it to the plotter and see the pattern emerge in a physical sense, taking home the printed version if they wish to do so.

Three distinct modes of operation are present in the interface:
- Altering the ruleset
- Printing the ruleset
- Transitioning between setting and printing

Figures 9 shows the final controller; a total of ten black and white tokens can be placed in the receptacle and scanned by pushing down an array of five infrared sensors that determine the current rule configuration. Holding the array in place for a short amount of time switches the rule state between on and off (see fig. 10). When a rule state is changed the pattern is updated on the LCD. This immediate feedback shows the rule has been read by the system, even if changing that particular rule has no effect on the pattern.

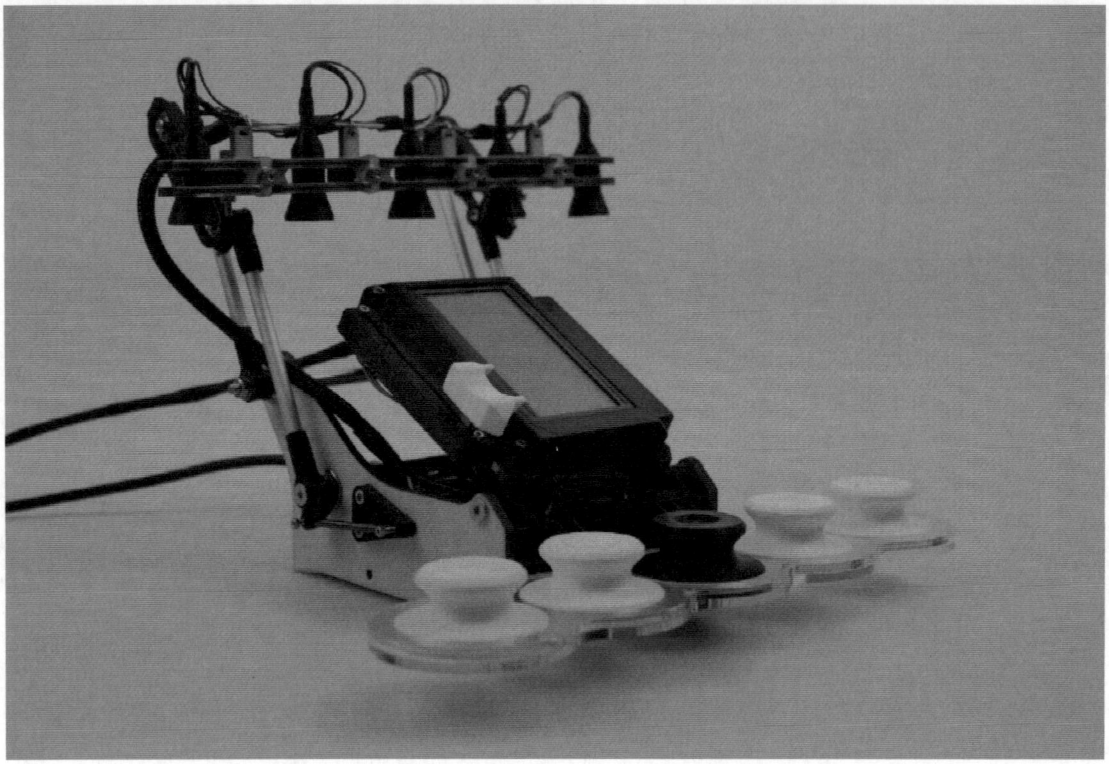

Fig. 9. Final iteration of the controller for interaction with cellular automata.

When the user is satisfied with the pattern they can instruct the plotter to print it on paper by pushing a knob mounted on a sliding potentiometer. Moving the knob upwards pushes the pattern out of the screen and into the printer; a section of whitespace is left between each printed pattern.

In previous iterations the ruleset was printed as long as no changes were made; following feedback from user tests the pattern matches exactly with what is shown on the LCD - after 124 lines the process is stopped. Users can cancel the printing and retract the pattern at any point by moving the slider back down: the pattern reappears on screen and can be altered

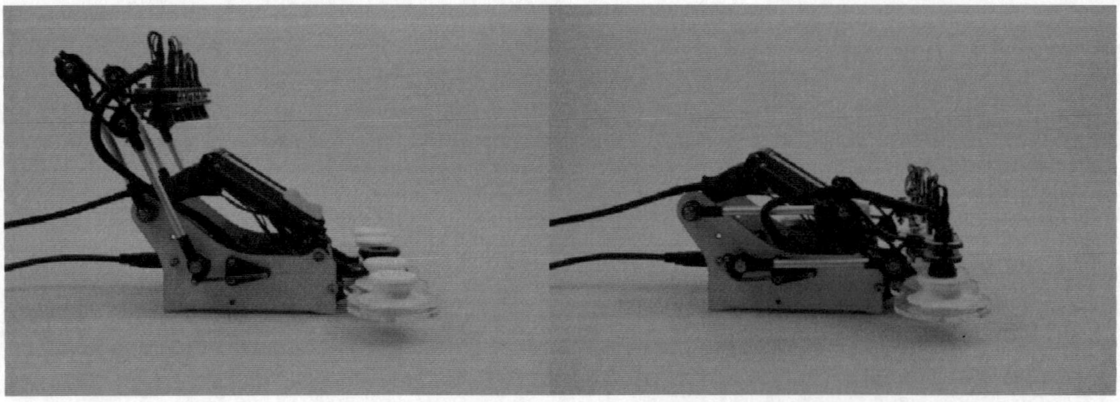

Fig. 10. The configuration of tokens is scanned by holding down the infrared sensors.

Fig. 11. photographs showing the transition from altering the ruleset to printing.

again by arranging and scanning tokens when the knob is fully lowered - fig. 11 shows the transitioning of the pattern between altering and printing. Figure 12 shows the arrangement of tokens to alter specific rules.

4.2 Conclusions

This projects attempts to create a meaningful interaction with cellular automata with the goal of increasing interest in STEM-related disciplines, in particular the elements of mathematics and engineering.

Different frameworks for interaction design are incorporated throughout iterations of the interface, starting out with a simple tangible interaction where users can directly manipulate the eight rules of an elementary cellular automaton. This iteration showed the need for a physical embodiment of the output, resulting in the creation of a dedicated plotter. In an attempt to make the automata clearer to the users, more feedback elements were added in the shape of rule indication lights and a control over the process speed, but trying these functions out with users lead to the conclusion that more feedback does not necessarily mean improved conceptual models.

These experiences lead to an iteration for slightly larger cellular automata with fewer controls.

Fig. 12. photographs showing arrangement of tokens - each of the 32 possible arrangements alters a rule of the automaton.

Users had indicated their expectation of receiving immediate feedback upon altering the ruleset; therefore an LCD was included to show the pattern resulting from iterating the current ruleset. Bluntly stated the version discussed in section 2.5 is a glorified scrolling function; the final interaction would have to be more meaningful as users were observed whizzing past many patterns without spending effort in trying to understand how the patterns are generated.

The final iteration combines experience from earlier explorations and a final literature research on interaction frameworks and research directed towards the context of a museum. Tangible interaction still forms the base of the interaction, but following research recommendations the switches operating the rule settings have been replaced by objects representing the automaton rules. In this way the 5-area automaton can be controlled without resorting to a switchboard containing 32 interaction modalities. Scanning the tokens is done in a way inspired by rich interaction - instead of pressing a button, the construction hints towards its functionality by transparently displaying the hinges and the resulting freedom of movement of the sensors. Another control from this framework is the transition slider: the wiring harness is located at the top of the screen, suggesting to the user that sliding the knob upwards transfers the data to the plotter, as if pushing the pattern from the screen into the printer.
From conversations with users, the most important feedback is that an interaction needs to have (nearly) immediate feedback on input; an unresponsive system is abandoned rather quickly. Printing the patterns has certain 'hypnotising' or 'fascinating' properties, lending credibility to the idea of placing a version of this prototype in a museum for science and technology. The resulting product shows the promise for object-based interactions with mathematical principles; the ten tokens can be combined to create over four billion different patterns, a compact and accessible way to navigate a large space of possibilities.

Acknowledgement
I would like to thank Loe Feijs for his supportive guidance throughout this project and his encouragement to write and submit this paper as well as the d.search lab at TU/e ID for their friendly help with technical difficulties throughout the project. I would also like to Bas and Pim for their insights on both engineering details and overarching design decisions.

References

[1] http://theinstitute.ieee.org/tech-history/technology-history/the-jacquard-loom-a-driver-of-the-industrial-revolution

[2] http://mathworld.wolfram.com/Rule30.html

[3] Reiter, C. (2004). With J: e Hodge Podge Machine. ACM SIGAPL APL Quote Quad, 34(4), 2-7.

[4] Feijs, L. M., & Toeters, M. J. (2017). A Cellular Automaton for Pied-de-poule (Houndstooth)

[5] Feijs, L. M., & Toeters, M. J. (2013). Constructing and applying the fractal pied de poule (houndstooth). Proceed- ings of Bridges, 27-31.

[6] P. Dourish (1999) Embodied Interaction: Exploring the Foundations of a New Approach to HCI

[7] Wensveen, S. A., Djajadiningrat, J. P., & Overbeeke, C. J. (2004, August). Interaction frogger: a design frame- work to couple action and function through feedback and feedforward. ACM.

[8] Djajadiningrat, T., Matthews, B., & Stienstra, M. (2007). Easy doesn't do it: skill and expression in tangible aesthetics. Personal and Ubiquitous Computing, 11(8), 657-676.

[9] Ciolfi, L., & Bannon, L. (2003, September). Learning from museum visits: Shaping design sensitivities. In *Proceedings of HCI international* (Vol. 1, pp. 63-67).

[10] Hornecker, E., & Stifter, M. (2006, November). Learning from interactive museum installations about interaction design for public settings. In *Proceedings of the 18th Australia conference on Computer-Human Interaction: Design: Activities, Artefacts and Environments* (pp. 135-142). ACM.

[11] Djajadiningrat, J. P., Overbeeke, C. J., & Wensveen, S. A. (2000, April). Augmenting fun and beauty: a pamphlet. In *Proceedings of DARE 2000 on Designing augmented reality environments* (pp. 131-134). ACM.

[12] G. Simondon (1982) On techno-aesthetics. Parrhesia journal, published 2012

[13] Norman, D. A. (1999). Affordance, conventions, and design. *interactions*, *6*(3), 38-43.

[14] Frens, J. J. (2006). Designing for rich interaction: Integrating form, interaction, and function (Doctoral dissertation, Technische Universiteit Eindhoven).

Current address

Haude L T H, BSc Eng.
Eindhoven University of Technology, Faculty of Industrial Design
Laplace 32, 5612 AZ Eindhoven, Netherlands
E-mail: l.t.h.haude@student.tue.nl

Slovak University of Technology in Bratislava
Faculty of Mechanical Engineering
17th Conference on Applied Mathematics **APLIMAT 2018**

Proceedings

PORTFOLIO OPTIMIZATION FOR RISK AVERSE AND RISK SEEKING INVESTOR

HECKENBERGEROVÁ Jana (CZ), ŠILD Petr (CZ)

Abstract. The goal of this work is to optimize portfolio for two types of investors: risk averse and risk seeking investor. The main purpose of optimization is to make the highest profit as possible in investing into the mutual funds with bearable risk corresponding to investor's profile. As input to this work is selection from equity, bond and commodity funds offered by ČSOB Asset management, because of the largest offer on the Czech market.

Keywords: portfolio optimization, profit, volatility, yield, simplex method

Mathematics Subject Classification: Primary 90C05.

1 Introduction

Investor who wants to invest on capital market has three ways how to pick securities into his portfolio. The first one is by stock picking which include many analyses such as technical analysis, fundamental analysis, macroeconomic analysis and other. The goal of stock picking is to buy securities which will bring better results than a market. The second one is to buy an index as DJIA, S&P 500 etc. Investor will reduce a risk but potential yield too and in the crisis he will report a massive loses. The third one is by using some algorithm such as Simplex method. [2] [7]

Collective investment is based on the principle of association of smaller investors, which brings more efficient asset management, cost reduction, risk diversification. Another benefit is investor's access to markets, which are not reserved to small investors, either due to high transaction costs or high risk for an inexperienced investor. The history of collective investment dates back to the time of the ancient Romans, who have been grouping their businesses to limit the loss of an individual. A similar principle is also applied, for example, in insurance companies where from collected assets from the group (premium) are covered individual damages. [5]

Risk assessment and risk evaluation stems from the entity's attitudes to risk. Different subjects can evaluate the same phenomenon differently. One will evaluate the phenomenon favourably, the other unfavourably. We can divide investors by maximum risk they are willing to take on risk seeking (dynamic) and risk averse (conservative) ones.

The goal of this work is to optimize portfolio for two types of investors: risk averse and risk seeking investor. Solution of portfolio optimization problem is provided by transformation into the linear programming task. Two different sets of limitations have been set for each type of investor. The final composition of portfolio is set by maximizing of profit function.

2 Financial Background

2.1 Mutual Funds

The operation of mutual funds is based on collective investment, which is based on the principle of the association of smaller investors that results in more efficient asset management, cost reduction and risk diversification. Another investor's benefit is an access to markets which are not reserved to small investors either because of high transaction costs or high risk for an inexperienced investor.

But how does such a mutual fund work in practice? Investors invest any amounts into the fund and buy the so-called units emitted by the fund. Therefore, the total assets of the fund increase, but the number of units is proportionally increased, so the price of the individual unit does not change by the transaction. If an investor wishes to withdraw invested money from a mutual fund, this process works exactly the opposite. The fund's total assets will be reduced and the units repurchased back by the fund will cease to exist again. For this purpose, the funds must hold part of their assets (around 5 %) in the deposits. [6]

2.2 Performance and Risk Analysis

If you are selecting a fund, it is important not to blindly choose the one fund, which an "investment adviser" recommends in the bank. Choosing fund only by historical yields is not the wisest idea, as the market does not have memory (past revenues will not secure future ones). It is important to objectively analyze the performance and risk of the fund, as described below.

Volatility

By using volatility, it is possible to measure the fund's risk. This value reflects the rate at which the fund's yield fluctuates around the mean value. In the case of mutual funds, it is often possible to find historical volatility in the monthly reports (calculated from historical data), which is calculated by using the equation:

$$\sigma = \sqrt{\frac{1}{n-1}\sum_{i=1}^{n}(\ln\frac{S_i}{S_{i-1}} - \mu)^2} \quad (1)$$

,

where $\mu = \frac{1}{n}\sum_{i=1}^{n} \ln\frac{S_i}{S_{i-1}}$, S_i is the value of the underlying asset (in our case, units of mutual fund) at time i and n is the number of observations. It is therefore a standard deviation from the logarithm of the fund's returns for a certain period. This value is further annualized due to better comparative capabilities by relation$\sigma_{p.a.} = \sigma\sqrt{h}$, where h is the number of trading periods per year. [7]

Sharpe ratio

Author of the Sharpe Index (William Sharpe) wanted to compare funds with the same yield by the risk. Thus he defined the Sharpe index, which is defined as the ratio of the yield of the asset over the safe interest rate and its risk:

$$S = (R_a - RF)/\sigma_a. \tag{2}$$

Funds with this index of 1 or more are very few and they made very good results. [7] [8]

2.3 Simplex Method

The Simplex method is an iterative computational procedure which can be used for finding an optimal solution for a linear programming task. The base principle is explained in the following diagram.

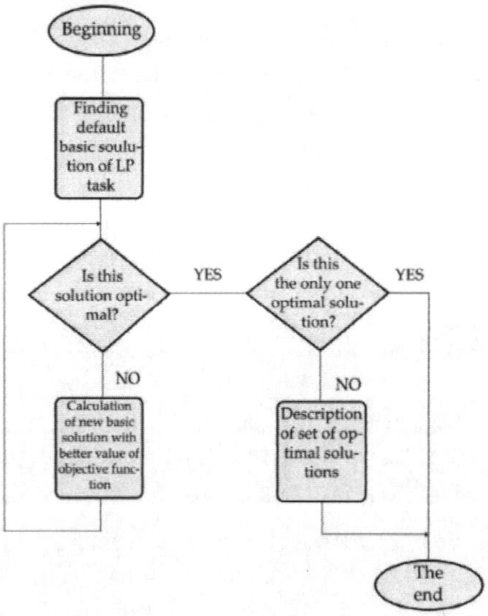

Fig. 1. Scheme of Simplex method.

In the first phase of this algorithm, is necessary to find the default base solution. Then the Simplex method in each step calculates a solution with better or at least the same value of an objective function. In case of maximizing the objective function, this value is higher, in case of minimalizing the objective function, the value is smaller. This method must lead to finding the base solution with the best value of the objective function, or to determine that there is no such solution. This solution must be optimal in compliance with the base sentence of linear programming. [3] [4]

In some cases, the first phase of finding the initial base solution is so easy that it is dropped and the algorithm continues with an iterative process that leads to the optimization of the objective z function. In this case the method is being called a one-phase simplex method. If non-trivial calculations are needed to find the initial base solution, it is a two-phase simplex method. [3]

3 Problem Formulation

Two investors are defined: conservative one and dynamic one. Both investors want to maximize their profits by investing in selected eleven mutual funds offered by ČSOB Asset Management under the following conditions:

Conservative investor

- In mutual funds, will be invested 100% of all money;
- Share of one mutual fund in portfolio can't exceed 25% of the whole portfolio;
- The weighted arithmetic mean of the standard deviations (volatility) of the individual funds in the portfolio must be less than or equal to 0.06;
- At least 50% of the portfolio must be composed of funds with a standard deviation of no more than 0.025;
- At least 20% of the portfolio must be composed of funds which have a market correlation close to zero.

Dynamic investor

- In mutual funds, will be invested 100% of all money;
- Share of one mutual fund in portfolio can't exceed 25% of the whole portfolio;
- The weighted arithmetic mean of the standard deviations (volatility) of the individual funds in the portfolio must be less than or equal to 0.12;
- At least 20% of the portfolio must be composed of funds with a standard deviation of no more than 0.05;
- At least 20% of the portfolio must be composed of funds which have a market correlation close to zero.

Index	Fund	Yield (c)	Standard deviation (σ)	Sharpe index
1	ČSOB Akciový	16,4 %	11,0 %	1,45
2	ČSOB Akciový Amerika	30,8 %	14,6 %	2,07
3	ČSOB Akciový realitní	6,1 %	14,3 %	0,39
4	KBC Equity Fund Food and Personal	5,6 %	10,6 %	0,49
5	ČSOB Komoditní	20,6 %	16,1 %	1,24
6	ČSOB Střednědobých dluhopisů	-1,0 %	0,8 %	-1,99
7	KBC Bonds High Interest	3,7 %	4,4 %	0,73
8	KBC Bonds Corporates EURO	4,5 %	2,2 %	1,87
9	KBC Multi Interest ČSOB CZK Medium	0,7 %	0,4 %	0,42
10	Index Fund Euroland	18,7 %	19,1 %	0,95
11	ČSOB Akciový BRIC	39,9 %	16,2 %	2,44

Tab. 1. Basic characteristics of selected funds.

Selected input funds are described in Table 1. The standard deviation is annualized over the preceding 12 months. For the purpose of this issue, yield of the fund is taken as the historical yield for the preceding 12 months. Fund of food, personal products and commodity fund are considered as funds with a very low correlation to the market.

4 Transformation into Linear Programming Model

First of all, it is necessary to define x_i as a share of the i-th fund in the portfolio. This variable is unknown and the goal of the task is to find all x_i that maximize profit of the portfolio and that is in compliance with investor's criteria. Since the goal is to maximize profit, the objective function will take the form of:

$$\text{maximize } \sum_{i=1}^{11} x_i \cdot c_i, \text{ for each } i \text{ applies } x_i \geq 0 \tag{3}$$

Restrictive conditions for a conservative user are:

$$\begin{gathered} \sum_{i=1}^{11} x_i = 1 \\ x_i \leq 0{,}25 \\ \sum_{i=1}^{11} x_i \cdot \sigma_i \leq 0{,}06 \\ x_6 + x_8 + x_9 \geq 0{,}5 \\ x_4 + x_5 \geq 0{,}2 \end{gathered} \tag{4}$$

For dynamic user, the restrictive conditions will be:

$$\begin{gathered} \sum_{i=1}^{11} x_i = 1 \\ x_i \leq 0{,}25 \\ \sum_{i=1}^{11} x_i \cdot \sigma_i \leq 0{,}12 \\ x_6 + x_7 + x_8 + x_9 \geq 0{,}2 \\ x_4 + x_5 \geq 0{,}2 \end{gathered} \tag{5}$$

5 Solution

The task was solved by using already mentioned Simplex method in the Solver tool of MS Excel 2016. The following table describes the portfolio of the conservative and dynamic investor.

Index	Fund	Conservative investor: share [x]	Dynamic investor: share [x]
1	ČSOB Akciový	0,0 %	4,9 %
2	ČSOB Akciový Amerika	0,0 %	25,0 %
3	ČSOB Akciový realitní	0,0 %	0,0 %
4	KBC Equity Fund Food and Personal	0,0 %	0,0 %
5	ČSOB komoditní	20,0 %	20,0 %
6	ČSOB Střednědobých dluhopisů	17,5 %	0,0 %
7	KBC Bonds High Interest	0,0 %	0,0 %
8	KBC Bonds Corporates EURO	25,0 %	25,0 %
9	KBC Multi Interest ČSOB CZK Medium	25,0 %	0,1 %
10	Index Fund Euroland	0,0 %	0,0 %
11	ČSOB Akciový BRIC	12,5 %	25,0 %

Tab. 2. Investor's portfolio.

In the case of a conservative investor, this portfolio has a standard deviation of 4.5 %, and in the period 23.2.2016 to 23.2.2017, the yield was 10.2%, resulting in a Sharpe index of 2.16. The dynamic investor portfolio has a standard deviation of 8.78%, reaching 23.7 % yield in the same period. Sharpe index is in this case 2.64. The following chart compares the earnings of a conservative and dynamic investor from February 23, 2016 to February 23, 2017.

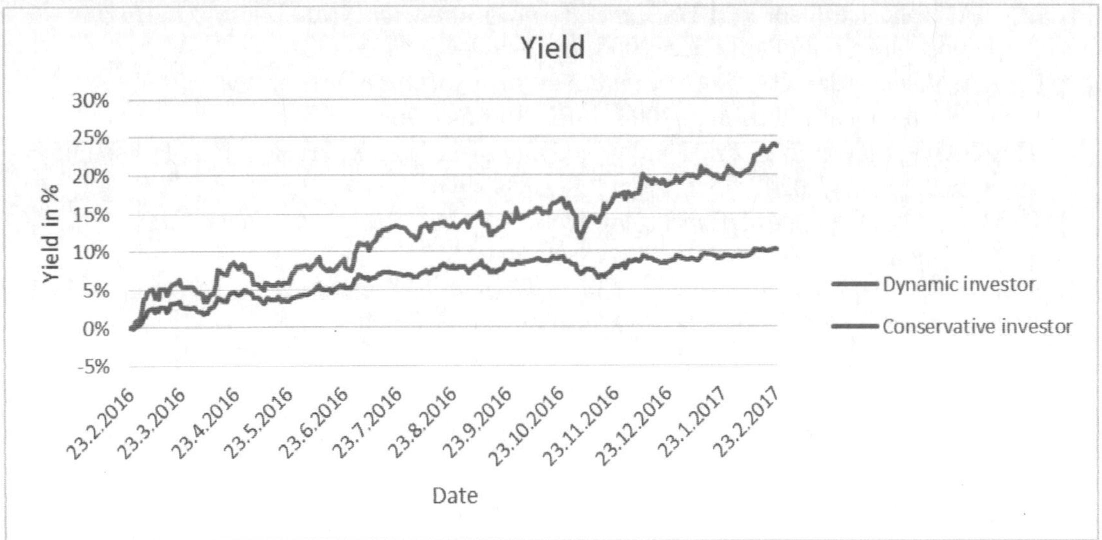

Fig. 2. Comparison of the conservative investor's and dynamic investor's portfolio.

6 Conclusion

The goal of the work was to create a portfolio of mutual funds which is in compliance with investor's criteria and which maximize profit. For the purpose of solving this problem was used the Simplex method. The data related to mutual funds has been obtained from the ČSOB company website.

The dynamic investor's portfolio achieved higher yield than conservative's portfolio, but with greater volatility. The dynamic investor portfolio shows even better Sharpe index than any of the funds in the portfolio. This is caused by diversification of nonsystematic risk.

Future improvement of this algorithm could be by adding an ability to change portfolio in real time which would depend on changes on market. This could protect investors' money in case of depression and make better profit in upturn.

Acknowledgement

This contribution was supported by the Student Grant of University of Pardubice.

References

[1] ČSOB *Všechny fondy* [online]. 2017 [cit. 2017-04-10]. https://www.csob.cz/portal/lide/produkty/investicni-produkty/podilove-fondy/vsechny-fondy

[2] INVESTOPEDIA *Guide to Stock-Picking Strategies* [online]. 2017 [cit. 2017-04-10]. http://www.investopedia.com/university/stockpicking/

[3] JABLONSKÝ, J. *Operační výzkum*. Professional Publishing, 2007, ISBN: 978-80-8694-644-3

[4] LINDA, Bohdan a Josef VOLEK. *Lineární programování*. Vyd. 2. Pardubice: Univerzita Pardubice, 2008. ISBN 978-80-7395-133-7.

[5] LIŠKA, Václav a Jan GAZDA. *Kapitálové trhy a kolektivní investování*. Praha: Professional Publishing, 2004. ISBN 80-864-1963-0.

[6] STEIGAUF, Slavomír. *FONDY jak vydělávat pomocí fondů*. Praha: Grada Publishing, 2003. ISBN 80-247-0247-9.

[7] TŮMA, Aleš. *Průvodce úspěšného investora: vše, co potřebujete vědět o fondech*. Praha: Grada, 2014. Partners. ISBN 978-80-247-5133-7.

[8] VALACH, Josef. *Investiční rozhodování a dlouhodobé financování*. 2., přeprac. vyd. Praha: Ekopress, 2006. ISBN 80-869-2901-9.

Current address

Heckenbergerová Jana, Mgr., Ph.D.
Institute of Mathematics and Quantitative Methods
Faculty of Economics and Administration
University of Pardubice
Studentska 95, 532 10 Pardubice, Czech Republic
E-mail: jana.heckenbergerova@upce.cz

Šild Petr, Bc.
Institute of Mathematics and Quantitative Methods
Faculty of Economics and Administration
University of Pardubice
Studentska 95, 532 10 Pardubice, Czech Republic
E-mail: sild.petr@gmail.com

Slovak University of Technology in Bratislava
Faculty of Mechanical Engineering
17th Conference on Applied Mathematics **APLIMAT 2018**

Proceedings

GRAPHLET COUNTING

HOČEVAR Tomaž (SI)

Abstract. Graphlet analysis describes a graph's topology by observing frequencies of 4- or 5-node induced subgraphs (graphlets) in the graph. We developed a graphlet counting approach based on a system of linear equations that relate the graphlet counts. For sparse large networks, such as those appearing in bioinformatics, the resulting algorithm is an order of magnitude faster than the existing approaches.

Keywords: graphlets, counting, network, bioinformatics

Mathematics subject classification: Primary 92C42, 05C60; Secondary 68R10, 68W01

1 Introduction

Networks are a very general modeling tool. All we need is a set of objects and a binary relation between them and we can build a network. Therefore, networks arise in a variety of research areas, from social sciences [2] and linguistics [3] to chemistry, pharmacology and bioinformatics [23]. We can analyze such networks to discover interesting properties that translate to new understandings of the underlying process.

One such approach is graphlet analysis. Graphlets are small connected patterns. Examples of 4-node graphlets are a path (P_4), a cycle (C_4), a cycle with a diagonal (called a diamond), etc. They are typically observed as induced subgraphs of the network. As small patterns present in the network, they represent a summary of the network's local structure. For example, denser networks would contain more diamonds and cliques, while sparser ones would contain a higher proportion of paths and cycles. Some other types of networks might differentiate themselves by a higher or lower presence of some other graphlet.

We can use graphlets to summarize the entire structure of the network by counting how many times each graphlet appears in it [22]. The other option is more node-centric and can be used to describe the structure around a particular node [16]. Let us describe a node by a vector of how many times it participates in each graphlet. To be even more precise, we can introduce the notion of orbits. We can say that there are two types or roles of nodes in a star graph: the central node and leaves. More formally, the orbits are equivalence classes of nodes under the action of the automorphisms of each graphlet. There are 11 orbits in 4-node graphlets. Therefore, we can assign a vector of orbit counts to each node that tells us, how many times does a node occur in the role of each orbit.

Usually, we observe graphlets consisting of up to 4 or 5 nodes. Because the number of graphlets of larger sizes grows very quickly (Table 1), it is not feasible to observe larger ones. Therefore, the graphlet frequency vectors would contain mostly zeros. Besides, larger graphlets already span a large portion of the network (small-world effect), which defeats the purpose of observing the local structure of a network.

k	2	3	4	5	6	7	8	9	10
graphlets	1	2	6	21	112	853	11117	261080	11716571

Tab. 1. The number of k-node graphlets.

If we use graphlet counts as a signature of a network's topology, how well does this actually describe the network? An interesting related problem is the reconstruction conjecture [13, 20]—are two graphs with the same multiset of vertex-deleted subgraphs isomorphic? A multiset of vertex-deleted subgraphs is equivalent to the frequencies of all $(n-1)$-node graphlets in a graph with n nodes. What about smaller graphlets? It is not hard to find a counterexample for 5-node graphlets. Graphs in Figure 1 have the same number of all 5-node graphlets but are not isomorphic.

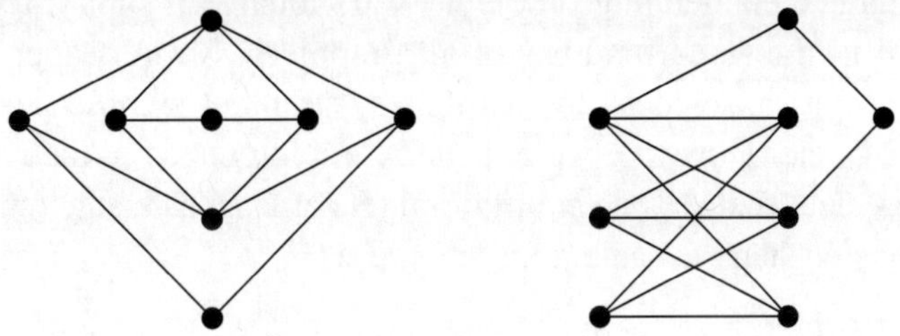

Fig. 1. Two nonisomorphic graphs with equal 5-node graphlet counts.

1.1 Applications

Graphlet analysis originates from bioinfomatics. There it was first used to propose a better random model for protein-protein interaction (PPI) networks. Pržulj et al. [22] proposed the relative graphlet frequency distance: a function that compares the local structure of two networks by comparing the relative frequency of each graphlet in the networks. They used their distance measure to show that the graphlet distribution in PPI networks is more similar to the graphlet distribution in random geometric graphs (induced by a set of randomly placed points in 3D space that are close enough) than to random networks that are modeling scale-free graphs.

Node-centric applications of graphlet analysis in bioinformatics are based on the assumption that the network topology of the modeled process is related to the function of objects corresponding to nodes. Let us consider the most prominent example of PPI networks. We model proteins with nodes and their interactions with edges. We assume that proteins with similar functions should form similar patterns of connections to neighbouring nodes in the network. Therefore, a similar local topology of two nodes is an indication that they might have similar protein functions or properties and should be investigated further. The goal of graphlet analysis in bioinformatics is often to narrow the search space of possible candidates that have to be further investigated experimentally; speeding-up the research process and reducing the expenses. Graphlets have been successfully employed to predict protein functions [16] and aid in discovery of cancer-related [14] or age-related genes [17].

Besides the already mentioned applications, they also have other, more indirect uses. One such example is their use in algorithms for network alignment. Given two similar networks, we would like to align or pair the nodes from both networks to maintain the adjacency. The alignment should be such that two nodes in the first network are adjacent exactly when their corresponding aligned nodes in the second network are adjacent. Because there is often no exact solution, we have to allow some mismatches. Exact penalties for such mismatches and the optimization function are a matter of application. In any case, we can make use of vectors of orbit counts for each node as a heuristic for aligning nodes that have a similar local topology. This led to development of several graphlet-based network alignment algorithms: GRAAL [11], H-GRALL [15], MI-GRAAL [12].

Finally, let us mention some other non-biological applications of graphlet analysis. Juszczyszyn et al. [9] observed graphlets in email-based social networks. Zhang et al. [28] used graphlets as an additional feature in the problem of aerial image categorization, while Mugan et al. [18] employed them for the task of entity resolution.

2 Graphlet Counting

Graphlet counting tools used in bioinformatics rely on exhaustive enumeration of all k-node graphlets. Enumeration of graphlets is a challenging problem itself. Ideally, it should require a computational time that is proportional to the actual number of all graphlets. Various approaches aim to achieve this by optimizing isomorphism testing [26], exploiting graphlet symmetries [27], simultaneously counting all graphlets [24], etc. However, we're usually not interested in their actual occurrences but only in their counts. It is clear that we can count graphlets at least as fast as we can enumerate them; but can we count them faster than we can enumerate them?

Let us consider the most simple case, a triangle. Surprisingly, we can count the number of triangles in an arbitrary graph with n nodes in $o(n^3)$ [8]. Let A represent a graph's adjacency matrix. Then, j-th element in i-th row of A^3 contains the number of paths of length 3 between from node i to j. It is well known that we can multiply two matrices in $o(n^3)$. The number of triangles is equivalent to $\frac{1}{6}\sum_{i=1}^{n}(A^3)_{i,i}$. Every cycle will be considered once for each of its three nodes and once for each direction, together six times. The solution for detecting larger cliques by Nešetřil and Poljak [19] reduces the problem to detecting a triangle in an auxiliary graph with $O(n^{k/3})$ nodes, resulting in a $o(n^3)$ algorithm.

We can count cliques faster than we can enumerate them. What about other graphlets? Kloks [10] presented a system of equations that relates all 4-node graphlets and can be set up and solved in $O(n^\omega)$, where ω represents the exponent in the time complexity of matrix multiplication. Eq. 1 is one of his equations (A denotes the adjacency matrix of graph $G = (V, E)$ and C the adjacency matrix of its complement).

$$\sum_{(x,y)\in E} (AC)_{x,y}(CA)_{x,y} = 4\#C_4 + \#P_4 \tag{1}$$

All mentioned approaches focus on solving a pattern counting problem in arbitrary dense graphs and rely on fast matrix multiplication techniques. However, in practical applications we encounter mostly sparse graphs, where it is often not possible to even construct an adjacency matrix, let alone multiply it.

2.1 ORCA

We developed ORCA (ORbit Counting Algorithm) [5] with the goal of outperforming existing enumeration techniques on sparse networks. It is based on a system of carefully chosen equations that can be set up faster than we can enumerate all k-node graphlets. In fact, it requires enumeration of only $(k-1)$-node graphlets. ORCA computes all orbit counts for every node in the network. From this more detailed statistic we can easily compute graphlet counts for each node and the total number of graphlets in the network.

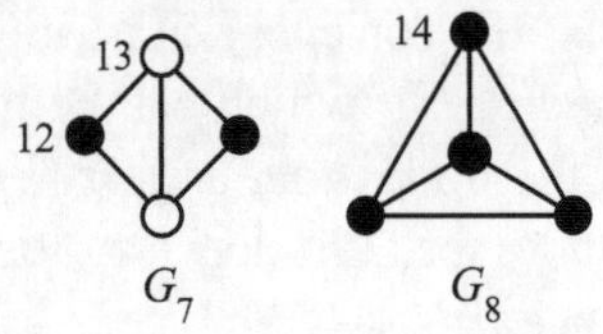

Fig. 2. Graphlets G_7 and G_8.

$$o_{12} + 3o_{14} = \sum_{y,z:\, y<z, G[\{x,y,z\}]\cong G_2} (c(y,z) - 1) \tag{2}$$

Equation 2 is an example of one such relation for a given node x in graph G that establishes a connection between orbits 12 and 14 of 4-node graphlets. The sum runs over all graphlets G_2 (C_3, a cycle with 3 nodes) in G that contain x as one of the nodes. We refer to the other two nodes as y and z. There are $c(y,z) - 1$ common neighbours of nodes y and z (without x), which together with x, y and z induce either a subgraph equivalent to graphlet G_7 with x in orbit 12 or graphlet G_8 with x in orbit 14. It turns out that we consider each occurrence of x in orbit 14 three times.

The system of equations for counting 4-node graphlets was designed on paper through trial and error of finding a system of independent equations that is efficient to set up. The rank of this system of equations is by 1 smaller than the number of orbits. Therefore, we have to know or compute one of the orbit counts. It turns out that we can do this efficiently (in sparse real-world network) for complete graphlets.

Moving on to 5-node graphlets required a more systematic approach (there are 58 orbits in 5-node graphlets). We observed possible extensions of 4-node graphlets by attaching a new node to some subset of its nodes. Each such extension leads to a candidate equation. We also showed that a similar approach as for counting node-orbits can be used for counting edge-orbits [6]. The developed algorithm has been used as a benchmark for other parallel [1] and approximate [4] graphlet counting methods.

A generalization to larger graphlets is of more interest from theoretical than practical perspective. We determined the conditions for construction of the required equations for the ORCA algorithm. For every node x in every graphlet G there has to exist another node y such that:

1. $d(y) \leq k-2$,

2. $G \setminus \{y\}$ is a connected graph,

3. if $d(y) = k-2$, the neighbours of y induce a connected graph,

where $d(y)$ represents the degree of node y. Such node y will represent the node by which we will try to extend a smaller graphlet with $k-1$ nodes. The first condition ensures that we are not concerned with common neighbours of too large sets of nodes, which are a part of precomputation. The second condition ensures that the algorithm will have to enumerate $(k-1)$-node graphlets and not some disconnected patterns. The last condition restricts the algorithm's space complexity and prevents memory accesses from dominating the run time. The algorithm's time complexity for counting k-node graphlets is $O(ed^{k-3}+T_k)$; e represents the number of edges, d is the maximum degree of a node and T_k is the time required to count cliques of size k.

We proved that such nodes indeed exist [7] for every noncomplete graphlet with at least 4 nodes. The only exception is a cycle with 4 nodes, C_4, which can be handled separately. There is a simple strategy that accomplishes this. From the nodes that are farthest away from x (suppose they are at distance u) pick the one with the lowest degree. If it doesn't satisfy the conditions, pick the node with the lowest degree among those at distance $u-1$.

Can we adapt this approach for the dynamic setting of the graphlet counting problem? We are given interleaved queries of three types: add an edge, remove an edge, compute orbit counts for a given node. The algorithm already handles the computation of orbit counts for each node independently, we just have to efficiently update the numbers of common neighbours (i.e. values $c(a,b)$), which are used in equations. This can be done and results in a dynamic version of ORCA which is as fast as the original version when used to add all edges into an empty graph and than compute the orbit counts for all nodes. Another version of dynamic graphlet counting dispenses with the last query type (obtaining orbit counts for a given node) and instead constantly maintains the number of graphlets in the network after each addition or removal of an edge.

2.2 Application: Generating Random Graphs

How do we describe a general topology of a real-life network? By picking a random model that generates similar networks; social networks, for instance, are described as scale-free. This also enables us to generate an arbitrary number of topologically similar networks. Such ensembles can be used, for example, to evaluate the performance of a newly designed method on a larger set of networks. Another related concept are network motifs [25], where an ensemble of random networks is used as a null model to discover over-represented patterns (motifs) in the original network. Because graphlet counts represent a summary of a network's topology, we attempted to generate random networks that have a prescribed (or at least very similar) number, X_i, of each 4-node graphlet.

$$F(G)=\sum_{i=0}^{8}|\log(N_i(G)+1)-\log(X_i+1)| \tag{3}$$

The process of generating a random network with a given graphlet distribution was formulated as an optimization procedure. We specified an optimization function, $F(G)$, based on the relative graphlet frequency distance, but removed the "relative" part (Eq. 3). A simple hill-climbing optimization picks random edges to add or remove from the network and checks the similarity of the modified network to the target distribution of graphlets. If the similarity improves, we keep the change, otherwise we revert it. Using the dynamic version of ORCA that is capable of maintaining graphlet counts under such modifications of the network makes the process feasible for the current size of PPI networks.

	$F(G)$	G_0	G_1	G_2	G_3	G_4	G_5	G_6	G_7	G_8
start	2.761	11000	42251	13492	167375	27308	1462	95264	25235	9321
finish	0.082	11626	418249	24796	11976090	9379476	387973	2749137	285672	22358
target	0.0	11626	418270	18225	11110782	9383348	445186	2216446	285736	22376

Tab. 2. Graphlet counts obtained by starting with a random geometric network.

The similarity of the obtained random network depends on the choice of the network that we start from. We attempted to approximate the graphlet counts observed in a PPI network of bacteria E. Coli by starting with a random geometric network (Table 2). The starting and target networks both contained around 3000 nodes and 11000 edges. The optimization process requires less than an hour for 1.5 million iterations. The resulting network that we obtained through optimization of 4-node graphlet counts had a surprisingly similar distribution of 5-node graphlet counts (Figure 3). This is a preliminary result that requires further investigation. Intuitively, the number of 2-node graphlets (the number of edges) does not tell us much about the number of 3-node graphlets. On the other hand, 4-node graphlets seem to restrict the space of 5-node graphlets much more. It would make sense that this influence grows with the size of graphlets.

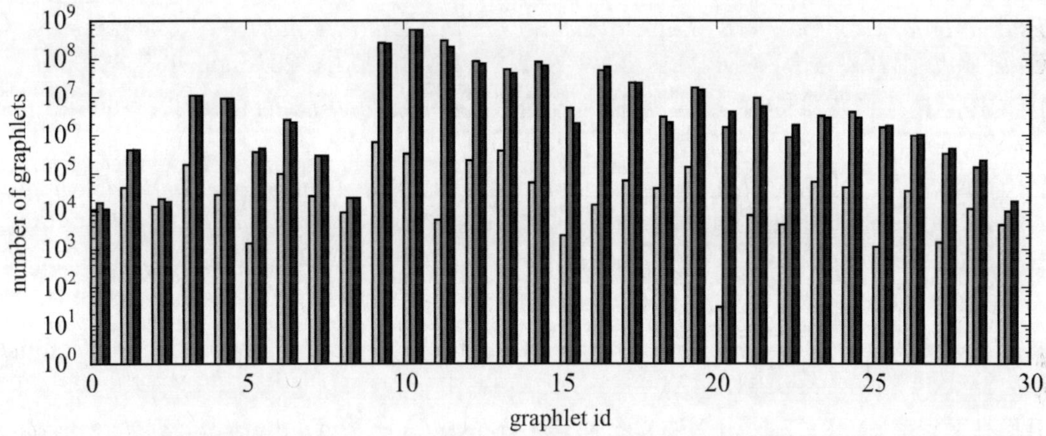

Fig. 3. The 5-node graphlet counts obtained by optimizing 4-node counts. White bars correspond to the distribution in the starting graph, gray bars corresponds to the obtained (generated) graph and black bars to the target graphlet distribution of a PPI network.

3 Conclusion

Bioinformatics is a rapidly evolving field with a growing amount of data due to recent technological advances. Simple improvements such as parallelization are often not enough; instead, new algorithmic and mathematical insights are required. We developed ORCA, which is based on combinatorial observations that establish relations between individual graphlet counts. These are employed to design an efficient graphlet counting algorithm that outperforms other approaches, which are based on exhaustive enumeration.

However, there is still a lot of room for improvement. Ortmann et al. [21] have already surpassed our method for counting 4-node graphlets. We know that the system of equations must have nonnegative integral solutions. Could we somehow incorporate this property into a more efficient algorithm?

Several applications would benefit from an efficient graphlet counting method in weighted graphs. Can we approach this with a similar method or does it require something new?

Acknowledgement

The research presented in this paper was funded by the Slovenian research agency grants J2-5480 and P2-0209.

References

[1] AHMED, N. K., NEVILLE, J., ROSSI, R. A., DUFFIELD, N.: *Efficient Graphlet Counting for Large Networks*, in *2015 IEEE International Conference on Data Mining*, IEEE, nov 2015, pp. 1–10.

[2] BORGATTI, S. P., MEHRA, A., BRASS, D. J., LABIANCA, G.: *Network Analysis in the Social Sciences*, Science, vol. 323, no. 5916, (2009), pp. 892–895.

[3] CANCHO, R. F. i., SOLE, R. V.: *The small world of human language*, Proceedings of the Royal Society B: Biological Sciences, vol. 268, no. 1482, (2001), pp. 2261–2265.

[4] ELENBERG, E. R., SHANMUGAM, K., BOROKHOVICH, M., DIMAKIS, A. G.: *Distributed Estimation of Graph 4-Profiles*, in *Proceedings of the 25th International Conference on World Wide Web - WWW '16*, ACM Press, New York, New York, USA, 2016, pp. 483–493.

[5] HOČEVAR, T., DEMŠAR, J.: *A combinatorial approach to graphlet counting*, Bioinformatics, vol. 30, no. 4, (2014), pp. 559–565.

[6] HOČEVAR, T., DEMŠAR, J.: *Computation of Graphlet Orbits for Nodes and Edges in Sparse Graphs*, Journal of Statistical Software, vol. 71, no. 10.

[7] HOČEVAR, T., DEMŠAR, J.: *Combinatorial algorithm for counting small induced graphs and orbits*, PLOS ONE, vol. 12, no. 2.

[8] ITAI, A., RODEH, M.: *Finding a Minimum Circuit in a Graph*, SIAM Journal on Computing, vol. 7, no. 4, (1978), pp. 413–423.

[9] JUSZCZYSZYN, K., KAZIENKO, P., MUSIAŁ, K.: *Local Topology of Social Network*, in *KES '08 Proceedings of the 12th international conference on Knowledge-Based Intelligent Information and Engineering Systems, Part II*, Zagreb, 2008, pp. 97–105.

[10] KLOKS, T., KRATSCH, D., MÜLLER, H.: *Finding and counting small induced subgraphs efficiently*, Information Processing Letters, vol. 74, no. 3-4, (2000), pp. 115–121.

[11] KUCHAIEV, O., MILENKOVIĆ, T., MEMISEVIĆ, V., HAYES, W., PRŽULJ, N.: *Topological network alignment uncovers biological function and phylogeny.*, Journal of the Royal Society, Interface / the Royal Society, vol. 7, no. 50, (2010), pp. 1341–1354.

[12] KUCHAIEV, O., PRŽULJ, N.: *Integrative network alignment reveals large regions of global network similarity in yeast and human.*, Bioinformatics (Oxford, England), vol. 27, no. 10, (2011), pp. 1390–1396.

[13] MCKAY, B. D.: *Small graphs are reconstructible.*, Australasian Journal of Combinatorics, vol. 15, (1997), pp. 123–126.

[14] MILENKOVIĆ, T., MEMISEVIĆ, V., GANESAN, A. K., PRŽULJ, N.: *Systems-level cancer gene identification from protein interaction network topology applied to melanogenesis-related functional genomics data.*, Journal of the Royal Society, Interface, vol. 7, no. 44, (2010), pp. 423–437.

[15] MILENKOVIĆ, T., NG, W. L., HAYES, W., PRŽULJ, N.: *Optimal network alignment with graphlet degree vectors.*, Cancer informatics, vol. 9, (2010), pp. 121–137.

[16] MILENKOVIĆ, T., PRŽULJ, N.: *Uncovering biological network function via graphlet degree signatures.*, Cancer informatics, vol. 6, (2008), pp. 257–273.

[17] MILENKOVIĆ, T., ZHAO, H., FAISAL, F. E.: *Global Network Alignment In The Context Of Aging*, in *Proceedings of the International Conference on Bioinformatics, Computational Biology and Biomedical Informatics - BCB'13*, ACM Press, New York, New York, USA, 2013, pp. 23–32.

[18] MUGAN, J., CHARI, R., HITT, L., MCDERMID, E., SOWELL, M., QU, Y., COFFMAN, T.: *Entity resolution using inferred relationships and behavior*, in *2014 IEEE International Conference on Big Data (Big Data)*, IEEE, 2014, pp. 555–560.

[19] NEŠETŘIL, J., POLJAK, S.: *On the complexity of the subgraph problem*, Commentationes Mathematicae Universitatis Carolinae, vol. 26, no. 2, (1985), pp. 415–419.

[20] O'NEIL, P. V.: *Ulam's Conjecture and Graph Reconstructions*, The American Mathematical Monthly, vol. 77, no. 1, (1970), pp. 35–43.

[21] ORTMANN, M., BRANDES, U.: *Quad census computation: Simple, efficient, and orbit-aware*, in *Lecture Notes in Computer Science*, vol. 9564, Springer International Publishing, 2016, pp. 1–13.

[22] PRŽULJ, N., CORNEIL, D. G., JURISICA, I.: *Modeling interactome: scale-free or geometric?*, Bioinformatics, vol. 20, no. 18, (2004), pp. 3508–3515.

[23] PUJOL, A., MOSCA, R., FARRÉS, J., ALOY, P.: *Unveiling the role of network and systems biology in drug discovery*, Trends in Pharmacological Sciences, vol. 31, no. 3, (2010), pp. 115–123.

[24] RIBEIRO, P., SILVA, F.: *G-tries: an efficient data structure for discovering network motifs*, in *Proceedings of the 2010 ACM Symposium on Applied Computing - SAC '10*, ACM Press, New York, New York, USA, 2010, pp. 1559–1566.

[25] SHEN-ORR, S. S., MILO, R., MANGAN, S., ALON, U.: *Network motifs in the transcriptional regulation network of Escherichia coli.*, Nature genetics, vol. 31, no. 1, (2002), pp. 64–8.

[26] STOICA, A., PRIEUR, C.: *Structure of Neighborhoods in a Large Social Network*, in *2009 International Conference on Computational Science and Engineering*, IEEE, 2009, pp. 26–33.

[27] WERNICKE, S.: *Efficient detection of network motifs.*, IEEE/ACM transactions on computational biology and bioinformatics / IEEE, ACM, vol. 3, no. 4, (2006), pp. 347–59.

[28] ZHANG, L., HAN, Y., YANG, Y., SONG, M., YAN, S., TIAN, Q.: *Discovering discriminative graphlets for aerial image categories recognition.*, IEEE transactions on image processing, vol. 22, no. 12, (2013), pp. 5071–5084.

Current address

Hočevar Tomaž, PhD.
Faculty of Computer and Information Science
University of Ljubljana
Večna pot 113, 1000 Ljubljana, Slovenia
E-mail: tomaz.hocevar@fri.uni-lj.si

Slovak University of Technology in Bratislava
Faculty of Mechanical Engineering
17th Conference on Applied Mathematics **APLIMAT 2018**

Proceedings

SERLIO´S, GUARINI´S AND MEYER´S CONSTRUCTIONS OF OVALS IN ARCHITECTURE

HOLEŠOVÁ Michaela (SK)

Abstract. When reconstructing some of the buildings with elliptical ground plan, it is often very difficult to unambiguously determine whether an ellipse or an oval was constructed to approximate this ellipse. We deal with Serlio´s and Guarini´s constructions of ovals. F. S. Meyer showed constructions of ovals whose gives very good approximation of the ellipse. We modify Serlio´s and Guarini´s constructions of ovals so that lengths of both axes of the ellipse are preserved.

Keywords: an oval, a construction, approximation of an ellipse

Mathematics Subject Classification: Primary 51E21, 51N05; Secondary 65D17

1 Introduction

Ellipse is a form that is found common in architecture. F. S. Meyer in his book A Hand Book of Ornament [6] said: "The Ellipse is of comparatively late appearance in art, the construction presupposing a certain knowledge of Geometry, which was not possessed by primitive peoples. The ellipse is a very popular shape for ceilings, panels, boxes and dishes."
Exact construction of ellipse is difficult because its radius of curvature is continually changing. In practice, therefore, it is approximated by circular arcs and the expression "Oval" is often used for the ellipse, but it is erroneous.
The word oval derived from the Latin word "ovus" for egg. Here it is used in terms of the shape of curves that resemble an egg. It is a curve for which strict mathematical definition is missing but has the common features: is a closed curve in a plane which "loosely" resembles the outline of an egg, is differentiable (smooth-looking), convex, closed, simple (not self-intersecting) plane curve, which has at least one axis of symmetry. In a special case and also in architecture, we see the oval as a curve, which is composed of circular arcs. We can find such an approach in books of authors as S. Serlio [1], G. Guarini [2] and F. S. Meyer [6].
We are interested in the study and geometric analysis of buildings with elements of ovals. These are buildings that have ground oval or elliptic plan or ovals have been used in the construction of theirs internal building elements. In Slovakia we can for example show buildings as Cathedral of St John of Matha in Bratislava and elliptic St. Ladislav Chapel in the former Ostrihom archbishop´s palace in Bratislava (Fig.1)

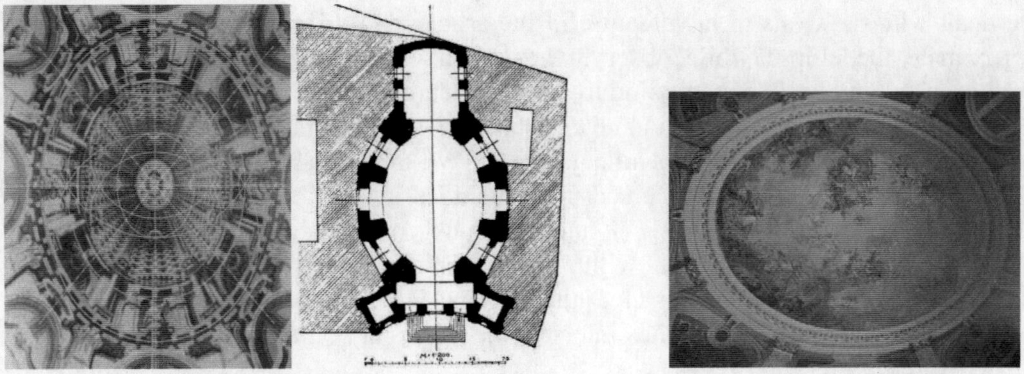

(a) (b) (c)

Fig. 1. Cathedral of St John of Matha (a), (b), Ostrihom archbishop´s palace (c).

Architects approximate the ellipse by circular arcs (oval) in practice. Due to the symmetry of an ellipse it is enough to use two different circles most often. We search the centers and radii of the two circles that are connected smoothly. The circles have a common tangent at the point of connection, which means that the centers *H, K* (Fig. 2) of these circles must belong on one line. Let $h = |SH|$ and $k = |SK|$ where *S* is the center of the oval. Found circles have radii $a - h, a - h + \sqrt{h^2 + k^2}$ where *a* is the length of semi-major axis of the ellipse.

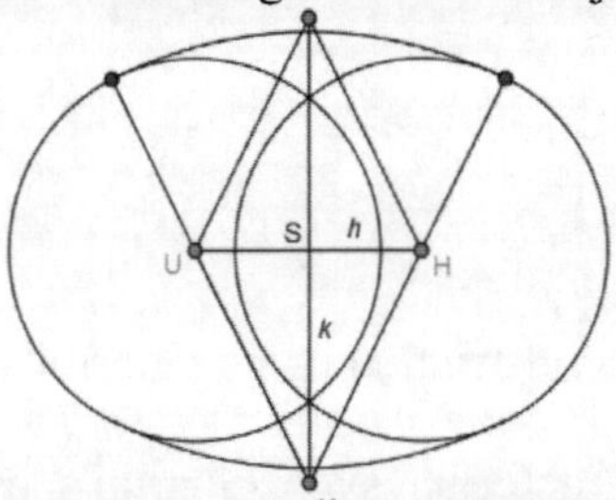

Fig. 2. Construction of two-center oval, based upon Serlio´s treatise.

In paper [3], P. L. Rosin dealt with Serlio´s constructions (Fig. 3). The construction of the Fig. 3(a) is in the case that ΔUHK is the equilateral triangle and $h = (a - b)/(\sqrt{3} - 1)$, $k = \sqrt{3}h$. The ratio of the radius of the circles, and the ratio a/b are not constant. These ratios are constant in other Serlio's construction. In case Fig. 3(b) is $h = k = a/(\sqrt{2} + 1)$, $a/b = (\sqrt{2} + 1)/(2\sqrt{2} - 1)$ and the ratio of the radii of the arcs is ½. In case Fig. 3(c) is $h = k = a/2$, $a/b = \sqrt{2}$ and the ratio of the radii of the arcs is $\sqrt{2} - 1$. In case Fig. 3(d) is $h = a/3$, $a/b = 3/(4 - \sqrt{3})$, $k = \sqrt{3}h$ and the ratio of the radii of the arcs is ½.

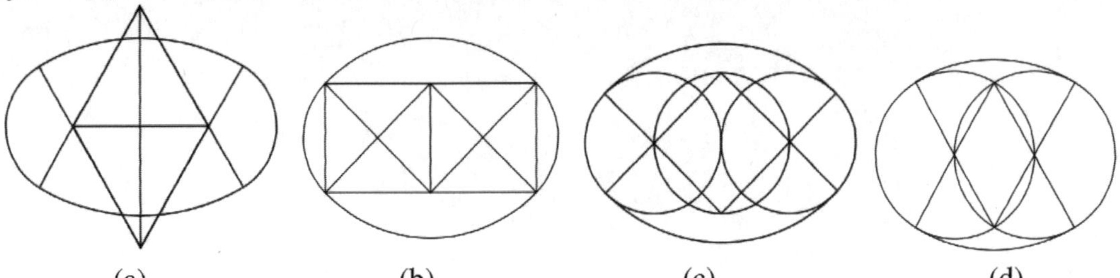

(a) (b) (c) (d)

Fig. 3. Oval constructions by S. Serlio.

We have dealt with the ovals in architecture in the articles [4], [5]. These are constructions that approximate the ellipse while observe the length of the major axis of the ellipse. G. Guarini is describing in [2] construction to an oval generally.

We can take two circles with centers *A* and *F*, which have any distance, non-intersecting or intersecting (each other), with equal or unequal radii. We take non-intersecting circles (Fig. 4(a)) with different radii. Line *AF* intersects these circles in points *I* and *C*. We construct points *G* and *O*, such that $|IO|$ equals $|CG|$ and it is equal to the length greater than half the line segment *CI* length. We have circles with centers *A* and *F* and radii *AO*, *GF*. These circles have common the points *M* and *H*. It is clear that the line *MH* is perpendicular to the line *AF*. We get lines *MF, HF, MA, HA*. The lines intersect the circles at points *S, R*. (Fig. 4), points *S* and *R* lie on a circle with center *H*.

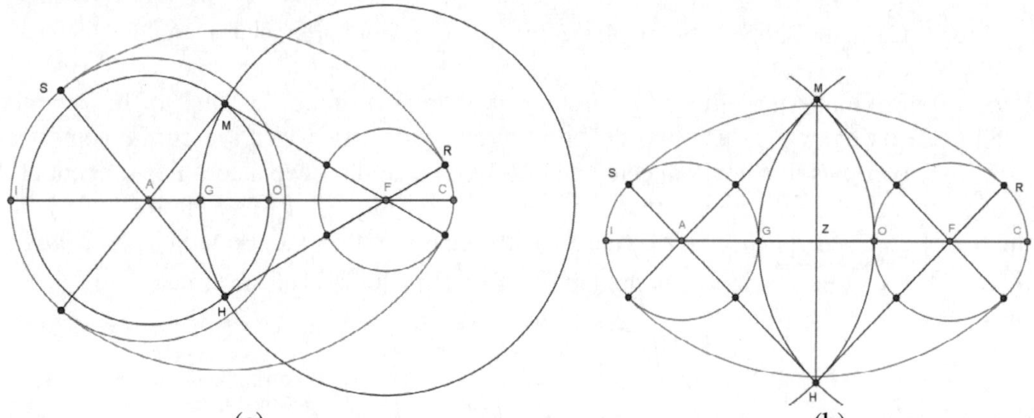

(a) (b)

Fig. 4. Ovals constructed using Guarini´s method.

By appropriate choice of the characteristics of Guarini's oval we can get Serlio's oval. For example, we get Serlio's oval in the construction in Fig. 3(d) so that we choose the radii of the circles as $2a/3$, the distance between centers of the circles is $2a/3$ and the point *G* is identical to the point *A*.

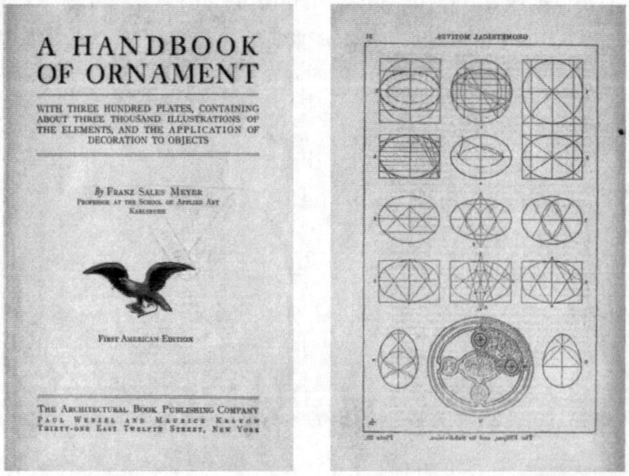

(a) (b)

Fig. 5. F. S. Meyer, A Handbook of Ornament (a), Plate 20, pp. 31 (b).

2 Meyer´s constructions

Franz Sales Meyer in his book A Handbook of Ornament [6] introduced the modified Serlio´s and Guarini´s constructions. Unlike these two authors, he showed constructions where not only the length of the major axis of the ellipse but also the length of minor axis of the ellipse is preserved.

First we will show the construction by which F.S. Meyer in [6, Plate 20, pp. 31, no. 8] modified the Serlio's construction (c) (see Fig. 5 (b)). One center of the approximation circle is point S_2. For Serlio's construction it is point K. In Fig. 6 it is a comparison of these two constructions.

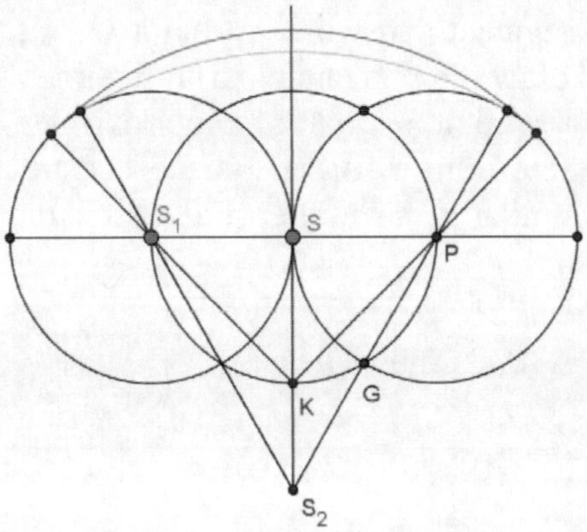

Fig. 6. Meyer´s construction (green colour), Serlio´s construction (c) (red colour).

It is clear that the triangle $\Delta\ S_2S_1P$ is equilateral and hence lengths are $h = a/2$, $k = \sqrt{3}a/2$, $a/b = (\sqrt{3}+3)/3$ and the ratio of radii of the circles is 1/3.

We have now the ellipse given by its axes. We are dealing with the construction, which is mentioned in [6, Plate 20, pp. 31, no. 11].

Point S is the center of the ellipse. We take half the difference between the semi-major axis and the semi-minor axis of the ellipse. From the center S we apply this distance three times to the major axis and four times to the minor axis. We obtain centers S_1, S_2 of circles that we will use to approximate the ellipse, see Fig. 7.

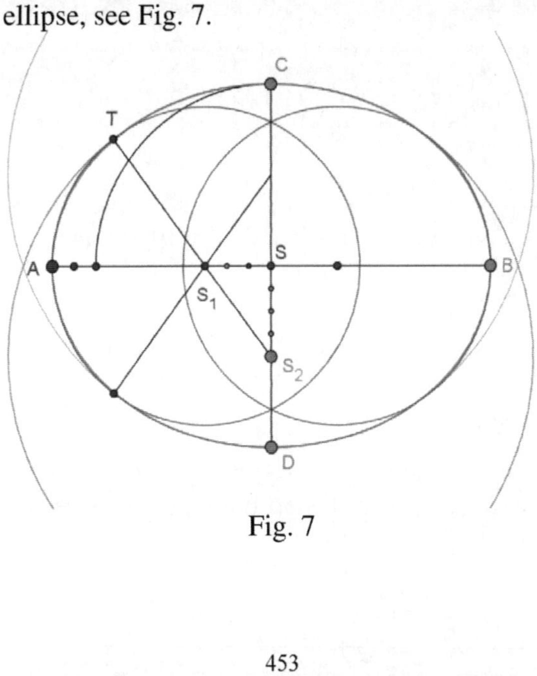

Fig. 7

We will denote a distance between a center of a circle on the major axis and the center of the ellipse as $h = |S_1S|$ and a distance of a center of a circle on the minor axis from the center of the ellipse as $k = |SS_2|$ as with Serlio´s and Guarini´s constructions.
In this case is $|SA| = a$, $|SC| = b$, $h = 3(a\text{-}b)/2$, $k = 2(a\text{-}b)$, $h = 3k/4$ and radii of circles are $r_1 = (3b\text{-}a)/2$, $r_2 = 2a\text{-}b$. We still have to show that $|S_2T| = |S_2C|$. Based on the construction is $|S_1S_2| = 5(a\text{-}b)/2$. $|S_2T| = |S_1S_2|+|S_1T| = 5(a\text{-}b)/2 + (3b\text{-}a)/2 = 2a\text{-}b = |S_2C|$. Ellipse (red colour) in Fig. 7 has $a/b = 5/4$ and its approximation according to the construction just described.

We construct the major and minor axes of the given ellipse as in [6, Plate 20, pp. 31, no. 10]. Point S is the center of the ellipse. We take the difference between the semi-major axis and the semi-minor axis of the ellipse. We construct a line segment AC and apply the length $a\text{-}b$ from the point C to this line segment and we get a point Q. We mark the center of the line segment AQ as a point L. We draw the perpendicular to the line AC through the point L. This line intersects the major axis at a point S_1 and the minor axis at a point S_2, which are centers of circles that approximate these ellipse. Based on the construction, the triangles $\Delta\, ALS_1$, $\Delta\, S_2LC$, $\Delta\, ASC$, $\Delta\, S_2SS_1$ are similar to each other (Fig. 8) and the length

$$|AL| = \frac{\sqrt{a^2+b^2}-(a-b)}{2}.$$

From the similarity $\Delta\, ALS_1 \approx \Delta\, ASC$ follows that

$$r_1 = |AS_1| = \frac{|AL||AC|}{|AS|} = \frac{\sqrt{a^2+b^2}\left(\sqrt{a^2+b^2}-(a-b)\right)}{2a}.$$

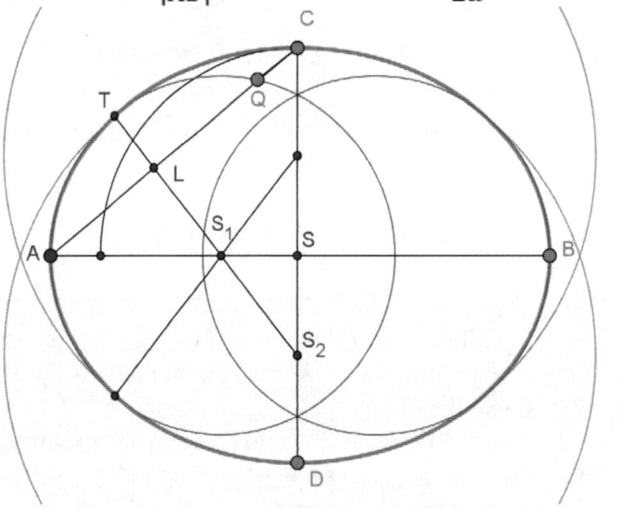

Fig. 8

We express length $h = |SS_1| = a - |AS_1| = (a-b)(\sqrt{a^2+b^2}+a+b)/(2a)$.
The similarity $\Delta\, ASC \approx \Delta\, S_2SS_1$ implies $|SS_1|/|SC| = |S_2S|/|AS|$ and $k = |S_2S| = a|SS_1|/b = ah/b$. From the same similarity, we have $|SS_1|/|SC| = |S_2S_1|/|AC|$, since $|S_2S_1| = \sqrt{a^2+b^2}h/b$. The similarity $\Delta\, S_2LC \approx \Delta\, ASC$ gives

$$r_2 = |CS_2| = \frac{|CL||AC|}{|CS|} = \frac{\sqrt{a^2+b^2}\left(\sqrt{a^2+b^2}+(a-b)\right)}{2b}.$$

We must prove that it is $|S_2T| = |S_2C|$. Based on the construction it is $|S_2T| = |S_1S_2|+|S_1T| = \sqrt{a^2+b^2}h/b + |AL|\sqrt{a^2+b^2}/a = \sqrt{a^2+b^2}(\sqrt{a^2+b^2}+(a-b))/(2b) = |S_2C|$. In Fig. 8 is the ellipse with $a/b = 5/4$ (red colour) and its approximation by circular arcs.

We see that both constructions very well approximate this ellipse. We now show constructions that use the osculating circles of the ellipse and the ellipse is approximated by an oval determined by three different circles.

F. S. Meyer showed in [6, Plate 20, pp. 31, no. 12] a smooth connecting of osculating circles of the ellipse with another circular arc.

We construct a rectangle *ORPQ* (Fig. 9) whose sides have a length equal to the major and minor axis of the ellipse and a center of the rectangle is the center *S* of the ellipse and the axes of its sides are line segments *AB*, *CD*. We draw a line segment *AC*. In the point *Q* draw a perpendicular line to the line *AC*. This line crosses the major axis in point S_1 and the minor axis in point S_2, which are the centers of osculating circles.

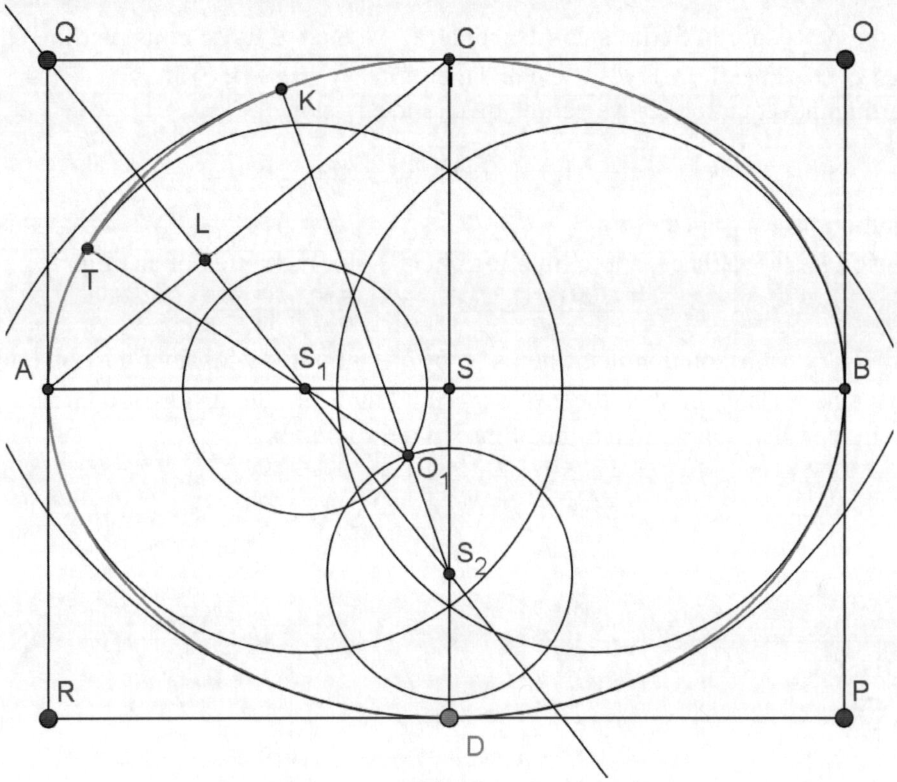

Fig. 9

The radius of the osculating circles is known, but we can easily deduce them from the similarity of triangles $\Delta\, ALS_1$, $\Delta\, S_2LC$, $\Delta\, ASC$, $\Delta\, S_2SS_1$, $r_1 = |AS_1| = b^2/a$, $r_2 = |CS_2| = a^2/b$. The similarity $\Delta\, ASC \approx \Delta\, S_2SS_1$ gives $|SS_1|/|SC| = |S_2S|/|AS|$ and it is $k = |S_2S| = a|SS_1|/b = ah/b$, where $h = a - r_1 = (a^2 - b^2)/a$.

We draw circles with centers at points S_1, S_2 and with the radius $(r_2 - r_1)/2$. In their intersection is the center O_1 of the circle (green colour), which smoothly feeds the osculating circles into the continuous curve - oval. The lines O_1S_1, O_1S_2 intersect osculating circles at points *T*, *K*. We must prove that $|TO_1| = |KO_1|$. By construction we have

$$|TO_1| = (r_2 - r_1)/2 + r_1 = (r_2 + r_1)/2 \text{ and } |KO_1| = r_2 - (r_2 - r_1)/2 = (r_2 + r_1)/2.$$

Although this construction is an approximate construction, the constructed oval is the best approximation of the ellipse from all of these constructions. In Fig. 9 is the ellipse with a/b = 5/4 (red colour).

3 Modification of Serlio´s and Guarini´s constructions

If we give the condition for constructions of S. Serlio and G. Guarini that we want to approximate the ellipse with known the major and the minor axes, then the following equality must be satisfied

$$r_1 + \sqrt{h^2 + k^2} = k + b, \qquad (1)$$

where $h = a - r_1$ and r_1 is a radius of a circle q_1, which approximates the ellipse in its vertex. If we calculate unknown k from the equation (1) we will get

$$k = \frac{(a-b)}{2}\frac{(a+b-2r_1)}{b-r_1} \qquad (2)$$

Based on the procedure in Serlio´s construction (a), where we try to construct the equilateral triangle $\Delta\, S_2S_1O_1$ (Fig. 10 (b)) is $k = \sqrt{3}h$ and therefore $r_1 = a - k/\sqrt{3}$.
After inserting into equation (2) we get the quadratic equation

$$k^2 - (a-b)\left(1+\sqrt{3}\right)k + \frac{\sqrt{3}}{2}(a-b)^2 = 0,$$

whose solutions are $k_1 = (a-b)\left(3+\sqrt{3}\right)/2$, $\bar{k}_1 = (a-b)\left(\sqrt{3}-1\right)/2$. From this we get the expression for the radius r_1 approximating circle q_1 for the vertex of an ellipse

$$r_1 = \frac{a(1-\sqrt{3})+b(1+\sqrt{3})}{2}, \quad \bar{r}_1 = \frac{a+b}{2} + \frac{(a-b)}{2\sqrt{3}}.$$

Clearly, r_1 is a possible solution just when $a < b\left(2+\sqrt{3}\right)$. If we consider the condition that it should be a smooth tangent, then the circle q_1 should tangent the inside of a circle q_2, where q_2 is the circle, which approximates the ellipse in the co-vertex C.

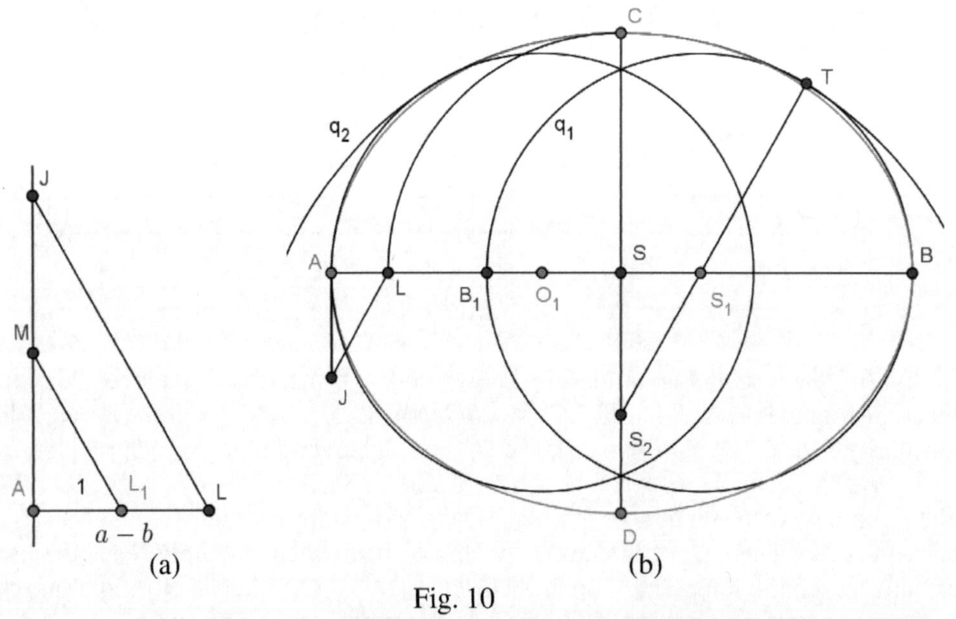

Fig. 10

It follows from this condition that $|S_1S_2| = k + b - r_1$ where $|S_1S_2| = 2h$ and we have $k = (a-b)\left(3+\sqrt{3}\right)/2$. The second solution $\bar{k}_1$ can also be excluded because in this case the co-vertex C of the ellipse would be the inside point of the circle q_1. When we know lengths of the major and the minor axes of the ellipse, for example, point S_1 must be constructed. If we express the radius as $r_1 = (a + b - \sqrt{3}(a-b))/2$, we can construct a line

segment of length $\sqrt{3}(a-b)$ as in Fig. 10 (a), where $|AL| = a - b$, $|AL_1| = 1$, $|M L_1| = 2$ and $|AJ| = \sqrt{3}(a-b)$.
We presented in Fig. 10 (b) the construction of the ellipse whose $a / b = 5/4$ (red colour). The point S_1 is the center of the line segment BB_1, where $|LB_1| = |AJ| = \sqrt{3}(a-b)$, $|LB| = a + b$, $|AL| = a - b = 1$.

Guarini introduced the constructions of ovals which approximate the ellipse with known major axis. We modify his construction so that the oval approximates the ellipse even in its co-vertex. If we choose the radius r_1 and hence the center F (Fig. 4 (b)), so $k = |ZH|$ is expressed in equation (2) and for the point G must already apply $|FG| = |FH|$, so we cannot choose it arbitrarily.

We have the major and minor axes of the ellipse. Let the circle q_1 be the osculating circle for the point A with the center at point S_1 (Fig. 11). For its radius $r_1 = b^2/a$ and the equation (2) shows that

$$|SS_2| = |SH| = k = \frac{(a-b)}{2}\frac{(a+2b)}{b}.$$

The point S_2 is the center of the circle q_2 which approximates the ellipse in the co-vertex C. Construction of line segment SH is based on the similarity of the triangles $\Delta\, SHG_1$, $\Delta\, SFC$, $\Delta\, SKC$, where $|SG_1| = (a-b)/2$, $|SE| = 1$, $|SK| = a + 2b$, $|SF| = (a + 2b)/b$.

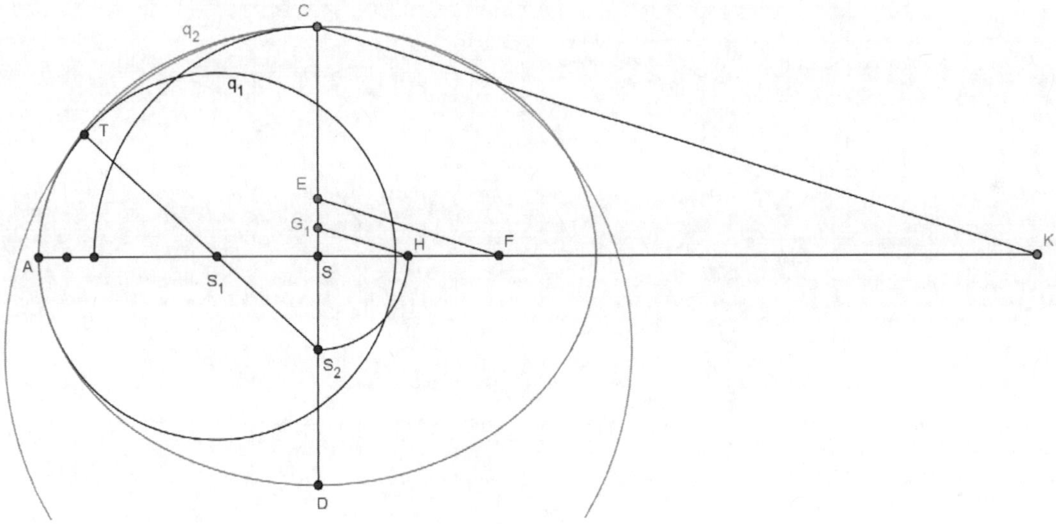

Fig. 11

We now compare our modified Guarini's construction with Meyer´s construction. It is clear that Meyer's construction is more precise (Fig. 12 and Fig. 13), because we use both osculating circles of the ellipse (green colour). In modified Guarini's construction we use the circle q_2 (blue colour in Fig. 12 or Fig. 13) which is not the osculating circle of the ellipse (red colour).

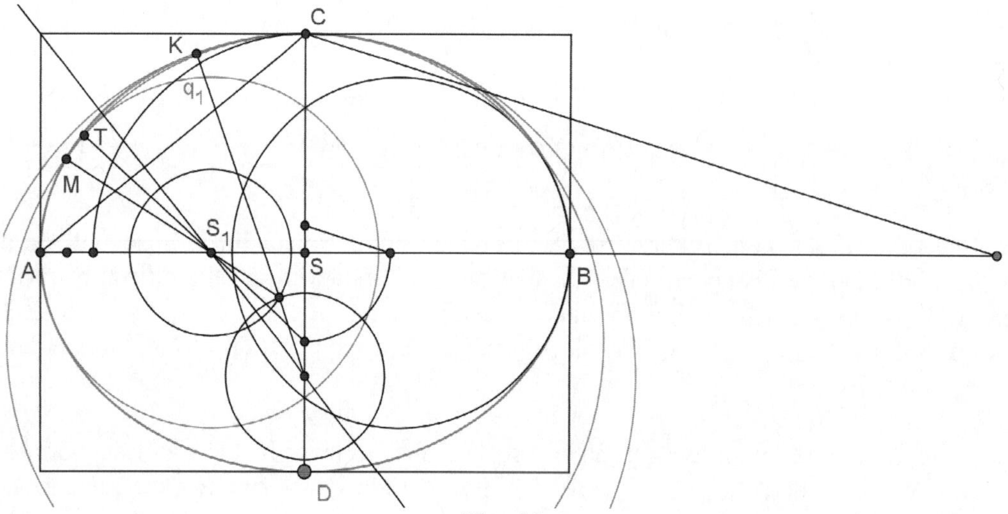

Fig. 12

Point T is the tangent point of the circles in the Guarini´s construction (Fig. 13), and points M, K are tangent points of the osculating arcs with a third circular arc in the Meyer´s construction.

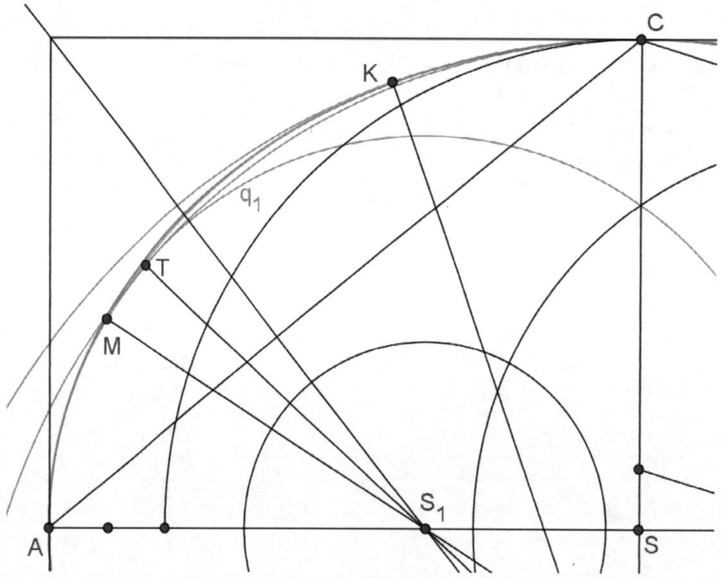

Fig. 13

We can say that given ellipses are the most precisely approximated by the Meyer´s constructions because he used osculating circles. Not even the modified Serlio´s and Guarini´s constructions do not give us more precise approximation of the ellipse than the Meyer´s construction in Fig. 9. However, we used only two circular arcs. The difference between the ovals presented in this article and the ellipse is minimal. Therefore, when we reconstruct some of buildings, we think that it is very difficult to determine unambiguously whether the ellipse was constructed or an oval that approximates this ellipse. Due to technical practice, we think that they were used ovals for approximation of the ellipse. For interpreting an elliptical shape, these constructions are, in our opinion, sufficient and practically usable. We believe they will continue to be used as they did in the past.

References

[1] SERLIO, S., *Il primo [-secondo] libro d'architettura*. Venice, 1545. (italian) https://archive.org/details/gri_33125008448702.

[2] GUARINI, G., GUJENOTTI, F., ABBIATI, G., *Architettura civile: opera postuma*. Published 1737. https://archive.org/details/gri_33125008641249.

[3] ROSIN, P. L., On Serlio's constructions of ovals. *The Mathematical Intelligencer*, 23(1), 2001, pp. 58-69. DOI: 10.1007/BF03024523.

[4] GRÚŇOVÁ, Z., HOLEŠOVÁ, M., Ellipse and oval in baroque sacral architecture in Slovakia. *Civil and environmental engineering*. Vol. 13, No. 1, 2017, pp. 30-41.

[5] GRÚŇOVÁ, Z., HOLEŠOVÁ, M., Geometrical analysis of Palesch family chapel in Kľačno, former Gaidel in Western Slovakia. *RSP 2017 – XXVI R-S-P Seminar 2017 Theoretical Foundation of Civil Engineering,* Vol. 117, No. 00060, 2017.

[6] MEYER, F. S., *A Hand Book of Ornament.* Published 1896. https://archive.org/details/handbookoforname00meyeuoft.

Current address

Holešová Michaela, RNDr., PhD.
Dept. of Structural Mechanics and Applied Mathematics, Faculty of Civil Engineering
University of Žilina
Univerzitná 8215/1, 010 26 Žilina, Slovak Republic
E-mail: michaela.holesova@fstav.uniza.sk

Slovak University of Technology in Bratislava
Faculty of Mechanical Engineering
17th Conference on Applied Mathematics **APLIMAT 2018**

Proceedings

THE GENERALIZED SOLUTION TO A LINEAR OPERATOR EQUATION. THE CASE OF FINITE DIMENSION

HORSKÝ Richard (CZ)

Abstract. It is well-known that a system of linear algebraic equations need not have a solution. There is a way allowing to define a solution to such a system which is not dependent on the properties of the linear operator (injective, surjective) represented by the matrix of the system. This leads in special case to the known Least Square Method, however here in rather unusual point of view.

Keywords: system of linear algebraic equations, normal solution, Moore Penrose pseudoinverse

Mathematics Subject Classification: Primary 15A09, 47A05; Secondary 15A18

1 Basic Notions and Facts

In this section we recall some basic notions and facts and accept suitable notation. Suppose we have a linear mapping (operator)

$$A : \Re^n \to \Re^m, \tag{1}$$

where $\Re^n$ is the n-th dimensional Euclidean space. This mapping is represented by a matrix $\mathbf{A} = (a_{ij})$ of the type (m, n). For our purposes it is not necessary to distinguish strictly between the linear operator and its matrix representation. So we will use the letter A instead of **A**. and the word operator may be exchanged by the word matrix whenever it occurs in the sequel. We will consider the operator equation

$$Ax = b, \tag{2}$$

which is in fact the system of m linear algebraic equations for n unknowns ξ_i, $x = (\xi_1, \ldots, \xi_n)$. Two important sets are connected with the operator (1). They are the *kernel* and the *range of the operator* (1):

$$\ker A = \{x \in \Re^n : Ax = o_m\}, \ \text{ran}\, A = \{y \in \Re^m : \exists (x \in \Re^n) y = Ax\}.$$

The symbol o_m is used for the zero vector of the space $\Re^m$. The Euclidean space is equipped by the inner product which is here taken in general sense, see [5] (page 110). It is not necessary to consider the standard Euclidean inner product. An inner product of the vectors x and y is usually denoted (x, y) and to distinguish among the inner products in different Euclidean spaces we write $(x, y)_k$ for the inner product in $\Re^k$. The vector norm is derived from the inner product in usual way: $\|x\| = \sqrt{(x, x)}$. We distinguish again among the norms in the different Euclidean spaces by the index. The inner product allows to define the so called *adjoint operator (conjugate transpose) of A*

$$A^* : \Re^m \to \Re^n \tag{3}$$

by the equality

$$(Ax, y)_m = (x, A^* y)_n$$

for any $x \in \Re^n, y \in \Re^m$. Its existence and uniqueness follow from Riesz representation theorem, see [5](page 237).

In the end of this section let us recall the relations among kernels and ranges of the operator (1) and its adjoint (3):

$$\Re^n = \ker A \oplus \text{ran}\, A^*, \quad \Re^m = \text{ran}\, A \oplus \ker A^*. \tag{4}$$

The direct decompositions (4) are even orthogonal. It means for example that any vector $x \in \Re^n$ can be expressed in the form $x = v + w$, where $Av = o_m$, $w = A^* u$ for some $u \in \Re^m$ and $(v, w)_n = 0$. The vectors v and w are determined by x uniquely.

2 Generalized Solution to the System of Linear Algebraic Equations

We define

$$F(x) = (Ax - b, Ax - b)_m = \|Ax - b\|_m^2. \tag{5}$$

The function (5) is called the *price functional* (terminology used in [1]). It is a nonnegative function $F : \Re^n \to \Re$. We will consider the problem of its minimum. The *generalized solution* to the system of linear algebraic equations (2) is any point at which the price functional (5) achieves its minimum.

Proposition 2.1
There is at least one point at which the price functional achieves its minimum.

Proof. Due to the orthogonal decompositions of Euclidean spaces (4) we may write

$$x = v + A^* u \,,\; Av = o_m \,,$$
$$b = Ac + d \,,\; A^* d = o_n \,.$$

Hence we obtain

$$F(x) = \left(A\left(v + A^* u\right) - Ac - d, A\left(v + A^* u\right) - Ac - d\right)_m = \\ = \left(A\left(A^* u - c\right), A\left(A^* u - c\right)\right)_m + \left(d, d\right)_m . \qquad (6)$$

Since the vector $d \in \Re^m$ is determined by the fixed vector $b \in \Re^m$ we may affect only the first term in (6). Thus we set $x = c$. Due to $Ac = AA^* u$ we obtain $F(x) = (d,d)_m$. □

The minimal value $F(x) = (d,d)_m$ is so called the *price of the generalized solution.* It is obvious that in the case there exists the classical solution to (2) the price is zero.

There may exist more than one generalized solution for the given operator A and the vector b in (2). Let us denote

$$S = \left\{x \in \Re^n : F(x) = (d,d)_m\right\}.$$

The set S is not empty as follows from the Proposition 2.1. Further S is a closed set because the function (5) is continuous. Finally S is convex. Suppose $x \in S,\ y \in S,\ t \in [0,1]$. If we set $z(t) = tx + (1-t)y$ then we obtain $F(z(t)) = (d,d)_m$ and hence $z(t) \in S$ for any $t \in [0,1]$. In the case that (2) has a classical solution the set S is a linear set of the form $c + \ker A,\ Ac = b$. However, if the price of the generalized solution is positive, the structure of S is the same as in the case of the classical solution.

Proposition 2.2
There is a unique solution among the generalized solutions which has the least norm. This solution is called normal solution.

Proof. The claim of this Proposition says that there is a unique vector $x^* \in S$ such that

$$\left\|x^*\right\|_n = \min\left\{\|x\|_n : x \in S\right\}.$$

However it follows from the fact that S is nonempty, closed and convex set, see [5](page 232). □

Remark 2.3
Let us notice that we have no requirements as to the inner products in Euclidean spaces. If we consider the standard inner product (Euclidean norm is derived from this product) we come to the well-known Least Square Method. Particularly, set $x = (\xi_1, \ldots, \xi_k),\ y = (\eta_1, \ldots, \eta_k)$. The

standard inner product in $\Re^k$ is $(x,y)_k = \sum_{i=1}^{k} \xi_i \eta_i$ and hence the price functional (5) is of the form

$$F(x) = \sum_{i=1}^{m} \left(\sum_{j=1}^{n} a_{ij} \xi_j - b_i \right)^2 .$$

3 Moore-Penrose Pseudoinverse Operator

Let us recall one concept of a generalization of the notion of the inverse operator (matrix). We consider the operator (1) and the following system of equations

(i) $AXA = A$,
(ii) $XAX = X$,
(iii) $(AX)^* = AX$,
(iv) $(XA)^* = XA$.

The operator $X : \Re^m \to \Re^n$ (the matrix is of the type (n,m)). However, it is not clear whether such an operator exists and whether it is unique.

Remark 3.1
If there is $m = n$ in (1) and the operator (1) is injective then $X = A^{-1}$. It is obvious and we can verify this by direct substitution.

Proposition 3.2
For any operator (1) there is unique operator X satisfying (i)-(iv).

Proof of uniqueness. Suppose *X* and *Y* are matrices satisfying (i)-(iv). It holds

$$AX = (AX)^* = X^* A^* = X^* (AYA)^* = X^* A^* (AY)^* = (AX)^* (AY)^* = AXAY = AY .$$

It also holds

$$XA = (XA)^* = A^* X^* = (AYA)^* X^* = (YA)^* A^* X^* = YA(XA)^* = YAXA = YA .$$

We can conclude

$$X = X(AX) = X(AY) = (XA)Y = (YA)Y = Y .$$

Proof of existence. Consider the operator

$$T = A^* A : \Re^n \to \Re^n . \quad (7)$$

The operator (7) is self-adjoint (its matrix is symmetric) and nonnegative in the sense that $(Tx,x)_n = \|Ax\|_n^2 \geq 0$ for any $x \in \Re^n$. It has all eigenvalues nonnegative. We denote each of the different positive eigenvalues of (7) by the symbol $\lambda_j,\ j=1,\ldots,s$. The operator (7) may have a non-trivial kernel (its matrix is singular) and then $\lambda_0 = 0$ is also its eigenvalue.
There are the spectral projections corresponding to positive eigenvalues $\lambda_j,\ j=1,\ldots,s$

$$P_j : \Re^n \to V_j,\ j=1,\ldots,s, \tag{8.1}$$

where V_j are the eigenspaces corresponding to λ_j. We define also the mapping

$$P_0 : \Re^n \to V_0, \tag{8.2}$$

which is the spectral projection for the eigenvalue $\lambda_0 = 0$ with eigenspace V_0 in the case (7) is singular and $P_0 = 0$ in the other case. We may write

$$T = \sum_{j=0}^{s} \lambda_j P_j. \tag{9}$$

The mappings (8.1) and (8.2) are orthogonal (symmetric) projections:

$$P_j P_k = 0,\ j \neq k,\ P_j^2 = P_j,\ P_j^* = P_j,\ I - P_0 = \sum_{j=1}^{s} P_j,\ j,k = 1,\ldots,s. \tag{10}$$

The expansion (9) is the *spectral decomposition of the operator* (7) and (10) is the well-known *spectral decomposition of unit.* Now we set

$$X = \sum_{j=1}^{s} \frac{1}{\lambda_j} P_j A^*. \tag{11}$$

We prove that (11) is desired pseudoinverse operator, i.e. it satisfies the properties (i)-(iv) above. Prove only (i). The remaining properties may be proved similarly.

$$AXA = A\sum_{j=1}^{s} \frac{1}{\lambda_j} P_j A^* A = A\sum_{j=1}^{s}\sum_{k=1}^{s} \frac{\lambda_k}{\lambda_j} P_j P_k =$$
$$= A\sum_{j=1}^{s} P_j = A(I - P_0) = A - AP_0.$$

It suffices to show that $AP_0 = 0$. However it is obvious since the range of P_0 is the kernel of the operator T and $\ker A \subseteq \ker T$. □

We have showed that there is a unique operator satisfying the properties (i)-(iv). This is operator is called *Moore-Penrose pseudoinverse operator (matrix)* and it is usually denoted by the symbol A^+. As we noted in Remark 3.1 if A is injective operator then $A^+ = A^{-1}$. In this case $A^+A = I$, however in general case it holds $A^+A = I - P_0$.

4 The normal solution and the Moore-Penrose Pseudoinverse Operator

The pseudoinverse operator (11) is in fact $A^+ : \Re^m \to \Re^n$. Let us denote

$$x^+ = A^+b. \tag{12}$$

We explain the relation between the vector (12) and the normal solution defined in Proposition 2.2.

Proposition 4.1
The normal solution of the system (2) is the vector (12).

Proof. Firstly we show that (12) is a generalized solution to (2). It means that the price functional (5) achieves its minimum at the point (12). Let us compute

$$F(x^+) = (Ax^+ - b, Ax^+ - b)_m = (AA^+b - b, AA^+b - b)_m = (AA^+Ac - Ac - d, AA^+Ac - Ac - d)_m =$$
$$= (A(I - P_0)c - Ac - d, A(I - P_0)c - Ac - d)_m = (d,d)_m.$$

The point of the minimum of (5) with the least norm has to be orthogonal to the kernel of the operator (1) because we observed that the set S of the minima of (5) is of the form $x^* + \ker A$. Using (11) we obtain

$$A^+b = \sum_{j=1}^{s} \frac{1}{\lambda_j} P_j A^* b \in \operatorname{ran}(T).$$

Using the relations (4) applied on the self-adjoint operator T we have that A^+b is orthogonal to any vector lying in $\ker T$. Since $\ker A \subseteq \ker T$ (in fact even equality holds) we have proved the claim of the proposition.

Example. The lag operator B is of the basic meaning in the time series analysis, e.g. see [2]. In the finite dimensional case this operator is represented by the matrix

$$\mathbf{B} = \begin{bmatrix} 0 & 1 & 0 & \dots & 0 \\ 0 & 0 & 1 & \dots & 0 \\ 0 & 0 & 0 & \ddots & 0 \\ \vdots & \vdots & \dots & \ddots & 1 \\ 0 & 0 & \dots & \dots & 0 \end{bmatrix} \in M(n),$$

where $M(n)$ is the space of all matrix of the order *n*. It is clear that 0 is the unique eigenvalue of *B*. Its adjoint operator B^* is obviously represented by the matrix

$$\mathbf{B}^* = \begin{bmatrix} 0 & 0 & 0 & \dots & 0 \\ 1 & 0 & 0 & \dots & 0 \\ 0 & 1 & 0 & \ddots & \vdots \\ \vdots & \vdots & \ddots & \ddots & 0 \\ 0 & 0 & \dots & 1 & 0 \end{bmatrix} \in M(n).$$

Finally the operator *T* defined by (7) has the matrix representation

$$\mathbf{T} = \begin{bmatrix} 0 & 0 & 0 & \dots & 0 \\ 0 & 1 & 0 & \dots & 0 \\ 0 & 0 & 1 & \ddots & 0 \\ \vdots & \vdots & \dots & \ddots & 0 \\ 0 & 0 & \dots & \dots & 1 \end{bmatrix} \in M(n).$$

Its eigenvalues are $\lambda_0 = 0$ with eigenspace $V_0 = \text{span}(e_1)$ and $\lambda_1 = 1$ with eigenspace $V_1 = \text{span}(e_2, \dots, e_n)$ respectively. Here $e_k = (0, \dots, 0, 1, 0, \dots, 0)$ with 1 on the *k*-th coordinate. In order to express the pseudoinverse operator B^+ in the form (11) it is necessary to find the projection P_1. However it is seen that $P_1 = T$. Hence we easily compute that $B^+ = B^*$.

5 Some notes to generalizations

The introduced approach to the generalized solution to the linear operator equation (2) is heavily dependent on the fact we consider the Euclidean space, i.e. the finite dimensional space equipped with an inner product. We may generalized in the direction of more generalized structure in a finite dimensional space: to take a normed linear space of the type $\ell_p(n)$, $p \in [1, +\infty]$, $p \neq 2$, see [4] or [5]. Another generalization is to relax the requirement of the finite dimension. The infinite dimensional analogue to the Euclidean space is the Hilbert space, i.e. the space of infinite dimension with an inner product which is complete with respect to the norm derived from this product. The typical example is the space of all square summable scalar sequences ℓ_2. This is the model of any separable Hilbert space. The price functional (5) corresponding to the linear operator $A : U \to H$, where *U* is a linear space and *H* is a Hilbert space achieves its minimum if *A* has the closed range. The proof is given in [4] (page 141). This proof is based again on Riesz projection theorem which is referred in the proof of Proposition 2.2 above. Unfortunately the range of *A* is not closed in many important situations.. It turns out that the sufficient and necessary condition for existence of the generalized solution to (2) is that $b \in \text{ran } A \oplus (\text{ran } A)^{\perp}$, where $(\text{ran } A)^{\perp}$ is the orthogonal complement of $\text{ran } A$.

The natural generalization of the operator (1) to infinite dimension is the compact operator, see [3](page 29). It is a consequence of the open mapping theorem that any compact operator of infinite rank between Hilbert spaces has non-closed range. Another example of an operator with non-closed range is given in [2]. It is the difference operator. In these cases there always exists vectors for which the price functional has no minimum and thus there is no generalized solution to the operator equation (2). Other generalizations and description of pseudoinverse operator can be found in [4].

References

[1] URL, http://mat.fsv.cvut.cz/marekivo/

[2] HORSKÝ, R., The lag operator and its spectral properties. *Mathematical Methods in Economics 2013.* Proceedings Part I (2013), 285-290

[3] LUKEŠ, J., Zápisky z funkcionální analýzy. Univerzita Karlova v Praze-Nakladatelství Karolinum, 2012, ISBN 978-80-246-2069-5

[4] NAIR, M. T., Linear operator equations; approximation and regularization. World Scientific Publishing Company, 2009

[5] TAYLOR, A. E., Úvod do funkcionální analýzy. Academia, Praha, 1973, ISBN 04-018-80, 216 s.

Current address

Horský Richard, RNDr., Ph.D.
Department of Mathematics , Faculty of Informatics and Statistics
University of Economics
Ekonomická 957, 148 00 Praha 4, Czech Republic
E-mail: rhorsky@vse.cz

Slovak University of Technology in Bratislava
Faculty of Mechanical Engineering
17th Conference on Applied Mathematics **APLIMAT 2018**

Proceedings

THE ACOUSTICS OF THE VANVITELLIANA HUNTING LODGE

IANNACE Gino (IT), TREMATERRA Amelia (IT)

Abstract. For many centuries, circular and vaulted structures were realised and used for aesthetic and constructive needs as buildings of worship, monumental buildings or courtyards. In vaulted environments, the effects of sound reflection focus on the central area due to the focus of multiple reflections. This paper presents the results of acoustic measurements carried out in a circular plant environment with a vaulted cover of the Vanvitelliana hunting lodge near Lake Fusaro in the north of the city of Naples. In the central area of the room, due to the geometry of the environment, there are focusing sound effects that generate a "flutter echo".

Keywords: acoustic measurements, impulse response, vault, reverberation time, flutter echo

Mathematics Subject Classification: Primary 68-04, 08A72; Secondary 28-04.

1 Introduction

For many centuries, circular and vaulted structures were realised and used, for aesthetic and constructive needs, as buildings of worship, monumental buildings or courtyards. In vaulted environments, the effects of sound reflection focus on the central area due to multiple reflections, generating special sound sensations for visitors listening to these acoustic phenomena. In the central area, underneath the highest part of the vault, there are phenomena of repeated echoes. Echoes are defined as a delayed repetition of sounds. Focusing effects can cause high sound pressure levels, echo and sound coloration [1]. The aural effects of concave curved surfaces have been studied by several authors [2, 3, 4]. Concave surfaces always create a convergence of sound energy and thus reflected sound amplification (compared to a flat surface) near the focal point, along with an attenuation of sound in other areas (far from the focal point). The echoes are often excluded from the acoustic design of halls due the acoustic risks associated with irregular sound distribution, fluctuating echoes and tone colouring; but the echoes can lead to positive acoustic effects [5].

In the 16th century, numerous works were published in which there were solutions and projects for the realisation of vaults and domes. Attanasio Kircher studied the circular and the vaulted rooms, and He demonstrated as the geometric shape influences acoustic behaviour of the halls [6, 7]. Kircher understood that in an room with the shape of ellipse, the people could communicate with each other when stay in the focus points. Wulfrank et al. [8] studied the sound focusing effect in the "Wigmore Hall" in London. While in the Royal Palace of Caserta, the double elliptical vault presents a great effect and inside, on the planking level, musicians played for the king and his guests during royal receptions [9, 10]. The architect Vanvitelli did not know the fundamentals of architectural acoustics and of the sound behaviour of large places when He designing the double vault. Several centuries prior to Vanvitelli, "De Architectura" by Vitruvio had been reprinted (written in 1st century B.C.) and described the fundamentals of theatrical acoustics [11]. However, most of the studies on architectural acoustics were carried out in the 19th century by Wallace Clement Sabine [12]. When sound meets a curved surface strange effects can be heard, the whispers reflected from a giant hemispherical ceiling were described by Wallace Sabine as having "... the effect of an invisible and mocking presence". In 1935, Alvar Aalto, a famous modernist architect, used an undulating ceiling for the Viipuri Library. The ceiling wood had many concave sections, each designed to amplify the sound for particular listeners [13].

2 The Acoustics in Vaulted Spaces

In the study of the acoustics of large rooms the main acoustic defect is the echo. For an attentive listener, small defects such as insufficient reverberation or low intensity can be unnoticed, a barely perceptible echo is annoying. An echo is a delayed repetition of a sound, and is a sound sensation in which, due to reflection, the listener distinctively perceives a replica, albeit distorted, of direct sound. Since the human ear can distinguish distanced sound impulses of at least 50 milliseconds, with the reflects sound being perceived as distinct from the direct one, it is necessary for the path of the two sound impulses to have a difference of at least 17 metres. This is not the only necessary condition for the echo to occur, in addition to an adequate delay, the reflected sound needs to be sufficiently intense with respect to the other sounds that the listener perceives. For this reason, an echo is manifested for short and particularly intense sound, or in environments with highly reflective walls and at a great distance from the source. In other cases the reflects sound that arrive with a delay are not always perceivable, because they are weakened by multiple reflections or covered by background noise. In large rooms, while realizing spatial conditions for this phenomenon, the presence of more reflective surfaces can cause an echo generated by a sound fail to emerge with respect to its reverberation and subsequent sounds. Thus, an echo manifests itself in particular cases. For example, when reflected energy is concentrated by large concave surfaces, such as domes and curved walls, which produce an increase in the intensity of reflections at certain points, or when the walls of the room are sufficiently absorbent except one which generates the sound reflection responsible for the echo. Curve surfaces can cause aural illusions, due to sound focusing. Standing the right distance in front of an appropriately curved surface and the sound will be focused and amplified. What is heard depends on how far the listener and the sound source are away from surface, the size, shape and extent of the curve as well as the type of sound being made. In an existing situation, an echo can be eliminated by making the reflecting surface sound absorbing and/or diffusing. Another form of echo is a flutter echo, this phenomenon occurs more often in not very large areas when the source is placed between two parallel and reflecting walls. Multiple reflections cause a sound tail similar to the flattering of bird's wings.

3 The Vanivitelliana hunting lodge

This work presents the study of the acoustics of a particular room of the Vanivitelliana hunting lodge near Lake Fusaro in the town of Bacoli, in the North of Naples. The city of Bacoli was a renowned resort for the nobles of the Roman Empire, but after the barbarian invasions and the phenomena of Bradyseism, it was abandoned. The city was reborn during the 17th century and became one of the favourite destinations of visitors to Northern Europe. From 1752 the Fusaro area became the hunting and fishing reserve of the Bourbons that ruled the city of Naples. It is a picturesque hunting lodge located on the Lake. The architect Carlo Vanvitelli built in 1782 the "Royal Hunting Lodge" then called "Casina Vanvitelliana" on the islet of Lake Fusaro, a short distance from the shore. The architect Carlo Vanvitelli was sensitive to the charm of Greek-Roman ruins, and He was inspired by the ruins of buildings and temples. Vanvitelli, inspired by the many examples of octagonal drums, designed a pier with three octagons inside each other, the outer ones being lower so that the tallest central so that it could emerge from the others. As a whole, the building has three octagonal bodies intersecting one at the top of the other with large windows arranged on two levels; a long wooden jetty connects the building to the shore of the lake. The care for of detail, elegance of the shapes and fusion with the lake environment give it a fairy-tale appearance. The building develops over two floors and in one of the two parts, there is the staircase that goes upstairs. Fig. 1 shows the external view of the Vanivitelliana hunting lodge on the Fusaro lake [14]. While Fig. 2 shows the prospectus of the Vanivitelliana hunting lodge. The central hall on the first floor has a circular plan and faces south and north with three openings per side. This circular room has a slightly concave dome ceiling. It has a 8.40 m diameter and minimum height to the side walls of 4.50 m and maximum at 5.30 m in the centre, the volume is about 270 m^3.

Fig. 1. External view of the Vanivitelliana hunting lodge.

This room is characterized by a reverberation time at medium frequencies of about 3.0 seconds. Sound impulses are prolonged by multiple reflections and accentuate in the central position giving rise to a particular flutter echo that captures the attention of visitors. Fig. 3 shows the dimensions of the circular room on the first floor. While the Fig. 4 shows the immage of the circular room.

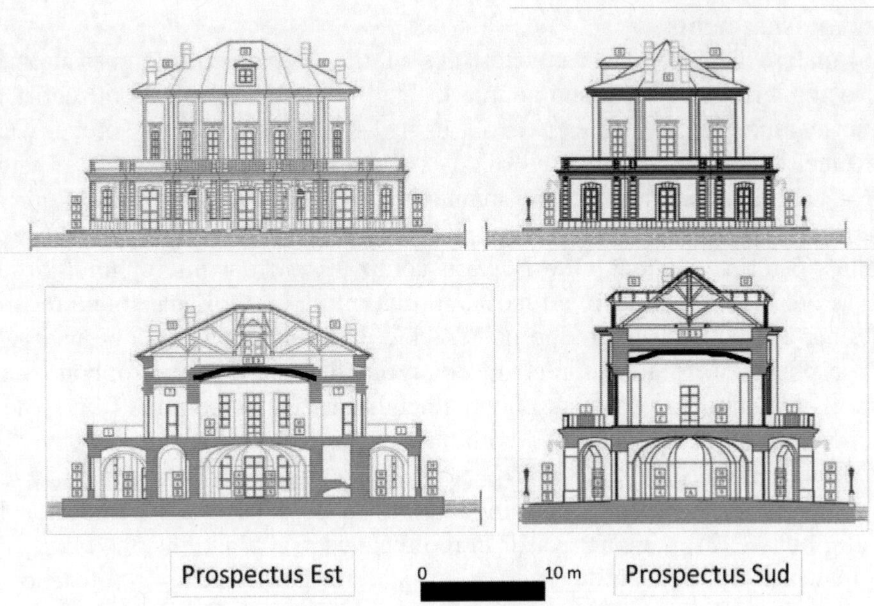

Fig. 2 Prospectus of the Vanivitelliana hunting lodge.

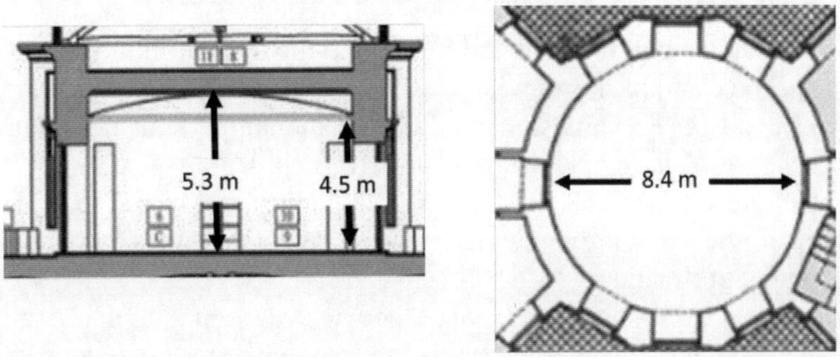

Fig. 3 Dimensions of the circular room on the first floor.

Fig. 4 Immage of the circular room on the first floor.

4 Acoustic measurements

In order to analyse the acoustic characteristics of the room, acoustic measurements were carried out using an impulsive sound source located in centre of the room, under the vault. The acoustic measurements were carried out using an impulsive sound source, which given the lack of electricity supply, consisted of toy balloons inflated with air. The height of the sound source and the microphones were maintained fixed at 1.6 m from the floor. The used microphone was a digital four channels equipment which also allowed to convert recording in an Ambisonics B-format. In fact, BRAHMA is composed of four microphones arranged in a tetrahedral array and allows to convert native format into an orthogonal sound records during post-processing. To reduce background noise, the measurements were done without visitors, so impulse responses were all recorded in empty conditions. The microphone was located along the radial direction at a constant pitch, at height of 1.60 m from the floor. The recorded impulse responses were then elaborated with the software Dirac 4.0, and the several acoustic parameters defined in the ISO 3382-1 [15], such as the early decay time (EDT), reverberation time (T30), the clarity (C80), the definition (D50), and the sound transmission index for speech intelligibility (STI) were analysed. In room acoustic evaluations, the clarity represents the degree to which different reflections arrives and are perceived by the listener, and it is assessed as an early-to-late arriving sound energy ratio. This ratio can be calculated for either a 50 ms or an 80 ms early time limit, depending on whether it relates to conditions for speech or music respectively. The definition considers the early arriving sound energy over the overall sound energy and, similarly to the clarity, it can be calculated for either a 50 ms or an 80 ms early time limit. The Speech Transmission Index (STI) represents the degree of amplitude modulation in a speech signal and it refers to the distortion in speech signals caused by reverberation, echoes, and background noise.

In particular, typical suggested values of the different monaural acoustic parameters for both the speech comprehension and music listening are [16, 17]:

- the reverberation time T30 should assume values below 1 second for a clearer perception of speeches, while it may assume greater values, around 2 seconds for music listening preference;
- the clarity C80, expressing the balance between the early and late arriving energy, should have a higher value if the goal is to separate the initial sounds from the diffuse ones and making the discrete sounds stand apart from each other. In a sound field which is not completely diffuse, the clarity C80 is uncorrelated to the reverberation time, and for the purposes of good listening conditions of music is generally reported that C80 should be in the range between -2 dB and 2 dB, while it is expected to be above 2 dB if speech perception is a priority;
- the definition D50 may assume values from 0 to 1, but for a good speech comprehension, it is often accepted that D50 should have values above 0.50;
- the indices STI can take values between 0 and 1, being greater than 0.5 for favourable speech conditions.

The acoustic procedure and post processing methodology were similar to those used in other spaces, such as the large theatre of Pompeii [18], the theatre of Benevento [19], as well as in many other ancient theatres [20, 21].

5 Discussion

Fig. 5 shows the position of the sound source in centre of the room, and the three receiver points disposed along the radial direction at the distance of 1 metre from each other, Fig. 5 shows the receiver point (1) that is at 1 metre from the sound source and it is close to the centre of the room. The receiver point (2) is at 2 metres from the sound source. While the receiver point (3) is at 3 metres from the sound source and it is close to the perimeter walls. Fig. 6 shows the impulse responses in the three different receiver points (1), (2) and (3), the figure shows the multi reflections that give the flutter echo effect in the receiver point (1); in the other points the effect of the flutter echo are very small. So the the effects of echo are concetrated in the center area of the room. Fig. 7 shows the measured values of different acoustic parameters, in the octave bands from 125 Hz to 4.0 kHz, for the three different measurement points (1), (2) and (3). The reverberation time (T30) at low frequencies exceeds the value of 10 seconds and decreases as the frequency increases, the reverberation time (T30) value is almost equal to the three measuring points. The early decay time (EDT) value at low frequencies is higher for the point located in the middle of the room, i.e. below the vault. For points (2) and (3) that are far from the centre the early decay time (EDT) values are almost equal. The clarity (C80) values reflect the trend of the reverberation time (T30), which is lower at low frequencies and increases as frequency increases; the farthest points from (2) and (3) centres have the best listening conditions for music. Parameter definition (D50) is always less than 0.5, in these environments, albeit small in size, the intelligibility of speech is always low.

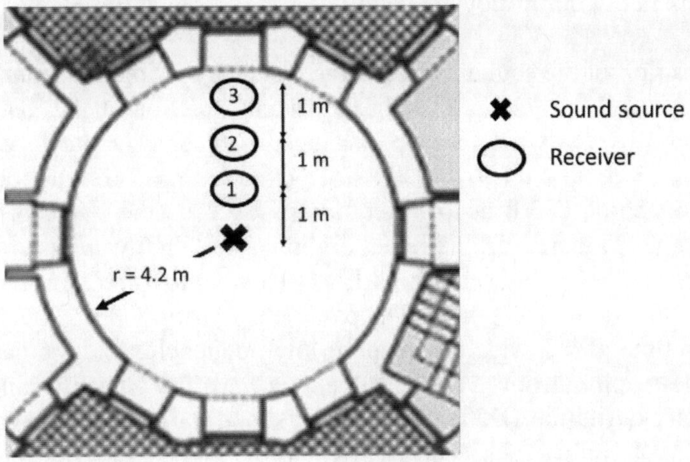

Fig.5 First measurement point.

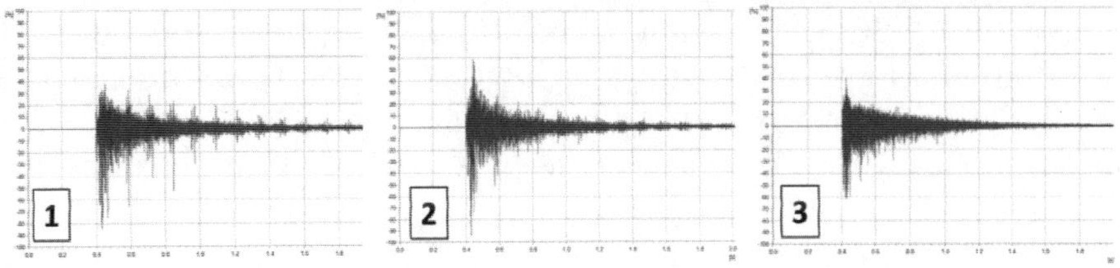

Fig.6 Impulse responses at the point 1, 2 and 3.

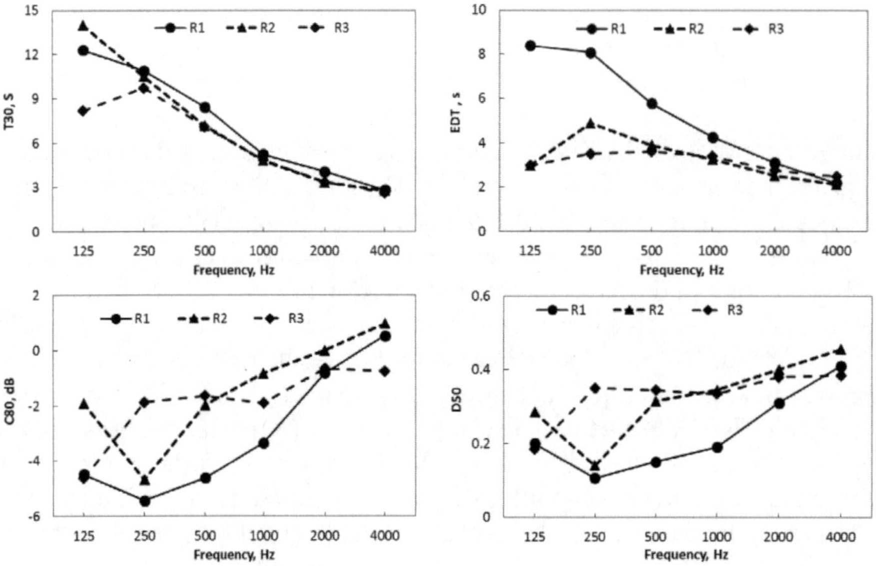

Fig.7 First measurement point. Acoustic parameters measured at the points (1), (2) and (3).

Fig. 8 shows the position of the sound source at the distance of 2 metres from the centre of the room, and the three receivers disposed along the radial direction at the distance of 1 metre from each other. Fig. 9 reports the measured values of different acoustic parameters in the octave bands, from 125 Hz to 4.0 kHz, for the three different measurement points (1), (2) and (3). The reverberation time (T30) at low frequencies exceeds the value of 10 seconds and decreases as the frequency increases, the reverberation time (T30) value is almost equal to the three measuring points. The early decay time (EDT) value at low frequencies is higher, but is similar for the three measurements points. The parameter clarity (C80) values reflect the trend of the reverberation time (T30), which is lower at low frequencies and increases as frequency increases; the farthest points from (2) and (3) centres have the best listening conditions for music. The parameter definition (D50) is always less than 0.5, in these environments, albeit small in size, the intelligibility of speech is always low. Fig. 10 shows the position of the sound source at the distance of 3 metres from the centre of the room, and the three receivers disposed along the radial direction at the distance of 1 metre from each other. Fig. 11 reports the measured values of different acoustic parameters in the octave bands, from 125 Hz to 4.0 kHz, for the three different measurement points (1), (2) and (3). The reverberation time (T30) at low frequencies exceeds the value of 10 seconds and decreases as the frequency increases. The reverberation time (T30) value is almost equal to the three measuring points. The early decay time (EDT) value at low frequencies is higher for the point located in the middle of the room, but the early decay time (EDT) values are almost equal, this valuea are under 4.0 seconds. The parameter clarity (C80) values reflect the trend of the reverberation time (T30), which is lower at low frequencies and increases as frequency increases; the farthest points from (2) and (3) centres have the best listening conditions for music. The parameter definition (D50) is always less than 0.5, in these environments, albeit small in size, the intelligibility of speech is always low.

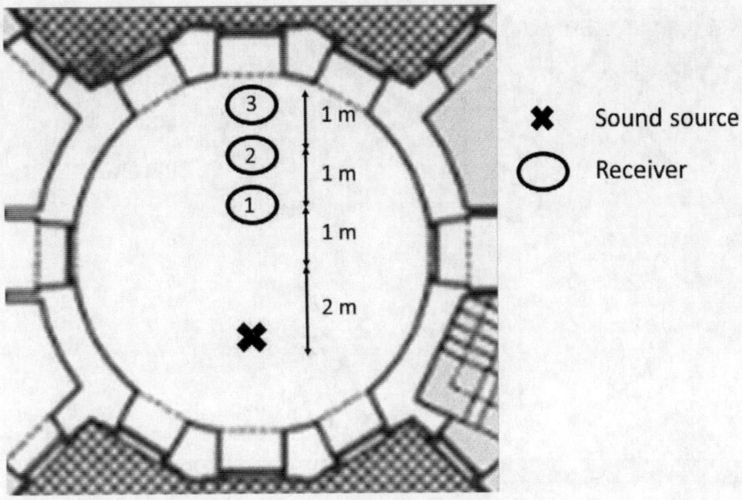

Fig.8 Second measurement point.

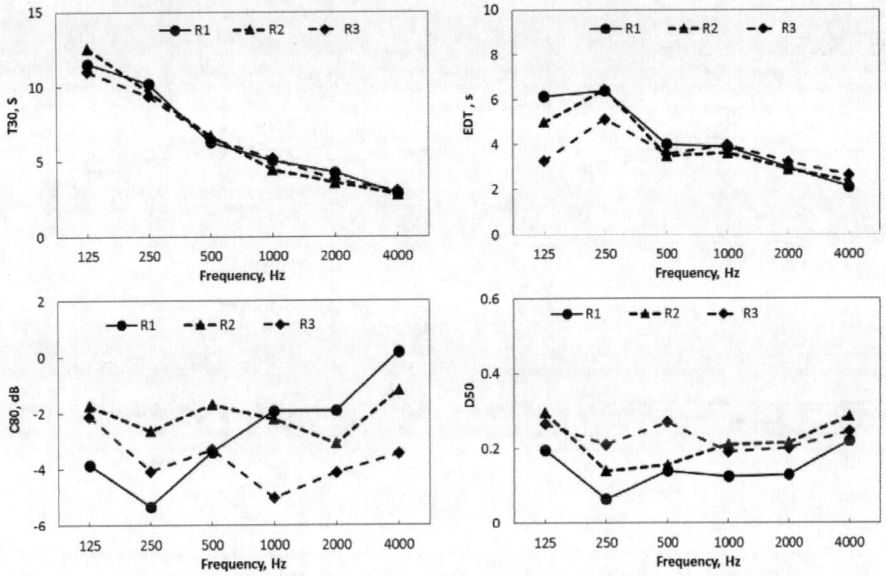

Fig.9 Second measurement point. Acoustic parameters measured at the points (1), (2) and (3).

6 Conclusions

This paper has discussed some of the aural effects that can be created by concave curved surfaces as well as the rooms where they can be heard. Vaulted rooms have particular acoustic features such as flutter echo. This effect is mainly manifested in the central area of the room and tends to decrease at the furthest points from the centre. In particular this paper has discussed the circular room located on the second floor of the Vanvitelliana hunting lodge on Lake Fusaro. The echoes can be considered acoustic faults, but in ancient and evocative buildings echoes can be an attraction for visitors.

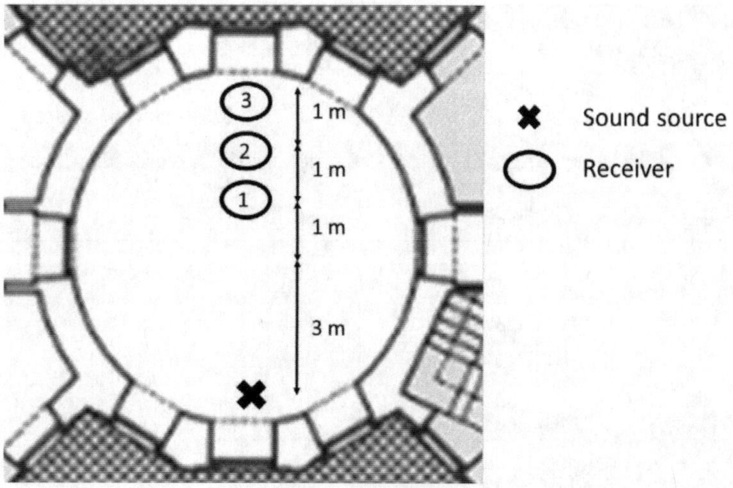

Fig.10 Third measurement point.

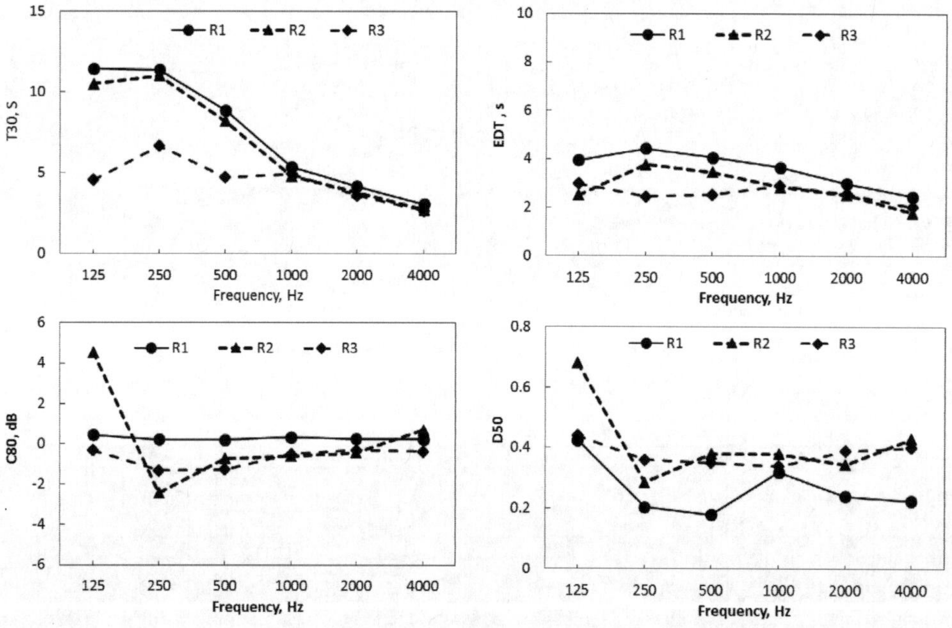

Fig.11 Third measurement point. Acoustic parameters measured at the points (1), (2) and (3).

References

[1] CREMER, L., MULLER, H. A. (translated by T. J. Schultz), *Principles and applications of room acoustics*, Applied Science Publishers, New York (1982)

[2] IANNACE, G., DI GABRIELE, M., SICURELLA, F., *Sound Focusing Effects in Horseshoe Plan Theatre*, Acoustics Australia, 44 (2), 2016, 359-368, doi: 10.1007/s40857-016-0059-2

[3] COX, T.J., *Sonic Wonderland: A Scientific Odyssey of Sound,* Bodley Head: Oxford, UK, 2014.

[4] IANNACE, G., IANNIELLO, E., *Sound-focusing effects in the plan of horse-shoe shaped opera theatres*. In: Proceedings of Acoustics '08, Paris, France, 2008, 1639–1643.
[5] COX, T.J., *Concave acoustics*. In: Proceedings of 5th International Symposium on Temporal Design, Sheffield, UK, 2011.
[6] KIRCHER, A., *Phonurgia Nova*; Kempten: Rudolph Dreherr, Germany, 1673.
[7] TRONCHIN, L., *The 'Phonurgia Nova' of Athanasius Kircher: The Marvellous sound world of 17th century*. Proc. Meet. Acoust. 4, 2008, 015002.
[8] WULFRANK, T., ORLOWSKI, R.J., *Acoustic analysis of Wigmore Hall, London, in the context of the 2004 refurbishment*. In: Proceedings of the Institute of Acoustic, 28(2), 2006.
[9] BERARDI, U., IANNACE, G., TREMATERRA, A., *The Acoustics of the Double Elliptical Vault of the Royal Palace of Caserta (Italy),* Buildings 7(1), 2017, doi:10.3390/buildings7010018
[10] BERARDI, U., IANNACE, G., IANNIELLO, C., *Acoustic intervention in a cultural heritage: The chapel of the Royal Palace in Caserta, Italy*, Buildings, 6(1) 2015, Article number 1. doi 10.3390/buildings6010001
[11] VITRUVIO, M.P., *De Architectura.*
[12] SABINE, W.C., *Collected Papers on Acoustics*, Cambridge, Mass, 1923.
[13] DECLERCQ, N. F, DEWIJNGAERT, K., VANDERHAEGHE, K., VERLEYSEN, P., *An acoustic diffraction study of a specifically designed auditorium having a corrugated ceiling : Alvar Aalto's lecture room,* Acta Acustica united with Acustica 97(4), 2011, 599-606, doi 10.3813/AAA.918440
[14] BARRELLA, G., TARÌ C., GARZONI, G.C., *Fusaro. Il restauro del complesso borbonico*, Giannini Editore, 2010.
[15] ISO 3382-1. Acoustics – Measurement of the reverberation time of rooms with reference to other acoustical parameters
[16] IANNACE, G., TREMATERRA, A., *Acoustic measurements and correction of a council room*, Noise and Vibration Worldwide, 45 (8), 2014, 12-16, doi: 10.1260/0957-4565.45.8.12
[17] IANNACE, G. Acoustic correction of monumental churches with ceramic material: The case of the Cathedral of Benevento (Italy), Journal of Low Frequency Noise Vibration and Active Control, 35(3), 2016, 230 – 239, doi: 10.1177/0263092316661028
[18] IANNACE, G., TREMATERRA, A., MASULLO, M., *The large theatre of Pompeii: Acoustic evolution*, Building Acoustics, 20(3), 2013, 215-227, doi: 10.1260/1351-010X.20.3.215
[19] IANNACE, G., TREMATERRA, A., *The rediscovery of Benevento Roman Theatre Acoustics*, Journal of Cultural Heritage, 15(6), 2014, 698, doi: http://dx.doi.org/10.1016/j.culher.2013.11.012
[20] BERARDI, U., IANNACE, G., MAFFEI, L. Virtual reconstruction of the historical acoustics of the Odeon of Pompeii, Journal of Cultural Heritage, 19, 2016, 555-566, doi: http://dx.doi.org/10.1016/j.culher.2015.12.004
[21] IANNACE G. The use of historical courtyards for musical performances. Building Acoustics 2016, 23(3-4), 207-222. DOI: https://doi.org/10.1177/1351010X16678219

Current address

Iannace Gino, PhD.
Università degli Studi della Campania Luigi Vanvitelli
Department of Architecture and Industrial Design Borgo San Lorenzo
81031 Aversa (Italy)
Tel. + 39.3315323642
gino.iannace@unicampania.it

Trematerra Amelia, PhD.
Università degli Studi della Campania Luigi Vanvitelli
Department of Architecture and Industrial Design Borgo San Lorenzo
81031 Aversa (Italy)
Tel. + 39.3402389882
amelia.trematerra@unicampania.it

Slovak University of Technology in Bratislava
Faculty of Mechanical Engineering
17th Conference on Applied Mathematics **APLIMAT 2018**

Proceedings

THE ACOUSTICS OF THE LONGOBARD CHURCH OF SANTA SOFIA IN BENEVENTO

IANNACE Gino (IT), TREMATERRA Amelia (IT)

Abstract. This paper presents the results of acoustic measurements carried out in the Longobard church of Santa Sofia in Benevento (Italy). The church is on a central plan, with a diameter of 23.5 m. In the centre there are six columns, placed at the vertexes of a hexagon and connected by arches which support the dome. Around this central hexagon there is a second decagonal ring, with eight pillars of white limestone blocks interspersed with layers of bricks; and two columns immediately after the entrance. The acoustic measurement were done with a sound source on the altar, and the microphone points in the area were seat people. The average value of reverberation time at the frequency of 1.0 kHz was about 1.5 seconds, so the church can be used for good listening of symphonic music.

Keywords: church, acoustic measurements, impulse response, reverberation time

Mathematics Subject Classification: Primary 68-04, 08A72; Secondary 28-04

1 Introduction

Music and architecture are linked by complex characteristics related to the concepts of space, and they are sciences that have evolved together [1, 2, 3]. The composers created their works even taking into consideration the acoustics of the environment in which the music was to be performed. Furthermore, the relationship between music and architecture has always been clear to musicians rather than architects. In the past, architects had no knowledge of indoor acoustics. Towards the end of the 1800s the notions of architectural acoustics began to be clearer thanks to the studies of W. C. Sabine, the father of modern acoustics, who considered the acoustics of the closed environments influenced the composition of the musical works [4]. From the origins of the Christian world, music has always been considered as a means of bringing together the values of faith and as a means of elevation and salvation. The first Christian songs were in Hebrew and Aramaic, then in Greek and finally in Latin. They formed a repertoire that was fully codified only between the 5th century and the end of the first millennium. The typical song was with a single voice sung by males without instruments (because the instruments were associated with pagan festivals). An important liturgical song is the Gregorian chant. The name of the "Gregorian chant" comes from Pope Gregory who included the accounts in the liturgical reform. Gregorian chants were best performed in the wide cathedrals, in fact the melody of Gregorian chants exploits the advantage of the long reverberation of the Romanesque and Gothic churches.

2 The acoustics of the churches

In the study of the acoustics of large environments, the focus is on aspects related to the physical and geometrical characteristics [5, 6, 7]. The churches are acoustically complex places because of the size of the volumes and the reflecting materials of which the walls are made and the absence of adequate sound-absorbing surfaces. Audience has the responsibility of most sound absorption, consequently acoustic conditions vary between the conditions of empty church and occupied church. One of the most used acoustical descriptors for evaluating the goodness of a room is the reverberation time (measured in seconds). This parameter was introduced at the end of the 1800s by Sabine and is defined as the time interval in which the sound energy decreases by 60 dB after switching off the source. In a room the longer the reverberation time, the greater the contribution of the components of the reflected sound compared to the direct one. The value of the reverberation time is a function of the volume of the room and of the total sound absorption of its internal surfaces. A short reverberation corresponds to a dull room, in which the surfaces absorb most of the sound energy and, as in the open air, the sound coming directly from the source is weakly reinforced. During the celebration of a liturgy different types of sound messages coexist, each of which requires different acoustic conditions. The organ music requires a long reverberation (> 3.0 seconds) but the instrumental music that sometimes requires a reverberation of 1.5 seconds. The readings and sermons require a short reverberation (< 1.0 seconds) to give good speech intelligibility while the Gregorian chants require reverberation 2.0 seconds to encourage a sense of community participation. To have good acoustics for all parts of the liturgy, a church should be transformed into a sophisticated hall of variable acoustics. The ideal values have been developed as indexes based of widely share perceptions. But derive from studies perfomed in concert halls, conference venues and theatres that do not have a high geometric variability. In addition, the churches can vary in size from a few hundred cubic meters to a small chapel; to several hundred thousand for a large basilica [8].

3 The church of Santa Sofia

The church of Santa Sofia was built by the Duke Gisulfo II and completed by Arechi II. It was built next to a Benedictine abbey, was completed in the year 762, as the Church of the Longobard people, and it was the most daring and imaginative construction of the high Middle age. Arechi II annexed a community of nuns, and entitled the Santa Sofia, that is, to the Holy Wisdom. The church of Santa Sofia is a building of exceptional interest in early medieval architecture. It is of modest dimensions, contained as it is within a circle of only 23.5 meters of diameter, and average height of 8.0 metres, the dome height is 14.0 metres. The general plan is original and entirely new for the time, not derived from Roman or Byzantine examples. The church has a central nucleus consisting of a hexagon whose vertices are placed six columns connected to each other with arches on which the dome develops. Around this central hexagon is a second decagonal ring, with eight pillars of white limestone blocks interspersed with layers of bricks and two columns immediately after the entrance. The pillars are radially arranged, each with the sides differently oriented, so as to make them parallel to the back walls of the perimeter. The course of the external walls is first circular, but at a certain point it is interrupted by star-shaped walls to return again to the circular at the entrance portal. All this creates games of perspectives, decompositions, closures of spaces coordinated with precise geometrical effects and based on reciprocal relationships, the result of an acute and original constructive intelligence. For example, the extraordinary variety of vaults, due to the unusual coupling of the hexagonal crown with the decagonal one: the succession of quadrangular, then rhomboidal and finally triangular folds is probably a reference to the shape of the tents used by the Lombard people during its long wanderings in Europe. The church of Santa Sofia has not always maintained the same appearance over the centuries. In the twelfth century, the church underwent a first restoration that, leaving intact the original plan, added a bell tower on the left side of the small facade and an elegant portico at the entrance, resting on four columns. The earthquake of 1688, which razed the city to the ground, caused enormous damage even in the church of Santa Sofia. The whole structure was damaged: the central hexagonal dome in wedges collapsed, much lower than the current one and without openings; the Romanesque bell tower overturned on the roof, completely destroying it. With the reconstruction in baroque forms of 1698 (and the further modifications following the subsequent earthquake of 1702) radical changes were made that led to the disappearance of the primitive Longobard configuration and caused the almost complete destruction of the precious frescoes of the IX century. The interventions consisted, among other things, in the transformation of the plant from star to circular, in the demolition and reconstruction in new forms of the central apse, in the tapering of the eight pillars and in the construction of the new façade, which still exists today. There were also two side chapels and the sacristy. The interior was completely plastered and furnished according to the Baroque taste. In 1951 the restoration work began, which, with scrupulous interventions, allowed to bring to light the original Lombard structural structure and then complete the demolished or tampered parts during the baroque transformation. In particular, the two chapels on the side of the façade, the central apse and the circular wall which had incorporated the outer edges of the star walls were eliminated and the walls were rebuilt following the indications provided by the archaeological research [9, 10, 11]. Fig. 1 shows the external view of the church. Fig. 2 shows the aerial view of the church. While Fig. 3 shows the plant and the section of the church with the main dimensions.

Fig. 1. External view of the church.

Fig. 2. Aerial view of the church

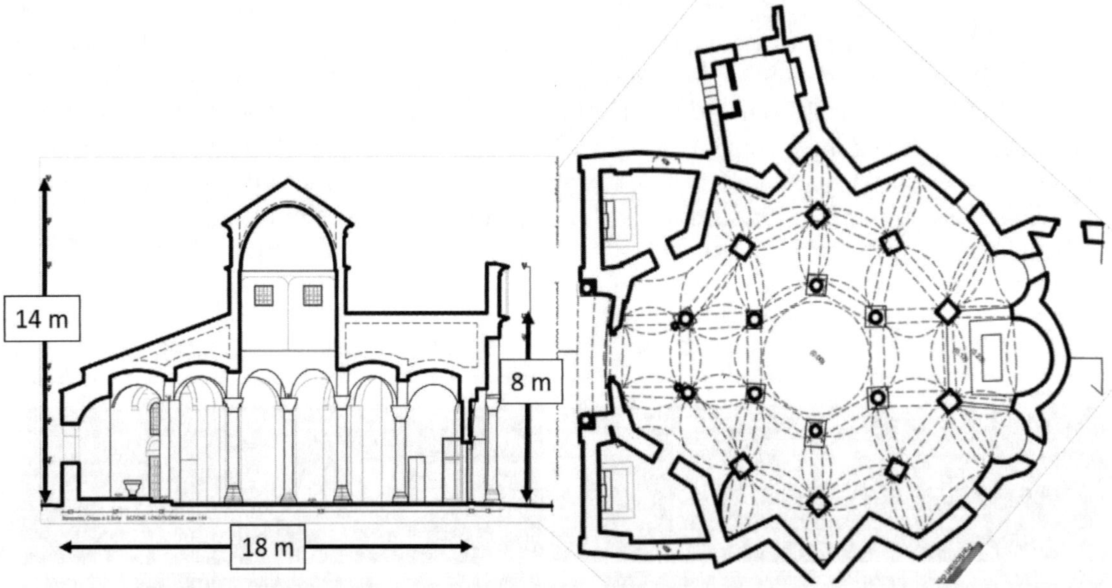

Fig. 3. Plant and section of the church with the main dimensions.

4 Acoustic measurements

The acoustic measurements were carried out with a spherical omnidirectional sound source located on the altar and 8 microphone points of measurement placed in different positions, in accordance with ISO 3382-1 [12]. The measuremt points were at height of 1.60 m from the floor, in an area occupied to the listeners. The sound source was powered by MLS signal, the processing of impulse responses had been analyzed by the Dirac 4.0 software. Fig. 4 shows the plan of the church with the indication of the source and of the receiver microphone points. The recorded impulse responses were then elaborated with the software Dirac 4.0, and the several acoustic parameters defined in the ISO 3382-1, such as the early decay time (EDT), reverberation time (T30), the clarity (C80), the definition (D50), and the sound transmission index for speech intelligibility (STI) were analysed. In room acoustic evaluations, the clarity represents the degree to which different reflections arrives and are perceived by the listener, and it is assessed as an early-to-late arriving sound energy ratio. This ratio can be calculated for either a 50 ms or an 80 ms early time limit, depending on whether it relates to conditions for speech or music respectively. The definition considers the early arriving sound energy over the overall sound energy and, similarly to the clarity, it can be calculated for either a 50 ms or an 80 ms early time limit. The Speech Transmission Index (STI) represents the degree of amplitude modulation in a speech signal and it refers to the distortion in speech signals caused by reverberation, echoes, and background noise. In particular, typical suggested values of the different monaural acoustic parameters for both the speech comprehension and music listening are [13]:

- the reverberation time T30 should assume values below 1 second for a clearer perception of speeches, while it may assume greater values, around 2 seconds for music listening preference;
- the clarity C80, expressing the balance between the early and late arriving energy, should have a higher value if the goal is to separate the initial sounds from the diffuse ones and making the discrete sounds stand apart from each other. In a sound field which is not completely diffuse, the clarity C80 is uncorrelated to the reverberation time, and for the purposes of good listening conditions of music is generally reported that C80 should be in the range between -2 dB and 2 dB, while it is expected to be above 2 dB if speech perception is a priority;
- the definition D50 may assume values from 0 to 1, but for a good speech comprehension, it is often accepted that D50 should have values above 0.50;
- the indices STI can take values between 0 and 1, being greater than 0.5 for favourable speech conditions.

The acoustic procedure and post processing methodology were similar to those used in other spaces, such as the large theatre of Pompeii [14], the theatre of Benevento [15], as well as in many other ancient theatres and places [16, 17, 18]. Table 1 shows the optimal values of the acoustic parameters for the different listening conditions, from which it is noted that for an environment intended for speech an optimal reverberation time of less than 1.0 seconds is required. The Fig. 5 shows the average values and standard deviation of the acoustic parameters in the octave band from 125 to 4.0 kHz of the eight measurement points.

Parameters	EDT, s	T_{30}, s	C_{80}, dB	D_{50}
Optimal values for musical performances	$1.8 < EDT < 2.6$	$1.6 < T_{30} < 2.2$	$-2 < C_{80} < 2$	< 0.5
Optimal values for speech performances	1.0	$0.8 < T_{30} < 1.2$	> 2	> 0.5

Tab. 1. Optimal acoustic parameters values for the different listening conditions

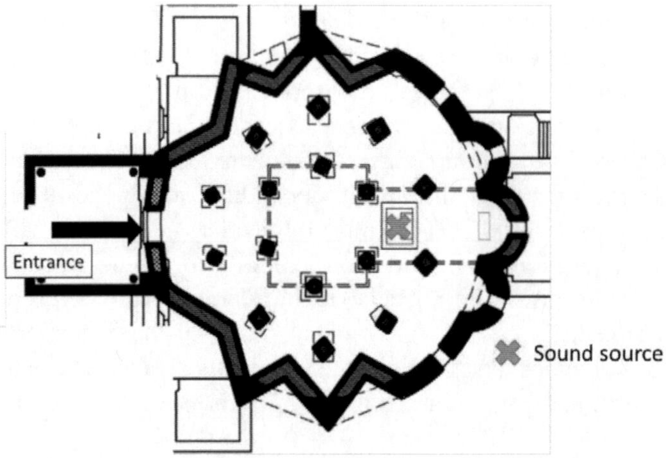

Fig. 4. Position of the sound source on the altar

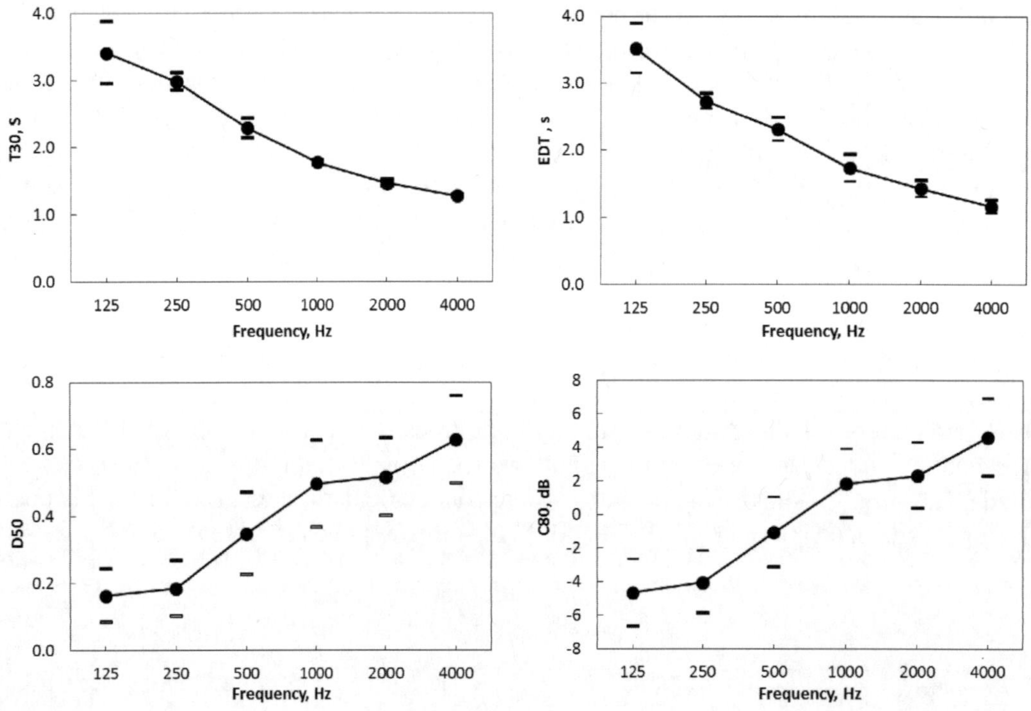

Fig.5 Average values of the acoustic parameters measured and standard deviations.

5 Discussion

The values of EDT and T30 at the low frequencies and around 3.5 seconds while at medium frequencies and on the order of 1.5 seconds. T30 and EDT decrease rapidly as the frequency increases. The values of the standard deviation are small at medium frequencies, while at 125 Hz the value of the standard deviation is greater, the spatial parameters at this frequency chenge with the measuremnt positions. So for D50 and C80 the values of the standard deviation vary significantly, this variation means that the acoustic parameters D50 and C80 change from point to point.The value of C80 reflects the trend of the reverberation time is less than - 2 dB at low frequencies at medium frequencies is around 0 dB. Similarly for the value of the definition which is greater than 0.5 for frequencies above 0.5. The average value of the parameter STI = 0.5. The room does meet the criteria of good listening for the music and the speech.

6 Conclusions

This paper has discussed the acoustic characteristic of the Longobard church of the Santa Sofia in Benevento. The church built as a place of prayer for the nearby monastery; the church has been rebuilt many times because of the damage due to the numerous earthquakes. This place has good characteristics for listening to music and for understanding speech. The modest volume makes the acoustic characteristics are optimal.

References

[1] CREMER, L., MULLER, H. A. (translated by T. J. Schultz), *Principles and applications of room acoustics*, Applied Science Publishers, New York (1982)

[2] IANNACE, G., Acoustic correction of monumental churches with ceramic material: The case of the Cathedral of Benevento (Italy), Journal of Low Frequency Noise Vibration and Active Control, 35(3), 2016, 230 – 239, doi: 10.1177/0263092316661028

[3] FORSYTH, M., *Buildings for Music: The Architect, the Musician, the Listener from the Seventeenth Century to the Present Day*, MIT Press, 1985

[4] SABINE, W.C., *Collected Papers on Acoustics*, Cambridge, Mass, 1923.

[5] BERARDI, U., IANNACE, G., TREMATERRA, A., *The Acoustics of the Double Elliptical Vault of the Royal Palace of Caserta (Italy),* Buildings 7(1), 2017, doi:10.3390/buildings7010018

[6] BERARDI, U., IANNACE, G., IANNIELLO, C., *Acoustic intervention in a cultural heritage: The chapel of the Royal Palace in Caserta, Italy*, Buildings, 6(1) 2015, Article number 1. doi 10.3390/buildings6010001

[7] CIRILLO, E, MARTELLOTTA, F., *Worship, Acoustics and Architecture*, Multi – Science Publishing Co. Ltd, 2006.

[8] BARRON, M., *Auditorium Acoustics and Architectural Design*, E&FN SPON, London, 1993.

[9] MEOMARTINI, A., *I monumenti e le opere d'arte della città di Benevento*, 1889.

[10] KUBACH, H., E., *Architettura Romanica*, Storia Universale dell'Architettura Electa, 1998.

[11] LAVAGNINA, E., *L'arte medievale*, U.T.E.T., Torino, 1949.

[12] ISO 3382-1. *Acoustics – Measurement of the reverberation time of rooms with reference to other acoustical parameters.*

[13] BREZINA, P., *Measurement of intelligibility and clarity of the speech in romanesque churches*, Journal of Cultural Heritage, 16 (3), 2015, 386–390, doi:10.1016/j.culher.2014.06.010

[14] IANNACE, G., TREMATERRA, A., MASULLO, M., *The large theatre of Pompeii: Acoustic evolution*, Building Acoustics, 20(3), 2013, 215-227, doi: 10.1260/1351-010X.20.3.215

[15] IANNACE, G., TREMATERRA, A., *The rediscovery of Benevento Roman Theatre Acoustics*, Journal of Cultural Heritage, 15(6), 2014, 698, doi: http://dx.doi.org/10.1016/j.culher.2013.11.012
[16] BERARDI, U., IANNACE, G., MAFFEI, L., *Virtual reconstruction of the historical acoustics of the Odeon of Pompeii*, Journal of Cultural Heritage, 19, 2016, 555-566, doi: http://dx.doi.org/10.1016/j.culher.2015.12.004
[17] IANNACE, G., *The use of historical courtyards for musical performances*, Building Acoustics 2016, 23(3-4), 207-222. doi: https://doi.org/10.1177/1351010X16678219
[18] IANNACE, G., TREMATERRA, A., *Acoustic measurements and correction of a council room*, Noise and Vibration Worldwide, 45 (8), 2014, 12-16, doi: 10.1260/0957-4565.45.8.12

Current address

Iannace Gino, phd
Università degli Studi della Campania "Luigi Vanvitelli"
Department of Architecture and Industrial Design
Borgo San Lorenzo
81031 Aversa (Italy)
Tel. + 39.3315323642
gino.iannace@unicampania.it

Trematerra Amelia, phd
Università degli Studi della Campania "Luigi Vanvitelli"
Department of Architecture and Industrial Design
Borgo San Lorenzo
81031 Aversa (Italy)
Tel. + 39.3402389882
amelia.trematerra@unicampania.it

Slovak University of Technology in Bratislava
Faculty of Mechanical Engineering
17th Conference on Applied Mathematics **APLIMAT 2018**

Proceedings

PARAMETER ESTIMATION FOR SCALAR STOCHASTIC DIFFERENTIAL EQUATION OF SECOND ORDER

JANÁK Josef (CZ)

Abstract. We introduce (scalar) stochastic differential equation of second order with two unknown parameters and its solution, which is Ornstein–Uhlenbeck process. Based on ergodicity, we derive two families of the strong consistent estimators of these parameters. Then we study their asymptotic normality and we present their implementation.

Keywords: Ornstein–Uhlenbeck process, parameter estimation, asymptotic normality

Mathematics subject classification: Primary 62M05; Secondary 60G35, 93E10.

1 Introduction

In this article, we are improving and extending some previous results on the parameter estimation for stochastic differential equation of second order, which were previously published in [3] and [2]. Namely, we have found another family of estimators $(\tilde{a}_t, \tilde{b}_t)$, we have proved their asymptotic normality and we have implemented them in the program R.

The following section introduces the stochastic differential equation for the harmonic oscillation, its rewriting and its solution, that is the Ornstein–Uhlenbeck process. Using the observation of the trajectory $\{X^{x_0}(t),\ 0 \leq t \leq T\}$ and ergodicity of the Ornstein–Uhlenbeck process, two types of strong constistent estimators of unknown parameters are proposed. The proof of Theorem 1 may be found in [4]. The proof of Lemma 1 is only a matter of computation. We also mention the asymptotic normality of the estimators $\hat{a}_T$ and $\hat{b}_T$, which was studied in [2].

The asymptotic normality of the estimators $\tilde{a}_T$ and $\tilde{b}_T$ is proved in the third section. We are using Itô's formula to obtain different formulae for the processes $Y_T = \int_0^T |X_1^{x_0}(t)|^2\,dt$ and $H_T = \int_0^T |X_2^{x_0}(t)|^2\,dt$ (on which the estimators $\tilde{a}_T$ and $\tilde{b}_T$ are based on) and then central limit theorem for the stochastic integral (which may be found in [5]). The results are summarized in Theorem 3.

In the section 4, we introduce the implementation of all estimators and on one particular example, we compare the two methods both graphically and numerically.

2 Parameter estimation and strong consistency

Consider the following (scalar) stochastic differential equation

$$\ddot{x} + 2a\dot{x} + bx = \sigma\dot{\beta}(t), \tag{1}$$

with initial values $X(0) = x_1$ and $\dot{X}(0) = x_2$. Let $a > 0$, $b > 0$ be unknown real parameters, $\sigma > 0$ is known and let $\dot{\beta}(t)$ be the formal time derivative of the standard Brownian motion.

We may rewrite this equation in the form

$$dX(t) = AX(t)\,dt + \Phi\,dB(t), \quad X(0) = x_0, \tag{2}$$

if we set

$$X(t) = \begin{pmatrix} X_1(t) \\ X_2(t) \end{pmatrix} \in \mathbb{R}^2,\ A = \begin{pmatrix} 0 & 1 \\ -b & -2a \end{pmatrix} \in \mathbb{M}_{2\times 2},\ \Phi = \begin{pmatrix} 0 & 0 \\ 0 & \sigma \end{pmatrix} \in \mathbb{M}_{2\times 2},$$

$$B(t) = \begin{pmatrix} 0 \\ \beta(t) \end{pmatrix} \in \mathbb{R}^2,\ x_0 = \begin{pmatrix} x_1 \\ x_2 \end{pmatrix} \in \mathbb{R}^2,$$

where $\mathbb{M}_{2\times 2}$ stands for the space of all 2×2 matrices with real entries, equipped with the operator norm $\|\cdot\|_{\mathcal{L}(\mathbb{R}^2)}$.

If we denote $S(t) := e^{At}$ a strongly continuous semigroup on $\mathbb{R}^2$, then the solution $(X^{x_0}(t), t \geq 0)$ to (2) is defined by the mild form

$$X^{x_0}(t) = S(t)x_0 + Z(t), \quad t \geq 0, \tag{3}$$

where $(Z(t), t \geq 0)$ is the convolution integral

$$Z(t) = \int_0^t S(t-u)\Phi\,dB(u). \tag{4}$$

The solution $(X^{x_0}(t), t \geq 0)$ to the equation (2) is called the Ornstein–Uhlenbeck process.

Theorem 1. *If the semigroup $(S(t), t \geq 0)$ is exponentially stable, i.e., there exist constants $M > 0$ and $\rho > 0$ such that for all $t \geq 0$*

$$\|S(t)\|_{\mathcal{L}(\mathbb{R}^2)} \leq Me^{-\rho t}, \tag{A1}$$

then there is a Gaussian centered limiting measure $\mu_\infty^{(a,b)} = N(0, Q_\infty^{(a,b)})$ for $(X^{x_0}(t), t \geq 0)$, such that

$$w^* - \lim_{t\to\infty} \mu_t^{x_0} = \mu_\infty^{(a,b)}$$

for each initial condition $x_0 \in \mathbb{R}^2$, where $\mu_t^{x_0} = \mathrm{Law}(X^{x_0}(t))$ and $\mathrm{Law}(\cdot)$ denotes the probability distribution. The covariance matrix $Q_\infty^{(a,b)}$ has the following form

$$Q_\infty^{(a,b)} = \int_0^\infty S(t)\Phi\Phi^\top S^\top(t)\,dt. \tag{5}$$

If $a^2 < b$, then the real parts of eigenvalues of matrix A are negative and the condition (A1) holds true. If we denote $\alpha = -a$, $\beta = \sqrt{b - a^2}$, then we may write $\lambda_1 = \alpha + i\beta$, $\lambda_2 = \alpha - i\beta$.

Lemma 1. The fundamental system $(S(t), t \geq 0)$ for the equation (1) has the following form

$$S(t) = \begin{pmatrix} e^{\alpha t}\left(\cos(\beta t) - \frac{\alpha}{\beta}\sin(\beta t)\right) & \frac{1}{\beta}e^{\alpha t}\sin(\beta t) \\ e^{\alpha t}\left(-\beta - \frac{\alpha^2}{\beta}\right)\sin(\beta t) & \frac{1}{\beta}e^{\alpha t}(\alpha\sin(\beta t) + \beta\cos(\beta t)) \end{pmatrix}.$$

The covariance matrix $Q_\infty^{(a,b)}$ of the limiting measure $\mu_\infty^{(a,b)}$ equals to

$$Q_\infty^{(a,b)} = \int_0^\infty S(t)\Phi\Phi^\top S^\top(t)\,dt = \int_0^\infty \frac{\sigma^2}{\beta^2}\begin{pmatrix} q_{11}(t) & q_{12}(t) \\ q_{21}(t) & q_{22}(t) \end{pmatrix} dt = \frac{\sigma^2}{\beta^2}\begin{pmatrix} \frac{\beta^2}{4ab} & 0 \\ 0 & \frac{\beta^2}{4a} \end{pmatrix}$$

$$= \sigma^2 \begin{pmatrix} \frac{1}{4ab} & 0 \\ 0 & \frac{1}{4a} \end{pmatrix}, \tag{6}$$

where

$$q_{11}(t) = e^{2\alpha t}\sin^2(\beta t),$$
$$q_{12}(t) = q_{21}(t) = e^{2\alpha t}\sin(\beta t)(\alpha\sin(\beta t) + \beta\cos(\beta t)),$$
$$q_{22}(t) = e^{2\alpha t}(\alpha\sin(\beta t) + \beta\cos(\beta t))^2.$$

Remark 1. In the cases of $a^2 > b$ and $a^2 = b$, the eigenvalues of matrix A are also negative and the condition (A1) also holds true. The formulae for the fundamental system $(S(t), t \geq 0)$ are different, but the resulting formulae for the covariance matrix $Q_\infty^{(a,b)}$ coincide. Hence the assumption $a^2 < b$ may be omitted.

For the parameter estimation, we are interested in the trace of the covariance matrix $Q_\infty^{(a,b)}$. From the formula (6), it follows that

$$\operatorname{Tr} Q_\infty^{(a,b)} = \frac{b+1}{4ab}\sigma^2. \tag{7}$$

Since the Ornstein–Uhlenbeck process $X(t)$ is ergodic in $\mathbb{R}^2$ (see Example 2.1. in [7]), we may use Birkhoff's theorem. Namely

$$\lim_{T\to\infty}\frac{1}{T}\int_0^T \|X^{x_0}(t)\|_{\mathbb{R}^2}^2\,dt = \int_{\mathbb{R}^2}\|y\|_{\mathbb{R}^2}^2\,d\mu_\infty^{(a,b)}(y) = \operatorname{Tr} Q_\infty^{(a,b)}, \tag{8}$$

for any initial contidion $x_0 \in \mathbb{R}^2$ (see Theorem 4.9. in [6]).

If we denote $I_T := \frac{1}{T}\int_0^T \|X^{x_0}(t)\|_{\mathbb{R}^2}^2\,dt$, then (based on (7)) some strongly constistent estimators of parameters a and b may be proposed. If we know the true value of the parameter b, then the strongly consistent estimator of the parameter a is

$$\hat{a}_T = \frac{b+1}{4bI_T}\sigma^2. \tag{9}$$

Similarly, if we know the true value of the parameter a, then

$$\hat{b}_T = \frac{\sigma^2}{4aI_T - \sigma^2} \tag{10}$$

is the strong consistent estimator of the parameter b.

In [2], we have also proved the asymptotic normality of these estimators.

Theorem 2. *The estimators $\hat{a}_T$ and $\hat{b}_T$ are asymptotically normal, i.e.,*

$$\text{Law}\left(\sqrt{T}\left(\hat{a}_T - a\right)\right) \to N\left(0, \frac{4a^3}{b(b+1)^2} + a\right), \quad T \to \infty, \tag{11}$$

$$\text{Law}\left(\sqrt{T}\left(\hat{b}_T - b\right)\right) \to N\left(0, 4ab + \frac{b^2(b+1)^2}{a}\right), \quad T \to \infty. \tag{12}$$

The estimators $\hat{a}_T$ and $\hat{b}_T$ are indeed very easily implemented, but they have one major disadvantage: We have to know the true value of the other parameter. However another family of estimators $(\tilde{a}_T, \tilde{b}_T)$ may be proposed, which do not possess this disadvantage. Since

$$\|x\|^2_{\mathbb{R}^2} = |x_1|^2 + |x_2|^2, \quad \forall x = \begin{pmatrix} x_1 \\ x_2 \end{pmatrix} \in \mathbb{R}^2,$$

the integral in (8) may be split into two parts

$$\begin{aligned} I_T &= \frac{1}{T}\int_0^T \|X^{x_0}(t)\|^2_{\mathbb{R}^2}\, dt \\ &= \frac{1}{T}\int_0^T |X_1^{x_0}(t)|^2\, dt + \frac{1}{T}\int_0^T |X_2^{x_0}(t)|^2\, dt \\ &=: Y_T + H_T, \end{aligned} \tag{13}$$

where $X^{x_0}(t) = (X_1^{x_0}(t), X_2^{x_0}(t))^\top \in \mathbb{R}^2$ is the solution to the equation (2). The formula (7) for the $\operatorname{Tr} Q_\infty^{(a,b)}$ may also be split into two parts. Indeed, the two parts in (13) converge to their corresponding limits individually, since it is just using the Birkhoff's theorem to the individual components, i.e.,

$$\lim_{T\to\infty} Y_T = \lim_{T\to\infty} \frac{1}{T}\int_0^T |X_1^{x_0}(t)|^2\, dt = \frac{\sigma^2}{4ab}, \tag{14}$$

$$\lim_{T\to\infty} H_T = \lim_{T\to\infty} \frac{1}{T}\int_0^T |X_2^{x_0}(t)|^2\, dt = \frac{\sigma^2}{4a}. \tag{15}$$

Based on these convergences, we may introduce new family of strongly consistent estimators $\tilde{a}_T$ and $\tilde{b}_T$

$$\tilde{a}_T = \frac{\sigma^2}{4H_T}, \tag{16}$$

$$\tilde{b}_T = \frac{H_T}{Y_T}. \tag{17}$$

3 Asymptotic normality of estimators $\tilde{a}_T$ and $\tilde{b}_T$

In this section, we show asymptotic normality of estimators (16) and (17), i.e., the weak convergence of $\sqrt{T}\left(\tilde{a}_T - a\right)$ or $\sqrt{T}\left(\tilde{b}_T - b\right)$ to a Gaussian distribution.

Let us start with the definition of operators $P_1 : \mathbb{R}^2 \to \mathbb{R}^2$ and $P_2 : \mathbb{R}^2 \to \mathbb{R}^2$:

$$P_1 x = P_1 \begin{pmatrix} x_1 \\ x_2 \end{pmatrix} = \begin{pmatrix} b & 0 \\ 0 & 1 \end{pmatrix}\begin{pmatrix} x_1 \\ x_2 \end{pmatrix}, \quad \forall x = \begin{pmatrix} x_1 \\ x_2 \end{pmatrix} \in \mathbb{R}^2, \tag{18}$$

$$P_2 x = P_2 \begin{pmatrix} x_1 \\ x_2 \end{pmatrix} = \begin{pmatrix} b + 4a^2 & 2a \\ 2a & 1 \end{pmatrix}\begin{pmatrix} x_1 \\ x_2 \end{pmatrix}, \quad \forall x = \begin{pmatrix} x_1 \\ x_2 \end{pmatrix} \in \mathbb{R}^2. \tag{19}$$

Let us denote $\langle \cdot, \cdot \rangle$ and $\| \cdot \|$, the Euclidean scalar product and the norm in $\mathbb{R}^2$. The properties of these two matrices are summarized in the following Lemma 2.

Lemma 2. The matrices P_1 and P_2 are symmetric and

$$\langle P_1 x, Ax \rangle = -2a|x_2|^2, \quad \forall x = \begin{pmatrix} x_1 \\ x_2 \end{pmatrix} \in \mathbb{R}^2, \tag{20}$$

$$\langle P_2 x, Ax \rangle = -2ab|x_1|^2, \quad \forall x = \begin{pmatrix} x_1 \\ x_2 \end{pmatrix} \in \mathbb{R}^2. \tag{21}$$

Proof. The symmetry of matrices P_1 and P_2 is evident. For every $x = (x_1, x_2)^\top \in \mathbb{R}^2$, we have

$$\begin{aligned} \langle P_1 x, Ax \rangle &= \left\langle \begin{pmatrix} b & 0 \\ 0 & 1 \end{pmatrix} \begin{pmatrix} x_1 \\ x_2 \end{pmatrix}, \begin{pmatrix} 0 & 1 \\ -b & -2a \end{pmatrix} \begin{pmatrix} x_1 \\ x_2 \end{pmatrix} \right\rangle \\ &= \left\langle \begin{pmatrix} bx_1 \\ x_2 \end{pmatrix}, \begin{pmatrix} x_2 \\ -bx_1 - 2ax_2 \end{pmatrix} \right\rangle \\ &= bx_1x_2 - bx_1x_2 - 2ax_2^2 \\ &= -2a|x_2|^2 \end{aligned}$$

and

$$\begin{aligned} \langle P_2 x, Ax \rangle &= \left\langle \begin{pmatrix} b + 4a^2 & 2a \\ 2a & 1 \end{pmatrix} \begin{pmatrix} x_1 \\ x_2 \end{pmatrix}, \begin{pmatrix} 0 & 1 \\ -b & -2a \end{pmatrix} \begin{pmatrix} x_1 \\ x_2 \end{pmatrix} \right\rangle \\ &= \left\langle \begin{pmatrix} bx_1 + 4a^2x_1 + 2ax_2 \\ 2ax_1 + x_2 \end{pmatrix}, \begin{pmatrix} x_2 \\ -bx_1 - 2ax_2 \end{pmatrix} \right\rangle \\ &= bx_1x_2 + 4a^2x_1x_2 + 2ax_2^2 - 2abx_1^2 - 4a^2x_1x_2 - bx_1x_2 - 2ax_2^2 \\ &= -2ab|x_1|^2. \end{aligned}$$

□

We will also need the alternative representations for the processes Y_T and H_T, which were defined by (13).

Lemma 3. The process Y_T admits the following representation

$$\begin{aligned} Y_T &= \frac{1}{T} \int_0^T |X_1^{x_0}(t)|^2 \, dt \\ &= -\frac{1}{4abT} \left(\langle P_2 X^{x_0}(T), X^{x_0}(T) \rangle - \langle P_2 x_0, x_0 \rangle \right) + \frac{1}{2abT} \int_0^T \langle P_2 X^{x_0}(t), \Phi \, dB(t) \rangle + \frac{\sigma^2}{4ab}. \end{aligned} \tag{22}$$

The process H_T admits the following representation

$$\begin{aligned} H_T &= \frac{1}{T} \int_0^T |X_2^{x_0}(t)|^2 \, dt \\ &= -\frac{1}{4aT} \left(\langle P_1 X^{x_0}(T), X^{x_0}(T) \rangle - \langle P_1 x_0, x_0 \rangle \right) + \frac{1}{2aT} \int_0^T \langle P_1 X^{x_0}(t), \Phi \, dB(t) \rangle + \frac{\sigma^2}{4a}. \end{aligned} \tag{23}$$

Proof. Define the function $g_1 : \mathbb{R}^2 \to \mathbb{R}$ by

$$g_1(x) = \langle P_1 x, x \rangle, \quad \forall x \in \mathbb{R}^2. \tag{24}$$

The application of Itô's formula to the function $g_1(X^{x_0}(t))$ yields

$$dg_1(X^{x_0}(t)) = 2\langle P_1 X^{x_0}(t), dX^{x_0}(t)\rangle + \frac{1}{2}\operatorname{Tr}\left(2P_1\Phi\Phi^\top\right) dt. \tag{25}$$

The second term may be simplified via following calculation

$$\frac{1}{2}\operatorname{Tr}\left(2P_1\Phi\Phi^\top\right) = \operatorname{Tr}\begin{pmatrix} 0 & 0 \\ 0 & \sigma^2 \end{pmatrix} = \sigma^2.$$

The expression (25) may be now written in the following way

$$\begin{aligned} dg_1(X^{x_0}(t)) &= 2\langle P_1 X^{x_0}(t), AX^{x_0}(t)\rangle\, dt + 2\langle P_1 X^{x_0}(t), \Phi\, dB(t)\rangle + \sigma^2\, dt \\ &= -4a|X_2^{x_0}(t)|^2\, dt + 2\langle P_1 X^{x_0}(t), \Phi\, dB(t)\rangle + \sigma^2\, dt. \end{aligned}$$

After integrating previous formula over the interval $(0, T)$ and after some algebraic operations, we will arrive at (23).

Similarly, if we define function $g_2 : \mathbb{R}^2 \to \mathbb{R}$ by

$$g_2(x) = \langle P_2 x, x \rangle, \quad \forall x \in \mathbb{R}^2 \tag{26}$$

and apply Itô's formula to the function $g_2(X^{x_0}(t))$, we will obtain

$$dg_2(X^{x_0}(t)) = 2\langle P_2 X^{x_0}(t), dX^{x_0}(t)\rangle + \frac{1}{2}\operatorname{Tr}\left(2P_2\Phi\Phi^\top\right) dt. \tag{27}$$

Since the second term equals to

$$\frac{1}{2}\operatorname{Tr}\left(2P_2\Phi\Phi^\top\right) = \operatorname{Tr}\begin{pmatrix} 0 & 2a\sigma^2 \\ 0 & \sigma^2 \end{pmatrix} = \sigma^2,$$

the formula (27) and Lemma 2 yield

$$\begin{aligned} dg_2(X^{x_0}(t)) &= 2\langle P_2 X^{x_0}(t), AX^{x_0}(t)\rangle\, dt + 2\langle P_2 X^{x_0}(t), \Phi\, dB(t)\rangle + \sigma^2\, dt \\ &= -4ab|X_1^{x_0}(t)|^2\, dt + 2\langle P_2 X^{x_0}(t), \Phi\, dB(t)\rangle + \sigma^2\, dt. \end{aligned}$$

After integrating previous formula over the interval $(0, T)$ and after some algebraic operations, we will arrive at (22). □

3.1 Asymptotic normality of the estimator $\tilde{a}_T$

Using formula (16) for the estimator $\tilde{a}_T$ and formula (23) for H_T, we are able to compute the following

$$\begin{aligned} &\sqrt{T}\left(\tilde{a}_T - a\right) = \sqrt{T}\left(\frac{\sigma^2}{4H_T} - a\right) = \sqrt{T}\,\frac{\sigma^2 - 4aH_T}{4H_T} \\ &= \frac{\sqrt{T}}{4H_T}\left(\frac{1}{T}\left(\langle P_1 X^{x_0}(T), X^{x_0}(T)\rangle - \langle P_1 x_0, x_0\rangle\right) - \frac{2}{T}\int_0^T \langle P_1 X^{x_0}(t), \Phi\, dB(t)\rangle\right) \\ &= \frac{1}{4H_T}\frac{1}{\sqrt{T}}\left(\langle P_1 X^{x_0}(T), X^{x_0}(T)\rangle - \langle P_1 x_0, x_0\rangle\right) - \frac{1}{2H_T}\frac{1}{\sqrt{T}}\int_0^T \langle P_1 X^{x_0}(t), \Phi\, dB(t)\rangle. \end{aligned} \tag{28}$$

Using Chebyshev's inequality, it is easy to show, that the first term $\frac{1}{\sqrt{T}}\left(\langle P_1 X^{x_0}(T), X^{x_0}(T)\rangle - \langle P_1 x_0, x_0\rangle\right) \to 0$ in probability as $T \to \infty$. We may also write

$$Z_1(T) := \frac{1}{\sqrt{T}} \int_0^T \langle P_1 X^{x_0}(t), \Phi\, dB(t)\rangle = \frac{1}{\sqrt{T}} \int_0^T \sum_{n=1}^{2} \langle P_1 X^{x_0}(t), e_n\rangle \, d\langle e_n, \Phi B(t)\rangle$$
$$= \frac{\sigma}{\sqrt{T}} \int_0^T \langle P_1 X^{x_0}(t), e_2\rangle \, d\beta(t) = \frac{\sigma}{\sqrt{T}} \int_0^T X_2^{x_0}(t)\, d\beta(t), \tag{29}$$

where we have used that $\Phi B(t) = (0, \sigma\beta(t))^\top$.

By the central limit theorem for the stochastic integral (see Proposition 1.22. in [5]), $Z_1(T)$ converges weakly to a Gaussian distribution with a zero mean and variance given by

$$\lim_{T\to\infty} \frac{\sigma^2}{T} \int_0^T \left(X_2^{x_0}(t)\right)^2 dt = \sigma^2\, \mathbb{E}\left(X_2(\infty)\right)^2 = \sigma^2\, \mathrm{Var}\left(X_2(\infty)\right) = \frac{\sigma^4}{4a}, \tag{30}$$

where $X(\infty) = (X_1(\infty), X_2(\infty))^\top$ is an $\mathbb{R}^2$–valued Gaussian random variable with zero mean and covariance matrix $Q_\infty^{(a,b)}$. (It has the invariant distribution $\mu_\infty^{(a,b)}$.)

Since the multiplicative factor $-\frac{1}{2H_T}$ of $Z_1(T)$ in (28) converges to $-\frac{2a}{\sigma^2}$ as $T \to \infty$, we have the following results

$$\mathrm{Law}\left(Z_1(T)\right) \to N\left(0, \frac{\sigma^4}{4a}\right), \, T \to \infty, \tag{31}$$
$$\mathrm{Law}\left(\sqrt{T}\left(\tilde{a}_T - a\right)\right) \to N\left(0,\, a\right), \, T \to \infty. \tag{32}$$

3.2 Asymptotic normality of the estimator $\tilde{b}_T$

Using formula (17) for the estimator $\tilde{b}_T$ and Lemma 3 for representation of Y_T and H_T, we are able to compute the following

$$\sqrt{T}\left(\tilde{b}_T - b\right) = \sqrt{T}\left(\frac{H_T}{Y_T} - b\right) = \frac{\sqrt{T}}{Y_T}\left(H_T - bY_T\right)$$
$$= \frac{\sqrt{T}}{Y_T}\Bigg(-\frac{1}{4aT}\left(\langle P_1 X^{x_0}(T), X^{x_0}(T)\rangle - \langle P_1 x_0, x_0\rangle\right) + \frac{1}{2aT}\int_0^T \langle P_1 X^{x_0}(t), \Phi\, dB(t)\rangle$$
$$+\frac{1}{4aT}\left(\langle P_2 X^{x_0}(T), X^{x_0}(T)\rangle - \langle P_2 x_0, x_0\rangle\right) - \frac{1}{2aT}\int_0^T \langle P_2 X^{x_0}(t), \Phi\, dB(t)\rangle\Bigg)$$
$$= \frac{1}{4aY_T}\frac{1}{\sqrt{T}}\left(\langle (P_2 - P_1) X^{x_0}(T), X^{x_0}(T)\rangle - \langle (P_2 - P_1) x_0, x_0\rangle\right)$$
$$- \frac{1}{2aY_T}\frac{1}{\sqrt{T}}\int_0^T \langle (P_2 - P_1) X^{x_0}(t), \Phi\, dB(t)\rangle . \tag{33}$$

Similarly as above, the term

$$\frac{1}{\sqrt{T}}\left(\langle (P_2 - P_1) X^{x_0}(T), X^{x_0}(T)\rangle - \langle (P_2 - P_1) x_0, x_0\rangle\right) \to 0$$

in probability as $T \to \infty$. We may also compute

$$\begin{aligned} Z_2(T) &:= \frac{1}{\sqrt{T}} \int_0^T \langle (P_2 - P_1) X^{x_0}(t), \Phi \, dB(t) \rangle \\ &= \frac{1}{\sqrt{T}} \int_0^T \sum_{n=1}^{2} \langle (P_2 - P_1) X^{x_0}(t), e_n \rangle \, d \langle e_n, \Phi B(t) \rangle \\ &= \frac{\sigma}{\sqrt{T}} \int_0^T \langle (P_2 - P_1) X^{x_0}(t), e_2 \rangle \, d\beta(t) = \frac{2a\sigma}{\sqrt{T}} \int_0^T X_1^{x_0}(t) \, d\beta(t), \end{aligned} \tag{34}$$

since $\langle (P_2 - P_1) X^{x_0}(t), e_2 \rangle = 2a X_1^{x_0}(t)$. Similarly to $Z_1(T)$, $Z_2(T)$ also converges weakly to a Gaussian distribution with a zero mean and variance given by

$$\lim_{T \to \infty} \frac{4a^2\sigma^2}{T} \int_0^T (X_1^{x_0}(t))^2 \, dt = 4a^2\sigma^2 \, \mathbb{E} \left(X_1(\infty) \right)^2 = 4a^2\sigma^2 \operatorname{Var} \left(X_1(\infty) \right) = \frac{a\sigma^4}{b}. \tag{35}$$

The multiplicative factor $-\frac{1}{2aY_T}$ of $Z_2(T)$ in (33) converges to $-\frac{2b}{\sigma^2}$ as $T \to \infty$, which brings us to the following results

$$\operatorname{Law} \left(Z_2(T) \right) \to N \left(0, \frac{a\sigma^4}{b} \right), \; T \to \infty, \tag{36}$$

$$\operatorname{Law} \left(\sqrt{T} \left(\tilde{b}_T - b \right) \right) \to N \left(0, 4ab \right), \; T \to \infty. \tag{37}$$

We may summarize the results in the following theorem.

Theorem 3. *The estimators $\tilde{a}_T$ and $\tilde{b}_T$ are asymptotically normal, i.e.,*

$$\operatorname{Law} \left(\sqrt{T} \left(\tilde{a}_T - a \right) \right) \to N \left(0, a \right), \; T \to \infty,$$

$$\operatorname{Law} \left(\sqrt{T} \left(\tilde{b}_T - b \right) \right) \to N \left(0, 4ab \right), \; T \to \infty.$$

The family of estimators $(\tilde{a}_T, \tilde{b}_T)$ is also better than the family $(\hat{a}_T, \hat{b}_T)$ in the sense that its limiting variances are smaller. Indeed, if we compare (32) to (11), we see that the limiting variance of $\hat{a}_T$ is $a + \frac{4a^3}{b(b+1)^2}$ and the limiting variance of $\tilde{a}_T$ is just a. Similarly, if we compare (37) to (12), we recognize that the limiting variance of $\hat{b}_T$ is $4ab + \frac{b^2(b+1)^2}{a}$ and the limiting variance of $\tilde{b}_T$ is just $4ab$. The estimators $(\tilde{a}_T, \tilde{b}_T)$ are strict upgrade to the estimators $(\hat{a}_T, \hat{b}_T)$.

4 Implementation and statistical evidence

We have generated a trajectory of the solution to the stochastic differential equation (1) by Euler's method (see for example [1]). We have chosen $T = 100$ (the length of the time interval), $\Delta t = 0,01$ (the mesh of the partition), $x_1 = 1$, $x_2 = 1$ (the initial values), $a = 1$, $b = 4$ and $\sigma = 1$. The implementation in R code is as follows.

```
T <- 100
N <- 10000
Delta <- T/N
t <- seq(0, T, length = N+1)
```

```
X <- numeric(N+1)
Y <- numeric(N+1)
X0 <- 1
Y0 <- 1
a <- 1
b <- 4
sigma <- 1
X[1] <- X0
Y[1] <- Y0
set.seed(123)
Z <- rnorm(N)
for (i in 2:(N+1)){
X[i] <- X[i-1] + Y[i-1]*Delta
Y[i] <- Y[i-1] + (- b*X[i-1] - 2*a*Y[i-1])*Delta + sigma*Z[i-1]*sqrt(Delta)
}
plot(t, X, type = "l")
plot(t, Y, type = "l")
```

The following figure shows the solution to the equation (1) (that is the process $X_1^{x_0}(t)$) and its derivative (that is the process $X_2^{x_0}(t)$).

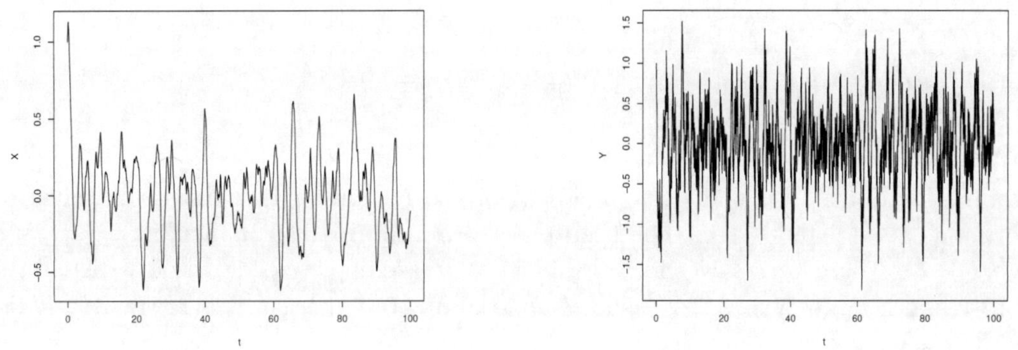

(a) The process $X_1^{x_0}(t)$.

(b) The process $X_2^{x_0}(t)$.

Fig. 1

The implementation of estimators $\hat{a}_T$ and $\hat{b}_T$ is described by the following code.

```
I <- numeric(N+1)
ahat <- numeric(N+1)
bhat <- numeric(N+1)
I[1] <- X[1]^2 + Y[1]^2
ahat[1] <- sigma^2*(b+1)/(4*b*I[1])
bhat[1] <- sigma^2/(4*a*I[1] - sigma^2)
for (i in 2:(N+1)){
I[i] <- (I[i-1]*(i-1) + X[i]^2 + Y[i]^2)/i
ahat[i] <- sigma^2*(b+1)/(4*b*I[i])
bhat[i] <- sigma^2/(4*a*I[i] - sigma^2)
}
```

The value of the statistic I_T (on which the estimators $\hat{a}_T$ and $\hat{b}_T$ are based on (see (8))) equals to $I_T = 0,3233$, while the trace of the matrix $Q_\infty^{(a,b)}$ equals to $Q_\infty^{(a,b)} = \frac{b+1}{4ab}\sigma^2 = 0,3125$. The estimators of a and b are $\hat{a}_T = 0,9667$, $\hat{b}_T = 3,4122$ and their time evolution is shown in the following Figure.

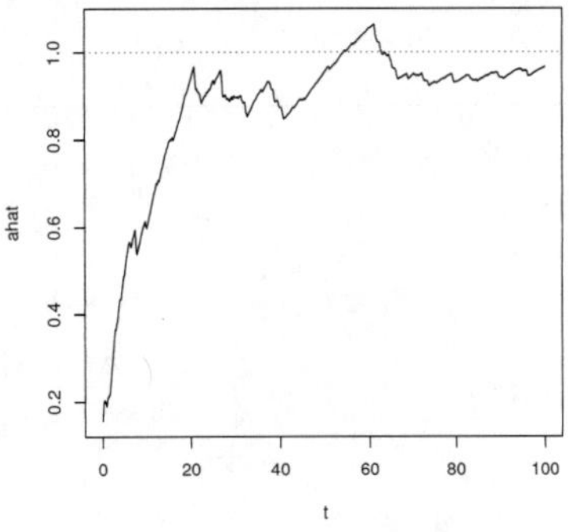

(a) The estimator $\hat{a}_t$.

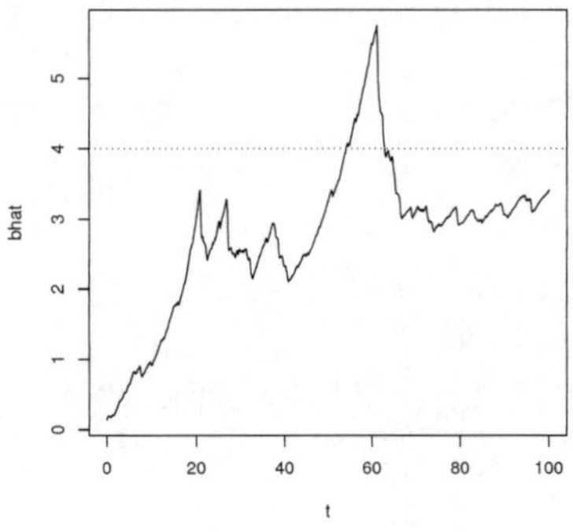

(b) The estimator $\hat{b}_t$.

Fig. 2

The implementation of formulae from Theorem 2 is as follows.

```
ahatnormal <- sqrt(t)*(ahat - a)
bhatnormal <- sqrt(t)*(bhat - b)
vara <- a + (4*a^3)/(b*(b+1)^2)
2*sqrt(vara)
varb <- 4*a*b + (b^2*(b+1)^2)/a
2*sqrt(varb)
```

The limiting variance from the formula (11) equals to 1,04 and the limiting variance from the formula (12) equals to 416. The Figure 3 shows the progress of variables $\sqrt{t}(\hat{a}_t - a)$ and $\sqrt{t}(\hat{b}_t - b)$, where boundaries for "the rule of 2σ" are depicted. (Normally distributed random variable $X \sim N(\mu, \sigma^2)$ is realized in the interval $(\mu - 2\sigma, \mu + 2\sigma)$ with 95,45% probability.)

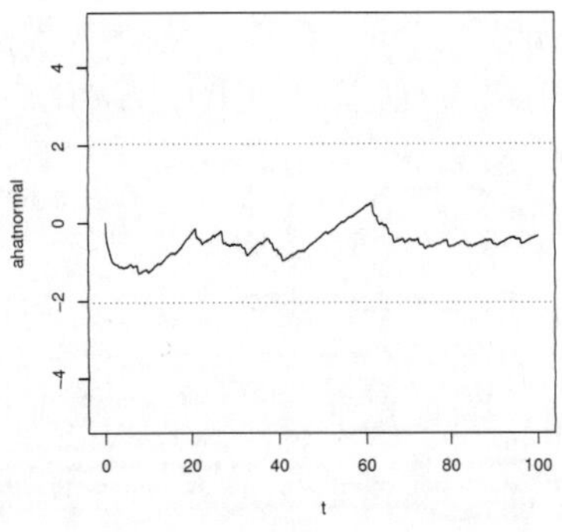

(a) The process $\sqrt{t}(\hat{a}_t - a)$.

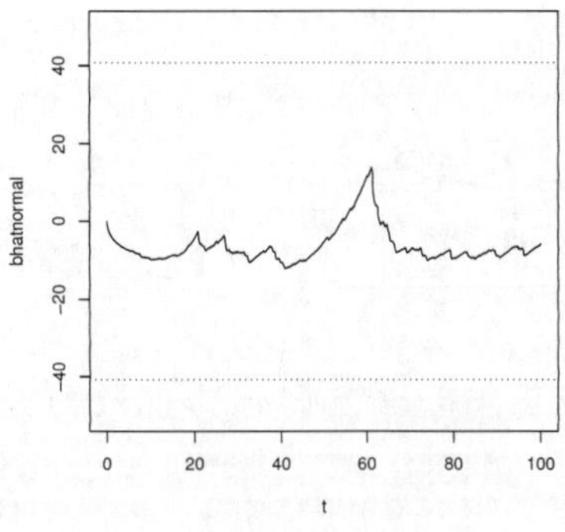

(b) The process $\sqrt{t}(\hat{b}_t - b)$.

Fig. 3

All the pictures seem very satisfactory, however let us introduce the estimators $\tilde{a}_T$ and $\tilde{b}_T$. Their implementation is also very simple.

```
I2X <- numeric(N+1)
I2Y <- numeric(N+1)
atilde <- numeric(N+1)
btilde <- numeric(N+1)
I2X[1] <- X[1]^2
I2Y[1] <- Y[1]^2
atilde[1] <- sigma^2/(4*I2Y[1])
btilde[1] <- I2Y[1]/I2X[1]
for (i in 2:(N+1)){
I2X[i] <- (I2X[i-1]*(i-1) + X[i]^2)/i
I2Y[i] <- (I2Y[i-1]*(i-1) + Y[i]^2)/i
atilde[i] <- sigma^2/(4*I2Y[i])
btilde[i] <- I2Y[i]/I2X[i]
}
```

The results are as follows

$$Y_T = 0,0622, \quad \frac{\sigma^2}{4ab} = 0,0625,$$
$$H_T = 0,2611, \quad \frac{\sigma^2}{4a} = 0,25,$$
$$\tilde{a}_T = 0,9576, \quad \tilde{b}_T = 4,1964.$$

Time evolution of the estimators $\tilde{a}_T$ and $\tilde{b}_T$ is shown in the following figure.

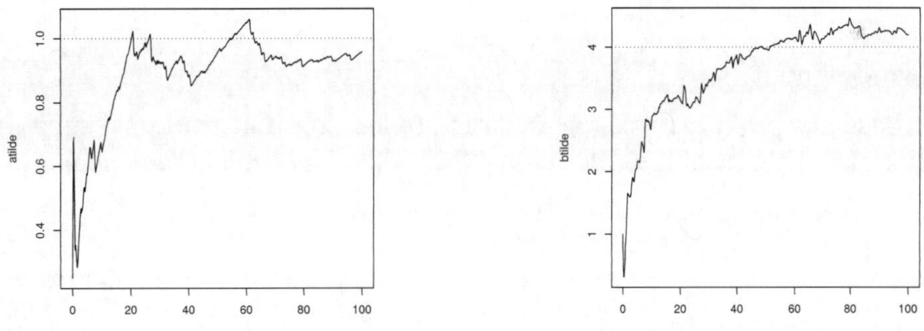

(a) The estimator $\tilde{a}_t$.

(b) The estimator $\tilde{b}_t$.

Fig. 4

The implementation of the formulae (32) and (37) is similar as above. Let us recall, that the limiting variance of the estimator $\tilde{a}_T$ is $a = 1$ and limiting variance of the estimator $\tilde{b}_T$ is $4ab = 16$. The following figure shows the progress of variables $\sqrt{t}(\tilde{a}_t - a)$ and $\sqrt{t}(\tilde{b}_t - b)$ as well as the $95,45\%$ confidence interval.

>From the previous (concrete) simulation it follows, that although the estimator $\tilde{a}_T$ is (locally) worse than the estimator $\hat{a}_T$, the estimator $\tilde{b}_T$ is much more better than $\hat{b}_T$. After running many and many simulations (also with different parameters a, b, σ, x_1, x_2, T, Δt), we claim that all estimators have their derived properties and that our implementation is correct and fully functional.

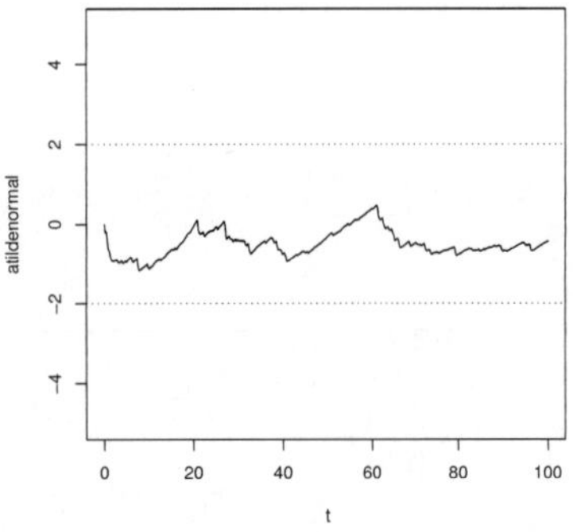

(a) The process $\sqrt{t}(\tilde{a}_t - a)$.

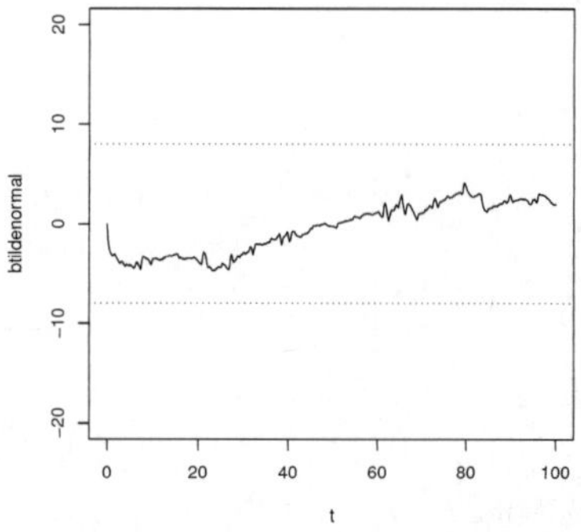

(b) The process $\sqrt{t}(\tilde{b}_t - b)$.

Fig. 5

5 Conclusion

We have introduced the stochastic differential equation of second order and its solution, which is Ornstein–Uhlenbeck process. Based on ergodicity, two families of the strong consistent estimators of unknown parameters have been derived.

In the second part of the paper, we have proved the asymptotic normality of the family $(\tilde{a}_T, \tilde{b}_T)$ and we have shown that this family of estimators is better than the family $(\hat{a}_T, \hat{b}_T)$ (e.g., it is possible to use them without any knowledge of the true value of the other parameter and their limiting variances are smaller).

The third part of the paper presents the implementation of used methods and their comparison on one concrete example.

6 Acknowledgement

This paper has been produced with contribution of long term institutional support of research activities by Faculty of Informatics and Statistics, University of Economics, Prague.

The paper was supported by the GAČR Grant no. 15–08819S entitled "Stochastic Processes in Infinite Dimensional Spaces".

References

[1] IACUS, S. M: *Simulation and Inference for Stochastic Differential Equations*, Springer Series in Statistics, 2008.

[2] JANÁK, J.: Asymptotic normality of parameter estimates for stochastic differential equation of second order, *Proceedings of the 16th Conference on Applied Mathematics Aplimat 2017*, Slovak University of Technology, Bratislava, 704–711, 2017.

[3] JANÁK, J.: Fractional Brownian Motion, *Proceedings of the 15th Conference on Applied Mathematics Aplimat 2016*, Slovak University of Technology, Bratislava, 557–567, 2016.

[4] JANÁK, J.: *Geometrical Fractional Brownian Motion*, Bachelor Thesis, Faculty of Mathematics and Physics, Charles University in Prague, 2007.

[5] KUTOYANTS, Y. A.: *Statistical Inference for Ergodic Diffusion Processes*, Springer London, 2004.

[6] MASLOWSKI, B., POSPÍŠIL, J.: Ergodicity and parameter estimates for infinite–dimensional fractional Ornstein–Uhlenbeck process, *Applied Mathematics and Optimization*, Vol. 57, Issue 3, 401–429, 2007.

[7] MASLOWSKI, B., SEIDLER, J.: Probabilistic approach to the strong Feller property, *Probability Theory and Related Fields*, Vol. 118, Issue 2, 187–210, 2000.

Current address

Janák Josef, Mgr.
Department of Mathematics, University of Economics in Prague
Ekonomická 957, 148 00 Prague 4, Czech Republic
E-mail: janj04@vse.cz

Slovak University of Technology in Bratislava
Faculty of Mechanical Engineering
17th Conference on Applied Mathematics **APLIMAT 2018**

Proceedings

RESPONSE DEVICES IN TERTIARY EDUCATION

JANČAŘÍK Antonín (CZ)

Abstract. The goal of this paper is to introduce the potential of the use of response devices in teaching at university level. The research question of this study is how response devices can be used in teaching at university level, what advantages and disadvantages its use brings and how pre-service teachers feel about the use of this technology both during their pre-service training and in their future teaching practice.

Keywords: response devices, clickers, teacher training, discrete mathematics, tertiary education, experiences

Mathematics Subject Classification: Primary 97B50, 97U70; Secondary 97B40

1 Introduction

The goal of this paper is to introduce the potential of the use of response devices in teaching at university level. It builds on experience with using this technology in teaching the subject Algebra in Computer Science and Discrete Mathematics at Faculty of Education, Charles University in Prague. Both of these subjects are taught as a part of bachelor studies of Mathematics for Education. Response devices have been used in these courses for several years both in regular and distance form of studies. The research question of this study is how response devices can be used in teaching at university level, what advantages and disadvantages its use brings and how pre-service teachers feel about the use of this technology both during their pre-service training and in their future teaching practice.

The paper first recapitulates research on the use of ICT and response devices in education and then presents examples of how response devices are used at the Faculty of Education, Charles University. This is followed by findings from of a questionnaire survey among pre-service teachers on their attitude to response devices and on the way they plan to use it in their teaching practice.

Thus effective ways as well as potential drawbacks of the use of this system are outlined as well as examples of efficient uses discussed.

2 ICT and teacher education

The role of technologies and ICT tools has been growing in importance ever since its appearance at schools. Technologies are nowadays easily available both to teachers and pupils. However, easy availability is not a synonym to quality and does not guarantee better results in education. The question that educators and teachers must naturally ask is how to use these technologies in education effectively for the benefit of pupils and the learning process. Current research in the Czech Republic, e.g. Robová, Vondrová ([1]), Novotná ([2]); Jančařík and Novotná ([3]) shows that Czech in-service teachers lack the confidence, efficacy and skills to be working with ICT to its full potential, feel to be at loss and often use it "for show". The conclusions speak of missed opportunities. Moraová ([4]) points out issues very important for teacher training – pre-service teachers need to be trained not only to be able to use technologies but also to consider how to use if effectively and for their pupils' benefit. They must be shown examples of good practice.

Without any doubt teachers must be able to use technologies meaningfully, must use them as a means of support, not burden. Having little or no experience with technology supported lessons from their school years and thus have nothing to build on, they have to learn how to use technologies efficiently during undergraduate studies as well as within in-service teacher training.

Mishra and Koehler ([5]) describe the set of knowledge a teacher must have (see Fig. 1.) and named it the Technological Pedagogical Content Knowledge (TPACK).

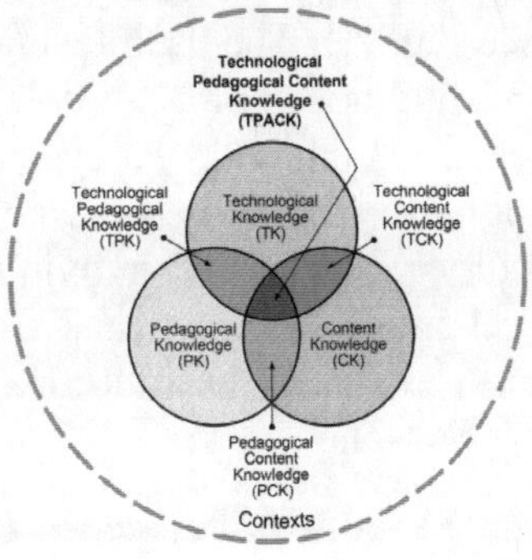

Fig. 1. TPACK (from http://matt-koehler.com/)

It is not enough for a teacher to have knowledge of technologies. They need specific knowledge on the relations and connections between technology and their own subject (e.g. of specific mathematical programmes) as well as pedagogical knowledge and skills connected to the use of technology in lessons. The intersection of these areas is the area of knowledge known as TPACK. In connection to mathematics, it is the subject didactic knowledge of the specifics of the use of ICT in teaching mathematics. That is why introduction to and work with modern technologies must be an integral part of pre-service teacher education (see [6]).

Learning on how to use modern technologies cannot be the matter of subjects where technology is a topic as such exclusively. It is essential that future teachers of mathematics get hands-on authentic experience with the use of ICT tools in mathematics lessons. That is why modern technologies are used in various courses of mathematics with the aim not only of improving teaching in the courses but also of giving pre-service teachers the chance to see and experience ICT supported teaching. The lecturers use computers, visualizers and projectors as well as less common technologies such as response devices.

3 Response devices in education

Response devices (Clickers) are a tool that allows learners to send answers to questions asked by the teacher electronically. Response devices are used especially for getting immediate feedback and for evaluation of students ([7]). Use of clickers in teaching mathematics has been the subject of several studies (e.g. [8], [9]), which include also studies focusing on university education (e.g. [10], [11]). Research shows that clickers boost students' motivation. They are also an efficient tool for getting feedback both from the students and from the teacher. In some cases clickers can be used for peer assessment of students' performance ([12]). Special hardware (clicker) may be substituted by mobile phones or tablets ([13]). It is also possible to get answers from students using a website ([14]).

3.1 Response devices on Faculty of Education, Charles University

Faculty of Education, Charles University, purchased response devices into the newly equipped multimedia seminar rooms at the Department of mathematics and didactics of mathematics. The number of the devices bought is such to ensure that the devices are available to all students in one year. Currently response devices are used regularly in the subjects Discrete Mathematics and Algebra in Computer Science. PowerPoint presentations were developed for these subjects, which involve study check questions to be answered by students using response devices (see Fig. 2.). The used technology allows to respond only by selecting from multiple choice items and does not allow to send e.g. a more digit number as the answer. When all the responses are submitted, the system offers immediate portrayal of the answers in the form of a histogram or by name of each student. The first possibility is used in the seminars at the Faculty, answers of individual students by names are not presented. However, all students get immediate feedback on what the correct answer is. In case of erroneous answers, the teacher analyses the solution and looks for sources of mistakes in the answers. Immediate remedy is possible.

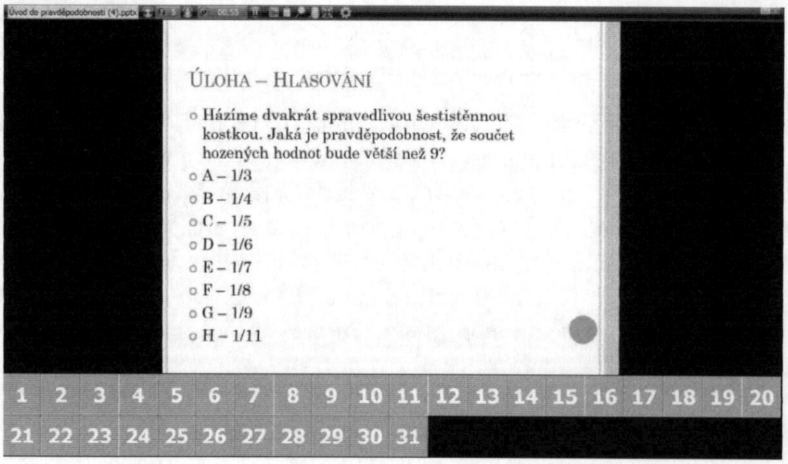

Fig. 2. Screenshot from the presentation with questions.

Results from students' responses are collected continually the whole semester. The system enables creation of various overviews (see Fig. 3.). These overviews are a useful basis for a student's evaluation at the end of the course.

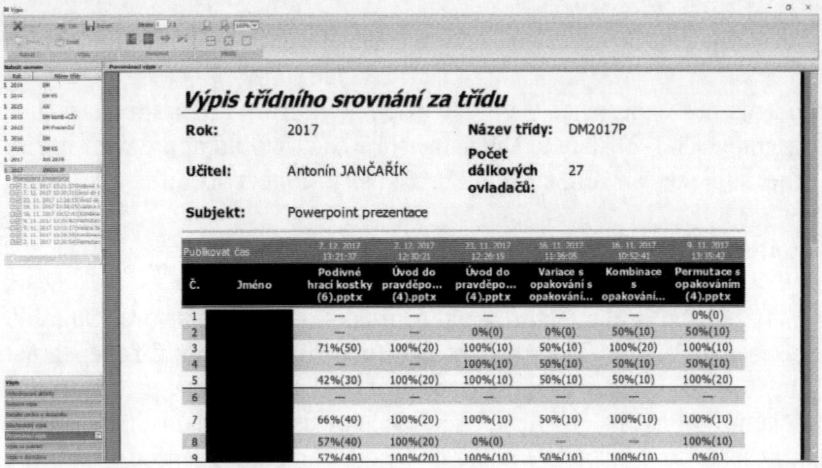

Fig. 3. Overview of the course.

4 Research methodology

For the needs of this study a questionnaire survey was conducted with 2nd year pre-service teachers after they had finished the lectures and seminars in the course of Discrete Mathematics – Combinatorics. What is specific for this course is that presentations for each of the topics are available to students in advance in the LMS system Moodle. The presentations contain study check questions to be answered by response devices. Within the block, seminars and lectures on 7 topics were supported by response device use. The course was finished by a credit test.

The questionnaire consisted of 6 closed and 3 open questions. Due to the low number of respondents (15) the results were evaluated only with descriptive statistical methods and qualitatively.

Questions were related to the following areas:

- Technical difficulties
- Experience with clickers from other subjects
- Preparation for education
- Advantages and disadvantages of the system
- Use in teaching practice

The results will be presented in the next section.

5 The findings from the study

5.1 Technical difficulties

Technical problems are usually quoted as one of the frequent obstacles that make use of ICT in lessons difficult. Technical problems may be of various nature, sometimes as simple as flat batteries in clickers (batteries in clickers should be changed once a year). However, in this survey only one of the respondents claimed he had encountered technical problems while using response devices in his lessons. The rest of the respondents (14) do not have this experience.

5.2 Use in other subjects

None of the students has experience with use of clickers from other subjects. A vast majority of students (see graph 12-1-2) state at the same time they would appreciate if this technology were used in other subjects as well, since its use is very subject specific.

5.3 Study of materials

Presentations that are used in the lectures and seminars in the course are available to students prior to each seminar or lecture. This enables the students to prepare for the lecture or seminar if interested. This possibility was used by one third of the students enrolled in the course, out of whom 80% claim that the use of response devices was the reason that made them want to go through the presentations prior to the lectures and seminars. In other words it is the use of response devices that seems to be the most powerful students' motive to prepare for seminars and lectures.

5.4 Advantages and disadvantages of the system

Open questions were used for assessment of the advantages and disadvantages of the system. The following are considered to be advantages of the use of response devices by the respondents:

- Active involvement in the lesson – the need to think deeper about the solution. This answer was given in some formulation by 60% of the respondents.
- "Anonymity" – the respondents appreciate they do not have to tell the results out loud and do not have to go to the board. The replies to this item imply that it is very stressful for students to present their solutions in front of their classmates.
- Immediate feedback – the respondents appreciate that they get immediate feedback. The respondents are also aware of the fact that teachers get information about the real level of understanding and thus can react to the situation. One of the students wrote the following: "This is followed by gratification from the correct result ➔motivation to further studies."

The respondents refer to much fewer disadvantages than advantages. The most often stated disadvantage is not enough time for some answers and unclarity in how to respond in Yes-No questions. Three students claim they select both possibilities in a yes-no question, one states that in case of any doubts he prefers not to respond at all.

5.5 Use in teaching practice

The last two questions in the questionnaire focus on the potential of the use of response devices in the respondents' future teaching career. All students stated they regarded response devices as a good tool in mathematics lessons on both primary and secondary school levels. The reasons for this belief were partly analogical to what was listed as advantages of response devices but several new ideas were presented as well.

- "Weaker pupils have the time to think before the better pupils give away the result."
- "We can get a reply for each pupil even in the case that we cannot ask each of them individually because there are too many in the group."
- "It forces everybody to join in and calculate."
- "It makes pupils pay attention."
- "It captures their attention."

6 Conclusion

The findings from the survey among pre-service teachers of mathematics show that their experience with the use of response devices is positive. Based on their personal experience they state they regard clickers as a useful tool that they will want to use in their future teaching practice. Thanks to the active use of the device in mathematical courses which are part of their undergraduate studies they gain hands-on authentic experience with this modern technology, which enables them to assess its advantages as well as limits of its use in lessons. Clickers are a simple tool that boosts pupils' motivation. Thanks to the immediate feedback and the necessity to pay attention to the subject matter clickers contribute to deeper understanding and performance.

Further research among in-service teachers and pupils should be conducted to see how efficiently and to what extent clickers are used in everyday school practice and how pupils assess lessons supported by response devices.

Acknowledgement

The paper was supported by grant Progres from Charles University no. Q17 entitled "Teacher and teacher training in the context of science and research".

References

[1] ROBOVÁ, J., VONDROVÁ, N. Missed learning opportunities in the teaching of mathematics with netbooks, In *Proceedings of the 10th International Conference on Efficiency and Responsibility in Education (ERIE 2013)*, Prague, 2013, pp. 524–533.

[2] NOVOTNÁ, J., Multimedia Teaching Materials for Development of Pupils' Culture of Problem Solving, In *Proceedings of the 13th European Conference on e-Learning ECEL2014*, Copenhagen: Aalborg University, 2014, pp. 370–377.

[3] JANČAŘÍK, A., NOVOTNÁ, J., "For show" or efficient use of ICT in mathematics teaching?, In *Proceedings of the 10th International Conference on Technology in Mathematics Teaching*, University of Portsmouth. Portsmouth, 2011, pp. 166-171.

[4] MORAOVÁ, H., Going interactive and multicultural in CLIL. In *Proceedings of the 12th International Conference Efficiency and Responsibility in Education 2015*, Praha: Česká zemědělská univerzita. 2015, pp. 377–384.

[5] MISHRA, P. KOHLER, M. Technological pedagogical content knowledge: A framework for teacher knowledge. *The Teacher College Record*, 108(6), 2006, p. 1017–1054.

[6] ADAMEC, M., JANČAŘÍK, A., STEHLÍKOVÁ, N. Podpora ICT v přípravě budoucích učitelů. In: *APLIMAT 2009: 8th International Conference, Proceedings*. Bratislava: Slovak university of technology, 2009, pp. 637–643. ISBN 978-80-89313-31-0.

[7] HAN, J. H., FINKELSTEIN, A., Understanding the effects of professors' pedagogical development with Clicker Assessment and Feedback technologies and the impact on students' engagement and learning in higher education. *Computers & Education*, 65, 2013, pp. 64–76.
[8] SUURTAMM, C., KOCH, M., ARDEN, A., Teachers' assessment practices in mathematics: Classrooms in the context of reform. *Assessment in Education: Principles, Policy & Practice*, 17(4), 2010, pp. 399–417.
[9] SIMELANE, S., SKHOSANA, P. M., Impact of clicker technology in a mathematics course. Knowledge Management & e-Learning: An International Journal (KM&EL), 4(3), 2012, pp. 279–292.
[10] STRASSER, N., Who wants to pass math? Using clickers in calculus. *Journal of College Teaching and Learning*, 7(3), 2010, 49.
[11] LIU, W. C., STENGEL, D. N., Improving Student Retention and Performance in Quantitative Courses Using Clickers. *International Journal for Technology in Mathematics Education*,18(1), 2011.
[12] BENNETT , S., BARKER, T. , THOMAS, P., LILLEY, M., Modelling and Motivating High Quality Academic Work With Live Peer Evaluation. In: Proceedings of the 14th European Conference On E-Learning (Ecel 2015). NR Reading, 2015, pp. 57–64. ISBN 978-1-910810-71-2.
[13] STAV, J., NIELSEN, K., HANSEN-NYGARD, G., THORSETH, T., Experiences obtained with integration of student response systems for iPod Touch and iPhone into e-learning environments. *Electronic Journal of e-learning*, 8(2), 2010, pp. 179–190.
[14] KUNDISCH, D., HERRMANN, P., WHITTAKER, M., NEUMANN, J., MAGENHEIM, J., REINHARDT, W., ZOYKE, A., Designing a web-based classroom response system. In *International Conference on Design Science Research in Information Systems*, Berlin, Heidelberg: Springer, 2013, pp. 425–431.

Current address

Jančařík Antonín, doc. RNDr., Ph.D.
Institute of Mathematics and Physics, Faculty of Education
Charles University
M. Rettigové 4, 116 39 Praha, Czech republic
E-mail: antonin.jancarik@pedf.cumi.cz

Slovak University of Technology in Bratislava
Faculty of Mechanical Engineering
17th Conference on Applied Mathematics **APLIMAT 2018**

Proceedings

GAGE STUDIES FOR VARIABLES AVERAGE AND RANGE METHOD

JANIGA Ivan (SK)

Abstract. There are several methods that can be used to measure gauge variability. The average and range method is widely used in industry because its calculations can be done by hand. The Average and Range method is an approach which will provide an estimate of both repeatability and reproducibility for a measurement system. This approach will allow the measurement system´s variation to be decomposed into two separate components, repeatability and reproducibility. However, variation due to the interaction between the appraiser and the part or gage is not accounted for in the analysis.

Keywords: gage, measurement system, repeatability, reproducibility, average, part to part variability, total variability

Mathematics Subject Classification: C1, C12, C6.

1 Introduction

There are several methods that can be used to measure gauge variability. In this study we will use the Average and Range method ($\bar{X} \& R$). This method is a statistical method that provides an estimate of the repeatability and reproducibility of the measurement system. Now we explain what is repeatability and reproducibility.

1.1 Repeatability

Repeatability is the variability of the measurements obtained by one person while measuring the same item repeatedly. This is also known as the inherent precision of the measurement equipment. Consider the probability density functions in Figure 1. The density functions were constructed from measurements of the thickness of a piece of metal with Gage A Gage B. The density functions demonstrate that Gage B is more repeatable than Gage A.

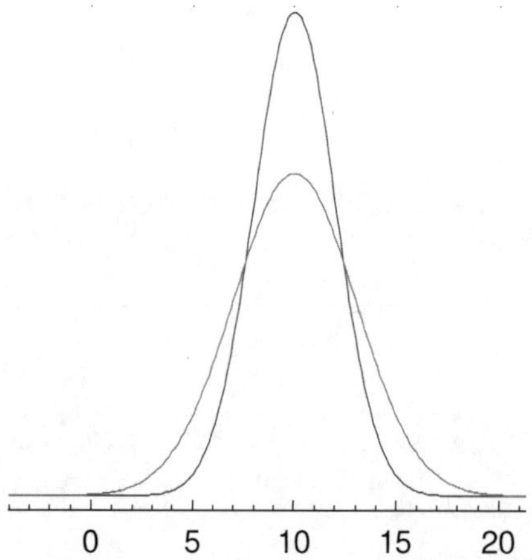

Fig. 1. Probability density functions for the thickness of 2 gages.

1.2 Reproducibility

Reproducibility is the variability of the measurement system caused by differences in operator behavior. Mathematically, it is the variability of the average values obtained by several operators while measuring the same item. Figure 2 displays the probability density functions of the measurements for three operators. The variability of the individual operators are the same, but because each operator has a different bias, the total variability of the measurement system is higher when three operators are used than when one operator is used. Figure 3 also displays the probability density functions of the measurements for three operators using the same scale as Figure 2. Notice that there is more difference in the means of the measurements shown in Figure 3 than those shown in Figure 2. The reproducibility of the system shown in Figure 3 is higher than the reproducibility of the system shown in Figure 2.

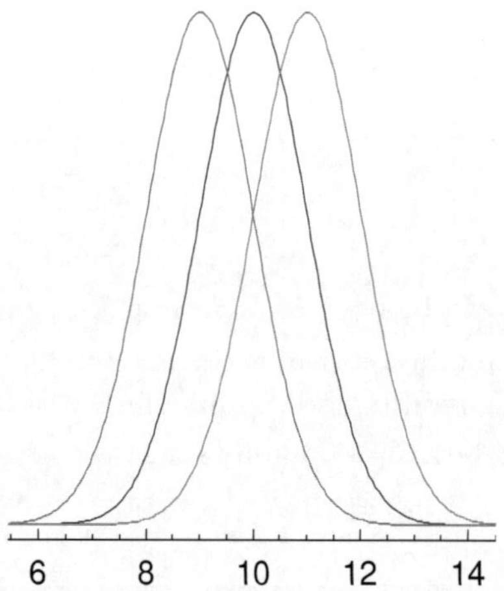

Fig. 2. Reproducibility demonstration.

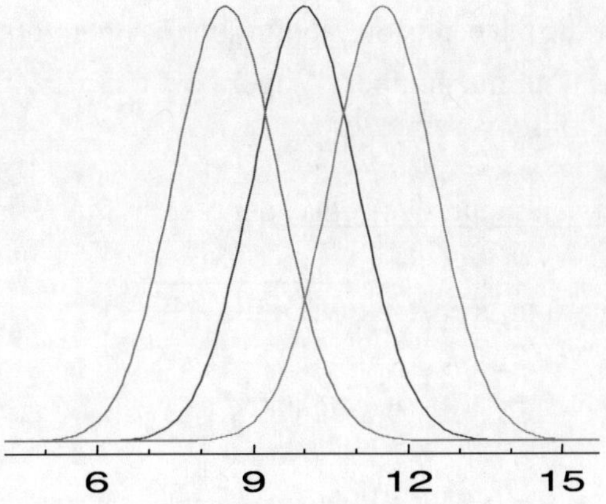

Fig. 3. Reproducibility demonstration.

The most commonly used method for computing repeatability and reproducibility is the Average and Range method.

2 Average and Range Method

All calculations described here are based upon a specified multiple of σ, where the multiple v can be 4; 5,15; or 6. The relevant quantities are calculated as follows:

The measure of repeatability (or equipment variation), denoted by *EV*, is calculated as

$$EV = \overline{R} \times K_1 \tag{1}$$

where $\overline{R}$ is the average range and $K_1 = v / d_2$ is the adjustment factor.

The quantity d_2 (Duncan A. J. 1986) depends on the number of trials used to calculate a single range. In the GAGE application, the number of trials can vary from 2 to 4. Use of d_2 is valid when "operators $\times$ parts $\geq$ 16"; otherwise, the GAGE application uses d_2^* (Duncan 1986), which is based on the number of ranges calculated from "operators $\times$ parts" and on the number of trials.

The measure of reproducibility (or appraiser variation), denoted by *AV*, is calculated as

$$AV = \sqrt{(\overline{X}_{\text{diff}} \times K_2)^2 - \frac{(EV)^2}{nr}} \tag{2}$$

where $\overline{X}_{\text{diff}}$ is the difference between the maximum operator average and the minimum operator average, $K_2 = v / d_2^*$ is the adjustment factor, n is the number of parts, and r is the number of trials. Reproducibility is contaminated by gage error and is adjusted by subtracting $(EV)^2 / nr$.

The quantity d_2^* (Duncan 1986) depends on the number of operators used to calculate a single range. In the GAGE application, the number of operators can vary from 1 to 4. When there is only one operator, reproducibility is set to zero.

The measure of repeatability and reproducibility, denoted by *R&R*, is calculated as

$$R\&R = \sqrt{(EV)^2 + (AV)^2} \tag{3}$$

Part-to-part variability, denoted by PV, is calculated as

$$PV = R_p \times K_3 \tag{4}$$

where R_p is the range of part averages and $K_3 = \nu / d_2^*$ is the adjustment factor.

Here the quantity d_2^* (Duncan 1986) depends on the number of parts used to calculate a single range. In the GAGE application, the number of parts can vary from 2 to 15.

Total variability, denoted by *TV*, is based on gage R&R and part-to-part variability.

$$TV = \sqrt{(R\&R)^2 + (PV)^2} \tag{5}$$

The percent of total variability accounted for by each factor is calculated as follows:

$$\%EV = 100\left[\frac{EV}{TV}\right] \tag{6}$$

$$\%AV = 100\left[\frac{AV}{TV}\right] \tag{7}$$

$$\%R\&R = 100\left[\frac{R\&R}{TV}\right] \tag{8}$$

$$\%PV = 100\left[\frac{PV}{TV}\right] \tag{9}$$

Note that the sum of these percentages does not equal 100%. You can use these percentages to determine whether the measurement system is acceptable for its intended application.

Instead of percent of process variation, your analysis may be based on percent of tolerance. For this you must specify a tolerance value. Then $\%EV$, $\%AV$, $\%R\&R$ and $\%PV$ are calculated by substituting the tolerance value for TV (the denominator) in the preceding formulas.

What is considered acceptable for *%R&R*? Barrentine (1991) gives the following guidelines:

10% or less	excellent
11% to 20%	adequate
21% to 30%	marginally acceptable
over 30%	unacceptable

3 Example

The thickness, in millimeters, of 10 parts have been measured by 3 operators, using the same measurement equipment. Each operator measured each part twice, and the data is given in Table 1.

	Operator					
	A		B		C	
Part	Trial 1	Trial 2	Trial 1	Trial 2	Trial 1	Trial 2
1	55,2	50,1	52,9	46,3	61,6	50,6
2	75,8	76,3	75,7	70,5	82,0	77,4
3	90,2	84,8	90,1	84,5	97,3	94,4
4	75,0	85,1	74,8	80,3	82,3	84,6
5	44,7	55,8	41,7	50,0	48,9	57,2
6	88,7	80,2	82,7	77,2	88,9	83,5
7	84,5	84,5	81,0	83,4	85,4	93,3
8	77,2	72,4	73,9	68,8	83,0	75,8
9	72,4	72,2	70,7	70,3	77,9	78,1
10	90,2	94,9	89,7	93,2	94,3	101,5

Tab. 1. Average & Range method example data.

Repeatability is computed using the average of the ranges for all appraiser and all parts. This data is given in Table 2.

	Operator								
	A			B			C		
Part	Trial 1	Trial 2	R	Trial 1	Trial 2	R	Trial 1	Trial 2	R
1	55,2	50,1	5,1	52,9	46,3	6,6	61,6	50,6	11,0
2	75,8	76,3	0,5	75,7	70,5	5,2	82,0	77,4	4,6
3	90,2	84,8	5,4	90,1	84,5	5,6	97,3	94,4	2,9
4	75,0	85,1	10,1	74,8	80,3	5,5	82,3	84,6	2,3
5	44,7	55,8	11,1	41,7	50,0	8,3	48,9	57,2	8,3
6	88,7	80,2	8,5	82,7	77,2	5,5	88,9	83,5	5,4
7	84,5	84,5	0,0	81,0	83,4	2,4	85,4	93,3	7,9
8	77,2	72,4	4,8	73,9	68,8	5,1	83,0	75,8	7,2
9	72,4	72,2	0,2	70,7	70,3	0,4	77,9	78,1	0,2
10	90,2	94,9	4,7	89,7	93,2	3,5	94,3	101,5	7,2

Tab. 2. Example problem range calculations.

The average of the 30 ranges, $\bar{R}$, is 5,18333. From Table 3, with $n = 30$ (10 parts multiplied by 3 appraisers) and $r = 2$ (2 trials), d_2 is 1,128. The constant K_1 is $5,15/1,128 = 4,5656$, then the repeatability is

$$EV = \bar{R} \times K_1 = 5,18333 \times 4,5656 = 23,665$$

d_2^*	Size of samples								
Samples	2	3	4	5	6	7	8	9	10
1	1,414	1,912	2,239	2,481	2,673	2,830	2,963	3,078	3,179
2	1,279	1,805	2,151	2,405	2,604	2,768	2,906	3,025	3,129
3	1,231	1,769	2,120	2,379	2,581	2,747	2,886	3,006	3,112
4	1,206	1,750	2,105	2,366	2,570	2,736	2,877	2,997	3,103
5	1,191	1,739	2,096	2,358	2,563	2,730	2,871	2,992	3,098
6	1,181	1,731	2,090	2,353	2,558	2,726	2,867	2,988	3,095
7	1,173	1,726	2,085	2,349	2,555	2,723	2,864	2,986	3,092
8	1,168	1,721	2,082	2,346	2,552	2,720	2,862	2,984	3,090
9	1,164	1,718	2,080	2,344	2,550	2,719	2,860	2,982	3,089
10	1,160	1,716	2,077	2,342	2,549	2,717	2,859	2,981	3,088
11	1,157	1,714	2,076	2,340	2,547	2,716	2,858	2,980	3,087
12	1,155	1,712	2,074	2,339	2,546	2,715	2,857	2,979	3,086
13	1,153	1,710	2,073	2,338	2,545	2,714	2,856	2,978	3,085
14	1,151	1,709	2,072	2,337	2,545	2,714	2,856	2,978	3,085
15	1,150	1,708	2,071	2,337	2,544	2,713	2,855	2,977	3,084
d_2									
> 15	1,128	1,693	2,059	2,326	2,534	2,704	2,847	2,970	3,078

Tab. 3. Values of d_2^* and d_2.

The average reading for appraiser A is 75,51; the average reading for appraiser B is 72,885 and the average reading for appraiser C is 79,9. To compute reproducibility, the average of

Part	Trial	Operator B	Operator C	R
1	1	52,9	61,6	8,7
2	1	75,7	82,0	6,3
3	1	90,1	97,3	7,2
4	1	74,8	82,3	7,5
5	1	41,7	48,9	7,2
6	1	82,7	88,9	6,2
7	1	81,0	85,4	4,4
8	1	73,9	83,0	9,1
9	1	70,7	77,9	7,2
10	1	89,7	94,3	4,6
1	2	46,3	50,6	4,3
2	2	70,5	77,4	6,9
3	2	84,5	94,4	9,9

4	2	80,3	84,6	4,3
5	2	50,0	57,2	7,2
6	2	77,2	83,5	6,3
7	2	83,4	93,3	9,9
8	2	68,8	75,8	7,0
9	2	70,3	78,1	7,8
10	2	93,2	101,5	8,3

Tab. 4. Reproducibility example computations.

the range between the appraiser with the smallest average reading (appraiser B in this example) and the appraiser with the largest average reading (appraiser C in this example) is needed. Table 4 shows this data.

The average of the ranges, $\bar{X}_{\text{diff}}$, is 7,015. From Table 3, with $n = 1$ and $r = 3$ for 3 appraisers, d_2^* is 1,912. The adjustment factor K_2 is equal to $5{,}15/1{,}912 = 2{,}6935$ Then the reproducibility is

$$AV = \sqrt{(7{,}015 \times 2{,}6935)^2 - \frac{(23{,}665)^2}{10 \times 2}} = 18{,}1388$$

The repeatability and reproducibility is

$$R\,\&\,R = \sqrt{(23{,}665)^2 + (18{,}1388)^2} = 29{,}8169$$

The part variability is computed using the difference between the largest and smallest part measurement, where the average is taken for all parts and appraisers. This data is shown in Table 5.

	Operator						
	A		B		C		
Part	Trial 1	Trial 2	Trial 1	Trial 2	Trial 1	Trial 2	Avg
1	55,2	50,1	52,9	46,3	61,6	50,6	52,78
2	75,8	76,3	75,7	70,5	82,0	77,4	76,28
3	90,2	84,8	90,1	84,5	97,3	94,4	90,22
4	75,0	85,1	74,8	80,3	82,3	84,6	80,35
5	44,7	55,8	41,7	50,0	48,9	57,2	49,72
6	88,7	80,2	82,7	77,2	88,9	83,5	83,53
7	84,5	84,5	81,0	83,4	85,4	93,3	85,35
8	77,2	72,4	73,9	68,8	83,0	75,8	75,18
9	72,4	72,2	70,7	70,3	77,9	78,1	73,60
10	90,2	94,9	89,7	93,2	94,3	101,5	93,97

Tab. 5. Example part variability computations.

The part with the largest average belongs to part 10 and is 93,97. The lowest average belongs to part 5 and is 49,72. This difference 44,25 is the range of part averages. From Table 3, with n = 1 and r = 10 for 10 parts, d_2^* is 3,179. The adjustment factor K_3 is 1,62. Then the part variability is

$$PV = 44,25 \times 1,62 = 71,685$$

The total measurement system variability is

$$TV = \sqrt{(29,8169)^2 + (71,685)^2} = 77,6388$$

The percent total variability is calculated as follows:

$$\%EV = 100\left[\frac{23,665}{77,6388}\right] = 30,48$$

$$\%AV = 100\left[\frac{18,1388}{77,6388}\right] = 23,363$$

$$\%R\&R = 100\left[\frac{29,8169}{77,6388}\right] = 38,4$$

$$\%PV = 100\left[\frac{71,6853}{77,6388}\right] = 92,33$$

Since $\%R\&R$ is greater than 30, namely 38,4, we consider the gauge to be unacceptable.

3 Conclusion

The Average and Range method will provide information concerning the causes of measurement system or gage variation. The repeatability ($\%EV = 30,48$) is large compared to reproducibility ($\%AV = 23,363$) it follows that the reasons may be:

- The instrument needs maintenance.
- The gage may need to be redesigned to be more rigid.
- The clamping or location for gaging needs to be improved.
- There is excessive within-part variation.

A fixture of some sort may be needed to help the appraiser use the gage more consistently.
Since $\%R\&R$ is greater than 30, namely 38,4, we consider the gauge to be unacceptable.

Acknowledgement

The paper was supported by grant from Grant Agency of VEGA no. 1/0544/16.

References

[1] DUNCAN, A. J., Quality Control and Industrial Statistics Appendix D3, 1986.

[2] BARRENTINE, L. B., Concepts for R&R Studies, 1991.

[3] ANALÝZA SYSTÉMŮ MĚŘENÍ, Česká společnost pro jakost 4/1999, 1999.

[4] DIETRICH, E., SCHULZE, A., *Statistical procedures for machine and process qualification*. ASQ Quality Press, 1999, ISBN 0-87389-447-2, 396 s.

[5] JANIGA, I., KULIFAJ, J. Overovanie kalibrácie prístroja Phascope. In: *Forum statisticum slovacum*, roč. VII, č. 3, 2011, s. 134-139, ISSN 1336-7420.

[6] PALENČÁR, R., RUIZ, J. M., JANIGA, I., HORNÍKOVÁ, A., *Štatistické metódy v skúšobných a kalibračných laboratóriach*, Bratislava: Grafické štúdio Ing. Peter Juriga, 2001, ISBN 80-968449-3-8, 380 s.

[7] BREYFOGLE, F. W., *Implementing Six Sigma: Smarter Solutions Using Statistical Methods*, John Wiley & Sons, 1999, ISBN 0-471-29659-7, 791 s.

[8] JANIGA, I., Measurement system studies using ANOVA. In *Aplimat 2018 [elektronický zdroj]: Proceedings of the 17th conference on Applied Mathematics. Bratislava, 6.2. -8.2. 2018*. 1. vyd. Bratislava : Vydavateľstvo Spektrum STU, 2018, (v tomto zborníku).

Current address

Janiga Ivan, doc. RNDr., PhD.
Institute of Mathematics and Physics, Faculty of Mechanical Engineering
Slovak University of Technology in Bratislava
Nám. slobody 17, 812 31 Bratislava, Slovak Republic
E-mail: ivan.janiga@stuba.sk

Slovak University of Technology in Bratislava
Faculty of Mechanical Engineering
17th Conference on Applied Mathematics **APLIMAT 2018**

Proceedings

MEASUREMENT SYSTEM STUDIES USING ANOVA

JANIGA Ivan (SK)

Abstract. Analysis of variance (ANOVA) is a standard statistical technique and can be used to analyze the measurement error and other sources of variability of data in a measurement systems study. In the analysis of variance, the variance can be decomposed into four categories: appraisers, parts, interaction between appraisers and parts and replication error due to the gage. In the paper it is studied measurement system with respect to repeatability and reproducibility. In contrast to the method of average and range, the ANOVA method allows to determine the variability of the interaction between the appraisers and the parts. The paper deals with the ANOVA method of computing repeatability and reproducibility.

Keywords: measurement system, repeatability, reproducibility, interaction, part to part variability, total variability, degrees of freedom, mean squares

Mathematics Subject Classification: C1, C12, C6.

1 Introduction

The analysis of variance method (ANOVA) is the most accurate method for quantifying repeatability and reproducibility (see [8]). In addition, the ANOVA method allows the variability of the interaction between the appraisers and the parts to be determined.

The advantages of ANOVA techniques as compared with Average and Range methods are:

- They are capable of handling any experimental set-up
- Can estimate the variances more accurately
- Extract more information (such as interaction between appraisers and parts effect) from the experimental data.

The disadvantages are that the numerical computations are more complex and users require a certain degree of statistical knowledge to interpret the results. The ANOVA method as described in the following sections is advised, especially if a computer is available.

2 ANOVA Method

The ANOVA method for measurement assurance is the same statistical technique used to analyze the affects of different factors in designed experiments. The ANOVA design used is a two-way, fixed effects model with replications. The ANOVA table is shown in Table 1.

$$SS_A = \sum_{i=1}^{a}\sum_{j=1}^{b}\sum_{k=1}^{n}(\bar{y}_{i..} - \bar{y}_{...})^2 = \sum_{i=1}^{a}\frac{y_{i..}^2}{bn} - \frac{y_{...}^2}{N} \quad (1)$$

$$SS_B = \sum_{i=1}^{a}\sum_{j=1}^{b}\sum_{k=1}^{n}(\bar{y}_{.j.} - \bar{y}_{...})^2 = \sum_{i=1}^{a}\frac{y_{.j.}^2}{an} - \frac{y_{...}^2}{N} \quad (2)$$

$$SS_{AB} = \sum_{i=1}^{a}\sum_{j=1}^{b}\sum_{k=1}^{n}(\bar{y}_{ij.} - \bar{y}_{...})^2 = \sum_{i=1}^{a}\sum_{j=1}^{b}\frac{y_{ij}^2}{n} - \frac{y_{...}^2}{N} - SS_A - SS_B \quad (3)$$

$$SS_T = \sum_{i=1}^{a}\sum_{j=1}^{b}\sum_{k=1}^{n}(\bar{y}_{ijk} - \bar{y}_{...})^2 = \sum_{i=1}^{a}\sum_{j=1}^{b}\sum_{k=1}^{n}y_{ijk}^2 - \frac{y_{...}^2}{N} \quad (4)$$

$$SS_\varepsilon = SS_T - (SS_A + SS_B + SS_{AB}) \quad (5)$$

where

a = number of appraisers,
b = number parts,
n = the number of trials and
N = total number of readings (abn).

Source of variation	Sum of Squares	Degrees of Freedom	Mean squares	*F* Statistic
Appraiser	SS_A	$a-1$	$MS_A = \frac{SS_A}{a-1}$	$F = \frac{SS_A}{MS_e}$
Parts	SS_B	$b-1$	$MS_B = \frac{SS_B}{b-1}$	$F = \frac{MS_B}{MS_e}$
Interaction (Appraisers, Parts)	SS_{AB}	$(a-1)(b-1)$	$MS_{AB} = \frac{SS_{AB}}{(a-1)(b-1)}$	$F = \frac{MS_{AB}}{MS_e}$
Gage Error	SS_ε	$ab(n-1)$	$MS_\varepsilon = \frac{SS_\varepsilon}{ab(n-1)}$	
Total	SS_T	$N-1$		

Tab. 1. Two-Way ANOVA Table.

When conducting a study, the recommended procedure is to use 10 parts, 3 appraisers and 2 trials, for a total of 60 measurements. Then the relevant quantities are calculated as follows:

The measurement system repeatability is

$$EV = 5{,}15\sqrt{MS_{\varepsilon}} \tag{5}$$

The measurement system reproducibility is

$$AV = 5{,}15\sqrt{\frac{MS_A - MS_{AB}}{bn}} \tag{6}$$

The interaction between the appraisers and the parts is

$$INT = 5{,}15\sqrt{\frac{MS_{AB} - MS_{\varepsilon}}{n}} \tag{7}$$

The measurement system repeatability and reproducibility is

$$R\&R = \sqrt{(EV)^2 + (AV)^2 + INT^2} \tag{8}$$

The measurement system part variability is

$$PV = 5{,}15\sqrt{\frac{MS_B - MS_{AB}}{an}} \tag{9}$$

The total measurement system variability is

$$TV = \sqrt{(R\&R)^2 + (PV)^2} \tag{10}$$

The percent of total variability accounted for by each factor is calculated as follows:

$$\%EV = 100\left[\frac{EV}{TV}\right] \tag{11}$$

$$\%AV = 100\left[\frac{AV}{TV}\right] \tag{12}$$

$$\%R\&R = 100\left[\frac{R\&R}{TV}\right] \tag{13}$$

$$\%PV = 100\left[\frac{PV}{TV}\right] \tag{14}$$

What is considered acceptable for $\%R\&R$, gives the following guidelines:

10% or less	excellent
11% to 20%	adequate
21% to 30%	marginally acceptable
over 30%	unacceptable

3 Example

The thickness [mm] of 10 parts have been measured by 3 operators, using the same measurement equipment. Each operator measured each part twice and the data is given in Table 2.

To compute the characteristics of this measurement system, the two-way ANOVA table must be completed. The sum of the 20 readings (10 parts multiplied by 2 trials) for appraiser A is 1510,2. The sum of the 20 readings for appraiser B is 1457,7. The sum of the 20 readings

	Operator					
	A		B		C	
Part	Trial 1	Trial 2	Trial 1	Trial 2	Trial 1	Trial 2
1	55,2	50,1	52,9	46,3	61,6	50,6
2	75,8	76,3	75,7	70,5	82,0	77,4
3	90,2	84,8	90,1	84,5	97,3	94,4
4	75,0	85,1	74,8	80,3	82,3	84,6
5	44,7	55,8	41,7	50,0	48,9	57,2
6	88,7	80,2	82,7	77,2	88,9	83,5
7	84,5	84,5	81,0	83,4	85,4	93,3
8	77,2	72,4	73,9	68,8	83,0	75,8
9	72,4	72,2	70,7	70,3	77,9	78,1
10	90,2	94,9	89,7	93,2	94,3	101,5

Tab. 2. ANOVA method example data.

for appraiser C is 1598,0. The sum of all 60 readings is 4565,9. The sum of squares for the appraisers is

$$SS_A = \frac{1810,2^2}{10\times 2} + \frac{1457,7^2}{10\times 2} + \frac{1598,0^2}{10\times 2} - \frac{4565,9^2}{60} = 502,486$$

The sum of the 6 readings for each part (3 appraisers multiplied by 2 trials) is given in Table 3 along with the square of this sum and the square of this sum divided by 6.

The sum of squares for the parts is

$$SS_B = 359\,002,9 - \frac{4565,9^2}{60} = 11\,545,52$$

Part	Sum	Sum Squared	Sum Squared/6
1	316,7	100298,9	16716,48
2	457,7	209489,3	34914,88
3	541,3	293005,7	48834,28
4	482,1	232420,4	38736,74
5	298,3	88982,89	14830,48
6	501,2	251201,4	41866,91
7	512,1	262246,4	43707,74
8	451,1	203491,2	33915,2
9	441,6	195010,6	32501,76
10	563,8	317870,4	52978,41
Total			359 002,9

Tab. 3. Part sum of squares computations.

The sum of the 2 trials for each combination of appraiser and part is given in Tab. 4 along with the square of this sum and the square of this sum divided by 2.

Part	Appraiser	Sum	Sum Squared	Sum Squared/2
1	A	105,3	11 088,1	5 544,0
2	A	152,1	23 134,4	11 567,2
3	A	175,0	30 625,0	15 312,5
4	A	160,1	25 632,0	12 816,0
5	A	100,5	10 100,3	5 050,1
6	A	168,9	28 527,2	14 263,6
7	A	169,0	28 561,0	14 280,5
8	A	149,6	22 380,2	11 190,1
9	A	144,6	20 909,2	10 454,6
10	A	185,1	34 262,0	17 131,0
1	B	99,2	9 840,6	4 920,3
2	B	146,2	21 374,4	10 687,2
3	B	174,6	30 485,2	15 242,6
4	B	155,1	24 056,0	12 028,0
5	B	91,7	8 408,9	4 204,4
6	B	159,9	25 568,0	12 784,0
7	B	164,4	27 027,4	13 513,7
8	B	142,7	20 363,3	10 181,6
9	B	141,0	19 881,0	9 940,5
10	B	182,9	33 452,4	16 726,2
1	C	112,2	12 588,8	6 294,4
2	C	159,4	25 408,4	12 704,2
3	C	191,7	36 748,9	18 374,4
4	C	166,9	27 855,6	13 927,8
5	C	106,1	11 257,2	5 628,6
6	C	172,4	29 721,8	14 860,9
7	C	178,7	31 933,7	15 966,8

8	C	158,8	25 217,4	12 608,7
9	C	156,0	24 336,0	12 168,0
10	C	195,8	38 337,6	19 168,8
Total				359 541

Tab. 4. Interaction sum of square computations.

The sum of squares for the interaction between the appraisers and the parts is

$$SS_{AB} = 359\,541 - \frac{4565{,}9^2}{60} - 502{,}5 - 11\,545{,}52 = 35{,}6$$

Squaring all 60 individual reading and summing the values gives 457 405,8. The total sum of squares is

$$SS_T = 359\,087{,}8 - \frac{4565{,}9^2}{60} = 12\,630{,}4$$

The sum of squares for the gage or error is

$$SS_\varepsilon = 12\,630{,}4 - 502{,}5 - 11\,545{,}5 - 35{,}6 = 546{,}8$$

There are 2 degrees of freedom for the appraisers, the number of appraisers minus one; 9 degrees of freedom for the parts, the number of parts minus one, 18 degrees of freedom for the interaction between the appraisers and the parts, the number of appraisers minus one multiplied by the number of parts minus one; 59 total degrees of freedom; the total number of readings minus one, and 30 degrees of freedom for the gage, total degrees of freedom minus the degrees of freedom for the appraisers minus the degrees of freedom for the parts minus the degrees of freedom for the interaction. Since the mean square error is the sum-of-square divided by degrees-of-freedom, the ANOVA table can be completed as shown in Table 5.

Source of variation	Sum of Squares	Degrees of Freedom	Mean squares	*F*-statistic	Significance
Appraiser	502,5	2	251,3	13,8	0,000057
Parts	11 545,5	9	1 282,8	70,5	0,000000
Interaction (Appraisers by Parts)	35,6	18	1,98	0,11	0,999996
Gage (Error)	546,8	30	18,2		
Total	12 630,4	59	214,1		

Tab. 5. Example ANOVA table.

The ANOVA table divides the total variability into several parts. The first part represents differences between the appraisers. The second part represents differences between the parts. The third part represents interactions between appraiser and parts. Since the P-value in the ANOVA table is greater than or equal to 0,05, there is not a statistically significant interaction between appraisers and parts at the 95,0% confidence level. The last part represents the residual error, which corresponds to Repeatability.

The significance listed in Table 5 represents the probability of Type I error. Stated another way, if the statement is made "the appraisers are a significant source of measurement variability", the probability of this statement being incorrect is 0,000057. This significance is the area under the F probability density function to the right of the computed F-statistic. This value can be found using the function FDIST(F-statistic, d_1, d_2) in Microsoft Excel, where d_1 and d_2 are the appropriate degrees of freedom.

Continuing with the example the relevant value are the following:

Repeatability is

$$EV = 5{,}15\sqrt{18{,}2} = 21{,}97$$

Reproducibility is

$$AV = 5{,}15\sqrt{\frac{251{,}3 - 1{,}98}{10 \times 2}} = 18{,}18$$

The interaction between the appraisers and the parts is

$$INT = 5{,}15\sqrt{\frac{1{,}98 - 18{,}2}{2}} = \text{cannot be computed}$$

Obviously the variability due to the interaction cannot be imaginary (the square root of a negative number is an imaginary number); what happened? Each mean square is an estimate subject to sampling error. In some cases the estimated variance will be negative or imaginary. In these cases, the estimated variance is zero.

The repeatability and reproducibility is

$$R\&R = \sqrt{21{,}97^2 + 18{,}18^2 + 0^2} = 28{,}517$$

The part variation is

$$PV = 5{,}15\sqrt{\frac{1\,282{,}8 - 1{,}98}{3 \times 2}} = 75{,}245$$

We now calculate confidence intervals. The results are shown in the Table 6.

	Lower Limit	5,15×Std. dev.	Upper Limit
Repeatability; EV	17,5701	21,987	29,3895
Reproducibility; AV	8,51443	18,1812	114,184
Interaction	0,0	0,0	0,0
R&R	18,8587	28,5304	115,867
parts	43,9656	75,2456	144,767

Tab. 6. 95% Confidence Intervals.

This table shows intervals equal to 5,15 times the standard deviations due to Repeatability, Reproducibility, combined R&R and variability between parts. These intervals can be expected to contain 98,9976% percent of the errors attributed to each source. For example, we would expect the measurements to deviate from the true values by +/-14,2652 due to combined R&R, an interval 28,5304 units wide. Since the estimates of variability are subject to sampling error, the confidence intervals show how precise these estimates are.

The total measurement system variability is

$$TV = \sqrt{28,517^2 + 75,245^2} = 80,468$$

Then the percent total variability is calculated as follows:

$$\%EV = 100 \times \sqrt{\frac{21,97}{80,468}} = 27,3$$

$$\%AV = 100 \times \sqrt{\frac{18,18}{80,468}} = 22,6$$

$$\%R\&R = 100 \times \sqrt{\frac{28,517}{80,468}} = 35,4$$

$$\%PV = 100 \times \sqrt{\frac{75,245}{80,468}} = 93,5$$

Since $\%R\&R$ is greater than 30, namely 35,4, we consider the gauge to be unacceptable.

4 Conclusion

ANOVA method will provide information concerning the causes of measurement system or gage variation. The repeatability is large compared to reproducibility it follows that the reasons may be:

- The instrument needs maintenance.
- The gage may need to be redesigned to be more rigid.

- The clamping or location for gaging needs to be improved.
- There is excessive within-part variation.

A fixture of some sort may be needed to help the appraiser use the gage more consistently. Since $\%R\&R$ is greater than 30, namely 35,4, we consider the gauge to be unacceptable.

Acknowledgement

The paper was supported by grant from Grant Agency of VEGA no. 1/0544/16.

References

[1] DUNCAN, A. J., Quality Control and Industrial Statistics Appendix D3, 1986.
[2] BARRENTINE, L. B., Concepts for R&R Studies, 1991.
[3] ANALÝZA SYSTÉMŮ MĚŘENÍ, Česká společnost pro jakost 4/1999, 1999.
[4] DIETRICH, E., SCHULZE, A., *Statistical procedures for machine and process qualification*. ASQ Quality Press, 1999, ISBN 0-87389-447-2, 396 s.
[5] JANIGA, I., KULIFAJ, J. Overovanie kalibrácie prístroja Phascope. In: *Forum statisticum slovacum*, roč. VII, č. 3, 2011, s. 134-139, ISSN 1336-7420.
[6] PALENČÁR, R., RUIZ, J. M., JANIGA, I., HORNÍKOVÁ, A., *Štatistické metódy v skúšobných a kalibračných laboratóriach*, Bratislava: Grafické štúdio Ing. Peter Juriga, 2001, ISBN 80-968449-3-8, 380 s.
[7] BREYFOGLE, F. W., *Implementing Six Sigma: Smarter Solutions Using Statistical Methods*, John Wiley & Sons, 1999, ISBN 0-471-29659-7, 791 s.
[8] JANIGA, I., Gage studies for variables - Average and Range Method. In *Aplimat 2018 [elektronický zdroj]: Proceedings of the 17th conference on Applied Mathematics. Bratislava, 6.2. -8.2. 2018*. 1. vyd. Bratislava : Vydavateľstvo Spektrum STU, 2018, (v tomto zborníku).

Current address

Janiga Ivan, doc. RNDr., PhD.
Institute of Mathematics and Physics, Faculty of Mechanical Engineering
Slovak University of Technology in Bratislava
Nám. slobody 17, 812 31 Bratislava, Slovak Republic
E-mail: ivan.janiga@stuba.sk

Slovak University of Technology in Bratislava
Faculty of Mechanical Engineering
17th Conference on Applied Mathematics **APLIMAT 2018**

Proceedings

GENERALIZED LINEAR MODEL WITH SOFTWARE MATHEMATICA®

JANIGA Ivan (SK), GABKOVÁ Jana (SK)

Abstract. Generalized linear model is used to model the functional dependencies, where one variable is a random variable and the second variable is non-random.

Keywords: generalized linear model, vector of measurements, vector of parameters, design matrix, covariance matrix.

Mathematics Subject Classification: C1, C12, C6.

1 Introduction

In practice, there are many technical problems that are mathematically modeled as **Model 1**

$$Y_i = Y(x_i) = a_1 f_1(x_i) + a_2 f_2(x_i) + \ldots + a_k f_k(x_i) + e_i, \quad i = 1,\ 2, \ldots,\ n \qquad (1)$$

where Y_i, $i = 1, 2, \ldots, n$ are random variables, whose realizations $y_1, y_2, \ldots, y_n$ are the measurement results.

These results depend on the function $a_1 f_1(x) + a_2 f_2(x) + \ldots + a_k f_k(x)$, known and non-random effects $x_1, x_2, \ldots, x_n$ and k parameters $a_1, a_2, \ldots, a_k$, whose values are unknown; e_i is a random error of the i-th measurement. Model 1 is called a **linear model** (linear in parameters).

Here are two examples from practice:

1. It is known from theory that in a certain range of temperatures x, the extension of the y copper pipe is a linear function of the temperature passing thru the null point. Measurements in this area can be written using a linear model

$$Y_i = a_1 x_i + e_i, \quad i = 1,\ 2, \ldots,\ n$$

2. Fuel consumption in a certain area of a car's speed is a quadratic function of speed. Therefore, the speed measurement x in this area can be described by a linear model

$$Y_i = a_1 + a_2 x_i + a_3 x_i^2 + e_i\,, \quad i = 1,\ 2, ...,\ n$$

2 Generalized linear model

Model 1 can be generalized and expressed in matrix form. We supposed, that

$$\boldsymbol{e} = (e_1,\ e_2,\ ...,\ e_n)^{\mathrm{T}} \tag{2}$$

is a vector of **random errors**, which is assumed to have the n-dimensional normal distribution with zero means:

$$E(e_i) = 0\,, \quad i = 1,\ 2, ..., n \tag{3}$$

and

$$\boldsymbol{\Sigma}_{\boldsymbol{e}} = \boldsymbol{\Sigma}_Y = \sigma^2 \boldsymbol{H} \tag{4}$$

is a symmetric matrix of the type $n \times n$ called **covariance matrix**. $\boldsymbol{H}$ is a **matrix of weights** with elements h_{ij}, $i, j = 1, 2, \ldots, n$. The **variance** of the random variable Y_i is $\sigma^2 h_{ii} = \mathrm{cov}(Y_i, Y_i) = \sigma_i^2$ and the **covariance** between variables Y_i and Y_j is $\sigma^2 h_{ij} = \mathrm{cov}(Y_i, Y_j)$, $i \neq j$. Now we can write a generalized linear model in matrix form as **Model 2** as follows:

$$\boldsymbol{Y} = \boldsymbol{A\beta} + \boldsymbol{e}\,, \quad \boldsymbol{\Sigma}_Y = \sigma^2 \boldsymbol{H} \tag{5}$$

where $\boldsymbol{Y} = (Y_1, Y_2, \ldots, Y_n)^{\mathrm{T}}$ is a known measurement vector,

$\boldsymbol{A}$ is known experimental design matrix of type $n \times k$,

$\boldsymbol{\beta} = (a_1,\ a_2, ...,\ a_k)^{\mathrm{T}}$ is a vector of unknown regression coefficients,

$\boldsymbol{e} = (e_1,\ e_2,\ ...,\ e_n)^{\mathrm{T}}$ is vector of random errors and

$\boldsymbol{\Sigma}_Y$ is a covariance matrix.

Model 2 can be written as an ordered triple as follows:

$$(\boldsymbol{Y}, \boldsymbol{A\beta}, \sigma^2 \boldsymbol{H}) \tag{6}$$

Point estimators in Model

An estimator $\hat{\boldsymbol{\beta}}$ of the regression coefficients vector $\boldsymbol{\beta}$ is obtain by the generalized least squares method as the minimum of quadratic form

$$e^{\mathrm{T}} H^{-1} e = (Y - A\beta)^{\mathrm{T}} H^{-1} (Y - A\beta) = f(\beta)$$

The quadratic form is adjusted to $f(\beta) = Y^{\mathrm{T}} H^{-1} Y - 2Y^{\mathrm{T}} H^{-1} A\beta + \beta^{\mathrm{T}} A^{\mathrm{T}} H^{-1} A\beta$, and by deriving is to get $\dfrac{\partial f(\beta)}{\partial \beta} = -2A^{\mathrm{T}} H^{-1} Y + 2A^{\mathrm{T}} H^{-1} A\beta$.

It can be proved that the **least squares estimator of** $\hat{\beta}$ is

$$\hat{\beta} = \; A^{\mathrm{T}} H^{-1} A \;^{-1} \; A^{\mathrm{T}} H^{-1} Y \tag{7}$$

This is the **minimum variance unbiased estimator (MVUE)** of each component a_i of the regression coefficients vector β. Furthermore, it can be proved that the **covariance matrix of** $\hat{\beta}$ is

$$\Sigma_{\hat{\beta}} = \sigma^2 \; A^{\mathrm{T}} H^{-1} A \;^{-1} \tag{8}$$

and the **unbiased estimator of variance** σ^2 is

$$\hat{\sigma}^2 = \frac{(Y - A\hat{\beta})^{\mathrm{T}} H^{-1} (Y - A\hat{\beta})}{n-k} \tag{9}$$

When (9) is inserted into (8) we get **the estimator of the covariance matrix of** $\hat{\beta}$

$$\hat{\Sigma}_{\hat{\beta}} = \hat{\sigma}^2 \; X^{\mathrm{T}} H^{-1} X \;^{-1} \tag{10}$$

It is often necessary to estimate the various linear combinations of the parameter vector, which we call **parametric functions** and write in matrix form

$$p^{\mathrm{T}} \beta = p_1 a_1 + p_2 \, a_2 + \; ... + p_k \, a_k \tag{11}$$

where $p^{\mathrm{T}} = (p_1, \, p_2, \, ..., \, p_k)$ and $\beta^{\mathrm{T}} = (a_1, a_2, \ldots, a_k)$.

Let us consider the linear model $(Y, X\beta, \sigma^2 H)$ and the parametric function $p^{\mathrm{T}} \beta$. It can be proved that **the minimum variance unbiased estimator of the parametric function** is

$$p^{\mathrm{T}} \hat{\beta} = p^{\mathrm{T}} \; X^{\mathrm{T}} H^{-1} X \;^{-1} \; X^{\mathrm{T}} H^{-1} Y \tag{12}$$

and **the variance of the parametric function** is

$$D(\boldsymbol{p}^{\mathrm{T}}\hat{\boldsymbol{\beta}}) = \sigma^2 \boldsymbol{p}^{\mathrm{T}} \;\; \boldsymbol{X}^{\mathrm{T}}\boldsymbol{H}^{-1}\boldsymbol{X} \;\;^{-1} \; \boldsymbol{p} \tag{13}$$

When (9) is inserted into (13) we get the estimator of $D(\boldsymbol{p}^{\mathrm{T}}\hat{\boldsymbol{\beta}})$

$$D(\boldsymbol{p}^{\mathrm{T}}\hat{\boldsymbol{\beta}}) = \hat{\sigma}^2 \boldsymbol{p}^{\mathrm{T}} \;\; \boldsymbol{X}^{\mathrm{T}}\boldsymbol{H}^{-1}\boldsymbol{X} \;\;^{-1} \; \boldsymbol{p} \tag{14}$$

Assume that the vector $\boldsymbol{Y}$ in the linear models $(\boldsymbol{Y}, \boldsymbol{X\beta}, \sigma^2\boldsymbol{H})$ has n-dimensional normal distribution then it is valid:

$$\frac{\boldsymbol{p}^{\mathrm{T}}\hat{\boldsymbol{\beta}} - \boldsymbol{p}^{\mathrm{T}}\boldsymbol{\beta}}{\sqrt{\sigma^2 \boldsymbol{p}^{\mathrm{T}} \;\; \boldsymbol{X}^{\mathrm{T}}\boldsymbol{H}^{-1}\boldsymbol{X} \;\;^{-1} \; \boldsymbol{p}}} \sim N(0,1) \tag{15}$$

$$\sigma^{-2}(\boldsymbol{Y} - \boldsymbol{X}\hat{\boldsymbol{\beta}})^{T}\boldsymbol{H}^{-1}(\boldsymbol{Y} - \boldsymbol{X}\hat{\boldsymbol{\beta}}) \sim \chi^2(n-k) \tag{16}$$

It can be proved that these two random variables are independent.

Then from the definition of t-distribution and relationship (9) it follows that

$$\frac{\boldsymbol{p}^{\mathrm{T}}\hat{\boldsymbol{\beta}} - \boldsymbol{p}^{\mathrm{T}}\boldsymbol{\beta}}{\sqrt{\hat{\sigma}^2 \boldsymbol{p}^{\mathrm{T}} \;\; \boldsymbol{X}^{\mathrm{T}}\boldsymbol{H}^{-1}\boldsymbol{X} \;\;^{-1} \; \boldsymbol{P}}} \sim t(n-k) \tag{17}$$

From the last statement, confidence intervals can be derived.

Confidence Intervals in Model 2

It is true that the $100(1-\alpha)$ % two-sided confidence interval for the parametric function $\boldsymbol{p}^{\mathrm{T}}\boldsymbol{\beta}$ is

$$\boldsymbol{p}^{\mathrm{T}}\hat{\boldsymbol{\beta}} \pm t(n-k,\alpha)\sqrt{\frac{(\boldsymbol{Y} - \boldsymbol{X}\hat{\boldsymbol{\beta}})^{\mathrm{T}}\boldsymbol{H}^{-1}(\boldsymbol{Y} - \boldsymbol{X}\hat{\boldsymbol{\beta}})}{n-k} \boldsymbol{p}^{\mathrm{T}} \;\; \boldsymbol{X}^{\mathrm{T}}\boldsymbol{H}^{-1}\boldsymbol{X} \;\;^{-1} \; \boldsymbol{p}} \tag{18}$$

Using (18), it can be specify the interval estimate of any parameter (e.g. $a_1 = \boldsymbol{p}^{\mathrm{T}}\boldsymbol{\beta}$, where $\boldsymbol{p}^{\mathrm{T}} = (1,\ 0,\ ...,\ 0)$) and the functional value at any point.

When calculating estimates of the functional values at several points, we get the **confidence band around (theoretical) function** of $a_1 f_1(x) + a_2 f_2(x) + ... + a_k f_k(x)$. However, it cannot be interpreted this band in such a way that it in confidence $1-\alpha$ covers the entire theoretical function.

The confidence band for the entire function $a_1 f_1(x) + a_2 f_2(x) + ... + a_k f_k(x)$ covering this function with confidence $1-\alpha$ is given by the relationship

$$\boldsymbol{p}^{\mathrm{T}}\hat{\boldsymbol{\beta}} \pm \sqrt{kF(k,n-k,\alpha)}\sqrt{\frac{(\boldsymbol{Y}-\boldsymbol{X}\hat{\boldsymbol{\beta}})^{\mathrm{T}}\boldsymbol{H}^{-1}(\boldsymbol{Y}-\boldsymbol{X}\hat{\boldsymbol{\beta}})}{n-k}\boldsymbol{p}^{\mathrm{T}}\ \boldsymbol{X}^{\mathrm{T}}\boldsymbol{H}^{-1}\boldsymbol{X}\ ^{-1}\ \boldsymbol{p}} \tag{19}$$

where $F(k,n-k,\alpha)$ is critical value of F–distribution with degrees of freedom k a $n-k$.

Tests of Hypotheses in Model 2

It is supposed to test the hypothesis H_0 that the actual value of the parametric function is equal to the real number q versus the alternative H_1 that it is not equal to this number. This hypothesis can be written as follows

$$H_0 : \boldsymbol{p}^{\mathrm{T}}\boldsymbol{\beta} = q \text{ vs. } H_1 : \boldsymbol{p}^{\mathrm{T}}\boldsymbol{\beta} \neq q \tag{20}$$

The test statistic, which has the t-distribution with $n-k$ degrees of freedom under the hypothesis H_0 validity, is given by the relation

$$T = \frac{\boldsymbol{p}^{\mathrm{T}}\hat{\boldsymbol{\beta}} - q}{\sqrt{\frac{(\boldsymbol{Y}-\boldsymbol{X}\hat{\boldsymbol{\beta}})^{\mathrm{T}}\boldsymbol{H}^{-1}(\boldsymbol{Y}-\boldsymbol{X}\hat{\boldsymbol{\beta}})}{n-k}\boldsymbol{p}^{\mathrm{T}}\ \boldsymbol{X}^{\mathrm{T}}\boldsymbol{H}^{-1}\boldsymbol{X}\ ^{-1}\ \boldsymbol{p}}} \tag{21}$$

The null hypothesis in (20) is rejected at the level of significance α when

$$|T| > t(n-k,\ \alpha) \tag{22}$$

These tests can be used to test arbitrary coefficient of the parameter vector, a functional value at any point, and so on. It is often used to test the statistical significance of individual regression coefficients. In this case, the hypothesis

$$H_0\!: \ a_i = 0 \tag{23}$$

is tested.

3 Example

It is given 4 independent measurements y_i with double precision at extreme points that are shown in Tab. 1.

x_i	2	3	4	5
y_i	3.5	1.7	1.3	2.6

Table 1. Independent measurements.

Assuming that the functional dependence between x, y is quadratic, we estimate the regression coefficients and their variance.

Solution

Since the functional dependence between x, y is quadratic, we are looking for the regression function $\hat{y} = \hat{a} + \hat{b}x + \hat{c}x^2 = \hat{a}x^0 + \hat{b}x + \hat{c}x^2$. It is necessary to calculate (estimate) unknown regression coefficients $\hat{a}, \hat{b}, \hat{c}$.

- For individual measured values y_i are :

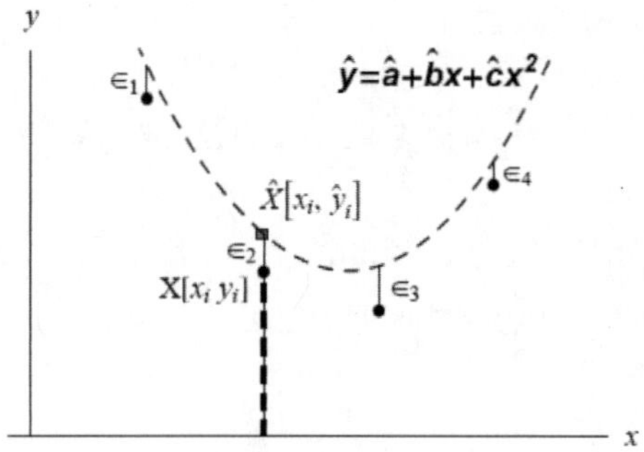

Fig. 1. Graphic individual measured values y_i.

in the form of equations:

$$y_i = \hat{y}_i + \varepsilon_i,\ i = 1, ..., 4$$

$$y_i = \hat{a} \cdot x_i^0 + \hat{b} \cdot x_i^1 + \hat{c} \cdot x_i^2 + \varepsilon_i, \quad i = 1, ..., 4$$

$$3{,}5 = \hat{a} \cdot 2^0 + \hat{b} \cdot 2^1 + \hat{c} \cdot 2^2 + \varepsilon_1$$

$$1{,}7 = \hat{a} \cdot 3^0 + \hat{b} \cdot 3^1 + \hat{c} \cdot 3^2 + \varepsilon_2$$

$$1{,}3 = \hat{a} \cdot 4^0 + \hat{b} \cdot 4^1 + \hat{c} \cdot 4^2 + \varepsilon_3$$

$$2{,}6 = \hat{a} \cdot 5^0 + \hat{b} \cdot 5^1 + \hat{c} \cdot 5^2 + \varepsilon_2$$

in matrix form:

$$\boldsymbol{y} = \boldsymbol{X} \times \boldsymbol{\beta} + \boldsymbol{\varepsilon}$$

$$\begin{pmatrix} 3{,}5 \\ 1{,}7 \\ 1{,}3 \\ 2{,}6 \end{pmatrix} = \begin{pmatrix} 1 & 2 & 4 \\ 1 & 3 & 9 \\ 1 & 4 & 16 \\ 1 & 5 & 25 \end{pmatrix} \times \begin{pmatrix} \hat{a} \\ \hat{b} \\ \hat{c} \end{pmatrix} + \begin{pmatrix} \varepsilon_1 \\ \varepsilon_2 \\ \varepsilon_3 \\ \varepsilon_4 \end{pmatrix}$$

where: $\boldsymbol{y}$ – measurement vector, $(\boldsymbol{Y}, \boldsymbol{X\beta}, \sigma^2 \boldsymbol{H})$,

$\boldsymbol{X}$ – design matrix (of experiment),

$\boldsymbol{\varepsilon}$ – vector of random errors (vector of deviations between measured and estimated

- Covariance matrix of vectors $\boldsymbol{y}$ or $\boldsymbol{\varepsilon}$ is : $\boldsymbol{\Sigma}_Y = \boldsymbol{\Sigma}_\varepsilon = \sigma^2 \boldsymbol{H} = \sigma^2 \cdot \begin{pmatrix} 2\,0\,0\,0 \\ 0\,1\,0\,0 \\ 0\,0\,1\,0 \\ 0\,0\,0\,2 \end{pmatrix}$

where: σ^2 is unknown dispersion factor,

$\boldsymbol{H}$ is matrix of weights ($h_{11} = h_{44} = 2$ – double inaccuracy at extreme points).

- Regression coefficients $\hat{a}, \hat{b}, \hat{c}$ are estimated by the generalized least squares method (GLSM) – the sum of squares of measurement errors is required to be minimal.

For random errors ε_i of measurement y_i is valid:

in the form of equations:

$$\varepsilon_i = y_i - \hat{y}_i,\ i = 1,\dots,4$$

$$\varepsilon_i = y_i - \ \hat{a}\cdot x_i^0 + \hat{b}\cdot x_i^1 + \hat{c}\cdot x_i^2\ ,\ i = 1,\dots,4$$

$$\varepsilon_1 = 3{,}5 - \ \hat{a}\cdot 2^0 + \hat{b}\cdot 2^1 + \hat{c}\cdot 2^2$$

$$\varepsilon_2 = 1{,}7 - \ \hat{a}\cdot 3^0 + \hat{b}\cdot 3^1 + \hat{c}\cdot 3^2$$

$$\varepsilon_3 = 1{,}3 - \ \hat{a}\cdot 4^0 + \hat{b}\cdot 4^1 + \hat{c}\cdot 4^2$$

$$\varepsilon_2 = 2{,}6 - \ \hat{a}\cdot 5^0 + \hat{b}\cdot 5^1 + \hat{c}\cdot 5^2$$

in matrix form:

$$\boldsymbol{\varepsilon} \ = \ \boldsymbol{y} \ - \ \boldsymbol{X} \ \times \ \boldsymbol{\beta}$$

$$\begin{pmatrix}\varepsilon_1\\ \varepsilon_2\\ \varepsilon_3\\ \varepsilon_4\end{pmatrix} = \begin{pmatrix}3{,}5\\ 1{,}7\\ 1{,}3\\ 2{,}6\end{pmatrix} - \begin{pmatrix}1 & 2 & 4\\ 1 & 3 & 9\\ 1 & 4 & 16\\ 1 & 5 & 25\end{pmatrix}\cdot\begin{pmatrix}\hat{a}\\ \hat{b}\\ \hat{c}\end{pmatrix}$$

It is required, that $\boldsymbol{\varepsilon}^{\mathrm{T}}\boldsymbol{H}^{-1}\boldsymbol{\varepsilon} = (\boldsymbol{Y} - \boldsymbol{X\beta})^{\mathrm{T}}\boldsymbol{H}^{-1}(\boldsymbol{Y} - \boldsymbol{X\beta}) = \min$.

Here it should be noted that

- known: $\boldsymbol{y}$, $\boldsymbol{X}$ – design matrix, $\boldsymbol{H}$ – matrix of weights,
- unknown: $\boldsymbol{\beta} = [a, b, c]^T$ – wants to be estimated.

Therefore

$$\varphi(\boldsymbol{\beta}) = (\boldsymbol{Y} - \boldsymbol{X\beta})^{\mathrm{T}}\boldsymbol{H}^{-1}(\boldsymbol{Y} - \boldsymbol{X\beta})$$

$$\varphi([\hat{a}, \hat{b}, \hat{c}]^T) = (\boldsymbol{Y} - \boldsymbol{X\beta})^{\mathrm{T}}\boldsymbol{H}^{-1}(\boldsymbol{Y} - \boldsymbol{X\beta})$$

Minimize $\boldsymbol{\varepsilon}^{\mathrm{T}}\boldsymbol{H}^{-1}\boldsymbol{\varepsilon}$ actual means to find a local minimum vector $[\hat{a}, \hat{b}, \hat{c}]^T$ of the function $\varphi(\boldsymbol{\beta})$.

The function $\varphi(\boldsymbol{\beta})$ acquires a minimum at a stationary point $[\hat{a}, \hat{b}, \hat{c}]^T$ that is a solution to a system of normal equations

$$\frac{\partial\ \varphi(\boldsymbol{\beta})}{\partial\ \boldsymbol{\beta}} = \boldsymbol{0} \Leftrightarrow \frac{\partial\varphi(\boldsymbol{\beta})}{\partial\hat{a}} = 0 \wedge \frac{\partial\varphi(\boldsymbol{\beta})}{\partial\hat{b}} = 0 \wedge \frac{\partial\varphi(\boldsymbol{\beta})}{\partial\hat{c}} = 0\,.$$

By solving the system of normal equations an estimate of the individual regression coefficients is obtained:

$$\boldsymbol{\beta} = \ \hat{a}, \hat{b}, \hat{c}\ ^T \ = \ \boldsymbol{X}^T\boldsymbol{H}^{-1}\boldsymbol{X}\ ^{-1}\cdot\ \boldsymbol{X}^T\boldsymbol{H}^{-1}\boldsymbol{y}$$

$$\hat{\beta}=\begin{pmatrix}\hat{a}\\ \hat{b}\\ \hat{c}\end{pmatrix}=\begin{pmatrix}\frac{1165}{22} & \frac{-1365}{44} & \frac{17}{4}\\ \frac{-1365}{44} & \frac{1649}{88} & \frac{-21}{8}\\ \frac{17}{4} & \frac{-21}{8} & \frac{3}{8}\end{pmatrix}\cdot\begin{pmatrix}6,05\\ 20,3\\ 75,6\end{pmatrix}=\begin{pmatrix}11,9136\\ -5,74318\\ 0,775\end{pmatrix}$$

Estimated regression function: $\hat{y}=\hat{a}+\hat{b}x+\hat{c}x^2=11,9136-5,74318x+0,775x^2$

Note.

The solution $\boldsymbol{\beta}$ exists only if the inverse matrix $(\boldsymbol{X}^T\boldsymbol{H}^{-1}\boldsymbol{X})^{-1}$ exists, that means the matrix $\boldsymbol{X}^T\boldsymbol{H}^{-1}\boldsymbol{X}$ is regular $\Rightarrow$ its determinant is different from zero ($\left|\boldsymbol{X}^T\boldsymbol{H}^{-1}\boldsymbol{X}\right|\neq 0$).

Auxiliary calculations (symmetric matrixes):

$$\boldsymbol{X}^T\boldsymbol{H}^{-1}\boldsymbol{y}=\begin{pmatrix}1 & 1 & 1 & 1\\ 2 & 3 & 4 & 5\\ 4 & 9 & 16 & 25\end{pmatrix}\cdot\begin{pmatrix}0,5 & 0 & 0 & 0\\ 0 & 1 & 0 & 0\\ 0 & 0 & 1 & 0\\ 0 & 0 & 0 & 0,5\end{pmatrix}\cdot\begin{pmatrix}3,5\\ 1,7\\ 1,3\\ 2,6\end{pmatrix}=\begin{pmatrix}6,05\\ 20,3\\ 75,6\end{pmatrix}$$

symmetric matrixes:

$$\boldsymbol{X}^T\boldsymbol{H}^{-1}\boldsymbol{X}=\begin{pmatrix}1 & 1 & 1 & 1\\ 2 & 3 & 4 & 5\\ 4 & 9 & 16 & 25\end{pmatrix}\cdot\begin{pmatrix}0,5 & 0 & 0 & 0\\ 0 & 1 & 0 & 0\\ 0 & 0 & 1 & 0\\ 0 & 0 & 0 & 0,5\end{pmatrix}\cdot\begin{pmatrix}1 & 2 & 4\\ 1 & 3 & 9\\ 1 & 4 & 16\\ 1 & 5 & 25\end{pmatrix}=\begin{pmatrix}3 & 10,5 & 39,5\\ 10,5 & 39,5 & 157,5\\ 39,5 & 157,5 & 657,5\end{pmatrix}$$

$$(\boldsymbol{X}^T\boldsymbol{H}^{-1}\boldsymbol{X})^{-1}=\begin{pmatrix}\frac{1165}{22} & \frac{-1365}{44} & \frac{17}{4}\\ \frac{-1365}{44} & \frac{1649}{88} & \frac{-21}{8}\\ \frac{17}{4} & \frac{-21}{8} & \frac{3}{8}\end{pmatrix}=\begin{pmatrix}52,9545 & -31,0227 & 4,25\\ -31,0227 & 18,7386 & -2,625\\ 4,25 & -2,625 & 0,375\end{pmatrix}$$

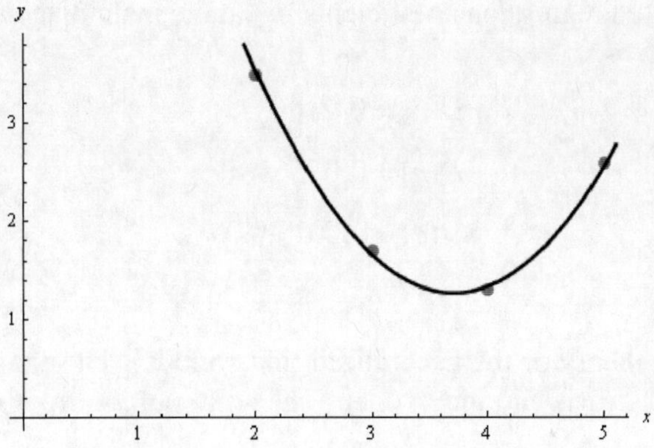

Fig. 2. Graphic verification of accuracy of the regression function found.

- To estimate the dispersion of regression coefficients, it is first necessary to calculate the dispersion estimate of σ^2:

$$\hat{\sigma}^2 = \frac{(\boldsymbol{y}-\boldsymbol{X\beta})^T \boldsymbol{H}^{-1}(\boldsymbol{y}-\boldsymbol{X\beta})}{n-k} = \frac{1}{n-k}(\boldsymbol{y}-\boldsymbol{X\beta})^T \boldsymbol{H}^{-1}(\boldsymbol{y}-\boldsymbol{X\beta})$$

where: $n = 4$ – number of measurements,
$k = 3$ – number of regression coefficient.

Then

$$\hat{\sigma}^2 = \frac{1}{4-3}(\boldsymbol{y}-\boldsymbol{X\beta})^T \boldsymbol{H}^{-1}(\boldsymbol{y}-\boldsymbol{X\beta}) = (\boldsymbol{y}-\boldsymbol{X\beta})^T \boldsymbol{H}^{-1}(\boldsymbol{y}-\boldsymbol{X\beta})$$

$$\hat{\sigma}^2 = \begin{pmatrix} -0,0272727 \\ 0,0409091 \\ -0,0409091 \\ 0,0272727 \end{pmatrix}^T \cdot \begin{pmatrix} 0,5 & 0 & 0 & 0 \\ 0 & 1 & 0 & 0 \\ 0 & 0 & 1 & 0 \\ 0 & 0 & 0 & 0,5 \end{pmatrix} \cdot \begin{pmatrix} -0,0272727 \\ 0,0409091 \\ -0,0409091 \\ 0,0272727 \end{pmatrix}$$

$$\hat{\sigma}^2 = 0,00409091$$

The estimator of the covariance matrix of $\hat{\boldsymbol{\beta}}$ (symmetric matrix) is:

$$\hat{\boldsymbol{\Sigma}}_{\hat{\beta}} = \hat{\sigma}^2 \left(\boldsymbol{X}^{\mathrm{T}} \boldsymbol{H}^{-1} \boldsymbol{X} \right)^{-1}$$

$$\hat{\boldsymbol{\Sigma}}_{\hat{\beta}} = 0,00409091 \cdot \begin{pmatrix} \frac{1165}{22} & \frac{-1365}{44} & \frac{17}{4} \\ \frac{-1365}{44} & \frac{1649}{88} & \frac{-21}{8} \\ \frac{17}{4} & \frac{-21}{8} & \frac{3}{8} \end{pmatrix} = \begin{pmatrix} 0,216632 & -0,126911 & 0,017386 \\ -0,126911 & 0,076658 & -0,010738 \\ 0,017386 & -0,010738 & 0,001534 \end{pmatrix}$$

Variances of the estimated regression coefficients lie on the main diagonal of the matrix and their values are:

$$D(\hat{a}) = 0,216632$$

$$D(\hat{b}) = 0,0766581$$

$$D(\hat{c}) = 0,00153409$$

4 Conclusion

The article presents the theory of the generalized linear model. The example shows the use of a diagonal covariance matrix in the model. The end points are measured with double precision.

Acknowledgement

The paper was supported by grant from Grant Agency of VEGA no. 1/0544/16.

References

[1] LAMOŠ, F., POTOCKÝ, R., *Pravdepodobnosť a matematická štatistika*. Vyd. ALFA, 1989. 342 s. ISBN 80-05-00115-0.

[2] WIMMER, G., *Štatistické metódy v pedagogike*. Nakl. GAUDEAMUS, 1993. 154 s. ISBN 80-7041-864-8.

[3] PALENČÁR, R., RUIZ, J. M., JANIGA, I., HORNÍKOVÁ, A., *Štatistické metódy v skúšobných a kalibračných laboratóriách*, Bratislava: Grafické štúdio Ing. Peter Juriga, 2001, ISBN 80-968449-3-8, 380 s.

[4] DIETRICH, E., SCHULZE, A., *Statistical procedures for machine and process qualification*. ASQ Quality Press, 1999, ISBN 0-87389-447-2, 396 s.

[5] MONTGOMERY, D.C., RUNGER, G.C., *Applied statistics and probability for engineers*. John Wiley & Sons, Inc., 2002. 706 p. ISBN 0 471-20454-4.

[6] HEBÁK, P. – HUSTOPECKÝ, J. 1987. *Vícerozměrné statistické metody s aplikacemi*. Vyd. SNTL/ALFA, 1987. 452 s.

[7] JANIGA, I., GABKOVÁ, J., OMACHELOVÁ, M., *Štatistické analýzy*. Vydavateľstvo Spektrum STU, 2017, ISBN 978-80-227-4704-2.

Current address

Janiga Ivan, doc., RNDr., PhD.
Institute of Mathematics and Physics
Faculty of Mechanical Engineering, Slovak University of Technology in Bratislava
Nám. slobody 17, 812 31 Bratislava, Slovakia
E-mail: ivan.janiga@stuba.sk

Gabková Jana, RNDr., PhD.
Institute of Mathematics and Physics
Faculty of Mechanical Engineering, Slovak University of Technology in Bratislava
Nám. slobody 17, 812 31 Bratislava, Slovakia
E-mail: jana.gabkova@stuba.sk

Slovak University of Technology in Bratislava
Faculty of Mechanical Engineering
17th Conference on Applied Mathematics **APLIMAT 2018**

Proceedings

DESTRUCTIVE R&R STUDY - EVALUATION PROBLEMS

JAROŠOVÁ Eva (CZ)

Abstract. The paper deals with the estimation of variance components which is part of a study to assess the capability of a measurement system. The moment method based on the ANOVA model with random effects can yield negative estimates which leads to ambiguous results. Apart from this, example from a wire drawing laboratory shows how the excessive sample-to-sample variation in a destructive gage study may distort the assessment of the system's capability.

Keywords: variance components, ANOVA, approximate confidence intervals

Mathematics Subject Classification: 62P30, 62-07

1 Introduction

Sufficient quality of measured data is an important prerequisite for their correct evaluation. The measurement quality is influenced both by the measuring instrument used and by the conditions under which the measurements take place, including the operator who performs the measurements. A set of methods known as MSA (Measurement System Analysis) are used to verify the capability of the whole measurement system. Through the repeatability and reproducibility (R&R) study, the variability of measurements is analysed. Repeatability refers to the variation in repeated measurements made on the same part under identical conditions including the operator, while reproducibility refers to the variation in measurements made on the part under changing conditions, such as the measurements taken by different operators. The total variability of the measurement system is then compared to the process variability or to the specification limits and the suitability of the measurement system for a given application is assessed.

Typically, several operators participate in the study, who repeatedly measure several selected parts using the same device. Factorial design is used, the factors being *Part* and *Operator*. Both the experimental design and evaluation methods are described in many publications, see [1], [2], [8], and in the manuals of various statistical software products. The problem occurs if the characteristic measured on the same part changes over time, if its value is influenced by the measurement or if the measurement is destructive. These cases are referred to as non-repeatable or destructive measurements. The basic designs of the experiment for destructive R&R are listed, for example, in [5]. Instead of a part being measured repeatedly, a batch of samples which are similar in terms of the measured characteristic is used in the experiment. It means that parts in the standard design are substituted with several such batches. The term repeatability becomes somewhat meaningless in this case, but it is still used. Then estimation of the variance representing repeatability reflects not only the variability due to the measuring device itself, but also the differences between samples. Some approaches aiming to reduce the impact of sample-to-sample differences are discussed in [4].

The paper deals with the problems associated with the design and evaluation of the R&R study. Only effects of two factors and their possible interaction are studied. Some methods for estimation of variance components and for constructing confidence intervals are described and the outputs of gage study analyses that are implemented in Minitab and Statgraphics are examined.

2 Experimental design for destructive gage study

The first problem concerns designs that are often referred to in connection with the destructive gage study - the crossed and nested designs [5]. Their schemes are illustrated in Fig. 1 and Fig. 2. In the illustrations two operators are participating in the experiment, each measuring nine samples. In the first case, there are three different batches of six samples and the samples from the same batch are measured by both operators. In the second case, the batches are six and the samples from one batch are measured by only one operator. An insufficient size of batches is stated as a reason for the nested design. Both Statgraphics and Minitab include nested designs on their gage study menu.

Without looking at the model equation, it is clear that in Fig. 2 the operator effect is confounded with differences between two groups of batches. Only the choice of similar batches could reduce this confounding, however, it would be contrary to the recommended procedure when the choice of parts (batches in the destructive study) should reflect the actual distribution of the measured characteristic. Consequently, only the crossed design is rationale for the destructive R&R study. The option of the nested design is the result of misunderstanding, see also [4]. Although the samples of the *j*th batch that are measured by the *i*th operator are nested within the *ij*th combination, batches are crossed with operators. Further, only the crossed design will be considered.

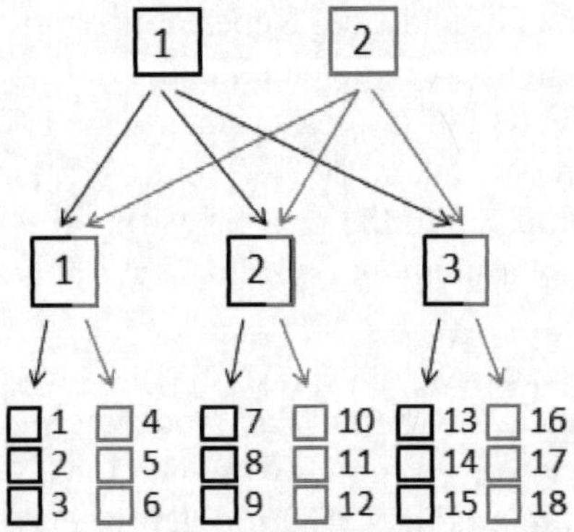

Fig. 1. Crossed design.

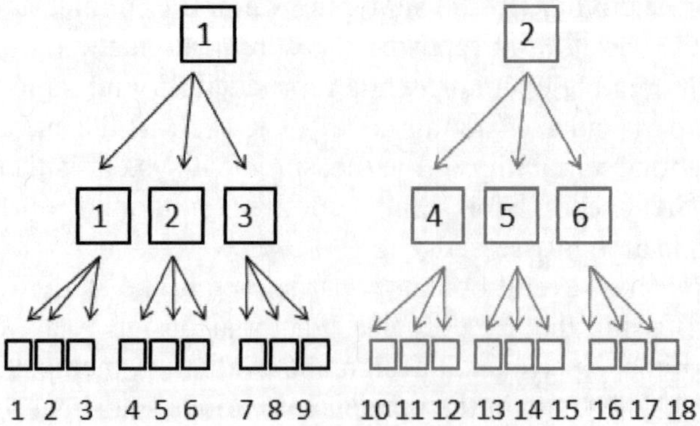

Fig. 2. Nested design.

By default, a complete factorial design with replications is used in non-destructive R&R studies and the factors being investigated are *Part* and *Operator*. The study includes I parts and J operators, each operator measures each of I parts r-times. Considering the recommended procedure for operators to proceed to the next measurement of the same part after all I parts have been measured, it would be appropriate to consider the layout of blocks made up of individual replicas [11] and the model should have the form

$$y_{ijk} = \mu + p_i + o_j + (po)_{ij} + b_k + e_{ijk} \qquad (1)$$
$$i = 1, 2, ..., I;\ j = 1, 2, ..., J;\ k = 1, 2, ..., r$$

where μ denotes a constant (overall mean), $p_i, o_j, (po)_{ij}, b_k$ denote random effects of factors *Operator*, *Part*, *Operator*Part* interaction and blocks, e_{ijk} represents an error component. It is commonly assumed that the result is not influenced by the time of measuring, and therefore the model without the blocking factor is used

$$y_{ijk} = \mu + p_i + o_j + (po)_{ij} + e_{ijk} \qquad (2)$$

The random effects of both factors correspond to the idea that both the parts and operators represent random samples. The aim of the study is not to compare the particular parts or the operators participating in the experiment; it is to measure the variability between different parts and different operators in general. It is assumed that $p_i, o_j, (po)_{ij}$ and e_{ijk} are mutually independent and normally distributed random variables with zero means and variances $\sigma_p^2, \sigma_o^2, \sigma_{po}^2$ and σ^2. Then the variance of response Y can be expressed in the form

$$Var(y_{ijk}) = \sigma_t^2 = \sigma_p^2 + \sigma_o^2 + \sigma_{po}^2 + \sigma^2 \qquad (3)$$

The aim of the R&R study is to determine the variance components due to the measurement system

$$\sigma_{R\&R}^2 = \sigma_o^2 + \sigma_{po}^2 + \sigma^2 \qquad (4)$$

and compare it with the total variance (3). Symbol σ^2 represents repeatability and $\sigma_o^2 + \sigma_{op}^2$ reproducibility.

In the destructive R&R study, where parts with repeated measurements are replaced by batches of similar samples, slight changes will be made in equations (2) to (4) - "p" will be replaced by "b".

3 Estimation of variance components

There are several methods to estimate variance components. ANOVA, maximum likelihood (ML) or restricted maximum likelihood (REML) methods belong to the best-known. For balanced data, ANOVA estimators of all variance components are unbiased and have the smallest variance of all estimators that are both quadratic functions of the observations and unbiased [10]. Under normality, they are minimum variance and unbiased. As the distribution of estimated variance components with exception of the estimate of σ^2 cannot be described by any theoretical model, the exact confidence limits for $\sigma^2_{R\&R}$ cannot be found. Three methods for constructing approximate confidence limits can be used (see 3.2). The disadvantage of ANOVA is that it can yield negative estimates of variance components.

ML estimates of variance components are never negative but they are negatively biased. Under certain conditions of regularity, which are normally met in practice, they are consistent and asymptotically normal and efficient [7]. Using the asymptotic variance-covariance matrix, which is the inverse of the information matrix, confidence intervals for individual components and their sum $\sigma^2_{R\&R}$ can be constructed.

REML estimates, which are based on the logarithm of the restrictive maximum likelihood, take account of the number of the fixed effect parameters estimated (there is only one in the case of model (1)) and are recommended for mixed-effect models. They are usually approximately unbiased. For balanced data and if the ANOVA estimates are nonnegative, the REML and ANOVA estimates are identical. The REML estimates usually have higher mean-squared error (MSE) than the ML estimates.

Only the ANOVA method and three approaches to construction of approximate confidence are described; explanation of other two methods goes beyond the scope of this article.

3.1 ANOVA method

For balanced data, it is common practice to estimate variance components by the moment method based on ANOVA model (2). The mean squares in the ANOVA table are equated to their expected values (Tab. 1) and the resulting equations are solved for the variance components. The method is usually referred to as the ANOVA method. Sums of squares, degrees of freedom, and average squares are determined as in fixed effects models, see for example [8].

Source	Sum of squares	Degrees of freedom	Mean square	Expected mean square
Part	SS_1	$I-1$	M_1	$\theta_1 = \sigma^2 + r\sigma^2_{po} + Jr\sigma^2_p$
Operator	SS_2	$J-1$	M_2	$\theta_2 = \sigma^2 + r\sigma^2_{po} + Ir\sigma^2_o$
Interaction	SS_3	$(I-1)(J-1)$	M_3	$\theta_3 = \sigma^2 + r\sigma^2_{po}$
Residual	SS_4	$IJ(r-1)$	M_4	$\theta_4 = \sigma^2$

Tab. 1. Expected mean squares, ANOVA, random effects model.

The expected mean squares are also used to construct F-statistics and test whether the variance components differ significantly from zero (see [8] for details).

Estimates of the variance components are

$$\hat{\sigma}_p^2 = \frac{M_1 - M_3}{Jr} \quad \hat{\sigma}_o^2 = \frac{M_2 - M_3}{Ir} \quad \hat{\sigma}_{po}^2 = \frac{M_3 - M_4}{r} \quad \hat{\sigma}^2 = M_4 \tag{5}$$

Using $\psi = \sigma_{R\&R}^2$, we can write

$$\psi = \frac{\theta_2 + (I-1)\theta_3 + I(r-1)\theta_4}{Ir} \tag{6}$$

and

$$\hat{\psi} = \frac{M_2 + (I-1)M_3 + I(r-1)M_4}{Ir} \tag{7}$$

A negative estimate in (5) indicates that the true value of the corresponding variance component is zero. Even if it is replaced by zero, some problems remain. Replacing the negative estimate by zero disturbs the properties of estimates - they are not unbiased but their mean squared error is smaller [10]. Another approach is to drop the corresponding effect from the model. This approach is commonly used in connection with the interaction term but sometimes the factor *Operator* should be omitted. However, some statistical packages do not allow omitting this main effect while leaving the interaction term in the model.

3.2 Confidence limits

It can be shown that under the normality assumptions, the following notation applies

$$\frac{f_q M_q}{E(M_q)} \sim \chi^2(f_q), \; q = 1, 2, 3, 4 \tag{8}$$

and M_q are independent. Consequently, the exact confidence intervals for $\theta_q = E(M_q)$ can be determined. However, no theoretical model of the distribution of $\hat{\sigma}_o^2$ or $\hat{\sigma}_{po}^2$, which are linear combinations of mean squares, exists and therefore only approximate confidence limits for σ_o^2 and σ_{po}^2, as well as for $\sigma_{R\&R}^2$ can be determined.

Three methods for constructing approximate confidence intervals are presented in the paper: modified large sample method [3], Satterthwaite method [9] and AIAG method [1]. Due to the specific application in which small values of variance components are desirable, only upper one-sided confidence limits are considered.

3.2.1 Modified large sample method

For $\hat{\psi} = \sum_q c_q M_q$, the approximate upper 100(1–α)% one-sided confidence limit is given by the formula [6]

$$U_{MLS} = \hat{\psi} + \sqrt{\sum_q H_q^2 c_q^2 M_q^2} \tag{9}$$

where

$$H_q = \frac{f_q}{\chi^2_\alpha(f_q)} - 1 \quad \text{and} \quad c_2 = \frac{1}{Ir} \quad c_3 = \frac{I-1}{Ir} \quad c_4 = \frac{I(r-1)}{Ir}$$

3.2.2 Satterthwaite method

The distribution of $\frac{m\hat{\psi}}{\psi}$ is approximated by the chi-square distribution with m degrees of freedom, where m is the highest integer for which

$$m \leq \frac{\hat{\psi}^2}{\sum_q \frac{c_q^2 M_q^2}{f_q}} \tag{10}$$

Consequently, the one-sided upper limit is

$$U_{satt} = \frac{m\hat{\psi}}{\chi^2_\alpha(m)} \tag{11}$$

3.2.3 AIAG method

The method recommended by AIAG is based on the exact confidence interval for $\theta_2 = E(M_2)$. The exact confidence limit for θ_2 or θ_2 / Ir is adjusted by adding the remaining terms in (6), with the expected mean squares θ_3 and θ_4 being replaced by their estimates M_3 and M_4. The approximate upper confidence limit is

$$U_{AIAG} = \frac{(H_2+1)M_2 + (I-1)M_3 + I(r-1)M_4}{Ir} \tag{12}$$

4 Example

In the R&R study samples of drawn wire were tested for tensile strength Rm. Due to the destructive nature of the measurements, twelve drawn wires with the final diameter of 2.5 mm were produced from 5.5 mm thick rolled rods by the drawing machine under different experimental conditions. From each wire, nine samples were prepared and divided between three operators.

4.1 Results in Minitab

Minitab uses the ANOVA method to estimate variance components and the modified large sample method to construct confidence limits. Two parts of the output yielded by Minitab are displayed in Tab. 2 and Tab. 3.

Source	Sum of Squares	Df	Mean Square	F-Ratio	P-Value
Batch	260975	11	23725.0	14.1000	0.000
Operator	126	2	63.2	0.0376	0.963
Batch*Operator	37017	22	1682.6	2.0719	0.011
Residual	58471	72	812.1		
Total (corrected)	356590	107			

Tab. 2. ANOVA for full factorial design, Minitab.

Source	Var Comp	95% Upper Bound	% Contribution (of Var Comp)	95% Upper Bound
Total Gage R&R	1102.27	1547.689	31.04	48.05
Repeatability	812.10	1093.688	22.87	37.21
Reproducibility	290.17	695.486	8.17	19.08
Operator	0.00	33.460	0.00	0.00
Operator*Batch	290.17	730.997	8.17	21.52
Batch -to- Batch	2449.15	6147.498	68.96	85.60
Total Variation	3551.42	7267.499	100.00	

Tab. 3. Gage R&R, Minitab.

The variance component estimates in Tab. 3 indicate that no significant differences between operators exist ($\hat{\sigma}_o^2 = 0$). Based on the corresponding P-value in Tab. 2, we can consider $\sigma_o^2 = 0$. However, the estimate of the interaction variance component $\hat{\sigma}_{bo}^2 = 290.17$ differs significantly from zero. Reproducibility is given only by $\hat{\sigma}_{bo}^2$. The contribution of repeatability ($\hat{\sigma}^2 = 812.10$) to the total gage variation, the estimate of which is 1102.27, is considerable. Based on this value, the measurement system is found inadequate since the contribution of R&R is greater than 30 %, which is the maximum acceptable value according to [1]. However, since the repeatability cannot be separated from the within-batch variation and the same can be said about the interaction, it is more likely that the large variation results from non-uniform mechanical properties along the wire length due to an imperfect technology of the wire production.

Let us now look at the results of the estimation in more detail, see Tab. 4. Together with the confidence limits obtained by the modified large sample method, which are further discussed, the results of other two methods are displayed for information. Using formulas (5) we get

$$\hat{\sigma}_o^2 = -44.984 \quad \hat{\sigma}_{bo}^2 = 290.171 \quad \hat{\sigma}^2 = 812.099 \quad \text{and} \quad \hat{\sigma}_{R\&R}^2 = 1057.286$$

The upper confidence limit according to (9) is $U_{MLS} = 1502.705$, Minitab yields 1547.689.

Differences between the values of $\hat{\sigma}^2_{R\&R}$ and the values of the confidence limits correspond to replacing the negative estimate by zero. The question is whether the confidence limit should not be based on the formula $\psi = \sigma^2_{bo} + \sigma^2$. In this case we get $\hat{\sigma}^2_{R\&R} = 1102.27$ and $U_{MLS} = 1579.928$. Searle et al. [10] suggest reducing the original model, as is commonly done in the case of insignificant interaction, however, even the GLM procedure does not allow it in Minitab.

Estimation method	Estimate	CL method	95% upper CL	Note
ANOVA	1057.286	MLS	1502.705	Model (2) Negative estimate not replaced
	1102.270	MLS	1547.689	Model (2) Negative estimate replaced by 0, Formulas for CL unchanged
		SATT	1474.831	
		AIAG	1089.750	
		MLS	1579.928	Model (2) Negative estimate replaced by 0 CL based on $\psi = \sigma^2_{bo} + \sigma^2$
		SATT	1506.040	
		AIAG	1541.492	
	1088.776	MLS	1564.615	Model without *Operator* CL based on $\psi = \sigma^2_{bo} + \sigma^2$
		SATT	1442.292	
		AIAG	1527.999	
ML	1057.287	-	-	Calculation of ASYCOV matrix failed
REML	1057.233	-	-	

Tab. 4. Differences in estimates and confidence limits.

4.2 Results in Statgraphics

The output of R&R procedure displays estimated standard deviations and two-sided confidence limits. The formulas for confidence limits are not available in the Statgraphics manual, though. Tab. 5 shows the recalculated values of estimated variances and confidence limits for variance components to compare the results with Minitab.

The estimates in Tab. 5 agree with the results in Minitab while the confidence limits except for repeatability are strikingly different. Especially the confidence interval for R&R is suspiciously narrow and if the level of confidence is changed to 90 %, the upper limit is even lower than the estimate.

GLM procedure in Statgraphics can be used to omit the main effect of *Operator* in (2). Resulting estimates are (Tab. 4)

$$\hat{\sigma}^2_{bo} = 296.918\ \hat{\sigma}^2 = 791.858 \text{ and } \hat{\sigma}^2_{R\&R} = 1088.776$$

The procedure does not enable to calculate confidence limits.

	95% confidence			90% confidence		
	Lower CL	Estimate	Upper CL	Lower CL	Estimate	Upper CL
Repeatability	600.608	812.102	1159.498	630.020	812.102	1093.691
Reproducibility	0	0	0	0	0	0
Interaction	155.169	290.171	615.789	171.505	290.171	543.389
R & R	1056.010	1102.273	1124.858	1056.120	1102.273	1089.753
Parts	925.310	2449.151	7830.055	1084.405	2449.151	6431.783

Tab. 5. Variance estimates and confidence limits, Statgraphics, recalculated.

5 Conclusion

Based on the study, following conclusions can be drawn:

1. The procedure for nested design implemented in Minitab or Statgraphics is not suitable for R&R study because of confounding effects of *Operator* and differences between batch groups assigned to them.
2. The analysis of destructive R&R study can yield an evidence of satisfactory system capability if the total gage variance is small. On the other side, nothing can be concluded if the gage variance is large, because its components can be influenced by sample-to-sample variation.
3. From the statistical point of view, the way of estimation is not unambiguous if some of the estimates are negative. This is especially true for construction of the approximate confidence limits. Moreover, the confidence limits yielded by Statgraphics are not usable.
4. The methods for CL constructions should be compared as to how the stated confidence level is maintained. For example, based on their simulation study, Burdick and Larsen [3] claim that only MLS intervals met the stated confidence level - simulated confidence levels for other methods were lower. The lower confidence limits for SATT and AIAG in Tab. 4 do not contradict these findings.
5. Although the way of ML and REML estimation excludes the possibility of negative estimates, the resulting estimates of the total gage variance almost coincide with the unbiased estimates obtained by ANOVA (without replacement by 0).
6. Asymptotic confidence limits based on ML or REML estimation and the asymptotic variance-covariance matrix of estimates could not be determined because the calculation of the asymptotic variance-covariance matrix failed in SPlus.
7. Another approach could make use of the Bayes mixed model and credible intervals. This approach will be the subject of further study.

References

[1] AUTOMOTIVE INDUSTRY ACTION GROUP, *Measurement System Analysis* (*MSA*), 4th ed. Detroit, MI, 2010

[2] BURDICK, R.K., BORROR, C.M., MONTGOMERY, D.C., A Review of Methods for Measurement Systems Capability Analysis. *Journal of Quality Technology*, 35(4), 2003, pp. 342 –354.

[3] BURDICK, R.K., LARSEN, G.A. Confidence Intervals on Measures of Variability in R&R Studies. *Journal of Quality Technology*, 29(3), 1997, pp. 261–273.

[4] DE MAST, J., TRIP, A., Gauge R&R Studies for Destructive Measurements, *Journal of Quality Technology*, 37(1), 2005, pp. 40–49.

[5] GORMAN, D., BOWER, K.M., Measurement System Analysis and Destructive Testing. *Six Sigma Forum Magazine,* 1(4), 2002. Avail. from https://www.minitab.com/uploaded Files/Content/News/Published_Articles/measurement_system_analysis_destructive_ testing.pdf

[6] GRAYBILL, F.A.WANG, C.M., Confidence intervals on non-negative linear combinations of variances, *J. Amer. Stat. Assoc.* 75, 1980, pp. 869–873.

[7] HARVILLE, D.A., Maximum-likelihood approaches to variance component estimation and to related problems *J. Amer. Stat. Assoc.* 72, 1977, pp. 320–340.

[8] MONTGOMERY, D. C., *Design and Analysis of Experiments*, 8th ed. New York: Wiley, 2012,741 p.

[9] SATTERTHWAITE, F.E., An Approximate distribution of estimates of variance components. *Biometrics Bull.* 2, 1946, pp. 110–114.

[10] SEARLE, S.R., CASELLA, G., Mc CULLOCH, C.E., *Variance components*. John Wiley & Sons, New York, 1992, 501 p.

[11] SENOL, S.: Measurement System Analysis Using Designed Experiments with Minimum α-β Risks and *n*. *Measurement*, 36, 2004, pp. 131–141.

Current address

Jarošová Eva, doc. Ing., CSc.
ŠKODA AUTO University
Na Karmeli 1457, 293 01 Mladá Boleslav, Czech Republic
E-mail: eva.jarosova@savs.cz

Slovak University of Technology in Bratislava
Faculty of Mechanical Engineering
17th Conference on Applied Mathematics **APLIMAT 2018**

Proceedings

LIENHARD INTERPOLATION METHOD, ITS GENERALISATION AND ADDITION

KAŇKA Miloš (CZ)

Abstract. This article is based on the primary publication LIENHARD, H.: Interpolation von Funktionswerten bei numerischen Bahnsteuerungen, furthermore from the publication MATUŠŮ, J. and MATUŠŮ, M.: The Lienhard Interpolation $L_{Q,p}$ – method (see the list of literature), in which the original structure of interpolation was extended. This presented article follows directly on $L_{Q,p}$ – method, which expands $L_{Q,p}$ – method in two ways. The first way relates to the enumeration of elements on the basic matrix of $L_{Q,p}$ – method, with its assistance it is possible to obtain parametric equations of interpolation curve. The other way contains parametric equations of interpolation curve in a situation in which the parameter for each part of the curve is changing at the given interval limited by the extreme (node) points.

Keywords: Lienhard interpolation method, elements of the matrix calculation, system of linear equation and its solution

Mathematics Subject Classification: Primary 65D05, 65D07, 65D15;
Secondary 65D10, 65D17, 41-00, 41-01, 68U05

1 Introduction

Let $Q \geq 1, n \geq 3$ be integers. In the Euclidean space $R^m (m > 1 \text{ integer})$, we think of n different points $P_i = x_j^{(i)} (i = 1, \ldots, n;\ j = 1, \ldots, m)$. We search for the polynomials of real variables t of the highest K-th degree

$$P_j^{(i)}(t) = \textstyle\sum_{k=0}^{K} a_{jk}^{(i)}\, t^k \tag{1}$$

for $i = 1, \ldots, n-1$ (the case of the opened interpolation curve), or for $i = 1, \ldots, n$ (the case of closed interpolation curve, where for the point $P_{n+1} = x_j^{(n+1)}$, we understand the point $P_1 = x_j^{(1)}$), and the way that applies

$$P_j^{(i)}(-1) = x_j^{(i)}, \qquad P_j^{(i)}(1) = x_j^{(i+1)} \tag{2}$$

and furthermore $(q = 1, \ldots, Q)$

$$\frac{\mathrm{d}^q}{\mathrm{d}t^q} P_j^{(i)}(1) = \frac{\mathrm{d}^q}{\mathrm{d}t^q} P_j^{(i+1)}(-1) \tag{3}$$

for $i = 1, \ldots, n-2$ (the opened interpolation curve), or for $i = 1, \ldots, n$ (closed interpolation curve, where for the function $P_j^{(n+1)}(t)$, we understand the function $P_j^{(1)}(t)$).

Conditions (2) guarantee, that the i-th segment of the curve parameterised by the function $P_j^{(i)}(t)$ $(j = 1, \dots, m)$, goes through points P_i, P_{i+1}; based on conditions (3) the transition from the i-th segment to $(i+1)$-st segment implements smoothly from 1st to Q-th derivation. To be brief we add

$$\frac{\mathrm{d}^q}{\mathrm{d}t^q} P_j^{(i)}(-1) = \mathrm{D}^q x_j^{(i)}, \ \frac{\mathrm{d}^q}{\mathrm{d}t^q} P_j^{(i)}(1) = \mathrm{D}^q x_j^{(i+1)} \quad . \tag{4}$$

We will talk later about the way to define values $\mathrm{D}^q x_j^{(i)}, \mathrm{D}^q x_j^{(i+1)}$.

By (2), (3) we have $2Q + 2$ conditions available, from their bases is each polynomial (1) of the highest degree $K = 2Q + 1$:

$$P_j^{(i)}(t) = \sum_{k=0}^{2Q+1} a_{jk}^{(i)} t^k \quad . \tag{5}$$

For q-th derivation applies

$$\frac{\mathrm{d}^q}{\mathrm{d}t^q} P_j^{(i)}(t) = \sum_{k=0}^{2Q+1} k(k-1)\dots(k-q+1) a_{jk}^{(i)} t^{k-q} = \tag{6}$$

$$= \sum_{k=0}^{2Q+1} q! \binom{k}{q} a_{jk}^{(i)} t^{k-q} \quad .$$

After setting values $t = -1,1$ to (5), (6), we get (while taking into account (2), (4)) the following system of $2Q + 2$ linear equation for $2Q + 2$ undefined coefficients $a_{jk}^{(i)}$:

$$\sum_{k=0}^{2Q+1} (-1)^k a_{jk}^{(i)} = x_j^{(i)} \quad , \tag{7}$$

$$\sum_{k=0}^{2Q+1} (-1)^{k-q} q! \binom{k}{q} a_{jk}^{(i)} = \mathrm{D}^q x_j^{(i)} \quad ,$$

$$(q = 1, \dots, Q) \, .$$

$$\sum_{k=0}^{2Q+1} a_{jk}^{(i)} = x_j^{(i+1)} \, ,$$

$$\sum_{k=0}^{2Q+1} q! \binom{k}{q} a_{jk}^{(i)} = \mathrm{D}^q x_j^{(i+1)} \quad ,$$

$$(q = 1, \dots, Q) \, .$$

We implement matrices

$$A_{ij} = (a_{j0}^{(i)}, a_{j1}^{(i)}, \dots, a_{j,2Q+1}^{(i)}) \, , \tag{8}$$

$$X_{ij} = (x_j^{(i)}, X_{ij}^{+}, x_j^{(i+1)}, X_{i+1,j}^{+}) \, , \tag{9}$$

where

$$X_{ij}^{+} = (\mathrm{D}^1 x_j^{(i)}, \mathrm{D}^2 x_j^{(i)}, \dots, \mathrm{D}^Q x_j^{(i)}) \tag{10}$$

$$X_{i+1,j}^{+} = (\mathrm{D}^1 x_j^{(i+1)}, \mathrm{D}^2 x_j^{(i+1)}, \dots, \mathrm{D}^Q x_j^{(i+1)}) \tag{11}$$

and let A_Q be the system matrix (7). We call it the basic matrix of method $L_{Q,p}$. If A_Q is regular, then the solution of the system (7): $A_Q \cdot A_{ij}^T = X_{ij}^T$, equals to

$$A_{ij}^T = A_Q^{-1} \cdot X_{ij}^T\,, \tag{12}$$

where A_{ij}^T, X_{ij}^T are transposed matrices to matrices (8), (9) and A_Q^{-1} is the inverse matrix of the matrix A_Q. Due to (12), polynoms (5) are:

$$P_j^{(i)}(t) = (1, t, t^2, \ldots, t^{2Q+1}) \cdot A_{ij}^T = \tag{13}$$

$$= (1, t, t^2, \ldots, t^{2Q+1}) \cdot (A_Q^{-1} \cdot X_{ij}^T)\,.$$

2 Grouping of Support Points

Let's start with the case of the opened interpolation curve $P_1P_2 \ldots P_n$. Let $W = 2n - 2$ and to the final sequence of support points $P_1P_2 \ldots P_n$, we add these points to the right

$$P_{n+1} = P_{n-1}, P_{n+2} = P_{n-2}, \ldots, P_W = P_{2n-W} \quad . \tag{14}$$

For the given integer k, we determine the smallest non-negative remainder r when dividing the number k by the number W and set

$$P_k = \begin{cases} P_r & for\, r > 0\,, \\ P_W & for\, r = 0\,. \end{cases} \tag{15}$$

Therefore, we get the infinite sequence of points

$$\ldots, P_{-5}, P_{-4}, P_{-3}, P_{-2}, P_{-1}, P_0, P_1, P_2, P_3, P_4, P_5, \ldots \tag{16}$$

For instance, for $n = 5$ we think of an opened interpolation curve $P_1P_2P_3P_4P_5$. Set $W = 8$ and to the sequence of points P_1, P_2, P_3, P_4, P_5 we add these points to the right $P_6 = P_4, P_7 = P_3, P_8 = P_2$. For instance, for $k = 9$ applies $9 = 1 \cdot 8 + 1$, and therefore $P_9 = P_1$, or for $k = 10$ applies $10 = 1 \cdot 8 + 2$, and therefore $P_{10} = P_2$.
Then $P_{11} = P_3, P_{12} = P_4, P_{13} = P_5, P_1 = P_6 = P_4, P_{15} = P_7 = P_3$. For $k = 0$ applies $0 = 0 \cdot 8 + 0$, therefore $P_0 = P_8 = P_2$, for $k = -1$ applies $-1 = (-1) \cdot 8 + 7$, therefore $P_{-1} = P_7 = P_3$. For instance, for $k = -315$ applies $-315 = (-40) \cdot 8 + 5$, therefore $P_{-315} = P_5$. The sequence (16) will then be composed of points

$$\ldots, P_{-2} = P_{-4}, P_{-1} = P_3, P_0 = P_2, P_1, P_2, P_3, P_4, P_5, P_6 = P_4, P_7 = P_3, P_8 = P_2 \ldots \tag{17}$$

In case of the closed interpolation curve $P_1P_2 \ldots P_nP_1$, we will proceed as follows. Set $W = n$. For the given integer k, we determine the smallest non-negative remainder r when dividing the number k by the number W, and set

$$P_k = \begin{cases} P_r & for\, r > 0\,, \\ P_W & for\, r = 0\,. \end{cases} \tag{18}$$

Thus, once again, we get the infinite sequence of points (16). For instance, $n = 4$ will be the closed interpolation curve $P_1P_2P_3P_4P_1$. For $k = 5$ applies $5 = 1 \cdot 4 + 1$, therefore $P_5 = P_1$, then $P_6 = P_2, P_7 = P_3, P_8 = P_4$. Furthermore $P_0 = P_4, P_{-1} = P_3, P_{-2} = P_2, P_{-3} = P_1, P_{-4} = P_4$. For instance, for $k = -1026$ applies $-1026 = (-257) \cdot 4 + 2$, therefore $P_{-1026} = P_2$. The sequence (16) is then composed of points

$$\ldots, P_4, P_1, P_2, P_3, P_4, P_1, P_2, P_3, P_4, P_1, P_2, P_3, P_4, \ldots \tag{19}$$

3 The Definition of Derivative Values of Parametric Functions

We now define values from 1st to Q-th derivation of function $P_j^{(i)}(t)$ at points $t = -1,1$. For that purpose, we display on the plane t, x_j points $(2h, x_j^{(i+h)})$, where $-Q + p \leq h \leq Q - p$, h integer; with the given Q representing p of any number fulfilling the inequality $0 \leq p \leq Q - 1$. The value $x_j^{(i+h)}$, we understand as the j-th coordinate of the point P_{i+h} in the sequence (16) while respecting the type of interpolation curve (opened or closed), see the figure 1.

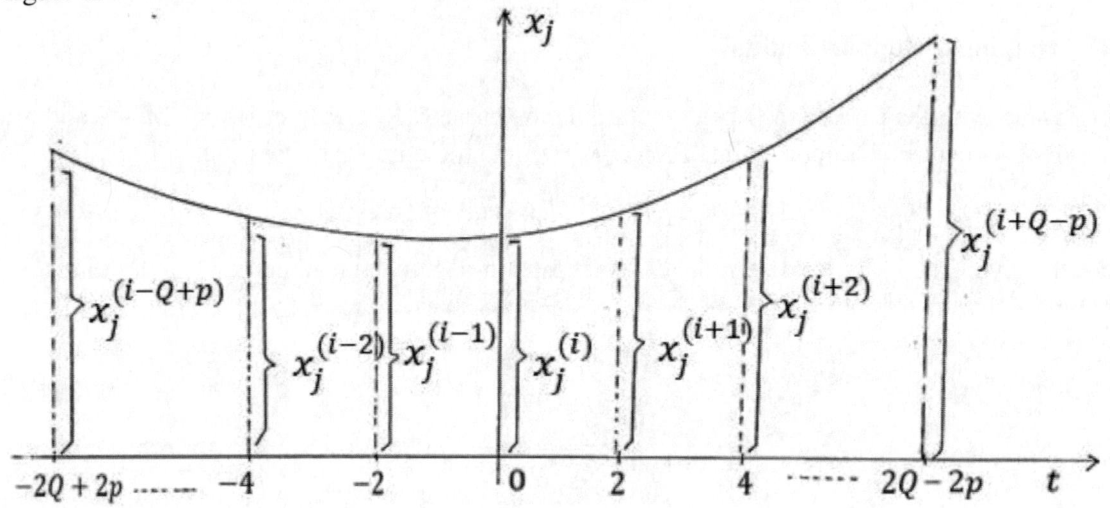

Fig. 1

Through these points, with their quantity equalled to the number $2Q - 2p + 1$, the polynomial of degree $2Q - 2p$ is then unambiguously determined

$$R_j^{(i)}(t) = \sum_{k=0}^{2Q-2p} b_{jk}^{(i)} t^k \,. \tag{20}$$

And just by using the polynomial (20), we define the derivation $\mathrm{D}^q x_j^{(i)}$ with relation

$$\mathrm{D}^q x_j^{(i)} = \frac{\mathrm{d}^q}{\mathrm{d}t^q} R_j^{(i)}(0) \tag{21}$$

$$(q = 1, \dots, Q).$$

Then we can easily find out that for $2Q - 2p \geq Q$, i.e. for $Q \geq 2p$, the following applies

$$\mathrm{D}^q x_j^{(i)} = q!\, b_{jq}^{(i)} \tag{22}$$

$$(q = 1, \dots, Q),$$

for $2Q - 2p < Q$, i.e. for $Q < 2p$, the following applies

$$\mathrm{D}^q x_j^{(i)} = q!\, b_{jq}^{(i)} \tag{23}$$

$$(q = 1, \dots, 2Q - 2p),$$

and

$$D^q x_j^{(i)} = 0 \quad (24)$$

$$(q = 2Q - 2p + 1, \ldots, Q).$$

3.1 Note

The derivative values of parametric functions $P_j^{(i)}(t)$ at points $t = -1{,}1$ is then equalled to the derivative values of the auxiliary polynomial (20) at the point zero. The meaning of the integer $p \in < 0, Q-1 >$ is that for making these derivations, we more or less need given support points; for $p = 0$ the quantity is maximum $(= 2Q + 1)$, for $p = Q - 1$ minimum $(= 3)$. With the given Q, the number p affects the shape of the resulting interpolation curve. For instance, if $Q = 3, p = 2$ then, according to (24), $D^3 x_j^{(i)} = 0$.

4 The Universal Matrix of Method $L_{Q,p}$

Because each coefficient of polynomial (20) is a linear combination of values $x_j^{(i-Q+p)}, x_j^{(i-Q+p+1)}, \ldots, x_j^{(i+Q-p)}$, also every derivation $D^q x_j^{(i)}$ $(q = 1, \ldots, Q)$ consists of some linear combination of those values. Therefore, there also exists a numerical matrix $B_{Q,p}$ of type $(Q, 2Q - 2p + 1)$ that is (see (10))

$$X_{ij}^{+} = \left(x_j^{(i-Q+p)}, x_j^{(i-Q+p+1)}, \ldots, x_j^{(i+Q-p)}\right) \cdot B_{Q,p}^{T}, \quad (25)$$

and similarly (see (11)),

$$X_{i+1,p}^{+} = \left(x_j^{(i-Q+p+1)}, x_j^{(i-Q+p+2)}, \ldots, x_j^{(i+Q-p+1)}\right). \quad (26)$$

By continuing in a similar way, to the one described in a publication [4], we get a matrix (8), described in a following way:

$$A_{ij}^{T} = C_{Q,p} \cdot \begin{pmatrix} x_j^{(i-Q+p)} \\ x_j^{(i-Q+p+1)} \\ \vdots \\ x_j^{(i+Q-p+1)} \end{pmatrix},$$

where

$$C_{Q,p} = A_Q^{-1} \cdot \left(\begin{array}{ccccccccc} & & & Q\text{-}p\text{+}1 & & Q\text{-}p\text{+}2 & & & \\ 0 & \dots & \dots & 1 & & 0 & \dots & \dots & 0 \\ & & & & & & & & 0 \\ & & & B_{Q,p} & & & & & \vdots \\ & & & & & & & & \vdots \\ & & & & & & & & 0 \\ 0 & \dots & \dots & 0 & & 1 & \dots & \dots & 0 \\ 0 & & & & & & & & \\ \vdots & & & & B_{Q,p} & & & & \\ \vdots & & & & & & & & \\ 0 & & & & & & & & \end{array} \right) \cdot Q+2 \tag{27}$$

is the matrix of type $(2Q + 2, 2Q - 2p + 2)$. We call it the universal matrix of method $L_{Q,p}$. The final form of interpolation polynomials is then

$$P_j^{(i)}(t) = (1, t, t^2, \dots, t^{2Q+1}) \cdot C_{Q,p} \cdot \begin{pmatrix} x_j^{(i-Q+p)} \\ x_j^{(i-Q+p+1)} \\ \vdots \\ \vdots \\ x_j^{(i+Q-p+1)} \end{pmatrix}. \tag{28}$$

5 The Calculatuion of Basic Matrix Elements of Method $L_{Q,p}$

If we mark elements of the matrix A_Q system of equation (7) with the symbol m_{rs}, then $A_Q = (m_{rs})_{1 \le r,s \le 2Q+2}$, the matrix A_Q is of the type $(2Q + 2, 2Q + 2)$.
Applies

$$m_{rs} = \tag{29}$$

$$= \begin{cases} (-1)^{s-1} & \text{for } r = 1 \text{ and } 1 \le s \le 2Q + 2 \\ 0 & \text{for } 2 \le r \le Q + 1 \text{ and } 1 \le s \le r - 1 \\ (-1)^{s-1}(r-1)!\binom{s-1}{r-1} & \text{for } 2 \le r \le Q + 1 \text{ and } r \le s \le 2Q + 2 \\ 1 & \text{for } r = Q + 2 \text{ and } 1 \le s \le 2Q + 2 \\ |m_{r-(Q+1),s}| & \text{for } Q + 3 \le r \le 2Q + 2 \text{ and } 1 \le s \le 2Q + 2 \end{cases} .$$

According to (32), we find out that e.g. the element m_{43} of the matrix A_1 is equal to

$$m_{43} = |m_{23}| = (-1)^1 1! \binom{2}{1} = 2 ,$$

the element $m_{33} = 1$ and

$$m_{24} = (-1)^2 1! \binom{3}{1} = 3 .$$

The whole matrix A_1 is

$$A_1 = \begin{pmatrix} 1 & -1 & 1 & -1 \\ 0 & 1 & -2 & 3 \\ 1 & 1 & 1 & 1 \\ 0 & 1 & 2 & 3 \end{pmatrix} \tag{30}$$

and the corresponding inverse matrix is

$$A_1^{-1} = \frac{1}{4} \begin{pmatrix} 2 & 1 & 2 & -1 \\ -3 & -1 & 3 & -1 \\ 0 & -1 & 0 & 1 \\ 1 & 1 & -1 & 1 \end{pmatrix} . \tag{31}$$

According to (29), we find out that e.g. the element m_{26} of the matrix A_2 equals to

$$m_{26} = (-1)^4 1! \binom{5}{1} = 5 ,$$

the element $m_{32} = 0$ and

$$m_{54} = |m_{24}| = \left|(-1)^2 1! \binom{3}{1}\right| = 3 .$$

The whole matrix A_2 is

$$A_2 = \begin{pmatrix} 1 & -1 & 1 & -1 & 1 & -1 \\ 0 & 1 & -2 & 3 & -4 & 5 \\ 0 & 0 & 2 & -6 & 12 & -20 \\ 1 & 1 & 1 & 1 & 1 & 1 \\ 0 & 1 & 2 & 3 & 4 & 5 \\ 0 & 0 & 2 & 6 & 12 & 20 \end{pmatrix} \tag{32}$$

and the corresponding inverse matrix is

$$A_2^{-1} = \frac{1}{16} \begin{pmatrix} 8 & 5 & 1 & 8 & -5 & 1 \\ -15 & -7 & -1 & 15 & -7 & 1 \\ 0 & -6 & -2 & 0 & 6 & -2 \\ 10 & 10 & 2 & -10 & 10 & -2 \\ 0 & 1 & 1 & 0 & -1 & 1 \\ -3 & -3 & - & 3 & -3 & 1 \end{pmatrix} . \tag{33}$$

According to (29), we find out that e.g. the element m_{37} of the matrix A_3 equals to

$$m_{37} = (-1)^4 2! \binom{6}{2} = 30 ,$$

the element $m_{83} = 0$ and

$$m_{88} = |m_{48}| = \left|(-1)^4 3! \binom{7}{3}\right| = 210 .$$

The whole matrix A_3 is

$$A_3 = \begin{pmatrix} 1 & -1 & 1 & -1 & 1 & -1 & 1 & -1 \\ 0 & 1 & -2 & 3 & -4 & 5 & -6 & 7 \\ 0 & 0 & 2 & -6 & 12 & -20 & 30 & -42 \\ 0 & 0 & 0 & 6 & -24 & 60 & -120 & 210 \\ 1 & 1 & 1 & 1 & 1 & 1 & 1 & 1 \\ 0 & 1 & 2 & 3 & 4 & 5 & 6 & 7 \\ 0 & 0 & 2 & 6 & 12 & 20 & 30 & 42 \\ 0 & 0 & 0 & 6 & 24 & 60 & 120 & 210 \end{pmatrix} \quad (34)$$

and the corresponding inverse matrix is

$$A_3^{-1} = \frac{1}{96} \begin{pmatrix} 48 & 33 & 9 & 1 & 48 & -33 & 9 & -1 \\ -105 & -57 & -12 & -1 & 105 & -57 & 12 & -1 \\ 0 & -45 & -21 & -3 & 0 & 45 & -21 & 3 \\ 105 & 105 & 30 & 3 & -105 & 105 & -30 & 3 \\ 0 & 15 & 15 & 3 & 0 & -15 & 15 & -3 \\ -63 & -63 & -24 & -3 & 63 & -63 & 24 & -3 \\ 0 & -3 & -3 & -1 & 0 & 3 & -3 & 1 \\ 15 & 15 & 6 & 1 & -15 & 15 & -6 & 1 \end{pmatrix} . \quad (35)$$

For interests, e.g. the element $m_{16,15}$ of the matrix A_9 which is the type (20,20), according to (29) equals

$$m_{16,15} = |m_{6,15}| = \left|(-1)^9 5! \binom{14}{5}\right| = 240240 .$$

We won't show the whole matrix A_9 and the corresponding inverse matrix A_9^{-1} either but they do exist.

5.1 Example

On the plane R^2, we think of support points $P_1 = (1,1), P_2 = (2,3), P_3 = (5,-1), P_4 = (2,-3), P_5 = (4,5)$. With the method $L_{3,1}$, we create an opened interpolation curve $P_1P_2P_3P_4P_5$. The matrix system of equations (7) is A_3 (which is the matrix (34)) with the according inverse matrix (35). For the purpose of getting the matrix $B_{3,1}$ of the type (3,5) (see (25)), we add to the polynomial (see (20))

$$R_j^{(i)}(t) = \sum_{k=0}^{4} b_{jk}^{(i)} t^k \quad (36)$$

points $(2h, x_j^{(i+h)}), -2 \le h \le 2, h$ integer. Polynomial coefficients (36) are calculated with a simple calculation

$$b_{j0}^{(i)} = x_j^{(i)} ,$$

$$b_{j1}^{(i)} = (-x_j^{(i+2)} + 8x_j^{(i+1)} - 8x_j^{(i-1)} + x_j^{(i-2)})/24 ,$$

$$b_{j2}^{(i)} = (-x_j^{(i+2)} + 16x_j^{(i+1)} - 30x_j^{(i)} + 16x_j^{(i-1)} - x_j^{(i-2)})/96 ,$$

$$b_{j3}^{(i)} = (x_j^{(i+2)} - 2x_j^{(i+1)} + 2x_j^{(i-1)} - x_j^{(i-2)})/96 ,$$

$$b_{j4}^{(i)} = (x_j^{(i+2)} - 4x_j^{(i+1)} + 6x_j^{(i)} - 4x_j^{(i-1)} + x_j^{(i-2)})/384 .$$

According to (22) which is $\mathrm{D}^1 x_j^{(i)} = b_{j1}^{(i)}, \mathrm{D}^2 x_j^{(i)} = 2b_{j2}^{(i)}, \mathrm{D}^3 x_j^{(i)} = 6b_{j3}^{(i)}$, so (see (25))

$$X_{ij}^{+} = \left(\mathrm{D}^1 x_j^{(i)}, \mathrm{D}^2 x_j^{(i)}, \mathrm{D}^3 x_j^{(i)}\right) =,$$

$$= (x_j^{(i-2)}, x_j^{(i-1)}, x_j^{(i)}, x_j^{(i+1)}, x_j^{(i+2)}) \cdot B_{3,1}^T ,$$

where

$$B_{3,1} = \frac{1}{48} \begin{pmatrix} 2 & -16 & 0 & 16 & -2 \\ -1 & 16 & -30 & 16 & -1 \\ -3 & 6 & 0 & -6 & 3 \end{pmatrix} . \tag{37}$$

Using (35), (37), we get (see (27))

$$C_{3,1} = \tag{38}$$

$$= \frac{1}{1536} \begin{pmatrix} 18 & -150 & 900 & 900 & -150 & 18 \\ -33 & 197 & -1194 & 1194 & -197 & 33 \\ -20 & 156 & -136 & -136 & 156 & -20 \\ 57 & -317 & 666 & -666 & 317 & -57 \\ 2 & -6 & 4 & 4 & -6 & 2 \\ -31 & 155 & -310 & 310 & -155 & 31 \\ 0 & 0 & 0 & 0 & 0 & 0 \\ 7 & -35 & 70 & -70 & 35 & -7 \end{pmatrix} .$$

With the symbol $\tilde{C}_{3,1}$, we mark the matrix which forms from the matrix $C_{3,1}$ (see (37)) by omitting the seventh row, composed of zeros.

Proceed to the calculation of parametric polynomials first and second segment of the opened interpolation curve $P_1P_2P_3P_4P_5$. According to (28) and (17), parametric equations of the first segment P_1P_2: $-1 \leq t \leq 1$ are:

$$x_1 = P_1^{(1)}(t) = \tag{39}$$

$$= (1, t, t^2, t^3, t^4, t^5, t^7) \cdot \tilde{C}_{3,1} \cdot (5,2,1,2,5,2)^T =$$

$$= \frac{1}{192}(222 + 63t + 68t^2 + 57t^3 - 2t^4 - 31t^5 + 7t^7) ,$$

$$x_2 = P_2^{(1)}(t) = \tag{40}$$
$$= (1, t, t^2, t^3, t^4, t^5, t^7) \cdot \tilde{C}_{3,1} \cdot (-1,3,1,3,-1,-3)^T =$$
$$= \frac{1}{768}(1614 + 1555t - 76t^2 - 1243t^3 - 2t^4 + 589t^5 - 133t^7)\,.$$

$(P_1^{(1)}(-1), P_2^{(1)}(-1)) = (1,1) = P_1, (P_1^{(1)}(1), P_2^{(1)}(1)) = (2,3) = P_2$ is in accordance with the primary requirement.

According to (28) and (17), parametric equations of the second segment $P_2P_3\colon -1 \le t \le 1$ are:

$$x_1 = P_1^{(2)}(t) = \tag{41}$$
$$= (1, t, t^2, t^3, t^4, t^5, t^7) \cdot \tilde{C}_{3,1} \cdot (2,1,2,5,2,4)^T =$$
$$= \frac{1}{1536}(5958 + 3451t - 604t^2 - 1795t^3 + 22t^4 + 837t^5 - 189t^7)\,,$$
$$x_2 = P_2^{(2)}(t) = (1, t, t^2, t^3, t^4, t^5, t^7) \cdot C_{3,1} \cdot (3,1,3,-1,-3,5)^T =$$
$$= \frac{1}{768}(1122 - 1961t - 372t^2 + 641t^3 + 18t^4 - 279t^5 + 63t^7)\,.$$

$(P_1^{(2)}(-1), P_2^{(2)}(-1)) = (2,3) = P_2, (P_1^{(2)}(1), P_2^{(2)}(1)) = (5,-1) = P_3$ is in accordance with the primary requirement.

Further applies

$$\frac{\mathrm{d}}{\mathrm{d}t} P_1^{(1)}(1) = \frac{\mathrm{d}}{\mathrm{d}t} P_1^{(2)}(-1) = \frac{4}{3}\,,$$
$$\frac{\mathrm{d}^2}{\mathrm{d}t^2} P_1^{(1)}(1) = \frac{\mathrm{d}^2}{\mathrm{d}t^2} P_1^{(2)}(-1) = \frac{2}{3}\,,$$
$$\frac{\mathrm{d}^3}{\mathrm{d}t^3} P_1^{(1)}(1) = \frac{\mathrm{d}^3}{\mathrm{d}t^3} P_1^{(2)}(-1) = -\frac{1}{2}$$

and

$$\frac{\mathrm{d}}{\mathrm{d}t} P_2^{(1)}(1) = \frac{\mathrm{d}}{\mathrm{d}t} P_2^{(2)}(-1) = -\frac{5}{12}\,,$$
$$\frac{\mathrm{d}^2}{\mathrm{d}t^2} P_2^{(1)}(1) = \frac{\mathrm{d}^2}{\mathrm{d}t^2} P_2^{(2)}(-1) = -\frac{45}{24}\,,$$
$$\frac{\mathrm{d}^3}{\mathrm{d}t^3} P_2^{(1)}(1) = \frac{\mathrm{d}^3}{\mathrm{d}t^3} P_2^{(2)}(-1) = -\frac{1}{8}\,,$$

thus, the transition from the first segment to the second segment of the constructed interpolation curve runs smoothly from the 1st to the 3rd derivation. This fact is also in accordance with the primary requirement.

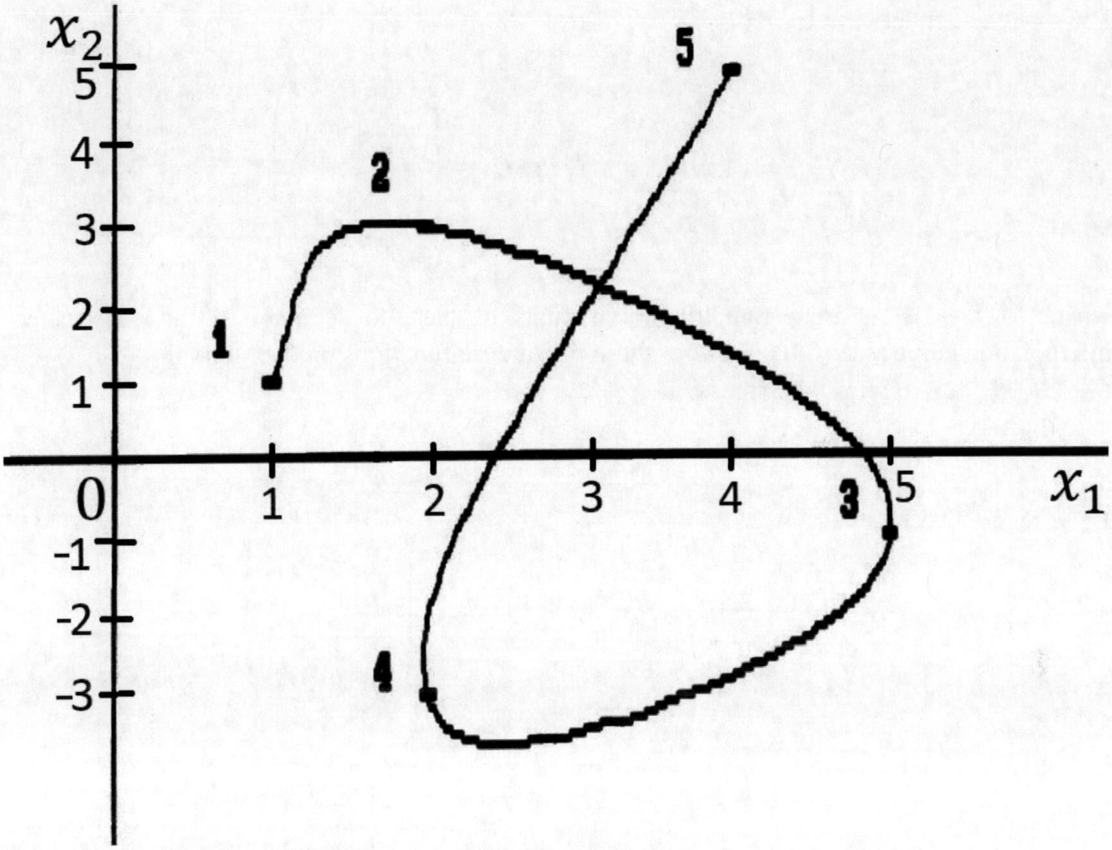

Fig. 2. The course of the opened interpolation curve $P_1P_2P_3P_4P_5$.

6 General Parameterisation of the Interpolation Curve

We recall that each segment of an opened interpolation curve $P_1P_2 \dots P_n$, or a closed interpolation curve $P_1P_2 \dots P_nP_1$, is parameterised on the same interval $< -1,1 >$. On the numerical axis, we now choose points $T_1 < T_2 < \dots < T_n$ in case of the opened curve $P_1P_2 \dots P_n$, or points $T_1 < T_2 < \dots < T_n < T_{n+1}$ in case of the closed curve $P_1P_2 \dots P_nP_1$, and the i-th interval $t \in < T_i, T_{i+1} >$ for $i = 1, \dots, n-1$ in case of the opened interpolation curve, or for $i = 1, \dots, n$ in case of the closed interpolation curve, we simply display it on the interval $< -1,1 >$ according to the formula

$$\frac{t-A_i}{B_i}\text{, where } A_i = \frac{T_i+T_{i+1}}{2}\text{, } B_i = \frac{T_{i+1}-T_i}{2}. \tag{42}$$

For $\lambda \in < 0,1 >$, the point $t = T_i + \lambda(T_{i+1} - T_i)$ from the interval $< T_i, T_{i+1} >$ displays on the number $2\lambda - 1$. For instance, for $\lambda = 0$, the point T_i displays on the number -1, for $\lambda = 1$, the point T_{i+1} displays on the number 1, for $\lambda = ½$, the centre of the interval $< T_i, T_{i+1} >$ displays at zero.

For $j = 1, \ldots, m$, let

$$P_j^{(i)}(t) = (1, t, t^2, \ldots, t^{2Q+1}) \cdot C_{Q,p} \cdot \begin{pmatrix} x_j^{(i-Q+p)} \\ x_j^{(i-Q+p+1)} \\ \vdots \\ x_j^{(i+Q-p+1)} \end{pmatrix},$$

where $t \in< -1,1 >$, are parametric polynomials of the i-th segment of the constructed interpolation curve (see (28)). Then for these parametric functions on the interval $t \in< T_i, T_{i+1} >$ applies the expression

$$P_j^{(i)}(t) = \tag{43}$$

$$= \left(1, \frac{t-A_i}{B_i}, (\frac{t-A_i}{B_i})^2, \ldots, (\frac{t-A_i}{B_i})^{2Q+1}\right) \cdot C_{Q,p} \cdot \begin{pmatrix} x_j^{(i-Q+p)} \\ x_j^{(i-Q+p+1)} \\ \vdots \\ x_j^{(i+Q-p+1)} \end{pmatrix}.$$

7 Conclusion

The originality of Lienhard interpolation method is solely based on the given support points and with its assistance it generates all other quantities appearing in the mathematical formulation of interpolation problem. In contrast with the original version of Lienhard method, in which every segment of the interpolation curve output is parameterised on the same interval <-1,1>, the author of this article came up with the process of individual belonging of the parametric interval for each part of the interpolation curve.

Acknowledgement

I would like to thank prof. Hindls for inspiring me to write this article.

References

[1] KAŇKA, M. Segmented Regression Based on B-Splines with solved Examples. STATISTIKA: Statistics and Economy Journal, 2015, Prague, pp. 47 - 66.

[2] KAŇKA, M. Segmented Regression Based on Cut-off polynomials. STATISTIKA: Statistics and Economy Journal, 2016, Prague, pp. 60 - 72.

[3] LIENHARD, H. Interpolation von Funktionswerten bei numerischen Bahnsteuerungen. Undated publication of CONTRAVES AG, Zürich.

[4] MATUŠŮ, J. - MATUŠŮ, M. The Lienhard Interpolation LQ,p - method. ACTA POLYTECHNICA, vol. 40, No. 5-6/2000, Prague, pp.15 - 21.

[5] MATUŠŮ, J. - NOVÁK, J. Constructions of interpolation curves from given supporting elements (I). APLIKACE MATEMATIKY, 30 (1985), No. 4, Prague, pp. 425 - 452.

[6] MATUŠŮ, J. - NOVÁK, J. Constructions of interpolation curves from given supporting elements (II). APLIKACE MATEMATIKY, 31 (1986), No. 2, Prague, pp. 141 - 162.

Current address

Kaňka Miloš, doc. RNDr., CSc.
Department of Mathematics, Faculty of Informatics and Statistics
University of Economics, Prague
Ekonomická 957, 140 00 Prague 4, Czech Republic
E-mail: kanka@vse.cz

AUTHOR INDEX

AUTHOR INDEX

17th Conference on Applied Mathematics (APLIMAT 2018)

Bratislava, Slovak Republic
6-8 February 2018

Volume 2 of 2

Editors:

Dagmar Szarkova
Peter Letavaj

Daniela Richtarikova
Jana Gabkova

ISBN: 978-1-5108-6248-7

Printed from e-media with permission by:

Curran Associates, Inc.
57 Morehouse Lane
Red Hook, NY 12571

Some format issues inherent in the e-media version may also appear in this print version.

Printed by Curran Associates, Inc. (2018)

For permission requests, please contact Institute of Mathematics and Physics, Faculty of Mechanical Engineering, Slovak University of Technology in Bratislava at the address below.

Institute of Mathematics and Physics, Faculty of Mechanical Engineering
Slovak University of Technology in Bratislava
Namestie Slobody 17
812 31 Bratislava 1
Slovak Republic

Phone: +421 2 572 96 111
Fax: +421 2 529 25 749

www.stuba.sk

Additional copies of this publication are available from:

Curran Associates, Inc.
57 Morehouse Lane
Red Hook, NY 12571 USA
Phone: 845-758-0400
Fax: 845-758-2633
Email: curran@proceedings.com
Web: www.proceedings.com

CONTENTS

Slovak University of Technology in Bratislava
Faculty of Mechanical Engineering
17th Conference on Applied Mathematics **APLIMAT 2018**

Proceedings

DATA SCIENCE AND STATISTICS WITH PYTHON

KASPŘÍKOVÁ Nikola (CZ)

Abstract. With Data Science as the recent trend, the tools for efficient manipulation and analysis of data have become important for (training of) professionals in data analysis and statistics. The Python programming language starts to be one of the most popular tools for data analysis, in many cases replacing well-established statistical software packages. We report on a short case study on elementary statistical evaluation of a real-world data on air quality, using Python.

Keywords: Data science, Python, statistics, software

Mathematics subject classification: Primary 97R30; Secondary 68N15

1 Introduction

There is a paper by John M. Chambers [1], the author of the S programming language, the language which can be considered the predeccessor of the widely used R software for the statistical computig [2]. The paper [1] is advocating the so-called greater statistics, which means everything related to learning from data – starting with the planning of the research, including collection of the data and organization of the data, to reporting the results. The paper [1] was published as early as in 1993, and it was probably not the first and definitely not the last such paper. The Data Science has recently become popular as a new discipline related to data analysis. This discipline, actually developing the ideas in [1], aims at the ability to efficiently operate with the data technologies and software tools, ask the right questions about the problem to be analysed and present the results appropriately. The Python programming language [3] is considered as one of the primary tools for Data Science. Many profiles of jobs related to quantitative methods and data analysis include Python as one of the required skills. This fact can be easily checked for example at the MathJobs website [4], run by the American Mathematical Society, and similarly at the well-known ResearchGate website [5].

These facts should be reflected in the training of professionals in mathematics, data analysis and statistics. And indeed many top-class universities have already started including the Data Science courses in the curriculum. There are textbooks like [6] (UC Berkeley), [7] (MIT) and many others and there are initiatives focused at including computational thinking in the curriculum, see e. g. [8].

The goal of this paper is to assess the suitability of using Python for data analysis and statistics courses. We report on a short case study on elementary statistical evaluation of a real-world data on air quality (the levels of PMO10) in Praha and in Moravskoslezský region in the Czech Republic, using Python. After having discussed the motivation in the introduction, the remaining parts of this paper briefly introduce the data science and the Python programming language for statistics, then the case study is reported, followed by some conclusions.

2 Data Science and Python

2.1 Data Science as the new trend

It is rather difficult to find an objective way of evaluating the popularity of some concept. We try using the Google Trends tool [9] to analyze the "Data Science" search query popularity between December 2012 and November 2017. The figure 1 is the visualization of the data obtained from [9] and it represents the relative interest in the search query at the Google search engine in comparison to the highest value in the time series (the value 100 represents the highest popularity). Obviously "Data Science" has been gaining popularity quite rapidly in the recent years.

The Data Science concept, which has recently started to be used much more often then earlier, may be still finding its precise meaning, but often it refers to (see [6]) parallel computing, dealing with non-standard data formats (such as access log files, e-mail messages), data technologies like XML, JSON, web scraping, NoSQL database technologies (including Map Reduce, Key - value database systems or graph database systems) and others, all aimed at dealing the data efficiently and ability to solve complex real world problems using data analytics.

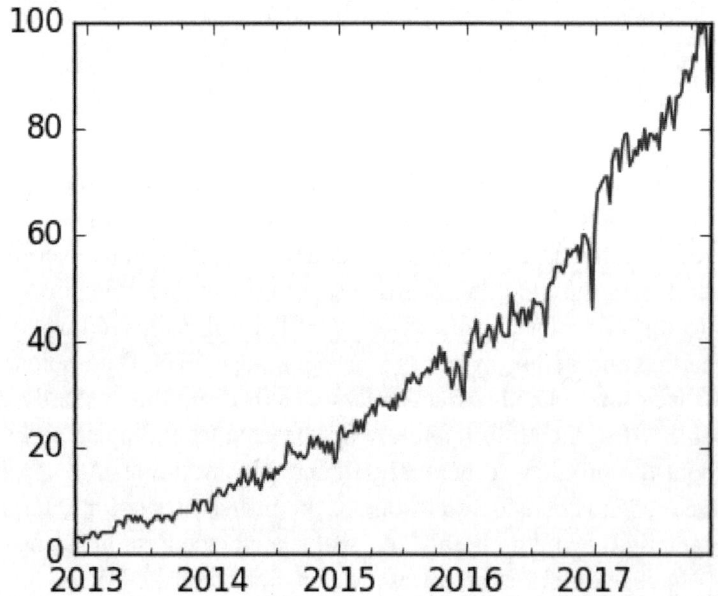

Fig. 1. The "Data Science" query popularity over time in Google Trends.

2.2 Python programming language and statistics

The Python [3] programming language is one of the most often used languages today and it has the following basic features:

- open source software
- interpreted language
- general purpose language
- there exist many extension packages designed for specific needs
- quite easy to learn

The Google Trends tool states "python for data science" as the top related query for the "data science" query. The Python extension packages often used for data analysis, plotting and statistical computing include SciPy [10], pandas [11] and seaborn [12]. There are many papers advocating using Python in data analysis and even in high performance computing, see e. g. [13].

3 Case study: the statistical analysis of air quality data

3.1 Material and methods

We analyze the yearly average levels of PMO10 (measured in $\mu g/m^3$) at the measurement stations in year 2016 in Praha and in the Moravskoslezský region in the Czech Republic. Let's assume that the locations of the stations (13 and 18 respectively) are representative. These two regions are known for troubles with the air quality, which can be generally attributed to heavy transportation, high industry-driven pollution and some geographical characteristics. The dataset has been obtained from the data published by the Czech Hydrometeorological Institute, available at [14].

We will do elementary exploratory statistical analysis - calculate basic descriptive statistical characteristics and obtain the probability density estimate plot. Then we evaluate the difference in the distributions of PM10 levels between the Praha and the Moravskoslezský region using the Wilcoxon rank-sum test.

3.2 Calculation in Python and results

First we read the dataset into the Python computing environment. We use the pandas package [11] for reading the csv file using the following statements.

```
>>> import pandas as pd
>>> dta2 = 'dataCHMU.csv'
>>> df2 = pd.read_csv(dta2)
```

Then, using the describe function, we calculate the basic descriptive statistics for each of the columns of the dataset, which stores the data of Praha and the Moravskoslezský (MS) region.

```
>>> df2.describe()
```

Which results in the following output:

```
            Praha          MS
count   13.000000   18.000000
```

```
mean    22.753846  31.694444
std      2.840368   4.332839
min     19.300000  24.700000
25%     20.400000  28.200000
50%     22.700000  32.900000
75%     25.200000  33.725000
max     26.900000  41.000000
```

The results of the calculations suggest that the levels of PMO10 are rather lower in Praha, with median at around 23 $\mu g/m^3$, than in Moravskoslezský region, with median at around 33 $\mu g/m^3$. The probability density estimate plot (shown in Fig. 2) supports this too. The probability density estimate plot can be conveniently produced using the functions available in the seaborn package. The statements for obtaining the plot are as follows:

```
>>> import seaborn as sns
>>> p1=sns.kdeplot(df2['Praha'], shade=True, color="r")
>>> p1=sns.kdeplot(df2['MS'], shade=True, color="b")
>>> sns.plt.show()
```

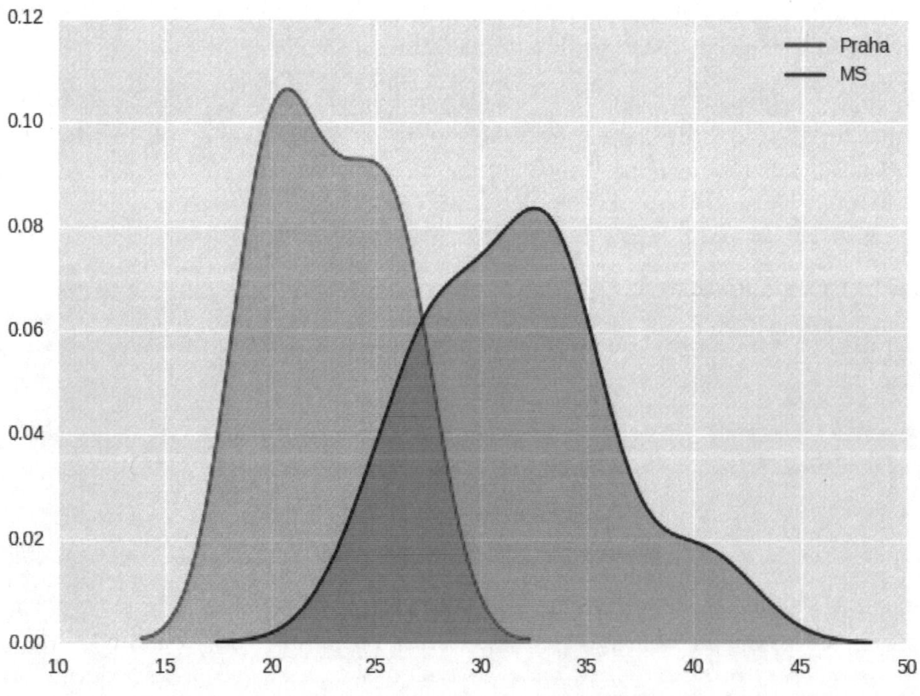

Fig. 2. Probability density estimate of PM10 levels [$\mu g/m^3$] by region in 2016.

Finally we run the Wilcoxon rank-sum test to detect formally if there is the difference in the distributions of PM10 levels between the Praha and the Moravskoslezský region. After importing the stats module from the scipy package, we retrieve directly the p-value of the test.

```
>>> from scipy import stats
>>> round(stats.ranksums(df2['Praha'],df2['MS'])[1],3)
```

Which gives 0.035. Based on this output, it may be concluded at 0.05 confidence level, that the PMO10 levels distribution is different in Praha and in Moravskoslezský region.

4 Conclusions

While perhaps still not offering the full comfort of some professional statistical packages, the Python programming language, with many data analysis related features under development, does already provide convenient tools for data analysis and common statistical calculations and may be easily used in courses on statistics. It has been shown that the syntax needed for elementary statistical analysis does not seem too complex and could easily be mastered, even by non-technical students.

References

[1] CHAMBERS, J. M.: Greater and Lesser Statistics: A Choice for Future Research, *Statistics and Computing*, 3(4), 1993, pp. 182-184.

[2] R CORE TEAM: *R: A language and environment for statistical computing.* R Foundation for Statistical Computing, Vienna, Austria, 2017. URL http://www.R-project.org.

[3] URL: https://www.python.org. Accessed 1/12/2017.

[4] URL: http://www.mathjobs.org. Accessed 1/12/2017.

[5] URL: http://www.researchgate.net. Accessed 1/12/2017.

[6] NOLAN D., LANG, D. T.: *Data science in R. A case studies Approach to computational reasoning and problem solving*, CRC Press, 2015, ISBN 978-1-4822-3481-7.

[7] GUTTAG, J. V.: *Introduction to computation and programming using Python*, The MIT Press, 2013, ISBN 978-0-262-52500-8.

[8] URL: http://www.computerbasedmath.org. Accessed 1/12/2017. x

[9] URL: https://trends.google.com/trends. Accessed 1/12/2017.

[10] URL: https://scipy.org. Accessed 1/12/2017.

[11] URL: http://pandas.pydata.org. Accessed 1/12/2017.

[12] URL: https://seaborn.pydata.org/. Accessed 1/12/2017.

[13] BILINA, R., LAWFORD, S.: Python for unified research in econometrics and statistics. *Econometric Reviews*, 31(5):558–591, ISSN: 0747-4938.

[14] URL: http://portal.chmi.cz/files/portal/docs/uoco/web_generator/tab_reports/automated/tab_2016_01_1Y_CZ.html. Retrieved 1/12/2017.

Current address

Kaspříková Nikola, Ph. D.
University of Economics in Prague, Department of Mathematics
Nám. W. Churchilla 4, 130 67 Praha, Czech Republic
E-mail: nb33@tulipany.cz

Slovak University of Technology in Bratislava
Faculty of Mechanical Engineering
17th Conference on Applied Mathematics **APLIMAT 2018**

Proceedings

MAPPINGS OF SPACES WITH AFFINE CONNECTION

KIOSAK Volodymyr (UA), LESECHKO Olexandr (UA), SAVCHENKO Olexandr (UA)

Abstract. By modification of A. Norden methodic, we have found the formulae connecting main tensors of spaces with affine connection A_n and $\bar{A}_n$, admitting mappings onto each other. We introduce the notion of shortened mapping and its particular case a half-mapping. When we turn our attention to covariant derivatives under medium connection, the main equations are simplified to a notable degree.

Keywords: mappings, spaces with affine connection, deformation tensor, Riemannian tensor

Mathematics subject classification: Primary 53B30, 53B35

1 Introduction

The notion of a connection was introduced to differential geometry in works of Tullio Levi-Civita (1917). Later H. Weyl (1920) invoked the spaces with affine connection while studying translation of vectors.

This paper treats the mappings of spaces with affine connection. Morphisms (mappings) of generalized geometric spaces are a subject of an up-to-date research field in modern differential geometry. There are three main directions of inquiry [1, 2, 3, 4]:

1. A study of general patterns of mappings.
2. Research on the problem whether a given generalized space admits a special mapping or not.
3. Search for a connecting mapping when a pair of spaces is given.

The scheme of investigation is as follows: study of mappings is reduced to a system of differential equations; the system of differential equations is reduced to an algebraic system, which represents conditions of integrability; these systems redefined by introduction of additional limitations, they are simplified or integrated [5].

A local solution of this problem is possible in general, meanwhile in practice it meets serious technical difficulties.

The paper suggests the method for construction of a simplified system of differential equations in covariant derivatives for research on mappings of spaces with affine connection.

2 Main results

A space with affine connection A_n of a given dimension n is a differential manifold with every curve having an affine connection defined. Or in other words: for any point M and for any vector field in the vicinity of this point, absolute differential of a vector belonging to this field, if calculated in the point M for any curve passing through it, is a linear function of a vector of elementary translation along the curve.

Authors treat spaces with affine connections A_n without torsion, as follows

$$\Gamma^h_{ij}(x) = \Gamma^h_{ji}(x). \tag{1}$$

The space A_n belongs to class C^r $(A_n \in C^r)$, if $\Gamma^h_{ij}(x) \in C^r$.

Here we treat two spaces with affine connection.

A one-to-one correspondence of points of spaces with affine connection A_n and $\bar{A}_n$ is called a mapping. Then in a system of coordinates common in respect to the mapping the following conditions exist:

$$\bar{\Gamma}^h_{ij}(x) - \Gamma^h_{ij}(x) = P^h_{ij}(x). \tag{2}$$

here $\Gamma^h_{ij}, \bar{\Gamma}^h_{ij}$ — objects with affine connection of spaces A_n and $\bar{A}_n$ respectively. In the following discussion the objects $\bar{A}_n$ will be denoted by a bar.

A system of curvilinear coordinates is called a system of coordinates common in respect to a mapping if the coordinates of respective points coincide.

The tensor $P^h_{ij}(x)$ is called a tensor of deformation of connection at a given mapping.

If

$$P^h_{ij}(x) \not\equiv 0, \tag{3}$$

then the mapping is called nontrivial.

Let us mention that the deformation tensor is symmetrical in sense of covariant indexes. Or $P^h_{ij} = P^h_{ji}$ for torsion-free spaces with affine connection.

Theorem 1. *When the space with affine connection A_n is mapped onto the space with affine connection $\bar{A}_n$, Riemannian tensors of spaces A_n and $\bar{A}_n$ are connected by an equation in a single coordinate system*

$$\bar{R}^h_{.\,ijk} = R^h_{.\,ijk} + \frac{1}{2}(\nabla_k P^h_{ji} - \nabla_j P^h_{ki} + \bar{\nabla}_k P^h_{ji} - \bar{\nabla}_j P^h_{ki}). \tag{4}$$

Here R^h_{ijk} — Riemannian tensor and ∇ — a symbol of covariant derivative.

While for purposes of definition of mapping, we introduce a one-to-one correspondence, in fact we order the given pair of spaces with affine connection A_n and $\bar{A}_n$ by ascribing a sign for a deformation tensor. Any pair of spaces A_n and $\bar{A}_n$ has a correspondence defined by objects of connections of these spaces. On the other hand, object of connection A_n and deformation tensor characterize connection of space $\bar{A}_n$. It permits to introduce a mapping that we propose to name shortened in relation to a given mapping. Further we will call it a shortened mapping.

Object $\overset{\lambda}{\Gamma}{}^h_{ij}$ is constructed following a rule

$$\overset{\lambda}{\Gamma}{}^h_{ij} = \Gamma^h_{ij}(x) + \frac{\lambda}{1+\lambda} P^h_{ij}(x), \qquad (5)$$
$$\lambda = const > 0.$$

It characterizes the connection of a given space with affine connection $\overset{\lambda}{A}_n$ [6].

A mapping of space with affine connection A_n onto a space with affine connection $\overset{\lambda}{A}_n$ is called a shortened mapping, when such an equation is true in a common system of coordinates (5).

Theorem 2. *When spaces A_n and $\bar{A}_n$ admit a mapping that corresponds to deformation tensor P^h_{ij}, then there exists a shortened mapping, in respect to which Riemannian tensor corresponds to limitations:*

$$\overset{\lambda}{R}{}^h_{.ijk} = R^h_{.ijk} + \frac{\lambda}{1+\lambda}(\nabla_k P^h_{ji} - \nabla_j P^h_{ki} + \overset{\lambda}{\nabla}_k P^h_{ji} - \overset{\lambda}{\nabla}_j P^h_{ki}). \qquad (6)$$

When $\lambda = 1$ then such a mapping is called shortened in half or a half-mapping and the connection is called medium.

Theorem 3. *When spaces A_n and $\bar{A}_n$ admit the mapping that corresponds to the deformation tensor P^h_{ij}, then there exists the half-mapping with Riemannian tensor that satisfies conditions:*

$$\bar{R}^h_{.ijk} = R^h_{.ijk} + \overset{c}{\nabla}_k P^h_{ji} - \overset{c}{\nabla}_j P^h_{ki}. \qquad (7)$$

Contracting the latter with respect to indices h and i, we get

$$\bar{R}^\alpha_{\alpha jk} = R^\alpha_{\alpha jk} + \overset{c}{\nabla}_k \tau_j - \overset{c}{\nabla}_j \tau_k, \qquad (8)$$

where $\tau_i \overset{def}{=} P^\alpha_{i\alpha} = \bar{\Gamma}^\alpha_{i\alpha} - \Gamma^\alpha_{i\alpha}$.

A manifold A_n with a symmetric affine connection is called equiaffine manifold if the Ricci tensor is symmetric.

Since in the case of symmetric connection

$$R_{ij} - R_{ji} = R^\alpha_{\alpha ji}. \qquad (9)$$

A manifold A_n with a symmetric affine connection is an equiaffine manifold if and only if in any coordinate system (x^i) there exists a function $f(x)$ satisfying

$$\Gamma^\alpha_{i\alpha} = \partial_i f(x). \qquad (10)$$

When the spaces are equiaffine and the Ricci tensors are symmetrical then we get the equation:

$$\overset{c}{\nabla}_k \tau_j - \overset{c}{\nabla}_j \tau_k = 0. \tag{11}$$

Let us formulate the theorem:

Theorem 4.

If the Weyl tensor is preserved in the course of the mapping of spaces with affine connection, the deformation tensor holds the conditions:

$$\begin{aligned} &\overset{c}{\nabla}_k P^h_{ji} - \overset{c}{\nabla}_j P^h_{ki} = \\ &= \frac{1}{n-1}\left(\delta^h_k(\overset{c}{\nabla}_\alpha P^\alpha_{ji} - \overset{c}{\nabla}_j P^\alpha_{\alpha i}) - \delta^h_j(\overset{c}{\nabla}_\alpha P^\alpha_{ki} - \overset{c}{\nabla}_k P^\alpha_{\alpha i})\right) - \\ &\quad - \frac{1}{n+1}(\delta^h_i(\overset{c}{\nabla}_j P^\alpha_{\alpha k} - \overset{c}{\nabla}_k P^\alpha_{\alpha j}) - \\ &\quad - \frac{1}{n-1}(\delta^h_k(\overset{c}{\nabla}_j P^\alpha_{\alpha i} - \overset{c}{\nabla}_i P^\alpha_{\alpha j}) - \delta^h_j(\overset{c}{\nabla}_k P^\alpha_{\alpha i} - \overset{c}{\nabla}_i P^\alpha_{\alpha k}))) \end{aligned} \tag{12}$$

and for equiaffine spaces

$$\begin{aligned} &\overset{c}{\nabla}_k P^h_{ji} - \overset{c}{\nabla}_j P^h_{ki} = \\ &= \frac{1}{n-1}\left(\delta^h_k(\overset{c}{\nabla}_\alpha P^\alpha_{ji} - \overset{c}{\nabla}_j P^\alpha_{\alpha i}) - \delta^h_j(\overset{c}{\nabla}_\alpha P^\alpha_{ki} - \overset{c}{\nabla}_k P^\alpha_{\alpha i})\right). \end{aligned} \tag{13}$$

For further investigation the methods developed by [7] and [8] can be applied.

3 Proofs for theorems

Theorem 1. Proof

In a case of tensor field S of type $\binom{p}{q}$ covariant derivative of connection A_n (it will be designated by ∇) is defined in any coordinate system $x^1, x^2, \ldots, x^n$ as follows:

$$\begin{aligned} \nabla_k S^{i_1 i_2 \ldots i_p}_{j_1 j_2 \ldots j_q}(x) &= \partial_k S^{i_1 i_2 \ldots i_p}_{j_1 j_2 \ldots j_q}(x) + \Gamma^{i_1}_{k\alpha}(x) S^{\alpha i_2 \ldots i_p}_{j_1 j_2 \ldots j_q}(x) + \ldots + \\ &+ \Gamma^{i_p}_{k\alpha}(x) S^{i_1 i_2 \ldots i_{p-1}\alpha}_{j_1 j_2 \ldots j_q}(x) - \Gamma^{\beta}_{kj_1}(x) S^{i_1 i_2 \ldots i_p}_{\beta j_2 \ldots j_q}(x) - \ldots - \\ &- \Gamma^{\beta}_{kj_q}(x) S^{i_1 i_2 \ldots i_p}_{j_1 j_2 \ldots j_{q-1}\beta}(x), \\ &(i_1, \ldots, i_p; \quad j_1, \ldots j_q; \quad k = 1, 2, \ldots, n). \end{aligned} \tag{14}$$

For the space $\bar{A}_n$ and covariant derivative $\bar{\nabla}$ in it the equation is true in a common coordinate system :

$$\begin{aligned} \bar{\nabla}_k S^{i_1 i_2 \ldots i_p}_{j_1 j_2 \ldots j_q}(x) &= \partial_k S^{i_1 i_2 \ldots i_p}_{j_1 j_2 \ldots j_q}(x) + \bar{\Gamma}^{i_1}_{k\alpha}(x) S^{\alpha i_2 \ldots i_p}_{j_1 j_2 \ldots j_q}(x) + \ldots + \\ &+ \bar{\Gamma}^{i_p}_{k\alpha}(x) S^{i_1 i_2 \ldots i_{p-1}\alpha}_{j_1 j_2 \ldots j_q}(x) - \bar{\Gamma}^{\beta}_{kj_1}(x) S^{i_1 i_2 \ldots i_p}_{\beta j_2 \ldots j_q}(x) - \ldots - \\ &- \bar{\Gamma}^{\beta}_{kj_q}(x) S^{i_1 i_2 \ldots i_p}_{j_1 j_2 \ldots j_{q-1}\beta}(x), \\ &(i_1, \ldots, i_p; \quad j_1, \ldots j_q; \quad k = 1, 2, \ldots, n). \end{aligned} \tag{15}$$

Subtracting from the latter (14) and taking into account (2) we will get

$$\begin{gathered}\bar{\nabla}_k S^{i_1 i_2 \dots i_p}_{j_1 j_2 \dots j_q}(x) - \nabla_k S^{i_1 i_2 \dots i_p}_{j_1 j_2 \dots j_q}(x) = P^{i_1}_{k\alpha}(x) S^{\alpha i_2 \dots i_p}_{j_1 j_2 \dots j_q}(x) + \dots + \\ + P^{i_p}_{k\alpha}(x) S^{i_1 i_2 \dots i_{p-1}\alpha}_{j_1 j_2 \dots j_q}(x) - P^{\beta}_{kj_1}(x) S^{i_1 i_2 \dots i_p}_{\beta j_2 \dots j_q}(x) - \dots - \\ - P^{\beta}_{kj_q}(x) S^{i_1 i_2 \dots i_p}_{j_1 j_2 \dots j_{q-1}\beta}(x), \\ (i_1, \dots, i_p; \quad j_1, \dots j_q; \quad k = 1, 2, \dots, n).\end{gathered} \tag{16}$$

The latter is true for any tensor and for deformation tensor (16). It takes a shape as follows:

$$\bar{\nabla}_k P^h_{ij}(x) - \nabla_k P^h_{ij}(x) = P^h_{k\alpha}(x) P^\alpha_{ij}(x) - P^\alpha_{ki}(x) P^h_{\alpha j}(x) - P^\alpha_{kj}(x) P^h_{i\alpha}(x). \tag{17}$$

Symmetrizing the latter we will get

$$\bar{\nabla}_k P^h_{ij} + \bar{\nabla}_j P^h_{ik} - \nabla_k P^h_{ij} - \nabla_j P^h_{ik} = -2 P^\alpha_{kj} P^h_{i\alpha}. \tag{18}$$

Alternating —

$$\bar{\nabla}_k P^h_{ij} - \bar{\nabla}_j P^h_{ik} - \nabla_k P^h_{ij} + \nabla_j P^h_{ik} = -2(P^h_{k\alpha} P^\alpha_{ij} - P^\alpha_{ki} P^h_{\alpha j}). \tag{19}$$

The law of change of curvature tensor that is defined as:

$$R^h_{.\,ijk} = \partial_j \Gamma^h_{ik} + \Gamma^\alpha_{ik} \Gamma^h_{j\alpha} - \partial_k \Gamma^h_{ij} - \Gamma^\alpha_{ij} \Gamma^h_{k\alpha}, \tag{20}$$

in the course of the mapping of space A_n onto $\bar{A}_n$ can be written as follows:

$$\bar{R}^h_{.\,ijk} = R^h_{.\,ijk} + \nabla_k P^h_{ji} - \nabla_j P^h_{ki} + P^h_{\alpha k} P^\alpha_{ji} - P^h_{\alpha j} P^\alpha_{ki} \tag{21}$$

or taking into account (19),

$$\bar{R}^h_{.\,ijk} = R^h_{.\,ijk} + \frac{1}{2}(\nabla_k P^h_{ji} - \nabla_j P^h_{ki} + \bar{\nabla}_k P^h_{ji} - \bar{\nabla}_j P^h_{ki}). \tag{22}$$

Thus, the theorem is proved.

Theorem 2. Proof

From equation (5), taking into account (2), we will obtain

$$\overset{\lambda}{\Gamma}{}^h_{ij} = \frac{\Gamma^h_{ij} + \lambda \bar{\Gamma}^h_{ij}}{1 + \lambda}. \tag{23}$$

For the deformation tensor the latter permits to write down:

$$\begin{gathered}\overset{\lambda}{\nabla}_k P^h_{ij}(x) - \nabla_k P^h_{ij}(x) = \\ = \frac{\lambda^2}{(1+\lambda)^2}\left(P^h_{k\alpha}(x) P^\alpha_{ij}(x) - P^\alpha_{ki}(x) P^h_{\alpha j}(x) - P^\alpha_{kj}(x) P^h_{i\alpha}(x)\right)\end{gathered} \tag{24}$$

and for Riemannian tensor — (6).

So the statement is proved.

Theorem 3. Proof

The reliability of theorem can easily be checked applying the definition of a medium connection and previous theorems.

Theorem 4. Proof

Weyl tensor W^h_{ijk} is defined as follows

$$W^h_{ijk} \stackrel{def}{=} R^h_{ijk} - \frac{1}{n-1}\left(\delta^h_k R_{ij} - \delta^h_j R_{ik}\right) +$$
$$+\frac{1}{n+1}\left(\delta^h_i R_{[jk]} - \frac{1}{n-1}\left(\delta^h_k R_{[ji]} - \delta^h_j R_{[ki]}\right)\right). \quad (25)$$

For equiaffine spaces the latter equation looks as follows

$$W^h_{ijk} \stackrel{def}{=} R^h_{ijk} - \frac{1}{n-1}(\delta^h_k R_{ij} - \delta^h_j R_{ik}). \quad (26)$$

In respect to the mapping Weyl tensors A_n and $\bar{A}_n$ are connected by

$$\bar{W}^h_{ijk} = W^h_{ijk} + \overset{c}{\nabla}_k P^h_{ji} - \overset{c}{\nabla}_j P^h_{ki} -$$
$$-\frac{1}{n-1}\left(\delta^h_k(\overset{c}{\nabla}_\alpha P^\alpha_{ji} - \overset{c}{\nabla}_j P^\alpha_{\alpha i}) - \delta^h_j(\overset{c}{\nabla}_\alpha P^\alpha_{ki} - \overset{c}{\nabla}_k P^\alpha_{\alpha i})\right) +$$
$$+\frac{1}{n+1}(\delta^h_i(\overset{c}{\nabla}_j P^\alpha_{\alpha k} - \overset{c}{\nabla}_k P^\alpha_{\alpha j}) - \quad (27)$$
$$-\frac{1}{n-1}(\delta^h_k(\overset{c}{\nabla}_j P^\alpha_{\alpha i} - \overset{c}{\nabla}_i P^\alpha_{\alpha j}) - \delta^h_j(\overset{c}{\nabla}_k P^\alpha_{\alpha i} - \overset{c}{\nabla}_i P^\alpha_{\alpha k}))),$$

or for equiaffine spaces

$$\bar{W}^h_{ijk} = W^h_{ijk} + \overset{c}{\nabla}_k P^h_{ji} - \overset{c}{\nabla}_j P^h_{ki} -$$
$$-\frac{1}{n-1}\left(\delta^h_k(\overset{c}{\nabla}_\alpha P^\alpha_{ji} - \overset{c}{\nabla}_j P^\alpha_{\alpha i}) - \delta^h_j(\overset{c}{\nabla}_\alpha P^\alpha_{ki} - \overset{c}{\nabla}_k P^\alpha_{\alpha i})\right). \quad (28)$$

Thus, the theorem is proved.

References

[1] BOLSINOV A., KIOSAK, V. A., MATVEEV, V. S.: *Fubini Theorem for pseudo-Riemannian metric*,Journal of the London Mathematical Society., v. 80, 2, 2009, pp. 341 - 356.

[2] KIOSAK, V., HINTERLEITNER, I.:*φ(Ric)–Vector Fields in Riemannian Spaces*, Archivum-mathematicum, Brno, v. 44, 2008, pp. 385 - 390.

[3] KIOSAK, V. A., MATVEEV, V. S.: *Proof of the Projective Lichnerowicz Conjecture for Pseudo-Riemannian Metrics with Degree of Mobility Greater than Two*, Communications in Mathematical Physics, Springer, v.297, 2010, pp. 401 - 426.

[4] MATVEEV, V. S., KIOSAK, V. A., FEDOROVA, A., ROSEMANN S.: *The only closed Kahler manifold with degree of mobility >2 is (CP(n), g-Fubini-Study*, Proc. London Math. Soc., V.105, 1, 2012, pp. 153 - 188.

[5] MIKES, J., KIOSAK, V., VANZUROVA, A.:*Geodesic mappings of manifolds with affine connecti*, Palacky University, Olomouc, 2008, 220 p.

[6] NORDEN, A. P.: *Spaces of affine connection*, Nauka, Moskow, 1976, 432 p.

[7] RODIONOV, E. D., SLAVSKI, V. V., CHIBRIKOVA, L. N.:*Locally conformally homogeneous pseudo-Riemannian spaces*, Siberian Advances in Mathematics, v.17, 3, 2007, pp. 186 - 212.

[8] YANO, K.:*The theory of Lie derivatives and its applications*, North-Holland publishing Co., Amsterdam, Groningen, New-York, 1957.

Current address

Kiosak Volodymyr, doc. RNDr., CSc.
Institute of Engineering
Odessa State Academy of Civil Engineering and Architecture
Didrihsona st., 4, Odessa, 65029, Ukraine
E-mail: kiosakv@ukr.net

Lesechko Olexandr, doc. RNDr., CSc.
Institute of Engineering
Odessa State Academy of Civil Engineering and Architecture
Didrihsona st., 4, Odessa, 65029, Ukraine
E-mail: a.lesechko@ukr.net

Savchenko Olexandr, prof. RNDr., DrSc.
Economic Departament
Kherson State Agricultural University
Sretenskaya st., 23, Kherson, 73006, Ukraine
E-mail: savchenko.o.g@ukr.net

Slovak University of Technology in Bratislava
Faculty of Mechanical Engineering
17th Conference on Applied Mathematics **APLIMAT 2018**

Proceedings

VISUALIZATION OF MATHEMATICAL CONCEPTS IN TEACHING LINEAR AND VECTOR ALGEBRA TO FIRST-YEAR ENGINEERING STUDENTS

KOLISKINA Valentina (LV), ILTINA Marija (LV)

Abstract. A method based on the use of dynamical features of PowerPoint to teach linear and vector algebra to first-year students at the Faculty of Mechanical Engineering (FME) at the Riga Technical University (RTU) is proposed in the paper. Power Point has built-in animation features which can be quite helpful for students. This approach is used to explain such topics as matrix multiplication, computation of determinants, vector coordinates in three-dimensional space. The use of dynamical features of PowerPoint is positively evaluated by students.

Keywords: visualization technique, PowerPoint, higher mathematics

Mathematics Subject Classification: Primary 97B40, 97D40; Secondary 97U70.

1 Introduction

Teaching mathematics to first-year engineering students is a challenging task nowadays. In the present paper some approaches to teaching linear and vector algebra to first-year students at the Faculty of Mechanical Engineering (FME) at the Riga Technical University (RTU) are discussed. In order to assess the knowledge of high school mathematics all first-year students of RTU complete a test during the first tutorial session at RTU. The test is given during the last five years. The results [1] show a rather poor knowledge of mathematics among many first-year students (note that only a minimal passing grade in mathematics is required in order to be admitted to RTU). Starting from the year 2016, RTU is even forced to offer an additional (two-credit point) course on basics of high school mathematics for those students who have low or failing grades on the test.

In addition, only two or three semesters (depending on students' specialization) are devoted to linear algebra, vector calculus, analytic geometry and calculus at the FME of RTU. Thus, the challenge for an instructor is to deliver the material required for other engineering courses in a relatively short time under the assumption that many students do not have a strong background in mathematics.

In the present paper, the authors share their experience in teaching first year students at the FME of RTU. The major emphasis of the paper is given to visualization of mathematical concepts (using software) in explaining such topics as linear algebra and vector calculus. Active use of animation features of PowerPoint allows the instructor to save time (no need to draw figures of the board). In addition, students can work at their own pace at home to analyze the material covered during the lectures since the slides are available on the Moodle platform (called ORTUS at RTU). Moreover, the instructor can control the speed at which the drawing appear on the screen during the lecture and give comments while students are following the presentation. Lecture notes developed by the instructor are also available on ORTUS so that students can use the combination of the learning tools to understand the material better.

There are different approaches to visualization in the literature. Sometimes professors use hand-made visual aids to explain the concepts. Two pieces of plexiglas with wooden dowels are used in [2] to visualize two planes with normal vectors. Modeling clay is used in [2] to clarify the concept of a function of two variables. Development of computer graphics generated interest in visualization tools for teaching and learning mathematics. One of the first sources in this area is the collection of papers addressing the role of visualization in three-dimensional geometry, calculus and other fields [3]. Nowadays the use of special software such as Autograph [4] or Geogebra [5] becomes quite popular. There are examples where special software is developed by professors (see [6]). Video presentations are also gaining popularity [7]. Applications of special dynamical software in teaching and learning mathematics are discussed in the recent book [8]. In the present paper the use of PowerPoint dynamical features is proposed to visualize some concepts in linear and vector algebra which are difficult to explain to students (especially if their knowledge of high school mathematics is quite poor).

2 Organization of the course

There are no separate courses on linear algebra, analytic geometry, vector calculus and calculus at RTU. All these topics are taught in one course "Higher mathematics". The first semester covers linear algebra, analytic geometry, vector algebra and calculus of a function of one and two variables. This is a five-credit point course with 3x45 min. of lectures and 3x45 min. of tutorials per week. The lectures are given to large classes (typically, to 100 students or more), but the tutorial sessions are given to smaller groups (about 25-30 students in each group).

The Department of Engineering Mathematics (DEM) of RTU uses the following grading system for many years: 50% of the total grade is based on the midterm and final exams, 40% - on tests given during the semester and 10% - on homework assignments. There are 7 tests (given during tutorial sessions) and 7 homework assignments. Two homework assignments are performed in the form of online tests (based on Moodle system), two homework assignments are graded by the instructor, and the remaining three homework assignments are partially performed in the form of online tests and partially are graded by the instructor. After each test students have a chance to retake it once. About 25% of all students use this option. The marking is done by the instructor and requires a lot of time. The DEM is a service department (there are no graduate students at the department who can help with marking). Assuming that the average time needed to grade one homework assignment (or a test) is 5 minutes per student (this is quite an optimistic estimate) the instructor devotes about 8 hours per week to grading only (this estimate is based on a class of 100 students). However, instructors have to do it, otherwise the results would be quite poor (a lot of students would fail the course). Thus, satisfactory results (in terms of gained knowledge) can be achieved only at the increase of the teaching load for the instructor. In summary, a closer collaboration between high school teachers and university professors is required in order to reduce dropout rate of first year students and raise the level of mathematical culture among the students who passed the course.

3 The use of dynamical features of PowerPoint

As the results of the survey among students have shown (see the next section) many first-year students of the FME at RTU use lecture notes and presentation slides as the primary sources of their gained knowledge in higher mathematics course. Knowing this trend, the instructors decided to use a combination of PowerPoint presentations and "classical blackboard" teaching to deliver lectures. PowerPoint has built-in animation features which can be quite helpful for students.

Let us consider how some topics covered in the first semester can be delivered with the help of PowerPoint. The first topic discussed below is matrix multiplication. The rule of matrix multiplication is dynamically explained as shown in Fig. 1. At first, the definition of matrix multiplication is presented (Fig. 1(a)). It is emphasized at the next step that the product exists if and only if the number of rows of the first matrix is equal to the number of columns of the second matrix (Fig. 1(b)). Next, a simple example is shown. The elements of two matrices A and B are displayed (Fig. 1(c)). It is checked that the product exists (Fig. 1(d)). Then the arrows appear on the screen showing the rows of matrix A and columns of matrix B (Fig. 1(e)). In addition, the 2x2 matrix C is also shown. Then the first row of A and second column of B is encircled with red and blue colors (Fig. 1(f)). Also, the product of Row 1 of A and Column 1 of B is explicitly shown. Similar procedure is repeated for other rows of A and columns of B. Note that all the information displayed on Figs. 1(a)-(g) appears on one slide when the example is solved. The instructor carefully measures the amount of time needed to explain a certain step in the calculation. The next feature on the screen appears only after time delay specified by the instructor. Thus, the process of matrix multiplication is visualized by students and allowed them to understand the concept better.

Another example of visualization is related to computation of determinants. First, the determinant of order three using Sarrus' rule is computed by visualization: the terms taken with the "+" sign are shown first and the terms taken with the "-" sign are shown later. Computation of determinants of higher order is also illustrated with the help of visualization. Similar visualization technique is used in the course to represent a system of linear algebraic equations in matrix form.

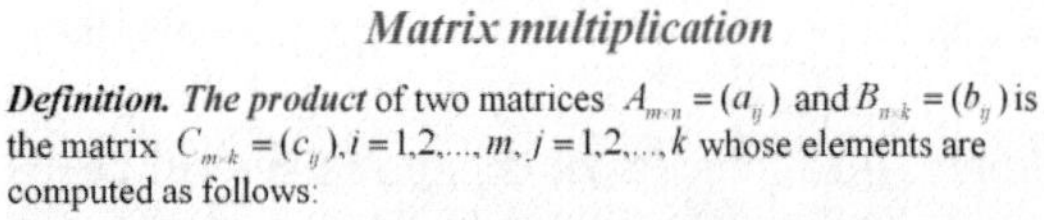

Fig. 1(a). The initial configuration.

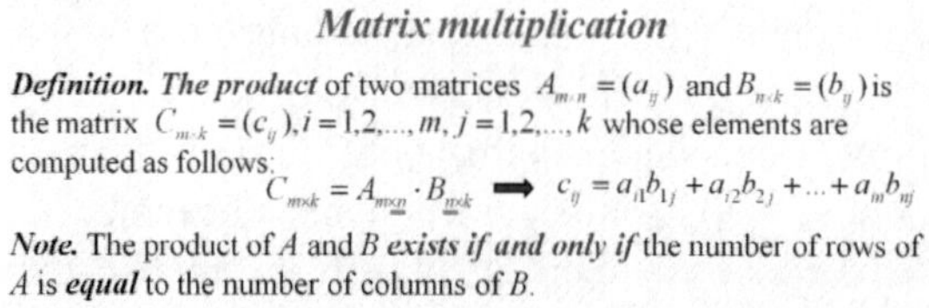

Fig. 1(b). The next step.

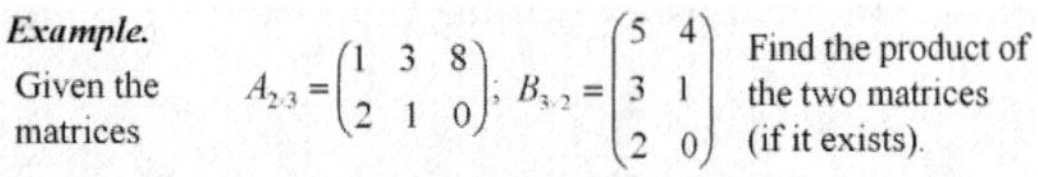

Fig. 1(c). Example.

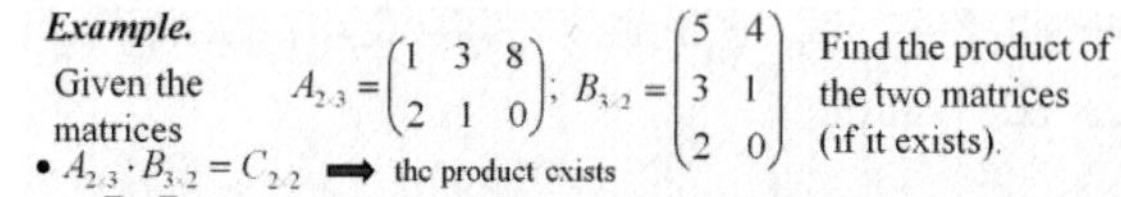

Fig. 1 (d). Check whether the product exists.

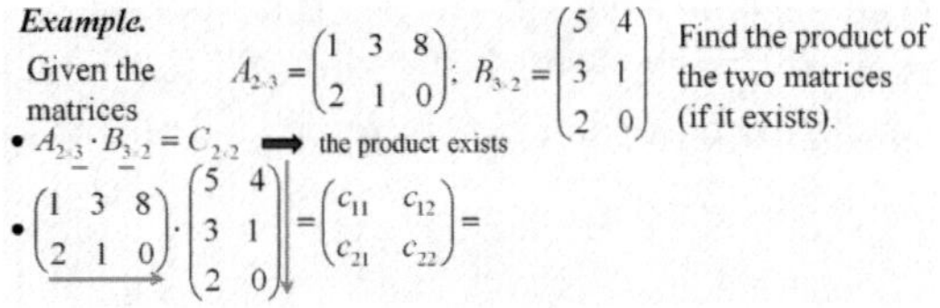

Fig. 1(e). Visual explanation of the rule of matrix multiplication.

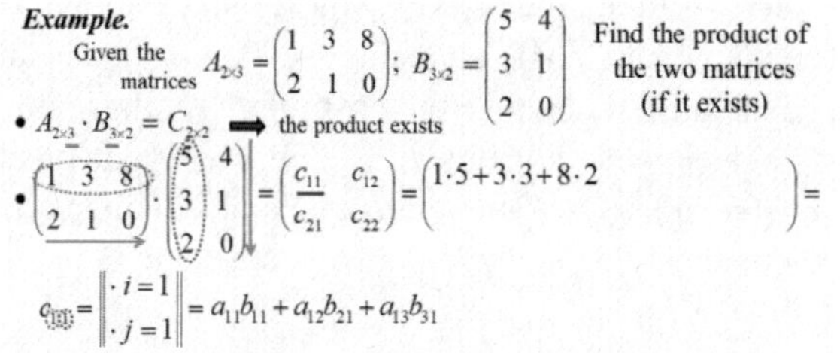

Fig. 1(f). Calculation of $\mathbf{c_{11}}$.

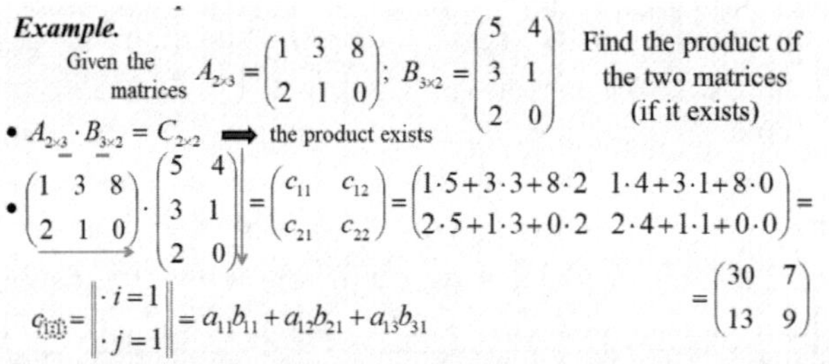

Fig. 1(g). Calculation of all elements of C.

Visualization technique is also actively used to explain elements of vector algebra (to clarify the concepts of vector coordinates and an arbitrary vector expansion in terms of base unit vectors). It usually takes time to draw all figures on board in case of "classical blackboard" teaching. The concept of orthogonal projections of a vector is explained as follows (using PowerPoint). The coordinate system with three mutually orthogonal unit vectors is shown first (Fig. 2(a)).

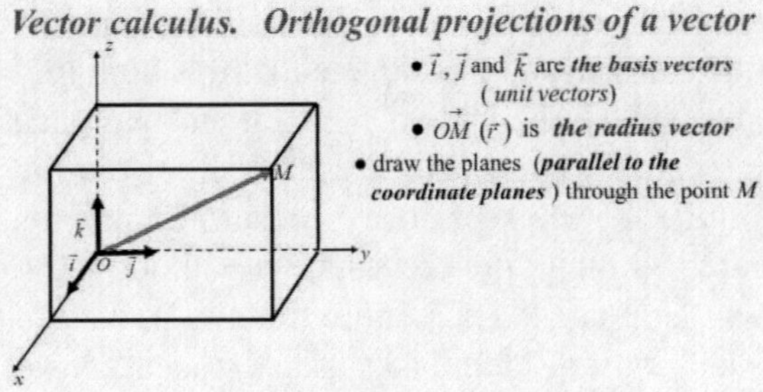

Fig. 2(a). The initial configuration.

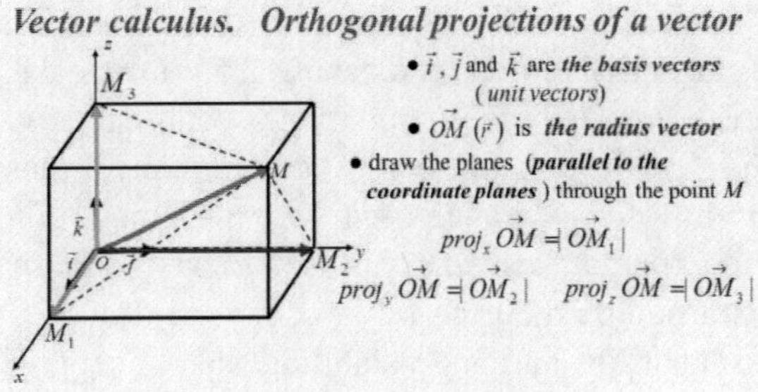

Fig. 2(b). The parallelepiped.

Radius vector is drawn next (Fig. 2(b)). The parallelepiped is constructed as follows. Three mutually perpendicular planes that are parallel to the coordinate planes are drawn through the point M (Fig. 2(c)). The projections of the given vector are shown in different colors (Fig. 2(d)).

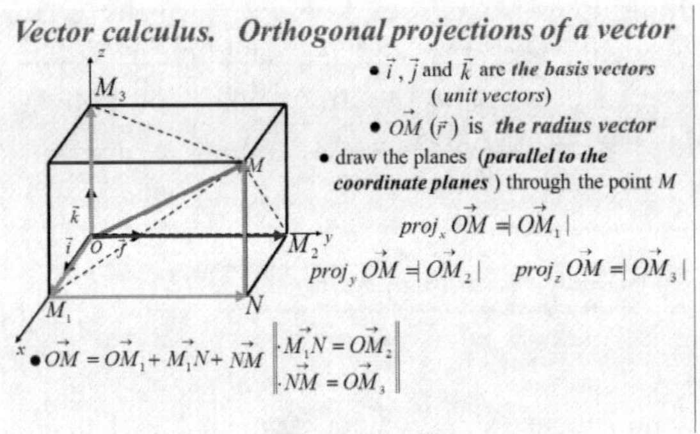

Fig. 2(c). The orthogonal projections.

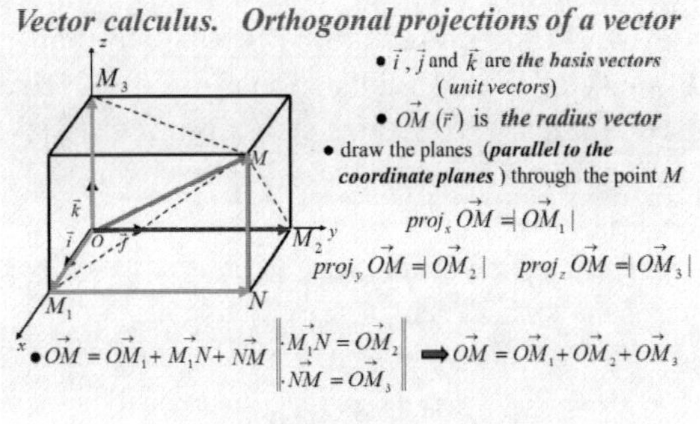

Fig. 2(d). Radius vector expansion.

The rule of vector addition is illustrated next (Fig. 2(e)). The next figure shows how to represent the given vector in the form of the sum of three mutually perpendicular vectors (Fig. 2(f)).

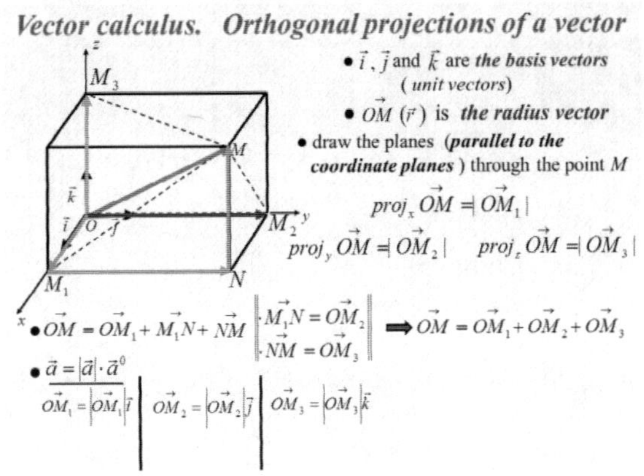

Fig. 2(e). Definition of orthogonal components.

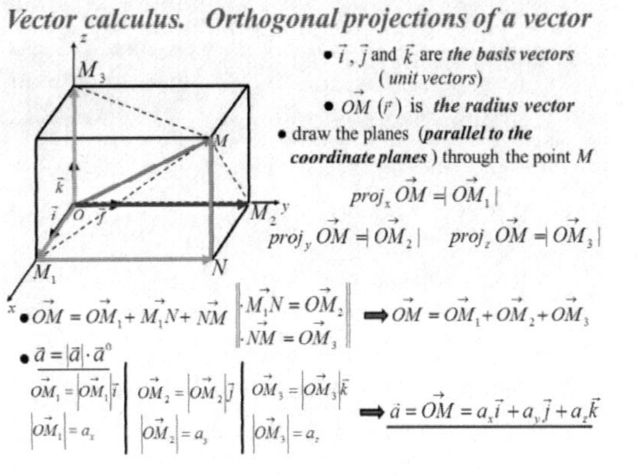

Fig. 2(f). The final step.

There are several advantages of using visualization techniques during lectures. First, students do not have to copy a drawing from the board after the instructor completed it (this certainly saves time for the instructor). Second, students can analyze the slides later at their own pace and go through the derivation as many times as needed. Third, the instructor controls the timing of appearance of different elements of a figure on the screen (as a result, comments and explanations are given to students while they are following the presentation without being disturbed by the necessity to copy the drawing from the board). Thus, more time is devoted to explanations and understanding the concept. This is very important since many students have rather poor background in math and they need more time to absorb all the aspects of the concept.

4 Results of students' survey

A survey is conducted among first-year students of the FME at RTU in the fall semester of 2016. The total number of participants was 62. The goal of the survey was the following: assess students' opinion on how mathematics is taught (for example, do lectures in the form of PowerPoint presentations are better than "classical" lectures with a blackboard?). The summary of the results is shown in Fig. 4 (five-point Likert scale ranging from "Completely agree" (label 1) to "Completely disagree (label 5) is used in the survey).

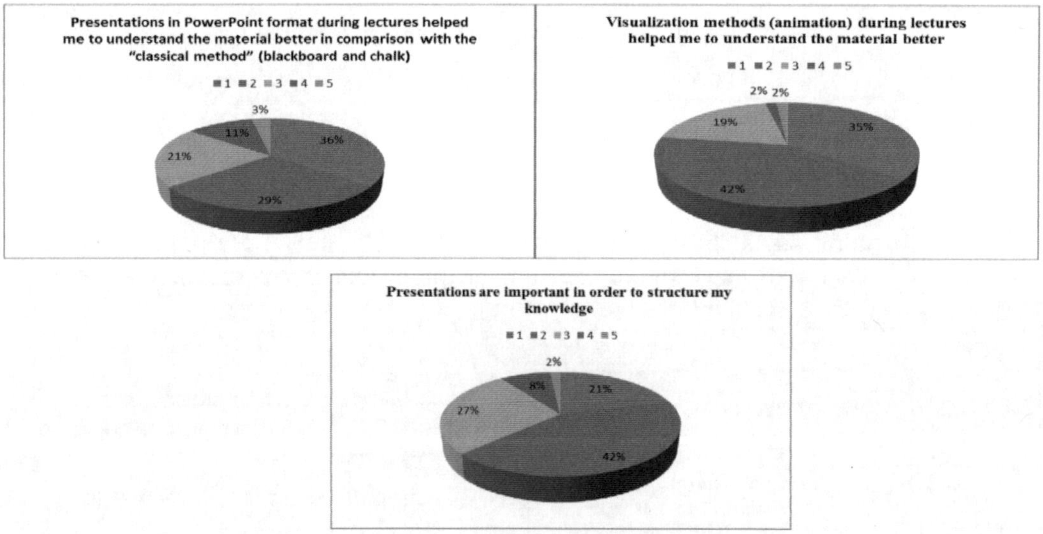

Fig.3. The results of the survey.

As can be seen from the results of the survey, students' responses to PowerPoint presentations and animations during lectures were quite positive.

5 Conclusions

1. A method based on animation features in PowerPoint is proposed in the paper to teach higher mathematics course to first-year students at the Faculty of Mechanical Engineering of RTU.
2. The use of dynamical features of PowerPoint during lectures on linear and vector algebra is positively evaluated by students.
3. The lecturer has a flexibility to control the time delay between appearances of different pieces of information on the slide. .
4. The author is currently developing the approach based on PowerPoint presentations to teach calculus of one variable and calculus of several variables.

References

[1] CERNJAEVA, S., VOLODKO, I. Improvement of teaching methodology of mathematics for students and pupils using the MOOC platform, In *Proceedings of International Conference "Engineering for Rural Development"*, Jelgava, Latvia, 2016, ISSN 16913043, pp. 1286-1290.

[2] BAILEY, D.E., KOBYLSKI, G. Interactive teaching ads for multivariable calculus. *The College Mathematics Journal*, vol. 32, 2001, pp. 283 – 287.

[3] ZIMMERMAN, W. *Visualization in teaching and learning mathematics: a project (MAA notes).* Mathematical Association of America, 1991.

[4] Software Autograph. http://www.autograph-maths.com

[5] Software Geogebra. http://www.geogebra.org

[6] HAUSNECHT, A.O., KOWALCZYK, R.E. Seeing is believing – visualizing calculus. In *Proceedings of the 12th Annual International Conference on Technology in Collegiate Mathematics*, Boston, USA, 1999, pp. 102 – 106.

[7] CHAMBEL, T., SANTOS, L., NAPOLES, S., RODRIGUES, J.F., APOSTOL, T. Visualization in learning mathematics with hypervideo. Encontro Nacional de Visualização CientÍfica, Centro Multimeios, Espinho, Setembro. http://cmaf.fc.ul.pt/arquivo/smwp%20docs/Pi_envc2005_paper_ACNRS.pdf

[8] HABRE, S. *Enhancing mathematics understanding through visualization: the role of dynamical software.* IGI Global, 2013.

Current address

Koliskina Valentina, Doc., Dr. Math.
Institute of Applied Mathematics, Department of Engineering Mathematics
Riga Technical University
Daugavgrivas street 2, LV 1007, Riga, Latvia
E-mail: v.koliskina@gmail.com

Iltina Marija, Asoc. Prof., Dr.Sc.Eng.
Institute of Applied Mathematics, Department of Engineering Mathematics
Riga Technical University
Daugavgrivas street 2, LV 1007, Riga, Latvia
E-mail: marijai@inbox.lv

Slovak University of Technology in Bratislava
Faculty of Mechanical Engineering
17th Conference on Applied Mathematics **APLIMAT 2018**

Proceedings

THE LEAVE-ONE-OUT MAXIMUM LIKELIHOOD METHOD FOR THE PRIESTLEY-CHAO ESTIMATOR OF CONDITIONAL DENSITY

KONEČNÁ Kateřina (CZ)

Abstract. The contribution is focused on a kernel estimation of conditional density. Kernel smoothing is still popular non-parametric method, in theory as well as in practice. The Priestley-Chao estimator of conditional density is introduced and the statistical properties of the estimator are given. The smoothing parameters called bandwidths play a significant role in kernel smoothing. This is the reason for suggesting the methods for their estimation. The typical approach - the cross-validation method - is supplemented with the leave-one-out maximum log-likelihood method. The performance of the suggested methods is compared via a simulation study and an application on a real data set.

Keywords: kernel smoothing, conditional density, bandwidths, Priestley-Chao estimator, leave-one-out maximum likelihood method, cross-validation method

Mathematics subject classification: Primary 62G07; Secondary 62G20, 62G08

1 Introduction

A conditional density provides a comprehensive information about the data set. It can be regarded as a generalization of regression: while regression models only the conditional expectation, conditional density models also an uncertainty. Thus, the conditional density estimator provides except the estimation of the distribution in fixed points of the independent variable also the estimation of a regression function.

The estimators of the conditional density depend on the smoothing parameters called bandwidths. The bandwidths control the smoothness in the both directions - in the direction of the independent and dependent variable. This is the reason why so much importance is given to their detection. The optimal values of the smoothing parameters generally depend on the true marginal and conditional density, that is why they can be used only for the simulated data. In the case of the real data sets, their values need to be estimated by any data-driven method, there are several methods for bandwidths selection in the literature.

Some methods proceed from the methods suggested for kernel density estimation or kernel regression, because the conditional density estimator is a combination of the density estimator and the estimator of a regression function. Thus, the bandwidths can be estimated separately - at first, by methods used for the bandwidth estimation of kernel density, for example the cross-validation method ([4]), the reference rule method ([17], [18]) or the maximum smoothing principle ([20], [19]) could be used. Further, the methods used in kernel regression can be applied for the estimation of the smoothing parameter h_x, for example the cross-validation method ([5]) or the method of penalizing functions (see [15], [10]).

The other methods for bandwidth selection are usually a generalization of the methods used in kernel density estimation and/or kernel regression. The most popular method is the cross-validation method, introduced by [3], which is based on a minimization of the global measure of the quality of the estimate. An iterative method is based on the minimization of the proper relation between the Asymptotic Integrated Square Bias (AISB) and Asymptotic Integrated Variance (AIV). The method is described in [12] and it is inspired by the iterative method for kernel density estimations (see [6] and [7]) and for kernel regression ([9]).

The expressions for the bandwidths estimations are given by a reference rule method, which assume uniform or normal marginal density and normal conditional density with linear mean and linear variance. This method was proposed by Bashtannyk and Hyndman in [1]. Further, a bootstrap method (see [1], [3]) and a method of penalizing functions (for detailed information see [1]) could be mentioned.

All the mentioned methods were derived for the widest used estimator of conditional density - for the Nadaraya-Watson estimator. In this contribution, the Priestley-Chao estimator of conditional density is focused on, the proposed methods - the cross-validation methods and the leave-one-out maximum log-likelihood method - will be derived for this estimator.

2 The Priestley-Chao estimator of conditional density and its statistical properties

In this section, the construction of the estimator is focused on and the statistical properties of the estimator are given. As the most of the methods for bandwidth detection is based on the minimization of a global measure of the quality of the estimator, the derivation of the statistical properties is very important.

In kernel smoothing generally, the elemental building block is the kernel function.

Definition 2.1 *[21] Let K be a real valued function satisfying:*

1. $K \in \mathrm{Lip}[-1,1]$, *i.e.* $|K(x) - K(y)| \leq L|x-y|$, $\forall x, y \in [-1,1]$, $L > 0$,
2. $\mathrm{supp}(K) = [-1,1]$,
3. *moment conditions:*

$$\int_{-1}^{1} K(x)\,\mathrm{d}x = 1, \quad \int_{-1}^{1} xK(x)\,\mathrm{d}x = 0, \quad \int_{-1}^{1} x^2 K(x)\,\mathrm{d}x = \beta_2(K) \neq 0.$$

Such a function K is called a kernel of order 2.

There are several examples of the kernel function - the Epanechnikov kernel, the uniform, the quartic, the triangular kernel, etc. Probably the best known kernel function is the Gaussian kernel, represented by the density of the random variable with standardized normal distribution. The disadvantage of this kernel is in its unconstrained support. On the other hand, this kernel is used in practice very often, especially because of its simple manipulation and computational aspect.

The estimation of conditional density with one independent variable is focused on, this approach can be generalized for multidimensional independent vector. In the most cases, the independent variable is considered to be a random variable X with a marginal density $g(x)$. Then, we talk about a model with random design.

Let (X, Y) be a random vector and (X_1, Y_1), (X_2, Y_2), $\ldots$, (X_n, Y_n) its observations. The kernel estimation of the distribution of a random variable Y given by a random variable X characterized by a fixed point x is generally given by the expression

$$\hat{f}\,(y|x) = \sum_{i=1}^{n} w_i(x) K_{h_y}\,(y - Y_i)\,,$$

where $w_i(x)$ is a weight function at the point x. In kernel conditional density estimates, there are several types of the estimates corresponding to the proper weight function.

Probably the widest used and best known estimator is the Nadaraya-Watson estimator of conditional density ([16]). Its name comes from the kernel estimator of the regression function - the Nadaraya-Watson estimator - due to the similarity between the weight functions of both estimators. The weight function takes the form

$$w_i^{NW}(x) = \frac{K_{h_x}\,(x - X_i)}{\sum\limits_{i=1}^{n} K_{h_x}\,(x - X_i)}.$$

The local linear estimator is another widely used estimator. The estimator is suggested in [2], its weight function of the estimator is expressed as

$$w_i^{LL}(x) = \frac{K_{h_x}\,(x - X_i)\,(\hat{s}_2(x) - (x - X_i)\,\hat{s}_1(x))}{\hat{s}_0(x)\hat{s}_2(x) - \hat{s}_1^2(x)},$$

$\hat{s}_r(x) = \frac{1}{n}\sum\limits_{i=1}^{n} (x - X_i)^r\, K_{h_x}\,(x - X_i)$ is the auxiliary function. The local linear estimator is more difficult, especially as statistical properties and the methods for bandwidths selection are concerned. On the other hand, the estimator is distinguished by better bias property and better boundary effects.

These two estimators mostly suppose a non-uniform distributed random design variable X. In practice, there are many situations when the conditional distribution has to be estimated, but the conditioning variable X is made by equally spaced values of X. Analysis of time series can be such an example in which the variable X is comprised by the time occasions. This is the reason for proposing a new type of estimator - the Priestley-Chao estimator of conditional density.

The Priestley-Chao estimator was originally suggested for the kernel estimation of a regression function. The original paper by Priestley and Chao ([14]) is followed and the generalization of the original estimator to the estimator of conditional density is made.

Suppose a fixed design, i.e. suppose the equally spaced design variable X with fixed values x_i, $i = 1,\ 2,\ \ldots,\ n$ and $\delta = x_{i+1} - x_i$. Although the fixed design in the form $x_i = \frac{i}{n}$, $i = 1,\ 2,\ \ldots,\ n$ is usually assumed, the design points can not be restricted only on the interval $[0, 1]$ but generally on $[a, b]\,,\ a < b$.

The Priestley-Chao estimator of conditional density is defined as

$$\hat{f}_{PC}\,(y|x) = \delta \sum_{i=1}^{n} K_{h_x}\,(x - x_i)\, K_{h_y}\,(y - Y_i)\,. \tag{1}$$

The similarity of the Priestley-Chao estimator of conditional density with the Priestley-Chao estimator of regression function is following. The conditional expectation of the estimator (1), denoted by $\widehat{m}_{PC}(x)$, takes the form

$$\widehat{m}_{PC}(x) = \delta \sum_{i=1}^{n} K_{h_x}\,(x - x_i)\, Y_i, \tag{2}$$

the estimator (2) is exactly the Priestley-Chao estimator of regression function, introduced in [14].

Now, the statistical properties of the estimator (1) are focused on.

Theorem 2.1 *[11] Let x be a fixed design, Y a random variable with conditional density $f\,(y|x)$ being at least twice continuously differentiable, and K be a kernel function satisfying Definition 2.1. For $h_x \to 0$, $h_y \to 0$ and $nh_xh_y \to \infty$ as $n \to \infty$, asymptotic bias (AB) and asymptotic variance (AV) of the Priestley-Chao estimator are given by the expressions*

$$\mathrm{AB}\left\{\hat{f}_{PC}\,(y|x)\right\} = \frac{1}{2}h_x^2\beta_2(K)\frac{\partial^2 f(y|x)}{\partial x^2} + \frac{1}{2}h_y^2\beta_2(K)\frac{\partial^2 f(y|x)}{\partial y^2}, \tag{3}$$

$$\mathrm{AV}\left\{\hat{f}_{PC}\,(y|x)\right\} = \frac{\delta}{h_xh_y}R^2(K)f(y|x), \tag{4}$$

where $R(K) = \int K^2(u)\,\mathrm{d}u$.

Sketch of the proof. At first, the property

$$\mathrm{E}\left\{\hat{f}_{PC}\,(y|x)\right\} = n\delta\mathrm{E}\left\{K_{h_x}\,(x - x_i)\, K_{h_y}\,(y - Y_i)\right\}$$

along with the symmetry of the kernel function, the Taylor expansion of the conditional density and $\mathcal{O}$-notation are used for deriving the expression (3).

A well known law of the total variance is the fundamental idea for deriving (4). Let U and V be a random variables, then variance of V is stated by

$$\mathrm{var}\,\{V\} = \mathrm{E}\left\{\mathrm{var}_{V|U}\,\{V|U\}\right\} + \mathrm{var}\left\{\mathrm{E}_{V|U}\,\{V|U\}\right\}.$$

Thus, the variance of i-th term of the estimator (1) is derived. Finally, using the stochastically independence of all the terms in (1), the final variance is expressed as the sum of the variances.

The complete proof can be found in [11]. □

For assessing the local measure of the quality of the estimator (1) at the point $[x, y]$, the Asymptotic Mean Square Error (AMSE) should be used. AMSE is defined as a sum of the the Asymptotic Variance and the Asymptotic Square Bias by the expression

$$\mathrm{AMSE}\left\{\hat{f}_{PC}\,(y|x)\right\} = \frac{\delta}{h_xh_y}R^2(K)f(y|x) + \left(\frac{1}{2}h_x^2\beta_2(K)\frac{\partial^2 f(y|x)}{\partial x^2} + \frac{1}{2}h_y^2\beta_2(K)\frac{\partial^2 f(y|x)}{\partial y^2}\right)^2.$$

As it was said earlier, the values of the smoothing parameters play a very important role in kernel smoothing. A lot of methods are based on a minimization of the global measure of the quality of the estimate, that is why this characteristics is needed. The global quality of the estimator (1) is given by the Asymptotic Mean Integrated Square Error (AMISE) in the form

$$\mathrm{AMISE}\left\{\hat{f}_{PC}\left(\cdot|\cdot\right)\right\} = \frac{\delta}{h_x h_y} c_1 + c_2 h_x^4 + c_3 h_y^4 + c_4 h_x^2 h_y^2, \tag{5}$$

where $c_1,\ c_2,\ c_3$ and c_4 are constants defined below by the integrals over the ranges of x and Y:

$$\begin{aligned}
c_1 &= \int R^2(K)\,\mathrm{d}x, \\
c_2 &= \frac{1}{4}\beta_2^2(K) \iint \left(\frac{\partial^2 f(y|x)}{\partial x^2}\right)^2 \mathrm{d}x\,\mathrm{d}y, \\
c_3 &= \frac{1}{4}\beta_2^2(K) \iint \left(\frac{\partial^2 f(y|x)}{\partial y^2}\right)^2 \mathrm{d}x\,\mathrm{d}y, \\
c_4 &= \frac{1}{2}\beta_2^2(K) \iint \frac{\partial^2 f(y|x)}{\partial x^2}\frac{\partial^2 f(y|x)}{\partial y^2}\,\mathrm{d}x\,\mathrm{d}y.
\end{aligned}$$

3 Methods for bandwidths detection

The smoothing parameters influence the final estimation of conditional density significantly. While choosing too great values of the bandwidths, the estimate will tend to oversmooth, on the other hand, the estimate will be undersmoothed in the choice of the small values. Our aim is to estimate the proper values of the smoothing parameters to get a satisfactory estimate.

As the values of the smoothing parameters depend on the unknown conditional density, their optimal values of them can be computed only in the case of the simulated data. While having the real data set, a data-driven method has to be used for the bandwidths detection.

At first, the attention is paid to the optimal values of the smoothing parameters. Further, the cross-validation method and the leave-one-out maximum log-likelihood method are described.

3.1 The optimal values of the smoothing parameters

The optimal values of the smoothing parameters can be used in the cases when the true conditional density is known. Their knowledge is used especially while implementing new methods for the bandwidths estimations to determine the suitability of the method.

The optimal values of the bandwidths are such values which minimize a global measure of the quality of the estimator. They are derived by differentiating AMISE with respect to both smoothing parameters, and by solving the system of two non-linear equations given by setting the derivatives to zero.

The optimal values of the smoothing parameters, denoted by h_x^* and h_y^*, are given by

$$\begin{aligned}
h_x^* &= \delta^{1/6} c_1^{1/6} \left(4\left(\frac{c_2^5}{c_3}\right)^{1/4} + 2c_4\left(\frac{c_2}{c_3}\right)^{3/4}\right)^{-1/6} \\
h_y^* &= \left(\frac{c_2}{c_3}\right)^{1/4} h_x^*.
\end{aligned}$$

3.2 The cross-validation method

The cross-validation method is a standard procedure widely used for estimating the smoothing parameters, not only in the kernel estimates of conditional density, but in kernel smoothing generally.

The main idea of the method is based on a minimization of the cross-validation function (6). The cross-validation function is associated with the global quality of the estimator - ISE (the Integrated Square Error) in this case. In this contribution, the leave-one-out cross-validation method is concerned, i.e. we use the estimation at the points (x_i, Y_i) using the points $\{(x_j, Y_j)\,,\ j \neq i\}$.

The leave-one-out cross-validation function is given in the form

$$\mathrm{CV}\,(h_x, h_y) = \delta^2 \sum_i \sum_{j \neq i} h_x h_y K_{h_x\sqrt{2}}\,(x_i - x_j)\, K_{h_y\sqrt{2}}\,(Y_i - Y_j) - 2\delta \sum_i \hat{f}_{PC}\,(Y_i | x_i)\,. \tag{6}$$

The values of the smoothing parameters using the leave-one-out cross-validation method are given by

$$\left(h_x^{CV}, h_y^{CV}\right) = \arg\min_{(h_x, h_y)} \mathrm{CV}\,(h_x, h_y)\,.$$

3.3 The leave-one-out log-likelihood method

The maximum likelihood method is a standard statistical procedure for selecting unknown parameters. In general, having an underlying density is supposed. In this case, the underlying unknown conditional density is substituted by the Priestley-Chao estimator (1), represented by given observations and unknown bandwidths.

This approach was proposed by Leiva-Murillo and Artes-Rodríguez for kernel density estimations (see the paper [13]). In this paper, their idea is followed and extended to conditional density estimations.

Assume the objective function

$$\mathcal{L}\,(h_x, h_y | \boldsymbol{x}, \boldsymbol{Y}) = \prod_{j=1}^{n} \hat{f}_{PC}\,(Y_j | x_j) \tag{7}$$

with the independent and identically distributed observations $(x_i, Y_i)\,,\ i = 1,\ 2,\ \ldots,\ n$ considered as the fixed parameters, and the smoothing parameters h_x, h_y as the unknown parameters.

In the case that all n observation is considered, the maximizing problem (7) has a trivial degenerate solution. In (7), if $i = j$, then

$$K_{h_x}\,(x_j - x_i) = K_{h_x}\,(0) \quad \text{and} \quad K_{h_x}\,(x_j - x_i) = K_{h_x}\,(0)\,.$$

Letting $h_x \to 0$ and $h_y \to 0$, then $\mathcal{L} \to \infty$, i.e. the objective function becomes unbounded and does not have a finite maximum. This is an undesirable property because of estimating the values of bandwidths tending to zero, thus the final estimate is undersmoothed, usually with abundance of information and high variability. This is a motivation for implementation of a modification of the classical maximum-likelihood approach.

The modification of the classical maximum likelihood approach lies in the leaving one observation out - here is the name of the method, the leave-one-out maximum likelihood. Thus, the modified likelihood function is expressed by

$$\mathcal{L}^*\left(h_x, h_y|\boldsymbol{x}, \boldsymbol{Y}\right) = \prod_{\substack{j=1\\ j\neq i}}^{n} \hat{f}_{PC}\left(Y_j|x_j\right). \tag{8}$$

As maximizing (8) is equivalent to maximizing the functional

$$\ell^*\left(h_x, h_y|\boldsymbol{x}, \boldsymbol{Y}\right) = \ln \prod_{\substack{j=1\\ j\neq i}}^{n} \hat{f}_{PC}\left(Y_j|x_j\right) = \sum_{\substack{j=1\\ j\neq i}}^{n} \ln \hat{f}_{PC}\left(Y_j|x_j\right), \tag{9}$$

the maximizing problem (9) will be used rather than (8), especially due to the computational aspects and computational time.

The estimations of the smoothing parameters are given by

$$\left(\hat{h}_x^{\ell^*}, \hat{h}_y^{\ell^*}\right) = \underset{(h_x,h_y)}{\arg\max}\, \ell^*\left(h_x, h_y|\boldsymbol{x}, \boldsymbol{Y}\right).$$

4 Simulation study

In this section, we conduct two simulation studies comparing the results given by the the cross-validation method and by the proposed leave-one-out maximum log-likelihood method. The simulation study involves two models defined as

$$M_1 : Y_i = \mathrm{e}^{x_i} + \varepsilon_i, \quad x_i = \frac{i}{n}, \quad i = 1\ldots, 100, \quad \varepsilon_i \sim N(0, 0.5^2)$$
$$M_2 : Y_i = \sin\left(3\pi x_i^2\right) + \varepsilon_i, \quad x_i = \frac{i}{n}, \quad i = 1\ldots, 100, \quad \varepsilon_i \sim N(1, 1)$$

At first, one hundred observations are generated from each model to apply the Priestley-Chao estimator for conditional density detection. For both simulation studies, an exactly given grid of 100 times 100 points is considered to construct an estimation and to measure the error term. The x grid is formed by the observations x_i, the y grid is formed by the exact equidistant points at the range of Y values.

The methods for bandwidth detection are performed from several points of view, the accuracy of the estimates of the smoothing parameters to the optimal bandwidths is assessed as well as the measure of the quality estimation is focused on. The measure of the quality of the estimate is given by the Integrated Square Error

$$\mathrm{ISE}\left\{\hat{f}_{PC}\left(y|x\right)\right\} = \iint \left\{\hat{f}_{PC}\left(y|x\right) - f\left(y|x\right)\right\}^2 \mathrm{d}x\, \mathrm{d}y.$$

Due to computational aspect, its estimation is used:

$$\widehat{\mathrm{ISE}}\left\{\hat{f}_{PC}\left(y|x\right)\right\} = \frac{\Delta}{n}\sum_{j=1}^{N}\sum_{i=1}^{n}\left(\hat{f}_{PC}\left(y_j|x_i\right) - f\left(y_j|x_i\right)\right)^2,$$

where $\boldsymbol{y} = (y_i, \ldots, y_N)$ is a vector of equally spaced values over the sample space of Y and Δ is the distance between two consecutive values of $\boldsymbol{y}$.

The observations were generated two hundred times in total. The values of the smoothing parameters, the ISE estimation and the computational time were received from every repetition of the simulation. The results of the estimates using the proposed methods are compared with the theoretical values and they are presented by boxplots.

The estimations of the smoothing parameters for the model M_1 are displayed in Fig. 1. The CV method gives more stable results than the LOO-MLln method in the estimation of h_x, on the other hand the median is slightly closer to the optimal value h_x^* (vertical line) while the LOO-MLln method is used. Both methods tend to underestimate the parameter h_x. As the smoothing parameter h_y is concerned, the CV method gives the underestimated values too, the median is located slightly above the optimal value h_y^* while using the LOO-MLln method.

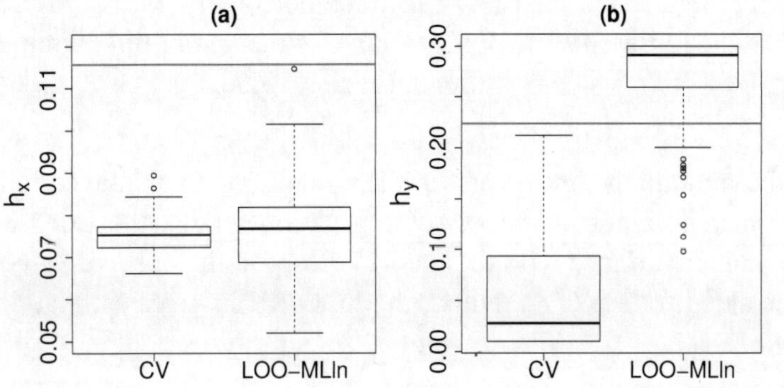

Fig. 1. The model M_1: The estimations of the smoothing parameters (a) h_x and (b) h_y for the CV and the LOO-MLln method. The red vertical line represents the optimal value of the smoothing parameter (a) $h_x^* = 0.116$, (b) $h_y^* = 0.224$.

In the model M_1, the estimations of the ISE values and the computational times are displayed in Fig. 2. The ISE values given by the CV and the LOO-MLln method are compared with the ISE values for the estimations constructed for the optimal values of the smoothing parameters (OPT).

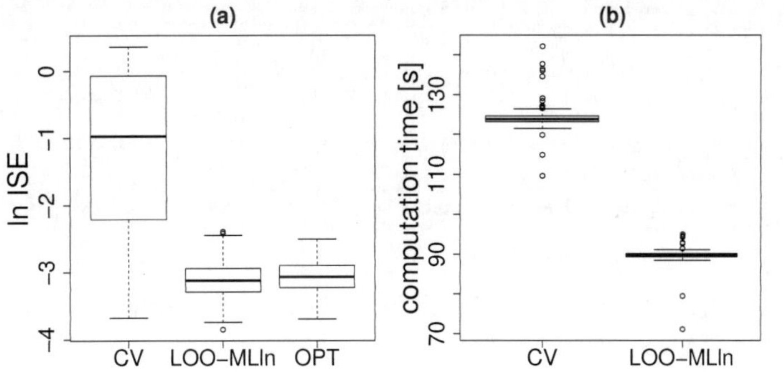

Fig. 2. The model M_1: The estimations of (a) the natural logarithms of the ISE values for the CV, the LOO-MLln method and OPT, (b) computational time.

The LOO-MLln method gives the estimations of the natural logarithms of the ISE error estimation very close to the optimal ISE. Other advantage of the LOO-MLln method lies in its computational difficulty - the computation using the LOO-MLln method takes about three quarters of the CV computational time.

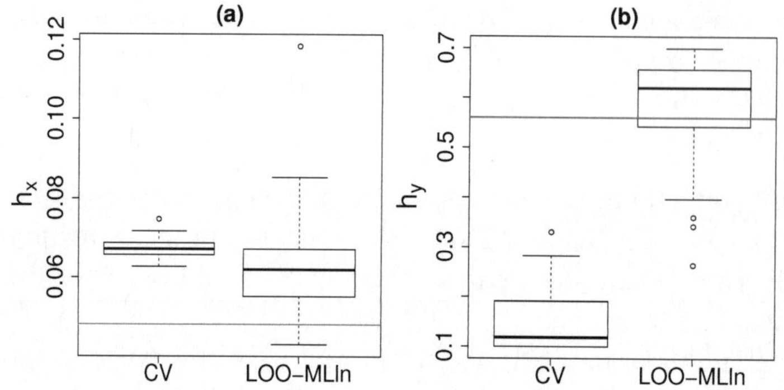

Fig. 3. The model M_2: The estimations of the smoothing parameters (a) h_x and (b) h_y for the CV and the LOO-MLln methods in the model M_2. The red vertical line represents the optimal value of the smoothing parameter (a) $h_x^* = 0.048$, (b) $h_y^* = 0.56$.

The estimations of the smoothing parameters for the model M_2 are displayed in Fig. 3. Both methods give slightly overestimated values of the smoothing parameter h_x, the LOO-MLln estimations are characterized by the higher variance with the median closer to the optimal value h_x^*. The estimations of h_y are very undervalued for the CV method, the LOO-MLln gives estimations well reflecting the optimal value h_y^*.

The estimations of the ISE values and the computational times are displayed for the model M_2 in Fig. 4. As it can be seen in Fig. 4 (a), the LOO-MLln method gives the ISE estimations very close to the optimal ISE values, while the estimations by the CV method are high and variable. The LOO-MLln method is distinguished by the lower computational time.

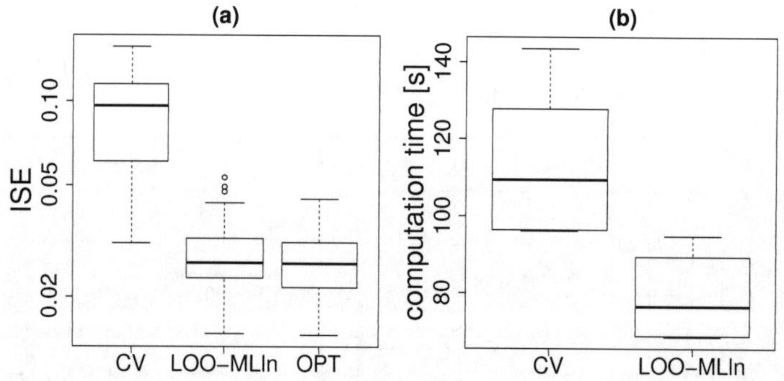

Fig. 4. The model M_2: The estimations of (a) the ISE values for the CV, the LOO-MLln method and OPT, (b) computational time.

5 Application to a temperature data

In this chapter, the comparison among the results given by the proposed leave-one-out maximum log-likelihood method and the cross-validation method is focused on. Both methods are implemented on a real data set provided by Berkeley Earth hosted in Kaggle [8].

The average annual temperatures in the Czech Republic during 1753 – 2012 are focused on. Early data was collected by technicians using mercury thermometers, while in the 1980's, there was a move to electronic thermometers that are said to have a cooling bias.

Our aim is to compare the estimations of the smoothing parameters by the cross-validation method (CV) and the leave-one-out log-likelihood method (LOO-MLln) and their influence on the final estimate of conditional density. Also the computation time and the conditional mean estimation are in our interest.

There are two variables in the data set:

- time factor (year), independent variable,
- the average annual temperature (measured in $^\circ$ C), dependent variable.

The temperature data is illustrated in Fig. 5. The results of the methods are presented in Tab. 1 and in Fig. 6 and Fig. 7.

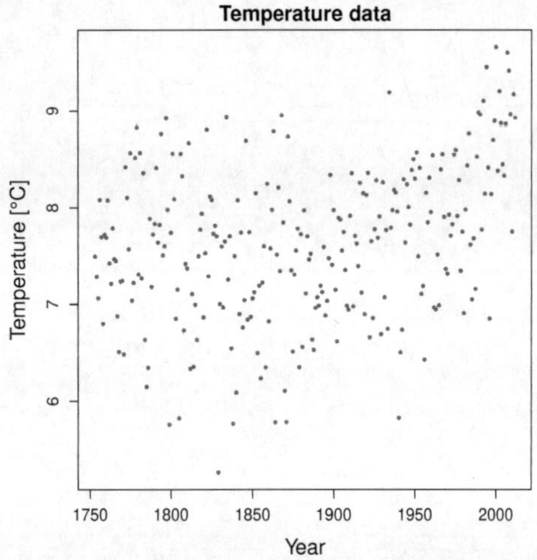

Fig. 5. A scatterplot of the temperature data.

At first, the estimates of the values of the smoothing parameters as well as the computational times were focused on. The results are brought in Tab. 1. Here, the most significant advantage, the computational time, of the LOO-MLln method over the CV method can be seen. The computational time for the LOO-MLln method is almost three times lower than the computational time for the CV method.

method	$\hat{h}_x$	$\hat{h}_y$	computational time $[s]$
CV	8.45	0.00656	1021
LOO-MLln	10.2	0.34	285

Tab. 1. Estimates of the smoothing parameters and computational times for methods used for estimating the temperature data.

Considering the estimates of the smoothing parameters, the differences among their estimations are distinguished by a significant impact on the final visual presentation of the estimations of conditional density. While the estimation using the CV method is characterized by the high-variable estimate

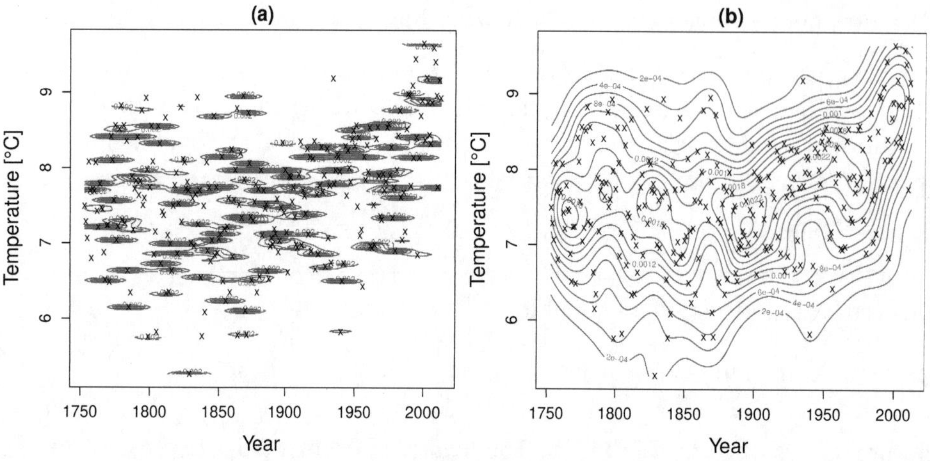

Fig. 6. Contour plots for conditional density estimations for two compared methods: (a) CV, (b) LOO-MLln.

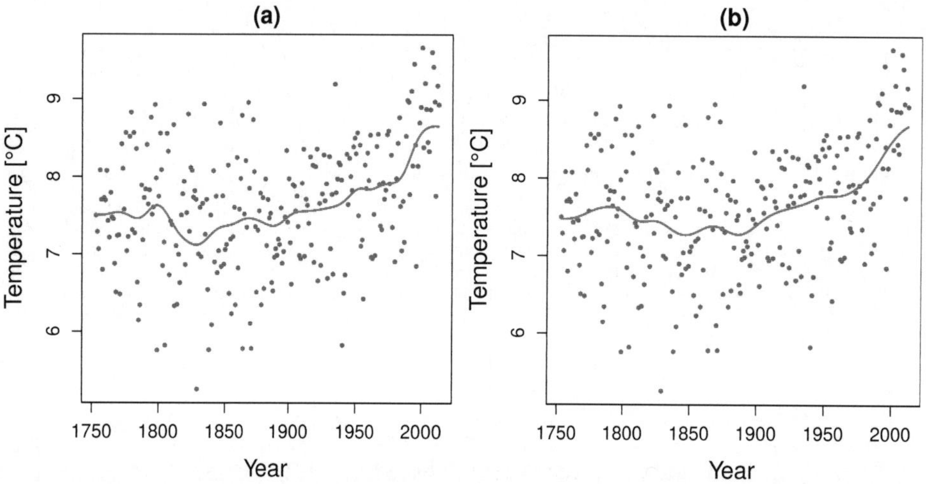

Fig. 7. Estimations of conditional expectation for two compared methods: (a) CV, (b) LOO-MLln.

with many peaks and unrecognizable distribution while time changing, the estimation using the LOO-MLln method is not so much fluctuating, it is much smoother with evident distribution.

The estimates of the conditional mean are displayed for each method in Fig. 7. Despite the very different conditional distribution, not so distinctive differences among the estimated conditional means can be seen. The estimation using the CV method is characterized by lower smoothness.

6 Conclusion

A new type of the kernel estimator of conditional density - the Priestley-Chao estimator - was focused on. Despite the "classical" approach consisted in the random design assumption, the fixed design is assumed in the Priestley-Chao estimator. The statistical properties including the local and the global measures of the quality of the estimate are included. As the smoothing parameters play a significant role in kernel smoothing, the optimal values of the smoothing parameters are supplemented with the data-driven methods. The typical approach, the cross-validation method, is mentioned and a new method, the leave-one-out log-likelihood method, is proposed.

The suggested methods are compared via a simulation study. Two simulation studies are conducted, the emphasis is placed on the comparisons of the estimations of the smoothing parameters to the theoretical values and of the estimations of ISE with its theoretical values. The computational time for both methods is also included.

The results show the unequivocal advantage of the maximum likelihood approach. Except for the lower computational time, the proposed method gives even more precise and stable estimations of the smoothing parameters. The estimations of ISE approach to the ISE values computed for the estimates with the optimal values of the smoothing parameters.

The CV and LOO-MLln are also applied to a real data set. The temperature data in the Czech Republic during 1753 – 2012 is focused on. The results show the differences between the estimations of conditional density. While the cross-validation method results in very undersmoothed estimates with very complicatedly recognizable distribution, the LOO-MLln gives satisfactory results.

The results show that the proposed maximum log-likelihood method can be a reasonable and reliable tool for bandwidth selection.

Acknowledgement

The research was supported by the project 22738 Institutional support for development of research organization (Brno University of Technology).

References

[1] BASHTANNYK, D. M., HYNDMAN, R. J., Bandwidth selection for kernel conditional density estimation. *Computational Statistics & Data Analysis*, vol. 36, no. 3, 2001: pp. 279–298, ISSN 0167-9473.

[2] FAN, J., YAO, Q., TONG, H., Estimation of conditional densities and sensitivity measures in nonlinear dynamical systems. *Biometrika*, vol. 83, no. 1, 1996: pp. 189–206, ISSN 00063444.

[3] FAN, J., YIM, T. H., A crossvalidation method for estimating conditional densities. *Biometrika*, vol. 91, no. 4, 2004: pp. 819–834, ISSN 0006-3444.

[4] HALL, P., Cross-validation in density estimation. *Biometrika*, vol. 69, no. 2, 1982: pp. 383–390, ISSN 0006-3444.

[5] HÄRDLE, W., *Applied nonparametric regression*. Cambridge university press, 1990.

[6] HOROVÁ, I., KOLÁČEK, J., VOPATOVÁ, K., Full bandwidth matrix selectors for gradient kernel density estimate. *Computational Statistics & Data Analysis*, vol. 57, no. 1, 2013: pp. 364–376, ISSN 0167-9473.

[7] HOROVÁ, I., ZELINKA, J., Contribution to the bandwidth choice for kernel density estimates. *Computational Statistics*, vol. 22, no. 1, 2007: pp. 31–47, ISSN 1613-9658.

[8] KAGGLE, *Climate change: Earth surface temperature data*, https://www.kaggle.com/berkeleyearth/datasets, accessed: 2017-09-13.

[9] KOLÁČEK, J., HOROVÁ, I., Iterative bandwidth method for kernel regression. *Journal of Statistics: Advances in Theory and Applications*, vol. 8, 2012: pp. 91–103, ISSN 0975-126.

[10] KOLÁČEK, J., Problems of automatic data-driven bandwidth selectors for nonparametric regression. *Journal of Electrical Engineering*, vol. 53, 2002: pp. 48–51, ISSN 1335-3632.

[11] KONEČNÁ, K., Priestley-chao estimator of conditional density, In *Mathematics, Information Technologies and Applied Sciences 2017, post-conference proceedings of extended versions of selected papers*, Brno: University of Defence, 2017, ISBN 978-80-7582-026-6, pp. 151–163.

[12] KONEČNÁ, K., HOROVÁ, I., Conditional Density Estimations. *International Society for the Advancement of Science and Technology (ISAST)*, Athens, 2014: pp. 15–31, ISBN 978-618-81257-7-3.

[13] LEIVA-MURILLO, J. M., ARTÉS-RODRÍGUEZ, A., Algorithms for maximum-likelihood bandwidth selection in kernel density estimators. *Pattern Recognition Letters*, vol. 33, no. 13, 2012: pp. 1717–1724, ISSN 0167-8655.

[14] PRIESTLEY, M. B., CHAO, M. T., Non-parametric function fitting. *Journal of the Royal Statistical Society. Series B (Methodological)*, vol. 34, no. 3, 1972: pp. 385–392, ISSN 00359246.

[15] RICE, J., Bandwidth choice for nonparametric regression. *The Annals of Statistics*, 1984: pp. 1215–1230.

[16] ROSENBLATT, M., Conditional probability density and regression estimators. *Multivariate analysis II*, vol. 25, 1969: p. 31.

[17] SILVERMAN, B. W., Algorithm as 176: Kernel density estimation using the fast fourier transform. *Journal of the Royal Statistical Society. Series C (Applied Statistics)*, vol. 31, no. 1, 1982: pp. 93–99, ISSN 00359254, 14679876.

[18] SILVERMAN, B. W., *Density estimation for statistics and data analysis*. CRC press, 1986.

[19] TERRELL, G. R., The maximal smoothing principle in density estimation. *Journal of the American Statistical Association*, vol. 85, no. 410, 1990: pp. 470–477, ISSN 01621459.

[20] TERRELL, G. R., SCOTT, D. W., Oversmoothed nonparametric density estimates. *Journal of the American Statistical Association*, vol. 80, no. 389, 1985: pp. 209–214, ISSN 01621459.

[21] WAND, M. P., JONES, M. C., *Kernel smoothing*. Crc Press, 1994, ISBN 9780412552700.

Current address

Konečná Kateřina, Mgr.
Institute of Mathematics and Descriptive Geometry
Faculty of Civil Engineering
Brno University of Technology
Žižkova 17, 602 00 Brno, Czech Republic
E-mail: konecna.k@fce.vutbr.cz

Slovak University of Technology in Bratislava
Faculty of Mechanical Engineering
17th Conference on Applied Mathematics **APLIMAT 2018**
Proceedings

STUDENTS' AUTOEVALUATION IN THE FIRST YEAR OF SCIENCE STUDY FIELDS ESPECIALLY IN MATHEMATICS (COMPLEX ANALYSIS)

KONEČNÁ Petra (CZ), HABIBALLA Hashim (CZ)

Abstract. This paper follows published results of the questionnaires that were done in 2010 and 2011. We continued in questionnaires next three years, however, collected data have not been published yet. We came back to them and decided to analyse them for the years 2011 – 2014 with purpose to check, if and how the observed parameters develop in a long-term horizon.

Keywords: mathematics, education, secondary school, university study

Mathematics Subject Classification: Primary 97A06; Secondary 97R06.

1 Introduction

In years 2009 – 2012, new Framework Educational Programmes (FEP) were implemented into School Educational Programmes (SEP) and schooling in the Czech Republic. Even though FEP describes compulsory and recommended extensional topics for each study field, it gives more freedom to schools in processing these topics into SEP in comparison to unified curriculum. Thus, huge differences in preliminary knowledge and skills occur between students, even at those from the same study field. This fact is visible in disunited knowledge in basic areas of Mathematics in first year students at universities, which influences the success in study. It was also the reason for making a questionnaire whose aim was to check the orientation in basic areas of Mathematics at students at the beginning of their studies and find, if there is any dependence of results on a type of absolved secondary school or if there are differences at students of different study fields.[4], [5]

Currently, thanks to amendment of Higher Educational Act, new study fields are being prepared at all Czech universities. It gives us unique opportunity to create study plans and edit content of subjects to make them correspond as much as possible with the profile of a graduate as well as with preliminary requirements on students. Thus, we decided to analyse data from questionnaires done in years 2011 – 2014[1] that have not been tested yet and use

[1] The questionnaire was slightly changed in 2011, it have not changed since then, thus we chose only data for four-year period.

these results in preparing new study programmes, when it is unique opportunity to create study plans and content of subjects to fit it the best to preliminary mathematical knowledge of students.

2 Description of the examination

Using the method Analysis of variance (ANOVA) we analysed results in 8 observed mathematical topics upon the secondary school attended and study fields:

- Sets and numerical fields (SET)
- Propositional logic and proofs (LOGIC)
- Functions (FUNC)
- In/equations (EQUAT)
- Fundamentals of the differential and integral calculus (MA)
- Combinatorics (COMB)
- Probability and statistics (STAT)
- Analytical geometry (GEOM)

Questionnaire is described in details in [4], [5].

Particular types of schools are the following: G - Gymnasium (preparation for university studies), SPS - industrial school (preparation for practice in several technical branches), SOS - integrated school (specialized branches for practice), OA - secondary business school (economically oriented for practice), SOU - secondary education for practice in a trade.

3 Results

Results in both evaluated years confirmed statistically significant differences in preliminary basic mathematical knowledge depending both on absolved type of school and chosen study field. [5]. We decided to check again, if these differences will be confirmed in a five-year term and we added to our analysis year-on-year comparison of results and mutual comparison of results between each areas. We used ANOVA, eventually MANOVA methods.

3.1 Results – dependence on a type of absolved secondary school

When evaluating ANOVA results from 2011 we found out statistically significant differences in total results (in all observed mathematical areas) depending on a type of absolved school [5]. Differences were also confirmed in 2014 with finding that results at SOU decreased much more. Also multiple comparison tests for all pairwise differences between the means identified statistically significant differences between types of school slightly differently.

Group	Mean 2011	Different from Groups 2011	Mean 2014	Different From Groups 2014
SOU	17,73214	SPS, G	14,5625	SPS, G
SOS	18,45821	SPS, G	18,79048	G
OA	20,44667		20,675	
SPS	22,10543	SOS, SOU	22,58833	SOU
G	23,71539	SOS, SOU	23,96364	SOU, SOS

Tab.1. Results of Tukey-Kramer Multiple Comparison test in 2011 and 2014 – differences between types of schools.

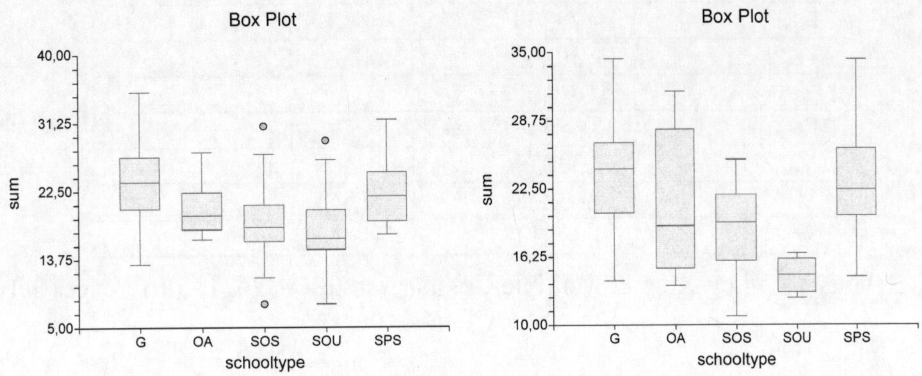

Fig. 1. Box plots in 2011 and 2014 – differences between types of schools.

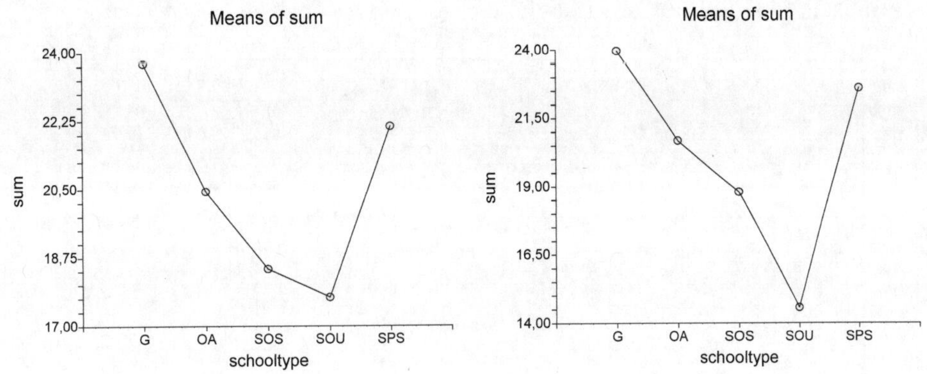

Fig. 2. Plots of mean in 2011 and 2014 – differences between types of school.

When we compared data for whole for-year period, differences between schools are visible again, collectively for all areas and in all areas as well. The best results were reached especially by graduates of G and SPS, namely in more areas together, however, in areas FUNC, SET and GEOM, students of G were slightly better, at areas STAT, COMB and MA, graduates of SPS reached same, even better results than graduates of G. Trend of giving more emphasis on differential calculus, combinatorics and statistics at industrial schools is being confirmed, which relates to the specialisations of study fields, that are taught there. In area EQUAT, students of OA approached in results to these two types of secondary schools. IN area LOGIC, results of students of G significantly outreached all other graduates, which relates to the fact, that only at this type of secondary school, the area of propositional logic and proofs is in compulsory topics in FEP.

3.2 Results – dependence on a chosen study field

With respect to the fact, that since 2012 data without geographical and biological fields were collected, in observed period 2011 – 2014 we focused only on comparison of fields, from which we have complete data, thus mathematical, informatics, physical, chemical and double-study field studies.

Group	Mean	Different from Groups
CH	19,23333	2ob, AME
AI	19,59242	2ob, AME
I	20,41167	
BF	21,6875	
AM	22,33077	
AME	22,63828	CH, AI
2ob	24,49211	CH, AI

Tab. 2. Results of Tukey-Kramer Multiple Comparison test in 2011 - differences between Study fields.

In 2011, ANOVA showed statistically significant differences in reached results of students of each study field (see Tab. 2), however in 2014 it confirmed zero hypothesis, which means that medians at each study fields are the same. We can say, that preliminary knowledge of applicants at observed study fields is in balance.

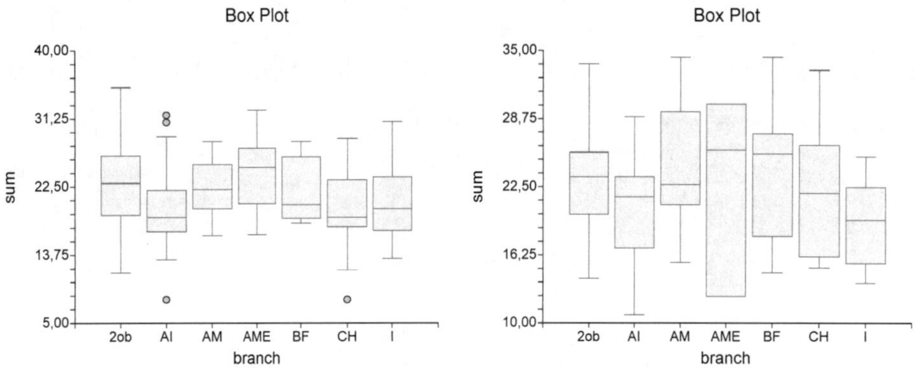

Fig. 3. Box plots in 2011 and 2014 – Differences between study fields.

3.3 Results in each observed mathematical areas

When analysing data in first two years, they were not yea-to-year compared as well as there was no comparison in achieved results mutually in each area. Thus, we focused on it now.

At first, we focused on year-to-year comparison. For each area, results from years 2011-2014 were compared and we observed, if there are statistically significant differences. We found out that results do not differ, except of two areas. Year-to-year results confirmed differences only in areas EQUAT and STAT.

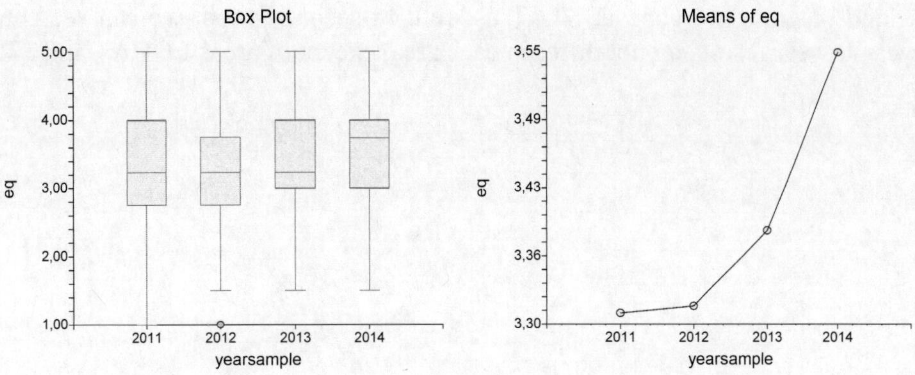

Fig. 4. Box plot and Plot of mean for the topic EQUAT.

Results in area EQUAT got significantly better in observed area, thus it is evident, that to this area, a great attention is paid at school. Tukey-Kramer Multiple-Comparison Test showed, that results differ conclusively in area EQUAT in years 2011, 2012 and 2014.

In area STAT, results in years 2011-2014 were also increasing, however, Tukey-Kramer Multiple-Comparison Test did not showed significant differences in comparison each pairs of years. Grouping was identified by different test, which identified differences between years 2011 and 2013, these tests are approximately accurate[2].

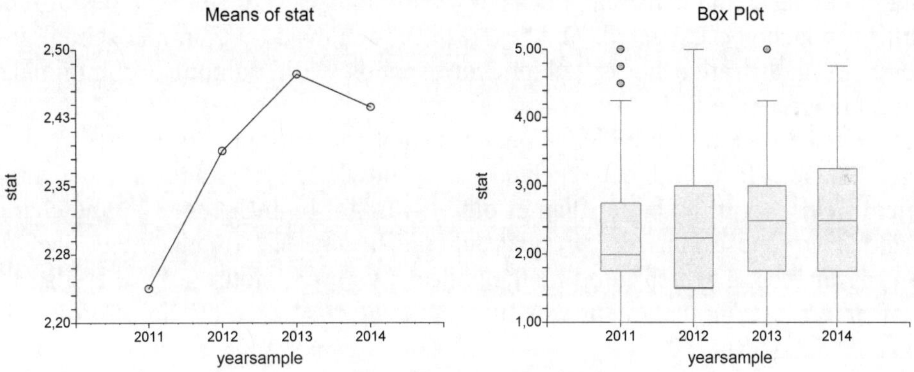

Fig. 5. Box plot and Plot of mean for the topic STAT.

At the end, we focused on comparison each area mutually. MANOVA showed, that there are statistically significant differences in knowledge at students in each area mutually. From graphs (Fig. 6) we can see, that the best results applicants show in area EQUAT, the worst in

[2] Each of the Multiple Comparison Tests has some different properties, differing primarily in how they handle the magnitude of error 1 of the α (level of significance of the test) when testing. Some of the tests are rather conservative, ie. maintain the required level of materiality throughout the experiment under fairly free assumptions, and by making the appropriate decisions as a rule at a lower level of materiality, do not allow the probability of error α to rise uncontrollably. Other tests are rather liberal, ie. they are very likely to reject the zero hypothesis on the compression of the compared pairs of mean values (in other words, we can easily obtain the statistical significance of the difference between the tested pairs of mean values). However, it should be remembered that these resulting significances may sometimes be false because the liberal tests do not adequately modify (ie reduce) the significance level in testing the differences in individual pairs of groups. Therefore, the error of the first type α in the entire experiment may increase disproportionately. [1]

area logic, and they are not good at STAT as well. In second graph, we can see confirmed fact, that year-to-year results do not differ in each area, except of areas EQUAT and STAT.

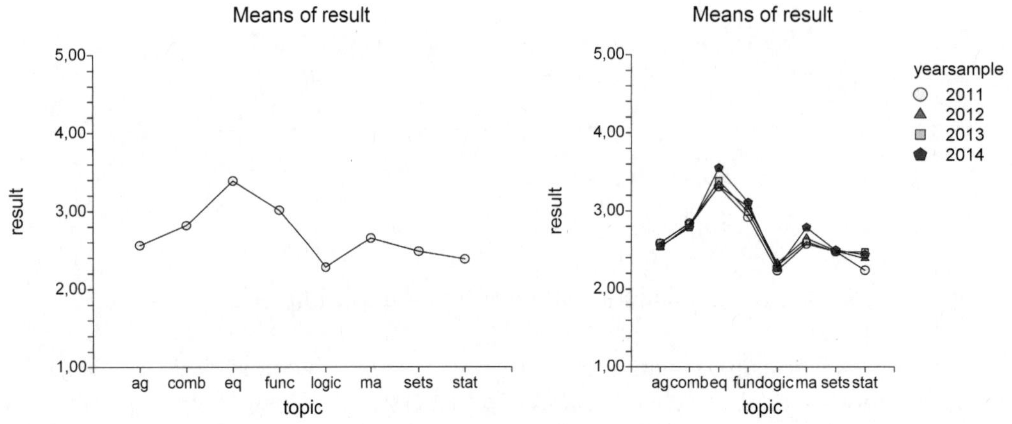

Fig. 6. Results of MANOVA – topics and years.

4 Conclusion

Evaluating the data from questionnaire for whole four-year period confirmed, that knowledge and orientation in basic mathematical areas between graduates of different types of secondary schools differs, which was expected. The best result is achieved by grammar school graduates and students of industrial schools, which corresponds with compulsory minimal lesson dotation of mathematics in FEP.

Less expected result was, that knowledge between students according to selected study field is getting into balance. It would be expected, that knowledge of students who enrolled at mathematical fields would be better than at other. It might be influenced by the character of fields included in questionnaire, physical, chemical and informatic fields belong into industrial programmes at which good mathematical literacy. Another effect is also the fact, that part of entrance examination on chemical and informatic study fields were requested entrance examination tests.

Results of third part were interesting. Knowledge of secondary school graduates do not differ in year-to-year comparison, with some exceptions, but the level of knowledge at each of eight areas of mathematics do differ.

Last two results can be used for the conception of new study programmes. E.g. where it is necessary for the next stage, it is necessary to include subjects containing the basics of mathematical statistics and propositional logic at the beginning of the higher education, while students in these areas cannot be expected to have sufficient secondary school base. With respect to the absence of differences between disciplines, such introductory subjects may be uniform for whole groups of disciplines and programs.

References

[1] HABIBALLA, H., *Aplikace analýzy rozptylu v pedagogickém výzkumu*. Ostravská univerzita v Ostravě, 2014.

[2] HABIBALLA, H.; KMEŤ, T. Theoretical branches in teaching computer science. In International Journal of Mathematical Education in Science and Technology. 6/2004(35), pp.829-841, Taylor and Francis : Great Britain, ISSN 0020-739X.

[3] MELOUN, M.MILITKÝ, J., *Kompendium statistického zpracování dat*. Academia, 2002, ISBN 80-200-1008-4, 764 s.

[4] KONEČNÁ P., HABIBALLA, H.: *Student's Autoevaluation in the First Year of Science Fields Especially Mathematics Abilities.* In Mathematics XVI, Scientivic Issue Jan Długosz University in Częstochowa, pp. 219-224, 2011. ISBN 978-83-7455-209-7.

[5] KONEČNÁ P., HABIBALLA, H.: *Student's Autoevaluation in the First Year of Science Fields Especially Mathematics Abilities – phase 2.* In *Proceedings of 11th International Conference of Applied Mathematics,* Bratislava, 2011, ISBN 978-80-89313-58-7, pp. 741-746.

Current address

Konečná Petra, RNDr., Ph.D.
Department of Mathematics, Faculty of Science University of Ostrava
30. dubna 22, Ostrava 701 03, Czech Republic
E-mail: petra.konecna@osu.cz.

Habiballa Hashim, doc. RNDr. PaeDr., Ph.D.
Department of Informatics and Computers, Faculty of Science University of Ostrava
30. dubna 22, Ostrava 701 03, Czech Republic
E-mail: hashim.habiballa@osu.cz

Slovak University of Technology in Bratislava
Faculty of Mechanical Engineering
17th Conference on Applied Mathematics **APLIMAT 2018**
Proceedings

AUGMENTED REALITY IN MATHEMATICS EDUCATION FOR PRE-SERVICE TEACHERS IN PRIMARY LEVEL

KOREŇOVÁ Lilla (SK), GUNČAGA Ján (SK)

Abstract. New technologies with modern method of teaching must be implemented in the university studies. Important part is teacher training study. These technologies make the study more attractive and also it brings bigger motivation an understanding of notions by students. We present in our contribution augmented reality technology as a part of using of mobile technologies in mathematics education of future primary school teachers.

Keywords: Augmented reality, constructivism, constructionism, solids, digital technology, teacher training of pre-service teachers at primary level

Mathematics Subject Classification: Primary 97C30, 97D40, 97G40; Secondary 97U60, 97U70

1 Mobile Technologies and Virtual Reality in Education

The technological and social changes that accompany the current ubiquitous use of mobile devices brings the question, how it is possible to implement these technologies into the educational process. We can find several approaches to defining mobile learning in professional literature: Mobile learning refers to the use of mobile or wireless devices for the purpose of learning while on the move. Typical examples of the devices used for mobile learning include cell phones, smartphones, palmtops, and handheld computers; tablet PCs, laptops, and personal media players can also fall within this scope (see [6]). Taylor in [17] has defined mobile learning as "learning mediated by mobile devices, or mobility of learners (regardless of their devices), or mobility of content/resources in the sense that it can be accessed from anywhere".

Sharples in [14] has made a good summary on different views of defining mobile learning. Current perspectives on mobile learning generally fall into the following four broad categories:

1. Technocentric. This perspective dominates the literature. Here mobile learning is viewed as learning using a mobile device, such as a PDA, mobile phone, iPod, PlayStation Portable etc.
2. Relationship to e-learnig. This perspective characterises mobile learning as an extension of e-learning.
3. Augmenting formal education.
4. Learner-centered.

What is needed is clarity: in agreement with Traxler in [18], the technocentric/e-learning based definitions only seek to place "mobile learning somewhere on eLearning's spectrum of portability". Augmenting formal education. In the mobile learning literature, formal education is often characterised as face-to-face teaching, or more specifically, as a stereotypical lecture. However, it is not at all clear that this perspective is wholly correct. Forms of distance education (for example, distance correspondence) have existed for over 100 years, leading to the questions regarding the place of mobile learning in relation to all forms of "traditional" learning, not only the classroom.

2 Augmented Reality in Education for pre-service teachers in primary level

The term Augmented Reality (AR) was created around the year 1990 and it translates the integration of virtual images in the real world, i.e. the reality is augmented of virtual elements. The integration of such images is made by the use of Information and Communication Technologies (ICT), through a mobile device with a camera (computer, tablet, mobile phone with android or iOS operating systems) which allows the access to the available contents with AR. Furthermore, the development of such contents encourages higher learning autonomy and the use of systems that support mobile-learning. Besides, the exploration of ICT by the students can promote collaboration, innovation and creativity skills. One characteristic that AR applications offer is the integration and interaction between the real and the virtual, allowing a huge versatility and creativity in applications (see Coimbra, Cardoso, Mateus in [4]).

The basic goal of an AR system is to enhance the users perception of and interaction with the real world through supplementing the real world with 3D virtual objects that appear to coexist in the same space as the real world. Many recent papers broaden the definition of AR beyond this vision, but in the spirit of the original survey we define AR systems to share the following properties (see Azuma et al. in [2]):
1) Blends real and virtual, in a real environment
2) Real-time interactive
3) Registered in 3D
Augmented reality – AR is the supplementation of the reality perceived by the user with virtual elements. The use of mobile AR technology applications brings advantage that the work of educators can be made more effective, and in addition it enables pupils to become actively involved in the education process. Besides all this it makes studying both in the lesson and outside the lesson more attractive and more experiential. It is also suitable for arousing interest in subject matters for which the generation socialising on TV, on the Internet and on the Facebook portal is less receptive.

3 Constructivism and constructionism in the teacher training at university

The inductive (constructivist) approach in teaching comparing to deductive approach is characterized by distinctly different characteristics from the deductive approach, cognitive development and the learning process are defined as follows (see Kostrub [10]):

- to base always on the achieved level of learner's development,
- to provide meaningful learning,
- to enable learners to realize their own meaningful learning process
- to have influence on learners in that way that they will modify their own knowledge schemes,
- to create and maintain rich relationship among new knowledge and already existing knowledge schemes.

The constructivist theory of learning assumes that each person creates himself (constructs) his own knowledge of the world in which he lives. Constructivism tries to overcome the transmissiveness of traditional teaching - the transfer of "teacher's" knowledge to the student. It deals with learning with understanding (see Stoffová [15]).

According Ackermann [1] the word constructionism is a mnemonic for two aspects of the theory of science education. From constructivist theories of psychology, we take a view of learning as a reconstruction rather than as a transmission of knowledge. Then we extend the idea of manipulative materials to the idea that learning is most effective when part of an activity the students´ experiences as constructing a meaningful product.

According Papert, Harel [12] the idea of the constructionism is thinking of it as "learning-by-making". It is nowadays in mathematics and informatics education used LEGO set and Logo program in primary and upper secondary level, which brings possibility for children to make some "cybernetic" activities, thus becoming as a part of the lives of young children as playing with toys and dolls. Future teachers for primary level need to have possibility of an experience with this play.

The opposite approach to constructionism is instructionism. There is typicall for this kind of teaching, that teacher makes instructions for children and they have small space for their own activity and personal way of thinking. Instructionism vs. conctructivism looks like a split about two strategies for education: two ways of thinking about transmission of knowledge. But behind this there is a split that goes beyond the acquisition of knowledge to touch on the nature of knowledge and nature of knowing.

An increasing number of researchers have come to the following view: The knowledge is essentially "situated" and it doesn't not be detached from the situations in which it is constructed and actualized. This growing interest in the idea of situated knowledge, or knowledge as it lives and grows in context, has lead many researchers to look closely at individual people's ways of knowing, or relating. Constructionism as an educational theory has many applications in the digital school environment (see Sabelli [13]).

The root for constructionism was constructivism. Constructivist instructional design according Kalaš in [8] aims to provide generative mental constructions embedded in relevant learning environments that facilitate knowledge construction by learners. Constructivists approach has many applications in different areas. A good example in the field of languages is possible to find in Šipošová [16] and in the field of science education for disable children in Vančová, Sulovská [19]. We used in our following research these pedagogical theories, because students – pre-service teachers for primary level works with different applications used augmented reality. These applications were tools for better understanding of the base mathematical notions. In the field of the functional thinking see Fulier, Ďuriš, Frantová in [7].

4 Using of applications with augmented reality for teaching of base properties of solids

We conducted the research in the group of 27 students of the first year of the bachelor study in the teacher training program for future teachers in primary level. The pedagogical experiment was realized in November 2017 at Faculty of Education of the Comenius University in Bratislava.

Our research focused on analysing the use of mobile technology and manipulation activities in teaching mathematics based on constructivist and constructionist concepts. We looked into the problem of how to incorporate the use of smartphones and tablets in education. We believe that when these two methods are combined, the students′ understanding is deeper, their motivation is greater, and, last but not least, their creativity is strongly supported.

There was installed the AR application "Augmented Polyhedra" with cards on tablets and smartphones (see Fig. 1). This application is free of charge and the markers can be also printed free of charge. It works in a similar way as the AR Flashcards Shapes. It is good for increasing users' motivation. Site of this application is: http://mirage.ticedu.fr/?p=2635
The task for the students was to find under the every card the solid, his properties, volume and surface.

Fig. 1. Application "Augmented polyhedra".

Some students used the lineal, they tried to measure some lengths of different solids (see Fig. 2). They used it for calculating the surface and volume.

Fig. 2. Student working with lineal.

Some solids in the apllication were prisms. Students tried to make network of the prism with the geometric manipulative construction product Polydron (see http://www.polydron.co.uk). This network also helped them by searching of the volume and area of the prism (see Fig. 3).

Fig. 3. Student working with Polydron and AR application.

Students worked in the groups of 2 – 4 persons. During the lesson using the constructive method. In the course of the qualitative research, we would have liked to observe the effect of the AR Polyhedron application on the learning in a constructivist environment.

At the beginning of the activity, the students select randomly a few cards with a code that the application recognizes. Using the Polyhedron application on the tablet or on the smartphone, they can use the augmented reality to study the geometric shape that appears in three dimensions.

During the research, the task of the students was to find the name of the solids, his base properties and find his volume and area. Every group obtains 10 cards. They work with these 10 tasks for 90 minutes. After the students asked if they could use the Internet, the teacher allowed to use it in the course of the problem solving.

A particularly interesting part of the research was to see how students tried to find those geometric shapes on the Internet which were unknown for them. They were looking for pictures to identify the shape. Occasionally, only the English name was found in case of truncated pyramid and truncated cones. Some of them find on the Internet the expressions for volume and area of the solid surface and also the art of dimensions needed. The students as a homework prepare presentations about solids, which they found through the cards with AR application.

Next block of teaching was oriented for presentation of the students' results. This block was one week after the first block. The students work at home with the material prepared during the first block. Every group choose 2 solids, which was presented for the whole class of students. Some students find wrong information on Internet and other students make corrections. If they cannot find the right answer, it was the case for the intervention of the teacher.

As a conclusion, students were motivated by the new method, they cooperated very well and learning was constructive. They gained new knowledge, collaborated, discussed their observations, argued and reasoned. The difficulties of problem solving were well managed, they had logical ideas. Our qualitative pedagogical research relies on the description of the teacher's observations and on the video recording of the students' work.

5 Conclusions

Our observations from the work of students we can conclude in the following table:

CATEGORY	SUBCATEGORY	EXPRESSIONS AND ACTIVITIES
1. Students´ mutual learning - constructionist approach	1. Students support each other	We are going to …, so that we can have, we will make, we have made, we have named, we have built, we are talking about, we are going to write, we are making a presentation
	2. Mutual learning helps them to understand terms and relations	We are creating together, to us, we will put, we are, let´s talk together, let´s take – so that we can, we are solving together
	3. They do not collaborate with each other	We were not able to explain it to each other, I am
2. Influence of mobile technology	1. It helps them when they can make several attempts thanks to the technology	Several tries, explanations, I have several attempts, several solutions, I can correct myself, with a simple click
	2. Mobile technology gives them feedback	We are verifying, she evaluated it on his/her own, so that it would be correct, I tried
	3. Thanks to mobile technology, they become aware of selected mathematical expressions and relations	We are thinking over, we are considering, we are thinking about, I can remember it
3. Manipulation activities	1. Discovery of new technologies	We are drawing, we are using a string, Polydron, cubes, we prepare a network, we calculate now the surface or volume
	2. Students are more motivated	We enjoy it, we are finding out, now we know, I am going to try it, we want to work
	3. Better awareness of mathematical knowledge and relations	A square, a rectangular, a triangle, a cube, construction, a diagram, a table, a network, expression for the volume or surface

Tab. 1. Categories of students activities.

The constructionist design of teaching involves constructionism, design, architecture. It is a conceptual art shared jointly and mutually by the students and their teachers as a result of the discursive and narrative essence of teaching in which they make proposals, discuss, describe, present, approve, evaluate, judge, agree, create and rework the non-material form of their products into a material form and vice versa (see Čavojský [5]). This takes place in the form of intentional but indirect student and teacher participation in teaching activities, controlled by the teacher, in which, however, the teacher acts as a consultant when he/she is invited by the students to comment on their ideas, while avoiding any reference to mistakes. Constructionist teaching takes place through didactically considered but conceptually open teaching activities, and through discourse (controlled argumentation, handling facts) in the form of individual as well as group exploration (learning groups), thanks to which common knowledge and understanding is established.

The students apply corresponding cognitive tools, such as thinking and speech in connection with the cognitive prostheses available in their surroundings. Their minds are thus formed in a different way, which means that digital technology delimits and structures cognitive schemes in a way that was unfamiliar to students of previous decades. The confidence arising from the availability of conventional information is pertinent; nevertheless, on the basis of such information, a conceptually reflective teacher supports the formation of unconventional knowledge which can be used for other teaching contexts. The students are able to perform both specific and unspecific transfers, which is a precondition for the teacher to lead the teaching process on the basis of the principles of constructionism (see also Kónya in [9]).

The goal of the activities for students from mathematical point of view was not to calculate the exact value of the surface area and volume of the solids, but to find the name and basic properties, network of the solid and all dimensions needed for the calculation of the surface area and volume of the solid. Students find that it was impossible to measure punctually dimensions of solids visible through AR application, because if they change the distance between smartphone, tablet and the card the solids change their dimensions. Only proportion of dimensions was constant. The advantage of this teaching was, that students must find which kind of dimensions they need for calculating of volume and surface area of solids. They use many modern sources and devices of information such AR application, Internet, smartphone, tablet, and computer, textbook. Group work of students' trains their collaborative abilities, inquiry-based teaching methods (see in Kovács [11]) which they can use in their future pedagogical practice. Similar collaborative activities is possible to find in the using of some educational software (see Beňačka, Čeretková in [3]).

Acknowledgment

The paper was written with the support of the grant KEGA 012UK-4/2018 "The Concept of Construcionism and Augmented Reality in the Field of the Natural and Technical Sciences of the Primary Education (CEPENSAR)".

References

[1] ACKERMANN, E., *Piaget's Constructivism, Papert's Constructionism:What's the difference?* Available from: http://learning.media.mit.edu/content/publications/EA.Piaget_Papert.pdf

[2] AZUMA, R., BAILLOT Y., BEHRINGER, R., FEINER, S., JULIER, S., MACINTYRE, B., Recent advances in augmented reality. *IEEE computer graphics and applications*, 21(6), 2001, pp. 34–47.

[3] BEŇAČKA, J., ČERETKOVÁ, S. Graphing functions and solving equations, inequalities and linear systems with pre-service teachers in Excel. *CERME 9 – Ninth Congress of the European Society for Research in Mathematics Education*, 2015, pp. 2311–2318.

[4] COIMBRA, M. T., CARDOSO, T., MATEUS, A., Augmented reality: an enhancer for higher education students in math's learning? *Procedia Computer Science*, 67, 2015, pp. 332–339.

[5] ČAVOJSKÝ, I., Digital technology in primary education. *Wyzwania i zaniechania w wychowaniu przedszkolnym i kształceniu wczesnoszkolnym*. Siedlce: Uniwersytet Przyrodniczo-Humanistyczny, 2014, pp. 271-281.

[6] FERKO, A., KOREŇOVÁ, L. Some possibilities for using mobile learning in mathematics. *Scientific Issues MATHEMATICA V*, Ružomberok: VERBUM, 2015, pp. 15–20.

[7] FULIER, J., ĎURIŠ, V., FRANTOVÁ, P. *Systémy počítačovej algebry (CAS) vo vyučovaní matematiky*. Nitra: FPV UKF, 2007.

[8] KALAŠ, I., *Integrácia informačných a komunikačných technológií do všeobecného vzdelania*. Bratislava: ŠPÚ, 2001.

[9] KÓNYA, H. E., How can high school students solve problems based on the concept of area measurement? *Problem solving in mathematics education (PROMATH)*, Budapest: Eötvös Loránd University, 2014, pp. 95–107.

[10] KOSTRUB, D., *Dieťa/žiak/študent-učivo-učiteľ, didaktický alebo bermundský trojuholník*, Prešov: Rokus, 2008.

[11] KOVÁCS, Z., Modelling with difference equations supported by GeoGebra: Exploring the Kepler Problem. *International Journal for Technology in Mathematics Education*, 17, 2010, pp. 141–146.

[12] PAPERT, S.; HAREL, I., *Constructionism*. Ablex Publishing Corporation, 1991.

[13] SABELLI, N., Constructionism: A New Opportunity for Elementary Science Education. *DRL Division of Research on Learning in Formal and Informal Settings*. 2008, pp. 193–206.

[14] SHARPLES, M., „Mobile learning," rev. *Technology-enhanced learning*. Springer Netherlands, 2009, pp. 233–249.

[15] STOFFOVÁ, V. Modelling and simulation as a recognizing method in education. Educational Media International, Vol. 4, Issue1, 2004, pp. 51–58.

[16] ŠIPOŠOVÁ, M., Constructivist Aspects in Developing Reading Comprehension Skills as a Means of Improving ESP Learners´ Competencies. *Aplikované jazyky v univerzitnom kontexte*. Zvolen, Technical University of Zvolen, 2017, pp. 27–42.

[17] TAYLOR, J., Towards a task model for mobile learning: a dialectical approach. *International Journal of Learning Technology 2*, 2006, pp. 138–158.

[18] TRAXLER, J., The evolution of mobile learning. *The evolution of mobile teaching and learning*, 2009, pp. 1–14.

[19] VANČOVÁ, A., SULOVSKÁ, M., Innovative trends in geography for pupils with mild intellectual disability. *CBU International Conference on Innovations in Science and Education (CBUIC).* Prague: Central Bohemia Univ, Unicorn College, 2016, pp. 392–398.

Current address

Gunčaga Ján, doc. PaedDr., PhD.
Department of Science Education, Faculty of Education
Comenius University in Bratislava
Šoltésovej 4, 813 34 Bratislava, Slovak republic
E-mail: guncaga@fedu.uniba.sk

Koreňová Lilla, doc. PaedDr., PhD.
Department of Science Education, Faculty of Education
Comenius University in Bratislava
Šoltésovej 4, 813 34 Bratislava, Slovak republic
E-mail: korenova@fedu.uniba.sk

THE USAGE OF MOBILE DEVICES IN THE STUDENTS' MATHEMATICS LEARNING

KOREŇOVÁ Lilla (SK), VERESS-BÁGYI Ibolya (HU)

Abstract. Mobile devices enables university students to develop innovative learning methods. The question is that can they exploit these potentials, do they know the relevant online curriculums, platforms and applications and do they have incentive for integrating mobile devices in their math learning process? In our survey, we searched the answers for these above issues. Our pilot survey suggests that the theme is very actual, arouses interest and we have to pay particular attention to the future research.

Keywords: math learning, mobile device, augmented reality, math application

Mathematics Subject Classification: Primary 97C30, 97C70, 97D40; Secondary 97U50, 97U70

1 Introduction

Nowadays, the Neumann Galaxy lives together with the Gutenberg Galaxy. Both printed books and mobile devices are applied. The question often arises: In which proportion should they be used by teachers and by students? We believe in the learner-centred pedagogy and we analyse this question from the student side. We are focusing on the types of mobile learning and talking about the change of mathematics learning.

In our research, we were interested in the quality of using mobile devices by students. The main questions in the questionnaire were: What do they generally use the smart tools for? What do they use the mobile technology for during their learning process? And in a new section we asked them about math learning with the help of mobile devices which the main topic of our survey was.

The balance between mobile learning and learning activities designed by a teacher should be found. The students of Neumann Galaxy are in a difficult situation if we think of the information flood but on the other side, they can reach easily that kind of information which was unimaginable and inaccessible earlier. We are convinced that pedagogy changes irreversibly. The term “students” refers to university level students in all cases.

2 Theoretical background

“If we want students to become smarter than a smartphone, we need to think harder about the pedagogies we are using to teach them. Technology can amplify great teaching but great technology cannot replace poor teaching” (OECD, 2015, p. 4).

Today’s students’ needs were identified in six areas. These are the followings: Sharism, Shifting identities, Border-crossing, Literacies beyond print, A culture of gaming and A culture of bricoleurs. In details, you don’t need to hide information. You do not have to be in a single role. You do not have to stay in one place. You do not have to learn the tools, it is enough to use them. There is no need to be motivated, only to give room for learning. There is no need for a rigid curriculum, but rather a tangible experience [1].

Mathematics is usually hated because it is difficult and incomprehensible. But if the education changes to the right direction and mobile devices get a little gamification in the math learning, this subject will become likeable.

It usually happens that both teachers and parents console their students by saying to them not to worry about doing badly in math because not everyone can excel in it. This resulted the appearance of the sentence: ‘I’m not a math person’. Although the sentence is still alive in these days, the new evidence of brain research tells the opposite. Everyone, with the right teaching and messages, can be successful in math. According to researchers, there are a few children who have very particular educational needs, but the vast majority of them – about 95% - any levels of school math are within their reach [5].

During mathematics courses, student can make use of digital technology in various ways:

- during numerical calculations so they can concentrate on the solution of the problem itself;
- for visualisation, modelling and simulation of problems and thus to obtain such a graphical representation of the problem which pushes them towards a solution;
- as a source of educational materials e.g. e-books or videos, interactive educational materials;
- drilling exercises, a student can make use of electronic working sheets or e-tests to evaluate himself.
- They can use applications, such as: Geogebra, GeoMatech, 3D Geometry, Wolfram Alpha, PhotoMath etc.
- Watching videos about problem solving [15]

Moreover, the Augmented Reality technology enables us to develop new learning methods for mathematics learning. We are talking about augmented reality when with the aid of an application on a mobile device virtual objects can be displayed in the space we can see in front of us. Thus, we create a bridge between reality and virtual reality with the help of which we can reach great influence on every generation and achieve extraordinary results.

The term Augmented Reality (AR) was created by Tom Caudell, in 1990, while he was working at Boing, and it translates the integration of virtual images in the real world, i.e. the reality is augmented of virtual elements. The integration of such images is made by the use of Information and Communication Technologies (ICT), through a mobile device with a camera (computer, tablet, mobile phone with android or iOS operating systems) which allows the access to the available contents with AR. Furthermore, the development of such contents encourages higher learning autonomy and the use of systems that support mobile-learning. Besides, the exploration of ICT by the students can promote collaboration, innovation and creativity skills. One characteristic that AR applications offer is the integration and interaction between the real and the virtual, allowing a huge versatility and creativity in applications [8].

With students it is extremely motivating, if we make fields, which are difficult to understand, such as geometry, more comprehensible for students with the aid of augmented reality.
The main goal of geometry learning is to improve the spatial abilities. Spatial skills present an important component of human intelligence. The term spatial abilities contain five components, spatial perception, spatial visualization, mental rotations, spatial relations and spatial orientation. These abilities can be improved by virtual and augmented reality [12].

Our earlier research [16] carried out with the Augmented Polyhedrons application deals with the introduction of an application into primary schools, which portrays polyhedra in 3D. In mathematics and geometry education at high school as well as university level the Construct3D is a development which allows us to observe the changes in various geometric bodies that we have drawn and their connections with each other, e.g. sections, all in a manually controlled virtual space on a display placed on the head.

We think that these alternatives will be fully utilised one day by students. Teachers will be only the mentor on students learning process.
Mike Sharples [20] has made a good summary on different views of defining mobile learning. Current perspectives on mobile learning generally fall into the following four broad categories:

- *Technocentric.* This perspective dominates the literature. Here mobile learning is viewed as learning using a mobile device, such as a PDA, mobile phone, iPod, PlayStation Portable etc.
- *Relationship to e-learning.* This perspective characterises mobile learning as an extension of e- learning. These definitions are often are all-inclusive and do not help in characterising the unique nature of mobile learning. What is needed is clarity: in agreement with Traxler [21], the technocentric/e-learning based definitions only seek to place "mobile learning somewhere on e-learning's spectrum of portability".
- *Augmenting formal education.* In the mobile learning literature, formal education is often characterised as face-to-face teaching, or more specifically, as a stereotypical lecture. However, it is not at all clear that this perspective is wholly correct. Forms of distance education (for example, distance correspondence) have existed for over 100 years, leading to the questions regarding the place of mobile learning in relation to all forms of "traditional" learning, not only the classroom.
- *Learner-centred.* Any sort of learning that happens when the learner is not at a fixed, predetermined location, or learning that happens when the learner takes advantage of learning opportunities offered by mobile technologies".

Nowadays, there are vast online math learning possibilities. In this chapter, we present some platforms and math applications which can be used for teaching and learning. These tools are used by a significant part of the respondents of the questionnaire.

Separated in two categories, in the first part, you can read about searching platforms such as GeogGebra Materials, GeoGebraBook, GEOMATECH project and Planeta Vedomosti project as well. In the second part, several of the students' favorite applications are presented.

2.1 Sites with searchable online math materials

1. GeoGebra is an open source and worldwide used dynamic mathematics software. It covers all levels of math learning in primary and secondary school. We can use it for learning geometry and algebra, for spreadsheets, graphing, statistic and calculus. It is the master software in learner and learning centered pedagogy in the territory of STEM (Science, Technology, Engineering, and Mathematics). The software can be used for desktop but there are applications for tablets, which runs with Windows, Android and iOS and for mobile phones as well.
GeoGebra Materials (https://www.geogebra.org/materials/) is the platform where the GeoGebra applets can be saved to. This is the official cloud service and repository of GeoGebra lessons. There are more than 1 million searchable materials. Teachers can prepare *GeoGebraBook* for their students. These materials can be viewed and used on mobile devices as well.

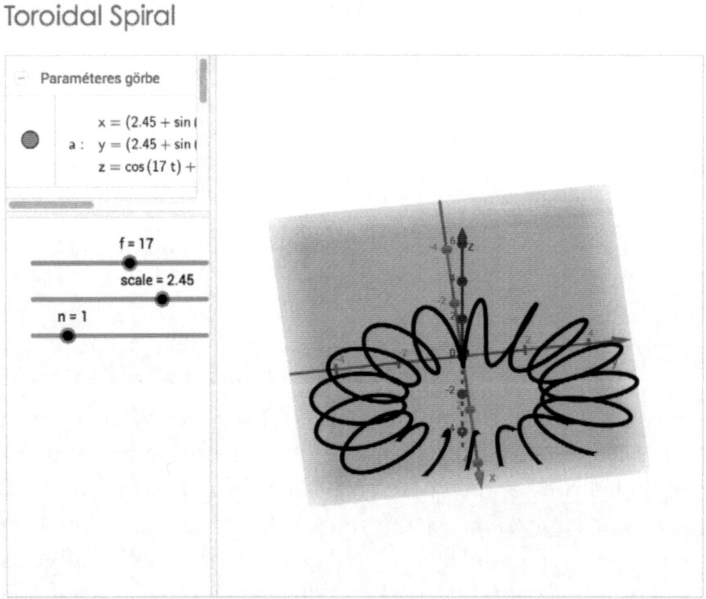

Fig. 1. Toroidal Spiral in GeoGebra https://www.geogebra.org/m/jy8mAzRk.

2. GEOMATECH (http://geomatech.hu) is a highlighted project in the Hungarian public education the goal of which is to have students like math and the other science subjects. This

is an online and interactive curriculum website, which can be used as digital math textbook too.
The curriculum portal is based on the GeoGebra software, which provides the potential for making interactive learning material. We can say that GEOMATECH is more than GeoGebra because it gives a frame to GeoGebra applet. Every exercise contains both the description of the problem and of the usage of GeoGebra applet.
We can easily search among subjects by choosing the class and the material, which we want to see.

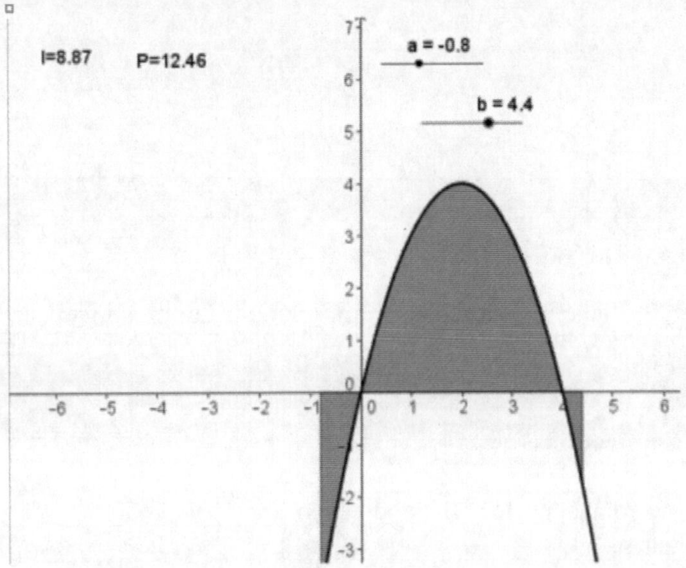

Fig. 2. Visualization of fractional calculus (Korenova, Takaci, 2014).

3. Planeta Vedomosti (http://planetavedomosti.iedu.sk) is a Slovakian, free of charge website. It contains a number of professionally made material for mathematics like videos, electronic worksheets and tests which are grouped together according to the curriculum topics for students and teachers alike. The content of this site covers a large part of both elementary and secondary school mathematics but also includes some topics for university students. It is accessible on computers and interactive boards but unfortunately only a part of these is available on smartphones.
Digital Library is the repository with lots of electronic materials and their methodological descriptions. They are categorized by the types of school, subjects and their use of ICT: https://www.modernizaciavzdelavania.sk/Digi/DigiLibrary.aspx

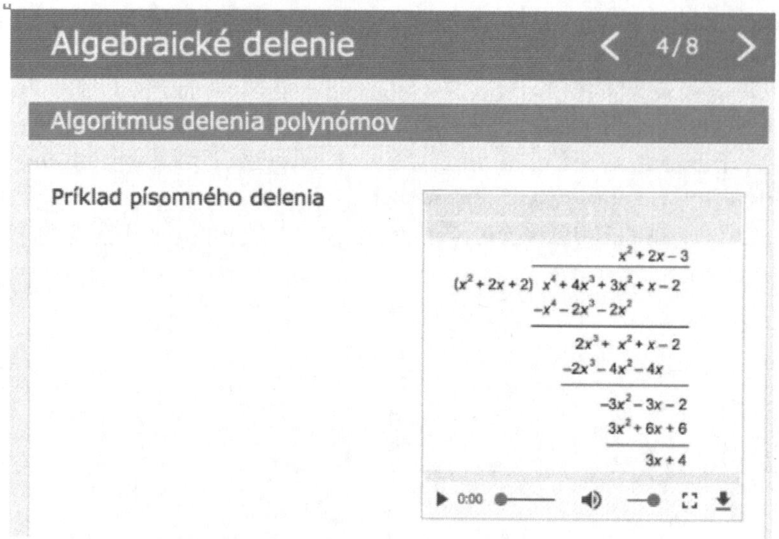

Fig. 3. Planeta Vedomosti - http://planetavedomosti.iedu.sk

2.2 Mathematics applications

1. *GeoGebra* has now four apps for iOS and seven apps for Android. We present the most often used ones of them.

GeoGebra Graphing Calculator easy to utilize for graph functions, solve equations, find special points of functions.

GeoGebra Geometry easily construct triangles, drag points, draw parallel lines, and intersect circles.

GeoGebra Classic is applicable for solving math problems, graph functions, create geometric constructions, do statistics and calculus.

A first version of *GeoGebra Augmented Reality* app was launched this year (2017) only for iOS platforms, but this is under construction. With this app, the developers would like to explore the potential of AR for learning and teaching mathematics.

2. *Wolfram Alpha* is an education tool, which can give answer for almost everything which is needed in the secondary school. This is a tool, which calculates the math problems and answers the questions. These possibilities are free of charge but there are some potential which you can get only in paid Pro version. For instance, if you want to upload a photo with math problem the solution of this is only possible in version Pro.

Wolfram Alpha has apps for iPad, iPhone, Android, Windows Phone and Tablet and Kindle fire.

3. *Photo Math* is a free 2D AR app for Android and iOS. We point our camera toward a math problem and Photomath will show the result with a detailed step-by-step instruction. What is really exciting about it is that this app contains handwriting recognition. We tried it and the result was excellent. PhotoMath, also named camera calculator, makes math problems easy to solve and everyone is motivated to continue this “game”.

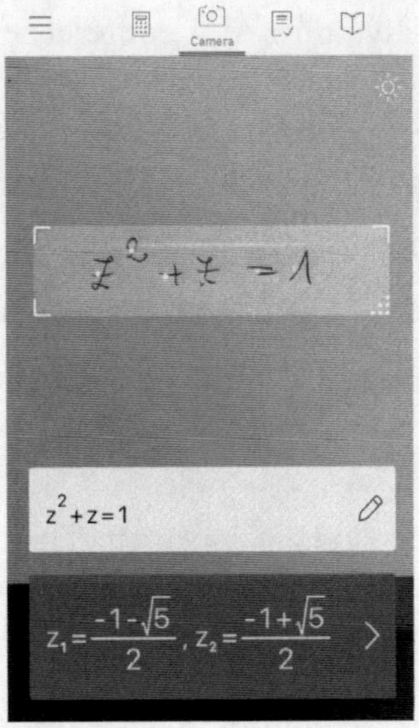

Fig. 4. Resolution of a handwritten equation ($z^2+z=1$).

There are many other applications in this segment of education but we just focused on the well-known and used ones.

3 M-learning in our region

We studied the M-learning current situation in Slovak, Hungarian and Romanian universities. At the present, nearly all schools make use of some kind of digital technologies in education. Digitaliada (http://digitaliada.ro) has been a special program for rural areas digital math education in Romania since 2016. The project covers 30 secondary schools from the countryside, this means 3400 students and 82 teachers. The site of Digitaliada is a platform for collecting new online materials in this theme as well.

The most active Hungarian blog is TanárBlog (http://tanarblog.hu) made by Prievara Tibor and Nádori Gergely. This blog is an almost daily updated site with actual and exciting content about the Hungarian digital education. Essentially it is a teacher blog as its name shows, where the authors share their teaching experience in M-learning too, but unfortunately only in Hungarian. Prievara has also created a new electronic evaluation system which is used by other teachers. The ICT Association in Hungary (http://ivsz.hu/en/) is concentrating on the education and digital skills in schools with Program for promoting digital education and The Digital Thematic Week.

There are many projects, domestic and foreign, which are focusing on pilot introduction of M-learning to schools. In Slovakia, there are currently two such projects that can be found under the following links: http://www.skolanadotyk.sk and http://www.digiskola.sk. The former project, "school by touch", was an initiative from private companies, which supplied ten schools with interactive whiteboards and tablets for every pupil in a classroom. Besides equipping schools with technology, teacher trainings were carried out as well and an empirical research was trying to determine the effectiveness of teaching using these technologies. Teachers were creating demonstrational materials suitable for M-learning, students created and shared videos. They are available at http://www.skolanadotyk.sk/materialy.html. At a conference that followed the project's end, the research team stated that in all subjects except mathematics, significant results were gathered which point out the effectiveness of such education. We presume that the case of mathematics classes was caused by insufficiently digitally literate teachers of mathematics. The latter project started in 2014. The Ministry of Education of the Slovak Republic, using EU funds, bought 5680 interactive whiteboards, 5680 notebooks, 2686 colour printers, 20000 tablets and 1000 Wi-Fi routers for schools all around Slovakia.

4 Research design

The usage of mobile devices in the students' mathematics learning research is based on an online questionnaire which were made, spread and analysed by the authors.

The purpose of this survey is to compare the characteristics of the use and adoption of mobile learning in higher education in three countries. These are Slovakia, Hungary and Romania. The main scope of this study is to survey the utilization of the mobile devices in the students' math learning and their opinion about the usage of mobile technologies in the learning process.

Research questions: In which proportion do university students use mobile devices in their learning process, especially in mobile math learning and what is their opinion about the utilisation of these devices in learning?

4.1 Survey

In order to get answers for our questions, we chose an online survey based on questionnaires. There were two questionnaires, one in Hungarian language and another in Slovakian language. Both the Slovak and Hungarian questionnaires were launched in Slovakia. The Hungarian language questionnaire was sent to almost all the universities in Hungary and Transylvania. Unfortunately, we caught only the Hungarian minority in Romania because we don't have enough connection with the Romanian colleagues.
The questionnaire needs only few minutes (approximately 5 minutes) to be filled. There are three sections, demographic and geographic characteristics, mobile devices in students learning and especially in math learning and finally it contains questions about their opinion in this theme.

Some of the most essential research questions

1. Are students utilising smartphones, tablets, e-Book Readers and other technologies in their learning? If yes, for what? Is there a difference between Slovakian, Hungarian and Romanian students regarding the utilisation of these technologies?
2. What technologies do Slovakian, Hungarian and Romanian students prefer for study purposes?
3. Are students utilising mobile devices for math learning? Which math applications do they know and which ones do they utilise?
4. Are their teachers encouraging the use of smart tools in the classroom?
5. What are the students' opinion about M-learning, about the mobile devices in math learning and about the degree of which their professors are using these tools?

6. What challenges are associated with mobile math learning in these three countries?

The filling period of the research called 'The usage of mobile devices in the students' mathematics learning' was between 27.11.2017. and 10.12.2017. On the first day, we sent a lot of emails to our teacher and researcher colleagues to Slovakia, Hungary and Romania. Altogether the questionnaires were sent for more than 50 colleagues in Hungary and 30 teachers both in Slovakia and Romania.
In the second round, we posted the questionnaire on the universities' students Facebook groups.
Almost all of our colleagues were very cooperative and the answers come one after the other. They said that they had sent it for their students or for the student council. There were a few cases when it was sent for the whole university on Neptun platform. Some students spread it on their own network.

The target group of the survey is the university students of the three countries studying in full time or distance learning, both from science and art territories.

There are 525 responses total, 391 responses arrived from the Hungarian language questionnaire and 134 from the Slovakian language one. The distribution of them by gender, kind of training and countries can be seen in the table below.

		Total responses by gender		Total responses by the type of learning process	
Country	Total responses	Male	Female	Full-time training	Distance learning
Hungary	187	88	99	104	81
Slovakia	141	20	121	90	36
Romania	194	92	102	128	65

Tab. 1. The distribution of responses.

These students were born into the digital world, and independently from their age and learning language and the kind of faculty where they are studying, they almost have smart tools in their pockets for sure. 98.2% of the respondents possess mobile devices. It is interesting that the over thirty year category gave the most active respondents and they are also the winner of the category containing respondents without mobile devices (Fig.5.).

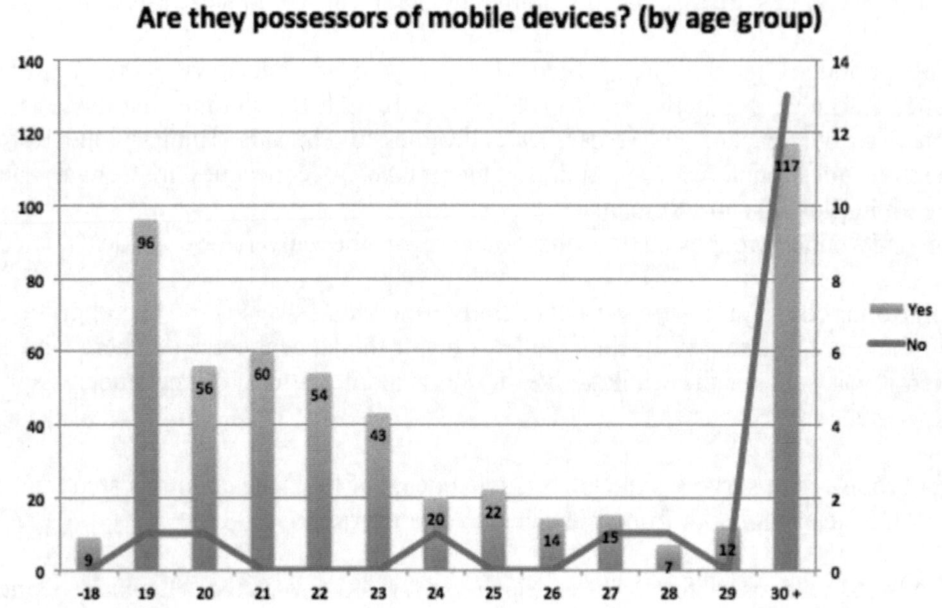

Fig. 5. Combination chart of age groups and their two categories to all respondents.

We were interested in seeing their habits in using mobile devices both in private life and at courses. As we thought, they use the mobile devices in their private life first of all for chatting, checking emails and social media platforms, making photographs and searching for information. The answers were similar in the three countries to this question, so we can say that there is not any difference between Slovakian, Hungarian and Romanian students regarding the utilisation of these technologies (Fig.6.).

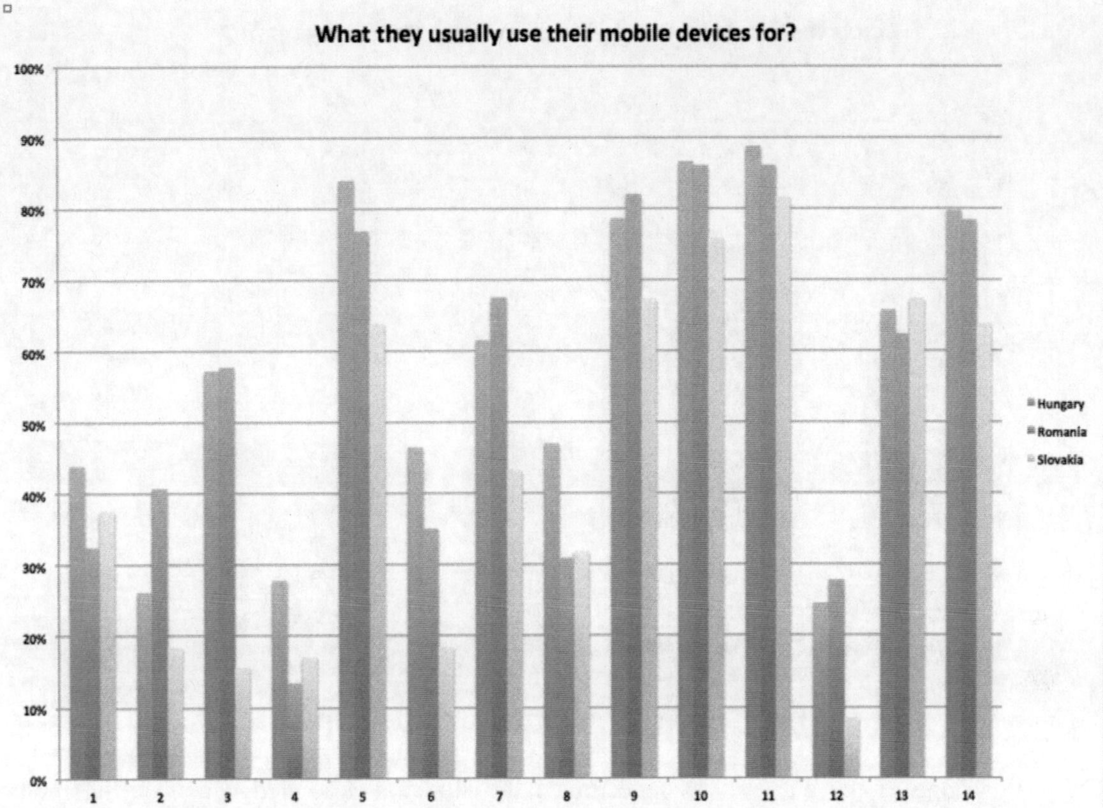

Fig. 6. The similar scores of using mobile devices in their private life.

1 - For online shopping, 2 - I use the Calculator, 3 - I use the online maps (e.g. Google Maps), 4 - For booking an appointment at different institution, 5 - For making photograph, 6 - I use the Cloud Systems (as Google Drive, OneDrive, Dropbox, iCloud), 7 - I use online dictionaries and translator programs as well (Google Translator, Glosbe, Dictzone), 8 - For mobile banking, SmartBanking, 9 - I use for checking my social media platforms (Facebook, Instagram, Twitter, SnapChat, Pinterest etc.), 10 - For chatting (Facebook Messenger, Viber, WhatsApp, Skype), 11 - Checking my emails, 12 - I use them during my sports training for checking my performance, 13 - I use the Calendar, 14 - For information searching

In the second section of the questionnaire, we asked them about the general usage of mobile technologies during their learning process. We can see that they rarely use them for solving math problems. (Fig. 7.).

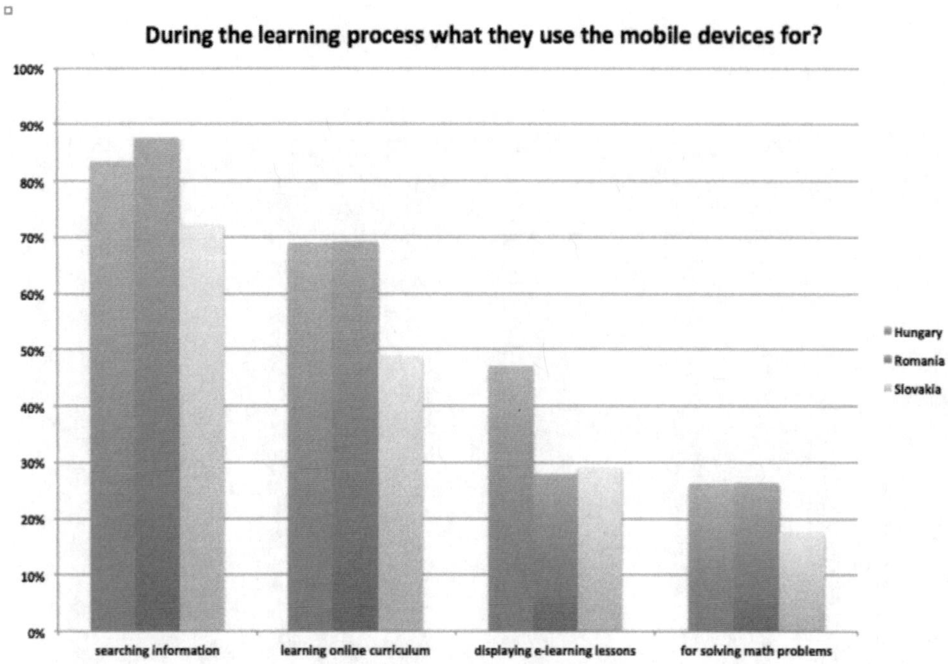

Fig. 7. Generally usage of mobile devices.

The main scope of the survey was to estimate the students' knowledge concerning math applications. This is a relatively new potential in education. Therefore we searched for the answer of the question which math applications they know and utilise. We can see above the top 7 applications used by students (Fig.8.).

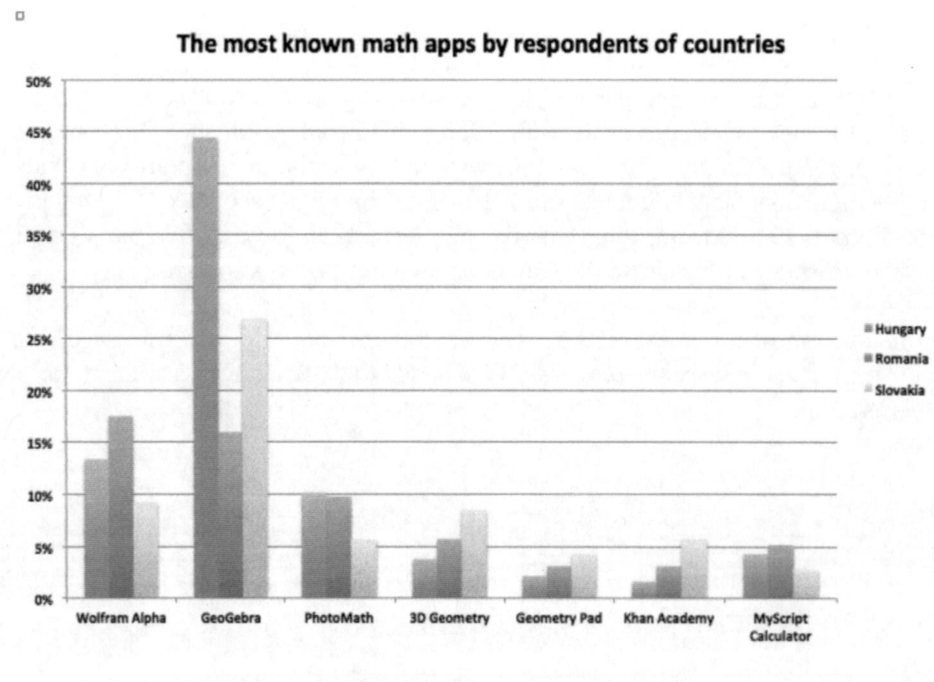

Fig. 8. Comparison of top 7 used application.

The quantitative research surpassed our waiting, because university students have other more interesting programs. We were satisfied with the number of the completed questionnaires. It can be said that Romanian students (who were in reality Transylvanian students) were the most active. We know that this generation was born into the digital world but that does not mean that they have highly developed digital competence. They are not native digital, but they are the children of Neumann Galaxy whose digital competence emerges only from their everyday usage. They are influenced very much by their cultural environment.

5 Conclusion

Today's students are always online, they live in the digital world. They can get the needed information it within a few seconds. At the same time, they have to learn the operation and the rules of the digital world because the information received is not a credible information in every case. They have to know some important searching rules, relevant sites, considerable applications, and they have to use the social media platforms intelligently if applicable.

In this article we searched the answer for the question that in which proportion do university students use mobile devices in their learning process, especially in mobile math learning and what is their opinion about the utilisation of these devices in learning?
For this purpose we establish a few lists about the university students' activities on their mobile devices. One list is about the possible online daily activities, the other is about online practices belong to learning and finally about math applications used by them. These lists were put in questionnaires made in Slovakian and Hungarian language and spread both in Hungary and Slovakia and in the Transylvanian part of Romania.
We obtained several interesting results. To the question that do their professors encourage the usage of mobile devices 40% of the respondents said yes. It is also interesting that the over thirty-year category gave the most active respondents and they are also the winner of the category containing respondents without mobile devices.
In our opinion, students do not know enough about the potential of mobile devices in learning. The solution is in the hand of the teachers. They can inspire students by changing the teaching method and allowing them to utilise mobile devices in their learning process. Additionally, it would be better if universities supported the development of more curriculums which can be used on these devices as well.
Our pilot project highlights this essential theme which naturally needs further survey. One possible future research can be about the influence of apps on the students' math learning.

Acknowledgment

The paper was written with the support of the grant KEGA 012UK-4/2018 "The Concept of Construcionism and Augmented Reality in the Field of the Natural and Technical Sciences of the Primary Education (CEPENSAR)".

References

[1] ACKERMANN, E., Minds in Motion, Media in Transition Growing up in the digital age: Areas of change, 2011. Website: http://www.childresearch.net/papers/pdf/digital_2011_01_ACKERMANN.pdf

[2] AZUMA, R. T., A survey of augmented reality. In *Presence: Teleoperators and virtual environments*, 1997, pp. 355-385.

[3] AZUMA, R., BAILLOT, Y., BEHRINGER, R., FEINER, S., JULIER, S., & MACINTYRE, B., Recent advances in augmented reality. In *IEEE computer graphics and applications*, 2001, pp. 34-47.

[4] BEŇAČKA, J., ČERETKOVÁ, S. Graphing functions and solving equations, inequalities and linear systems with pre-service teachers in Excel. *CERME 9 – Ninth Congress of the European Society for Research in Mathematics Education*, 2015, pp. 2311–2318.

[5] BOALER, J., *Mathematical Mindsets, Unleashing Students' POTENTIAL Through Creative Math, Inspiring Messages and INNOVATIVE TEACHING*, , 2016.

[6] ČAVOJSKÝ, I., Digital technology in primary education. *Wyzwania i zaniechania w wychowaniu przedszkolnym i kształceniu wczesnoszkolnym.* Siedlce: Uniwersytet Przyrodniczo-Humanistyczny, 2014, pp. 271-281.

[7] COCHRANE, T., NARAYAN, V., *Design Considerations for Mobile Learning,* Chapter: 15., 2017.

[8] COIMBRA, M. T., CARDOSO, T., & MATEUS, A., Augmented reality: an enhancer for higher education students in math's learning?. In *Procedia Computer Science*, 2015, pp. 332-339.

[9] FERKO, A., KOREŇOVÁ, L. Some possibilities for using mobile learning in mathematics. In *MATHEMATICA V*, Ružomberok: VERBUM, 2015, pp. 15–20.

[10] FULIER, J., ĎURIŠ, V., FRANTOVÁ, P. *Systémy počítačovej algebry (CAS) vo vyučovaní matematiky.* Nitra: FPV UKF, 2007, ISBN 978-80-8094-139-0

[11] KALAŠ, I., *Integrácia informačných a komunikačných technológií do všeobecného vzdelania.* Bratislava: ŠPÚ, 2001, ISBN: 8085756552

[12] KAUFMANN, H., SCHMALSTIEG, D., & WAGNER, M. Construct3D: a virtual reality application for mathematics and geometry education. In *Education and information technologies*, 2000, pp. 263-276.

[13] KHALIL, A., Mobile Learning Technologies. In *International Journal of Electrical and Computer Engineering*, Vol. 7, No. 5, pp. 2833-2837, 2017.

[14] KÓNYA, H. E., How can high school students solve problems based on the concept of area measurement? *Problem solving in mathematics education (PROMATH)*, Budapest: Eötvös Loránd University, 2014, pp. 95–107.

[15] KOREŇOVÁ, L. Mobile learning in elementary and secondary school mathematics in Slovakia. In *Electronic Journal of Mathematics & Technology* 9 (3), 259-268., 2015.

[16] KOREŇOVÁ, L., VERESS-BÁGYI, I., Augmented reality in mathematics education in primary, 2017.

[17] KOSTRUB, *D., Dieťa/žiak/student-učivo-učiteľ, didaktický alebo bermundský trojuholník*, Prešov: Rokus, 2008.

[18] KOVÁCS, Z., Modelling with difference equations supported by GeoGebra: Exploring the Kepler Problem. In *International Journal for Technology in Mathematics Education*, 17, 2010, pp. 141–146.

[19] NAGYOVÁ, A., Analysis of possibilities of the digital technologies use in teaching foreign languages Slavonic Pedagogical Studies. In *The Scientific Educational Journal.* - Roč. 5, č. 1 (2016), s. 101-112, ISSN 1339-8660
[20] SHARPLES, M., *Big Issues in Mobile Learning, in Report of a workshop by the Kaleidoscope. Network of Excellence Mobile Learning Initiative*, 2006.
[21] TRAXLER, J., Defining, Discussing and Evaluating Mobile Learning: the Moving Finger Writes, In *The International Review of Research in Open and Distance Learning.*vol 8, no 2., 2007.
[22] TRAXLER, J., *The evolution of mobile teaching and learning*, pp. 1-14, 2009.
[23] ZÁHOREC, J., [et al.] Results of a research evaluating quality of computer science education. In *Informatics in Education.* - Vol. 11, No. 2 (2012), s. 283-300 ISSN 1648-5831

Current address

Koreňová Lilla, doc. PaedDr., PhD.
Faculty of Education, Department of Pre-Primary and Primary Education
Comenius University in Bratislava
Račianska 59, 813 34 Bratislava, Slovak Republic
Email: korenova@fedu.uniba.sk

Veress-Bágyi Ibolya
Doctoral School of Mathematics and Computational Science
University of Debrecen, Hungary
4010 Debrecen, Pf. 12, Hungary
Email: veressbibolya@gmail.com

Slovak University of Technology in Bratislava
Faculty of Mechanical Engineering
17th Conference on Applied Mathematics **APLIMAT 2018**

Proceedings

THE THREE-MEASURABLE PROBLEM CHANGE OF COORDINATES FUNCTIONS OF FLOATER SUSPENDED IN ACOUSTIC ENVIRONMENT

KOROBIICHUK Igor (PL), KARACHUN Volodimir (UA), MEL'NICK Viktorij (UA), ASAFTEI Olena (UA), SZEWCZYK Roman (PL)

Abstract. The constructed analytical model of elastic-stressed state of a floated gyroscope gimbal. Clarified the pattern of elastic displacement of the gyroscope gimbal under the action of the penetrating acoustic radiation. Analyzed the possibility of manifestation of local features. The obtained values of the coordinate functions of the floated gyroscope gimbal provide the possibility of manifestation, in the operating conditions, of the effect of selectivity of the angular motion of an aircraft, components of acoustic vibration of its surface in the form of the formation of a systematic measurement error.

Keywords: hypersonic technology, gyroscope gimbal, float-shell gimbal, acoustic emission, floated gyroscope

Mathematics Subject Classification: Primary 74J20, 74J25; Secondary 70Q05

1 Introduction

Nowadays, a country's borders are defended by intercontinental ballistic and space missiles of different classes and deployment types, nuclear submarines and strategic missile forces. The present-day launch vehicles (LV) are known to be able to deliver the means of destruction from the country's continental territory to any part of the globe not only with a great accuracy but also within the shortest time allowed.

Ballistic missiles (BM), while in flight, are controlled by inertial or radio inertial systems. The first-mentioned type should be considered the most reliable one it is only these systems which offer the decisive advantage —self-containment. This feature makes it possible to implement a simultaneous launch of a large number of missiles, provides a sufficient degree of independence of the starting positions and, in addition, a high readiness and reliability of the on-alert means of tactical carrier-based aircraft (TPA), strategic bombers (SBA), the ship-to-air missile systems (for example, class "Fort" with 64 C-300 missiles), which are deployed with cruisers that have gas turbine engines, missile launchers "Basalt" class with cruise missiles P-500 (superprecision shooting to a distance (radius) of 500 km), as well as unmanned aircrafts (autonomous robots) having freedom of movement, the explosive ordnance disposal (EOD) robots (Mini-Andros) and, finally, remotely controlled vehicles. Robots are finding the ever widening applications for military intelligence purposes, demining of the water area, destruction of troops in a zone of military conflicts, and others.

1.1 Diffraction of N-waves on the impedance surface of gimbal

Analyzing the interaction of an N-wave and mechanical systems of inertial autonomous positioning, we will consider, to be more specific, a commercially available single-axis gyroscope with a liquid-static gimbal. Assuming that the oscillation generated in the gimbal is not transmitted to the mating surfaces, the consideration can be limited to merely the shell part, while for sake of completeness we will construct a three-dimensional model. The results obtained can be fully used, inter alia, for the analysis of dynamics of a two-axis gyro case.

2 Analysis of the problem state

It is the spectral density of the process of energy distribution, which provide the most complete picture of the sonic boom [1, 2]. During the sound–barrier breaking, the level of N-waves may exceed ten times the level of starting from open positions [3, 4].

Acoustic radiation that penetrates an aircraft generates numerous vibration modes, including resonance one, in the hardware of inertial sensors [5, 6]. Together, they produce disturbance torques of Euler inertial forces, which give rise to errors (or drift) of inertial sensor output signals [7-14].

The simplest while also the least labor-intensive methods to eliminate this phenomenon are the methods of design and technological solutions, i.e., passive methods [15-17].

We consider partial cases of the analytical model of the influence of geometry of the floating gyroscope gimbal on the gimbal coordinate functions.

The **object** of the research is the process of elastic interaction of penetrating radiation and mechanical systems of on-board equipment.

The purpose of the investigations is to build an accurate computational model of elastic interaction of acoustic emission and the mechanical system of floated gyroscope under hypersonic flight conditions.

To this end, the following should be accomplished:

1. Build an analytical model of the movable portion of the gyroscope gimbal.
2. Clarify the pattern of elastic displacement of the gyroscope gimbal under the action of the penetrating acoustic radiation.
3. Analyze the possibility of manifestation of local features.

3 Gimbal with arbitrary outline of the meridian line

Boundary conditions of mathematical model under study are defined in the assumption that there is no transfer of flexural shell surface oscillation energy to the end surface and vice versa. This is observed in the hinged connection of these surfaces, which is assumed in the manuscript materials. In this case, simulation model of the shell and the flat end can be described as an infinite in length element – the shell and plates.

Initial conditions. Suppose the shell belongs to the curvilinear orthogonal coordinates α_1 and α_2. They are regarded as lines of curvature with a radius R_1 and R_2.

Denote Lame parameters of the middle surface of the π shell by A_1 and A_2. Then, adding inertial forces, we can use the shell equilibrium equations, which in expanded form are written as follows:

$$\frac{\partial A_2 T_1}{\partial \alpha_1}+\frac{1}{A_1}\frac{\partial A_1^2 S}{\partial \alpha_2}-\frac{\partial A_2}{\partial \alpha_1}T_2+\frac{1}{R_1}\left(\frac{\partial A_2 M_1}{\partial \alpha_1}+\frac{1}{A_1}\frac{\partial A_1^2 H}{\partial \alpha_2}-\frac{\partial A_2}{\partial \alpha_1}M_2\right)+$$
$$+\frac{\partial}{\partial \alpha_2}\left(\frac{A_1}{R_1}H\right)+\frac{1}{R_2}\frac{\partial A_1}{\partial \alpha_2}H=-A_1 A_2 q_1+\rho A_1 A_2 h\frac{\partial^2 U_1}{\partial t^2}\;;$$

$$\frac{\partial A_1 T_2}{\partial \alpha_1}+\frac{1}{A_2}\frac{\partial A_2^2 S}{\partial \alpha_1}-\frac{\partial A_1}{\partial \alpha_2}T_1+\frac{1}{R_2}\left(\frac{\partial A_1 M_2}{\partial \alpha_2}+\frac{1}{A_2}\frac{\partial A_2^2 H}{\partial \alpha_1}-\frac{\partial A_1}{\partial \alpha_2}M_1\right)+$$
$$+\frac{\partial}{\partial \alpha_1}\left(\frac{A_2}{R_2}H\right)+\frac{1}{R_1}\frac{\partial A_2}{\partial \alpha_1}H=-A_1 A_2 q_2+\rho A_1 A_2 h\frac{\partial^2 U_2}{\partial t^2}\;;$$

$$\frac{T_1}{R_1}+\frac{T_2}{R_2}-\frac{1}{A_1 A_2}\left\{\frac{\partial}{\partial \alpha_1}\frac{1}{A_1}\left(\frac{\partial A_2 M_1}{\partial \alpha_1}+\frac{1}{A_1}\frac{\partial A_1^2 H}{\partial \alpha_2}-\frac{\partial A_2}{\partial \alpha_1}M_2\right)+\right.$$
$$\left.+\frac{\partial}{\partial \alpha_2}\frac{1}{A_2}\left(\frac{\partial A_1 M_2}{\partial \alpha_2}+\frac{1}{A_2}\frac{\partial A_2^2 H}{\partial \alpha_1}-\frac{\partial A_1}{\partial \alpha_2}M_1\right)\right\}=q_n+\rho h\frac{\partial^2 W}{\partial t^2}\;,\;(1)$$

where $q_1=p_1+\frac{m_1}{R_1}\approx p_1$; $\quad q_2=p_2+\frac{m_2}{R_2}\approx p_2$;

$$q_n=p_n+\frac{1}{A_1 A_2}\left(\frac{\partial A_2 m_1}{\partial \alpha_1}+\frac{\partial A_1 m_2}{\partial \alpha_2}\right)\approx p_n\,,$$

as in most cases quantities m_i are of the hp order, so that by identifying q_i and p_i, the terms of order $\frac{h}{R}$ are thus dropped out in comparison with unity; T_1, T_2 are normal forces, and S is tangential force; M1, M2 are bending moments; H is torque; ρ is density of shell material; h is shell thickness; u_i is elastic displacement of points of π surface towards the coordinate α_i.

In the form presented, equations (1) are inconvenient to use. Therefore, a series of transformations shall be done over them, after which they should be written in a form that is acceptable for integration.

In dimensionless form, the differential equations of a shell with an arbitrary outline of the meridian line are written as [18]

$$\frac{\partial^2 U_z}{\partial z^2}-(1+2\nu)\xi'(z)\frac{\partial U_z}{\partial z}+\left[(1+\nu\mu)\xi'^2(z)-\nu\xi''(z)\right]U_z+\frac{1}{R(1+\zeta)}\times$$

$$\times\left[\frac{1+\nu}{2}+\nu(1+\mu)\xi(z)\right]\frac{\partial^2 U_\phi}{\partial z\partial\phi}-$$

$$\frac{1}{R(1+\zeta)}\left\{-\left[(1+\nu)\mu+3\mu^2\right]\xi'(z)W+(\mu+\nu)\frac{\partial W}{\partial z}\right\}=$$

$$=\frac{1-\nu^2}{Eh}\left[1+2\mu\xi(z)\right]\left(-q_1+\rho h\frac{\partial^2 U_z}{\partial t^2}\right); \tag{2}$$

$$\frac{\partial^2 U_\phi}{\partial s^2}+\frac{1}{2}(1+\nu)\left[1-(1+\mu)\xi(z)\right]\frac{\partial^2 U_z}{\partial z\partial s}-\frac{1}{2}(1-\nu)\left[1-2(1+\mu)\xi(z)\right]\frac{\partial^2 U_\phi}{\partial z^2}-$$

$$-\frac{1}{2}(1-\nu)(1+\mu)\xi'(z)\frac{\partial U_\phi}{\partial z}-\frac{1}{2}(3-\nu)\xi'(z)\frac{\partial U_z}{\partial s}+\frac{1}{2}(1-\nu)\xi''(z)U_\phi-$$

$$-\frac{1}{R(1+\zeta)}\{1+\nu\mu-\left[(1+\nu)\mu+3\nu\mu^2\right]$$

$$\xi(z)+\left[(1+3\nu)\mu^2+\frac{15}{2}\nu\mu^3\right]\xi^2(z)\}\frac{\partial W}{\partial s}=$$

$$=\left[1-\xi(z)\right]^2\left(-q_2+\rho h\frac{\partial^2 U_\phi}{\partial t^2}\right)\frac{1-\nu^2}{Eh}; \tag{3}$$

$$\frac{\partial^2}{\partial z^2}\left\{-\left[1-(1+2\mu)\xi(z)\right]\frac{\partial^2 W}{\partial z^2}+(\nu+\mu)\xi'(z)\frac{\partial W}{\partial z}-\nu\frac{\partial^2 W}{\partial s^2}\right\}-$$

$$-\frac{\partial}{\partial z}\left\{(1-\nu)\left[2+\xi(z)\right]\frac{\partial^3 W}{\partial z\partial s^2}+\right.$$

$$\left[1+2(1-\nu)\xi'(z)\right]\frac{\partial^2 W}{\partial s^2}+\nu\xi'(z)\frac{\partial^2 W}{\partial z^2}+\frac{1-\nu}{R(1+\zeta)}\times$$

$$\times\left[1+\mu(\mu-2)\xi(z)\right]\frac{\partial^2 U_\phi}{\partial z\partial s}\Bigg\}-\frac{\partial^4 W}{\partial s^4}-\nu\frac{\partial^4 W}{\partial z^2\partial s^2}+\mu\xi'(z)\frac{\partial^3 W}{\partial z^3}+$$

$$+\left[1+\mu(1-\nu)\right]\xi'(z)\frac{\partial^3 W}{\partial z\partial s^2}-\mu(\nu+\mu)\xi'^2(z)\frac{\partial^2 W}{\partial z^2}-(\nu+\mu)\mu\xi'(z)\xi''(z)\frac{\partial W}{\partial z}-$$

$$-\frac{1}{R(1+\zeta)}\left[1+\left(2+\mu^2\right)\xi(z)\right]\frac{\partial^3 U_\phi}{\partial s^3}-\frac{\nu\mu}{R(1+\zeta)}\frac{\partial^3 U_z}{\partial z\partial s^2}+\frac{\mu^2\xi'(z)}{R(1+\zeta)}\frac{\partial^2 U_z}{\partial z^2}+$$

$$+\frac{\mu(1-\nu)}{R(1+\zeta)}\xi'(z)\frac{\partial^2 U_\phi}{\partial z\partial s}$$

$$+\frac{12}{h^2}\left[\frac{\mu+\nu}{R(1+\zeta)}\frac{\partial U_z}{\partial z}+\frac{1+\mu\nu}{R(1+\zeta)}\frac{\partial U_\phi}{\partial s}-\frac{1+\mu\nu}{R(1+\zeta)}\xi'(z)U_z\right]=$$

$$=-\frac{1}{D}\left[1-(1-\mu)\xi(z)\right]\left(q_3+\rho h\frac{\partial^2 W}{\partial t^2}\right), \qquad (4)$$

where U_z, U_φ, W are the elastic displacements of the shell surface along the generatrix, along the parallel, and in the transverse plane; h is the shell thickness; ρ is the density of the material; E is the Young modulus; ν is the Poisson ratio; $R=f(0)=f(\ell)=const$ is the radius at the edges; ℓ is the shell length; $r=f(z)$ is the distance from the axis of rotation to the point M; $f(z)$ is the rotation curve (meridian line); $\frac{\partial}{\partial S}=\frac{1}{R(1+\zeta)}\frac{\partial}{\partial\varphi}$; $\eta=\frac{R}{\ell}$ $\zeta=\frac{\delta}{R}<1$; $\xi=\frac{\delta}{\ell}<1$; $\mu=8\zeta(1+\zeta)\eta^2$; $2\mu\ll1$; δ is the ascent of the meridian line; provided that $\delta\to0$, of course, we will have $\xi\to0$ and $\mu\to0$; ω_0 is the natural frequency.

Equations (2) provide an opportunity to further carry out a comparative analysis of the four types of float-shell gimbal: with an arbitrary outline of the meridian line; convex (Figure 1, *a*.); concave or *catenoid*, from Lat. *catena* (Fig. 1, *b*), and (iv) circular cylinder (Fig. 1, *c*). In all the cases, the curve $f(z)$ that generates the shell is assumed to be symmetrical relative to the line *CM*.

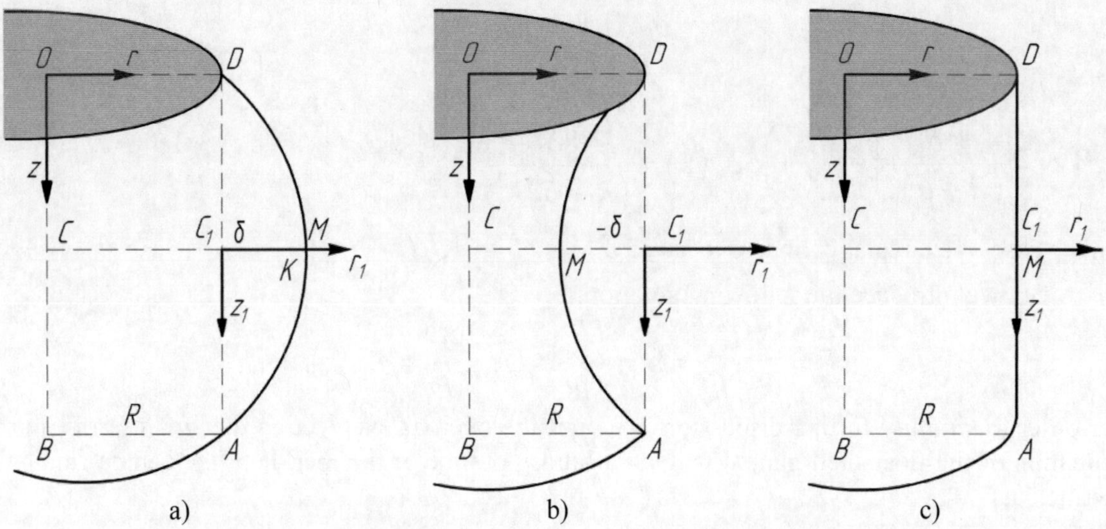

Fig. 1. Float shell of special form: *a)* a convex shell of rotation; *b)* a concave shell; *c)* a circular cylinder

The coordinate systems $C_1 z_1 r_1$ and Ozr are related by the analytical expressions

$$r = r_1 + R; \quad z = z_1 + \frac{1}{2}\ell .$$

In the coordinate system $C_1 z_1 r_1$ the meridian line is given by

$$r_1 = \pm f(z_1),$$

and the "+" and "–" signs refer to the schemes shown in Fig. 1, *a* and Fig. 1, *b*, respectively. Specify the class of the curves $f_1(z_1)$ for the implementation of the required shell. First of all, the following conditions must be necessarily met:

$$f_1(-z_1) = f_1(z_1);$$

$$f_1\left(\pm\frac{1}{2}\ell\right) = 0.$$

The function $\left[+f_1(z_1)\right]$ is considered to be strictly convex, and the function $\left[-f_1(z_1)\right]$ strictly concave. The function $f_1(z_1)$ is descending if $z_1 \in \left(0;\frac{\ell}{2}\right)$ (Fig. 1, *a*) and ascending if $z_1 \in \left(0;\frac{\ell}{2}\right)$ (Fig. 1, *b*). Obviously,

$$f_1(-z_1) = f_1(z_1).$$

It follows from equations (1), (2), and (3) that the elastic displacement of gimbal surface in all three directions, influence each other to a certain extent. The degree of this influence will be determined later on.

For the convenience of integration we should turn to dimensionless coefficients.

Considering that

$$\zeta = \frac{\delta}{R}; \qquad \eta = \frac{R}{\ell}; \qquad \mu = 8\zeta(1+\zeta)\eta^2; \qquad \xi(z) = \frac{\zeta}{1+\zeta}\left(\frac{2z}{\ell}\right)^2;$$

$$\xi'(z) = \frac{4\delta}{(R+\delta)\ell}\left(\frac{2z}{\ell} - 1\right); \; \xi''(z) = \frac{8\delta}{(R+\delta)\ell^2},$$

in the case of a circular cylinder we have: $\delta = 0$; $\zeta = 0$; $\mu = 0$; $\xi(z) = 0$.

So, we introduce the following notation:

$$\frac{z}{\ell} = \bar{z}; \; \frac{U_z}{h} = \bar{U}_z; \; \frac{U_\varphi}{h} = \bar{U}_\varphi; \; \frac{W}{h} = \bar{W}; \; \omega_0 t = \bar{t}. \tag{4}$$

In order to simplify further discussion, we omit the overbar. Neglecting the small terms, the equation of the float shell gimbal with an arbitrary outline of the meridian line can be finally written as:

$$\frac{\partial^2 U_z}{\partial z^2} - a_1(2z-1)\frac{\partial U_z}{\partial z} - a_2 U_z + a_3\frac{\partial^2 U_\varphi}{\partial z \partial \varphi} - a_4\frac{\partial W}{\partial z} =$$

$$= \left[1 + \alpha_1(2z-1)^2\right]\left(-q_1^* + \alpha^{*2}\frac{\partial^2 U_z}{\partial t^2}\right); \tag{5}$$

$$\frac{\partial^2 U_\phi}{\partial \phi^2} + b_1\left[1 - \beta_1(2z-1)^2\right]\frac{\partial^2 U_z}{\partial z \partial \phi} - b_2\left[1 - \beta_2(2z-1)^2\right]\frac{\partial^2 U_\phi}{\partial z^2} - b_3(2z-1)\frac{\partial U_\phi}{\partial z} -$$

$$-b_4(2z-1)\frac{\partial U_z}{\partial \varphi} + b_5 U_\varphi - b_6\frac{\partial W}{\partial \varphi} = \left[1 - \beta_3(2z-1)^2\right]\left(-q_2^* + \beta^{*2}\frac{\partial^2 U_\varphi}{\partial t^2}\right); \tag{6}$$

$$\left[-1 + \beta_4(2z-1)^2\right]\frac{\partial^4 W}{\partial z^4} - c_1\frac{\partial^4 W}{\partial z^2 \partial \varphi^2} - c_2\frac{\partial^4 W}{\partial \varphi^4} + c_3(2z-1)\frac{\partial^3 W}{\partial z^3} - c_4\frac{\partial^3 W}{\partial z\, \partial \varphi^2} +$$

$$+c_5\frac{\partial^2 W}{\partial z^2} - -c_6\frac{\partial^2 W}{\partial \varphi^2} - c_7(2z-1)\frac{\partial W}{\partial z} - c_8\frac{\partial^3 U_\varphi}{\partial \varphi^3} - c_9\frac{\partial^3 U_\varphi}{\partial z^2 \partial \varphi} - c_{10}\frac{\partial^3 U_z}{\partial z\, \partial \varphi^2} +$$

$$+c_{11}(2z-1)\frac{\partial^2 U_z}{\partial z^2} + +c_{12}(2z-1)\frac{\partial^2 U_\varphi}{\partial z\, \partial \varphi} + c_{13}\frac{\partial U_z}{\partial z} + c_{14}\frac{\partial U_\varphi}{\partial \varphi} - c_{15}(2z-1)U_z =$$

$$= \left[1 - \beta_5(2z-1)\right]\left(q_3^* + \gamma^{*2}\frac{\partial^2 W}{\partial t^2}\right), \tag{7}$$

where

$$a_1 = 4(1+2\nu)\frac{\delta}{R(1+\zeta)}; \; a_2 = 8\nu\frac{\delta}{R(1+\zeta)}; \; a_3 = \frac{1}{2}\frac{1+\nu}{1+\zeta}\frac{l}{R}; \; a_4 = \frac{\nu+\mu}{1+\zeta}\frac{h}{R};$$

$$q_1^* = (1-\nu^2)\frac{l^2}{E h^2} q_1 ; \quad \alpha^{*2} = (1-\nu^2)\frac{\rho \omega_0^2 l^2}{E}; \quad \alpha_1 = 2\mu \frac{\delta}{R(1+\zeta)}.$$

$$b_1 = \frac{1}{2}(1+\nu)(1+\zeta)\left(\frac{R}{l}\right); \qquad b_4 = 2(3-\nu)\frac{\delta}{l};$$

$$b_2 = \frac{1}{2}(1-\nu)(1+\zeta)^2\left(\frac{R}{l}\right)^2; \qquad b_5 = 4(1-\nu)(1+\zeta)\frac{\delta R}{l^2};$$

$$b_3 = 2(1-\nu)(1+\mu)(1+\zeta)\frac{\delta R}{l^2}; \qquad b_6 = 1+\nu\mu ;$$

$$\beta_1 = \frac{1+\mu}{1+\zeta}\frac{\delta}{R}; \quad \beta_2 = 2\frac{1+\mu}{1+\zeta}\frac{\delta}{R}; \quad \beta_3 = \frac{\delta}{R(1+\zeta)}; \quad \beta_2 = 2\beta_1 ;$$

$$q_2^* = (1-\nu^2)\frac{R^2(1+\zeta)^2}{E h^2} q_2 ; \quad \beta^{*2} = (1-\nu^2)\frac{\rho\omega_0^2}{E} R^2 (1+\zeta)^2 ;. \ \beta_4 = \frac{1+2\mu}{1+\zeta}\left(\frac{\delta}{R}\right);$$

$$c_1 = \frac{2}{(1+\zeta)^2}\left(\frac{l}{R}\right)^2; c_2 = \frac{1}{(1+\zeta)^4}\left(\frac{l}{R}\right)^4; c_3 = 8\frac{1+3\mu}{1+\zeta}\left(\frac{\delta}{R}\right);$$

$$c_4 = 4\frac{(1-\nu)(3-\mu)}{(1+\zeta)^3}\frac{\delta}{R}\left(\frac{l}{R}\right)^2;$$

$$c_5 = 8\frac{(1+\nu+4\mu)}{1+\zeta}\left(\frac{\delta}{R}\right); \; c_6 = 16\frac{(1-\nu)}{(1+\zeta)^3}\left(\frac{\delta}{R}\right)\left(\frac{l}{R}\right)^2; c_7 = \frac{32\mu(\nu+\mu)}{(1+\zeta)^2}\left(\frac{\delta}{R}\right)^2;$$

$$c_8 = \frac{1}{(1+\zeta)^4}\left(\frac{l}{R}\right)^4; c_9 = \frac{1-\nu}{(1+\zeta)^2}\left(\frac{l}{R}\right)^2; c_{10} = \frac{\nu\mu}{(1+\zeta)^3}\left(\frac{l}{R}\right)^3; c_{11} = \frac{4\mu^2}{(1+\zeta)^2}\left(\frac{\delta}{R}\right)\left(\frac{l}{R}\right);$$

$$c_{12} = \frac{4\mu\,(1-\nu)(3-\mu)}{(1+\zeta)^3}\left(\frac{\delta}{R}\right)\left(\frac{l}{R}\right)^2; c_{13} = 12(\nu+\mu)\frac{l^3}{Rh^2}; c_{14} = 12\frac{1+\nu\mu}{(1+\zeta)^2}\frac{l^4}{R^2 h^2};$$

$$c_{15} = \frac{4(1+\nu\mu)}{(1+\zeta)^2}\left(\frac{\delta}{R}\right)\frac{12 l^3}{Rh^2}; \; \gamma^{*2} = 12(1-\nu^2)\left(\frac{l}{h}\right)^4 \frac{\rho h \omega_0^2}{E}; \beta_5 = \frac{1-\mu}{1+\zeta}\left(\frac{\delta}{R}\right).$$

4. Conclusions

The constructed analytical model of elastic-stressed state of a floated gyroscope gimbal provides the following:

Conduct a qualitative and quantitative analyses of the elastic displacements of the shell surface of the float in three directions - along the meridian lines, along the perimeter of the frame, and in the radial direction.

The constructed analytical model enables one not only to analyze the structure of the coordinate functions as a function of time and geometrical parameters of the surface, but also to solve the optimization problem of the gyroscope gimbal for the minimum surface acoustic vibration generated in the material.

By determining the value of sound transmission in the shell can clarify the causes and development of technological risks hypersonic aircraft in time, in particular the manifestation of the effect of "acoustic transparency" of the object as a result of the wave match.

The obtained values of the coordinate functions allow one to find the values of Euler inertial forces acting on the surface float under flight conditions, determine the value of additional errors of floated gyroscope to determine the magnitude of errors due to direct action of moments of couple of Euler inertial forces in accordance with the Rezal theorem, when velocity of vector end of gyroscope's angular moment acquires velocity directed in the same way as the vector of perturbing moment of couple of Euler inertial forces and movable part of the device would rotate relative to the output axis as long as there is parallelism between them. Additional error of a gyroscope, besides the mentioned one, would contain one more component, which is caused by the indirect action of perturbing moments of Euler inertia forces, more precisely, Coriolis inertia forces due to elastic displacement velocities of float surface ($\dot{U}_\varphi, \dot{U}_z, \dot{W}$) on the aircraft body rotating relative to the cross axes of gyroscope suspension ($\omega_y, \omega_x, \omega_z$). This action is carried out through gyroscopic moments.

Numerical values of additional gyro errors in the operating conditions can thus be determined, as said above, on the basis of the known kinematics of aircraft body in the form of three angular velocities with respect to cross axes of the float and three velocities $\dot{U}_\varphi(t), \dot{U}_z(t), \dot{W}(t)$ of elastic deformation (surface coordinate functions), $(U_\varphi(t), U_z(t), W(t))$ of float surface under the influence of sound waves.

The obtained values of the coordinate functions of the floated gyroscope gimbal provide the possibility of manifestation, in the operating conditions, of the effect of selectivity of the angular motion of an aircraft, components of acoustic vibration of its surface in the form of the formation of a systematic measurement error.

Acknowledgement

The funds for covering the costs to publish were received from the Warsaw University of Technology Grant No. 504/03274

References

[1] FEODOSJEV, V.I., SINJAROV, G.B., Introduction to rocketry, *Russia, Moscow: Oboroniz*, 1969, 506. [in Russian]

[2] KRENDELL, S., Accidental fluctuations. *Russia, Moscow: Mir,* 1967, 356. [in Russian]

[3] KARACHUN, V.V., MELNIC, V.N., Hypersonic technology. *Prospects and problems. Gyro technologies, navigation, traffic control and design of aerospace engineering materials IX Intern. scientific and practical. conf.,* Kyiv, NTU "KPI", 2013, 38-44. [in Ukrainian]

[4] KARACHUN, V.V., SHIBECKIY, V.U., The action of the shock N-wave at the float suspension in circulation. *Integrated robotehnical complex materials Sixth Intern. scientific and practical. conf.*, Kyiv, NAU, 2013, 32-33. [in Ukrainian]

[5] KUZMENKO,V.A., MATOHNUK, L.E., PISARENKO, G.G, TROJAN, I.A., SHEVCHUK, A.D., Fatigue testing at higher loading frequencies. Ukraine, Kiev: *Sciences opinion,* 1976, 336. [in Russian]

[6] MELNIC, V.N., KARACHUN, V.V., The main causes of noise carrier rockets. *Gyro technologies, navigation, traffic control and design of aerospace engineering materials IX Intern. scientific and practical. conf.*, Kyiv, NTU "KPI", 2013, 98-104. [in Ukrainian]
[7] MELNIC, V.N., KARACHUN, V.V., Axleunsimetrical hanging floating deformation under the action of the shock wave. *The first approximation. Education and Science of the XXI Century materials of VIII International Scientific practical conference*, Sofia, 17-25 October 2012, 5-7.
[8] KARACHUN V.V., MELNIC V.N., Lame parameters and the curvature of the surface of floating hanging. *Scientific Industries of the European continent: Materials VII international scientific-practical conference, Prague, Publishing House "Science and Education"*, 2011, 38-42.
[9] KARACHUN V., MEL'NICK V., KOROBIICHUK I., NOWICKI M., SZEWCZYK R., KOBZAR S., The Additional Error of Inertial Sensor Induced by Hypersonic Flight Condition. *Sensors.* Volume 16, Issue 3, 2016. doi:10.3390/s1603029
[10] KOROBIICHUK I., KARACHUN V., MEL'NICK V., KACHNIARZ M., Modeling of the hypersonic conditions influence on the gyroscopic inertial navigation sensor suspension. *Metrol. Meas. Syst.,* 2017, 2, pp. 357–368. DOI: 10.1515/mms-2017-0024
[11] KOROBIICHUK I., BEZVESILNA O., TKACHUK A., CHILCHENKO T., NOWICKI M., SZEWCZYK R., Design of piezoelectric gravimeter for automated aviation gravimetric system. *Journal of Automation, Mobile Robotics & Intelligent Systems (JAMRIS)*, 2016, vol. 10, I. 1, pp. 43-47. DOI: 10.14313/JAMRIS_1-2016/6
[12] KOROBIICHUK I., NOWICKI M., SZEWCZYK R., Design of the novel double-ring dynamical gravimeter. *Journal of Automation, Mobile Robotics and Intelligent Systems*, 2015, vol. 9, I. 3, p. 47-52. DOI: 10.14313/JAMRIS_3-2015/23
[13] KOROBIICHUK I., Analysis of Errors of Piezoelectric Sensors Used in Weapon Stabilizers. *Metrol. Meas. Syst.*, 2017, Vol. 24, No. 1, pp. 91–100. DOI: 10.1515/mms-2017-0001.
[14] Korobiichuk I., Experimental Investigations of a Precision Sensor for an Automatic Weapons Stabilizer System. *Sensors*, 2017, 17(1), 23. DOI:10.3390/s17010023
[15] FOMENKOVA, A.O., KARACHUN, V.V., Elastic-stress state as a factor in hanging gyro additional errors in the operational use of aircraft. *Scientific session of the NUAP: Fri. Reports.: in 3h. Part I. Technical sciences.* Russia, St. Petersburg, NUAC, 14-17 April 2012, 55-57. [in Russian]
[16] KARACHUN, V.V., MELNIC, V.N., KOSOVA, V.P., Fourier series to determine the compliance of the elastic shell. *European science XXI century: materials ix international scientific-practical conference*, Prremyśl, 07-15 May 2013, 57-59.
[17] MEL'NICK, V.N., TRIVAILO, M.S., KARACHUN, V.V., The loss of sound waves. *Nat. Tech. Univ. Ukraine "KPI":Korniychuk*, 2010, 120 p.
[18] KARACHUN V.V., KAYUK YA.F., MEL'NYK V.N., Khvyl'ovi zadachi poplavkovoho hiroskopa. *Nats. tekhn. un-t Ukrayiny «KPI».* Kyyiv: «Korniychuk», 2007, pp. 217 - 228.

Current Address

Korobiichuk Igor, doc., PhD.
Institute of Automatic Control and Robotics, Warsaw University of Technology,
Boboli 8, 02-525, Warsaw, Poland;
E-mail: igor@mchtr.pw.edu.pl

Karachun Volodimir, Professor, DSc.
National Technical University of Ukraine "Igor Sikorsky Kyiv Polytechnic Institute"
37, Avenue Peremogy, Kyiv, Ukraine, 03056
E-mail: karachun11@i.ua

Mel'nick Viktorij, Professor, DSc.
National Technical University of Ukraine "Igor Sikorsky Kyiv Polytechnic Institute"
37, Avenue Peremogy, Kyiv, Ukraine, 03056
E-mail: vmm71@i.ua

Asaftei Olena, MSc., PhD student
National Technical University of Ukraine "Igor Sikorsky Kyiv Polytechnic Institute"
37, Avenue Peremogy, Kyiv, Ukraine, 03056
E-mail: bti@fbt.ntu-kpi.kiev.ua

Szewczyk Roman, Professor, DSc.
Industrial Research Institute for Automation and Measurements PIAP
Jerozolimskie 202, 02-486 Warsaw, Poland
E-mail: rszewczyk@piap.pl

Slovak University of Technology in Bratislava
Faculty of Mechanical Engineering
17th Conference on Applied Mathematics **APLIMAT 2018**

Proceedings

VERIFYING THE AUTHENTICITY OF THE ACCOUNTING DATA USING THE BENFORD'S ALGORITHM

KOŠŤÁLEK Josef (CZ), KOŤÁTKOVÁ STRÁNSKÁ Pavla (CZ)

Abstract. This article describes examples of such algorithms, which is basing on the knowledge of American physics Benford. Benford's law describes the behavior of empirical values, when the size of the relative frequency digits forms a numeric or financial data. The aim of the article is a description of the model that is capable of decomposition on single digit figures. The digits are calculating relative to frequency and those are then comparing to the probabilistic model respectful of Benford's law.

Keywords: analysis of the accounting data, Benford's distribution, Benford's law, first digit distribution

Mathematics Subject Classification: Primary 62E99; Secondary 62P99, 62P05.

1 Literature review

The initial scientific studies dealing with the application the Benford's law in the area of accounting and taxation issues come from authors [3,11], and the collective [4,14]. They followed the articles on mathematics [5,12]. At present, the scientific studies focus on the application of the law in the area of Benford macro-economic data, and government economic data, especially in studies of [12] will be used in the article named On the Application of Benford's Law's International Macroeconomic Statistics published in the magazine of the German Economics. Finally, it is necessary to mention also the credibility of scientific results, for example the regression coefficients. In scientific study [13] dealt with the place of convergence test data to evaluate the Benford distribution of national according to the width of the data from the Benford distribution of variances, respectively, the average result of χ^2 test in the field of national accounting. The results of the scientific research of the 27 member States of the EU have shown that the Czech Republic showed the smallest diameters of the χ^2 test. The research focused on the 27 Member States. The resulting findings have been that of all the EU Member States showed a significant deviation from the Benford distribution, namely Greece and Ireland.

1.1 Theoretical basis of Benford´s law

The model uses Benford's law of frequency distributions of numbers in the first place of the numeric data. The likelihood that the figure will begin with the digit 1 is not as high as the probability that will begin with the digit 9. This probability according to [1,5,6,7,8] (see Tab. 1. and formula (1)).

Digits	1	2	3	4	5	6	7	8	9
Probability (%)	30.10	17.61	12.49	9.69	7.92	6.70	5.80	5.11	4.58

Tab. 1. Benford's probability distribution of first digits.

The values (of Tab. 1.) are calculated using (1), in [2,8].

$$P(\%) = log\left(1+\frac{1}{d_1}\right)\times 100 \text{, where } d_1 \in \{1, 2, 3, \ldots, 8, 9\} \quad (1)$$

In addition to the likelihood of the digit in the first place can be formula (1) extend to formula (2), which calculates the probability of occurrence of the first two digits. We calculate so eg. What is the probability that a numeric value begins with the value 11 or 36.

$$P(\%) = log\left(1+\frac{1}{10\times d_1 + d_2}\right)\times 100 \text{, where} d_1 \in \{1, 2, 3, \ldots, 8, 9\} \text{ a } d_2 \in \{1, 2, 3, \ldots, 8, 9\} \quad (2)$$

These are similar s for determining the probability of the first three digits, etc., but here the differences in probabilities are no longer so much contrast. In formula (3) it used to calculate the probability of the occurrence of digits in second place. The probability that it will be the second, for example the number 3 is the sum of the probabilities of scenarios: 13, 23, 33... 83, 93.

$$P(\%) = \sum_{d_1=1}^{9} log\left(1+\frac{1}{10\times d_1 + d_2}\right)\times 100 \text{, } d_1 \in \{1, 2, 3, \ldots, 8, 9\} \text{ a } d_2 \in \{1, 2, 3, \ldots, 8, 9\} \quad (3)$$

The use of these patterns has its limits. Since the probability is necessary to ensure the most figures. This rule States that the basic statistical (4). To prove the measurability and Benford's distribution recommended at least 200 figures, in [2,4,5,6].

$$\lim \frac{n_i}{n} \to p \quad (4)$$

The input condition is that the figures include at least three digits.

1.2 Principle of functioning of the model

The model created in SW MS Excel in which accounting data entered. Then automatically verify the accuracy of data according to Benford's laws. The first digit validates of the frequency, the frequency of the first two digits and the frequency of the second digit. The Benford's distribution verified with of statistical tests: χ^2 test and Kolmogorov-Smirnov test when the specified confidence level "α", which can specified as a parameter in the model. [9]

Created by the program must first take the figure and divide it into individual digits, in order to calculate the frequency of these digits. If the input entered is a negative number, the absolute value created from it. If the specified entry has only one number, it is automatically skip. Decimal places deleted. First, it is necessary to eliminate incorrect values from the input data. The following algorithm is in a form (5).

IF(size of the input sample of data; S VYHLEDAT (serial number of the input data; search area; column to select rounded data; untruth); empty cell) (5)

After the elimination of incorrect values from the input data are calculated the frequency digits. It shows (see Fig. 1.) the result of the frequency digits in first place compared with theoretical – Benford´s values (according to formula (1)).

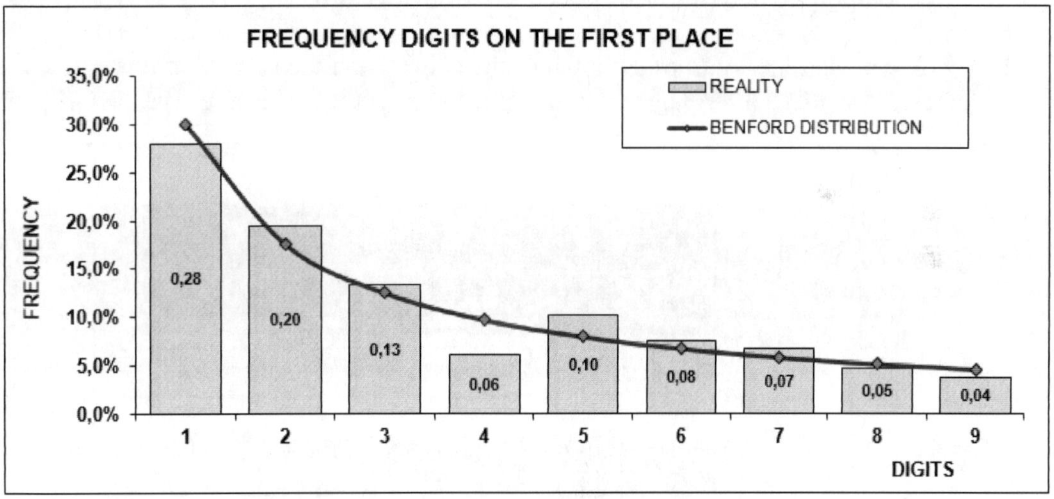

Fig. 1. Frequency digits on the first place.

Further statistical tests performed to verify the conformity of the Benford´s frequency distributions with frequency digits of empirical data used for testing the first digit χ^2 test. When testing the second digit and the first two digits is used Kolmogorov-Smirnov test, due to the existence of small probability and failure to comply with conditions for χ^2 test, and that the theoretical frequency must have a minimum value "5", in [5,8, 9].

All calculations and statistical tests created automatically in the model (see Fig. 2.). Just insert the data and test result appears with "OK" or "NOK".

The tests could also do with the frequencies of digits in third and fourth place, etc. Why is it usually does not do is that while the first digits are the differences of frequency (likely) very contrasting (see Tab. 1. or Fig. 2.) (frequency number 1 is 30.1%, the frequency of the number 9 is 4.6%) and the additional digits, these contrasts gradually reduced.

If there is a small difference of real numbers, and frequency numbers Benford´s breakdown (model), is a testing criterion of small value. Then the reality matches the model and the statistical test to apply. If the difference is greater than the critical value of the statistical test does not apply and there is no consensus.

The validity of the test is according to the parameter α (level of significance). The normal value is 0.05. The critical value is calculating according to statistical functions CHISQ.INV.

2 Result of the model

The model created in MS Excel, easily accessible for everyone. The data we need to verify is enough to insert into the green entry column. Automatic calculations of three statistical tests to verify compliance with the Benford´s distribution performed automatically.

In Fig. 2., you can see a part of the input of the values, and the parameter α is entered, eg 0.05. If the test is positive, "OK" appears in the green box, if the test is negative, "NOK" appears in the red box.

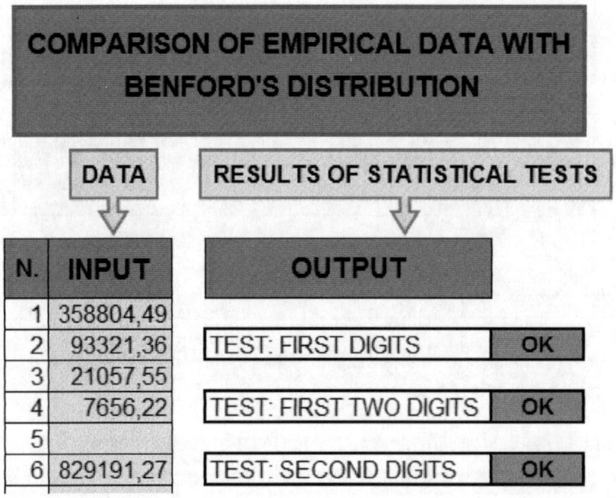

Fig. 2. Demonstration of work in the model.

The real data entered in to the model. Then there was a validation when α (level of significance) of 0.05. The result indicates that the frequency digits on the first two points correspond to the Benford's laws. Therefore, there is no reason to believe that the data tampered with and falsified.

3 Conclusion

The accounts of any company gets special characteristics, in terms of frequency distributions of digits. Quantity of digits that occurs in the accounting respects the Benford's law. Data input assumption is realistic data. Therefore, you cannot take advantage of Benford's law for the so-called "random digits". As mentioned above, Benford's law used to detect falsified data, in particular falsification of data in accounting. If the handling of a digits in accounting (small, medium or large business), then cease to satisfy Benford's law. In the case that occurs to manipulate the digits, it is with a certain probability followed the premise of the so-called statistical distribution. The statistical breakdown achieved by efforts to ensure the same likelihood of digit 1 or 2 or ... 9 at the beginning of the accounting data. This is a big mistake. With real data, the occurrence of the digit "1" at the beginning of the accounting data according to the Benford's law about 30%, rather than about 11% in the case of compliance with the so-called statistical distribution.

As has already been mentioned earlier in the text, the data file contained over 300 figures from the accounts of the real business. The results confirmed the validity of the Benford's distribution. It is to demonstrate the functioning this law. It also includes a confirmation of the fact that, in the accounting of this business not to manipulate the data.

The described model can very quickly and easily, whether the data are in some way suspicious. It used here to describe some properties of the Benford´s laws of the empirical data. This model is well suited especially for the verification of accounting data as first degree of control that identifies which of the accounts is in order and where values are suspicious and possibly rigged. This selection of the accounting unit is then to be exam such as dale using the audit of the accounts.

References

[1] BARROW, D. J. Benford's very strange law. *Lecture on Gresham College* [online]. 2011.25. August 2011. [vid. 18. Mai 2017]. Available at: https://www.youtube.com/watch?v=4iz4EHriYz0&feature=related.

[2] BERGER, A., HILL, P. T. A Basic Theory of Benford's Law. *Probability Surveys,* vol. 8, no. 1. 2011, ISSN 1549-5787. DOI: 10.1214/11-PS175 php?id=696&article.

[3] CARSLAW, C. A. P. Anomalies in Income Numbers: Evidence of Goal Oriented Behaviour. *Accounting Review*, vol. 63, no. 2, 1988: pp. 321–327, ISSN 0001-4826.

[4] DURTSCHI, C., HILLISON, W., PACINI, C. The effective use of Benford's law to assist in detecting fraud in accounting data. *Journal of Forensic accounting*, vol. 5, no. 1, 2004: pp.17–34. ISSN 1524-5586.

[5] HILL, T. P. The First Digit Phenomenon. *American Scientist*, vol. 86, no. 1, 1998: pp. 358–363, ISSN 0003-0996.

[6] HILL, T. P. A Statistical Derivation of the SignificantDigit Law. *Statistical Science*, vol. 10, no. 4, 1995b: pp. 354–363, ISSN 0883-4237.

[7] HILL, T. P. Base-Invariance Implies Benford's Law. *Proceedings of the American Mathematical Society*, vol. 123, no. 3, 1995b: pp. 887–895, ISSN 0894-0347.

[8] CHO, W. K. T., GAINES, B. J. Breaking the (Benford) Law: Statistical Fraud Detection in Campaign Finance. *The American Statistician*, vol. 61, no. 3, 2007: pp. 218-223, ISSN 0003-1305. DOI 10.1198/000313007X223496.

[9] NIGRINI, M. *Forensic Analytics: Methods and Techniques for Forensic Accounting Investigation.* NY: John Wiley&Sons Inc., 2011. ISBN 9781118087633. 352 s.

[10] NIGRINI, M. J. *Benford's law: applications for forensic accounting, auditing, and fraud detection.* NY: John Wiley&Sons Inc., 2012. ISBN 978-1-118-15285-0. 352 s.

[11] NIGRINI, M., MITTERMAIER, L. Audit Sampling Using Benford's Law: A Review of the Literature With Some New Perspectives. *Journal of Emerging Technologies in Accounting*. DOI 10.2308/jeta-51783.

[12] NYE, J., MOUL, C. The Political Economy of Numbers: On the Application of Benford's Law to International Macroeconomic Statistics. *Journal of Macroeconomics,* vol. 7, no. 1, 2007: pp. 1–14, ISSN 0164-0704.

[13] RAUCH, B., GÖTTSCHE, M., BRÄHLER, G. AND ENGEL, S. Fact and Fiction in EU-Governmental Economic Data. *German Economic Review*, vol. 12, no. 1, 2011: pp. 243–255, ISSN 1468-0475. DOI 10.1111/j.1468-0475.2011.00542.x.

[14] WATRIN, CH. Benford's Law: An Instrument for Selectin Tax Audit Targets?. *Review of Managerial Science*, vol. 2, no. 3, 2008: pp. 219–237, ISSN 1863-6683. DOI 0.1007/s11846-008-0019-9 1863-6683.

Current address

Košťálek, Josef, Ing.
Department of Management and Economics
Faculty of Mechanical Engineering
Czech Technical University in Prague
Karlovo náměstí 13, 121 35 Prague 2, Czech Republic
E-mail: josef.kostalek@fs.cvut.cz

Koťátková Stránská Pavla, Ing., PhD.
Department of Economic studies
Masaryk Institute of Advanced Studies, MIAS School of Business Institution
Czech Technical University in Prague
Kolejní 2637/2A, 160 00 Praha 6, Czech Republic
E-mail: pavla.kotatkova.stranska@cvut.cz

Slovak University of Technology in Bratislava
Faculty of Mechanical Engineering
17th Conference on Applied Mathematics **APLIMAT 2018**

Proceedings

REGULARITY CRITERION FOR LOCAL IN TIME EXISTENCE OF STRONG SOLUTIONS TO THE NAVIER-STOKES EQUATIONS WITH NAVIER'S TYPE BOUNDARY CONDITIONS

KUČERA Petr (CZ), PÍŠOVÁ Jitka (CZ), VACKOVÁ Petra (CZ)

Abstract. In this contribution we deal with the system of the Navier-Stokes equations with boundary conditions of the Navier's type on the bounded smooth convex domain. The main theorem gives condition for local in time existence of strong solutions to this system. This result is modification of results of Robinson et al. for solutions of the Navier-Stokes system on the whole space or on the space-periodic domains.

Keywords: Navier-Stokes equations, Navier's type boundary conditions, regularity

Mathematics subject classification: Primary 35Q30, 35B35; Secondary 76D05, 76E09.

1 Introduction

Let $\beta > 0$, $0 < T \leq \infty$ and $\Omega \subset \mathbb{R}^3$ be a convex bounded domain with the boundary $\partial\Omega$ of the class $C^{2+\beta}$. We denote $Q_T = \Omega \times (0, T)$. We study the following initial–boundary value Navier–Stokes problem

$$\partial_t \boldsymbol{u} + \nu\, \mathbf{curl}^2 \boldsymbol{u} + \boldsymbol{u} \cdot \nabla \boldsymbol{u} + \nabla P = \mathbf{0} \qquad \text{in } Q_T, \tag{1.1}$$

$$\mathbf{div}\, \boldsymbol{u} = 0 \qquad \text{in } Q_T, \tag{1.2}$$

$$\text{a) } \boldsymbol{u} \cdot \boldsymbol{n} = 0, \quad \text{b) } \mathbf{curl}\, \boldsymbol{u} \times \boldsymbol{n} = \mathbf{0} \qquad \text{on } \partial\Omega \times (0, T), \tag{1.3}$$

$$\boldsymbol{u}(\,.\,, 0) = \boldsymbol{u}_0 \qquad \text{in } \Omega. \tag{1.4}$$

Here, $\boldsymbol{u} = (u_1, u_2, u_3)$ and P denote the velocity of motion and the associated pressure. By $\boldsymbol{n}$ and ν we denote the outer normal vector field on $\partial\Omega$ and the kinematic coefficient of viscosity, respectively. For simplicity we suppose $\nu = 1$. Equations (1.1), (1.2) describe the motion of a viscous incompressible fluid in domain Ω.

H. Navier formulated so called Navier's boundary conditions in 1824. They takes the form

$$\boldsymbol{u} \cdot \boldsymbol{n} = 0, \quad [\mathbb{T} \cdot \boldsymbol{n}]_\tau + \gamma \boldsymbol{u} = 0. \tag{1.5}$$

The symbols $\mathbb{T}$ and the subscript τ denote the stress tensor and the tangential component, respectively. The first Navier's condition is the condition of impermeability of the wall. The second condition expresses the requirement that the tangential component of the stress, with which the fluid acts on the boundary, is proportional to the velocity. The stress tensor has the form $\mathbb{T} = -p\mathbb{I} + \nu\,[\nabla \boldsymbol{u} + (\nabla \boldsymbol{u})^T]$. Consequently, the second Navier condition (1.5b) can be written in the form

$$\nu\,\mathbf{curl}\,\boldsymbol{u} \times \boldsymbol{n} - 2\nu\,\boldsymbol{u}\cdot\nabla\boldsymbol{n} + \gamma\,\boldsymbol{u} \;=\; 0.$$

If we suppose that coefficient γ of friction between the fluid and the boundary is zero (the case of the so called perfect slip) and the curvature of the wall is neglected then (1.5) is equivalent to our boundary condition (1.3b). The Navier–Stokes equations with the boundary conditions (1.3) are studied, e.g., in papers [1], [2], [3], [4], [13], [16].

1.1 Notation of function spaces and operators.

We denote vector–valued functions and spaces of such functions by boldface letters. We use these function spaces and operators:

- c is a generic constant, i.e. a constant whose value may vary from line to line. On the other hand, numbered constants have fixed values throughout the whole paper.
- $\mathbf{C}^\infty_{0,\sigma}(\mathbb{R}^3) = \{\boldsymbol{\phi} \in \mathbf{C}^\infty(\Omega);\ supp\,\phi \subset \Omega,\ \mathbf{div}\,\phi \equiv 0\}$.
- $\mathbf{L}^s_\sigma(\Omega)$ is a closure of $\mathbf{C}^\infty_{0,\sigma}(\Omega)$ in $\mathbf{L}^s(\Omega)$.
- The Hemholtz projection of $\mathbf{L}^s_\sigma(\Omega)$ onto $\mathbf{L}^s(\Omega)$ is denoted by $\mathbf{P}_\sigma$. The following holds

$$\mathbf{L}^s(\Omega) = \mathbf{L}^s_\sigma(\Omega) \oplus G(\Omega) \tag{1.6}$$

  where $G(\Omega) = \{\nabla p;\ p \in \mathbf{W}^{1,s}(\Omega)\}$, see, e.g., [6] for the proof.
- $\mathbf{L}^2_\sigma(\Omega)^\perp = \{\nabla\varphi;\ \varphi \in W^{1,2}(\Omega)\}$ (The symbol $\mathbf{L}^2_\sigma(\Omega)^\perp$ denotes the orthogonal complement to $\mathbf{L}^2_\sigma(\Omega)$ in $\mathbf{L}^2(\Omega)$).
- $\mathbf{W}^{1,2}_\sigma(\Omega) := \mathbf{W}^{1,2}(\Omega) \cap \mathbf{L}^2_\sigma(\Omega)$. (Functions in this space are divergence–free in Ω and have the normal component on $\partial\Omega$ equal to zero in the sense of traces.)
- $\|\,.\,\|_k$ (respectively $\|\,.\,\|_{l,k}$) denotes the L^k–norm (respectively the $W^{l,k}$–norm) of a scalar–valued or vector–valued or tensor–valued function in Ω. The scalar product in $\mathbf{L}^2(\Omega)$ is denoted by $(\,.\,,\,.\,)_2$.
- The dual space to $\mathbf{W}^{1,2}_\sigma(\Omega)$ is denoted by $\mathbf{W}^{-1,2}_\sigma(\Omega)$. The duality between the elements of $\mathbf{W}^{-1,2}_\sigma(\Omega)$ and $\mathbf{W}^{1,2}_\sigma(\Omega)$ is denoted by $\langle\,.\,,\,.\,\rangle_\Omega$ and the norm in $\mathbf{W}^{-1,2}_\sigma(\Omega)$ is denoted by $\|\,.\,\|_{-1,2}$.
- $\mathbf{W}^{2,2}_{\sigma,nc}(\Omega) := \big\{\boldsymbol{u} \in \mathbf{W}^{2,2}(\Omega) \cap \mathbf{W}^{1,2}_\sigma(\Omega);\ \mathbf{curl}\,\boldsymbol{u} \times \boldsymbol{n} = \mathbf{0} \text{ on } \partial\Omega\big\}$. (The subscript „nc" indicates the Navier–type boundary conditions.)
- $\mathbf{H}(\Omega)$ is the space of all divergence–free functions $\boldsymbol{u} \in \mathbf{W}^{1,2}(\Omega)$ such that $\boldsymbol{u} \times \boldsymbol{n} = \mathbf{0}$ on $\partial\Omega$.
- $\mathbf{T}_1 := \mathbf{curl}$ is the operator with the domain $D(\mathbf{T}_1) = \mathbf{W}^{2,2}_{\sigma,nc}(\Omega)$.
- $\mathbf{T}_2 := \mathbf{curl}$ is the operator with the domain $D(\mathbf{T}_2) = \mathbf{H}(\Omega)$. The kernel of both the operators $\mathbf{T}_1$ and $\mathbf{T}_2$ is trivial because domain Ω is simply connected.

- $\mathbf{S} := \mathbf{T}_2 \circ \mathbf{T}_1$ is an operator in in $\mathbf{L}^2_\sigma(\Omega)$. $\mathbf{S}$ is one of the concrete realizations of the so called Stokes operator. The domain of $\mathbf{S}$ is $D(\mathbf{S}) = D(\mathbf{T}_1) = \mathbf{W}^{2,2}_{\sigma,nc}(\Omega)$.

Below we list some properties of operators $\mathbf{T}_1$, $\mathbf{T}_2$ and $\mathbf{S}$, see [13] and [16].

- $\mathbf{T}_1$ is a linear bijective operator from $\mathbf{W}^{2,2}_{\sigma,nc}(\Omega)$ onto $\mathbf{H}(\Omega)$. The inverse operator is bounded as an operator from $\mathbf{W}^{1,2}(\Omega)$ into $\mathbf{W}^{2,2}(\Omega)$.
- $\mathbf{T}_2$ is a linear bijective operator from $\mathbf{H}(\Omega)$ onto $\mathbf{L}^2_\sigma(\Omega)$. The inverse operator is bounded as an operator from $\mathbf{L}^2(\Omega)$ into $\mathbf{W}^{1,2}(\Omega)$.
- Operator $\mathbf{S}$ is a linear bijective operator from $\mathbf{W}^{2,2}_{\sigma,nc}(\Omega)$ onto $\mathbf{L}^2_\sigma(\Omega)$. The inverse operator $\mathbf{S}^{-1} = \mathbf{T}_1^{-1} \circ \mathbf{T}_2^{-1}$ is a bounded operator from $\mathbf{L}^2_\sigma(\Omega)$ onto $\mathbf{W}^{2,2}_{\sigma,nc}(\Omega)$ and $\mathbf{S} = -\Delta$.
- Operator $\mathbf{S}$ commutes with the projection $\mathbf{P}_\sigma$, i.e.

  $$\mathbf{P}_\sigma \mathbf{S} = \mathbf{S}\mathbf{P}_\sigma \boldsymbol{u}$$

  for $\boldsymbol{u} \in D(\mathbf{S})$. Consequently

  $$-\mathbf{P}_\sigma \Delta = -\Delta. \tag{1.7}$$

- Operator $\mathbf{S}$ is positive and selfadjoint in $\mathbf{L}^2_\sigma(\Omega)$. The eigenvalues of $\mathbf{S}$ form a non–decreasing sequence $\{\lambda_i\}$ of positive real numbers and they have the same algebraic and geometric multiplicity. Corresponding eigenfunctions $\{\mathbf{e}^i\}$ can be chosen so that they form a complete orthonormal system in $\mathbf{L}^2_\sigma(\Omega)$.
- Although the operator $\mathbf{S}^{1/2}$ is different from $\mathbf{T}_1$, one can easily check the identities: $\|\mathbf{S}^{1/2}\boldsymbol{u}\|_2^2 = (\mathbf{S}\boldsymbol{u}, \boldsymbol{u})_2 = \|\mathbf{curl}\,\boldsymbol{u}\|_2^2 = \|\mathbf{T}_1\boldsymbol{u}\|_2^2$ for $\boldsymbol{u} \in D(\mathbf{S})$. Consequently, there exist positive constants c_1 and c_2 so that for $\boldsymbol{u} \in D(\mathbf{S})$, we have

  $$c_1 \|\mathbf{S}^{1/2}\boldsymbol{u}\|_2 \leq \|\boldsymbol{u}\|_{1,2} \leq c_2 \|\mathbf{S}^{1/2}\boldsymbol{u}\|_2. \tag{1.8}$$

  Inequalities (1.8) show that the norms $\|\mathbf{S}^{1/2}.\|_2$ and $\|\,.\,\|_{1,2}$ are equivalent in $\mathbf{W}^{2,2}_{\sigma,nc}(\Omega)$. Using the Friedrichs–type inequality

  $$\|\boldsymbol{u}\|_2 \leq C \|\nabla \boldsymbol{u}\|_2 \tag{1.9}$$

  (where $C = C(\Omega)$), satisfied by vector functions whose normal component is zero on $\partial\Omega$ (see [7, Exercise II.5.15]), we deduce that the norm $\|\mathbf{S}^{1/2}.\|_2$ is also equivalent to $\|\nabla.\|_2$ in $\mathbf{W}^{2,2}_{\sigma,nc}(\Omega)$.
- If $\boldsymbol{u} \in \mathbf{W}^{1,2}(\Omega)$ and $\boldsymbol{v} \in \mathbf{H}(\Omega)$ then $(\boldsymbol{u}, \mathbf{T}_2\boldsymbol{v})_2 = (\boldsymbol{u}, \mathbf{curl}\,\boldsymbol{v})_2 = (\mathbf{curl}\,\boldsymbol{u}, \boldsymbol{v})_2$.
- If $\boldsymbol{u} \in \mathbf{W}^{1,2}(\Omega)$ and $\boldsymbol{v} \in D(\mathbf{S})$ then $(\boldsymbol{u}, \mathbf{S}\boldsymbol{v})_2 = (\boldsymbol{u}, \mathbf{T}_2 \circ \mathbf{T}_1\boldsymbol{v})_2 = (\boldsymbol{u}, \mathbf{curl}^2\boldsymbol{v})_2 = (\mathbf{curl}\,\boldsymbol{u}, \mathbf{curl}\,\boldsymbol{v})_2$.

Definition 1 *Let $\boldsymbol{u}_0 \in \mathbf{L}^2_\sigma(\Omega)$. A function $\boldsymbol{u} \in L^\infty\big(0,T;\, \mathbf{L}^2_\sigma(\Omega)\big) \cap L^2\big(0,T;\, \mathbf{W}^{1,2}_\sigma(\Omega)\big)$ is said to be a weak solution of the problem (1.1)–(1.4) if*

$$\int_0^T \int_\Omega \big[-\boldsymbol{u} \cdot \partial_t \boldsymbol{\phi} + \nu\, \mathbf{curl}\,\boldsymbol{u} \cdot \mathbf{curl}\,\boldsymbol{\phi} + (\boldsymbol{u} \cdot \nabla)\boldsymbol{u} \cdot \boldsymbol{\phi}\big] = \int_\Omega \boldsymbol{u}_0 \cdot \boldsymbol{\phi}(\,.\,,0)$$

for all infinitely differentiable functions $\boldsymbol{\phi}$ in $\overline{Q_T}$ such that $\boldsymbol{\phi}(\,.\,,t) \in \mathbf{W}^{1,2}_\sigma(\Omega)$ for all $0 \leq t \leq T$ and $\boldsymbol{\phi}(\,.\,,T) = \mathbf{0}$.

The boundary condition (1.3b) does not explicitly appear in the weak formulation of the problem (1.1)–(1.4). However, it can be shown by standard methods that if the weak solution is smooth then it satisfies (1.3b).

Definition 2 *Let $\boldsymbol{u}$ be a weak solution of the problem (1.1)–(1.4). We say that $\boldsymbol{u}$ is a strong solution if $\boldsymbol{u}(0) \in \mathbf{W}^{1,2}_{\sigma}(\Omega)$ and $\boldsymbol{u} \in L^r\big(0,T';\, \mathbf{L}^s(\Omega)\big)$ for all $0 < T' \le T$, $T' < \infty$ and for some $r,\ s \in \mathbb{R}$, satisfying $2 < r < \infty$, $3 < s < \infty$ and $2/r + 3/s = 1$.*

The following lemma is proved in [10].

Lemma 1 *If $\boldsymbol{u}$ is a strong solution of the problem (1.1)–(1.4) then $\boldsymbol{u} \in L^2\big(0,T';\, \mathbf{W}^{2,2}_{\sigma,nc}(\Omega)\big)$, $\partial_t \boldsymbol{u} \in L^2\big(0,T';\, \mathbf{L}^2_{\sigma}(\Omega)\big)$ for every $0 < T' \le T$, $T' < \infty$. Moreover, $\boldsymbol{u}$ is a continuous function from $[0,T']$ into $\mathbf{W}^{1,2}_{\sigma}(\Omega)$ (after an eventual change on a set of measure zero).*

2 Local in time existence of strong solutions

In this section we prove the main result of the paper, providing sufficient condition for existence of strong solutions which satisfy the boundary conditions (1.3) on given time interval. Similar results for solutions which satisfy the boundary condition (1.9) were published, e.g., in [5]. J.C. Robinson et al. proved similar results in [14], in which they studied local in time existence of strong solutions with initial data in L^3. They proved sufficient conditions for local in time existence of strong solutions of the Navier-Stokes on the whole space and for local in time existence of strong solutions which satisfy the space-periodic boundary conditions. Our result which is formulated in Theorem 1 is a modification of their result for problem with boundary conditions 1.3 on some smooth bounded convex domain.

Very similar problems were solved in [9], [10], [11] and [12]. In these papers authors studied problems of robustness and stability of strong solutions with respect to perturbations of initial velocities.

In this section we deal with the non-steady Stokes system.

$$\partial_t \boldsymbol{\varphi} + \nu\, \mathbf{curl}^2 \boldsymbol{\varphi} = \mathbf{0} \qquad \text{in } Q_T, \tag{2.1}$$

$$\operatorname{\mathbf{div}} \boldsymbol{\varphi} = 0 \qquad \text{in } Q_T, \tag{2.2}$$

$$\text{a) } \boldsymbol{\varphi} \cdot \boldsymbol{n} = 0, \quad \text{b) } \mathbf{curl}\, \boldsymbol{\varphi} \times \boldsymbol{n} = \mathbf{0} \qquad \text{on } \partial\Omega \times (0,T), \tag{2.3}$$

$$\boldsymbol{\varphi}(\,.\,,0) = \boldsymbol{\varphi}_0 \qquad \text{in } \Omega. \tag{2.4}$$

Weak solutions of the problem (2.1)–(2.4) are defined the following way:

Definition 3 *Let $\boldsymbol{\varphi}_0 \in \mathbf{L}^2_{\sigma}(\Omega)$. A function $\boldsymbol{\varphi} \in L^\infty\big(0,T;\, \mathbf{L}^2_{\sigma}(\Omega)\big) \cap L^2\big(0,T;\, \mathbf{W}^{1,2}_{\sigma}(\Omega)\big)$ is said to be a weak solution of the problem (2.1)–(2.4) if*

$$\int_0^T \int_\Omega \big[-\boldsymbol{\varphi} \cdot \partial_t \boldsymbol{\phi} + \nu\, \mathbf{curl}\, \boldsymbol{\varphi} \cdot \mathbf{curl}\, \boldsymbol{\phi}\big] = \int_\Omega \boldsymbol{\phi}_0 \cdot \boldsymbol{\varphi}(\,.\,,0)$$

for all infinitely differentiable functions $\boldsymbol{\phi}$ in $\overline{Q_T}$ such that $\boldsymbol{\phi}(\,.\,,t) \in \mathbf{W}^{1,2}_{\sigma}(\Omega)$ for all $0 \le t \le T$ and $\boldsymbol{\phi}(\,.\,,T) = \mathbf{0}$.

2.1 The main result

Theorem 1 *There exists an absolute contant* $\epsilon > 0$ *with the following property: Let* $\boldsymbol{u}_0 \in \mathbf{W}^{1,2}_{\sigma}(\Omega)$, $\tau \in \mathbb{R}$, $0 < \tau \leq T$ *and* φ *be a solution of (2.1)–(2.4) with initial velocity* $\boldsymbol{u}_0$ *such that*

$$\|\boldsymbol{u}_0\|_3 \int_0^T \int_\Omega |\nabla\varphi|^2|\varphi| < \epsilon. \tag{2.5}$$

Then there exists a strong solution $\boldsymbol{u}$ *of the problem (1.1)–(1.4) on the time interval* $(0, \tau)$.

2.2 The mathematical model of behaviour of fluid

We study qualitative properties of solutions of the Navier-Stokes equations with the boundary conditions of the Navier's type in this contribution. We derive a criterion for local in time existence of strong solution. Although this result looks to be a theoretical one, it has a very specific relationship to fluid dynamics. The system of the Navier-Stokes equations which is solved in this contribution is a model of behaviour of incompressible fluid in bounded domains. Equations (1.1) which are called the Navier-Stokes equations are derived from the momentum conservation law. Equation (2.2) which is called the continuity equation expresses the fact that liquid is incompressible. Conditions (1.3) describe interaction between fluid and fixed wall. We are supposing density and viscosity of fluid are constant, in this model.

We can say that the system (1.1), (1.2), (1.4) with various types of boundary conditions (we are studying this system with boundary conditions (1.3) here) is the most commonly used model of behaviour of incompressible fluid. This model is studied in at least two areas of mathematics. Many mathematicians dealing with the theory of partial differential equations publish papers in which the qualitative properties of solutions of this model are studied. At the same time, many mathematicians working in the field of numerical mathematics publish papers that deal with the approximate numerical solutions of this model. These approximate solutions are then applied to solve various technical problems in different areas, mechanical engineering, civil engineering, hydrodynamics. It should be said that the results dealing with the qualitative properties of this system often help to optimize work of numerical mathematicians.

Unfortunately, mathematicians and physicists are not yet sure if this mathematical model corresponds to the real fluid behavior. In the qualitative theory of the Navier-Stokes equations there are still some open problems that are of crucial importance. Suppose we have a weak solution to the problem with a smooth enough initial velocity. It is well known that there is a weak solution of this problem, satisfying the energy inequality in addition. This means that its kinetic energy is uniformly bounded over the entire time interval. We do not know whether this solution satisfies energy identity (energy identity is a special case of energy inequality). This means we do not know whether or not the fluid behavior corresponds with the energy conservation law. The next open question remains whether or not the dissipative energy is uniformly limited over the entire time interval. If we knew that there exists a weak solution of the problem which is also a strong solution, all of the above mentioned questions would be positively answered. Such a solution would satisfy energy identity and its dissipative energy would be uniformly bounded over the time interval. We also know that if there exists a strong solution, then this solution is the unique strong solution. In addition, it is the only solution that satisfies energy equality. It is even known that such a solution would be then the only solution that satisfies the energy inequality. It is therefore reasonable to think that such a solution would be a mathematical model that actually describes the behavior of the fluid.

The open problem mentioned here is one of the most well-known open problems of applied mathematics, included in the so-called millennium problems, and the Clay Institute has issued a financial reward for solution of this problem.

The main result of this paper (Theorem 1) provides a partial contribution to the solution of this problem. We know that there is a time interval on which the solution is strong, and we know how to determine the length of this interval. This means that we know that at this time interval the solution of the Navier-Stokes problem describes the behavior of a fluid that complies with the law of conservation of energy.The question of whether or not this solution is strong outside this time interval remains open.

2.3 Proof of the main result

At first we mention some estimates and lemmas which we apply in order to prove Theorem 1. The following estimate is proved in [11, Estimate (2.1)]. If $\boldsymbol{v} \in \mathbf{W}^{1,3}(\Omega)$, $\boldsymbol{v} \cdot \boldsymbol{n} = 0$ on $\partial(\Omega)$, then

$$\|\boldsymbol{v}\|_9^3 \leq c_3 \int_\Omega |\nabla \boldsymbol{v}|^2 |\boldsymbol{v}|. \tag{2.6}$$

Let $\boldsymbol{v} \in \mathbf{W}^{1,2}_\sigma(\Omega)$. Then there exists $q \in \mathbf{W}^{1,2}(\Omega)$ such that

$$\boldsymbol{v}|\boldsymbol{v}| = \mathbf{P}_\sigma(\boldsymbol{v}|\boldsymbol{v}|) + \nabla q. \tag{2.7}$$

Applying operator $\mathbf{div}$ to 2.7 and using the fact that $\boldsymbol{v} \cdot \boldsymbol{n} \equiv 0$ on $\partial\Omega$ we obtain

$$\Delta q = \mathbf{div}\,(\boldsymbol{v}|\boldsymbol{v}|) \quad \text{in } \Omega \qquad \frac{\partial q}{\partial \boldsymbol{n}} = 0 \quad \text{in } \partial\Omega \tag{2.8}$$

Therefore ,

$$\left\|D^2 q\right\|_{3/2} \leq c_4 \left\| \nabla \boldsymbol{v}|\boldsymbol{v}| \right\|_{3/2} \leq c_5 \left\|\nabla \boldsymbol{v}|\boldsymbol{v}|\right\|_2 \left\|\boldsymbol{v}\right\|_3^{1/2}. \tag{2.9}$$

The following inequalities are proved in [10, Lemma 2.] where assumption of convexity of Ω has been applied. Let $\boldsymbol{w} \in \mathbf{W}^{2,2}_{\sigma,nc}(\Omega)$. Then

$$\int_\Omega \mathbf{S} \cdot \mathbf{P}_\sigma(\boldsymbol{v}|\boldsymbol{v}|) \equiv -\int_\Omega \Delta \boldsymbol{v} \cdot \mathbf{P}_\sigma(\boldsymbol{v}|\boldsymbol{v}|) \geq \int_\Omega |\nabla \boldsymbol{v}|^2 |\boldsymbol{v}| + \frac{4}{9}\int_\Omega |\nabla |\boldsymbol{v}|^{3/2}|^2. \tag{2.10}$$

The following lemma is proved in [14].

Lemma 2 *Let $c_6 > 0$, θ, ϑ and ς are real-valued, non-negative functions which are continuous on $[0, \tau)$, $\theta \in C(0, \tau)$, $\theta(0) = 0$ and the inequality*

$$\theta'(t) + \vartheta(t) \leq c_6 \theta(t)\vartheta(t) + \varsigma(t) \tag{2.11}$$

holds on $(0, \tau)$. Put $D = \int_0^\tau \varsigma(t)\, dt$. If $D < \frac{1}{4c_6}$ then

$$\sup_{t \in (0,\tau)} \theta(t) \leq 2D < \frac{1}{2c_6} \tag{2.12}$$

and

$$\int_0^\tau \vartheta(t)\, dt \leq 2D < \frac{1}{2c_6}. \tag{2.13}$$

Integrating (2.1) and using (2.10) and embedding $\mathbf{W}^{1,2}_\sigma(\Omega) \hookrightarrow \mathbf{L}^3(\Omega)$ we prove the following lemma.

Lemma 3 *Let $\varphi_0 \in \mathbf{W}^{1,2}_\sigma(\Omega)$, φ be a solution of (2.1)–(2.4). Then the inequality*

$$\|\varphi(T')\|_3^3 + \int_0^{T'} \int_\Omega |\nabla\varphi|^2|\varphi| \le \|\varphi_0\|_3^3 \tag{2.14}$$

holds for every T', $0 < T' \le T$.

Now we prove Theorem 1:

It is well known that if $\boldsymbol{u}_0 \in \mathbf{W}^{1,2}_\sigma(\Omega)$ then $\varphi \in L^2\big(0,T';\, \mathbf{W}^{2,2}_{\sigma,nc}(\Omega)\big)$. Multiplying (2.1) by $\mathbf{P}_\sigma(\varphi|\varphi|)$, integrating it over Ω and $(0,T')$ and using (2.10) we obtain (2.14) with $\boldsymbol{u}_0$ instead of φ_0. Let

$$\varsigma = \sup\{\varsigma';\, \varsigma',\ \text{such that } \boldsymbol{u} \text{ is a strong solution on } (0,\varsigma')\}. \tag{2.15}$$

Since $\boldsymbol{u}_0 \in \mathbf{W}^{1,2}_\sigma(\Omega)$ then $\varsigma > 0$. This statement can be proven by analogy with [8, Theorem 6.1]. We prove that $\varsigma \ge \tau$. Suppose by contradiction that $\varsigma < \tau$. It is sufficient to prove that

$$\boldsymbol{u} \in C([0,\varsigma];\, \mathbf{W}^{1,2}_\sigma(\Omega)). \tag{2.16}$$

By (2.16) $\boldsymbol{u}(\varsigma) \in \mathbf{W}^{1,2}_\sigma(\Omega)$. Consequently, there exists $\varsigma^* > \varsigma$ such that $\boldsymbol{u}$ is a strong solution on $(0,\varsigma^*)$ and we obtain the contradiction with (2.15).

Let $\boldsymbol{u} = \varphi + \psi$ such that ψ is solution of the system

$$\psi' - \Delta\psi + (\psi\cdot\nabla)\psi + (\varphi\cdot\nabla)\psi + (\psi\cdot\nabla)\varphi + (\varphi\cdot\nabla)\varphi + \nabla P = 0, \tag{2.17}$$

$$\psi(0) = 0 \tag{2.18}$$

and φ is solution of the problem (2.1)–(2.4) with initial velocity $\boldsymbol{u}_0$. By (1.6) there exists $q \in \mathbf{W}^{1,3/2}(\Omega)$ such that

$$\psi|\psi| = \mathbf{P}_\sigma(\psi|\psi|) + \nabla q. \tag{2.19}$$

Multiplying (2.17) by $\mathbf{P}_\sigma(\psi|\psi|)$ on the time interval $(0,\varsigma)$, integrating it over Ω and using (2.7), (2.10), (2.19) and a well-known identity (see, e.g., [15, Lemma 3.2.1])

$$(((\psi\cdot\nabla)\psi\,,\,\psi|\psi|)) = (((\varphi\cdot\nabla)\psi\,,\,\psi|\psi|)) = 0$$

we obtain

$$\begin{aligned}
&\frac{1}{3}\frac{d}{dt}\|\psi\|_3 + \|\nabla\psi|\psi|^{1/2}\|_2^2 + \frac{4}{9}\|\nabla|\psi|^{3/2}\|_2^2 \le \\
&\quad |(((\psi\cdot\nabla)\psi + (\varphi\cdot\nabla)\psi + (\psi\cdot\nabla)\varphi + (\varphi\cdot\nabla)\varphi\,,\,\psi|\psi|))| + \\
&\quad |(((\psi\cdot\nabla)\psi + (\varphi\cdot\nabla)\psi + (\psi\cdot\nabla)\varphi + (\varphi\cdot\nabla)\varphi\,,\,\nabla q))| \le \\
&\quad |(((\psi\cdot\nabla)\varphi\,,\,\psi|\psi|))| + |(((\varphi\cdot\nabla)\varphi\,,\,\psi|\psi|))| + |(((\psi\cdot\nabla)\psi\,,\,\nabla q))| + \\
&\quad |(((\varphi\cdot\nabla)\varphi\,,\,\nabla q))| + |(((\varphi\cdot\nabla)\psi\,,\,\nabla q))| + |(((\psi\cdot\nabla)\varphi\,,\,\nabla q))| = \\
&\quad I_1 + I_2 + I_3 + I_4 + I_5 + I_6
\end{aligned} \tag{2.20}$$

Integrating by parts and applying (2.6) and (2.9) we infer

$$I_1 \le c\left\|\nabla\psi|\psi|\right\|_{3/2}\|\psi\|_6\|\varphi\|_6 \le c\left\|\nabla\psi|\psi|\right\|_2\|\psi\|_3^{1/2}\|\psi\|_9^{3/4}\|\psi\|_3^{1/4}\|\varphi\|_9^{3/4}\|\varphi\|_3^{1/4} \le$$
$$c\left\|\nabla\psi|\psi|\right\|_2^{3/2}\|\psi\|_3^{3/4}\|\varphi\|_9^{3/4}\|\varphi\|_3^{1/4} \le c_7\left\|\nabla\psi|\psi|\right\|_2^2\|\psi\|_3^3 + \frac{1}{2}\left\|\nabla\varphi|\varphi|\right\|_2^2\|\boldsymbol{u}_0\|_3 \qquad (2.21)$$

$$I_2 \le c\left\|\nabla\psi|\psi|\right\|_{3/2}\|\varphi\|_6^2 \le c\left\|\nabla\psi|\psi|\right\|_2\|\psi\|_3^{1/2}\|\varphi\|_9^{3/2}\|\varphi\|_3^{1/2} \le$$
$$\frac{1}{8}\left\|\nabla\psi|\psi|\right\|_2^2 + c_8\left\|\nabla\psi|\psi|\right\|_2^2\|\psi\|_3^3 + \frac{1}{2}\left\|\nabla\varphi|\varphi|\right\|_2^2\|\boldsymbol{u}_0\|_3 \qquad (2.22)$$

$$I_3 \le c\left\|D^2 q\right\|_{3/2}\|\psi\|_6^2 \le c\left\|\nabla\psi|\psi|\right\|_{3/2}\|\psi\|_9^{3/2}\|\psi\|_3^{1/2} \le$$
$$c\left\|\nabla\psi|\psi|\right\|_2^2\|\psi\|_3 \le \frac{1}{8}\left\|\nabla\psi|\psi|\right\|_2^2 + c_9\left\|\nabla\psi|\psi|\right\|_2^2\|\psi\|_3^3 \qquad (2.23)$$

$$I_4 \le c\left\|D^2 q\right\|_{3/2}\|\varphi\|_6^2 \le c\left\|\nabla\psi|\psi|\right\|_{3/2}\|\varphi\|_9^{3/2}\|\varphi\|_3^{1/2} \le$$
$$\frac{1}{8}\left\|\nabla\psi|\psi|\right\|_2^2 + c_{10}\left\|\nabla\psi|\psi|\right\|_2^2\|\psi\|_3^3 + \left\|\nabla\varphi|\varphi|\right\|_2^2\|\boldsymbol{u}_0\|_3 \qquad (2.24)$$

$$I_5 + I_6 \le c\left(I_3 + I_4\right) \le \frac{1}{8}\left\|\nabla\psi|\psi|\right\|_2^2 + c_{11}\left\|\nabla\psi|\psi|\right\|_2^2\|\psi\|_3^3 + \left\|\nabla\varphi|\varphi|\right\|_2^2\|\boldsymbol{u}_0\|_3 \qquad (2.25)$$

Set $c_{12} = c_7 + c_8 + c_9 + c_{10} + c_{11}$. Using estimates (2.20)–(2.25) we obtain

$$\frac{d}{dt}\|\psi\|_3^3 + \left\|\nabla\psi\,|\psi|^{1/2}\right\|_2^2 \le 3\,c_{12}\left\|\nabla\psi\,|\psi|^{1/2}\right\|_2^2\|\psi\|_3^3 + 3\,\|\boldsymbol{u}_0\|_3\left\|\nabla\varphi\,|\varphi|^{1/2}\right\|_2^2.$$

Put $\theta(t) = \|\psi(.,t)\|_3^3$, $\vartheta(t) = \left\|\nabla\psi(.,t)|\psi|^{1/2}\right\|_2^2$, $\varsigma(t) = 3\,\|\boldsymbol{u}_0\|_3\left\|\nabla\varphi\,|\varphi|^{1/2}\right\|_2^2$ and $\epsilon = \frac{1}{12c_{12}}$. Let $0 < \tau < T$ such that

$$\|\boldsymbol{u}_0\|_3\int_0^\tau\left\|\nabla\varphi\,|\varphi|^{1/2}\right\|_2^2 < \frac{1}{12c_{12}}.$$

If $\tau \le \varsigma$ the theorem is proved. Suppose by contradiction $\varsigma < \tau$. Since

$$3\,\|\boldsymbol{u}_0\|_3\int_0^\varsigma\left\|\nabla\varphi\,|\varphi|^{1/2}\right\|_2^2 < \frac{1}{4c_{12}}.$$

we get by Lemma 2 and by (2.6)

$$\int_0^\varsigma\|\psi\|_9^3 \le c_3\int_0^\varsigma\left\|\nabla\psi\,|\psi|^{1/2}\right\|_2^2 < \frac{1}{2c_{12}} < \infty.$$

The last inequalities, (2.6) and (2.14) imply that $\boldsymbol{u} \in L^3(0,\varsigma;\,\mathbf{L}^9(\Omega))$. Then, $\boldsymbol{u} \in C([0,\varsigma];\,\mathbf{W}_\sigma^{1,2}(\Omega))$. Since (2.16) holds we obtain the contradiction with (2.15). Consequently, $\tau \le \varsigma$ and theorem is proved.

ACKNOWLEDGEMENTS: The research was supported by the project SGS17/116/OHK1/2T/11.

References

[1] BEIRAO DA VEIGA, H., CRISPO, F.: *Sharp inviscid limit results under Navier type boundary condition. An L^p theory.* J. Math. Fluid Mech., 12, 2010, 397–411.

[2] BELLOUT, H., NEUSTUPA, J., PENEL, P.: *On viscosity–continuous solutions of the Euler and Navier–Stokes equations with a Navier–type boundary condition.* Comptes Rendus Math., 347, 2009, 1141–1146.

[3] BELLOUT, H., NEUSTUPA, J., PENEL, P.: *On a ν–continuous family of strong solutions to the Euler or Navier–Stokes equations with the Navier–type boundary condition.* Discr. and Cont. Dyn. Systems–A, 27, no. 4, 2010, 1353–1373.

[4] CHEN, G.Q., OSBORNE, D., QIAN, Z.: *The Navier–Stokes equations with the kinematic and vorticity boundary conditions on non–flat boundaries.* Acta Mathematica Scientia, 29, no. B(4), 2009, 919—948.

[5] FARWIG, R., SOHR, H., VARNHORN, V.: *On optimal initial value conditions for local strong solutions of the Navier–Stokes equations.* Ann. Univ. Ferrara, 55, 2009, 89–110.

[6] FUJIWARA, D., MORIMOTO, H.: An L^r-theorem of the Helmholtz decomposition of vector fields. *J. Fac. Sci. Univ. Tokyo Sect. IA Math.*, 24, 1977, 685–700.

[7] GALDI, G.P.: *An Introduction to the Mathematical Theory of the Navier–Stokes Equations. Vol. I: Linearized Steady Problems.* Springer–Verlag, New York–Berlin–Heidelberg, 1994.

[8] GALDI, G.P.: *An Introduction to the Navier–Stokes initial–boundary value problem.Fundamental Directions in Mathematical Fluid Mechanics,* editors G. P. Galdi, J. Heywood, R. Rannacher, series "Advances in Mathematical Fluid Mechanics", Birkhauser–Verlag, Basel 2000, 1–98.

[9] KUČERA, P., NEUSTUPA, J.: *On perturbations of solutions to the Navier–Stokes equations with large initial data and their dynamics.* Nonlinear Analysis, 71, no. 12, 2009, e2690—e2695.

[10] KUČERA, P., NEUSTUPA, J.: *On L^3-stability of strong solutions of the Navier-Stokes equations with the Navier-type boundary conditions.* Journal of Mathematical analysis and applications, 405, 2013, 731–737.

[11] KUČERA, P., NEUSTUPA, J.: *On robustness of a strong solution to the Navier-Stokes equations with Navier's boundary conditions in the L 3 -norm.* Nonlinearity, 30, no. 4, 2017, 1564—1583.

[12] KUČERA, P.: *Small perturbation of initial conditions of solutions of the Navier–Stokes equations in the L^2–norm and applications.* Advances in Mathematical Fluid Mechanics, ed. R. Rannacher, A. Sequeira, Springer, Berlin, 2010, 319—328.

[13] MAHALOV, A., NICOLAENKO, B.: *Global solvability of three-dimensional Navier–Stokes equations with uniformly high initial vorticity.* Russian Math. Surveys, 58, No.2, 2003, 287–318.

[14] ROBINSON, J.C., RODRIGO, J.L., SADOWSKI, W.: *The three-dimensional Navier-Stokes equations. Classical theory.* Cambridge Studies in Advanced Mathematics, No.157. Cambridge University Press, Cambridge, 2016.

[15] SOHR, H.: *The Navier–Stokes Equations. An Elementary Functional Analytic Approach.* Birkhäuser Advanced Texts, Basel–Boston–Berlin 2001.

[16] XIAO, Y.L., XIN, Z.P.: *On the vanishing viscosity limit for the 3D Navier–Stokes equations with a slip boundary condition.* Comm. on Pure and Appl. Math., 60, 2007, 1027—1055.

Current address

Kučera Petr, doc. RNDr., CSc.
Department of Mathematics
Faculty of Civil Engineering
Czech Technical University in Prague
Thákurova 7, 166 29 Prague, Czech Republic
E-mail: petr.kucera@cvut.cz

Píšová Jitka, Ing. Arch.
Department of Mathematics
Faculty of Civil Engineering
Czech Technical University in Prague
Thákurova 7, 166 29 Prague, Czech Republic
E-mail: marguete@seznam.cz

Vacková Petra, Mgr.
Department of Mathematics
Faculty of Civil Engineering
Czech Technical University in Prague
Thákurova 7, 166 29 Prague, Czech Republic
E-mail: petra.vackova@fsv.cvut.cz

Slovak University of Technology in Bratislava
Faculty of Mechanical Engineering
17th Conference on Applied Mathematics **APLIMAT 2018**

Proceedings

NATURE INSPIRED ARCHITECTURES AND JEWELS PARAMETRIC DESIGN

LORENZI Marcella Giulia (IT), SCHIAVELLO Francesco (IT)

Abstract. The concept of "geometric form" aroused from the observation of forms already existing in Nature. These can be sources of inspiration for artists. We present some design projects in which architectures and jewels are created in this framework, modelled using parametric processes and softwares.

Keywords: geometric forms, inspiration from nature, parametric design, math&art, architecture; jewel design

Mathematics Subject Classification: Primary 97M80; Secondary 00A67

0. Introduction. Mathematics for Parametric Design

New (and old) Mathematics have gradually become instruments and sources of inspiration for new forms of Art that aim at transcending the experiential world, rather than trying to reproduce it in a stereotypical way. In order to understand most of the new forms of Contemporary Art, one needs to understand their mathematical and technological roots. It is well known that Geometry was born out of much older knowledge as that part of Mathematics explicitly devoted to investigate shapes in Space and measure their "extension"; developed as a practical discipline since the Palaeolithic age [12], during the Greek age it became a formalized elegant tool of theoretical thought. The concept of "geometric form" did in fact slowly arise from the observation of forms already existing in Nature and more or less hidden in the structures through which the human mind tends to perceive the "Order of Nature". [19] The language that we use to understand "Kosmos" is indeed formed by a subtle intertwining of Numbers (the act of "counting", i.e. "Arithmetic") and Forms (the core of Geometry. [11] An infinite family of "geometrical shapes" that we like to call "forms without an age" since they have crossed the ages, from Prehistory to now, giving rise to what we call the "persistence of forms" [9]. Along with experimental developments that are still under their way, for many artists of XX Century the simple forms of Geometry – and more generally, the entire field of Mathematics - have again become a fresh source of inspiration for Art [19], in a kind of "travel backward in time", towards a renewed aesthetics of "simple forms": circles, straight lines, triangles ("without which we cannot intend human words" as Galileo said), as well as squares, pentagons, hexagons and other geometric constructions often found in Nature [6;7].
On the other hand, we have examined, during the previous Aplimat conferences, some examples in which, starting from Nature and simple geometric forms, contemporary artists and especially architects designed complex surfaces, often using a parametric approach [6; 16; 17].

The aesthetics related to the use of geometry is a topic in the field of theories of architecture gradually applied to the compositional process to draw plans and elevations according dimensional relationships, hidden but perceived at a deeper level, probably by virtue of a rooted cultural conditioning; also applied to polygonal modularity, up to the representation of buildings according to new geometries, to the topological transformations of surfaces and volumes, to the dynamic and kinematic parametric representations, and even with reference to René Thom catastrophe theory [6]. Mathematics and geometry represent the core of the architectural design process, in the age of digital parametric non-standard architecture. They have a central role from the initial stage of finding form, shaping form, generating form, to the process of manufacturing architectural elements. Contemporary computer technology allows the application of a number of tools for the design, analysis, simulation and manufacturing of complex architectural forms. In the process of design, today's leading architects as well as leading schools of architecture use different software packages and digital technologies, thus contributing to the formation of a new aesthetics of digital architecture. On the one hand, technical possibilities open up new horizons in architecture, while on the other, they give rise to new issues related to the disciplines of mathematics and geometry [21].
Parametric design is a process based on algorithmic thinking that enables the expression of parameters and rules that, together, define, encode and clarify the relationship between design intent and design response [13; 22]. Parametric design is a paradigm in design where the relationship between elements is used to manipulate and inform the design of complex geometries and structures. The term parametric originates from mathematics (parametric equation) and refers to the use of certain parameters or variables that can be edited to manipulate or alter the end result of an equation or system [4].
While today the term is used in reference to computational design systems, there are precedents for these modern systems in the works of architects such as Antoni Gaudí, who used analog models to explore design space [1; 18].
Parametric thinking introduces the shift in the mindset between the search for a specific static and defined formal solution, and the design of the specific stages and factors used to achieve it. It is the use of algorithms and advanced computational techniques not for the sake of drawing shapes, but rather for creating formal possibilities. It is not about producing a solution, but a family of possible outcomes [14].
Today many topics from different fields influence the design and mathematics can be considered as a tool of governance of finite element calculation. Former mathematical and geometrical algorithms, forms and structures are now visible and spatial understandable for architects and, therefore, usable. Using new technique architectural design has established computational design concepts such as: topological space (topological architectures), isomorphic surfaces (isomorphic architectures), motion kinematics and dynamics (animate architectures), keyshape animation (metamorphic architectures), parametric design (parametric architectures), genetic algorithms (evolutionary architectures) or fractal geometry (fractal architecture) [15].
In conceptual parametric design, it is the parameters of a particular design that are declared, not its shape. By assigning different values to the parameters different objects or configurations can be easily created. They are powerful owing to their ability to capture a high degree of variation in a few numerical values. Software like Rhinoceros offers such script editors for parametric design. Many architects have used it for conceptual design [21].

1. Projects

Within this framework, we will briefly present some of the projects realized during the last ten years by the SAO Schiavello Architects Office, in which nature and mathematics lay at the basis of the inspirational as well as design processes. SAO rely on principles such as Shape and Patterns and Form generation process, starting from the observation of nature and, in particular, on how form and its varieties in nature are found. Patterns organize and define relationships in nature and can be integrated into design to substantiate and support visual communication. Since design's purpose is to create a relationship with the viewer, the language of pattern helps to frame what that relationship is before the message is read or even consciously processed [20].

1.1 The Earthquake Museum

The Earthquake Museum (acronym "MuTerr", from the Italian words Museo and Terremoto) is located in the major hypogean spaces of the San Domenico Convent in Soriano, in the Province of Vibo Valentia (Calabria Region, in the South of Italy), a religious XV Century complex which was partially destroyed by the earthquake of 1783. The Convent, formerly called "Casa Santa" (i.e. "Holy House"), has historically been the focus of the "Dominican Universe", having had the important support of many popes, cardinals and kings of Spain and Naples, ruling in the south of Italy (Fig. 1.). Since 1654, the "Holy House" has administered the County of Soriano and therefore its sphere of economical, cultural and social influence has greatly contributed to the development of the surrounding areas and settlements. Internally there were a guesthouse, a pharmacy, a print house and a well-stocked library, which still exists and that contains many prestigious volumes. For many years, the Dominican complex has been one of the most important symbols of the earthquake history and today it is reborn as a museum center, second only to that of Reggio Calabria, the biggest in the region. The complex also includes the Museum of Marbles (Italian acronym "Mumar") where of particular value is the Head of Saint Catherine of Siena, in Carrara marble, probably attributed to Gian Lorenzo Bernini. Besides the "MuTerr", other three museums are under construction in the same site: those of Ceramics, Sacred Arts and Contemporary Art Gallery.
MuTerr represents an attractive meeting and information place on earthquakes, not only of the Calabria Region, but of the entire Italian country. The earthquake phenomena are analyzed from different perspectives: historical, anthropological and architectural.
Thanks to this restoration, another important part of this complex, which remained so far "hidden" and never used, has been brought to light. All the ancient architectural spaces of the southwestern wing, commonly called the "Darks", testimony of what was once the most "deadly catastrophe" caused by the earthquake of 1783, are now accessible in the "MuTerr" and can be used for temporary exhibitions, different cultural events or conventions.

Fig. 1. The Earthquake Museum, San Domenico Convent in Soriano, Vibo Valentia (Italy).

The Museum has two distinct environments: the visitor is welcomed in a modern area, marked by the installation of a suspended monumental wooden ribs ceiling, created with digital processes of

parametric design and cut with CNC machines. The ceiling is a true piece of art that recalls the wave and fleeting movement of the earthquake and which visually envelops the visitor in an aggressive and dynamic place.
The ceiling was designed (Fig. 2.) by sectioning a "seismic wave" (Fig. 3.) in more than thousand pieces (Fig. 4.), snap-fit assembled by attacks and steel cables to the IPE beams which are anchored to the perimeter walls (Fig. 5.).

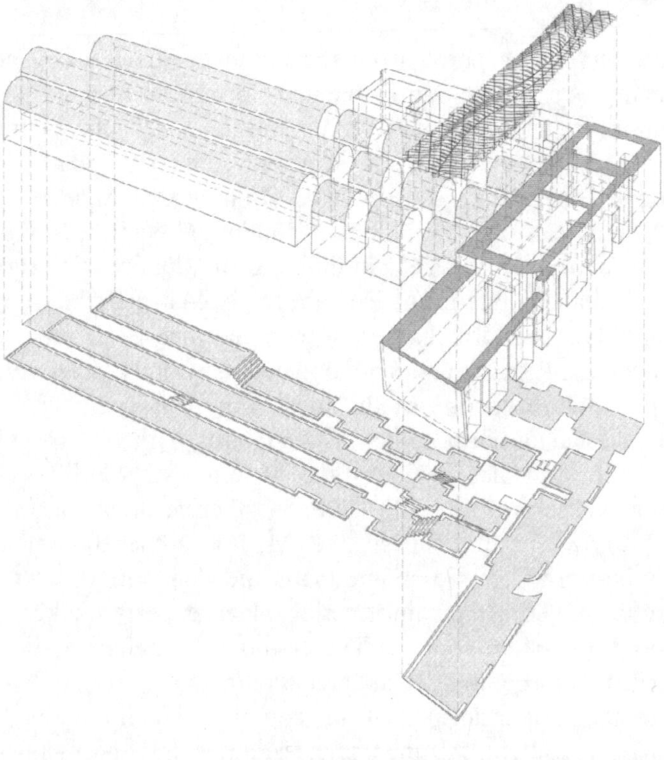

Fig. 2. Design of the ceiling

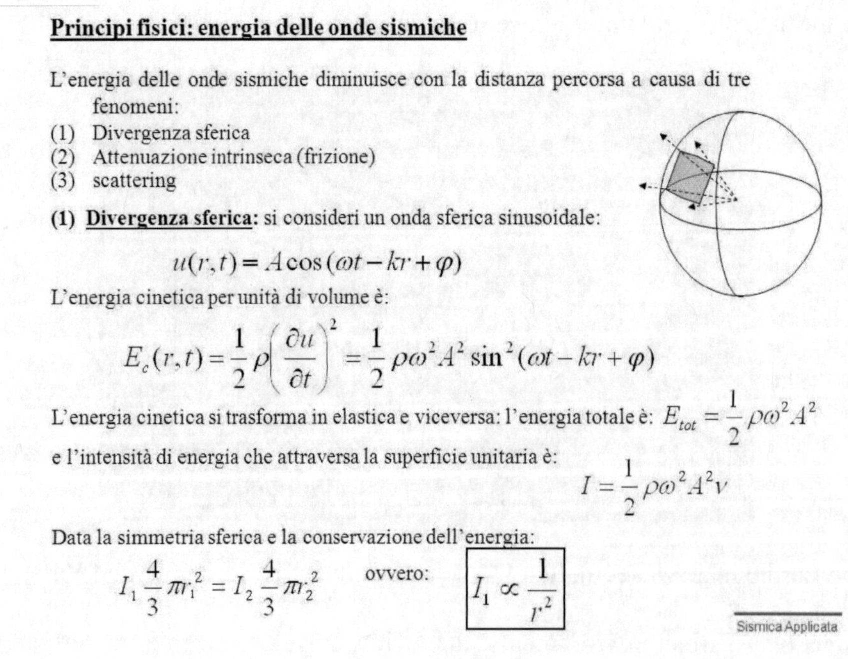

Fig. 3. Seismic wave.

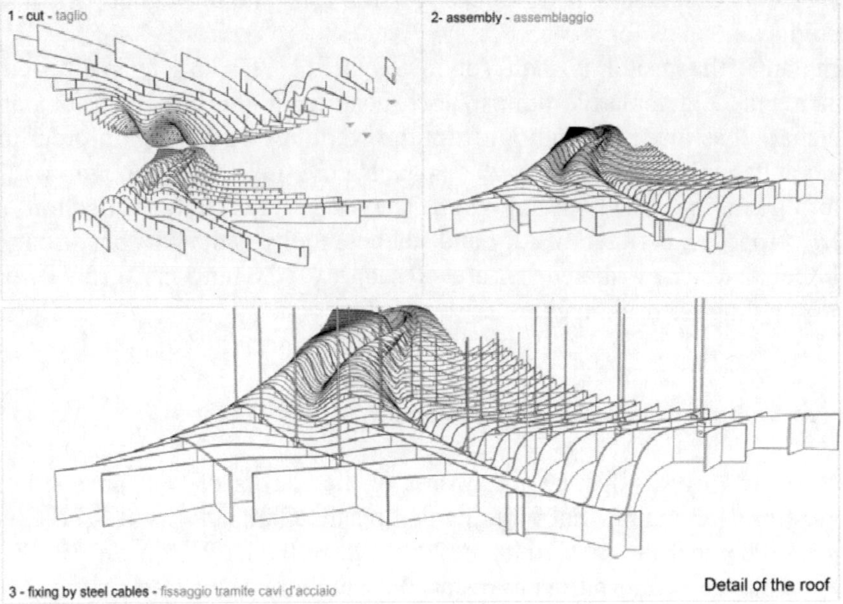

Fig. 4. Detail of the roof design.

Fig. 5. Installation of the ceiling.

Cutting and preparation for digital manufacturing have been created with Rhino / Grasshopper parametric software. Grasshopper (originally Explicit History) is a plug-in for Rhinoceros 3D that presents the users with a visual programming language interface to create and edit geometry [2]. Components or nodes are dragged onto a canvas in order to build a grasshopper definition. Grasshopper is based on graphs (see Graph (discrete mathematics)) that map the flow of relations from parameters through user-defined functions (nodes), resulting in the generation of geometry.

Changing parameters or geometry causes to changes to propagate throughout all functions, and the geometry to be redrawn.
From this bright area, the visitor is then conducted in the "Darks", characterized instead by a charming, calmer and contemplative atmosphere. Feature of this second room are the vaulted ceilings, whose charm has made it stand out from the white resin flooring. In order to highlight the direction of the vaults and of the opaque glass panels covering the channeling of the installation system, the flooring continues without any gaps. The long and complex reflection on earthquakes that the "MuTerr" proposes, between history and science, looks at the past but strongly interests the future, in the belief that a new shared culture of security of tenure can grow on these bases of knowledge on the earthquake phenomena.

1.2 Portonave Auditorium in Navegantes harbor

The "Portonave" (Company has opened a new area of the Harbor of Navegantes, in Brazil, one of the nation's largest private commercial ports owned by the Company itself. The Company bought the plot of land in 2005, starting its activity in 2007. Since then, "Portonave" has been employing more than 1,100 employees, contributing to restart the economy of the city.
In 2008, a new project planning has started. The aim of the company was to create a building hosting, on the ground floor, a company canteen, a library, a relaxation area and a dressing room serving its workers and, on the first floor, an auditorium and a large raised and covered square for events. Although they took a long time to start, works have been concluded in just 14 months.
The goal was to create a building acting as a new landmark in the area and open to the local community as well. As a matter of fact, the Company has started a partnership with the public administration aimed at promoting both cultural and social projects (health promotion, child and elderly care) and environmental initiatives (requalification of different areas of the city, awareness projects on environmental issues and other initiatives focused on sustainable development).
The "Portonave Project" is extended over a surface of 5,600 square meters and it has been carried out thanks to parametric processes.
The idea of the "Schiavello Architects Office" and the "Canhadas Bot Tarelli Arquitetura" studios was to give a strong, simple and innovative sign to the Project. The building shape reminds the form of a ship ready to sail and its surface was inspired by the geometrical texture of the so-called butterfly fish, belonging to the Chaetodontidae family and that can be found also in the Atlantic Ocean (Fig. 6.).

Fig. 6. Geometrical texture of the so-called butterfly fish (Chaetodontidae family).

The large, custom-made aluminum panels of different sizes cover an area of 90 x 40 meters; the high facade has been conceived as an urban and panoramic "promenade", shining at night. The building creates a sort of "path", structured on various levels and connected with ramps, staircases and a raised square, with a view on the harbor. The external green areas are also distinctive elements of the Project (Fig. 7.).

Fig. 7. Portonave Auditorium in Navegantes harbor.

From a functional point of view, the working spaces are located on the ground floor while the upper part of the building hosts the most representative ones as well as the relaxation area. The connection elements - ramps, stairs and corridors, that link the different functional areas of the building, have been designed to have both a functional and an expressive and representative effect. The building is well placed in the context it belongs to.

1.3 Cheong-na City Tower: geometrical equilibrium

Project for the Cheong-na City Tower (Korea) International competition by the team Francesco Schiavello, Junseung Woo, Yeol Park is for an observatory of approx. 450m (1,480 ft) height, 200000 m^2, including exhibition, shopping area and entertainment areas. It is intended to be the sixth largest tower in the world, and the third highest observatory in the world.

This project aimed at achieving the representative aspiration and symbolic meaning of Cheong-na City and satisfy people's stable and grand visual experience to appreciate. Cheong-na is continuously changing, evolving to become a technological and cultural centre in the world, also through a precise strategy for an economically and structurally safe and sustainable development.

The Cheong-na City Tower is based on the mathematical principle of turning curves, which design results in woven structural conditions, in order to deal with its own stress and load. Mathematically generated circulating structure is related both to structural aspects than to symbolic meanings.

Its design has a reference to the Korean *Taegeuk* graphic symbol [3]: the interlocking-spirals, historically associated with Korean Taoism philosophical values, permanent circulation of cosmos, and is used in the flag of South Korea (Fig. 8.). It represents here the city tower idea as a symbol of a sustainable, economical and cultural agreement.

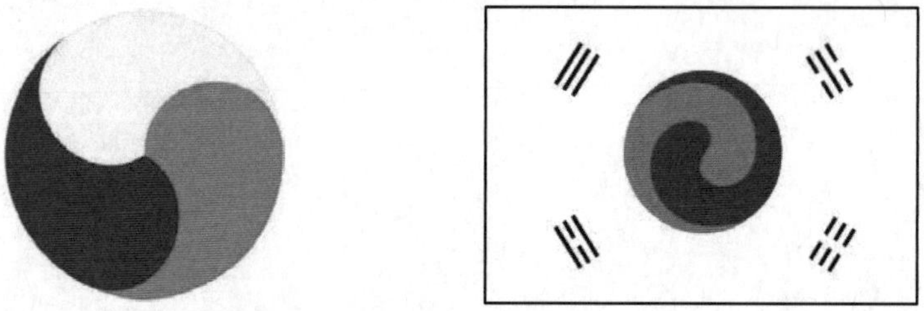

Fig. 8. Left: Korean *Taegeuk* graphic symbol; right: the flag of South Korea.

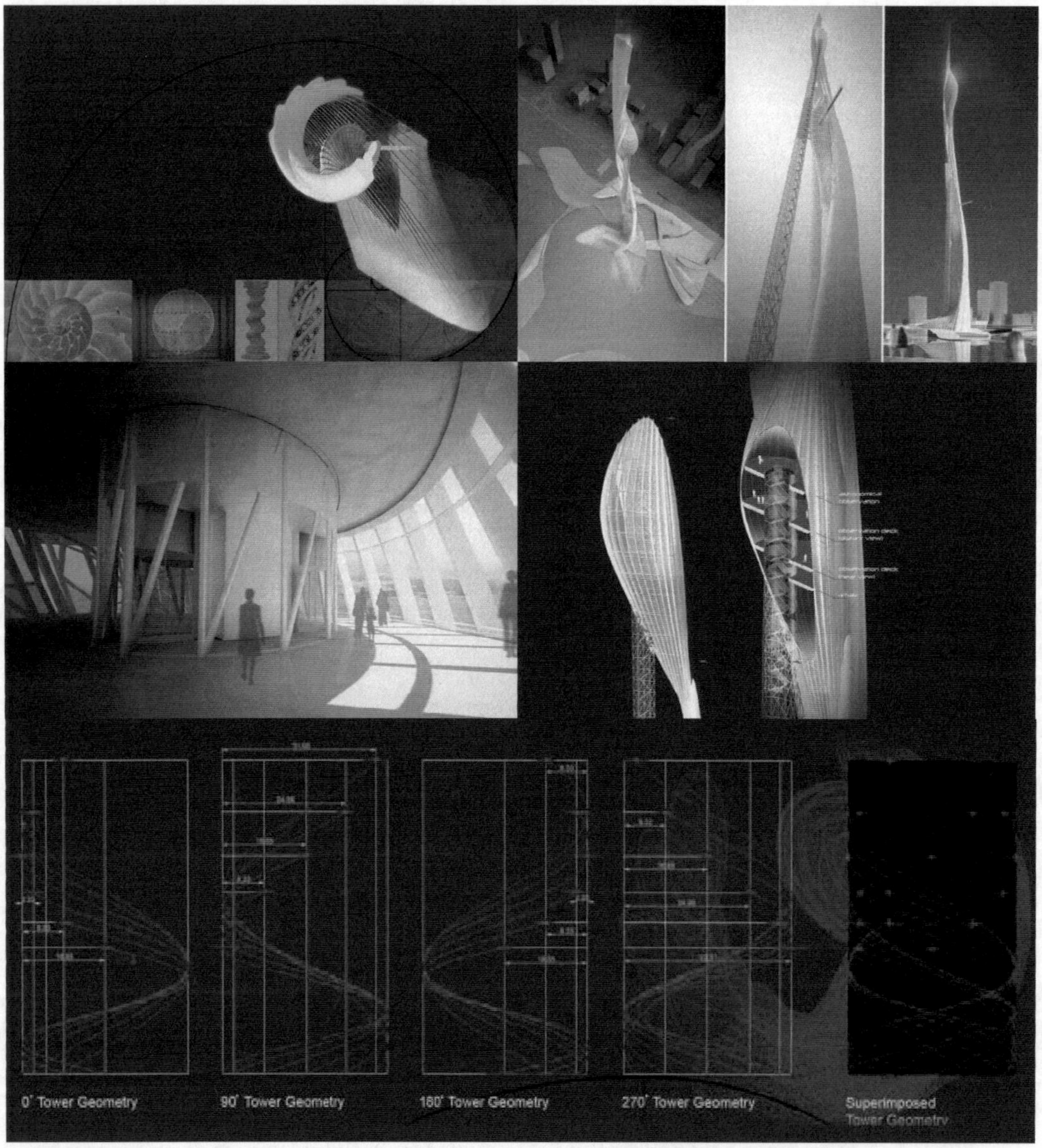

Fig. 9. Design and geometry of the Cheong-na City Tower

In the project, composite steel and concrete are provided along the perimeter of the Cheong-na City Tower at regular spacing. The bundle of the six tube structure is provided to maintain equilibrium under vertical loads restrain capability. Each of the tube structure is composted with embedded steel trusses. The tower is designed to meet three parts of major functions, corresponding to a turning angle, by steps of 90°: Observation level (90°), Entertainment level (180°) and Retail & Cultural (270°) (Fig. 9.). At a very simple inspirational level, and at the core of the project, there is an helicoidal staircase, a form that persists over the centuries [9], for example, in staircases and ancient cochlear columns, which also where used for visual storytelling purposes [8].

1.4 Metro station covering in Turin

Nature is always an infinite source of inspiration, especially when its forms reflect a mathematical harmony. In this case the inspiration came from the wings of the dragonfly. They show a particular texture and their functional morphology, with a considerable wingspan, has the property of not bending under considerable stress, despite their length [5].
The project by Francesco Schiavello and Yeol Park for the International contest for the design of covering of the Metro stations in Turin (Italy) indeed followed these principles. In addition, the pattern recalls the internal cupola of the mole Antonelliana Tower, the symbol of the City, which hosts the National Museum of Cinema.[1]
Turin, as a post-industrial city, has been trying to find a way to develop the city from the cultural point of view and, after hosting the winter Olympic games, has undergone a continuous transformation.. The main character of Turin is that it is surrounded by nature: the Alps mountains, covered with snow during winter, are a natural and suggestive frame to the city. The proposal for the roof intended to recall a pattern from nature. To obtain the fluid forms for the roof we generated a new geometry, which is based on an hexagonal geometry pattern, subsequently deformed parametrically (Fig. 10.).

Fig. 10. Dragonly wings inspiration for the design of the metro station in Turin.

[1] Turin was the first Italian city in which the industry of cinema was born and developed, in 1896.

1.5 Be[e] Italy and Radiolaria Jewelry

A jewelry collection was presented for the first time during the second edition of the Trieste Mini Maker Fair, under the premises of the International Center for Theoretical Physics (ICTP), a research center that works on the basis of an agreement between the Italian Government and two UN Agencies and where studies and researches in physics and mathematics are carried out.
This project is indeed linked to the worlds of physics and mathematics; it also combines biology, these all merged through parametric design.
The goal is creating unique design jewels, starting from the study of natural phenomena, then using models that we can find in nature, and by means of software which create codes based on parametric processes. This process emulates biological models of construction, so far indecipherable codes, putting digital manufacturing at the service of architecture and design. The design of this creations in particular is inspired to forms and geometries which bees are able to create: endless configurations, characterized by the alternation of solids and voids and by the irregular Combination of 110 modules (honeybees parametric). This design process has been used to represent the Italian region's forms. Jewels which are dynamic by their very nature (Fig. 11-12.).
On the occasion of the Trieste Mini Maker Fair, SAO has therefore decided to present a project of innovation and design, while denouncing the problem of the widespread bees death. Beyond its aesthetic function design becomes thus also a sensitization and awareness raising tool.

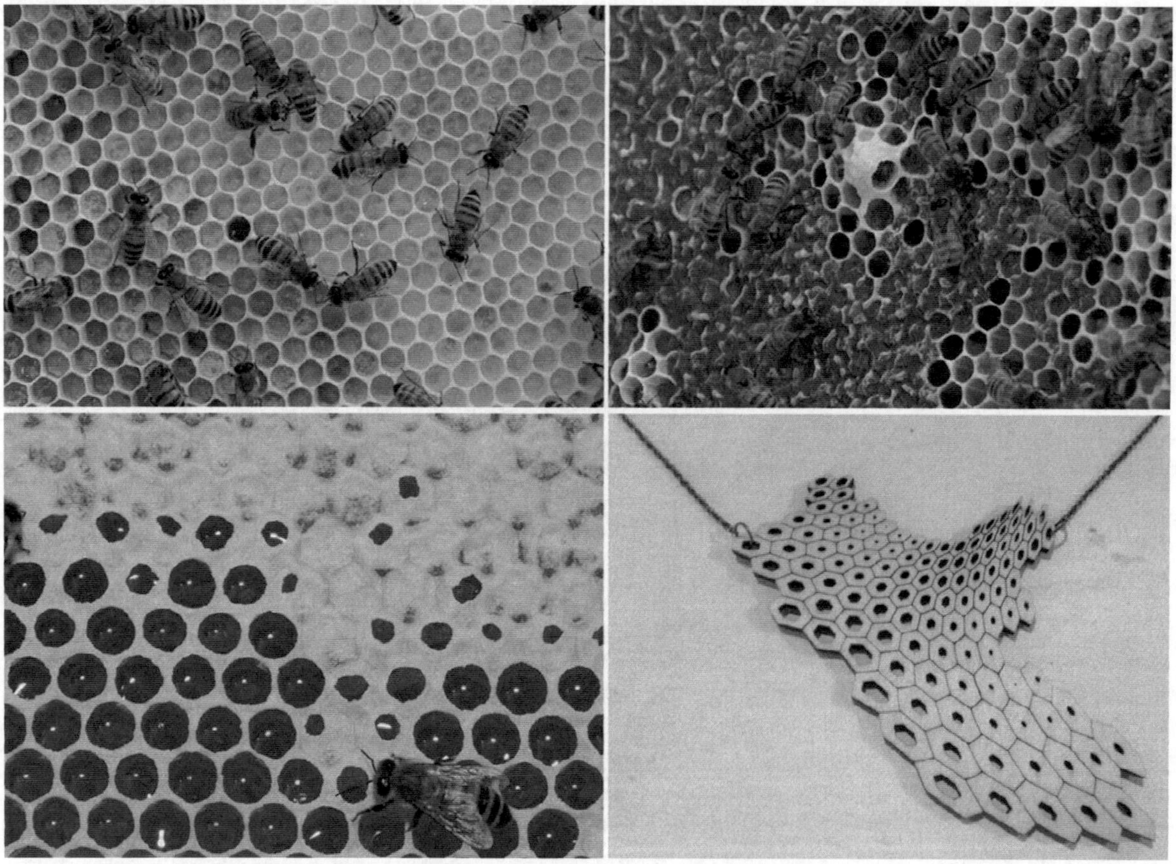

Fig. 11. Inspired by nature Be[e] Italy Jewelry.

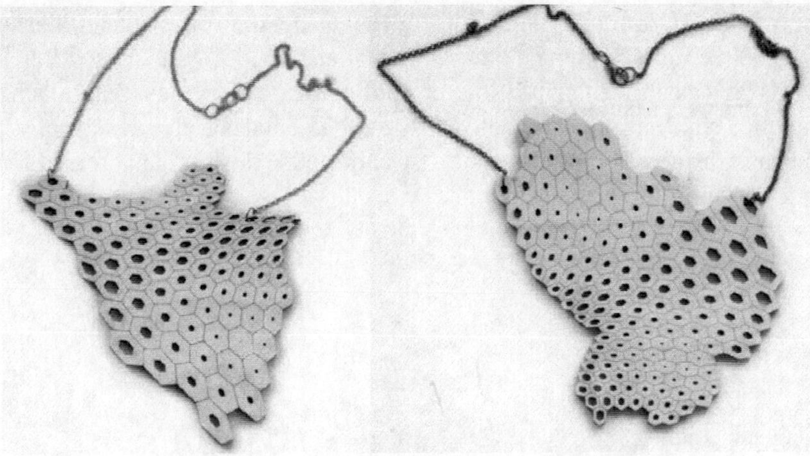

Fig. 12. Be[e] Italy Jewelry.

Fig. 13. Radiolaria Jewelry inspiration.

The necklace comes from the decomposition of the design of the radiolaria: a protozoon characterized by a siliceous skeleton. The linear dimensions of this protozoon are most often between hundredths and tenths of a millimeter, but there are species that reach or exceed the millimeter. Schiavello, combining design and biology, has enlarged the design of a protozoon and, by modifying its shape through digital processes of parametric design, has created a unique piece of jewel, with a fluid, harmonious and dynamic shape.

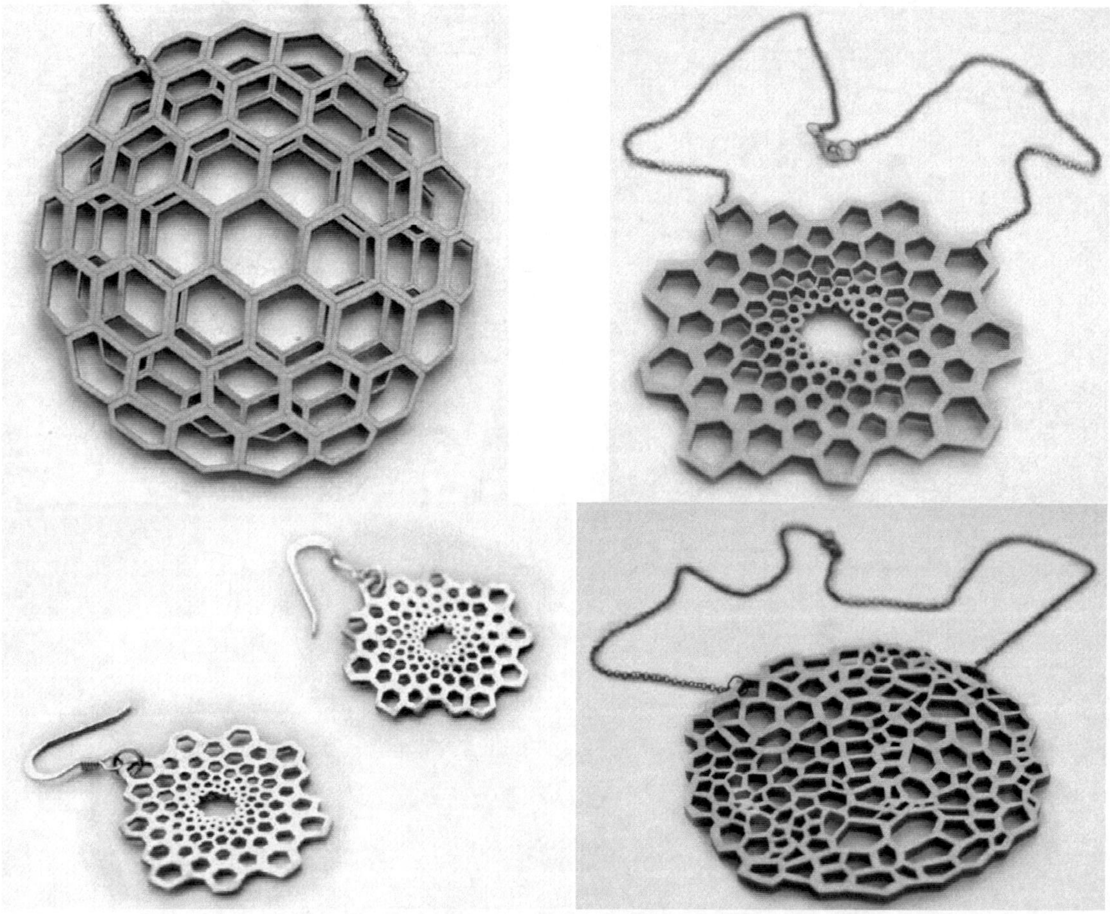

Fig. 14. Radiolaria Jewelry.

2.0 Conclusions

Mathematics pervades nature in its various forms, which can become an infinite source of inspiration for the artist, the architect, the designer. The process of design itself benefits from various parametric software that have a strong mathematical basis. Machines for cutting and 3d printers make these design become real objects.

Epilogue

In the brief epilogue, Thompson D'Arcy [10] writes that he will have succeeded "if I have been able to shew [the morphologist] that a certain mathematical aspect of morphology ... is ... complementary to his descriptive task, and helpful, nay essential, to his proper study and comprehension of Form." More lyrically, he writes that "For the harmony of the world is made manifest in Form and Number, and the heart and soul and all the poetry of Natural Philosophy are embodied in the concept of mathematical beauty" and quotes Isaiah 40:12 on measuring out the waters and heavens and the dust of the earth. He ends with a paragraph praising the French entomologist Jean-Henri Fabre who "being of the same blood and marrow with Plato and Pythagoras, saw in Number 'la clef de voute' [the key to the vault (of the universe)] and found in it 'le comment et le pourquoi des choses' [the how and the why of things]".

Acknowledgement

We are very grateful to Prof. Laura Tedeschini Lalli for her precious encouragement and kind suggestions, which we treasured.
This article and the entire session within the Aplimat conference is dedicated to the memory of Prof. Mauro Francaviglia (1953-2013), first co-organizer of the Math&Art session.
Work performed under the moral auspices of the National research group INdAM-GNSAGA (M. G. Lorenzi).

References

[1] URL, FRAZER, J., "Parametric Computation: History and Future". Architectural Design (March/April), 2016. http://www.danieldavis.com/a-history-of-parametric/
[2] URL, Grasshopper http://www.grasshopper3d.com/
[3] URL, Korea's Sam-Taegeuk Symbol, http://san-shin.org/Sacred-Mtns-01.html
[4] URL, Parametric Design https://en.wikipedia.org/wiki/Parametric_design
[5] URL, The geometry of the dragonfly wing, www.insects-en-maquettes.over-blog.com
[6] CAPANNA Alessandra (I), LORENZI, Marcella Giulia (I), EXPLORING EXPO GEOMETRIC EXPLOITS, Aplimat 2016
[7] CAPANNA A LORENZI M.G., Memories of lost cubes and domes self-similarity in modular contemporary buildings, the persistence of form between math and architecture, Aplimat 2014 proceedings, STU Bratislava, 2014.
[8] CARLINI, A., TEDESCHINI-LALLI L. , Imperial roman cochlear columns and their historical narrative freizes. Aplimat 2017 proceedings, STU Bratislava, 2017.
[9] CONVERSANO E., FRANCAVIGLIA M., LORENZI M.G., TEDESCHINI-LALLI L., Persistence of form in art and architecture: catenaries, helicoids and sinusoids. Aplimat - Journal of applied mathematics, vol. IV, n. IV, pp. 101-112, 2011.
[10] D'ARCY W. T., On Growth and Form, Cambridge University Press, 1917.
[11] FRANCAVIGLIA M., LORENZI M.G., Dal Cosmo al Numero ed alla Geometria Euclidea, Technai, 1, 23-37, 2009.
[12] GHEORGHIU D., The Decoration of Prehistoric Ceramic Vases with Bezier Curves, APLIMAT - Journal of Applied Mathematics, 4, 2011.
[13] JABI, W., Parametric Design for Architecture. London: Laurence King, 2013. ISBN 9781780673141.
[14] JANUSZKIEWICZ K., KOWALSKI K. G., Parametric Architecture in the Urban Space, IOP Conf. Ser.: Materials Scienceand engineering., 245, 052082, 2017. Doi:10.1088/1757-899X/245/5/052082
[15] KOLAREVIC, B (ed.), Architecture in the Digital Age: Design and Manufacturing, Spon Press, London, 2003.
[16] LALATTA COSTERBOSA Cecilia Parametric hybrid wall: a responsive surface for exhibition design, Aplimat 2014 proceedings, STU Bratislava, 2014.

[17] LALATTA COSTERBOSA Cecilia, (Italy IT), ROCCETTI CAMPAGNOLI Beatrice, (Italy IT), THE SMART GREENHOUSE, Responsive surfaces for agriculture, Aplimat 2017 proceedings, STU Bratislava, 2017.
[18] LORENZI M.G., FRANCAVIGLIA M., Art & Mathematics in Antoni Gaudí's Architecture: "La Sagrada Familía", APLIMAT Journal Applied Mathematics, 3 (1), 125-145, 2010. – ISSN 1337-6365
[19] LORENZI M., FRANCAVIGLIA M., Geometry as a Source of Inspiration in Contemporary Art. In: Bridges (Mathematics, Music, Art, Architecture, Culture), Coimbra Conference Proceedings; R. Sarhangi & C. Sequin Ed.s, with P. Machado as co-Editor; Coimbra, Phoenix: Tessellation Publishing, p. 365-372, ISBN/ISSN: 978-0-9846042-6-5, 2011. Also available online at url: http://archive.bridgesmathart.org/2011/bridges2011-365.html
[20] MACNAB M., Design by Nature: Using Universal Forms and Principles in Design, New Riders, 2012.
[21] STAVRIC M. AND MARINA O., Parametric Modeling for Advanced Architecture, International journal of applied mathematics and informatics, Issue 1, Volume 5, 2011.
[22] WOODBURY, R., Elements of Parametric Design. Routledge, 2010. ISBN 0415779871.

Current address

Lorenzi Marcella Giulia, Dr. PhD
University of Calabria
Via P.Bucci, 7-11, 87036 Arcavacata di Rende (CS)- ITALIA
Tel. +39 (0)984-49.3703
E-mail: marcella.lorenzi@unical.it

Schiavello Francesco, Dr. PhD
SAO Schiavello Architects Office
Via Luciano Manara, 57, 00153, Roma, Italia
Tel. +39 (0)6 68 300 199
E-mail: office@schiavello.it
website: www.schiavello.it

Slovak University of Technology in Bratislava
Faculty of Mechanical Engineering
17th Conference on Applied Mathematics **APLIMAT 2018**

Proceedings

COMPARISON OF RESULTS IN DETERMINING THE OPTIMAL NUMBER OF CLUSTERS IN CLUSTER ANALYSIS

LÖSTER Tomáš (CZ)

Abstract. Cluster analysis is a multivariate statistical method which main aim is to classify objects into groups called clusters. Classification of objects into groups is a natural requirement of various scientific disciplines. There are a lot of methods which can be used to classification of objects into clusters, in the current scientific literature. The main aim of this paper is to compare results of cluster number determination in using of different clustering methods. To comparison were used 19 selected real datasets from *The UCI Machine Learning Repository* (http://archive.ics.uci.edu/ml/datasets.html). On the basis of these analyzes, it can be said, that in case of the high level of variable variability, in situation, that the individual clusters are overlaped each other, it is better to use the Mahalanobis distance measure in process of cluster number determination. The best results were obtained using the CHF coefficient and Davides Bouldin index. Their success rate was, especially when using the centroid method, 54, 55 % in the case of the CHF coefficient (54, 55%) and 59, 09 % in case of the Davies-Bouldin index (59, 09%).

Keywords: clustering, evaluating of clustering, methods, coefficients for determining the optimal number of clusters, real datasets

Mathematics Subject Classification: Primary 62H30; Secondary 91C20.

1 Introduction

The basic issue of many scientific disciplines is the modernly so called taxonomy: to organize objects in groups. They may be represented by customers, patients, cars, documents, etc. To create groups there could be used different statistical methods and procedures. If the object (observation) is included into the existing group, there is used the discrimination analysis, if the object is classified into classes that may not be known in advance, there is used the cluster analysis. This is a multivariate statistical method that aim is to create groups of objects, which are called clusters. The objective of cluster analysis is to find objects within one cluster that are as similar as possible and objects from two different clusters should be as least similar as possible. Cluster analysis is very often used statistical method, see e.g. [3], [4], [8], [11], [12]. Very often it is used to regions classification. Authors very often used wages to describe regions. The problem of wages and poverty is described e.g. in [1], [2]. Other demographic variables, which are very often considered in cluster analysis, are described in [10], etc. In the current scientific literature, there are many cluster analysis methods that are used to classify objects. These methods and procedures can be categorized according to various criteria see e.g. [3], [12].They can be divided into hierarchical and non-hierarchical. They can also be divided into traditional or new approaches.

2 Clustering methods

Among the best known clustering methods can be included the nearest neighbour, furthest neighbour, centroid method, the average distance and also the Ward's method. These methods are included for example in the SYSTAT software and the researcher has also the option to apply the coefficients to determine the optimal number of clusters in connection with these methods, see below.

Nearest neighbour method

Nearest neighbour is the oldest and simplest method. Its origin dates back in 1957 and is associated with the name of P. H. A. Sneath, see [3]. The principle of this method is based on the idea that there is always looking for a pair of the most similar (closest) objects and those objects are associated. Firstly there are found two objects whose distance is the shortest and there is created the first cluster from them. Subsequently there is looking for other object whose distance is the smallest to this cluster. If we analyse the distances between clusters, we have to find the minimum distance of any object in the cluster in relation to any object in another cluster, see [12]. According to professional literature, this method is the easiest, but unfortunately has the disadvantage, which is also known as pipelining. This means the fact that there can be included two objects into one cluster, which are really the closest, however, they are not the closest to the most of other objects. This is due to the fact that a larger number of other objects between them creates the "chain" (also bridge). Another disadvantage is, that this method produces relatively elongate clusters. Formula for adjusting the distance matrix can be found e.g. in [3].

Furthest neighbour method

The author of the furthest neighbour method is Sörensen, see [3]. This method is based on the opposite principle than nearest neighbour. There are connected that two clusters (objects) into one cluster, which have a minimum distance between the most distant objects. The advantage of this method especially is [12], that the resulting clusters are small, consistent, compact, and well separated clusters, and that there is not arises the problem of chaining of the objects of clusters. The detailed formulas for adjusting the distance matrix can be found e.g. in [3].

Average distance method

The average distance method is sometimes called as a compromise method between Closest and Furthest neighbour, see [12]. Their priority is that the results are not affected by the extreme values. The criterion for the formation of clusters is defined as the average distance of all objects in one cluster to all of the objects in the second cluster. In this case, the criterion on whose basis become to the creation of new clusters, is influenced by the values of all the objects in the cluster. There are connected such two clusters, whose average distance is minimal. The formulas for the formal form of the distance matrix can be found e.g. in [3].

Centroid method

Centroid method is associated with the names of Sokal and Michener and it was published under the title "weighted group method", see [3]. The idea of this method is based on the centres of gravity (centroids). There are linked such two clusters whose centroid distance is minimal. In this case the "centroid" is defined as the average of the values of variables in the cluster. As an important advantage of this method is reported that the results are not affected by outlying objects. On the other hand, as a disadvantage of this method is reported the "inversion", see [3]. There may also arise the "confusing clusters", which means that the distance between the centroid of one pair of clusters may be smaller than the distance between the centroid of another pair, created in the previous step. The formulas for the formal form of the distance matrix can be found e.g. in [3].

Ward's method

Ward's method is associated with two names - Ward and Wishart. Author Ward designed a method for measuring of similarities / dissimilarities of clusters, the author Wishart designed the calculation of Ward's coefficient. The principle of clustering is different from the above mention methods of clustering, in which the distance of clusters were optimized. Ward's method is a process to minimize the heterogeneity of clusters, it means that the clusters are created by maximizing within-groups homogeneity. With this method are small clusters removed and as the result of the clustering process are the clusters with approximately the same number of objects. As noted [3], this method is most commonly used in practice. The criterion of homogeneity of the clusters is the within-group sum of squared deviations from the average (centroid) of the cluster. When connecting the clusters there is the main issue based on the idea, that in every step of clustering must be the smallest increment of Ward's G criterion, see [3].

More detailed description of these methods can be found, for example in [3]. Methods differ, apart from other facts, in the time of formation and also in the way of clustering. Some of them minimize the distance between objects, which means the distance among the closest and furthermost objects. Ward's method solves the principle of clustering based on minimizing the intra-group variability, see [3]. Clustering methods may be combined with various measures of distances. The most famous include Euclidean distance measure or Mahalanobis distance measure, see for example [3]. The result of these combinations is a large number of options that can be used to classify objects.

Part of the cluster analysis is very often the number of clusters determination in which the objects may be distributed. To determine the number of clusters (groups of objects), there is a number of criteria and procedures. As noted above, various techniques for clustering may involve a different distribution of objects into clusters. The issues of determining the number of clusters, in the case of quantitative variables, are the subject of many scientific papers, including for example [4], [5], [6], [7], [8], [9] etc. However, in the current literature there is no clear rule to determine the use of concrete coefficients in different conditions. Various authors create and modify their coefficients and sometimes the coefficients are compared with the existing ones. Among the selected criteria for determining the number of clusters can be included Davies-Bouldin's (DB) index, Dunn's index, RMSSTD index, CHF index and PTS index. These coefficients are already applied e.g. in SYSTAT software and they are commonly used for determining the number of clusters.

For clustering can be used a lot of software products. Among the best-known and most frequently used can be included e.g. IBM SPSS, SAS, STATISTICA, S-PLUS, SYSTAT, STATGRAPHICS, etc. In these are implemented especially the traditional clustering methods, including possible methods of decomposition. Some of them, such as the SAS system do not allow to user to select the appropriate combination of clustering methods and distance measures. The system offers a combination like that. When determining the number of clusters can be used only selected software products. But only the selected coefficients are there implemented. Another widespread software products, such as STATISTICA or STATGRAPHICS does not contain coefficient for determining the number of clusters. Table 1 summarizes the evaluation and the location of selected coefficients listed above for determining the number of clusters.

Coefficient	Founded extreme	Software
CHF index	maximum	SAS system LE, SYSTAT
PTS index	maximum	SAS system LE, SYSTAT
RMSSTD	minimum	SYSTAT
Davies-Bouldin (DB)	minimum	SYSTAT
Dunn's index	maximum	SYSTAT

Tab. 1. Overview of selected coefficients in software products.

3 Evaluation of coefficients for determining the number of clusters on real data sets

In this part of the paper there are described and evaluated the results of clustering in the existing socio-economic data files. All datasets come from The UCI Machine Learning Repository database (web address: *http://archive.ics.uci.edu/ml/datasets.html*). For the purposes were selected 19 files. Some files are very large (millions of objects). For this reason, were produced individual subgroups, and those were evaluated separately. In this random selection the objects were removed (after the inclusion in a sample file) from the original database, for the reason that the object could not be analysed twice in different random samples. Random sampling was carried out using the SPSS version 20. There were evaluated 22 data files in total, some of which are conventional and well-known classification files. Individual existing files are different in terms of number of clusters, the number of objects and the number of variables. The files also differ by separation of objects, and so in some cases, clusters are significantly overlapped. When selecting files from the database it was controlled, that the files are intended for classification of objects with a possibility to analyse all variables simultaneously, i.e. that to the selection must not enter the files that contain qualitative variables. To subsequent evaluation there were subjected those files for which the number of clusters (classification of object to the cluster) is known, (because of comparison). It is obvious that the result (the number of clusters), which is written in the database, is not always the only one possible classification of objects, but for the purposes of evaluation is the value from the database used. If, for example, will be selected only some variables, it could lead to other initial classification of objects. In the case that in some datasets occurred missing values for some variables, the objects were discarded from further analyses. In the case of unequal units it was carried out standardization using by z-scores. Analysed files (sorted alphabetically):
Abalone, Banknote Authentication, Blood Transfusion Service Center, Cardiotocography, Connectionist Bench (Vowel Recognition - Deterding Data), Energy Efficiency, Glass, Indian Liver Patient, Ionosphere, Iris, Musk (Version 1) QSAR Biodegradation, Statlog (Vehicle Silhouettes) a+b, Susy (extensive file from which was conducted random selection), Vertebral Column 2c, Vertebral Column 3c, Wall-Following Robot Navigation Data, Wholesale Customers, Wine.

Table 2 shows the number of correctly set clusters (in %) at each of the method using Euclidean distance measure. The table shows that the success rate of most coefficients is very low and does not reach 50%.

Method/coefficient	RMSSTD	CHF	PTS	D-B	Dunn
Nearest neighbour	9,09	45,45	40,91	36,36	50,00
Farthest neighbour	22,73	18,18	18,18	59,09	36,36
Centroid method	27,27	36,36	22,73	45,45	50,00
Average distance	22,73	31,82	22,73	45,45	45,45
Ward's method	18,18	31,82	31,82	18,18	40,91

Tab. 2. Number of correctly set clusters (in %) – Euclidean distance measure.

Table 3 shows the number of correctly set clusters (in %) at each of the method using Mahalanobis distance measure. It can be seen from the table 3 that the success rate of most of coefficients is higher than the Euclidean distance measure was used.

Method/coefficient	RMSSTD	CHF	PTS	D-B	Dunn
Nearest neighbour	4,55	45,45	50,00	50,00	45,45
Farthest neighbour	22,73	31,82	36,36	36,36	36,36
Centroid method	0,00	54,55	40,91	59,09	36,36
Average distance	9,09	50,00	50,00	59,09	50,00
Ward's method	22,73	40,91	22,73	4,55	59,09

Tab. 3. Number of correctly set clusters (in %) – Mahalanobis distance measure.

3 Conclusion

Cluster analysis is a multivariate statistical method which aim is to classify objects into groups called clusters. The objective of cluster analysis is to find objects within one cluster that are as similar as possible and objects from two different clusters should be as least similar as possible. The output of cluster analysis very often contains the number of clusters into which the objects are classified. This information is not known in advance in most real cases. To determine the optimal number of clusters there are many ways (coefficients) that again can be combined with different methods and different measures of distances.

Based on the results were compiled conclusins to Table 4, which contains differences betwwn average succes rate (in %) for both distances. It is evident that success rate is higher for almost of coefficients in using of Mahalanobis distance measure.

Method/coefficient	RMSSTD	CHF	PTS	D-B	Dunn
Nearest neighbour	4,55	0,00	-9,09	-13,64	4,55
Farthest neighbour	0,00	-13,64	-18,18	22,73	0,00
Centroid method	27,27	-18,18	-18,18	-13,64	13,64
Average distance	13,64	-18,18	-27,27	-13,64	-4,55
Ward's method	-4,55	-9,09	9,09	13,64	-18,18

Tab. 4. Difference Average success rate (in %) of Euclidean and Mahalanobis distances.

Graphical comparison of differences between success rate of Euclidean and Mahalanobis distance measure is evident from the figure 1.

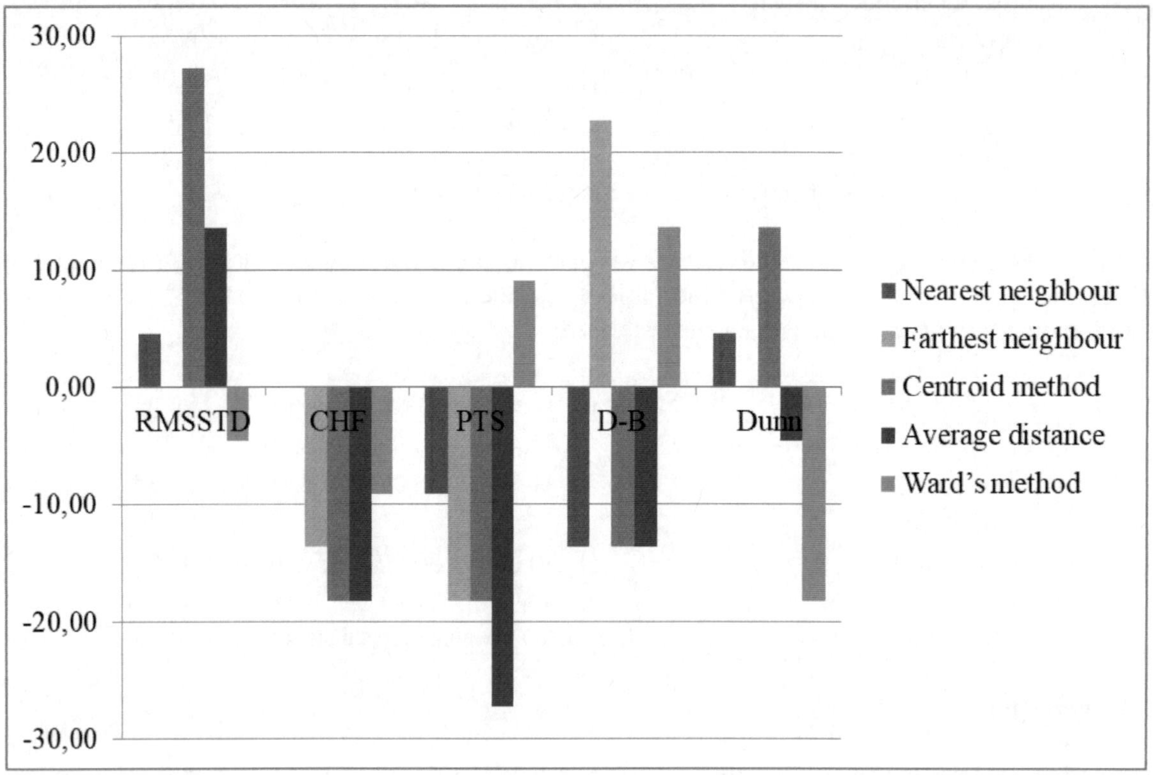

Fig.1. Comparison of differences between success rate of Euclidean and Mahalanobis distance (Eucledian – Mahalanobis).

In conclusion we can say, that the CHF and Davies-Bouldin coefficient are more successful in determining the number of clusters compared with the other coefficients, which are noted e.g. in [6]. The more are the individual clusters overlapped, (i.e. that the separation rate decreases), the less its success rate. Best results are achieved in connection with the average distance method, and Ward's method. The success rate in use of other methods is low. The lowest success of CHF coefficient was achieved in connection with centroid and average distance method. When we compare coefficients in connection with the Mahalanobis distance measure o at the same group of files, it is obvious that the use of Mahalanobis distances extent always leads to a higher success of this coefficient, rather than using Euclidean distance measure. This statement is particularly valid for significantly overlapping clusters.

Acknowledgement

This paper was processed with contribution of long term institutional support of research activities number IP400040 by Faculty of Informatics and Statistics, University of Economics, Prague, Czech Republic.

References

[1] BÍLKOVA, D. *Modelling of income and wage distribution using the method of l-moments of parameter estimation* . In Loster Tomas, Pavelka Tomas (Eds.), *International Days of Statistics and Economics* (pp. 40-50), 2011, ISBN 978-80-86175-77-5.

[2] BÍLKOVA, D. *Development of wage distribution of the czech republic in recent years by highest education attainment and forecasts for 2011 and 2012.* In Loster Tomas, Pavelka Tomas (Eds.), *6th International Days of Statistics and Economics* (pp. 162-182), 2012, ISBN 978-80-86175-86-7.

[3] GAN, G., MA CH., WU J.: *Data Clustering Theory, Algorithms, and Applications*, ASA, 2007, Philadelphia.

[4] HALKIDI, M., VAZIRGIANNIS, M.: *Clustering validity assessment: Finding the optimal partitioning of a data set*, Proceedings of the IEEE international conference on data mining, 2001, (pp. 187-194).

[5] LÖSTER, T. *The Evaluation of CHF coefficient in determining the number of clusters using Euclidean distance measure.* In: The 8th International Days of Statistics and Economics., 2014, (pp. 858–869), ISBN 978-80-87990-02-5.

[6] LÖSTER, T. *The Evaluation of CHF coefficient in determining the number of clusters using Mahalanobis distance measure.* In: *14th Conference on Applied Mathematics – Aplimat 2015 [CD].* Bratislava, 03.02.2015 – 05.02.2015. Bratislava : Slovak University of Technology, 2015, s. 546–554. ISBN 978-80-227-4314-3.

[7] LÖSTER, Tomáš. Evaluation of Coefficients for Determining the Optimal Number of Clusters in Cluster Analysis on Real Data Sets. In: The 9th International Days of Statistics and Economics (MSED 2015) [online]. Praha, 10.09.2015 – 12.09.2015. Slaný : Melandrium, 2015, s. 1014–1023. ISBN 978-80-87990-06-3.

[8] LÖSTER, Tomáš. Evaluation of Coefficients for Determining the Optimal Number of Clusters in Cluster Analysis on Real Data Sets. In: *The 9th International Days of Statistics and Economics (MSED 2015).* [online] Praha, 10.09.2015 – 12.09.2015. Slaný : Melandrium, 2015, s. 1014–1023. ISBN 978-80-87990-06-3.

[9] MAKHALOVÁ, E. 2015. The Fuzzy Clustering Problems and Possible Solutions *The 9th International Days of Statistics and Economics.* Praha, 2015. s. 1052–1061. ISBN 978-80-87990-06-3

[10] MEGYESIOVA, S., & LIESKOVSKA, V. *Recent population change in europe*. In Loster Tomas, Pavelka Tomas (Eds.), *International Days of Statistics and Economics*, (pp. 381-389), 2011, ISBN 978-80-86175-77-5.

[11] MELOUN, M., MILITKÝ, J., HILL, M.: Počítačová analýza vícerozměrných dat v příkladech, Academia, 2005, Praha.

[12] ŘEZANKOVA, H., HÚSEK, D., SNÁŠEL, V.: *Shluková analýza dat,* 2. vydání, Professional Publishing, 2009, Praha.

[13] ŘEZANKOVA, H., & LOSTER, T. Shlukova analyza domacnosti charakterizovanych kategorialnimi ukazateli. *E+M. Ekonomie a Management, 16*(3), (pp 139-147), 2013, ISSN: 1212-3609.

[14] STANKOVIČOVÁ, I., VOJTKOVÁ, M.: Viacrozmerné štatistické metódy s aplikáciami, Ekonómia, 2007, Bratislava.

Current address

Löster Tomáš, Ing., Ph.D.
University of Economics, Prague, Dept. of Statistics and Probability
W. Churchill sq. 4, 130 67 Prague 3, Czech Republic
E-mail: tomas.loster@vse.cz

Slovak University of Technology in Bratislava
Faculty of Mechanical Engineering
17th Conference on Applied Mathematics **APLIMAT 2018**

Proceedings

TWO NUMERICAL APPROACHES TO THE NON-LINEAR LEAST-SQUARES METHOD VIA PRACTICAL EXAMPLES

LUDVÍK Pavel (CZ), MORÁVKOVÁ Zuzana (CZ)

Abstract. Our contribution is meant to be an example of multidisciplinary approach to the teaching of numerical methods. By employing the Michaelis-Menten's model for enzyme kinetics, we show a practical application of numerical methods in biochemistry. Given the experimental data, we find a dependence of a reaction rate on a concentration of a substrate. In a preliminary section, we derive and explain The Michaelis-Menten kinetics from behaviour of biochemical reactions. In a first part, the problem is linearized and then solved by the least squares method. In a second part, we do not use linearization and solve the original problem by the Newton's method for systems of non-linear equations. We conclude our contribution with a comparison of both approaches and results. We also offer several problems for students to clarify and deepen their understanding of the linearization. Our solution is provided in a form of a thoroughly commented MATLAB Code.

Keywords: teaching of numerical methods, non-linear least-squares method, Newton's method for systems, linearization

Mathematics subject classification: Primary 97N50; Secondary 49M15, 93E24

1 Introduction

We put the strong accent on the theoretical aspects of engineering which allows our graduates to develop an interdisciplinary way of thinking and to be able to handle engineering problems with efficiency and bold creativity. For this purpose we have developed a collection of solved exercises where (besides other things) we show usage of numerical methods in applied sciences (see [4]). We strongly believe that the students of engineering are capable of comprehension and even appreciation of the abstract concepts of the pure mathematics. Nevertheless, the first contact should be carried out under the cover of a real tangible problem. In what follows you can taste a flavour of our blend of biochemistry, numerical mathematics and MATLAB coding.

2 Michaelis-Menten kinetics

In 1913, Michaelis and Menten proposed a model which is one of the best-known for enzyme kinetics ([2, p. 7]). They assumed a reaction scheme where the enzyme E converts the substrate S into the product P through a two-step process with a complex C being an intermediate. The reaction scheme is represented by

$$S + E \underset{k_{-1}}{\overset{k_1}{\rightleftharpoons}} C \underset{k_{-2}}{\overset{k_2}{\rightleftharpoons}} P + E.$$

Bothe reactions must be reversible but reaction rates are typically measured under conditions where P is continually removed which prevents the reverse reaction k_{-2} to happen. Hence, it is reasonable to assume no reverse reaction occurs:

$$S + E \underset{k_{-1}}{\overset{k_1}{\rightleftharpoons}} C \overset{k_2}{\rightarrow} P + E.$$

Let us denote the quantities $s = [S]$, $c = [C]$, $e = [E]$ and $p = [P]$. The *law of mass action* then gives four differential equations for the rates of change of s, c, e and p,

$$\begin{aligned}
\frac{ds}{dt} &= k_{-1}c - k_1 se,\\
\frac{de}{dt} &= (k_{-1} + k_2)c - k_1 se,\\
\frac{dc}{dt} &= k_1 se - (k_2 + k_{-1})c,\\
\frac{dp}{dt} &= k_2 c.
\end{aligned}$$

Let us remark that p can be found by direct integration and also $\frac{de}{dt} + \frac{dc}{dt} = 0$ which means that $e + c$ is constant. We denote the constant by e_0 and it has meaning of the total amount of available enzyme.

Michaelis and Menten made an additional assumption that the substrate s is in instantaneous equilibrium with the complex c, which means

$$k_1 se = k_{-1}c.$$

Since $e + c = e_0$, we obtain a system of two equations, we consider c and e as unknowns and solve it for c:

$$c = \frac{k_1 e_0 s}{k_{-1} + k_1 s} = \frac{e_0 s}{\frac{k_{-1}}{k_1} + s}.$$

Finally, the velocity v of the reaction is given by

$$v = \frac{dp}{dt} = k_2 c = \frac{k_2 e_0 s}{\frac{k_{-1}}{k_1} + s}.$$

3 Introducing the problem

Michaelis-Menten model takes the form of an equation

$$v(x) = \frac{a_1 x}{a_2 + x},$$

relating a reaction rate v to a concentration of a substrate denoted by x. Constants a_1 and a_2 are unknown.

Analyze the experimental data (see Table 1) and find appropriate values of constants a_1 and a_2.

x_i [10^{-4} mol $\cdot$ dm^{-3}]	2.267	2.703	3.575	4.098	4.098	3.974	4.211
v_i [10^{-8} mol $\cdot$ dm^{-3} $\cdot$ s^{-1}]	1.220	2.451	4.897	9.780	14.688	19.576	24.483

Tab. 1. The experimental data

4 Solution with linearization

The original equation

$$v = \frac{a_1 x}{a_2 + x}$$

can be rewritten into an equivalent form

$$\frac{1}{v} = \frac{a_2 + x}{a_1 x} \qquad \Rightarrow \qquad \frac{1}{v} = \frac{a_2}{a_1} \cdot \frac{1}{x} + \frac{1}{a_1}.$$

After substitutions $V := \frac{1}{v}$, $X := \frac{1}{x}$, $A := \frac{1}{a_1}$ and $B := \frac{a_2}{a_1}$, it becomes

$$V = BX + A.$$

A manoeuvre we have just described is called a **linearization** (see [1, p. 201]) of a problem. Considering the new variables X and V we obtain a transformed data, see Table 2.

X_i	0.441	0.370	0.280	0.244	0.244	0.252	0.238
V_i	0.820	0.408	0.204	0.102	0.068	0.051	0.041

Tab. 2. The linearized data

To acquire reasonable values of constants A and B we employ the least square method (see [3, p. 85] or [5, p. 431]) for the modified problem. Hence, our goal is finding a minimum of the function

$$\Psi(A, B) = \sum_i \left[(BX_i + A)^2 - V_i\right]^2.$$

First, we compute the partial derivatives of Ψ and we set both of them equal zero.

$$\frac{\partial \Psi}{\partial A} = 2 \sum_i (BX_i + A - V_i) = 0,$$

$$\frac{\partial \Psi}{\partial B} = 2 \sum_i (BX_i + A - V_i) X_i = 0.$$

After straightforward modification we get the **normal equation** in a matrix form

$$\begin{pmatrix} \sum_i 1 & \sum_i X_i \\ \sum_i X_i & \sum_i X_i^2 \end{pmatrix} \begin{pmatrix} A \\ B \end{pmatrix} = \begin{pmatrix} \sum_i V_i \\ \sum_i X_i V_i \end{pmatrix}.$$

The normal equation is easily solvable with use of MATLAB.

```
>> x = [2.267 2.703 3.575 4.098 4.098 3.974 4.211];
>> v = [1.220 2.451 4.897 9.780 14.688 19.576 24.483];
>> X = 1./x;
>> V = 1./v;
>> [7 sum(X); sum(X) sum(X.^2)]\[sum(V); sum(X.*V)]
ans =
   -0.8054
    3.5455
```

Therefore $A = -0.8054$ and $B = 3.5455$. If we go back to the original unknowns, we end up with

$$a_1 = \frac{1}{A} = -1.2416, \quad a_2 = \frac{B}{A} = -4.4022,$$
$$\text{hence} \quad v^L(x) = \frac{-1.2416\,x}{-4.4022 + x}.$$

5 Solution without linearization

The linearization leads to a great simplification of the problem. Nevertheless, we pay a price for this simplicity as the resulting function is not necessarily the best fit in the least-squares sense. Now, we carry out a computation without the linearization trick. We are looking for a minimum of a function

$$\Psi(a_1, a_2) = \sum_i \left(\frac{a_1 x_i}{a_2 + x_i} - v_i \right)^2 .$$

We compute the partial derivatives of Ψ and we set both of them equal zero.

$$\frac{\partial \Psi}{\partial a_1} = 2 \sum_i \left(\frac{a_1 x_i}{a_2 + x_i} - v_i \right) \frac{x_i}{a_2 + x_i} = 2 \sum_i \left(\frac{a_1 x_i^2}{(a_2 + x_i)^2} - \frac{x_i v_i}{a_2 + x_i} \right) = 0,$$
$$\frac{\partial \Psi}{\partial a_2} = 2 \sum_i \left(\frac{a_1 x_i}{a_2 + x_i} - v_i \right) \frac{-a_1 x_i}{(a_2 + x_i)^2} = -2 \sum_i \left(\frac{a_1^2 x_i^2}{(a_2 + x_i)^3} - \frac{a_1 x_i v_i}{(a_2 + x_i)^2} \right) = 0.$$

We adjust the equations by cancelling the constants 2 and -2. Moreover, we denote the functions by f_1 and f_2 and rewrite the equations as

$$f_1(a_1, a_2) = \sum_i \left(\frac{a_1 x_i^2}{(a_2 + x_i)^2} - \frac{x_i v_i}{a_2 + x_i} \right) = 0,$$
$$f_2(a_1, a_2) = \sum_i \left(\frac{a_1^2 x_i^2}{(a_2 + x_i)^3} - \frac{a_1 x_i v_i}{(a_2 + x_i)^2} \right) = 0.$$

The computation is carried out in MATLAB. We begin with definitions of vectors x, v and functions f_1, f_2. Take notice of a vector `A` as an input variable of the functions.

```
>> x = [2.267 2.703 3.575 4.098 4.098 3.974 4.211];
>> v = [1.220 2.451 4.897 9.780 14.688 19.576 24.483];
>> f1 = @(A)sum(A(1)*x.^2./(A(2)+x).^2-x.*v./(A(2)+x));
>> f2 = @(A)sum(A(1).^2*x.^2./(A(2)+x).^3-A(1)*x.*v./(A(2)+x).^2)
   ;
```

We have obtained a rather difficult system of equations but with the assistance of the **Newton's method** (see [3, p. 31] or [5, p. 283])may gain at least an approximate solution. We need to find a Jacobi matrix of the mapping $f(a_1, a_2) = (f_1(a_1, a_2), f_2(a_1, a_2))^\top$ which is

$$\boldsymbol{J_f}(a_1, a_2) = \begin{pmatrix} \frac{\partial f_1}{\partial a_1}(a_1, a_2) & \frac{\partial f_1}{\partial a_2}(a_1, a_2) \\ \frac{\partial f_2}{\partial a_1}(a_1, a_2) & \frac{\partial f_2}{\partial a_2}(a_1, a_2) \end{pmatrix} =$$

$$= \begin{pmatrix} \sum_i \frac{x_i^2}{(a_2 + x_i)^2} & \sum_i \frac{-2a_1 x_i^2}{(a_2 + x_i)^3} + \frac{x_i v_i}{(a_2 + x_i)^2} \\ \sum_i \frac{2a_1 x_i^2}{(a_2 + x_i)^3} - \frac{x_i v_i}{(a_2 + x_i)^2} & \sum_i \frac{-3a_1^2 x_i^2}{(a_2 + x_i)^4} + \frac{2a_1 x_i v_i}{(a_2 + x_i)^3} \end{pmatrix}.$$

Assuming $\boldsymbol{J_f}$ is regular at points where necessary, the Newton's method defines a sequence $(\boldsymbol{a}^k)$ such that

$$\boldsymbol{a}^0 = (-1.2416, -4.4022)^\top \quad \text{(i.e., the result of linearized problem),}$$

$$\boldsymbol{a}^{k+1} = \boldsymbol{a}^k - \boldsymbol{J_f}(\boldsymbol{a}^k)^{-1}\, \boldsymbol{f}(\boldsymbol{a}^k).$$

Then, we define the Jacobian matrix in MATLAB.

```
>> Jf = @(A)[sum(x.^2./(A(2)+x).^2),
       sum(-2*A(1)*x.^2./(A(2)+x).^3+x.*v./(A(2)+x).^2);
       sum(2*A(1)*x.^2./(A(2)+x).^3-x.*v./(A(2)+x).^2),
       sum(-3*A(1)^2*x.^2./(A(2)+x).^4+2*A(1)*x.*v./(A(2)+x).^3)
   ];
```

We pick a result of the linearized problem as an initial approximation and accomplish the first step of the Newton's method. We also save the computed value into the variable X.

```
>> X = [-1.2416;-4.4022];
>> Xnew = X-inv(Jf(X))*[f1(X);f2(X)], X=Xnew;
Xnew =
   -1.2411
   -4.4224
```

We repeat the last command while the visible digits of the approximation vector `Xnew` change.

```
% After several repetitions:
>> Xnew = X-inv(Jf(X))*[f1(X);f2(X)], X = Xnew;
Xnew =
   -1.6433
   -4.5168
```

We may conclude that

$$a_1 = -1.6433, \qquad a_2 = -4.5168$$

is a good approximation of the problem and hence

$$v^N(x) = \frac{-1.6433\,x}{-4.5168 + x}.$$

6 Solution by MATLAB

We may also use a non-linear least-squares solver implemented in MATLAB. It's application is simple but it uses internal numerical methods which are sensitive on a choice of initial approximation. Since we are furnished with knowledge of a very good approximation of solution, we can choose a very precise initial approximation here.

```
>> f = @(a)a(1)*x./(a(2)+x)-v;
>> a0=[-1 -5]; % the initial approximation
>> aM = lsqnonlin(f,a0)
aM =

   -1.6430    -4.5168
```

MATLAB generates a solution

$$v^M(x) = \frac{-1.6430\,x}{-4.5168 + x}.$$

Notice that the parameters are almost the same as in case of the previous solution without linearization.

7 Comparison of results

Let us summarize our results

$$v^L(x) = \frac{-1.2416\,x}{-4.4022 + x}, \qquad v^N(x) = \frac{-1.6433\,x}{-4.5168 + x}.$$

The linearization is a very useful tool which may help us handle problems that would be unsolvable otherwise. Nevertheless, as we observe in the Table 3 and in the Figure 1, the results obtained by linearization may be distorted in general.

x_i	v_i	$v^L(x_i)$	$v^N(x_i)$	$(v^L(x_i) - v_i)^2$	$(v^N(x_i) - v_i)^2$
2.2670	1.2200	1.3182	1.6559	0.0097	0.1900
2.7030	2.4510	1.9751	2.4489	0.2265	0.0000
3.5750	4.8970	5.3660	6.2378	0.2199	1.7979
4.0980	9.7800	16.7261	16.0799	48.2482	39.6882
4.0980	14.6880	16.7261	16.0799	4.1538	1.9373
3.9740	19.5760	11.5229	12.0311	64.8519	56.9257
4.2110	24.4830	27.3451	22.6290	8.1915	3.4375

Tab. 3. Comparison of results

Finally, let us show the sums of squares of both results which allows to measure an accuracy in sense of least squares ("the smaller the better"). Their difference is less negligible than one would probably expect

$$\sum_i (v^L(x_i) - v_i)^2 = 125.9015, \qquad \sum_i (v^N(x_i) - v_i)^2 = 103.9764 .$$

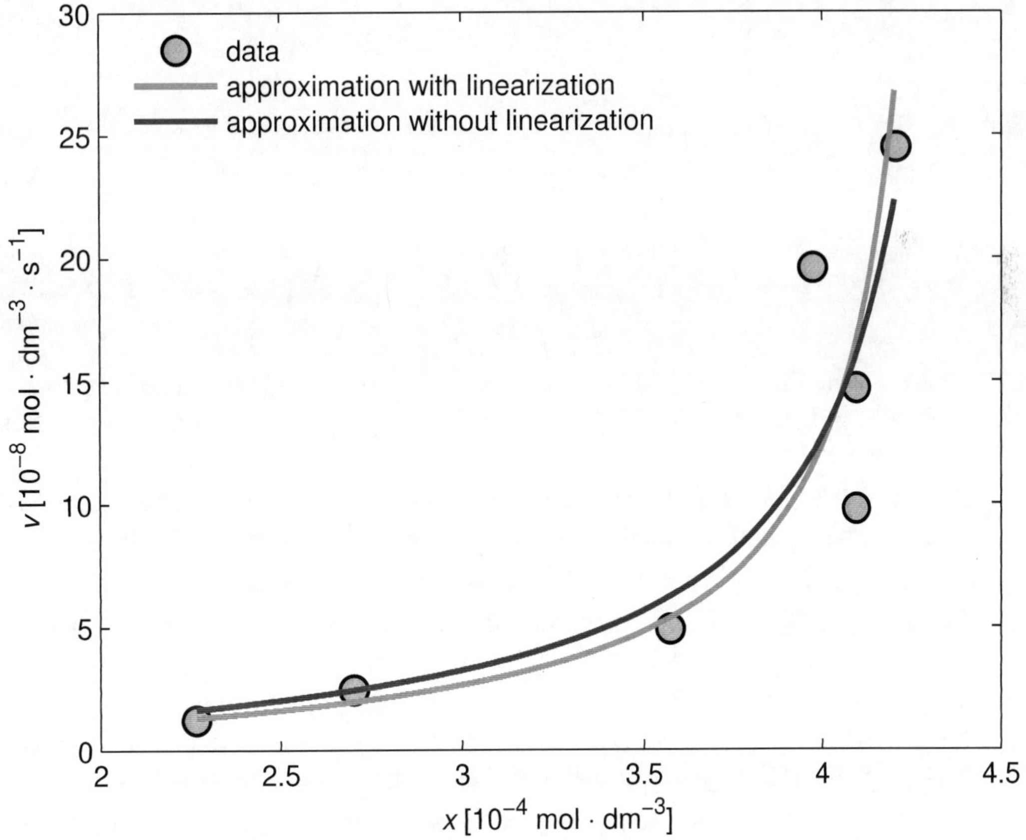

Fig. 1. Comparison of results and data

8 The examples of linearization

Linearization is a simple tool to transform a non-linear approximation problem to a linear one. We can apply a linear least-squares method afterwards. The quality of a resulting approximation is usually satisfying even though a routine linearization might add an error which is difficult to control. On the other hand, mathematically rigid usage of a non-linear least-squares method leads to a non-linear system of equations which is often solvable only by a numerical (and therefore only approximate) method.

We append a table of some of the most useful linearization formulas.

$v = a_1 x^{a_2}$	$V := \ln v$, $X := \ln x$, $A := \ln a_1$, $B := a_2$	$V = A + BX$
$v = a_1 \mathrm{e}^{a_2 x}$	$V := \ln v$, $X := x$, $A := \ln a_1$, $B := a_2$	$V = A + BX$
$v = \mathrm{e}^{a_1 + \frac{a_2}{x}}$	$V := \ln v$, $X := \frac{1}{x}$, $A := \ln a_1$, $B := a_2$	$V = A + BX$
$v = \ln(a_1 + a_2 x)$	$V := \mathrm{e}^v$, $X := x$, $A := a_1$, $B := a_2$	$V = A + BX$
$v = a_1 x e^{a_2 x}$	$V := \ln \frac{v}{x}$, $X := x$, $A := \ln a_1$, $B := a_2$	$V = A + BX$
$v = a_1 + a_2 \ln x$	$V := v$, $X := \ln x$, $A := a_1$, $B := a_2$	$V = A + BX$

Conclusion

We hope this exercise teaches a lesson that every problem in numerical mathematics permits many solutions. It can be very useful to use a simplified model of the situation. Nevertheless, you cannot expect to obtain the same result as if you do all the hard work.

References

[1] GILAT, Amos. a Vish. SUBRAMANIAM. *Numerical methods for engineers and scientists: an introduction with applications using Matlab*. Third edition. Hoboken, NJ: John Wiley & Sons, Inc., 2014. ISBN 978-1-118-55493-7.

[2] KEENER, James P. a James SNEYD. *Mathematical physiology*. 2nd ed. New York, NY: Springer, c2009. Interdisciplinary applied mathematics, 8. ISBN 978-0-387-75846-6.

[3] KUČERA, Radek a Zuzana MORÁVKOVÁ. *Numerická matematika* [online]. 1. Ostrava: Vysoká škola báňská-Technická univerzita Ostrava, 2016. ISBN 978-80-248-3893-9. Available online: http://mdg.vsb.cz/portal/nm/nm.pdf

[4] LUDVÍK, Pavel a Zuzana MORÁVKOVÁ. Numerická matematika: Řešené příklady s Matlabem a aplikované úlohy [online]. 1. Ostrava: Vysoká škola báňská-Technická Ostrava, 2016. ISBN 978-80-248-3892-2. Available online: http://mdg.vsb.cz/portal/nm/nmsm.pdf

[5] QUARTERONI, Alfio, Riccardo SACCO and Fausto SALERI. *Numerical mathematics*. New York: Springer, c2000, xx, 654 p. ISBN 0-387-98959-5.

Current address

Ludvík Pavel, RNDr., Ph.D.
Department of Mathematics and Descriptive Geometry
VŠB – Technical University of Ostrava
17. listopadu 15, 708 33 Ostrava – Poruba
Czech Republic
E-mail: pavel.ludvik@vsb.cz

Morávková Zuzana, Mgr., Ph.D.
Department of Mathematics and Descriptive Geometry
VŠB – Technical University of Ostrava
17. listopadu 15, 708 33 Ostrava – Poruba
Czech Republic
E-mail: zuzana.moravkova@vsb.cz

Slovak University of Technology in Bratislava
Faculty of Mechanical Engineering
17th Conference on Applied Mathematics **APLIMAT 2018**

Proceedings

GRAVITATIONAL MUSIC

LUMINET Jean-Pierre (FR)

Abstract. Since the 1960s a growing number of composers have engaged with scientific research and have tried to incorporate their understanding of various models and theories into their musical works. Among them, Hèctor Parra (b. 1976) has been particularly impressed by the recent developments of gravitational physics and astrophysics, namely the part of astronomy which deals with gravity rather than light. Black holes, gravitational waves (first detected in September 2015), cosmology, or quantum gravity models belong to such fields of intensive research, bringing surprising new concepts such as the coarse-graining of space-time, multiverse or holographic principle. In this framework we have collaborated in the conception of a large piece for soloist ensemble, orchestra and electronic devices, which tries to transpose gravitational phenomena into a new form of contemporary music.

Keywords: contemporary music, black holes, gravitational waves, holography, quantum gravity

Mathematics Subject Classification: Primary 83C35, 83C57A72; Secondary 83E99.

1 Science and Contemporary Music

Modern science upsets our vision of the world and influences the theory and the musical practice on an often-metaphoric mode. A recent, striking illustration was the workshop *L'influence des théories scientifiques sur le renouvellement des formes musicales dans la musique contemporaine*, held in November 2014 at the Centre de Musique Contemporaine in Paris [3]. As noticed by the organizer of the conference Mártha Grabócz, in the years 1970-1980 major advances in the field of physics, such as the catastrophe theory by R. Thom, the chaos theory by I. Prigogine, the fractal geometry by B. Mandelbrot, the big bang models of cosmology exposed in a best-seller book by S. Weinberg or the fascinating black hole studies popularized by S. Hawking, R. Penrose, J. A. Wheeler or myself, have led many composers to study scientific works and try to transpose them into musical creations [4].

As I had the privilege to give the opening lecture of the workshop [8], I began to emphasize the fact that the recent deep-sky observations by large ground-based and space telescopes, as well as by neutrino detectors, gravitational interferometers, particle accelerators, and the astrophysical modeling aimed to interpret such experimental data, revealed that space was no more the harmonious, peaceful and unchanging place dreamt by generations of scholars from Pythagoras to Kepler and Newton, but the arena of discordant, arrhythmic and utterly violent phenomena. Taking this for granted, a number of creations in contemporary music have tried to reflect this new, disturbing picture of cosmic frenzy. I gave several examples of contemporary works that used mathematical and/or astrophysical structures to create new forms of musical language. I had myself the chance to have specific collaborations with leading composers, such as Gérard Grisey for *Le Noir de l'Étoile* (1990), written for six percussionists, magnetic tape and on-site broadcast of astronomical signals emitted by pulsars [6]. Pulsars are rapidly rotating neutron stars, i.e. compact remnants of supernovae explosions that emit electromagnetic radiation in the form of periodic radio pulses. It is technically possible to monitor the pulses and figure out a way to translate them live, into sound, as the genesis of a piece of music. This was done in a striking way in Grisey's masterpiece.

The next speaker at the CDMC workshop was Hèctor Parra. The first time I had met him was in 2011, when he visited me in my office of Paris-Meudon Observatory, and I was delighted to hear that the reading of my book *Le destin de l'univers, trous noirs et énergie sombre* [7] had played a role in the composition of his work of 2011 *Caressant l'horizon* [1]. Before this, Hèctor had composed *Hypermusic Prologue* (2009), a chamber opera written in collaboration with the Harvard physicist Lisa Randall, a specialist of the so-called string theory [11]. Indeed these two pieces emerged from the keen interest of Hèctor (whose father was an engineer, as he told me later) for theoretical physics, and more particularly on the new vision of the world offered by Einstein's theory of general relativity, with such spectacular consequences as space-time curvature, time elasticity, gravity waves, black holes, wormholes or cosmological big bang models. As Hèctor explained [10], in *Caressant l'horizon* he tried to transpose the physical feeling one could have by being crossed by strong gravity waves generated by the coalescence of a binary black hole.

2 From black holes to quantum gravity

Science often creates its own language full of technical words, but also uses terms that belong to the layman vocabulary, although with a very different meaning that may be confusing for the non-informed reader. For instance, the nice and poetic title of Hèctor's piece *Caressant l'horizon* refers to usual terms such as "caresser" (to caress) and "horizon". However, in the vocabulary of black hole physics, "horizon" refers to the surface of a black hole (more precisely called the *event horizon*), viewed as a purely geometric border between the external universe and the inner zone of no return. And the seemingly attractive word "caress" refers to the fact that any object doomed to graze the event horizon of a black hole will suffer a violent and destructive gravitational pull, not at all a pleasant loving touch! Twenty-five years ago I myself studied how a star grazing the event horizon of a giant black hole undergoes huge tidal forces and is transformed into a stellar pancake before being fully destroyed in a titanic explosion [2]. Our telescopes have since captured such scenes.

Nevertheless, beyond such possible confusions of terms, it is important to feed the numerous gateways between science and art, even when we deal with technical concepts hardly comprehensible to the layman. Hèctor is fascinated by the relationship between the notions of a strongly distorted, curved or "atomized" space-time in the neighborhood of black hole and big bang singularities, and the nature of life itself, taking place in the apparently peaceful oasis of our planet Earth.

Einstein's theory of general relativity provides a consistent picture of the propagation of gravitation. One century ago, he discovered special solutions of the gravitational field equations representing waves of space-time curvature, travelling at the velocity of light. The major difficulty for detecting gravitational waves is their extreme weakness. The best astrophysical generators of gravitational waves are closely coupled pairs of black holes, that radiate enough gravitational energy for its effects to be directly detectable, at least in the last phase of coalescence of the binary system and the merger into a single black hole. The gravitational luminosity from the collision of two stellar mass black holes is expected to be more than the electromagnetic luminosity of the full universe filled with its hundreds of billion galaxies! However, the amplitude of the signal received on Earth is so weak that it is only in September 2015 that two detectors in the United States, achieved the required sensitivity to catch several signals of black hole coalescences, whose waveforms strikingly corresponded to the theoretical calculations previously performed by supercomputers [5]. Thus gravitational astronomy was born, opening a new window on the universe and one of the most promising developments of astrophysics for the decades to come.

Another spectacular prediction of general relativity (although not yet checked by astronomical observations) is the possible existence of "wormholes", long a staple of science fiction, as they arise as solutions to Einstein's equations describing the inner structure of rotating black holes. Such "bridges" or "throats" could join two different regions of space-time, that in some cases one is at liberty to interpret also as "parallel universes". In the upper universe, the throat has the appearance of a black hole consuming matter, light and energy; below, it appears as a white fountain expelling matter, light and energy. Thus a wormhole can be thought of as a tunnel with two ends at separate points in space-time, and the tunnel would function as a shortcut offering fantastic possibilities for interstellar travel.

Now, when the curvature radius of space-time reaches huge values, as it is the case near the center of a black hole or very close to the big bang, then matter, energy, and the gravitational field must all be quantized, and this requires a comprehensive treatment of new theories called « quantum gravity ».

Quantum mechanics governs the evolution of the microscopic world. It describes local interactions between elementary particles, which are considered to be point-like, within the geometry of a fixed background, namely, the flat space-time of Newtonian mechanics or of special relativity. Throughout the twentieth century, developments in quantum mechanics provided a unified description of electromagnetic and short-range nuclear interactions, leading to the powerful "Standard Model" of particle physics. Its technological successes are innumerable, including lasers, transistors, supercomputers, etc. At the astronomical scale, however, quantum effects play no role at all, and physics is reduced to its classical version. Gravity reigns supreme, and this is described by general relativity, a geometric theory of space-time that models gravity, not as a force acting in a space with a fixed background, but instead as the curvature of space-time, whose value at each point changes when matter is displaced. It too has shown remarkable power in describing astrophysical and cosmological phenomena.

Faced with these two domains of quite different scales and theoretical frameworks—which, through an abuse of the language, are commonly called the “infinitely small” and the “infinitely large”—the physicists may apply either quantum theory or relativity depending on whether they are dealing with small or large-scale phenomena. If this strategy is perfectly appropriate for a practical approach, it is not at all satisfactory for a theoretical one. One wants physics as a whole to be coherent. From this perspective, there must be a theory of a higher order of which quantum mechanics and general relativity are but approximations, valid only in their respective domains. This theory is called quantum gravity.

True comprehension of the Big Bang and of black holes precisely requires a theory of quantum gravity. Given the technical difficulty of the problem, it was necessary to wait until the beginning of the 1990s for physicists to learn how to transform classical theories into quantum theories. For this they needed to develop complex mathematical tools, such as knot theory, gauge group theory, topology, and so on. For a half-century, two main routes to quantifying gravity have been followed.

The first privileges the geometric approach of general relativity, and seeks thus to quantify space and time themselves. The fundamental idea is that there exist "atoms of space", namely entities at a level in which quantum effects radically change the texture of space-time itself, making it granular. The second route privileges the approach of quantum field theory, and thus considers gravitation as a force field between the objects, which it seeks to quantify in terms of an exchange of particles within a passive, exterior space-time. The fundamental ingredient is that the particles are extended objects, not points, with the radical consequence that space has more than three dimensions. These two methods have followed twisted and difficult paths. The formalism of loops for the first approach, and that of strings for the second represent their current state.

In these two cases as well, quantum gravity theories — which are still under construction — put in question the entire foundation of our understanding of the universe. That is to say, they both imply, in very different ways, the fascinating concept of a "multiverse". This term refers to a hypothetical group containing a very large, perhaps infinite, number of “possible” universes, of which ours would just be one individual example. The multiverse—the collection of all the universes in existence—would represent all that exists: space in the fullest sense of the term, time, matter, energy, and all the possible physical laws, together with the fundamental constants of nature which determine them.

Eventually, in some approaches (e.g. string theory), it is hypothesized that all the information about the matter and energy within the three-dimensional volume of a black hole is encoded as a hologram upon its two-dimensional surface, the event horizon. Such an idea, initially applied to black holes only in the hope to solve the so-called information problem, has been studied by hundreds of physicists, with hopes that it could be generalized for *any* physical system occupying a finite volume of space-time. According to this so-called "holographic principle", all the physics within a N-dimensional volume, including gravitational phenomena, could be entirely described by a simpler quantum theory without gravitation that operates only on the (N-1)-dimensional border of the volume [9].

3 The Inscape Project

It is fascinating to see how the most difficult and speculative concepts of modern physics, such as quantum gravity, holography or multiverse models briefly mentioned above, can be transposed into artworks. Of course it is not about strict equivalences, it has more to do with metaphors and analogies.

The work of the artist often translates the fragility of human existence, its extreme limitation in time. When we begin thinking of our existence in a cosmic perspective, we end inevitably in a kind of derisory status of human condition, so limited in space and time. Nevertheless, it is in this so short lifetime that our brain seems to be able to decipher the hidden laws of our immense universe, which normally should overtake us completely. Thus our "cosmic feeling", if it can generate some anxiety or more simply humility, may also provide some pride coming precisely from the fact that, in spite of our short-lived and infinitely modest dimension, we may succeed in achieving some big things.

Physics is precisely such a big thing, but for me another very big thing is the musical composition, that I place over any other kind of artistic creation, because it is the one which seems to me the most elaborate, the most emotional also although the most abstract, and it is probably my training in mathematics and theoretical physics that feeds my fascination for the musical language, that I had the opportunity to acquire very early, in my childhood.

Anyway, the various feelings generated by our own condition of limited human beings extend far beyond our practice as a musician or a physicist. Could "being human" just mean to be moved by the always-deeper knowledge of the surrounding nature, what we are, where we are, what we see? It seems to me that great music can give and amplify a sensation of emotional, physical and intellectual self-fulfillment, at the same time as it can give us the sensation to be overtaken. It is partly on this psychological basis that with Hèctor Parra we undertook to build together a narrative structure as a basis for an orchestral composition entitled *Inscape,* planned for 2018.

We had a first working session in January 2016, at my home place in Marseille. The musicologist José Luis Besada, a post-doc researcher at IRCAM whose project was to study *in vivo* the genesis of the work, emphasizing on the cognitive sharing and transfers between the composer and his scientific collaborators, accompanied us.

Hèctor had the idea to transpose into sounds a realistic journey through the mysteries of space-time curved by gravity, taking account of the various related physical phenomena. Another key idea was to create an acoustic illusion in which the audience would have the feeling to be in a hologram. Indeed, using the sophisticated tools of electroacoustic music, we wanted to look for a sound equivalent of a hologram, which is of course primarily visual. As holography assumes that the information contained in a volume can be entirely encoded on its border, we began to think about various ways to create acoustic holograms. This could be achieved for instance by a specific arrangement of the orchestra and space position of the instruments. We could also have musical sequences, which resume a former sequence already played by the orchestra in the full volume of the concert hall, but replayed only on the edges although displaying the same quantity of information.

How to realize that? Hèctor proposed that the acoustic superstructure played on the edge could echo the breath of the public. All the sound information generated by the spectators, their breaths, whispers, even coughing, would be encoded in some way on the edge of the concert hall - Universe.

I suggested that it could be also interesting to transpose the idea that the universe can be perceived at three very different scales, from the very small (micro scale) to the very large (macro scale), passing by the intermediate human scale (meso scale), and to make listen how we can acoustically travel through these scales.

After a night of reflection, the next morning Hèctor and I exchanged our ideas, and we noticed that we had, independently one of the other, ended in a common structure that followed a sequence of physical events associated to a gravitational journey through a giant black hole. Here are some elements of the (very tentative) scenario:

Start from the meso scale by hearing some biological activity (human noise, whispers, etc.) with sounds confined within a small volume. Then the music grows and amplifies at the macro-scale to simulate the genesis of a massive star. At this moment an instability occurs and gravitational collapse begins. Acceleration, transposition of relativistic effects such as time dilation and space contraction, emission of gravitational waves and formation of the event horizon of a black hole. Once the black hole formed, astrophysical processes take place around: formation of a turbulent gaseous accretion disk, orthogonal jets of gases at relativistic speeds, slow growth of the black hole by accretion of matter and energy. It becomes supermassive. Destruction of full stars by tidal effects and star-star collisions: flambéed stellar pancakes, accidental supernovae. Approach and collision of two galaxies, each harboring a supermassive black hole. Ballets of their central black holes. Loss of progressive energy by emission of gravitational waves. Acceleration, orbital spiraling. Huge burst of gravitational energy when the two black holes merge into a single one. Plunging into the giant black hole. All the information of the matter and energy, which formed the black hole, is deposited on the event horizon, in the form of bits. Entering the wormhole: passage through an annular singularity, taking out of the wormhole by a white fountain. Birth of a "baby-universe" which a dodecahedral topology of space. Expansion of the new universe, at first slowed down, then accelerated by dark energy. Decreasing of dark energy, general collapse. Atomization of the space, inversion of gravitational collapse and Big Bounce.

4 Work in Progress

A second working session took place in Paris one year later, in January 2017, at IRCAM, with Thomas Goepfer as the computer music director of the project. We worked in a room equipped with a system of spatialization of sound able to create the illusion of an acoustic hologram. With Thomas and Hèctor, we looked for new electronic treatments in order to transpose acoustically gravity waves and the idea of holography. Eventually Hèctor asked me to draft a text to be recorded in various ways, speeds and tones that would be distorted by electronic treatment. I provided (in French) the following text (in a very approximate English translation).

"The star which was light became dark, silent, fathomless. Black hole, funnel of cold hells. Once crossed its horizon, an endless fall towards a bottomless center. Mixed, inverted, time and space tighten and collide. The first state of the world vaporizes into interlaced elementary grains. What become matter, energy, waves that fall? Are there a bottom, a final stop of fall,

and a crushing singularity? Yet an absolute end cannot exist. An inexhaustible blooming is the only possible fruit. At the bottom of the black hole a tunnel now opens, a shortcut leading to somewhere else in the universe, or even in other universes. When any border vanishes in all directions, there is no more remedy to dizziness. New universes mature, as full and delicious as newborn fruits. Our Big Bang is just the moment when occurred a bounce. From then on, the metamorphosis of the many-worlds is faster than we can think. When sprouting out of their quantum matrix, baby-universes are endowed with inconceivable shapes. The ignorant considers them flat, Nature thickens their curvatures. And more and more widely space extends, expanding, always expanding, still and forever beyond. The sky, brimming with dark energy, becomes of a frightening transparency."

From now on, Hèctor Parra is working on the musical composition and I am eager to be in the concert hall for the world creation of *Inscape* in Barcelona on 19 may 2018, followed in june by performances in Francfort, Lille and Paris…

References

[1] BESADA, J. L., Musique, science et crise, quel rapport? In A. Mora Luna & P. Ordónez-Eslava, *Les arts en [temps de] crise* (pp. 113-129). Paris: L'Harmattan, 2000.

[2] CARTER, B., LUMINET, J.-P., Pancake Detonation of Stars by Black Holes in GalacticNuclei. *Nature, 296*, 211, 1982.

[3] Cdmc, http://www.cdmc.asso.fr/fr/actualites/saison-cdmc/influence-theoriesscientifiques-sur-renouvellement-formes-dans-musique, 2014.

[4] GRABÓCZ, M., *Entre naturalisme sonore et synthèse en temps réel: Images et formes expressives dans la musique contemproaine.* Paris: Éditions des archives contemporaines, 2013.

[5] LIGO COLLABORATION, http://www.ligo.catlech.edu/news/ligo20160211, 2016.

[6] LUMINET, J.-P., Musique avec pulsar obligé. *Les Cahiers de l'IRCAM / Recherche et Musique 4*, 133-143, 1993.

[7] LUMINET, J.-P., *Le Destin de l'Univers : trous noirs et énergie sombre.* Paris : Fayard, 2006.

[8] LUMINET, J.-P., Astrophysique et musique contemporaine [Video]. Retrieved from https://www.youtube.com/watch?v=ECd15UOIh20, 2014.

[9] LUMINET, J.-P., The Holographic Universe. *Inference : The International Review of Science,* 2(1), 2016 [arXiv : 1602.07258].

[10] LUMINET, J.-P., PARRA, H., Caressant l'horizon : entretien entre Jean-Pierre Luminet et le compositeur Hèctor Parra. *Revue de l'Ensemble Intercontemporain*, Saison 2011-2012, 14-17, 2012.

[11] PARRA, H., Hypermusic Prologue. Òpera projectiva en set plans. In C. Massip & J. Perelló (Eds), *En ressonància. Encontre Noves Fronteres de la Ciència, l'Art i el Pensament*

Current address

Luminet Jean-Pierre, Prof. doc.
Laboratoire d'Astrophysique de Marseille, UMR 7326 Aix-Marseille Université/CNRS
38 rue Frédéric Joliot-Curie, 13388 Marseille Cedex 13, France
E-mail: jean-pierre.luminet@lam.fr

Slovak University of Technology in Bratislava
Faculty of Mechanical Engineering
17th Conference on Applied Mathematics **APLIMAT 2018**

Proceedings

GEOMETRICAL ANALYSIS OF A DESIGN ARTWORK COFFEE TABLE DESIGNED BY THE ARCHITECT AUGUSTO MAGNAGHI-DELFINO

MAGNAGHI-DELFINO Paola (IT), NORANDO Tullia (IT)

Abstract. For his home (Caboto Street, in Milan) the architect Augusto Magnaghi-Delfino designed a prototype of coffee table in marble and iron and assigned the construction to his trusted artisan in Carrara. The unusual shape of the top table and the single leg and the lack of the original drawing stimulated our interest in trying to recreate the draw of this coffee table. We create two templates, and then we fitted points with approximating curves. For the components of the leg and the table's top we concluded that the shapes are similar respectively to branches Lamé curve and superellipse, a curve used by the Danish writer and inventor Piet Hein in 1959, which started in Architecture and Industrial Design the use of mathematical curves.

Keywords: geometry, design, rational architecture

Mathematics Subject Classification: Primary 51M15; Secondary 51N20

Art is solving problems that cannot be formulated before they have been solved.
The shaping of the question is part of the answer.
Piet Hein

1 Introduction

Italian design refers to all forms of design in Italy, including interior design, urban design, fashion design and architectural design. Italy is recognized as being a worldwide trendsetter and leader in design: Italian architect Achille Castiglioni, Pier Giacomo Castiglioni and Luigi Caccia claims that *"Quite simply, we are the best"* and that *"We have more imagination, more culture, and are better mediators between the past and the future"*. After World War II, however, there was a period in which Italy had a true *avant-garde* in interior design. With the fall of Fascism, rise of Socialism and the 1946 RIMA exhibition, Italian talents in interior decorating were made evident, and with the Italian economic miracle, Italy saw a growth in industrial production and mass-made furniture. Yet, the 1960s and 1970s saw Italian interior design reach the top of stylishness and, from that time, the phrases *"Bel Design"* and *"Linea Italiana"* entered into the vocabulary of furniture design.

Born from an idea of Giò Ponti and Alberto Rosselli, in 1954 *La Rinascente* company established a prize, named *Compasso d'Oro* that now is the highest honour in the industrial designing circle in Italy, as famous as the first grade international awards like the Red Dot Award. It was the first award of its kind in Europe and soon took on an international dimension and relevance. Augusto Magnaghi-Delfino won the first edition with SAFFA Modular Kitchen designed for the SAFFA firm (Società Anonima Fabbriche Fiammiferi e Affini). Then he studied for FGB, the company Busnelli of Meda, numerous furniture models: the sofa bed Relax, the Maxia armchair, lounge chairs and the living room Talia, exposed to the *Triennale* exhibits.

2 Cultural Environment and career of the architect Augusto Magnaghi-Delfino

The Italian Rationalism (Architettura Razionale) had its maximum exponents in the Gruppo 7, founded in 1929 by a group of young architects: Sebastiano Larco, Guido Frette, Carlo Enrico Rava, Adalberto Libera, Luigi Figini, Gino Pollini and Giuseppe Terragni. Gruppo 7 declared that their intent was to strike a middle ground between the classicism and the industrially inspired architecture.

Giuseppe Terragni, born on 1904 in a family of builders, in 1921 enrolled at the Superior School of Architecture of the Polytechnic of Milan, where he studied with particular passion the architectures of the past and appreciated the new rationalist architecture encoded by Le Corbusier. Obtained the degree in 1926, he engaged himself in the study of the theoretical premises of the new architecture, still in elaboration. Elaborating a personal rationalist language, marked by the search of classical proportions, the architecture of Giuseppe Terragni starts from a pure geometric shape, that corresponds to the square or to the rectangle and for the plan to a solid form that coincides with the parallelepiped.

Le Corbusier, master of the modern movement, in relation to the new way of conceiving the furnishings wrote "*Equipping is to classify, through the analysis of the problem, the various elements necessary for domestic use. Substituting the countless furniture dressed by different forms and names, the "cases-standard" are incorporated in the walls or supported by them [...] equipped inside according to their exact destination, they are only the furniture of a house. Only the chairs and tables remain.*"

Chairs and armchairs, therefore, are the master. The famous Sant'Elia armchair by Giuseppe Terragni of the 1936, the Modern Armchair by Le Corbusier of 1928 and the Chaise Lounge; the "Barcelona Armchair" by the German architect Mies Van Der Rohe are widely known. All design works characterized by the combination of chrome-plated steel-leather and simple and essential lines. In the modern armchair of Le Corbusier, for example, the separation between the cage of the structure and the cushions expresses in full the logic of rationalism.

One of the famous phrases of Mies Van Der Rohe "*Less is more*" is a "brand" rationalist. That is, all that does not serve the function ("the most", see for example the decoration) becomes "less". Another famous phrase of the famous architect is "*God is in the details*" so we understand that the true point of architecture and rationalist design is not the excess of decoration but the care of the small detail.

Augusto Magnaghi-Delfino, born in Milan in 1914, graduated in 1939 at the Faculty of Architecture of the Politecnico. During his studies, he knew Mario Terzaghi, with whom he started a lasting professional partnership and a solid friendship, and Luigi Comencini and Alberto Lattuada who later will became famous filmmakers. Still a student, he began to attend the study of Terragni and Lingeri, where he also brought his friend Terzaghi as a further practitioner to help the drafting of the contest project for the Palazzo del Littorio in Rome. In

Terragni's architectural firm, he met Filippo Tommaso Marinetti, Lucio Fontana and Adone Asinari.
In 1939, the two young architects built the "Casa dei nidi" (house of the nests) in Fino Mornasco, a residential building with rationalist lines and they won, with Lingeri and Cattaneo, a competition for the Palace of Trade Unions in Como and in the 1941 participate in the primordial Futurist group Sant'Elia.
From 1948 to 1952, Augusto worked for the firm SAFFA. The firm produced matches and derived from cellulose and asked to Magnaghi to draw the first modular kitchen. On 1954, he won the prize "Compasso d'Oro "for this kitchen.
Magnaghi-Delfino, very attentive to public relations, wrote some articles for the magazine Domus, on which, in the fifties, some of their projects of furniture and design objects appeared. They worked fruitfully together in all fields, separating themselves seldom to follow their own personal interests. Magnaghi-Delfino designed for the company Olivari the handles BICA that were then placed on the doors of the transatlantic *Raffaello* and *Michelangelo* and later exhibited at the Museum of Modern Art in New York. After almost 50 years, they are still in production.
The two architects dedicated themselves with particular commitment to public housing, school and hospital. In 1947, they studied the application of prefabricated construction systems to the using popular building in some buildings in the QT8 district in Milan. In 1950, they signed together the *Village for workers* of the paper mill Vita & Mayer in Cairate and the small quarter in Lonate Pozzolo. They designed numerous complexes for the INA-CASA, in Gemonio and Orago in the 1951, in Magenta in 1953 and in Monza in 1955. After the happy realization of the elementary school in Jerago with Orago, followed interventions in the Feltre district and Comasina (Milan) and the INAM clinics of Laveno Mombello (1957-58), Cremona (1960) and Milan, among which the beautiful structure in Piazzale Accursio (1957-60). They collaborated with Ignazio Gardella at the design of the Ivrea Hospital and with Vittorio Gandolfi for the project of the new airport in Lagos (Nigeria). In 1959 they built the house in Caboto street where Augusto Magnaghi-Delfino lived till 1963 when, undermined by a serious illness, get off. This closed the brilliant activity of a knit professional team, while the projects remained entrusted to Mario Terzaghi, which was the only one owner of the architectural firm.

3 Furnishing works: right lines or curved lines?

Civilized man is surrounded on all sides, indoors and out, by a subtle, seldom-noticed conflict between two ancient ways of shaping things: the orthogonal and the round. Cars on circular wheels, guided by hands on circular steering wheels, move along streets that intersect like the lines of a rectangular lattice. Buildings and houses are made up mostly of right angles, relieved occasionally by circular domes and windows. At rectangular or circular tables, with rectangular napkins on our laps, we eat from circular plates and drink from glasses with circular cross sections... [7].

In 1953, architect Saarinen had the idea for the design of his furniture with a single leg. Because the irrational overlap of legs disturbed him, he made of the chair and the table a clear and distinct unity. Dozens of designs will follow the stories of different models: the first in the 1:4 scale, then in proportions corresponding to those of a room of the size of a dollhouse, until it reaches a good effect in space. Knoll International realized the Saarinen collection, which is still in production today and is part of the permanent collection at the Museum of Modern Art in New York. The work on the tables began after 1956 and was the last series of furniture designed by Saarinen before his premature death occurred in 1961. Saarinen continued to work on the problem of "organic mobile", that is to say the furniture in a single form and in a single material. Visually, the pedestal furniture group has an organic unit. However, given the technical limits of the period, Saarinen failed to make this series with a single mould and a single printed material. The base was made of aluminium casting because the plastic printed in such a form was not sufficiently resistant to hold a tabletop or the weight of a seated person. Saarinen's coffee tables of the Pedestrial Collection have tabletops in a wide range of woods, marbles and granite, features with bevelled edges; the top attaches to base with threaded rod and base is heavy moulded cast aluminium with white or black paint.

The architect Arne Jacobsen, born in 1902 in Copenhagen, is one of the grandfathers of modern Danish furniture and the minimalist Danish style. He is remembered for his elegant and functional chair design. From 1934, he cooperated with Fritz Hansen and in 1952, they made the Jacobsen Ant Chair and the Jacobsen Series 7 Chair. At the end of the 50s, Jacobsen designed the famous Egg chair, Swan sofa and Series 3300 chairs, in which the curved lines are an admired and outstanding design.

3.1 Particular furnishing works: coffee tables

The coffee table can have many uses: a spot for a cocktail or the centre of a conversation area. We have many styles of coffee tables of different forms and materials. Traditional coffee tables tend to have ornate detailing, such as curved legs, carvings or inlaid tops and be made of rich woods. Transitional furniture falls between traditional and contemporary/modern furniture. It has a more casual, functional feel. Transitional coffee tables will suit most decor styles. Though contemporary and modern may seem interchangeable, contemporary style refers to what is currently popular. Contemporary designs often feature stark interiors with bold, bright accents. Glass tops and metal bases are common features of contemporary coffee tables. Mid-century modern refers to the interior design style that was popular between the 1950s and 1960s. Mid-century modern coffee tables are characterized by clean, simple lines and bentwood or moulded plastic construction. Most of the time, wood furniture has a simple finish to show off its natural beauty.

4 The Magnaghi-Terzaghi coffee table

The Magnaghi-Terzaghi architectural firm from 1950 designed sophisticated lounge set and coffee tables with oval top for industrial production For example, we consider the coffee table with top painted etched glass top and brass legs (38.7cm high, 120cm principal diameter).

Fig. 1. Coffee Table (1955).

For his home (Caboto Street, in Milan) Augusto Magnaghi-Delfino preferred to assign to his trusted artisan in Carrara the construction of a prototype in marble and iron. The unusual shape of the top table and the single leg and the lack of the original drawing stimulated our interest in trying to recreate the draw of the top and the legs.

(a) (b)

Fig. 2. Caboto home's coffee table (a), (b).

Description of the coffee table:
The top is made in Carrara marble, its measures are:
Length: 116.5cm. Width: 79.5cm. Thickness: 2cm.
Specific weight of the marble: 2.5 Kg/dm^3 Total weight: about 46Kg.
The top is supported by a single iron leg and is removable.
The single leg is composed by two iron pieces jointed, perfectly symmetrical.

Fig. 3

The measures of the legs are:
Length: 48.5cm.
Width: 69cm.
Thickness: 1.5cm.
Height: 21cm.

4.1 Coffee table study

We made templates for both elements, top and leg, and then we tried to recreate the shapes, using GeoGebra. In Fig. 4 we see the templates and the paper with plotted points.

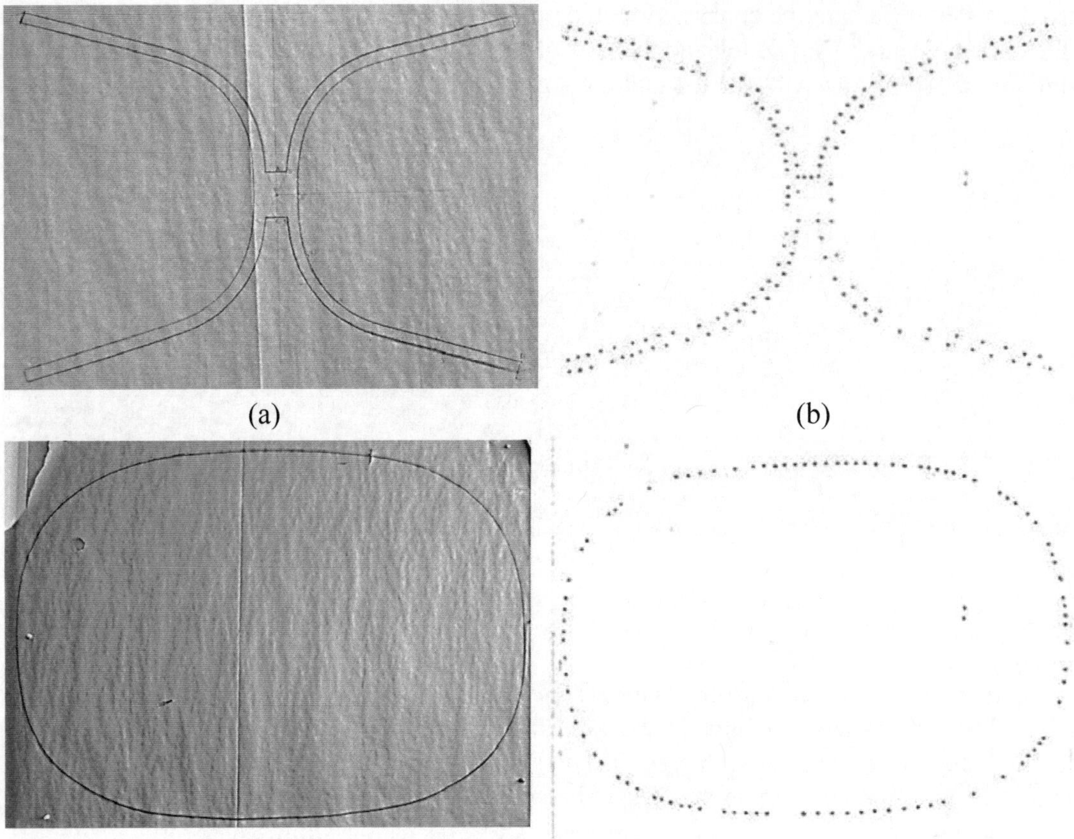
(a) (b) (c) (d)

Fig. 4. Figures (a), (b), (c), (d).

4.1.1 Leg's study

For the components of the leg, we try with catenaries, Musmeci's curves, generalized parabolas and hyperbolas (not Apollonius' conic sections), superellipses.
The catenary is the curve that an idealized hanging chain or cable assumes under its own weight when supported only at its ends. The catenary curve has a U-like shape, superficially similar in appearance to a parabolic arch. Mathematically, the catenary curve is the graph of the hyperbolic cosine function.
The equation of a catenary in Cartesian coordinates has the form

$$y = cosh\left(\frac{x}{a}\right) \tag{1}$$

Catenaries and related curves are used in architecture and engineering, in the design of bridges and arches, so that forces do not result in bending moments.
Over any horizontal interval, the ratio of the area under the catenary to its length equals *a*, independent of the interval selected. The catenary is the only plane curve other than a horizontal line with this property. In addition, the geometric centroid of the area under a stretch of catenary is the midpoint of the perpendicular segment connecting the centroid of the curve itself and the

x-axis [5]. In our opinion, the curves designed by Augusto Magnaghi-Delfino are not catenaries, because the curvature decreases too quickly.

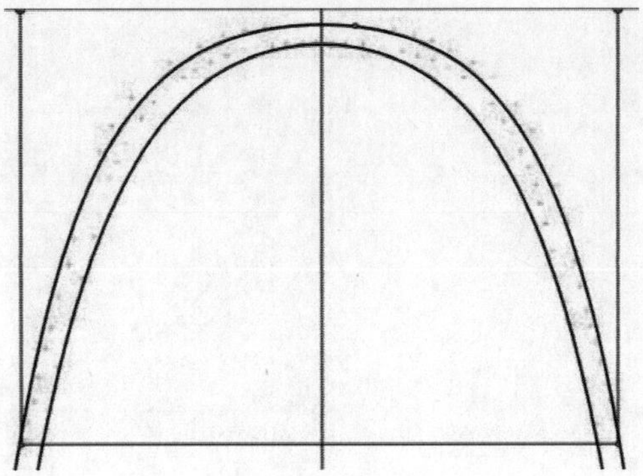

Fig. 5

The curves designed by Augusto Magnaghi-Delfino are also similar to Sergio Musmeci's logarithmic curve. Sergio Musmeci (1926-1981) was a Roman engineer, who worked in the Riccardo Morandi e Pier Luigi Nervi firm and designed many famous bridges, in collaboration with his wife, the architect Zenaide Zanini. The Musmeci's curve has the form

$$y = ln(cos(ax)) \quad (2)$$

There is a particular geometric property of this curve: the angle that the axis form with the x-axis is proportional to x-coordinate. Musmeci wrote "*A continuous form is an organic form; it is modern because it represents the solution, in a single structure, of forces and balances that descend from a complex much wider of events*".
But, in our opinion, the curves designed by Augusto Magnaghi-Delfino do not agree with Musmeci curves.

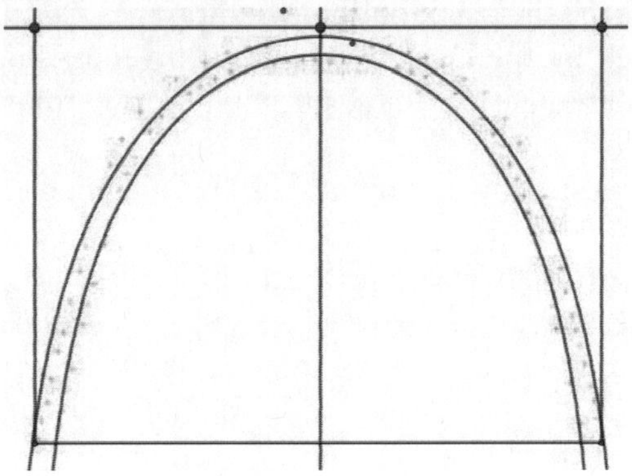

Fig. 6

Generalized parabolas and hyperbolas have respectively the form

$$\left(\frac{x}{a}\right)^p - \left(\frac{y}{b}\right)^q = 1 \text{ with } p=1 \text{ or } q=1 \quad (3)$$

$$\left(\frac{x}{a}\right)^p - \left(\frac{y}{b}\right)^q = 1 \text{ with } p \text{ and } q\text{, positive real numbers} \quad (4)$$

None of these curves fitted well the points of our template. In the parabolic case we have an error of 5.1% that correspond with 3.5cm in the hyperbolic case we have an error of 4.7% that correspond with 3.24cm.

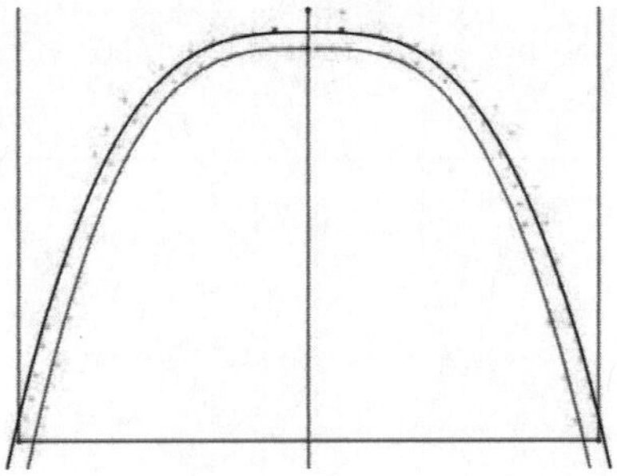

Fig. 7. Generalized parabolas.

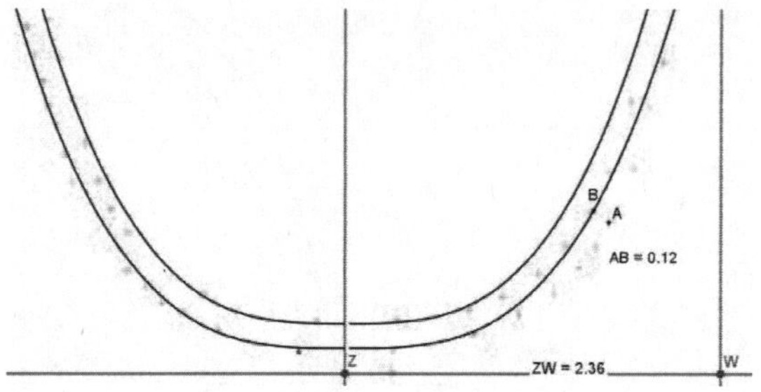

Fig. 8. Error evaluation.

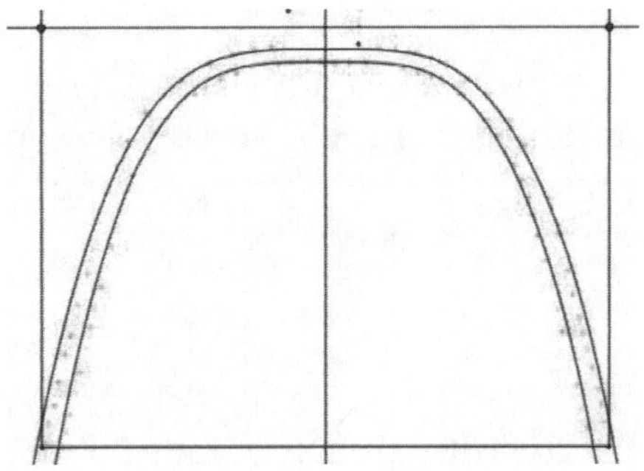

Fig. 9. Generalized hyperbolas.

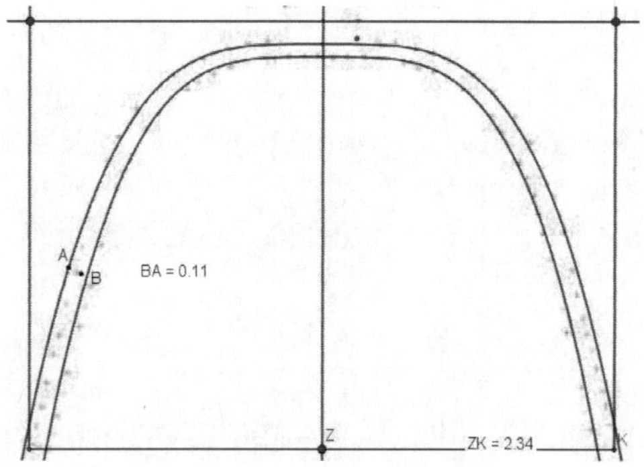

Fig. 10. Error evaluation.

So we considered to approximate with branches of quartic curves. The quartic we considered is a particular case of superellipses, also known as Lamé curves, after the French mathematician Gabriel Lamé, who generalized the equation for the ellipse.
The superellipse was named by the Danish writer and inventor Piet Hein (1905-1996), that considered this curve in order to answer to the question to fit harmoniously an oval shape into a rectangular space.
The question had been suggested by a complicated city-planning problem that first arose in 1959 in Sweden. Many years earlier Stockholm had decided to raze and re-build a congested section of old houses and narrow streets in the heart of the city. Two broad new traffic arteries running north south and east west have been cut through the centre of the city. At the intersection of these avenues, a large rectangular space about 183m long is being laid out. At its centre, would be construct an oval basin with a fountain surrounded by a large oval pool containing several hundred smaller fountains.
Daylight would filter through the pool's translucent bottom into an oval self-service restaurant, below street level, surrounded by oval rings of pillars and shops.
In planning the exact shape of this centre, the Swedish architects ran into unexpected problems. The city planners next tried a curve made up of eight circular arcs, but it had a patched-together look with ugly "jumps" of curvature in eight places. In addition, plans called for nesting different sizes of the oval shape, and the eight-arc curve refused to nest in a pleasing way.
The architectural team in charge of the project consulted Piet Hein. To understand Piet Hein's novel answer, we must first consider the ellipse, as he did, as a special case of a more general family of curves with the following formula in Cartesian coordinates

$$\left(\frac{x}{a}\right)^p + \left(\frac{y}{b}\right)^p = 1 \text{ with } p \text{ is a positive real number} \quad (5)$$

where a and b are unequal parameters (arbitrary constants) that represent the two semi axes of the curve. When $p>2$, the oval develops flatter and flatter sides, becoming like a rectangle; indeed, the rectangle is its limit as p approaches infinity.

Moreover, such curves could be nested, to give a strong feeling of harmony and parallelism between the concentric ovals. Piet Hein called all such curves with exponents greater than 2 "super-ellipses."

Bruno Mathsson, a well-known Swedish furniture designer, has enthusiastically adopted the Piet Hein's superellipse. He first produced a variety of superelliptical desks, followed by superelliptical tables, chairs and beds. Industries in Denmark, Sweden, Norway and Finland have turned to Piet Hein for solutions to various orthogonal-v. -circular problems and he has been working on superelliptical furniture, dishes, coasters, lamps, silverware, textile patterns and so on.

We thought to generalize the Hein's curves, creating a non-Hein superellipse

$$\left(\frac{x}{a}\right)^p + \left(\frac{y}{b}\right)^q = 1 \text{ with } p = 1 \text{ and } q = 4 \quad (6)$$

We obtained a pleasant shape, approximately near to Magnaghi's.

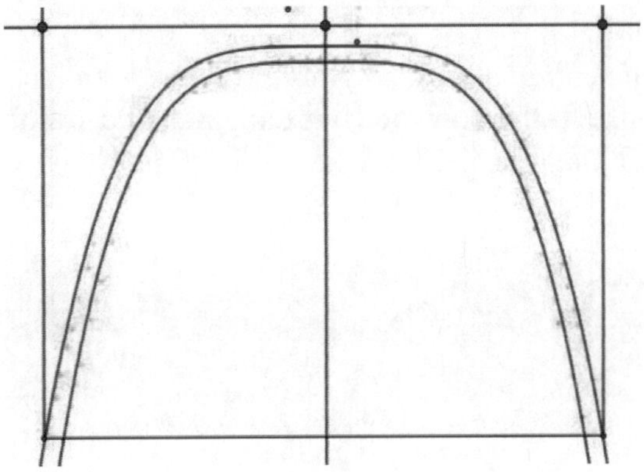

Fig. 11

4.1.2 Top table's study

For the top table, we have considered alternative solutions: ellipse, polycentric curves, quartic curves. We rejected ellipse because its pointed ends would interfere with smooth flow around it; moreover, it did not fit harmoniously into a rectangular space. We rejected polycentric curves although they look like more simple to carve in the marble. Mathematically the polycentric curves are curves made of circle arcs subsequently connected at a point where they share a common tangent. Their use in architecture has been and still is widespread. An application of the polycentric curves is to drawing a class of egg-shaped curves, see [8]. A simple conjecture about ovals, which he also applied to eggs, showed how it is possible to decide, among an infinite number of possibilities, which oval to draw once the measures of the two axes are given. Franco Ghione in the same paper gave a rigorous mathematical proof of this conjecture. Ragazzo's conjecture on how to draw all possible ovals inside a rectangle reduces the problem to *quarter* ovals inside rectangles. In any rectangle, there is only one inscribed ellipse and

infinite number of four centre ovals. The connection's problem in polycentric construction is the curvature's discontinuity.
We drew polycentric curve with eight centres.

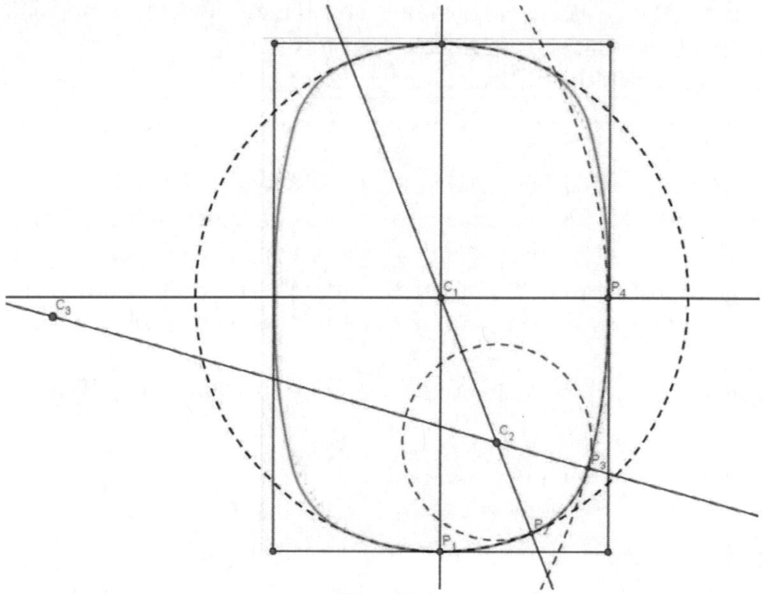

Fig. 12

Then we tried to use quartic curve, but we had more success with Hein superellipse.
For $p = 2.78$ the superellipse fit the points of our template, and we obtained a pleasant shape, approximately very near to Magnaghi's.

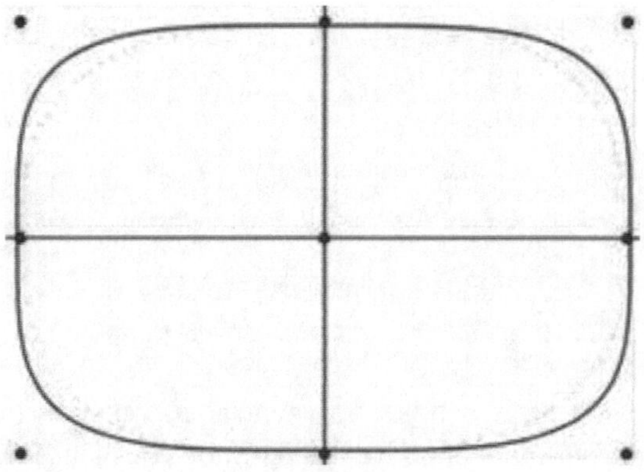

Fig. 13. Quartic curve.

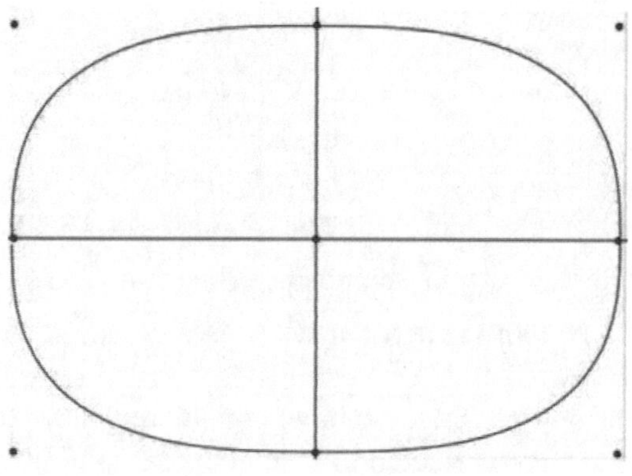

Fig. 14. Superellipse.

5 How Augusto Magnaghi-Delfino did sketch the table?

In 60's the designers did not have computers or electronic devices but only French Curves. French curve is a flat drafting instrument with curved edges and several scroll shaped cut-outs, used as a guide in drawing curves. In the 20th century, among mathematicians and engineers, there was a renewed interest in kinematics, the geometry of pure motion. Study of these mechanisms can be used as a way to show interconnections between mathematics and technology.

Descartes in the 17th Century considered only those curves that could be drawn with mechanical devices. Curves were constructed from geometrical actions, many of which were pictured as mechanical tools. After curves had been drawn, Descartes introduced coordinates and then analysed the curve-drawing actions in order to arrive at an equation that represented the curve. Equations did not create curves; curves gave rise to equations.
A common set of curves is the Burmester set [4] displayed here. The first item is very handy for ellipses, the second very often fits large parts of parabolas and the third item is used most for hyperbolas.

(a) Ellipses (b) Parabolas (c) Hyperbolas

Fig. 16. Figures (a), (b), (c).

French curves, that are flat drafting instruments with curved edges and several scroll shaped cut-outs, were used in manual drafting to draw curves of various shapes. In general you would plot various known points on the paper, and seek to join them up on the edge of a French curve. The one on the top left is meant to be good for hyperbolas, top right for ellipses, and the big one for parabolas (all the conic sections except the circle, which can be drawn easily anyway).
They can make smooth complicated curves. If you have a bunch of points you want to fit a curve to you sort of pick a piece of French curve that fits through the three points. You then pick another slice of curve that fits enough the adjacent points and so on.
There's no standard for French curve shapes. The complicated shapes are just a means to get all kinds of curves on one piece of plastic.
We considered the templates and we tried to join the plot points with French curves, obtaining

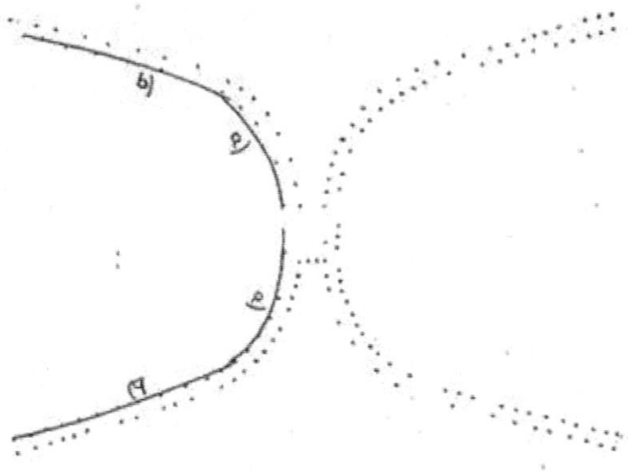

Fig. 17

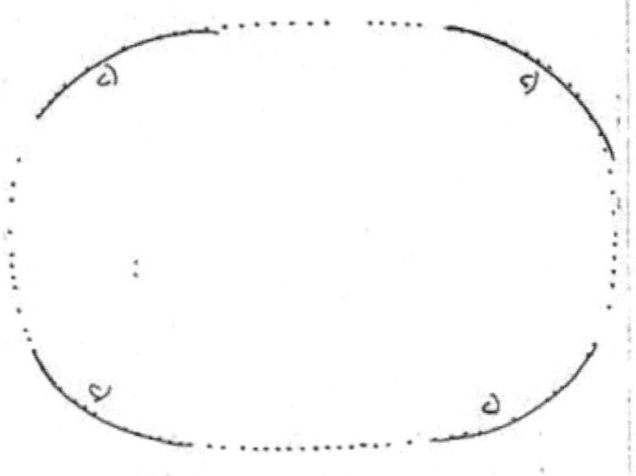

Fig. 18

In this approach we used 16 (a) and 16 (b) for the leg and 16 (c) for the top.

6 Conclusions

In all the works of the architect Magnaghi-Delfino, both of architecture and design, we find strongly a geometric component typical of the works of those years. In the absence of the original drawings, we tried to reconstruct the coffee table prototype. We think that the original shapes of the top table and leg and the unusual materials were design experiments to evaluate market trends.
We propose to look into other achievements in the future such as the BICA handle of the company Olivari and the armchair prototype in Magnaghi-Delfino's home.

Acknowledgements

The authors thank

- Historical archive of the Politecnico di Milano
- Comune di Milano, **CASVA** (**C**entro di **A**lti **S**tudi sulle **A**rti **V**isive) in Milan

for giving us the opportunity to consult archives and publish their documents.

References

[1] URL, https://architextsassociation.wordpress.com/2012/09/19/giuseppe-terragni-and-the-rationalist-architecture-in-italy/

[2] URL, http://www.piethein.com/page/superellipse-24/

[3] BEVILACQUA, F., Augusto Magnaghi, In *E. Godoli (a cura di), Dizionario del futurismo*,
vol I, Firenze: Vallecchi, 2001, p. 675

[4] CECCARELLI, M., KOETSIER, T., *Burmester and Allievi: A Theory and Its Applications for Mechanism Design at the End of 19th Century*. Journal of Mechanical Design, vol. 130, 2008, downloaded from: https://mechanicaldesign.asmedigitalcollection.asme.org

[5] DODD, J., *Characteristic Average Property of the Catenary*. V. Coll, The American Mathematical Monthly, August 2016, pp.683-689

[6] FERABOLI, M.T., Augusto Magnaghi e Mario Terzaghi, in *G.L. Ciagà (a cura di), Gli archivi di architettura e in Lombardia.* Comune di Milano, Milano, 2003

[7] GARDINER, M., *The superellipse: a curve that lies between the ellipse and the rectangle*. Scientific American, 21, 1965, pp. 222-234

[8] RAGAZZO, F., *Geometria delle figure ovoidali*. Disegnare idee immagini VI (11), 1995, pp.17–24.

Current address

Magnaghi-Delfino Paola, Researcher
Dipartimento di Matematica del Politecnico di Milano
Piazza Leonardo da Vinci 32, 20133 Milan, Italy
E-mail:paola.magnaghi@polimi.it

Norando Tullia, Associate Professor
Dipartimento di Matematica del Politecnico di Milano
Piazza Leonardo da Vinci 32, 20133 Milano, Italy
E-mail: tullia.norando@polimi.it

Slovak University of Technology in Bratislava
Faculty of Mechanical Engineering
17th Conference on Applied Mathematics **APLIMAT 2018**

Proceedings

TOWARDS A DIFFUSED (OPEN AIR) MUSEUM IN THE HISTORICAL CENTER OF ROME. UNIVERSITY THIRD MISSION INVOLVING ART, ARCHITECTURE, MATHEMATICS

MAGRONE Paola (IT), BRANCALEONI Fabio (IT), CARLINI Alessandra (IT), FALCOLINI Corrado (IT), GARGANO Maurizio (IT), TEDESCHINI LALLI Laura (IT)

Abstract. Among third mission of universities is to nurture scientific citizenship. Our *scientific trips* initiated (see [1]) on occasion of the 2014 European Science Week. Here we describe two such trips: *Modularity and symmetries in the city of Rome*, observing the "cosmatesque" geometric compositions in the medieval Roman churches. The second, *Under the bridges of Rome more than one story*, is a walk along the banks of the Tiber, providing a picture about historical and structural aspects affected by the presence of the River Tiber and its bridges.

Keywords: mathematics, architecture, early structural engineering, bridges, cultural heritage, Rome.

Mathematics Subject Classification: Primary 97A30, Secondary 00A67

1 Introduction

Diffusion of scientific culture aims at fostering and updating scientific citizenship, and is encompassed in the “third mission” of Universities, as defined in the Horizon2020 European program. We experienced a new way of spreading abstract scientific culture, by exploiting artefacts of Cultural Heritage widely available in Rome, and promoting *scientific trips*. Such field-trips, open to a large public, draw on the beauty of these artefacts and the sense of belonging they instill, to motivate a large public towards a scientific aspect that can be ascertained by direct observation. We initiated Scientific trips [1, 4] on the occasion of 2014 European Science Week, and then maintained them regularly under different institutional settings.

Focussing on the experiential component is crucial to a scientific approach, together with its assessment and possible critique, shared by other scientific disciplines. Given the great amount of available digital resources, the experiential approach risks to be overshadowed. Our experience then merged into the larger multidisciplinary project entitled "From the cosmos to the earth, to life, to culture. A resilient strategy for the dissemination of natural, physical and mathematical sciences: a proposal for a spread museum for the city of Rome", funded by the Italian Ministry of Education (see [2] for details about planners participants, developed material etc.) for 2016. The goal of the project this time went further, as it was to be a pilot experience, so as to develop material useful in the future for other teachers, classes, cultural associations or public libraries, to perform the activities by themselves. The experiences in Astronomy, Botany, Geology and Mathematics have therefore been collected on the website [2], together with scholarly material especially developed, and material realized by the participating classes. In this paper we report on the mathematical activities, delving into the mathematics implicitly weaved into the artefacts.

We describe two scientific trips: *Modularity and symmetries in the city of Rome*, where we observe the geometric compositions of the Roman Marmorari, or "cosmati", in the medieval Roman churches, and *Under the bridges of Rome more than one story*, a walk along the banks of the Tiber, from the bridges around Tiberina isle to Ponte Elio, accompanied by the narration of historical facts, geological evolution, evolution of constructive techniques. These walks offer the opportunity to enjoy monuments and resources such as riverfronts, churches, and bridges, observing them through the eye of a mathematician, an historian of architecture and a civil engineer.

Fig. 1. Some of our scientific trips in previous years, from left to right: Surveys and mathematical models of Titus's arch at Circo Massimo; The Aurelian Wall at Porta Latina, the science of restoration; Geometric flash mob in St. Peter's Square.

The entire realm of activities follows in the path of "diffused museums", much advocated in the 1970's to treasure the available Cultural Heritage on a widespread scale [7].
In Italy there are many examples of widespread museums. The idea of a diffused open air museum is often associated to the idea of virtual and augmented reality. Our work adds the direction of direct experience, of embodied learning and knowledge: direct experience can be assisted by the help of technology, but not replaced. Younger generations, who are often too immersed in technology, can benefit from the direct on-field experience, not mediated by technology, involving not only sight but also the senses of touch and hearing.

These particular trips were aimed at high schools. The structure of the project was planned to provide a project presentation meeting in January, a scientific introduction to the visits in March, the guided visits in April. Then some time was left to the teachers in order to organize some original material to be developed with their classes. Some of them left this as a task to the students for the summer holidays. In october the products were ready, they all can be seen on the website of Museodiffuso [2]. In November we gathered all the teachers and part of students for a final general meeting at our University together with the botanical, geological and astronomical trips, where selected products developed by students togheter with their teachers where shown.

2 Modularity and symmetries in the city of Rome

Modularity procedures go through the whole history of art and architecture. The theoretical-scientific systematization of tessellations was born in the 18th century, with the research on the crystallographic groups in the geological field, and finds its definite structure in Polya's Theorem of algebraic geometry of 1926 [9, 15]. On the other hand, discovering repetitions and modules assists in the survey activity, so much that the first Escher's studies about tessellations were born out of his architectural surveys in Alhambra [8]. With this point of view we led our groups of students to observe the geometrical compositions of the Roman Marmorari, or "Cosmati", in some medieval Roman churches and to perform surveys. This scientific trip was planned by Alessandra Carlini and Laura Tedeschini Lalli [3, 5], the trips were guided by the young architect Monica Del Grasso. To stay in the theme of diffused museum, the team developed scientific records such as for museum catalogues, and made them available on the website.

The diffused museum of cosmatesque floors way organized around an itinerary touching five churches: San Clemente, Santi Quattro Coronati, San Lorenzo out of the walls, Santa Maria in Cosmedin, Santa Croce in Gerusalemme.

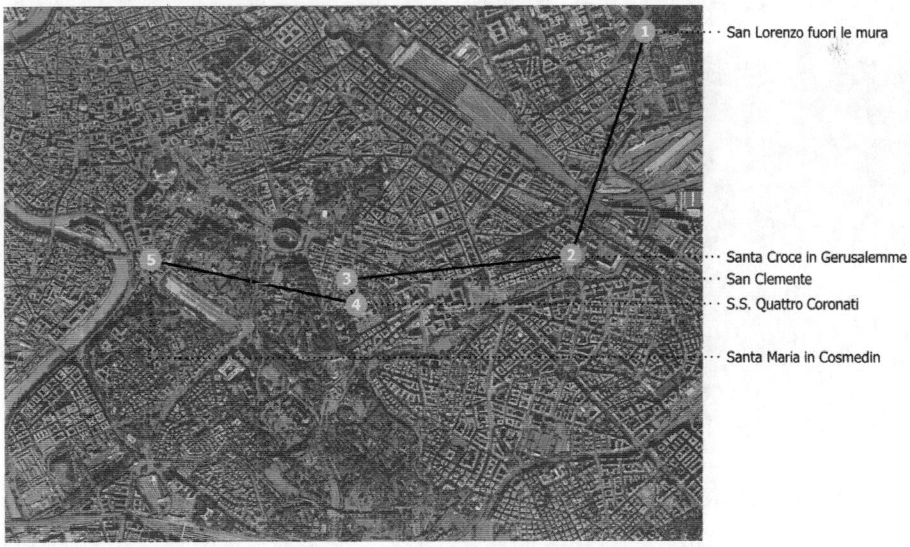

Fig. 2. The map of the itinerary "Modularity and symmetries in the city of Rome"

The guided tours expect the group to perform a real field experience, for this purpose the leader carries mirrors and tissue paper, to survey correctly. Looking for the fundamental domain of the symmetry group of each piece of floor, will save time, redundant information, and will avoid errors in the survey. A symmetry for a mathematician is the invariance of the graphical motive, under rigid transformations of the plane. So mirrors are used to verify axes of symmetry of the observed motifs, and ascertain when there are no mirror symmetries. i.e. when the composition is in fact not invariant when looked at into the mirror. In medieval churches often the immediate intuition is of greater symmetry, so this is crucial, and medieval floors are good grounds. Tissue paper is used to make a contact survey (fig. 3-4) of the pattern from the ground, and then verify symmetries by actually moving the drawing with one of the rigid movements of the plane. We have proposed to survey mainly the symmetrical motifs in what are called *tappetini*, rectangular areas, much recalling muslim prayer carpets, usually placed in the left and right sides of the churches, where the devoutees would stand. They are decorated with geometric motifs made of stone tiles of various sizes, mostly in red, white, green, yellow.

So the students can use the survey drawn by themselves on tissue paper, to analyze the rigid movements that would leave the pattern unaltered, by posing it on the floor and verify that the superposition of motives is still exact after the movements. To find the fundamental domain, i.e. the minimum region necessary to regenerate the whole pattern under the action of the isometries we similarly invite the students to use their drawing on the tissue paper.

Participating schools and teachers (67 students in total): Prof. Simonetta Martignani (mathematics), Prof. Paolo La Civita (history of art), Liceo Peano in Rome: Prof. Lidia Ansini (mathematics), Istituto Einaudi, Roma; Prof. Letizia Sassolino, Prof. Caterina Palla (mathematics), I.I.S. Federico Caffè in Rome.

Fig. 3. Surveys of the floors, students of Einaudi Institute in Santa Croce in Gerusalemme

Fig. 4. Surveys of the floors, students of Einaudi Institute in Santa Croce in Gerusalemme

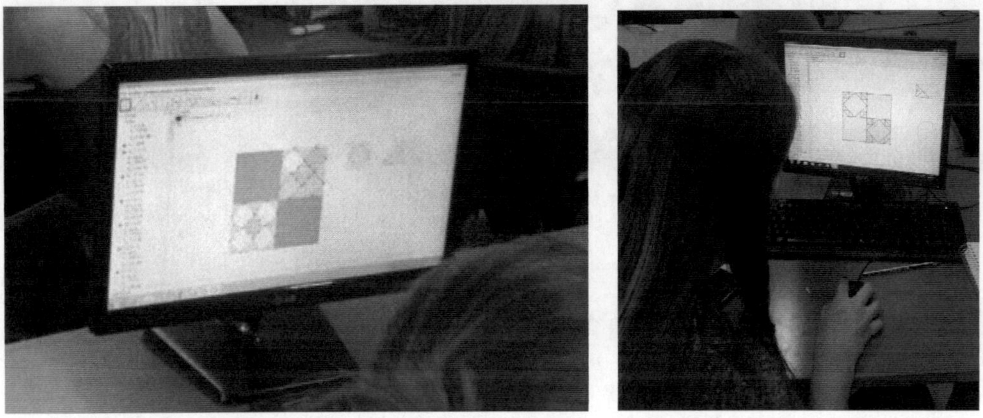

Fig.5. Students of Liceo Peano working at school to reconstruct the symmetries of patterns

3 Under the bridges of Rome more than one story

How was the river of Rome formed, the relationship with the city and the challenge of the bridges, from the Roman antiquity to the nineteenth century. We proposed a walk along the banks of the Tiber river, from the bridges around Tiberina to Ponte Elio, in front of Castel Sant' Angelo, fig.6. The activity was first conceived and activated by Fabio Brancaleoni and Maurizio Gargano (respectively structural engineer and historian of architecture of the Department of Architecture of Roma Tre University) in their University courses. In the diffusion project, the school groups were led by the young engineer Marco Mininni.

The main objective of this activity was to provide a picture, as complete as possible, with outstanding features of interdisciplinarity, about several aspects conditioned by the presence of the River Tiber. In particular, its bridges [13]. To this purpose we told about the evolution of constructive techniques, the floods and overflows of the Tiber River and the great change of the landscape around the river due to the construction of the walls and all the related works. Providing a scientific explanation for the presence of a number of architectural and structural elements the visit aims to give the participants the necessary tools to allow them to understand and interpret by themselves other similar elements in different places.

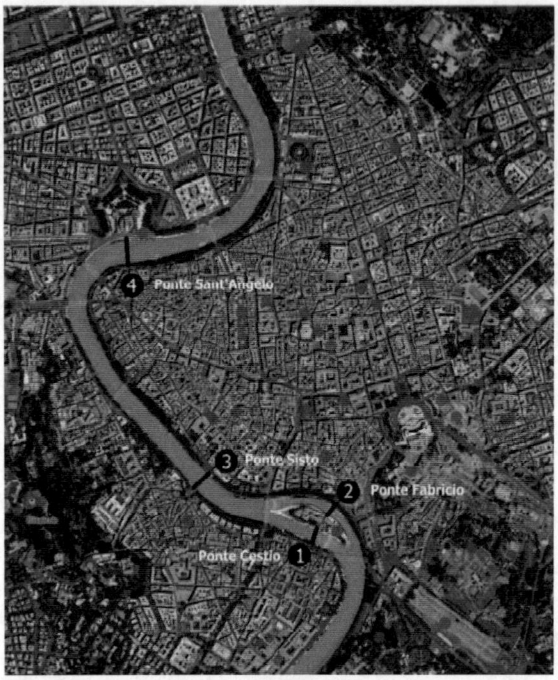

Fig. 6. The map of the itinerary "Under the bridges of Rome more than one story"

During the visit the guide explains how some issues were solved time by time, for example about the construction of ancient bridges in masonry [10], such as Ponte Cestio. Because of the circular shape of the round arch, as the bridge increases in size, also its height grows, therefore there is a difference in height between the top of the arch and the banks of the river. This drawback was overcome with the construction of two inclined access ramps. The base of the roadway was completed with concrete reinforcements necessary to balance the lateral thrust (fig. 7, historical picture taken from [6]). The arches were built with the aid of "centrings", i.e. temporary wooden structures whose goal was to support the arch during its construction. The phase of removing the centrings was called "striking" of the arc. It was a very delicate phase: the ribs had to be removed in a sudden manner so that each and all the segments of the arch took the load at the same time, settling into equilibrium.

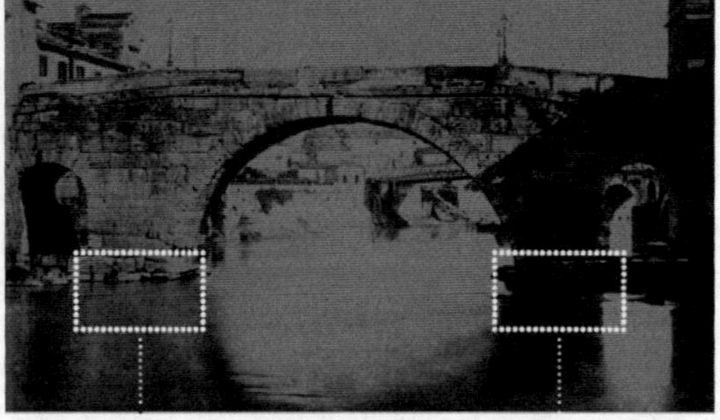

Fig. 7. Ponte Cestio before the construction of the walls, marking structural elements. On the left, the white dots point towards the arcades, on the right the dots mark the piers.

The piers ere equipped with cutwaters, or protrusions (figure 8) necessary to facilitate the outflow of the current reducing the vortices and thus limiting the possible impact damage.

Fig. 8. Ponte Sisto, the cutwaters. Photo: Marco Mininni

The foundations, obviously immersed in water, were made by means of a *tura*. The tura is a parallelepiped-shaped structure, with a double wall made from oak poles. The in-gap between the two walls is emptied from the silt and re-filled with matte and quick-setting tufa scales. Then the tura constituted a watertight chest, which was emptied of lime and water by means of Archimedes' screws, and was then filled with concrete. The tura thus allowed to improve the quality of the ground on which to then realize the piers.

Historical aspects are the key to understand differences in material and in position of the bridges [11, 12] therefore much time is spent in front of Ponte Sisto, one of the most important historically.

Fig. 9. Three-dimensional maquette of Ponte Sisto, students of Liceo Cavour

We realized a three-dimensional portable maquette, showing the orography of the center of Rome (fig. 10). This model, realized in forex at the Laboratory of Models and Prototypes, Department of Architecture Roma Tre, is a scale model, whose ratio between the horizontal and the vertical scale is approximately equal to 1/4, in order to make the differences in level more evident. The model has proved to be an excellent tool for the concentration of attention, and it allows also to focus the area of the historical center of the city with respect to the different hills that constitute the orography of the city. Moreover it inspires new projects that schools can develop in the neighborhood of their own premises, drawing on the complicated orography of the town: the famous seven hills count only those in the initial district. All around the land is hilly and at time with deep gorges (forre). This activity implies searching the data and then implementing a model to be cut digitally.

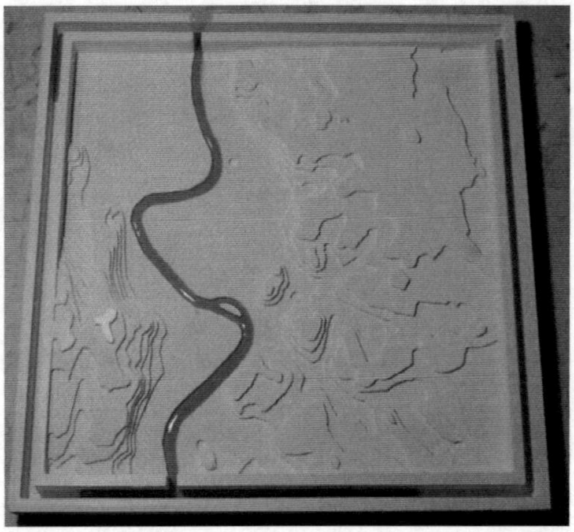

Fig. 10. Three-dimensional portable maquette showing the orography of the center of Rome, Designed by Alessandra Carlini.

Participating schools and teachers (79 students in total): Prof. Alessandra Carlini, Teresita D'agostino, Liceo Scientifico Cavour. Prof. Maria Grazia Nuzzo, Liceo Linguistico Margherita di Savoia. Prof. Maria Teresa Califano I.I.S. Luigi Einaudi.

Acknowledgement
This paper refers to the project: Progetto MIUR per la Diffusione della Cultura Scientifica (Legge 113/91) D.D. 1524/2015 - Titolo 3 PANN15T3_00048, title: "Dal Cosmo alla Terra alla Vita alla Cultura. Una strategia resiliente per la diffusione e la disseminazione della Scienze Naturali, Fisiche e Matematiche: proposta di Museo Diffuso per la città di Roma", P.I. Prof. Settimio Mobilio, Dipartimento di Scienze, Roma Tre University.

References

[1] URL: http://www.formulas.it/?page_id=7902 Our Scientific trips for European Reasearchers' week
[2] URL: http://museodiffuso.uniroma3.it/index.php/it/
[3] CARLINI A., CONVERSANO E., TEDESCHINI LALLI L.,Mathematics and Archeology, *APLIMAT – Journal of Applied Mathematics*, Vol. 1 (2), pp. 61-68, 2008.
[4] CARLINI A., MAGRONE P. Ellipses and ovals in the physical space of St. Peter's square in Rome, *Proceedings Aplimat 2017*, Slovak University of Technology Bratislava, 2017 pp. 672-685.

[5] CARLINI A., TEDESCHINI LALLI L., Sierpinski Triangles in Stone on Medieval Floors in Rome, *Aplimat Journal of Applied Mathematics*, Vol. 4, pp. 113-122, 2011.
[6] D'ONOFRIO C. Il Tevere : l'Isola tiberina, le inondazioni, i molini, i porti, le rive, i muraglioni, i ponti di Roma, *Editore: Romana Società editrice,* 1980
[7] DRUGMAN F., Il Museo diffuso, in *"Lo specchio dei desideri. Antologia sul museo*", a cura di M. Brenna, 2010, pp. 65-70.
[8] ESCHER M.C, The Regular Division of the Plane, *De Roos Foundation*, Utrecht 1958.
[9] FEDOROV ZAPISKI E.S., Imperatorskogo S. Peterburgskogo Mineralogicheskogo Obshchestva, (2), 28 (1891)
[10] GALLIAZZO V., *I ponti romani*, Treviso 1995.
[11] GARGANO M., Sisto IV.Il 'Pons Ruptus': tecniche, simbologie e finalità di una ricostruzione, in AA.VV., *Il modo di costruire*, Edilstampa s.r.l., Roma 1990, pp. 219-238;
[12] GARGANO M., Nuove acquisizioni su Ponte Sisto (1473-75), in *Quaderni dell'Istituto di Storia della Architettura* (Università degli Studi di Roma "La Sapienza"), n.s., 1993 (ma 1994), 21, pp. 29-38.
[13] GARGANO M., Note sul gettar-ponti a Roma nel XV secolo. Ponte Sisto: tra Leon Battista Alberti e Leonardo da Vinci, in *Rassegna di Architettura e Urbanisti*ca, Anno XXVII, n. 84-85 (Settembre 1994-Aprile 1995), 1996, pp. 15-27.
[14] PIRANESI G.B., *Le antichità romane*, 1784.
[15] POLYA G., NIGGLI P., Zeitschrift fur Kristallographie und Mineralogie 60 (1924), pp. 278-298

Current address

Carlini Alessandra, Ph.D., architect
Liceo Scientifico Cavour
Via delle Carine 1, I-00146 Rome, Italy
E-mail: alessandracarlini@yahoo.it

Brancaleoni Fabio, Professor of Structural Engineering
Dipartimento di Architettura, Università Roma Tre
Via Madonna dei Monti 40, I-00146 Rome, Italy
E-mail: fabio.brancaleoni@uniroma3.it

Falcolini Corrado, Professor of Mathematical Physics
Dipartimento di Architettura, Università Roma Tre
Via Madonna dei Monti 40, I-00146 Rome, Italy
E-mail: falco@mat.uniroma3.it

Gargano Maurizio, Professor of History of Architecture
Dipartimento di Architettura, Università Roma Tre
Via Madonna dei Monti 40, I-00146 Rome, Italy
E-mail: maurizio.gargano@uniroma3.it

Magrone Paola, Assistant Professor of Mathematical Analysis
Dipartimento di Architettura, Università Roma Tre
Via Madonna dei Monti 40, I-00146 Rome, Italy
E-mail: magrone@mat.uniroma3.it

Tedeschini Lalli Laura, Professor of Mathematical Physics
Dipartimento di Architettura, Università Roma Tre
Via Madonna dei Monti 40, I-00146 Rome, Italy
E-mail: tedeschi@mat.uniroma3.it

Slovak University of Technology in Bratislava
Faculty of Mechanical Engineering
17th Conference on Applied Mathematics **APLIMAT 2018**

Proceedings

WHEN TURIN CAME TO ROME: A WALK IN THE PRATI DISTRICT AND ITS SYMMETRIES

MAGRONE Paola (IT), SCIROCCHI Antonio (IT), TEDESCHINI-LALLI Laura (IT), TRICARICO Daniele (IT)

Abstract. Symmetry, in mathematics today, is the property of invariance with respect to any "isometry", i.e. with respect to any transformation leaving distances unchanged. Thanks to isometries it is possible reduce complex shapes into primary forms for the purpose of classification, for example. The buildings of the Prati district of Rome, show symmetries at various spatial scales (urban scale, front, decorations). We analyze them and propose some significance to their urban scale symmetry.

Keywords: symmetry, urban scale, architecture, friezes, Rome 20th century

Mathematics Subject Classification 00A67, 58D19, 05B45

1 Introduction

Symmetry, in mathematics today, is the property of invariance with respect to any transformation that transforms an object while leaving distances unchanged. Such a transformation of a space is called "isometry". Thanks to the isometries it is possible reduce complex shapes into primary forms for the purpose of classification, for example. In particular, all information about a symmetrical object can be reduced to the description of a portion of the object, together with a transformation of the space; for instance, a motive with rotational symmetry of angle $2\pi/n$ can be memorized by the graphic information contained in a circular sector of angle $2\pi/n$, and the instructions to rotate it n times of that angle, thus filling the entire circle. It is obvious that this description is shorter, at least from a computational point of view, as it occupies less memory, and therefore it can be legitimately called "less complex". A very symmetrical object is also a simpler one, because it is invariant with respect to many isometries so that the graphic information needed to memorize it is less and less. Under another viewpoint, if you look at an object, close your eyes, and while your eyes are shut somebody performs an isometry characteristic of the object, the fact that the object is "invariant" means exactly that you will not be able to see any difference. The architecture has exploited the symmetries also to speed the design and realization processes of artifacts since antiquity. In this study we show how in the early twentieth century buildings of

the Prati district of Rome, isometries have been used systematically on the large urban scale and on smaller on the scale of individual buildings decorations. In [1,2] we performed studies using planar symmetry in architecture and Cultural heritage in Rome, yielding to hypothesis of reconstruction of patchy floors from existing fragments: the same was later done also in Villa Adriana [3].
Moreover symmetries on urban scale usually suggest some privileged directions, for instance as vanishing point of a large-scale perspective. Prati district was built, as Rome became the Capital of the new Kingdom of Italy, to host its functionaries and employees. We show its urban symmetries are never aligned on the view of the adjacent Vatican City, itself the capital, of another State. We think the choice of orientation of the symmetry is significant.
On the scale of the individual buildings, all buildings of Prati show a highly symmetric front. Many of the buildings also carry decorative friezes. We will look in particular at the front of the Courthouse.

1.1 Mathematical friezes

Given a pattern, the set of isometries leaving it unchanged, forms a group under the operation of composition of transformations. Any such group contains finite and infinite Subgroups (possibly void). The finite subgroup is generated by transformations, which under a finite number of compositions with themselves end up at the identity (mirrors and rotations by a divisor of 2π). If the infinite part of the group is generated by the translations along two independent directions, the group is considered a "tessellation group", or a "wallpaper group", It is called a "frieze group" if its infinite part is generated by one translation. There exist only 7 frieze groups [7]. The repetition patterns of the seven frieze groups can be summarized graphically in the scheme of fig.1

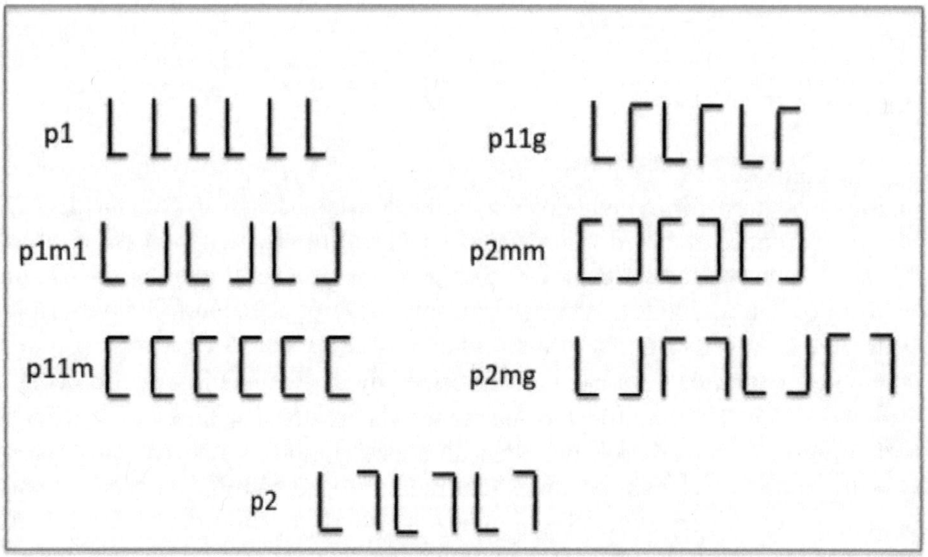

Fig. 1. The seven frieze groups: graphical repetition patterns, labeled by international classification of the groups.

2. Prati District in Rome, some history

The Prati district's territory consisted of fields that during the reign of Nero (54-68 A.D.), were called Prata Neronis and later were part of the properties of Domitia, wife of Domitian (81-96 A.D). During the Middle Ages, the area was called Prata Sancti Petri for its proximity to St Peter's Basilica. [4]
The king Umberto I (1844-1900), first king of a united Italy, decided to place in the Prati neighborhood the housing for the new administrative class that would move to Rome after it was proclaimed the capital of the Kingdom of Italy. [4]
In fact, until the 19th century the area was used for agriculture and only in 1883 it was urbanized in the part that was intended for military exercises, called the parade ground.

2.1 Symmetries at urban scale: orientation of Prati district in the surrounding town

The new district was planned in a very symmetric distribution, much as it happens with the plan of Turin, from where the King and his architects came. When a plan has symmetries and large straight streets, it necessarily also permits long scale perspectives. However at this time the climates between the Kingdom and the adjacent Vatican City was still politically very tense and, probably for this reason, the urban plan was designed in such a way that none of the new streets has as a vanishing point the dome of St. Peter's Basilica. Vatican City was to become the capital of a separate state: the Holy Siege. In addition the names of the streets in the new neighborhood were chosen among all historical characters belonging either to the pagan world of ancient Rome, or else to the lay world of the Italian unification. The main street of the neighborhood was named after Cola di Rienzo, nickname of Nicola Gabrini, a Roman tribune who in the 15th century rebelled against the papal power trying to restore the republic of Rome [8]. This marks even more the separation of the district from the adjacent capital of the Holy Siege.

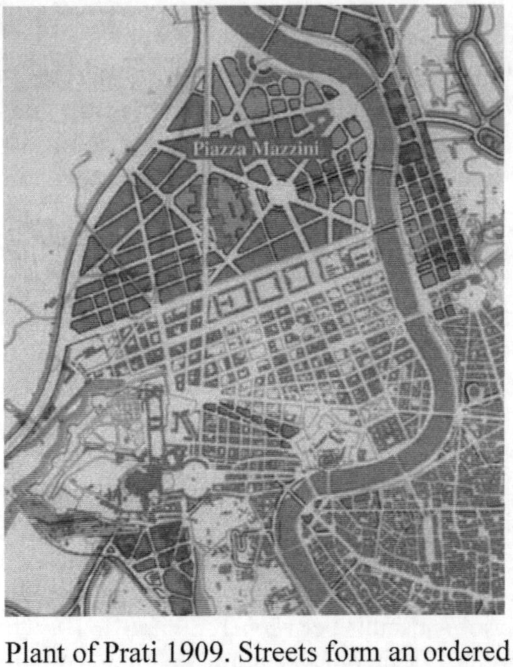

Fig. 2. Plant of Prati 1909. Streets form an ordered lattice.
Rome Town plan 1909, Edmondo Sanjust.

In the 1929 the Lateran Treaty between the Holy Siege and the State of Italy was finally signed, conciliating the two. In fact in 1933 via della Conciliazione was achieved, demolishing the neighborhood "Spina di Borgo" that did impede the view of the basilica from the center of Rome. This road was built as a visible urban sign of reconciliation. The new street, much photographed today, allows a new perspective with a St. Peter's in its vanishing point; as a consequence the Vatican can in fact be seen from several viewpoints in Rome.

Fig. 3. The three adjacent centers of power in Rome: religious, Military, Judiciary. (Aerial photo from Google earth).

2.2 Piazza Cavour

Piazza Cavour is the heart of the Prati district. The square was designed by Savoy planners as a symbol of laicism and for this reason they decided to place in this area the new Courthouse and a theatre: two public places, highly significant for the administration of a new state; the independent action and administration of culture, art, law. The Adriano theatre was inaugurated in 1898.

The planners made impossible the view of Saint Peter's Basilica from any point of the square and they named it to Camillo Benso Conte di Cavour, the Piedmontese statesman who worked to bring the union of the Italian kingdom, to emphasize the separation of the Kingdom of Italy by the Papal States [8]. One of his most famous quotes, for all Italian citizens, is "libera Chiesa in libero Stato", i.e. "a free Church in a free State". He advocated separation of Church and State.

In the middle of the square there is a garden designed by Nicodemus Severi and at the center of which stands a bronze statue of Camillo Benso Conte di Cavour by the sculptor Stefano Galletti between 1885 and 1895.

The square is rectangular. Standing in it, one perceives the sides as theatrical wings.

The wing towards the river Tiber consists of the Justice palace, or Courthouse designed by architect Guglielmo Calderini and built between 1889 and 1910, while one of the short sides hosts the Waldensian Church (Evangelica Valdese) church built in 1911, in Liberty style.

Waldensian is another religion, whose followers had had quite some difficulties with the ruling Catholic one in France and in Italy.

3. The Courthouse

The Courthouse hosts currently the headquarter of the Supreme Court of Cassation, the Lawyers Bar Council of Rome and the central Law library. The people of Rome call it "palazzaccio" or ugly palace, probably due to its role.
Built between 1889 and 1911 by the architect Guglielmo Calderini, it is one of the major works created after the proclamation of Rome as capital of the Kingdom of Italy. The prime minister Zanardelli of that era, an illustrious jurist, wanted the new Courthouse to reunify the various Courts, or Tribunals, of the capital in prestigious headquarters, in the Prati neighborhood that was then being built. [5]
The Courthouse was designed following the principles of "eclectic style", then in vogue, combining elements of Renaissance to Baroque. The building has an extraordinary size, competing in stateliness with more ancient buildings present in town. Its plant is a rectangle 170x155 m.; its fronts are entirely covered with travertine blocks. [6]
On the sides of the entrances of the building have been placed the travertine statues of the great Latin jurists and on top of the facade towards the Tiber is placed in a large bronze Quadriga by the sculptor Ettore Ximenes [8]. The quadriga can actually be seen from the entire Prati district.

Fig. 4. Front of Courthouse facing Piazza Cavour; notice the bronze quadriga on top. (Photo Tricarico)

3.1 The friezes of the Courthouse

Anybody who has performed this kind of mathematical analysis on an actual architectural or art piece, knows that symmetry groups are "mathematical models" to such end, and therefore applying the models requires some modelling assumptions. We think this is one of the instructive things to do. At times we consider various possibilities.

Fig. 5. The friezes of the Courthouse.

In fig. 5 we marked in red the friezes we are going to classify, on the front of the Courthouse in Rome. They will be shown ordered from bottom to top.

In Figure 6, a classification of three friezes.

Motive A: if we consider half of the strip, the frieze with waves is invariant only under a horizontal translation, (group p1 in the scheme of fig.1). On the other hand, if we take into consideration the mirror axis in the center, the strip pattern cannot be considered invariant under translation anymore, and it can be considered as a single element, with an axis of symmetry. (So it is not really a frieze group).

Motive B in the balustrade, shows two vertical and one horizontal mirror axis (group p2mm).

Motive C shows a vertical axis of symmetry and a rotation of π, corresponding to group p2mg.

Fig. 6. Three friezes belonging to the front of the Courthouse. (Photo Scirocchi)

Motive D, pictured in fig.7, is architecturally positioned on a drainer. It can be analyzed either as frieze or as planar group, depending on what you presume to allow being infinite.
As a frieze, pattern D displays a vertical mirror axis (group p1m1), and then repeated along horizontal translation. If we take into account the obvious repetition of the pattern, the planar group sees a glide-reflection, with axis parallel to the mirror axis.

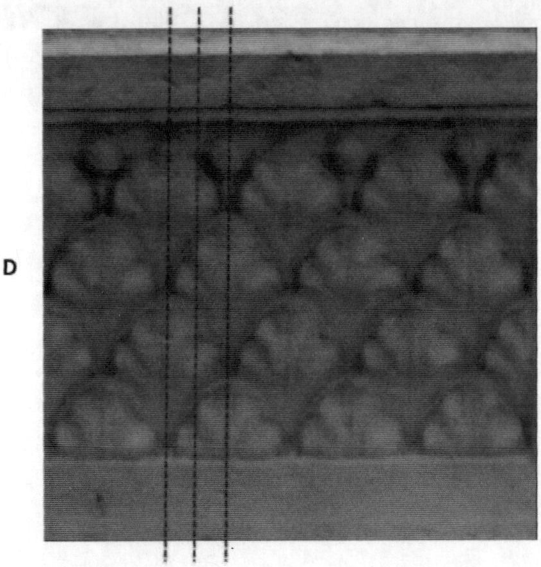

Fig. 7. The drainer. The red dotted lines mark the reflection axis; the blue dotted line marks the glide reflection axis. (Photo Tricarico)

Motive E, the beautiful and complex frieze of the summit cornice in Figure 8 shows a vertical mirror axis, going through the head of the oxen (group p1m1). We could also think to analyze some of its element separately.
From top: the first cornice has two mirror axes. If we extend it we notice that it passes in the middle of the oxen heads, in other words this axis belongs to the group of symmetry of the whole pattern.

Motive F: Flowers below the first cornice. The axis passing through the flowers is not part of the symmetries, which leave the top cornice unvaried. The flower as a single pattern is invariant under rotation of $2\pi/6$, and has six axes of symmetry.
In any case, the entire frieze and the single cornice, belong to group p1m1.

E

Fig. 8. The frieze of the summit cornice, and the axis of its symmetries. (Photo Scirocchi)

F

Fig. 9. The flower in the top cornice, the axis of symmetries, the center of rotation. (Photo Scirocchi)

Acknowledgements

LTL and PM wish to thank our colleague Luca Montuori, professor of urban scale architecture, for useful discussions on the topic of orientation of Prati. Also our thankful thoughts to Domenico Cecchini who made all cartographic historic documentation available.

References

[1] CARLINI, A., CONVERSANO, E., TEDESCHINI LALLI, L., Matematica per l'archeologia*, Dialoghi di archeologia e architettura, cura scientifica*, D.Manacorda, R.Santangeli Valenziani, L.Franciosini, E.Pallottino, R.Volpe, S.Picciola, A.Carlini, P.Porretta, edizioni Quasar, Roma, 2009, pp.168-181

[2] CARLINI A., CONVERSANO E., TEDESCHINI LALLI L. Mathematics and Archeology, *Aplimat – Journal of Applied Mathematics*, 1 (2), 61-68 (2008), ISSN 1337-6365

[3] CARLINI A., TEDESCHINI LALLI L., *Interrogare lo spazio. Esperienze di matematica ad architettura*, Roma, Cangemi Editore, 2012.

[4] CUCCIA, G., Roma. Prati di Castello. Dai Romani ai Barbari ai Piemontesi, *Roma, Gangemi Editore*, 2007, ISBN 978-88-492-1144-3.

[5] FABBRI, M., (a cura di), Il Palazzo di Giustizia di Roma, *Gangemi Editore*, Roma, 1988, ISBN 978-88-492-0231-1.

[6] MANODORI SAGREDO A., (a cura di), *La Corte di cassazione. Le opere d'arte del palazzo di giustizia di Roma*, Roma, Gangemi Editore, 2007, ISBN 978-88-492-1311-9.

[7] NIKULIN,V., SHAFAREVICH, I. Geometries and Groups, *Springer V.* 1987 Armstrong, M. Groups and Symmetry Sprinter 1988

[8] RENDINA C., PARADISI D., Le strade di Roma, vol. 1, Roma, *Newton Compton Editori*, 2004, ISBN 88-541-0208-3.

Current address

Magrone Paola, Assistant Professor of Mathematical Analysis
Dipartimento di Architettura, Università Roma Tre
Via Madonna dei Monti 40, I-00146 Rome, Italy
E-mail: magrone@mat.uniroma3.it

Scirocchi Antonio, Architect
Via Fabio Filzi 21, Anguillara Sabazia, Rome, IT 00061, Italy
E-mail: antonio.scirocchi@gmail.com

Tedeschini Lalli Laura, Professor of Mathematical Physics
Dipartimento di Architettura, Università Roma Tre
Via Madonna dei Monti 40, I-00146 Rome, Italy
E-mail: tedeschi@mat.uniroma3.it

Tricarico Daniele, Architect
Via Avanzini 79, Roma, IT 00163, Italy
E-mail: danieletricarico91@gmail.com

Slovak University of Technology in Bratislava
Faculty of Mechanical Engineering
17th Conference on Applied Mathematics **APLIMAT 2018**

Proceedings

HANKEL TRANSFORM AND FREE VIBRATION OF A LARGE CIRCULAR MEMBRANE

MALACKÁ Zuzana (SK)

Abstract. Integral transforms are a powerful apparatus for solving initial value and boundary value problems for linear differential equations. Paper is primarily attended to Hankel integral transform and shows a utilization of the integral transforms for solving partial differential equation- the equation of vibration of circular membrane.

Keywords: integral transform, Hankel transform, partial differential equation

Mathematics Subject Classification: 35L10, 35L20 , 58J45.

„One cannot understand ... the universality of laws of nature, the relationship of thinngs, withouth an understanding of mathematics. There is no other way to do it."

Richard P. Feynman

The theory of integral transorms in the classical sence can be traced back to about two hundred years. But the theory of distributions is only ninety five years old. Some elements of the theory of distributions may be found in [7]. The author actually tried to generalize to concept and classical operations to a larger field in which the Cauchy problem can be solved.

The papers [3],[4],[5] gave a firm footing to the theory of distributions. It has been rightly stated by a reputed mathematician that „twentieth century may appropriatety be called the century of Functional Analysis". It may be pointed out that the theory of distributions occupies a very important role in the field of application of functional analysis. During the last seventy years the phenomenal growth in the theory of partial differential equations is the result of theory of distributions which has acted as a catalytic agent.

In [6], [8], [2], authors gave other approaches to distribution theory.

It all started with giving mathematically accepted definition to Dirac´s Delta function defined by

$$\delta(x) = 0 \text{ for } x \neq 0 \text{ and}$$

$$\int_{-\infty}^{\infty} \delta(x)dx = 1.$$

This function was used to represent mathematically the concept of a shock, like the impact of hammer in mechanics or a large voltage of very small duration in electrical engineering. But it is not a function in the strict sence. For in the theory of integrals if a function is zero almost everywhere in an interval, its integral over that interval is also zero. Actually Delta function is a functional.

The Dirac delta function, $\delta(x)$ is defined so that for any good function $\Phi(x)$,

$$\int_{-\infty}^{\infty} \delta(x)\Phi(x)dx = \Phi(0).$$

Herman Hankel (1839-1873), a German mathematician, is remembered for his numerous contributions to mathematical analysis including the Hankel transformation, which occurs in the study of functions which depend only on the distance from the origin. He also studied Bessel functions. The Hankel transform involving Bessel functions as the kernel arises naturally in axisymmetric problems formulated in cylidrical polar coordinates.

Bessel function

Function $J_v(x)$ is a solution of a Bessel equation

$$x^2y'' + xy' + (x^2 - v^2)y = 0,$$

where $v \geq 0$ is called index of Bessel equation.

If $v = n$

$$J_n(x) = \frac{1}{\pi}\int_0^{\pi} \cos(n\theta - x sin\theta)\, d\theta,$$

or

$$J_n(x) = \frac{1}{2\pi}\int_{\phi_0}^{2\pi+\phi_0} e^{i(n\theta - x sin\theta)} d\theta \text{ [1].}$$

We introduce the definition of the *Hankel transform* from the two-dimensional Fourier transform and its inverse given by

$$\mathcal{F}\{f(x,y)\} = F(k,l) = \frac{1}{2\pi}\int_{-\infty}^{\infty}\int_{-\infty}^{\infty} e^{-i(\kappa,r)} f(x,y)dx\, dy, \quad (1)$$

$$\mathcal{F}^{-1}\{F(k,l)\} = f(x,y) = \frac{1}{2\pi}\int_{-\infty}^{\infty}\int_{-\infty}^{\infty} e^{i(\kappa,r)} F(k,l)dk\, dl, \quad (2)$$

where $\boldsymbol{r} = (x,y)$, $\boldsymbol{\kappa} = (k,l)$.

Introducing polar coordinates

$$x = r.cos\theta, \qquad k = \kappa.\cos\phi,$$

$$y = r.sin\theta, \qquad l = \kappa.\sin\phi,$$

where $r, \kappa \in \langle 0, \infty)$ a $\theta, \phi \in \langle 0, 2\pi)$ and $J = r$.

Than

$$\boldsymbol{r}.\boldsymbol{\kappa} = \kappa.r(\cos\phi\ cos\theta - \sin\phi\ sin\theta) = \kappa.rcos(\theta - \phi).$$

We denote $F(\kappa.\cos\phi, \kappa.\sin\phi) = G(\kappa, \phi)$,

where $G(\kappa,\phi) = \frac{1}{2\pi}\int_0^\infty \int_0^{2\pi} e^{-i\kappa.rcos(\theta-\phi)} f(r,\theta) r\, d\theta dr$ (3)

We next assume $f(r,\theta) = e^{in\theta} f(r)$, which is not a very severe restriction, and make a change of variable $\theta - \phi = \alpha - \frac{\pi}{2}$ to reduce (3) to the form

$$G(\kappa,\phi) = \frac{1}{2\pi}\int_0^\infty \int_{\frac{\pi}{2}-\phi}^{2\pi+\frac{\pi}{2}-\phi} e^{-i\kappa.rcos\left(\alpha-\frac{\pi}{2}\right)} e^{in.rcos\left(\alpha+\phi-\frac{\pi}{2}\right)} f(r) r\, d\alpha dr.$$

Because $cos\left(\alpha - \frac{\pi}{2}\right) = \sin\alpha$, than we have

$$G(\kappa,\phi) = e^{in\left(\phi-\frac{\pi}{2}\right)} \int_0^\infty \left[\frac{1}{2\pi}\int_{\frac{\pi}{2}-\phi}^{2\pi+\frac{\pi}{2}-\phi} e^{i(n\alpha-\kappa r\sin\alpha)} d\alpha\right] f(r) r dr.$$

Using the integral representation of the Bessel function of order *n*, we have

$$G(\kappa,\phi) = e^{in\left(\phi-\frac{\pi}{2}\right)} \int_0^\infty r J_n(\kappa r) f(r) r dr = e^{in\left(\phi-\frac{\pi}{2}\right)} \tilde{f}_n(\kappa),$$

where $\tilde{f}_n(\kappa)$ is called the *Hankel transform* of $f(r)$ defined formally by

$$\tilde{f}_n(\kappa) = \mathcal{H}_n\{f(r)\} = \int_0^\infty r J_n(\kappa r) f(r) dr.$$

Similarly, in terms of the polar variables with the assumption $f(x,y) = f(r,\theta) = e^{in\theta} f(r)$, $J = \kappa$, the inverse Fourier transform becomes

$$e^{in\theta} f(r) = \frac{1}{2\pi}\int_0^\infty \int_0^{2\pi} e^{i\kappa.rcos(\theta-\phi)} G(\kappa,\phi)\ \kappa d\phi d\kappa,$$

which is, by the change of variables $\theta - \phi = -\alpha - \frac{\pi}{2}$, we have

$$e^{in\theta} f(r) = \frac{1}{2\pi}\int_0^\infty \int_{-\theta-\frac{\pi}{2}}^{2\pi+\left(-\theta-\frac{\pi}{2}\right)} e^{i\kappa.rcos\left(-\alpha-\frac{\pi}{2}\right)} e^{in(\theta+\alpha)} \tilde{f}_n(\kappa)\ \kappa d\alpha d\kappa =$$

$$= e^{in\theta} \int_0^\infty \left[\frac{1}{2\pi}\int_{-\theta-\frac{\pi}{2}}^{2\pi+\left(-\theta-\frac{\pi}{2}\right)} e^{i(n\alpha-\kappa r sin\alpha} d\alpha\right] \tilde{f}_n(\kappa)\kappa d\kappa =$$

$$= e^{in\theta} \int_0^\infty \kappa J_n(\kappa r) \tilde{f}_n(\kappa) d\kappa.$$

Thus, the inverse Hankel transform is defined by

$$f(r) = \mathcal{H}_n^{-1}\{\tilde{f}_n(\kappa)\} = \int_0^\infty \kappa J_n(\kappa r) \tilde{f}_n(\kappa) d\kappa.$$

Free Vibration of a Large Circular Membrane

Obtain the solution of the free vibration of a large circular elastic membrane governed by the initial value problem

$$c^2\left(\frac{\partial^2 u(r,t)}{\partial r^2}+\frac{1}{r}\frac{\partial u(r,t)}{\partial r}\right)=\frac{\partial^2 u(r,t)}{\partial t^2}, 0<r<\infty, t>0, \tag{4}$$

$$u(r,0)=f(r), \qquad \left.\frac{\partial u(r,t)}{\partial t}\right|_{t=0}=g(r), \quad 0\le r<\infty \tag{5}$$

where $c^2=\frac{T}{\rho}$ =constant, T is the tension in the membrane, and ρ is the surface density of the membrane.

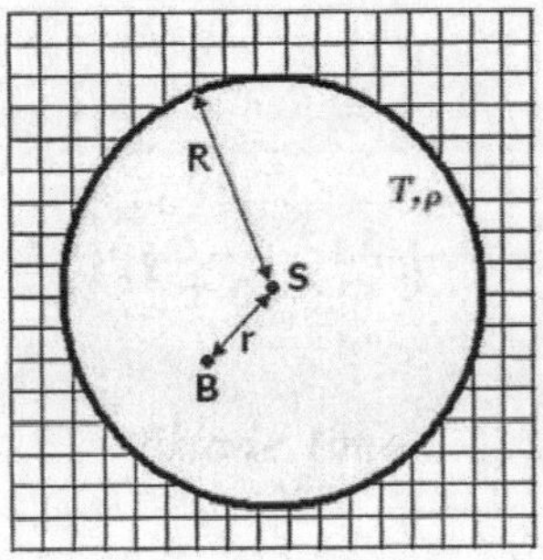

Fig. 1. Circular elastic membrane.

At the beginning, the membrane deflects from the equilibrium position.

Solution:

Equation (4) is a hyperbolic type, because for the matrix

$$\begin{pmatrix} c^2 & 0 \\ 0 & -1 \end{pmatrix}$$

$$\lambda_1 = c^2, \qquad \lambda_2 = -1.$$

Application of the zero-order Hankel transform with respect to r

$$\tilde{u}(\kappa,t) = \int_0^\infty rJ_0(\kappa r)u(r,t)dr,$$

to (4) with conditions (5) gives

$$\frac{d^2\tilde{u}}{dt^2} = -c^2\kappa^2\tilde{u}(\kappa,t), \tag{6}$$

$$\tilde{u}(\kappa,0)=\tilde{f}(\kappa), \qquad \left.\frac{d\tilde{u}}{dt}\right|_{t=0}=\tilde{g}(\kappa) \tag{7}$$

We got a homogenne ordinar linear differential equation of second order with constant coeficients. Characteristic equation is

$$\lambda^2 + c^2\kappa^2 = 0,$$

where

$$\lambda_1 = ic\kappa\,, \qquad \lambda_2 = -ic\kappa.$$

Solution of equation (6) is in the form

$$\tilde{u}(\kappa, t) = c_1 \cos(c\kappa t) + c_2 \sin(c\kappa t).$$

From (7) we got, that $c_1 = \tilde{f}(\kappa),\ c_2 = (c\kappa)^{-1}\tilde{g}(\kappa),$

and $\tilde{u}(\kappa, t) = \tilde{f}(\kappa)\cos(c\kappa t) + (c\kappa)^{-1}\tilde{g}(\kappa)\sin(c\kappa t).$

The inverse Hankel transformation leads to the solution

$$u(r,t) = \int_0^\infty \kappa J_0(\kappa r)\tilde{f}(\kappa)\cos(c\kappa t)\,d\kappa + \frac{1}{c}\int_0^\infty J_0(\kappa r)\tilde{g}(\kappa)\sin(c\kappa t)\,d\kappa\,. \qquad (8)$$

In particular, we consider

$$f(r) = Aa(r^2 + a^2)^{-\frac{1}{2}},\quad a\,, A \in R,\ \ g(r) = 0.$$

So that

$$\tilde{g}(\kappa) = \int_0^\infty rJ_0(\kappa r)0dr \equiv 0\,,$$

$$\tilde{f}(\kappa) = \int_0^\infty rJ_0(\kappa r)Aa(r^2 + a^2)^{-\frac{1}{2}}\,dr = \frac{Aa}{\kappa}e^{-a\kappa}\,.$$

From (8) we have

$$u(r,t) = \int_0^\infty \kappa J_0(\kappa r)\frac{Aa}{\kappa}e^{-a\kappa}\cos(c\kappa t)\,d\kappa = Aa\int_0^\infty J_0(\kappa r)e^{-a\kappa}\cos(c\kappa t)\,d\kappa =$$

$$= Aa\mathrm{Re}\int_0^\infty J_0(\kappa r)e^{-\kappa(a+ict)}d\kappa = Aa\,Re\left\{\int_0^\infty \kappa J_0(\kappa r)\frac{e^{-\kappa(a+ict)}}{\kappa}d\kappa\right\} = Aa\,Re\left\{\frac{1}{\sqrt{r^2+(a+ict)^2}}\right\}.$$

We will graphically illustrate the solution. The figure shows the vibration of the circular membrane with the center at the point $S = [0,0]$ and the radius $R = 4\,m$ from $t = 0\,s$ to $t = 20\,s$, where $T = 2{,}43\ N\,m^{-1}$, and $\rho = 4{,}29\ kg\,m^{-2}$, $A = 4{,}5$, $a = 1{,}3$.

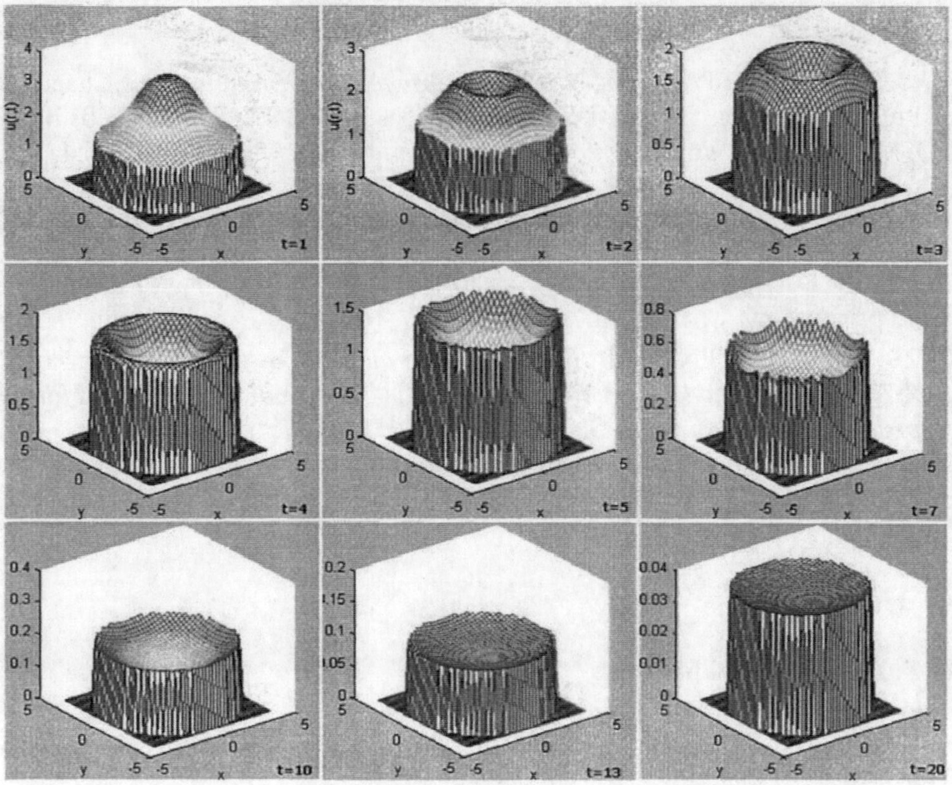

Fig. 2. Vibration of circular elastic membrane.

Conclusion

"A teacher can never truly teach unless he is still learning himself. A lamp can never light another lamp unless it continues to burn its own flame. The teacher who has come to the end of his subject, who has no living traffic with his knowledge but merely repeats his lessons to his students, can only load their minds, he cannot quicken them. "

Rabindranath Tagore

The importance of integral transforms is that they provide powerful operational methods for solving initial value problems and initial-boundary value problems for linear differential and integral equations. Integral transforms have many mathematical and physical applications, their use is still predominant in advanced study and research. The paper was developed as a result of experience in a teaching advanced undergraduates students in mathematics and engineering. Main feature is a systematic mathematical treatment of the theory and method of integral transforms that gives the students a clear understanding of the subject and its varied applications.

References

[1] DEBNATH, L., BHATTA, D., *Integral Transform and Their Applications,*Chapman CRC Press, Boca Raton, Florida, USA, 2007. 315-339, ISBN 1-58488-575-0, 700 p.

[2] SAFO, M., *On a generalization of concept of function,* Proc. Japan Acad., Vol. 34, 1958, pp.126-130.

[3] SCHWARTZ, L., *Generalizations de la notion de function et de derivation, Theorie des distributions,* Ann Telecommunum, No 3,1945, pp.135-140.

[4] SCHWARTZ, L., *Theorie des distributions,* Vol,I, 1950.

[5] SCHWARTZ, L., *Theorie des distributions,* Vol,II, 1951.

[6] SIKORSKI, R., MIKUSINSKI, J., *The elementary theory of distributions*, Warsaw, 1957

[7] SOBOLEV, S. L., *Methode nouvelle le probleme de cauchy pourles equations lineares hyperbeliques,* Rec. Math. [Mat. Sbornik] N.S., **1(43)**:1, 1936, pp. 39-72.

[8] TEMPLE, G., *The Theory of generalized functions,* Proc. Roy. Soc., London, Series A, 228, 1955, pp. 175-190.

Current address

Malacká Zuzana, RNDr., PhD.
Department of Applied Mathematics, Faculty of Mechanical Engineering
University of Žilina
Univerzitná 8215/1, 010 26 Žilina, Slovak Republic
E-mail: zuzana.malacka@fstroj.uniza.sk

Slovak University of Technology in Bratislava
Faculty of Mechanical Engineering
17th Conference on Applied Mathematics **APLIMAT 2018**

Proceedings

COMPARISON OF HOME TEAM ADVANTAGE IN ENGLISH AND SPANISH FOOTBALL LEAGUES

MAREK Patrice (CZ), VÁVRA František (CZ)

Abstract. Home team advantage in sports is widely analysed phenomenon. This paper builds on results of recent research that – instead of points gained – uses goals scored and conceded to describe home team advantage. Using this approach, the home team advantage is random variable that can be described by trinomial distribution, and it is possible to use Jeffrey divergence and test for homogeneity of parallel samples to compare and test home team advantage of different leagues. The paper also introduces procedure that is based on identification of all distances between home advantages of leagues. These distances are later used to construct disconnected graph with components that contain leagues with similar home team advantage. Procedures are demonstrated on five top level English football leagues and two top level Spanish leagues from the 2007/2008 season to the 2016/2017 season.

Keywords: Home team advantage, Combined measure, Jeffrey divergence, Test for homogeneity of parallel samples, Premier League, La Liga

Mathematics subject classification: Primary 62P99; Secondary 62F03, 62F10

Home team advantage in sports is widely analysed phenomenon. Usually, the home team advantage is analysed by comparing the number of points that a team earned at its home ground with the total number of points that the team earned. This paper builds on results of recent research that – instead of points gained – uses goals scored and conceded to describe home team advantage.

We define random variable C *combined measure of home team advantage* that – according to results between teams – can take values -1 (away team advantage), 0 (no advantage), and 1 (home team advantage). Random variables $Z_r, r = -1, 0, 1$ describe number of cases (in a season) where it is possible to identify home team advantage ($r = 1$), away team advantage ($r = -1$), and no advantage ($r = 0$), i.e. Z_r sums number of cases where $C = r$. Vector (Z_{-1}, Z_0, Z_1) follows trinomial distribution.

To compare leagues, we use *Jeffrey divergence* – symmetric version of Kullback-Leibler distance – that allows to measure distances between probability distributions that describe home team advantage. Next, approach based on the *Test for homogeneity of parallel samples* allows to test hypothesis H_0 : *Two samples come from the same population* (or in our case: that home advantage in both tested leagues is described by the same distribution) against alternative hypothesis H_1 : *Two samples do not come from the same population* (or in our case: that home advantage in both tested leagues come from different distributions). Moreover, χ^2 statistic used in this test is also used to measure distances between probability distributions that describes home team advantage. The paper also introduces procedure that is based on identification of all distances between home advantages of leagues. These distances are later used to construct disconnected graph with components that contain leagues with similar home team advantage. Procedures are demonstrated on five top level English football leagues and two top level Spanish leagues from the 2007/2008 season to the 2016/2017 season.

1 Introduction

Home team advantage in sports was investigated by many researchers, for exhaustive study see Pollard and Pollard [6] and Allen and Jones [1]. Home team advantage in these papers is based on comparing the number of points that a team earned at its home ground with the total number of points that the team earned. Marek and Vávra [5] used approach where – instead of points – home team advantage is based on number of goals scored and conceded. This approach is used in this paper because of its advantage that can be illustrated by the following example. Let us assume that two teams A and B played each other exactly two times in a season, and team A won both matches: 4–1 at the home ground and 2–1 at the home ground of team B. Using points in this situation will result in conclusion that team A won 50% of points at the home ground, i.e. that there is no home team advantage. However, using goals and their differences – for team A: 3 goals at the home ground and 1 goal at the away team ground – shows that team A played better at the home ground. The same holds true for the team B who lost by 1 goal at the home ground and by 3 goals at the away team ground, i.e. the better result was recorded at the home ground.

Marek and Vávra [5] defined three types of measures: *Active measure of home team advantage* that takes into account difference between goals scored at home and away; *Passive measure of home team advantage* that takes into account difference between goals conceded at home and away; and *Combined measure of home team advantage* – used in this paper and specified in Definition 1 – that takes into account both goals scored and conceded.

Definition 1 Combined measure of home team advantage *is random variable C that can take values $-1, 0,$ and 1. $C = -1$ for team T_1 if two matches between teams T_1 and T_2 in a season ended with a better result – measured by goal differences in matches – for team T_1 away from home (at T_2's ground). $C = 0$ for team T_1 if goal differences in both matches were exactly the same from T_1's point of view, and $C = 1$ for team T_1 if this team recorded better result – measured by goal differences in matches – at its own ground. With results $h_{T_1} : a_{T_2}$ at the home ground of team T_1 and $h_{T_2} : a_{T_1}$ at the home ground of team T_2 the value of random variable C for team T_1 is determined as*

$$C = \text{sgn}((h_{T_1} - a_{T_2}) - (a_{T_1} - h_{T_2})). \tag{1}$$

2 Data and Methods

This section describes data and methods used for the estimation of parameters. It also introduces approach to compare home team advantage of different leagues.

2.1 Compared Leagues

Home team advantage in five top level English football leagues and two top level Spanish football leagues are compared by the methods described in the following subsections. Results of ten latest seasons – i.e. from the 2007/2008 season to the 2016/2017 season – used in the comparison are obtained from [3]. The following leagues are used for England:

- Premier League (level 1), ENG_1,
- English Football League Championship (level 2), ENG_2,
- English Football League One (level 3), ENG_3,
- English Football League Two (level 4), ENG_4,
- National League (level 5), ENG_5.

Next two leagues are used for Spain:

- La Liga (level 1), ESP_1,
- Segunda Divisíon (level 2), ESP_2.

Websites [7] and [4] were used for basic control of all data, e.g., total number of goals scored by each team in the whole season, and to fill in results that were occasionally not recorded in the original source. Following matches were excluded (and consequently the opposite matches): Blackpool vs. Huddersfield in 2014/2015 season of ENG_2 (because the match was abandoned after 48 minutes of play, and the final result 0–0 reflects only these 48 minutes) and all matches of Chester City in 2009/2010 season of ENG_5 (because Chester City was expelled during the season).

2.2 Parameters estimation method

A balanced schedule was used in all seasons, i.e. each team played each other team exactly two times, once as a home team and once as a visiting team. For example, the English Premier League consisted of 20 teams in all seasons and total number of matches in one season is always 380. The combined measure of home team advantage joins results of the first part of the season (where team A plays at its home ground against team B) with results of the second part of the season (where team A plays against team B at B's home ground). Therefore, 190 observations of combined measure are available for one season in the English Premier League.

Let $Z_r, r = -1, 0, 1$ are random variables which describe number of cases (in a season) where it is possible to identify home team advantage ($r = 1$), away team advantage ($r = -1$) and no advantage ($r = 0$), i.e. Z_r sums number of cases where $C = r$. Vector (Z_{-1}, Z_0, Z_1) follows trinomial distribution with parameters p_{-1}, p_0, p_1, and K with probability function

$$P(k_{-1}, k_0, k_1) = \frac{K!}{k_{-1}!k_0!k_1!} p_{-1}^{k_{-1}} p_0^{k_0} p_1^{k_1}, \tag{2}$$

where K is total number of observations of combined measure in a season, p_{-1}, p_0, p_1 are probabilities of occurring home team advantage ($r = 1$), away team advantage ($r = -1$) and no advantage ($r = 0$). k_{-1}, k_0, k_1 ($k_{-1} + k_0 + k_1 = K$) are observations of appropriate advantage.

Maximum likelihood estimator of parameters $p_r, r = -1, 0, 1$ is

$$\hat{p}_r = \frac{k_r}{K}, \quad r = -1, 0, 1. \tag{3}$$

The results of the highest leagues in England and Spain in the 2016/2017 season are used to illustrate previous procedure. Observed counts of combined measure of home team advantage are in Table 1, and estimated parameters are later presented in Table 2

League	k_{-1}	k_0	k_1
ENG_1	56	34	100
ESP_1	52	41	97

Tab. 1. Observed counts of combined measure in 2016/2017 season.

League	$\hat{p}_{-1}$	$\hat{p}_0$	$\hat{p}_1$
ENG_1	0.295	0.179	0.526
ESP_1	0.274	0.216	0.511

Tab. 2. Estimated probabilities in 2016/2017 season.

2.3 Methods of Comparison

Two approaches were adopted to compare home team advantage in different leagues. The first one is *Jeffrey divergence* (JD), i.e. symmetric version of *Kullback-Leibler distance*, which allows to measure distance between two probability distributions P and Q

$$JD(P,Q) = \sum_x \left(p(x) \ln \frac{p(x)}{q(x)} + q(x) \ln \frac{q(x)}{p(x)} \right). \tag{4}$$

More about Jeffrey divergence can be found in Deza and Deza [2] (page 185). Equation 2 allows to measure distance between probability distributions of (Z_{-1}, Z_0, Z_1) for each two leagues. These distances can be compared to find leagues with similar, or, on the contrary, different home team advantages. Values of JD are frequently small, therefore – for better readability in the following parts – we multiply them by 100. Jeffrey divergence (multiplied by 100) of data presented in Table 2 is 0.894.

The second approach is based on the *Test for homogeneity of parallel samples* that allows to test hypothesis H_0 : *Two samples come from the same population* (or in our case: that home advantage in both tested leagues is described by the same distribution) against alternative hypothesis H_1 : *Two samples do not come from the same population* (or in our case: that home advantage in both tested leagues come from different distributions). χ^2 statistic can be – for the case with two leagues – computed as

$$\chi^2 = \sum_{r=-1}^{1} \frac{\left(n_{1,r} - \frac{n_{1,.}n_{.,r}}{n_{.,.}}\right)^2}{\frac{n_{1,.}n_{.,r}}{n_{.,.}}} + \sum_{r=-1}^{1} \frac{\left(n_{2,r} - \frac{n_{2,.}n_{.,r}}{n_{.,.}}\right)^2}{\frac{n_{2,.}n_{.,r}}{n_{.,.}}}, \tag{5}$$

	$C = -1$	$C = 0$	$C = 1$	Total
Sample 1	$n_{1,-1}$	$n_{1,0}$	$n_{1,1}$	$n_{1,.}$
Sample 2	$n_{2,-1}$	$n_{2,0}$	$n_{2,1}$	$n_{2,.}$
Total	$n_{.,-1}$	$n_{.,0}$	$n_{.,1}$	$n_{.,.}$

Tab. 3. Observed numbers in given sample and class.

where appropriate observed numbers ($n_{1,-1}, n_{1,0}$, etc.) are obtained according to Table 3.

Assymptotical distribution of χ^2 statistic for the data in Table 3 is χ^2 distribution with two degrees of freedom (i.e. number of classes reduced by one). More about this test and its more general form can be found in [8] (pages 398–402).

χ^2 statistic for the data presented in Table 1 is 0.847, and corresponding p-value is 0.655. The hypothesis H_0 that home team advantage in ENG_1 (Premier League) and ESP_1 (La Liga) in the season 2016/2017 is described by the same distribution cannot be rejected, i.e. we cannot reject that there is no difference in home team advantage (measured by combined measure) in these two leagues in the 2016/2017 season.

Method based on χ^2 statistic can be used to test whether the home team advantage in two leagues differ or not. Next, both methods – the first based on JD and the second based on χ^2 – can be used to identify groups of leagues with similar home team advantage. In this case, we will proceed according to these steps:

- construct graph, where nodes represent leagues, and edges represent distance (JD or χ^2) between leagues,
- remove the edge with the highest distance from the graph,
- repeat the previous step until the original full graph becomes disconnected, i.e. until two components are obtained.

Previous procedure will identify two groups of leagues where distance (JD or χ^2) between any league from the first group and the second group is always equal or greater than the last removed distance of the graph. This procedure can be further used to obtain more than two components.

3 Results

The first part will present full results for data summed over all ten seasons for each league. Next part will present major findings for single seasons.

3.1 Categorization by Leagues

In this part, we do not distinguish among seasons; therefore, the findings are valid for leagues in general (or, more precisely, for the last ten seasons).

Table 4 presents observed counts of combined measure for all leagues over all seasons, i.e. from the 2007/2008 season to the 2016/2017 season. ENG_4 recorded the lowest value of $\hat{p}_1$ and ESP_1 recorded the highest value. The second highest value of $\hat{p}_1$ was recorded by ESP_2. Both ESP_1 and ESP_2 recorded two lowest values of $\hat{p}_{-1}$. Based on these numbers, we can formulate hypothesis that home team advantage in Spain is higher than in England.

JD divergence (multiplied by 100) is used to measure distances between each two leagues, and results are presented in Table 5. For example, the distance between the English Premier League and Spanish

League	$k=-1$	$k=0$	$k=1$	Total	$\hat{p}_{-1}$	$\hat{p}_0$	$\hat{p}_1$
ENG_1	520	346	1 034	1 900	0.274	0.182	0.544
ENG_2	796	510	1 453	2 759	0.289	0.185	0.527
ENG_3	845	503	1 412	2 760	0.306	0.182	0.512
ENG_4	922	518	1 320	2 760	0.334	0.188	0.478
ENG_5	845	494	1 398	2 737	0.309	0.180	0.511
ESP_1	484	304	1 112	1 900	0.255	0.160	0.585
ESP_2	600	438	1 272	2 310	0.260	0.190	0.551

Tab. 4. Observed counts of combined measure for all ten seasons with appropriate estimated probabilities.

La Liga is quantified by $100 \cdot JD(ENG_1, ESP_1) = 0.721$. This gives us brief look at difference between home team advantage in studied leagues. The procedure described in section 2.3 is used to obtain disconnected graph (where nodes are leagues, and edges are distances between leagues). After removing edges – the last removed edge is between ENG_1 and ESP_1 – we obtain six leagues in one group (white colour in Table 5) and ESP_1 in the second group (grey colour in Table 5).

League	ENG_1	ENG_2	ENG_3	ENG_4	ENG_5	ESP_1	ESP_2
ENG_1		0.140	0.566	2.072	0.636	0.721	0.111
ENG_2	0.140		0.152	1.138	0.196	1.398	0.421
ENG_3	0.566	0.152		0.484	0.004	2.226	1.080
ENG_4	2.072	1.138	0.484		0.442	4.752	2.892
ENG_5	0.636	0.196	0.004	0.442		2.299	1.191
ESP_1	0.721	1.398	2.226	4.752	2.299		0.724
ESP_2	0.111	0.421	1.080	2.892	1.191	0.724	

Tab. 5. JD divergence for leagues over all ten seasons.

Last view is based on χ^2 statistic. Table 6 presents p-values for leagues over all ten seasons. All p-values are based on χ^2 distribution with two degrees of freedom; therefore, values of χ^2 are not presented, and values in Table 6 can be easily used to identify which leagues are more distant (the lower p-value, the higher distance). The procedure to obtain disconnected graph was also used for distances based on χ^2. The result is exactly the same as in the previous case, i.e. ESP_1 belongs to one group and the rest of leagues to the other group.

League	ENG_1	ENG_2	ENG_3	ENG_4	ENG_5	ESP_1	ESP_2
ENG_1		0.456	0.042*	<0.001*	0.029*	0.033*	0.561
ENG_2	0.456		0.350	<0.001*	0.260	<0.001*	0.071
ENG_3	0.042*	0.350		0.036*	0.973	<0.001*	0.001*
ENG_4	<0.001*	<0.001*	0.036*		0.048*	<0.001*	<0.001*
ENG_5	0.029*	0.260	0.973	0.048*		<0.001*	0.001*
ESP_1	0.033*	<0.001*	<0.001*	<0.001*	<0.001*		0.024*
ESP_2	0.561	0.071	0.001*	<0.001*	0.001*	0.024*	

Tab. 6. p-values for leagues over all ten seasons.

Asterisk in Table 6 is used to identify p-values lower that 0.05, i.e. pairs of leagues where – on 5% level of significance – the hypothesis H_0 that home team advantage in these two leagues is described by the same distribution can be rejected, and the alternative H_1 that home team advantage in these two leagues is described by different distributions can be accepted. Test of ESP_1 with any other league results in rejection of H_0 and acceptance of H_1, i.e. that there is difference in home team advantage. We can interpret this result – taking into account values from Table 4 – that home team advantage in ESP_1 is significantly stronger than in other tested leagues.

Next league with all p-values lower that 0.05 is ENG_4. In this case we can interpret this result – taking into account values from Table 4 – that home team advantage in ENG_4 is significantly weaker than in other tested leagues. This result is supported by the fact that if the procedure of finding components in Table 5 and Table 6 is be used to find three components then one component is ESP_1 (higher home team advantage), the second component is ENG_4 (lower home team advantage), and the third component is group of rest leagues.

3.2 Categorization by Leagues and Seasons

Previous part suggests that the highest difference is between ESP_1 and ENG_4 (for example, the highest JD divergence in Table 5); therefore, these two leagues are chosen to present evolution of $\hat{p}_1$ in Figure 1. Moreover, this figure contains evolution of $\hat{p}_1$ for the English Premier League ENG_1. Other leagues are omitted to maintain readability of the figure. Point estimate of p_1 for ESP_1 is in all seasons above point estimate of p_1 for ENG_4 and usually above point estimate of p_1 for ENG_1.

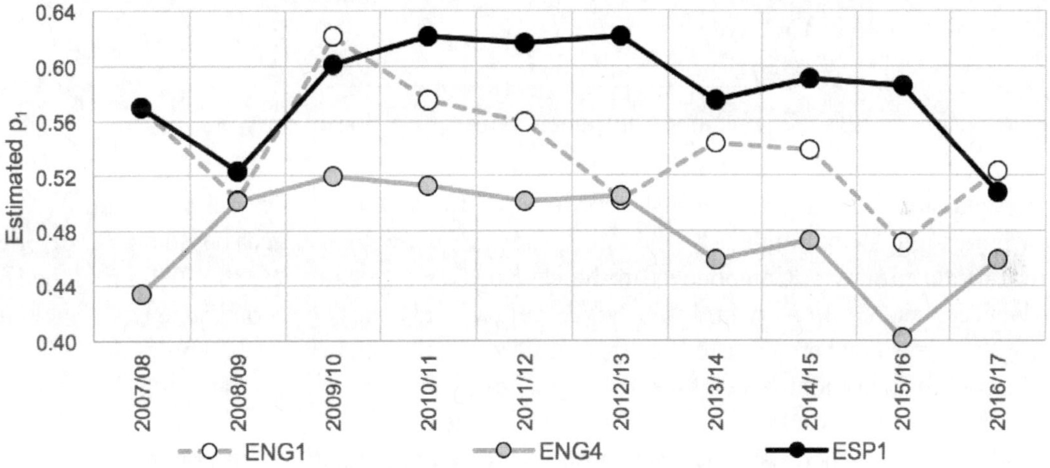

Fig. 1. Evolution of $\hat{p}_1$ for ENG_1, ENG_4, and ESP_1.

The test for homogeneity of parallel samples (see Section 2.3) can be used identify differences in home team advantage between leagues in single seasons. Here, we will present only results for the test of ESP_1 against ENG_1 and ENG_4. Hypothesis H_0 that home team advantage in ESP_1 and ENG_4 is described by the same distribution cannot be rejected in three seasons: 2008/2009, 2009/2010, and 2016/2017. The hypothesis H_0 can be rejected in the rest of tested seasons, and its alternative H_1 that home team advantage in ESP_1 and ENG_4 is described by different distributions can be accepted. The same test where ESP_1 and ENG_1 are used is not able to reject H_0 in all tested seasons. These results suggest that when single seasons are used then – to identify difference in home team advantage – the difference has to be strong. Therefore, we will use the procedure based on searching components of graph (see section 2.3).

Components obtained in every season are presented in Table 7. Results are based on distance measured by χ^2 statistic that was defined in Equation 5. For example, a graph with two components is obtained in the 2010/2011 season. The first component contains ENG_2, ENG_3, ENG_4, and ENG_5, and the second component contains ENG_1, ESP_1, and ESP_2. The interpretation is based on values of $\hat{p}_r, r = -1, 0, 1$. The first component contains leagues where all leagues have value of $\hat{p}_{-1}$ higher than leagues of the second component. The situation in the case of $\hat{p}_1$ is the opposite, and all leagues of the second component have value of $\hat{p}_1$ higher than leagues of the first component. We emphasize that words *higher* and *lower* are here used only to distinguish between two components (i.e. the meaning is relative), and it should not be interpreted as components with high or low home team advantage. Other leagues – except ENG_3 – are consequently classified as leagues with higher home team advantage.

Season	Lower home adv.	Higher home adv.
2007/2008	ENG_4	all others
2008/2009	all others	ESP_2
2009/2010	all others	ENG_1
2010/2011	all others	$ENG_1, ESP_{1,2}$
2011/2012	all others	ESP_1
2012/2013	all others	$ESP_{1,2}$
2013/2014	$ENG_{2,3,4}$	all others
2014/2015*	ENG_3	all others
2015/2016**	ENG_4	$ENG_{1,2,5}, ESP_{1,2}$
2016/2017	all others	ESP_2

Tab. 7. Obtained components in seasons (based on χ^2).

Table 7 contains two remarks. The first remark (*) is for the 2014/2015 season which is the only season where components obtained by the same procedure but with distances based on JD divergence (Equation 4) are different. Components in the case of JD divergence are ESP_1 with higher home team advantage and *all others* with lower home team advantage. The second remark (**) belongs to the 2015/2016 season where obtained components are ENG_3 and *all others*. This is the only season where the conclusion is not made in favour of higher or lower home team advantage, and it is possible to observe both situations at the expense of the lower value of $\hat{p}_0$. ENG_3 has the second highest value of $\hat{p}_{-1}$ and also of $\hat{p}_1$. Therefore, the procedure continued to obtain third component (ENG_4) which can be classified as a league with lower home team advantage.

4 Conclusion

This paper introduced procedure for comparison of the home team advantage in different leagues. The procedure uses combined measure of home team advantage. This measure is based on both goals scored and conceded; and therefore, it is able to detect home team advantage in a better way than usual approach based on points gained. Two procedures were proposed – the first is based on determining distances between leagues, and the second uses classical statistical hypothesis testing. Distances can be further used to obtain groups of leagues with similar home team advantage based.

Procedures applied on five top level English football leagues and two top level Spanish leagues suggest that home team advantage in Spain is higher that in England. More specifically, La Liga (level 1 Spanish league) has the highest home team advantage among the analysed leagues, and English

Football League Two (level 4 English league) has the lowest home team advantage among analysed leagues. More convincing results are obtained when all ten analysed seasons are used together.

Acknowledgement

This publication was supported by the project LO1506 of the Czech Ministry of Education, Youth and Sports.

References

[1] M. S. ALLEN and M. V. JONES. The home advantage over the first 20 seasons of the English Premier League: Effects of shirt colour, team ability and time trends. *International Journal Of Sport And Exercise Psychology*, 12(1):10–18, 2014.

[2] E. DEZA and M. M. DEZA. *Dictionary of Distances*. Elsevier, 2006.

[3] England Football Results and Betting Odds. Premiership Results and Betting Odds, 2017. http: // www.football-data.co.uk/englandm.php.

[4] Liga de Fútbol Profesional. The official website for Laliga, 2017. http://www.laliga. es/en/ laliga-santander.

[5] P. MAREK and F. VÁVRA. Home Team Advantage in English Premier League. In *Mathsport International 2017 Conference Proceedings*, pages 244–254, 2017.

[6] R. POLLARD and G. POLLARD. Long-term trends in home advantage in professional team sports in North America and England (1876–2003). *Journal of Sports Sciences*, 23(4):337–350, 2005.

[7] Premier League Football News, Fixtures, Scores & Results. Premier League Football Scores, Results & Season Archives, 2017. https://www.premierleague.com/results?co= 1&se=1&cl=-1.

[8] C. R. RAO. *Linear Statistical Inference and its Applications*. John Wiley & Sons, 2002.

Current address

MAREK Patrice, Ing., Ph.D.
European Centre of Excellence NTIS – New Technologies for Information Society
Faculty of Applied Sciences, University of West Bohemia
Univerzitní 8, 301 00 Plzeň, Czech Republic
E-mail: patrke@kma.zcu.cz

VÁVRA František, doc. Ing., CSc.
European Centre of Excellence NTIS – New Technologies for Information Society
Faculty of Applied Sciences, University of West Bohemia
Univerzitní 8, 301 00 Plzeň, Czech Republic
E-mail: vavra@kma.zcu.cz

Slovak University of Technology in Bratislava
Faculty of Mechanical Engineering
17th Conference on Applied Mathematics **APLIMAT 2018**

Proceedings

NETWORK MODEL INSPIRED BY THE REAL DATA – FUNCTIONAL BRAIN NETWORKS

MARKOŠOVÁ Mária (SK)

Abstract. In this paper we studied degree distributions of functional brain networks, which are extracted from the real fMRI measurements of young healthy participants at three different correlation thresholds of voxel activity. To explain how the degree distribution changes with the decreasing threshold, we created a dynamic network model. The model reflects how initially scale-free networks change their structure, manifested in the degree distribution, due to the processes of the network growth and edge addition.

Keywords: Dynamics, Brain, fMRI, Functional brain networks, Dynamic network model

Mathematics Subject Classification: Primary 68R01; Secondary 68R10

1 Introduction

1.1 fMRI

Functional Magnetic Resonance Imaging (fMRI) is a technique capturing high-resolution images of the brain neural activity, based on the blood-oxygen-level-dependent (BOLD) signal [4]. Images in the fMRI method are captured in a series of 3-d slices. A single slice is comprised of a grid of discrete 3D regions known as voxels. Spatial resolution of a single fMRI image is around 3 mm^3 per each voxel. fMRI is an ideal technique for deriving functional brain connectivity, defined by the extent to which spatially distinct regions of the brain exhibit similar behavior over time. By modelling this functional connectivity as a network (functional brain network), we can explore the ways in which regions of the brain interact, and use techniques from the graph theory to evaluate their topological characteristics. Voxels, are represented in a network as nodes, and interaction between voxels as edges [3, 8, 13].

1.2 Graph theoretical concepts

Graph $G(V,E)$ consists of $N=|V|$ nodes, connected by $M=|E|$ edges [18], while V and E denote node and edge sets, respectively. Network is a graph that has some dynamics. Binary network is a network having undirected unweighted edges. If there are no multiple edges and no node is connected to itself in a binary network, it can be represented as a simple graph [18]. All the following definitions are for simple, binary networks. The node degree is the number of edges incident with the node. The degree distribution shows how many nodes in the network has certain degree. To measure clustering of nodes clustering coefficient of the node i is defined as:

$$C_i = \frac{2m_i}{k_i\,(k_i - 1)} \,, \tag{1}$$

where m_i denotes the number of edges among neighbors of the node i and k_i is its degree. Average network clustering coefficient is denoted as C. Average shortest path l, is defined as an average of shortest paths (measured in the number of edges) between all pairs of nodes [18]. As shown by [1] and others [5, 9], real-world networks developed by self-organized processes, have usually scale-free structure. Scale-free property is reflected in the power law degree distribution (2). The normalized number of nodes with the degree k, $P(k)$, decreases as a power law with the scaling exponent γ,

$$P(k) \propto k^{-\gamma} \,. \tag{2}$$

As shown in [1], the scale-free structure develops due to a preferential node linking.

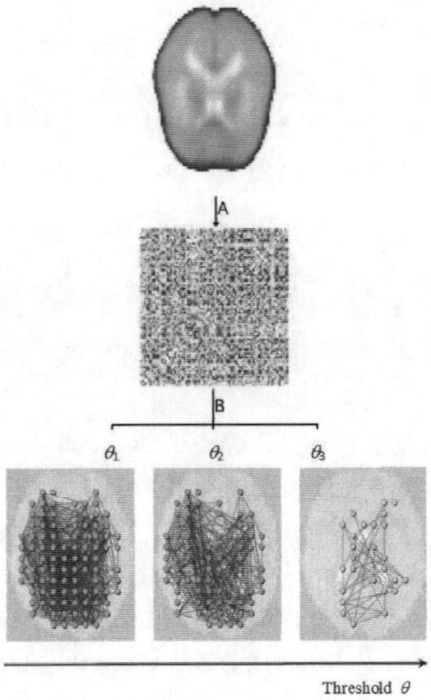

Fig. 1. Steps in functional network creation. A: Correlation matrix for the fMRI volume of each participant. B: From each correlation matrix, three unweighted, undirected networks were created, using three correlation thresholds $\theta_1 \;\pi\; \theta_2 \;\pi\; \theta_3$. So the resulting networks contain different number of nodes and edges.

1.3 The goals

The aim of this paper is to study and model real world networks by means of the graph theoretical methods. For this purpose, the functional brain networks derived from the brain fMRI data of healthy young (HY) participants collected by Buckner et al. [2] were used. We analyze how degree distributions of functional brain networks change due to the correlation threshold changes. Based on this analysis we developed a network model that links the underlying network dynamics with the network structure manifested as the degree distribution at different thresholds. We test how the model fits the data by its numerical simulation.

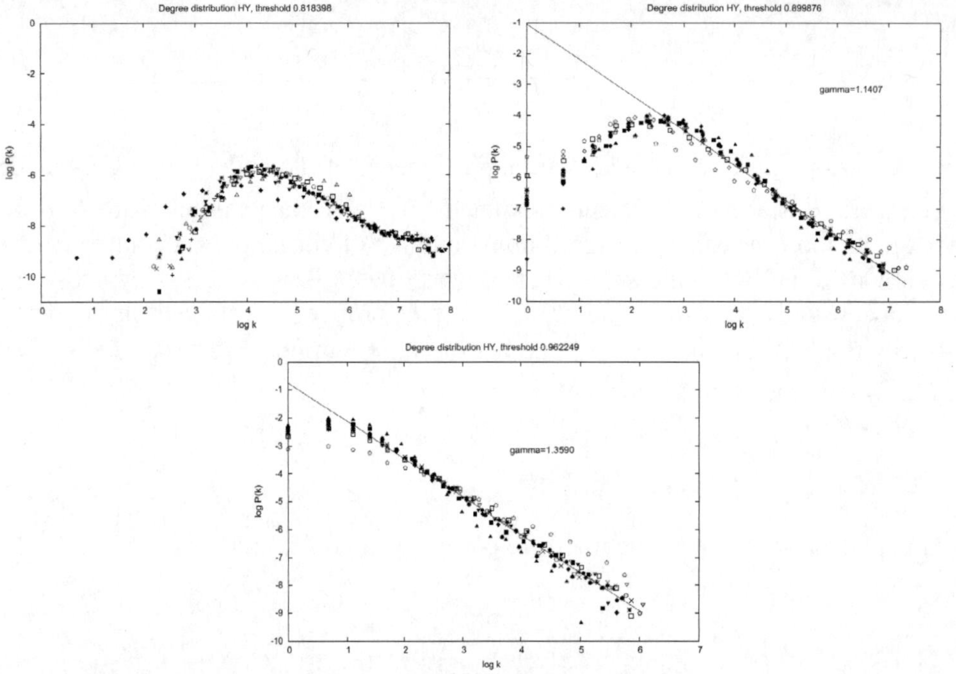

Fig. 2. Degree distributions from left to right: distribution at the threshold $\theta_1 = 0.818398$, the threshold $\theta_2 = 0.899876$, and the threshold $\theta_3 = 0.962249$. See how the scale-free network structure is destroyed, as the correlation threshold gets lower and lower.

2 The data

2.1 The fMRI datasets

The raw fMRI data were collected by Buckner et al. [2] and are publicly available (data set no. 2-2000-118W from the fMRI Data Center: http://www.fmridc.org). Structural and functional MRI data were acquired from 41 subjects in total. The participants were divided into three groups: healthy young (HY) participants, healthy elderly (HE) participants and elderly participants with diagnosed Alzheimer's disease (AE). The HY group used for our studies has 14 subjects (9 females/5 males) with the mean age 21.1 years. The data are preprocessed before being used in this and other studies [13]. Preprocessing applied to the data is described in detail in [11].

2.2 Functional brain networks

For each participant in each group the prepossessed fMRI data were used to create functional brain networks [11]. Functional brain networks, contrary to the neuronal brain networks, are based on the temporal correlation of signals between the voxels over some time. Thus, the functional brain network reflects the functional cooperation of different brain areas. Because the smallest unit of the measured fMRI signal is an integrated signal of the neurons contained in one voxel, voxels are natural candidates for the nodes of the functional brain network. If the two voxels functionally cooperate, the measured BOLD signal is highly correlated over time. The Pearson correlation coefficient is calculated for all the voxel pairs:

$$r(i,j)=\frac{\langle V(i,t)V(j,t)\rangle-\langle V(i,t)\rangle\langle V(j,t)\rangle}{\sigma(V(i))\sigma(V(j))}, \quad (3)$$

where $r(i,j)$ is the correlation coefficient between the pair of nodes, $V(m,t)$, $m=i,j$ is the measured activity in the m-th voxel at time t, $\langle . \rangle$ denotes the time average, and $\sigma^2(V(m))=\langle V^2(m,t)\rangle-\langle V(m,t)\rangle^2$. A link between the voxel pair (nodes) is established, if $|r(i,j)| \phi\ \theta$, where θ is a prescribed correlation threshold (Fig.1). We opted for an absolute value of correlation that is both strongly positively and strongly negatively correlated voxels to be included in the functional network. Thus, the networks are unweighted and undirected, because correlation $r(i,j)$ is a symmetric function.

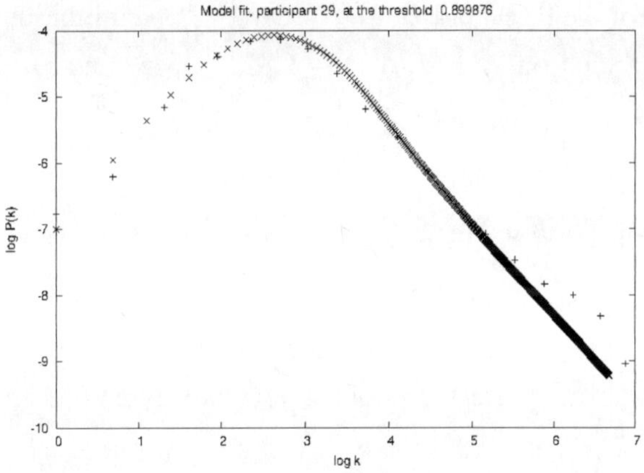

Fig. 3. The best fit for the threshold θ_2. Model fit for the participant No.29, `+' -- data, `x' -- simulation. 800 iterations, b=3.7479, a=1.3958, $b_1=243.14$, $a_1=0.0$, mse=0.086.

2.3 Towards the model of functional network dynamics

We used the degree distributions extracted from the HY functional networks for different correlation thresholds, in order to get an insight into the network changes as the value of the threshold changes from high to low. The data for HY group were chosen because of their coherence and the smallest differences among subjects. Based on this analysis we built our dynamic network model. We are aware, that to speak about "dynamics" in the context of our model is not quite correct, because the independent variable is not time. Nevertheless, we believe, to use this term with respect to the model iteration steps, is not misleading.

Character of the degree distribution is an indicator whether the functional brain networks are scale-free or not. The scale-free property means that the model should incorporate preferential linking, while the loss of the scale-free structure with the changing correlation threshold, means that there are random link additions present too. As the power law degree distribution gradually changes, each intermediate stage reflects the ratio of preferential versus random linking. All of this should be incorporated into the formal mathematical model.

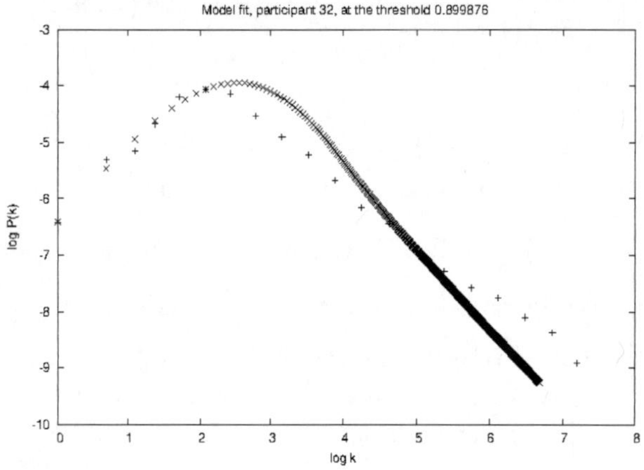

Fig. 4. The worst fit for the threshold θ_2, `+' -- data, `x' -- simulation. Model fit for the participant No. 32 , 800 iterations, $b = 3.61$, $a = 1.87$, $b_1 = 528,52$, $a_1 = 0.0$, $mse = 0.28$.

3 Results

3.1 Data analysis – whole brain network

In this section, a basic network analysis of the measured data is presented. The networks are constructed at the three different correlation thresholds $\theta_1 = 0.819398$, $\theta_2 = 0.899876$ and $\theta_3 = 0.962249$. The reason for these threshold values are explained in [11]. For the lowest correlation threshold $\theta_1 = 0.819398$, the average number of network nodes is $N_1^{HY} = 10487$ (interval of the individual values is $[9850, 11140]$). The average group clustering coefficient is $C_1^{HY} = 0.4594$; and the average group shortest path is $l_1^{HY} = 2.5677$. All individual values of the average clustering coefficients are in the interval $[0.4029, 0.5593]$ and that of the shortest paths are in the interval $[2.3519, 2.6877]$, respectively. For the lowest correlation threshold, the degree distribution does not have a power law character (Fig. 2).

For the threshold $\theta_2 = 0.899876$, the group average number of nodes is somewhat lower, i.e. $N_2^{HY} = 10426$ with the individual values in the interval $[9842, 11140]$. The average group clustering coefficient is $C_2^{HY} = 0.3982$, with individual values in the interval $[0.36128, 0.4822]$ and the average group shortest path is $l_2^{HY} = 3.7981$ with the individual values in the interval $[3.3168, 4.015]$. The degree distribution reveals a more pronounced power law tail, with the average group exponent $\gamma_2^{HY} = 1.41$ (see Fig. 2).

The situation changes for the correlation threshold $\theta_3 = 0.962249$. The functional brain networks have a well-defined scale-free structure, reflected in the power law degree distribution with the average scaling exponent $\gamma_3^{HY} = 1.36$ (Fig. 2) with the individual values in the interval $[0.9346,\ 1.7812]$. Other characteristics are $N_3^{HY} = 7280$, $C_3^{HY} = 0.3616$ and $l_3^{HY} = 7.3596$, with the individual values in the intervals $[6482, 8582]$, $[0.3366, 0.4049]$, and $[5.6129,\ 9.6049]$, respectively.

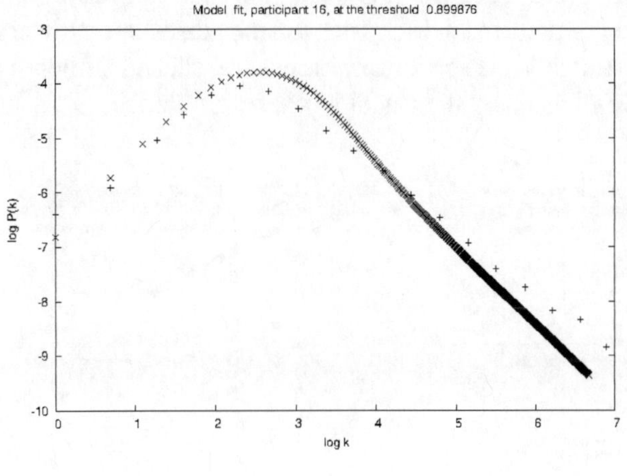

Fig. 5. Typical fit for the threshold θ_2, `+' -- data, `x' -- simulation. Model fit for the participant No. 16, 800 iterations, $b = 4.9$, $a = 0.0$, $b_1 = 255.48$, $a_1 = 0.0$, $mse = 0.19$.

4 The model

Recently, several papers proposed models of functional brain networks using various principles [7, 14, 16]. Our data analysis has shown that the properties of the functional brain networks depend mainly on the correlation threshold. For the lower correlation thresholds, the degree distributions do not have a power law character, but rather have a shape more resembling the degree distribution of the random graph [18]. As the threshold increases, the distributions develop power law tails (Fig. 2). Similar scenario is described in Scholz et al. [15] for the noisy scale-free networks. The authors started from an initially scale free network, disturbed by several types of noise: i.e. random link removal, random link exchange and random link addition. They have studied how the degree distribution drifts from the power law character with the increasing noise.

We adopted the idea of the random link addition noise to reflect degree distribution changes while lowering the threshold of correlation. Unlike the original model, we allow the network to grow and we also allow preferential link addition. We suppose, it is possible to lower the threshold by such a slow way, that at each threshold jump one new node comes to the system. Thus, each network growth step or iteration is marked by an addition of a new node. The situation, can be described as follows: One starts at the highest threshold θ_3, where the network is scale-free having N_0 nodes, L_0 edges and the power law degree distribution. Then the threshold is lowered. New nodes and edges are added to the network by both -

preferential and random linking. New node is added due to the fact, that the signal of such a new node (voxel) is now correlated with some node (voxel) signal already present in the network, which were not previously correlated at the higher threshold. Each new node brings a_1 new edges, which are linked preferentially and a edges linked randomly. Simultaneously another process takes place. As the threshold decreases, some correlations between the pairs of nodes already present in the network become significant. Therefore new edges are distributed randomly (b) and preferentially (b_1) among the nodes being already in the network. Thus, our network grows in the number of nodes and edges. Because the network at the highest threshold is scale-free, we suppose, that the real correlations between voxels construct the scale-free structure, which is, as the threshold lowers, disturbed by the accidental correlations (edges). These correlations are caused either by the real similarity between the two voxel signals or by their accidental resemblance.

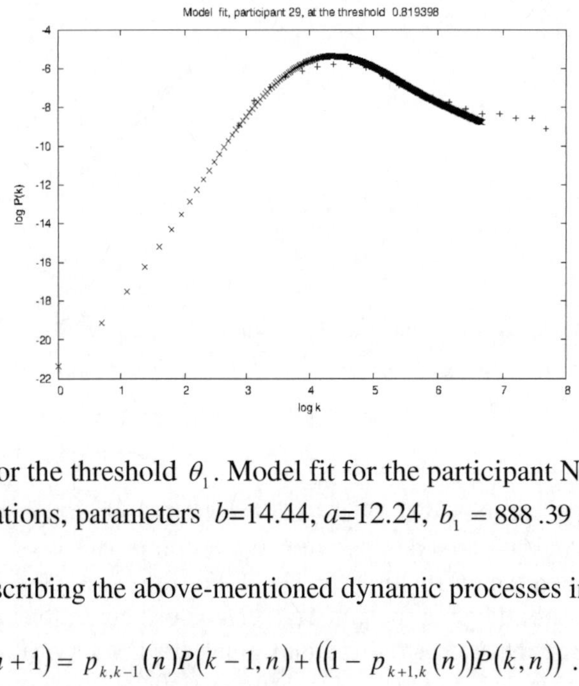

Fig. 6. The best fit for the threshold θ_1. Model fit for the participant No. 29, `+' -- data, `x' – simulation, 800 iterations, parameters *b*=14.44, *a*=12.24, $b_1 = 888.39$, $a_1 = 1.77$, *mse*=0.29.

The rate equation describing the above-mentioned dynamic processes in the model is:

$$P(k,n+1) = p_{k,k-1}(n)P(k-1,n) + ((1 - p_{k+1,k}(n))P(k,n)) . \qquad 4$$

The transition term $p_{k,k-1}(n)$ is defined as

$$p_{k,k-1}(n) = \frac{a+2b}{N_0+n} + \frac{(a_1+2b_1)(k-1)}{2L_0 + A(n)} , \qquad 5$$

where $A(n) = 2n(a+b+a_1+b_1)$. In equation (4), $P(k,n)$ is the normalized number of nodes having the degree *k* at the iteration *n*. At each iteration, this number changes, due to the fact, that some nodes having at the previous iteration *n* the degree *k-1*, gain a new edge. This is expressed in the first term of the equation (4). The second term expresses, that some nodes have the degree *k* already at iteration *n* and with the probability $1 - p_{k+1,k}(n)$ no new edges are added. In equation (5), N_0, L_0 denote initial number of nodes and edges, a, b are the number of randomly added edges per iteration, where a is the number of edges fetched by a new coming node and b is the number of edges added between older network nodes. Similarly, a_1 denotes the number of edges by which a new node links preferentially to the network and b_1 is the number of edges linking older nodes preferentially. Factor two denotes, that some edges are linked by both of their ends, unlike those edges, which have one edge end linked to the new coming node. The transition term $p_{k+1,k}(n)$ describes how the number of nodes having the degree k changes due to the above-mentioned dynamic processes. The first term of $p_{k+1,k}(n)$ says that $a+2b$ edge ends are linked randomly, with equal probability $\frac{1}{N_0+n}$ to the node in the network. The number of nodes in the network at iteration n is N_0+n, because, as stated before, at each iteration (threshold step) exactly one new node appears. The second term of $p_{k+1,k}(n)$ describes preferential attachment of a_1+2b_1 edge ends with the probability proportional to the node degree. The normalization factor is a sum of all node degrees $2L_0 + A(n)$.

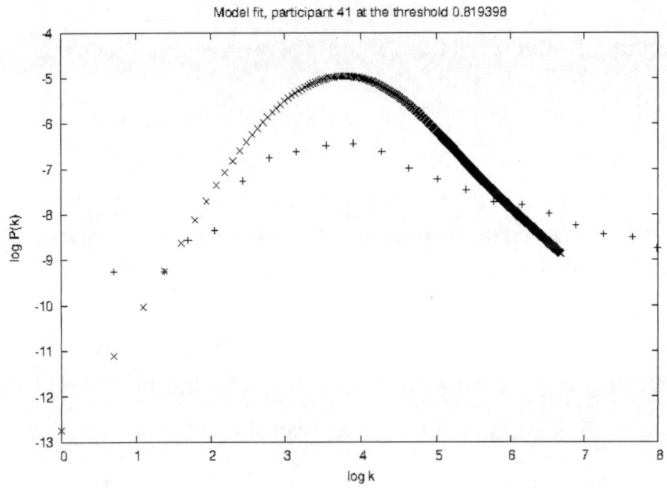

Fig. 7. The worst fit for the threshold θ_1. Model fit for the participant No 41, `+' -- data, `x' -- simulation. 800 iterations, $b = 6.97$, $a = 6.88$, $b_1 = 824.46$, $a_1 = 0.06$, $mse = 1.38$.

5 Results of numerical simulations

We simulated the proposed model (4, 5) numerically. Each simulation was done for all appropriate functional brain HY networks and compared to the data at the thresholds θ_2 and θ_1. Here we present the best, the worst and the typical fit for the threshold θ_2 (Fig 3, Fig 4, Fig 5) and the same for the threshold θ_1 (Fig 6, Fig 7, Fig 8).

First, we used the experimental data to find the parameters in the power law distributions at threshold θ_3 (6). This threshold corresponds to the initial number of iterations $n_0 = 0$. Thus, the initial degree distribution is power law i.e.:

$$P(k) = ck^{-\gamma}. \tag{6}$$

Both parameters c and γ are derived from the data. The power law distribution at the highest threshold was normalized by the constant z calculated from the equation

$$z = \int_{1}^{\infty} P(k)dk, \tag{7}$$

and it was checked whether the sum of all probabilities of the initial distribution is close to one after the normalization. We excluded the networks, for which the integral (7) does not converge (case $0.0 \pi \gamma \pi 1.0$). The model (4, 5) was first applied to model the transition between the two highest correlation thresholds θ_3 and θ_2. Each model has been iterated $N_2 - N_0$ times for the defined set of parameters a, a_1, b, b_1. N_0, N_2 denote the number of nodes at the initial network state (threshold θ_3) and at the network state corresponding to the correlation threshold θ_2, respectively. In each network growth step (a discrete small threshold change), a fixed number of edges is added, namely $\frac{L_2 - L_0}{N_2 - N_0}$, where L_2 is the number of network edges gained from the measured data at the threshold θ_2 and L_0 is the initial number of edges. To find the best set of parameters a, b, a_1, b_1 we used the hill climbing algorithm, in which the mean square error between the measured and simulated data is calculated. From the best fit parameters in the current simulation, fifteen new sets of parameters were derived by the slight perturbations of the currently best fit parameter set. This is a standard procedure in the hill climbing algorithm. The hill climbing algorithm was iterated 800 times. Second, the same job as before has been done to model the data at the threshold θ_1. The only difference is, that the model was iterated $N_1 - N_0$ times, where N_1 is the number of nodes at the threshold θ_1. Also the number of edges added in each threshold (network growth) step is different, namely $\frac{L_1 - L_0}{N_1 - N_0}$, where L_1 is the number of edges in the functional brain network created at the lowest threshold θ_1.

5. 1 Results for the HY group

In the HY group we excluded one participant (number 39), since the γ exponent is such that the normalization (7) does not converge. In Fig 3, the best fit at the θ_2 threshold is presented. The participant number is 29 and for this participant the overall best fits have been achieved. We got the worst fit at the θ_2 threshold for the participant number 32 (Fig.4). For illustration also a typical results (Fig.5) in the HY group are presented. Next results are for the lowest threshold θ_1 (Fig.6). We show the best fits again for the participant (No. 29). The worst fit has been achieved for the participant 41 (Fig. 7). The typical results at this threshold are presented at Fig. 8.

6 Summary

In this paper, the degree distributions of functional brain networks were analyzed. We found that:

1. At the threshold θ_3 the degree distributions are scale-free (Fig.2). This is in agreement with previous studies [10, 6] on different fMRI data sets.

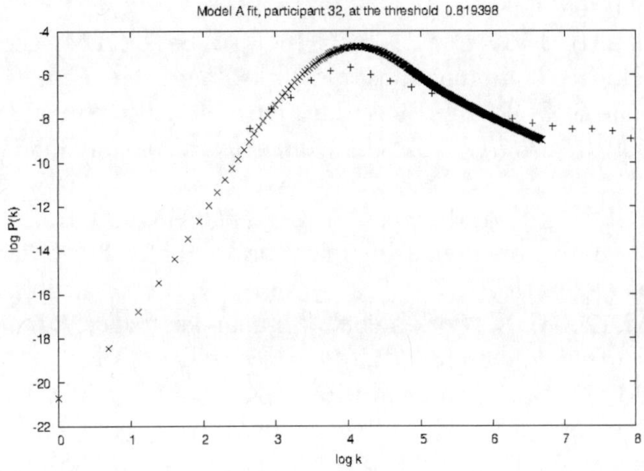

Fig. 8. The typical fit for the threshold θ_1. Model fit for the participant No. 32, `+' -- data, `x' -- simulation. 800 iterations, $b = 22.43$, a=5.34, $b_1 = 867.64$, $a_1 = 0.97$, *mse*=0.87.

2. The character of the degree distribution changes with the threshold. As the threshold decreases, the degree distributions are losing their scale-free character.
3. There are individual differences in the degree distributions at each threshold.
4. Our model is successful in fitting the different correlation thresholds.

The distribution analysis formed a prerequisite for the model addressing the question which dynamical processes causes the scale-free functional brain networks to change their structure.

In conclusion, we created a mathematical models of noisy growing networks inspired by the behavior of functional brain networks derived from the fMRI brain scans of healthy young participants. This research was supported by VEGA grant 1/0039/17. Preliminary work was done at the University of Otago, Dunedin, New Zealand, at the Department of Computer Science. We are grateful for their financial and technical support. I thank Paul McCarthy for providing functional brain networks used in this study and for the parts of Fig. 1. I am grateful to Ľ. Beňušková and B. Rudolf for careful reading of the manuscript.

References

[1] BARABÁSI A. L. and ALBERT. R. (1999). Emergence of scaling in random networks. *Science* 286 (5439), 509--512.

[2] BUCKNER R. L., SNYDER A. Z., SANDERS A. L., RAICHLE M. E., and MORRIS J. C. (2000). Functional brain imaging of young, nondemented, and demented older adults. *Journal of Cognitive Neuroscience* 12 (Supplement 2), 24--34.

[3] BULLMORE E., BRAMMER M., WILLIAMS S.C., RABE-HESKETH S., JANOT N., DAVID A., MELLERS J., HOWARD R., SHAM P. ,(1996). Statistical methods of estimation and inference for functional MR image analysis. *Magn. Reson. Med.* 35(2), 261--277.

[4] BUXTON R. B. (2009). *Introduction to Functional Magnetic Resonance Imaging: Principles and Technique (second edition).* Cambridge Univ. Press, Cambridge.

[5] DOROGOVTSEV S. N., and MENDES J. F. F. (2003). *Evolution of Networks*, Oxford University Press, Oxford U.K.

[6] EGUÍLUZ V. M., CHIALVO D. R., CECCHI G. A., BALIKI M., and APAKARIAN, A. V. (2005). Scale-free brain functional networks. *Phys. Rev. Lett.* 94, 018102.

[7] GLEISER P. M., and SPOORMAKER V. I. (2010). Modelling hierarchical structure in functional brain network. *Philos. Trans. A Math. Phys. Eng. Sci.* 368 (1933), 5633--5644.

[8] HORWITZ B., MCINTOSH A. R., HAXBY J. V., and GRADY C. L. (1995). Network analysis of brain cognitive function using metabolic and blood flow data. *Behavioural Brain Research* 66 (1–2), 187–193.

[9] MARKOŠOVÁ M. (2008). Network model of human language. *Physica A: Statistical Mechanics and its Applications* 387 (2), 661--666.

[10] MARKOŠOVÁ M., FRANZ E. A., and BEŇUŠKOVÁ L. (2009). Topology of brain functional networks: towards the role of genes. In: *Advances in Neuro-Information Processing,* Volume 5506, Chapter Lecture Notes in Computer Science, eds M. Koppen, N. Kasabov,and G. Coghill (Berlin: Springer-Verlag, 111--118.

[11] MCCARTHY P., BEŇUŠKOVÁ L., FRANZ E. A. (2013). Functional network analysis of aging and Alzheimer's disease: Results. *Technical Report OUCS-2013-12*, University of Otago, New Zealand.

[12] MCCARTHY P., BEŇUŠKOVÁ L., FRANZ E.A. (2014). The age-related posterior-anterior shift as revealed by voxel wise analysis of functional brain networks. *Frontiers in Aging Neuroscience* 6:301. doi: 10.3389/ fnagi.2014.00301

[13] MCINTOSH A. R., and GONZALES-LIMA F. (1994). Structural equation modeling and its application to network analysis in functional brain imaging. *Human Brain Mapping* 2 (2), 2-22.

[14] PORTILLO I. J. G., and GLEISER P. M. (2009). An adaptive complex network model for brain functional networks. *PLoS One* 4 (9), e6863. doi:10.1371, 0006863.

[15] SCHOLZ J., DEJORI M., STETTER M., GREINER M. (2005). Noisy scale-free networks. *Physica* A 350 (2-4), 622--642.

[16] VÉRTES P. E., ALEXANDER- BLOCH A. F., GOGTAY N., GIEDD J. N., RAPPORT J. L., and BULLMORE E. T. (2012). Simple models of human brain functional networks. *Proc. Natl. Acad. Sci. USA* 109 (15), 5868--5873.

[17] WATTS D. J. and STROGATZ S. H. (1998). Collective dynamics of "small world" networks. *Nature* 393, 440--442.

[18] WEST D. B. (2001). *Introduction to Graph Theory*. Prentice Hall UK, London.

Markošová Mária, doc. RNDr., PhD.
Department of Applied Informatics, Faculty of Mathematics, Physics and Informatics
Comenius University in Bratislava,
Mlynská dolina, 842 48 Bratislava, Slovak Republic
E-mail: markosova@ii.fmph.uniba.sk

Slovak University of Technology in Bratislava
Faculty of Mechanical Engineering
17th Conference on Applied Mathematics **APLIMAT 2018**

Proceedings

SQUARE AND TRIANGULAR MATRICES IN MAGNAGHI-TERZAGHI'S ARCHITECTURES

MELE Giampiero (IT), CAPANNA Alessandra (IT)

Abstract. The encounter, in the past editions of Aplimat Conference, with Paola Magnaghi Delfino, daughter of Augusto Magnaghi who, together with his friend and fellow student Mario, founded the Studio Magnaghi-Terzaghi, was an opportunity to start a systematic research on the drawings and projects carried out by these two architects, collaborators of Pietro Lingeri and Giuseppe Terragni, that were two of the most important exponents of Modern Architecture in Milan. The modern architecture created by the two architects is strongly mathematical in its conception.

Keywords: modern architecture, geometry, geometrical layout, square, equilateral triangle, architectural shape

Mathematics Subject Classification: Primary 00A05, 51F05, 97G99; Secondary 97F90

1 Introduction

The encounter, in the past editions of Aplimat Conference, with Paola Magnaghi Delfino, daughter of Augusto Magnaghi who, together with his friend and fellow student Mario, founded the Studio Magnaghi-Terzaghi, was an opportunity to start a systematic research on the drawings and projects carried out by these two architects, collaborators of Pietro Lingeri and Giuseppe Terragni, that were two of the most important exponents of Modern Architecture in Milan.

What at first came to light, in addition to the quality in itself of their architectural works, is a systematic use of the geometry as a basic layout on which they structured their projects. As a matter of facts, architectural structures based on a geometric form have ancient roots, adopted with continuity throughout history up to modern times.

The rational composition of the geometric elements comes from the evolution of the Vitruvian precepts in which we read the thesis endorsed also by Le Corbusier for which the beauty of a building consists in the rational integration of all the parts, such that nothing can be added or eliminated without destroying the harmony of the whole.

If, however, in the past, the composition of codified orders and ornaments was defined by interpreting geometrical objects as linguistic elements, Le Corbusier's system from which a large part of the so-called rational architecture is born, sets the limits of a compositional process regulating the possibility to order and precisely proportion the project as indoor and outdoor spaces as a whole. Consequently, in the architectures we are analyzing geometry more than a language, is a method.

2 Square Matrices

In the year of their graduation (1939), the architects Magnaghi and Terzaghi received the assignment from Mr Cicogna to design a residential building from then on known as the Casa dei Nidi [2] (House of the nests).
Various History of Architecture books quote this building as one of the first examples of Italian Modern Architecture, yet the writings have always returned to a single article written by Giuseppe Pagano in 1940 and published in Casabella n. 150 [4] (Fig. 1).
The drawings published in this article have been the starting point for an accurate redrawing, followed by a comparison with the actual survey. The obtained model was thereafter analyzed with the aim to identify the geometric matrices on which the architects based the project of the building.
The 80 cm-sided square is the module that the architects used to design this architecture. The plan is incorporated into a double square made by 11,5 x 22 modules and the elevation is 22 x 16 modules. The 0.80 m module has an obvious convenience for the control of the façade because the distance from the storey of the floor below to the floor above is 3.20 m, equal to 4 modules.
The control of the project is given both by the tessellation of the plan with squares and of the space with cubes of side equal to 0.80 m. The general measures of the plan are 9.20 x 17.60 = 161.92 square meters (11.5 x 22 = 253 modules). The total volume of the building, all inclusive (calculated empty to full) is 161.92 x 12.80 = 2072.576 cubic meters (253 x 16 = 4048 modules). The gross surface area of 6 apartments from the first to the third floor is 66,24 m (9 x 11.5 = 103.5 modules). A semi-open staircase connecting the different levels of the building is placed in a space of 6.5 x 4 = 26 modules equal to 3.20 x 5.20 = 16.64 square meters. The apartments consist of two rooms and a kitchen. The rationality of the scheme is the essential tool to control the project in all its parts.
The "ribbons" of load-bearing masonry on which the slabs lie are ½ module wide (40 cm) and allow to have apartments free from load-bearing structures visible in the interiors, as well as the possibility of having proportioned openings measured according to a coherent scheme that does not betray the quality of the design philosophy.
The openings on the main façade are equal to 3 modules (2 + 1) with sliding shutters fitting exactly in the filled part of the wall. The ground floor originally had two single-span porches on the front side of building, in plan composed of 9 x 6.5 modules (7.20 x 5.20 = 37.44 square meters), which almost immediately have been closed to obtain two additional housing units. In the lower area of the back side of the building, 6 garages were placed, one for each original housing unit.
As Pagano wrote «from this healthy realism, this very first construction by Magnaghi and Terzaghi was born, and it is thanks to their faith and their intelligence if this house is not an "ordinary house"» [4]. The simple volumetric value of this composition guarantees quality and cost effectiveness for the intervention. Starting from the square, vertical and horizontal rectangles are obtained. In this way the architects managed a sequence of light-dark, full and empty relationships that make this building one of the examples of Italian rationalism. The square grid is the tool used by the two Milanese architects to generate an innovative

architecture that is perfectly in line with their times or even, for the period in which it was built, a cutting-edge project. In fact, the aesthetic rhythm of this architecture is based on a geometric grid that generates pure forms. This grid is referred to the classic *ad quadratum* background, so deep-rooted in the Italian tradition to constitute the basis of the new Italian Rationalist Architecture as theorized by the "Gruppo 7".

Fig. 1. Plans, elevations, sections [4], and photos of Casa dei Nidi at Fino Mornasco (CO) Italy.

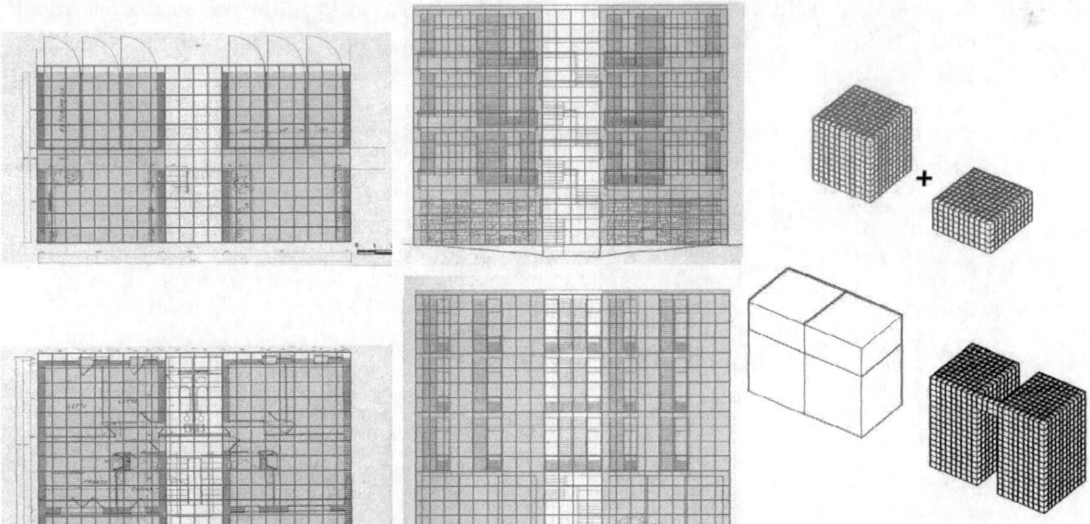

Fig. 2. Grid and geometric schemes for the definition of Casa dei Nidi.

Also in the Bica-Montecatini office building in Via San Giovanni sul Muro – Milano [2], designed in 1955, the two architects chose the square as a basic module by adopting a 3,34 x 3,34 meter grid that generates the shape in plan and scans the rhythm of the façade. The plan of the office building is set on a 5:14 ratio grid and the main elevation (front view from Via San Giovanni sul Muro) is a rectangle of 8:14 ratio. As already mentioned, the starting module for a first sizing of the building is the square with side equal to 3.34 meters which is reduced by a few centimeters following the adjustments due to the presence of the via Giulini passage that crosses the building and in particular interrupts the rhythm of the ground floor and of the first floor. In this two levels are located a number of commercial activities.
The structure of the main façade towards Via San Giovanni sul Muro presents a protrusion of 0.3 meters wide that emphasizes the verticality of the building and of the grid. The measurement of the span rating of the pillars is the same as the side of the module chosen to draw the building (3.34 meters). The number of pillars is not equal to the number of modules: they are 13 instead of 15 because of the presence of the span without pillars under which the road passes through. This reduction has been wisely incorporated by the measured drawing of the façade. Indeed, 3.34 x 14 would give a total length of 46.76 meters, but due to the lack of the two pillars the real length of the façade is 46.16 meters (2 x 0.3 = 0.60; 46.76 - 0, 60 = 46.16 meters). The size of the aluminum window-frames that characterize the façade is a consequence of the choice of the module and the dimension of the pillars. The size of the window frames, equal to 1,033 meters, derives from the subdivision of the measure of the module minus that of the pillar (3.34 - 0.30 = 3.04 meters, 3.04 / 3 = 1.033). Therefore the window frame has a ratio of 1: 3. In height the square shape of the module is emphasized by the non-transparent window part that covers the sides of the slabs on each different floor. As previously mentioned, the size of the window and the lack of two pillars cause a reduction in the overall length of the project. This length has been cleverly rationalized, in fact, its calculation is simple when the structure of the matrix that generated the project is known. If we consider, in fact, the number of spaces between the pillars (11) and the measure of the span (3.34 mt), the total length is 3.34 x 11 = 36.74. To this measure we should add the span of 9.42 meters and the width of a pillar 0.30 meters (0.15 x 2, i.e. ½ pillar per side). The result is equal to 46.46 meters. It could be precisely 46.76 if the width of a pillar was that theorized in the drawing.

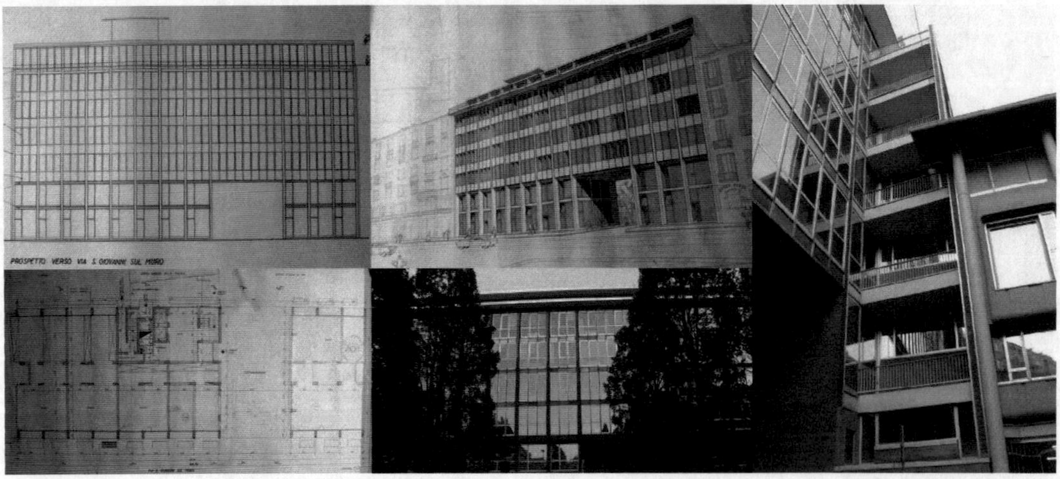

Fig. 3. Plan, elevation, perspective view [2], and photos of BICA Montecatini office building, via San Giovanni sul Muro – Milano.

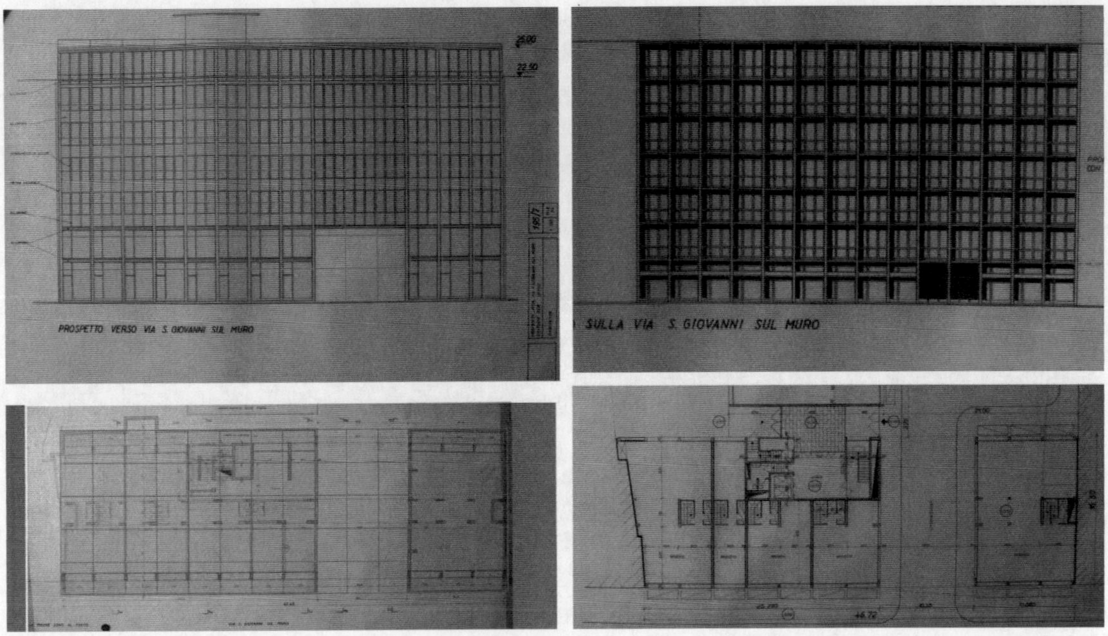

Fig. 4. Square grids generating the general shape of BICA Montecatini office building [2].

The aim of the present description is to explain the method used by Magnaghi and Terzaghi to outline the project of this building, starting from the square grid. It is a process of rationalization of the measure that contributes to the construction of the shape of the building for the Beni Immobili Civili Agricoli Montecatini company whose offices are located from the second to the seventh floor. The top floor is set back on all sides of the building and its height reaches the value identified in the grid. The façade on Via Polezza has the same characteristics of the front on Via San Giovanni sul Muro with two different aspects: the insertion of the glass block in correspondence of the stairwell - a material that the two architects had never used before this period - and the advanced position of the stairway, with respect to the façade, memory of a previous unrealized project that included a building orthogonal to the constructed one. This office building as well can be defined as rational because the method for its design is highly similar to that of the house in Fino Mornasco. The collaboration, when still students, with the architects Lingeri, Cattaneo and Origoni and the awards obtained in 1936 and 1938 [2] are evidence that even at that time Magnaghi and Terzaghi mastered this method, learned from the rationalist masters.

3 Triangular Matrices

On the other hand, in the projects for some religious buildings, (the Cagnola villa church in Gazzada and San Filippo Neri in Bovisasca district in Milan, designed respectively in 1959 and 1961), instead of the geometry of the square, they used the equilateral triangle at the basis of the matrix generating the shape.

This regular geometric figure, closely linked to the number three, represents both the concept of stability and the religious concept of the Trinity,

In fact, the triangle, in the biblical symbolic language, represents the transcendence of God and the Trinity. The latter should be considered the absolute postulate of the New Testament that professes one God in three persons. This is why, as an elementary geometric figure, the equilateral triangle is often used to generate the form of sacred spaces and, above all in the Baroque period, a composition of triangles becomes a generator of planimetric schemes with a high symbolic value. In the form of the intersection of two equilateral triangles which determines the figure of the Star of David, it generated a whole series of important monuments: among them it is enough to remember the Borromini's church of S. Ivo alla Sapienza, where the concept of wisdom (Sapienza means Wisdom and is the name of the ancient academic institution), inherent in the geometrical figure of the 6-pointed star, becomes the basic form of the plant. The triangle therefore, and in particular the equilateral triangle, is considered the symbol of an absolute perfection, both material and spiritual, and has been found since ancient times in all religions. With the vertex upwards or downwards it indicates the four elements (fire, air, water, earth), which once again, in the fusion of the two symbols in the one quoted above, with the meaning of "wisdom" is in the ancient iconography as a seal in the ring of King Solomon.

As regards instead the hexagonal plan, it had the precise symbolic function related to the creative force of God: the hexagon is, in fact, composed of six triangles that allude to the Trinity and to the number of days of creation. Hexagonal plans in ancient times were adopted for central-plan systems, typical of mausoleums, baptisteries, imperial chapels or temples and churches of great importance, all declined with a close symbolic relationship with the shape of the plan, which was more often octagonal or circular.

In the modern design of the churches the questions of setting up the plan and related spaces, maintained an underlying continuity with the classical symbology, but subordinating it to functional choices and of architectural language. From the latter, an important component was the renewed interest in the adoption of geometric and arithmetic proportional relationships that derive from rationalist theories, but also from the organic architecture. While the rationalist research developed within the Modern Movement get back to the golden ratio as the basis of Le Corbusier's Modulor or of the research on the measure of the Raumpan of Loos or of the existenzminimum and of the researches born within the Bauhaus, the organic architects, in Europe and America (from Alvaar Aalto to Frank Lloyd Wright) adopted a formative process that instead considered the creative event as an extension of the action of nature, which works according to the same geometric rules. The subtle difference compared to the method previously described concerns the fact that in order to reach the completed work no rules defined a priori are applied, but the discovery of its own is identified from time to time; only in this way creativity and production go hand in hand shaping the composition "according to such a practice that while it does, it invents the way of doing" [7].

Both, therefore, refer to an ordering principle as an indispensable condition of the creative moment, which remains, at times, linked to the a priori search for creative action as the reason for the composition.

The organization of the modern project spaces, abandoned the safe way of the systematic adoption of the proportional rules, that guaranteed a "beauty-just-like-nature architecture", is subjected to the analysis and the "a posteriori" interpretation; a literary, narrative, sometimes post-planning exercise, which is the principle of a self-reproducing architecture that ends up identifying its model in similar architectural objects, in the questionable logic of references and citations. It is therefore necessary to reason on the compositional structure as a place of regulated expression not so much and not more by the use of data elements, linguistic codes, prefixed forms, but by the mathematical concepts that regulate the making and, if we accept the pregnant image of "aesthetics of the formativity", of the constitutive simultaneity of idea and form, the act of thinking [7].

For the purpose of our research it is important to set some dates of the revival of the theory of proportions applied to architecture: in particular the decade of the 1940s was crucial. The discourse emanated from two different sources: one established by art and architectural historians, the other by practicing architects, the former with analytical goals, the latter with creative purposes. Wittkower's theory of harmonic proportions was first published in 1949, in a chapter of his seminal Architectural Principles in the Age of Humanism [6] and in 1951 he participated in the IX Triennale in Milan, entitled 'Divina Proporzione'. The projects we are analyzing are born in this context. For the two churches, which were built between 1959-63 and 1961-64, at the turn of the liturgical reform of the Second Vatican Council that changed the functional and spatial relationship of the altar and the assembly, the adopted geometric system allows us to make further reflections.

We have seen that the hexagonal plan by its nature generates a central symmetry system. Still typical of the Orthodox Christian rite or of different religions such as the Muslim, the Catholic churches over time have adopted a basilica system, apparently in contrast with a tessellation of the plan with the exact triangular shape. It should be noted, however, that many post-conciliar churches designed on the basis of a composition of triangles also have an organization of the basilical liturgical action within a space that is geometrically similar to that generating a central symmetry. After careful analysis, many of these architectures adopt isosceles triangles, derived from the intersection of squares along the diagonal, then with the base angles of 45 degrees and the consequent rotation of the axis joining the altar with the entry of the church that is no longer parallel to the perimeter of the building, but diagonal. This produces a dynamic effect that represents, once again in a symbolic and metaphorical form, the changed relationship of man towards God, which is no longer passive and static, but active.

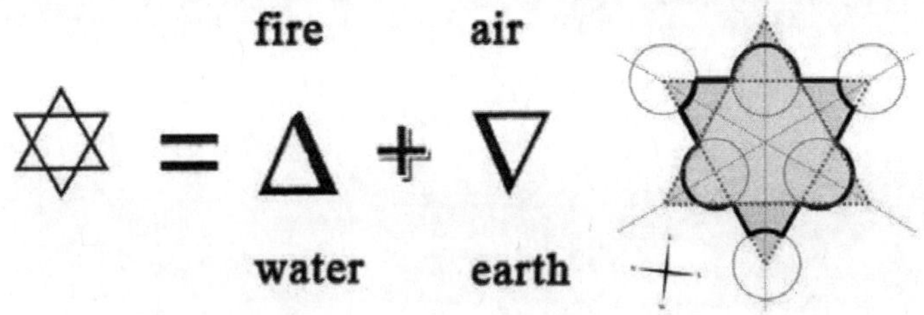

Fig. 5. The symbols related to the equilateral triangles and a compositional scheme generating the plan of Borromini's Sant'Ivo alla Sapienza in Rome.

In the churches we are analyzing this character is also evident in the adoption of a tessellation of the plan made with the equilateral triangle. In the case of the church of Villa Cagnola which is of basilica type, the covering of the plan, as we shall see, determines the architectural-landscaping solution described below, with the exception of mediating the rectilinear perimeter of the wall with the need to choose how to finish the composition arriving at the edge of the rectangular hall: that is, to choose to halve the series of triangles along the height intercepted by the long side of the rectangle, or to use volumetrically this exception. For the large hexagonal hall of the church of San Filippo Neri, however, it will be precisely the position of the accessory spaces to entail a sort of hybridization of the distributional-functional system with the geometrical-spatial scheme.

The church of Villa Cagnola has a hall-plan with a fully windowed wall that opens the view from the altar to the pre-existing historical park with a secular Lebanon cedar. The project turns its attention to the old tree and does it by opening one of the long sides of the church. A series of equilateral triangles, in plan, have one of the sides consisting of a full-height wall and the other consisting of a stained glass window, which allows you to see in transparency the magnificent tree that in some way reminds the biblical tree of the life. On the opposite side of the church, the continuous wall is composed as a succession of full vertical panels, positioned across the reinforced concrete beams that support the roof and overhang from the façade. Under the beams there are long windows that reach the floor. The walls are completely covered with brown tesserae. The plan of the church is rectangular. The ratio of the rectangle is 24:43. The side of the equilateral triangles of the open wall towards the garden is 8 modules. The height of each equilateral triangle of side 8 is 6,9282. The thickness of the walls is equal to 1 module. Also in this case the number 8 is linked to 3 and 6. In the Christian religion these three numbers are meaningful from the symbolic point of view: 6 is the number of creation, 8 is related to the resurrection and 3 to the Trinity.

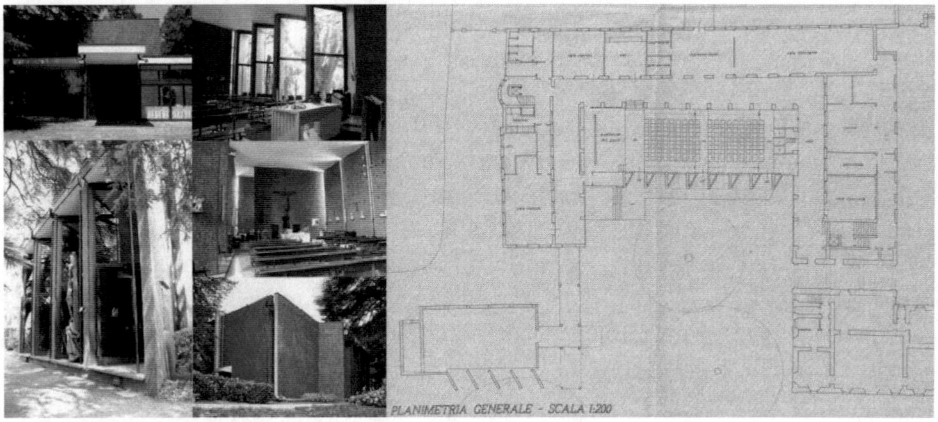

Fig. 6. Photos of the church of villa Cagnogla at Gazzada (VA) and general plan.

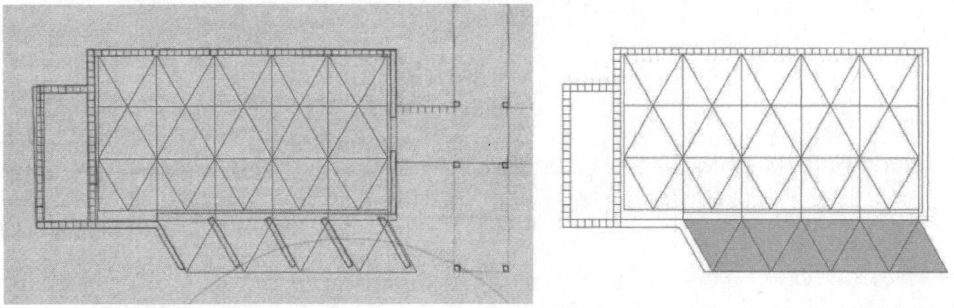

Fig. 7. Geometric studies of the church of villa Cagnogla.

In fact, if we divide the short side of the rectangle of the church plan into three, we get the side of the square generating the plan of the portico connecting the church to the remaining part of the structure of the villa Cagnola. The number of equilateral triangles that generates the open wall is 8. The ratio of the rectangle that composes the main façade is 16:24. The sacred building is a single hall, and is developed like a simple volume at a constant height with a small sacristy on the back of the same height as the church with a rectangular plan of ratio 8:19. Outside, the long façade is punctuated by a series of gargoyles of 1 module that are separated from each other by 7 modules. The church, even in the simplicity of the lines, shows a rational geometry that binds number, form and measure to create that ascensional path that brings the spirit to God. The interior space is rich in symbols while reflecting simplicity and geometric purity.

The church of San Filippo Neri has a central hexagonal plan with a series of annexes generated by the same geometrical shape based as well on the geometry of the equilateral triangle. The reference to regular polygons as the principle of the form and the related traditional value of geometric "perfection," becomes a model of symbolic rationality, and the courtly instrument of conception of the sacred space.

The church built in 1963 presents a uniform height with annexed hexagonal spaces around the perimeter. On the right of the entrance is the weekday chapel, on the left the baptistery and another little chapel with a secondary altar. Natural light pours from above through 6 semi-transparent plastic domes with a slight underwater twilight effect. A further series of 13 stained glass windows are situated in the upper part of the vertical structure representing the stations of the Via Crucis.

The basic module that allows the description of the project is the equilateral triangle of side equal to 4.66 meters. The relationship between the side of the basic module and that of the large hexagon of the church's hall plan is 1:4. The weekday chapel and the entrance porch are generated by the composition of 18 triangular modules (6x3). The chapel and the baptistery are made out of 6 modules that generate the hexagons of each plan and are connected by a space composed of 3 modules which generates a half hexagon. The church's hexagon consists of 144 triangular modules. The height of the church is two modules. Also in the case of the church of San Filippo Neri, number, form and size contribute to generate a shape with great symbolic value. The 3, 4, 6, and 8 are contained in this scheme; about the symbolic meaning of these numbers we have already spoken.

Fig. 8. Plans, sections, photos and schemes of the church dedicated to San Filippo Neri in Milano.

About the 144 it is important to underline that it is a number that belongs to the sphere of the 60 divisors of 5040 (number of the ideal city described in Plato - Book V, Le Leggi) [3] [6]. This number has an important reference also in the description of the Heavenly Jerusalem: is clear in fact 144 cubits is the measure of the side of the cube of the walls of the holy city described in the Apocalypse: "*15. And he who was talking with me had a gold measuring-rod to take the measure of the town, and of its doors, and its wall. 16. And the town is square, as wide as it is long; and he took the measure of the town with the rod, one thousand and five hundred miles: it is equally long and wide and high. 17. And he took the measure of its wall, one hundred and forty-four cubits, after the measure of a man, that is, of an angel.*"[1].
So 144 is a symbolic number and its geometric arithmetic quality is a perfect concept to base the design of a sacred building. The design of the church of San Filippo Neri despite the rationality of the scheme, has a high symbolic value as well, just like the historical churches. Our question is: what is the genesis of this design? As a matter of fact, the purpose for this scheme is not only symbolic but also practical, in fact, the hexagon with the same perimeter covers a larger surface than the square or the equilateral triangle. The hexagon is therefore a convenient form for a large church like this one in the Bovisasca district of Milan, which can house more than 1000 faithful, because it allows to cover, with a defined perimeter, a larger surface.

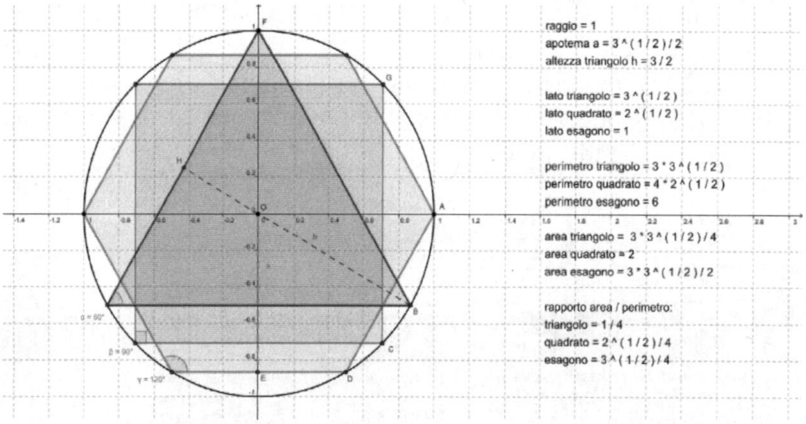

Fig. 9. Comparison of the ratio perimeter-area among equilateral triangle, square and regular hexagon.

4 Conclusions

The modern architecture created by the two architects is strongly mathematical in its conception.
The square and the equilateral triangle at the base of these architectures show a rationality that binds number, shape and size, generating a modern form that bases its rationalist roots in the classical tradition of the History of Architecture of the past.
The reference to the periodic tessellation of the plan is evident and guarantees a strong and rational design method. The geometric figures that generate it are precisely the square, the equilateral triangle and the hexagon. In fact, a tessellation of the plane is a set of limited plane figures and a way of placing them on the **R2** plane, so as to cover it without overlapping and without empty spaces. A tessellation of the plane is periodic, if there are two independent translations **v**, **w** and a finite portion **U** of the tessellation, such that all the drawing is obtained

by copying U and its translations with all the translations generated by **v w**, i.e. all the drawings of the type

$$\mathbf{U + mv + nw}$$

with **m**, **n** part of **Z**. Translated from the mathematical language, it means that when we say that we are tiling a plane we mean that we are covering it completely (without leaving holes) with figures that repeat and do not overlap, and are periodic if these figures are always the same ones next to the others.

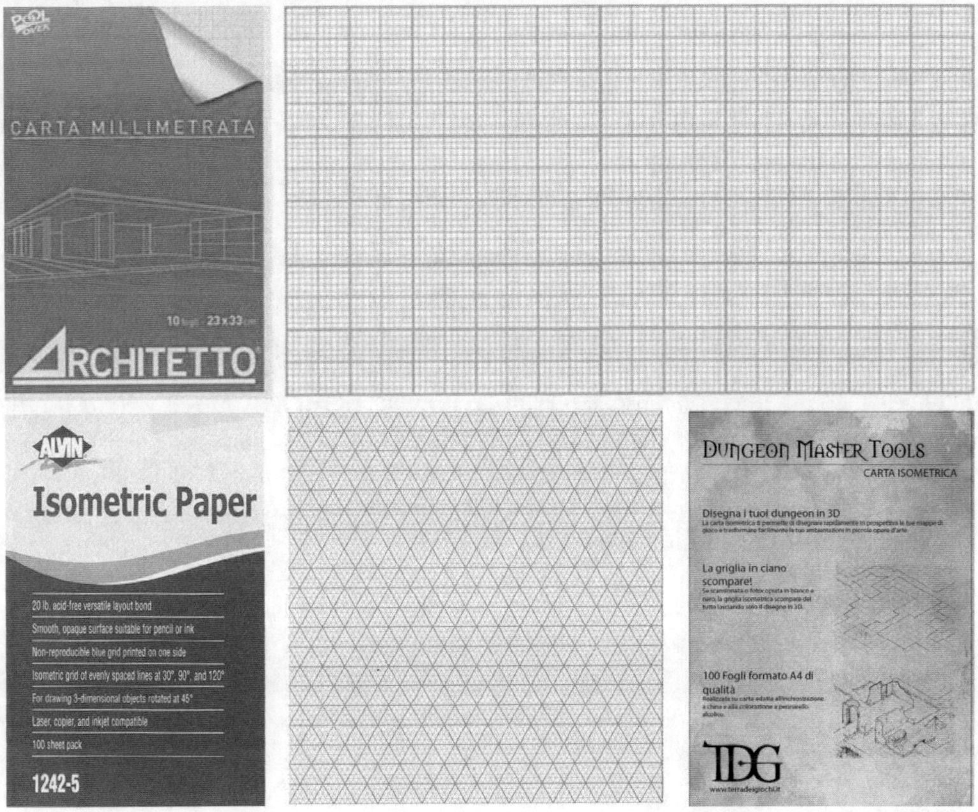

Fig. 10. Cartesian graph paper and isometric or triangular graph papers, irreplaceable tools in the Magnaghi-Terzaghi architectural firm.

Moreover, the Cartesian graph paper and the isometric, trilinear or triangular papers were irreplaceable tools never missing in an architectural office, useful for the first layout of the project based on regular grids arising from the periodic tessellations of the plan. Even though nowadays the Cartesian graph papers are still widespread and known, the triangular graph paper or isometric papers that were usually employed for trilinear diagrams or for quickly obtaining isometric axonometric projections and planivolumetric drawings, are less known. Triangular graph paper, in particular is useful to graph the relationships between three variables on an equilateral triangle. This type of two-dimensional graph can be made when the sum of the three variables always adds up to a constant - usually 100% or 1. The charts produced from this method of graphing are generally referred to as "ternary plots" or "ternary diagrams." They are frequently used to plot the compositions of mixtures in physical

chemistry, petrology, mineralogy, metallurgy and other physical sciences. Ternary plots are also used in genetics and game theory.
In conclusion, the tools and rational methodology described are at the basis of the architectural composition and in particular in Augusto Magnaghi Delfino's and Mario Terzaghi's design method. They handled a logical deductive approach with practical purposes that in geometry found an important tool to generate the architectural form.

References

[1] AA.VV, *Bibbia - Apocalisse*, 21.15-17 – (English version: *Holy Bible - Revelation,* 21.15-17 http://www.transcripture.com/italiano-inglese-apocalisse-21.html)

[2] FERABOLI M.T., edited by, *Archivio dell'architetto Terzaghi*, CASVA, Grafiche Serenissima, Milano, 2006, pp. 70–75.

[3] MELE G., *A geometrical analysis of the layout of Acaya, Italy*, in Nexus Network Journal, vol.14, No 2, Birkhauser, 2012 pp. 385–386

[4] PAGANO G., *Casa a Fino Mornasco*, in Casabella n. 150, giugno 1940, p. 25

[5] PAREYSON L., *Estetica*, Milano 1988, pp. 278-279 (first edition *Estetica. Teoria della formatività* Torino, Edizioni di «Filosofia», 1954)

[6] PLATONE *Le leggi*, a cura di F. Ferrari e S. Poli, Rizzoli, Milano 2005; in english language a commented version of Plato's Law's: BOBONICH C., *Plato's Laws: a Critical Guide*, Cambridge: Cambridge University Press, 2011

[7] WITKOWER R., *Architectural Principles in the Age of Humanism*, London: Warburg Institute, 1949 (first edited in Italy *Principi architettonici nell'Età dell'Umanesimo*, trad. Renato Pedio, Torino, Giulio Einaudi Editore, 1964).

NB. In the present research Alessandra Capanna's contribution is in the analysis of the projects and Giampiero Mele's is in the study of the geometric matrices.

Current address

Mele Giampiero, Ass. Prof., PhD.
Dipartimento eCampus
Università degli Studi eCampus
Via Isimbardi, 10, 22060 Novedrate (Co), Italia
E-mail: giampiero.mele@uniecampus.it

Capanna Alessandra, Res. PhD
Dipartimento di Architettura e Progetto
Università degli Studi di Roma La Sapienza
Via Flaminia 359, 00196 Roma, Italia
E-mail: alessandra.capanna@uniroma1.it

Slovak University of Technology in Bratislava
Faculty of Mechanical Engineering
17th Conference on Applied Mathematics **APLIMAT 2018**

Proceedings

SOME ALGEBRAIC PROPERTIES OF TENSORS

MIKEŠ Josef (CZ)

Abstract. This paper is devoted to the study of some algebraic properties of tensors of type (1,3) which have the properties similar to Riemannian tensor on manifolds with affine connection. The investigations are related to certain decompositions of these tensors. The criteria are connected with systems of coordinates and some components of the tensors mentioned above.

Keywords: algebraic properties, tensors, Riemannian tensor

Mathematics subject classification: Primary 53A45; Secondary 15A72, 53B05.

1 Introduction

In the works by I. P. Egorov [1] and N. S. Sinyukov [9] were used criteria of spaces with non constant curvature connected with existence of non zero components of Riemannian tensor in certain special coordinates. These authors applied the criteria for studying of motions and geodesic mappings of Riemannian spaces and spaces with affine connection. The above mentioned criteria were formulated in the monograph by I. A. Schouten and J. J. Struik [8].

These criteria have been developed in the paper [6] by J. Mikeš and D. Moldobayev. Moreover, this problem is closely connected with decompositions of tensors, see [2, 4, 3].

This study is focused on detailed description of these criteria for tensors which have the algebraic properties of Riemannian tensor.

2 Tensors which have the algebraic properties of Riemannian tensor

Let manifolds M is assigned on some of coordinate neighborhood with coordinates $x^1, x^2, \dots, x^n$ and let in M we determine the tensor T^h_{ijk} which has the algebraic properties of Riemannian tensor. We can write

$$T^h_{ijk} + T^h_{ikj} = 0, \qquad (1)$$

$$T^h_{ijk} + T^h_{jki} + T^h_{kji} = 0. \qquad (2)$$

These properties (1) and (2) have, for example [4, 7, 9], *Weyl tensor of projective curvature* on spaces with affine connection

$$W^h_{ijk} = R^h_{ijk} - \frac{1}{n+1}\,\delta^h_i(R_{jk} - R_{kj}) + \frac{1}{(n+1)\,(n-1)}\,[(nR_{ij} + R_{ji})\,\delta^h_k - (nR_{ik} + R_{ki})\,\delta^h_j)],$$

where R^h_{ijk} and $R_{ik} = R^\alpha_{i\alpha k}$ are the Riemannian and Ricci tensors, respectively. Weyl tensor of projective curvature in equiaffine spaces (for which $R_{ik} = R_{kj}$) has the following simply form

$$W^h_{ijk} = R^h_{ijk} + \frac{1}{n-1}\,(R_{ij}\,\delta^h_k - R_{ik}\,\delta^h_j),$$

and, for $n > 2$, *Weyl tensor of conformal curvature* on (pseudo-) Riemannian spaces

$$C^h_{ijk} = R^h_{ijk} - \delta^h_j L_{ik} + \delta^h_k L_{ij} - L^h_j g_{ik} + L^h_k g_{ij},$$

where

$$L_{ij} = \frac{1}{n-2}\,(R_{ij} - \frac{R}{2(n-1)}\,g_{ij})$$

is the Brinkmann tensor, $L^h_i = g^{h\alpha}L_{\alpha i}$, $R = R_{ij}g^{ij}$ is a scalar curvature, g_{ij} is a metric tensor, and g^{ij} are components of its inverse matrices.

These properties yield also for *Yano tensor of concircular curvature*

$$Y^h_{ijk} = R^h_{ijk} + \frac{R}{n\,(n+1)}\,(g_{ij}\,\delta^h_k - g_{ik}\,\delta^h_j).$$

3 Tensor transformation law

During the transformation of the coordinate system

$$x'^h = x'^h(x^1, x^2, \ldots, x^n),$$

components of tensor T are changed according to the tensor law

$$T'^h_{ijk}(x') = T^\alpha_{\beta\gamma\delta}(x)\,A^h_\alpha\,B^\beta_i\,B^\gamma_j\,B^\delta_k, \tag{3}$$

where

$$A^h_i = \frac{\partial x'^h}{\partial x^i}$$

and $\|B\|$ is inverse matrix of $\|A\|$.

Furthermore, the same is valid for a given point on these spaces. Here the tensor T is considered to be given in this point. Hence, the formulae analogous to (3) is valid.

The components of tensor T are changed according to the law

$$T'^h_{ijk} = T^\alpha_{\beta\gamma\delta}\,A^h_\alpha B^\beta_i\,B^\gamma_j\,B^\delta_k, \tag{4}$$

where A^h_i is real regular matrix, and $\|B\|$ is inverse matrix of $\|A\|$.

These changes of tensors components satisfy linear transformations

$$x'^h = A^h_i\,x^i.$$

4 The Main Theorem of special tensors decomposition

We proved the following Main Theorem.

Theorem 1 *Let in each coordinates is valid one of the conditions*

$$a) \qquad T^H_{IIK} = 0 \tag{5}$$

or

$$b) \qquad T^H_{IKM} = 0 \tag{6}$$

where H, I, K, M are fixed commonly different indices.

Then tensor T has the following decomposition

$$T^h_{ijk} = \delta^h_i\,(p_{jk} - p_{kj}) + \delta^h_k\,p_{ji} - \delta^h_j\,p_{ki} \tag{7}$$

where

$$p_{kj} = \frac{n\,T_{jk} + T_{kj}}{n^2 - 1} \quad \textit{and} \quad T_{ij} = T^\alpha_{ij\alpha}.$$

Proof Let us assume that the conditions mentioned above in the Theorem are true. Firstly, we prove that formula (5) or (6) are true in each coordinates, so consequently, it follows that

$$a) \qquad T^h_{iik} = 0 \tag{8}$$

or

$$b) \qquad T^h_{ijk} = 0 \tag{9}$$

in each coordinates for any commonly different indices h, i, j, k.

Let us consider the following linear transformations

$$x'^p = x^p + r \cdot x^q, \quad x'^s = x^s, \quad s \neq p, \tag{10}$$

where r is a certain constant, p and q are commonly different fixed indices.

Then A^h_i and B^h_i in formula (4) have the following form $A^h_i = B^h_i = \delta^h_i$ and $A^p_q = -B^p_q = r$ for $h \neq p$ or $h \neq q$.

Let us compute the following components of tensor T in a new coordinate system determined by transformation (10)

$$T'^h_{pqk} = T^h_{pqk} - r \cdot T^h_{ppk}, \tag{11}$$

$$T'^h_{qqk} = T^h_{qqk} - r \cdot (T^h_{pqk} + T^h_{qpk}) + r^2 \cdot T^h_{ppk}, \tag{12}$$

$$T'^p_{qjk} = T^p_{qjk} + r \cdot (T^q_{qjk} - T^p_{pjk}) - r^2 \cdot T^q_{pjk}, \tag{13}$$

$$T'^p_{iqk} = T^p_{iqk} + r \cdot (T^q_{iqk} + T^p_{ipk}) - r^2 \cdot T^q_{ipk}, \tag{14}$$

$$T'^p_{qqk} = T^p_{qqk} + r \cdot (T^q_{qqk} - T^p_{pqk} - T^p_{qpk}) - r^2 \cdot (T^q_{pqk} + T^q_{qpk} - T^p_{ppk}) - r^3 \cdot T^q_{ppk}. \tag{15}$$

In the last formulas all different kind indices are not equivalent. Here and further by p and q no summing is provided.

First of all considering (11) it is easy to find out that from (9) follows (8).

Let us prove the contrary. Let the conditions (8) are valid, then $T^h_{iik} = 0$ in any coordinate system, under commonly different indices h, i, j. So from (14) we obtain $T^h_{ijk} + T^h_{jik} = 0$ where $h, i, j, k \neq$. The above mentioned indices are commonly different.

After alternation the last formula under j, k and considering the properties (1) and (2) of the tensor T we obtain that $T^h_{ijk} = 0$ for $h, i, j, k \neq$.

Finally we get

$$T^h_{ijk} = 0 \tag{16}$$

in any coordinate system for all indices $h, i, j, k \neq$.

Furthermore, from (13) and (14) we consequently obtain

$$T^q_{qjk} = T^p_{pjk} \quad \text{and} \quad T^q_{iqk} = T^p_{ipk} \tag{17}$$

for $p, q \neq j, k$.

Basing on (17), we can come to the following conclusion

$$T^p_{pjk} = A_{jk} \quad \text{and} \quad T^p_{ipk} = B_{jk} \tag{18}$$

for all indices p, j, k bat $p \neq j, k$, and A_{jk} and B_{jk} are certain geometric objects.

Considering (16) and (18) and that r, generaly speaking, is any constant, then from (15) we get

$$T^p_{ppk} = A_{pk} + B_{pk} \tag{19}$$

for any indices p and k, $p \neq k$.

From (16), (18), (19) we can write the following $T^h_{ijk} = \delta^h_i A_{jk} + \delta^h_j B_{ik} - \delta^h_k B_{ij}$, and by graduate contracting, taking into consideration the algebraic properties of tensor T, we get the formula (7). The theorem is proved.

5 Applications of the Main Theorem

Basic on Theorem 1 the following theorem is true.

Theorem 2 *Let in each coordinate system of space with affine connection A_n for the components of Riemannian tensor is valid one of the conditions*

$$a) \quad R^H_{IIK} = 0 \quad or \quad b) \quad R^H_{IKM} = 0$$

where H, I, K, M are fixed commonly different indices. Then A_n is projectively Euclidean space.

Proof of this theorem follows from the structure of the curvature tensor R, which we obtain from the Main Theorem. Analog of Theorem 2 is true also for the Weyl tensor of projective curvature W^h_{ijk}, whereas, if we use the traceless property [3, 5].

This Theorem generalizes the results obtained by I. A. Schouten and D. J. Struik [8, p. 197] and shortens I. P. Egorov computing [1] assessment of the order of the complete groups of moving in Riemannian spaces and spaces with affine connections.

In (pseudo-) Riemannian spaces there are the following theorems

Theorem 3 *Let in each coordinate system of (pseudo-) Riemannian space V_n for the components of Riemannian tensor is valid one of the conditions $R^H_{IIK} = 0$ or $R^H_{IKM} = 0$, where H, I, K, M are fixed commonly different indices. Then V_n has constant curvature.*

Analog of Theorem 3 is true also for the Weyl tensor of projective curvature W^h_{ijk}, and for the Yano tensor of concircular curvature Y^h_{ijk}. Let as note that for dimension $n = 2$ these tensors are vanishing. The Weyl tensor of conformal curvature C^h_{ijk} is vanishing for dimension three. For dimension $n > 3$ tensor C^h_{ijk} is vanishing if and only if V_n is conformally (pseudo-) Euclidean. Because this tensor is traceless then, from mentioned above, the following theorem holds.

Theorem 4 *Let in each coordinate system of (pseudo-) Riemannian space V_n for the components of Riemannian tensor is valid one of the conditions $C^H_{IIK} = 0$ or $C^H_{IKM} = 0$, where H, I, K, M are fixed commonly different indices. Then V_n is conformally (pseudo-) Euclidean.*

References

[1] EGOROV, A. I., EGOROV, I. P.: *Motion in generalized spaces*, Ryazan. Gos. Ped. Inst., Ryazan, 1986, 86 pp.

[2] JUKL, M., JUKLOVÁ, L., MIKEŠ, J.: *Applications of local algebras of differentiable manifolds*, J. Math. Sci., New York, vol. 207, no. 3, (2015), pp. 485–511.

[3] MIKEŠ, J.: *On the general trace decomposition problem*, in *Differential geometry and applications. Proceedings of the 6th international conference, Brno, Czech Republic, August 28–September 1, 1995*, Brno: Masaryk University, 1996, ISBN 80-210-1369-9, pp. 45–50.

[4] MIKEŠ, J., et al.: *Differential geometry of special mappings*, Palacký University Olomouc, Faculty of Science, Olomouc, 2015, ISBN 978-80-244-4671-4, 568 pp.

[5] MIKEŠ, J., JUKL, M., JUKLOVÁ, L.: *Some results on traceless decomposition of tensors*, J. Math. Sci., New York, vol. 174, no. 5, (2011), pp. 627–640.

[6] MIKEŠ, J., MOLDOBAEV, D.: *On certain algebraic properties of tensors*, Developed of non Eucleidian Geometry and Topology, (1982), pp. 58–62.

[7] MIKEŠ, J., VANŽUROVÁ, A., HINTERLEITNER, I.: *Geodesic mappings and some generalizations*, Palacký University Olomouc, Faculty of Science, Olomouc, 2009, ISBN 978-80-244-2524-5, 304 pp.

[8] SCHOUTEN, I. A., STRUIK, J. J.: *Introduction in the new methods in differential geometry. II. Geometry*, Gosudarstv. Izdat. Inostr. Lit., Moscow, 1948, 348 pp.

[9] SINYUKOV, N. S.: *Geodesic mappings of Riemannian spaces*, Nauka, Moscow, 1979, 256 pp.

Current address

Mikeš Josef, prof. RNDr., DrSc.
Department Algebra and Geometry
Faculty of Science
Palacky University in Olomouc
17. listopadu 12, 77146 Olomouc, Czech Republic
E-mail: josef.mikes@upol.cz

Slovak University of Technology in Bratislava
Faculty of Mechanical Engineering
17th Conference on Applied Mathematics **APLIMAT 2018**

Proceedings

USE OF STRUCTURAL DYNAMIC MODIFICATION TO REDUCING VIBRATION LEVEL

MUSIL Miloš (SK), HAVELKA René (SK), HAVELKA Ferdinand (SK), MIHELOVÁ Silvia (SK)

Abstract. Dynamic parameters tuning is a required operation when significant vibration of a construction component is expected. Structural dynamic modification represents modal-spectral parameters changes by adding substructures to the original to avoid resonances and reduce the level of vibration. Those substructures, which influent mass, stiffness and damping matrices of the original system, can increase or decrease natural frequencies and influents corresponding mode shapes according to the desire behaviour of the structure. Additional tuning of dynamic parameters performed by structural dynamic modification is a useful operation to adapt the structure to changed operation conditions. Modal and spectral parameters of the modified structure can be determined by analytical approach using structural coefficient matrices of modifying substructures and modal parameters of the original structure, that can be obtained experimentally

Keywords: structural dynamic modification, modal synthesis, system DOF reducing, dynamic parameters tuning

Mathematics Subject Classification: Primary 15A18, 65E99

1 Introduction

A common occurrence in engineering practice are undesirable levels of vibration in the structures of machinery, which decrease their functionality, safety, reliability and service life. Current trends in the dynamic operation of machinery inherently generate such undesirable effects. That is to say, increasing the operational capacity of the machine (higher speeds, higher loads, more changes in operational regimes, etc...) are financially counterproductive to any desired savings in the material/technological realization of such structures. The dynamic properties of the machines structure itself, to an extent, affects the level of vibration of each of its individual parts. In the research stage, it is now necessary to extensively analyse/synthesize the dynamic properties of the machine and its structure followed by the optimization of significant parameters. Therefore an effective methodology to push (decrease) any high level

of vibration in critical parts of the machines structure is becoming increasingly more interesting to satisfy more demanding technological and economic design requirements.
In the design of a structure which must vibrate within acceptable levels during operation, an suitable (effective) concept must be chosen. For example, on the basis of numeric analysis and optimization of individual structural components, it is possible to create a real structure which satisfies the operational conditions set upon it. But in general, the real structure partially exhibits differing properties than those predicted computationally, either due to inappropriate simplification or inaccurate physical or geometrical parameters. It is therefore necessary to modify critical structural elements in order to fulfil the desired properties. To do this it is possible to use some correction method for the mathematical models. With them it is possible to refine the parameters of critical nodes in the structure to reflect the operation of the real structure. Afterwards it is necessary to repeat the optimization procedure of the revised structural parameters such that, based on them, the built structure reflects the required properties.

Another possible approach to modify the dynamic properties of the structure is through modal synthesis. This approach combines the modal properties of the real structure obtained through measurements and the modal properties of additional components obtained computationally or through measurements. This approach is particularly effective if the computational model of the built structure is incorrect. Through optimization of the additional components it is possible to obtain the desired properties of a modified structure while reducing the computational requirements and increasing accuracy of the results. The aforementioned approach is valid for a variety of applications, most notably the automotive and aerospace industries amongst many others.

2 Modelling a structure with a vibroinsulating layer made of aluminium foam

To reduce unacceptable levels of vibration, layers of different material are usually used to reduce the levels of vibration. Materials continue to evolve. New materials have not only good damping properties, but their behaviour also show other favourable characteristics. Such materials are typically difficult and expensive to produce, however the economic benefits from improving the properties if a structure far outweigh the initial cost for their production [1], [3].

2.1 Properties of aluminium foam

Metal and aluminium foams are amongst one group of materials with special properties. Their combination of beneficial physical properties allows them to be useful for a wide variety of applications in practice. The main benefit is in their low weight, high stiffness, and good damping properties. They are materials with many beneficial properties and wide spectrum of application. Even though these materials are relatively expensive the appropriate design of geometrical parameters and applying them in suitable locations, can effectively ensure the required stiffness and damping parameters to be achieved with a minimal use of material. A large part of practical calculations lays in the determination of frequency transfer (modal properties). Determining the modal properties for a specific position and geometrical parameter of the vibro-isolating layers subsequently followed by the processing of the results, yields optimal parameters.

As was discussed, aluminium foams are used also as a material with properties of vibro-isolation. Tab. 1 shows a comparison of material properties between conventional materials and aluminium foam (Alulight from Alulight International GmbH) [1], [18], [4]. In respect the properties of Alulight, it is possible to apply where relatively stiff structural components are needed with a minimum of added weight, or where heat resistant, high stiffness, low density structures are required.

Material	Density ρ [kg/m^2]	Modulus E [GPa]	E 10^{-5} [$GPa.kg^2/m^6$]
Alulight	500	5	2,0
Epoxy	1300	3	0,3
steel	7800	210	0,4
aluminium	2700	69	1,0
glass	2500	70	1,1
concrete	2500	50	0,8

Tab 1. Comparison of physical properties of aluminium foam and conventional materials.

2.2 Modelling of layered structures

Currently, the finite element method is used when modelling vibration of a machine structures [4], [5], [19], [13]. Using this method, it is possible to solve for layered materials and directly use elements respecting the required inertial, stiffness and damping properties. In the case of a thin, layered beam and considering the bending vibration, the element matrices are written in the form:

$$\mathbf{M}_E = m_E \begin{bmatrix} \frac{13}{35} & & & \\ \frac{11l}{420} & \frac{l^2}{105} & sym & \\ \frac{9}{700} & \frac{13l}{420} & \frac{13}{35} & \\ \frac{-13l}{420} & \frac{-l^2}{140} & \frac{-11l}{210} & \frac{l^2}{105} \end{bmatrix} \quad (1)$$

$$\mathbf{K}_E = \frac{2EJ}{l^3} \begin{bmatrix} 6 & & & \\ 2l & 2l^2 & sym & \\ -6 & -2l & 6 & \\ 2l & l^2 & -2l & 2l^2 \end{bmatrix}$$

where the displacement and rotation within nodes of the element are represented by the vector: $[y_j, \varphi_K, y_j, \varphi_K]^T$, where m_E is the mass, E is the equivalent module of elasticity, J is the equivalent moment of inertia of the cross-section area and l is the length of the layered element (Fig. 1). The equivalent modulus of elasticity is the beams module of elasticity $E = E_1$. The moment of inertia of the cross-sectional area is the moment of inertia of the transformed cross-sectional area of the added layer [13], [6].

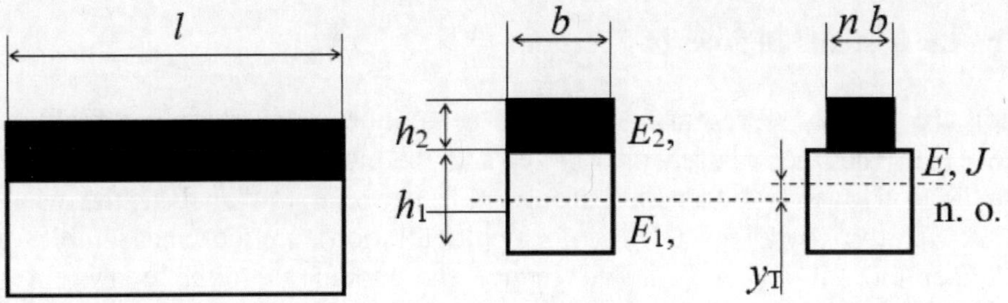

Fig. 1. Layered beam element with cross section and transformed cross section.

The element damping matrix is expressed in proportional terms $\mathbf{B}_E = \beta\,\mathbf{K}_E$, where $\beta = \eta / \omega$, η is the loss coefficient, and ω is the angular excitation frequency. The loss coefficient is determined experimentally and is given by the manufacturer. Tab. 2 shows the necessary material properties of aluminium foam to calculate the element matrix [1], [18], [2].

Density ρ	kg/m^3	500	750	1000
Module of Elasticity E	GPa	5	9	14
Loss coefficient η	-	0,003	0,004	0,004

Tab 2. Comparison of physical properties of aluminium.

The complete FEM beam model is created by connecting elements in corresponding node points (see Fig. 2).

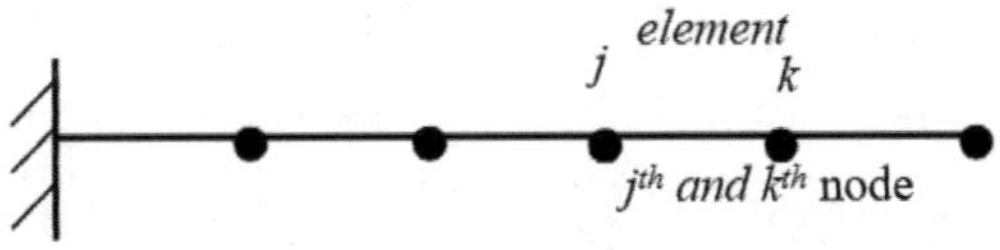

Fig. 2. FE model of the beam.

Its mathematical model has the form:

$$\mathbf{M}\,\ddot{\mathrm{q}} + \mathbf{B}\,\dot{\mathrm{q}} + \mathbf{K}\,\mathbf{q} = \mathbf{f} \tag{2}$$

where matrices for mass **M**, stiffness **K** and damping **B** have a striped character and make up an overlapping element matrix in the rows and columns where elements have a common displacement and rotation (the same node). Simultaneously in the the element matrices the first two rows are tied to the displacement and rotation in the jth node and the rest are tied to the displacement and rotation of the kth node. The same applies to the columns of the element matrices.

3 Reducing the Systems Degrees of Freedom

FEM models that closely represent the dynamic behaviour within the given frequency range tend to have many degrees of freedom (DOF) and therefore the coefficient matrices of the system are large. However, when measuring, it is necessary to analyse the measurement points of the structure whose DOF (for example rotation) are not measurable or poorly measured. Therefore, it is beneficial to reduce the model as to be consistent with the experiment. This means that the computational model will be represented by movements which are measured and are true to the tested results within the given frequency range. There are many methods which can be used to reduce such a model [7], [9] – [12], [16].
Methods for reduction, in principle, are distinguished by the approximation of insignificant common motions in the system ("insignificant – omitted - secondary" DOF) within the considered range of frequencies, and the significant common dynamic motion ("dominant - primary - main" DOF) determined based on physical/modal properties of the system (Fig. 3). In this way the insignificant (secondary) DOF are omitted based on the approximation function chosen based on identifying the (primary) DOF.

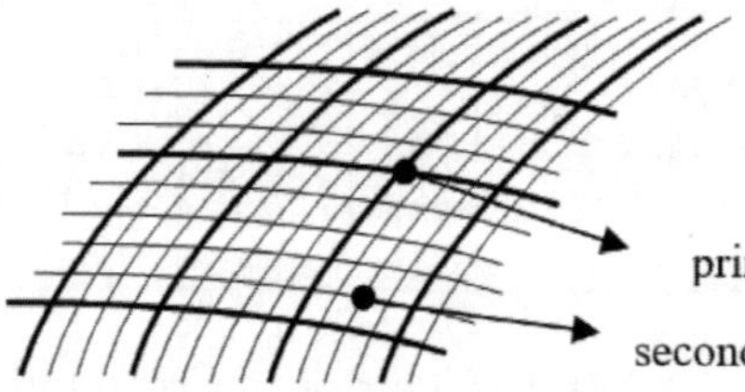

Fig. 2. FE model with primary and secondary nodes.

In principle, the shown method is based on the election of the transformation matrix **T** of type *n, r* and ranking *m*, for the transformation of coordinates $\mathbf{q}(t)$ of the model by dimension *n* to the coordinate $\mathbf{q}_R(t)$ of the model with dimension *r*, where:

$$\mathbf{q}(t) = \mathbf{T}\,\mathbf{q}_R(t). \tag{3}$$

Such a transformation matrix characterizes the relationship between secondary and primary coordinates of the system. Modifying relation (2) using this transformation will yield the reduced model in the form:

$$\mathbf{M}_R\,\ddot{\mathbf{q}}_R(t) + \mathbf{B}_R\,\dot{\mathbf{q}}_R(t) + \mathbf{K}_R\,\mathbf{q}_R(t) = \mathbf{f}_R\,(t), \tag{4}$$

where

$$\mathbf{M}_R = \mathbf{T}^T\,\mathbf{M}\,\mathbf{T}\,;\ \mathbf{B}_R = \mathbf{T}^T\,\mathbf{B}\,\mathbf{T}\,;\ \mathbf{K}_R = \mathbf{T}^T\,\mathbf{K}\,\mathbf{T}\,;\ \mathbf{f} = \mathbf{T}^T\,\mathbf{f}(t). \tag{5}$$

The accuracy of any method depends on the choice of a suitable transformation matrix **T**, which should ensure that the reduced model to the degree r remains modally and spectrally (in r selected coordinates, $m \leq r$ selected Eigen modes) true to the original model to the nth degree.
It follows from the modal analysis of such a system that the reduction will be more accurate depending on how correctly the transformation equation between elements of the structures Eigen modes $\mathbf{v}_j$ (n^{th} amount) and from them the selected components $\mathbf{v}_{Rj}$ (r amount) [9], [11]. Afterward, if:

$$\mathbf{v}_j = \mathbf{T}\,\mathbf{v}_{Rj}, \tag{6}$$

then both of the following relations apply:

$$(\mathbf{K} + s_j\,\mathbf{B} + s_j^2\,\mathbf{M})\,\mathbf{v}_j = \mathbf{0},\ (\mathbf{K}_R + s_j\,\mathbf{B}_R + s_j^2\,\mathbf{M}_R)\,\mathbf{v}_{Rj} = \mathbf{0}, \tag{7}$$

where s_j (in amount $m \leq r$) are the values from the considered frequency range.
In cases where, for example, the system with transformrmation matrix **T** (System Equivalent Reduction and Expansion Process – SEREP method) in the form:

$$\mathbf{T} = \begin{bmatrix} \mathbf{I} \\ \mathbf{V}_{00}\,\mathbf{V}_{0R}^{+} \end{bmatrix},\ \mathbf{V}_0 = \mathbf{T}\,\mathbf{V}_{0R}, \tag{8}$$

where $\mathbf{V}_0 \in R^{n,m}$ is a matrix with m – significant Eigen modes , which are applicable for specific directional actions and excitation frequency ranges. Its submatrix $\mathbf{V}_{0R} \in R^{r,m}$ is composed of elements characterized by r – most significant common deviations of the structure (usually $m \leq r << n$) and its sub matrix $\mathbf{V}_{00} \in R^{o,m}$ is composed of elements characterized by $o = n - r$ remaining elements of this set.
Using the transformation matrix **T,** the relationship for orthonormality are satisfied only in the considered frequency range. In the experimental verification of the aforementioned reduction method, if the sub matrix $\mathbf{V}_{0R}$ is determined experimentally and sub matrix $\mathbf{V}_{00}$ is determined analytically, then the given relations are satisfied in respect to the minima of the quadratic standards of such a regression model. This problem, in general, occurs as a result of an inconsistent analytical model to the r - degree with the nth degree of the model (that is with errors in modelling and measuring) [10, 8, 17, 14]. Such a transformation matrix must represent the real properties of the structure, which is used for example for the correction of the mathematical models of the systems based on measured data.

Inconsistencies often occur due to errors in the analytical modelling of the structure and in measurements during experimental tests. Therefore, in order to minimize computational errors in terms of problem verification, identification, system corrections (modal synthesis), it is necessary to choose the elements of the modal vector which dominantly represent the orthonormal properties of the system. This will guarantee the rank m of the transformation matrix **T**. Their selection can be objectivised by a singular or QR (Gramm-Schmidt orthogonalization process) decomposition of the modal matrix $\mathbf{V}_0$. Any sudden decrease (large difference) between j and k - th singular values of the modal matrix ($\sigma_1 > \sigma_2 \ldots \sigma_j >> \sigma_k\ \sigma_{k+1} > \ldots$) [7], [9], [12] will not occur in the case where the analytically obtained submatrices are

given because in this case the errors from measurements do not occur, however matrix $\mathbf{V}_0$ must be regular (its rank must be m).
Since the modal analysis is relatively taxing computationally, the transformation matrix **T** can be expressed by the already computed coefficient matrices of the system. The following is true for the system (2) (matrices assumed to be symmetric and **B=0** with homogeneous solution), expanded to block form:

$$\left(\begin{bmatrix}\mathbf{K}_{11} & \mathbf{K}_{12}\\ \mathbf{K}_{21} & \mathbf{K}_{22}\end{bmatrix} - \omega_{0j}^2 \begin{bmatrix}\mathbf{M}_{11} & \mathbf{M}_{12}\\ \mathbf{M}_{21} & \mathbf{M}_{22}\end{bmatrix}\right)\begin{bmatrix}{}^r\mathbf{v}_{0j}\\ {}^o\mathbf{v}_{0j}\end{bmatrix} = \mathbf{0}\,,\ \mathbf{V}_0 = \{\mathbf{v}_{0j}\},\ \mathbf{v}_{0j} = \begin{bmatrix}{}^r\mathbf{v}_{0j}\\ {}^o\mathbf{v}_{0j}\end{bmatrix} \tag{9}$$

where $\mathbf{K}_{11}, \mathbf{M}_{11} \in R^{r,r}$, $\mathbf{K}_{12} = \mathbf{K}_{21}^{\mathrm{T}}$, $\mathbf{M}_{12} = \mathbf{M}_{21}^{\mathrm{T}} \in R^{r,o}$, $\mathbf{K}_{22}, \mathbf{M}_{22} \in R^{o,o}$, ${}^r\mathbf{v}_{0j} \in R^r$, ${}^o\mathbf{v}_{0j} \in R^o$.
then for the transformation matrix the following applies:

$${}^o\mathbf{v}_{0j} = \mathbf{T}_{\mathrm{D}}\, {}^r\mathbf{v}_{0j} = -(\mathbf{K}_{22} - \omega_{0j}^2\, \mathbf{M}_{22})^{-1}\, (\mathbf{K}_{21} - \omega_{0j}^2\, \mathbf{M}_{21})\, {}^r v_{0j},\ \mathbf{T} = \begin{bmatrix}\mathbf{I}\\ \mathbf{T}_{\mathrm{D}}\end{bmatrix} \tag{10}$$

Relation (9) approximates the static condition – negligible inertial effects (Guayan reduction method) in the form [6]:

$${}^o\mathbf{v}_{0j} = -\mathbf{K}_{22}^{-1}\, \mathbf{K}_{21}\, {}^r\mathbf{v}_{0j}\,,\ \mathbf{T} = \mathbf{T}_{\mathrm{G}} = -\mathbf{K}_{22}^{-1}\, \mathbf{K}_{21}, \tag{11}$$

where the reduced matrix of the model is then:

$$\mathbf{K}_{\mathrm{R}} = \mathbf{K}_{11} - \mathbf{K}_{12}\, \mathbf{K}_{22}^{-1}\, \mathbf{K}_{21},\ \mathbf{B}_{\mathrm{R}} = \mathbf{B}_{11} - \mathbf{K}_{12}\, \mathbf{K}_{22}^{-1}\, \mathbf{B}_{21} - \mathbf{B}_{12}\, \mathbf{K}_{22}^{-1}\, \mathbf{K}_{21} + \mathbf{K}_{12}\, \mathbf{K}_{22}^{-1}\, \mathbf{B}_{22}\, \mathbf{K}_{22}^{-1}\, \mathbf{K}_{21}$$
$$\mathbf{M}_{\mathrm{R}} = \mathbf{M}_{11} - \mathbf{K}_{12}\, \mathbf{K}_{22}^{-1}\, \mathbf{M}_{21} - \mathbf{M}_{12}\, \mathbf{K}_{22}^{-1}\, \mathbf{K}_{21} + \mathbf{K}_{12}\, \mathbf{K}_{22}^{-1}\, \mathbf{M}_{22}\, \mathbf{K}_{22}^{-1}\, \mathbf{K}_{21}, \tag{12}$$

This approximation becomes more accurate the lower the natural angular frequencies $\omega_{\mathrm{Red}1}$ if

$$(\mathbf{K}_{22} - \omega_0^2\, \mathbf{M}_{22}) = \mathbf{0} \tag{13}$$

is greater than the maximum angular frequency of the considered excitation frequency range. It is therefore obvious that the selection of coordinates, which will be considered in further relations (primary), effect the accuracy of this method. Whereas the residual (secondary) coordinates are from the area of high stiffness and low mass. The stated selection can be objectively expressed in cases where the coefficient matrices have a dominant diagonal character. Any *j-th* coordinate which satisfies the following condition can be considered as residual (secondary) coordinates:

$$\frac{k_{jj}}{m_{jj}} > \omega_{\max}. \tag{14}$$

The static condition (11) can be precisely satisfied in cases where, for example, the model is obtained by FEM with diagonal mass matrices $\mathbf{M}_{12} = \mathbf{M}_{21}^{\mathrm{T}} = \mathbf{0}$, whereas the mass of the system is concentrated only at specific nodes, while the coordinates of the residuals are represented by the secondary coordinates connected with $\mathbf{M}_{22} = \mathbf{0}$.

In more complex systems, where it is necessary to reduce the dynamic properties of individual connecting between components a combination of this and the previously mentioned (Craig – Bampton method) can be used [7], [12] . In the system where the force effects act on the boundary of individual substructures $\mathbf{q}_R$ (in the primary coordinates) and the dynamic properties of the individual substructure $\mathbf{q}_O$ are represented by their modal properties (secondary coordinates), then for a single j-th substructure [10, 14] it then applies:

$$\mathbf{q}_{Oj} = \mathbf{V}_{0j}\, \tilde{\mathbf{q}}_{O_j} + \mathbf{T}_{Gj}\, \mathbf{q}_{Rj}, \tag{15}$$

$$\mathbf{q}_j = \begin{bmatrix} \mathbf{q}_R \\ \mathbf{q}_O \end{bmatrix}_s = \begin{bmatrix} \mathbf{0} & \mathbf{I} \\ \mathbf{V}_0 & \mathbf{T}_G \end{bmatrix}_s \begin{bmatrix} \tilde{\mathbf{q}}_O \\ \mathbf{q}_R \end{bmatrix}_s = \mathbf{T}_{Hj}\, \mathbf{q}_{Hj}, \tag{16}$$

$$\begin{bmatrix} \mathbf{I} & \mathbf{M}_{P1} \\ \mathbf{M}_{P2} & \mathbf{M}_{P3} \end{bmatrix}_j \begin{bmatrix} \ddot{\tilde{\mathbf{q}}}_O \\ \ddot{\mathbf{q}}_R \end{bmatrix}_j + \begin{bmatrix} 2\boldsymbol{\Delta} & \mathbf{0} \\ \mathbf{0} & \mathbf{B}_P \end{bmatrix}_j \begin{bmatrix} \dot{\tilde{\mathbf{q}}}_O \\ \dot{\mathbf{q}}_R \end{bmatrix}_j + \begin{bmatrix} \boldsymbol{\Omega}_0^2 & \mathbf{0} \\ \mathbf{0} & \mathbf{K}_P \end{bmatrix}_j \begin{bmatrix} \tilde{\mathbf{q}}_O \\ \mathbf{q}_R \end{bmatrix}_j = \begin{bmatrix} \mathbf{0} \\ \mathbf{f} \end{bmatrix}_j \tag{17}$$

$$\mathbf{M}_{Rj}\, \ddot{\mathbf{q}}_{R_j} + \mathbf{B}_{Rs}\, \dot{\mathbf{q}}_{R_j} + \mathbf{K}_{Rj}\, \mathbf{q}_{Rj} = \mathbf{f}_{Rj}, \quad \mathbf{M}_{Rj} = \mathbf{M}_{P3j}, \ \mathbf{B}_{Rj} = \mathbf{B}_{Pj}, \quad \mathbf{K}_{Rj} = \mathbf{K}_{Pj}, \tag{18}$$

where $\mathbf{V}_{0j}$, Δ_j, $\Omega_0{}^2{}_j$ are modal-spectral properties of the jth substructures and matrix $\mathbf{T}_{Gj}$ results from the Guayan reduction method (11). The modal properties of the substructure are commonly obtained experimentally or on the basis of modal analysis of the FEM model of the relatively complex structure. Penetration of the common coordinates of individual substructures into one global vector $\mathbf{q}$ can later form the complete structure of the model. In this way it is possible to significantly reduce the DOF of the systems FEM model without compromising its dynamic properties.

4 Design of the added structural element by method of modal synthesis

Often times it is necessary to change the dynamic p properties of an existing structure. If the modal properties are known, the modal synthesis method can be used to obtain the desired modal – spectral properties. In such a case, the modal properties of the original system are virtually realized, for example, through FEM modeling of the added substructures in such a way that the modal properties of the modified system change in the desired way. The added substructures must be described through the same coordinates of the original (existing) structure, for which any of the above mentioned methods for reduction can be used and thus ensure the connection between the original and added structures [7], [9], [12].

If the existing proportionally damped system with the measured modal – spectral properties $\mathbf{V}_{0P}$, Ω_{0P}, $2\Delta_{0P}$, is added to by the substructure represented by the reduced (to common coordinates) FEM model with coefficient matrices $\mathbf{M}_A$, $\mathbf{K}_A$, $\mathbf{B}_A$, where the damping matrix $\mathbf{B}_A$ is also proportional, the resulting system obtains the modal – spectral properties $\mathbf{V}_{0L}$, Ω_{0L}, $2\Delta_{0L}$. Usually the added component is located in a defined area and not evenly across the original structure, the resulting structure then becomes non-proportionally damped.

4.1 Modal analysis of the non-porportionally damped system

To determine the modal and frequency properties of the disproportionately damped system, the system from n dimensional space represented by coefficients matrices **M, B, K** of order n is transformed to a system of $2n$ dimensional space with the coefficient matrices **N**, **P** [15] , [16] , [17]. Then

$$\mathbf{N}\,\dot{\mathbf{x}} - \mathbf{P}\mathbf{x} = \mathbf{r}, \tag{19}$$

where

$$\mathbf{P} = \begin{bmatrix} -\mathbf{K} & \mathbf{0} \\ \mathbf{0} & \mathbf{M} \end{bmatrix}, \quad \mathbf{N} = \begin{bmatrix} \mathbf{B} & \mathbf{M} \\ \mathbf{M} & \mathbf{0} \end{bmatrix} \in R^{2n,2n}, \quad \mathbf{x} = \begin{bmatrix} \mathbf{q} \\ \dot{\mathbf{q}} \end{bmatrix}, \mathbf{r} = \begin{bmatrix} \mathbf{f} \\ \mathbf{0} \end{bmatrix} \in R^{2n}. \tag{20}$$

From the solution of the eigenvalues for the symmetric coefficient matrix it also applies:

$$(\mathbf{P} - s_j\,\mathbf{N})\,\mathbf{w}_j = \mathbf{0}, \tag{21}$$

$$(\mathbf{K} + s_j\,\mathbf{B} + s_j^2\,\mathbf{M})\,\mathbf{v}_j = \mathbf{0}, \tag{22}$$

where vector $\mathbf{v}_j$ represents its jth Eigen mode and $s_j = -\,\delta_j \pm i\,\omega_{Dj}$ is its jth eigenvalue (assuming subcritical damping of the system). The real part of the eigenvalues represent the damping properties and the imaginary part represents the frequency properties. The conditions for orthonormality can be expressed by:

$$\mathbf{W}^{T}\,\mathbf{P}\,\mathbf{W} = \mathbf{S},\ \mathbf{W}^{T}\,\mathbf{N}\,\mathbf{W} = \mathbf{I}, \tag{23}$$

$$\mathbf{S}\,\mathbf{V}^{T}\,\mathbf{M}\,\mathbf{V}\,\mathbf{S} - \mathbf{V}^{T}\,\mathbf{K}\,\mathbf{V} = \mathbf{S},\ \mathbf{V}^{T}\,\mathbf{B}\,\mathbf{V} + \mathbf{V}^{T}\,\mathbf{M}\,\mathbf{V}\,\mathbf{S} + \mathbf{S}\,\mathbf{V}^{T}\,\mathbf{M}\,\mathbf{V} = \mathbf{I} \tag{24}$$

where for the spectral matrix **S** and modal matrix **W** and **V**:

$$\mathbf{S} = diag(s_j) \in C^{2n}, \quad \mathbf{W} = \{\mathbf{w}_j\} = \begin{bmatrix} \mathbf{V} \\ \mathbf{V}\mathbf{S} \end{bmatrix} \in C^{2n,2n},\ \mathbf{V} = \{\mathbf{v}_j\} \in C^{n,2n}. \tag{25}$$

In cases where the substructure with the proportional damping matrix is for example $\mathbf{B} = \alpha\,\mathbf{M} + \beta\,\mathbf{K}$, the solution of the eigenvalues is simplified to [16]:

$$(\mathbf{K} - \mathbf{M}\,\omega_{0j}^2)\,\mathbf{v}_{0j} = \mathbf{0}, \quad \mathbf{K}\,\mathbf{V}_0 - \mathbf{M}\,\mathbf{V}_0\,\Omega_0^2 = \mathbf{0}, \tag{26}$$

$$\Omega_0 = diag(\omega_{0j}) \in R^{n}, \quad \Delta = diag(\delta_j) \in R^{n},\ \mathbf{V}_0 = \{\mathbf{v}_{0j}\} \in R^{n,n},$$
$$\mathbf{V}_0^{T}\,\mathbf{K}\,\mathbf{V}_0 = \Omega_0^2,\ \mathbf{V}_0^{T}\,\mathbf{M}\,\mathbf{V}_0 = \mathbf{I},\ \mathbf{V}_0^{T}\,\mathbf{B}\,\mathbf{V}_0 = 2\,\Delta = 2\,(\alpha\,\mathbf{I} + \beta\,\Omega_0^2). \tag{27}$$

The diagonal matrix Ω_0 contains elements of natural angular frequencies of the non-damped system ω_{0j} and the matrix of constant decay Δ contains elements of constant decay for a proportionally damped system $\delta_j = 2\ \xi\ \omega_{0j}$, where ξ is the comparative damping of the analyzed substructure. The modal matrix $\mathbf{V}_0$ is composed of vectors $\mathbf{v}_{0j}$ representing the Eigen modes of this system. For the eigenvalues s_j and spectral matrix **S** of the proportional system the following applies:

$$s_j = - \delta_j \pm i\,(\omega_{0j}^2 - \delta_j^2)^{0,5} = - \delta_j \pm i\,\omega_{Dj}, \tag{28}$$

$$\mathbf{S} = - \boldsymbol{\Delta} \pm i\,(\boldsymbol{\Omega}_0^2 - \boldsymbol{\Delta}^2)^{0,5} = - \boldsymbol{\Delta} \pm i\,(\boldsymbol{\Omega}_0^2 - \boldsymbol{\Delta}^2)^{0,5} = - \boldsymbol{\Delta} \pm i\,\boldsymbol{\Omega}_D. \tag{29}$$

4.2 Modal synthesis of non-proportionally damped system

The above mentioned relations can be used to determine the modal and spectral properties of the original substructures modified by means of added substructures. The modal and spectral properties of the original proportionally damped system $\mathbf{V}_{0P}$, $\boldsymbol{\Omega}_{0P}$, $2\,\boldsymbol{\Delta}_{0P}$ are determined experimentally. The added, also proportionally damped system, is represented by the FEM model with coefficient matrices $\mathbf{M}_A$, $\mathbf{B}_A$, $\mathbf{K}_A$. These are composed so that the FEM model of the added structure reduces to the measured coordinates and forms elements of the matrix that are responsible for the measured coordinates. Elements of this matrix contain coordinates for which the added structures do not belong because it is localized only at predetermined areas equal to 0. This way, the experimentally and analytically obtained matrices have the same dimension.

The properties of such a modified system represent the modal and spectral matrix $\mathbf{V}_L$ and $\mathbf{S}_L$. From the conditions of orthonormality affected from the modal analysis of the modified system in *2n* space (23), it follows that:

$$[\mathbf{V}_L^T, \mathbf{S}_L \mathbf{V}_L^T] \begin{bmatrix} -(\mathbf{K}_P + \mathbf{K}_A) & \mathbf{0} \\ \mathbf{0} & \mathbf{M}_P + \mathbf{M}_A \end{bmatrix} \begin{bmatrix} \mathbf{V}_L \\ \mathbf{V}_L \mathbf{S}_L \end{bmatrix} = \mathbf{S}_L \tag{30}$$

$$[\mathbf{V}_L^T, \mathbf{S}_L \mathbf{V}_L^T] \begin{bmatrix} \mathbf{B}_P + \mathbf{B}_A & \mathbf{M}_P + \mathbf{M}_A \\ \mathbf{M}_P + \mathbf{M}_A & \mathbf{0} \end{bmatrix} \begin{bmatrix} \mathbf{V}_L \\ \mathbf{V}_L \mathbf{S}_L \end{bmatrix} = \mathbf{I}, \tag{31}$$

If transformation is used

$$\mathbf{V}_L = \mathbf{V}_{0P}\, \mathbf{T}_L, \tag{32}$$

$$[\mathbf{T}_L^T, \mathbf{S}_L \mathbf{T}_L^T] \begin{bmatrix} -(\boldsymbol{\Omega}_{0P}^2 + \mathbf{V}_{0P}^T \mathbf{K}_A \mathbf{V}_{0P}) & \mathbf{0} \\ \mathbf{0} & \mathbf{I} + \mathbf{V}_{0P}^T \mathbf{M}_A \mathbf{V}_{0P} \end{bmatrix} \begin{bmatrix} \mathbf{T}_L \\ \mathbf{T}_L \mathbf{S}_L \end{bmatrix} = \mathbf{S}_L \tag{33}$$

$$[\mathbf{T}_L^T, \mathbf{S}_L \mathbf{T}_L^T] \begin{bmatrix} 2\boldsymbol{\Delta}_P + \mathbf{V}_{0P}^T \mathbf{B}_A \mathbf{V}_{0P} & \mathbf{I} + \mathbf{V}_{0P}^T \mathbf{M}_A \mathbf{V}_{0P} \\ \mathbf{I} + \mathbf{V}_{0P}^T \mathbf{M}_A \mathbf{V}_{0P} & \mathbf{0} \end{bmatrix} \begin{bmatrix} \mathbf{T}_L \\ \mathbf{T}_L \mathbf{S}_L \end{bmatrix} = \mathbf{I} \tag{34}$$

These relations represent conditions for orthonormality of the transformed system with spectral matrix $\mathbf{S}_L$ and modal matrix:

$$\mathbf{W}_T = \{\mathbf{w}_{Tj}\} = \begin{bmatrix} \mathbf{T}_L \\ \mathbf{T}_L\, \mathbf{S}_L \end{bmatrix}. \tag{35}$$

and with the coefficient matrices

$$\mathbf{P}_T = \begin{bmatrix} -(\mathbf{\Omega}_{0P}^2 + \mathbf{V}_{0P}^T\, \mathbf{K}_A\, \mathbf{V}_{0P}) & \mathbf{0} \\ \mathbf{0} & \mathbf{I} + \mathbf{V}_{0P}^T\, \mathbf{M}_A\, \mathbf{V}_{0P} \end{bmatrix}, \tag{36}$$

$$\mathbf{N}_T = \begin{bmatrix} 2\mathbf{\Delta}_P + \mathbf{V}_{0P}^T\, \mathbf{B}_A\, \mathbf{V}_{0P} & \mathbf{I} + \mathbf{V}_{0P}^T\, \mathbf{M}_A\, \mathbf{V}_{0P} \\ \mathbf{I} + \mathbf{V}_{0P}^T\, \mathbf{M}_A\, \mathbf{V}_{0P} & \mathbf{0} \end{bmatrix}. \tag{37}$$

$\mathbf{P}_T$ and $\mathbf{N}_T$ are known elements of the coefficient matrix. the unknown spectral and modal matrices $\mathbf{S}_L = diag(s_{Lj})$ and $\mathbf{W}_T$ can be determined by solving the eigenvalues in the form (21):

$$(\mathbf{P}_T - s_{Lj}\, \mathbf{N}_T)\, \mathbf{w}_{Tj} = \mathbf{0}. \tag{38}$$

Directly from this the modal matrix of the modified system $\mathbf{S}_L$ emerges and therefore obtaining the modal matrix of the modified system $\mathbf{V}_L$ can be achieved using the transformation relation in (32)

4.3 Modal synthesis of the non-proportionately damped layered beam

The above mentioned modal synthesis method can be used also in determining the modal and spectral properties of beam structures with added layers of vibroisolation at specific points. A very simple illustration of such a system can be explained on an existing cantilever beam with a connected (added) beam which creates the modified beam seen in Fig. 4. also shows the schematic illustration of the above mentioned modal synthesis methodology.

From the schematic, coefficient matrices $\mathbf{M}_N$, $\mathbf{K}_N$ of the added beam reduces to the measurement coordinates y_2, y_3, which are shared with the original beam. The reduced coefficient matrices $\mathbf{M}_R$, $\mathbf{K}_R$ obtained in this way create a non-zero sub matrix expanded by the coefficient matrices $\mathbf{M}_A$, $\mathbf{K}_A$ of the added beam such, that it also represents the extra unmeasured coordinates y_1, y_4. This way the matrices have the same dimension as those measured by natural angular frequency, constant decay and with the modal matrix $\mathbf{\Omega}_{0P}$, $\mathbf{\Delta}_P$, $\mathbf{V}_{0P}$.

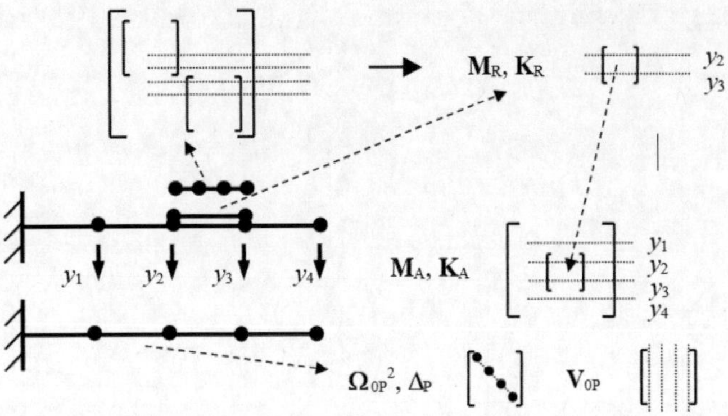

Fig. 4. Schematic methodology of modal synthesis.

The presented method can be automated and used for the parameter optimization of vibroisolating layers (orientation, geometry, material properties, etc...) [14]. The schematic representation of optimizing position a and thickness h of the vibroisolating layer with respect to the maximum ratio of damping ξ in the second Eigen mode is shown in Fig. 5. (c). From this dependence it is obvious that by increasing the thickness of the isolating layer, the ratio of damping also gradually increases. More complex characteristics have values of proportional damping dependent on the location of the isolating layer, which directly depends on the corresponding eigen modes in the presented system. Fig. 6. depicts the dependence of proportional damping of the 3rd and 4th modes and the position of the vibroisolating layer. The steel beam with an aluminium foam layer was considered (Tab. 1 and Tab. 2 – 500kg/m^3).

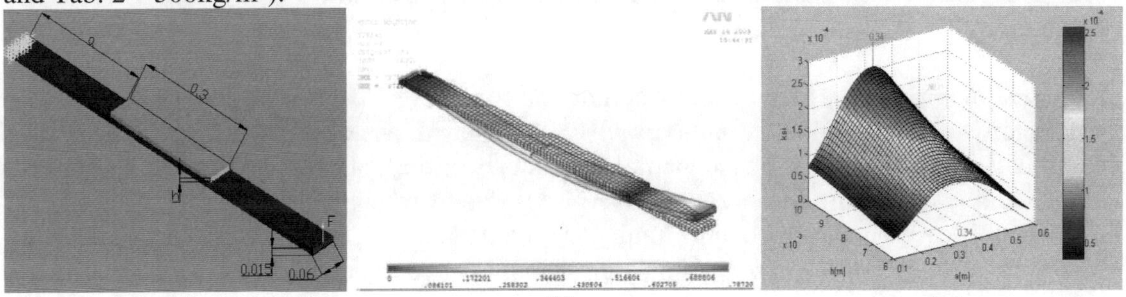

(a) (b) (c)

Fig. 5. Hierarchy (left to right) for the optimization of position "a" and thickness "h" of the vibro-isolating layer.

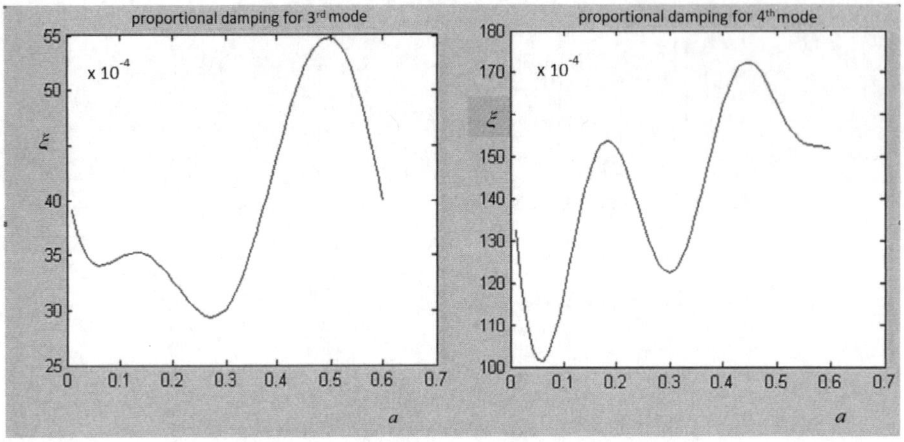

Fig. 6. Dependence of the proportional damping (3rd and 4th modes) on the position of the vibroisolating layer.

From the presented properties, it can be seen how it is possible to achieve the desired damping of some Eigen modes by choosing the appropriate position and thickness of the vibroisolating layer. Therefore, on the basis of modal synthesis it is possible, without any time-consuming calculations, to determine how effective the chosen parameters of the vibroisolating layer are and in what manner do the dynamic properties of the modifies structure change.

4.4 Structural dynamic modification of the beam by aluminium foams demonstrated on FE model

Structural modification of the beam by aluminium foam layers causes the change of dynamic parameters of the original structure, which is known as well as structural dynamic modification. It is possible to tune natural frequencies of the beam structure by adding one or more additive aluminium layers on suitable places due to which natural frequency change is required. Effective natural frequency tuning can be done by adding of modifying structure to places of anti-nodes of the appropriate mode shape. Desired dynamic parameters of the beam structure can be reached using aluminium layers with corresponding parameters that can be calculated by mentioned modal synthesis method. Next figure shows first three natural frequencies of bending vibration of the cantilever beam.

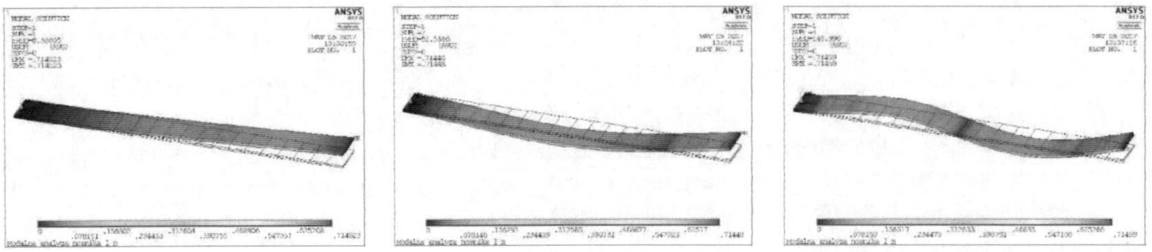

8.3881 Hz 52.539 Hz 147.00 Hz

Fig. 7. First three natural frequencies and mode shapes.

The picture bellow demonstrates structural dynamic modifications of the mentioned beam to reach desire modal and spectral properties.

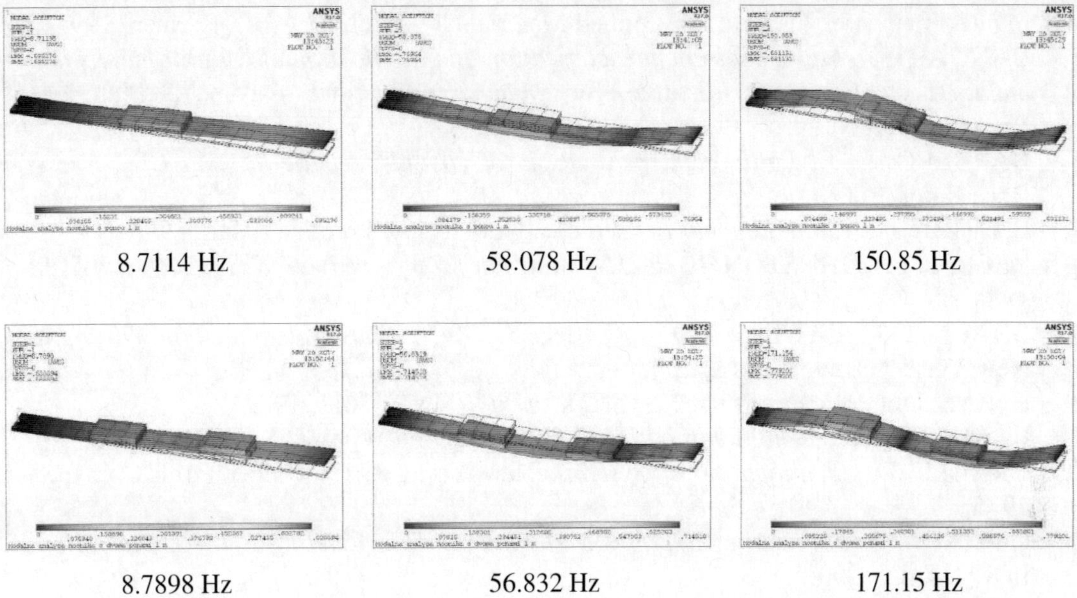

Fig. 8. Structural dynamic modifications of the beam to change its natural frequencies.

As is shown in pictures, modification with the one layer mostly influenced the second mode shape, while the two-layer modification is appropriate for the increasing of the third natural frequency.

5 Conclusion

This work presents methodology which allows for the purposeful change in dynamic properties of existing machine structures by adding additional components. This method is based on modal synthesis and is particularly effective in cases where it is not possible to obtain the reliable and standalone FEM model of the structure but measured modal properties of the system are available. At the same time, it is considered that the added components can be modelled utilizing FEM. Based on modal synthesis, it is possible to determine the modal properties with added component, using the methodology presented in this paper.
The presented methodology is applied to a situation where the added component is a layer of aluminium foam and is illustrated on a beam of a structure.

Acknowledgement

This work was supported by the grant from the Research and Development Agency under the contract no. APVV-15-0630

References

[1] URL: http://www.alulight.com

[2] BANHART, J., BAUMEISTER, J., WEBER, M., *Damping properties of aluminium foams*. Fraunhojer-Institute for Applied Materials Research (IFA M), Bremen, 1995

[3] BANGART, J., *Manufacture, characterisation and application of cellular metals and metal flaks*. Fraunhofer-Institute for Manufacturing and Advanced Materials, Bremen, 2000

[4] BARBERO, E. J., *Finite element analysis of composite materials*, In *CRC Press*, Boca Raton, 2008

[5] BATHE, K. J., *Finite Element Procedures*. New Jersey: Prentice - Hall, 1996

[6] BEER, F. P., JOHNSTON, E. R., *Mechanics of Materials*. New York: McGraw-Hill, 1985

[7] BRAUN, S., EWINS, D., RAO, S. S., *Encyclopedia of Vibration*. New York: Academic Press, 2002

[8] EWINS, D. J., *Modal testing*. New York: John Wiley & Sons, 1987

[9] EYKHOFF, P., *System Identification*. New York: John Wiley & Sons, 1974

[10] GRAUPE, D., *Identification of Systems*. New York: R. E. Krieger Publ. Co., Inc., 1976

[11] GYSIN H., *Comparison of Expansion Method for FE Modeling Error Localization*. In 8th IMAC, 1990

[12] HARRIS, C. M., PIERSOL, A. G., *Harris' Shock and Vibration Handbook*. New York: McGraw - Hill, 2002

[13] INMAN, D. J., *Engineering Vibration*. New Jersey: Prentice-Hall, Upper Saddle River, 2001

[14] KIRCH, U., *Structural Optimization*. Berlin: Springer-Verlag, 1993

[15] LANCASTER P., *Theory of Matrices*. New York: Acad. Press, 1969

[16] STAREK, L., MUSIL, M., INMAN, D. J., *Updating of a Nonconservative System Using Inverse Methods*. In *ASME Design Technical Conference*, Sacramento, 1997

[17] STAREK, L., INMAN, D. J., MUSIL, M., *Updating of Non-Conservative Structure via Inverse Methods with Parameter Subset Selection*. In 5th *International Conference on Sound and Vibration 3*, USA, 1997

[18] URL: http://www.reade.com

[19] ZIENKIEVICZ, O. C.: *The Finite Element Method*. New York: McGraw-Hill, 1977

Current address

Musil Miloš, Prof., Ing., CSc.
Slovak University of Technology in Bratislava, Faculty of Mechanical Engineering
Nám slobody 17, 812 31 Bratislava, Slovak Republic
E-mail: milos.musil@stuba.sk

Havelka René, Ing.
Slovak University of Technology in Bratislava, Faculty of Mechanical Engineering
Nám slobody 17, 812 31 Bratislava, Slovak Republic
E-mail: rene.havelka@stuba.sk

Havelka Ferdinand, Ing., PhD.
Slovak University of Technology in Bratislava, Faculty of Mechanical Engineering,
Nám slobody 17, 812 31 Bratislava, 811 06
E-mail: ferdinand.havelka@stuba.sk

Mihelová Silvia, Ing.
Slovak University of Technology in Bratislava, Faculty of Mechanical Engineering,
Nám slobody 17, 812 31 Bratislava 811 06
E-mail: silvia.mihelova@stuba.sk

Slovak University of Technology in Bratislava
Faculty of Mechanical Engineering
17th Conference on Applied Mathematics **APLIMAT 2018**

Proceedings

COMPARING EEG SIGNALS AND EMOTIONS PROVOKED BY IMAGES WITH DIFFERENT AESTHETIC VARIABLES USING EMOTIVE INSIGHT AND NEUROSKY MINDWAVE

NEDVĚDOVÁ Marie (CZ), MAREK Jaroslav (CZ)

Abstract. The paper is part of a broader study to verify some hypotheses about the relationship between the aesthetic characteristics and the emotions that an image provokes. Here we will not go through a survey where respondents would communicate their emotions, but we will use EEG signal monitoring devices. The main objective is to compare options and outputs from the Brain devices when examining emotions in images with different aesthetic characteristics.

Keywords: brain computer interfaces, aesthetic characteristics, electroencephalogram, EEG signals

Mathematics Subject Classification: Primary 00A66; Secondary 68U35.

1 Introduction

An instrument for recording the electrical activity of the brain, usually by means of electrodes distributed on the scalp, is called an electroencephalograph. The recording produced by such an instrument is called an electroencephalogram, commonly abbreviated EEG. The measurement of neurological signals was first used by Vladimir Pravdich-Neminsky, who measured the activity of a dog's brain in 1912. He used the term "electrocerebrogram". Hans Berger first measured human brain signals 10 years later. The electrode positioning system was designed by Herbert Jasper in the 1950s.
In this paper, we will try to introduce cheap devices to monitor the EEG signal. These devices are known as Brain-Computer Interface (Furthermore, only BCI). These devices enable measuring of human brain activity. At the same time, we give an overview of software for processing the obtained EEG signal. There is a device designed by an Australian entrepreneur Tan Le called Emotiv Insight for the price of 299 USD. Another option is a device called NeuroSky mindwave for the price of 79.99 USD, which was developed in California. These headsets differ in the number of sensors and the possibility of getting raw data. Of course, these systems do not achieve the quality of professional medical devices. Both headsets are wireless. For the Emotiv Insight device, wave measurement is a paid service (99 USD/month)

and only the characteristics derived from them can be obtained directly. These characteristics are: Engagement, Excitement, Interest, Relaxation, Stress, Focus. The Neurosky mindwave offers delta, theta, alpha, beta, and gamma wave tracking. These waves are part of the electroencephalogram.

The Emotiv Insight headset offers five EEG sensors, and two reference sensors. The Neurosky mindwave offers 2 EEG sensors. These headsets are shown in Fig. 1.

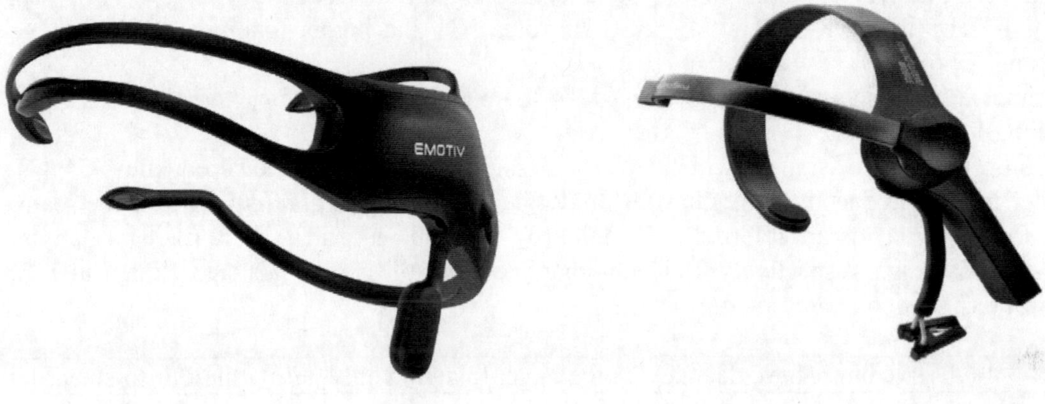

Fig. 1. Fotography of headsets: Emotiv Insight and Neurosky Mindwave. Source: [7], [8]

Affective computing is a very modern interdisciplinary field that tries to study and develop systems and devices that can recognize, interpret, process, and simulate human affects. The main objective of our paper is to compare options and outputs from the BCIs when examining emotions in images with different aesthetic characteristics. BCIs use different measures for a description of the user's emotions. Researches also developed other approaches for measuring, cf. [4].

Now, we will describe the concept of realizing a sample survey in which we will measure the emotions of viewers watching the picture. Klinger and Salingaros stated that the aesthetic impression of an image was determined by the variables Life and Complexity. They proposed the hypothesis; that by measurements L and C for patterns in the image, it is possible to determine the overall impression of the image. Klinger and Salingaros in article [5] considered these emotions: relaxing–distressing; pleasant–unpleasant; exciting–gloomy; arousing–sleepy.
We will try to measure these emotions using the above BCIs. In paper [5] Klinger and Salingaros form the hypothesis that aesthetic parameters can be used for explanation of emotion provoked by images. Much more information can be found in paper [6]. Using of brain-computer interfaces, opens space for testing of Klinger-Salingaros hypothesis.
Of course, it is possible to test the hypothesis even based on a direct description of emotions from the audience, but the headset provides an interesting option for verifying the hypothesis. If the hypothesis is confirmed, it will allow us, in the future, to create an image causing the selected emotion.
Based on the Klinger-Salingars approach, cf. [5], several images with different values of aesthetic characteristics L and C were generated. From these images, we select several paintings to analyze the suitability of using EEG devices to measure audience emotions.

The viewer will be looking at the painting "Colors on a grid" by Ellsworth Kelly, the painting "Echiquier I" by Aurélie Nemours and another six of our paintings. Each painting was viewed for one minute, and the EEG feature vector will be measured.

1.1 Electrophysiological features

With its detection of brain electrical activity over a wide frequency range, the brain computer interfaces detect several electrophysiological features. These include the alpha, beta and gamma rhythms. Devices are capable of detecting signals at frequency range 0.1 – 4 Hz (delta), 4-7 Hz (theta), 8-12 (alpha), 12-30 Hz (beta), 31 and higher (gamma). Cf. [1] (page 1216) and [2] (page 3).

Sometimes a more detailed division is used, so the alpha band can, for example, be divided into alpha 1 and alpha 2.

EEG comprises a set of signals, which may be classified according to their frequency. Cf. [1], [2]. Well-known frequency ranges have been defined according to distribution over the scalp. These frequency bands are referred to as delta (δ), theta (θ), alpha (α), beta (β), and gamma (γ) from low to high, respectively. Relevant characteristics of these bands are detailed in Tab. 1, their interpretation is presented in Tab. 2.

Delta	Observed in babies, decrease as they age. In adults only during their deep sleep.
Theta	Observed in young and older children, and in drowsy, meditative or sleep state adults.
Alpha	Primarily reflect visual processing in the occipital brain region. Increases when the eyes close and the body relaxes.
Beta	Associated with motor activities. During real movement or motor imagery are desynchronized.
Gamma	The presence in the brain activity of a healthy adult is related to certain motor functions or perceptions.

Tab 1. The characteristics of the frequency bands.

	Small value	Large value	Optimal value
Delta	Inability to revitalize the brain, poor sleep	Brain injuries, learning problems, inability to think	Natural healing, restorative/deep sleep
Theta	Anxiety, stress	Depression, hyperactivity, impulsivity, inattentiveness	Creativity, intuition, relaxation
Alpha	Anxiety, high stress	Daydreaming, inability to focus, too relaxed	Relaxation
Beta	Daydreaming, depression	Adrenaline, anxiety, high arousal, stress	Conscious focus, memory, problem solving
Gamma	Depression	Anxiety, high arousal, stress	Information processing, learning, REM sleep

Tab 2. The amount of the measured values of EEG waves and their most common causes.

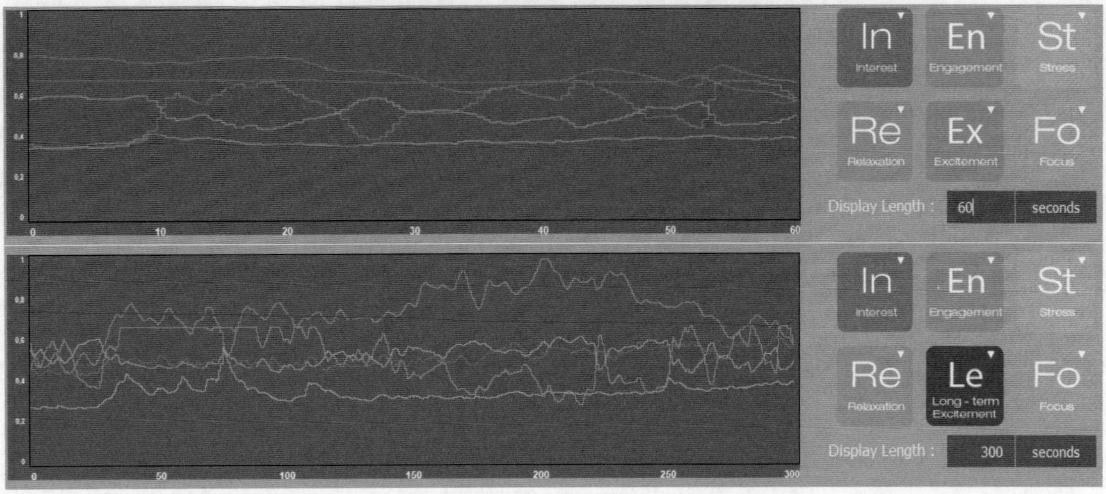

Fig. 2. Emotiv Xavier Control Panel: Detections performance metrix.

2 Brain Computer Interfaces

Now, we will summarize the properties of used BCI's, and their software. Cf. [3].

2.1 Emotiv Insight

This BCI monitors and records the degree of six key cognitive and emotional metrics. Those measures are focus, stress, excitement, relaxation, interest and engagement.
For using of this BCI a long calibration of the sensors is needed, and it requires application of saline substance to capture brain signals with high precision.
The application Emotiv Xavier ControlPanel allows you to view a chart of emotional measures. See Fig. 2.

2.2 Neurosky Mindwave

User mental activity in this BCI is measured by the following waves: delta, theta, low alpha 8-9Hz (alpha 1), high alpha 10-12Hz (alpha 2), low beta 13-17Hz (beta 1), high beta 18-30Hz (beta 2) and low gamma 31-40Hz (gamma 1), high gamma 41-50Hz (gamma 2) waves. See Fig. 3 and Fig. 4. This headset is simple, and it does not require callibration of sensors and application of a saline substance.

However, with the current software NeuroExperimenter, it is possible to recognize brain waves. Mental states, such as attention and meditation, are evaluated in percentages. See Fig. 3.
Attention and Meditation are calculated directly by the device and it is not possible to obtain the formula used in their calculation.
Based on information captured from user's brain, this software can be used to draw graphs of brain waves.

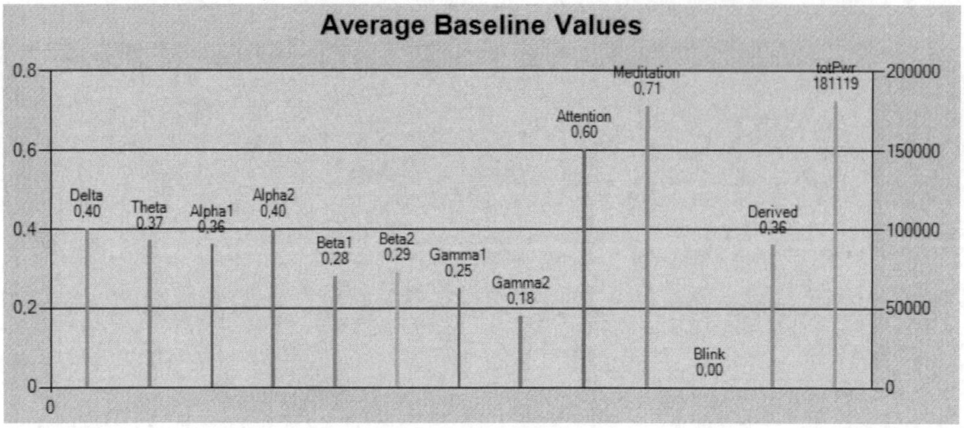

Fig. 3. NeuroExperimenter: derived variables.

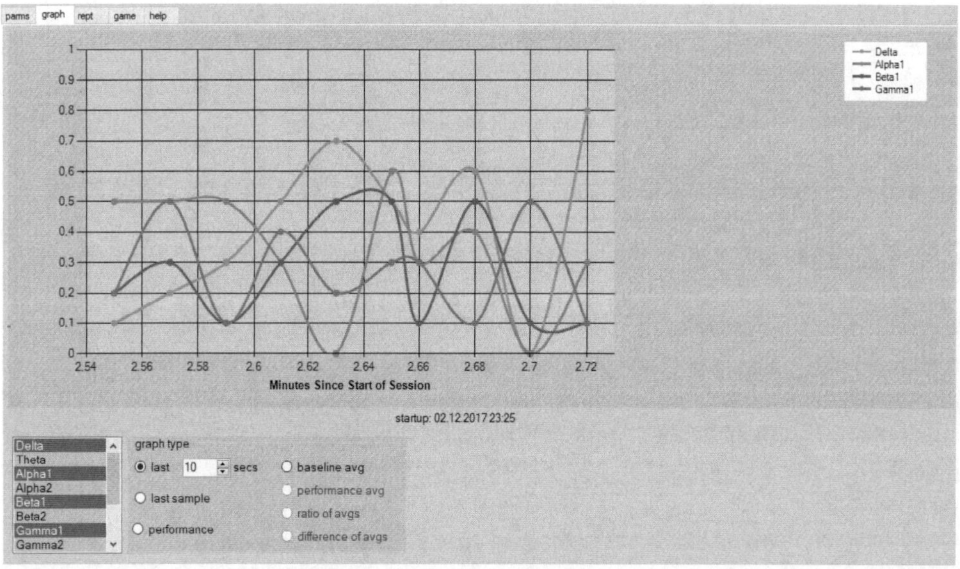

Fig. 4. NeuroExperimenter: measured signals.

3 Experiment

3.1 Choice of test images

Methods for constructing square images with different aesthetic features of Life L and Complexity C were presented in paper [6]. The algorithm for calculating these aesthetic characteristics is described in [5] and [6].
To monitor emotions using the brain computer devices, we select eight images (see Fig. 5). Figure (g) was created by Ellsworth Kelly and picture (h) was painted by Aurélie Nemours.

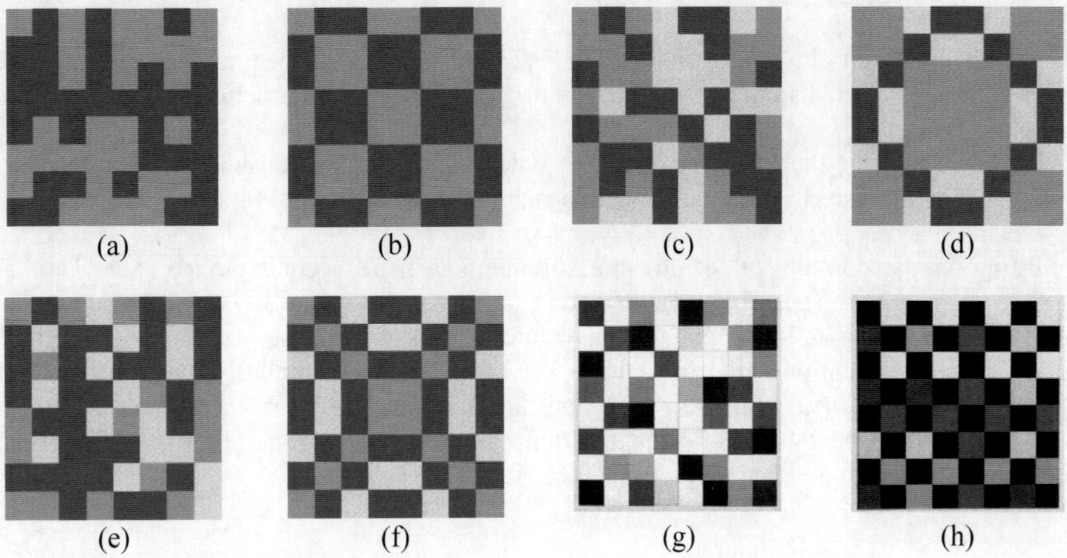

Fig. 5. Testing paintings (a), (b), (c), (d), (e), (f), (g), (h).

The other six paintings are our own and have been designed so that their L and C characteristics are as extreme as possible. In Tab. 3 the aesthetic characteristics of paintings, according to paper [5] are calculated.

Painting/characteristic	(a)	(b)	(c)	(d)	(e)	(f)	(g)	(h)
L	0.839	7.0	0.458	9.75	0.0	14.0	1.288	2.594
C	7.786	2.0	16.042	3.75	25.125	7.0	68.274	28.531

Tab 3. The aesthetic characteristics of paintings.

3.2 Measured signals

For testing of those BCIs, an experiment was made with one person. He needed to look at paintings from Fig. 5. Our all measurements were performed during one hour. Each session involved measuring variables for every image over a 60-second period.

Painting/characteristic	(a)	(b)	(c)	(d)	(e)	(f)	(g)	(h)	avg
Engagement (En)	73	66	71	65	67	73	70	55	67.500
Excitement (Ex)	73	70	69	68	68	66	62	63	67.375
Interest (In)	47	51	53	54	54	54	55	58	53.250
Relaxation (Re)	27	32	34	31	33	27	32	34	31.250
Stress (St)	58	58	58	58	58	58	58	58	58.000
Focus (Fo)	70	66	66	66	65	66	62	62	65.375

Tab 4. Measuring of emotions by Emotiv Xavier Control Panel.

We present the results in Tab. 4 and Tab. 5. We will particularly comment on whether the EEG signals are within a similar shape for paintings with different aesthetic characteristics.

From Tab. 4, it seems that Emotive Insight is unable to distinguish differences in perception of images. The measured values are almost constant (for example Stress).

Interesting is the comparison of the measurements by the second device. See Tab. 5. Symmetric paintings (i.e. (b), (d), (f)) had an average value of attention equal to (0.68+0.81+0.72)/3 = 0.74. While the unsymmetrical paintings (i.e. (a), (c), (e)) had an average value of attention equal to (0.46+0.58+0.43)/3 = 0.49. Similarly, we can see, that symmetric paintings (i.e. (b), (d), (f)) had an average value of meditation equal to (0.39+0.59+0.45)/3 = 0.48. While the unsymmetrical paintings (i.e. (a), (c), (e)) had an average value of meditation equal to (0.81+0.69+0.78)/3 = 0.76.

Painting/characteristic	(a)	(b)	(c)	(d)	(e)	(f)	(g)	(h)	avg
Delta	0.44	0.46	0.55	0.48	0.37	0.40	0.49	0.40	0.44875
Theta	0.38	0.38	0.35	0.38	0.39	0.38	0.43	0.37	0.38250
Alpha 1	0.36	0.34	0.27	0.32	0.42	0.34	0.31	0.36	0.34000
Alpha 2	0.35	0.33	0.35	0.35	0.42	0.35	0.35	0.40	0.36250
Beta 1	0.27	0.27	0.26	0.27	0.29	0.29	0.27	0.28	0.27500
Beta 2	0.31	0.30	0.26	0.30	0.29	0.33	0.26	0.29	0.29250
Gamma 1	0.23	0.26	0.22	0.24	0.21	0.27	0.19	0.25	0.23375
Gamma 2	0.17	0.19	0.14	0.17	0.15	0.19	0.14	0.18	0.16625
Attention	0.46	0.68	0.58	0.81	0.43	0.72	0.31	0.60	0.57375
Meditation	0.81	0.39	0.69	0.59	0.78	0.45	0.61	0.71	0.62875

Tab 5. Measuring of emotions by Neurosky mindwave.

3.3 Dependency between EEG based measures, emotions and aesthetic characteristics

First, we examine the correlation between Emotiv Insight variables and aesthetic characteristics. Then we determine the correlation between Neurosky Mindwave values and L, C characteristics. Values of Pearson's correlation coefficient

$$r_{x,y} = \frac{\sum (x_i - \bar{x})(y_i - \bar{y})}{\sqrt{\sum (x_i - \bar{x})^2 \sum (y_i - \bar{y})^2}} \qquad (1)$$

are calculated in Tab. 6.

Emotiv	Engagement	Excitement	Interest	Relaxation	Stress	Focus
L	0.0894	-0.0577	0.0973	-0.4982	NaN	0.1107
C	-0.0812	-0.7358	0.4591	0.3605	NaN	-0.7196

Neurosky	Delta	Theta	Alpha 1	Alpha 2	Beta 1	Beta 2
L	-0.16	-0.08	-0.12	-0.41	0.31	0.70
C	0.089	0.73	-0.08	0.25	-0.02	-0.67

Neurosky	Gamma 1	Gamma 2	Attention	Meditation
L	0.75	0.70	0.79	-0.77
C	-0.74	-0.62	-0.77	0.23

Tab. 6. Correlation coefficients.

We can see that dependence between Emotiv Insight variables and Life and Complexity characteristics have very small correlation coefficient values. When using the Neurosky variables, great absolute values of correlation are between L and Attention, L and Meditation, C and Attention. These correlation coefficients are close to 0.7 and -0.7 respectively. From this, we can conclude that images with larger values of the L characteristic and small values of the C characteristic were observed with greater attention. However, watching pictures with a low L characteristic had greater calming and relaxing effects, so that pictures had a greater value of the Meditation variable.

3 Conclusion

It is now possible to purchase many brain computer interfaces (EEG metering devices) for which emotional measurement software is available. In this paper, we compared outputs and inputs of two devices: Neurosky Mindwave and Emotiv Insight. We analysed the possibilities of using them to compare emotions that images, with different aesthetic characteristics, provoke. For widespread use of those BCIs, it is necessary to explore the current limitation and precision of obtained signals. Our recent research was focused to testing of the Klinger-Salingaros hypothesis. We tried to analyze a dependency of the human psyche frame on aesthetical characteristics. When positive emotional paintings were presented the viewer EEG characteristics were higher for attention. We found that the Neurosky Mindwave is more suitable for verifying the effect of aesthetic characteristics on the emotions of the viewer.

Acknowledgement

The paper was supported by grant SGS 2018 of the University of Pardubice. We would like to thank the reviewers for their detailed comments. Their suggestions for the design of experiment (with BCIs and paintings with different aesthetic characteristics) will be exploited in future works.

References

[1] NICOLA-ALONSO, L. F., GOMEZ-GIL, J., Brain Computer Interfaces, a Review. *Sensors*, vol. 12, no. 2, 2012: pp. 1211–1279, doi: 10.3390/s120201211.

[2] MOAZZAMI, M.-M., *EEG Signal Processing in Brain-Computer interface*. Michigan: Michigan State University, Master thesis. 2012

[3] VOURVOPOULOS, A., LIAROKAPIS, F., Evaluation of commercial brain–computer interfaces in real and virtual world environment: A pilot study. *Computers & Electrical Engineering*, vol. 40, no. 2, 2014: pp. 714–729, doi: 10.1016/j.compeleceng.2013.10.009.

[4] SORINAS, J., MURCIA, M.D.G., MINGUILLON, J., SÁNCHEZ-FERRER, F., VAL-CALVO, M., FERRÁNDEZ, J.M., and FERRÁNDEZ, E., Setting the Parameters for an Accurate EEG (Electroencephalography) - Based Emotion Recognition System. In 7th International Work-Conference on the Interplay Between Natural and Artificial Computation IWINAC 2017, June 19-23, 2017, Coruna, Spain, 2017, ISBN 978-3-319-59772-0, pp. 265–226.

[5] KLINGER, A., SALINGAROS, N.A., A pattern measure. *Environment and Planning B: Planning and Design*, vol. 27, no. 4, 2000: pp. 537-547, doi: 10.1068/b2676.

[6] NEDVĚDOVÁ, M., MAREK, J., Mathematical aesthetics: Domain of variables Life and Temperature. In 2017 IEEE 14th International Scientific Conference on Informatics, November 14-16, 2017, Poprad, Slovakia, 2017, ISBN 978-1-53860888-3, pp. 221–226.

[7] URL, https://www.emotiv.com/product/emotivpro/#

[8] URL, https://store.neurosky.com

Current address

Nedvědová Marie, Ing.
Department of Mathematics and Physics, Faculty of Electrical Engineering and Informatics, University of Pardubice
Studentská 95, 530 02 Pardubice, Czech Republic
E-mail: st36806@student.upce.cz

Marek Jaroslav, Mgr., PhD.
Department of Mathematics and Physics, Faculty of Electrical Engineering and Informatics, University of Pardubice
Studentská 95, 530 02 Pardubice, Czech Republic
E-mail: jaroslav.marek@upce.cz

Slovak University of Technology in Bratislava
Faculty of Mechanical Engineering
17th Conference on Applied Mathematics **APLIMAT 2018**

Proceedings

ACTIVE-LEARNING TECHNIQUES IN MATH COURSES IN ENGINEERING DEGREES: THE PURSUIT FOR A NEW ENGINEER PROFILE

NICOLA Susana (PT), MENDONÇA Jorge (PT), PINTO Carla (PT)

Abstract. In this work, we present results of application of new teaching methodologies for Math courses in Bachelor Engineering Degrees. The traditional way of teaching mathematics is based on a 'teaching by telling' approach, especially in the early years of the courses. It is characterized by classes with many students, where they are exposed to the contents of a single curricular unit. Recently, there has been an increasing interest by engineering professionals and bodies to accredit engineering degrees in promoting change in this paradigm. The new teaching paradigm proposed here consists on the implementation of eduScrum and Jigsaw in a Differential and Integral Calculus I course in a Bachelor Engineering Degree at the School of Engineering of the Polytechnic of Porto. EduScrum creates an environment where students are masters of their knowledge, increasing their initiative to learn and entrepreneurship. Jigsaw develops more responsibility/autonomy and notion of belonging in students, hardly achieved when working alone. Some results of the implementation of these methods are presented and discussed.

Keywords: eduScrum, Jigsaw, Active-learning, Novel teaching methodologies.

Mathematics subject classification: Primary 65D17; Secondary 53A04, 14Q05.

1 Introduction

The world of education encourages, inspires and helps students to achieve great things in life. Educators are faced with a myriad of individual learners' needs. The traditional classroom encompasses a range of mixed abilities among students, some struggling to meet grade level standards, and some performing above grade level. Educators must, thus, be able to adjust to this immense heterogeneity. Moreover, to engage and develop the new skills of the 3rm industrial revolution, namely, problem solving, critical thinking, communication, collaboration and self-management, educators need to think new pedagogical methods, new ways of teaching, and novel instructional management methods.

In this paper, we introduce results of application of new teaching methodologies, namely eduScrum [9] and Jigsaw Method Teaching Strategy [6], in a Differential an Integral Calculus I course of a Bachelor Engineering degree. This type of methodology engages students in what constitutes real life-learning experiences, and enhances their knowledge. EduScrum creates an environment where students are the owners of their knowledge, increasing the sense of self-initiative and entrepreneurship. Knowledge's responsibility is thus transferred from teachers to students. Applying the eduScrum framework enhances collaborative learning, emphasizing the students' interactions with each other. With Jigsaw, students learn course's contents cooperatively. It is a strategy in which students become experts in specific topics and teach their topic to their peers. This promotes students' engagement, learning and team group. Students interaction is thus highlighted, developing a sense of belonging to a community and value for each other contributions. *'Some teachers see themselves as the designated expert whose role is to impart their knowledge to students who are empty vessels. That's the wrong metaphor'* - Dr William Rando, Director of Chicago Center for Teaching at the University of Chicago.

In 2007, Shekar [13], underlines manners in which active-learning (AL) can raise up the students learning outcomes in a project based learning methodology. AL has been recognized 'as a means of skills transfer through apprenticeship' and it is a tool to integrate in the student's mind the culture practiced in the entrepreneurial/business world. This author presents a successful case study of a product development applying the PBL approach. Edström *et al* [2] point out the differences and similarities between problem/project-based learning (PBL), and conceive–design–implement–operate (CDIO) teaching methodologies in engineering education. They emphasize that the primordial principle of PBL is that students are responsible for their learning process. CDIO is described as an alternative way to the traditional engineering education, postulating an 'engineering culture' rather than an 'engineering science'. Bottom line, the authors conclude that PBL is a benchmark of student-centered education, paving the way to the implementation of CDIO. Gonçalves *et al* [5], apply a Problem-Based-Learning (PBL) approach to teaching and assessment of a Statistics chapter of the curricular unit of Computational Mathematics in an ISEP Bachelor Engineering degree. The authors implemented the PBL approach at a practical work on 'Simple Regression and Linear Correlation', using MS Excel, for computational support, to 394 students. A questionnaire was elaborated, focusing on the understanding and acquisition of knowledge using PBL. Authors conclude that the new method develops critical spirit, contributes to individual training, namely sense of responsibility, capacity and dis-inhibition in group working, analysis and decision. Pinto Ferreira *et al* [11] describe an eduScrum application, used as a pedagogical approach, on mathematics courses of the ISEP Engineering bachelor program. They provide and discuss results of application.

The goal of the present study is to reply to the following question: how does the implementation of these new pedagogical methodologies, namely Jigsaw and eduScrum, on the Differential and Integral Calculus I Course affect the students' own learning process?

With the aforementioned ideas in mind, the outline of the paper is as follows. In Section 2, we describe the implementation of the new teaching methodologies in the Differential and Integral Calculus I Course in an Engineering Bachelor Course at the School of Engineering of the Polytechnic of Porto (ISEP). In Section 3, we present and discuss the results of a questionnaire posed to students. In the last section we conclude our work and highlight *pros* and *cons* of this pilot-implementation of these novel methodologies.

2 Teaching methodologies

In this section, we describe in more detail the teaching techniques Jigsaw and eduScrum, and their implementation in a Differential and Integral Calculus I of a Bachelor Course at ISEP.

Fig. 1. Flowchart of the Jigsaw methodology [15].

2.1 Jigsaw technique

Jigsaw is a cooperative technique in which the teacher facilitates students' learning. His role is not to teach but to easy students' learning process. Students learn and teach in groups, ie., among their peers. Biggs says that 'Most people learn 95% of what they explain to others' [1]. All students in the groups must learn and teach their subject to others. The later increases responsibility and students' social skills, namely communication and teamwork. In the end, the teacher must check if students have adequately understand the contents of that class. The later may be done with some quiz questions in the end of the class, or any form of evaluation considered valid by the teacher. An implementation of Jigsaw in a class in which 4 concepts must be apprehended by the students is generically designed as follows [10]. The teacher must prepare notes for each of the topics, and these notes are made available to all students. Then students are gathered in groups of 4. These are called the *Home groups*. Each student in these groups learns one of the 4 concepts individually. This task may take 10 minutes. Then, all students who have learned topic i, $i = 1, \cdots, 4$, move to an *Expert group* for that specific topic i. There, they take 10 minutes to discuss between peers the key points of that topic and how they will instruct it to the other members of their *Home groups*, when they return. There, they have 10 minutes to teach their peers. In the end of this activity, the teacher verifies the degree of understanding of the 4 concepts by all students.

2.2 EduScrum methodology

Scrum is a framework for project management in which teamwork, accountability and iterative progress, towards a well-known objective, are highlighted. The main foundations of Scrum are transparency, inspection and adaptation [14]. EduScrum is an adaptation of Scrum to education. Schools are using Scrum to boost students' learning, in an enjoyable team group way. Application of Scrum is associated with a higher quality of education, better grades and higher motivated students [14]. The applicability of eduScrum is done by pre-defining events, the so-called time-boxed events, to generate

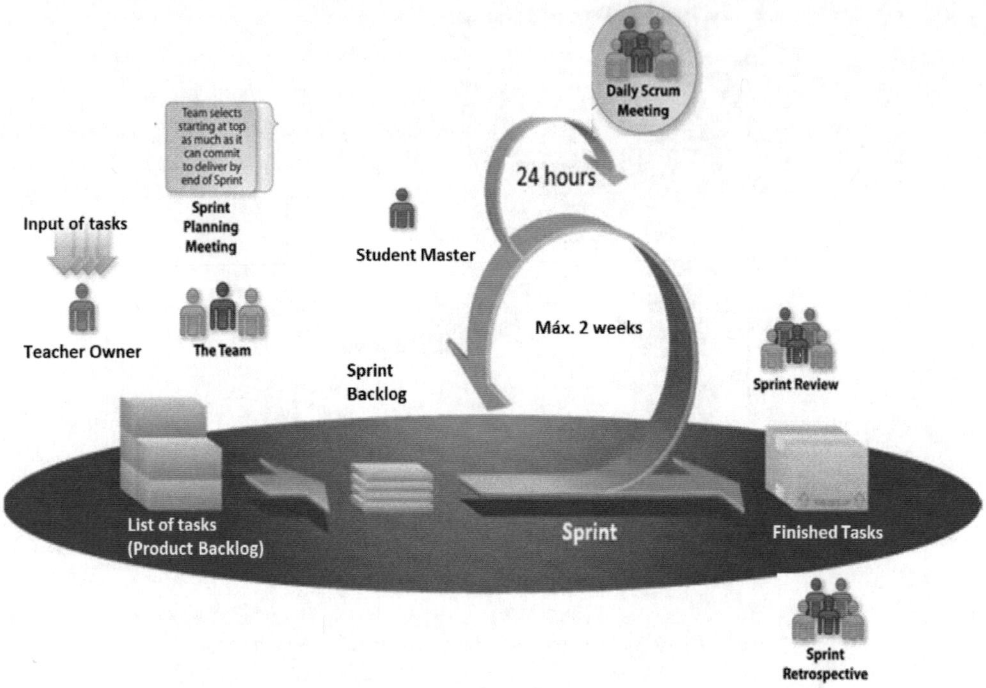

Fig. 2. Flowchart of the eduScrum methodology [12].

a sense of regularity and predictability. Each event in Scrum is an opportunity to inspect and adapt something, which enables critical transparency. The core of eduScrum is the Sprint. A Sprint is a logical set of learning material, designed to accomplish a certain learning goal. The time to execute a Sprint is pre-set, something up to two months, so student teams plan well and manage the complexity of the task. The timetable of the Sprint starts with the Sprint Planning Meeting, followed by the Team Formation. The Student Teams plan their actions independently during the Sprint. Stand ups allow teams to discuss assignments and tasks within the Sprint, at the beginning of each class. The Sprint ends with a review and retrospective, students inspect their final work and enumerate possible ameliorations. In our case-study, we implement eduScrum as follows. We (teacher owner) will propose a list of exercises (input tasks) in the practical classes every two weeks (sprints). Students in class are asked to form groups of 4/5 members each (the team). After forming the groups, students will choose a Student Master, who is responsible for distributing the tasks to each member of the group. This procedure ends with a Sprint Review, where it is performed the sprint assessment. The later consists of 3 components: 1) assessment of tasks performed - usually calculating the weighted average of accepted tasks; 2) activities not accepted have a 0; 3) assessing students' individual contribution by analyzing the team's Scrum board. In the Sprint retrospective students discuss key points of the solved tasks and make a brief report of what went well; what went wrong; what should be improved in the next Sprint. Moreover, students' will be evaluated in two written tests, one approximately in the middle of the semester and one in the end of the semester.

3 Discussion of the results

In this section we highlight the results, interpreting the findings and outlining what they mean. A questionnaire was done to 48 students attending the course of Differential and Integral Calculus I, at a Bachelor Engineering Degree at ISEP, in the first semester of 2017-2018. Were analyzed the following questions about JigSaw (JGi) and eduScrum (EDSi) methodologies in comparison with traditional method(TMi). Also, some benefits (Soft Skills) linked to the different methodologies were analyzed to emphasized the performance of those methods.

Questions:
Q1: JIG1- In your opinion, does working in groups enhances your learning skills?
Q2: JIG2 - Do you consider Jigsaw a useful learning tool?
Q3: JIG3 - Did you feel more motivated to learn with the Jigsaw methodology?
Q4: EDS1 - In your opinion, can the eduScrum methodology help you to develop soft skills for your future professional life?
Q5: EDS2 - Do you agree with the learning by doing method in the practical classes?

Benefits (Soft Skills):

Eduscrum methodology – EDSM

EDSM1- Communication; EDSM2 -Self-motivation; EDSM3 -Leadership ;
EDSM4 -Responsibility; EDSM5 -Teamwork; EDSM6 -Problem solving
EDSM7 -Ability to work under pressure and time management; EDSM8 -Flexibility
EDSM9 -Negotiation and Conflict resolution

Traditional methods of teaching - TM
TM1- Communication; TM2 -Self-motivation;
TM3 -Leadership; TM4 -Responsibility; TM5 -Teamwork
TM6 -Problem solving; TM7 -Ability to work under pressure and time management
TM8 -Flexibility; TM9 -Negotiation and Conflict resolution

We observe that these 9 soft skills are the same for eduScrum and the Traditional Method. The reasoning for the choice of these skills is as follows. Self-motivation improves team and class dynamics, by helping to build trust between peers and with the teacher. Students are motivated to take responsibility for their own and for their peers' learning. Communication helps to build cohorts (i.e., within majors) by enabling students to really get to know each other in class. The later can translate into more interactions outside class, in departmental and campus activities. Teamwork fosters student engagement through peer-learning and more equal participation in the group, by empowering individual students to share their own 'expertise'.

In the following tables are presented the correlation coefficients and chi-square tests for the results of the questionnaire to the questions above. At bold are marked the significant results. In fact, the correlations between JigSaw and EDSM questions show positive coefficients, against negative coefficients when considering the TM.

Table 1 shows the correlation between Jigsaw and the benefits intrinsic to the Edusrum methodology. In both tables 1,2 variables are related to each other. As an example, in table1,is rejected the null hypothesis with no relationship between JIG1 and EDSM1 ($p=0.027$) and also between JIG1 and EDSM4 ($p=0.006$) and EDMS5 ($p=0.008$). The same conclusion between JIG2 and EDMS2 ($p=0.021$), EDMS4 ($p=0.001$) and EDMS6 ($p=0.04$) were observed. The previous results were supported by chisquare test result as presented in table 2. Although JIG3 in table 2 show some association with more benefits.

Table 3 shows the benefits associated with the Jigsaw methodology. These conclusions are also

	EDSM1	EDSM2	EDSM3	EDSM4	EDSM5	EDSM6	EDSM7	EDSM8	EDSM9
JIG1	0,32	0,13	-0,17	0,39	0,38	0,15	0,14	0,16	0,08
	0,027	*0,382*	*0,253*	***0,006***	***0,008***	*0,321*	*0,329*	*0,27*	*0,599*
JIG2	0,15	0,33	0,05	0,45	0,09	0,3	0,28	0,29	0,2
	0,308	***0,021***	*0,727*	***0,001***	*0,533*	***0,04***	*0,052*	*0,048*	*0,164*
JIG3	0,2	0,32	0,25	0,11	-0,1	0,15	-0,05	-0,03	0,06
	0,811	***0,029***	*0,092*	*0,439*	*0,479*	*0,301*	*0,716*	*0,856*	*0,709*

Tab. 1. Correlations JIG vs EDSM

	EDSM1	EDSM2	EDSM3	EDSM4	EDSM5	EDSM6	EDSM7	EDSM8	EDSM9
JIG1	17,233	18,401	14,549	21,446	22,306	11,900	9,197	11,389	8,139
	0,045	*0,104*	*0,267*	*0,011*	*0,008*	*0,454*	*0,686*	*0,496*	*0,774*
JIG2	16,121	17,461	18,386	17,496	9,906	31,735	36,780	20,594	10,807
	0,186	*0,356*	*0,302*	*0,132*	*0,624*	*0,011*	*0,002*	*0,195*	*0,821*
JIG3	21,538	22,856	22,219	22,300	13,167	34,442	43,109	61,868	34,054
	0,043	*0,118*	*0,136*	*0,034*	*0,357*	*0,005*	*0,000*	*0,000*	*0,005*

Tab. 2. Chisquare tests and p-value JIG vs EDSM

	Benefits
JIG1	EDMS1- Communication
	EDMS4- Responsibility
	EDSM5 -Teamwork
JIG2	EDSM2 -Self-motivation
	EDMS4- Responsibility
	EDSM6 -Problem solving
JIG3	EDSM2 -Self-motivation

Tab. 3. Benefits associated to the Jigsaw methodology

confirmed in the literature [7], [8]. According to Murphy [7], the benefits of communication (EDMS1) in Jigsaw technique "develops communication and teamwork skills". The authors [8] argue that students become more responsible (EDMS4) by helping building cohorts "enabling students to really get to know each other in one class, which can translate to more interactions outside of class in departmental and campus activities", [8] . "Fosters student engagement through peer learning and more equal participation by everyone in the group by empowering individual students to share their own expertise [8], improve teamwork (EDMS5) and self-motivation (EDMS2), facilitating interaction between students [7], to achieve a common goal. Self-motivation is a benefit, also mentioned in other works, as linked with jigsaw technique: "Improves team and class dynamics by helping to build trust", [8]; "Students are motivated to take responsibility for their own and each others", [7]. Problem-solving (EDMS6) and critical thinking are other benefits related with Jigsaw "The students were able to answer questions by critically analyzing their papers" [4].

	TM1	TM2	TM3	TM4	TM5	TM6	TM7	TM8	TM9
JIG1	0,04	-0,04	-0,08	0,07	0,08	-0,09	0,06	-0,09	0,14
	0,8	*0,811*	*0,6*	*0,623*	*0,603*	*0,53*	*0,662*	*0,522*	*0,337*
JIG2	-0,12	-0,2	-0,13	-0,06	-0,08	-0,26	0,09	-0,01	0
	0,421	*0,163*	*0,363*	*0,696*	*0,577*	*0,075*	*0,549*	*0,966*	*0,977*
JIG3	0,18	-0,27	0,25	-0,13	0,13	-0,27	-0,1	0,11	0,06
	0,214	*0,063*	*0,086*	*0,378*	*0,381*	*0,061*	*0,511*	*0,475*	*0,701*

Tab. 4. Correlations JIG vs TM

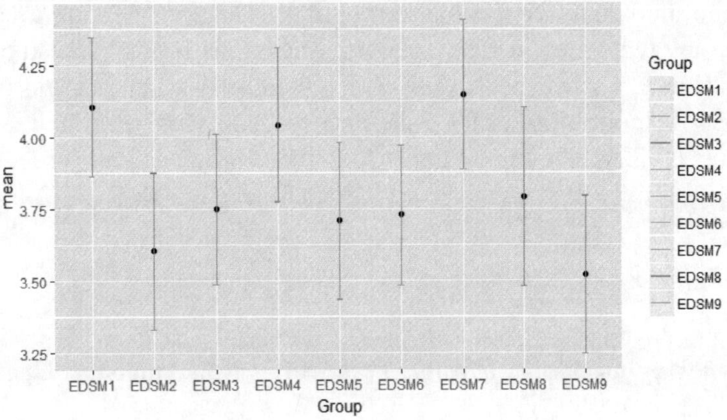

Fig. 3. Eduscrum Methodology

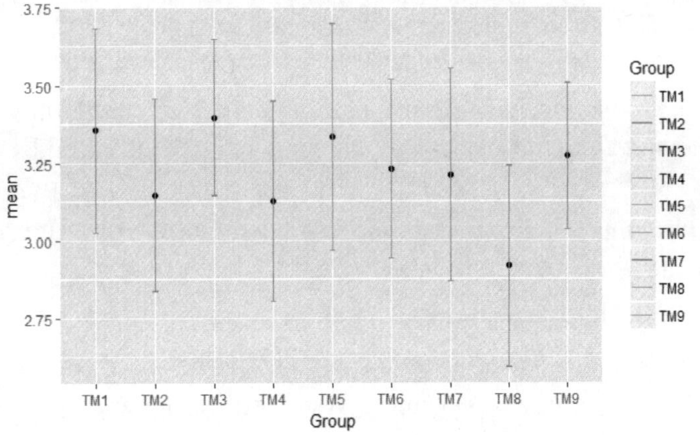

Fig. 4. Traditional Methods

These two figures 3, 4 allow conclude that overall results from traditional methods are below those eduScrum. So, there is a clear that new methodologies promote these skills.

4 Conclusions

In this paper, we focus on some preliminary results of application of Jigsaw and eduScrum in a Bachelor Degree Course at the School of Engineering of the Polytechnic of Porto. The authors presented new teaching-learning methodologies of Mathematics in Engineering Degree courses. We focus on two AL methodologies, namely the eduScrum method [8] and Jigsaw [7]. AL revealed itself useful by helping students to be involved in their knowledge and help them recognize how dynamic a math class could be. Students never felt bored, attending classes till the end, and learning contents happily. According to the results, one could argue, that instead of creating a class of competitive study we contributed to create a cooperative learning math classroom. The students could achieve success by paying attention to their peers (EDMS5 - Teamwork), helping and teaching each other (EDMS5 and EDMS4-Responsibility), they were encouraged to embrace the knowledge from each student around them (EDMS2- Self-Motivation). Also they enjoy a higher sense of ownership themselves (EDMS4-Responsibility) and a greater trust in their peers (EDMS5- Teamwork and EDMS6 - Problem Solving).

References

[1] BIGGS, J.B.: *Teaching for quality learning at university*. Buckingham: Open University Press, 1999.

[2] EDSTRÖM K., KOLMOS, A.: *PBL and CDIO: complementary models for engineering education development*, European Journal of Engineering Education, vol.39, no. 5, pp. 539–555, 2014.

[3] FREEMAN, S.: *Active learning increases student performance in science, engineering, and mathematics*. Proceedings of the National Academy of Sciences, vol.111, no. 23, pp. 8410-8415, 2014.

[4] GÖÇER, A. (2010). *A comparative research on the effectivity of cooperative learning method and jigsaw technique on teaching literary genres*. Educational Research and Reviews, v5(8), 439.

[5] GONÇALVES, G., NICOLA, S., AFONSO, L., FERREIRA, M., FERRO, T., MENDONÇA, J.: *Problem Based Learning: An Experience in Computer Engineering*, Journal of Education and Human Development, vol.3, no.4, pp. 271-281, 2014.

[6] Jigsaw classroom, URL: https://www.jigsaw.org/.

[7] Jigsaw-CDIO, URL: http://slideplayer.com/slide/9344337/

[8] Jigsaw-Sesync, URL: https://www.sesync.org/system/tdf/resources/ses case study best practices jigsaw and cooperative learning.pdf

[9] KINGSTON, J.: *Doing Twice as Much Maths in Half the Time: Implementing Scrum Methodology in a Year 7 Mathematics Classroom*. TEACH Journal of Christian Education, vol. 9, no.2, pp. 4-7, 2015.

[10] MURPHY, M., MACCARTAN, C.: *THE CDIO APPROACHTO ENGINEERING EDUCATION:3. Engaging Students in Their Learning*, URL: http://slideplayer.com/slide/9344337/, accessed at november 22 2017.

[11] PINTO FERREIRA, E., MENDONÇA, J., NICOLA, S.:*'Eduscrum methodology in mathematical engineering education*, INTED2017 Proceedings, pp. 2994-3000. 2017.

[12] Scrum Methodology, URL: https://www.inloox.com/company/blog/articles/the-scrum-methodology/

[13] SHEKAR, A.: *Active learning and reflection in product development engineering education.* European Journal of Engineering Education, vol.32, no.2, pp. 125–133, 2007.

[14] SUTHERLAND, J.: Scrum: *The Art of Doing Twice the Work in Half the Time*, Crown Business, ISBN 978-0385346450. 2014.

[15] Yotam's Courses. Accessed June 29th 2017. Retrieved from URL: https://sites.google.com/a/leobaeck.net/courses/grade-8/13-base-groups/activity2-jigsaw-23102010.

Current address

Nicola Susana, PhD
School of Engineering, Polytechnic of Porto
Rua Dr António Bernardino de Almeida, 431, 4249–015 Porto, Portugal
E-mail: sca@isep.ipp.pt

Mendonça Jorge, PhD
School of Engineering, Polytechnic of Porto
Rua Dr António Bernardino de Almeida, 431, 4249–015 Porto, Portugal
E-mail: jpm@isep.ipp.pt

Pinto Carla, PhD
School of Engineering, Polytechnic of Porto
Rua Dr António Bernardino de Almeida, 431, 4249–015 Porto, Portugal
E-mail: cap@isep.ipp.pt

Slovak University of Technology in Bratislava
Faculty of Mechanical Engineering
17th Conference on Applied Mathematics **APLIMAT 2018**

Proceedings

APPLICATION OF SHAPE DISTRIBUTIONS TO COMPARE TRIANGULAR MESHES

PAJEROVÁ Nikola (CZ), LINKEOVÁ Ivana (CZ)

Abstract. In computer graphics and its applications where shape-based recognition problem is solved, a shape distributions are used to calculate a similarity measure and discriminate among similar and dissimilar shapes. In this paper, we apply the shape distributions method to five triangular meshes obtained by optical scanning of the same object. The aim is to find relation between form deviations of individual meshes and similarity measures obtained by shape distributions approach.

Keywords: shape recognition, shape function, shape distribution, triangular mesh

Mathematics Subject Classification: Primary 68U05, 68U10; Secondary 62-07.

1 Introduction

Shape-matching or shape-based recognition problem is widely solved in computer graphics, computer vision, mechanical engineering, molecular biology, etc. Application of shape distributions represents an efficient approach to compute similarity measures of 3D shapes. The main idea is to represent a digital shape signature for a 3D model as a probability distribution sampled from a suitable shape function measuring geometric properties of the 3D model. This generalization of geometric histograms is so called shape distribution [1], [2]. This approach reduces the shape matching problem to the relatively simpler problem, i.e. comparison of two probability distributions instead of traditional shape matching methods, e.g. parametrization, feature correspondence and model fitting. The method for construction of shape distributions from 3D polygonal models to compute a measure of their dissimilarities consists in the following steps: shape function selection, random points sampling, shape distributions calculation and shape distributions comparison. Shape function is usually based on simple measurements of geometrical features, e.g. angles, distances, areas, volumes, [1] to [4]. Shape distributions calculated for a sufficiently large number of random sample points of the surface are compared using well-known curve matching techniques, e.g. Minkowski L_N norm.

In this paper, the shape distributions method is used to determine similarity of five triangular meshes in stl format obtained by optical scanning of ball-bar standard with two precise spheres connected by a cylinder (Fig. 1). This standard is used for calibration of optical scanners when the diameters of the two spheres (obtained by least squares fitting to the corresponding parts of the triangular mesh) and the distance of their centres are measured. The goal of this paper is to find the correspondence between the form deviations of individual meshes and similarity measures obtained by the shape distributions approach.

Fig. 1. Ball-bar standard.

2 Triangular meshes pre-processing

The standard was scanned by portable handheld optical scanner separately five times. Thus, five different triangular meshes, each of them in general position with respect to the coordinate system, were obtained. The meshes contain different parts of scanned handles which have to be removed before evaluation because they can distort the results, see example in Fig. 2.

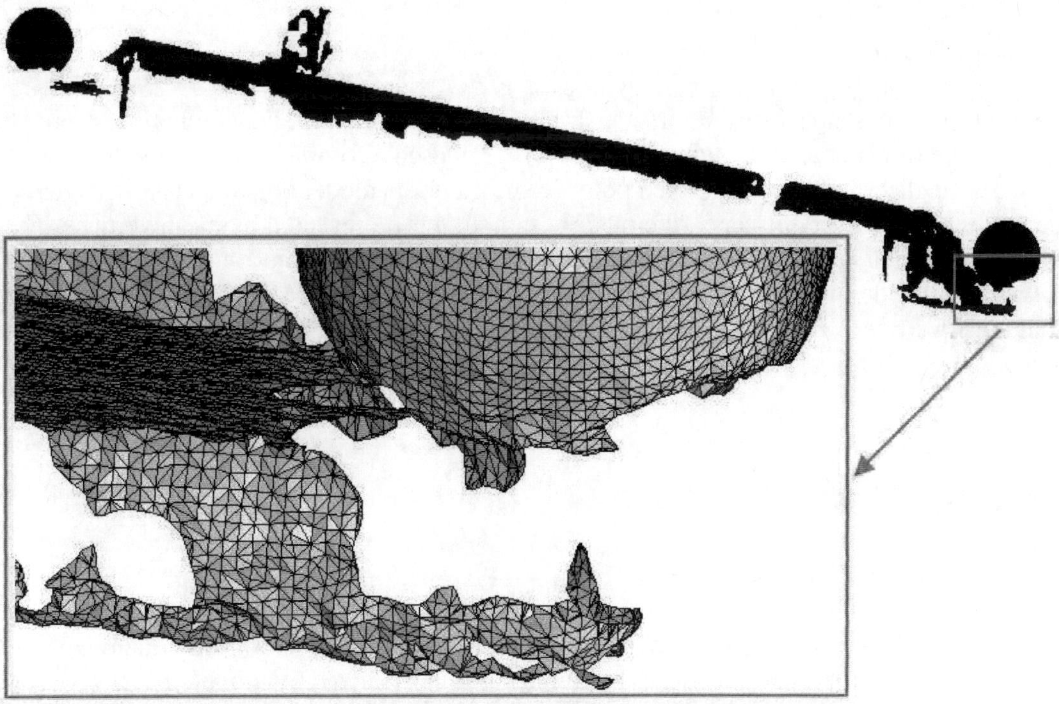

Fig. 2. Ball-bar standard and a detail of triangular mesh.

To remove the same part corresponding to the handles, the meshes were aligned with respect to the coordinate system in the following way. The least square method was used to associate two spheres to the corresponding parts of the triangular mesh. Denoting S_1 and S_2 the centres of the spheres, the centre of straight line segment S_1S_2 lies at origin O and $S_1S_2 \subset x$. Now, it is possible to choose suitable position of section planes and trim inappropriate parts of meshes, see Fig. 3. Vertices of meshes depicted in Fig. 3 represent the input data for further processing.

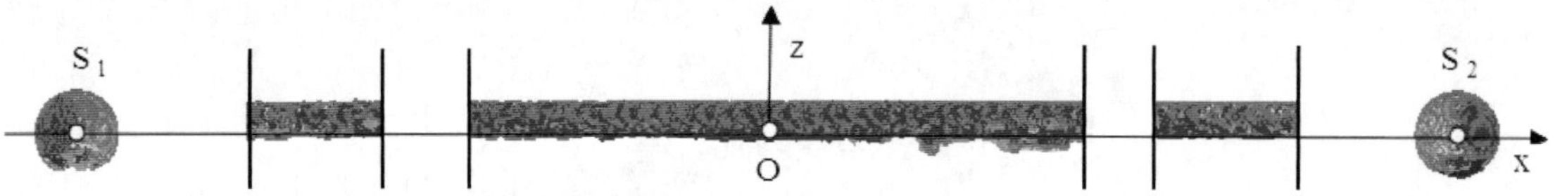

Fig. 3. Five triangular meshes aligned with respect to the coordinate system.

In addition to the five scanned triangular meshes, a triangular mesh of the nominal CAD model, see Fig. 4, of the standard was generated (nominal mesh). The position of the nominal CAD model with respect to the coordinate system is identical to the position of meshes in Fig. 3.

3 Shape distributions

In this section, a detailed description of the method used for construction of shape distribution based on triangular mesh is given.

3.1 Shape function

The first step of shape distribution construction is to choose a shape function. Here, the modification of $D1$ shape function described in [1] is used. Originally, the shape function $D1$ measures the distance between a fixed point (centroid of the model boundary) and one random point on the surface. In this paper, the random sampling is represented by mesh vertices. Then, the shape function is a function measuring the oriented distance of mesh vertex from origin O. As the sampling density strongly influences the precision of the shape distribution, the distances of all mesh vertices $V_i = (x_i, y_i, z_i)$ from origin, i.e.

$$f_i = \mathrm{sign}(x_i)\sqrt{x_i^2 + y_i^2 + z_i^2}, i = 0,1,\cdots,n, \quad (1)$$

are taken into consideration.

3.2 Shape distribution construction

The second step is a frequency histogram calculation, i.e. determination how many values f_i fall into each of k fixed sized classes. The frequency is normalized by number of mesh vertices n to eliminate the influence of different number of meshes vertices. An example of histogram for nominal mesh is shown in Fig. 4. Due to the mutual position of the nominal CAD model and the histogram for nominal mesh depicted in Fig. 4, it is obvious that the oriented distance reflects the shape characteristic of the mesh very well.

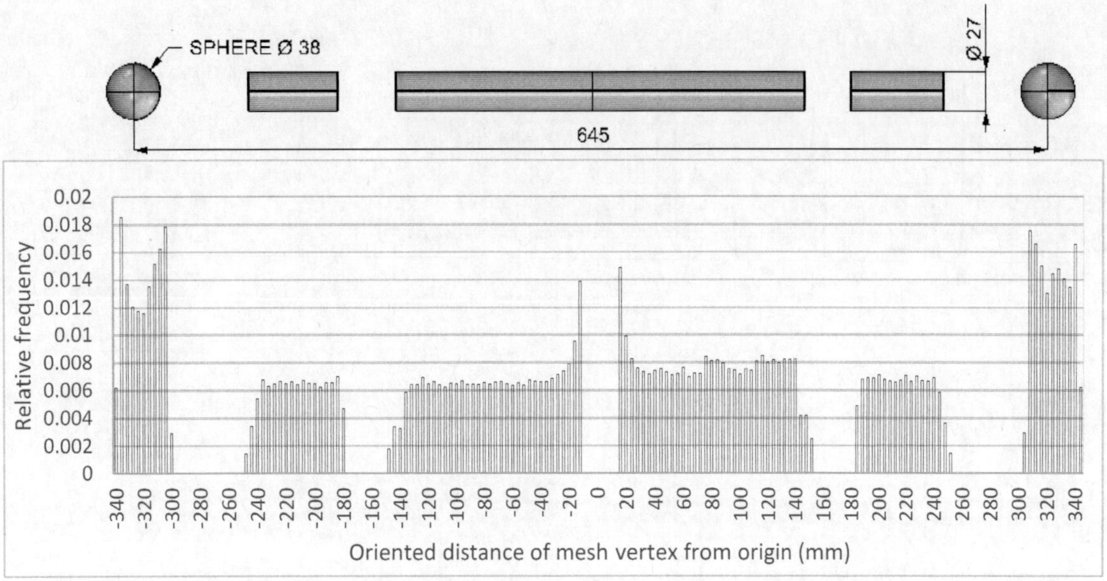

Fig. 4. Nominal CAD model and relative frequency histogram of nominal mesh.

A piecewise linear function constructed from the relative frequency histogram forms the representation for the shape distribution, see charts in Fig. 5.

Additionally, it is possible to express shape distribution by a cumulative frequency distribution, i.e. number of values f_i which are less than or equal to a reference value, see charts in Fig. 6.

3.3 Shape distribution comparison

In case the shape distribution is represented by the relative frequency histogram, the similarity measurement is based on Minkowski L_1 norm

$$D(f,g) = \sum_{i=1}^{k} |f_i - g_i|, \tag{2}$$

where f_i and g_i calculated according to (1) represent the relative frequency of two triangular meshes. In case the shape distribution is represented by the cumulative frequency distribution, the Minkowski L_1 norm is given by

$$D(\hat{f},\hat{g}) = \sum_{i=1}^{k} |\hat{f}_i - \hat{g}_i|, \tag{3}$$

where $\hat{f}_i$ and $\hat{g}_i$ are cumulative frequency of two triangular meshes

4 Experimental results

The shape distributions in form of relative frequency histograms calculated for all triangular meshes $M1, M2, \cdots, M5$ and the Nominal mesh are depicted in Fig. 5. The resemblance of characteristics is obvious. The resulting shape distributions for all triangular meshes are plotted in Fig. 6. The more the curves overlap the greater their similarity.

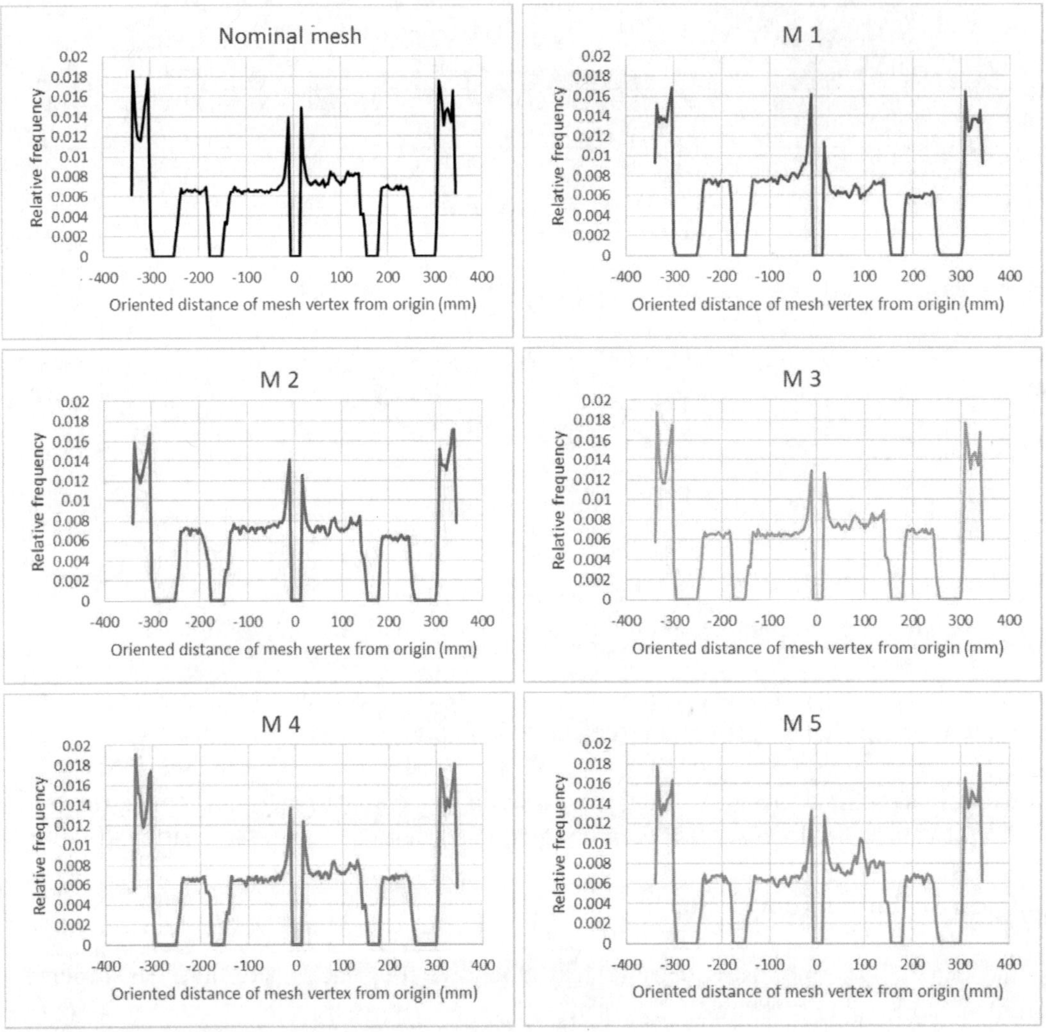

Fig. 5. Shape distributions (relative frequency histogram).

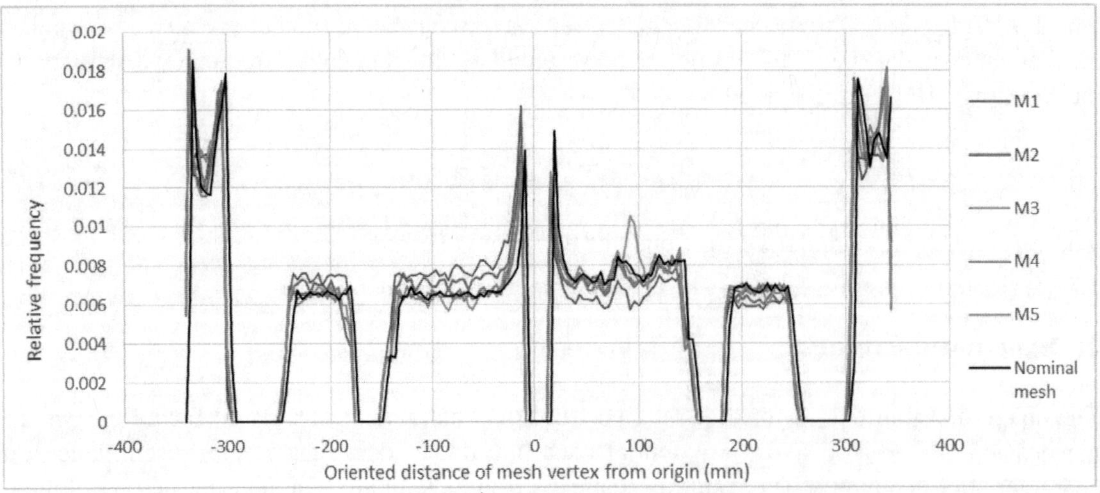

Fig. 6. Shape distributions (relative frequency histogram) of 6 triangular meshes.

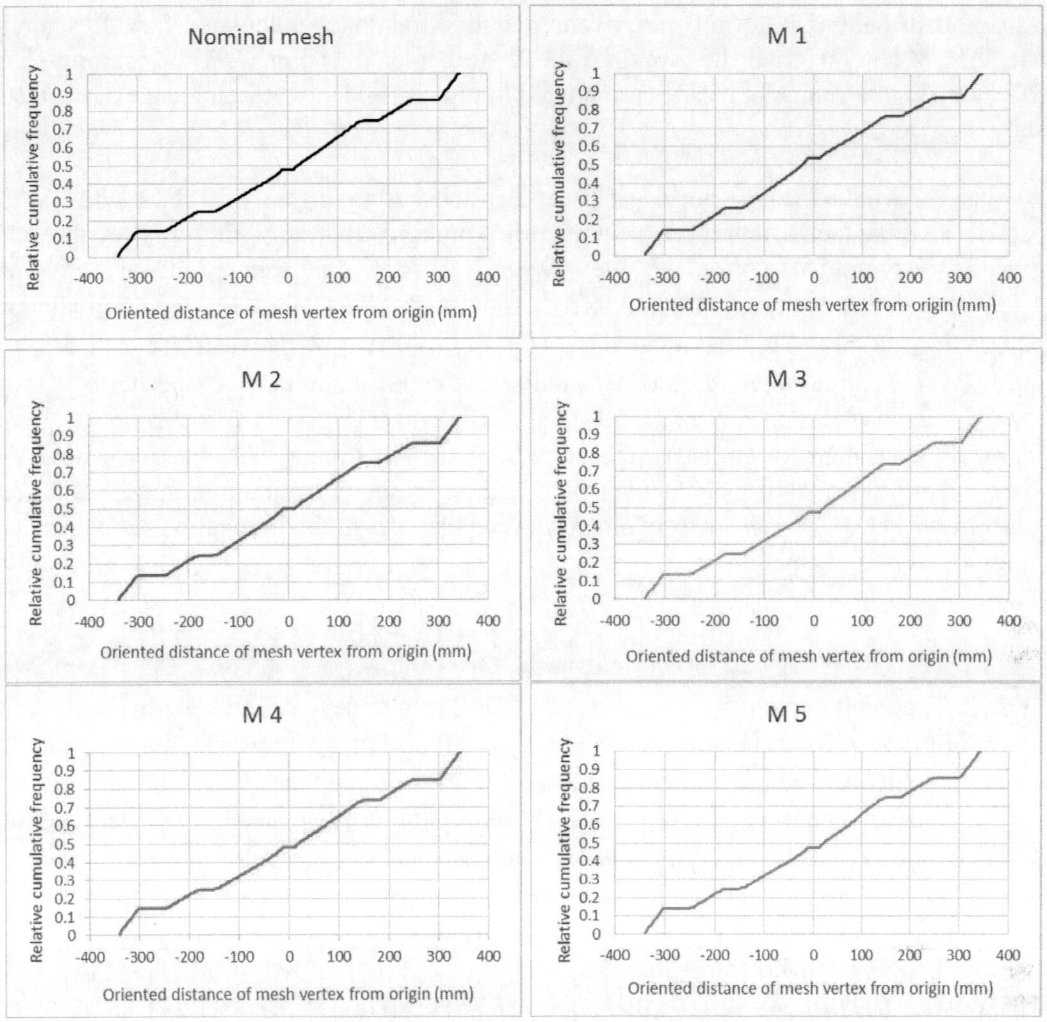

Fig. 7. Shape distributions (relative cumulative frequency).

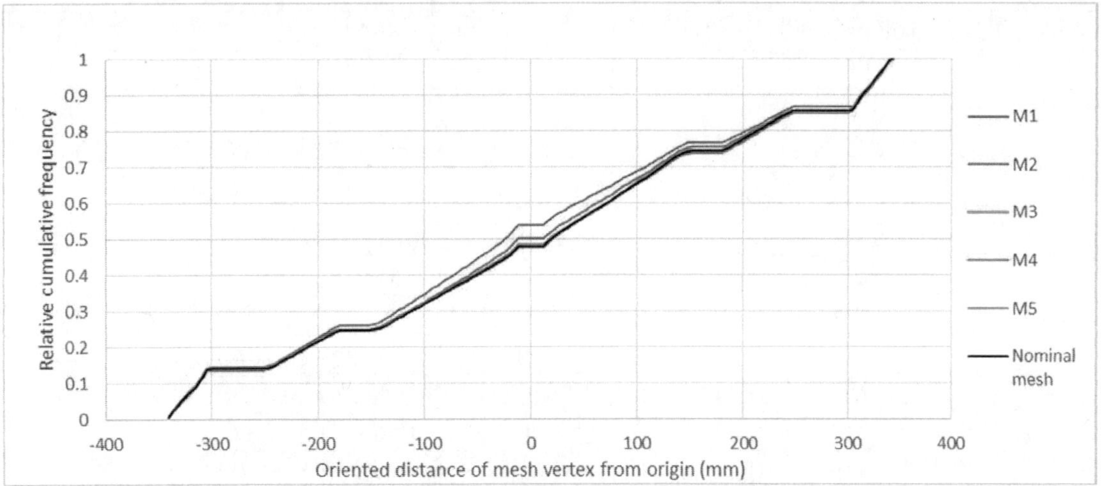

Fig. 8. Shape distributions (relative cumulative frequency) of 6 triangular meshes.

The shape distributions in form of relative cumulative frequency calculated for all triangular meshes $M1, M2, \cdots, M5$ and the Nominal mesh are depicted in Fig. 7. The resulting shape distributions for all triangular meshes are plotted in Fig. 8. Again, the resemblance of the tested meshes is obvious.

The values of Minkowski L_1 norm given by (2) and (3) are listed in Tab. 1 and Tab. 2, respectively. Both tables represent a symmetric similarity matrix with zeros on the main diagonal. The minimal nonzero value ($D_{min}(f,g)$ =0.05 between $M3$ and $M4$ and $D_{min}(\hat{f},\hat{g}) = 0.416$ between $M3$ and $M5$) indicates the best similarity between the two corresponding meshes. The maximal value ($D_{max}(f,g)$ =0.153 between $M1$ and $M5$ and $D_{max}(\hat{f},\hat{g}) = 3.834$ between $M1$ and $M4$) indicates the worst similarity of meshes.

To compare the obtained similarity measure L_1 with real dimensions, the form deviation $d_1, d_2, \cdots, d_5$ (the normal distance between the farthest mesh vertex and nominal CAD model) for each mesh $M1, M2, \cdots, M5$ was evaluated. After that, the mutual deviations between each two meshes were calculated by

$$d_{i,j} = |d_i - d_j|, \ i,j = 1,2,\cdots,5, \tag{4}$$

see Tab. 2. The minimum nonzero value $d_{3,5} = 0.062$ mm represents the best shape similarity between $M3$ and $M5$, the maximum value $d_{1,4} = 0.582$ mm represents the worst shape similarity not only between $M1$ and $M4$ but also among all triangular meshes $M1, M2, \cdots, M5$. Moreover, when comparing the values of all three similarity measures $d_{i,j}$, $D(f,g)$ and $D(\hat{f},\hat{g})$, it is obvious that $D(\hat{f},\hat{g})$ and $d_{i,j}$ indicate the same results, i.e. the best shape similarity between $M3$ and $M5$ and the worst shape similarity between $M1$ and $M4$.

L_1	M1	M2	M3	M4	M5
M1	0.000	0.093	0.146	0.147	0.153
M2	0.093	0.000	0.087	0.083	0.093
M3	0.146	0.087	0.000	0.050	0.059
M4	0.147	0.083	0.050	0.000	0.069
M5	0.153	0.093	0.059	0.069	0.000

(a) $D(f,g)$

L_1	M1	M2	M3	M4	M5
M1	0.000	2.798	3.823	3.834	3.701
M2	2.798	0.000	1.268	1.587	1.329
M3	3.823	1.268	0.000	0.753	0.416
M4	3.834	1.587	0.753	0.000	0.819
M5	3.701	1.329	0.416	0.819	0.000

(b) $D(\hat{f},\hat{g})$

Tab. 1. L_1 Similarity matrices for triangular meshes $M1, M2, \cdots, M5$.

$d_{i,j}$	M1	M2	M3	M4	M5
M1	0.000	0.112	0.253	0.582	0.315
M2	0.112	0.000	0.141	0.470	0.204
M3	0.253	0.141	0.000	0.329	0.062
M4	0.582	0.470	0.329	0.000	0.267
M5	0.315	0.204	0.062	0.267	0.000

Tab 2. Mutual form deviation similarity matrix for triangular meshes $M1, M2, \cdots, M5$.

5 Conclusion

In this paper, a shape similarity of five triangular meshes obtained by portable handheld optical scanner was investigated. The method was based on application of shape distributions commonly used in computer vision for recognition of shape of 3D models. The shape function measuring the oriented distance of each mesh vertex from suitable chosen reference point was defined, its function values were evaluated. The shape disributions in the form of relative frequency histogram and relative cumulative frequency were constructed. Three different similarity measures were evaluated – the Minkowski L_1 norm for the relative frequency distribution and for the relative cumulative frequency distribution as well as the difference of form deviations between each two tested meshes. The obtained results were compared and the relation between L_1 norm and real form deviation of scanned meshes was determined.

Acknowledgement

The paper was supported by grant from Grant Agency of CTU in Prague no. SGS16/206/OHK2/3T/12 entitled "Numerical solution of fluid flow in engineering practice".

References

[1] OSADA, R., FANKHOUSER, T., CHAZELLE, B., DOBKIN, D., *Matching 3D Models with Shape Distributions*. In *Proceedings International Conference on Shape Modeling and Applications*, Genova, 2001, ISBN 0-7695-0853-7

[2] OSADA, R., FANKHOUSER, T., CHAZELLE, B., DOBKIN, D., *Shape Distributions*, ACM Transactions on Graphics, Vol. 21, No. 4, 2002, Pages 807–832.

[3] YU I.C., LAPADAT, D., SIEGER, L., REGLI, W.C., *Using shape distributions to compare solid models*. In *Proceedings of the seventh ACM symposium on Solid modeling and applications* (SMA '02). ACM, New York, NY, USA, 273-280.

[4] MONTEVERDE L.C., RUIZ C.R., HUANG Z., *A Shape Distribution for Comparing 3D Models*. In: *Advances in Multimedia Modeling*. MMM 2007. Lecture Notes in Computer Science, vol 4351. Springer, Berlin, Heidelberg

Current address

Pajerová Nikola, Mgr.
Department of Technical Mathematics, Faculty of Mechanical Engineering
Czech Technical University in Prague
Karlovo nám. 13, 121 35 Praha 2, Czech Republic
E-mail: Nikola.Pajerova@fs.cvut.cz

Linkeová Ivana, doc. Ing., Ph.D.
Department of Technical Mathematics, Faculty of Mechanical Engineering
Czech Technical University in Prague
Karlovo nám. 13, 121 35 Praha 2, Czech Republic
E-mail: Ivana.Linkeova@fs.cvut.cz

Slovak University of Technology in Bratislava
Faculty of Mechanical Engineering
17th Conference on Applied Mathematics **APLIMAT 2018**

Proceedings

MODELING AND SYNTHESIS OF CONTROL THE PROCESS OF CASTING

PALENČÁR Jakub (SK), KUBIŠ Milan (SK), PALENČÁR Rudolf (SK)

Abstract. This article is focused on modeling and design of thermal control of distributed parameter systems. Partial differential equations describe the dynamics of these systems. Currently, to solve these equations using numerical methods, concrete the finite elements method. The numerical model was created based on the finite elements method using software COMSOL Multiphysics, which is adequate for modeling the process of casting. This software was linked with MATLAB & Simulink environment, where control of temperature fields in casting mold was simulated. Linked software provides for foundry practice to include the possibility of controlling the process at the very beginning of the design and by the can reduce the cost of casting production.

Keywords: distributed parameter systems, finite element method, simulation, control, COMSOL Multiphysics, MATLAB & Simulink

Mathematics Subject Classification: Primary 60A05, 08A72; Secondary 28E10

1 Introduction

Most of the dynamical systems analyzed in engineering practice, including the casting process, are distributed parameter systems (DPS) defined on complex-shape 3D definition domains. In order to obtain the desired structure during solidification in casting, the process requires a specific temperature field of the mold. Modeling, simulation and evaluation of real-time experiments in this area is now widely accepted as an important tool in product design and process development to improve productivity and casting quality.

Techniques of finite elements method (FEM) based modelling and design of control synthesis methods of DPS is demonstrated on modelling and simulation of control of temperature fields in casting mold. For control synthesis purpose, numerical models in the form of a lumped-input/distributed-output system (LDS) were created by means of numerical dynamical – FEM solution. Analysis of formulated models, simulation of temperature fields, but also design of control synthesis was carried out using the Distributed Parameter Systems Blockset for MATLAB & Simulink - Third-party software product of The MathWorks, developed at the

Institute of Automation, Measurement and Applied Informatics, Faculty of Mechanical Engineering, Slovak University of Technology in Bratislava.

2 Modeling systems with distributed parameters

In general, DPS are systems whose state or output variables, X(x,y,z,t)/Y(x,y,z,t) are distributed variables or fields of variables, where (x,y,z) are spatial coordinates in 3D. These systems are often considered as systems whose dynamics is described by partial differential equations (PDE) as infinite-dimensional systems, [1]. [2]. However, from point of view of implementation of DPS control in engineering practice, where a finite number of sensors and actuators for practical sensing and control is at disposal, such infinite-dimensional systems need to be approximated by finite-dimensional systems. There are many dimension reduction methods, which can be used to solve this problem, e.g. continuous and approximation theories aimed to control of parabolic systems presents monograph [3].

2.1 LDS representation of distributed parameter systems

In the input-output relation, PDE define distributed-input/distributed-output systems (DDS) between distributed input, U(x,y,z,t) and distributed output variables, Y(x,y,z,t), at initial and boundary conditions given. Distributed parameter systems very frequently are found in various technical and non-technical branches with limited number of manipulated input variables, or actuators and may be interpreted as LDS, Fig. 1, [4]. These lumped input variables by means of interaction of fields and variables generate distributed output of real DPS.

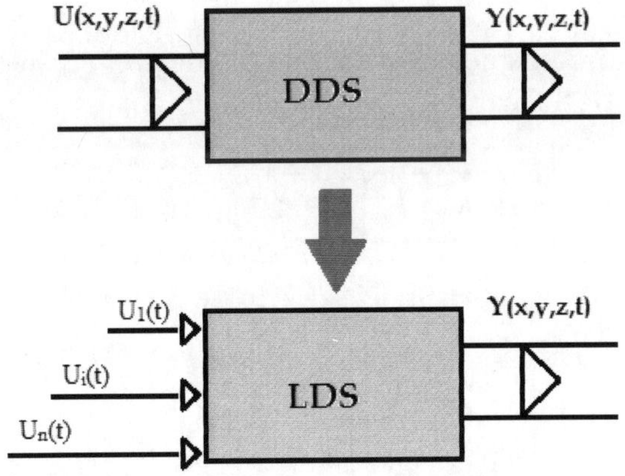

Fig. 1. Lumped-input/distributed-output system representation of distributed parameter systems.

2.2 Dynamics of lumped-input/distributed output systems

Output of the linear LDS in time domain, or in s-domain is in the form:

$$Y(\bar{x},t)=\sum_{i=1}^{n}Y_i(\bar{x},t)=\sum_{i=1}^{n}\mathcal{G}_i(\bar{x},t)\otimes U_i(t) \quad (1)$$

$$Y(\bar{x},s)=\sum_{i=1}^{n}Y_i(\bar{x},s)=\sum_{i=1}^{n}\mathcal{S}_i(\bar{x},s)\otimes U_i(s) \quad (2)$$

where $\bar{x}=(x,y,z)$ is position vector in 3D, $U_i(t)$ - lumped input quantity, $\mathcal{G}_i(\bar{x},t)$ - i-th distributed impulse response, $\otimes$ - denotes convolution product, $Y_i(\bar{x},t)$ - system response to the i-th input, $\mathcal{S}_i(\bar{x},s)$ - i-th transfer function. When $U_i(t)$ is a unit-step (Heaviside) function, s in the form of distributed transient response function $\mathcal{H}_i(\bar{x},t)$. For a discrete-time system considering zero-order hold (ZOH) units, the overall distributed output variable of LDS and ZOH can be expressed in the form:

$$Y(\bar{x},k)=\sum_{i=1}^{n}Y_i(\bar{x},k)=\sum_{i=1}^{n}\mathcal{GH}_i(\bar{x},k)\oplus U_i(k) \quad (3)$$

where $\oplus$ denotes convolution sum. For points $\bar{x}_i=(x_i,y_i,z_i)$ located in surroundings of lumped input variables, $U_i(t)$, where partial distributed transient responses $\mathcal{H}_i(\bar{x}_i,t)$ attain maximal amplitudes, partial distributed output variables are obtained in time-domain, or in z-domain respectively:

$$\{Y_i(\bar{x}_i,k)=\mathcal{GH}_i(\bar{x}_i,k)\oplus U_i(k)\}_{i=1,n} \quad (4)$$

$$\{Y_i(\bar{x},z)=\mathcal{SH}_i(\bar{x}_i,z)U_i(z)\}_{i=1,n} \quad (5)$$

2.3 DPS feedback control loop based on LDS structure

Decomposition of dynamics enables also to decompose the control synthesis to time and space control tasks in the DPS feedback control loop, Fig. 2.

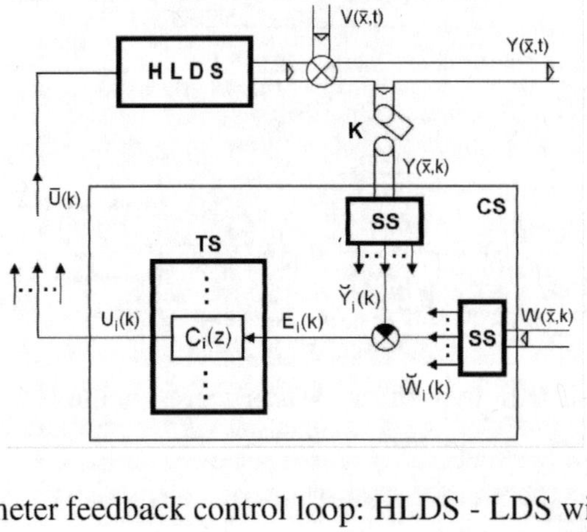

Fig. 2. Distributed parameter feedback control loop: HLDS - LDS with zero-order holds $\{\mathcal{H}_i\}_i$ on the input, CS - control synthesis, TS - control synthesis in time domain, SS - control synthesis in space domain, K - time/space sampling, $V(\bar{x},t)$ - disturbance variable, $Y(\bar{x},t)$ - distributed controlled variable, $Y(\bar{x},k)$ - sampled distributed controlled variable, $\breve{Y}_i(k)$ - approximation parameters of controlled variable, $W(\bar{x},k)$ - reference variable, $\breve{W}_i(k)$ -

approximation parameters of reference variable, $E_i(k)$ - control errors, $C_i(z)$ - controllers, $U_i(k)$ - lumped control variables

Let us consider a step change of distributed parameter control variable $W(\overline{x},k)=W(\overline{x},\infty)$ and $V(\overline{x},t)=0$. The goal of the control synthesis is to generate a sequence of control inputs $\overline{U}(k)$ in such manner, that in the steady-state, for $k\rightarrow\infty$, the control error $E(\overline{x},k)$ will approach its minimal value $\left\|\breve{E}(\overline{x},\infty)\right\|$ in the quadratic norm:

$$\min\left\|E(\overline{x},\infty)\right\| = \min\left\|W(\overline{x},\infty)-Y(\overline{x},\infty)\right\| = \left\|\breve{E}(\overline{x},\infty)\right\| \tag{6}$$

First, in the SS blocks, the approximation both of sampled distributed controlled variable $Y(\overline{x},k)$ and reference variable $W(\overline{x},\infty)$, on the set of reduced steady-state distributed step responses $\left\{\mathcal{H}HR_i(\overline{x},\infty)\right\}_i$, are solved in following form:

$$\min_{\breve{Y}_i}\left\|Y(\overline{x},k)-\sum_{i=1}^{n}Y_i(\overline{x}_i,k)\mathcal{H}HR_i(\overline{x},\infty)\right\| = \left\|Y(\overline{x},k)-\sum_{i=1}^{n}\breve{Y}_i(k)\mathcal{H}HR_i(\overline{x},\infty)\right\| \tag{7}$$

$$\min_{\breve{W}_i}\left\|W(\overline{x},\infty)-\sum_{i=1}^{n}W_i(\overline{x}_i,\infty)\mathcal{H}HR_i(\overline{x},\infty)\right\| = \left\|W(\overline{x},\infty)-\sum_{i=1}^{n}\breve{W}_i\mathcal{H}HR_i(\overline{x},\infty)\right\|. \tag{8}$$

3 Numerical modeling of temperature fields of casting mold in COMSOL Multiphysics

For study of the physical phenomena occurring during casting solidification of aluminium-alloy in the permanent mould, from a DPS control point of view, control system development as well as mathematical model validation, a benchmark of the casting processes with a two-part steel mould of a complex-shape has been made, Fig 3. Inside of the mould are built-in 26 electric heating elements in the cylinder shape, each with maximal heating power 400 W. Heating elements are grouped to 5 zones and their heating power is actuated by the input voltage range of (0–10) V. In the body of the mould is also placed 7 water-cooled copper chills and 11 thermocouples, Fig. 4.

Fig. 3. Construction of casting die.

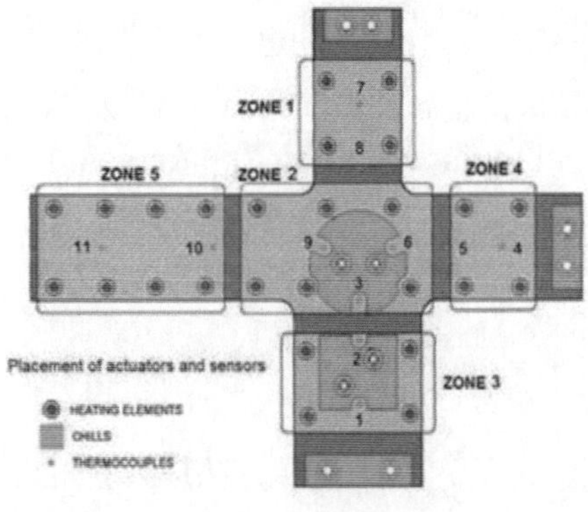

Fig. 4. Casting die with integrated action ang measuring members.

The casting mould is one of the key components of a casting process. It is well known, that the quality of the castings is affected strongly by the surface quality and the distribution of temperature in the mould, which has both time and space dependence. Influence of fundamental input factors to the results properties of casting products is depicted in Fig. 5. For various control methods during of casting process is necessary also modelling of temperature fields in casting processes. There exist many methods of control according given requirements, as predictive control, robust control and others, [5], [6], [7], [8]. Numerical modeling of technological processes is also in the frame of concept Idustry 4.0, [9].

Fig. 5. Schematic entry-exit models casting system.

CAD model of casting die was imported into COMSOL Multiphysics graphical environment to create finite element network that is avoidable for further numerical calculations. Were defined boundary conditions with environment temperature of 20 °C. The heating zones were defined to match the physical connection of the experimental facility. To solve the model contains several types of COMSOL Multiphysics solver, dealing with linear and nonlinear problems and a task in the frequency and time domain or with the tasks of the selected parameter. When solving systems of linear equations solver can be used directly as a UMPACK spools, GMRES iterative solver, solver with complex gradients, or with geometric multigrid. After finding performance ratio simulations were performed stepwise increase in performance in individual areas. Return to particular zones was made of the performance increase of 30%. The results of these simulations were nuts transient responses measuring 11x61. The number of lines is 11, because the measurement points have eleven. Number of pillars 61 tells us that the length of the simulation was 18,000 seconds with sampling plus 300 seconds is added to the first sample at the beginning of the simulation. For modelling of temperature fields as DPS based on FEM, wide possibilities are offered by COMSOL Multiphysics with heat transfer module, [10], [11]. Distribution of temperatures $T(\overline{x},t)$ in the casting die over the definition domain $\Omega \in E_3$ for lumped input variables $\{U_i(\overline{x},t)\}_{i=1,5}$ in the form of heat sources (W/m^3) is modelled by PDE of parabolic type with boundary conditions (BC) and initial condition T_{init} for heat flux:

$$\begin{aligned} &\frac{\partial T(\overline{x},t)}{\partial t} - a\nabla^2 T(\overline{x},t) = \sum_{i=1}^{5} U_i(\overline{x},t) \\ &-\mathrm{n}(-\lambda \nabla T) = h(T_{ext} - T) \\ &T(\overline{x},0) = T_{init} \end{aligned} \tag{9}$$

Parameters of PDE and BC are in Tab. 1

Quantity	Value	Unit	Description
λ	76.2	W/(m.K)	Thermal conductivity
ρ	7870	kg/m^3	Density
C_p	440	J/(kg.K)	Heat capacity
h	10	W/(m^2.K)	Heat transfer coefficient

Tab. 1 Parameters of PDE for Iron

In Fig. 6 we can see distributed transient responses in steady-state and steady temperature at 11 measuring points for actuating of the 5. zone, transient responses of systems after the power increase of the 5. zone are depicted in Fig.7.

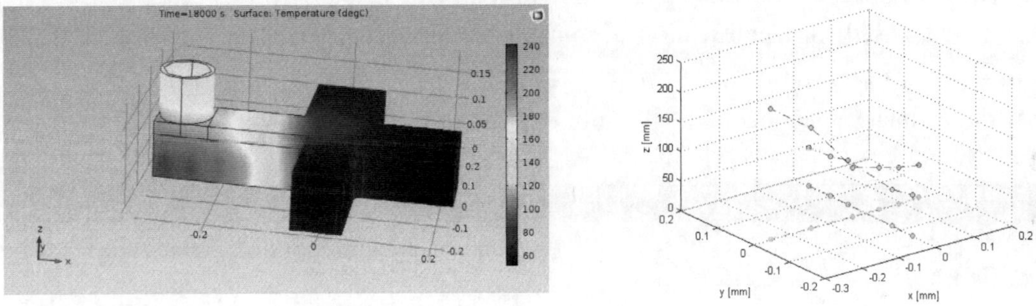

Fig. 6. Distributed transient responses in steady-state and steady temperature at 11 measuring points for actuating of the 5. zone.

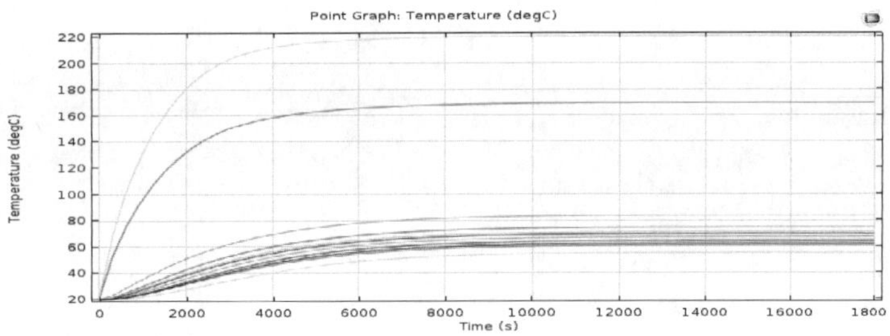

Fig. 7. Transient responses of system after the power increase of the 5. zone.

3.1 Identification of transient responses

After performing simulations in COMSOL Multiphysics data were downloaded using COMSOL LiveLink with MATLAB to the MATLAB, see Fig. 8. Subsequently, the step responses were identified using Identification Toolbox in MATLAB [12]. The role of the Identification Toolbox was to find the coefficients of the transfer system. System represents a step response obtained from simulations in COMSOL Multiphysics. Mathematical models in the form of PDE with given BC will be associated with distributed-input/distributed output system.

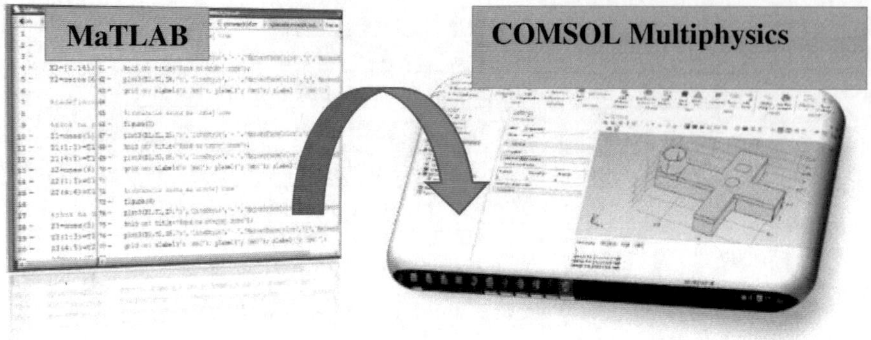

Fig. 8. Communication between software MATLAB and COMSOL Multiphysics.

Next, transient responses with maximal amplitude (e.g. for actuating in 5th zone it is position 11) were used for identification of transfer functions $\{S_i(\overline{x}_P, s)\}_{i=1,5}$. In Tab. 2 we see the setting for the estimates of our transmission. The following Fig. 9 it can be seen that the estimated parameters of the model is almost identical to our model. The conformity of those two models is 97.6%.

Zone	1	2	3	4	5
Position of thermocouple	7	4	1	9	11
$\{S_i(\overline{x}_P, s)\}_{i=1..5}$	$\frac{140,61}{1424,2s+1}$	$\frac{187,71}{1307,4s+1}$	$\frac{114,57}{1828s+1}$	$\frac{86,87}{2410,2s+1}$	$\frac{220,24}{1230,5s+1}$

Tab. 2. Position of thermocouples and identified transfer functions.

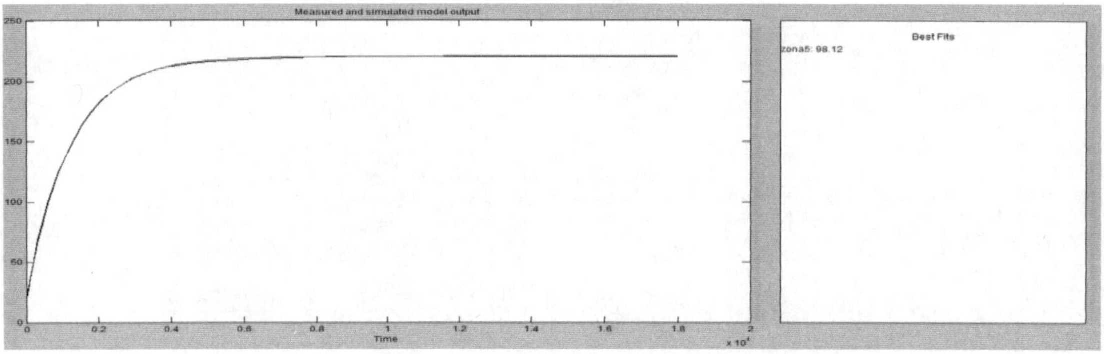

Fig. 9. Comparison of the original and the estimated model.

4 Temperature control of the casting mold

In MATLAB & Simulink we made synthesis of control with help of software packages DPS Blockset, [13], [14]. DPS Blockset is a special software product, which was developed at the Institute of automation, measurement and applied informatics In the Fig. 10 we can see scheme of control with using DPS Blockset in graphical environment MATLAB & Simulink.

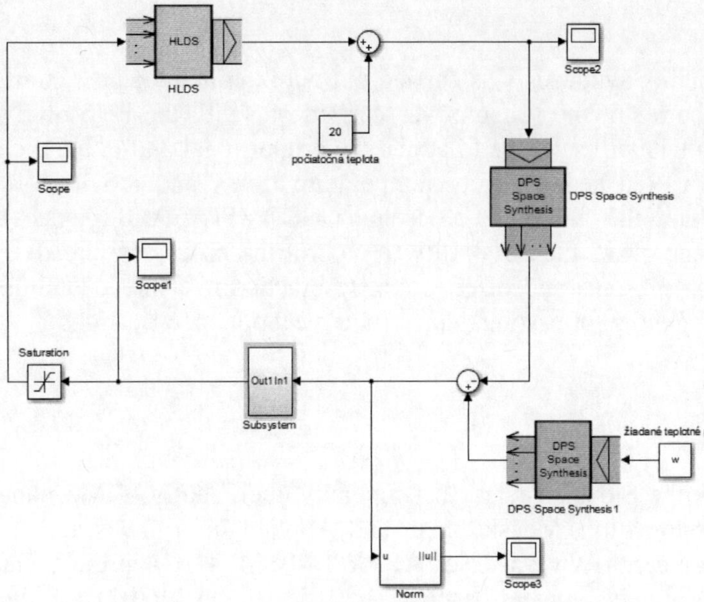

Fig. 10. Scheme of control loop with using DPS Blockset.

This product is used to control synthesis of distributed parameter systems. The next important block id DPS Space Synthesis in this block runs space synthesis, because we have 5 PI controllers in the time synthesis. Block of DPS Space synthesis adjust the regulation aberrance to the require measure. The transmission parameters derived from the System Identification Toolbox was applied to a transfer function block. This will serve to controller parameter settings. For adjusting the controller parameters will serve us PID Controller block in which the possibility of tuning parameters automatically. This toolbox then gives the user the result of parameter estimation or it is tuned properly remains on the users. Placed to make the final effect of the regulator is correct, basic knowledge of automatic control. The following Fig. 11 see the progress resulting from temperature control to the reference variables required by us.

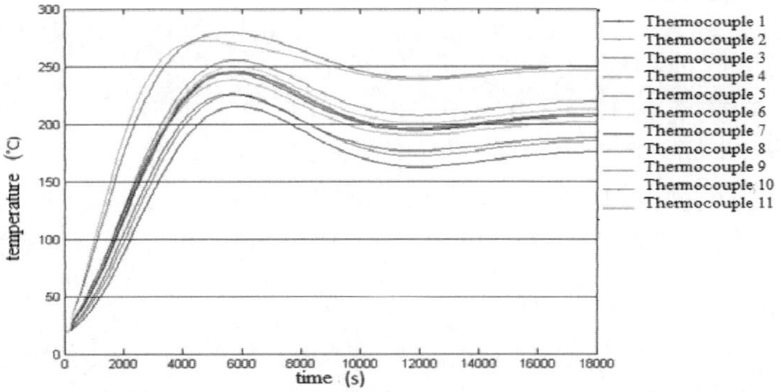

Fig. 11. Controlled variables at given position of the casting mold.

5 Conclusion

MATLAB & Simulink synthesis was carried out with the help of post control software tools DPS Blockset. Control of the scheme was carried out with the help of five PI controllers. Block PID Controllers will serve us to adjust the proportional and integral components of the various controllers. Reached a steady temperature limits and moving around the desired temperature. Linking the graphical environment COMSOL Multiphysics with MATLAB & Simulink package offers the possibility to control the processes already in their proposal. This option, which presents and offers foundry practice to achieve optimal control, which would increase production of castings that would meet our specifications.

Acknowledgement

The authors thank the Slovak Technical University, the Faculty of Mechanical Engineering, the Institute of Automation, Measurement and Applied Informatics and the grant agencies APVV projects number APVV 15-0295, APVV 15-0164, VEGA projects number 1/0604/15, 1/0748/15, KEGA projects number 014STU-4/2015, 039STU-4/2017 for their support during writing of this article.

References

[1] BUTKOVSKIJ, A. G.: *Optimal control of distributed parameter systems*. Nauka, Moscow, 1965. (in Russian)

[2] LIONS, J.-L.: *Optimal control of systems governed by partial differential equations*. Springer-Verlag, Berlin - Heidelberg - New York, 1971.

[3] LASIECKA, I. - TRIGGIANI, R.: *Control Theory for Partial Differential Equations: Continuous and approximation theories I. Abstract parabolic systems*. Cambridge UK: Cambridges University Press, 2000. 644 pgs. ISBN 0-521-43408-4

[4] HULKÓ, Gabriel et al.: *Modeling, Control and Design of Distributed Parameter Systems with Demonstrations in MATLAB. Bratislava*: Publishing House of STU Bratislava, 1998. 265 pgs. ISBN 80-227-1083-0

[5] MAIJER, D., M., OWEN, W., S., VETTER, R., A.: An investigation of predictive control for aluminum wheel casting via a virtual process model. *Journal of Materials Processing Technology*. ISSN 0924-013. Iss. 209, pp. 1965-1979. 2009.

[6] METAN, V., EIGENFELD, K., RABIGER, R., LEONHARDT, M., ECKER, S.: Grain size control in Al–Si alloys by grain refinement and electromagnetic stirring. *Journal of Alloys and Compounds*. Vol. 487, pp. 163-172. 2009.

[7] BELAVÝ, C., VLČEK, M., VITÁLOŠ, F., BARBOLYAS, B.: Modelling and robust control of temperature fields in casting mould. In *Strojárstvo - Strojírenství*. Roč. 18, č. 7-8, príloha Extra (2014), s. 76-77. ISSN 1335-2938.

[8] BELAVÝ, Cyril - VITÁLOŠ, Filip - VLČEK, Marcel - ŠIŠMIŠOVÁ, Dana - MICHALEČKO, Marek. Robust control synthesis for a class of distributed parameter systems. In *Proceedings of the 15th International Carpathian Control Conference: ICCC 2014; Velké Karlovice, Czech Republic, May 28-30, 2014*. [s.l.] : IEEE- Czechoslovakia Section of IEEE, 2014, CD-ROM, p. 40-45. ISBN 978-1-4799-3527-7.

[9] VACHÁLEK, Ján - LOKŠÍK, Milan - MORHÁČ, Martin - BARTALSKÝ, Lukáš - ROVNÝ, Oliver - ŠIŠMIŠOVÁ, Dana. The digital twin of an industrial production line within the Industry 4.0 concept. In *21st International Conference on Process Control (PC) 2017: June, 6-9, 2017, Štrbské Pleso, Slovakia*. 1. vyd. [s.l.] : IEEE, 2017, S. 258-262, ISBN 978-1-5386-4010-4.

[10] BELAVÝ, C., HULKÓ, G., BUČEK, P., LEĽO, S.: Control synthesis for distributed parameter systems modeled by FEM in COMSOL Multiphysics. In: *COMSOL Conference 2010: Proceedings*, Boston, USA, October 7-9, 2010.

[11] SLADKÁ, K., BABIUCH, M. Application of control algorithms in.NET Micro Framework Technology. In Proceedings of 17th International Carpathian Control Conference ICCC´2016. Tatranská Lomnica, Slovakia, May 29- June 01, 2016, pp. 684-687. ISBN: 978-146738606-7, DOI: 10.1109/CarpathianCC.2016.7501182.

[12] TAKÁCS, Gergely - VACHÁLEK, Ján - ROHAĽ-ILKIV, Boris. *Identifikácia sústav*. 1. vyd. Bratislava Nakladateľstvo STU 2014. 281 s., 100 obr., 5 tab. ISBN 978-80-227-4288-7.

[13] HULKÓ, G. et al.: *Distributed Parameter Systems Blockset for MATLAB & Simulink - Third-Party MathWorks product*. Bratislava - Natick, MA, USA, 2004.

[14] VASEK, V., KOLOMAZNIK, K., JANACOVA, D. *Tannery Wastes Processing Distributed Control System*. In Proceedigs of 17th International Conference on CAD/CAM, Robotics and Factories of the Future. Durban : MRRG University of Natal, 2001, s. 894;900. ISBN 1-86840-432-3.

Current address

Palenčár Jakub, Ing., PhD.
Institute of automation, measurement and applied informatics
Slovak University of Technology in Bratislava
Nám. slobody 17, 812 31 Bratislava, Slovak Republic
E-mail: jakub.palencar@stuba.sk

Kubiš Milan, Ing.,
Institute of automation, measurement and applied informatics
Slovak University of Technology in Bratislava
Nám. slobody 17, 812 31 Bratislava, Slovak Republic
E-mail: milan.kubis@stuba.sk

Slovak University of Technology in Bratislava
Faculty of Mechanical Engineering
17th Conference on Applied Mathematics **APLIMAT 2018**

Proceedings

WEAK PROPERTY B AND PROPERTY B OF SYSTEM OF FOUR NONLINEAR DIFFERENCE EQUATIONS

PASÁČKOVÁ Jana (CZ)

Abstract. We investigate a system of the four nonlinear difference equations, where the first equation obtain a neutral term. We state sufficient conditions for system to have the weak property B and property B.

Keywords: Property B, nonoscillatory solutions, oscillatory solution, difference equation

Mathematics subject classification: Primary 39A10; Secondary 39A21.

1 Introduction

In this paper, we study asymptotic behavior of solutions of a four-dimensional system

$$\begin{aligned}\Delta\left(x_n + p_n x_{n-\sigma}\right) &= A_n\, f_1\left(y_n\right)\\ \Delta y_n &= B_n\, f_2\left(z_n\right)\\ \Delta z_n &= A_n\, f_3\left(w_n\right)\\ \Delta w_n &= D_n\, f_4\left(x_{\gamma_n}\right),\end{aligned} \tag{S}$$

where $n \in \mathbb{N}_0 = \{n_0, n_0+1, ...\}$, n_0 is a positive integer, σ is a nonnegative integer, $\{A_n\}, \{B_n\}, \{D_n\}$ are positive real sequences defined for $n \in \mathbb{N}_0$. Δ is the forward difference operator given by $\Delta x_n = x_{n+1} - x_n$.

The sequence $\gamma : \mathbb{N} \to \mathbb{N}$ satisfies

$$\gamma_n \geq n + \sigma. \tag{H1}$$

The most common form of this sequence is $\gamma_n = n \pm \tau$, where $\tau \in \mathbb{N}$.

The sequence $\{p_n\}$ is a sequence of the real numbers and it satisfies

$$\lim_{n\to\infty} p_n = P, \text{ where } |P| < 1. \tag{H2}$$

Functions $f_i : \mathbb{R} \to \mathbb{R}$ for $i = 1, .., 4$ satisfy

$$\frac{f_i(u)}{u} \geq 1,\, u \in \mathbb{R}\backslash 0. \tag{H3}$$

Nonlinear difference systems or difference equations are often studied when

$$\sum_{n=n_0}^{\infty} A_n = \infty, \quad \sum_{n=n_0}^{\infty} B_n = \infty, \tag{H4}$$

which is called that the system (S) is in the canonical form. In this paper, we study the system under the following conditions

$$\sum_{n=n_0}^{\infty} A_n = \infty, \quad \sum_{n=n_0}^{\infty} B_n < \infty, \quad \sum_{n=n_0}^{\infty} B_n \left(\sum_{i=n_0}^{n-1} A_i \right) = \infty. \tag{H5}$$

By a solution of the system (S) we mean a vector sequence (x, y, z, w) which satisfies the system (S) for $n \in \mathbb{N}_0$. We investigate oscillatory or nonoscillatory solutions.

Definition 1. *The component x is said to be* ***nonoscillatory*** *if there exists $n_1 \geq n_0$ such that $x_n \geq 0$ (respectively $x_n \leq 0$) for all $n \geq n_1$. A solution of (S) is said to be nonoscillatory if all of its components x, y, z, w are nonoscillatory.*

Definition 2. *The component x is said to be* ***oscillatory*** *if for any $n_1 \geq n_0$ there exists $n \geq n_1$ such that $x_{n+1}x_n < 0$. A solution of (S) is said to be oscillatory if all of its components x, y, z, w are oscillatory.*

We study when (S) has a Property B or Weak property B. Property B is defined in accordance with those for the higher-order differential equations or for the system of differential equations, see [9] and references therein.

Definition 3. *The system (S) has* ***weak property B*** *if every nonoscillatory solution of (S) satisfies*

$$x_n z_n > 0 \quad \text{and } y_n w_n > 0 \quad \text{for large } n. \tag{1}$$

Definition 4. *The system (S) has* ***property B*** *if any of its solutions either is oscillatory or satisfies either*

$$\lim_{n\to\infty} |x_n| = \lim_{n\to\infty} |y_n| = \lim_{n\to\infty} |z_n| = \lim_{n\to\infty} |w_n| = \infty, \tag{2}$$

or

$$\lim_{n\to\infty} x_n = \lim_{n\to\infty} y_n = \lim_{n\to\infty} z_n = \lim_{n\to\infty} w_n = 0. \tag{3}$$

Solutions satisfying (1) and $x_n y_n > 0$ are called *strongly monotone solutions*, while solutions satisfying (1) and $x_n y_n < 0$ are called *Kneser solutions*.

The system (S) can be easily rewritten as a fourth-order nonlinear neutral difference equation. Equations with quasi-differences have been widely studied in the literature; see, for example, [1]–[7], [10]. Equations of this form appear in the discretization process for solving models concerning physical, biological, and chemical henomena, such as, for instance, problems of elasticity, deformation of structures, or soil settlement.

The aim of this paper is to extend our results about asymptotic behavior of nonoscillatory solutions of (S). We try to extend our results from [8]. We give sufficient conditions that (S) has weak property B and property B for the system (S) with the different assumptions then in previous. In [8], we establish sufficient conditions for the system (S) to have weak property B and we suppose that (S) is in the canonical form with positive sequence $\{p_n\}$. In this paper, we have the symetric operator of (S) and we modify the conditions (H4) into (H5) and sequence γ_n in order to have more general conditions.

2 Types of nonoscillatory solutions

If the system (S) has a solution (x, y, z, w), then it has the solution $(-x, -y, -z, -w)$ as well. Thus, throughout the paper, we can focus on solutions whose first component is eventually positive for large n.

We use the notation

$$s_n = x_n + p_n x_{n-\sigma}, \tag{4}$$

where $n \in \mathbb{N}_0$.

In proofs of our theorems we use the following lemma, which was proved in [8] for p positive, but it is easy to see that it holds for (H2) as well.

Lemma 1. *[8, Lemma 1] Let $\{x_n\}$ be eventually positive sequence and $\{p_n\}$ satisfies* (H2), *$n \in \mathbb{N}_0$. Let $\{s_n\}$ be the sequence defined by* (4). *Then $\{x_n\}$ is bounded if and only if $\{s_n\}$ is bounded. Moreover, if $\{s_n\}$ is positive and increasing for large n, then*

$$x_n \geq s_{n-\sigma}(1 - p_n) \quad \textit{for large } n. \tag{5}$$

The following lemma describes the possible types of nonoscillatory solutions.

Lemma 2. *Assume* (H4). *Then any nonoscillatory solution (x, y, z, w) of (S) with eventually positive x is one of the following types:*

type (a)	$x_n > 0$	$y_n > 0$	$z_n > 0$	$w_n > 0$	*for large n,*
type (b)	$x_n > 0$	$y_n > 0$	$z_n > 0$	$w_n < 0$	*for large n,*
type (c)	$x_n > 0$	$y_n < 0$	$z_n > 0$	$w_n < 0$	*for large n,*
type (d)	$x_n > 0$	$y_n < 0$	$z_n < 0$	$w_n < 0$	*for large n,*
type (e)	$x_n > 0$	$y_n > 0$	$z_n < 0$	$w_n < 0$	*for large n.*

Proof. Let (x, y, z, w) be a nonoscillatory solution of (S) such that $x_n > 0$ for large n. There are eight possible types of these solutions. We prove that solutions of the following types do not exist.

type (i)	$x_n > 0$	$y_n > 0$	$z_n < 0$	$w_n > 0$	for large n,
type (ii)	$x_n > 0$	$y_n < 0$	$z_n < 0$	$w_n > 0$	for large n.
type (iii)	$x_n > 0$	$y_n < 0$	$z_n > 0$	$w_n > 0$	for large n.

Assume that there exist $n_1 \in \mathbb{N}_0$ and a solution such that $z_n < 0$, $w_n > 0$ for $n \geq n_1 \geq n_0$. From the fourth equation of (S) we have $\Delta w_n > 0$ and this implies that there exists $k > 0$ such that $w_n \geq k$ for large n. Using (H3) we have $f_3(w_n) \geq w_n \geq k$. By the summation of the third equation of (S) we have

$$z_n - z_{n_0} = \sum_{i=n_0}^{n-1} A_i f_3(w_i) \geq k \sum_{i=n_0}^{n-1} A_i.$$

Passing $n \to \infty$, we get a contradiction with the fact that $z_n < 0$. This excludes solutions of types (i) and (ii).

Assume that (x, y, z, w) is a type (iii) solution. Since w is positive and increasing, there exists $k > 0$ such that $w_n \geq k$ for large n. By the summation of the third equation of (S) we get

$$z_n - z_{n_0} = \sum_{i=n_0}^{n-1} A_i f_3(w_i) \geq \sum_{i=n_0}^{n-1} A_i w_i \geq k \sum_{i=n_0}^{n-1} A_i.$$

Using the summation of the second equation of (S) we have

$$y_n - y_{n_0} = \sum_{i=n_0}^{n-1} B_i f_2(z_i) \geq \sum_{i=n_0}^{n-1} B_i z_i \geq k \sum_{i=n_0}^{n-1} B_i \left(\sum_{j=n_0}^{i-1} A_j \right).$$

Passing $n \to \infty$ we get the contradiction with the negativity of y.

□

By Definition 3, the system (S) has the weak property B if there exist only nonoscillatory solutions of type (a) and (c). Solutions of type (a) are called strongly monotone and solutions of type (c) are called Kneser solutions.

Lemma 3. *Any solution of type (a) satisfies*

$$\lim_{n\to\infty} x_n = \infty, \quad \lim_{n\to\infty} z_n = \infty. \tag{6}$$

Proof. Let (x, y, z, w) be a solution of type (a). Since y is positive and increasing, there exists $k > 0$ such that $y_n \geq k$ for large n. By the summation of the first equation of (S) we get

$$s_n - s_{n_0} = \sum_{i=n_0}^{n-1} A_i f_1(y_i) \geq \sum_{i=n_0}^{n-1} A_i y_i \geq k \sum_{i=n_0}^{n-1} A_i. \tag{7}$$

Passing $n \to \infty$ we get $s_n \to \infty$. Lemma 1 implies that s is unbounded if and only if x is unbounded. Therefore $\lim_{n\to\infty} x_n = \infty$.

Since w is eventually positive increasing, thus there exists $l > 0$ such that $w_n \geq l$ for large n. By the summation of the third equation of (S) we obtain that $z_n \to \infty$ for $n \to \infty$.

□

Lemma 4. *Any solution of type (c) satisfies*

$$\lim_{n\to\infty} w_n = 0. \tag{8}$$

Proof. Assume that the solution (x, y, z, w) is of type (c). Since w is eventually negative increasing, there exists $\lim_{n\to\infty} w_n = l \leq 0$. Suppose $l < 0$. By the summation of the third equation of (S) we obtain a contradiction with the positivity of z. Therefore $\lim_{n\to\infty} w_n = 0$. □

3 Weak property B and property B

The first theorem gives the simple criterion that the system (S) has property B, the proof of the theorem is omitted, we can proceed the similar way as in [8].

Theorem 1. *Assume* (H5). *If*

$$\sum_{n=n_0}^{\infty} D_n = \infty \tag{9}$$

holds, then the system (S) has property B.

In view of Theorem 1, in the sequel, we assume $\sum_{n=n_0}^{\infty} D_n < \infty$.

Theorem 2. *Let* (H1) - (H3) *hold. If*

$$\sum_{i=n_0}^{\infty} D_i \left(\sum_{j=n_0}^{i-1} A_j \right) = \infty, \tag{10}$$

then the system (S) has weak property B.

Proof. The weak property B means that there are no solutions of type (b), (d) and (e). Therefore we have to exclude these solutions. Assume that (x, y, z, w) is a type (b) solution. By (7) we get $\lim_{n\to\infty} x_n = \infty$, thus there exists $k > 0$ such that $x_n \geq k$ for large n. By the summation of the fourth equation of (S) we get

$$w_\infty - w_n = \sum_{i=n}^{\infty} D_i f_4(x_{\gamma_i}) \geq \sum_{i=n}^{\infty} D_i x_{\gamma_i} \geq k \sum_{i=n}^{\infty} D_i,$$

Using the summation of the third equation of (S) we have

$$z_n - z_{n_0} = \sum_{i=n_0}^{n-1} A_i f_3(w_i) \leq \sum_{i=n_0}^{n-1} A_i w_i,$$

$$-z_n + z_{n_0} \geq \sum_{i=n_0}^{n-1} A_i (-w_i) \geq k \sum_{i=n_0}^{n-1} A_i \left(\sum_{j=i}^{\infty} D_j \right).$$

Passing $n \to \infty$ and using the change of summation

$$\sum_{i=n_0}^{\infty} A_i \left(\sum_{j=i}^{\infty} D_j \right) = \sum_{i=n_0}^{\infty} D_i \left(\sum_{j=n_0}^{i-1} A_j \right) = \infty,$$

we get the contradiction with the boundedness of z. Thus, solutions of type (b) do not exist.

Assume that (x, y, z, w) is a type (d) solution. Since y is negative and decreasing, there exists $k < 0$ such that $y_n \leq k$ for large n. By the summation of the first equation of (S) we get

$$s_n - s_{n_0} = \sum_{i=n_0}^{n-1} A_i f_1(y_i) \leq \sum_{i=n_0}^{n-1} A_i y_i \leq k \sum_{i=n_0}^{n-1} A_i. \tag{11}$$

Using the summation of the fourth equation of (S) we have

$$w_n - w_{n_0} = \sum_{i=n_0}^{n-1} D_i f_4(x_{\gamma_i}) \geq \sum_{i=n_0}^{n-1} D_i x_{\gamma_i} \tag{12}$$

From (H2) and (4) we have $s_n \geq x_n - x_{n-\sigma} \geq -x_{n-\sigma}$ for large n. Thus

$$x_n \geq -s_{n+\sigma}. \tag{13}$$

Using this and (12) and (11) we get

$$w_n - w_{n_0} \geq \sum_{i=n_0}^{n-1} D_i (-s_{\gamma_i+\sigma}) \geq -k \sum_{i=n_0}^{n-1} D_i \left(\sum_{j=n_0}^{\gamma_i+\sigma-1} A_j \right).$$

Passing $n \to \infty$ we get from (H1) that (10) implies

$$\sum_{i=n_0}^{\infty} D_i \left(\sum_{j=n_0}^{\gamma_i+\sigma-1} A_j \right) = \infty.$$

Thus, we get the contradiction with the negativity of w.

Assume that (x, y, z, w) is a type (e) solution. Since z is negative and decreasing, there exists $h < 0$ such that $z_n \leq h$ for large n. By the summation of the second equation of (S) we get

$$y_\infty - y_n = \sum_{i=n}^{\infty} B_i f_2(z_i) \leq \sum_{i=n}^{\infty} B_i z_i \leq h \sum_{i=n}^{\infty} B_i.$$

Using the summation of the first equation of (S) we get

$$s_n - s_{n_0} = \sum_{i=n_0}^{n-1} A_i f_1(y_i) \geq \sum_{i=n_0}^{n-1} A_i y_i \geq -h \sum_{i=n_0}^{n-1} A_i \left(\sum_{j=i}^{\infty} B_j \right). \tag{14}$$

In case $p_n \leq 0$, we get $s_n \leq x_n$ from (4). Using this fact, the summation of the fourth equation of (S) and the estimation (14) we obtain

$$w_n - w_{n_0} = \sum_{i=n_0}^{n-1} D_i f_4(x_{\gamma_i}) \geq \sum_{i=n_0}^{n-1} D_i x_{\gamma_i} \geq \sum_{i=n_0}^{n-1} D_i s_{\gamma_i},$$

$$w_n - w_{n_0} \geq -h \sum_{i=n_0}^{n-1} D_i \left(\sum_{j=n_0}^{\gamma_i-1} A_j \left(\sum_{k=j}^{\infty} B_k \right) \right). \tag{15}$$

In case $p_n > 0$ we use (5) and we get

$$w_n - w_{n_0} \geq \sum_{i=n_0}^{n-1} D_i x_{\gamma_i} \geq \sum_{i=n_0}^{n-1} D_i s_{\gamma_i-\sigma}(1 - p_{\gamma_i}),$$

$$w_n - w_{n_0} \geq -h(1-P) \sum_{i=n_0}^{n-1} D_i \left(\sum_{j=n_0}^{\gamma_i-\sigma-1} A_j \left(\sum_{k=j}^{\infty} B_k \right) \right). \tag{16}$$

Passing $n \to \infty$ we get the contradiction with the negativity of w in both cases (15), (16). □

Theorem 3. *Let* (10) *hold. In addition, if*

$$\sum_{i=n_0}^{\infty} A_i \left(\sum_{j=i}^{\infty} D_j \right) = \infty \tag{17}$$

holds, then the system (S) has property B.

Proof. By Theorem 2, the system (S) has only solutions of type (a) and (c). First, assume that (x, y, z, w) is a solution of type (a). Thus, w is positive increasing and there exists a constant $k_1 > 0$ such that $w_n \geq k_1$ for large n. From the third equation of (S) we get

$$z_n - z_{n_0} = \sum_{i=n_0}^{n-1} A_i f_3(w_i) \geq \sum_{i=n_0}^{n-1} A_i w_i \geq k_1 \sum_{i=n_0}^{n-1} A_i.$$

Substituting this into the second equation of (S) we obtain

$$y_n - y_{n_0} = \sum_{i=n_0}^{n-1} B_i f_2(z_i) \geq \sum_{i=n_0}^{n-1} B_i z_i \geq k_1 \sum_{i=n_0}^{n-1} B_i \left(\sum_{j=n_0}^{i-1} A_j \right).$$

Passing $n \to \infty$ we have $y_n \to \infty$.

Taking into account $\lim(1 - p_n) = 1 - P > 0$, there exists $p > 0$ such that $1 - p_n \geq p$, for large n. Using the summation of the fourth equation of (S) and (5) we have

$$w_n - w_{n_0} = \sum_{i=n_0}^{n-1} D_i f_4(x_{\gamma_i}) \geq \sum_{i=n_0}^{n-1} D_i x_{\gamma_i} \geq \sum_{i=n_0}^{n-1} D_i s_{\gamma_i - \sigma}(1 - p_{\gamma_i}) \geq \tag{18}$$

$$\geq p \sum_{i=n_0}^{n-1} D_i s_{\gamma_i - \sigma} \geq kp \sum_{i=n_0}^{n-1} D_i \left(\sum_{j=n_0}^{\gamma_i - \sigma - 1} A_j \right).$$

From (18) we get $w_n \to \infty$ passing $n \to \infty$.

Now, assume that (x, y, z, w) is a solution of type (c). Assume that $\lim x_n = t_1 \geq 0$. First, assume that $t_1 > 0$. Using the summation of the third and the fourth equation of (S) we have

$$z_n - z_{n_0} = \sum_{i=n_0}^{n-1} A_i f_3(w_i) \leq -t_1 \sum_{i=n_0}^{n-1} A_i \left(\sum_{j=i}^{\infty} D_j \right).$$

Passing $n \to \infty$ we get the contradiction with the boundedness of z, therefore $\lim_{n\to\infty} x_n = 0$.

Now, assume that $\lim y_n = t_2 \leq 0$. First, assume that $t_2 < 0$. Using the summation of the first and the fourth equation of (S) we obtain

$$w_n - w_{n_0} = \sum_{i=n_0}^{n-1} D_i f_4(x_{\gamma_i}) \geq \sum_{i=n_0}^{n-1} D_i x_{\gamma_i} \geq \sum_{i=n_0}^{n-1} D_i(-s_{\gamma_i + \sigma}) \geq -t_2 \sum_{i=n_0}^{n-1} D_i \left(\sum_{j=n_0}^{\gamma_i + \sigma - 1} A_j \right).$$

Passing $n \to \infty$ we get the contradiction with the boundedness of w, therefore $\lim_{n\to\infty} y_n = 0$.

Finally, assume that $\lim z_n = t_3 \geq 0$. First, assume that $t_3 > 0$. Using the summation of the first and the second equation of (S) we get

$$s_n - s_{n_0} = \sum_{i=n_0}^{n-1} A_i f_1(y_i) \leq -t_3 \sum_{i=n_0}^{n-1} A_i \left(\sum_{j=i}^{\infty} B_j \right).$$

Since

$$\sum_{i=n_0}^{\infty} A_i \left(\sum_{j=i+1}^{\infty} B_j \right) = \sum_{i=n_0}^{\infty} B_i \left(\sum_{j=n_0}^{n-1} A_j \right) = \infty,$$

then we get the contradiction with the boundedness of s. Since x is bounded, then s is bounded as well. Therefore $\lim_{n\to\infty} z_n = 0$.

Now, we get the assertion by Lemma 3, Lemma 4 and Definition 4.

□

4 Conclusion

We found conditions for the system (S) to have proberty B. We extended the conditions from [8]. Now, we can investigate the system with the different conditions or it could be interesting to find the sufficient conditions for system (S) to have any type of the solutions.

Acknowledgements

This paper was processed with contribution of long term institutional support of research activities by Faculty of Informatics and Statistics, University of Economics, Prague.

References

[1] DIBLÍK, J. et. al: *Bounded and unbounded non-oscillatory solutions of a four-dimensional nonlinear neutral difference system*, Adv. Difference Equ., 2015, 319, 1–11.

[2] DOŠLÁ, Z., KREJČOVÁ, J.: *Nonoscillatory solutions of the four-dimensional differences system*, Electron. J. Qual. Theory Differ. Equ., 2011, 4, 1–11.

[3] DOŠLÁ, Z., KREJČOVÁ, J.: *Oscillation of a class of the fourth-order nonlinear difference equations*, Adv. Difference Equ., 2012, 99, 1–14.

[4] DOŠLÁ, Z., KREJČOVÁ, J.: *Asymptotic and oscillatory properties of the fourth-order nonlinear difference equations*, Appl. Math. Comput., 2014, 249, 164-173.

[5] DOŠLÁ, Z., KREJČOVÁ, J., MARINI, M.: *Minimal and maximal solutions of the fourth-order nonlinear difference equations*, Difference Equ. Appl., 2014, 20, 12, 1–14.

[6] JANKOWSKI, R., SCHMEIDEL, E., ZONENBERG, J.: *Oscillatory properties of solutions of the fourth order difference equations with quasidifferences*, Opuscula Math., 2014, 34(4).

[7] KREJČOVÁ, J.: *Oscillation of fourth-order difference equations*, 14th Conference on Applied Mathematics, Aplimat, 2015.

[8] KREJČOVÁ, J.: *Asymptotic properties of nonoscillatory solutions of neutral difference systems*, 15th Conference on Applied Mathematics, Aplimat, 2016.

[9] KIGURADZE, I.T., CHANTURIA, A.: *Asymptotic Properties of Solutions of Nonautonomous Ordinary Differential Equations*, Kluwer Acad. Publ. G., Dordrecht, 1993.

[10] MIGDA, M., SCHMEIDEL, E.: *Convergence of solutions of higher order neutral difference equations with quasi-differences*, Tatra Mt. Math. Publ., 2015, 63, 205-213.

Current address

Pasáčková Jana, RNDr., Ph.D.
University of Economics
Faculty of Informatics and Statistics
W. Churchill Sq. 4, 130 67 Prague, Czech Republic
E-mail: jana.pasackova@vse.cz

Slovak University of Technology in Bratislava
Faculty of Mechanical Engineering
17th Conference on Applied Mathematics **APLIMAT 2018**
Proceedings

PARALLEL ILLUMINATION IN GEOGEBRA

PIRKLOVÁ Petra (CZ), BÍMOVÁ Daniela (CZ)

Abstract. The paper introduces several parallel illumination problems solved using geometric GeoGebra software. The solution of the problem both in 3D and in 2D can be identified in GeoGebra at the same time. As a result it provides students better imagination of the problem submission and furthermore its solution.

Keywords: descriptive geometry, parallel illumination, GeoGebra, Monge projection shadow, moveable

Mathematics Subject Classification: Primary 97B50, 68U05; Secondary 51N05

1 Introduction

The descriptive geometry is very demanding in terms of spatial imagination. Students have serious difficulties with solving spatial problems by using different projecting methods. They are often able to memorize the individual constructs, but they are not able to imagine the constructs in space. If the solution of the problem is non-standard, the students are uncertain of the solution. GeoGebra can additionally be used for parallel illumination.

Geometric software GeoGebra provides a better idea of the exercise and its solution as it displays spatial exercises both in space and their projection onto the plane. Therefore it is very helpful to the students. GeoGebra is also dynamic software, which means it allows to shift certain elements and to study various solutions of the problems (including non-standard problems). The dynamical applets facilitate better comprehension of the issues.

GeoGebra workspace can be divided into one, two or three parts. One part is for a 3D view and two others for a 2D view. We use one of them for the "Check Boxes" that allow to show and hide one or more parts of the constructions. All examples introduced in the paper are solved in 3D and in 2D together. The solution is displayed in Monge projection on the plane. All applets are dynamic and linked. If some points are shifted in 3D, subsequently the same points have been shifted in 2D. Moveable points are always marked blue. If we decide to change placement of the points, we move them by cursor.

Points in Monge projection in a 2D window are given by using coordinates of the same points in a 3D view. However, the constructions in 2D and in 3D windows need to be done separately.

2 Some Exercises solved in GeoGebra

2.1 The Parallel Illumination of Quadrilateral Prism

First problem to be solved is the illumination of the quadrilateral prism. The initial submission sets the prism with one base in the horizontal plane. Vertices *A, B, C, D* are chosen on the horizontal plane and the vertex *E* on the upper base. Their location is random, therefore these vertices are moveable. The base and the height of prism can be modified. These points are marked blue in the picture. Other vertices are dependent on the moveable vertices (these points are marked red). The direction of the illumination is also given by two points in 3D and these points are moveable (blue). The shape of the prism and the direction of the illumination can be modified (see Fig. 1).

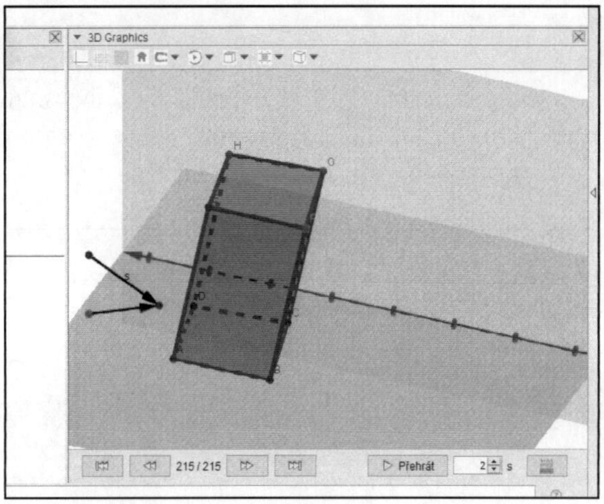

Fig. 1

Vertices A_1, A_2, B_1, B_2, …, H_1, H_2 of the prism and points of direction in Monge projection in 2D window are given by using coordinates of the points *A, B, C, ..., H* in a 3D view. If the blue vertices in a 3D window have been shifted, then the points in a 2D window have also moved. The folding line is set by two points (see Fig. 2.)

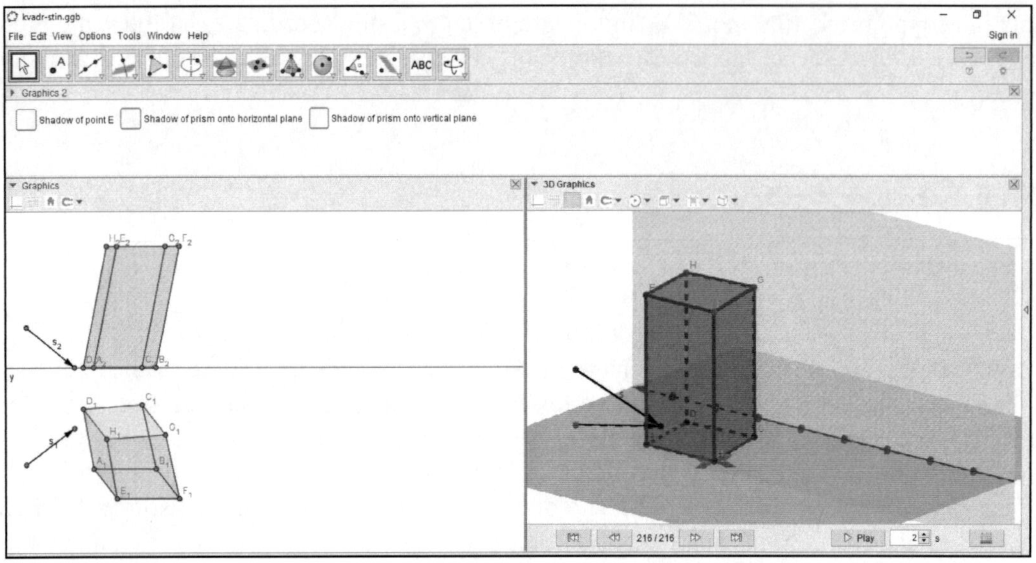

Fig. 2

Three check boxes are placed into one 2D window. The first check box shows and hides construction of the shadow of the point *E*. It is evident that the shadow of point *E* on the vertical and horizontal plane are vertical and horizontal trace points *E´´, E´* of parallel line intersecting point *E* with direction of the illumination. See Fig. 3.

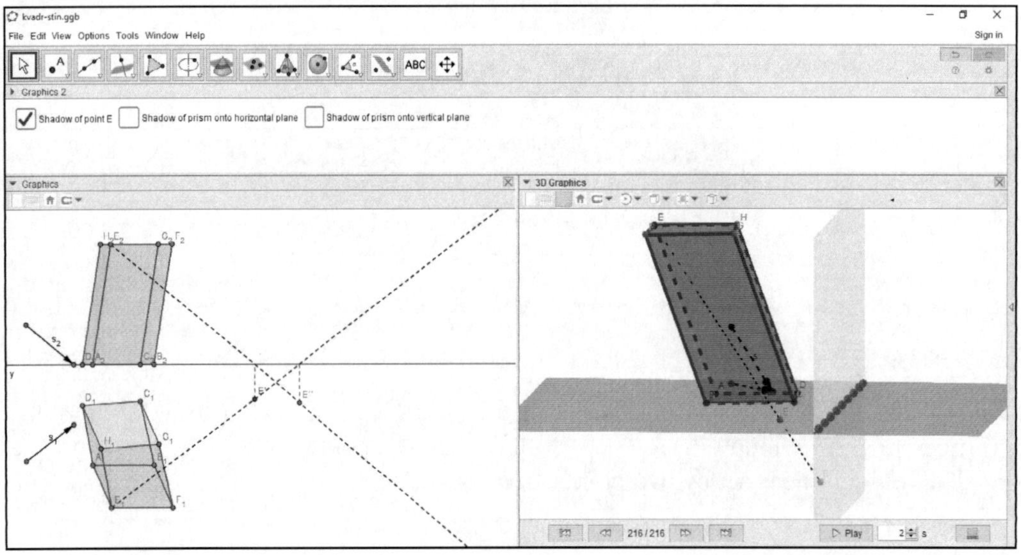

Fig. 3

The second check box "Shadow of the prism onto the horizontal plane" shows and hides the complete construction of the shadow in a 3D and a 2D window. The shadow of the prism on the horizontal plane is marked blue. We aim to construct shadows of all vertices and then to connect them. The lower base of the prism is on the horizontal plane, therefore we can create shadows of the vertices of the upper base only. However, we can also place vertices of the lower base to other points above and below on the horizontal plane.

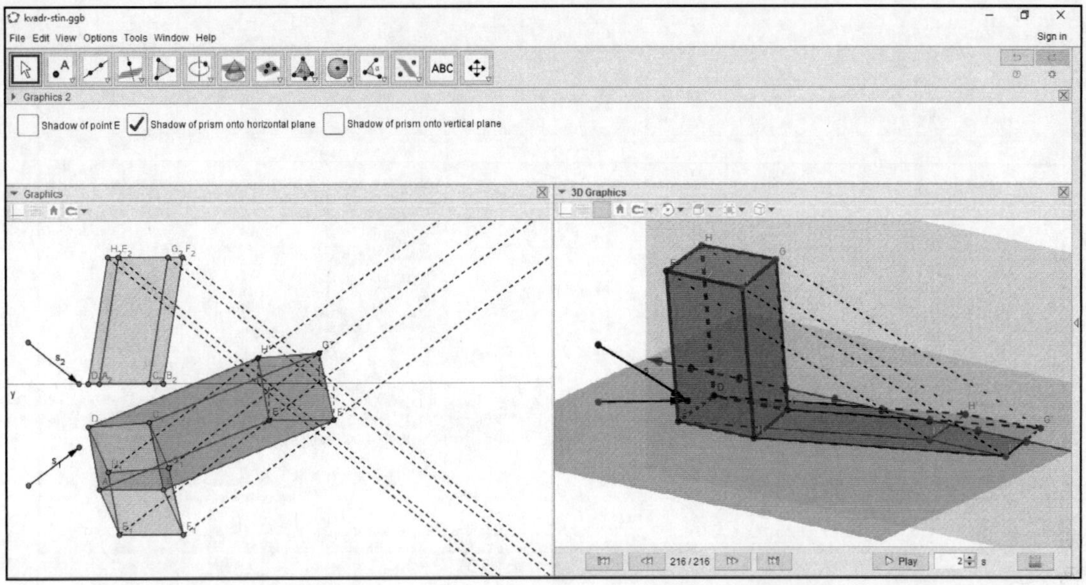

Fig. 4

The third check box displays the illumination of the prism on the vertical plane. The shadow is marked green, and we have to create the shadow of all vertices of the lower and upper base of the prism. In a 3D window the shadow in the vertical plane is displayed only above the folding line for clarity (see Fig. 5.).

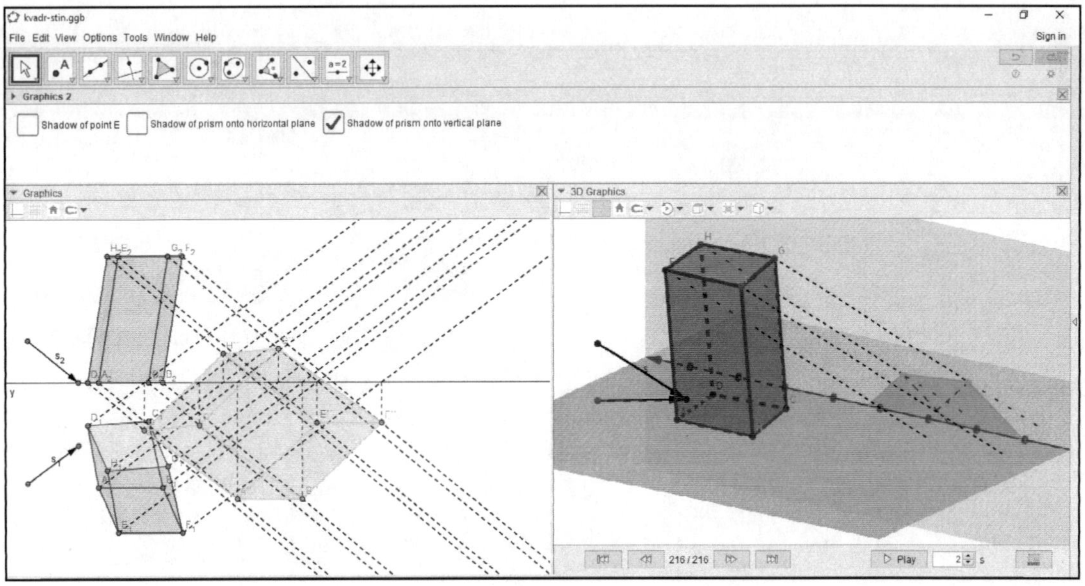

Fig. 5

If the second and third check boxes are displayed together, it is noticeable that the shadows on the vertical and horizontal plane intersect on the folding line (see Fig. 6).

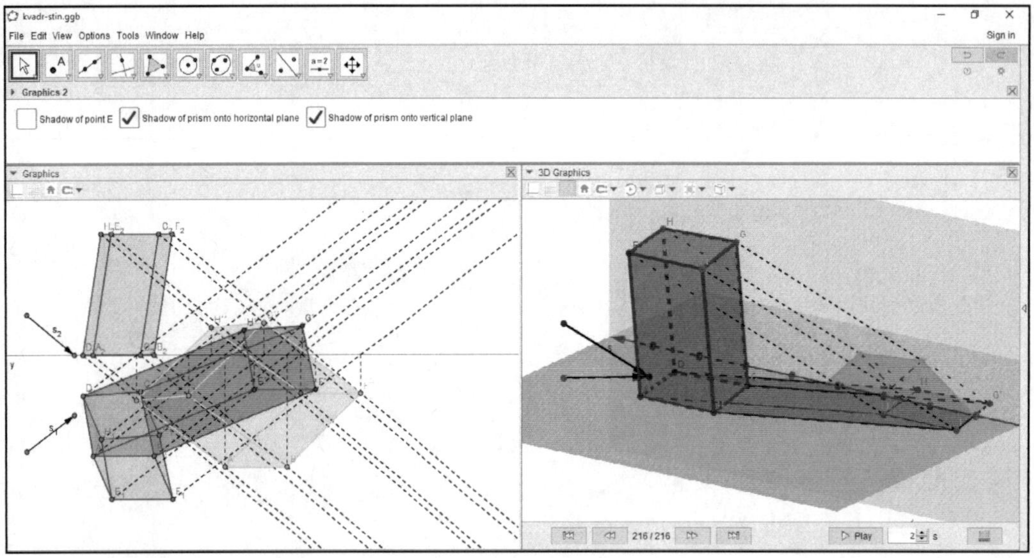

Fig. 6

2.2 The illumination of the Pyramid and the Line

The base of pyramid *ABCDV* is set on the horizontal plane. Vertices and the apex are moveable. Line *p* is given by two points that are also moveable. Direction of the illumination is also given by two points. The moveable points are marked blue and the dependent points are marked red (see Fig. 7.).

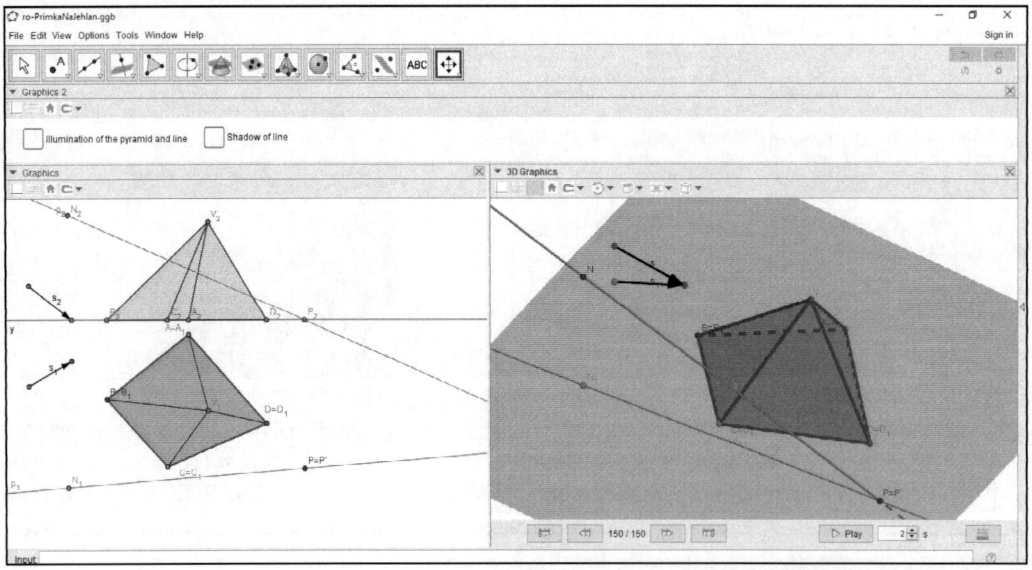

Fig. 7

The first check box displays the illumination of the pyramid and the line on the horizontal plane. The shadow on the vertical plane is not constructed for simplicity and clarity. The shadow of line *p* is line *p′* and the shadow of apex *V* is point *V′* (see Fig. 8.). Vertices *A, B, C, D* are on the horizontal plane, their shadows are the same points, however, moving of the vertices above or under the horizontal plane is possible. Subsequently the shadows of the vertices are to be found.

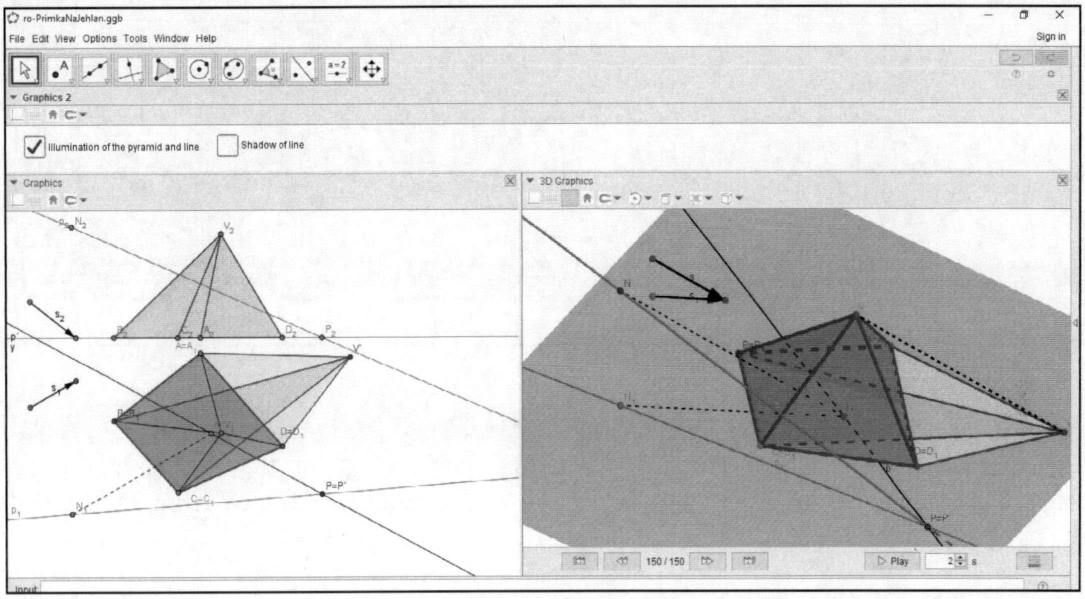

Fig. 8

Following the previous constructions, the shadow of line *p*, which is projected on the pyramid (polyline) is to be found. The shadow *p′* intersects the shadows of edges *AV′, BV′, CV′, DV′* in the points 1′, 2′, …, 4′ or on the base edges in the points 1*, 4* (depending on the position of the pyramid and the line). Points 1′, …, 4′ represent the shadows of points 1*, 2*, 3*, 4* displayed on the shadow line *p* projected on the pyramid (see Fig. 9.).

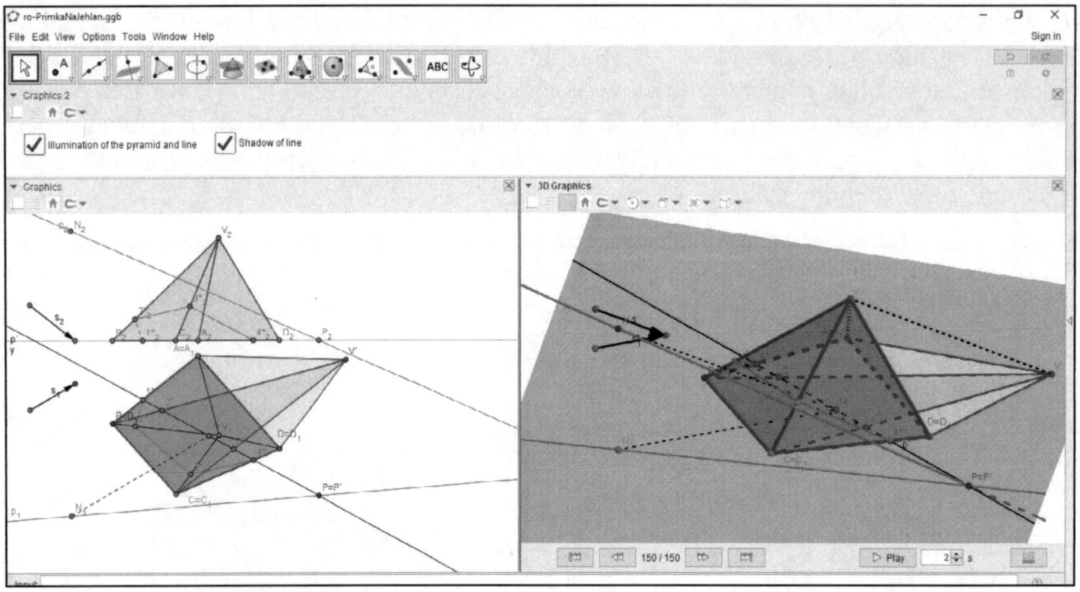

Fig. 9

When the points *P* or *N* of the line or the points on the direction of the illumination are shifted, the shadow of line *p* on the pyramid has also been changed. GeoGebra indicates all the possible positions of the shadow.

Line *p* can intersect the pyramid. The intersection points *X, Y* are marked green. The shadow of the line ends in the above mentioned points. Unfortunately, it cannot be displayed in GeoGebra (see Fig. 10.). Therefore the students need to be made aware of the deficiency.

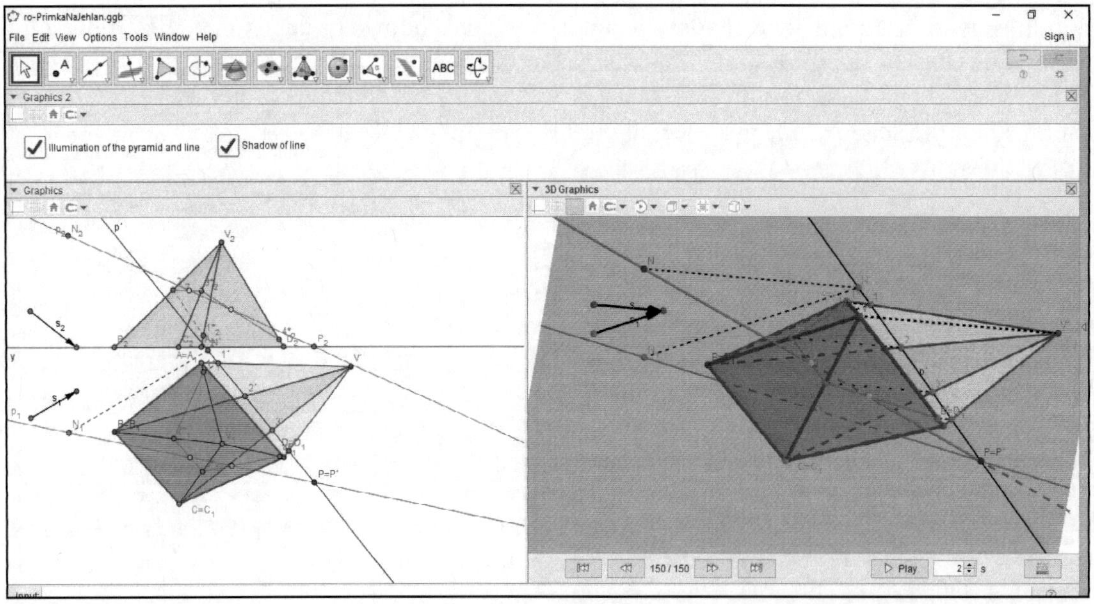

Fig. 10

2.3 The Illumination of the Hollow Cylinder

In this exercise the shadow of the hollow cylinder thrown inside is aimed to be found. Another task is to learn what curve the shadow creates.

The cylinder has the first base on the horizontal plane and the cylinder is perpendicular. An oblique cylinder cannot be created in GeoGebra. We choose the centre and the radius of the lower circle and the height of the cylinder. The radius and height can be changed. A tool slider can be used for this. The direction of the illumination is given by two moveable points (see Fig. 11.).

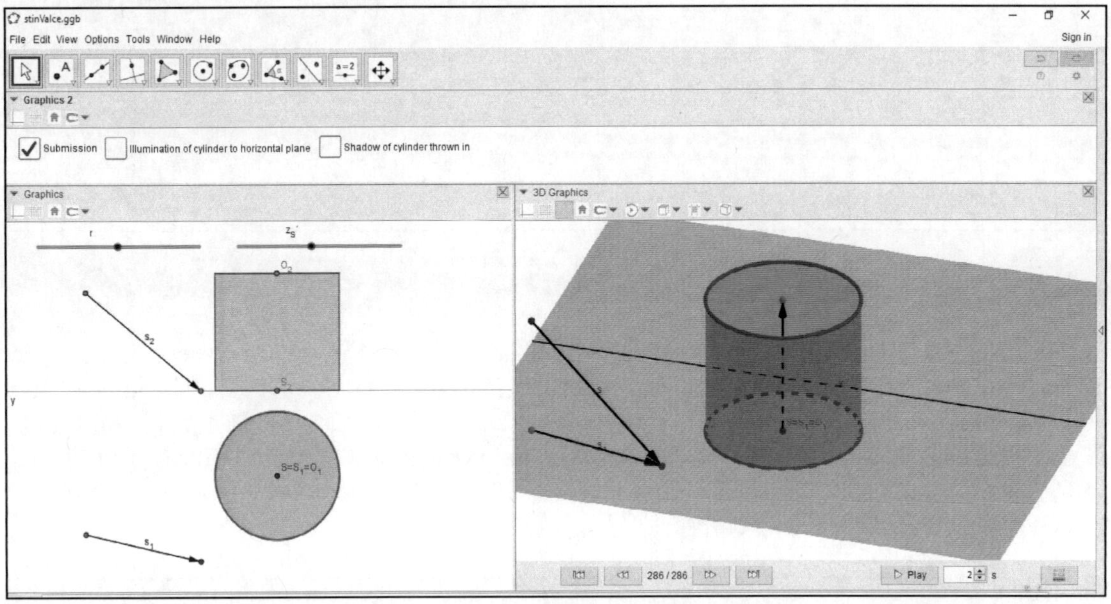

Fig. 11

The second checkbox displays the shadow of the cylinder on the horizontal plane. The shadow on the vertical plane is not constructed in this exercise. At first the shadow of the centre of the upper circle is to be found. Then the shadow of the whole circle is the identical circle because the plane of the circle is parallel with the horizontal plane (see Fig. 12.).

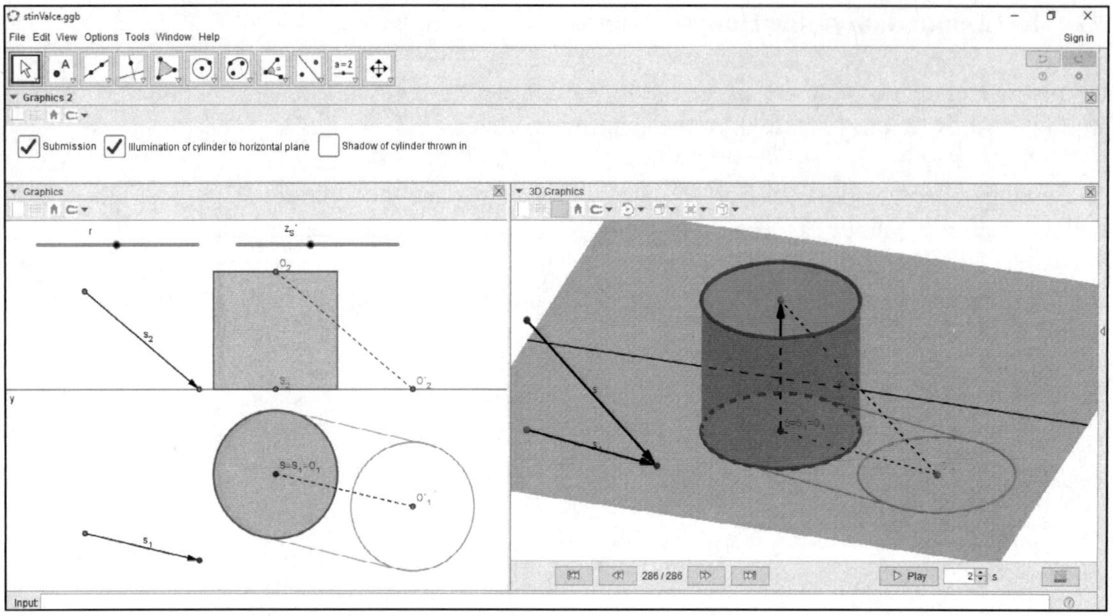

Fig. 12

The third check box displays the shadow of the upper circle thrown into the cylinder. We choose a few points 2′, 3′, …, 7′ on the inner circular arc. These points are shadows of the points in the part of the upper circle, which also create shadows 2*, …, 7* thrown in. Points 1′ and 7′ are the shadows of points 1, 7 of the circle in which the shadow thrown in begins and ends (see Fig. 13.). When the points 1, 2*, 3*, …, 7 are connected by the curve, the GeoGebra names this curve an ellipse.

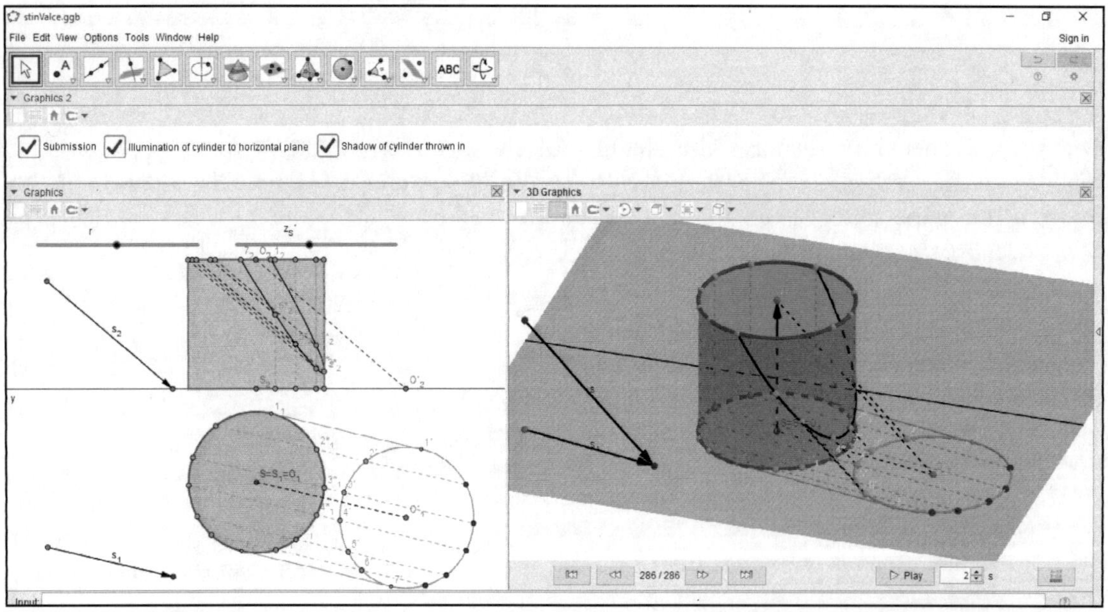

Fig. 13

If the direction of the illumination or the cylinder are moved, the shadow of the upper circle can intersect the lower circle. The intersection points are marked green (see Fig. 14.). Then the shadow thrown in intersects the lower circle in these points and the upper circle has the shadow on the lower circle (yellow on the figure). However, the ellipse cannot be ended in the green points in GeoGebra.

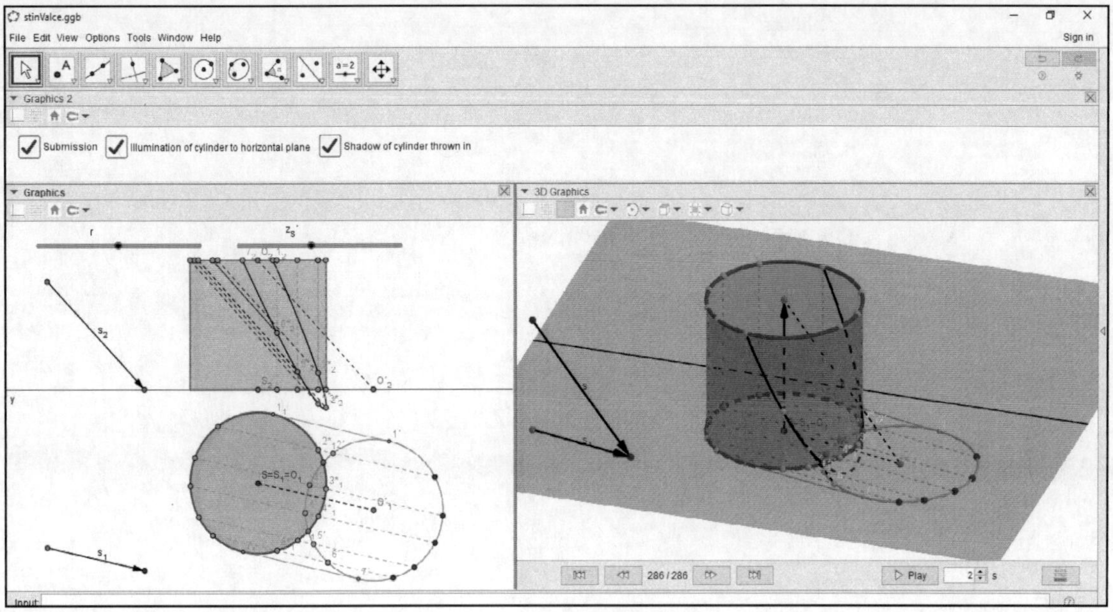

Fig. 14.

3 Conclusion

Teaching of descriptive geometry, not only parallel illumination using software GeoGebra is more visual. Students obtain a better idea of the spatial situation and the solution of the exercises. It is also possible to show different submissions and their solutions in one exercise.

References

[1] Manual:Main Page – GeoGebraWiki, https://wiki.geogebra.org/en/Manual.
[2] URBAN, A., *Deskriptivní geometrie II*. SNTL Praha, 1967.

Current address

Pirklová Petra, Mgr., Ph.D.
Department of Mathematics and Didactics of Mathematics
Faculty of Science, Humanities and Education
Technical University of Liberec
Studentská 1402/2, 461 17 Liberec, Czech Republic
E-mail: petra.pirklova@tul.cz

Bímová Daniela, Mgr., Ph.D.
Department of Mathematics and Didactics of Mathematics
Fakulty of Science, Humanities and Education
Technical University of Liberec
Studentská 1402/2, 461 17 Liberec, Czech Republic
E-mail: daniela.bimova@tul.cz

Slovak University of Technology in Bratislava
Faculty of Mechanical Engineering
17th Conference on Applied Mathematics **APLIMAT 2018**

Proceedings

IMPROVING THE SENSITIVITY OF TRACTOGRAPHY USING MICROSTRUCTURE MODELING AND PSEUDO-INVERSION

PISKOŘOVÁ Zuzana (CZ), BAŠTINEC Jaromír (CZ), LABOUNEK René (CZ)

Abstract. Tractography is a method to map white matter bundles within brain tissue based on diffusion MRI techniques. It seems to be a promising way of detecting white matter structure in vivo, however, method seems to be prone to false positives if standard methods of tractography are used. The paper promises to explore the possibilities of optimization techniques which would enable to include more informative models to the tractography procedure by filtering out false positives from a set of candidate tracts. To detect anatomically plausible tracts, inverse problem from matrix equation $\mathbf{y} = \mathbf{A}\mathbf{x}$ is solved with help of pseudo-inversion technique. However, pseudo-inversion seems to be inappropriate for system of such dimensions and conditioning.

Keywords: tractography, MRI, diffusion, optimization, pseudo-inversion, inverse problem

Mathematics subject classification: 49M04

1 Introduction

Tractography is a MRI based procedure which aims to map the structure of brain white matter in vivo. White matter is a special kind of neuronal tissue formed by axons, longitudinal cells which tend to cluster into bundles. Since the watter diffusion within white matter follows tissue structure, and since it is possible to detect the directionality of diffusive motion by diffusion MRI techniques, we can derive strucutral information about tissue architecture from MRI data. Such assumption works well for white matter bundles with single population of parallel axons within voxel. However, experiments shown that voxels with strictly homogeneous fiber population are rather rare than in majority [1].

Modeling of the diffusive motion within white matter tissue is a process of per-voxel prediction of the diffusion signal by fitting some tissue model. The underlying architecture of the white matter can be suprisingly complex (multiple fiber populations within voxel etc.). If the more complex tissue model is applied, the more trustworthy the fit may be, however computational burden of such procedures critically grows.

Oversimplification of the real situation within neuronal tissue may often lead to untrustworthy results of tractography since the tractography builds the tracts on a basis of information about direction of the diffusive motion within each voxel. False positive results may be detected. Improving reliability of tractography may be done by using more complex models and smart tracking algorithm (such algorithm appeared in recent literature [2]) or filter out false positive tracts built by standard simple tractography methods [6].

This paper deals with the option of determining anatomically plausible tracts by incorporating microstructural models to the analysis, as shown in [6], but via inverse problem solution.

2 Theoretical background

2.1 COMMIT

COMMIT, acronym for convex optimization modeling for microstructure informed tractography, is a post-processing tractography framework working in a top-down fashion. Given a tractogram computed by any of the conventional tractography methods, it selects the set of fibers describing the acquired DMRI data the best. The only prerequisite for the candidate set is that it contains true positives, possibly includes many false positives [6].

Once the candidate pathways have been estimated, DMRI signal contribution of every tract is mapped to each voxel of the image. For this purpose, multi-compartment model containing hindered, restricted and free diffusion compartment is included to characterize neuronal tissue [6].

For restricted compartment, each fiber has one column. Hypothetically, each fiber can be described by more columns if more fiber populations are considered. In case of hindered compartment, only one extraaxonal compartment is described for every fiber population, aiming to avoid redundancy.

The predicted signal $S(\mathbf{q})$ at q-space location $\mathbf{q}$, in a voxel can be expressed as a linear combination of the subsets of tracts, equation (1). In equation (1) standard biomedical labeling of the variables is used.

$$S(\mathbf{q}) = \sum_{\mathcal{F}_\rangle \in \mathcal{F}} \mathbf{f}_i^{IC} R_i^{IC}(\mathbf{q}) + \sum_{\mathcal{F}_\rangle \in \mathcal{F}} \mathbf{f}_i^{EC} R_i^{EC}(\mathbf{q}) + f^{ISO} R^{ISO}(\mathbf{q}) \tag{1}$$

Where R_i^{IC}, R_i^{EC}, R_i^{ISO} are the predicted signal profiles for each compartment, these can be estimated from the data or synthetically generated with analytic models. It is assumed that microstructure parameters remain the same along the bundle. Variables f^{IC}, f^{EC} and f^{ISO} are the corresponding fractal volumes. $\mathcal{F}_\rangle$ is the ith fiber, in equation (1) meaning global contribution of the diffusion arising from all traced fibers. In other words, if the fiber does not cross a voxel there is no signal contribution from it in that voxel [6].

$$y = \mathbf{A}\mathbf{x} + \eta \tag{2}$$

Figure 1 pictures the COMMIT model, which can be also expressed by equation (2). Where $y \in R_+^{n_d n_v}$ is the vector containing the n_d q-space samples acquired in all n_v voxels, η means acquisition noise and modeling errors, $\mathbf{A} \in R^{n_d n_v \times n_c}$ is the observation matrix, a result of multi-compartment model in every voxel, $\mathbf{x} \in R^{n_c}$ are the contributions of the n_c basis functions in $\mathbf{A}$.

Schematic representation of COMMIT formulation. Vector $\mathbf{y}$ is the set of acquired data, every colored block means a different voxel (the number of rows of each block depends on a number of diffusion measurements). $\mathbf{A}$ is a block matrix filled with data computed by microstructure model. Each column

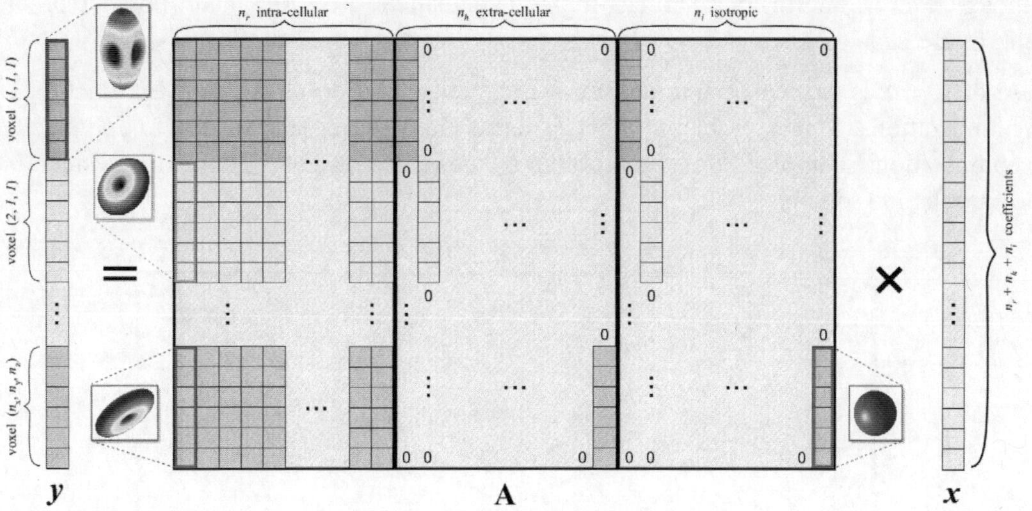

Fig. 1. Schematic representation of COMMIT formulation [6].

for every compartment in matrix $\mathbf{A}$ corresponds to the fiber from the set of candidates. If the fiber does not cross the voxel, there is no signal contribution from it. The single fibre crosses only few voxels within the dataset, hence the matrix $\mathbf{A}$ is sparse. Vector $\mathbf{x}$ is a positive weights vector and is solved by optimization method, non-negative least squares in this case [6].

The goal is to find the optimal vector x, hence to weight fibers with respect to how much their features are consistent alongside each fiber. In other words the goal is to detect anatomically correct fibers, if the assumption that fiber parameters are consistent alongside anatomically valid fibers is made.

3 Solution of linear systems

Task described by COMMIT algorithm can be understood as a typical inverse problem. Inverse problem is mathematical framework that is used to obtain infromation about a physical object or system from observed measurements. It is useful because it provides information about physical parameter that we can not measure directly (anatomical correctness of the axon fiber, parameters consistency, in this case).

If the linear operator $\mathbf{A}$ was regular matrix, finding of solution vector $\mathbf{x}$ would not be problematic. However, for singular systems (which arise often from the practical applications), finding the solution becomes an optimization problem.

To find optimal vector $\mathbf{x}$ from a solution space, we tend to minimize the equation (3),

$$\underset{x \geq 0}{\operatorname{argmin}} ||\mathbf{Ax} - \mathbf{y}||, \tag{3}$$

where $|| \cdot ||$ is a l_p norm defined as

$$||\mathbf{x}|| = \sqrt[p]{\sum_i |x_i|^p}, \; p\epsilon\mathbb{R}. \tag{4}$$

Even though every l_p norm look similar, their mathematical properties and applications are very different, figure 2.

The case of $p = 2$ is the most important norm in practice. All norms are convex functions, which makes optimization a convex problem. Other l_p norms are however problematic to optimize because of the computational complexity of the problem (l_0 optimization gives interesting results, but it is NP-hard problem to solve).

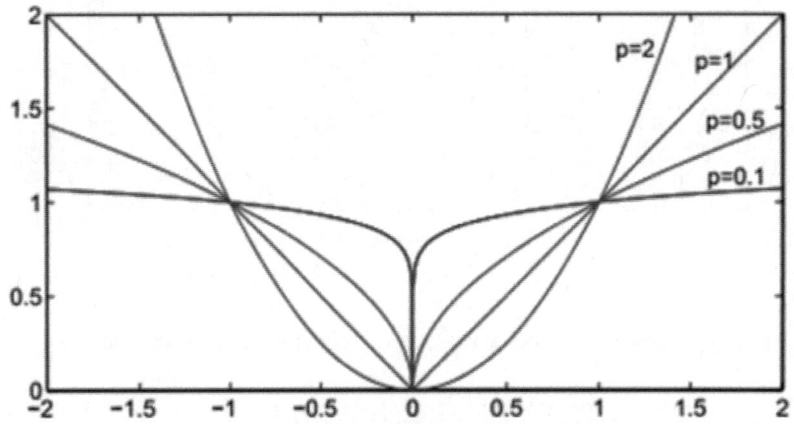

Fig. 2. The functions of various p norms.

The pursuit for solutions of

$$\mathbf{Ax} = \mathbf{y} \tag{5}$$

is then the minimization of error $||\mathbf{r}||_p = \mathbf{Ax} - \mathbf{y}$ for some suitable choice of p. For example l_2 norm is computationaly the easiest to reach. Different norms render different optimum solutions.

Minimization of the l_2 norm is widespread in various fields of engineering and is called least squares problem. It is preferred due to uniqueness of the solution (in contrast with other p-norms).

4 Problem formulation

The problem to be optimized is expressed by figure 1. The linear operator $\mathbf{A}$ can be understood as a dictionary of atoms (columns). The aim is to produce representation of the signal $\mathbf{y}$ by expressing it as a linear combination of atoms from $\mathbf{A}$ by retrieveing the best solution vector $\mathbf{x}$ by minimizing the support of

$$\mathbf{x} \text{ subject to } \mathbf{y} = \mathbf{Ax}. \tag{6}$$

Than it is possible to use the indices of nonzero components of $\mathbf{x}$ to identify the atoms (fibers) that are significant for reproducing the signal.

The shape of the linear system strongly depends on the size of the candidate set, model used and even on the acquisition scheme. In vast majority of applications, we are dealing with rectangular matrix whith tremendous dimensions (it is problematic to store the linear operator if no special format exploting sparseness of the matrix is used). Fortunatelly, since each tract crosses only few voxels of the full volume, the matrix is sparse.

Linear operator $\mathbf{A}$ based on a real data is also extremely ill-conditioned. The condition number of the system is close to infinity (computed by numpy.linalg.cond method on the most dense fragment of the linear operator, size (1000 x 1000).

5 Solution design

Problem solved is known to be large, sparse, singular (or close to singular) and ill-conditioned. Since it is known that singular problems are prone to roundoff errors when iterative methods are used, we decided to choose direct method of computation. Direct methods also enables to reach optimal solution without iterative procedure. Pseudo-inversion provides the optimal solution in means of l_2 optimization.

5.1 Pseudo-inversion by normal equations

In pursuit of finding a solution vector $\mathbf{x}$, it is very intuitive to rearrange the inverse problem equation as in (7). This solution was proposed because it is theoretically the simplest one. Also, the solution theoretically guarantees reaching global optimum in terms of $||\mathbf{Ax} - \mathbf{y}||_2$, hence least squares, minimization [4].

$$\begin{aligned} \mathbf{Ax} &= \mathbf{y} \\ \mathbf{x} &= \mathbf{A}^{-1}\mathbf{y} \end{aligned} \tag{7}$$

Since we are dealing with rectangular matrix, it is not possible to compute $\mathbf{A}^{-1}$ by inversion, but it is necessary to use pseudo-inversion, $\mathbf{A}^{+}$. More approaches how to reach pseudo-inverse of the matrix $\mathbf{A}$ exist. First, lets mention normal equations (8).

$$\begin{aligned} \mathbf{Ax} &= \mathbf{y} \\ \mathbf{A}^{\mathbf{T}}\mathbf{Ax} &= \mathbf{A}^{\mathbf{T}}\mathbf{y} \\ \mathbf{x} &= (\mathbf{A}^{\mathbf{T}}\mathbf{A})^{-1}\mathbf{A}^{\mathbf{T}} \end{aligned} \tag{8}$$

Normal equations are no doubt elegant analysis, since the product of $\mathbf{A}^{\mathbf{T}}\mathbf{A}$ is allways symmetric matrix. However, we are dealing with badly ill-posed system, and the condition number of $\mathbf{A}^{\mathbf{T}}\mathbf{A}$ product is the condition number of the original matrix $\mathbf{A}$ squared. Such process make the solution of the system even more prone to the roundoff errors [5].

The computational instability of the pseudo-inversion is shown on well-conditioned and ill-conditioned problem by comparison of defined vector of solution and vector of solution obtained by pseudo-inverse. If we define matrices $\mathbf{A}$ and $\mathbf{y}$, process is as in (9).

$$\begin{aligned} \mathbf{y} &= \mathbf{Ax} \\ \mathbf{x} &= \mathbf{A}^{+}\mathbf{y} \end{aligned} \tag{9}$$

First of all, analysis is shown on a well-conditioned matrix $\mathbf{A1}$ (condition number = 7.3, with solution vector $\mathbf{x}$.

$$\mathbf{A1} = \begin{pmatrix} 1 & 2 & 3 \\ 3 & 2 & 1 \\ 1 & 2 & 1 \end{pmatrix}, \mathbf{x} = \begin{pmatrix} 3 \\ 2 \\ 1 \end{pmatrix} \tag{10}$$

As shown in figure 3, solution vector $\mathbf{x}$ reached by pseudoinversion is identical with the original one.

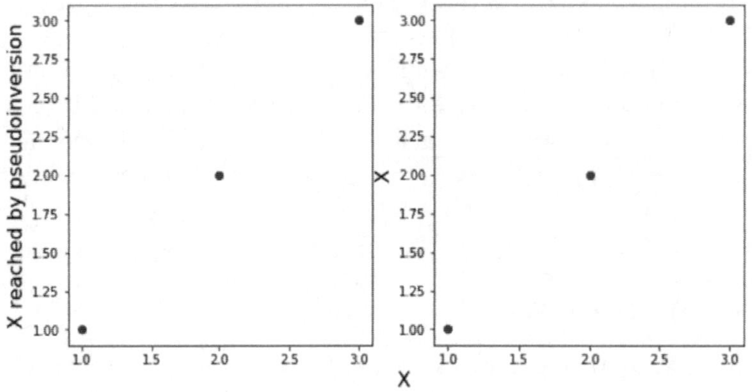

Fig. 3. Pseudo-inverse of the well-conditioned system.

Now, lets analyze ill-conditioned matrix $\mathbf{A2}$ (conditional number $2 * 10^{16}$) with solution vector $\mathbf{x}$.

$$\mathbf{A2} = \begin{pmatrix} 1 & 0 & 1 \\ 1 & 1 & 1 \\ 1 & 2 & 1 \end{pmatrix}, \mathbf{x} = \begin{pmatrix} 3 \\ 2 \\ 1 \end{pmatrix} \tag{11}$$

In figure 4 comparison of vector of solution $\mathbf{x}$ reached by pseudo-inversion is not identical to the vector $\mathbf{x}$ defined previously. Such error is caused by roundoff errors during the process of computation.

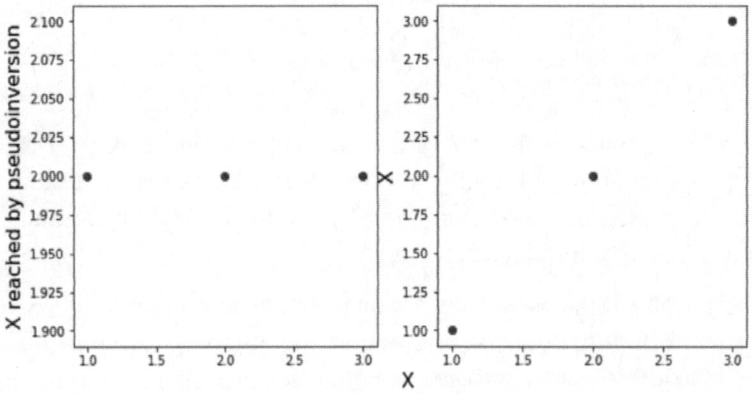

Fig. 4. Pseudo-inverse of the ill-conditioned system.

Note that the condition number of matrix $\mathbf{A2}$ is still very low in comparison with condition number of system built from neurological data. What is more, the pseudo-inverse of the sparse system is a very dense matrix, which throws off potentially beneficial feature of the matrix in terms of data storage and computational demands. In fact, solution of large ill-posed systems is very challenging task, sparseness brings a dramatic simplification if used in a sophisticated way.

6 Conclusion

It is not possible to use direct inversion matrix method because linear operator $\mathbf{A}$ is not square.

The method of pseudo-inversion is suitable for rectangular matrices, however it seems to be extremeley prone to rounoff errors and comes with inneglilible computational and memory burden if standard methods [3]) are used.

Method of pseudo-inversion also is not suitable for such task because it does not exploit the sparsenest of the linear operator, which comes with significant memory issues. What is more, let be kept in mind that COMMIT algorithm is just a postprocessing step of improving the sensitivity of the modelling. Incorporation of the more complex models could be computationally more feasible task, if the candidate dataset would be build based on more precise information. Hence, in COMMIT case, the task lays a great deal on how the computation is time and space consuming.

Using pseudo-inversion without exploiting the sparsness of the linear operator leads at least to very multi-dimensional problem resulting in ridiculous time estimations if standard methods and material was used (months of computation time if standard PC used). What is more, results reached by pseudo-inversion seems to be non-valid due to ill-posed nature of the problem.

Optimization of a matrix equation $\mathbf{y} = \mathbf{Ax}$ for real neurological data is challenging task due to dimensions of the data and also due to errors which arises from measurement itself. Let assume this problematics as open problem and find the right way to solve it.

Acknowledgement

The work was supported by project FEKT-S-17-4225 of Brno University of Technology.

References

[1] JEURISSEN B., A. LEEMANS, J.D. TOURNIER, D.K. JONES a J. SIJBERS. *Investigating the prevalence of complex fiber configurations in white matter tissue with diffusion magnetic resonance imaging.* Human Brain Mapping [online]. 2013, 34(11), 2747-2766. DOI: 10.1002/hbm.22099. ISSN 10659471.

[2] FAROOQ H., J. XU, J. W. NAM, D. F. KEEFE, E. YACOUB, T. GEORGIOU , CH. LENGLET. *Microstructure Imaging of Crossing (MIX) White Matter Fibers from diffusion MRI.* Scientific Reports [online]. 2016, 6, 38927. DOI: 10.1038/$srep$38927. ISSN 2045-2322.

[3] FILLARD P., Descoteaux M., Goh A., et al *Quantitative Evaluation of 10 Tractography Algorithms on a Realistic Diffusion MR Phantom*NeuroImage [online]. 2011, vol. 56, issue 1, s. 220-234 [cit. 2015-05-13]. DOI: 10.1016/$j.neuroimage$.2011.01.032.

[4] CHAPRA, Steven C. a Raymond P. CANALE. *Numerical methods for engineers* . Seventh edition. New York, NY: McGraw-Hill Education, 2015. ISBN $978 - 0073397924$.

[5] PRESS, William H. *Numerical Recipes in C: the art of scientific computing.* 2nd ed. Cambridge: Cambridge University Press, c1992. ISBN $0 - 521 - 43108 - 5$. .

[6] DADUCCI A., A. DAL PALU, A. LEMKADDEM, J.P. THIRAN. *COMMIT: Convex Optimization Modeling for Microstructure Informed Tractography.* IEEE TRANSACTIONS ON MEDICAL IMAGING,. 2015, 2015(34), 246-257. DOI:10.1109/TMI.2014.2352414.

[7] PISKOŘOVÁ, Z. *Multi-tensor model based tractography of axonal bundles* Brno: Brno University of Technology, Faculty of Electrical Engineering and Communication, Department of Biomedical Engineering, Master thesis, 2017. 74 p. Supervised by Ing. René Labounek

Current address

Piskořová Zuzana, Ing
Department of Mathematics, Faculty of Electrical Engineering and Communication, Brno University of Technology, Technická 10, 616 00 Brno, Czech republic
E-mail: xpisko01@stud.feec.vutbr.cz

Baštinec Jaromír, doc. RNDr., CSc.
Department of Mathematics, Faculty of Electrical Engineering and Communication, Brno University of Technology, Technická 10, 616 00 Brno, Czech republic
E-mail: bastinec@feec.vutbr.cz

Labounek René, Ing.
University Hospital Olomouc, I. P. Pavlova 6, 779 00 Olomouc, Czech Republic
E-mail: rene.labounek@gmail.com

Slovak University of Technology in Bratislava
Faculty of Mechanical Engineering
17th Conference on Applied Mathematics APLIMAT 2018

Proceedings

COMPARATIVE STUDY OF SEVEN METHODS FOR ESTIMATING THE WEIBULL DISTRIBUTION PARAMETERS FOR WIND SPEED IN BRATISLAVA - MLYNSKÁ DOLINA

POBOČÍKOVÁ Ivana (SK) , SEDLIAČKOVÁ Zuzana (SK), ŠIMON Ján (SK)

Abstract. In this paper the Weibull distribution is used to describe the wind speed data collected in Bratislava-Mlynská dolina in the period of the years 2005-2009. The parameters are estimated using seven methods: the maximum likelihood, method of moments, empirical method, power density method, least squares method and weighted least squares method with two weight factors. Their performance is compared using the root mean square error and coefficient of determination. It was observed that the weighted least squares method perform the best, followed by the maximum likelihood method and the method of moments.

Keywords: Weibull distribution, parameter estimation, wind speed

Mathematics Subject Classification: Primary 6207, Secondary 62P30, 62P12.

1 Introduction

Wind power belongs to among promising energy sources with minimal environmental impact and huge energetic potential. However, attention has to be paid to the optimal locality selection to optimize cost/benefit ratio (Morgan et al., 2011).

Many conditions have to be satisfied in the process of locality selection, but one of the most important is a suitable wind speed spectrum. The literature contains several distributions that can be applied to calculate the wind speed distribution (Morgan et al., 2011; Celik, 2003; Carta et al., 2009; Lun and Lam, 2000; Gryning, 2016). By far the most widely-used distribution for the characterization of average wind speeds is the two-parameter Weibull distribution. Most likely it is a consequence of relative simplicity and flexibility of Weibull distribution, as well as its good ability to fit measured data. The way the distribution parameters are estimated is at least of the same importance as the correct distribution choice. There are several methods for estimating the Weibull distribution parameters (Justus et al., 1978; Stevens et al., 1979; Conradsen et al., 1984; Seguro and Lambert, 2000; Dorvlo, 2002; Wu et al., 2006; Akdağ and Dinler, 2009). In the recent literature these methods are compared several times and in different ways (Kantar and Senoglu, 2008; Chang, 2011; Kantar et al., 2011; Costa Rocha et al., 2012).

This paper uses the two-parameter Weibull distribution to describe the wind speed in Bratislava-Mlynská dolina for given wind speed data that cover the period of years 2005 - 2009. The presented paper therefore examines seven different methods for estimating Weibull distribution parameters. Namely: the maximum likelihood method (MLM), the method of moments (MOM), the empirical method (EM), the power density method (PDM), the least squares method (LSM) and the weighted least squares method (WLSM) with two weight factors. The comparison of these methods is based on the monthly, seasonal and yearly wind speed data. Finally the root mean square error ($RMSE$) and the coefficient of determination (R^2) are used to measure the performance of the estimating methods. All the numerical computations are performed in Matlab 7.1.

2 Methodology

2.1 Description of wind speed data

The wind speed data processed in the presented paper were measured at the Meteorological observatory Bratislava-Mlynská dolina, situated on the campus of the Faculty of Mathematics, Physics and Iformatics, Comenius University in Bratislava, within time frame January 2005 to December 2009. The sampling site Mlynská dolina is situated at the SW boundary of Bratislava (N 48°09'08", E 17°04'13", 195 m.a.s.l). It is situated approximately 700 m to the north from the Danube river, on the university campus. The experimental setup is installed on the roof of the university building. The data are collected 7 m above the roof and 25 m above ground. The wind speed has been measured by universal anemograph METRA, working on Pitot tube principle. Data are recorded on the paper tape and manually transcripted once a day. Hourly averages are evaluated, rounded to the nearest integer and stored in the electronic form. The sample size to be researched is 43 824.

2.2 Methods for estimating the Weibull distribution parameters

The probability density function of the two-parameter Weibull distribution is given by

$$f(v) = \frac{k}{c^k} v^{k-1} \exp\left(-\left(\frac{v}{c}\right)^k \right) \tag{1}$$

and the cumulative distribution function is given by

$$F(v) = 1 - \exp\left(-\left(\frac{v}{c}\right)^k \right), \tag{2}$$

where $v > 0$, $k > 0$ and $c > 0$. Here v is the wind speed, k is the dimensionless shape parameter and c is the scale parameter in units of the wind speed. The mean wind speed $E(V)$ and the variance of the wind speed $Var(V)$ are given by

$$E(V) = c\Gamma\left(1 + \frac{1}{k}\right), \tag{3}$$

$$Var(V) = c^2\left[\Gamma\left(1+\frac{2}{k}\right)-\Gamma^2\left(1+\frac{1}{k}\right)\right], \tag{4}$$

where $\Gamma(a), a > 0$, is the gamma function.

Let $V_1, V_2, \ldots, V_n$ be a random sample of size n from the Weibull distribution with parameters k and c and let $v_1, v_2, \ldots, v_n$ be a realization of a random sample. Let $V_{(1)} < V_{(2)} < \ldots < V_{(n)}$ be the order statistics of $V_1, V_2, \ldots, V_n$ and let $v_{(1)} < v_{(2)} < \ldots < v_{(n)}$ be the observed ordered observations.

There are several methods described in the literature for estimating the Weibull distribution parameters. This paper presents seven methods for estimating the parameters k and c.

Maximum likelihood method

The likelihood function of the Weibull distribution is given by

$$L(k,c) = \prod_{i=1}^{n} \frac{k}{c^k} v_i^{k-1} \exp\left(-\left(\frac{v_i}{c}\right)^k\right). \tag{5}$$

By differentiating the logarithm of the function (5) with respect to k and c, respectively, and equating them to zero, we obtain

$$\begin{aligned} \frac{\partial \ln L(k,c)}{\partial k} &= \frac{n}{k} - n\ln c - \frac{\sum_{i=1}^{n} v_i^k \ln v_i - \ln c \sum_{i=1}^{n} v_i^k}{c^k} + \sum_{i=1}^{n} \ln v_i = 0, \\ \frac{\partial \ln L(k,c)}{\partial c} &= -\frac{nk}{c} + \frac{k}{c^{k+1}} \sum_{i=1}^{n} v_i^k = 0. \end{aligned} \tag{6}$$

The parameters k and c can be estimated by the following equations (Carta et al., 2009)

$$\frac{1}{k} - \frac{\sum_{i=1}^{n} v_i^k \ln v_i}{\sum_{i=1}^{n} v_i^k} + \frac{1}{n}\sum_{i=1}^{n} \ln v_i = 0, \tag{7}$$

$$c = \left(\frac{1}{n}\sum_{i=1}^{n} v_i^k\right)^{1/k}. \tag{8}$$

The estimate $\hat{k}$ of the parameter k can be obtained by solving (7) with respect to k. The estimate $\hat{c}$ of the parameter c can be obtained using equation (8). The equation (7) can be solved iteratively with respect to k, after which equation (8) can be solved analytically. The Newton method was used for the numerical computations.

Method of moments

The estimates of the parameters k and c can be obtained by equating the moments of the Weibull distribution with the corresponding sample moments. The parameters k and c can be estimated by the following equations (Carta et al., 2009)

$$c\Gamma\left(1+\frac{1}{k}\right) = \bar{v}, \tag{9}$$

$$c^2\Gamma\left(1+\frac{2}{k}\right)=\frac{1}{n}\sum_{i=1}^{n}v_i^2, \tag{10}$$

where $\bar{v}=\frac{1}{n}\sum_{i=1}^{n}v_i$ is the sample mean wind speed. By dividing (10) by the square of (9) one obtains

$$\frac{\Gamma\left(1+\frac{2}{k}\right)}{\Gamma^2\left(1+\frac{1}{k}\right)}=\frac{\frac{1}{n}\sum_{i=1}^{n}v_i^2}{\bar{v}^2}. \tag{11}$$

The estimate $\hat{k}$ of the parameter k can be obtained by solving (11) with respect to k. This equation does not have an analytical solution and can be solved iteratively with respect to k. The estimate $\hat{c}$ of the parameter c can be obtained using equation (9).

Empirical method

The empirical method can be used as a special case of the MOM. The parameter k can be estimated by following equation (Morgan et al., 2011; Akdağ and Dinler, 2009)

$$\hat{k}=\left(\frac{\bar{v}}{s_v}\right)^{1.086}, \tag{12}$$

where $s_v=\sqrt{\frac{1}{n-1}\sum_{i=1}^{n}\left(v_i-\bar{v}\right)^2}$ is the sample standard deviation. The parameter c can be estimated using (9).

Power density method

The energy pattern factor is defined as (Akdağ and Dinler, 2009)

$$E_{pf}=\frac{\overline{v^3}}{\bar{v}^3}, \tag{13}$$

where $\overline{v^3}=\frac{1}{n}\sum_{i=1}^{n}v_i^3$ is the sample mean of the wind speed cubes. The parameter k can be estimated by the following equation

$$\hat{k}=1+\frac{3.69}{\left(E_{pf}\right)^2}, \tag{14}$$

and the parameter c can be estimated using (9).

Least squares method

The cumulative distribution function (2) can be linearized as follows

$$\ln\left(-\ln\left(1-F(v)\right)\right)=k\ln(v)-k\ln(c). \tag{15}$$

Let $Y=\ln\left(-\ln\left(1-F(v)\right)\right)$, $X=\ln v$, $b=k$ and $a=-k\ln c$. Then the equation (15) can be rewritten as

$$Y=bX+a. \tag{16}$$

The parameters in equation (15) can be estimated using the least squares method. The estimates of the regression parameters a and b minimize the function

$$Q(a,b)=\sum_{i=1}^{n}(Y_i-a-b\ln v_{(i)})^2. \tag{17}$$

The parameters k and c can be estimated by the following equations

$$\hat{k}=\frac{n\sum_{i=1}^{n}\ln v_{(i)}\ln\left[-\ln\left(1-\hat{F}\left(v_{(i)}\right)\right)\right]-\sum_{i=1}^{n}\ln v_{(i)}\sum_{i=1}^{n}\ln\left[-\ln\left(1-\hat{F}\left(v_{(i)}\right)\right)\right]}{n\sum_{i=1}^{n}\ln^2 v_{(i)}-\left(\sum_{i=1}^{n}\ln v_{(i)}\right)^2}, \tag{18}$$

$$\hat{c}=\exp\left(-\frac{\sum_{i=1}^{n}\ln\left[-\ln\left(1-\hat{F}\left(v_{(i)}\right)\right)\right]-\hat{k}\sum_{i=1}^{n}\ln v_{(i)}}{\hat{k}n}\right). \tag{19}$$

To estimate the values of the cumulative distribution function, the mean rank is used

$$\hat{F}\left(v_{(i)}\right)=\frac{i}{n+1}, \tag{20}$$

where i denotes the ith smallest value of $v_{(1)},v_{(2)},\ldots,v_{(n)}$, $i=1,2,\ldots,n$.

Weighted least squares method

The estimates of the regression parameters a and b minimize the function

$$Q(a,b)=\sum_{i=1}^{n}w_i(Y_i-a-b\ln v_{(i)})^2, \tag{21}$$

where w_i is the weight factor, $i=1,2,...,n$. Two weight factors are used in this paper. The first weight factor was proposed in Bergman (1986)

$$w_i=\left[\left(1-\hat{F}(v_{(i)})\right)\ln\left(1-\hat{F}(v_{(i)})\right)\right]^2,\quad i=1,2,...,n. \tag{22}$$

and the second weight factor was proposed in Faucher and Tyson (1988)

$$w_i=3.3\,\hat{F}(v_{(i)})-27.5\left[1-\left(1-\hat{F}(v_{(i)})\right)^{0.025}\right],\quad i=1,2,...,n. \tag{23}$$

The parameters k and c can be estimated by the following equations

$$\hat{k}=\frac{\sum_{i=1}^{n}w_i\sum_{i=1}^{n}w_i\ln v_{(i)}\ln\left[-\ln\left(1-\hat{F}\left(v_{(i)}\right)\right)\right]-\sum_{i=1}^{n}w_i\ln v_{(i)}\sum_{i=1}^{n}w_i\ln\left[-\ln\left(1-\hat{F}\left(v_{(i)}\right)\right)\right]}{\sum_{i=1}^{n}w_i\sum_{i=1}^{n}w_i\ln^2 v_{(i)}-\left(\sum_{i=1}^{n}w_i\ln v_{(i)}\right)^2}, \tag{24}$$

$$\hat{c}=\exp\left(-\frac{\sum_{i=1}^{n}w_i\ln\left[-\ln\left(1-\hat{F}\left(v_i\right)\right)\right]-\hat{k}\sum_{i=1}^{n}w_i\ln v_i}{\hat{k}\sum_{i=1}^{n}w_i}\right). \tag{25}$$

2.3 Statistical analysis

The wind speed data were divided into subsets with respect to the months, four seasons and years, respectively. The months were divided as follows: winter (December, January, February), spring (March, April, May), summer (June, July, August) and autumn (September, October, November). The estimates of the Weibull distribution parameters were calculated for each month, season, year and the whole monitored period of 5 years, respectively.

To find the best method to estimate the Weibull distribution parameters the coefficient of determination (R^2) and the root mean square error ($RMSE$) were used. These parameters can be calculated using

$$R^2 = 1 - \frac{\sum_{i=1}^{n} (v_i - x_i)^2}{\sum_{i=1}^{n} (v_i - \overline{v})^2}, \tag{26}$$

$$RMSE = \sqrt{\frac{1}{n} \sum_{i=1}^{n} (v_i - x_i)^2}, \tag{27}$$

where n is the number of the wind speed data, x_i is the ith predicted data calculated using the Weibull distribution, v_i is the ith ordered observed wind speed.

The coefficient of determination R^2 ranges from 0 to 1. The ideal value of R^2 is equal to 1. The root mean square error $RMSE$ ranges from 0 to infinity. The ideal value of $RMSE$ is close to zero. The best method for estimating the Weibull distribution parameters can be determined according to the highest values of R^2 and the lowest values of $RMSE$. R^2 and $RMSE$ were calculated for each month, season, year and the whole monitored period of 5 years, respectively.

4 Results and discussion

Tables 1, 2 and 3 show the estimates of the Weibull distribution parameters and the results of the statistical analysis of all of the examined methods.

The comparison of seasonal Weibull probability density distributions with the observed seasonal probability density distributions of the wind speed are illustrated in Figure 1. The comparison of the Weibull probability density distributions with the observed probability density distribution of the wind speed for the whole monitored period of 5 years is illustrated in Figure 2. The comparison of the yearly Weibull probability density distributions for the years 2005-2009 with the observed yearly probability density distributions of the wind speed is illustrated in Figure 3.

The comparison of the results in Table 1 and Figure 1 shows that the seasonal value of $RMSE$ ranges from 0.00198 to 0.00939. The seasonal value of R^2 ranges from 0.81982 to 0.99559 which means that all the presented methods fit the data successfully. Each approach has R^2 value better than 0.96777 in spring, summer, autumn and better than 0.81982 in winter. The WLSMa seems to have the best performance for all seasons in terms of R^2 and $RMSE$. The second best method is the WLSMb, except spring, when the second best performance can be assigned to the LSM in terms of R^2 and $RMSE$.

	Winter				Spring			
Method	$\hat{k}$	$\hat{c}$	*RMSE*	R^2	$\hat{k}$	$\hat{c}$	*RMSE*	R^2
MLM	1.7169	12.6399	0.00878	0.84264	1.9022	12.0985	0.00427	0.96971
MOM	1.7502	12.6802	0.00902	0.83388	1.9073	12.1044	0.00429	0.96946
EM	1.7745	12.6700	0.00921	0.82671	1.9309	12.1088	0.00440	0.96777
PDM	1.7969	12.6981	0.00939	0.81982	1.9235	12.1075	0.00437	0.96835
LSM	1.6193	12.7254	0.00804	0.86787	1.8687	12.1287	0.00414	0.97158
WLSMa	1.6119	13.0680	0.00791	0.87232	1.8765	12.2167	0.00411	0.97198
WLSMb	1.6244	12.8472	0.00803	0.86823	1.8829	12.1384	0.00417	0.97113
	Summer				**Autumn**			
Method	$\hat{k}$	$\hat{c}$	*RMSE*	R^2	$\hat{k}$	$\hat{c}$	*RMSE*	R^2
MLM	1.9502	10.4870	0.00219	0.99457	1.8100	11.1920	0.00313	0.98530
MOM	1.9323	10.4670	0.00213	0.99487	1.8065	11.1848	0.00309	0.98561
EM	1.9557	10.4703	0.00216	0.99470	1.8307	11.1915	0.00335	0.98309
PDM	1.9233	10.4657	0.00214	0.99482	1.8233	11.1895	0.00327	0.98392
LSM	2.0365	10.4303	0.00271	0.99172	1.8123	11.1768	0.00315	0.98503
WLSMa	1.9303	10.3838	0.00198	0.99559	1.7457	11.2540	0.00262	0.98965
WLSMb	1.9478	10.4060	0.00201	0.99543	1.7640	11.2115	0.00273	0.98879

Tab. 1. The estimates of the Weibull distribution parameters and the results of the statistical analysis for four seasons of composite year.

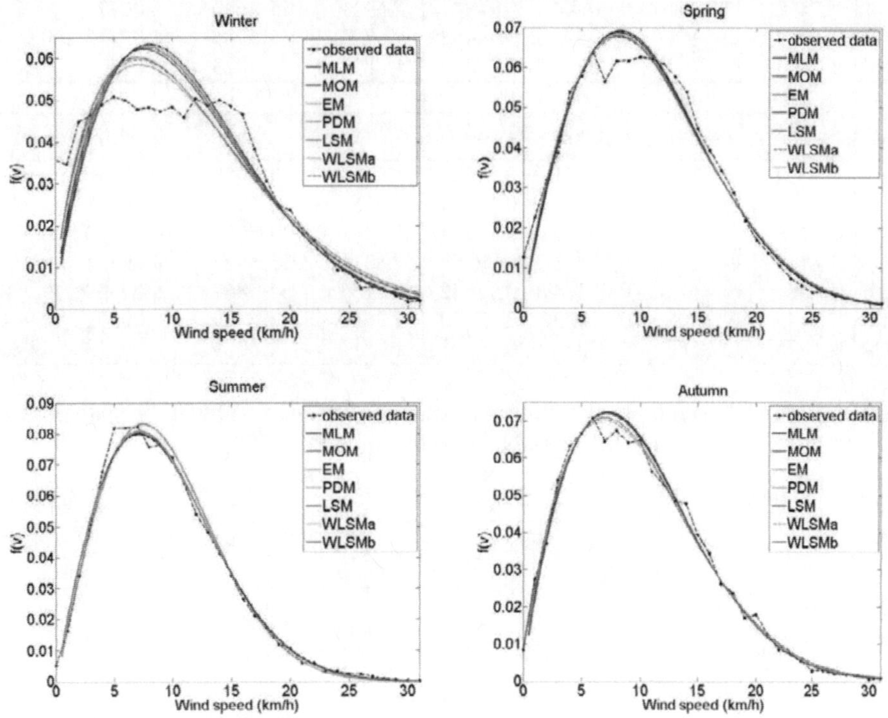

Fig. 1. Comparison of Weibull probability density distributions for four seasons of composite year

The analyses based on Table 2 and Figure 2 further show that the value of *RMSE* for the whole monitored period of 5 years ranges from 0.00302 to 0.00360 and the value of R^2 ranges from 0.97851 to 0.98482. The WLSMa performs better than the other methods in terms of R^2 and *RMSE*. The second best method is the WLSMb.

From Table 2 and Figure 3 comparison shows that the yearly value of *RMSE* ranges from 0.00223 to 0.00828 and the yearly value of R^2 ranges from 0.88641 to 0.99230. The WLSMa performs better than any other method for all the years in terms of R^2 and *RMSE*. The second best method is the WLSMb.

	2005				2006			
Method	$\hat{k}$	$\hat{c}$	*RMSE*	R^2	$\hat{k}$	$\hat{c}$	*RMSE*	R^2
MLM	1.8516	11.7633	0.00321	0.98358	1.8153	11.3147	0.00253	0.99047
MOM	1.8442	11.7510	0.00316	0.98412	1.8071	11.3032	0.00249	0.99081
EM	1.8682	11.7570	0.00335	0.98209	1.8313	11.3100	0.00264	0.98962
PDM	1.8608	11.7552	0.00329	0.98277	1.8111	11.3044	0.00251	0.99065
LSM	1.8692	11.7280	0.00337	0.98188	1.8378	11.2898	0.00270	0.98919
WLSMa	1.7727	11.8314	0.00282	0.98730	1.7780	11.3352	0.00239	0.99153
WLSMb	1.8017	11.7756	0.00290	0.98656	1.7951	11.2991	0.00243	0.99119
	2007				**2008**			
Method	$\hat{k}$	$\hat{c}$	*RMSE*	R^2	$\hat{k}$	$\hat{c}$	*RMSE*	R^2
MLM	1.8008	11.5603	0.00625	0.93528	1.7284	11.4348	0.00802	0.89352
MOM	1.8127	11.5720	0.00634	0.93339	1.7375	11.4436	0.00809	0.89168
EM	1.8369	11.5787	0.00654	0.92902	1.7618	11.4528	0.00828	0.88641
PDM	1.8437	11.5805	0.00660	0.92771	1.7525	11.4494	0.00821	0.88845
LSM	1.7458	11.6021	0.00584	0.94341	1.6929	11.4630	0.00774	0.90063
WLSMa	1.7419	11.7171	0.00578	0.94458	1.6783	11.5986	0.00763	0.90349
WLSMb	1.7433	11.6462	0.00581	0.94404	1.6859	11.5023	0.00769	0.90202
	2009				**2005-2009**			
Method	$\hat{k}$	$\hat{c}$	*RMSE*	R^2	$\hat{k}$	$\hat{c}$	*RMSE*	R^2
MLM	1.9024	11.8970	0.00251	0.99027	1.8183	11.5958	0.00340	0.98077
MOM	1.8876	11.8769	0.00239	0.99114	1.8165	11.5912	0.00339	0.98091
EM	1.9113	11.8816	0.00257	0.98975	1.8408	11.5979	0.00360	0.97851
PDM	1.8928	11.8780	0.00243	0.99087	1.8311	11.5953	0.00351	0.97952
LSM	1.9581	11.8375	0.00304	0.98572	1.8165	11.5867	0.00339	0.98089
WLSMa	1.8445	11.8798	0.00223	0.99230	1.7649	11.6729	0.00302	0.98482
WLSMb	1.8715	11.8586	0.00230	0.99182	1.7811	11.6189	0.00313	0.98377

Tab. 2. The estimates of the Weibull distribution parameters and the results of the statistical analysis for individual years and for years 2005-2009.

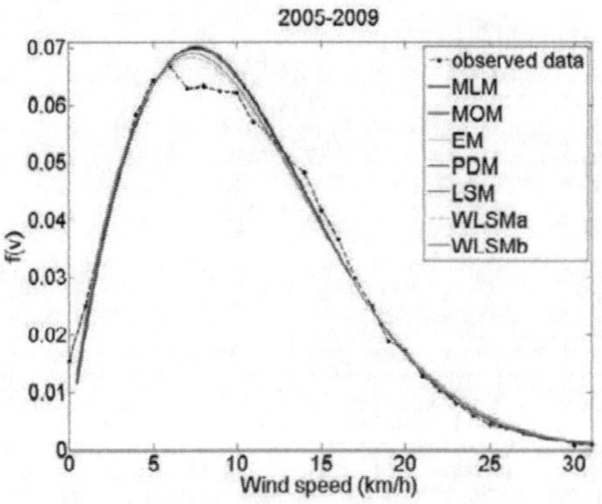

Fig. 2. Comparison of Weibull probability density distributions for the period 2005-2009.

	January				**February**			
Method	$\hat{k}$	$\hat{c}$	*RMSE*	R^2	$\hat{k}$	$\hat{c}$	*RMSE*	R^2
MLM	1.6900	13.3453	0.00903	0.81423	1.7669	12.2243	0.01584	0.56197
MOM	1.7331	13.4025	0.00930	0.80283	1.8089	12.2746	0.01616	0.54401
EM	1.7573	13.4134	0.00947	0.79540	1.8329	12.2819	0.01636	0.53275
PDM	1.7842	13.4244	0.00967	0.78671	1.8694	12.2915	0.01666	0.51514
LSM	1.5740	13.4581	0.00825	0.84487	1.6234	12.3657	0.01467	0.62399
WLSMa	1.5644	13.9283	0.00805	0.85222	1.6594	12.6871	0.01486	0.61416
WLSMb	1.5821	13.6107	0.00823	0.84575	1.6630	12.4637	0.01493	0.61069
	March				**April**			
Method	$\hat{k}$	$\hat{c}$	*RMSE*	R^2	$\hat{k}$	$\hat{c}$	*RMSE*	R^2
MLM	1.7926	13.3675	0.00675	0.89492	1.9366	11.2466	0.00598	0.94583
MOM	1.8152	13.3979	0.00684	0.89182	1.9433	11.2532	0.00603	0.94495
EM	1.8392	13.4056	0.00698	0.88756	1.9664	11.2564	0.00624	0.94114
PDM	1.8472	13.4080	0.00703	0.88604	1.9788	11.2579	0.00635	0.93890
LSM	1.7032	13.4657	0.00637	0.90632	1.8755	11.2831	0.00552	0.95384
WLSMa	1.7619	13.6143	0.00649	0.90264	1.8443	11.4240	0.00528	0.95778
WLSMb	1.7479	13.4989	0.00648	0.90294	1.8709	11.3296	0.00546	0.95495
	May				**June**			
Method	$\hat{k}$	$\hat{c}$	*RMSE*	R^2	$\hat{k}$	$\hat{c}$	*RMSE*	R^2
MLM	2.1266	11.6513	0.00532	0.96344	1.9643	10.0331	0.00358	0.98717
MOM	2.1258	11.6523	0.00532	0.96348	1.9448	10.0127	0.00357	0.98725
EM	2.1466	11.6526	0.00538	0.96258	1.9678	10.0155	0.00354	0.98744
PDM	2.1487	11.6526	0.00539	0.96248	1.9196	10.0091	0.00366	0.98661
LSM	2.0886	11.6708	0.00523	0.96464	2.0764	9.9750	0.00397	0.98426
WLSMa	2.0605	11.7785	0.00515	0.96578	1.9710	9.9094	0.00337	0.98865
WLSMb	2.0850	11.6944	0.00520	0.96499	1.9933	9.9249	0.00341	0.98838
	July				**August**			
Method	$\hat{k}$	$\hat{c}$	*RMSE*	R^2	$\hat{k}$	$\hat{c}$	*RMSE*	R^2
MLM	1.9115	10.9645	0.00286	0.98969	1.9994	10.4444	0.00312	0.98931
MOM	1.8932	10.9424	0.00279	0.99015	1.9869	10.4305	0.00306	0.98975
EM	1.9168	10.9466	0.00285	0.98979	2.0096	10.4327	0.00316	0.98903
PDM	1.8926	10.9422	0.00279	0.99015	1.9886	10.4307	0.00306	0.98971
LSM	1.9827	10.9057	0.00320	0.98705	2.0516	10.4086	0.00347	0.98680
WLSMa	1.8796	10.8608	0.00270	0.99079	1.9474	10.4005	0.00300	0.99011
WLSMb	1.8892	10.8964	0.00273	0.99059	1.9782	10.3997	0.00301	0.99008
	September				**October**			
Method	$\hat{k}$	$\hat{c}$	*RMSE*	R^2	$\hat{k}$	$\hat{c}$	*RMSE*	R^2
MLM	1.9773	9.8233	0.00326	0.98897	1.7302	10.8425	0.00475	0.96516
MOM	1.9747	9.8203	0.00324	0.98912	1.7319	10.8387	0.00477	0.96484
EM	1.9975	9.8226	0.00347	0.98748	1.7560	10.8476	0.00508	0.96024
PDM	1.9975	9.8226	0.00347	0.98747	1.7573	10.8480	0.00509	0.95998
LSM	1.9570	9.8309	0.00307	0.99019	1.7179	10.8313	0.00461	0.96717
WLSMa	1.9043	9.8800	0.00280	0.99186	1.6448	10.9560	0.00392	0.97630
WLSMb	1.9248	9.8479	0.00287	0.99145	1.6616	10.8925	0.00405	0.97475
	November				**December**			
Method	$\hat{k}$	$\hat{c}$	*RMSE*	R^2	$\hat{k}$	$\hat{c}$	*RMSE*	R^2
MLM	1.9012	12.9339	0.00519	0.94647	1.7220	12.3016	0.00656	0.91430
MOM	1.9086	12.9415	0.00523	0.94561	1.7423	12.3228	0.00674	0.90963
EM	1.9320	12.9461	0.00539	0.94218	1.7665	12.3325	0.00696	0.90338
PDM	1.9434	12.9481	0.00548	0.94033	1.7821	12.3383	0.00712	0.89909
LSM	1.8469	12.9660	0.00488	0.95267	1.6575	12.3427	0.00600	0.92829

WLSMa	1.8115	13.1461	0.00467	0.95662	1.6294	12.5881	0.00572	0.93488
WLSMb	1.8373	13.0319	0.00480	0.95424	1.6404	12.4448	0.00583	0.93228

Tab. 3. The estimates of the Weibull distribution parameters and the results of the statistical analysis for composite months for years 2005-2009.

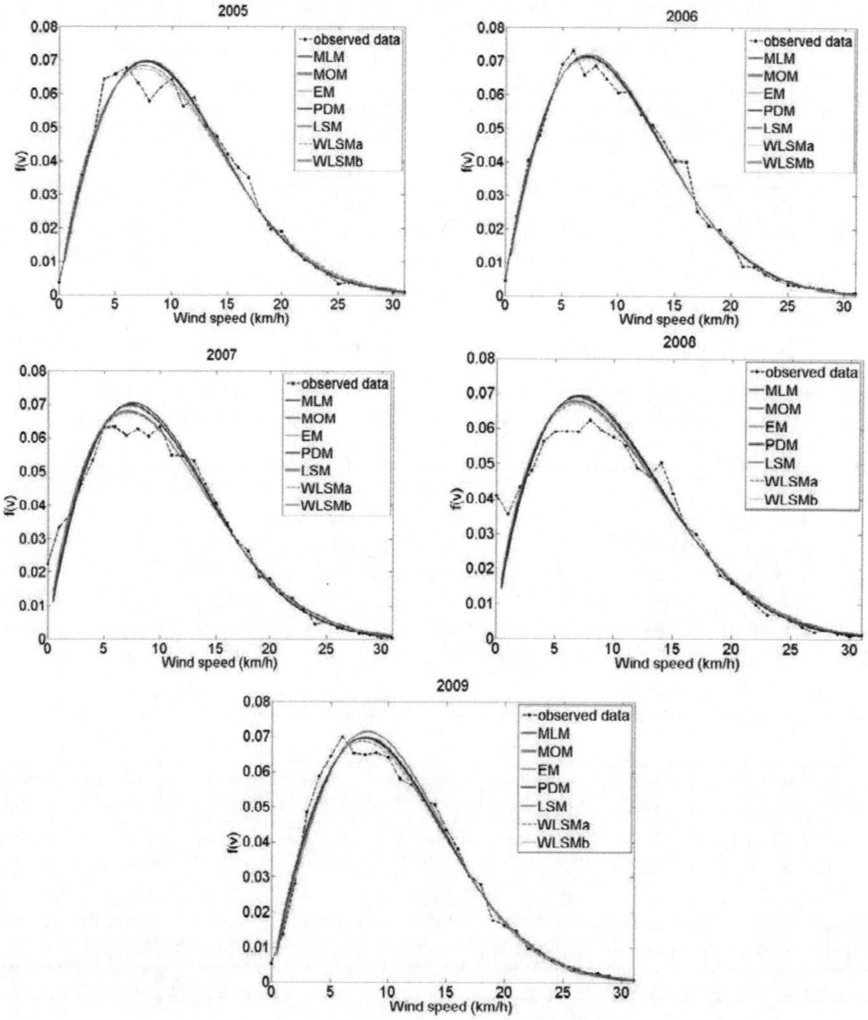

Fig. 3. Comparison of Weibull probability density distributions for individuals years.

Analyzing the results in Table 3 month by month one can find that the WLSMa performs better than other methods in January and in April - December in terms of R^2 and $RMSE$. In February and March the LSM performs best of all methods. The second best method is the WLSMb, except for February, when the second best method is the WLSMa in terms of R^2 and $RMSE$.

The comparison shows that all seven methods are applicable for estimating the Weibull distribution parameters for all seasons and for the whole monitored period of 5 years with a sufficient level of performance.

The comparison shows that the WLSMa and the WLSMb perform better than the other methods in terms of R^2 and $RMSE$ in the majority of the cases.

The MLM and the MOM give comparable values of R^2 and $RMSE$. These two methods perform well in terms of R^2 and $RMSE$. However, these methods perform worse than the WLSMa and the WLSMb.
The EM and the PDM are comparable methods in terms of R^2 and $RMSE$. These two methods give the largest values of $RMSE$ and the smallest values of R^2 in the majority of the cases. These methods perform worse than the others methods in terms of R^2 and $RMSE$.

5 Conclusion

In this paper the wind speed data from Bratislava-Mlynská dolina were fitted using the two-parameter Weibull distribution. The seven methods: the maximum likelihood method, the method of moments, the empirical method, the power density method, the least squares method and the weighted least squares method with two weight factors were used for estimating the Weibull distribution parameters. The coefficient of determination and the root mean square error were used to evaluate the performance of the considered methods. The following results can be concluded.
All seven methods perform well and are applicable for estimating the Weibull distribution parameters for all seasons, years and for the whole monitored period of 5 years as is indicate by high values of R^2 and low values of $RMSE$.
The EM and the PDM perform poorer compared to the other methods. Both methods are comparable in terms of R^2 and $RMSE$.
The MLM and the MOM that use numerical methods for estimating the Weibull distribution parameters perform a little worse than the WLSMa and the WLSMb in terms of R^2 and $RMSE$. These two methods are comparable.
The LSM perform worse than the WLSM, MLM and the MOM in terms of R^2 and $RMSE$. The study showed that the weight factor has an effect on the performance of the parameter estimation.
Statistical analysis showed that the values of R^2 and $RMSE$ are similar for the WLSMa and the WLSMb for the months, seasons, years and the whole monitored period in the majority of the cases. It can be concluded that these two methods have the best performance in estimating the Weibull distribution parameters. The WLSMa ranked first followed by the WLSMb in the majority of the cases. These methods provide the best performance in fitting the wind speed data measured in Bratislava – Mlynská dolina by Weibull distribution and are recommended. Another useful feature of these methods is the relative simplicity of the required numerical computations.

Acknowledgement

The authors gratefully acknowledge the Meteorological observatory Mlynská dolina and Faculty of mathematics, physics and informatics Comenius University in Bratislava for providing us with measured wind speed data.
The paper was supported by grant from Grant Agency of VEGA no. 1/0812/17.

References

[1] AKDAĞ, S. A., DINLER, A. A new method to estimate Weibull parameters for wind energy application. *Energy Conversion and Managemet, 50,*2009, p. 1761-1766.

[2] BERGMANN, B. Estimation of Weibull parameters using a weight function. *Journal of Materials Science Letters, 5,* 1986, p. 611-614.

[3] CARTA, J. A., RAMÍREZ, P., VELÁSQUEZ, S. A review of wind speed probability distributions used in wind energy analysis Case studies in the Canary Island. *Renewable and Sustainable Energy Rewiews, 13*, 2009, p. 933-955.

[4] CELIK, A.N. A statistical analysis of wind power density based on the Weibull and Rayleigh models at the southern region of Turkey. *Renewable Energy, 29,* 2003, p. 593-604.

[5] CHANG, T. P. Performance comparison of six numerical methods in estimating Weibull parameters for wind energy application. *Applied Energy, 88, 2011. P.* 272-282.

[6] CONRADSEN, K., NIELSEN, L. B., PRAHM, L. P. Review of Weibull Statistics for Estimation of Wind Speed Distributions. *Journal of Climate and Applied Meteorology 23,* 1984, p. 1173-1183.

[7] COSTA ROCHA, P. A., DE SOUSA, R. C., DE ANDRADE, C. F., DA SILVA, M. E. V. Comparison of seven numerical methods for determining Weibull parameters for wind energy generation in the northeast region of Brazil. *Applied Energy, 89,*2012, p. 395-400.

[8] DORVLO, A. S. S. Estimating wind speed distribution. *Energy Conversion and Management 43*, 2002, p. 2311-2318.

[9] FAUCHER, B., TYSON, W. R. On the determination of Weibull parameters. *Journal of Materials Science Letters*, 7, 1988, p. 1199-1203.

[10] GRYNING, S. E., FLOORS, R., PENA, A., BATCHVAROVA, E., BRÜMMER, B. Weibull Wind-Speed Distribution Parameters Derived from a Combination of Wind-Lidar and Tall-Mast Measurements Over Land, Coastal and Marine Sites. *Boundary-Layer Meteorology*, Volume 159, Issue 2, 2016, p. 329-348.

[11] JUSTUS, C. G., HARGRAVES, W. R., MIKHAIL, A., GRABER, D. Methods for estimating wind speed frequency distributions. *Journal of Applied Meteorology, 17*, 1978, p. 350-353.

[12] KANTAR, Y. M., SENOĞLU, B. (2008). A comparative study for the location and scale parameter of the Weibull distribution with given shape parameter. *Computers & Geosciences, 34*, 2008, p. 1900-1909.

[13] KANTAR, Y. M., KURBAN, M., HOCAOGLU, F. O. Comparison of six different parameter estimation methods in wind power applications. *Scientific Research and Essays, vol. 6(32)*, 2011, p. 6594-6604.

[14] LUN, I. Y. F., LAM, J. C. A study of Weibull parameters using long-term wind observation. *Renewable Energy, 20*, 2000, p. 145-153.

[15] MORGAN, M. C., LACKNER, M., VOGEL, R. M., BAISE, L.G. Probability distributions for offshore wind speeds. *Energy Conversion and Management, 52*, 2011, p. 15-26.

[16] POBOČÍKOVÁ, I., SEDLIAČKOVÁ, Z. The least square and the weighted least square methods for estimating the Weibull distribution parameters-a comparative study. *Communications: Scientific Letters of the University of Žilina*, vol. *14, no. 4*, 2012, p. 88-93.

[17] SEGURO, J. V., LAMBERT, T. W. Modern estimation of the parameters of the Weibull speed distribution for wind energy analysis. *Journal of Wind Engineering and Industrial Aerodynamics 85*, 2000, p. 75-84.

[18] STEVENS, M. J. M., SMULDERS, P. T. The estimation of the parameters of the Weibull wind speed distribution for wind energy utilization purposes. *Wind Engineering*, vol *3(2),* 1979, p. 132-45.

[19] WU, D., ZHOU, J., LI, Y. Methods for estimating Weibull parameters for brittle materials. *Journal of Materials Science 41*, 2006, p. 5630-5638.

Current address

Pobočíková Ivana, Mgr., PhD.
Department of Applied Mathematics , Faculty of Mechanical Engineering, University of Žilina
Univerzitná 1, 010 26 Žilina, Slovak Republic
Tel. number: +421 41 513 4967, e-mail: ivana.pobocikova@fstroj.uniza.sk

Sedliačková Zuzana, Mgr., PhD.
Department of Applied Mathematics , Faculty of Mechanical Engineering, University of Žilina
Univerzitná 1, 010 26 Žilina, Slovak Republic
Tel. number: +421 41 513 4963, e-mail: zuzana.sedliackova@fstroj.uniza.sk

Šimon Ján, Mgr., PhD.
Department of Applied Mathematics , Faculty of Mechanical Engineering, University of Žilina
Univerzitná 1, 010 26 Žilina, Slovak Republic
Tel. number: +421 41 513 4966 , e-mail: jan.simon@fstroj.uniza.sk

Slovak University of Technology in Bratislava
Faculty of Mechanical Engineering
17th Conference on Applied Mathematics **APLIMAT 2018**

Proceedings

THE SUM OF THE SERIES OF RECIPROCALS OF THE QUADRATIC POLYNOMIALS WITH INTEGER ROOTS

POTŮČEK Radovan (CZ)

Abstract. This contribution is a follow-up and a completion to the previous author's papers dealing with the sums of the series of reciprocals of quadratic polynomials with non-zero roots. We summarize these results and deal with the sum of the series of reciprocals of the quadratic polynomials with zero roots. We derive the formulas for these sums and verify them by some examples evaluated using the computer algebra system Maple 16.

Keywords: Sum of the series, telescoping series, harmonic number, computer algebra system Maple

Mathematics subject classification: Primary 40A05; Secondary 65B10

1 Introduction and basic notions

Let us recall the basic terms. For any sequence $\{a_k\}$ of numbers the associated *series* is defined as the sum

$$\sum_{k=1}^{\infty} a_k = a_1 + a_2 + a_3 + \cdots .$$

The *sequence of partial sums* $\{s_n\}$ associated to a series $\sum_{k=1}^{\infty} a_k$ is defined for each n as the sum

$$s_n = \sum_{k=1}^{n} a_k = a_1 + a_2 + \cdots + a_n .$$

The series $\sum_{k=1}^{\infty} a_k$ *converges* to a limit s if and only if the sequence of partial sums $\{s_n\}$ converges to s, i.e. $\lim_{n\to\infty} s_n = s$. We say that the series $\sum_{k=1}^{\infty} a_k$ has a *sum* s and write $\sum_{k=1}^{\infty} a_k = s$.

The *nth harmonic number* is the sum of the reciprocals of the first n natural numbers:

$$H_n = \sum_{k=1}^{n} \frac{1}{k} = 1 + \frac{1}{2} + \frac{1}{3} + \cdots + \frac{1}{n},$$

H_0 being defined as 0. The *generalized harmonic numbers* of order n in power r is the sum

$$H_{n,r} = \sum_{k=1}^{n} \frac{1}{k^r},$$

where $H_{n,1} = H_n$ are harmonic numbers. Generalized harmonic number of order n in power 2 can be written as a function of harmonic numbers using formula (see [6])

$$H_{n,2} = \sum_{k=1}^{n-1} \frac{H_k}{k(k+1)} + \frac{H_n}{n}.$$

From formulas for $H_{n,r}$, where $r = 1, 2$ and $n = 1, 2, \ldots, 10$, we get the following Table 1:

n	1	2	3	4	5	6	7	8	9	10
H_n	1	$\frac{3}{2}$	$\frac{11}{6}$	$\frac{25}{12}$	$\frac{137}{60}$	$\frac{49}{20}$	$\frac{363}{140}$	$\frac{761}{280}$	$\frac{7129}{2520}$	$\frac{7381}{2520}$
$H_{n,2}$	1	$\frac{5}{4}$	$\frac{49}{36}$	$\frac{205}{144}$	$\frac{5269}{3600}$	$\frac{5369}{3600}$	$\frac{266681}{176400}$	$\frac{1077749}{705600}$	$\frac{771817}{352800}$	$\frac{1968329}{1270080}$

Tab. 1. Ten first harmonic numbers H_n and generalized harmonic numbers $H_{n,2}$.

The *sum of the reciprocals* of some positive integers is generally the sum of unit fractions. For example the sum s of the reciprocals of all the nonzero triangular numbers

$$T_n = \sum_{k=1}^{n} k = 1 + 2 + 3 + \cdots + n = \frac{n(n+1)}{2} = \binom{n+1}{2},$$

which create the sequence $\{1, 3, 6, 10, 15, 21, 28, 36, 45, 55, 66, 78, 91, \ldots\}$, is

$$s = \sum_{n=1}^{\infty} \frac{1}{(n^2+n)/2} = 2\sum_{n=1}^{\infty} \frac{1}{n^2+n} = 2\sum_{n=1}^{\infty} \frac{1}{n(n+1)} = 2.$$

This can be shown by using the following sum of a telescoping series.

The *telescoping series* is a series whose partial sums eventually only have a fixed number of terms after cancellation. For example, the series $\sum_{n=1}^{\infty} \frac{1}{n(n+1)}$ simplifies as

$$\begin{aligned}
\sum_{n=1}^{\infty} \frac{1}{n(n+1)} &= \sum_{n=1}^{\infty} \left(\frac{1}{n} - \frac{1}{n+1}\right) = \lim_{N\to\infty} \sum_{n=1}^{N} \left(\frac{1}{n} - \frac{1}{n+1}\right) \\
&= \lim_{N\to\infty} \left[\left(1 - \frac{1}{2}\right) + \left(\frac{1}{2} - \frac{1}{3}\right) + \cdots + \left(\frac{1}{N} - \frac{1}{N+1}\right)\right] \\
&= \lim_{N\to\infty} \left[1 + \left(-\frac{1}{2} + \frac{1}{2}\right) + \left(-\frac{1}{3} + \frac{1}{3}\right) + \cdots + \left(-\frac{1}{N} + \frac{1}{N}\right) - \frac{1}{N+1}\right] \\
&= \lim_{N\to\infty} \left[1 - \frac{1}{N+1}\right] = 1.
\end{aligned}$$

2 The sum of the series of reciprocals of the quadratic polynomials with non-zero integer roots

In the papers [1], [2], [3], [4], and [5], the following results concerning the quadratic polynomials with non-zero integer roots were derived.

In the paper [1] it was derived that the sum $s(a,b)^{++}$ of the series $\sum_{\substack{k=1\\k\neq a,b}}^{\infty} \frac{1}{(k-a)(k-b)}$ of reciprocals of the quadratic polynomials with different positive integer roots a and b, $a < b$, is given by the formula

$$s(a,b)^{++} = \frac{1}{b-a}\left(H_{a-1} - H_{b-1} + 2H_{b-a} - 2H_{b-a-1}\right), \tag{1}$$

where H_n is the nth harmonic number.

In the paper [2] it was shown that the sum $s(a,b)^{--}$ of the series $\sum_{k=1}^{\infty} \frac{1}{(k-a)(k-b)}$ of reciprocals of the quadratic polynomials with different negative integer roots a and b, $a < b$, is given by the formula

$$s(a,b)^{--} = \frac{1}{b-a}\left(H_{-a} - H_{-b}\right). \tag{2}$$

In the paper [3] we derived the sum $s(a,a)^{--}$ of the series $\sum_{k=1}^{\infty} \frac{1}{(k-a)^2}$ of reciprocals of the quadratic polynomials with double negative integer root a is given by the formula

$$s(a,a)^{--} = \frac{\pi^2}{6} - H_{-a,2}, \tag{3}$$

where $H_{-a,2}$ is the generalized harmonic number of order $-a$ in power 2.

The formula for the sum $s(a,a)^{++}$ of the series $\sum_{\substack{k=1\\k\neq a}}^{\infty} \frac{1}{(k-a)^2}$ of reciprocals of the quadratic polynomials with double positive integer root a was derived in the paper [4] and has the form

$$s(a,a)^{++} = \frac{\pi}{2} + H_{a-1,2}. \tag{4}$$

The sum $s(a,b)^{-+}$ of the series $\sum_{\substack{k=1\\k\neq b}}^{\infty} \frac{1}{(k-a)(k-b)}$ of reciprocals of the quadratic polynomials with integer roots $a < 0$ and $b > 0$ was derived in the paper [5] and is given by the formula

$$s(a,b)^{-+} = \frac{(b-a)\left(H_{-a} - H_{b-1}\right) + 1}{(b-a)^2}. \tag{5}$$

3 The sum of the series of reciprocals of the quadratic polynomials with zero roots

Now, we analyze three cases of zero roots of the quadratic polynomials – the case of two zero roots, the case of one zero and one negative integer root, and the case of one zero and one positive integer root. We shall use the following notation: $s(0,0)$ instead of $s(a,b)^{00}$, $s(a,0)$ for negative integer a instead of $s(a,b)^{-0}$, and $s(0,b)$ for positive integer b instead of $s(a,b)^{0+}$.

3.1 The sum of the series of reciprocals of the quadratic polynomials with two zero roots

A problem to determine the sum $s(0,0) = \sum_{k=1}^{\infty} \frac{1}{(k-0)^2} = \frac{1}{1^2} + \frac{1}{2^2} + \frac{1}{3^2} + \frac{1}{4^2} + \cdots$ is so called *Basel problem.* This problem was posed by *Pietro Mengoli* (1625–1686) in 1644. In 1737 *Leonhard Euler* (1707–1783) showed his famous result

$$s(0,0) = \sum_{k=1}^{\infty} \frac{1}{k^2} = \frac{\pi^2}{6} = 1.644\,934\,066\ldots .$$

This sum presents the value $\zeta(2)$ of the *Riemann zeta function* $\zeta(s) = \sum_{k=1}^{\infty} k^{-s} = \frac{1}{1^s} + \frac{1}{2^s} + \frac{1}{3^s} + \cdots$. Note that the values of the nth partial sum

$$s_n(0,0) = \frac{1}{1^2} + \frac{1}{2^2} + \frac{1}{3^2} + \frac{1}{4^2} + \cdots + \frac{1}{n^2}$$

correspond to the values $H_{n,2}$, so their first ten values are presented in Table 1.

3.2 The sum of the series of reciprocals of the quadratic polynomials with one zero and one negative integer root

The sum $s(a,0)$ of the series of reciprocals of the quadratic polynomials with one zero and one negative integer root a, where $a = -A$, $A > 0$, whence $k - a = k + A$, is given by the following simple theorem:

Theorem 1 *Let $a = -A$ be a negative integer. Then it holds* $\sum_{k=1}^{\infty} \frac{1}{k(k+A)} = \frac{H_A}{A}$, *where H_A is the Ath harmonic number, so*

$$s(a,0) = \sum_{k=1}^{\infty} \frac{1}{k(k-a)} = \frac{H_{-a}}{-a}. \tag{6}$$

Proof: By using the partial fraction decomposition of the telescoping series above we get

$$\sum_{k=1}^{\infty} \frac{1}{k(k+A)} = \frac{1}{A} \sum_{k=1}^{\infty} \left(\frac{1}{k} - \frac{1}{k+A} \right) = \frac{1}{A} \lim_{N\to\infty} \sum_{k=1}^{N} \left(\frac{1}{k} - \frac{1}{k+A} \right) =$$
$$= \frac{1}{A} \lim_{N\to\infty} \left[\left(1 - \frac{1}{1+A} \right) + \left(\frac{1}{2} - \frac{1}{2+A} \right) + \cdots + \left(\frac{1}{A} - \frac{1}{2A} \right) + \left(\frac{1}{A+1} - \frac{1}{2A+1} \right) + \cdots \right.$$
$$\left. \cdots + \left(\frac{1}{N-A-1} - \frac{1}{N-1} \right) + \left(\frac{1}{N-A} - \frac{1}{N} \right) + \cdots + \left(\frac{1}{N} - \frac{1}{N+A} \right) \right].$$

After cancellation all of the inner terms starting from $\frac{1}{A+1}$ up to $\frac{1}{N}$ and considering the facts that $\lim_{N\to\infty}\frac{1}{\ell}=\frac{1}{\ell}$ and $\lim_{N\to\infty}\frac{1}{N+\ell}=0$ for any positive integer ℓ we get

$$\sum_{k=1}^{\infty}\frac{1}{k(k+A)}=\frac{1}{A}\lim_{N\to\infty}\left(1+\frac{1}{2}+\cdots+\frac{1}{A}-\frac{1}{N+1}-\frac{1}{N+2}-\cdots-\frac{1}{N+A}\right)=$$
$$=\frac{1}{A}\left(1+\frac{1}{2}+\cdots+\frac{1}{A}\right)=\frac{H_A}{A}.$$

3.3 The sum of the series of reciprocals of the quadratic polynomials with one zero and one positive integer root

The sum $s(0,b)$ of the series of reciprocals of the quadratic polynomials with one zero and one positive integer root b is given by the following simple theorem:

Theorem 2 *Let b be a positive integer. Then it holds*

$$s(0,b)=\sum_{\substack{k=1\\k\neq b}}^{\infty}\frac{1}{k(k-b)}=\frac{1-bH_{b-1}}{b^2}. \tag{7}$$

The formula (7) *can be also written in the equivalent form* $s(0,b)=\dfrac{H_b-2H_{b-1}}{b}$.

Proof: Obviously, by using the partial fraction decomposition, we can write

$$\sum_{\substack{k=1\\k\neq b}}^{\infty}\frac{1}{k(k-b)}=\sum_{k=1}^{b-1}\frac{1}{k(k-b)}+\sum_{k=b+1}^{\infty}\frac{1}{k(k-b)}=$$
$$=\frac{1}{b}\left[\sum_{k=1}^{b-1}\left(\frac{1}{k-b}-\frac{1}{k}\right)+\lim_{N\to\infty}\sum_{k=b+1}^{N}\left(\frac{1}{k-b}-\frac{1}{k}\right)\right]=$$
$$=\frac{1}{b}\left\{\left(\frac{1}{1-b}-1\right)+\left(\frac{1}{2-b}-\frac{1}{2}\right)+\cdots+\left(-\frac{1}{2}-\frac{1}{b-2}\right)+\left(-1-\frac{1}{b-1}\right)+\right.$$
$$+\lim_{N\to\infty}\left[\left(1-\frac{1}{b+1}\right)+\left(\frac{1}{2}-\frac{1}{b+2}\right)+\cdots+\left(\frac{1}{b+1}-\frac{1}{2b+1}\right)+\left(\frac{1}{b+2}-\frac{1}{2b+2}\right)+\cdots\right.$$
$$\left.\left.\cdots+\left(\frac{1}{N-2b}-\frac{1}{N-b}\right)+\left(\frac{1}{N-2b+1}-\frac{1}{N-b+1}\right)+\cdots+\left(\frac{1}{N-b}-\frac{1}{N}\right)\right]\right\}.$$

After summing up the counts before the limit, cancellation all of the inner terms beyond the limit starting from $\frac{1}{b+1}$ up to $\frac{1}{N-b}$ and considering the facts that $\lim_{N\to\infty}\frac{1}{\ell}=\frac{1}{\ell}$ and $\lim_{N\to\infty}\frac{1}{N+\ell}=0$ for any positive integer ℓ we get

$$\sum_{\substack{k=1\\k\neq b}}^{\infty}\frac{1}{k(k-b)}=\frac{-2}{b}\left(\frac{1}{b-1}+\frac{1}{b-2}+\cdots+\frac{1}{2}+1\right)+$$

$$+\frac{1}{b}\lim_{N\to\infty}\left(1+\frac{1}{2}+\cdots+\frac{1}{b}-\frac{1}{N-b+1}-\frac{1}{N-b+2}-\cdots-\frac{1}{N}\right)=$$

$$=\frac{-2}{b}\left(1+\frac{1}{2}+\cdots+\frac{1}{b-2}+\frac{1}{b-1}\right)+\frac{1}{b}\left(1+\frac{1}{2}+\cdots+\frac{1}{b}\right)=\frac{H_b-2H_{b-1}}{b}=$$

$$=\frac{-1}{b}\left(1+\frac{1}{2}+\cdots+\frac{1}{b-2}+\frac{1}{b-1}-\frac{1}{b}\right)=\frac{1-bH_{b-1}}{b^2}.$$

4 Numerical verification

We solve two problems – to determine the values of the sums

$$s(a,0)=\sum_{k=1}^{\infty}\frac{1}{k(k-a)}\qquad\text{and}\qquad s(0,b)=\sum_{\substack{k=1\\k\neq b}}^{\infty}\frac{1}{k(k-b)}$$

for negative integer a and positive integer b. For the numerical verification of the derived formulas (6) and (7) we choose these values: $a=-1,-2,-3,-10,-100$ and $b=1,2,3,10,100$. We use on the one hand an approximative direct evaluation of the sums

$$s(a,0,t)=\sum_{k=1}^{t}\frac{1}{k(k-a)}\qquad\text{and}\qquad s(0,b,t)=\sum_{\substack{k=1\\k\neq b}}^{t}\frac{1}{k(k-b)},$$

where $t=10^8$, using the computer algebra system Maple 16, and on the other hand the formula (6) for evaluation the sum $s(a,0)$ and the formula (7) for evaluation the sum $s(0,b)$.

For direct evaluation of the sums $s(a,0,10^8)$ and $s(0,b,10^8)$ we use the Maple commands illustrated by the following two examples, where $a=-10$ and $b=10$:

```
> evalf[8](sum(1/(k*(k+10)),k=1..10^8));
                                   0.29289682
> evalf[8](harmonic(10)/10);
                                   0.29289683
> evalf[8](sum(1/(k*(k-10)),k=1..9)+sum(1/(k*(k-10)),k=11..10^8));
                                  -0.27289684
> evalf[8]((1-10*harmonic(9))/100);
                                  -0.27289683
```

The approximative values of the sums $s(a,0,10^8)$ and $s(a,0)$ for $a=-1,-2,-3,-10,-100$ and the sums $s(0,b,10^8)$ and $s(0,b)$ for $b=1,2,3,10,100$, rounded to 8 decimals and obtained by the calculations made in the CAS Maple are written into the following Table 2, where the absolute errors of the calculations are all approximately equal to $1\cdot10^{-8}$:

a	-1	-2	-3	-10	-100
$s(a,0,10^8)$	0.99999999	0.74999999	0.61111110	0.29289682	0.05187377
$s(a,0)$	1.00000000	0.75000000	0.61111111	0.29289683	0.05187378
b	1	2	3	10	100
$s(0,b,10^8)$	0.99999999	−0.25000001	−0.38888890	−0.27289684	−0.05167379
$s(0,b)$	1.00000000	−0.25000000	−0.38888889	−0.27289683	−0.05167378

Tab. 2. Approximate values of the sums $s(a,0,10^8)$, $s(a,0)$, $s(0,b,10^8)$, $s(0,b)$.

5 Examples

Now, let us give some examples illustrating using formulas (1) to (7).

$$s(3,5)=\sum_{\substack{k=1\\k\neq 3,5}}^{\infty}\frac{1}{(k-3)(k-5)}=\frac{1}{5-3}(H_{3-1}-H_{5-1}+2H_{5-3}-2H_{5-3-1})=\frac{53}{24}\doteq 2.208,\text{ by (1).}$$

$$s(-5,-3)=\sum_{k=1}^{\infty}\frac{1}{(k+5)(k+3)}=\frac{1}{-3-(-5)}\left(H_{-(-5)}-H_{-(-3)}\right)=\frac{9}{40}\doteq 0.225,\text{ by (2).}$$

$$s(-5,-5)=\sum_{k=1}^{\infty}\frac{1}{(k+5)^2}=\frac{\pi^2}{6}-H_{-(-5),2}\doteq 1.645-\frac{5269}{3600}\doteq 0.181,\text{ by (3).}$$

$$s(5,5)=\sum_{\substack{k=1\\k\neq 5}}^{\infty}\frac{1}{(k-5)^2}=\frac{\pi}{2}+H_{5-1,2}\doteq 1.571+\frac{205}{144}\doteq 2.994,\text{ by (4).}$$

$$s(-3,5)=\sum_{\substack{k=1\\k\neq 5}}^{\infty}\frac{1}{(k+3)(k-5)}=\frac{(5+3)\big(H_{-(-3)}-H_{5-1}\big)}{(5+3)^2}=\frac{8\left(\frac{11}{6}-\frac{25}{12}\right)}{64}=\frac{-1}{32}\doteq -0.031,\text{ by (5).}$$

$$s(-5,0)=\sum_{k=1}^{\infty}\frac{1}{k)(k+5)}=\frac{H_{-(-5)}}{-(-5)}=\frac{\frac{137}{60}}{5}=\frac{137}{300}\doteq 0.457,\text{ by (6).}$$

$$s(0,5)=\sum_{\substack{k=1\\k\neq 5}}^{\infty}\frac{1}{k(k-5)}=\frac{1-5H_{5-1}}{5^2}=\frac{1-5\cdot\frac{25}{12}}{25}=\frac{-113}{300}\doteq -0.377,\text{ by (7).}$$

6 Conclusion

We deal with the sum $s(a,b)$ of the series of reciprocals of the quadratic polynomials with two integer roots a, b, i.e. with the series

$$\sum_{k=1}^{\infty}\frac{1}{(k-a)(k-b)},$$

where the index of summation $k\neq a$, if $a>0$, or $k\neq b$, if $b>0$.

This contribution pick up the threads of author's previous results stated in the papers [1], [2], [3], [4], [5] concerning non-zero integer roots and complete them with the case of zero roots. At first we remember the Euler's result

$$s(0,0)=\sum_{k=1}^{\infty}\frac{1}{(k-0)^2}=\frac{1}{1^2}+\frac{1}{2^2}+\frac{1}{3^2}+\frac{1}{4^2}+\cdots=\frac{\pi^2}{6}$$

and then derived the formulas for the sums $s(a,0)$, $a<0$, and $s(0,b)$, $b>0$:

$$s(a,0)=\sum_{k=1}^{\infty}\frac{1}{k(k-a)}=\frac{H_{-a}}{-a} \quad \text{and} \quad s(0,b)=\sum_{\substack{k=1\\k\neq b}}^{\infty}\frac{1}{k(k-b)}=\frac{1-bH_{b-1}}{b^2},$$

where H_n is the nth harmonic number. We verified these results by computing 10 sums (for $a=-1,-2,-3,-10,-100$ and $b=1,2,3,10,100$) by using the CAS Maple 16. We also stated that the last formula above can be written in the form

$$s(0,b)=\frac{H_b-2H_{b-1}}{b}.$$

The series of reciprocals of the quadratic polynomials with integer roots so belong to special types of infinite series, such as geometric and telescoping series, which sums are given analytically by means of a simple formula.

References

[1] POTŮČEK, R.: *The sum of the series of reciprocals of the quadratic polynomial with different positive integer roots*. In: Matematika, informační technologie a aplikované vědy (MITAV 2016). Univerzita obrany, Brno, 2016, p. 32-43. ISBN 978-80-7231-464-5.

[2] POTŮČEK, R.: *The sum of the series of reciprocals of the quadratic polynomial with different negative integer roots*. In: Ratio Mathematica - Journal of Foundations and Applications of Mathematics, 2016, vol. 30, no. 1/2016, p. 59-66. ISSN 1592-7415.

[3] POTŮČEK, R.: *The sum of the series of reciprocals of the quadratic polynomials with double non-positive integer root*. In: Proceedings of the 15th Conference on Applied Mathematics APLIMAT 2016. Faculty of Mechanical Engineering, Slovak University of Technology in Bratislava, 2016, p. 919-925. ISBN 978-80-227-4531-4.

[4] POTŮČEK, R.: *The sum of the series of reciprocals of the quadratic polynomials with double positive integer root*. In: MERAA – Mathematics in Education, Research and Applications, 2016, vol. 2, no. 1. Slovak University of Agriculture in Nitra, 2016, p. 15-21. ISSN 2453-6881.

[5] POTŮČEK, R.: *The sum of the series of reciprocals of the quadratic polynomials with one negative and one positive integer root*. In: Proceedings of the 16th Conference on Applied Mathematics APLIMAT 2017. Faculty of Mechanical Engineering, Slovak University of Technology in Bratislava, 2017, p. 1252-1253. ISBN 978-80-227-4650-2.

[6] WIKIPEDIA CONTRIBUTORS: *Harmonic number*. Wikipedia, The Free Encyclopedia, [online], [cit. 2017-11-30]. URL: https://en.wikipedia.org/wiki/Harmonic_number.

Current address

Potůček Radovan, RNDr. Ph.D.
Department of Mathematics and Physics
Faculty of Military Technology, University of Defence in Brno
Kounicova 65, 662 10 Brno, Czech Republic
Tel. number: + 420 973 443 056, e-mail: Radovan.Potucek@unob.cz

Slovak University of Technology in Bratislava
Faculty of Mechanical Engineering
17th Conference on Applied Mathematics **APLIMAT 2018**

Proceedings

TEMPERAMENT THEORY AND STATISTICAL ANALYSIS OF TEMPERAMENTAL DEPENDENCE ON OTHER FACTORS – EMPIRICAL STUDY

POZDÍLKOVÁ Alena (CZ), HEDVIČÁKOVÁ Martina (CZ)

Abstract. The aim of this article is an empirical study of a person's temperamental dependence on other factors. Belov's temperament test and Eysenck's lability and stability test were used for research. A questionnaire survey was conducted at the University of Hradec Králové using first and third year students. A correlation analysis and χ^2 – test of independence in a PivotTable will be used to show independency of temperament on other factors; and both tests between themselves. Results will be discussed and questions for further results will be formulated.

Key words: Belov's temperament test, Eysenck's lability and stability test, correlation analysis, χ^2 – test of independence

Mathematics Subject Classification: Primary 62J02, 62J05; Secondary 62B15

1 Introduction

In the first half of the twentieth century, E. Krečmer expanded somatotyping to three basic types according to body structure - pyknic, leptos, athletics. "Pyknics are usually cooperative and cheerful, an occasional transition to depression, and have a cycling temperament. In the case of leptosomes, we usually find a schismatic temperament: they are usually closed, either hypersensitive ("romantic") or cold without a lively expression of emotions. Athletes have a viscous temperament (corresponding approximately to phlegmatic), characterized by indifference, inertia, and which sometimes alternates with explosiveness " [Čáp, Mareš, 2001].

People have always responded differently to certain stimuli. The attempts have been made to seal the individual into a certain type. The best known is the typology based on the teachings of Hippocrates and Galena. It distinguishes four basic types and defines their characteristics [Poláková, 2007]:

Sanguine - cheerful, optimistically tuned. Their experiences are rather shallow and short. They respond weakly to strong stimuli. They are dominated by a "straw fire" reaction. This name was invented by J. Strelau and marks a quick dying experience and a rapid change of focus. This type of person adapts quickly and is emotionally balanced. In motion, the course is unbalanced, and the duration is short, the speed component is fast and strong. We will classify them as emotionally stable extroverts.

Phlegmatic - on the outside, has a quiet and lazy impression. They appear calm or cold or apathetic. Only exciting stimuli are excited. They choose friends and only have a deep relationship with them. They are rather passive, without great ambitions and demands. The motion response method is level, slow and weak. Emotionally we classify them as stable introverts.

Melancholic - a melancholic man approaching life is usually pessimistic with fear, and fear of the future. They love peace, their own well-being, and are characterized by profound experiences. They hate rage, noise and new people. They do not show their feelings outward, but live more internally. The relationships they establish are permanent and deep. The movement has a regular course; the duration is permanent. Emotionally labile introvert.

Choleric - often impulsive, irritable, impatient and overwhelming. They are prone to explosions of anger. Coexistence with them is difficult, they are often single minded. They are easily affected by emotions. Outwardly, they respond quickly, strongly, often without hindrance. Emotionally labile extrovert.

2 Data and methodology

The 2 tests of temperaments from Belov (1972) and from Eysenck (1960) were used for research. A questionnaire survey was conducted at the University of Hradec Králové for the first and third year students of the Faculty of Informatics and Management. The aim was to get at least 600 completed questionnaires from both University students, and other respondents over the age of 18.

The questionnaire was distributed to students in person throughout September 2016. During the submission of the questionnaires, students were trained to complete and disseminate the questionnaires. The data collection period ran from October to December 2016. In the following months January to May 2017, an evaluation of the questionnaires was carried out. In total, 650 questionnaires were distributed, of which 571 questionnaires were returned. 38 of them were poorly or incompletely filled. Altogether, 533 questionnaires were used for research. Returnability was 82%. Such a high return on the questionnaire was due to the fact that 315 questionnaires were distributed to students during class at the University of Hradec Králové, and the remaining 335 questionnaires were distributed outside the University.

The first three questions of the questionnaire concerned the input characteristics of the respondents. The following question dealt with net income for employees or monthly financial requirements for students (this area is planned for future research).

This is a case study of exploratory character. As a method, we chose quantitative research that uses random selections, experiments, and highly structured data collection using tests, questionnaires, or observations. Quantitative research uses experiment, observation, or polls to capture data. Structured, standardized questionnaires were selected for our research. Eysenck's Temperament Questionnaire [Eysenck, 1960] and temperament test by Belov [Čáp, 2001] were used. The Eysenck's Temperament questionnaire consists of 57 questions relating to behavior, behavior and feelings in different situations. For each of the questions, only YES or NO could be answered. [Bílý, Suss, 2006]

3 Temperament tests

The Belov's temperament test focuses on the characteristics, features and behavior of individuals. We use the test to determine the percentage of temperament components (sanguine, choleric, phlegmatic, melancholic). There are twenty questions for each type, for which the individual answers yes or no. Some questions are repeated in these four sections (A, B, C, D). The individual completing the questionnaire should respond spontaneously without long delay. Then an evaluation is made. [Poláková, 2007]

From each section the positive answers are added according to the formula:

$$\boxed{V_{t1} = \frac{Ch_a * 100}{a} + \frac{S_a * 100}{a} + \frac{F_a * 100}{a} + \frac{M_a * 100}{a}}$$

$$where:\ a = Ch_a + S_a + F_a + M_a$$

The second test from Eysenck determines the lability and stability, as we mentioned above. In other words, extrovert and neuroticism.

Sixty yes or no questions are asked. The evaluation is performed as follows: To determine positive lability we required a "yes" response to questions 2, 5, 7, 10, 13, 15, 18, 21, 23, 26, 29, 31, 34, 37, 39, 42, 45, 47, 50, 52 54, 56, 58, 60. A positive answer to questions 1, 3, 9, 11, 14, 17, 19, 22, 25, 27, 30, 35, 57 showed stability. In this case, we included questions 6, 33, 51, 55, 59, where the respondents should answer "no". Calculation:

$$\boxed{V_{t2} = \frac{N_a * 100}{a} + \frac{E_a * 100}{a}}$$

$$where:\ a = N_a + E_a$$

4 Correlation analysis and χ^2 – test of independence in a PivotTable

In the beginning of the first part a correlation analysis will be performed (correlation matrix will be composed). This will establish a correlation between both temperament tests, and age, and gender of respondents. [Hedvičáková, Pozdílková, 2015]

A Spearman correlation coefficient is an important characteristic in evaluating the validity of tests, because it determines how close together two related phenomena are captured. Thus, it allows quantitative determination of how far the two similar orders are created. For the calculation, it is necessary to have a table in which you can specify individual correlated pairs, which are compared to the individual components of the correlation, overall index and the basic form of vector analysis. The result is a dimensionless number, which indicates the degree of correlation between individual freedom and the steam created for each pair of correlation. [Marek, Svoboda, Brožek, 2016]

In the second part test of an independence in PivotTable will be used. If we examine the dependence of two variables X and Y, where variable X is not numeric, the ANOVA method is not appropriate to use. The better test is χ^2 – test of independence of two variables, which assumes that they are

detected
We arrange the data of the ordinal or nominal type into the PivotTables. [Hebák, 2006]

So we will test the hypothesis H: variables X and Y are independent, against the alternative hypothesis that they are dependent.

For independent phenomena A, B:

$$P(A \cap B) = P(A).P(B)$$

We will compare empirically established frequencies

$$n.\pi_{ij} = n.\pi_{i.}.\pi_{.j}$$

Estimates of theoretical frequencies are

$$\hat{\pi}_{i.} = \frac{n_{i.}}{n}$$
$$\hat{\pi}_{.j} = \frac{n_{.j}}{n}$$

an estimate of the theoretical combined probability is

$$\hat{\pi}'_{ij} = \hat{\pi}_{i.}\hat{\pi}_{.j} = \frac{n_{i.}}{n}\frac{n_{.j}}{n} = \frac{n_{i.}.n_{.j}}{n^2}$$

Then an estimate of the theoretical frequency is

$$n'_{ij} = n.\hat{\pi}'_{ij} = n.\frac{n_{i.}.n_{.j}}{n^2} = \frac{n_{i.}.n_{.j}}{n}$$

For test H: variables X and Y are independent → A: variables X and Y are dependent we use test statistics

$$\chi^2 = \sum_{i=1}^{r}\sum_{j=1}^{s}\frac{\left(n_{ij} - n'_{ij}\right)^2}{n'_{ij}},$$

provided that the characters X and Y are independent for a sufficiently large n Pearson distribution $\chi^2(\nu)$ with $\nu = (r-1)(s-1)$ degrees of freedom.

The hypothesis on the independence of variables X and Y is denied if

$$\chi^2 \geq \chi^2_{1-\alpha}(\nu),$$

where $\nu = (r-1)(s-1)$.

5 Experiments and results

Using Spearman coefficient, it was proved, that Belov's temperament test and Eysenck's lability and stability test are independent of age and gender.

The correlation matrix and independency is illustrated on the following graph:

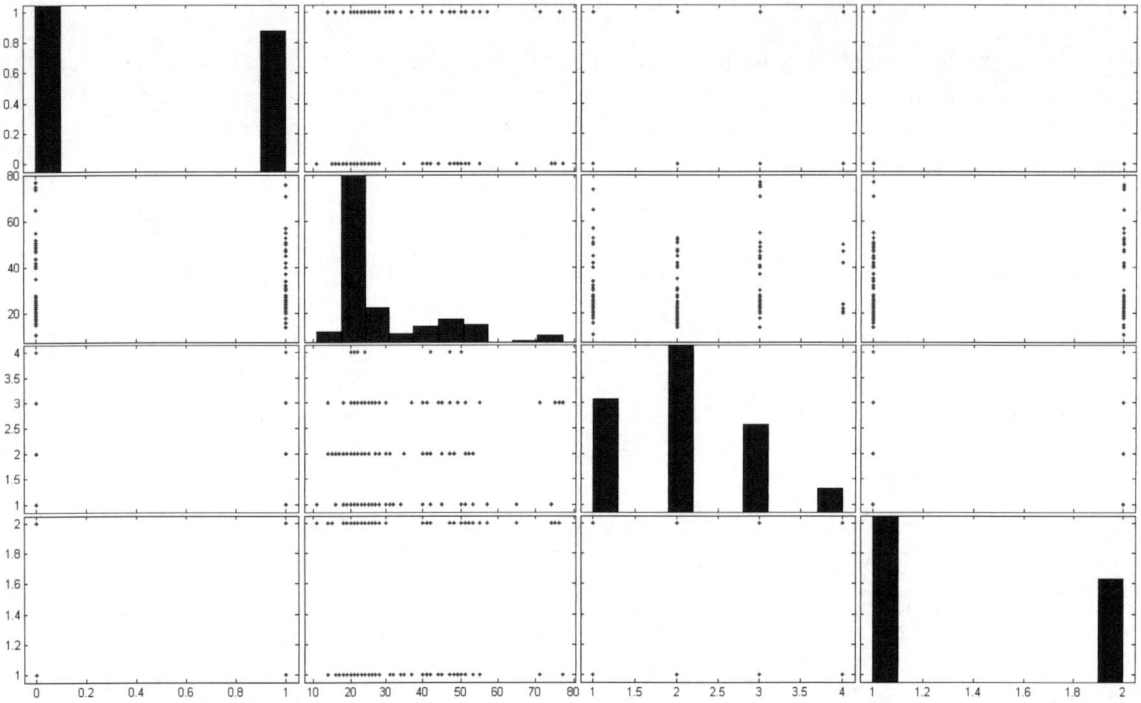

Fig. 1. Independency of test results on variables age and gender

Due to the fact that the correlation of Belov's temperament test and Eysenck's lability and stability test is very low (between 0.2 and 0.3), the test of independence in PivotTable was realized to confirm this hypothesis.

Hypothesis H: tests are independent, was tested using the test of independence in the PivotTable. The resulting test statistic χ^2 was computed as 6,24. Comparison with $\chi^2(\nu)$ with $\nu = (r-1)(s-1)$ degrees of freedom, showed, that the hypothesis of independence cannot be dismissed.

$$\chi^2 = \sum_{i=1}^{r}\sum_{j=1}^{s}\frac{\left(n_{ij} - n'_{ij}\right)^2}{n'_{ij}},$$

$$\chi^2 = 6{,}24 < 7{,}81 = \chi^2(3)$$

6 Discussion and conclusions

The results of the correlation analysis demonstrated the independence of both tests from other variables, such as age and gender, which is as expected. This result has also been confirmed by the test of independence in the PivotTable, which confirmed this hypothesis.

The reason for the small interdependence of both tests is the way they are evaluated. Belov's temperament test results for a given variant, are determined by the most questions answered in this variant. A better way would be to determine the standard of weights and evaluation with regard to all the answers. Alternatively, another option is to use fuzzy sets or graphical processing using 4 coordinates. This is the topic of our further research.

The other question in the discussion is whether the questionnaires are distorted because most of the respondents do not respond as they really feel, but rather as they would like to be seen. [Mummendey, 2006] Therefore, it is necessary to place emphasis on the truthfulness of the respondents and, if necessary, to supplement the questionnaire with further tests.

Acknowledgements

The paper was written with the support of the specific project 6/2017 grant "Determinants affecting job satisfaction" granted by the University of Hradec Králové, Czech Republic and thanks to help from student Veronika Domšová.
This paper was also supported by internal support of University of Pardubice.

References

[1] ČÁP, J. and MAREŠ, J. *Psychologie pro učitele.* Vyd. 1. Praha: Portál, 2001. 655 s. ISBN 80-7178-463-X.
[2] POLÁKOVÁ, B. *Výzkum vybraných dimenzí osobnosti sportovce ve volejbalu*, Bakalářská práce. Masarykova Univerzita, 2007.
[3] MAREK, J., SVOBODA, V. and BROŽEK, J. *Software for determining the quality of an exam test.* Konferencia ELEKTRO 2016, FEI TU Košice, 2016.
[4] HEDVIČÁKOVÁ, M. and POZDÍLKOVÁ, A. *Development of Mortgage Loans with Using of Regression Analysis.* Advances Science Letters 2015, American Scientific Publishers, 2015.
[5] HEBÁK, P. *Vícerozměrné statistické metody 1., 2., 3.* Informatorium, Praha, 2006. ISBN 978-80-7333-056-9.
[6] BÍLÝ, M., SÜSS, V. *Temperamentové vlastnosti a výkonnová motivace závodníků ve vodním slalomu.* Studia Kinanthropologica, 2006.
[7] EYSENCK, H. J. *Behaviour Therapy and Neuroses.* London Pergamon Press, 1960.

Current address

Pozdílková Alena, Mgr. PhD.
Department of Mathematics and Physics, University of Pardubice
Faculty of Electrical Engineering and Informatics
Studentská 85, 530 02 Pardubice, Czech Republic
E-mail: alena.pozdilkova@upce.cz

Hedvičáková Martina, Ing. PhD.
Department of Economics, University of Hradec Králové
Faculty of Informatics and Management
Rokitanského 62, 500 03 Hradec Králové, Czech Republic
E-mail: martina.hedvicakova@uhk.cz

Slovak University of Technology in Bratislava
Faculty of Mechanical Engineering
17th Conference on Applied Mathematics **APLIMAT 2018**

Proceedings

SPATIAL LAG MODEL FOR APARTMENT PRICES IN PARDUBICE REGION

POZDÍLKOVÁ Alena (CZ), MAREK Jaroslav (CZ)

Abstract. The article is devoted to modelling the relationship of apartment prices between neighbouring small municipalities within the Pardubice region. A spatial lag model will be used for the calculation, and the unknown parameters will be estimated using the least squares method. This model is often used to describe geo-informational phenomena.

Keywords: spatial modelling, spatial lag model, factor model, weights matrix, estimation of unknown parameters, exploratory visualization, spatial lag vs spatial error models

Mathematics Subject Classification: 90J15

1 Introduction

The aim of the article is modeling the relationship between housing prices in small municipalities within the Pardubice region. Specifically, it models the average price of apartments in the municipalities of Pardubice (Česká Třebová, Heřmanův Městec Hlinsko, Holice, Choceň Chrudim, Chvaletice, Lázně Bohdaneč, Letohrad, Litomyšl, Moravská Třebová, Opatovice nad Labem, Pardubice, Polička, Přelouč, Skuteč, Králíky, Svitavy, Ústí nad Orlicí, Vysoké Mýto). The "explaining variable" in this model is the average advertised price of residential units. The "explanatory variables" are the population, the unemployment rate in the district, and the distance from the regional capital. Several of the proposed factor models will be tested for their quality by using a modified index of determination.

It will also be tested to see if the model must also include a spatial factor. This spatial factor, as well as the availability matrix (known from graph theory), allow us to grasp the issue of neighbourhood regions.

2 Housing prices modelling

Econometrics attempts to measure some economic quantities based on the analysis of the real data by using various mathematical-statistical methods. The object of great interest is the measurement of supply and demand using the factor models. The first aim of this paper is to assemble a model for describing the relationships between housing prices and suitable explained variables. The estimators are computed by using a two-stage Least squares method. The second aim is to explore spatial relationship of housing prices throughout the Pardubice region municipalities. A similar problem is solved in [Meen, 1996], where the author studied spatial interactions in UK regional house prices.

The reason for modelling is to determine if the price can be explained by the unemployment rate and the distance from the regional center. Furthermore, we want to prove that the spatial lag model has greater ability to grasp the problem of apartment price modelling. That means whether or not the index of determination is greater.

The non-spatial model estimated by conventional regression procedures is not a reliable representation and should be avoided when there is a spatial phenomenon to be analyzed. [Augustin, 1998]

Special attention must be given to the possibility that errors and variables in the model show spatial dependence in data analysis. The influences of spatial autocorrelation could be drawn from statistical tests. If the interest focuses on obtaining proper statistical inference (estimates of unknown parameters, statistical hypothesis testing, forecasting) from the dependent data, spatial autocorrelation can be considered a nuisance.

2.1 Formulation of regression model

Firstly, we propose a simple regression model, which will feature spatial coefficient. We will work with tools from graph theory, which allow us to describe the existence or nonexistence of a boundary between the municipalities. [Anselin, Rey, 1991], [Augustin, 1998], and [LeSage, 2009] show formulas for the spatial lag model that is given by:

$$\boldsymbol{Y} = \lambda \boldsymbol{W}\boldsymbol{Y} + \boldsymbol{X}\boldsymbol{\gamma} + \boldsymbol{\varepsilon}, \boldsymbol{\varepsilon} \sim N(0, \sigma^2 \boldsymbol{I}) \,. \tag{1}$$

Here the symbols denote:

- λ… coefficient in a spatial autoregressive structure for the disturbance,
- $\boldsymbol{\gamma}$ … vector of unknown parameters,
- **X** … matrix of explanatory variables,
- $\boldsymbol{W}$ ... known spatial weight matrix, where $W_{ij} = 1$, if boundary between i-th and j-th region exists, $W_{ij} = 0$ if there is no boundary,
- $\boldsymbol{\varepsilon}$ … vector of independent possibly heteroscedastic spatial errors.

Explained variables

In [Kotatkova, 2015], we can see that suitable model for housing prices use three explained variables: distance from the regional center, population, and unemployment rate.

We will use following notation:

- **Y**... the average advertised price of 1 m² of apartment. Price per 1m² is determined as a proportion of the purchase price and floor area (except of balkonies, and terraces). [Reality, 2017] Data obtained automatically – by parsing of real estate webside.
- $\boldsymbol{x}_1$ ... distance from the regional center, cf. [Routeplanner, 2017],
- $\boldsymbol{x}_2$ ... average wage throughout the whole municipality, cf. [Payroll, 2017],
- $\boldsymbol{x}_3$ ... unemployment rate (in the selected municipality), cf. [Unimployment, 2017].

2.2 Estimation in spatial lag model

The residual regression estimation is divides into two steps [Bhatacharjee, 2006], [LeSage, 2009]:

First stage - Ordinary least squares regression of parameters γ in model (2) with estimate (3):

$$\boldsymbol{Y} = \boldsymbol{X\gamma} + \boldsymbol{u}, \boldsymbol{u} \sim \boldsymbol{N}(\boldsymbol{0}, \boldsymbol{\sigma^2 V}), \tag{2}$$

$$\widehat{\boldsymbol{\gamma}} = (\boldsymbol{X^T V^{-1} X})^{-1} \boldsymbol{X^T V^{-1} Y}, \tag{3}$$

$$\hat{u}_t = Y_t - \boldsymbol{X}_{t.} \widehat{\boldsymbol{\gamma}}. \tag{4}$$

Second stage – spatial lag parameter estimation:

$$\hat{u}_{kt} = \rho_k \sum_{j=1, j\neq k}^{k} w_{kj} \hat{u}_{jt} + \varepsilon_{kt}, \tag{5}$$

where the second stage regressions are estimated separately for each region.

2.3 Measuring of model quality

In this section, we describe characteristics for measuring the quality of the regression model. Let $\bar{y}$ be the mean estimate based on measurements $y_1, \ \ldots \ , y_n$ and let s_y^2 be the estimate of dispersion, $s_y = \frac{1}{n}\sum(y_i - \bar{y})^2$. Following quantities express variation of the dependent variable. The total sum of squares SS_T of Y: $SS_T = \sum(y_i - \bar{y})^2$, the sum of squares due to regression SS_R:

$$SS_R = \sum(\hat{y}_i - \bar{y})^2.$$

The goodness of fit can be defined by

$$R^2 = \sqrt{\frac{SS_R}{SS_T}}. \tag{6}$$

The square root of goodness of fit is the correlation coefficient $R = \sqrt{R^2}$. In the case of several explanatory variables, an adjusted measure of R^2 is used usually:

$$R^2 = 1 - (1 - R^2)\left(\frac{N-1}{N-k-1}\right). \tag{7}$$

2.4 Moran's I test

The Moran's statistic has been used as a way of spatial dependency description. Calculations for Moran's I are based on a weighted matrix W and measured values Y:

$$I = \frac{N}{S_0}\left(\frac{e^T We}{e^T e}\right), \tag{8}$$

with e as a vector of OLS residuals and $S_0 = \sum_i \sum_j w_{ij}$, a standardization factor that corresponds to the sum of the weights for the nonzero cross-products. The statistic shows a striking similarity to the familiar Durbin–Watson test. [Anselin, 2001]

Those interested in derivation of previous formula are referred to paper. [Moran, 1950]

2.5 Test statistic for spatial lag model: Lagrange multiplier test for spatial lag model

If we would like to explore in our model existence of spatial dependency, then the problem of lag coefficient statistical significance appears.

Our null hypothesis is that the spatial arrangement does not affect to value of dependent variable, i.e. the spatial coefficient is equal to zero. For detail see [Anselin, 2001], [Anselin, Rey, 1991].

The Lagrange Multiplier (or Score) test statistic is calculated using the following formula:

$$LM_{err} = [\boldsymbol{e}^T \boldsymbol{W}\boldsymbol{e}/(\boldsymbol{e}^T \boldsymbol{e}/N)]^2/[tr(\boldsymbol{W}^2 + \boldsymbol{W}^T\boldsymbol{W})]. \tag{9}$$

This statistic has an asymptotic $\chi^2(1)$ distribution and, apart from a scaling factor, corresponds to the square of Moran's I. From several simulation experiments [Anselin, Rey, 1991]; it follows that Moran's I has slightly better power than the LM_{err} test in small samples, but the performance of both tests becomes indistinguishable in medium and large size samples. The LM/RS test against a spatial lag alternative was outlined in and takes the form [Anselin, 2001]:

$$LM_{lag} = [\boldsymbol{e}^T \boldsymbol{W}\boldsymbol{e}/(\boldsymbol{e}^T \boldsymbol{e}/N)]^2/D, \tag{10}$$

where

$$D = [(\boldsymbol{W}\boldsymbol{X}\boldsymbol{\beta})^T(\boldsymbol{I} - \boldsymbol{X}(\boldsymbol{X}^T\boldsymbol{X})^{-1}\boldsymbol{X}^T)(\boldsymbol{W}\boldsymbol{X}\boldsymbol{\beta})/\sigma^2] + tr(\boldsymbol{W}^2 + \boldsymbol{W}^T\boldsymbol{W}).$$

3 Numerical studies

Example 1. The aim of our study will be to find a suitable model for describing the price of 1m^2 by using some / some variables: distance, population, unemployment. At the same time, we investigate whether spatial dependence exists between variables. This study is realized in the municipalities of Pardubice: Česká Třebová, Heřmanův Městec Hlinsko, Holice, Choceň Chrudim, Chvaletice, Lázně Bohdaneč, Letohrad, Litomyšl, Moravská Třebová, Opatovice nad Labem, Pardubice, Polička, Přelouč, Skuteč, Králíky, Svitavy, Ústí nad Orlicí, Vysoké Mýto. Data are available from [Reality, 2017], [Routeplanner, 2017], [Payroll, 2017], [Unemployment, 2017].

Table 1 shows the regions characteristics. The explanatory data consists of three variables. We will consider a lag model with explanatory variables $\mathbf{x_1}$, $\mathbf{x_2}$, $\mathbf{x_3}$, described in section 1.1. First we try to explain the dependent variable in models 1, 2, 3 using only one variable $\mathbf{x_1}$, $\mathbf{x_2}$ or $\mathbf{x_3}$. We determine the indices determination and decide which variable in the regression model (1) is best explained-cost apartments. Then we build a 4 model using the two best variables. From this model will be model with spatial coefficient created – model 5, see Table 3 in Results. Measured data are shown in table 1. In our case the number of measurement is N=15. Data was obtained on April 1st 2017.

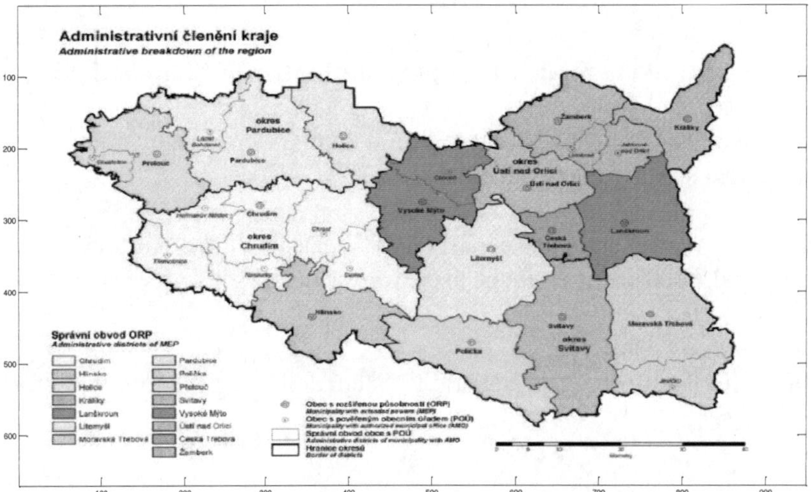

Fig. 1. Map of Pardubice Region.

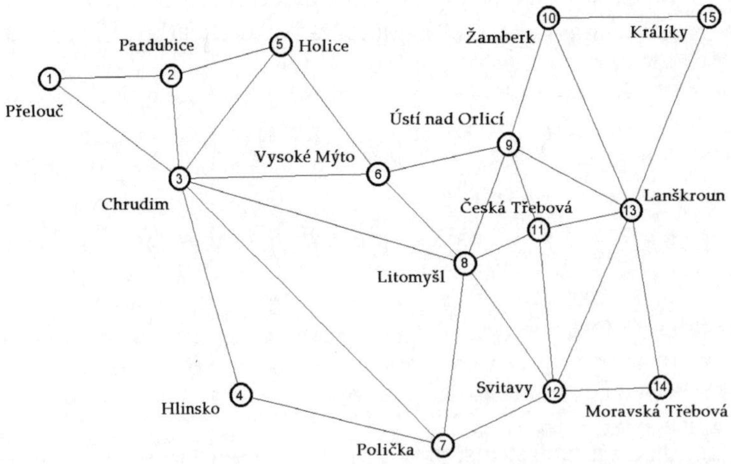

Fig. 2. Pardubice Region – description by graph.

No.	Region	Distance x_1	No. of inhabitants x_2	Unemploity x_3	Prize Y
1.	Přelouč	0.36	9 127	6.13 %	33 113 CZK
2.	Pardubice	0.00	89 638	6.13 %	37 501 CZK
3.	Chrudim	0.25	23 061	8.39 %	31 386 CZK
4.	Hlinsko	0.70	9 831	8.39 %	14 558 CZK
5.	Holice	0.32	6 514	6.13 %	28 041 CZK
6.	Vysoké Mýto	1.46	12 404	6.95 %	21 144 CZK
7.	Polička	1.06	8 783	9.29 %	17 269 CZK
8.	Litomyšl	0.83	10 043	9.29 %	32 101 CZK
9.	Ústí nad Orlicí	1.00	14 226	6.95 %	19 563 CZK
10.	Žamberk	1.04	6 062	6.95 %	21 104 CZK
11.	Česká Třebová	1.05	15 710	6.95 %	18 756 CZK
12.	Svitavy	1.05	17 005	9.29 %	26 279 CZK
13.	Lanškroun	1.30	10 031	6.95 %	24 771 CZK
14.	Moravská	1.23	10 267	9.29 %	19 947 CZK
15.	Králíky	1.42	4 300	6.95 %	9712 CZK

Tab. 1. Prices of housing in selected regions with explanatory variables.

The method used to compute unknown parameters, described in section 2.2, is based on the two step strategy. The existence of the boundaries of Figures 1 and 2 expresses the matrix W. Fig. 1 and Fig. 2 show municipalities in Pardubice region. From these figures weights matrix can be obtained:

$$\boldsymbol{W} = \begin{pmatrix} 0&1&1&0&0&0&0&0&0&0&0&0&0&0&0 \\ 1&0&1&0&1&0&0&0&0&0&0&0&0&0&0 \\ 1&1&0&1&1&1&1&1&0&0&0&0&0&0&0 \\ 0&0&1&0&0&0&1&0&0&0&0&0&0&0&0 \\ 0&1&1&0&0&1&0&0&0&0&0&0&0&0&0 \\ 0&0&1&0&1&0&0&1&1&0&0&0&0&0&0 \\ 0&0&1&1&0&0&0&1&0&0&0&1&0&0&0 \\ 0&0&1&0&0&1&1&0&1&0&1&1&0&0&0 \\ 0&0&0&0&0&1&0&1&0&1&1&0&1&0&0 \\ 0&0&0&0&0&0&0&0&1&0&0&0&1&0&1 \\ 0&0&0&0&0&0&0&1&1&0&0&1&1&0&0 \\ 0&0&0&0&0&0&1&1&0&0&1&0&1&1&0 \\ 0&0&0&0&0&0&0&0&1&1&1&1&0&1&1 \\ 0&0&0&0&0&0&0&0&0&0&0&1&1&0&0 \\ 0&0&0&0&0&0&0&0&0&1&0&0&1&0&0 \end{pmatrix}$$

Given that $\boldsymbol{y} = \boldsymbol{\lambda} \boldsymbol{W} \boldsymbol{y}$, we obtain a set of equations that we will render in relation to the unknown λ. [LeSage, 2009] And here we have the ability of each equation to get $\lambda_1, \dots, \lambda_{15}$. These values can be redirected to get one λ.

4 Results

In the first model, we use x_1 as the explanatory variable, in the second model we use the variable x_2 and in the third model we use the variable x_3. The fourth model is based on two best explanatory variables from previous models, such as distance and population. The fifth model is based on model 4 by adding a spatial expression $\boldsymbol{\lambda W y}$. By application of least squares method, following results were achieved:

in model 1: $\hat{\boldsymbol{\gamma}} = (34539, -12459)$, $R^2 = 0.7318$, $LM_{err} = 13.29$, $LM_{lag} = 41.93$,
in model 2: $\hat{\boldsymbol{\gamma}} = (20303, 0)$, $R^2 = 0.5561$, $LM_{err} = 33.40$, $LM_{lag} = 42.21$,
in model 3: $\hat{\boldsymbol{\gamma}} = (31625, -1045)$, $R^2 = 0.1694$, $LM_{err} = 86.89$, $LM_{lag} = 52.33$,
in model 4: $\hat{\boldsymbol{\gamma}} = (31545, -10460, 0.076)$, $R^2 = 0.7512$, $LM_{err} = 24.60$, $LM_{lag} = 40.54$,
in model 5: $\hat{\boldsymbol{\gamma}} = (31545, -10460, 0.076)$, $R^2 = 0.8053$., estimates of spatial coefficients $\boldsymbol{\lambda}$ are given in Tab. 3.

No.	Model 1 $\hat{Y}$	Model 1 $\hat{\varepsilon}$	Model 2 $\hat{Y}$	Model 2 $\hat{\varepsilon}$	Model 3 $\hat{Y}$	Model 3 $\hat{\varepsilon}$
1	30053.65	3059	22176.22	10937	25220.88	7892
2	34538.83	2962	38704.53	-1203	25220.88	12280
3	31424.12	-38	25036.77	6349	22859.77	8526
4	25817.64	-11260	22320.74	-7763	22859.77	-8302
5	30552.00	-2512	21639.79	6401	25220.88	2820
6	16348.92	4795	22848.96	-1705	24364.20	-3220
7	21332.46	-4064	22105.60	-4837	21919.50	-4651
8	24197.99	7903	22364.27	9737	21919.50	10182
9	22079.99	-2517	23223.00	-3660	24364.20	-4802
10	21581.63	-478	21547.00	-443	24364.20	-3260
11	21457.04	-2701	23527.66	-4772	24364.20	-5608
12	21457.04	4822	23793.51	2486	21919.50	4360
13	18342.33	6428	22361.80	2409	24364.20	406
14	19214.45	733	22410.25	-2463	21919.50	-1972
15	16847.27	-7135	21185.27	-11473	24364.20	-14652

Tab. 2. Results.

Obtained results illustrates the differences between models. We can see that model 3 is very poor. The best results are calculated in model 4. In this model null hypothesis is rejected, e.g. spatial dependency is significant.

No.	Model 4 $\hat{Y}$	Model 4 $\hat{\varepsilon}$	Model 5 $\hat{Y}$	Model 5 $\hat{\varepsilon}$	λ
1	28473.63	4639	28453	4660	0.13
2	38362.04	-861	38221	-720	-0.03
3	30683.92	702	30722	664	0.03
4	24970.61	-10413	26327	-11769	-0.43
5	28693.34	-653	28618	-577	-0.02
6	17216.34	3928	18045	3099	0.14
7	21125.15	-3856	20647	-3378	-0.15
8	23626.88	8474	23472	8629	0.38
9	22166.71	-2604	20468	-905	-0.11
10	21127.44	-24	21127	-23	-0.00
11	21756.54	-3001	19794	-1038	-0.12
12	21855.02	4424	23488	2791	0.20
13	18709.55	6061	16142	8629	0.32
14	19459.73	488	19669	278	0.02
15	17018.47	-7306	15086	-5374	-0.32

Tab. 3. Results.

5 Conclusions

Paper presented to the readers the solution of one type of spatial regression problem. This is a situation where it should be a spatial dependence of measured data to consider. To using this model, we are inspired by studies from [Kotatkova, 2015] and [Bhatacharjee, 2006]. First paper is devoted to regression model for description of flat prices in Czech Republic by suitable variables. Second paper studied spatial diffusion in housing starts across regions in England by regressing OLS residuals of the regression relationship for each region on residuals from all the other regions. Our results are analogous like in these studies. We proofed that use of spatial model is necessary.

Acknowledgement

This paper was supported by internal support of University of Pardubice.

References

[1] ANSELIN, L., *Spatial Econometrics*, 14th chapter, A Companion to Theoretical Econometrics, Copyright by Blackwell Publishing Ltd, 2001, 310-330.

[2] ANSELIN, L., REY, S., Properties of Tests for Spatial Dependence in Linear Regression. *Geographical analysis* 23, 1991, 112–131.

[3] AUGUSTIN, N. H. et al.: The role of simulation in modelling spatially correlated data. *Environmetrics* 9, 1998, 175–196.

[4] BHATACHARJEE, A., JENSEN-BUTLER, C.: *Estimation of Spatial Weights Matrix, with an Application to Diffusion in Housing Demand.* Glasgow: School of Economics and Finance, University of St. Andrews, U.K., 2006.

[5] KOTATKOVA STRANSKA, P., MAREK, J.: Simultaneous equations and modelling of new flat demand. In: *Mezinárodní vědecká konference Hradecké ekonomické dny 2015* (Jedlička, P., ed.). Univerzita Hradec Králové, Fakulta informatiky a managementu, Hradec Králové, 2015, díl I, 11–16.
[6] LESAGE, J., PACE, K. R.: *Introduction to spatial Econometrics.* CRC Press/Taylor & Francis Group, Boca Raton, 2009.
[7] MEEN, G.: Spatial Aggregation, Spatial Dependence and Predictability in the UK Housing Market. *Housing Studies* 11, 1996, 345–372.
[8] MORAN, P. A. P. : Notes on Continuous Stochastic Phenomena. *Biometrika* 37 (1), 1950, 17–23. DOI:10.2307/2332142. JSTOR 2332142. http://www-hsc.usc.edu/~mereditf/Spatial%20Statistics%206.pdf
[9] Reality: Real estate data – prize per 1 m2 of flat. [online], [cit. 8. 5. 2017]. URL: http://www.sreality.cz/hledani/prodej /byty/
[10] Routeplanner: Routeplanner SKODA auto. [online], [cit. 8. 5. 2017]. URL: www.planovac-tras-mapy.cz/
[11] Payroll. [online]. [cit. 8. 5. 2017]. URL: http://www/brno.czso.cz/xb/redakce.nsf/i/mesta_a_obce
[12] Unemployment. ČSÚ. [online]. [cit. 8. 5. 2017]. URL: http://www.czso.cz

Current address

Pozdílková Alena, Mgr. Ph.D.
Department of Mathematics and Physics, University of Pardubice
Faculty of Electrical Engineering and Informatics
Studentská 85, 530 02 Pardubice, Czech Republic
E-mail: alena.pozdilkova@upce.cz

Marek Jaroslav, Mgr. Ph.D.
Department of Mathematics and Physics, University of Pardubice
Faculty of Electrical Engineering and Informatics
Studentská 85, 530 02 Pardubice, Czech Republic
E-mail: jaroslav.marek@upce.cz

Slovak University of Technology in Bratislava
Faculty of Mechanical Engineering
17th Conference on Applied Mathematics **APLIMAT 2018**

Proceedings

NOTES ON ITERATIVE CLOSEST POINT ALGORITHM

PROCHAZKOVA Jana (CZ), MARTISEK Dalibor (CZ)

Abstract. Point clouds are commonly used in many areas of technical practice. Nowadays, the common applications are in area of autonomous cars and 3D terrain or object modelling. During the scanning process, the scanning device or object is moving. So, the problem of precise point cloud registration arises. In this article, we describe Iterative Closest Point (ICP) algorithm that is suitable for fine registration. We introduce some of the improvements that leads to higher precision, speed computation and efficiency. We also tested the ICP algorithm on the noisy point clouds to process the registration.

Keywords: Iterative closest point, point cloud, least square method

Mathematics subject classification: Primary 65D18; Secondary 68U05

1 Introduction

We currently see substantia arise of point clouds applications in a large number of technical disciplines. The most significant applications include the use of 3D scanning technology to describe the environment (roads, city parts, forests, etc.). Nowadays, we can see the progress of scanning and processing technologies in area of autonomous vehicles [20]. For this purpose is often used LiDAR technology (Light Detection and Ranging, [3] [12]). It is a surveying method that measures distance to a target by illuminating that target with a pulsed laser light, and measuring the reflected pulse with a sensor. In engineering applications, the optical 3D scanners are widely employed. The result of both scanning principles is a point cloud that can be described as a set of data points in some coordinate system, usually defined by x, y, z coordinates.

When a sensing device or a scanned object moves, the resulting cloud of points is partially overlaid. The main task is to assemble these partial scans together to form the final object. This process is usually called point registration.

Due to [4], we can divide the registration process into two main parts: coarse and fine matching. In the coarse matching part, we find the corresponding points with detectors and (or) descriptors. There is a huge number of detector and descriptor methods, e.g. Maximally Stable Volumes [14], MeshDoG, Instrinsic Shape Signatures [22], Point Signature descriptor [9] and Spin Image [10]. This part determines the coarse matching of the scans. In our article, we deal with to second part – fine matching that align these two point clouds precisely. The most powerful algorithm Iterative Closest Points is presented in Sec. 2 and the results are described in Sec. 3 in details.

Fig. 1. Point cloud registration, source [4].

In our article, we introduce Iterative Closest Point (ICP) algorithm that is one of the common used algorithms in practice. The algorithm was firstly described in [8], [1]. The basis of the algorithm is to find an optimal mapping consisting of a translation and a rotation that minimizes the distance between the corresponding points in the least squares sense.

The authors in [1] formulated the problem as:

"Given 3-D data in a sensor coordinate system, which describes a data shape that may correspond to a model shape, and given a model shape in a model coordinate system in a different geometric shape representation, estimate the optimal rotation and translation that aligns, or registers, the model shape and the data shape minimizing the distance between the shapes and thereby allowing determination of the equivalence of the shapes via a mean-square distance metric."

2 ICP algorithm

2.1 Theoretical background

In this section, we briefly describe the mathematical background of the ICP algorithm. For more details, see e.g. [2]. This algorithm is very popular because of its simplicity and clarity. However, the basic algorithm works only in ideal cases, leading to its further modifications and improvements, which will be described in 2.3.

Generally, let $\mathcal{M}, \mathcal{S}$ are two finite sets in finite dimensional real vector space R^d that contains M, resp. N points. We want to determine the rigid or non-rigid transformation $T : R^d \longrightarrow R^d$ that minimize the distance between given sets:

$$\mathtt{dist}_g(T(\mathcal{M}), \mathcal{S}) = \sum_{m \in T(\mathcal{M})} \sum_{s \in \mathcal{S}} g((m-s)^2) \tag{1}$$

Also Root Mean Square Distance (RMSD) can be used to comparison of $T(\mathcal{M})$ and $\mathcal{S}$:

$$RMSD(T(\mathcal{M}), \mathcal{S}) = \frac{\sum_{m \in T(\mathcal{M})} \sum_{s \in \mathcal{S}} |m-s|^2}{n} \tag{2}$$

In the case that the sets $\mathcal{M}, \mathcal{S}$ contain the outliers, i.e. points that are not in both sets, we can compute partial overlapping only. In these cases, it is suitable to set the threshold parameter thr that eliminates the mistakes. For a set of pairs of points $C = (m_i, s_j)$, where $m_i \in \mathcal{M}$, $s_j \in \mathcal{S}$ must hold:

$$\forall s_k \in \mathcal{S} : dist(m_i, s_k) \geq dist(m_i, s_j), dist(m_i, s_j) < thr \tag{3}$$

The Eq. (2) can be written as:

$$RMSD(T(\mathcal{M}), \mu(\mathcal{S})) = \frac{\sum_C dist(m_i, T(s_j)^2}{|C|}, m_i, s_j \in C \tag{4}$$

where distance $T(m_j)$ and s_i is minimal and $|C|$ is cardinality of the set C.

2.2 ICP overview

Let $\mathcal{S}$ and $\mathcal{M}$ are input point clouds where source point cloud is $\mathcal{S}$ and $\mathcal{M}$ is model point cloud. We are looking for the rigid transformation which satisfies the best correspondence between given point clouds. The following steps describes the ICP algorithm:

1. compute the nearest point in the set $\mathcal{S}$ for every point (part of the points) in the set $\mathcal{M}$ using e.g. Euclidean distance:

$$d_i = \min\{\sqrt{m_i^2 - s_j^2}\} \quad j = 1, \ldots, N_S \tag{5}$$

2. if the distance $d_i > treshold$, remove a given pair of points
3. add the weights to pairs of points: implicitly $w_{ij} = 1$, we can set the weights in the dependence of the direction of normal vectors as their dot product:

$$w_{ij} = n_i n_j \tag{6}$$

4. compute the rotation matrix R and translation vector $\mathbf{t}$ using the least square method for distance minimization that is described in Eq. (9)-(13)
5. compute the transformation of the set $\mathcal{S}$ using computed values – $Rs_j + \mathbf{t}$,
6. compute the error $E(R, \mathbf{t})$ using Eq.(7), iterate until we reach to required accuracy.

The pseudocode of the algorithm ICP:

$$
\begin{aligned}
&AlgorithmICP(\mathcal{M}, \mathcal{S}) \\
&\quad \theta := \theta_0 \\
&\text{while not registered:} \\
&\quad X := \emptyset \\
&\quad \text{for } \; m_i \in T(\mathcal{M}, \theta) \\
&\qquad \hat{s}_j := \text{closest point in } \mathcal{S} \text{ to } m_i \\
&\qquad X := X + \langle m_i, \hat{s}_j \rangle \\
&\qquad \theta = \; \text{least squares}(X) \text{ using SVD} \\
&\text{return } \theta
\end{aligned}
$$

In the pseudocode the symbol θ means the rotation matrix R and translation vector $\mathbf{t}$. The initial values θ_0 are: identity matrix R and zero translation vector $\mathbf{t}$. The set X contains all the pairs that cover the closest points in set $\mathcal{S}$ to points in the set $\mathcal{M}$.

In the next step, we calculate the new transformation $\theta = (R, \mathbf{t})$ using SVD and Eq.(9)-(13). Then, the transfromation θ is applied to the model points $\mathcal{M}$ and the error is computed. We iterate this setup until required accuracy.

2.2.1 Mathematical description of the ICP

Let $\mathcal{M}, \mathcal{S}$ are the finite sets with N_M and N_S points. Firstly, we compute the nearest point in the set $\mathcal{S}$ for every points in the set $\mathcal{M}$. It is possible to use all of the points or we can choose the points randomly. The ICP algorithm suppose that corresponding points are the nearest ones. For these pairs, we compute the rotation matrix and translation. We try to minimize the error:

$$E(R,t) = \sum_{i=1}^{N_M} \sum_{i=1}^{N_S} w_{ij} \|m_i - (Rs_j + t)\|^2 \tag{7}$$

where w_{ij} are the weights for corresponding points. If point m_i is the nearest point to d_j, we set the weights as $w_{ij} = 1$, otherwise the weight is equal to zero.

The Eq. 7 can be rewritten as:

$$E(R,t) \propto \frac{1}{N} \sum_{i=1}^{N} \|m_i - (Rs_i + t)\|^2 \tag{8}$$

where

$$N = \sum_{i=1}^{N_M} \sum_{i=1}^{N_S} w_{ij}$$

and corresponding matrix of w_{ij} is substituted by the sum of corresponding points.

The optimization step (find $(R, \mathbf{t})$) is based on Singular Value Decomposition (SVD). For the computation of SVD, we need centroid alignment. We compute the centroids of both point sets as:

$$C_M = \frac{1}{N} \sum_{i=1}^{N} m_i \tag{9}$$

$$C_S = \frac{1}{N} \sum_{i=1}^{N} s_i \tag{10}$$

Subsequently, we align all of the points of $\mathcal{M}, \mathcal{S}$ using corresponding centroid and we get:

$$\mathcal{M}' = \{m_i' = m_i - C_M\}_{1,\dots,N}$$

$$\mathcal{S}' = \{s_i' = s_i - C_S\}_{1,\dots,N}$$

For these values, it is easy to compute covariance matrix $H = \mathcal{M}'\mathcal{S}'^T$ and make the SVD as:

$$H = U\Lambda V^T \tag{11}$$

The matrix of rotation is given by:

$$R = VU^T \tag{12}$$

and vector of translation can be computed by:

$$\mathbf{t} = C_S - RC_M \tag{13}$$

First iteration consists of the transformation of the cloud points using rotation matrix R and translation vector $\mathbf{t}$. Subsequently, we compute the error and if the error is higher then given threshold, the iteration continues.

2.3 The improvement methods of ICP

The previous section outlines the basic steps and theoretical background of the ICP algorithm. Nevertheless, there is the amount of the improvements that try to satisfy better convergence and improve the computation time. The overview can be find in e.g. [17] in details. In this part, we show some of the possible way how to optimize the ICP algorithm.

k-d tree K-d trees are a special case of binary space partitioning trees that are often used as suitable structure in many applications, such as searches involving a multidimensional search key (e.g. range searches and nearest neighbor searches). ICP algorithm computation can be efficiently accelerated with its usage. Authors [11] declare that the usage of k-d tree can improve the speed of registration 10 times.

Corresponding points The original work [1] works with all the points in both cloud points. Certain enhancements have been made to the use of uniform sub-sampling of given points and random sampling [13]. The problem occurs in the case of noise or high curvature shapes, where these methods do not select enough reference points (see Figure 2, left). The solution is to use the normals for sampling [6]. The detection of corresponding points using normals is on Figure 2, right.

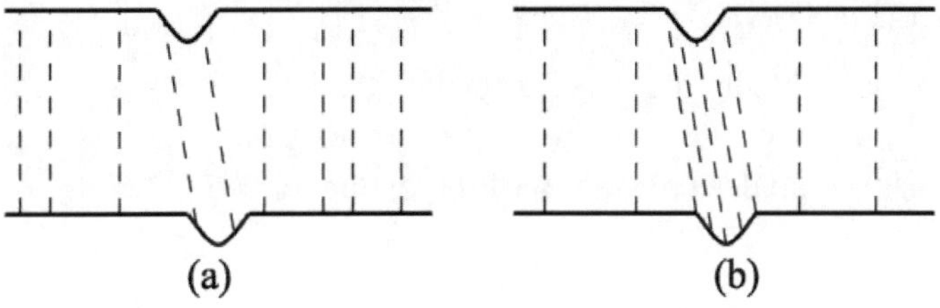

Fig. 2. Random sampling (left) and sampling with normals (right), source [17].

Point matching The algorithm [1] is based on the calculation of the Euclidean distance between points. However, this leads to significant computational demands, in the worst case, the algorithm has order $O(N_M N_S)$ of time complexity and average time complexity is $O(N_M\ logN_S)$. The use of point sorting in *kd*-tree leads to time complexity equal to $O(\ logN)$.

There are a number of methods to speed up the distance computation. The work [18] covers the most of possibilities of point matching methods. The normal method [8] is based on minimizing the distance not only of two points, but the approximate distance of the point from the plane. This computation reduces the number of iteration but the method is not so robust.

Another group of methods is based on projection of points from the original cloud (model) to the second cloud. The center of the projection can be given as the position of the camera that scans or generally. Then you can select a point of intersection or work with to points depending on the distance or intensity [19] or the color [16]. It is also possible to use the Hausdorff distance as in [5] that studies some theoretical results using Hausdorff distance in ICP.

It is also possible to use the SIFT feature point is used as corresponding point in the process of the improved ICP algorithm in order to reduce the error corresponding points. The reducing of the closest point search improves the accuracy and efficiency of the ICP algorithm as described the authors in [21].

Point weighting The trivial weighting is that every point has same weight. However, it is more appropriate to use the distance of points as a measure of suitability. Depending on the distance of the points, we can either totally eliminate the pair if it exceeds a predetermined threshold or add a weight to the pair according to the formula [7]:

$$Weight = 1 - \frac{\text{dist}(P_1 P_2)}{\text{dist}_{max}} \tag{14}$$

We can also assign the weights in dependence to the direction of normal vectors. If they are parallel at both points, the scalar product is large number. With increasing size of angle-normal we get the scalar product smaller (the perpendicular vectors have a scalar product equal to zero). The weight assignment can be performed as:

$$Weight = \mathbf{n}_1 \mathbf{n}_2 \tag{15}$$

Correspondence between points The point clouds often contain the noise points and outliers. To avoid misalling of the scans, criteria are added to eliminate or at least decrease this problem. The usual recommendation is to remove pairs of points that contain points at the borders if one of the skins exceeds the other (Fig. 3).

Another option is to remove points that have a greater distance than a given threshold, to remove a certain percentage of points depending on the metric, or to remove points whose distance is greater than a given multiple of the standard deviation of the distance.

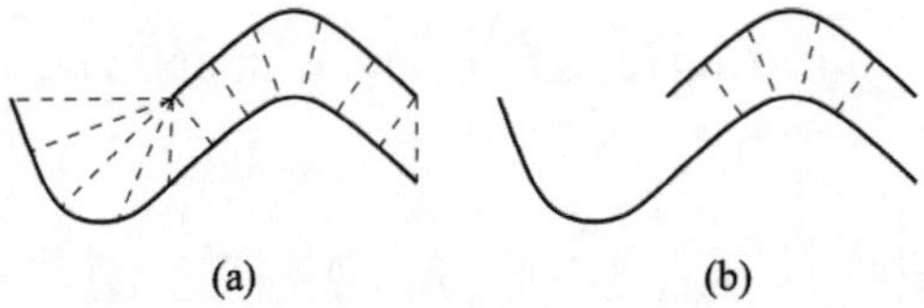

Fig. 3. The elimination of overlapping pairs, source [17].

3 Results

We tested the algorithm on the model point cloud in software Matlab, version 2017b [15]. We generated point clouds with random sets and we also added a noise. Tested point clouds have 3 sizes: 1000, 10 000 and 25 000 points. Then, we applied our algorithm ICP in Matlab. In Fig. 4, we can see the initial position of input point clouds and the registered position after application of ICP algorithm. The computation time and number of iterations is in Table 3.

Number of points	1000	10 000	25 000
Number of iteration	22	32	35
Time (s)	1.552	6.348	22.502

Tab. 1. Point cloud registration on tested point clouds.

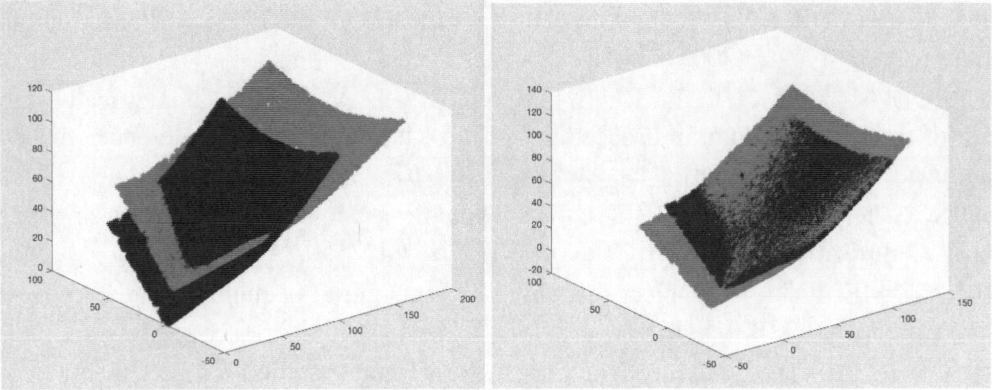

Fig. 4. Testing point cloud before (left) and after registration (right).

The implementation in Matlab consists of the steps described in 2.2. Matlab is suitable tool for ICP programming because of the easy use of matrices. We can also use the implemented functions as *svd*, *pdist* (Euclidean distance), etc. In implementation, we can choose the distance metrics (Euclidean, normal), threshold for ejecting noise points, type of weighting. The comparison of different parameters is difficult. It depends on the type of point clouds – their size, noise, overlapping region and shape. We also have different requirements on the result, we can prefer computation speed, precision or robustness. For example, the usage of normal distance computing increase the precision but it also decrease the robustness. Only the usage of k-d tree storage is strongly recommended. It can speed up the computation more than 10 times. Other parameters are optional and it depends on the input point clouds and future use of results.

4 Conclusion

Iterative closest point algorithm is the essential method of the point cloud registration. It is used for fine matching to ensure the best results. We present the overview of this method and also described the possible improvements of the algorithm. We tested the basic algorithm on the data. In future work, we would like to implement the possible improvements and compare the results.

Acknowledgment

This work was supported by Project LO1202 by financial means from the Ministry of Education, Youth and Sports under the National Sustainability Programme I.

References

[1] BESL, P. J., MCKAY, N.D.: A Method for Registration of 3-D Shapes. IEEE Transactions on Pattern Analysis and Machine Intelligence. 14 (2) (1992), 239–256.

[2] JIAN, B., VEMURI, B. C.: Robust Point Set Registration Using Gaussian Mixture Models. IEEE Transactions on Pattern Analysis and Machine Intelligence. 33(8), (2011), 1633–1645.

[3] CRACKNELL, A. P.: HAYES, L.:*Introduction to Remote Sensing* (2 ed.) London: Taylor and Francis. (2007).

[4] DIEZ, Y., ROURE, F., LLADO, X., SALVI J.: A Qualitative Review on 3D Coarse Registration Methods. ACM Comput. Surv. 47, 3, (2015), 54:1-45:36.

[5] EZRA, E., SHARIR, M., EFRAT, A.: On the performance of the ICP algorithm, Computational Geometry, vol. 41, 1–2, (2008), 77-93

[6] GELFAND, N., IKEMOTO, L., RUSINKIEWICZ, S. LEVOY, M.: Geometrically stable sampling for the ICP algorithm, in Proceedings of the 4th International Conference on 3-D Digital Imaging and Modeling, Banff, Canada, (2003) 260–67.

[7] GODIN, G. RIOUX, M. BARIBEAU, R.: Three-dimensional Registration Using Range and Intensity Information. Proc. SPIE: Videometrics III, vol. 2350, (1994), 279-290.

[8] CHEN, Y., MEDIONI, G.: Object modelling by registration of multiple range images. Image Vision Comput. (1992), 145–155.

[9] CHUA, C.S., JARVIS, R.: Point Signatures: A New Representation for 3D Object Recognition. International J. Computer Vision 25, 1, (1997), 63–85

[10] JOHNSON, A.E.: Spin-Images: A Representation for 3-D Surface Matching, (1997), Carnegie Mellon University Pittsburgh, Pennsylvania, Ph.D. thesis.

[11] LI, S., WANG, J., LIANG, Z., SU, L.: Tree point clouds registration using an improved ICP algorithm based on kd-tree. In Proceedings of 2016 IEEE International Geoscience and Remote Sensing Symposium (IGARSS). 4545-4548.

[12] National Oceanic and Atmospheric Administration (NOAA) Coastal Services Center. Lidar 101: An Introduction to Lidar Technology, Data, and Applications. Revised. Charleston, SC: NOAA Coastal Services Center, (2012)

[13] TAKESHI, M.: Generation of geometric model by registration and integration of multiple range images. In Proceedings of the IEEE International Conference on 3D Digital Imaging and Modelling. (2001), 254–261.

[14] MATAS, J., CHUM, O., URBAN, M., PAJDLA, T. : Robust wide baseline stereo from maximally stable extremal regions. Proc. of British Machine Vision Conference, (2002), 384-396.

[15] Matlab software, Math Works, https://www.mathworks.com/products/matlab. html

[16] PULLI, K.: Surface Reconstruction and Display from Range and Color Data, (1997), University of Washington, Ph.D. thesis.

[17] RUSINKIEWICZ, S. LEVOY, M.: Efficient variants of the ICP algorithm. In Proceedings of the IEEE International Conference on 3D Digital Imaging and Modeling, (2001), 145–152.

[18] WANG, F., ZHAO, Z.: A survey of iterative closest point algorithm, In Procceding of Chinese Automation Congress (CAC), Jinan, China, (2017), 4395-4399.

[19] WEIK, S.: Registration of 3-D Partial Surface Models Using Luminance and Depth Information, In Proceeding of International Conference on Recent Advances in 3D Digital Imaging and Modeling, (1997), 93-100.

[20] ZHENG, Z., LI, Y.: LIDAR Data Registration for Unmanned Ground Vehicle Based on Improved ICP Algorithm. In Proceedings of 7th International Conference on Intelligent Human-Machine Systems and Cybernetics, Vol. 2 (2015), 554-558.

[21] ZHENG, Z., LI, Y., JUN, W.: LiDAR point cloud registration based on improved ICP method and SIFT feature, In Proceeding o IEEE International Conference on Progress in Informatics and Computing (PIC), Nanjing, (2015), 588-592.

[22] ZHONG, Y.: Intrinsic Shape Signatures: A Shape Descriptor for 3D Object Recognition, IEEE, 12th International Conference on Computer Vision Workshops, ICCV Workshops, (2009), 689-696.

Current address

Procházková Jana, Mgr. Ph.D.
Institute of Materials Science and Engineering
NETME centre
Brno University of Technology
Technická 2896/2 616 69 Brno, Czech Republic
E-mail: prochazkova.j@fme.vutbr.cz

Martišek, Dalibor, doc., PaedDr.
Institute of Materials Science and Engineering
NETME centre
Brno University of Technology
Technická 2896/2 616 69 Brno, Czech Republic
E-mail: martisek@fme.vutbr.cz

Slovak University of Technology in Bratislava
Faculty of Mechanical Engineering
17th Conference on Applied Mathematics **APLIMAT 2018**

Proceedings

NUMERICAL SOLUTION OF A SPECIAL SINGULAR INTEGRAL EQUATION OF THE SECOND KIND

RAK Josef (CZ)

Abstract. This paper shows the method for numerical solution of a Fredholm integral equation of the second kind which has kernel function with a special type of singularity. The numerical solution is based on a Nyström method with a singularity subtraction technique. It requires a singular integral to be computed. Here the singular integral is converted to a non-singular one and a standard numerical quadrature is used.

Keywords: integral equations, Nyström method, numerical quadrature, singular integral

Mathematics Subject Classification: Primary 65R20, 45B05; Secondary 65Z05

1 Introduction

Fredholm integral equations of the second kind describe many physical phenomena. A numerical solution of integral equations leads to a solution of a system of linear equations with fully populated matrices. Many of them are described in [1]. The speed of finding a numerical solution depends on the speed of filling the matrix of a system of linear equations. One way to find the numerical solution is based on approximating the integral by numerical quadrature. Such methods are called Nyström methods. Let us have an integral equation of the following form:

$$y(x) - \int_0^1 \frac{h(x,t)y(t)}{|x-t|^\alpha} dt = f(x), \tag{1}$$

where $h(x,t) \in \mathcal{C}[0,1]$ and $\alpha \in (0,1)$. Here we integrate a singular function. So the Nytröm method can not be used directly. The Nyström method with singularity subtraction will be described in this paper. Its idea is to weaken the singularity in the integral.

2 Singularity subtraction

First let us weaken the singularity of the kernel function as in [2].

$$\frac{h(x,t)y(t)}{|x-t|^\alpha} = \frac{h(x,t)[y(t)-y(x)]}{|x-t|^\alpha} + f(x)\frac{h(x,t)}{|x-t|^\alpha}. \tag{2}$$

Substituting (2) to (1) we get

$$y(x) - \int_0^1 \frac{h(x,t)[y(t)-y(x)]}{|x-t|^\alpha}dt - y(x)\int_0^1 \frac{h(x,t)}{|x-t|^\alpha}dt = f(x). \tag{3}$$

The first integrand on the left hand side of (3) is singular when $x = t$. To apply numerical quadrature, we need to approximate it by a finite function. One approximation can be

$$\frac{1}{|x-t|^\alpha} \approx s_\varepsilon(x,t) = \begin{cases} \frac{1}{|x-t|^\alpha}, |x-t| \geq \varepsilon \\ \frac{1}{\varepsilon^\alpha}, |x-t| < \varepsilon. \end{cases} \tag{4}$$

We get

$$y(x) - \int_0^1 h(x,t)[y(t)-y(x)]s_\varepsilon(x,t)dt - y(x)\int_0^1 \frac{h(x,t)}{|x-t|^\alpha}dt = f(x). \tag{5}$$

Now the first integral on the left hand side of (5) is not singular and we can use standard numerical quadrature. The problem is with the second integral. It is singular, but its values are finite. It can be converted to a non-singular one.

3 Application of the Nyström method

Now let us take a numerical quadrature

$$\int_0^1 g(t)dt \approx \sum_{i=0}^{n} \omega_i g(x_i). \tag{6}$$

Points $x_i \in [0,1]$ are called the node points and numbers ω_i are called weights of the numerical quadrature.

Now we can approximate the first integral in (5) by (6) and run x through the node points. We get a system of linear equations for a solution in the node points

$$y(x_i) - \sum_{j=0}^{n} \omega_j h(x_i,x_j)[y(x_j)-y(x_i)]s_\varepsilon(x_i,x_j) - y(x_i)\int_0^1 \frac{h(x_i,t)}{|x-t|^\alpha}dt = f(x_i). \tag{7}$$

For sufficiently small ε is (7) equivalent to

$$y(x_i) - \sum_{j=0, j\neq i}^{n} \omega_i h(x_i,x_j)\frac{[y(x_j)-y(x_i)]}{|x_i-x_j|^\alpha} - y(x_i)\int_0^1 \frac{h(x_i,t)}{|x-t|^\alpha}dt = f(x_i). \tag{8}$$

We can see that the exact formula for construction of approximation (4) is not needed. We only need finite approximation of $|x-t|^{-\alpha}$ outside a certain neighborhood of $x = t$.

4 Computation of improper integral

The last problem is to compute the last integral on the left hand side of (8).

$$I(x) = \int_0^1 \frac{h(x_i, t)}{|x-t|^\alpha} dt \tag{9}$$

It is an improper integral but it reaches finite values. We will use an action as in [3] for a similar function. Since $x > 0$ we can split the integral $I(x)$ into a sum of two integrals

$$I(x) = \int_0^1 \frac{h(x,t)}{|x-t|^\alpha} dt = \int_0^x \frac{h(x,t)}{(x-t)^\alpha} dt + \int_x^1 \frac{h(x,t)}{(t-x)^\alpha} dt. \tag{10}$$

Using substitution $u = x - t$ the left integral in (10) can be written as

$$\int_0^x \frac{h(x,t)}{(x-t)^\alpha} dt = \int_0^x \frac{h(x, x-u)}{u^\alpha} du \tag{11}$$

Using substitution $u = t - x$ the right integral in (10) can be written as

$$\int_0^x \frac{h(x,t)}{(x-t)^\alpha} dt = \int_0^{1-x} \frac{h(x, x+u)}{u^\alpha} du. \tag{12}$$

Finally using substitution $\tau = u^{1-\alpha}$, (11) can be written as a non-singular integral

$$\int_0^x \frac{h(x, x-u)}{u^\alpha} du = \frac{1}{1-\alpha} \int_0^{x^{1-\alpha}} h\left(x, x - \tau^{\frac{1}{1-\alpha}}\right) d\tau \tag{13}$$

and (12) can be written as a non-singular integral

$$\int_0^{1-x} \frac{h(x, x+u)}{u^\alpha} du = \frac{1}{1-\alpha} \int_0^{(1-x)^{1-\alpha}} h\left(x, x + \tau^{\frac{1}{1-\alpha}}\right) d\tau. \tag{14}$$

From (11), (12), (13) and (14) the integral $I(x)$ has the non-singular form

$$I(x) = \frac{1}{1-\alpha} \left[\int_0^{x^{1-\alpha}} h\left(x, x - \tau^{\frac{1}{1-\alpha}}\right) d\tau + \int_0^{(1-x)^{1-\alpha}} h\left(x, x + \tau^{\frac{1}{1-\alpha}}\right) d\tau. \right] \tag{15}$$

and the common numerical integration rule can be used.

5 Error estimation

By using methods of functional analysis the error estimate can be derived. The error has two factors. First is the factor of singular function $|x-t|^{-\alpha}$ and the second is the error on the numerical integration rule. All important facts are summarized by the following theorem. It is proved in [4].

Theorem 5.1. Let y be the solution of (1) with $h(x,t) \in \mathcal{C}([0,1] \times [0,1])$, and $\gamma \in (0,1)$. Let y_n be the solution of (8), where the quadrature is a compound midpoint rule. Then for a sufficiently large n, there exist constants c_0 and c_1 such that

$$\|y - y_n\| \leq c_0 \left[\frac{2}{1-\gamma} \max_{x\in[0,1]} \omega\left(h_x, \frac{1}{n}\right) + \frac{2c_1}{n^{1-\gamma}} \right], \tag{16}$$

where ω is the modulus of continuity and $h_x(t) = y(t)h(x,t)$. Furthermore, if $y \in \mathcal{C}^2[0,1]$ and $h_x(t) \in \mathcal{C}^2([0,1] \times [0,1])$ for each $x \in [0,1]$, then for a sufficiently large n there exists c_2 such that

$$\|y - y_n\| \leq \frac{c_2}{n^{2-\gamma}}. \tag{17}$$

For other quadrature rules, similar theorems can be proved. Singularity of the kernel function is a very important factor which affects error behavior. So the error behavior is not better for more precise numerical quadratures. So it is not necessary to take a higher order integration rule than $\mathcal{O}(n^{-2})$, where n is the number of the node points.

6 Numerical tests

In this section let us verify the error estimation, which is described by the theorem 5.1. Let us take the compound mid-point rule. It has error $\mathcal{O}(n^{-2})$. Let us choose $\alpha = \frac{1}{2}$. It is the most common singular factor because it is the Euclidean distance. Let us choose the right hand side function $f(x)$ such that the exact solution is e^x. The right hand side function $f(x)$ is computed symbolically by Maple [5]. The same software is used for solution of the system of linear equations (8).

Let us verify the error behavior for exact solution $y(x) = e^x$ and $h(x,t) = x+t$. The columns labeled "ratio" in following tables give the ratio of successive errors. Since $h_x \in \mathcal{C}^2([0,1])$ the column ratio is expected to be approximately $\sqrt{8} \approx 2.83$.

n	error	ratio	n	error	ratio
10	0.0124685	-	20	0.0050904	2.45
40	0.0016737	3.04	80	0.0004682	3.57
160	0.0001236	3.78	320	0.0000318	3.88

Tab. 1. Midpoint rule, exact solution $y(x) = e^x$, $h(x,t) = (x+t)$

As we can see in table 1, the result is consistent to the theory. From table 2 we can see that the error is not worse if we take the function $h(x,t) = \ln(x+t)$. This is despite the fact that that h_x does not have continuous derivatives when $x = 0$. The reason is that the error (16) is too pessimistic for only continuous functions.

n	error	ratio	n	error	ratio
10	0.0244284	-	20	0.0059047	4.14
40	0.0014633	4.04	80	0.0003745	3.91
160	0.0000992	3.78	320	0.0000272	3.64

Tab.2. Midpoint rule, exact solution $y(x) = e^x$, $h(x,t) = \ln(x+t)$

Now let us take the exact solution $y(x) = \sqrt{x}$. This function is Hölder continuous with constant $\frac{1}{2}$. So the ratio should be approximately $\sqrt{2} \approx 1.41$. Tables 3 and 4 show error behavior for the same functions $h(x,t)$.

n	error	ratio	n	error	ratio
10	0,0009572	-	20	0,0004938	1.94
40	0,0002767	1.78	80	0,0001400	1.98
160	0,0000583	2.40	320	0,0000239	2.44

Tab. 3. Midpoint rule, exact solution $y(x) = \sqrt{x}$, $h(x,t) = x+t$

n	error	ratio	n	error	ratio
10	0.0104430	-	20	0.0052149	2.00
40	0.0033060	1.57	80	0.0020864	1.58
160	0.0013055	1.60	320	0.0008075	1.61

Tab. 4. Midpoint rule, exact solution $y(x) = \sqrt{x}$, $h(x,t) = \ln(x+t)$

From tables 3 and 4 we can see, that the theorem is also consistent to the results. Now let use choose $\alpha = \frac{1}{4}$. The function $f(x)$ is chosen such that the exact solution is $y(x) = e^x$. The ratio should decrease to the value $\sqrt[4]{2^7} \approx 3.36$ for continuous $h(x,t) \in C[0,1]$. From tables 5 and 6 we see the consistency with the theory.

n	error	ratio	n	error	ratio
10	0.0054453	-	20	0.0013761	3.96
40	0.0003450	3.99	80	0.0000865	3.99
160	0.0000216	4.00	320	0.0000054	4.00

Tab. 5. Midpoint rule, exact solution $y(x) = e^x$, $h(x,t) = x+t$, $\alpha = \frac{1}{4}$

n	error	ratio	n	error	ratio
10	0.0036882	-	20	0.0009687	3.80
40	0.0002566	3.78	80	0.0000704	3.64
160	0.0000252	2.79	320	0.0000088	2.86

Tab. 6. Midpoint rule, exact solution $y(x) = e^x$, $h(x,t) = \ln(x+t)$, $\alpha = \frac{1}{4}$

7 Conclusion

For the solution of integral equations many methods have been developed. Each of them lead to the solution of a system of linear equations with fully populated matrices. Let us compare this method to piecewise linear collocation. The error is $\mathcal{O}(n^{-2})$ (see [1] for details) for continuous $h(x,t)$ and $y(x) \in \mathcal{C}^2([0, 1])$. So the label ratio is expected to be 4. If we compare the column ratio in the tables above, we can see that we have very good results.

But this method has a big advantage. Piecewise linear collocation and other methods (for example product integration methods) have big computing time. The reason is that every element in the matrix of the system of linear equations is an integral which needs to be calculated. In case of the Nyström method with the singularity subtraction, only diagonal elements of the matrix are integrals. This fact rapidly decreases computing time.

References

[1] ATKINSON K.E. The Numerical Solution of Integral Equations of the Second Kind, Cambridge University Press, Cambridge 1997, ISBN: 978-0-521-58391-6.

[2] KANTOROVICH L. V. , KRYLOV V. I. Approximate Methods of Higher Analysis. Interscience, New York, 1958,

[3] AMARI S., BORNEMANN J. Efficient Numerical Computation of Singular Integrals with Applications to Electromagnetics, 1958, IEE Trans. on Atennas and Propag, vol 43, no 11, pp-1342-1348, 1995

[4] RAK J. Numerical Solution of a Fredholm Integral Equation of the Second Kind Related to Induction Heating, Praha 2012. Doctoral thesis. Charles University Prague, Faculty of Mathematics and Physics, 27.9.2012

[5] Maple. http://www.maplesoft.com.

Current Address

Rak Josef, RNDr., Ph.D.
Univerzita Pardubice
Fakulta elektrotechniky a informatiky
Studentská 95, 532 10 Pardubice, Czech Republic
Phone: +420 466 036 726, email: josef.rak@upce.cz

Slovak University of Technology in Bratislava
Faculty of Mechanical Engineering
17th Conference on Applied Mathematics **APLIMAT 2018**

Proceedings

AN APPLICATION OF BELL POLYNOMIALS IN NUMERICAL SOLVING OF NONLINEAR DIFFERENTIAL EQUATIONS

REBENDA Josef (CZ)

Abstract. Partial ordinary Bell polynomials are used to formulate and prove a version of the Faà di Bruno's formula which is convenient for handling nonlinear terms in the differential transformation. Applicability of the result is shown in two examples of solving the initial value problem for differential equations which are nonlinear with respect to the dependent variable.

Keywords: Faà di Bruno's formula, Bell polynomials, differential transformation, nonlinear differential equations

Mathematics subject classification: Primary 34A45; Secondary 34A34, 05A19, 26E05, 34A25

1 Introduction

The ability to find a numerical approximation of solution of a differential equation is important in particular practical applications where it is difficult or even impossible to find analytical solution of the given problem. Plenty of well-established and verified numerical methods for various kinds of problems involving differential equations can be found for instance in monographs [1], [2], [5], [6].

Among others, semi-analytical methods convenient for solving differential equations are in the forefront of study in the last two decades. However, the calculations and results are often expressed in a complicated way. We propose an easily applicable approach in this paper.

The differential transformation is closely related to Taylor expansion of real analytic functions with applications to different types of problems of solving differential equations. To indicate recent development in the field we mention several papers from the last three years, e.g. [9], [10], [11], [12], [15].

The paper is organized as follows. First we recall basic definitions and formulas of the differential transformation and give a brief overview of Bell polynomials and the Faà di Bruno's formula in Section 2. In Section 3 we develop the theory and prove the main result. Application of the results is shown in Section 4.

Convergence, error estimates and stability of Taylor series based methods is thoroughly discussed in literature on numerical methods, therefore we do not include such topics in the paper. However, interesting results on a-priori error bounds are published in recent paper [14] and the references cited therein.

2 Preliminaries

In this section we recall basic definitions and formulas of the differential transformation as well as notions and results related to partial Bell polynomials.

Definition 1 The differential transformation of a real function $u(t)$ at a point $t_0 \in \mathbb{R}$ is $\mathcal{D}\{u(t)\}[t_0] = \{U(k)[t_0]\}_{k=0}^{\infty}$, where $U(k)[t_0]$, the differential transformation of the $k-$th derivative of the function $u(t)$ at t_0, is defined as

$$U(k)[t_0] = \frac{1}{k!}\left[\frac{d^k u(t)}{dt^k}\right]_{t=t_0}, \tag{1}$$

provided that the original function $u(t)$ is analytic in some neighbourhood of t_0.

Definition 2 The inverse differential transformation of $\{U(k)[t_0]\}_{k=0}^{\infty}$ at t_0 is defined as

$$u(t) = \mathcal{D}^{-1}\Big\{\ \{U(k)[t_0]\}_{k=0}^{\infty}\ \Big\}[t_0] = \sum_{k=0}^{\infty} U(k)[t_0](t-t_0)^k. \tag{2}$$

In real applications the function $u(t)$ is expressed by a finite sum

$$u(t) = \sum_{k=0}^{N} U(k)[t_0](t-t_0)^k. \tag{3}$$

Plenty of transformation formulas can be derived from Definitions 1 and 2. We recall the following relations which will be used later in illustratory examples.

Lemma 1 Assume that $\{F(k)\}_{k=0}^{\infty}$, $\{G(k)\}_{k=0}^{\infty}$, $\{H(k)\}_{k=0}^{\infty}$ and $\{U_i(k)\}_{k=0}^{\infty}$, $i = 1,\dots,m$, are differential transformations of analytic functions $f(t)$, $g(t)$, $h(t)$ and $u_i(t)$, $i = 1,\dots,m$, at $t_0 \in \mathbb{R}$, respectively. Let $n \in \mathbb{N}$ and $\lambda \in \mathbb{R}$. Then

$i)$ If $f(t) = \dfrac{d^n g(t)}{dt^n}$, then $F(k) = \dfrac{(k+n)!}{k!}G(k+n)$.

$ii)$ If $f(t) = g(t)\,h(t)$, then $F(k) = \sum_{l=0}^{k} G(l)\,H(k-l)$.

$iii)$ If $f(t) = t^n$, then $F(k) = \delta(k-n), t_0 = 0$, where δ is the Kronecker delta symbol.

$iv)$ If $f(t) = e^{\lambda t}$, then $F(k) = \dfrac{e^{a\lambda}\,\lambda^k}{k!}, t_0 = a$.

$v)$ If $f(t) = \ln t$, then $F(0) = 0$ and $F(k) = \dfrac{(-1)^{k+1}}{k}$ for $k \geq 1, t_0 = 1$.

$vi)$ If $f(t) = (1+t)^{\lambda}$, then $F(k) = \dbinom{\lambda}{k} = \dfrac{\lambda(\lambda-1)\dots(\lambda-k+1)}{k!}, t_0 = 0$.

$vii)$ If $f(t) = \prod_{i=1}^{m} u_i(t)$, then

$$F(k) = \sum_{s_1=0}^{k}\ \sum_{s_2=0}^{k-s_1} \cdots \sum_{s_{m-1}=0}^{k-s_1-\dots-s_{m-2}} U_1(s_1)\dots U_{m-1}(s_{m-1})\,U_m(k-s_1-\dots-s_{m-1}).$$

The main disadvantage of most papers with applications of the differential transformation is that there is lack of direct applications on equations with nonlinear terms containing unknown function $u(t)$, e.g. $f(u) = \sqrt{1+u^2}$ or $f(u) = e^{\sin u}$. Usually, the "nonlinearity" is represented by terms u^n, $n \in \mathbb{N}$, and formula $vii)$ in Lemma 2 is used to transform such terms.

In paper [13] the differential transformation of components containing nonlinear terms is calculated using the so-called Adomian polynomials A_n in which each solution component u_i is replaced by the corresponding differential transformation component $U(i)$, $i = 0, 1, 2, \ldots$. The formula for the differential transformation $F(k)$ of a nonlinear term $f(u)$ is

$$\begin{aligned} F(k) &= \sum_{n=0}^{\infty} A_n(U(0), U(1), \ldots, U(n))\delta(k-n) = A_k(U(0), U(1), \ldots, U(k)) \\ &= \frac{1}{k!}\frac{d^k}{dt^k}\left[f\left(\sum_{i=0}^{\infty} U(i)t^i \right) \right]_{t=0}, \quad k \geq 0. \end{aligned}$$

The first four terms are:

$$\begin{aligned} F(0) &= f(U(0)), \\ F(1) &= U(1)f'(U(0)), \\ F(2) &= U(2)f'(U(0)) + \frac{1}{2!}U^2(1)f''(U(0)) \\ F(3) &= U(3)f'(U(0)) + U(1)U(2)f''(U(0)) + \frac{1}{3!}U^3(1)f'''(U(0)). \end{aligned}$$

As we can observe, the formula with the so-called Adomian polynomials is in fact the well-known Faà di Bruno's formula generalizing the chain rule to higher derivatives. However, there are derivatives of the function f contained in the formula, which means that symbolic derivatives of f need to be calculated and then evaluated. In such situation, one of the big advantages of the differential transformation is lost.

Fortunately, the differential transformation of components containing nonlinear terms can be easily found without calculating and evaluating symbolic derivatives. We will utilize a slightly modified Faà di Bruno's formula with not exponential but ordinary Bell polynomials.

For this purpose we recall some necessary notions and results in combinatorics. The proofs are omitted since they can be found in the cited literature [3] and [4].

Definition 3[[4], p. 133] The partial exponential Bell polynomials are the polynomials $B_{k,l}(x_1, \ldots, x_{k-l+1})$ in an infinite number of variables $x_1, x_2, \ldots$, defined by the series expansion

$$\sum_{k \geq l} B_{k,l}(x_1, \ldots, x_{k-l+1})\frac{t^k}{k!} = \frac{1}{l!}\left(\sum_{m \geq 1} x_m \frac{t^m}{m!} \right)^l, \quad l = 0, 1, 2, \ldots \tag{4}$$

Lemma 2[[4], p. 134] The partial exponential Bell poynomials have integer coefficients, are homogeneous of degree l and weight k, and their exact expression is:

$$B_{k,l}(x_1, \ldots, x_{k-l+1}) = \sum \frac{k!}{j_1! j_2! \cdots j_{k-l+1}!}\left(\frac{x_1}{1!}\right)^{j_1}\left(\frac{x_2}{2!}\right)^{j_2} \cdots \left(\frac{x_{k-l+1}}{(k-l+1)!}\right)^{j_{k-l+1}}, \tag{5}$$

where the summation takes place over all sequences $j_1, j_2, \ldots, j_{k-l+1}$ of non-negative integers such that

$$j_1 + j_2 + \ldots + j_{k-l+1} = l, \tag{6}$$
$$j_1 + 2j_2 + \ldots + (k-l+1)j_{k-l+1} = k. \tag{7}$$

Lemma 3[[3], p. 415] The partial exponential Bell polynomials $B_{k,l}(x_1, \ldots, x_{k-l+1})$, $l = 1, 2, \ldots$, $k \geq l$, satisfy the recurrence relation

$$B_{k,l}(x_1, \ldots, x_{k-l+1}) = \sum_{i=1}^{k-l+1} \binom{k-1}{i-1} x_i B_{k-i,l-1}(x_1, \ldots, x_{k-i-l+2}), \tag{8}$$

where $B_{0,0} = 1$ and $B_{k,0} = 0$ for $k \geq 1$.

Theorem 1[[4], p. 138-139] Let two functions $f(y)$ and $g(x)$ of a real variable be given, $g(x)$ of class C^∞ at $x = t_0$, and $f(y)$ of class C^∞ at $y = s_0 = g(t_0)$, and let $h(x) = (f \circ g)(x) = f(g(x))$. If we put $g_0 = g(t_0)$, $f_0 = f(s_0) = h_0 = h(t_0) = f(g(t_0))$, $g_m = \left.\frac{d^m g}{dx^m}\right|_{x=t_0}$, $f_l = \left.\frac{d^l f}{dy^l}\right|_{y=s_0}$, $h_k = \left.\frac{d^k h}{dx^k}\right|_{x=t_0}$, then the k-th order derivative of h at $t = t_0$ for $k \geq 1$ equals

$$h_k = \left.\frac{d^k h}{dx^k}\right|_{x=t_0} = \sum_{l=1}^{k} f_l B_{k,l}(g_1, g_2, \ldots, g_{k-l+1}), \tag{9}$$

where $B_{k,l}$ are explicitly given by (5).

Definition 4[[4], p. 136] The partial ordinary Bell polynomials are the polynomials $\hat{B}_{k,l}(\hat{x}_1, \ldots, \hat{x}_{k-l+1})$ in an infinite number of variables $\hat{x}_1, \hat{x}_2, \ldots$, defined by the series expansion

$$\sum_{k \geq l} \hat{B}_{k,l}(\hat{x}_1, \ldots, \hat{x}_{k-l+1}) t^k = \left(\sum_{m \geq 1} \hat{x}_m t^m \right)^l, \quad l = 0, 1, 2, \ldots \tag{10}$$

3 Results

Before we formulate the main theorem, we prove several auxiliary results. The following lemma is crucial in the proof of the main result.

Lemma 4 The relation between the partial exponential Bell polynomials $B_{k,l}$ and the partial ordinary Bell polynomials $\hat{B}_{k,l}$ is

$$B_{k,l}(x_1, \ldots, x_{k-l+1}) = \frac{k!}{l!} \hat{B}_{k,l} \left(\frac{x_1}{1!}, \frac{x_2}{2!}, \ldots, \frac{x_{k-l+1}}{(k-l+1)!} \right). \tag{11}$$

Proof. If we denote $\hat{x}_i = \frac{x_i}{i!}$ for all $i = 0, 1, 2, \ldots$, and substitute in (10), we obtain

$$\sum_{k \geq l} \hat{B}_{k,l} \left(\frac{x_1}{1!}, \frac{x_2}{2!}, \ldots, \frac{x_{k-l+1}}{(k-l+1)!} \right) t^k = \left(\sum_{m \geq 1} \frac{x_m t^m}{m!} \right)^l, \quad l = 0, 1, 2, \ldots \tag{12}$$

Multiplying both sides by $\frac{1}{l!}$ and each summand on the left side by $\frac{k!}{k!}$, we get

$$\sum_{k\geq l} \frac{k!}{l!} \hat{B}_{k,l}\left(\frac{x_1}{1!}, \frac{x_2}{2!}, \dots, \frac{x_{k-l+1}}{(k-l+1)!}\right) \frac{t^k}{k!} = \frac{1}{l!}\left(\sum_{m\geq 1} x_m \frac{t^m}{m!}\right)^l, \quad l = 0, 1, 2, \dots \tag{13}$$

Equating the coefficients of t^k in (13) and (4) we get the formula (11).

In calculation of the partial ordinary Bell polynomials, the following lemma can be useful.

Lemma 5 The partial ordinary Bell polynomials $\hat{B}_{k,l}(\hat{x}_1, \dots, \hat{x}_{k-l+1})$, $l = 1, 2, \dots$, $k \geq l$, satisfy the recurrence relation

$$\hat{B}_{k,l}(\hat{x}_1, \dots, \hat{x}_{k-l+1}) = \sum_{i=1}^{k-l+1} \frac{i \cdot l}{k} \hat{x}_i \hat{B}_{k-i,l-1}(\hat{x}_1, \dots, \hat{x}_{k-i-l+2}), \tag{14}$$

where $\hat{B}_{0,0} = 1$ and $\hat{B}_{k,0} = 0$ for $k \geq 1$.

Proof. Using the relation between the exponential and ordinary Bell polynomials (11), the formula (8) changes to

$$\frac{k!}{l!} \hat{B}_{k,l}\left(\frac{x_1}{1!}, \dots, \frac{x_{k-l+1}}{(k-l+1)!}\right) = \sum_{i=1}^{k-l+1} \binom{k-1}{i-1} x_i \frac{(k-i)!}{(l-1)!} \hat{B}_{k-i,l-1}\left(\frac{x_1}{1!}, \dots, \frac{x_{k-i-l+2}}{(k-i-l+2)!}\right).$$

After rearranging the expression and multiplying ith term in the sum by $\frac{i!}{i!}$, we obtain

$$\hat{B}_{k,l}\left(\frac{x_1}{1!}, \dots, \frac{x_{k-l+1}}{(k-l+1)!}\right) =$$
$$= \sum_{i=1}^{k-l+1} \frac{(k-1)!}{(i-1)!(k-i)!} \frac{l!}{k!} \frac{(k-i)!}{(l-1)!} i! \frac{x_i}{i!} \hat{B}_{k-i,l-1}\left(\frac{x_1}{1!}, \dots, \frac{x_{k-i-l+2}}{(k-i-l+2)!}\right).$$

Now we cancel all possible factors and factorials and we get

$$\hat{B}_{k,l}\left(\frac{x_1}{1!}, \dots, \frac{x_{k-l+1}}{(k-l+1)!}\right) = \sum_{i=1}^{k-l+1} \frac{i \cdot l}{k} \frac{x_i}{i!} \hat{B}_{k-i,l-1}\left(\frac{x_1}{1!}, \dots, \frac{x_{k-i-l+2}}{(k-i-l+2)!}\right). \tag{15}$$

Denoting $\hat{x}_i = \frac{x_i}{i!}$ for all $i = 1, \dots, k-l+1$, gives the formula (14).

The main result of the paper is formulated in the following theorem:

Theorem 2 Let g and f be real functions analytic near t_0 and $g(t_0)$ respectively, and let h be the composition $h(t) = (f \circ g)(t) = f(g(t))$. Denote $\mathcal{D}\{g(t)\}[t_0] = \{G(k)\}_{k=0}^{\infty}$, $\mathcal{D}\{f(t)\}[g(t_0)] = \{F(k)\}_{k=0}^{\infty}$ and $\mathcal{D}\{(f \circ g)(t)\}[t_0] = \{H(k)\}_{k=0}^{\infty}$ the differential transformations of functions g, f and h at t_0, $g(t_0)$ and t_0 respectively. Then the numbers $H(k)$ in the sequence $\{H(k)\}_{k=0}^{\infty}$ satisfy the relations $H(0) = F(0)$ and

$$H(k) = \sum_{l=1}^{k} F(l) \cdot \hat{B}_{k,l}\big(G(1), \dots, G(k-l+1)\big) \quad \text{for } k \geq 1. \tag{16}$$

Proof. The assumption of analyticity of functions g, f and h guarantees that Theorem 1 is valid. Applying Definition 1 on formula (9), we obtain

$$k!H(k) = \sum_{l=1}^{k} l!F(l) \cdot B_{k,l}\big(1!G(1), \ldots, (k-l+1)!G(k-l+1)\big). \tag{17}$$

If we divide both sides by $k!$ and apply formula (11) in Lemma 3, we have

$$H(k) = \sum_{l=1}^{k} \frac{l!}{k!} F(l) \cdot \frac{k!}{l!} \hat{B}_{k,l}\left(\frac{1!G(1)}{1!}, \ldots, \frac{(k-l+1)!G(k-l+1)}{(k-l+1)!}\right) \tag{18}$$

for $k \geq 1$. Cancelling all possible factorials gives the result.

4 Applications

To show efficiency of the derived algorithm, we apply the differential transformation to two differential equations nonlinear with respect to the dependent variable $u(t)$.

Example 1 Let us consider equation

$$u'(t) = u(t) - t + \ln(u(t)) \tag{19}$$

with initial condition

$$u(0) = 1. \tag{20}$$

Here we denote $h(t) = f(g(t))$, where $g(t) = u(t)$ and $f(x) = \ln(x)$. We are looking for an analytic solution in a neighbourhood of $t_0 = 0$. Then the righthand side of (19) is analytic too. The differential transformation turns the equation (19) into

$$(k+1)U(k+1)[0] = U(k)[0] - \delta(k-1) + H(k)[0], \tag{21}$$

with transformed initial condition $U(0)[0] = 1$.

To find the coefficients $H(k)[0]$, we use Theorem 2. First of all, we recall Lemma 2, formula v), to see that the differential transformation of $f(x) = \ln(x)$ at $x_0 = u(0) = 1$ is $F(0)[1] = 0$ and $F(k)[1] = \dfrac{(-1)^{k+1}}{k}$ for $k \geq 1$. Theorem 2 gives $H(0)[0] = F(0)[1] = 0$ and $H(k)[0] = \sum_{l=1}^{k} F(l)[1] \cdot \hat{B}_{k,l}\big(U(1)[0], \ldots, U(k-l+1)[0]\big)$ for $k \geq 1$. We calculate

$$\begin{aligned}
H(1)[0] &= F(1)[1] \cdot \hat{B}_{1,1}\big(U(1)[0]\big) = U(1)[0], \\
H(2)[0] &= \sum_{l=1}^{2} F(l)[1] \cdot \hat{B}_{2,l}\big(U(1)[0], U(2)[0]\big) = F(1)[1]\hat{B}_{2,1}\big(U(1)[0], U(2)[0]\big) + \\
&+ F(2)[1]\hat{B}_{2,2}\big(U(1)[0]\big) = U(2)[0] + \frac{(-1)}{2}\big(U(1)[0]\big)^2 = U(2)[0] - \frac{1}{2}\big(U(1)[0]\big)^2, \\
H(3)[0] &= \sum_{l=1}^{3} F(l)[1] \cdot \hat{B}_{3,l}\big(U(1)[0], U(2)[0], U(3)[0]\big) \\
&= F(1)[1]\hat{B}_{3,1}\big(U(1)[0], U(2)[0], U(3)[0]\big) + F(2)[1]\hat{B}_{3,2}\big(U(1)[0], U(2)[0]\big) + \\
&+ F(3)[1]\hat{B}_{3,3}\big(U(1)[0]\big) = U(3)[0] - \frac{1}{2} 2U(1)[0]U(2)[0] + \frac{1}{3}\big(U(1)[0]\big)^3.
\end{aligned}$$

Now we are prepared to calculate the coefficients $U(k)[0]$, $k = 1, 2, \ldots$.

$$\begin{aligned}
U(1)[0] &= U(0)[0] - \delta(-1) + H(0)[0] = 1, \\
U(2)[0] &= \frac{1}{2} = \left(U(1)[0] - \delta(0) + H(1)[0]\right) = \frac{1}{2}(1 - 1 + 1) = \frac{1}{2}, \\
U(3)[0] &= \frac{1}{3}\left(U(2)[0] - \delta(1) + H(2)[0]\right) = \frac{1}{3}\left(\frac{1}{2} + U(2)[0] - \frac{1}{2}\left(U(1)[0]\right)^2\right) = \frac{1}{3 \cdot 2} = \frac{1}{3!}, \\
U(4)[0] &= \frac{1}{4}\left(U(3)[0] - \delta(2) + H(3)[0]\right) = \frac{1}{4}\left(\frac{1}{3!} + U(3)[0] - \frac{1}{2}2U(1)[0]U(2)[0] + \right. \\
&+ \left. \frac{1}{3}\left(U(1)[0]\right)^3\right) = \ldots = \frac{1}{4!}, \\
&\vdots
\end{aligned}$$

It is not difficult to verify that the kth coefficient $U(k)[0]$ has the value $\frac{1}{k!}$. Using the inverse differential transformation (Definition 2) we obtain

$$u(t) = \mathcal{D}^{-1}\left\{ \{U(k)[0]\}_{k=0}^{\infty} \right\}[0] = \sum_{k=0}^{\infty} \frac{1}{k!} t^k = \mathrm{e}^t, \tag{22}$$

which is the exact solution of the initial value problem (19), (20).

Example 2 Let us solve the following equation

$$u'(t) = 2\sqrt{1 - u^2(t)} \tag{23}$$

with initial condition

$$u(0) = 0. \tag{24}$$

We denote $h(t) = f(g(t))$, where $g(t) = u(t)$ and $f(x) = \sqrt{1 - x^2}$. We want to find an analytic solution in a neighbourhood of $t_0 = 0$. The righthand side of (23) is analytic near t_0. The differential transformation of the equation (23) is

$$(k + 1)U(k + 1)[0] = 2H(k)[0], \tag{25}$$

with transformed initial condition $U(0)[0] = 0$.

We use Theorem 2 to find the coefficients $H(k)[0]$. First of all, we use Lemma 2, formula $vi)$ to derive the differential transformation of $f(x)$ at $x_0 = u(0) = 0$. Since $f(x) = \sqrt{1 - x^2} = (1 + (-x^2))^{\frac{1}{2}}$, Taylor series of $f(x)$ at 0 is $f(x) = \sum_{k=0}^{\infty} \binom{\frac{1}{2}}{k} (-x^2)^k$. Consequently, $F(2k)[0] = \binom{\frac{1}{2}}{k} (-1)^k$ and $F(2k + 1)[0] = 0$ for $k \in \mathbb{N}_0$. Theorem 2 gives $H(0)[0] = F(0)[0] = 1$ and

$$H(k)[0] = \sum_{l=1}^{k} F(l)[0] \cdot \hat{B}_{k,l}\left(U(1)[0], \ldots, U(k - l + 1)[0]\right) \text{ for } k \geq 1. \tag{26}$$

Now we take (25) and (26) in turns to obtain the coefficients.

$$\begin{aligned}
U(1)[0] &= \frac{2}{1}H(0)[0] = 2,\\
H(1)[0] &= F(1)[0]\cdot \hat{B}_{1,1}(U(1)[0]) = 0\cdot U(1)[0] = 0,\\
U(2)[0] &= \frac{2}{2}H(1)[0] = 0,\\
H(2)[0] &= \sum_{l=1}^{2} F(l)[0]\cdot \hat{B}_{2,l}(U(1)[0],U(2)[0]) = F(1)[0]\hat{B}_{2,1}(U(1)[0],U(2)[0]) +\\
&+ F(2)[0]\hat{B}_{2,2}(U(1)[0]) = 0\cdot U(2)[0] + \frac{1}{2}(-1)(U(1)[0])^2 = -\frac{1}{2}2^2,\\
U(3)[0] &= \frac{2}{3}H(2)[0] = \frac{2}{3}\left(-\frac{1}{2}\right)2^2 = -\frac{2^3}{3!},\\
H(3)[0] &= \sum_{l=1}^{3} F(l)[0]\cdot \hat{B}_{3,l}(U(1)[0],U(2)[0],U(3)[0])\\
&= F(1)[0]\hat{B}_{3,1}(U(1)[0],U(2)[0],U(3)[0]) + F(2)[0]\hat{B}_{3,2}(U(1)[0],U(2)[0]) +\\
&+ F(3)[0]\hat{B}_{3,3}(U(1)[0]) = 0 - \frac{1}{2}2U(1)[0]U(2)[0] + 0 = 0,\\
U(4)[0] &= \frac{2}{4}H(3)[0] = 0,\\
H(4)[0] &= \sum_{l=1}^{4} F(l)[0]\cdot \hat{B}_{4,l}(U(1)[0],U(2)[0],U(3)[0],U(4)[0])\\
&= F(1)[0]\hat{B}_{4,1}(U(1)[0],U(2)[0],U(3)[0],U(4)[0]) +\\
&+ F(2)[0]\hat{B}_{4,2}(U(1)[0],U(2)[0],U(3)[0]) + F(3)[0]\hat{B}_{4,3}(U(1)[0],U(2)[0]) +\\
&+ F(4)[0]\hat{B}_{4,4}(U(1)[0]) = 0 - \frac{1}{2}\left(2U(1)[0]U(3)[0] + (U(2)[0])^2\right) + 0 + \binom{\frac{1}{2}}{2}(U(1)[0])^4\\
&= -\frac{1}{2}(2\cdot 2\cdot \frac{-2^3}{3!} + 0) + \frac{1}{2}\cdot\left(-\frac{1}{2}\right)\cdot\frac{1}{2!}\cdot 2^4 = \frac{2^4}{6} - \frac{2^4}{8} = \frac{2^4}{24} = \frac{2^4}{4!},\\
U(5)[0] &= \frac{2}{5}H(4)[0] = \frac{2}{5}\cdot\frac{2^4}{4!} = \frac{2^5}{5!},\\
&\vdots
\end{aligned}$$

Also in this case, it is possible to see the pattern for the kth coefficient $U(k)[0]$, which is $U(k)[0] = 0$ if k is even and $U(k)[0] = (-1)^{\frac{k-1}{2}}\frac{2^k}{k!}$ if k is odd. The inverse differential transform gives

$$u(t) = \mathcal{D}^{-1}\left\{\{U(k)[0]\}_{k=0}^{\infty}\right\}[0] = \sum_{k=0}^{\infty}\frac{(-1)^k\cdot 2^{2k+1}}{(2k+1)!}t^{2k+1} = \sum_{k=0}^{\infty}\frac{(-1)^k}{(2k+1)!}(2t)^{2k+1} = \sin 2t.$$

Indeed, $u(t) = \sin 2t$ is the exact unique solution of the given initial value problem (23), (24).

Remark In recent works [7], [8], a more general discretization technique involving Hilbert spaces is introduced. Although the approach is different, the transformation rules are identical to the differential transformation formulas.

We can summarize the main advantages of the presented approach as follows:

- Using the presented algorithm, we are able to obtain approximate solution of the initial value problem. However, in some cases, there is the possibility to identify the unique solution in closed form.
- We do not need initial guess approximation and symbolic computation of multiple integrals or derivatives, hence less calculations are demanded compared to other popular semi-analytical methods (the variational iteration method, the homotopy perturbation method, the homotopy analysis method, the Adomian decomposition method).
- There is no need for numerical integration or differentiation either. Only arithmetical operations are used.
- In comparison to any purely numerical method, a specific advantage of this technique is that the approximate solution is always a function analytic near t_0.
- The algorithm is recurrent, so we use values computed in previous steps. Suitable arrangements can be made to reduce the necessary computational work. This fact was demonstrated in Example 2.

5 Conclusion

Results of this paper are based on combinatorial properties of Bell polynomials applied in the differential transformation theory. Modified version of the Faà di Bruno's formula with partial ordinary Bell polynomials was proved. Applicability of the algorithm was demonstrated on two particular examples of the initial value problem for differential equations with nonlinearity containing the unknown function. The algorithm can be generalized to other types of problems, e.g. boundary value problems.

6 Acknowledgements

The author was supported by the Grant FEKT-S-17-4225 of Faculty of Electrical Engineering and Communication, Brno University of Technology. This support is gratefully acknowledged.

References

[1] BELLEN A., ZENNARO M., Numerical Methods for Delay Differential Equations, Oxford University Press, Oxford, 2003, ISBN 0-19-850654-6.

[2] BUTCHER J.C., Numerical Methods for Ordinary Differential Equations, John Wiley & Sons, Chichester, 2003, ISBN 0-471-96758-0.

[3] CHARALAMBIDES Ch. A., Enumerative Combinatorics, Chapman & Hall, Boca Raton, 2002, ISBN 1-58488-290-5.

[4] COMTET L., Advanced Combinatorics - The Art of Finite and Infinite Expansions, D. Reidel, Dordrecht, 1974, ISBN 90-277-0380-9.

[5] HAIRER E., NØRSETT S.P., WANNER G., Solving ordinary differential equations I: Nonstiff problems, 2nd edition, Springer Verlag, Berlin, 1993, ISBN 3-540-56670-8.

[6] HAIRER E., WANNER G., Solving ordinary differential equations II: Stiff and differential-algebraic problems, second edition, Springer Verlag, Berlin, 1996, ISBN 3-540-60452-9.

[7] PETROPOULOU E.N., SIAFARIKAS P.D., TZIRTZILAKIS E.E., A "discretization" technique for the solution of ODEs, J. Math. Anal. Appl. 331 (2007), 279–296.

[8] PETROPOULOU E.N., SIAFARIKAS P.D., TZIRTZILAKIS E.E., A "discretization" technique for the solution of ODEs II, Numer. Funct. Anal. Optim. 30 (2009), 613–631.

[9] REBENDA, J., ŠMARDA, Z., A differential transformation approach for solving functional differential equations with multiple delays, Commun. Nonlinear Sci. Numer. Simul. 48 (2017), 246–257.

[10] REBENDA, J., ŠMARDA, Z., A semi-analytical approach for solving nonlinear systems of functional differential equations with delay, AIP Conference Proceedings 1863 (2017), art. nr. 530003, 4 pages. DOI: 10.1063/1.4992673

[11] REBENDA, J., ŠMARDA, Z., KHAN Y., A New Semi-analytical Approach for Numerical Solving of Cauchy Problem for Differential Equations with Delay, Filomat 31 (2017), 4725–4733.

[12] ŠAMAJOVÁ H., LI T., Oscillators near Hopf bifurcation, Communications : Scientific Letters of the University of Žilina 17 (2015), 83–87.

[13] ŠMARDA Z., KHAN Y., An efficient computational approach to solving singular initial value problems for Lane-Emden type equations, J. Comput. Appl. Math. 290 (2015), 65–73.

[14] WARNE P.G., POLIGNONE WARNE D.A., SOCHACKI J.S., PARKER G.E., CAROTHERS D.C., Explicit *A-Priori* Error Bounds and Adaptive Error Control for Approximation of Nonlinear Initial Value Differential Systems, Comput. Math. Appl. 52 (2006), 1695–1710.

[15] YANG X.-J., TENREIRO MACHADO J.A., SRIVASTAVA H.M., A new numerical technique for solving the local fractional diffusion equation: Two-dimensional extended differential transform approach, Appl. Math. Comput. 274 (2016), 143-–151.

Current address

Rebenda Josef, Mgr., PhD.
Department of Mathematics
Faculty of Electrical Engineering and Communication
Brno University of Technology
Technická 8, 616 00 Brno, Czech Republic
E-mail: josef.rebenda@ceitec.vutbr.cz

Slovak University of Technology in Bratislava
Faculty of Mechanical Engineering
17th Conference on Applied Mathematics **APLIMAT 2018**

Proceedings

SINUSOID CHARACTERISTICS IN APPLETS

RICHTARIKOVA Daniela (SK)

Abstract. The paper deals with visualisation of sinusoid characteristics. It discusses the importance of visualisation and interactive manipulations in the process of understanding the heart of the matter as the essential stage to acquire higher levels of knowledge and skills. Combining different teaching and learning aids and methods helps not only to supply but also develop various styles of cognition.

Keywords: visualisation, mathematical competencies, sinusoid characteristic

Mathematics Subject Classification: Primary 42B05; Secondary 33B10, 42B05

1 Introduction

In mathematical courses curriculum, there are some entries, which make a clue for substantial technical phenomena. They enable to describe them and even control through engineering processes. Such are sinusoids that are indisputably widely used, mostly for depiction of signals, distributed periodically, within various media. One can speak about electromagnetic waves, e.g. radio or TV waves; various radiations, e.g. light; mechanical waves – propagation of sound, corn waves, or water surface waves; not leaving out vibration of rotors or other mechanical systems; or one can speak even about gravitational waves, seismic waves or others. From mathematical point of view, signals can be generally expressed as real functions $f(t)$. Being elementary orthogonal functions they satisfy the conditions of the simplest mathematical approximation model $f(t)=\sum_k a_k \vartheta_k(t)$ with superposition property providing signal decomposition into $a_k \vartheta_k(t)$ parts. The response of system to the applied input function can be then explored by individual responses of parts, producing the net response in the sum. The study of vibrations usually begins with the key chapter on periodic signals that are articulated by means of Fourier series of harmonic motions. Harmonic motion as a function of time is sinusoidal; it is characterized by period T, angular frequency Ω, amplitude A and phase φ what allows to transform it into a function of frequency and analyse the signal by means of amplitude and phase spectra. Since relatively large variety of its possible analytical formulations, the understanding of each element in goniometric function representation is very important and leads to the success in the study and following engineering practice.

2 Sinusoids – Ambiguity and Pitfalls

2.1 Ambiguity Problem

Mathematically, sinusoids are the graphs of sine or cosine functions. As they have the same shape but a shift from the origin of the Cartesian coordinate system in basic position, it leads to more possibilities of their anallytical expression and gives rise to become a stumbling block for students. For instance the sinusoid in the figure 1 can be expressed as

$$\text{a) } 2\sin(t-\frac{\pi}{4})\text{; b) } 2\sin(t+(-\frac{\pi}{4}))\text{; c) } 2\cos(t-\frac{3\pi}{4})\text{; d) } 2\cos(t+(-\frac{3\pi}{4})),$$

where the value of the curve shift is restricted to the interval $\langle -\pi, \pi \rangle$. Moreover, taking into account the orthogonality of (cos *t*, sin *t*) system, it can be also expressed as $\sqrt{2}\sin(t)-\sqrt{2}\cos(t)$.

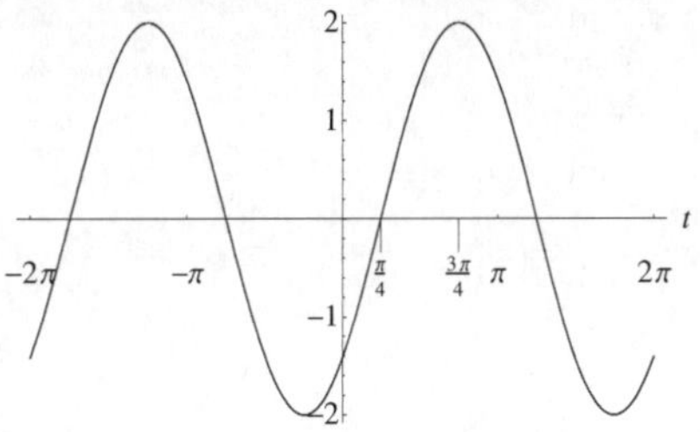

Fig. 1. Sinusoid

$$2\sin(t-\frac{\pi}{4})=2\sin(t+(-\frac{\pi}{4}))=2\cos(t-\frac{3\pi}{4})=2\cos(t+(-\frac{3\pi}{4}))=\sqrt{2}\sin(t)-\sqrt{2}\cos(t)$$

One posssibility, how to eliminate the ambiguity is to prefer only one form. Nevertheless, each representation has its advantages, what greatly facilitates the work with curve, so nobody should want to get rid of it. For instance, sine representation has the intuitive non-confusable position of starting point [0,0], and the shift could be easily recognized. On the other hand, the cosine repesentation could stand for signals that start at full power, and moreover, they correspond to exponentional (complex) representation $a\cos(\omega t+\varphi)=c\,e^{\omega t}+\overline{c}\,e^{\omega t}$, where c and $\overline{c}$ are complex conjugates and φ is also the argument of complex number *c* in polar form.

Going to simple and for students more transparent difference, in representation $a\sin(t-\varphi)$, φ stands directly for intersection point of the sinusoid with axis *t*, while taking into account the representation $a\sin(t+\varphi)$, the intersection point has to have the opposite sign, $-\varphi$.

What can help is the fact, that the sine and cosine graphs are shifted by $\pi/2$. What complicates the situation is that although a student is concious of the shift, she/he has to be aware of the shift direction.

2.2 Applets

To reveal this situation we decided to prepare applets as learning aids for students to help them better understand the role of used coefficients in the formulas.

Usually freshmen come to study at technical universities with relatively good familiarity with two concepts of sine and cosine functions. First, it is a practical concept of side ratios in right-angled triangle, sometimes also demonstrated within the unit circle. The second one is the approach through a standard sinusoidal graph of periodical real function in one variable, however perceived isolated, not in relation to the first concept, and in more abstract way. In calculus course, students work with graphs, they learn to understand the meaning, and changes of the graph with respect to coefficients *a, b, c* in expressions $a\sin x$, $a\sin(x+c)$, $a\sin(bx)$ (the same for cosine). The most strange for students is to understand the coefficient *b* responsible for the length of periodicity interval. A problem also occurs when putting all coefficients together in $a\sin(bx+c)$, especially in connection to shift of the starting/reference point of sinusoid (fig.2).

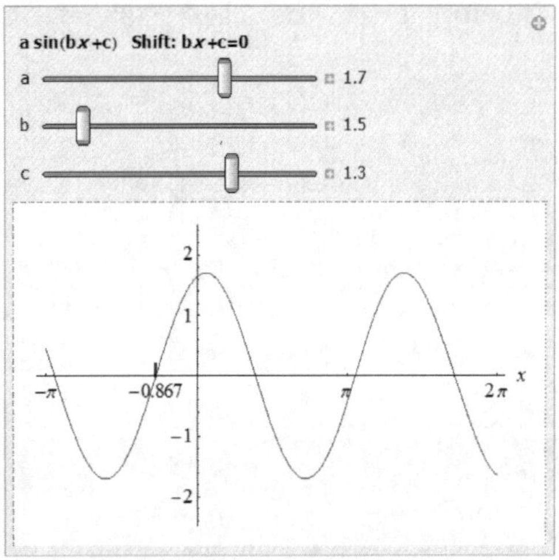

Fig. 2. Applet: $a\sin(bx+c)$, playing with coefficients.

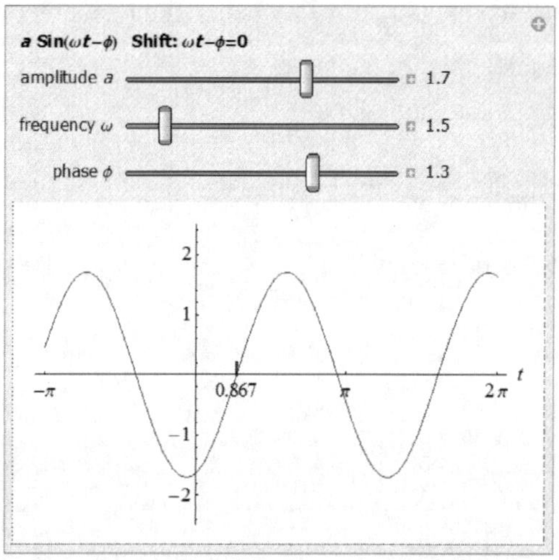

Fig. 3. Applet: harmonic signal $a\sin(\omega t-\varphi)$, reference point shift calculation.

Studying the elementary theory of signals, students of higher years can directly apply their previous skills. General coefficients *a, b, c* are substituted by technical terms: amplitude *a*, frequency ω, and phase φ (fig. 3).

Usually, there are no difficulties until the moment of harmonic analysis, when one has to decompose the periodic signal onto harmonic components, to calculate their frequencies, amplitudes and phases. The most problematic is calculation of phase, when a student is confronted with different formulas, even producing the different results. Alleviating the confusion, we have very good experience with kinematic way of sinusoid creation for sine and cosine representation, helping to achieve the necessary control. In presented applets, a periodical movement is represented by the movement along the unit circle. Simultauosly, the corresponding sinusoid is recorded in the Cartesian coordinate system. Guide angle *t* of the moving point is at each position transformed into the independent variable *t* of sine function. The value sin *t* is simply transferred horizontally from the vertical axis (fig. 3), while the value cos t is transferred vertically from the horizontal axis onto the vertical axis of the sinusoid graph (fig. 4). Recording the corresponding values we obtain desired sinusoids: sin t, and the rotated graph cos t, that can be then turned into its traditional position. For better recognition, the sine and cosine relations in the unit circle can be supported by image of the right angle triangle (seen in fig. 7)

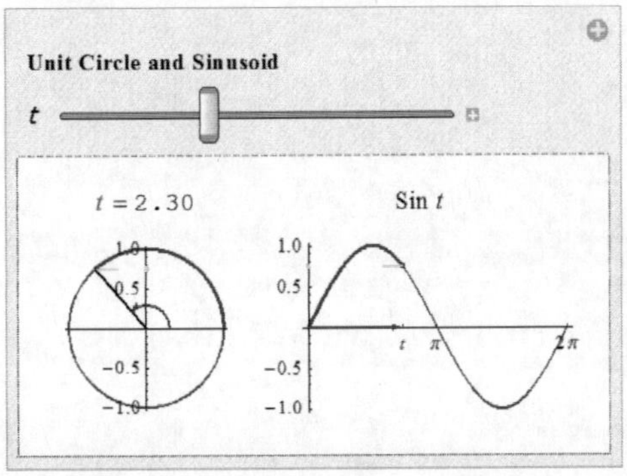

Fig. 4. Applet: sin t; kinematic approach.

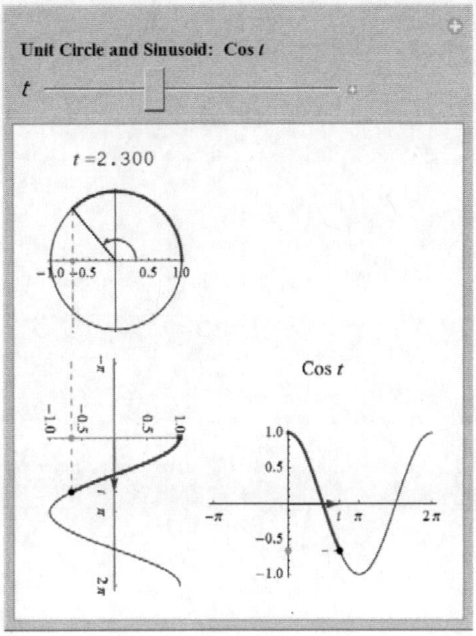

Fig. 5. Applet: cos t; kinematic approach.

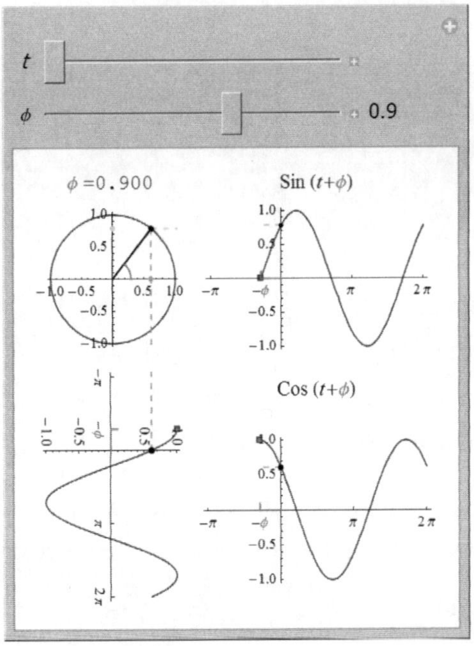

Fig. 6. Applet: $\sin(t+\varphi)$ & $\cos(t+\varphi)$,

one angle φ *2 different curves* $\sin\varphi$, $\cos\varphi$.

Innitial position of the movement.

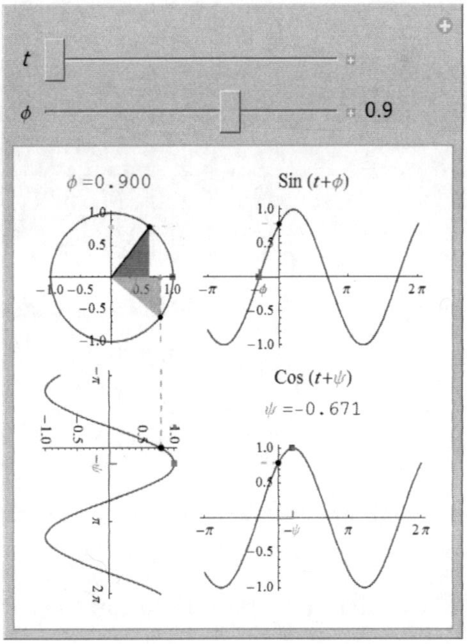

Fig. 7. Applet: $\sin(t+\varphi)$ & $\cos(t+\psi)$,

one curve 2 different angles φ, ψ.

Innitial position of the movement.

The shift of the sinusoidal initial reference point by the value φ can be demonstrated on the same principle. Starting the movement from the angle φ (fig. 6) on the unit circle produces two different curves, where the initial reference point on each curve is shifted by the same value φ or $-\varphi$ (depends on argument formulation). On the other hand, if we want to obtain the same sinusoid (fig. 7), it is necessary to start from different angle for each representation. As sine and cosine curves are shifted by $\pi/2$, the initial values of angles differ also by $\pi/2$.

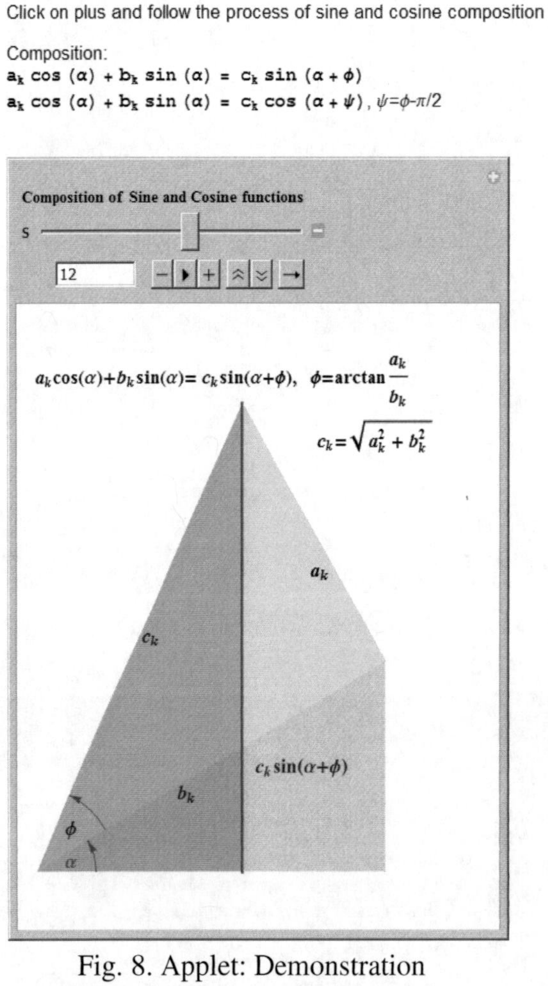

Fig. 8. Applet: Demonstration

$$c_k \sin(\alpha + \varphi) = a_k \cos\alpha + b_k \sin\alpha$$

$$c_k = \sqrt{a_k^2 + b_k^2}, \quad \tan\varphi = a_k / b_k .$$

While relations on composition of sine and cosine waves in trigonometric form of harmonic motion can be demonstrated by composition of right-angled triangles (fig. 8), phase counting itself holds another pitfall: from $\tan\varphi = a_k / b_k$ basically $\varphi \in \left(-\frac{\pi}{2}, \frac{\pi}{2}\right)$ ($\varphi \in \left(-\frac{\pi}{2} + \mathrm{k}\pi, \frac{\pi}{2} + \mathrm{k}\pi\right), k \in N$), being the value of shifting sinusoid, basically $\varphi \in (-\pi, \pi)$, what always requires to consider the current part of the interval.

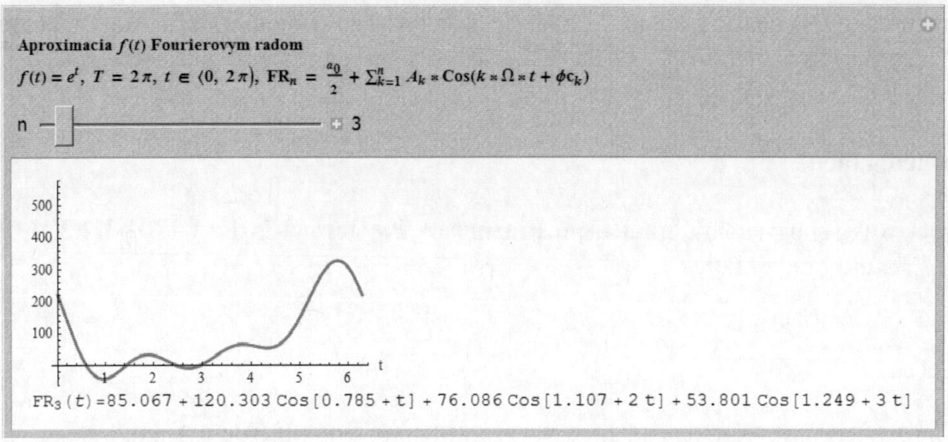

Fig. 9. Applet: Approximation $f(t)$ with Fourier series. The sum of first 4 items.

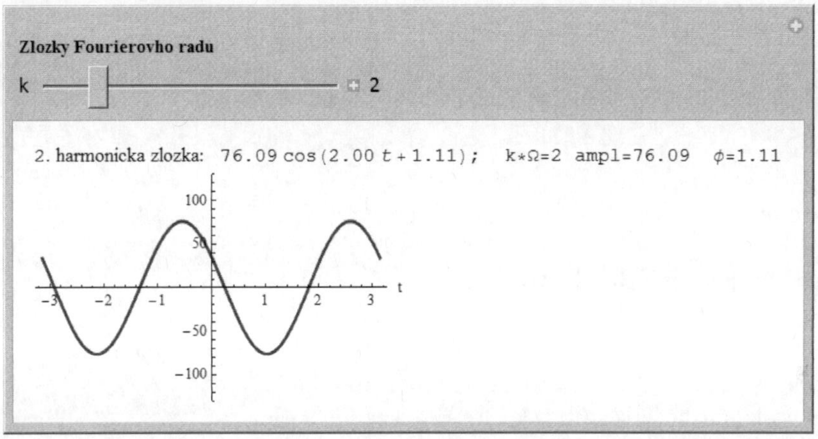

Fig. 10. Applet: Terms of Fourier series. Second harmonic and its characteristics.

3 Decisive Mathematical Competency

Elementary theory on signals is usually taught by technicians using technical textbooks, where mathematical details are rarely explained. Due to multiformity of analytical expressions, usage of formulas is not always consistent and frequently results in confusions. In this context, mathematical knowledge is inevitable. Presented applets can be highly conducive to deep comprehension. On the other side, engineering context directly motivates to study and gives actual background.

Nowadays, overwhelmed by modern technology, physical models have been gradually replaced by virtual teaching and learning aids, which allow to combine visual, read/(formulas) and manipulative approach, often equipped also by auditory output, attacking all senses not only supplying preferred learning styles but as well, involving and developing not preferred ones, with strong potential to affect the cognition. However, to aquire this, they do not work autonomly. Carefully formulated tasks, Socratic questioning, provoking to critical thinking, supplemented by necessary facts, actual background contribute to building mathematical competency part by part and step by step. However, working in passive mode is not sufficient.

Acknowledgement

The paper was supported by grant from Erazmus + Project no. 2015-1-FI01-KA203-009044 entitled " Future Mathematics".

References

[1] ONDRÁČEK, O. Signály a sústavy. STU Bratislava, 2008. ISBN 978-80-227-2956-7

[2] ALPERS, B., DEMLOVA, M., FANT, C.-H., GUSTAFSSON, TH., LAWSON, D., MUSTOE, L., OLSSON-LEHTONEN, B., ROBINSON, C., VELICHOVA, D. A Framework for Mathematics Curricula in Engineering Education, SEFI, Brussels, 2013. ISBN: 978-2-87352-007-6.

[3] KRATHWOHL D.R. *A Revision of Bloom's Taxonomy: An Overview, Theory Into. Practice*, 2002. https://teaching.yale-nus.edu.sg/wp-content/uploads/sites/25/2016/02/A-Revision-of-Blooms-Taxonomy_An-Overview.pdf.

Current address

Richtáriková Daniela, RNDr., PhD.
Slovak University of Technology
Faculty of Mechanical Engineering, Institute of Mathematics and Physics
Námestie slobody 17, 812 31 Bratislava, Slovak Republic
E-mail: daniela.richtarikova@stuba.sk

Slovak University of Technology in Bratislava
Faculty of Mechanical Engineering
17th Conference on Applied Mathematics **APLIMAT 2018**

Proceedings

FUZZY C-MEANS CLUSTERING. TECHNIQUE AND EVALUATING RESULTS.

RIHOVA Elena (CZ), MALEC Miloslav (CZ)

Abstract. The present study aims to provide a better understanding of the fuzzy C-means clustering, their technique and evaluating their results. In order to do so, the fuzzy C-means clustering technique was described in detail, then those technique was applicated on the data sets, as generated, as real. In doing so, novel insights into the key drivers of fuzzy C-means clustering and evaluating of fuzzy clustering results. For determining the right number of clusters, were used indices based not only on membership function, but indices based on membership function and data sets. In most cases, those type of indices were able to recognize the right number of clusters. Indices involving the membership values and data sets were detected as more successful indices, although not always successful.

Keywords: fuzzy clustering, C-means, number of clusters

Mathematics Subject Classification: 62H30, 62A86

1 Possibilistic Fuzzy C-means Clustering

Clustering is an unsupervised process and can be classified into two categories: hard and fuzzy clustering. Although those two different clustering categories, they have the common goal. The task of clustering is to divide the set in to the optimal number of groups. The objects in the same group (this group is called a cluster) must be more similar to each other than to those objects in other clusters. In this article will deal with the second category of clustering - fuzzy clustering.

Fuzzy clustering is a form of clustering in which each object can belong to more than one cluster. The main concept in fuzzy clustering is based on membership degrees. These membership grades indicate the degree to which data points belong to each cluster. A better reading of the memberships, avoiding misinterpretations, would be (Höppner, Klawonn, Kruse and Runkler, 1999): If the object x_i has to be assigned to a cluster, then with the probability u_{ij} to the cluster j. However, the normalization of u_{ij} can lead to unexpected bad result in finding and discovering outliners. The membership values affects the clustering results.

By dropping the normalization constraint (1) in the following definition one tries to achieve a more intuitive assignment of degrees of membership and to avoid undesirable normalization effects (Oliveira, 2007).

Let $X = \{x_1; \ldots ; x_n\}$ is a data set with k clusters, where number of clusters is $1 < k < n$ and represented by the fuzzy sets μ_{Ch}. Hence, $U_p = (u_{ij}) = \mu_{Ch}(x_i)$ is a possibilistic partition of X if:

$$\sum_{i=1}^{n} u_{ij} > 0 \text{, } \forall j \in \{1,...,k\}, \tag{1}$$

holds. The degree of the object x_i to cluster C_h is u_{ij} and achieves value [0, 1]. The membership degrees for one datum now resemble the possibility (in the sense of possibility theory (Dubois and Prade, 1988) of its being a member of the corresponding cluster (Daver and Krishnapuram,1997; Krishnapuram and Keller, 1992).Consequently, J would not be appropriate for this type of fuzzy clustering. The normalization term leads to following problem: J would reach its minimum for $u_{ij} = 0$ for all objects in data set, it means no one object is not assigned to cluster. Consequently, clusters are empty. According Krishnapuram and Keller (Krishnapuram and Keller, 1992) to avoid this problem the objective function must be modified to:

$$J = \sum_{j=1}^{k}\sum_{i=1}^{n} u_{ij}^{q} D_{ij}^{2} + \sum_{j=1}^{k} \eta_i \sum_{i=1}^{n} (1-u_{ij})^{q} \text{ ,} \tag{2}$$

where $\eta_i > 0$ $(i = 1; \ldots ; k)$.

The first part of this objective function leads to a minimization of the weighted distances. The second part puts down the first part of this function: when the first part leads to 1, the second part suppresses it: $(1-u_{ij})^q$. In tandem with the first term the high membership can be expected especially for data that are close to their clusters, since with a high degree of belonging the weighted distance to a closer cluster is smaller than to clusters further away (Oliveira, 2007). The updating the membership degrees that is derived from J by setting its derivative to0 is (Krishnapuram and Keller, 1992):

$$u_{ij} = \frac{1}{1+\left(\frac{D_{ij}^2}{\eta_i}\right)^{\frac{1}{q-1}}} \quad (3)$$

Eq. 3.8 shows that the membership u_{ij} (belonging the object $x_i$to cluster C_h) depends on the distance from this object to cluster. Small value of the distance (strong similarity) leads to high membership degree, and the large value of distance means to low membership value. And the other one parameter is η_i - the distance from object x_j to the cluster C_h ,which membership degree should be 0,5.

Since that value of membership can be seen as definite assignment to a cluster, the permitted extension of the cluster can be controlled with this parameter (Oliveira, 2007), but the parameter η_i may have the different geometrical interpretation, this interpretation depends on the cluster shape. In case of the possibilistic *k*-means, the clusters diameter is $\sqrt{\eta_i}$ (Höppner, Klawonn, Kruse and Runkler 1999). If a kind of information about clusters is known a prior, η_i can be set to any value. In case the same optionalities of all clusters this parameter can be the same for all clusters. But in the real world this information about cluster optionalities is unknown in advance. Hence, parameter η_i should be calculated. To calculate the optional value of η_i can be used a probabilistic clustering model. The parameters η_i are then estimated by the fuzzy intra-cluster distance using the fuzzy memberships matrix U_f as it has been determined by the probabilistic counterpart of the chosen possibilistic algorithm (Krishnapuram and Keller, 1992):

$$\eta_i = \frac{\sum_{i=1}^{n} u_{ij}^q D_{ij}^2}{\sum_{i=1}^{n} u_{ij}^q} \quad (4)$$

2 Validation indices

The problem for finding an optimal number of clusters C^* is usually called cluster validity problem. In order to solve the cluster validity problem, validity indices must enclose, take into account, some specific are as which enable to solve this problem successfully. Those areas are: compactness, separation, noise and overlap. A lot of validity indices for fuzzy clustering are. Early indices such as the partition coefficient and classification entropy make use only of membership values. The main advantage of those indices they are easy to compute. Now, it is widely accepted that a better definition of a validity index always consider both partition matrix U and the data set itself. In this paper we will go with the classification of indices by Wang (Wang, 2007). The first group of those indices is indices involving only the membership values.

Partition coefficient (*PC*). *PC* index is based on minimizing the overall content of pair wise fuzzy intersection in *U*, the partition matrix**.** The index was defined as

$$PC = \frac{1}{n}\sum_{j=1}^{k}\sum_{i=1}^{n} u_{ij}^{2} \tag{5}$$

Those index indicates the average relative amount of membership sharing done between pairs of fuzzy subsets in *U*, by combining into a single number, the average contents of pairs of fuzzy algebraic products. In general, we find an optimal cluster number *C** by solving $\max_{2 \le C \le n-1} PC$ to produce the best clustering performance for the data set *X*.¨

Modified *PC* (*PC$_{mod}$*) index (Dave, 1992) as a modification of the previous one and can take values ⟨0,1⟩:

$$PC_{\text{mod}}(C) = 1 - \frac{C}{C-1}(1 - PC(C)). \tag{6}$$

The optimal number of cluster *C** is defined by solving of:

$$\max_{2 \le C \le n=1} PC_{\text{mod}}(C). \tag{7}$$

When the variability in clusters is small, this modified Dunn's coefficient PC_{mod} usually determined the number of clusters correctly (Řezanková, Húsek, 2012).When the cluster variability is greater, the normalized Dunn's coefficient usually achieved its highest value for the highest possible number of clusters. (Řezanková, Húsek, 2012) Validity indices involving the membership values and the data set are more successful compared with validity indices involving the only membership values. (Wang, 2007). The most succesfull index from those group is **Xie and Beni (*XB*) validity index.** *XB* index was proposed in by Xie and Beni (Xie, Beni, 1991) with $q = 2$ and modified by Pal and Bezdek (Pal, Bezdek, 1995) was defined as:

$$XB = \frac{J_m(u,v)/n}{Sep(v)} = \frac{\sum_{i=1}^{k}\sum_{j=1}^{n} u_{ij}^{2}\left\|x_j - x_i\right\|^2}{n.\min_{i,j}\left\|v_i - v_j\right\|^2} \tag{8}$$

XB includes two components: compactness in the numerator and separation, which is represented in denominator. The optimal number of clusters C^* can be find by solving $\min_{2 \leq C \leq n-1} XB$ to produce the best clustering performance for the data set X. XB index has tendency to monotonically decrease with increasing number of clusters.

3 Case Study

The main objective of third section is to compare the performance of those indices in evaluating the optimal number of clusters. In the following experiments presented here, were tested the cluster validity indices for 5 well-known data set from *UCI Machine Learning Repository* and for 5 generated data sets.

The data set Abalone:

This data set comes from an original, non-machine study. Those data was made in 1994 by Warwick J Nash and collective. Data set Abalone contains 8 variables:

1. length (continuous in mm) Longest shell measurement,
2. diameter (continuous in mm) perpendicular to length,
3. height (continuous in mm) with meat in shell,
4. whole weight (continuous in grams) whole abalone,
5. shucked weight (continuous in grams) weight of meat,
6. viscera weight (continuous in grams) gut weight (after bleeding),
7. shell weight (continuous in grams) after being dried,
8. rings (integer) +1.5 gives the age in years.

Total number of samples is 4177. Abalone data set contains two clusters. Data set samples are highly overlapped and none correlated.

The data set Breast Tissue:

This database includes 106 instances. Six classes of freshly excised tissue were studied using electrical impedance measurements:

1. carcinoma,
2. fibro-adenoma,
3. mastopathy,
4. glandular,
5. connective,
6. adipose.

The data set can be used for predicting the classification of either the original 6 classes or of 4 classes by merging together the fibro-adenoma, mastopathy and glandular classes whose discrimination is not important (they cannot be accurately discriminated anyway). All those data were divided into 6 groups with help of 9 variables:

1. impedivity (ohm) at zero frequency,
2. phase angle at 500 khz,
3. high-frequency slope of phase angle,

4. impedance distance between spectral ends,
5. area under spectrum,
6. area normalized,
7. maximum of the spectrum,
8. distance between impedivity and real part of the maximum frequency object,
9. length of the spectral curve.

The data set Cardiography:

The represented data set obtains 2126 samples and 21 variables, which are divided on 3 clusters by fetal state class code: normal, suspect and pathologic. Data set samples are middling overlapped and none correlated.

The data set Connectionist Bench:

For this data set 15 people were gradually microphones into 6 series, where each series has 11 vowels pronounced. The data file contains of 990 subjects (voiced vowels), where for each vowel recorded 10 characteristics. Because the multicollinearity is not present in the data file, all of the variables will be enter into the analysis.

The data set Wholesale customers

Current data set contains 440 objects, divined into 2 clusters by following attributes:

1. fresh: annual spending (m.u.) on fresh products,
2. milk: annual spending (m.u.) on milk products,
3. grocery: annual spending (m.u.)on grocery products,
4. frozen: annual spending (m.u.)on frozen products,
5. detergents paper: annual spending (m.u.) on detergents and paper products,
6. delicatessen: annual spending (m.u.)on and delicatessen products (continuous).

3.2 Generated Data Sets

The data set with 8 clusters (Example_2):

Data set contains 320 objects, which were put in 8 clusters. The number of variables is 8.All data sets have the normal distribution.

The data set Example_2, obtains overlapped clusters.

The data set with 9 clusters (Example_7):

Current data set contains 360 objects, which were put in 9 clusters. The number of variables is 8. Data set have the normal distribution. The data set Example_7 obtains middling - overlapped clusters.

The data set with 10 clusters (Example_11):

The data set contains 400 objects, which were put in 10 clusters. The number of variables is 8 with normal distribution. The data set Example_11obtain middling - overlapped clusters.

The data set with 11 clusters (Example_19):

The data set contains 440 objects, which were put in 9 clusters. The number of variables is 8 with normal distribution. The data set Example_19 obtain middling - overlapped clusters.

The data set with 12 clusters (Example_25):

The data set contains 480 objects, which were put in 9 clusters. The number of variables is 8 with normal distribution. The data set Example_25 obtain middling - overlapped clusters.

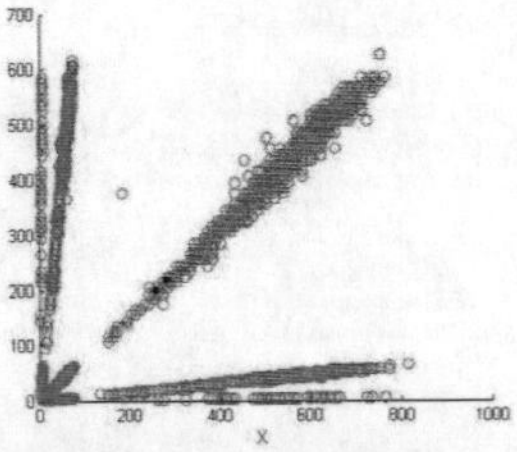

Fig. 1. The data set Abalone.

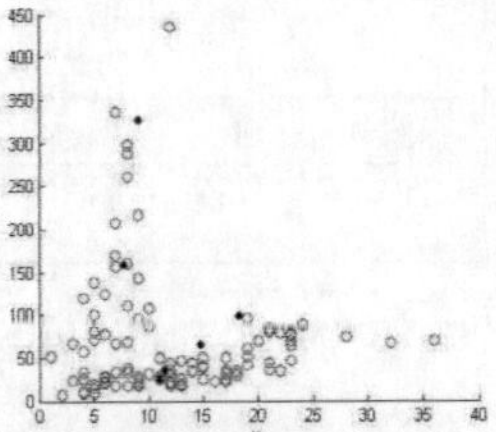

Fig. 2. The data set Breast Tissue.

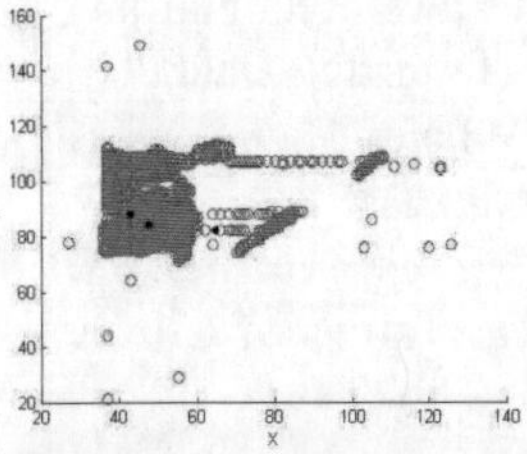

Fig. 3. The data set Cardiography.

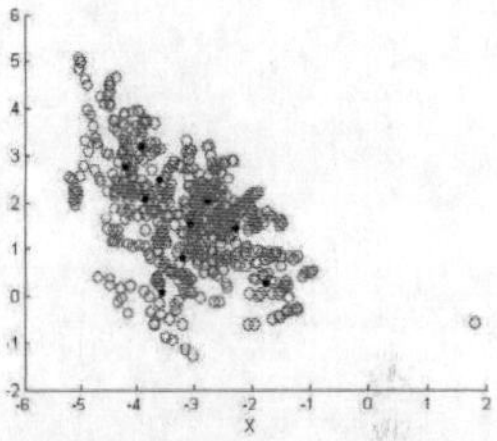

Fig. 4. The data set Connectionist Bench (Vowel Recognition – Deterring Data).

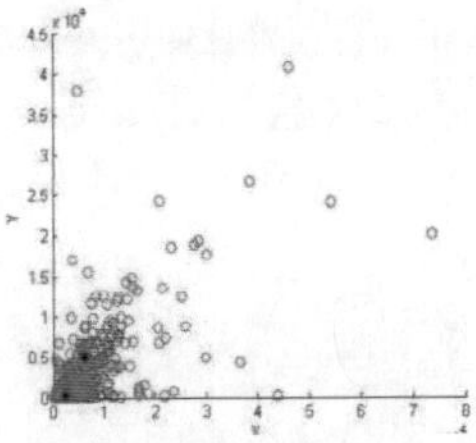

Fig. 5. The data set Wholesale customers.

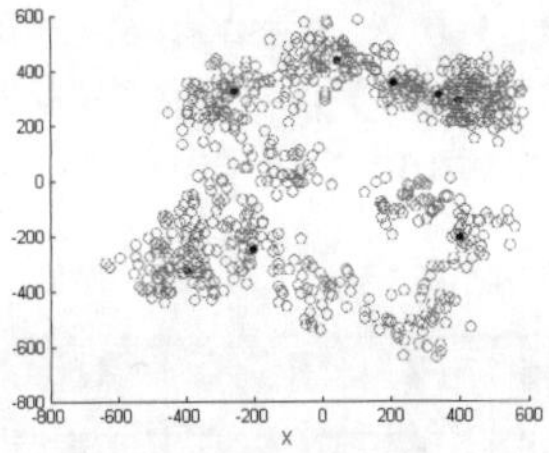

Fig. 6. Example_2: The data set with 8 clusters (obtaining 320 objects); overlapped clusters with normal distribution.

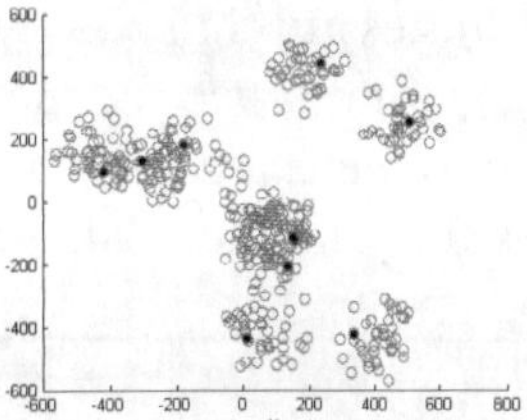

Fig. 7. Example_7: The data set with 9 clusters (obtaining 360 objects);middling - overlapped clusters with normal distribution.

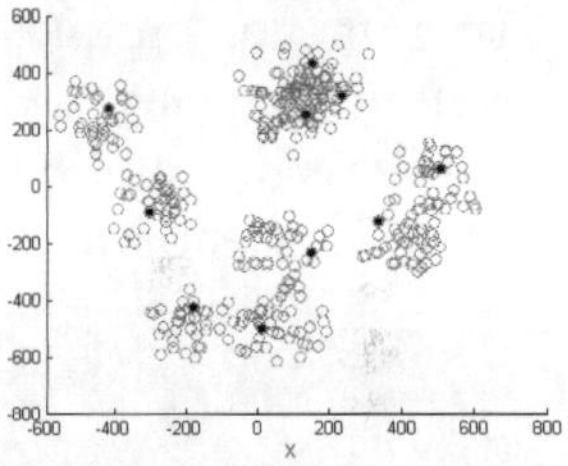

Fig. 8. Example_11: The data set with 10 clusters (obtaining 400 objects) middling - overlapped clusters with normal distribution.

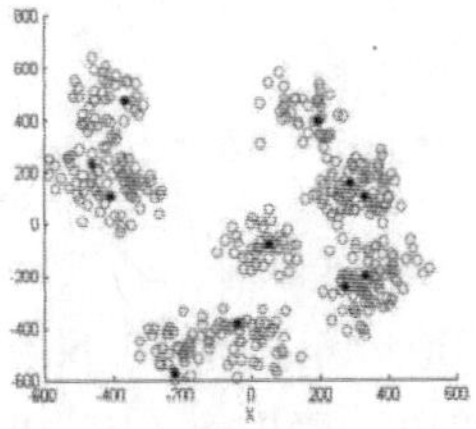

Fig. 9. Example_19: The data set with 11 clusters (obtaining 440 objects) middling - overlapped clusters with normal distribution.

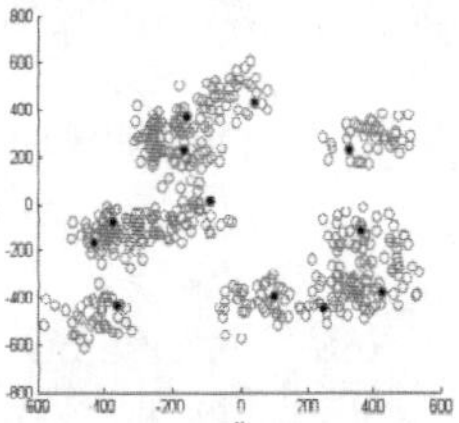

Fig. 10.: Example_25: The data set with 12 clusters ; (obtaining 480 objects); middling - overlapped clusters with normal distribution.

4 Results and Discussion

The result of evaluating fuzzy C-means clustering results are shown in Tab. 1. and 2. In Tab. 1. are shown the number of clusters, which are calculated for *PC*, PC_{mod}, and *XB* index for real data sets.

Data set	*C**	*PC*	PC_{mod}	*XB*
Abalone	2	2	2	2
Breast Tissue	6	2	2	2
Cardiotocography	3	2	2	3
Connectionist Bench (Vowel Recognition - Deterring Data)	11	3	2	8
Wholesale customers	2	2	3	2
Successfulness,%	-	40%	20%	60%

Tab. 1. Values of *C* by validity indices for real data sets.

Summing up, we can determine in which of the abovementioned cases the indices worked incorrectly, in other words, to find out what affected it.

To sum up, the most successful indexes are *XB* with success rates of 60%.The worst-performing index is PC_{mod} with a success rate of 20%. However the *PC* index was better with a success rate of 40%. In Table 2 are shown the number of clusters, which are calculated for *PC*, PC_{mod}, and *XB* index for generated data sets.

Data set	*C**	*PC*	PC_{mod}	*XB*
Example_2	8	3	8	8
Example_7	9	2	2	9
Example_11	10	2	2	2
Example_19	11	4	4	11
Example_25	12	2	6	12
Successfulness, %	-	0%	20%	60%

Tab. 2. Values of *C* by validity indices for generated data sets.

The number of clusters ranged from 8 to 12. As Bezdek (Bezdek, 1987) pointed out: the number of variables does not affect the fuzzy *C*-means clustering results. For this reason, the same number of variables was chosen for all generated data sets – eight. However, the degree of overlapping is different for each data set as shown in Fig 1 – 10. Fuzzy *C*-means clustering with Euclidean distance was applied to each data set. To sum up, the most successful indexes are *XB* with success rates of 60%.The worst-performing index is *PC* with a success rate of 0%. However the modified *PC* index was better with a success rate of 20%.

The validation of clustering structures is the most difficult and frustrating part of cluster analysis. That's why the issue of the definition of the indexes, which would be good for data with large variability and a large number of clusters, has not yet been resolved. As shown by the results of the approach, which we suggest, this modification can increase the efficiency of the correct determination of the number of clusters.

From the experiment's results, it can be drawn that the modification method determines the number of clusters correctly. The next part of current study is represented in following subchapter.
Study the data can help to define what behavior we can expect from the clusters with different overlap but with normal distribution. Let's observe how the behavior of those indices changes with an increasing number of clusters.

References

[1] URL, http://www.ics.uci.edu/~mlearn/databases.html

[2] BEZDEK, J.C., HATHAWAY, R.J., SABIN, M.J. (1987). Convergence theory for fuzzy *c*-means: counter-examples and repairs. *IEEE Transaction on Systems*, vol.17, no.5, p.: 873–877.

[3] DAVE, R.N., BHASWAN, K. (1992) Adaptive fuzzy *c*-shells clustering and detection of ellipses. *IEEE Transactions of Neural Networks,* vol.3, no.5, p: 643–662.

[4] DAVEГ, R., KRISHNAPURAM, R. (1997) Robust clustering methods: a unified view. *IEEE Transactions on Fuzzy Systems*, vol. 5, no.2, p: 270–293.

[5] DUBOIS, D., PRADE, H., (1988). *Possibility Theory: An Approach to Computerized Processing of Uncertainty*. New York: Plenum Press.

[6] HÖPPNER, F., KLAWONN, F., KRUSE, R., AND RUNKLER, T. (1999). *Fuzzy Cluster Analysis*. John Wiley & Sons, Inc., New York

[7] KRISHNAPURAM, R.,NASRAOUI, O., KELLER, J. (1992). The fuzzy *c* spherical shells algorithm: a new approach. *IEEE Transactions of Neural Networks,* vol.3, no. 5, p.: 663–671.

[8] OLIVEIRA, J.W. (2007) *Advances in Fuzzy Clustering and its Applications*. London: John Wiley & Sons, Ltd.

[9] PAL, N.R., BEZDEK, J.C. (1995). On cluster validity for fuzzy *C*-means model. *IEEE Transactions of Fuzzy Systems*, vol.3, no.3, p.: 370–379.

[10]REZANKOVA, H., HUSEK, D.(2012): Fuzzy Clustering: Determining the Number of Clusters. In: *Computational Aspects of Social Networks*. Sao Carlos: Research Publishing Services, p.: 277--282.

[11]WANG, W., ZHANG,Y. (2007). On fuzzy cluster validity indices. *Fuzzy Sets and Systems.* vol. 158, no. 19. P.: 2095 – 2110.

[12]XIE, X.L., BENI, G. (1991). A validity measure for fuzzy clustering. *IEEE Transactions of Pattern Analalysis and Machine Intelligence*, vol.13, p.: 841–847.

Current address

Rihova Elena, Mgr., Ph.D.
University of Business in Prague
Department of Information Technologies and Analytical Methods,
Spalená 76/14 110 00, Praha, Czech Republic
E-mail: rihova.elena@gmail.com

Malec Miloslav, doc. RNDr., CSc.
University of Business in Prague
Department of Information Technologies and Analytical Methods,
Spalená 76/14 110 00, Praha, Czech Republic
E-mail: malecm@vso-praha.eu

Slovak University of Technology in Bratislava
Faculty of Mechanical Engineering
17th Conference on Applied Mathematics **APLIMAT 2018**

Proceedings

RUBRICS AS SELF-ASSESSMENT TOOL OF MATHEMATICAL EDUCATION

ŠABÍKOVÁ Henrieta (SK), JUDINOVÁ Lenka (SK) , ČERETKOVÁ Soňa (SK)

Abstract. The source of the majority of disputes, misunderstandings and disagreements in a mathematics teacher-pupil relationship is testing, evaluation and classification of pupils. These parts of education are nowadays marked by defects which are an outcome of constant changes in lives of teachers, pupils and their parents. The focal point of education is no longer just acquirement of facts and but also development of teaching quality, creation of a good base for lifelong learning, increase of creativity, communication, digital literacy, the ability to resolve problems. The feedback accommodated to these demands on learning in a 21st century school should be new trend in university education of next generation mathematics teachers. According to a conventional approach, evaluation is solely an assessment of learning results, but evaluation ought to support the progress of the learner, as well as the process of learning itself. At any rate, the teacher should evaluate pupils' outputs while providing them with a space for self-assessment. Our experiment gives answers to fundamental questions: Are pupils currently able to evaluate their own knowledge and skills? Do they underestimate or overestimate themselves? Is self-assessment affected by their age or gender? Results of the experiment could be the initial motion for changing the current approach of teachers, for preparing next generation of mathematics teachers to evaluation of pupils and pointing it to the right direction of change – to the developing assessment and to new assessment tools, e.g. Assessment

Keywords: evaluation, self-assessment, experiment, assessment rubrics, mathematical education

Mathematics Subject Classification: Primary 97B10; Secondary 97C60

1 Introduction

The source of the majority of disputes, misunderstandings and disagreements in a mathematics teacher-pupil relationship, sometimes involving parents as well, is testing, evaluation and classification of pupils. These parts of education are nowadays marked by certain defects which are an outcome of constant changes in lives of teachers, pupils and their parents. The focal point of education is no longer just acquirement of facts and but also development of teaching quality, creation of a good base for lifelong learning, increase of creativity, communication, digital literacy, the ability to resolve problems and the skill of working with information. The feedback accommodated to these demands on learning in a 21st century school should be new trend in university education of next generation mathematics teachers. According to a conventional approach, evaluation is solely an assessment of learning results, however it should convey a more important message. Evaluation ought to support the progress of the learner, as well as the process of learning itself. At any rate, the teacher should evaluate pupils' outputs while providing them with a space for self-assessment. After all, it is self-assessment that is a tool for improvement of learning feedback. The new generation mathematics teachers should be notified with requirement and possibility of assessment changes and should be prepared to manage pupils′ self-assessment.

2 The level of pupils' self-assessment skills

This experiment gives answers to fundamental questions: Are pupils currently able to evaluate their own knowledge and skills? Do they underestimate or overestimate themselves? Is self-assessment affected by their age or gender? Results of the experiment, statistically very easily derived into graphs, could be the initial motion for changing the current approach of teachers and for preparing next generation of mathematics teachers to evaluation of pupils and pointing it to the right direction of change – to the developing assessment and to new assessment tools, e.g. Assessment Rubrics.

2.1 Rubric as a self-assessment tool

The aim of experiment is to determine the level of pupils' self-assessment skills. The most appropriate instrument for determining the level of pupils' self-assessment skills is Assessment Rubric. Whilst assembling the Rubric we first set categories, i.e. mathematical task and after that, we elaborated individual criteria of the categories, i.e. problems of varying difficulty. Thus, our Assessment Rubric contained four criteria levels in four categories. According to this Rubric, we prepared an exam with assignments that matched each problem, that is, the criteria of each type of mathematical task, that is, the category. Categories represented a type of mathematical task and each one was made up of four problems varying in difficulty – evaluation criteria.

2. 2 Hypothesis

Pupils of the first grade up to the fourth grade of High school of business were given an Assessment Rubric containing categories and evaluation criteria during a mathematics class just before an exam. The pupils were asked to assess their abilities and current knowledge. That essentially meaning to guess what grade they would get from the exam if the problems on the exam were picked from the Assessment Rubric they had seen. Founded on pedagogical experience, we assumed that pupils were not capable of objective assessing of their knowledge, and therefore unable to classify themselves into a fitting category. After they had guessed, the pupils took the exam. We subsequently graded and classified the pupils based on their filled exam sheets.

2. 3 Results of the experiment

Based on the acquired entries from pupils' self-assessment task and from grading their filled exam sheets, we drew a comparison between the actual results of pupils' examining and the results that pupils had expected before being handed the exam.

2.3.1 Pupils in the age category of 15 years old

Graph 1. contains the results of self-assessment of pupils in the age category of 15 years old. It shows that, in percentage representation, 30.8% of pupils had guessed correctly. The same percentage of pupils had assumed they would get a mark two grades worse than what they actually received. 23% had assumed to receive a mark one grade worse. Only 7.7% had assumed a mark one or two grades better than what they received.

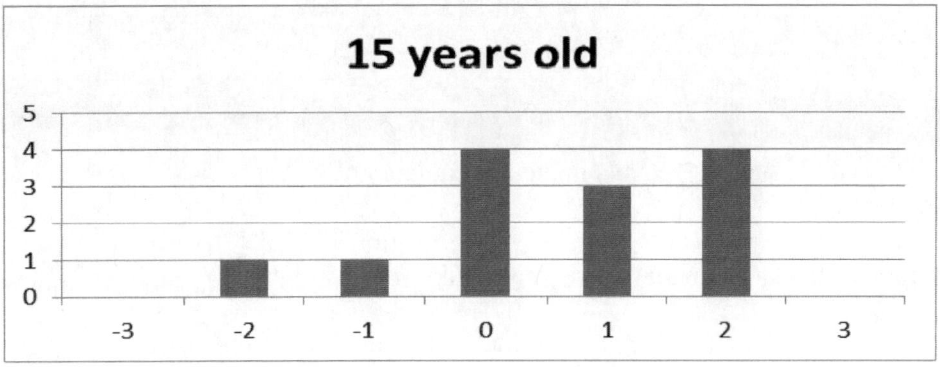

Graph 1. Pupils in the age category of 15 years old.

2. 3. 2 Pupils in the age category of 16 years old

Graph 2. demonstrates pupils in the age category of 16 years old. It points out that 39.1% of pupils had guessed their mark accurately, 34.8% had guessed they would get a mark one grade worse, 17.4% had assumed to get a mark one grade better and the last 8.7% of pupils had expected a mark two grades worse than the actual one they received.

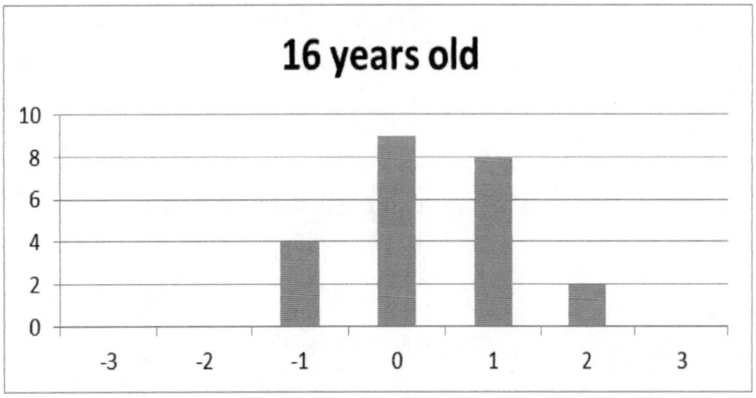

Graph 2. Pupils in the age category of 16 years old.

2. 3. 3 Pupils in the age category of 17 years old

Graph 3. indicates that almost a half of 17-years-old pupils, i.e. 49.94%, had guessed their mark correctly. 24.97% had guessed to get a mark one grade better, 18.16% had expected a mark one grade worse, 4.54% had predicted a mark two grades worse and only a small percentage of just 2.27% had expected to get a mark two grades better.

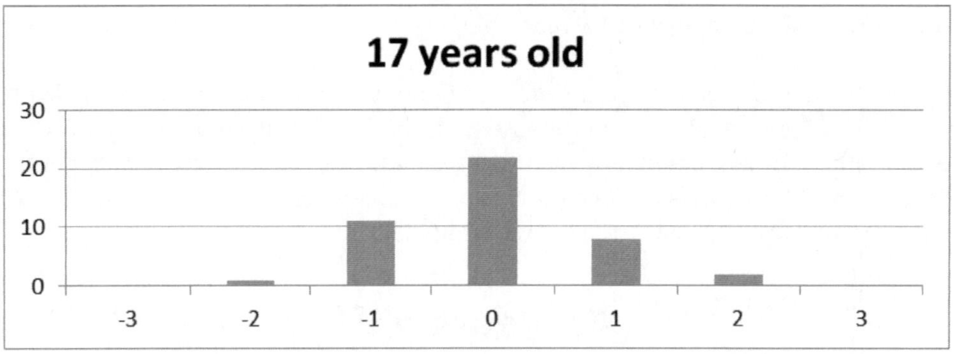

Graph 3. Pupils in the age category of 17 years old.

2. 3. 4 Pupils in the age category of 18 years old

Graph 4. shows that 36.4% of pupils had not made a mistake in their guessing, 22.75% had expected a mark one grade better, but the equal percentage had expected a mark one grade worse. A two grades better mark had been anticipated by 13.65% and only a single pupil representing the last 4.55% had expected a mark two grades worse.

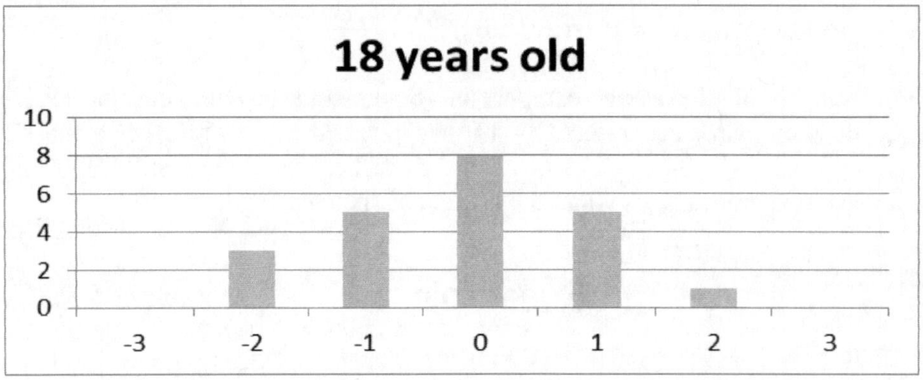

Graph 4. Pupils in the age category of 18 years old.

2. 3. 5 Gender differences in self-assessment

From the final graph, Graph 5., it follows that both genders had mostly guessed their mark correctly. According to the results, boys (45,45%) in our school show a tendency to predict the results of their exams more accurately than girls (43,53%), but the difference is minimum.

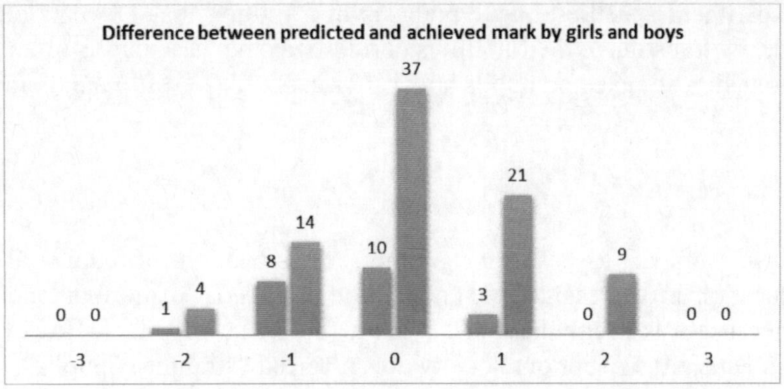

Graph 5. Gender differences of self-assessment.

2. 3. 6 Pupils´ of High school of business

The following Graph 6. summarizes the overall results of all the pupils who participated in the experiment. It indicates that 44.65% of pupils, that is almost a half, had guessed their mark exactly, 22.8% had expected a mark one grade worse, 20.9% had assumed to get a mark one grade better, 8.55% had expected a mark two grades worse and 4.75% a mark two grades better.

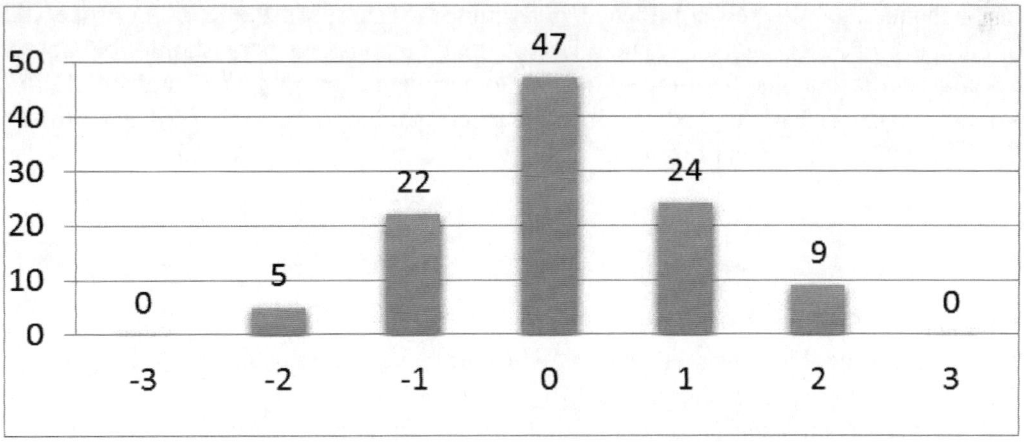

Graph 6. The level of pupils' self-assessment skills.

2. 4 Conclusion of the experiment

On each of the previous graphs, we could observe one common sign. No matter how many categories we divided the pupils into, the pupils who correctly estimated their knowledge never dominated. From the charts, it might appear as if the majority of the pupils had adequately assessed themselves, but the exact self-assessment that occurred in the total number of pupils (i.e. 107) is less than a half. Only 43% of pupils had correctly assessed their knowledge and skills. This shows that 57%, meaning more than a half, had wrongly assessed their abilities. With this simple pedagogical experiment, we have discovered that the ability of pupils to self-assess themselves is low and they tend to underestimate their abilities. In addition, the experiment specified our hypothesis in a finding that 17-years-old pupils have the highest tendency to estimate their abilities correctly as they are, pupils of an age below 17 years old tend to underestimate and pupils above 17 years old overestimate themselves.

3. Conclusion

Examining of the subject, assessment, grading, classification of pupils all represent a feedback in education in a present-day school. The main goal of current pupils is to get a "good" mark, because it is everywhere presented as the most important. However, what is to be understood in educating is that our society doesn't need "straight-A pupils", but especially people who are able to communicate, to problem-solve, are digitally literate, people willing to continue learning. To obtain these qualities, education and, of course, evaluation should be adjusted. We should abandon the old way of classifying pupils to proficient and less skilled ones, and instead opt for the developing assessment with its integral part, the self-assessment. Our experiment has proven that the ability of pupils to evaluate themselves and their knowledge is low, at the level of forty-three percent. The resulting situation is also due to inadequate feedback tools, with one of the solutions being the usage of the developing assessment. Feedback in the mathematics education of 21st century should be enriched with a possibility of self-assessment. Self-evaluation brings a number of benefits such as higher pupil's engagement and eagerness in learning, better communication between a pupil and a teacher, increased pupil's motivation, providing of valuable information about a pupil to a teacher, enhanced pupil's self-confidence, activating the learner in the learning process. The important part of learning next generation of mathematics teachers should be not only mathematics skills, but also teaching skills. Our experiment shows the need and the ability for assessment change. University education of teachers should contain the self-assessment of pupils and creating assessment rubrics as a new trend in mathematics education.

References

[1] URL, http://pblchecklist.4teachers.org/
[2] URL, http://school.discoveryeducation.com/schrockguide/assess.html
[3] URL, http://www.teach-nology.com/web_tools/rubrics/
[4] URL, http://www.teacherplanet.com/calendar/calendar.php?op=cal&month=1&year=2010
[5] URL, http://www.vazka.sk/site/site.php?id=169
[6] GAVORA, P.: Akí sú moji žiaci? Bratislava: Práca 1999. 5s. ISBN 80-7094-335-1
[7] PORUBSKÝ, Š.: Prezentácia o rozvíjajúcom hodnotení. Žiar nad Hronom: Nadácia Škola dokorán, 2000
[8] TUREK, I.: Úvod do didaktiky vysokej školy. Komárno: Univerzita J. Selye 2005, 318s. ISBN 80-8073-301-5

[9] TUREK, I.: Zvyšovanie efektívnosti vyučovania. Bratislava: Metodické centrum, 1997. 310s. ISBN 80-88796-49-0

Current address

Šabíková Henrieta, PaedDr.
Department of Mathematics
Faculty of Natural Sciences, Constantine the Philosopher University
Tr. A. Hlinku 1, 949 74 Nitra, Slovak Republic
E-mail: henrieta.sabik@gmail.com

Judinová Lenka
High School of Business
Bernolákova 26, 953 20 Zlaté Moravce, Slovak Republic
E-mail: lenkajudinova000@gmail.com

Čeretková Soňa, doc. PaedDr., PhD.
Department of Mathematics
Faculty of Natural Sciences, Constantine the Philosopher University
Tr. A. Hlinku 1, 949 74 Nitra, Slovak Republic
E-mail: sceretkova@gmail.com

Slovak University of Technology in Bratislava
Faculty of Mechanical Engineering
17th Conference on Applied Mathematics **APLIMAT 2018**

Proceedings

NORMALIZATION TECHNIQUES FOR UNIVARIATE BIOSTATISTICS ANALYSIS

ŠEDIVÁ Blanka (CZ), PELANTOVÁ Helena (CZ), BUGÁŇOVÁ Martina (CZ)

Abstract. The biostatistic processing of metabolomic data includes a number of mathematical and statistical methods. The series of preprocessing steps is necessary applied before application of the univariate or multivariate approaches for identifications of statistical significant factor.
The goal of this article is using simulation Monte Carlo approach for analysis of influence of normalizations processes on the results of univariate statistical methods. In our study four methods of normalization are compared - normalization by the area under the curve (AUC), normalization to creatinine, quantile normalization and probabilistic quotient normalization (PQN). The results of simulation experiments studies have shown that PQN, quantile normalization and creatinine normalization are more robust than AUC normalization, especially in case a small number of metabolites with a large fold change is presented. From a practical point of view PQN method is recomended as the robust normalization procedure with the broad application and easy data interpretation.

Keywords: data normalization, preprocessing, simulation, metabolomics

Mathematics subject classification: Primary 65C05; Secondary 62P10, 62H30

1 Introduction

The problem of multidimensional correlated data sets is very common in biostatistics. In these situations, mathematical statistical procedures focus on suitable data preprocessing methods, in particular on data standardization problems and other preprocessing methods that enable us to eliminate the effect of the large variability of biological samples on the one hand and to identify significant difference factors on the other. This problem is common for all so called "-omics" methods and metabolomics is no exception.

Metabolomics is focused on the comprehensive characterization of metabolites in biological systems (humans, animals, plants, bacteria, etc.); in humans all biological materials from biofluids (blood, urine) up to tissues can be analyzed (Alonso 2015 [1]). Therefore, it is increasingly being used in almost all fields of health science including pharmacology, pre-clinical drug trials, toxicology, newborn screening and many others. The metabolome of being as reflected in biofluids is defined by genetic factors, but it can also be affected by diet, age, disease, etc. One of the most prominent analytical methods, which is applied for analysis of biological samples in metabolomic studies is nuclear magnetic resonance (NMR) spectroscopy.

It can provide one- or multidimensional spectra. For metabolomics are predominately used proton NMR spectra (^{1}H NMR), which consist of all hydrogen resonances of metabolites present in analyzed sample. The position and shape of detected signals reflect the structure of molecules and their intensity is proportional to concentration in the sample. Thus, we can both identify molecules constituting the sample (usually by comparison with databases) and directly determine their concentration.

The NMR spectra require extensive data pre-processing prior to data analysis. It starts with the Fourier transformation, baseline correction, binning, which allows to transform the continuous signal to discrete datasets and the normalization of the dataset. Once the complete metabolic sets are generated, the univariate or multivariate statistical methods focused directly on the identification of differences between the groups of samples can then be applied.

In Fig. 1 displays representative urine spectra of three samples. The result of the univariate and multivariate analysis can be presented, for example, as a boxplot for selected metabolites or 2D score plots of the first and second components after Principal Component Analysis (PCA) or Partial Least Squares - Discriminant Analysis (PLS-DA). In Fig. 2, we show the visualization of the final statistical results of the effects of liraglutide in mice with diet-induced obesity studied by metabolomics (Buganova 2017 [2]).

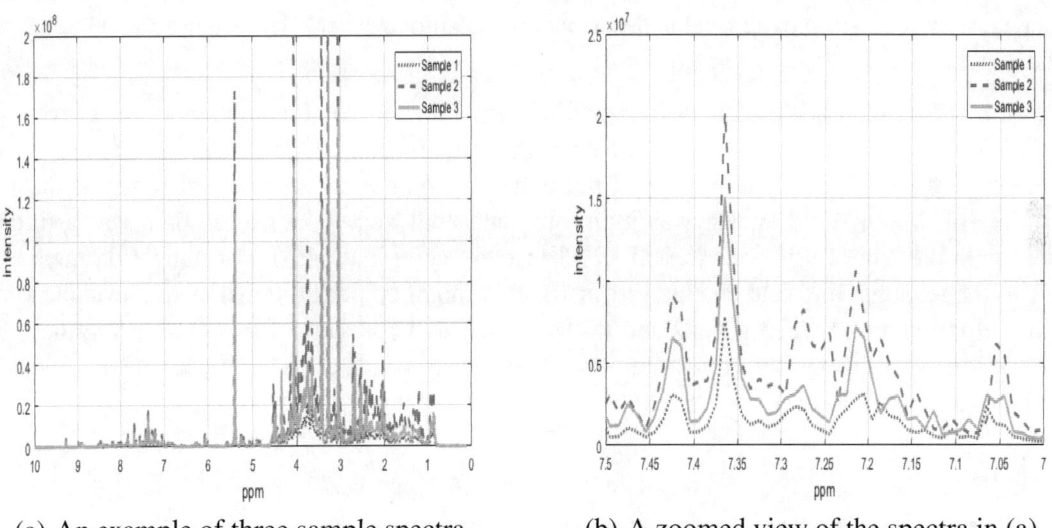

(a) An example of three sample spectra.

(b) A zoomed view of the spectra in (a).

Fig. 1. Examples of spectra obtained with NMR spectroscopy technologies.

Urine is one of the most common biofluid used for metabolomics. Unfortunately, urine samples can substantially vary in their dilution, which is caused by different water consumption and other physiological factors. Consequently, the concentrations of metabolites are influenced. Therefore, proper normalization for these effects prior to further statistical processing is recommended (Alonso 2015 [1]).

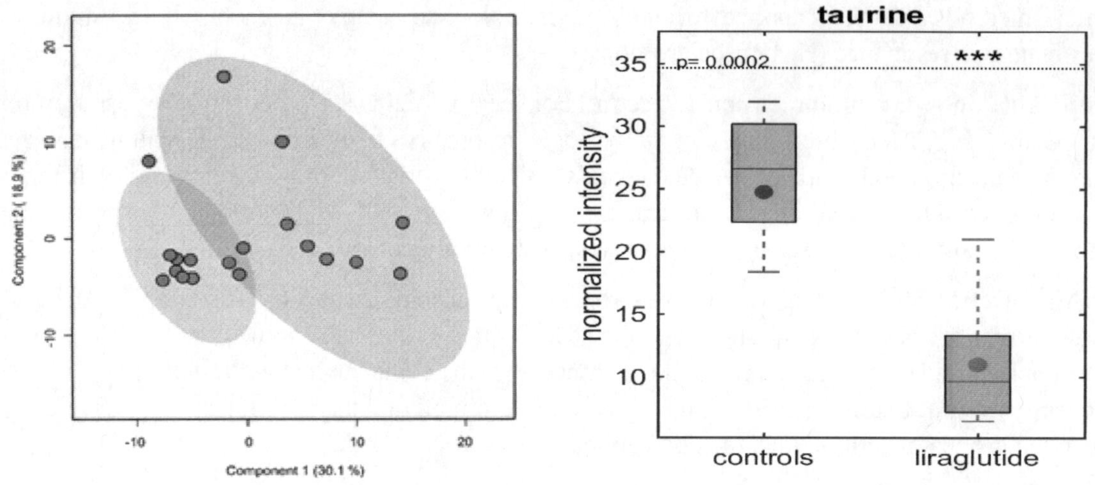

(a) 2D scores plots as a result of PLS-DA.

(b) Boxplots for intensity of selected metabolite in control group and treated group.

Fig. 2. The visualization of univariate and multivariate statistical results.

The aim of this article was to study the influence of the normalization methods of raw datasets on the results and conclusions based on univariate statistics of identified significant features. Raw data of ^{1}H NMR spectra of urine from experiments described in (Pelantova 2016 [9]) were used as a background of the dataset.

We performed four methods of data normalization - normalization by the area under the curve (AUC normalization), normalization to creatinine, quantile normalization and probabilistic quotient normalization (PQN). The effect of these types of normalization on the results of the parametric t-test and non-parametric Mann-Whitney test was monitored and analyzed. The source data was artificially modified with the aim of assessing the effect of different combinations of normalization methods in relation to the response and fold change. We artificially modified peak intensities to introduce known metabolic differences between groups and this provided us with a target for discovering these artificial peak intensities using both univariate and multivariate statistics. The effects of the processing steps on our ability to re-discover these known metabolic differences were evaluated.

2 Methods

2.1 Data set, raw data preprocessing and simulations

Real datasets from studies (Pelantova 2016 [9]) were used as a background for simulation studies. In order to get as close as possible to real-world experiments, two sets (the number of samples in each set is $n = 8$) of observation were always generated. For the first step of the preprocessing approach, we applied the binning with the fix bin step 0.01 ppm.

The structure of generated samples was derived from real binned spectra using bootstrapping methods. For the second set (simulated treated group), spectral values in the area from 6.445 ppm to 6.485 ppm were artificially increased. The additional increasing of intensities was based on the intensity values from 10^5 to 10^8 and in percentages from 1% to 150%.

The resulting generated dataset was arranged into a data matrix with rows $1-8$ corresponding to unmodified samples and rows $9-16$ to an included part with artificial values. The amount of analyzed bins (corresponding to the number of columns of the data array) was 922. The simulation processes were repeated $1,000$ times.

2.2 Normalizations

The generated datasets were normalized by four methods recommended for normalizations of spectral datasets in the literature Di Guida 2016 [4], Euceda 2012 [5] or Kohl 2012 [8].

(a) AUC normalization (area under curve normalization, normalization by sum) is a method where each value in a row (sample) is divided by the total sum of the row (sample) and multiplied by 100; the unit is $\%$.

(b) Normalization to creatinine is related to the very specific metabolite, called creatinine, that is presented in all urine samples. Creatinine is a chemical waste product in blood (by-product of normal muscle contractions) that passes through the kidneys to be filtered and eliminated in urine. Under normal conditions, urinary creatinine output is relatively constant and measurable.

(c) Quantile normalization consists of two steps: in the first step a mapping between ranks and values is created. For rank 1, find the n values, one per array, that are the smallest value on the array, and save their averages. Similarly, for rank 2 and the second smallest values, and on up to the largest n values, one per array. In the second step, the actual values for each array is replaced with these averages.

(d) Probabilistic quotient normalization (Dieterle 2016 [3]) assumes that biologically interesting concentration changes influence only parts of the NMR spectrum, while dilution effects will affect all metabolite signals. PQN begins with an AUC normalization of each spectrum, followed by the calculation of a reference spectrum such as a median or mean spectrum. Next, for each variable of interest, the quotient of a given test spectrum and reference spectrum is calculated and the median of all quotients is estimated. Finally, all variables of the test spectrum are divided by the median quotient.

2.3 Statistical analyses

After the application of the preprocessing approach with one of the selected normalization procedures, two univariate methods of identification differences between metabolites in the "control" group and "simulated treated" group were calculated. The parametric t-test and the nonparametric robust test, the Mann-Whitney test, were chosen for testing the differences between groups, for more information about the advantages and disadvantages of univariate methods see Saccenti 2014 [10].

We focused on identifying the differences in the area of the added metabolite (area 1) in the area of common metabolite occurrences (area 2) as well as in the area of the noise (area 3). Appropriate standardization should reveal the diffusion in the area of the artificially enhanced spectrum (area 1) but at the same time not identify the differences in other areas (especially in area 2) because it would be the differences caused by the inappropriately chosen standardization.

3 Results and discussion

Our result from the influence of several types of standardization processes is confirmed by other simulation studies, Gardlo 2016 [6] and Hertel 2017 [7]. In Fig. 3 and Fig. 4 we show the typical results for artificial metabolite simulation with gradual increments of 1% to 150%. The sensitivity of the results is dependent on the intensity value of bins that is modified. To illustrate the impact of the intensity value, two situations are presented - the intensity value of bins (in the modified area 1) is equal to 10^6 or 10^7.

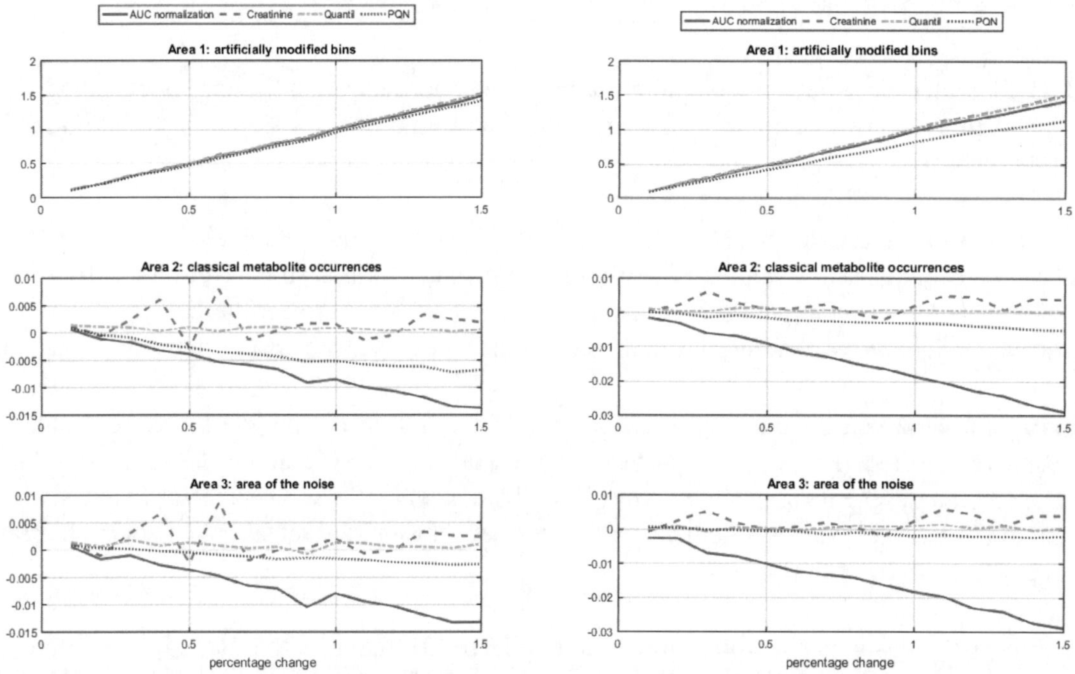

(a) The value of intensities of modified bins - 10^6 (b) The value of intensities of modified bins - 10^7

Fig. 3. The dependencies of changing values of the bins for four methods of the normalization and for three types of bin areas.

The x-axis in all graphs describes the percentage of the artificial bin changes. In Fig. 3, the dependencies of the changing values of the bins for four types of normalizations and for three types of bin areas are presented. It is clear that the changing of the values of bins in the modified area is proportionally increasing for all types of normalizations. Larger differences between normalization methods were observed in the second and third areas. The changes in area 1 also greatly influenced the bins after normalizations in area 2 and area 3. This effect is the biggest one for AUC normalization and PQN.

In Fig. 4, the results of parametric (left) and nonparametric tests (right) are presented. The curves capture the percentage of tests in which the hypothesis on the equality of the expected value in both groups of samples was rejected.

The AUC normalization shows the greatest problems in the field of the identification of fake differences in in all three areas under study. The problematic influence of the AUC normalization increases with the absolute value of modified bins before the normalization process. The percentage of false significant differences grew rapidly for this method of normalization. The PQN increases the number

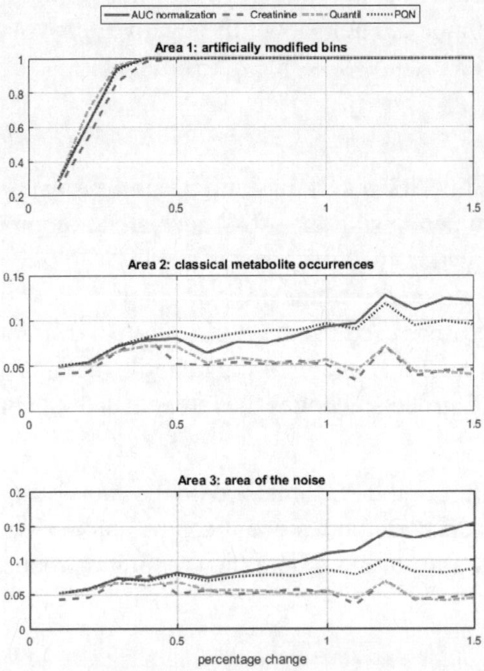

(a) T-test for the intensity value 10^6.

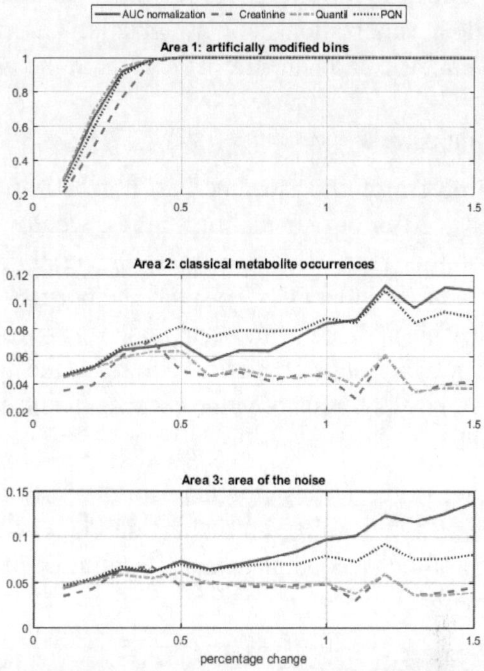

(b) M-W test for the intensity value 10^6.

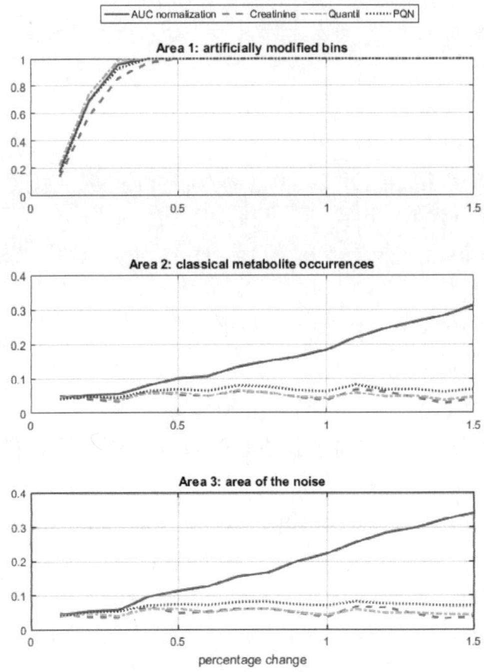

(c) T-test for the intensity value 10^7.

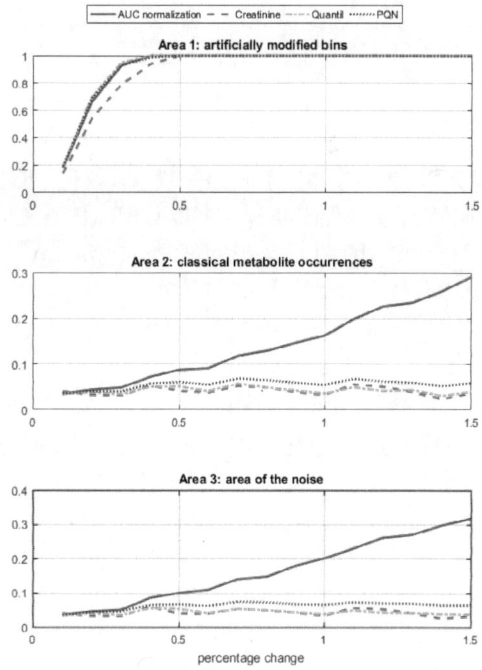

(d) M-W test for the intensity value 10^7.

Fig. 4. The visualization of results for simulation additive artificial metabolite for addition value of intensities 10^6 and 10^7 (y-axis represents the percentage of simulations for which the null hypothesis of the equality of the mean is rejected).

of false identified differences for the smaller absolute value of modified bins. On the contrary, the quantile normalization and normalization based on creatinine did not show this negative behavior and the parametric and nonparametric tests in the area 2 and 3 did not cross the 5% limit.

4 Conclusions

The problem of choosing the best normalization procedure for a certain dataset can be solved empirically. After our simulation studies, we do not recommend using the AUC normalization process because this approach can increase the number of falsely identified differences of bins. On the other hand, we found that PQN, quantile and normalization based on creatinine provide comparable results. Datasets normalized by these approaches are robust and can be used for the explanation of differences between two groups of spectra. However, the considerable disadvantage of the quantile method is the difficult interpretation of value because the normalization process changes the intensities of the bin to the rank.

Thus, we suggest using creatinine normalization procedures or the PQN method on the data. However, compared to creatinine normalization PQN is more universal method, since it can be applied even in other biofluids (blood plasma and serum) or in metabolomic studies where the creatinine production is impaired.

Acknowledgement

The authors gratefully acknowledge the project LO1509 of the Ministry of Education, Youth and Sports of the Czech Republic.

References

[1] ALONSO A., MARSAL S., and JULIÀ A.. Analytical methods in untargeted metabolomics: State of the art in 2015. *Frontiers in Bioengineering and Biotechnology*, 3:23, 2015.

[2] BUGÁŇOVÁ, M., PELANTOVÁ, H., HOLUBOVÁ, M., ŠEDIVÁ, B., MALETÍNSKÁ, L., ŽELEZNÁ, B., KUNEŠ, J., KAČER, P., KUZMA, M., HALUZÍK, M. The effects of liraglutide in mice with diet-induced obesity studied by metabolomics. *Journal of Endocrinology*, 2017, 233:1, pp. 93-104. ISSN: 0022-0795

[3] DIETERLE F., ROSS A., SCHLOTTERBECK G., SENN, H.: Probabilistic Quotient Normalization as Robust Method to Account for Dilution of Complex Biological Mixtures. Application in 1H NMR Metabonomics. *Analytical Chemistry* 78 (13), pp. 4281–4290, 2006.

[4] DI GUIDA R., ENGEL J., ALLWOOD J.W., WEBER R.J.M., JONES M.R., SOMMER U., VIANT M.R., and DUNN W.B.. Non-targeted UHPLC-MS metabolomic data processing methods: a comparative investigation of normalisation, missing value imputation, transformation and scaling. *Metabolomics*, 12(5):93, Apr 2016.

[5] EUCEDA L.R., GISKEØDEGÅRD G.F., and BATHEN F.T.. Preprocessing of NMR metabolomics data. *Scandinavian Journal of Clinical and Laboratory Investigation*, 75(3):193–203, 2015. PMID: 25738209.

[6] GARDLO A., SMILDE A.K., HRON K., HRDÁ M., KARLÍKOVÁ R., FRIEDECKÝ D., and ADAM T.. Normalization techniques for PARAFAC modeling of urine metabolomic data. *Metabolomics*, 12(7):117, Jun 2016.

[7] HERTEL J., VAN DER AUWERA S., FRIEDRICH N., WITTFELD K., PIETZNER M., BUDDE K., TEUMER A., KOCHER T., NAUCK M., and GRABE H.J. Two statistical criteria

to choose the method for dilution correction in metabolomic urine measurements. *Metabolomics*, 13(4):42, Feb 2017.

[8] KOHL S.K., KLEIN M.S., HOCHREIN J., OEFNER P.J., SPANG R., and GRONWALD W.. State-of-the art data normalization methods improve NMR-based metabolomic analysis. *Metabolomics*, 8(1):pp. 146–160, Jun 2012.

[9] PELANTOVÁ, H., BUGÁŇOVÁ, M., HOLUBOVÁ, M., ŠEDIVÁ, B., ZEMENOVÁ, J., SÝKORA, D., KAVÁLKOVÁ, P., HALUZÍK, M., ŽELEZNÁ, B., MALETÍNSKÁ, L., KUNEŠ, J., KUZMA, M. Urinary metabolomic profiling in mice with diet-induced obesity and type 2 diabetes mellitus after treatment with metformin and vildagliptin and their combination. *Molecular and Cellular Endocrinology*, 2016, 431:15 August 2016, pp. 88-100. ISSN: 0303-7207

[10] SACCENTI E., HOEFSLOOT H.C.J., SMILDE A.K., WESTERHUIS J.A., and HENDRIKS M.M.W. B. . Reflections on univariate and multivariate analysis of metabolomics data. *Metabolomics*, 10(3): pp.361–374, Jun 2014.

[11] XIA J., SINELNIKOV I.V., HAN B., and WISHART D.S.. Metaboanalyst 3.0—making metabolomics more meaningful. *Nucleic Acids Research*, 43(W1):W251–W257, 2015.

Current address

Šedivá Blanka, RNDr., Ph.D.
Institute of Microbiology of the Czech Academy of Sciences
Vídeňská 1083, 142 20, Prague 4, Czech Republic
E-mail: blanka.sediva@gmail.com

Pelantová Helena, RNDr., Ph.D.
Institute of Microbiology of the Czech Academy of Sciences
Vídeňská 1083, 142 20, Prague 4, Czech Republic
E-mail: pelantova@biomed.cas.cz

Bugáňová Martina, Mgr.
Institute of Microbiology of the Czech Academy of Sciences
Vídeňská 1083, 142 20, Prague 4, Czech Republic
E-mail: martina.buganova@gmail.com

Slovak University of Technology in Bratislava
Faculty of Mechanical Engineering
17th Conference on Applied Mathematics **APLIMAT 2018**

Proceedings

ON THE HOSOYA INDEX FOR THE MOLECULAR GRAPHS OF HELICENES

SEIBERT Jaroslav (CZ), ZAHRÁDKA Jaromír (CZ)

Abstract. The matching of an undirected graph $G = (V, E)$ is a subset M of E such that no two edges of M are adjacent in G. The Hosoya index of a graph G is given by the number of all matching of G. This graph invariant is one of the most interesting topological index in chemistry. Helicenes are extremal hexagonal chains with a simple graph representation as an important subclass of benzenoid molecules. We obtain the exact formula for the Hosoya index of the molecular graphs of helicenes as a function of the number of hexagons in it.

Key words. matching of a graph, Hosoya index, molecular graph, decomposition theorem, difference equation, helicene

Mathematics Subject Classification: Primary 12H10, 39A10; Secondary 11B39

1 Introduction

In this contribution, we consider undirected simple graphs (without loops of multiple edges). A subset M of the edge set of a graph G is called the matching if none two edges of M are adjacent in G. Other graph terminology and notation is taken from the book [1]. Denote by $m(G, k)$ the number k-matching, which means the number of k mutually independent edges can be selected in G.

Definition 1. Let $G = (V, E)$ be a simple connected graph. By definition, $m(G,0) = 1$ for all graphs, and $m(G,1) = |E|$. The Hosoya index of G is given by $Z(G) = \sum_{k \geq 0} m(G, k)$ as the number of all matching in G.

The chemist Haruo Hosoya introduced in 1971 a molecular graph based structure descriptor, which he named the topological index $Z(G)$. He showed that certain physico-chemical properties of saturated hydrocarbons, in particular their boiling points, are well correlated with $Z(G)$. The molecular structural description Z was soon re-named into Hosoya index, whereas the name topological index is used for any of the countless invariants that are found to have some chemical applicability [11].

Many authors found the Hosoya index for various classes of graphs or solved the problem of extremal values of the Hosoya index in special cases. LV and YU [6] investigated the Hosoya index of trees with a given maximum vertex degree. Hamzeh et al. [4] obtained exact formulas of the Hosoya index for the set of bicyclic graphs, caterpillars and dual star. Huang, Kuang and Deng [5] found the average values of the Hosoya index with respect to the set of all polyphenylene chains. Wagner and Gutman [11] collected and classified the results on the extremal values of the Hosoya index and also provided some useful tools and techniques that are frequently used in the study of this type of problem.

A hexagonal system is a connected plane graph without cut-vertices in which all inner faces are the cycles of the length 6 (hexagons), such that two hexagons are either disjoint or have exactly one common edge, and none three hexagons share a common edge. Two hexagons of a system may have either two common vertices (if they are adjacent) or none (if they are not adjacent). A hexagonal chain with n hexagons, $n \geq 2$, possesses two terminal hexagons and $n-2$ hexagons that have two neighbors. A hexagon adjacent to exactly two other hexagons possesses two vertices of degree 2.

Denote B_n an arbitrary chain with n hexagons and V_3 the set of its vertices of degree 3. The subgraph B_n' of B_n generated by V_3 is an acyclic graph [13]. If the subgraph B_n' is a matching with $n-1$ edges, then B_n is called a linear chain. If the subgraph B_n' is a path, then B_n is called a zigzag chain. Shiu [10] showed that the linear hexagonal spider and zig-zag hexagonal spider attain the extremal values of the Hosoya index. Gutman [3] reported some results on extremal hexagonal chains.

If the subgraph B_n' is a comb, then B_n is called a helicene chain H_n (Fig.1.), where a comb is a graph obtained by joining a single pendant edge to each vertex of a path (see e.g. [9]).

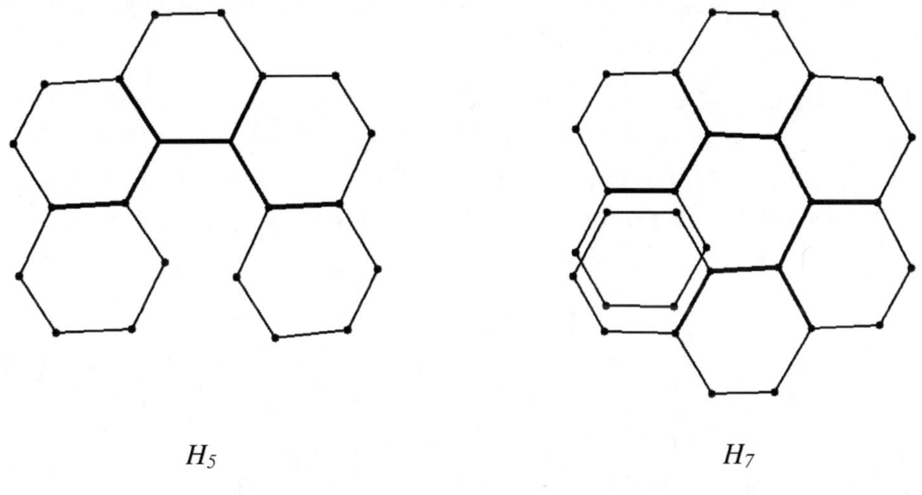

Fig. 1. The molecular graphs of [5]helicene and [7]helicene (the combs are drawn by bold lines).

Carbohelicenes belong to a class of fascinating, chiral, and helicoidal molecules, which have a rich history in chemistry since the very beginning of the 20th century. Helicene chemistry is being considered as an expanding and modern field, leading to several applications in supramolecular chemistry, in nanosciences, in chemical-biology, in polymers and materials science. A comprehensive report on non-stereoselective reactions and methods for producing helicenes can be found in [2].

The main aim of this contribution is to derive the exact formula for the Hosoya index of the molecular graphs of helicenes H_n as a function of n.

2 Preliminary results

In this section, we review some basic and general properties of the Hosoya index (e.g. [4], [11]).

Theorem 1. If $G_1, G_2, \ldots, G_m$ are the connected components of a graph G, then $Z(G)=\prod_{i=1}^{m} Z(G_i)$.

Theorem 2. Let G be a graph with at least two vertices and v be its arbitrary vertex. Then

$$Z(G)=Z(G-v)+\sum_{w} Z(G-\{v,w\}),$$

where the sum is taken over all the vertices w adjacent to v.

Theorem 3. Let vertices u,v be the ends of an edge e in a graph G. Then

$$Z(G)=Z(G-e)+Z(G-\{u,v\}).$$

Using Theorem 2 or Theorem 3 it is easy to obtain the following values of the Hosoya index for special graphs. We just recall that the Fibonacci numbers F_n are defined by the second order recurrence $F_{n+2}=F_{n+1}+F_n$, with $F_0=0, F_1=1$ and the Lucas numbers L_n satisfy the same recurrence but with the initial terms $L_0=2, L_1=1$.

Theorem 4.

(a) $Z(P_n)=F_{n+1}$, where P_n is a path with n vertices,

(b) $Z(C_n)=L_n$, where C_n is a cycle with n vertices.

Now, we will derive some auxiliary results to finding the Hosoya index for helicenes. Consider therefore the molecular graph H_n of $[n]$helicene and its subgraphs I_n, J_n, K_n and M_n (Fig.2).

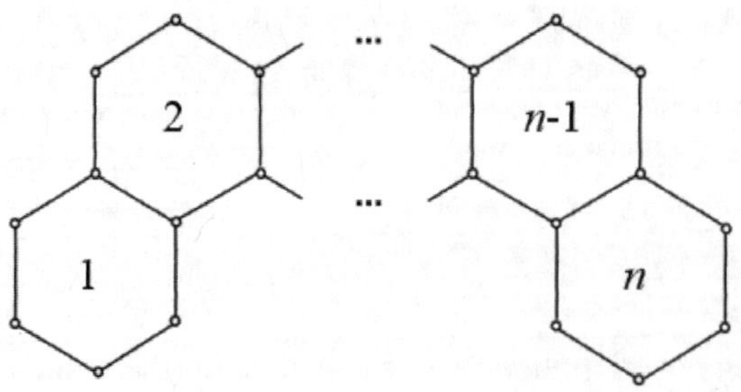

H_n

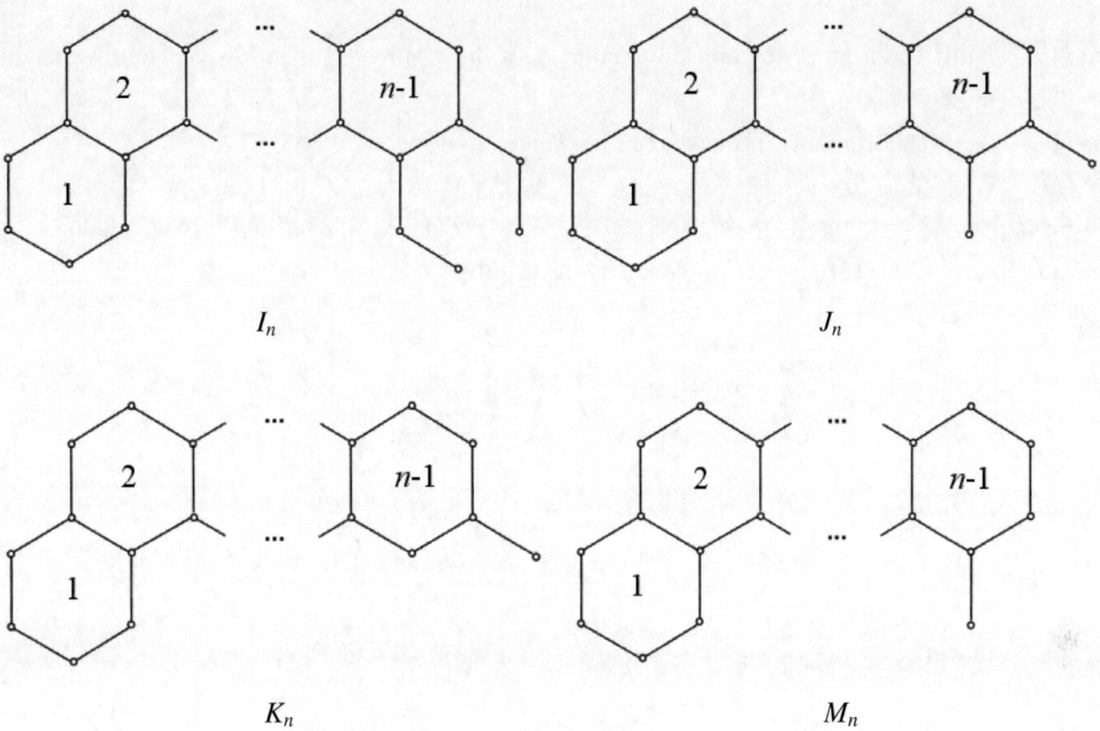

Fig. 2. The molecular graph H_n of $[n]$helicene and its subgraphs I_n, J_n, K_n and M_n.

Let us denote their Hosoya indices $Z(H_n)=h_n$, $Z(I_n)=i_n$, $Z(J_n)=j_n$, $Z(K_n)=k_n$ and $Z(M_n)=m_n$, for short. By direct using of Theorems 1-4 we can obtain the values of these indices for the small numbers of n. These values are collected in Tab. 1.

n	h_n	i_n	j_n	k_n	m_n
1	18	13	5	3	3
2	148	109	39	26	26
3	1233	906	327	213	218
4	10244	7531	2713	1778	1807
5	85169	62605	22564	14764	15033

Tab. 1. The Hosoya index of the graphs H_n, I_n, J_n, K_n, M_n for $1 \le n \le 5$.

Lemma 1. The terms of sequences $\{h_n\}$, $\{i_n\}$, $\{j_n\}$, $\{k_n\}$ and $\{m_n\}$ satisfy the following difference equations for any $n \ge 2$

$$h_n = i_n + j_n \quad (1)$$

$$i_n = j_n + k_n + m_n + h_{n-1} \quad (2)$$

$$j_n = k_n + 3k_{n-1} + 2h_{n-2} \quad (3)$$

$$k_n = h_{n-1} + j_{n-1} + m_{n-1} \quad (4)$$

$$m_n = h_{n-1} + 2k_{n-1} + h_{n-2} \quad (5)$$

Proof.

Relation (1) – Using Theorem 3 we choose the edge e of the graph H_n as in Fig. 3. Then we have directly $Z(H_n) = Z(I_n) + Z(J_n)$.

Relation (2) – Using Theorem 2 two times (see Fig. 3) we have successively

$Z(I_n) = Z(I_n - v_1) + Z(I_n - \{v_1, w_1\}) =$

$= Z(I_n') + Z(I_n'') = Z(I_n' - v_2) + Z(I_n' - \{v_2, w_2\}) + Z(I_n'' - v_2) + Z(I_n'' - \{v_2, w_2\})$, but it means that $Z(I_n) = Z(J_n) + Z(K_n) + Z(M_n) + Z(H_{n-1})$ which was to prove.

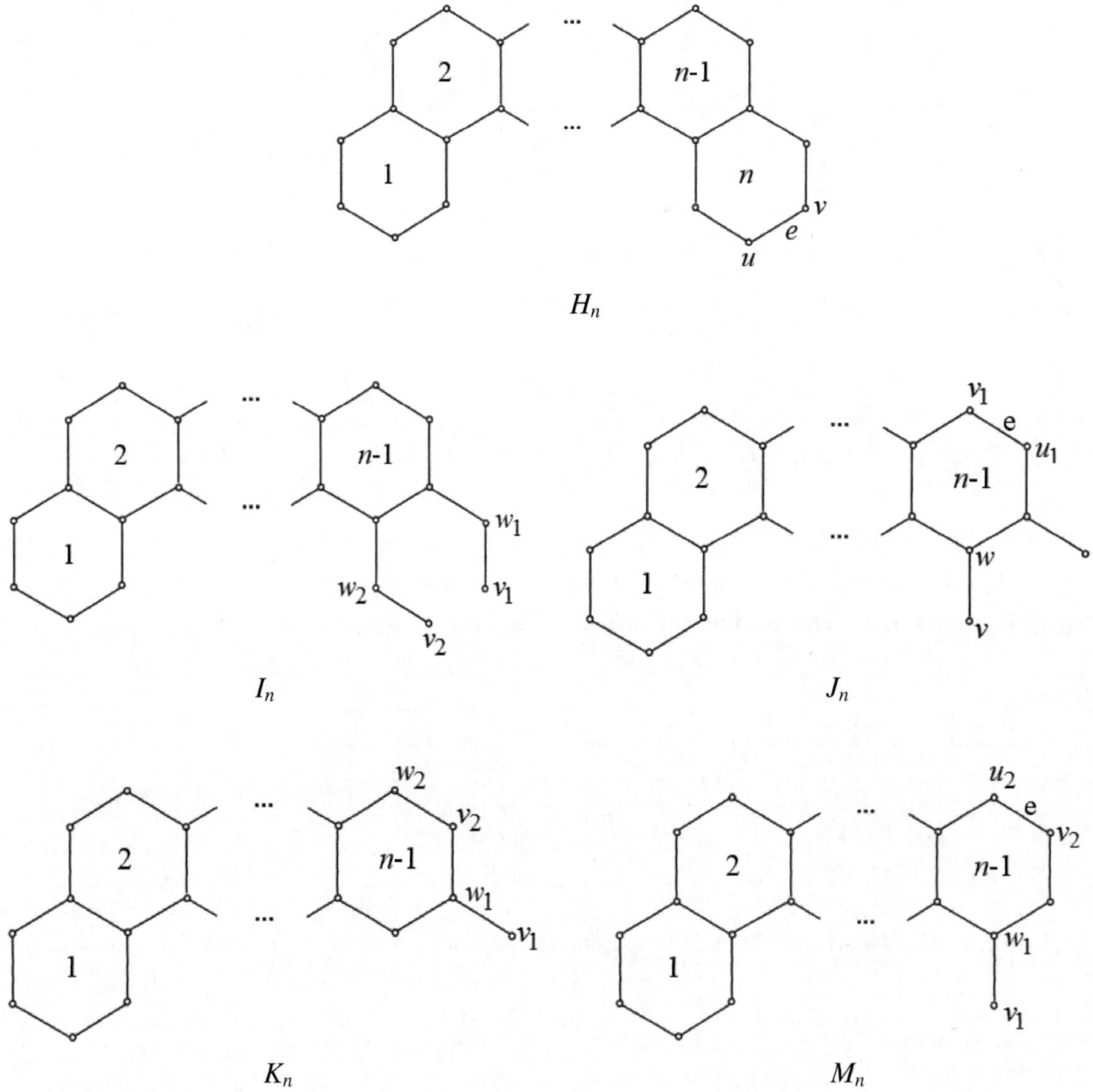

Fig. 3.

Relation (3) – First, we will use Theorem 2 so that we choose the vertex v of the graph J_n as in Fig. 3. Then $Z(J_n) = Z(K_n) + Z(J_n - \{v, w\})$. Now, with the respect of Theorem 3 by using the edge

$e = u_1 v_1$ we obtain $Z(J_n - \{v, w\}) = Z(P_3 \cup K_{n-1}) + Z(P_2 \cup H_{n-2})$. Using Theorem 1 we have $Z(J_n) = Z(K_n) + Z(P_3) Z(K_{n-1}) + Z(P_2) Z(H_{n-2}) = Z(K_n) + F_4\, Z(K_{n-1}) + F_3\, Z(H_{n-2})$, which gives the proved relation.

Relation (4) – Now, we will again use Theorem 2 two times (see Fig. 3). Thus

$$Z(K_n) = Z(K_n - v_1) + Z(K_n - \{v_1, w_1\}) = Z(H_{n-1}) + Z(K_n') = Z(H_{n-1}) + Z(K_n' - v_2) + Z(K_n' - \{v_2, w_2\}) = $$
$$= Z(H_{n-1}) + Z(J_{n-1}) + Z(M_{n-1}).$$

Relation (5) – Using Theorem 2 we first obtain

$Z(M_n) = Z(M_n - v_1) + Z(M_n - \{v_1, w_1\}) = Z(H_{n-1}) + Z(M_n')$. If we want to use Theorem 3 for the graph M_n' it is suitable to choose the edge e as in Fig. 3. Then

$Z(M_n) = Z(H_{n-1}) + Z(M_n' - e) + Z(M_n' - \{u_2, v_2\}) = Z(H_{n-1}) + Z(P_2) Z(K_{n-1}) + Z(P_1) Z(H_{n-2})$ and since $Z(P_2) = F_3 = 2$, $Z(P_1) = F_2 = 1$ the proof is over.

It is easy to see that $h_0 = Z(P_2) = F_3 = 2$ and therefore relations (1) and (2) are also valid for $n = 1$.

3 Main results

Theorem 5. For the Hosoya index k_n of the graphs K_n the following linear difference equation

$$k_{n+4} - 6k_{n+3} - 19k_{n+2} - 2k_{n+1} + k_n = 0 \qquad (7)$$

holds for each positive integer n.

Proof. First, we reduce the number of variables in the system relations from Lemma 1 for any $n \geq 2$

$$j_n = k_n + 3k_{n-1} + 2h_{n-2} \qquad (3)$$
$$j_n = k_{n+1} - 2k_{n-1} - h_n - h_{n-1} - h_{n-2} \qquad (8)$$
$$2j_n = -k_n - 2k_{n-1} + h_n - 2h_{n-1} - h_{n-2} \qquad (9)$$

when we obtain relation (8) from relations (4), (5) and relation (9) from relations (1), (2), (5).

We can eliminate j_n by a suitable way and then we have two relations for the members of the sequences $\{h_n\}$ and $\{k_n\}$

$$k_{n+2} - 4k_{n+1} - 13k_n = 3h_n + 8h_{n-1} \qquad (10)$$
$$k_{n+3} - 6k_{n+2} - 14k_{n+1} + 2k_n = 8h_n - h_{n-1} \qquad (11)$$

and the equality
$$h_n = \frac{1}{67}\left(8k_{n+3} - 47k_{n+2} - 116k_{n+1} + 3k_n\right) \qquad (12)$$

is directly obtained from the previous system.

Finally, we substitute (12) into (10) and after simplification we obtain difference equation (7).

Remark. It is known fact that each of the sequences from Lemma 1 satisfy the same linear difference equation as the sequence $\{k_n\}$. Therefore the general solution for these sequences are the same, but the concrete function expression (particular solution) of the members of sequences will be different with respect to the various initial members of sequences.

Theorem 6. The Hosoya index of the molecular graph H_n of $[n]$helicene can be written for any positive integer n in the form $h_n = c_1\alpha^n + c_2\beta^n + c_3\gamma^n + c_4\delta^n$, where the constants c_1, c_2, c_3, c_4 can be uniquely expressed by the numbers $\alpha, \beta, \gamma, \delta$.

Proof. The members of the sequence $\{h_n\}$ satisfy the homogeneous linear difference equation of the fourth order with constant coefficients

$$h_{n+4} - 6h_{n+3} - 19h_{n+2} - 2h_{n+1} + h_n = 0.$$

Its characteristic equation is an algebraic equation of the fourth order

$$x^4 - 6x^3 - 19x^2 - 2x + 1 = 0$$

with the different roots $\alpha, \beta, \gamma, \delta$. It means that $h_n = c_1\alpha^n + c_2\beta^n + c_3\gamma^n + c_4\delta^n$ and with help of the initial members $h_0 = 2,\ h_1 = 18,\ h_2 = 148,\ h_3 = 1233$ we obtain the following system of four linear equations for unknown c_1, c_2, c_3, c_4

$$\begin{aligned} c_1 + c_2 + c_3 + c_4 &= 2 \\ c_1\alpha + c_2\beta + c_3\gamma + c_4\delta &= 18 \\ c_1\alpha^2 + c_2\beta^2 + c_3\gamma^2 + c_4\delta^2 &= 148 \\ c_1\alpha^3 + c_2\beta^3 + c_3\gamma^3 + c_4\delta^3 &= 1233. \end{aligned}$$

It is easy to find the solution of this system in the symbolic form

$$c_1 = \frac{1233 - 148(\beta+\gamma+\delta) + 18(\beta\gamma+\beta\delta+\gamma\delta) - 2\beta\gamma\delta}{(\alpha-\beta)(\alpha-\gamma)(\alpha-\delta)}$$

$$c_2 = \frac{1233 - 148(\alpha+\gamma+\delta) + 18(\alpha\gamma+\alpha\delta+\gamma\delta) - 2\alpha\gamma\delta}{(\beta-\alpha)(\beta-\gamma)(\beta-\delta)}$$

$$c_3 = \frac{1233 - 148(\alpha+\beta+\delta) + 18(\alpha\beta+\alpha\delta+\beta\delta) - 2\alpha\beta\delta}{(\gamma-\alpha)(\gamma-\beta)(\gamma-\delta)}$$

$$c_4 = \frac{1233 - 148(\alpha+\beta+\gamma) + 18(\alpha\beta+\alpha\gamma+\beta\gamma) - 2\alpha\beta\gamma}{(\delta-\alpha)(\delta-\beta)(\delta-\gamma)}.$$

Then the proof is finished.

4 Numerical results

We also had to calculate the previous results using Symbolic Math Toolbox of MATLAB.

The roots of the characteristic equation $x^4 - 6x^3 - 19x^2 - 2x + 1 = 0$ can be found by MATLAB in the numerical form through the command

```
>> syms x, R=solve(x^4-6*x^3-19*x^2-2*x+1)
R =
  8.3128260775610574875139058467921
 0.17898242638989329333757646913108
```

-0.3077323367373107047565344171426
-2.1840761672136400760949478987805

The roots are four irrational numbers, which were denoted by $\alpha, \beta, \gamma, \delta$ in Theorem 6. It is obvious that the above mentioned values are only approximations these numbers.

The system of four linear equations for unknown c_1, c_2, c_3, c_4 was solved by using commands of MATLAB. Instead of the variables $\alpha, \beta, \gamma, \delta$ are used in MATLAB the symbolic variables a, b, c, d and instead c_1, c_2, c_3, c_4 are used the identifiers C1, C2, C3, C4.

```
>> syms a b c d , A = [1,1,1,1; a,b,c,d; a^2, b^2, c^2, d^2 ; a^3, b^3, c^3, d^3];
>> AC1=A; AC1(:,1)=[2; 18; 148; 1233]; C1=factor(det(AC1))/factor(det(A))
C1 =
-(148*b + 148*c + 148*d - 18*b*c - 18*b*d - 18*c*d + 2*b*c*d - 1233)/((a - b)*(a - c)*(a - d))
>> AC2=A; AC2(:,2)=[2; 18; 148; 1233]; C2=factor(det(AC2))/factor(det(A))
C2 =
(148*a + 148*c + 148*d - 18*a*c - 18*a*d - 18*c*d + 2*a*c*d - 1233)/((a - b)*(b - c)*(b - d))
>> AC3=A; AC3(:,3)=[2; 18; 148; 1233]; C3=factor(det(AC3))/factor(det(A))
C3 =
-(148*a + 148*b + 148*d - 18*a*b - 18*a*d - 18*b*d + 2*a*b*d - 1233)/((a - c)*(b - c)*(c - d))
>> AC4=A; AC4(:,4)=[2; 18; 148; 1233]; C4=factor(det(AC4))/factor(det(A))
C4 =
(148*a + 148*b + 148*c - 18*a*b - 18*a*c - 18*b*c + 2*a*b*c - 1233)/((a - d)*(b - d)*(c - d))
```

Calculation of approximations of the irrational constants c_1, c_2, c_3, c_4 can be realized by command

```
>> c1=subs(C1,[a,b,c,d],R), c2=subs(C2,[a,b,c,d],R), c3=subs(C3,[a,b,c,d],R), c4=subs(C4,[a,b,c,d],R)
c1 =
2.1454822181534150647686192500525
c2 =
0.046858209049296998559385477378165
c3 =
-0.14042920424633548952728375521586
c4 =
-0.051911222956376573800720972214777 .
```

The values of Hosoya index calculated by using MATLAB through the recurrence formula $h_{n+4} = 6h_{n+3} + 19h_{n+2} + 2h_{n+1} - h_n$ for the positive integers $1 \le n \le 20$ are

```
>> h=[sym(18) sym(148) sym(1233) sym(10244)];
>> for  n=1:16, h(n+4)= 6*h(n+3)+19*h(n+2)+2*h(n+1)-h(n); end, h
h =
[ 18, 148, 1233, 10244, 85169, 707968, 5885274, 48923130, 406689753, 3380740568,
28103509701, 233619585374, 1942038987946, 16143832328616, 134200870403717,
```

1115588495056524, 9273693133679053, 77090598116627092, 640840734359373890, 5327197568144459670] .

The values of Hosoya index can also be calculated through the previously derived formula $h_n = c_1\alpha^n + c_2\beta^n + c_3\gamma^n + c_4\delta^n$ for the positive integers $1 \le n \le 20$.

```
>> syms n, H=c1*R(1)^n+ c2*R(2)^n+ c3*R(3)^n+ c4*R(4)^n;
>> for  k=1:20, h(k)=subs(H,n,sym(k)); end, h
 h =
 [ 18.0, 148.0, 1233.0, 10244.0, 85169.0, 707968.0, 5885274.0, 48923130.0, 406689753.0,
3380740568.0, 28103509701.0, 233619585374.0, 1942038987946.0, 16143832328616.0,
134200870403717.0, 1115588495056524.0, 9273693133679053.0, 77090598116627092.0,
640840734359373890.0, 5327197568144459670.0]
```

It is easy to see that both used methods of calculating of the Hosoya index of the graphs H_n for the positive integers $1 \le n \le 20$ give identical results with respect to integers. The Hosoya index of $[n]$helicenes for $3 \le n \le 10$ is presented in Tab. 2.

n	3	4	5	6	7	8	9	10
h_n	1 233	10 244	85 169	707 968	5 885 274	48 923 130	406 689 753	3 380 740 568

Tab. 2.

5 Concluding remarks

The Hosoya index of a graph, also known as Z-index, is one of the large family of topological indices of graphs which are closely connected with selected physico-chemical characteristics of the respective compounds. Similar connections are also known for the Merrifield-Simmons index of a graph, which is defined via the number of ways in which mutually independent vertices can be selected in a graph. The two indices do not only have very similar definitions, they are also quite related in another respect. Mostly it is true that the graph which minimizes the Merrifield-Simmons index is also the one which maximizes the Hosoya index, and vice versa.

In the previous years we have calculated the Merrifield-Simmons index for special classes of hydrocarbons. In this contribution we have turned attention to the Hosoya index for one interesting class of molecular graphs. The methods of calculation are possible to use for other classes of the molecular graphs. The main problem is again in a rather complicated solution of system of difference equations which often leads only to numeric results.

Acknowledgment. The authors gratefully acknowledge the anonymous referee for his useful remarks and comments.

References

[1] DIESTEL, R. *Graph Theory*, Springer, 2006.

[2] GINGRAS, M. *One hundreds years of helicene chemistry. Part 1: non-stereoselective syntheses of carbohelicenes.* Chem. Soc. Rev., **42** (2013), 968-1006.

[3] GUTMAN, I. *Extremal hexagonal chains*. Journal of Mathematical Chemistry, **12** (1993), 197–210.

[4] HAMZEH, A., IRANMANESH, A., HOSSEIN-ZADEH, S. et al. *The Hosoya index and the Merrifield-Simmons index of some graphs.* Transactions on Combinatorics, **1** (2012), 51-60.

[5] HUANG, G., KUANG, M., DENG, H. *The expected values of Hosoya index and Merrifield-Simmons index in a random polyphenylene chain.* J. Comb. Optim., **12** (2015), 55-67.

[6] LV, X., YU, A. *The Merrifield-Simmons indices and Hosoya indices of trees with a given maximum degree.* MATCH Commun. Math. Comput. Chem., **56** (2006), 605-616.

[7] MERRIFIELD, R. E., SIMMONS, H. E. *Topological Methods in Chemistry*, Wiley, New York, 1989.

[8] PAN, X.F., XU, J.M., YANG, Ch. et al. *Some graphs with minimum Hosoya index and maximum Merrifield-Simmons index.* MATCH Commun. Math. Comput. Chem., **57** (2007), 235-242.

[9] SANDHYA, S. S., RAY, C. D., JAYASEKARAN, C. *Some new results on harmonic mean graphs.* International Journal of Mathematical Archive, **4** (2013), 240-245.

[10] SHIU, W. C. *Extremal Hosoya index and Merrifield-Simmons index of hexagonal spiders.* Discrete Applied Mathematics, **156** (2008), 2978-2985.

[11] VAGNER, S., GUTMAN, I. *Maxima and minima of the Hosoya index and the Merrifield-Simmons index. A survey of results and techniques*. Acta Appl. Math., **112** (2010), 323-346.

[12] YEH, L. *Fibonacci numbers of product graphs.* The Journal of Combinatorial Mathematics and Combinatorial Computing, **32**(2000), 223-229.

[13] ZHANG, L. *Extremal hexagonal chains concerning largest eigenvalue.* Science of China (Series A), **44** (2001), 1089-1097.

Current address

Seibert Jaroslav, doc. RNDr., CSc.
Institute of Mathematics and Quantitative Methods, Faculty of Economics and Administration
University of Pardubice
Studentská 95, 532 10 Pardubice, Czech Republic
E-mail: jaroslav.seibert@upce.cz

Zahrádka Jaromír, RNDr., PhD.
Institute of Mathematics and Quantitative Methods, Faculty of Economics and Administration
University of Pardubice
Studentská 95, 532 10 Pardubice, Czech Republic
E-mail: jaromir.zahradka@upce.cz

Slovak University of Technology in Bratislava
Faculty of Mechanical Engineering
17th Conference on Applied Mathematics **APLIMAT 2018**

Proceedings

STOCHASTIC EXTRAPOLATION OF MORTALITY RATES IN THE CZECH REPUBLIC WITH AN IMPACT ON PROBABILITY AND NUMBER OF SURVIVING

ŠIMPACH Ondřej (CZ), DOTLAČILOVÁ Petra (CZ)

Abstract. The aim of this paper is to compare the results of stochastic (autoregressive integrated moving averages) and deterministic (linear regression) extrapolations of logs of age-and-sex-specific mortality rates in the Czech Republic with impact on selected characteristics of mortality tables – probability and number of surviving. Research shows that selected ARIMA models provide more precise extrapolation of the analysed characteristics mainly in advanced ages and they are much less biased. Deterministic models are useful mainly in lower ages and up to 65 years.

Keywords: stochastic modelling, mortality rates, probability of surviving, projection

Mathematics Subject Classification: Primary 90C30; Secondary 62H12.

1 Introduction

Intensity of mortality (denoted $\mu_{x,t}$) at the advanced ages (65 years and above, which is mostly acknowledged as retirement age in developed populations) becomes important topic for demographers and for analysts in the field of health, pension insurance and for public policy planning. The countries of Central and Eastern Europe are aging as it is in the case of developed Western European countries. Extensions of length of human life in Central and Eastern Europe is influenced mainly by improvement in medicine and in healthcare, which has been running since the early 90s of the last century. At first, there was a significant improvement in a care of live born persons and infants, which caused the decrease of infant mortality and mortality rates during childhood. Later, mortality began to improve even at the advanced ages. Among reasons for this evolution could be included higher level of health care, standard of living, information technologies and scientific progress, better housing conditions or functional health, pension and social systems. Another equally important reason could be more interest in a healthy lifestyle of the people and better environment in most countries (especially in industrial cities with a heavy load transport). This slow prolongation of human life means for our future, that the proportion of persons at higher ages will continue to increase, while the proportion of young people is decreasing.

1.1 Current state and literature review

According to Finkelstein [11], "mortality rates of human populations in developed countries are declining with time". With increasing life expectancy (denoted $e_{x,t}$) improving long-term care and sustaining the pension system are becoming an important issue. Besides, establishing methodologically sound longitudinal data sets is necessary to examine the phenomena (Andel [1]). For the purposes of actuarial demography and life insurance, it is not sufficient to construct deterministic models only, but it is necessary to combine them and to compare the results also which probabilistic and stochastic ones. Finding optimal form of the model for age-and-sex specific mortality rates (denoted $m_{x,t}$) (respectively probability of surviving (denoted $p_{x,t}$) after recalculation according to mortality table algorithm) is a complicated process, because it is necessary to test not only statistical significance of the parameters, but also the significance of the whole models and to respect the results of the diagnostics tests, which include tests of autocorrelation, heteroskedasticity and normality of a non-systematic (residual) component (ε_t). "The actuarial and demographic literature has introduced a myriad of (deterministic and stochastic) models to forecast mortality rates of single populations" (Antonion, Bardoutsos and Ouburg [2]). Work of Gompertz [14] played a key role in shaping the emerging statistical science. Gompertz model provided a powerful stimulus to examine the patterns of death ("law of human mortality") across the life course not only in humans but also in a wide range of other organisms (Makeham [22], Kirkwood [17]). Since that many of models have been developed. Lee and Carter [19] published a new statistical method for forecasting US mortality in 1992. Since that it has been applied on many real data of the populations. For example, Li and Lee [21] applied the Lee-Carter model to a group of populations, allowing each its own age pattern and level of mortality but imposing shared rates of change by age. However, there are more models used. For example, Gogola [13] used stochastic mortality modelling approach – Cairns, Blake and Dowd model that is well suited at very high ages to calculate mortality rates of age categories from 85 to 115 for selected countries. Antonion, Bardoutsos and Ouburg [2] presented in their paper a Bayesian analysis of two related multi-population mortality models of log-bilinear type, designed for two or more populations. Godunov [12] compared several models applied on particular populations and found that „the most appropriate models of smoothing mortality curve are Kannistö-Thatcher (United Kingdom) Martinell (Sweden) and Kannistö (Canada). On the other side, the least suitable models are Coale-Kisker (Singapore), Gompertz-Makeham and modified Gompertz-Makeham (Czech Republic, Slovakia, Germany)." Mortality rates in the Czech Republic was calculated for example by Jindrová and Slavíček [15]. They applied Lee-Carter model on Czech population in the period of 1950–2009 and predicted the development of age-and-sex specific mortality rates and consequently life expectancy for period 2010–2029. Followed Arltová, Langhamrová and Langhamrová [3], who used the same data base, but the life expectancy was predicted up to the year 2050, because unlike Jindrová and Slavíček [15] they tried to find and use the cointegration relationship between male's and female's parameters (denoted k_t) of the Lee-Carter model. This long-term relationship was later used for a better and more significant prediction of life expectancy until 2050. In the broader context, Fiala and Langhamrová [10] analysed of the development of the sex-and-age structure of the Czech population of productive age based on the latest population projection of the Czech Statistical Office (CZSO). Probabilistic projections of age-specific mortality and fertility rates were done for example by Ševčíková et al. [23] to apply probabilistic population projections on United Nations (UN) countries. This important research is best suited to international comparison, because it is constructed with the same approach for all populations.

1.2 Problem formulation and goal of the paper

Finding a suitable model for extrapolation of $\ln(m_{x,t})$ to the future raises a problem that there are about 2 × 101 or more models in advance populations[1], because each gender and age[2] is necessary to model. Therefore, the aim of the paper is to compare selection of ARIMA models towards the linear deterministic models based on the results of the mortality projection that are recalculated on the probabilities of surviving ($p_{x,t}$) of x-year old persons and number of surviving persons (tabular $l_{x,t}$), where the base at age of 0 years is set on 100 000 persons. Paper shows that process of user selection of partial appropriate models is although slower to process, but the results of projections are more credible in comparison with linear deterministic models (see also paper by Šimpach and Langhamrová [26]).

2 Materials and methods

For the description of mortality development is most often used an indicator known as life expectancy ($e_{x,t}$), (as a key output of the mortality tables), but for analytical purposes may be used other characteristics of mortality, such as e.g. age-and-sex specific mortality rates ($m_{x,t}$), probability of surviving ($p_{x,t}$) and tabular number of surviving ($l_{x,t}$). These characteristics are obtained as an output indicator from mortality tables. The calculation is carried out in several steps. First, we calculate the age-and-sex specific mortality rates as

$$m_{x,t} = \frac{M_{x,t}}{\overline{S}_{x,t}}, \tag{1}$$

where $M_{x,t}$ is the number of deaths x years old and $\overline{S}_{x,t}$ is the mid-year number of living persons (exposure to risk). Between the mortality rate and the intensity of mortality (which is considered in similar studies and analysis, see e.g. paper by Bogue, Anderton, Arriaga [4], Fiala [9] or Dotlačilová, Šimpach and Langhamrová [8]) is valid the followed relationship

$$m_{x,t} \approx \mu\left(x + \frac{1}{2}\right). \tag{2}$$

Next, we will calculate the probability of death for 0-year-old person as

$$q_{0,t} = \frac{M_{0,t}}{\alpha N_t^v + (1-\alpha) N_{t-1}^v} \quad \text{for } x = 0, \tag{3}$$

where $M_{0,t}$ is the number of deaths at the age 0, α is the proportion of lower elementary file of died persons from Lexis [20] diagram (approximately 0.85–0.91) and N_t^v, respectively N_{t-1}^v is the number of live births in year t, respectively in year t-1 (see Keyfitz [16]). The probability of dying is given by formula

[1] As stated in subchapter 1.1, Gogola [13] used stochastic mortality modelling approach for persons in very high ages to calculate mortality rates of age categories from 85 to 115 years. It is understandable that at the age of 115, almost nobody lives from an empirical point of view the intensity of mortality at this age must be adjusted.
[2] Age range is usually surveyed by statistical offices in the intervals 0–100+ completed years of life.

$$q_{x,t} = 1 - p_{x,t}, \tag{4}$$

and is valid for age $x > 0$. The calculation of the probability of surviving is given by formula

$$p_{0,t} = 1 - q_{0,t} \text{ for age } x = 0, \tag{5}$$

and by formula

$$p_{x,t} = e^{-m_{x,t}} \text{ for age } x > 0. \tag{6}$$

Next part of the calculation relates to tabular (i.e. imaginary) population. First, we select the initial number of live births in tabular population: $l_{0,t}$ = 100 000. Based on knowledge of the probability of surviving, we can calculate the number of surviving in the further exact ages by

$$l_{x+1,t} = l_{x,t} p_{x,t}, \tag{7}$$

where $l_{x,t}$ is the number of surviving at the exact age x from the default file of 100 000 live births of tabular population.

Extrapolation approaches applied on the mortality development, which are based on deterministic principles are still used in a lot of countries, including the Czech Republic. We can mention e.g. study by Fiala [9] or Koschin et al. [18]. Especially the study by Fiala [9] provides information that $\ln(m_{x,t})$ is for each age x approximately linear and can be modelled by linear function. According to this evidence we can conclude that if we find the appropriate constant (intercept) (β_0) and slope (β_1) of the regression function

$$\ln(m_{x,t}) = \beta_0 + \beta_1 t + \varepsilon_t, \tag{8}$$

we can easily perform extrapolation. This approach is simple, however, because the empirical data of the Czech Republic is quite variable, and the nature of individual time series of mortality is volatile, residuals of these regressions could be autocorrelated and the slope of the regression line (especially in the lowest and highest age groups) may not be optimal for achieving the desired results. Another approach which was used e.g. by Arltová, Langhamrová and Langhamrová [3], Šimpach [24], [25] or Šimpach, Pechrová [27] is based on Box and Jenkins [5] methodology, where the linear regression model (8) is replaced by ARIMA models, which can be generally written as

$$\ln m_{x,t} = \beta_0 + \sum_{i=1}^{p} \phi_i \ln m_{x,t-i} + \sum_{j=1}^{q} \theta_j \varepsilon_{t-j}. \tag{9}$$

where ϕ_i (i = 1, ..., p), and θ_j (j = 1, ..., q) are the parameters of AR and MA process, β_0 represents the constant and ε_t is a pure white noise error term, where $E(\varepsilon_t) = 0$, $D(\varepsilon_t) = \sigma^2$, $cov(\varepsilon_t\, ; \varepsilon_t') = 0$ and $\varepsilon_t \approx$ N (normal) distribution.

Followed by this procedure it is obvious, that in the first step we estimate 101 regression models for males (age 0–100+ years of life) and 101 models for females with use of database of logarithms of age-specific mortality rates. Then, these models are used for extrapolation up to the year 2065, (in the same way as it performed e.g. CZSO [7]). In the second step there are calculated another 2 × 101 models using ARIMA approach with optimally selected parameters. Resulting forms are used to extrapolate the logarithms of age-specific mortality rates, also up to the year 2065. In the third step, there are the empirical and predicted mortality rates recalculated to probability ($p_{x,t}$) and number of surviving ($l_{x,t}$) using an algorithm of mortality tables. For mutual comparison, the differences were calculated as

$$diff\left(p_{x,t}\right) = p_{x,t}^{arima} - p_{x,t}^{regression}, \tag{10}$$

respectively

$$diff\left(l_{x,t}\right) = l_{x,t}^{arima} - l_{x,t}^{regression}, \tag{11}$$

and these results are subsequently presented in charts. The deviations are commented, and the advantages and disadvantages of both approaches are subsequently discussed.

3 Results

The empirical values of male's and female's age-and-sex specific mortality rates are shown in Fig. 1, logarithmic values in Fig. 2. Due to the nature of the data and a small number of living people, high variability is evident in the highest age groups.

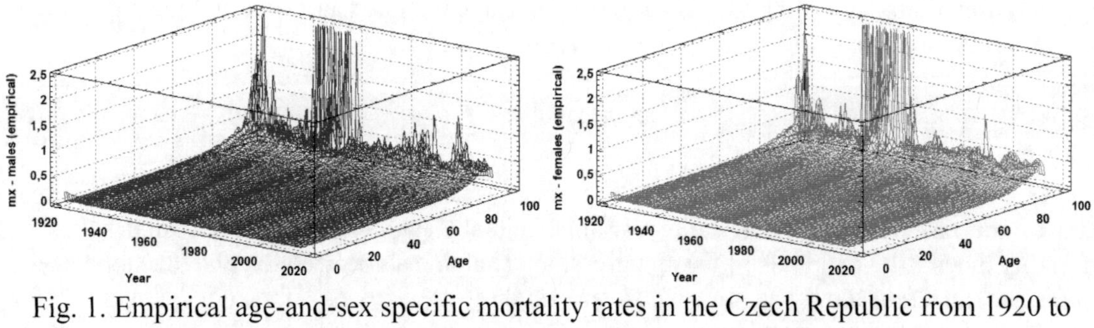

Fig. 1. Empirical age-and-sex specific mortality rates in the Czech Republic from 1920 to 2016, males – left chart, females – right chart.
Data source: CZSO [6], author's calculation and illustration.

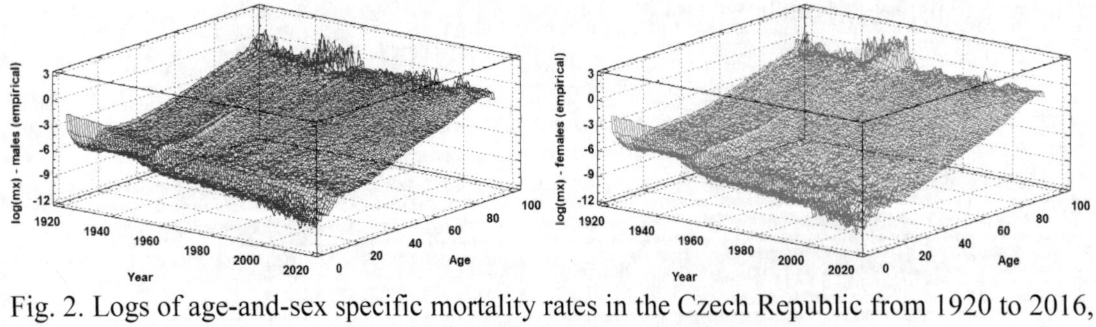

Fig. 2. Logs of age-and-sex specific mortality rates in the Czech Republic from 1920 to 2016, males – left chart, females – right chart.
Data source: CZSO [6], author's calculation and illustration.

From the mathematical nature of the indicator $m_{x,t}$ is clear, that its maximum value can be 1. Higher values presented in the graph are a systematic error which can be eliminated by smoothing in the highest ages. In this article, smoothing will not be performed to prevent data in front of external interference. Results will show how the used methodology can respond to variation of the Czech data.

According to mortality tables algorithm we recalculated $m_{x,t}$ using formulas (3), (4), (5) and (6) into $p_{x,t}$, which are presented in Fig. 3. The root of tabular number of 0-year-old persons is set on 100 000 every year and using by formula (7) we estimated tabular number of surviving. Results are presented in Fig. 4, from which it is clearly visible the change towards better mortality development.

3.1 Modelling

Optimized forms of ARIMA models for logarithms of age-and-sex specific mortality rates are written in Tab. 1 and 2 in Annex of this paper. We respected the residual diagnostic of these models. Forecast was calculated for all 101 males' and 101 females' time series and graphic presentation is shown in Fig. 5 (top charts).

Estimates of intercepts (β_0) and slopes (β_1) of the regression lines for logarithms of age-and-sex specific mortality rates were calculated as well. Extrapolation is presented in Fig. 5 (bottom charts). For better comparison of the results, Fig. 6 also shows the values $e^{\ln(m_{x,t})}$, i.e. classical mortality rates usable for mortality tables algorithm.

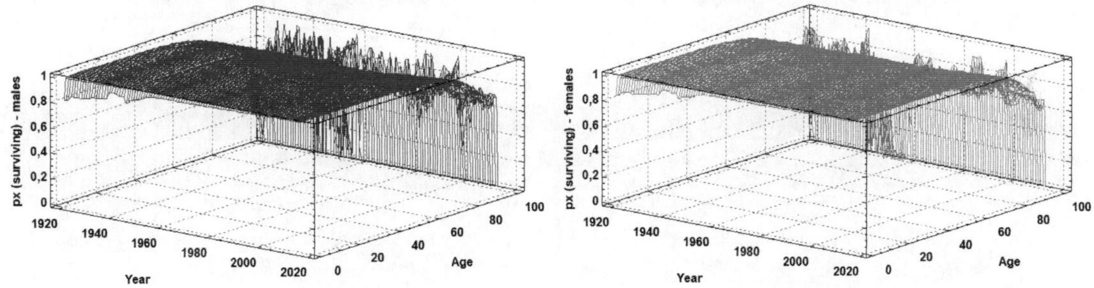

Fig. 3. Calculated age-and-sex specific probability of surviving in the Czech Republic from 1920 to 2016, males – left chart, females – right chart.
Data source: CZSO [6], author's calculation and illustration.

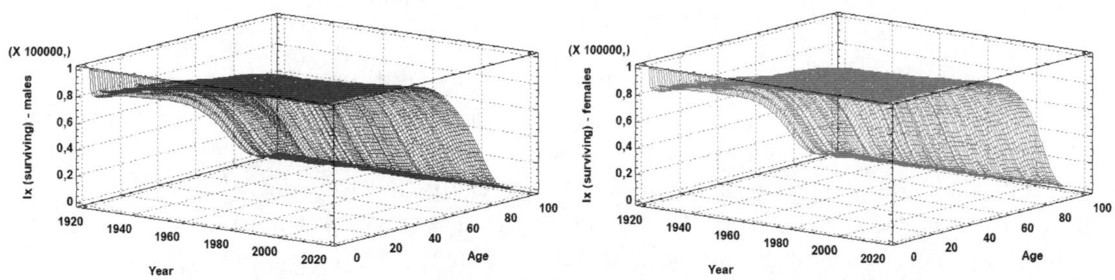

Fig. 4. Calculated age-and-sex specific tabular number of surviving in the Czech Republic from 1920 to 2016, males – left chart, females – right chart.
Data source: CZSO [6], author's calculation and illustration.

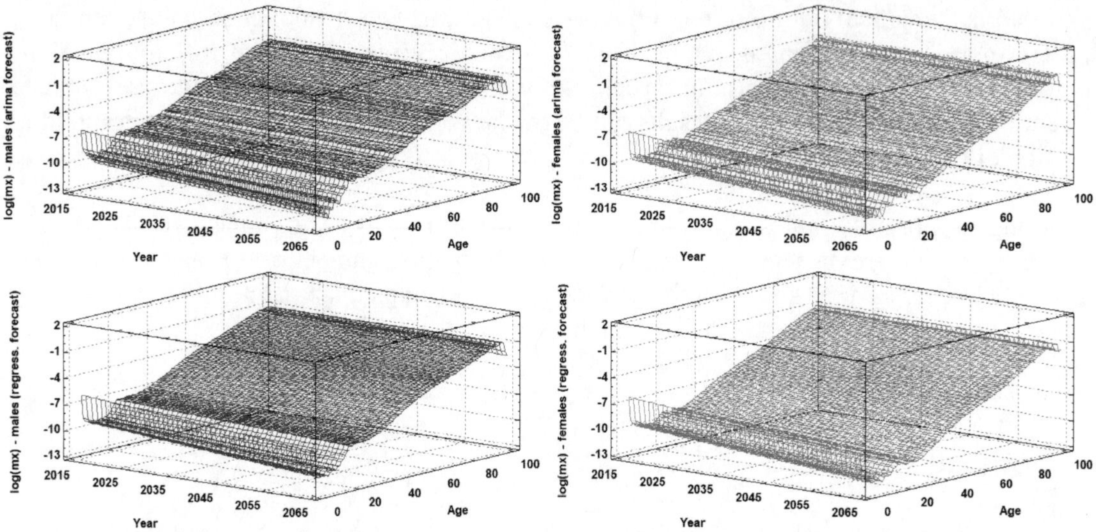

Fig. 5. Forecasted logs of age-and-sex specific mortality rates in the Czech Republic from 2016 to 2065 by selected ARIMA models (top) and by linear regression models (bottom), males – left chart, females – right chart. Source: author's calculation and illustration.

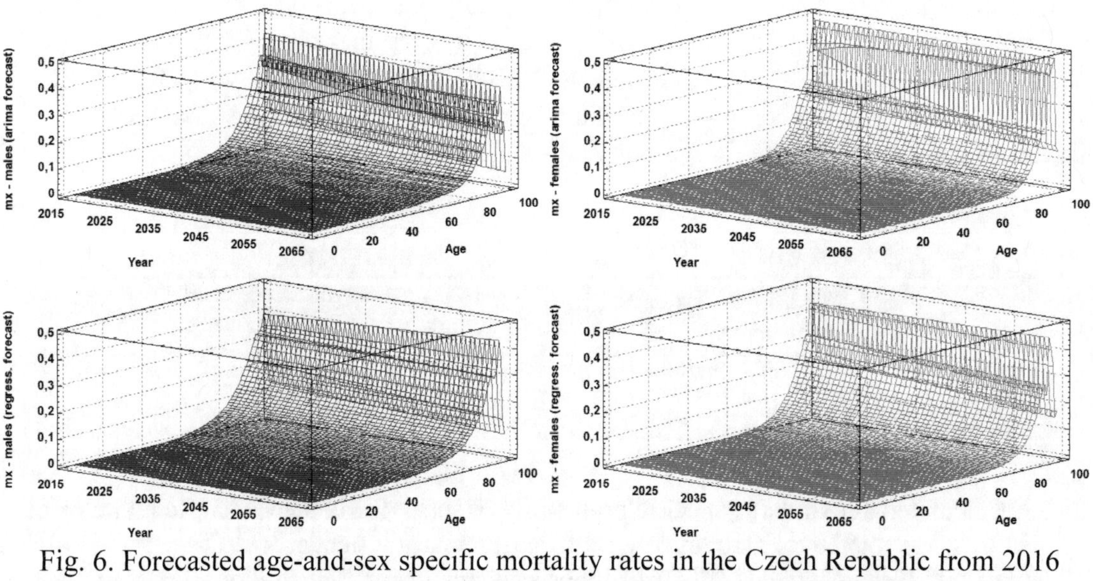

Fig. 6. Forecasted age-and-sex specific mortality rates in the Czech Republic from 2016 to 2065 by selected ARIMA models (top) and by linear regression models (bottom), males – left chart, females – right chart. Source: author's calculation and illustration.

It is clear that results by ARIMA models show higher variability in forecast, but it does not mean that they are wrong. On the contrary, they respond better to the past development of the time series and do not act so "artificially". In the case of regression approach and male population is obvious an error in the form of increasing development of mortality in the highest age groups. This error is due to the high variability in empirical data, and the parameters of linear regression were therefore deflected.

Forecast of probability ($p_{x,t}$) and tabular number ($l_{x,t}$) of surviving up to the year 2065 was calculated on the basis of mortality tables algorithm and results are displayed in Fig. 7 and 8.

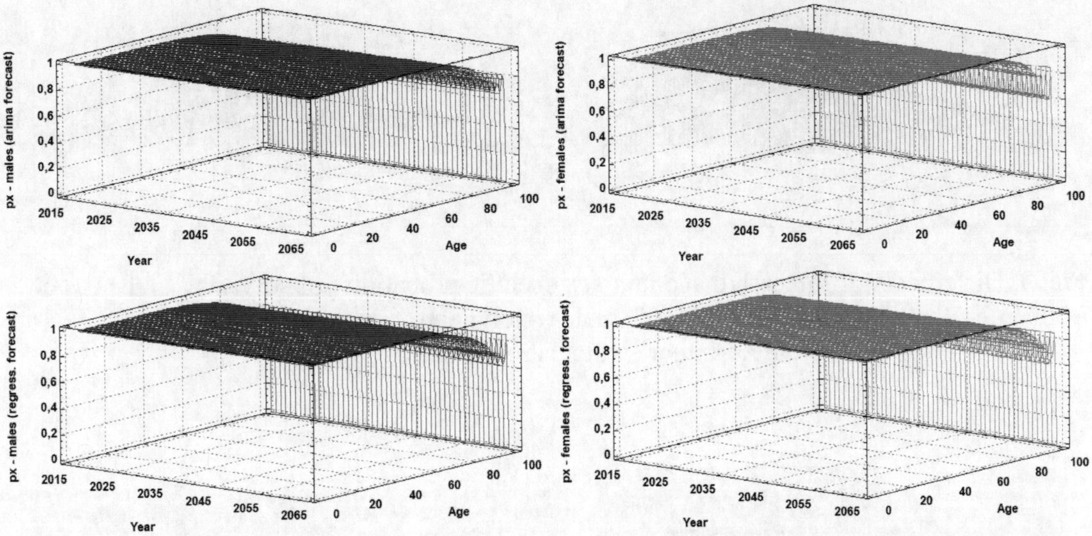

Fig. 7. Forecasted age-and-sex specific probability of surviving in the Czech Republic from 2016 to 2065 by selected ARIMA models (top) and by linear regression models (bottom), males – left chart, females – right chart. Source: author's calculation and illustration.

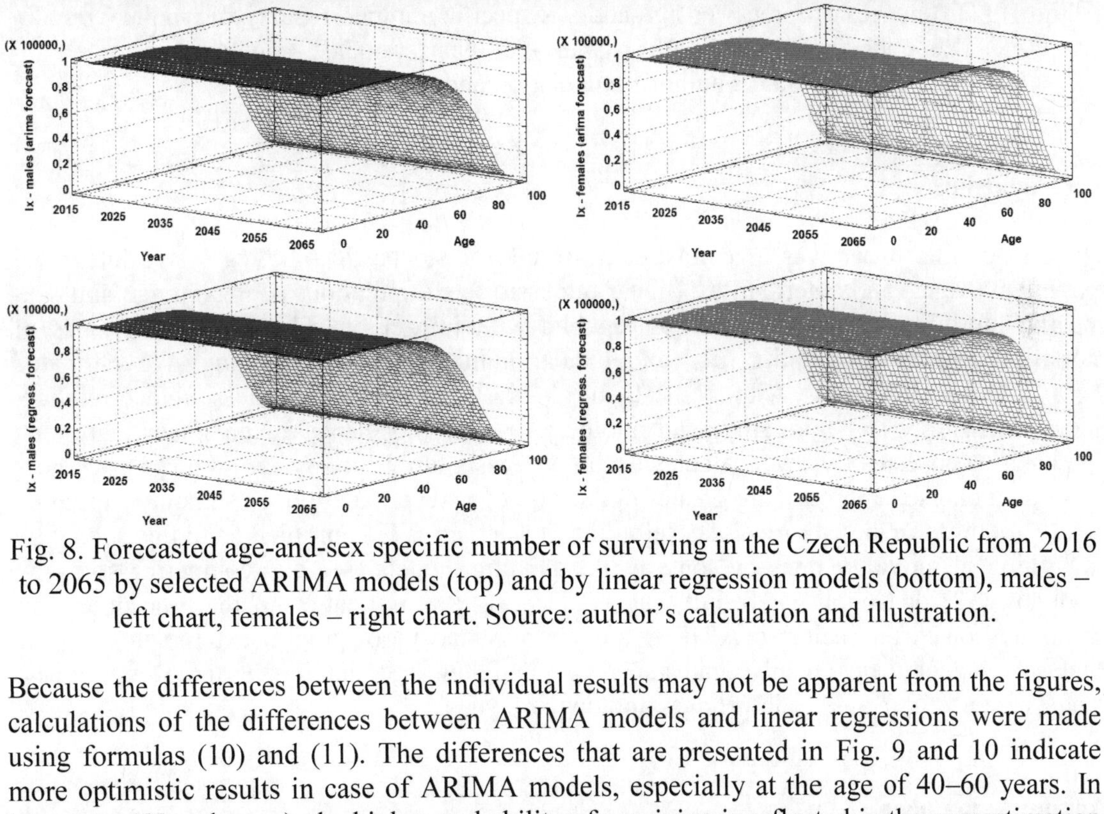

Fig. 8. Forecasted age-and-sex specific number of surviving in the Czech Republic from 2016 to 2065 by selected ARIMA models (top) and by linear regression models (bottom), males – left chart, females – right chart. Source: author's calculation and illustration.

Because the differences between the individual results may not be apparent from the figures, calculations of the differences between ARIMA models and linear regressions were made using formulas (10) and (11). The differences that are presented in Fig. 9 and 10 indicate more optimistic results in case of ARIMA models, especially at the age of 40–60 years. In later ages (65 and more), the higher probability of surviving is reflected in the overestimation of the table number of surviving persons, which is approximately 15,000 in case of male population and 8,000 in case of female population at the end of forecast in 2065. These values were observed in the mode of the distribution of $l_{x,t}$ differences. Increased variability in the case of probability of surviving in the highest age groups (95 and more) is caused by an error in linear regression models that deviated the prediction.

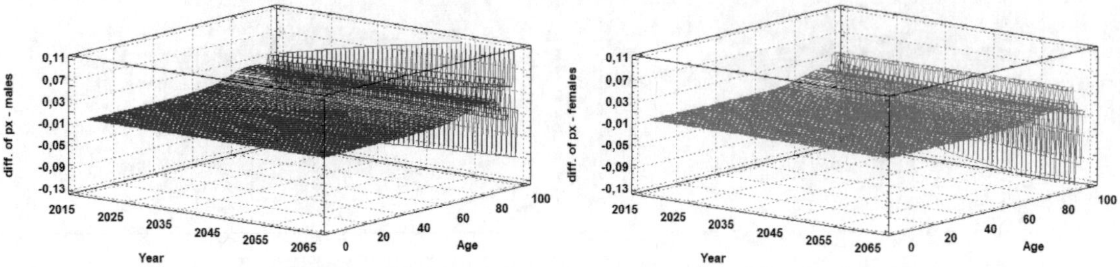

Fig. 9. Differences of forecasted age-and-sex specific probability of surviving in the Czech Republic from 2016 to 2065, males – left chart, females – right chart.
Source: author's calculation and illustration.

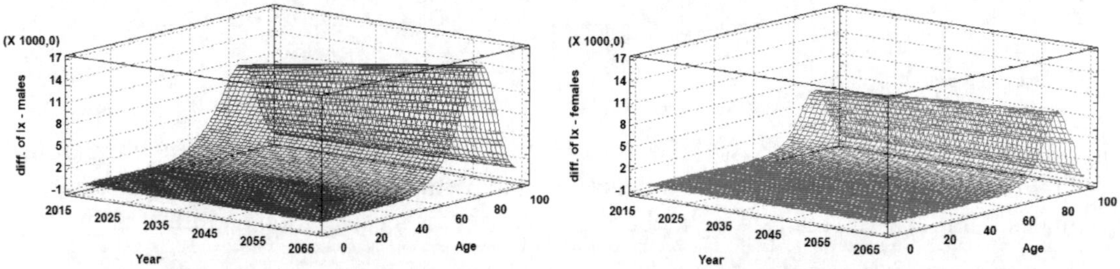

Fig. 10. Differences of forecasted age-and-sex specific number of surviving in the Czech Republic from 2016 to 2065, males – left chart, females – right chart.
Source: author's calculation and illustration.

4 Conclusion

The aim of this paper was to compare the results of stochastic (autoregressive integrated moving averages) and deterministic (linear regression) extrapolations of logs of age-and-sex-specific mortality rates in the Czech Republic with impact on selected characteristics of mortality tables – probability ($p_{x,t}$) and tabular number ($l_{x,t}$) of surviving. We estimated 2×101 ARIMA models with respect to diagnostic tests of residues (autocorrelation, heteroskedasticity and normality) and another 2×101 models based on linear regression approach. Consequently, we calculated the forecasts up to the year 2065. Differences depending on used approach are greater in the case of male population. It is mainly caused by great variability in the empirical dataset at the highest ages. This problem could be solved by smoothing of mortality rates by some of the existing models (see e.g. comparing study by Godunov [12] or Gogola [13]), but that would bring a prolongation of the procedure. Our research showed that selected ARIMA models provided more precise extrapolation of the analysed characteristics mainly in advanced ages and they are much less biased. Deterministic models are useful mainly in lower ages and up to 65 years.

Acknowledgement

The paper was supported from the Internal Grant Agency of University of Economics Prague under no. 35/2017 "Demographic models in R Software" and also from the Czech Science Foundation Project no. P402/12/G097 "Dynamic Models in Economics".

References

[1] ANDEL, R. Aging in the Czech Republic. *Gerontologist*, vol. 54, no. 6, 2014, pp. 893–900.

[2] ANTONION, K., BARDOUTSOS, A., and OUBURG, W. Bayesian Poisson log-bilinear models for mortality projections with multiple populations. *European Actuarial Journal*, vol. 5, no. 2, 2015, pp 245–281.

[3] ARLTOVÁ, M., LANGHAMROVÁ, JI., LANGHAMROVÁ, JA. Development of life expectancy in the Czech Republic in years 1920-2010 with an outlook to 2050. *Prague Economic Papers*, vol. 22, no. 1, 2013, pp. 125–143.

[4] BOGUE, D. J., ANDERTON, D. L., ARRIAGA E. E. Population models, Projections and Estimates. *Readings in Population Research Methodology*, vol. 5, 1993, United Nations Social Development Center, Chicago.

[5] BOX, G. E. P., JENKINS, G. *Time series analysis: Forecasting and Control*. San Francisco: Holden-Day, 1970.

[6] CZSO. Obyvatelstvo podle Sčítání lidu, domů a bytů, Česká republika a kraje, *[DVD] Sčítání lidu, domů a bytů 2011*, Praha: Český statistický úřad, 2012.

[7] CZSO. *Projekce obyvatelstva České republiky do roku 2100*, [on-line]. Available at: http://www.czso.cz/csu/2013edicniplan.nsf/publ/4020-13-n_2013, [cited: 10.11.2015].

[8] DOTLAČILOVÁ, P., ŠIMPACH, O., LANGHAMROVÁ, J. DERAS Versus MS EXCEL Solver in Levelling the Life Expectancy at Birth. *In: APLIMAT 2014: 13th Conference on Applied Mathematics*. Bratislava: Publishing House of Slovak University of Technology, 2014, pp. 108–114.

[9] FIALA, T. Reducing the Consequences of Population Ageing for the Pension System by Raising the Retirement Age in the Czech Republic. *In: The 7th International Days of Statistics and Economics*. Slaný: Melandrium, 2013, pp. 398–407.

[10] FIALA, T., LANGHAMROVÁ, J. Increase of Labor Force of Older Age – Challenge for the Czech Republic in Next Decades. *Procedia Economics and Finance*, vol. 12, 2014, pp. 144–153.

[11] FINKELSTEIN, M. S. Lifesaving explains mortality decline with time. *Mathematical Biosciences*, vol. 196, no. 2, 2005, pp. 187–197.

[12] GODUNOV, I. *Comparison of official life tables construction in selected countries [Bachelor's thesis]*. Charles University in Prague, Faculty of Science, Department of Demography and Geodemography, 2015.

[13] GOGOLA, J. Modelling mortality rate of the very old. *In: 9th Professor Aleksander Zelias International Conference on Modelling and Forecasting of Socio-Economic Phenomena*, Cracow: Cracow University of Economics, 2015, pp. 58–67.

[14] GOMPERTZ, B. On the Nature of the Function Expressive of the Law of Human Mortality, and on a New Mode of Determining the Value of Life Contingencies. *Philosophical Transactions of the Royal Society of London*, vol. 115, 1825, pp. 513–585.

[15] JINDROVÁ, P., SLAVÍČEK, O. Life Expectancy Development and Prediction for Selected European Countries *In: 6th International Scientific Conference Managing and Modelling of Financial Risks*, Ostrava: VŠB-TU Ostrava, 2014, pp. 303–312.

[16] KEYFITZ, N. Experiments in the projection of mortality. *Canadian Studies in Population*, vol. 18, no. 2, 1991, pp. 1–17.

[17] KIRKWOOD, T. B. L. Deciphering death: a commentary on Gompertz (1825) 'On the nature of the function expressive of the law of human mortality, and on a new mode of determining the value of life contingencies'. *Philosophical Transactions of the Royal Society, B-Biological Sciences*, vol. 370, no. 1666, 2015, UNSP 20140379.

[18] KOSCHIN, F. et al. *Prognóza lidského kapitálu obyvatelstva České republiky do roku 2050*. Praha: Oeconomica. 2007, 105 p.

[19] LEE, R. D. CARTER, L. Modeling and Forecasting the Time Series of U.S. Mortality. *Journal of the American Statistical Association*, vol. 87, 1992, pp. 659–671.

[20] LEXIS, W. *Einleintung in die Theorie der Bevölkerungs-Statistik [Introduction to the Theory of Population Statistics]*. Strasbourg: Trubner, 1875.

[21] LI, N., LEE, R. D. Coherent mortality forecasts for a group of populations: An extension of the Lee-Carter method. *Demography*, vol. 42, no. 3, 2005, pp. 575–594.

[22] MAKEHAM, W. M. On the Law of Mortality and the Construction of Annuity Tables. *The Assurance Magazine, and Journal of the Institute of Actuaries*, vol. 1860, no. 8, pp. 301–310.

[23] ŠEVČÍKOVÁ, H., LI, N., KANTOROVÁ, V., GERLAND, P., and RAFTERY, A. E. *Age-Specific Mortality and Fertility Rates for Probabilistic Population Projections*. Center for Statistics and the Social Sciences, University of Washington, Working Paper no. 150, March 17, 2015, pp. 1–31.

[24] ŠIMPACH, O. Application of the Modern Approach to Age-specific Fertility Rates Stochastic Modelling in the Czech Republic. *In: Mathematical Methods in Economics 2015*. Plzeň: University of West Bohemia, 2015, pp. 799–804.

[25] ŠIMPACH, O. Possible Structure Development of Job Seekers by the Highest Level of Education Attained. *In: Proceedings of the 21st International Conference Current Trends in Public Sector Research 2017*. Brno: Masaryk University, 2017, pp. 387–393.

[26] ŠIMPACH, O., LANGHAMROVÁ, J. Stochastic Modelling of Age-specific Mortality Rates for Demographic Projections: Two Different Approaches. *In: 32nd International Conference on Mathematical Methods in Economics (MME)*, Olomouc: Palacký University in Olomouc, 2014, pp. 890–895.

[27] ŠIMPACH, O., PECHROVÁ, M. Development of the Czech Farmers' Age Structure and the Consequences for Subsidy Policy. *Agris on-line Papers in Economics and Informatics*, vol. 7, no. 3, 2015, pp. 57–69.

Annex

Age	Model	Age	Model	Age	Model	Age	Model	Age	Model	Age	Model	Age	Model	Age	Model
0	(0,2,1) c	13	(0,1,1) c	26	(0,1,1) c	39	(0,1,1) c	52	(0,1,1) c	65	(0,1,1) c	78	(0,1,1) c	91	(0,1,2) c
1	(0,1,1) c	14	(0,1,1) c	27	(0,1,1) c	40	(2,2,1)	53	(0,1,1) c	66	(0,1,1) c	79	(1,1,1) c	92	(2,1,0) c
2	(2,2,1)	15	(0,1,1) c	28	(0,1,1) c	41	(0,1,1) c	54	(0,1,1) c	67	(0,1,1) c	80	(0,1,1) c	93	(0,1,1) c
3	(0,1,1) c	16	(0,1,1) c	29	(0,1,1) c	42	(0,1,1) c	55	(0,1,1) c	68	(0,1,1) c	81	(0,1,1) c	94	(0,1,2) c
4	(0,1,1) c	17	(0,1,1) c	30	(0,1,1) c	43	(0,1,1) c	56	(2,1,0) c	69	(0,1,1) c	82	(0,1,1) c	95	(1,1,1) c
5	(0,1,1) c	18	(0,1,1) c	31	(2,2,1)	44	(0,1,1) c	57	(0,1,1) c	70	(0,1,1) c	83	(1,1,1) c	96	(0,1,2) c
6	(0,1,1) c	19	(0,1,1) c	32	(0,1,1) c	45	(0,1,1) c	58	(0,1,1) c	71	(0,1,1) c	84	(0,1,1) c	97	(0,1,1) c
7	(0,1,1) c	20	(0,1,2) c	33	(0,1,1) c	46	(0,1,1) c	59	(0,1,1) c	72	(0,1,1) c	85	(0,1,1) c	98	(0,1,1) c
8	(0,1,1) c	21	(0,1,1) c	34	(0,1,1) c	47	(2,1,0) c	60	(0,1,1) c	73	(1,1,0) c	86	(0,1,1) c	99	(0,1,2) c
9	(0,1,1) c	22	(0,1,1) c	35	(0,1,1) c	48	(1,1,0) c	61	(0,1,1) c	74	(1,1,0) c	87	(0,1,1) c	100+	(0,1,1) c
10	(0,1,1) c	23	(0,1,1) c	36	(0,1,1) c	49	(0,1,1) c	62	(0,1,1) c	75	(2,1,0) c	88	(0,1,1) c		
11	(0,1,1) c	24	(0,1,1) c	37	(0,1,1) c	50	(2,1,0) c	63	(0,1,1) c	76	(0,1,1) c	89	(0,1,2) c		
12	(0,1,1) c	25	(2,1,0) c	38	(0,1,1) c	51	(1,1,0) c	64	(0,1,1) c	77	(0,1,1) c	90	(2,1,0) c		

Tab. 1. ARIMA (*p*, *d*, *q*) models with or without constant (*c*) for logs of male's age-specific mortality rates in range of 0–100+ years in the Czech Republic.
Source: author's calculation and illustration.

Age	Model	Age	Model	Age	Model	Age	Model	Age	Model	Age	Model	Age	Model	Age	Model
0	(0,2,1) c	13	(0,1,1) c	26	(0,1,1) c	39	(0,1,1) c	52	(0,1,1) c	65	(1,1,0) c	78	(2,1,2) c	91	(1,1,1) c
1	(0,1,1) c	14	(1,1,1) c	27	(0,1,1) c	40	(2,1,0) c	53	(0,1,1) c	66	(1,1,0) c	79	(0,1,1) c	92	(2,1,0) c
2	(1,1,1) c	15	(0,1,1) c	28	(0,1,1) c	41	(1,1,0) c	54	(0,1,1) c	67	(0,1,1) c	80	(0,1,1) c	93	(0,1,1) c
3	(2,1,0) c	16	(2,1,0) c	29	(0,1,1) c	42	(0,1,1) c	55	(0,1,1) c	68	(0,1,1) c	81	(0,1,1) c	94	(2,1,1) c
4	(2,1,0) c	17	(0,1,1) c	30	(0,1,1) c	43	(0,1,1) c	56	(0,1,1) c	69	(0,1,1) c	82	(0,1,1) c	95	(1,0,0) c
5	(0,1,1) c	18	(0,1,2) c	31	(0,1,1) c	44	(0,1,1) c	57	(0,1,1) c	70	(0,1,1) c	83	(0,1,1) c	96	(1,0,0) c
6	(1,1,1) c	19	(0,1,1) c	32	(0,1,1) c	45	(0,1,1) c	58	(2,1,0) c	71	(1,1,0) c	84	(0,1,1) c	97	(0,0,2) c
7	(0,1,1) c	20	(0,1,1) c	33	(0,1,1) c	46	(0,1,1) c	59	(1,1,1) c	72	(2,1,0) c	85	(0,1,1) c	98	(0,0,2) c
8	(2,1,0) c	21	(0,1,1) c	34	(0,1,1) c	47	(0,1,1) c	60	(0,1,1) c	73	(1,1,0) c	86	(0,1,1) c	99	(1,1,0) c
9	(2,1,0) c	22	(2,1,0) c	35	(2,1,0) c	48	(0,1,1) c	61	(0,1,2) c	74	(2,1,0) c	87	(0,1,1) c	100	(0,1,1) c
10	(0,1,1) c	23	(0,1,1) c	36	(0,1,1) c	49	(1,1,1) c	62	(1,1,0) c	75	(2,1,0) c	88	(0,1,1) c		
11	(0,1,1) c	24	(1,1,0) c	37	(0,1,1) c	50	(0,1,1) c	63	(1,1,0) c	76	(0,1,1) c	89	(1,1,1) c		
12	(1,1,1) c	25	(0,1,1) c	38	(0,1,1) c	51	(0,1,1) c	64	(0,1,1) c	77	(0,1,1) c	90	(2,1,0) c		

Tab. 2. ARIMA (*p*, *d*, *q*) models with or without constant (*c*) for logs of female's age-specific mortality rates in range of 0–100+ years in the Czech Republic.
Source: author's calculation and illustration.

Current address

Šimpach Ondřej, Ing., Ph.D.
University of Economics in Prague, Faculty of Informatics and Statistics
Department of Statistics and Probability
W. Churchill sq. 4, 130 67 Prague 3, Czech Republic
E-mail: ondrej.simpach@vse.cz, Tel.: +420 737 665 461, +420 224 095 273

Dotlačilová Petra, Ing.
University of Economics in Prague, Faculty of Informatics and Statistics
Department of Mathematics
W. Churchill sq. 4, 130 67 Prague 3, Czech Republic
E-mail: petra.dotlacilova@vse.cz, Tel.: +420 224 095 273

Slovak University of Technology in Bratislava
Faculty of Mechanical Engineering
17th Conference on Applied Mathematics **APLIMAT 2018**

Proceedings

MODELS USED IN FUZZY LINEAR REGRESSION

ŠKRABÁNEK Pavel (CZ), MAREK Jaroslav (CZ)

Abstract. The paper summarizes fuzzy linear models that were introduced in fuzzy regression analysis. The exhaustive review is extended with new models. This contribution is intended as a guide for implementing known fuzzy regression methods. However, it can be used as an inspiration for researchers developing new approaches.

Keywords: fuzzy regression, possibilistic regression, fuzzy linear model, fuzzy observations

Mathematics subject classification: Primary 62E86, 03E72.

Fuzzy linear regression analysis is a vital alternative to commonly used statistics-based regression methods. In the fuzzy linear regression analysis, a wide variety of fuzzy linear models can be used for approximation of a linear dependence, according to a set of observations. To orient the user in selection of an appropriate model, we issue an exhaustive review of fuzzy linear models that were expected while developing model parameter estimators. We provided the models with references on parameter estimators. We also introduced new fuzzy linear models that can be used in fuzzy regression. We classified the models into three categories according to datatype they are intended for. We also provided remarks on model features from the perspective of model predictions. The review is intended as a guide for implementing fuzzy regression estimators as well as an inspiration for researchers developing new parameter estimation methods aimed at fuzzy data.

1 Introduction

Fuzzy regression analysis is an alternative to the commonly used statistical regression analysis. Both in fuzzy regression analysis and in statistical regression analysis, a regression model is used to describe a functional relationship between a dependent y and independent variables x. Parameters of the model are estimated utilizing a set of observations of the variables x and y. With estimated parameters, the regression model can be used for prediction of the dependent variable value for a specific combination of the independent variable values. While any deviation of the prediction from the corresponding observation is supposed to be due to a random error or measurement errors in the statistical regression, the fuzzy regression expects the deviations due to fuzziness. The fuzziness can be inherently given by the nature of observations (e.g. observations described by linguistic terms) [31, 27, 24]; however, it can also be due to the imprecise observed data or the indefiniteness of the system structure and parameters [3]. The fuzzy regression is also a viable alternative to statistical regression when the dataset is insufficient to support statistical regression analysis [20].

Probably the most commonly used model in statistical regression analysis is a linear model. In fuzzy regression analysis, the linear dependence of y on x is also most often expected. However, the fuzzy regression analysis offers several fuzzy models that can be used to approximate the expected dependence. The observations of the dependent and independent variables can be either real value numbers x, y or fuzzy numbers $\tilde{X}, \tilde{Y}$. In this paper, we bring an exhaustive review of fuzzy models that were expected during development of model parameter estimators. We provide the models with references on parameter estimators. We also present new fuzzy linear models that can be used in fuzzy regression. The referred fuzzy models are classified into three categories according to the type of observations they are intended for. We further provide remarks on model features from a perspective of model predictions.

2 Preliminaries

This section originates from [2, 36].

2.1 Fuzzy numbers

Definition 1. A *fuzzy set* A defined on a universe X of elements x is defined by a mapping $\mu_A : X \to [0, 1]$, where $\mu_A(x)$ is the membership degree of x to the fuzzy set A, $\mu_A(x) = 1$ means full membership of x in A, and $\mu_A(x) = 0$ expresses non-membership. If a fuzzy set has a positive degree of membership to a single element $x \in X$, then the set is called *fuzzy singleton.*

Definition 2. Let A be a fuzzy set defined on the universe X of elements x. The set of elements that belong to the fuzzy set A at least to a degree α is called *α-level set* and it is given as

$$A_\alpha = \{x \in X | \mu_A(x) \geq \alpha\},$$

where $0 < \alpha \leq 1$. The α-level set of A for $\alpha = 1$ is called a *core* of the fuzzy set A.

Definition 3. A fuzzy subset $\tilde{A}$ defined on $\mathbb{R}$ with membership function $\mu_{\tilde{A}} : \mathbb{R} \to [0, 1]$ is called a *fuzzy number* if

(a) $\mu_{\tilde{A}}$ is normal, i.e. $\exists x_0 \in \mathbb{R}$ with $\mu_{\tilde{A}}(x_0) = 1$,
(b) $\mu_{\tilde{A}}$ is fuzzy convex, (i.e. $\mu_{\tilde{A}}(tx + (1-t)y) \geq \min\{\mu_{\tilde{A}}(x), \mu_{\tilde{A}}(y)\}, \forall t \in [0, 1], x, y \in \mathbb{R}$),
(c) $\mu_{\tilde{A}}$ is upper semi-continuous on $\mathbb{R}$ (i.e. $\forall \varepsilon > 0 \exists \delta > 0$: $\mu_{\tilde{A}}(x) - \mu_{\tilde{A}}(x_0) < \varepsilon, |x - x_0| < \delta$),
(d) $\mu_{\tilde{A}}$ is compactly supported, i.e. $\mathrm{cl}\{x \in \mathbb{R}; \mu_{\tilde{A}}(x) > 0\}$ is compact, where $\mathrm{cl}(A)$ denotes the closure of the set A.

Let us denote by $\mathbb{R}_\mathcal{F}$ the space of fuzzy numbers.

Remark. *A real number $a \in \mathbb{R}$ is also a fuzzy number $\tilde{A} \in \mathbb{R}_\mathcal{F}$ which has the positive degree of membership only for the element a, i.e. the fuzzy number $\tilde{A}$ is the fuzzy singleton. Such numbers are called crisp numbers.*

Definition 4. Let L and R be continuous decreasing functions $L, R : [0, +\infty) \to [0, 1]$ fulfilling $L(0) = R(0) = 1$ and $L(1) = R(1) = 0$, invertible on $[0, 1]$. Moreover, let $m^L_{\tilde{A}}, m^R_{\tilde{A}} \in \mathbb{R}$, where

$m^L_{\tilde{A}} \leq m^R_{\tilde{A}}$, and $\alpha_{\tilde{A}}, \beta_{\tilde{A}} \in \mathbb{R}^+$; then a fuzzy set $\tilde{A}$ is said to be an *L-R type fuzzy number*, if its membership function is

$$\mu_{\tilde{A}} = \begin{cases} L\left(\frac{m^L_{\tilde{A}} - x}{\alpha_{\tilde{A}}}\right), & x < m^L_{\tilde{A}}, \\ 1, & m^L_{\tilde{A}} \leq x \leq m^R_{\tilde{A}}, \\ R\left(\frac{x - m^R_{\tilde{A}}}{\beta_{\tilde{A}}}\right), & x > m^R_{\tilde{A}}, \end{cases}$$

where $m^L_{\tilde{A}}$ and $m^R_{\tilde{A}}$ are the *left* and the *right points of the core*, and $\alpha_{\tilde{A}}$ and $\beta_{\tilde{A}}$ are the *left* and the *right spread* of the fuzzy number $\tilde{A}$. The L-R type fuzzy number $\tilde{A}$ can be written as

$$\tilde{A} = \left(m^L_{\tilde{A}}, m^R_{\tilde{A}}, \alpha_{\tilde{A}}, \beta_{\tilde{A}}\right)_{LR}.$$

2.2 Fuzzy algebra

Using the extension principle [34], fuzzy algebra can be developed for the L-R type fuzzy numbers.

Definition 5. The *r-level set* of the fuzzy number $\tilde{A}$ is defined as $\tilde{A}_r = \{x \in \mathbb{R} | \mu_{\tilde{A}}(x) \geq r\}$. The set $\tilde{A}_r$ is a closed interval $\tilde{A}_r = [\tilde{A}^-_r, \tilde{A}^+_r]$.

Definition 6. *Scalar multiplication* $(\lambda \cdot \tilde{A})$ is multiplication between the real number $\lambda \in \mathbb{R}$ and the fuzzy number $\tilde{A} \in \mathbb{R}_{\mathcal{F}}$ where $(\lambda \cdot \tilde{A}) \in \mathbb{R}_{\mathcal{F}}$. This operation is defined as

$$(\lambda \cdot \tilde{A})_r = \left\{\lambda x \middle| x \in \tilde{A}_r\right\} = \lambda \tilde{A}_r, \forall r \in [0, 1].$$

Note that $\lambda \tilde{A}_r$ is the usual product of a number and a subset of $\mathbb{R}$. For any $\lambda, \kappa \in \mathbb{R}$ and $\tilde{A} \in \mathbb{R}_{\mathcal{F}}$, it holds that

$$(\lambda \kappa) \cdot \tilde{A} = \lambda \cdot (\kappa \cdot \tilde{A}).$$

Remark. *It holds that* $((\lambda \cdot \tilde{A})^+_r - (\lambda \cdot \tilde{A})^-_r) \geq (\tilde{A}^+_r - \tilde{A}^-_r)$ *for* $|\lambda| \geq 1, \forall r \in [0, 1]$; *and* $((\lambda \cdot \tilde{A})^+_r - (\lambda \cdot \tilde{A})^-_r) < (\tilde{A}^+_r - \tilde{A}^-_r)$ *for* $|\lambda| < 1, \forall r \in [0, 1]$.

Definition 7. Sum of two fuzzy numbers $\tilde{A}$ and $\tilde{B}$, $\tilde{A} \oplus \tilde{B}$, where $\tilde{A}, \tilde{B}, (\tilde{A} \oplus \tilde{B}) \in \mathbb{R}_{\mathcal{F}}$, is given as

$$(\tilde{A} \oplus \tilde{B})_r = \left\{x + y \middle| x \in \tilde{A}, y \in \tilde{B}\right\} = \tilde{A}_r + \tilde{B}_r, \forall r \in [0, 1].$$

(a) The sum of fuzzy numbers is commutative and associative, i.e.

$$\tilde{A} \oplus \tilde{B} = \tilde{B} \oplus \tilde{A}$$

and

$$\tilde{A} \oplus (\tilde{B} \oplus \tilde{C}) = (\tilde{A} \oplus \tilde{B}) \oplus \tilde{C}, \forall \tilde{A}, \tilde{B}, \tilde{C} \in \mathbb{R}_{\mathcal{F}}.$$

(b) The fuzzy singleton $\tilde{0} \in \mathbb{R}_{\mathcal{F}}$ (i.e. $\mu_{\tilde{0}}(x) = 1$ for $x = 0$ and $\mu_{\tilde{0}}(x) = 0$ for $x \neq 0$) is the neutral element with respect to (w.r.t.) $\oplus$, i.e.

$$\tilde{A} \oplus \tilde{0} = \tilde{0} \oplus \tilde{A} = \tilde{A}$$

for any $\tilde{A} \in \mathbb{R}_{\mathcal{F}}$.

(c) None of $\tilde{A} \in \mathbb{R}_{\mathcal{F}} \setminus \mathbb{R}$ has an opposite in $\mathbb{R}_{\mathcal{F}}$ (w.r.t. $\oplus$).

(d) For any $\lambda, \kappa \in \mathbb{R}$ with $(\lambda\kappa) \geq 0$ and any $\tilde{A} \in \mathbb{R}_{\mathcal{F}}$, we have the distributive law

$$(\lambda + \kappa) \cdot \tilde{A} = \lambda \cdot \tilde{A} \oplus \kappa \cdot \tilde{A}.$$

This property does not hold for any $\lambda, \kappa \in \mathbb{R}$.

(e) For any $\lambda \in \mathbb{R}$ and $\tilde{A}, \tilde{B} \in \mathbb{R}_{\mathcal{F}}$, the distributive law is fulfilled:

$$\lambda \cdot (\tilde{A} \oplus \tilde{B}) = \lambda \cdot \tilde{A} \oplus \lambda \cdot \tilde{B}.$$

Remark. *$\tilde{A}_r + \tilde{B}_r$ is the sum of two intervals (as subsets of $\mathbb{R}$). It means that* $((\tilde{A}\oplus\tilde{B})_r^+ - (\tilde{A}\oplus\tilde{B})_r^-) \geq \max\{(\tilde{A}_r^+ - \tilde{A}_r^-), (\tilde{B}_r^+ - \tilde{B}_r^-)\}, \forall r \in [0,1]$.

Definition 8. Product of two fuzzy numbers $\tilde{A}$ and $\tilde{B}$, $\tilde{A} \otimes \tilde{B}$, where $\tilde{A}, \tilde{B}, (\tilde{A} \otimes \tilde{B}) \in \mathbb{R}_{\mathcal{F}}$, is defined as

$$(\tilde{A} \otimes \tilde{B})_r^- = \min\left\{\tilde{A}_r^-\tilde{B}_r^-, \tilde{A}_r^-\tilde{B}_r^+, \tilde{A}_r^+\tilde{B}_r^-, \tilde{A}_r^+\tilde{B}_r^+\right\},$$

and

$$(\tilde{A} \otimes \tilde{B})_r^+ = \max\left\{\tilde{A}_r^-\tilde{B}_r^-, \tilde{A}_r^-\tilde{B}_r^+, \tilde{A}_r^+\tilde{B}_r^-, \tilde{A}_r^+\tilde{B}_r^+\right\}, \forall r \in [0,1].$$

(a) The fuzzy singleton set $\tilde{1} \in \mathbb{R}_{\mathcal{F}}$ (i.e. $\mu_{\tilde{1}}(x) = 1$ for $x = 1$ and $\mu_{\tilde{1}}(x) = 0$ for $x \neq 1$) is the neutral element w.r.t. $\otimes$, i.e.

$$\tilde{A} \otimes \tilde{1} = \tilde{1} \otimes \tilde{A} = \tilde{A},$$

for any $\tilde{A} \in \mathbb{R}_{\mathcal{F}}$.

(b) None of $\tilde{A} \in \mathbb{R}_{\mathcal{F}} \setminus \mathbb{R}$ has an opposite in $\mathbb{R}_{\mathcal{F}}$ (w.r.t. $\otimes$).

(c) For any $\tilde{A}, \tilde{B}, \tilde{C} \in \mathbb{R}_{\mathcal{F}}$ we have

$$((\tilde{A} \oplus \tilde{B}) \otimes \tilde{C})_r \subseteq (\tilde{A} \otimes \tilde{C})_r + (\tilde{B} \otimes \tilde{C})_r, \forall r \in [0,1].$$

and, in general, distributivity does not hold.

(d) For any $\tilde{A}, \tilde{B}, \tilde{C} \in \mathbb{R}_{\mathcal{F}}$, where none of the supports of $\tilde{A}, \tilde{B}, \tilde{C}$ contain 0, we have

$$\tilde{A} \otimes (\tilde{B} \otimes \tilde{C}) = (\tilde{A} \otimes \tilde{B}) \otimes \tilde{C}.$$

3 Fuzzy linear models

Fuzzy linear models describe linear relations between the independent variables x and the dependent variable y. The models differ in expected datatypes of the variables x, y, in datatypes of model parameters, as well as in number of the parameters. Since the model selection is primarily datatype driven, we used the datatype as the feature for classification of the models.

3.1 Models for fuzzy dependent and fuzzy independent variables

Fuzzy models belonging to this category, expect that observations of the dependent y, as well as observations of all m independent variables x are fuzzy numbers, i.e. $\tilde{X}_i, \tilde{Y} \in \mathbb{R}_{\mathcal{F}}, \forall i \in \{1, \ldots, m\}$. The extensively used model of this category [9, 29, 28, 10, 15, 23, 30, 18, 19, 21] is given as

$$\tilde{Y} = \tilde{A}_0 \oplus \left(\tilde{A}_1 \otimes \tilde{X}_1\right) \oplus \ldots \oplus \left(\tilde{A}_m \otimes \tilde{X}_m\right), \tag{1}$$

where $\tilde{A}$ denotes fuzzy parameters of the regression model, and $\tilde{A}_i \in \mathbb{R}_{\mathcal{F}}, \forall i \in \{0, \ldots, m\}$.

Other fuzzy linear models of this category can be inferred from the model (1) through specifying some model parameters as real value numbers. In our search, we found three such models that were used in development of parameter estimators. The first model expects that all parameters, except the intercept, are real value numbers a (i.e. $a_i \in \mathbb{R}, \forall i \in \{1, \ldots, m\}$ and $\tilde{A}_0 \in \mathbb{R}_{\mathcal{F}}$). The model is given as

$$\tilde{Y} = \tilde{A}_0 \oplus a_1 \cdot \tilde{X}_1 \oplus \ldots \oplus a_m \cdot \tilde{X}_m. \tag{2}$$

Parameter estimators for this model were introduced in [8, 32, 1].

The second model considers all parameters as real value numbers (i.e. $a_i \in \mathbb{R}, \forall i \in \{0, \ldots, m\}$). The model is given as

$$\tilde{Y} = a_0 \oplus a_1 \cdot \tilde{X}_1 \oplus \ldots \oplus a_m \cdot \tilde{X}_m. \tag{3}$$

Parameter estimators for this model were introduced in [22, 1, 14, 25].

The third model represents an extension of the fuzzy linear model (3) due to a fuzzy error term $\tilde{\delta}$. The model is given as

$$\tilde{Y} = a_0 \oplus a_1 \cdot \tilde{X}_1 \oplus \ldots \oplus a_m \cdot \tilde{X}_m \oplus \tilde{\delta}, \tag{4}$$

where $a_i \in \mathbb{R}, \forall i \in \{0, \ldots, m\}$ and $\tilde{\delta} \in \mathbb{R}_{\mathcal{F}}$. Parameter estimators for this model were introduced in [16, 17, 5, 7, 6, 18].

The idea to extend fuzzy linear models with the fuzzy error term $\tilde{\delta}$ can be applied to models (1) and (2). By adding the fuzzy error term, new fuzzy linear models can be obtained. They are given as

$$\tilde{Y} = \tilde{A}_0 \oplus \left(\tilde{A}_1 \otimes \tilde{X}_1\right) \oplus \ldots \oplus \left(\tilde{A}_m \otimes \tilde{X}_m\right) \oplus \tilde{\delta}, \tag{5}$$

where $\tilde{\delta}, \tilde{A}_i \in \mathbb{R}_{\mathcal{F}}, \forall i \in \{0, \ldots, m\}$; and

$$\tilde{Y} = \tilde{A}_0 \oplus a_1 \cdot \tilde{X}_1 \oplus \ldots \oplus a_m \cdot \tilde{X}_m \oplus \tilde{\delta}, \tag{6}$$

where $\tilde{\delta}, \tilde{A}_0 \in \mathbb{R}_{\mathcal{F}}$ and $a_i \in \mathbb{R}, \forall i \in \{1, \ldots, m\}$. Note that these two models were not considered yet by developers of the parameter estimators. Considering [7], we hypothesize that adding the fuzzy error term might improve precision of the fuzzy linear regression.

3.2 Models for fuzzy dependent and real value independent variables

Fuzzy models belonging to this category expect that observations of all independent variables x are real value numbers ($x_i \in \mathbb{R}, \forall i \in \{1, \ldots, m\}$), and observations of the dependent variable y are fuzzy numbers $\tilde{Y} \in \mathbb{R}_{\mathcal{F}}$. For such data, most of parameter estimators [12, 26, 4, 8, 33, 11, 15, 7, 1, 13, 18, 35] expect the linear dependence described by the model

$$\tilde{Y} = \tilde{A}_0 \oplus \tilde{A}_1 \cdot x_1 \oplus \ldots \oplus \tilde{A}_m \cdot x_m, \tag{7}$$

where $\tilde{A}_i \in \mathbb{R}_{\mathcal{F}}, \forall i \in \{0, \ldots, m\}$.

Derived from the model (7), new models can be obtained when some model parameters would be real value numbers. Such a way, a parallel of the model (2) can be obtained. The new fuzzy linear model is given as

$$\tilde{Y} = \tilde{A}_0 \oplus (a_1 x_1 + \ldots + a_m x_m), \tag{8}$$

where $a_i \in \mathbb{R}, \forall i \in \{1, \ldots, m\}$ and $\tilde{A}_0 \in \mathbb{R}_{\mathcal{F}}$. According to our search, no parameter estimator is available for this model yet.

Parameter estimator published in [7] considered an extension of the model (7) with the fuzzy error term $\tilde{\delta}$. The extended model is given as

$$\tilde{Y} = \tilde{A}_0 \oplus \tilde{A}_1 \cdot x_1 \oplus \ldots \oplus \tilde{A}_m \cdot x_m \oplus \tilde{\delta}, \tag{9}$$

where $\tilde{\delta}, \tilde{A}_i \in \mathbb{R}_{\mathcal{F}}, \forall i \in \{0, \ldots, m\}$.

The extended model (9) can be simplified, when all parameters of the regression model are considered as real value numbers, i.e.

$$\tilde{Y} = (a_0 + a_1 x_1 + \ldots + a_m x_m) \oplus \tilde{\delta}, \tag{10}$$

where $\tilde{\delta} \in \mathbb{R}_{\mathcal{F}}$, and $a_i \in \mathbb{R}, \forall i \in \{0, 1, \ldots, m\}$. An estimator based on this model was presented in [18].

3.3 Models for real value dependent and fuzzy independent variables

The opposite of the previous category are fuzzy models that are intended for datasets where the observations of the independent variables x are fuzzy numbers ($\tilde{X}_i \in \mathbb{R}_{\mathcal{F}}, \forall i \in \{1, \ldots, m\}$), while the observation of the dependent variable y are real value numbers $y \in \mathbb{R}$. To ensure the requirement on the model output y, all parameters related to the fuzzy independent variables $\tilde{X}$ must be also fuzzy numbers. The basic model of this category is given as

$$y = \tilde{A}_0 \oplus \left(\tilde{A}_1 \otimes \tilde{X}_1\right) \oplus \ldots \oplus \left(\tilde{A}_m \otimes \tilde{X}_m\right), \tag{11}$$

where $\tilde{A}_i \in \mathbb{R}_{\mathcal{F}}, \forall i \in \{0, \ldots, m\}$. A parameter estimator for this model was presented in [10].

A new model can be obtained with the intercept represented as a real value number. Thus, the model is given as

$$y = a_0 \oplus \left(\tilde{A}_1 \otimes \tilde{X}_1\right) \oplus \ldots \oplus \left(\tilde{A}_m \otimes \tilde{X}_m\right), \tag{12}$$

where $a_0 \in \mathbb{R}$, and $\tilde{A}_i \in \mathbb{R}_{\mathcal{F}}, \forall i \in \{1, \ldots, m\}$. As far as we know, no parameter estimator was designed for this model.

4 Model attributes

Since each real value number can be expressed as a fuzzy number, we can say that the fuzzy linear model (5), respectively the model (1), are the general fuzzy linear models, and the remaining models are their simplifications. However, existence of the simplified fuzzy linear models is well founded. The simplified models have naturally implemented assumptions on model prediction in their structures. Let us demonstrate the importance of this fact on several examples.

Significance of prior assumptions on model predictions is best evident in models for real value dependent and fuzzy independent variables. Models of this class are intended for applications where real value predictions of the models are required. Parameter estimators based on such models provide parameters that guarantee the desired model predictions, while parameter estimators based on the general model (1) do not.

Another good example are models for fuzzy dependent and real value independent variables. Predictions of these models are fuzzy numbers $\tilde{Y} = \left(m^L_{\tilde{Y}}, m^R_{\tilde{Y}}, \alpha_{\tilde{Y}}, \beta_{\tilde{Y}}\right)_{LR}$, where $m^L_{\tilde{Y}}, m^R_{\tilde{Y}}$ are left and right cores of the predictions $\tilde{Y}$, and $\alpha_{\tilde{Y}}, \beta_{\tilde{Y}}$ are their left and right spreads. The fuzzy linear models (7) and (9) are known for dependence of the model prediction spreads $\alpha_{\tilde{Y}}$ and $\beta_{\tilde{Y}}$ on absolute values of the independent variables x [26, 7]. The fuzzy linear models (8) and (10), which belong to the same class, have different properties. The spreads $\alpha_{\tilde{Y}}$ and $\beta_{\tilde{Y}}$ are independent of x for these models [18]. Thus, the models (7) and (9) allow to model relationships where the prediction spreads are dependent on x, while the models (8) and (10) are appropriate for relations with no correlation of the spreads and x.

In this context, limitations of the models (7) and (9), arising from the essence of the fuzzy arithmetic, should be mentioned. As stated above, the model prediction spreads $\alpha_{\tilde{Y}}$ and $\beta_{\tilde{Y}}$ are dependent on x for these two models. For positive values of x, the spreads increase with increasing values of x. For negative values of x, the spreads decrease with increasing values of x. This limitation can be overcome by translating the observations along x [3]. Using axes translations, decreasing spreads $\alpha_{\tilde{Y}}, \beta_{\tilde{Y}}$ with increasing values of x for $x \in \mathbb{R}^+$; and increasing spreads $\alpha_{\tilde{Y}}, \beta_{\tilde{Y}}$ with decreasing values of x for $x \in \mathbb{R}^-$ can be achieved with these two models.

5 Conclusion

Here we provided user guidelines for selection of fuzzy linear models. From the wide variety of fuzzy linear models, the appropriate model for a given dataset should be selected according to datatype of the observations. However, the spreads of model predictions should respect the nature of the approximated relation, and this fact should be reflected in the model selection as well. We hope that the review will provide an inspiration for researchers developing new parameter estimators. To support further development of the fuzzy regression, we proposed four new fuzzy linear models. However, other new models can be introduced for fuzzy dependent and fuzzy independent variables, and for fuzzy dependent and real value independent variables. Considering some model parameters to be real value numbers, new fuzzy linear models can be formulated for these two model categories.

Acknowledgement

The work was supported by the University of Pardubice.

References

[1] ARABPOUR, A. R., TATA, M.: *Estimating the parameters of a fuzzy linear regression model*, Iranian Journal of Fuzzy Systems, vol. 5, no. 2, (2008), pp. 1–19, doi:10.22111/IJFS.2008.322.

[2] BEDE, B.: *Mathematics of Fuzzy Sets and Fuzzy Logic*, Studies in Fuzziness and Soft Computing, 1st edn., Springer-Verlag Berlin Heidelberg, 2013, ISBN 978-3-642-35220-1.

[3] BISSERIER, A., BOUKEZZOULA, R., GALICHET, S.: *A revisited approach to linear fuzzy regression using trapezoidal fuzzy intervals*, Information Sciences, vol. 180, no. 19, (2010), pp. 3653 – 3673, ISSN 0020-0255, doi:10.1016/j.ins.2010.06.017.

[4] CHANG, P.-T., LEE, E. S.: *A generalized fuzzy weighted least-squares regression*, Fuzzy Sets and Systems, vol. 82, no. 3, (1996), pp. 289 – 298, ISSN 0165-0114, doi:10.1016/0165-0114(95)00284-7.

[5] CHEN, L.-H., HSUEH, C.-C.: *A mathematical programming method for formulating a fuzzy regression model based on distance criterion*, IEEE Transactions on Systems, Man, and Cybernetics, Part B (Cybernetics), vol. 37, no. 3, (2007), pp. 705–712, doi: 10.1109/TSMCB.2006.889609.

[6] CHEN, L.-H., HSUEH, C.-C.: *Fuzzy regression models using the least-squares method based on the concept of distance*, IEEE Transactions on Fuzzy Systems, vol. 17, no. 6, (2009), pp. 1259–1272, doi:10.1109/TFUZZ.2009.2026891.

[7] CHOI, S. H., BUCKLEY, J. J.: *Fuzzy regression using least absolute deviation estimators*, Soft Computing, vol. 12, no. 3, (2008), pp. 257–263, doi:10.1007/s00500-007-0198-3.

[8] DIAMOND, P.: *Fuzzy least squares*, Information Sciences, vol. 46, no. 3, (1988), pp. 141–157, ISSN 00200255, doi:10.1016/0020-0255(88)90047-3.

[9] DIAMOND, P., KÖRNER, R.: *Extended fuzzy linear models and least squares estimates*, Computers & Mathematics with Applications, vol. 33, no. 9, (1997), pp. 15 – 32, ISSN 0898-1221, doi:10.1016/S0898-1221(97)00063-1.

[10] D'URSO, P.: *Linear regression analysis for fuzzy/crisp input and fuzzy/crisp output data*, Computational Statistics & Data Analysis, vol. 42, no. 1, (2003), pp. 47–72, doi:10.1016/S0167-9473(02)00117-2.

[11] D'URSO, P., GASTALDI, T.: *A least-squares approach to fuzzy linear regression analysis*, Computational Statistics & Data Analysis, vol. 34, no. 4, (2000), pp. 427–440, ISSN 0167-9473, doi:10.1016/S0167-9473(99)00109-7.

[12] H. TANAKA, K. A., S. UEJIMA: *Linear regression analysis with fuzzy model*, Systems, Man and Cybernetics, IEEE Transactions on, vol. 12, no. 6, (1982), pp. 903–907, ISSN 0018-9472, doi:10.1109/TSMC.1982.4308925.

[13] HAO, P.-Y., CHIANG, J.-H.: *Fuzzy regression analysis by support vector learning approach*, IEEE Transactions on Fuzzy Systems, vol. 16, no. 2, (2008), pp. 428–441, doi: 10.1109/TFUZZ.2007.896359.

[14] HASSANPOUR, H., MALEKI, H., YAGHOOBI, M.: *Fuzzy linear regression model with crisp coefficients: a goal programming approach*, Iranian Journal of Fuzzy Systems, vol. 7, no. 2, (2010), pp. 1–153, doi:10.22111/IJFS.2010.168.

[15] HOJATI, M., BECTOR, C., SMIMOU, K.: *A simple method for computation of fuzzy linear regression*, European Journal of Operational Research, vol. 166, no. 1, (2005), pp. 172–184, ISSN 0377-2217, doi:10.1016/j.ejor.2004.01.039.

[16] KAO, C., CHYU, C.-L.: *A fuzzy linear regression model with better explanatory power*, Fuzzy Sets and Systems, vol. 126, no. 3, (2002), pp. 401 – 409, ISSN 0165-0114, doi:10.1016/S0165-0114(01)00069-0.

[17] KAO, C., CHYU, C.-L.: *Least-squares estimates in fuzzy regression analysis*, European Journal of Operational Research, vol. 148, no. 2, (2003), pp. 426–435, ISSN 0377-2217, doi: 10.1016/S0377-2217(02)00423-X.

[18] KELKINNAMA, M., TAHERI, S.: *Fuzzy least-absolutes regression using shape preserving operations*, Information Sciences, vol. 214, no. 0, (2012), pp. 105–120, ISSN 0020-0255, doi: 10.1016/j.ins.2012.04.017.

[19] KHAN, U. T., VALEO, C.: *A new fuzzy linear regression approach for dissolved oxygen prediction*, Hydrological Sciences Journal, vol. 60, no. 6, (2015), pp. 1096–1119, doi: 10.1080/02626667.2014.900558.

[20] KIM, K. J., MOSKOWITZ, H., KOKSALAN, M.: *Fuzzy versus statistical linear regression*, European Journal of Operational Research, vol. 92, no. 2, (1996), pp. 417–434, doi:10.1016/0377-2217(94)00352-1.

[21] LI, J., ZENG, W., XIE, J., YIN, Q.: *A new fuzzy regression model based on least absolute deviation*, Engineering Applications of Artificial Intelligence, vol. 52, (2016), pp. 54 – 64, ISSN 0952-1976, doi:10.1016/j.engappai.2016.02.009.

[22] MING, M., FRIEDMAN, M., KANDEL, A.: *General fuzzy least squares*, Fuzzy Sets and Systems, vol. 88, no. 1, (1997), pp. 107 – 118, ISSN 0165-0114, doi:10.1016/S0165-0114(96)00051-6.

[23] NASRABADI, M. M., NASRABADI, E., NASRABADY, A. R.: *Fuzzy linear regression analysis: a multi-objective programming approach*, Applied Mathematics and Computation, vol. 163, no. 1, (2005), pp. 245 – 251, ISSN 0096-3003, doi:10.1016/j.amc.2004.02.008.

[24] PAN, N.-F., LIN, T.-C., PAN, N.-H.: *Estimating bridge performance based on a matrix-driven fuzzy linear regression model*, Automation in Construction, vol. 18, no. 5, (2009), pp. 578 – 586, ISSN 0926-5805, doi:10.1016/j.autcon.2008.12.005.

[25] ROH, S.-B., AHN, T.-C., PEDRYCZ, W.: *Fuzzy linear regression based on polynomial neural networks*, Expert Systems with Applications, vol. 39, no. 10, (2012), pp. 8909–8928, doi: 10.1016/j.eswa.2012.02.016.

[26] TANAKA, H., HAYASHI, I., WATADA, J.: *Possibilistic linear regression analysis for fuzzy data*, European Journal of Operational Research, vol. 40, no. 3, (1989), pp. 389–396, ISSN 0377-2217, doi:10.1016/0377-2217(89)90431-1.

[27] TOYOURA, Y., WATADA, J., KHALID, M., YUSOF, R.: *Formulation of linguistic regression model based on natural words*, Soft Computing, vol. 8, no. 10, (2004), pp. 681–688, ISSN 1433-7479, doi:10.1007/s00500-003-0326-7.

[28] WU, H.-C.: *Fuzzy estimates of regression parameters in linear regression models for imprecise input and output data*, Computational Statistics & Data Analysis, vol. 42, no. 1, (2003), pp. 203 – 217, ISSN 0167-9473, doi:10.1016/S0167-9473(02)00116-0.

[29] WU, H.-C.: *Linear regression analysis for fuzzy input and output data using the extension principle*, Computers & Mathematics with Applications, vol. 45, no. 12, (2003), pp. 1849–1859, ISSN 08981221.

[30] WU, H.-C.: *The construction of fuzzy least squares estimators in fuzzy linear regression models*, Expert Systems with Applications, vol. 38, no. 11, (2011), pp. 13 632 – 13 640, ISSN 0957-4174, doi:10.1016/j.eswa.2011.04.131.

[31] YAGER, R. R.: *Fuzzy prediction based on regression models*, Information Sciences, vol. 26, no. 1, (1982), pp. 45 – 63, ISSN 0020-0255, doi:10.1016/0020-0255(82)90043-3.

[32] YANG, M.-S., KOL, C.-H.: *On cluster-wise fuzzy regression analysis*, IEEE Transactions on Systems, Man, and Cybernetics, Part B (Cybernetics), vol. 27, no. 1, (1997), pp. 1 – 13, doi: 10.1109/3477.552181.

[33] YEN, K., GHOSHRAY, S., ROIG, G.: *A linear regression model using triangular fuzzy number coefficients*, Fuzzy Sets and Systems, vol. 106, no. 2, (1999), pp. 167 – 177, ISSN 0165-0114, doi:10.1016/S0165-0114(97)00269-8.

[34] ZADEH, L.: *The concept of a linguistic variable and its application to approximate reasoning-I*, Information Sciences, vol. 8, no. 3, (1975), pp. 199 – 249, ISSN 0020-0255, doi:10.1016/0020-0255(75)90036-5.

[35] ZENG, W., FENG, Q., LI, J.: *Fuzzy least absolute linear regression*, Applied Soft Computing, vol. 52, no. Supplement C, (2017), pp. 1009 – 1019, ISSN 1568-4946, doi: 10.1016/j.asoc.2016.09.029.

[36] ZIMMERMANN, H. J.: *Fuzzy Set Theory-and Its Applications*, 4th edn., Springer Netherlands, 2001, ISBN 978-0-7923-7435-0.

Current address

Škrabánek Pavel, Ing., Ph.D.
Department of Process Control
Faculty of Electrical Engineering and Informatics
University of Pardubice
Studentská 95, 532 10 Pardubice, Czech Republic
E-mail: pavel.skrabanek@upce.cz

Marek Jaroslav, Mgr., Ph.D.
Department of Mathematics and Physics
Faculty of Electrical Engineering and Informatics
University of Pardubice
Studentská 95, 532 10 Pardubice, Czech Republic
E-mail: jaroslav.marek@upce.cz

Slovak University of Technology in Bratislava
Faculty of Mechanical Engineering
17th Conference on Applied Mathematics **APLIMAT 2018**

Proceedings

SECONDARY MATHEMATICS MISCONCEPTION AS A MAIN OBSTACLE IN SOLVING HIGHER MATHEMATICS PROBLEMS

SLAVÍČKOVÁ Mária (SK), VARGOVÁ Michaela (SK)

Abstract. Paper deals with the most common misconception in secondary mathematics and its influence on solving tasks of higher mathematics. We provide several examples from our pedagogical experiences, main misconceptions connected to the mentioned examples and our solutions for possible re-education of mentioned misconceptions.

Keywords: concept of function, mathematics analysis, basic mathematics misconceptions

Mathematics Subject Classification: Primary 97I20, Secondary 26A09

1 Introduction

Mathematics knowledge of university students seems to be lower and lower year by year. They struggle with many misconceptions concerning secondary level mathematics. These misconceptions cause freshman's poor performance in university mathematical courses.

As long as the lack of the secondary level mathematical concepts and misconceptions about basic mathematical objects, concepts and operations continue, it is impossible to learn and understand new concepts on university level.

It is well known that students can master various formalized rules or computational algorithms without a deeper understanding of the issue. This fact leads students to formal and incomplete knowledge. Students are able to learn some techniques to solve limits, derivatives, integrals, but there may be differences between their concept image[1] of the considered notion and formal definition of this mathematical notion. Such reduced and restricted understanding of the content of fundamental mathematical concepts can cause the inability to understand the proofs of some (often basic) mathematical theorems. Formal knowledge is a major obstacle to solving the nonstandard problems of higher mathematics, requiring a deep comprehension of the concept, not only routine algorithmic skills.

[1] The set of all the mental pictures associated in the student´s mind with the concept name, together with all the properties characterizing them. [1]

All the mentioned is manifested especially in mathematical analysis courses. The limit process, derivatives, integrals, all the "epsilon - delta" definitions and their applications are used without a concept understanding. Formal knowledge is enlarged by every other mathematical concept that is learned. It is a pity, because without a good mathematical background there are no good engineers, economists, computer scientists, and representatives of applied sciences.

We distinguish between random mistake (it may be the result of faulty memory, cognitive overload etc.) which can be corrected by a little bit warning and misconception. There are some definitions of misconception in current literature. According to [5] misconceptions are the wrong concepts or conceptions that a person assumes it as true and uses as a habit. In [4] is misconception defined as the inconsistency between the concept that we want students to learn and the mental model that they build in their minds.
We understand the misconception about mathematical concept in sense of [7] as a set of mathematical objects considered by students to be examples of the concept which is not necessarily the same as the set of mathematical object determined by the definition.

2 Function concept misconceptions

The concept of function is probably one of the most important concepts in contemporarily mathematics; it is included in the content of analysis (calculus), algebra, and geometry. Therefore, this concept is also important in secondary school mathematics and current curriculum emphasizes the importance of functions. In spite of that, a lot of freshman university students obtained the misconceptions about this concept.
The emergence of function concept can be dating to the end of 17th century, to the beginnings of infinitesimal calculus. Many mathematicians deal with the notion of function in their researche (for example, Newton (1642-1727) was one of the first mathematicians to show how functions could be developed in infinite power series, Leibniz (1646-1716) was the first, who used the term "function" in 1673, etc.) and the definition of this notion has changed over time. Jean Bernoulli in 1718 defined a function of a variable as a quantity that is composed in some way from that variable and constant. His student, Leonard Euler, later changed the term quantity by the term analytical expression. Finally, Dirichlet separated the concept of notion from its analytical representation and defined function as an arbitrary correspondence between two variables, so that to any value of the independent variable it is associated one and only one value of the dependent variable. The modern concept of functions is based on the set theory (in the literature also referred to as Dirichlet – Bourbaki concept) – the function is a correspondence between two nonempty sets that assigns to every element in the first set (the domain) exactly one element in the second set (the codomain). Instead of the term correspondence, one can talk about set of ordered pairs that satisfies a certain condition.
Current definition includes some correspondences that are not recognized as a function by mathematicians of previous two centuries. For example: discontinuous functions, function defined on split domain (by different rules or formulas) on different subdomains (Leonard Euler did not consider this correspondence as function) or functions by means of a graph.

The formal definition of function is introduced in Slovakia to high school students. In spite of that many high school and university students can memorize definition of the function in the form mentioned above, but their understanding of this notion is restricted the same way as it has been in history of mathematics and concepts of previous generations of mathematicians.

A similar opinion is given by Walton, who claims that many students make comparable definitions of function but they are not aware of what these definitions represent [8]. Another study set forth that the students cannot understand the concepts that are featured especially within operations in analysis courses [2].

During our pedagogical practise we often meet the students whose concept of the function is built on the formula conception (like Euler). The parallel of phylogeny and ontogenesis of the concept of function is very obvious in this case. It can be caused by examples of functions used in high school textbooks to illustrate this concept. These examples are often (sometimes exclusively) functions whose rule of correspondence is given by a formula. This fact can lead to students´ misconception or formulating the incorrect definition like "A function is a formula, an algebraic expression, etc" or "It is an equation expressing a certain relation between two objects".

Especially, we can observe the biggest problem with non-polynomial functions. Students are able to understand somehow the expression $y = x^2 + 5x + 6$. Problems with calculating and solving particular tasks start with function with "alphabetical name" like logarithm, sinus, cosine etc. Let's have a look on some examples of students operations with these functions:

Example 2.1: Logarithmic functions:

a) $\ln x(\ln x - 1) = 0 \Rightarrow x = \dfrac{\ln x - 1}{\ln}$

b) $\ln x - \ln a = \ln(x - a)$

Example 2.2: Goniometric functions

a) $\sin(\sin x) = \sin^2 x$

b) $\dfrac{\sin^2 x + \cos^2 x}{\sin^2 x} = x + \dfrac{\cos x}{\sin x}$

As we can see, the problem is very serious and its remedy is not easy. It will take quite a lot of time to make changes in students' conceptual maps. We cannot be sure about the results of our re-educating process.

3 Secondary mathematics misconceptions in mathematical analysis

We will provide brief summary of the most crucial problems based on secondary mathematics misconceptions and its influence on solving problems of mathematical analysis (or calculus).

3.1 Limits process

It is quite an easy algorithm to calculate limit of sequence or a function. The main problem is misunderstanding of the concept of function on secondary school. There are few examples of students' work which can demonstrate this situation (see also example 2.1 and 2.2):

Example 3.1: $\lim\limits_{n\to\infty} \dfrac{2^n - 1}{2^n + 1} \bullet \dfrac{\log_2}{\log_2} = \lim\limits_{n\to\infty} \dfrac{\log_2 2^n - \log_2 1}{\log_2 2^n + \log_2 1} = \lim\limits_{n\to\infty} \dfrac{n - 0}{n - 0} = 1$

Example 3.2: $\lim_{x\to 0}\frac{\sin(x+2\pi)}{\sin(x+3\pi)}=\frac{x+2\pi}{x+3\pi}=\frac{2}{3}$

On these two examples we can observe the wrong understanding of the function mentioned in paragraph 2. Not only mentioned students have no idea what the specific mathematical notation means. We think that the main problems are:

- Misconception of the function concept
- Insufficient training of calculation with other than polynomial functions
- Students' opinion – everything is on Wikipedia, why we should learn something?

Problems like these are more and more common in our lessons of mathematics. It might take a lot of time and energy to demonstrate the problem with such calculations, to argue why specific ways of solving is not correct and the necessity to re-educate all the misconceptions. The results are not pleasing. To correct so strong misconception of function on University level is almost impossible.

Example 3.3: $\lim_{n\to\infty}\frac{(n+1)^n}{n^n}=\lim_{n\to\infty}\frac{n^n+2n+1^n}{n^n}$

This example is very typical misunderstanding of basic algebra identities. Once they finally can correctly expand $(a+b)^2$ we can find out very formal knowledge of it: "there is always $2ab$ middle term". If we asked them about binomial theorem, they know about it and some of them start to think about how it could be connected to our specific problem (and when teacher asked about it, it has to be connected). Counting with fraction is well known problem in school mathematics. Unfortunately, this problem survives very long time. Let's have a look on next two examples:

Example 3.4: $\lim_{x\to 0}\frac{\tan x-\sin x}{x^3}=\lim_{x\to 0}\left(\frac{\tan x}{x^2}-\frac{\sin x}{x}\right)$

Example 3.5: $\lim_{n\to\infty}\left(\frac{1}{\alpha+1}-\frac{1}{n+\alpha+1}\right)=\lim_{n\to\infty}\left(\frac{1}{\alpha+1}-\frac{1}{n}-\frac{1}{\alpha+1}\right)=\lim_{n\to\infty}\left(-\frac{1}{n}\right)=0$

Misconception concerning fractions is also very common and questions concerning "when we can stroke out some numbers" or "why it was possible now and not before" sounds incredible on the university students of (applied) mathematics (yes, we are talking about students of mathematics). Problems with calculation with fractions are very frequent in calculations of integrals (see paragraph 3.3).

3.2 Derivative of function

We think that there is no easier algorithm for solving problems in mathematics analysis than finding the derivatives of functions. It is such routine that everyone should handle it. By everyone we mean someone who already has managed concept of function and operations with them (addition, multiplication, composing). If there is missing this essential knowledge we can find several kinds of problems:

Example 3.6: $\left(5\sin^2(\cos 3x)\right)' = 5\cos^2 x - \sin x + 3$

Example 3.7: $\left(\dfrac{\sin x + x^2}{5^x}\right)' = \dfrac{\cos x + 2x}{x.5^{x-1}}$

We have to say that students are allowed to use the derivative rules on the test. Example 6 shows us misconception of derivative of compound function. Example 7 represents the most common mistake – quotient rule for derivative and again misconception of the function. Students often do not recognize polynomial functions from exponential ones.

3.3 Integral and primitive function

To master integration of the function is much harder than master derivatives or other concept of mathematical analysis. There are not such easy rules. Students have to make sense for using correct substitution or method of solving. Besides that, there is still the biggest obstacle in solving the integrals - "lower mathematics" – fractions and concept of functions.

Example 3.8: $\int_0^{\pi} \sin^2 x\, dx = \int_0^{\pi} \dfrac{1}{\left(\sin^2 x\right)^{-1}} dx = \left[\left(-\operatorname{cotg} x\right)^{-1}\right]_0^{\pi}$

Example 3.9:

a) $\int t^2 \dfrac{t^2}{t^3+1} dt = \int t^2 dt - \int \dfrac{t^2}{t^3+1} dt$

b) $\int \dfrac{1}{t^2+3t-1} dt = \int \dfrac{1}{t^2} dt + \int \dfrac{1}{3t} dt + \int (-1) dt$

c) $\int \dfrac{dx}{\sin x + \cos x + 5} = \int \dfrac{dx}{\sin x} + \int \dfrac{dx}{\cos x} + \int \dfrac{dx}{5}$

d) $\int \dfrac{dx}{3x^2+2x+2} = \dfrac{1}{3} \int \dfrac{dx}{x^2+2x+2}$

Some of these problems remain unchanged since the first semester of university studies. Students seem to be not able to connect knowledge from algebra (division by x means multiplication by 1 over x in some contexts). As we mentioned in Section 3.1 of Function Limits, the misconception of basic algebraic identities and function concepts is a problem leading to an inappropriate solution to higher mathematical problems. In parallel with higher mathematics, the exaltation of secondary mathematics appears in lessons. Nevertheless their knowledge of basic schemes and concepts are very formal and our effort to change it has very weak effect.

4 Suggested solutions and their results

Our experiences as university mathematics teachers show that students need longer time period to reconsider their conceptual schemas but they do not have so much time. They did have it on secondary school. Unfortunately, this time has been wasted. Some of the students find out that mathematics is not good study program for them and they leave the study. The rest still remain and suffer.

We have to handle with this situation as good as possible. We'd like to encourage students to learn harder on their chosen study program. Therefore we started to prepare better environment for obtaining necessary mathematical skills:

- 2 lessons per week from mathematical analysis (voluntary subject),
- Enlarged consultation time for the groups,
- Using ICT on mathematical analysis lessons:
    - Demonstration tool (teachers work),
    - Tool for modelling, observation and practising (students work),
    - Preparing e-learning support for the subject,
- Preparing special voluntary courses focused on secondary mathematics.

The results are not very encouraging. The more we try the less the students try. We tried to help them more than students wanted. Therefore, this academic year we stopped some activities and remain only voluntary lessons and e-learning support. The observed results so far are:

- Students started searching more information (there are only very basic information in e-learning course)
- Many students realized their weaknesses and they started to ask more on lessons or they ask to have more consultation time

5 Conclusion

We have attempted to summarize the most important misconceptions we have encountered in our pedagogical practice. We think many university mathematics teachers have seen the same or very similar mistakes. According to the constructivist approach and its rules, the teacher should perceive the error as the developmental stage of the pupil's understanding of mathematics and the impulse for further work [4]. Unfortunately, some teachers of mathematics are not sure how to use error as an impulse for further work. Therefore we think that to solve problem of misconceptions (not only of function concept) we should do the following:

- To educate future mathematics teachers how to identify misconceptions and suggest an effective teaching strategy that helps their students to develop a correct understanding of these concepts.
- To encourage in-service teachers to be interested to take into consideration the misconceptions about basic mathematical concepts and implement an effective teaching strategy that helps their students to develop a correct understanding of these concepts.
- To consider 4 years of bachelor study (some of the technical universities already have it) or to implement into accreditation materials obligatory subject "The foundation of mathematics" focused on the most critical parts of mathematics on primary and secondary school.

Since, every learned concept in mathematics is closely related with previous or upcoming concepts, a difficulty experienced in learning a certain concept causes difficulties in learning many further concepts and paves the way for misconceptions (Altun, In: [1]). Therefore, building correct conceptual maps is necessary for further education.

References

[1] CANSIZ, S., KŰCŰK, B, ISLEYEN, T. Identifying the secondary school student´s misconceptions about functions. *Procedia Social and Behavioral Sciences*, vol. 15, 2011: ISSN 1877-0428

[2] GRAVEMEIJER, K., DOORMAN, M. Context Problems in Realistic Mathematics Education: A Calculus Course as an Example. *Educational Studies in Mathematics*, Vol. 39, 1999: pp. 111-129, ISSN 0013-1954

[3] KLEIN, F. *Elementary mathematics from an advanced standpoint*. (E. R. Hedrick & C. A. Noble, trans.). New York, NY: Dover, 1945. (Original work published 1908)

[4] MICHAEL, J. Misconceptions – What Students Think They Know, *Advanced in Physiology Education*, Vol. 26, No. 1, 2002: pp. 4 -6, ISSN 1043-4046

[5] ÖZKAN, C. M. Misconceptions in radicals in high school mathematics. *Procedia Social and Behavioral Sciences*, vol. 15, 2011: pp. 120 – 127, ISSN 1877-0428

[6] STEHLÍKOVÁ, N., CACHOVÁ, J. Konstruktivistické přístupy k vyučování a praxe. Studijní materiály k projektu Operační program Rozvoj lidských zdrojů), JČMF, Praha, 2006

[7] VINNER, S., DREYFUS, T. Images and definitions for the concept of function. *Journal for Research in Mathematical Education*, Vol. 20, No. 4, 1989: pp. 356 – 366, ISSN 00218251

[8] WALTON, K. D., Examining Functions in Mathematics and Science Using Computer Interfacing. *School Science and Mathematics*, Vol. 88, 1988: pp. 604-609, ISSN 1949-8594

Current address

Slavíčková Mária, PaedDr., PhD.
Faculty of Mathematics, Physics and Informatics, Comenius University in Bratislava
Mlynská dolina, 842 48 Bratislava, Slovak Republic
E-mail: maria.slavickova@fmph.uniba.sk

Vargová Michaela, Mgr., PhD.
Faculty of Mathematics, Physics and Informatics, Comenius University in Bratislava
Mlynská dolina, 842 48 Bratislava, Slovak Republic
E-mail: michaela.vargova@fmph.uniba.sk

Slovak University of Technology in Bratislava
Faculty of Mechanical Engineering
17th Conference on Applied Mathematics **APLIMAT 2018**

Proceedings

COOPERATION IN TEACHING AND LEARNING ANALYTIC GEOMETRY

ŠOVČÍKOVÁ Petronela (SK)

Abstract. Cooperative or small group learning is widely recognized as pedagogical practise that promotes learning and socialisation across a range of curriculum areas from primary school to high school and college. Our research was realized in two classes. There were 42 pupils (18 pupils in the first class and the 24 pupils in the second class). The main aim of this research was to verify our hypothesis: Can pupils accomplish better results when they cooperate with someone similar?.

Keywords: cooperation, teaching, learning, analytic geometry, teaching method

Mathematics Subject Classification: Primary 97G70; Secondary 97D40

1 Introduction

Cooperation is the basis of human lives. We assume that everybody have cooperated in his/her life – in childhood, at school, at work, etc. Cooperation is related to everybody with no difference. From this reason we decided to study cooperation in teaching and learning analytic geometry.

Cooperative learning has been well documented in the educational research as a successful pedagogy to improve academic achievements of students. It is a fundamental principle of cooperative learning that group members are linked together in such a way that they can succeed, they will actively assist each other to make sure that the assignment is done and the purpose of the group achieved. [1]

Numerous studies have been published over the past four decades that demonstrate the benefits of cooperative learning. These benefits include academic gains across different curriculum domains [2], improved participation in school-based learning [3] and enhanced socialisation among peers [4], including more cross-ethnic and cross-sex relationships [5].

Collaborative learning in the form of groups has been used in teaching for a long time. Many research indicated that a teaching effect can be greatly improved through collaborative learning. [6]

In our research we have chosen the problematics of cooperation as a teaching method. According to the results we have reached we would like to confirm the reliability of cooperation as a well - chosen teaching tool for pupils at grammar school also because of their age and period of life when they create new relationships, meet new people and it is a natural way of being for them.

2 Theoretical background

Cooperative learning is one of the two ways of organizing the learning environment of a classroom, the other being competitive. In cooperative learning environment, the goals of a separate individual become so linked that there is a positive correlation between them; on the contrary, in a competitive conventional environment, the goals of the students are so linked that there is a negative correlation between their goal attainments. [8]

Cooperative learning is a pedagogical practise strongly supported by many state because of the well-documented benefits that accrue to children who experience cooperative learning. [9]

One of the most influential perspectives was developed by [10] who proposed that children´s knowledge, ideas, attitudes, and values develop trough interactions with others. In fact, when children interact with adults or more able peers, children´s learning is mediated or reinforced so they can often complete tasks that they would not be able to do by themselves.

[10] recognized that children´s learning is mediated by adults and more capable peers who teach the knowledge and skills of their culture. This process of mediation of scaffolding, enables children to complete tasks they would not be able to do themselves. When children work cooperatively together, the group creates a zone of proximal development enabling members to be successful at tasks they would be unable to do alone. Instruction within the group is both explicit and implicit, enabling children to engage in more opportunities for developing understanding and meaning.

[11] found that cooperative learning can be enhanced when group size does not exceed four members, instructions are adapted to the needs of the group, and teachers are trained to implement small group work in their classroom.

According to [12] cooperative learning is the "instructional use of small groups so that students work together to maximize their own and each other's learning." In order to work well, cooperative learning needs to be planned, with consideration given to the appropriate size if the group, to each student´s role within the group, and to how the results will be evaluated and used in the class session.

The analytic geometry is a very problematic part of education. Pupils are focused on analytic geometry in the third year at grammar school. They learn about coordinate system, vectors and lines. [13]

3 Methodology

The research sample consisted of 42 pupils who study at grammar school. These pupils filled in worksheets which were oriented on solving some problems of analytic geometry especially analytic geometry of circle. Analytic geometry is a content of education in the third year at grammar school.

Pupils learned analytic geometry during the four weeks and then they filled in pre-test. This pre-test consisted of 4 tasks which were oriented on mentioned analytic geometry of circle.
The first task was oriented on a drawing circle in coordinate system when the equation of circle was given. They were able to find out coordinates of centre of circle and radius of circle from mentioned equation of circle. Than drawing circle in coordinate system was not complicated. Other tasks were aimed on acquisition information from a graph of a circle which was drawn in previously task. Pupils had to find points where the graph crosses the y-axis, then points where the graph crosses the x-axis and some more points on the graph.
In the next step pupils were divided into small groups. The positioning of pupils in group was deliberate. The starting point of dividing them into small groups was the result of pre-test from our previous step. Members of each group had one common thing – they had the same or very similar outcome of their pre-test because we wanted to verify our hypothesis: Can pupils accomplish better results when they cooperate with someone similar (someone who had the same or very similar results from his/her pre-test)?.

In each group pupils worked together during the two weeks. Then they filled in the post-test which was oriented on analytic geometry of circle, as well. This post-test consisted on four circle puzzles. The first circle puzzle was oriented on working out coordinates of the points A, B, C and D (Fig 1.) where the centre of the circle and the radius of the circle were known.

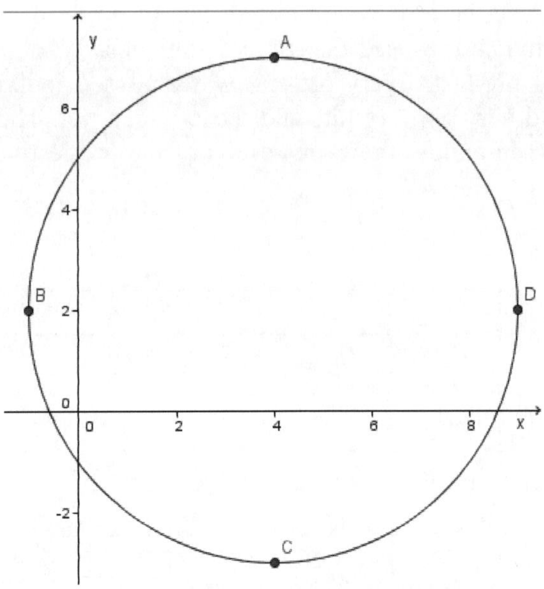

Fig. 1. The circle in the coordinate system.

In the next task there were the equations of two tangents to a circle. Pupils should find out the radius and the coordinates of the centre of the circle. The picture in the third task shows them the location of three circles in the area. They knew the centre of the circle and the radius of the circle. Pupils had to follow the instructions and answer the questions: Draw the y-axis and the x-axis; What is the radius of the smaller circles? Write down the coordinates of the centres of the two smaller circles. The situation was illustrated in the picture (Fig 2.).

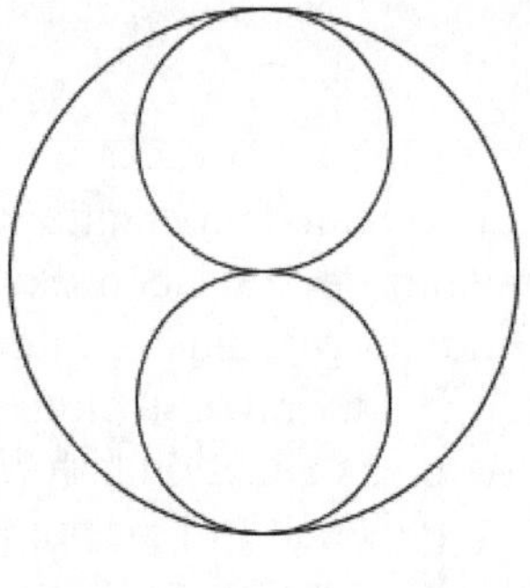

Fig. 2. Two circles inside the third circle.

The last task gave them information about the circle position according to which pupils had to draw it. They knew the radius of the circle and the circle has one intersection point with y-axis and one intersection point with x-axis.

4 Findings

The main aim of this research is to verify our hypothesis: Can pupils accomplish better results when they cooperate with someone similar?

In the methodology we describe how we collected data to prove it. When we had collected data, we could analyse them. We used descriptive statistics. When we sorted pupils in to groups we used result of pre-test. Every pre-test was analysed individually. Of course, we found pre-tests which had very good results and those which not. These results helped us to divide pupils into groups. In table 1 there are results of the pre-test which was sat by pupils from the first class.

the first class									
pupil	1	2	3	4	5	6	7	8	9
result	93%	69%	13%	78%	15%	91%	19%	72%	98%
pupil	10	11	12	13	14	15	16	17	18
result	89%	81%	22%	64%	32%	24%	95%	44%	55%

Tab. 1. Pre-tests´ results in the first class.

Subsequently we made this groups - group 1 (pupil 1, pupil 9 and pupil 16), group 2 (pupil 10, pupil 11 and pupil 6), group 3 (pupil 4, pupil 8 and pupil 2), pupil 4 (pupil 13, pupil 17 and pupil 18), group 5 (pupil 12, pupil 15 and pupil 14), group 6 (pupil 3, pupil 5 and pupil 7).

In table 2 there is the same situation than in the table 1 but there are results of the pre-test which was done by pupils from the second class.

the second class								
pupil	1	2	3	4	5	6	7	8
result	10%	5%	46%	13%	100%	60%	24%	57%
pupil	9	10	11	12	13	14	15	16
result	8%	63%	74%	79%	99%	50%	29%	49%
pupil	17	18	19	20	21	22	23	24
result	15%	81%	99%	76%	31%	89%	6%	84%

Tab. 2. Pre-tests´ results in the second class.

In the second class we made this groups - group 1 (pupil 5, pupil 13 and pupil 19), group 2 (pupil 2, pupil 9 and pupil 23), group 3 (pupil 4, pupil 7 and pupil 1), pupil 4 (pupil 15, pupil 7 and pupil 21), group 5 (pupil 3, pupil 16 and pupil 14), group 6 (pupil 8, pupil 6 and pupil 10), group 7 (pupil 20, pupil 11 and pupil 12), group 6 (pupil 24, pupil 22 and pupil 18).

This division into groups was deliberate because we were afraid of a situation in which pupils work with classmates with different school results. Then we would not have relevant information because in the group there could be pupils with excellent results and pupils with not so good results. Subsequently it could happen that their way of doing the tests would not be cooperation but only work one pupil or some pupils. Results from post-tests are in table 3 (the first class) and table 4 (the second class).

the first class									
pupil	1	2	3	4	5	6	7	8	9
result	93%	73%	26%	80%	19%	96%	25%	75%	99%
pupil	10	11	12	13	14	15	16	17	18
result	96%	84%	28%	69%	43%	35%	100%	54%	60%

Tab. 3. Post-tests´ results in the first class.

the second class								
pupil	1	2	3	4	5	6	7	8
result	13%	19%	53%	23%	93%	65%	29%	63%
pupil	9	10	11	12	13	14	15	16
result	14%	76%	98%	89%	89%	54%	31%	56%
pupil	17	18	19	20	21	22	23	24
result	15%	81%	99%	83%	50%	96%	26%	94%

Tab. 4. Post-tests´ results in the second class.

When we compare results of pre-test and the post-test (Tab 1., Tab 2., Tab 3., Tab 4.) we can notice some obvious facts:

- some pupils have better results of post-test,
- some pupils attain not any improvement (result of their pre-test and result of their post-test are similar),
- some pupils have worse results of post-tests.

The first finding can be considered as a validation of our supposition. In the second finding we see some pupils are not better or worse. They stay on the same level than before they fill in post-test. The third finding shows us some pupils (small group) have worse results of post-test. It can be caused by a lot of factors that we have not researched so far.
In general our hypothesis about cooperation in teaching and learning in analytic geometry is true in this case. We confirmed our hypothesis that pupils have better results of post-test than results of pre-test. We can say that cooperation in teaching and learning in analytic geometry is a good way for pupils at grammar school.

5 Conclusion

Cooperative learning establishes a community in which pupils can get help and support from other group members immediately in a non-competitive learning environment, just raising their hands and waiting for the right answers to be given.
We showed cooperation has its own place in teaching and learning in general but also especially in analytic geometry.
This study is not only a confirmation and a proof of our hypothesis, we also showed pupils how they can be better and reach greater results when they learn something – in our case not only new knowledge but also new social interactions – making friends, giving and getting support, helping one another, talking to someone with a purpose, etc. which are very important in everyday life in any period of life.
This study shows teacher they can be successful in teaching when they use cooperation in teaching pupils. Pupils are oriented on common work and they work as team then they feel their own need. These facts can help them to be better in learning problematic. We think the method of cooperation can be used in teaching and learning each problematic.

Acknowledgement

This paper was created with support of projects APVV-14-0446 - Hodnotenie kompetencií učiteľa and UGA-VII/17/2017– Aktivizujúce metódy a dynamické prostriedky vyučovania analytickej geometrie na gymnáziách.

References

[1] DEUTSCH, M., A theory of cooperative and completion. *Human Relation.* 1949. pp. 129-152.
[2] FALL, R., WEBB, N., Group discussion and large-scale language arts assessment: Effects on students' comprehension. *American Educational Research Journal, 37l.* 2000. pp. 911-941.

[3] STEVENS, R., SLAVIN, R., The cooperative elementary school. *American Educational Research Journal, 32.* 1995. pp.321-351.
[4] Slavin, R., Cooperative learning: Theory research, and practise (2nd ed.). *Boston: Allyn and Bacon.*1995.
[5] SHARAN, S., Cooperative learning and helping behaviour in the multi-ethnic classroom. *New York: Praeger.* 1990. Pp. 151-175.
[6] DAVIDSON, N., MAJOR, C. H., MICHAELSEN, L. K., Small-group learning in higher education – cooperative, collaborative, problem-based, and team-based learning: An introduction by the guest editors. *Journal on Excellence in College Teaching, 25.* Pp. 1-6.
[7] JOHNSON, D. W., JOHNSON, R. T., Learning together and alone. *London: Allyn and Bacon.* 1994.
[8] GILLIES, R. M., Structuring cooperative group work in classrooms. *International Journal of Educational Research, 39.* 2003. pp. 35-49.
[9] VYGOTSKY, L. Mind in society: The development of higher psychological processes. *Cambridge, MA: Harvard University Press.*
[10] LOU, Y., ABRAMI, P., SPENCE, J., POULSEN, C., CHAMBERS, B., D'APOLLONIA, S., Within-class grouping: A meta-anaysis. *Review of Educational Research, 66.* 1996. pp. 423-458.
[11] JOHNSON, D. W., JOHNSON, R. T., SMITH, K. A., Active Learning: Cooperation in the College Classroom. *Edina, MN: Interaction Book.* 1991.
[12] URL, http://www.statpedu.sk/files/articles/dokumenty/inovovany-statny-vzdelavaci-program/matematika_g_4_5_r.pdf.

Current address

Šovčíková Petronela, Mgr.
Constantine the Philosopher University in Nitra
Tr. A. Hlinku 1, 949 74 Nitra, Slovak Republic
E-mail: petronela.sovcikova@ukf.sk

Slovak University of Technology in Bratislava
Faculty of Mechanical Engineering
17th Conference on Applied Mathematics **APLIMAT 2018**

Proceedings

CYCLIC PERMUTATIONS: CROSSING NUMBERS OF THE JOIN PRODUCTS OF GRAPHS

STAŠ Michal (SK)

Abstract. The crossing number $\mathrm{cr}(G)$ of a graph G is the minimal number of edge crossings over all drawings of G in the plane. In the paper, we extend known results concerning crossing numbers for join of graphs of order six. We give the crossing number of the join product $G + D_n$, where the graph G consists of one 4-cycle and two leaves, and D_n consists on n isolated vertices. The proof is done with the help of software that generates all cyclic permutations for a given number k, and creates a graph for a calculating the distances between all $(k-1)!$ vertices of the graph. Finally, by adding some edges to the graph G, we are able to obtain the crossing numbers of the join product with the discrete graph D_n for other graphs.

Keywords: graph, drawing, crossing number, join product, cyclic permutation

Mathematics subject classification: Primary 05C10, 05C38

1 Introduction

Let G be a simple graph with the vertex set V and the edge set E. A *drawing* of the graph G is a representation of G in the plane such that its vertices are represented by distinct points and its edges by simple continuous arcs connecting the corresponding point pairs. In such a drawing, the intersection of the interiors of the arcs is called a *crossing*. A drawing is *good* if each two edges have at most one point in common, which is either a common end-vertex or a crossing. Moreover, no three edges cross in a point. It is easy to see that a drawing with minimum number of crossings (an optimal drawing) is always a good drawing. The *crossing number* $\mathrm{cr}(G)$ of a simple graph G is defined as the minimum possible number of edge crossings in a good drawing of G in the plane. Let G_1 and G_2 be simple graphs with vertex sets $V(G_1)$ and $V(G_2)$ and edge sets $E(G_1)$ and $E(G_2)$, respectively. The join product of two graphs G_1 and G_2, denoted by $G_1 + G_2$, is obtained from the vertex-disjoint copies of G_1 and G_2 by adding all edges between $V(G_1)$ and $V(G_2)$. For $|V(G_1)| = m$ and $|V(G_2)| = n$, the edge set of $G_1 + G_2$ is the union of disjoint edge sets of the graphs G_1, G_2, and the complete bipartite graph $K_{m,n}$. Let D $(D(G))$ be a good drawing of the graph G. We denote the number of crossings in D by $cr_D(G)$.

Let G_i and G_j be edge-disjoint subgraphs of G. We denote the number of crossings between edges of G_i and edges of G_j by $\mathrm{cr}_D(G_i, G_j)$, and the number of crossings among edges of G_i in D by $cr_D(G_i)$. In the paper, some proofs are based on the Kleitman's result on crossing numbers of the complete bipartite graphs [12]. More precisely, he proved that

$$\mathrm{cr}(K_{m,n}) = \left\lfloor \frac{m}{2} \right\rfloor \left\lfloor \frac{m-1}{2} \right\rfloor \left\lfloor \frac{n}{2} \right\rfloor \left\lfloor \frac{n-1}{2} \right\rfloor, \quad \text{if} \quad \min\{m,n\} \le 6.$$

2 The crossing number of $G + D_n$

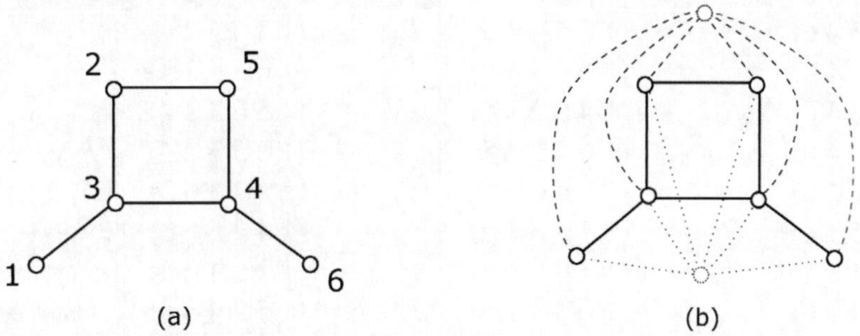

Fig. 1. Drawing of the graph G with the vertex notation and the graph $G + D_2$.

In the paper, we extent these results [4], [6], [7], [9], [10], [11] for another four graphs. Let G be the graph consisting of one 4-cycle and of two leaves. We consider the join product of G with the discrete graph on n vertices denoted by D_n. The graph $G + D_n$ consists of one copy of the graph G and of n vertices $t_1, t_2, \ldots, t_n$, where any vertex t_i, $i = 1, 2, \ldots, n$, is adjacent to every vertex of G. Let T^i, $1 \le i \le n$, denote the subgraph induced by the six edges incident with the vertex t_i. Thus, $T^1 \cup \cdots \cup T^n$ is isomorphic with the complete bipartite graph $K_{6,n}$ and

$$G + D_n = G \cup K_{6,n} = G \cup \left(\bigcup_{i=1}^{n} T^i \right). \tag{1}$$

2.1 Cyclic permutations and configurations

Let D be a good drawing of the graph $G + D_n$. The *rotation* $\mathrm{rot}_D(t_i)$ of a vertex t_i in the drawing D is the cyclic permutation that records the (cyclic) counter-clockwise order in which the edges leave t_i, see [3]. We emphasize that a rotation is a cyclic permutation. Hence, for $i, j \in \{1, 2, \ldots, n\}$, $i \ne j$, every subgraph $T^i \cup T^j$ of the graph $G + D_n$ is isomorphic with the graph $K_{6,2}$. In the paper, we will deal with the minimum necessary number of crossings between the edges of T^i and the edges of T^j in a subgraph $T^i \cup T^j$ induced by the drawing D of the graph $G + D_n$ depending on the rotations $\mathrm{rot}_D(t_i)$ and $\mathrm{rot}_D(t_j)$.

Let D be a good drawing of the graph $K_{6,n}$. Woodall [13] proved that in the subdrawing of $T^i \cup T^j \cong K_{6,2}$ induced by D, $\mathrm{cr}_D(T^i, T^j) \ge 6$ if $\mathrm{rot}_D(t_i) = \mathrm{rot}_D(t_j)$. Moreover, if $Q(\mathrm{rot}_D(t_i), \mathrm{rot}_D(t_j))$ denotes the minimum number of interchanges of adjacent elements of $\mathrm{rot}_D(t_i)$ required to produce the inverse cyclic permutation of $\mathrm{rot}_D(t_j)$, then $Q(\mathrm{rot}_D(t_i), \mathrm{rot}_D(t_j)) \le \mathrm{cr}_D(T^i, T^j)$. We will separate the subgraphs T^i, $i = 1, \ldots, n$, of the graph $G + D_n$ into three subsets depending on haw many the considered T^i crosses the edges of G in D. For $i = 1, 2, \ldots, n$, let $R_D = \{T^i : \mathrm{cr}_D(G, T^i) = 0\}$ and $S_D = \{T^i : \mathrm{cr}_D(G, T^i) = 1\}$. Every other subgraph T^i crosses G at least twice in D. Moreover, let F^i denote the subgraph $G \cup T^i$ for $T^i \in R_D$, where $i \in \{1, \ldots, n\}$. Thus, for a given drawing of

Name	Cyclic perm.	Name	Cyclic perm.	Name	Cyclic perm.	Name	Cyclic perm.
$P_1 \rightarrow$	(1 2 3 4 5 6)	$P_{31} \rightarrow$	(1 2 3 6 4 5)	$P_{61} \rightarrow$	(1 2 5 6 3 4)	$P_{91} \rightarrow$	(1 4 5 6 2 3)
$P_2 \rightarrow$	(1 3 2 4 5 6)	$P_{32} \rightarrow$	(1 3 2 6 4 5)	$P_{62} \rightarrow$	(1 5 2 6 3 4)	$P_{92} \rightarrow$	(1 5 4 6 2 3)
$P_3 \rightarrow$	(1 2 4 3 5 6)	$P_{33} \rightarrow$	(1 2 6 3 4 5)	$P_{63} \rightarrow$	(1 2 6 5 3 4)	$P_{93} \rightarrow$	(1 4 6 5 2 3)
$P_4 \rightarrow$	(1 4 2 3 5 6)	$P_{34} \rightarrow$	(1 6 2 3 4 5)	$P_{64} \rightarrow$	(1 6 2 5 3 4)	$P_{94} \rightarrow$	(1 6 4 5 2 3)
$P_5 \rightarrow$	(1 3 4 2 5 6)	$P_{35} \rightarrow$	(1 3 6 2 4 5)	$P_{65} \rightarrow$	(1 5 6 2 3 4)	$P_{95} \rightarrow$	(1 5 6 4 2 3)
$P_6 \rightarrow$	(1 4 3 2 5 6)	$P_{36} \rightarrow$	(1 6 3 2 4 5)	$P_{66} \rightarrow$	(1 6 5 2 3 4)	$P_{96} \rightarrow$	(1 6 5 4 2 3)
$P_7 \rightarrow$	(1 2 3 5 4 6)	$P_{37} \rightarrow$	(1 2 4 6 3 5)	$P_{67} \rightarrow$	(1 3 5 6 2 4)	$P_{97} \rightarrow$	(1 3 4 5 6 2)
$P_8 \rightarrow$	(1 3 2 5 4 6)	$P_{38} \rightarrow$	(1 4 2 6 3 5)	$P_{68} \rightarrow$	(1 5 3 6 2 4)	$P_{98} \rightarrow$	(1 4 3 5 6 2)
$P_9 \rightarrow$	(1 2 5 3 4 6)	$P_{39} \rightarrow$	(1 2 6 4 3 5)	$P_{69} \rightarrow$	(1 3 6 5 2 4)	$P_{99} \rightarrow$	(1 3 5 4 6 2)
$P_{10} \rightarrow$	(1 5 2 3 4 6)	$P_{40} \rightarrow$	(1 6 2 4 3 5)	$P_{70} \rightarrow$	(1 6 3 5 2 4)	$P_{100} \rightarrow$	(1 5 3 4 6 2)
$P_{11} \rightarrow$	(1 3 5 2 4 6)	$P_{41} \rightarrow$	(1 4 6 2 3 5)	$P_{71} \rightarrow$	(1 5 6 3 2 4)	$P_{101} \rightarrow$	(1 4 5 3 6 2)
$P_{12} \rightarrow$	(1 5 3 2 4 6)	$P_{42} \rightarrow$	(1 6 4 2 3 5)	$P_{72} \rightarrow$	(1 6 5 3 2 4)	$P_{102} \rightarrow$	(1 5 4 3 6 2)
$P_{13} \rightarrow$	(1 2 4 5 3 6)	$P_{43} \rightarrow$	(1 3 4 6 2 5)	$P_{73} \rightarrow$	(1 2 4 5 6 3)	$P_{103} \rightarrow$	(1 3 4 6 5 2)
$P_{14} \rightarrow$	(1 4 2 5 3 6)	$P_{44} \rightarrow$	(1 4 3 6 2 5)	$P_{74} \rightarrow$	(1 4 2 5 6 3)	$P_{104} \rightarrow$	(1 4 3 6 5 2)
$P_{15} \rightarrow$	(1 2 5 4 3 6)	$P_{45} \rightarrow$	(1 3 6 4 2 5)	$P_{75} \rightarrow$	(1 2 5 4 6 3)	$P_{105} \rightarrow$	(1 3 6 4 5 2)
$P_{16} \rightarrow$	(1 5 2 4 3 6)	$P_{46} \rightarrow$	(1 6 3 4 2 5)	$P_{76} \rightarrow$	(1 5 2 4 6 3)	$P_{106} \rightarrow$	(1 6 3 4 5 2)
$P_{17} \rightarrow$	(1 4 5 2 3 6)	$P_{47} \rightarrow$	(1 4 6 3 2 5)	$P_{77} \rightarrow$	(1 4 5 2 6 3)	$P_{107} \rightarrow$	(1 4 6 3 5 2)
$P_{18} \rightarrow$	(1 5 4 2 3 6)	$P_{48} \rightarrow$	(1 6 4 3 2 5)	$P_{78} \rightarrow$	(1 5 4 2 6 3)	$P_{108} \rightarrow$	(1 6 4 3 5 2)
$P_{19} \rightarrow$	(1 3 4 5 2 6)	$P_{49} \rightarrow$	(1 2 3 5 6 4)	$P_{79} \rightarrow$	(1 2 4 6 5 3)	$P_{109} \rightarrow$	(1 3 5 6 4 2)
$P_{20} \rightarrow$	(1 4 3 5 2 6)	$P_{50} \rightarrow$	(1 3 2 5 6 4)	$P_{80} \rightarrow$	(1 4 2 6 5 3)	$P_{110} \rightarrow$	(1 5 3 6 4 2)
$P_{21} \rightarrow$	(1 3 5 4 2 6)	$P_{51} \rightarrow$	(1 2 5 3 6 4)	$P_{81} \rightarrow$	(1 2 6 4 5 3)	$P_{111} \rightarrow$	(1 3 6 5 4 2)
$P_{22} \rightarrow$	(1 5 3 4 2 6)	$P_{52} \rightarrow$	(1 5 2 3 6 4)	$P_{82} \rightarrow$	(1 6 2 4 5 3)	$P_{112} \rightarrow$	(1 6 3 5 4 2)
$P_{23} \rightarrow$	(1 4 5 3 2 6)	$P_{53} \rightarrow$	(1 3 5 2 6 4)	$P_{83} \rightarrow$	(1 4 6 2 5 3)	$P_{113} \rightarrow$	(1 5 6 3 4 2)
$P_{24} \rightarrow$	(1 5 4 3 2 6)	$P_{54} \rightarrow$	(1 5 3 2 6 4)	$P_{84} \rightarrow$	(1 6 4 2 5 3)	$P_{114} \rightarrow$	(1 6 5 3 4 2)
$P_{25} \rightarrow$	(1 2 3 4 6 5)	$P_{55} \rightarrow$	(1 2 3 6 5 4)	$P_{85} \rightarrow$	(1 2 5 6 4 3)	$P_{115} \rightarrow$	(1 4 5 6 3 2)
$P_{26} \rightarrow$	(1 3 2 4 6 5)	$P_{56} \rightarrow$	(1 3 2 6 5 4)	$P_{86} \rightarrow$	(1 5 2 6 4 3)	$P_{116} \rightarrow$	(1 5 4 6 3 2)
$P_{27} \rightarrow$	(1 2 4 3 6 5)	$P_{57} \rightarrow$	(1 2 6 3 5 4)	$P_{87} \rightarrow$	(1 2 6 5 4 3)	$P_{117} \rightarrow$	(1 4 6 5 3 2)
$P_{28} \rightarrow$	(1 4 2 3 6 5)	$P_{58} \rightarrow$	(1 6 2 3 5 4)	$P_{88} \rightarrow$	(1 6 2 5 4 3)	$P_{118} \rightarrow$	(1 6 4 5 3 2)
$P_{29} \rightarrow$	(1 3 4 2 6 5)	$P_{59} \rightarrow$	(1 3 6 2 5 4)	$P_{89} \rightarrow$	(1 5 6 2 4 3)	$P_{119} \rightarrow$	(1 5 6 4 3 2)
$P_{30} \rightarrow$	(1 4 3 2 6 5)	$P_{60} \rightarrow$	(1 6 3 2 5 4)	$P_{90} \rightarrow$	(1 6 5 2 4 3)	$P_{120} \rightarrow$	(1 6 5 4 3 2)

Tab. 1. Names of Cyclic Permutations of 6-elements set.

G, any F^i is exactly represented by $\mathrm{rot}_D(t_i)$. All cyclic permutations of six elements can be generated by the algorithm [2], and they are collected in Tab. 1.

We will dealt with only drawings of the graph G with a possibility of an existence of a subgraph $T_i \in R_D$ because of arguments in the proof of the main Theorem 1. Assume a good drawing D of the graph $G + D_n$ in which the edges of G does not cross each other. In this case, without loss of generality, we can choose the vertex notations of the graph in such a way as shown in Fig. 1(a). It is easy to see that, in D, there are only four different possible configurations of F^i summarized in Tab. 2. In the rest of the paper, each cyclic permutation will be represented by the permutation with 1 in the first position. As for our considerations does not play role which of the regions is unbounded, assume the drawings shown in Figure 2. In a fixed drawing of the graph $G + D_n$, some configurations from the set $\mathcal{M} = \{A_1, A_2, A_3, A_4\}$ do not must appear. We denote by $\mathcal{M}_D$ the set of all configurations that exist in the drawing D belonging to $\mathcal{M}$.

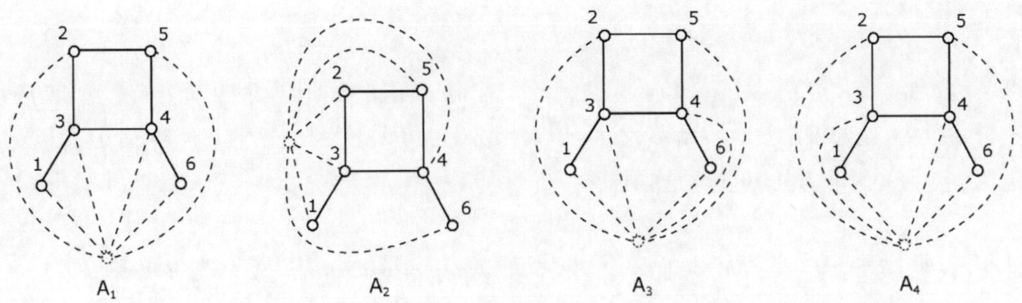

Fig. 2. Drawings of four possible configurations of graph F^i with the vertices of G denoted as in Fig. 1(a).

$A_1 : (125643)$	$A_2 : (132546)$
$A_3 : (125463)$	$A_4 : (132564)$

Tab. 2. Configurations of graph F^i with the vertices of G denoted as in Fig. 1(a).

$-$	A_1	A_2	A_3	A_4
A_1	6	4	5	5
A_2	4	6	5	5
A_3	5	5	6	5
A_4	5	5	5	6

Tab. 3. Lower-bounds of numbers of crossings for two configurations from $\mathcal{M}$.

Let X, Y be the configurations from $\mathcal{M}_D$. We shortly denote by $\mathrm{cr}_D(X,Y)$ the number of crossings in D between T^i and T^j for different $T^i, T^j \in R_D$ such that F^i, F^j have configurations X, Y, respectively. Finally, let $\mathrm{cr}(X,Y) = \min\{\mathrm{cr}_D(X,Y)\}$ over all good drawings of the graph $G + D_n$. In the next statements we are able to use the possibilities of the algorithm of the cyclic permutations of 6-elements set, see [2]. By $\overline{P_i}$ we will understand the inverse cyclic permutation to the permutation P_i, for $i = 1, \ldots, 120$. Woodall [13] defined the cyclic-ordered graph COG with the set of vertices $V = \{P_1, P_2, \ldots, P_{120}\}$, and with the set of edges E, where two vertices are joined by the edge if the vertices correspond to the permutations P_i and P_j, which are formed by the exchange of exactly two adjacent elements of the 6-tuple (i. e. an ordered set with 6 elements). Hence, if $d_{COG}(\text{"rot}_D(t_i)\text{"}, \text{"rot}_D(t_j)\text{"})$ denotes the distance between two vertices correspond to the cyclic permutations $\mathrm{rot}_D(t_i)$ and $\mathrm{rot}_D(t_j)$ in the graph COG, then

$$d_{COG}(\text{"rot}_D(t_i)\text{"}, \text{"}\overline{\mathrm{rot}_D(t_j)}\text{"}) = Q(\mathrm{rot}_D(t_i), \mathrm{rot}_D(t_j)) \leq \mathrm{cr}_D(T^i, T^j)$$

for any two different subgraphs T^i and T^j. The configurations A_1 and A_2 are represented by the cyclic permutations $P_{85} = (125643)$ and $P_8 = (132546)$, respectively. Using $\overline{P_8} = (164523) = P_{94}$ and $d_{COG}(\text{"}P_{85}\text{"}, \text{"}P_{94}\text{"}) = 4$ we obtain $\mathrm{cr}(A_1, A_2) \geq 4$. The same reason gives $\mathrm{cr}(A_1, A_3) \geq 5$, $\mathrm{cr}(A_1, A_4) \geq 5$, $\mathrm{cr}(A_2, A_3) \geq 5$, $\mathrm{cr}(A_2, A_4) \geq 5$ and $\mathrm{cr}(A_3, A_4) \geq 4$. Moreover, by a discussion of possible subdrawings, we can verify that $\mathrm{cr}(A_3, A_4) \geq 5$. Clearly, also $\mathrm{cr}(A_k, A_k) \geq 6$ holds for any $k = 1, \ldots, 4$. Thus, all lower-bounds of number of crossing of configurations from $\mathcal{M}$ are summarized in Tab. 3.

2.2 Main results

Lemma 1 *Let D be a good drawing of $G + D_n$, $n > 2$, in which $\mathrm{cr}_D(T^i, T^j) \neq 0$ for any different subgraphs T^i and T^j. Let $2|R_D| + |S_D| > 2n - 2\left\lfloor \frac{n}{2} \right\rfloor$ and let $T^n, T^{n-1} \in R_D$ be different subgraphs with $\mathrm{cr}_D(T^n \cup T^{n-1}) \geq 4$. If both conditions*

$$\mathrm{cr}_D(G \cup T^n \cup T^{n-1}, T^i) \geq 10 \qquad \text{for any } T^i \in R_D \setminus \{T^n, T^{n-1}\}, \tag{2}$$

$$\mathrm{cr}_D(G \cup T^n \cup T^{n-1}, T^i) \geq 7 \qquad \text{for any } T^i \in S_D \tag{3}$$

hold, then there are at least $6\left\lfloor \frac{n}{2} \right\rfloor \left\lfloor \frac{n-1}{2} \right\rfloor + 2\left\lfloor \frac{n}{2} \right\rfloor$ crossings in D.

Proof. We denote by $r = |R_D|$ and $s = |S_D|$. By the assumption of lemma, any $T^i \notin R_D \cup S_D$ satisfies the condition $\mathrm{cr}_D(G \cup T^n \cup T^{n-1}, T^i) \geq 4$, and the number of T^i that cross the graph G at least two times is equal to $n - r - s$. By fixing of the graph $G \cup T^n \cup T^{n-1}$ we have

$$\mathrm{cr}_D(G + D_n) = \mathrm{cr}_D(K_{6,n-2}) + \mathrm{cr}_D(K_{6,n-2}, G \cup T^n \cup T^{n-1}) + \mathrm{cr}_D(G \cup T^n \cup T^{n-1}) \geq$$

$$\geq 6\left\lfloor \frac{n-2}{2} \right\rfloor \left\lfloor \frac{n-3}{2} \right\rfloor + 10(r-2) + 7s + 4(n-r-s) + 4 = 6\left\lfloor \frac{n-2}{2} \right\rfloor \left\lfloor \frac{n-3}{2} \right\rfloor + 6r + 3s + 4n - 16 \geq$$

$$\geq 6\left\lfloor \frac{n-2}{2} \right\rfloor \left\lfloor \frac{n-3}{2} \right\rfloor + 3\left(2n - 2\left\lfloor \frac{n}{2} \right\rfloor + 1\right) + 4n - 16 \geq 6\left\lfloor \frac{n}{2} \right\rfloor \left\lfloor \frac{n-1}{2} \right\rfloor + 2\left\lfloor \frac{n}{2} \right\rfloor.$$

This completes the proof. □

Lemma 2 *Let D be a good drawing of $G + D_n$ with the vertex notations of the graph G as in Fig. 1(a), $n > 2$. If $T^n \in R_D$ such that F^n has configuration $A_i \in \mathcal{M}_D$, for $i = 1, 3, 4$, then*

$$\mathrm{cr}_D(T^n, T^k) \geq 3 \qquad \text{for any } T^k \in S_D. \tag{4}$$

Proof. Let, in D, the graph F^n has configuration A_1. If $T^k \in S_D$ with $\mathrm{cr}_D(T^n, T^k) = 2$, then the vertex t_k must be placed in a region with at least three vertices of G on its boundary, see Fig. 2. Since $T^k \in S_D$, the vertex t_k cannot be placed in the region bounded by 4-cycle of the graph G. Moreover, if t_k is placed in another regions, then $\mathrm{cr}_D(F^n, T^k) > 3$. The same idea can be used for configurations A_3 and A_4. This completes the proof. □

Remark that the property (4) is not true for configuration A_2, see the proof of the following statement.

Collorary 1 *Let D be a good drawing of $G + D_n$ with the vertex notations of the graph G as in Fig. 1(a), $n > 2$, in which $\mathrm{cr}_D(T^i, T^j) \neq 0$ for any different subgraphs T^i and T^j. If $T^n, T^{n-1} \in R_D$ such that F^n, F^{n-1} have configurations A_1, A_2, respectively, then*

$$\mathrm{cr}_D(G \cup T^n \cup T^{n-1}, T^k) \geq 7 \qquad \text{for any } T^k \in S_D. \tag{5}$$

Proof. Let, in D, the graphs F^n, F^{n-1} have configurations A_1, A_2, respectively. The configurations A_1 and A_2 are represented by the cyclic permutations $P_{85} = (125643)$ and $P_8 = (132546)$, respectively.

- If there is a subgraph $T^k \in S_D$ with $\mathrm{cr}_D(T^{n-1}, T^k) = 2$, then the vertex t_k must be placed in the region with four vertices of G and one vertex t_{n-1} on its boundary, see Fig. 2. Thus, the graph $F^k = G \cup T^k$ can be represented only by two possible cyclic permutations $P_{81} = (126453)$ and $P_{95} = (156423)$. By the above mentioned algorithm we have

$$d_{COG}("P_{26}", "P_{85}") = d_{COG}("P_{99}", "P_{85}") = 4,$$

where $\overline{P_{81}} = (135462) = P_{99}$ and $\overline{P_{95}} = (132465) = P_{26}$. By the properties of the cyclic permutations we have $\mathrm{cr}_D(T^n, T^k) \geq 4$. Thus, $\mathrm{cr}_D(G \cup T^n \cup T^{n-1}, T^k) \geq 1 + 4 + 2 = 7$.

- If $\mathrm{cr}_D(T^{n-1}, T^k) \geq 3$ for any subgraph $T^k \in S_D$, then $\mathrm{cr}_D(G \cup T^n \cup T^{n-1}, T^k) \geq 1+3+3 = 7$.

□

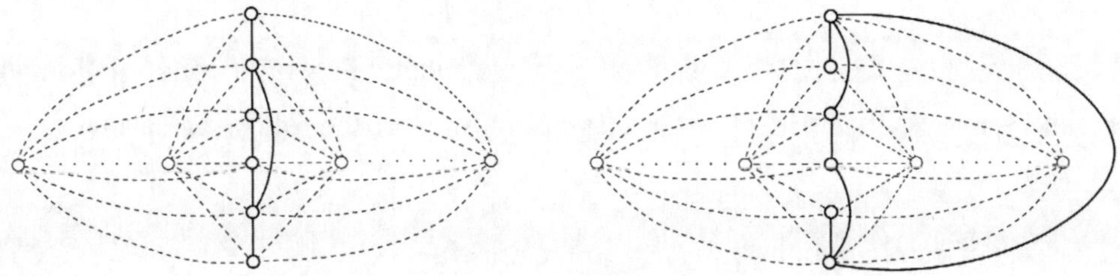

Fig. 3. Two good drawings of $G + D_n$.

Theorem 1 $\mathrm{cr}(G + D_n) = 6\left\lfloor \frac{n}{2} \right\rfloor \left\lfloor \frac{n-1}{2} \right\rfloor + 2\left\lfloor \frac{n}{2} \right\rfloor$ *for* $n \geq 1$.

Proof. In Fig. 3 there are the drawings of $G + D_n$ with $6\left\lfloor \frac{n}{2} \right\rfloor \left\lfloor \frac{n-1}{2} \right\rfloor + 2\left\lfloor \frac{n}{2} \right\rfloor$ crossings. Thus, $\mathrm{cr}(G + D_n) \leq 6\left\lfloor \frac{n}{2} \right\rfloor \left\lfloor \frac{n-1}{2} \right\rfloor + 2\left\lfloor \frac{n}{2} \right\rfloor$. We prove the reverse inequality by induction on n. The graph $G + D_1$ is planar, hence $\mathrm{cr}(G + D_1) = 0$. It is clear from Fig. 1(b) that $\mathrm{cr}(G + D_2) \leq 2$. The graph $G + D_2$ contains a subdivision of $K_{3,4}$, and therefore $\mathrm{cr}(G + D_2) \geq 2$. So, $\mathrm{cr}(G + D_2) = 2$ and the result is true for $n = 1$ and $n = 2$.

Suppose now that, for $n \geq 3$, there is a drawing D with

$$\mathrm{cr}_D(G + D_n) < 6\left\lfloor \frac{n}{2} \right\rfloor \left\lfloor \frac{n-1}{2} \right\rfloor + 2\left\lfloor \frac{n}{2} \right\rfloor, \tag{6}$$

and let

$$\mathrm{cr}(G + D_m) \geq 6\left\lfloor \frac{m}{2} \right\rfloor \left\lfloor \frac{m-1}{2} \right\rfloor + 2\left\lfloor \frac{m}{2} \right\rfloor \qquad \text{for any } m < n. \tag{7}$$

The drawing D has the following property:

$$\mathrm{cr}_D(T^i, T^j) \neq 0 \qquad \text{for all } i, j = 1, 2, \ldots, n,\ i \neq j. \tag{8}$$

To prove it assume that there are two different subgraphs T^i and T^j such that $\mathrm{cr}_D(T^i, T^j) = 0$. Without loss of generality let $\mathrm{cr}_D(T^{n-1}, T^n) = 0$. One can easy to verify that $\mathrm{cr}_D(G, T^{n-1} \cup T^n) \geq 2$.

As $\mathrm{cr}(K_{6,3}) = 6$, we have $\mathrm{cr}_D(T^k, T^{n-1} \cup T^n) \geq 6$ for $k = 1, 2, \dots, n-2$. So, for the number of crossings in D holds

$$\mathrm{cr}_D(G+D_n) = \mathrm{cr}_D\,(G + D_{n-2}) + \mathrm{cr}_D(T^{n-1} \cup T^n) + \mathrm{cr}_D(K_{6,n-2}, T^{n-1} \cup T^n) + \mathrm{cr}_D(G, T^{n-1} \cup T^n) \geq$$

$$\geq 6\left\lfloor\frac{n-2}{2}\right\rfloor\left\lfloor\frac{n-3}{2}\right\rfloor + 2\left\lfloor\frac{n-2}{2}\right\rfloor + 6(n-2) + 2 = 6\left\lfloor\frac{n}{2}\right\rfloor\left\lfloor\frac{n-1}{2}\right\rfloor + 2\left\lfloor\frac{n}{2}\right\rfloor.$$

This contradicts (6), and therefore $\mathrm{cr}_D(T^i, T^j) \neq 0$ for all $i, j = 1, 2, \dots, n$, $i \neq j$. Our assumption on D together with $\mathrm{cr}(K_{6,n}) = 6\left\lfloor\frac{n}{2}\right\rfloor\left\lfloor\frac{n-1}{2}\right\rfloor$ implies that

$$\mathrm{cr}_D(G) + \mathrm{cr}_D(G, K_{6,n}) < 2\left\lfloor\frac{n}{2}\right\rfloor.$$

Hence, if we denote $r = |R_D|$ and $s = |S_D|$, then

$$0r + 1s + 2(n - r - s) < 2\left\lfloor\frac{n}{2}\right\rfloor.$$

Thus, $r \geq 1$ and $2r + s > 2n - 2\left\lfloor\frac{n}{2}\right\rfloor$. We will fix one or two subgraphs with a contradiction with the assumption that there are less than $6\left\lfloor\frac{n}{2}\right\rfloor\left\lfloor\frac{n-1}{2}\right\rfloor + 2\left\lfloor\frac{n}{2}\right\rfloor$ crossings in the following cases:

Case 1: $\mathrm{cr}_D(G) = 0$.

We will deal with the sets of configurations $\{A_1, A_2\}$ in the drawing D.

1) $\{A_1, A_2\} \not\subseteq \mathcal{M}_D$.

   a) Let $A_2 \notin \mathcal{M}_D$ and $A_i \in \mathcal{M}_D$ for some $i \in \{1, 3, 4\}$, or let $A_2 \in \mathcal{M}_D$ and $A_i \in \mathcal{M}_D$ for some $i \in \{3, 4\}$. Without lost of generality, we can assume that $T^n \in R_D$ with F^n having configuration A_i. Thus, by fixing of the graph F^n using Lemma 2 we have

$$\mathrm{cr}_D(G+D_n) = \mathrm{cr}_D(K_{6,n-1}) + \mathrm{cr}_D(K_{6,n-1}, G \cup T^n) + \mathrm{cr}_D(G \cup T^n) \geq 6\left\lfloor\frac{n-1}{2}\right\rfloor\left\lfloor\frac{n-2}{2}\right\rfloor +$$

$$+5(r-1) + 4s + 3(n-r-s) = 6\left\lfloor\frac{n-1}{2}\right\rfloor\left\lfloor\frac{n-2}{2}\right\rfloor + 2r + s + 3n - 5 \geq 6\left\lfloor\frac{n-1}{2}\right\rfloor\left\lfloor\frac{n-2}{2}\right\rfloor +$$

$$+2n - 2\left\lfloor\frac{n}{2}\right\rfloor + 1 + 3n - 5 \geq 6\left\lfloor\frac{n}{2}\right\rfloor\left\lfloor\frac{n-1}{2}\right\rfloor + 2\left\lfloor\frac{n}{2}\right\rfloor.$$

   b) Let $\mathcal{M}_D = \{A_2\}$ and, without lost of generality, let $T^n \in R_D$.
   If there is no $T^k \in S_D$ with $\mathrm{cr}_D(T^n, T^k) = 2$, then we fix the graph F^n having configuration A_2 and we obtain the same inequalities as in the previous case. So, assume that there is a subgraph $T^k \in S_D$ with $\mathrm{cr}_D(T^n, T^k) = 2$. We can easily verify that $\mathrm{cr}_D(G \cup T^n \cup T^k, T^i) \geq 6 + 2 = 8$ for any $T^i \in R_D$, because both F^n and F^i have configuration A_2. Similarly by a discussion for two possible drawings of the graph T^k, see the proof of Corollary 1, we can verify that $\mathrm{cr}_D(G \cup T^n \cup T^k, T^i) \geq 7$ for any $T^i \in S_D$ and $\mathrm{cr}_D(G \cup T^n \cup T^k, T^i) \geq 6$ for any $T^i \notin R_D \cup S_D$. Thus, by fixing of the graph $G \cup T^n \cup T^k$ we have

$$\mathrm{cr}_D(G + D_n) = \mathrm{cr}_D(K_{6,n-2}) + \mathrm{cr}_D(K_{6,n-2}, G \cup T^n \cup T^k) + \mathrm{cr}_D(G \cup T^n \cup T^k) \geq$$

$$\geq 6\left\lfloor\frac{n-2}{2}\right\rfloor\left\lfloor\frac{n-3}{2}\right\rfloor + 8(r-1) + 7s + 6(n-r-s) + 3 = 6\left\lfloor\frac{n-2}{2}\right\rfloor\left\lfloor\frac{n-3}{2}\right\rfloor + 2r + s +$$

$$+6n - 12 \geq 6\left\lfloor\frac{n-2}{2}\right\rfloor\left\lfloor\frac{n-3}{2}\right\rfloor + 2n - 2\left\lfloor\frac{n}{2}\right\rfloor + 1 + 6n - 12 \geq 6\left\lfloor\frac{n}{2}\right\rfloor\left\lfloor\frac{n-1}{2}\right\rfloor + 2\left\lfloor\frac{n}{2}\right\rfloor.$$

2) $\{A_1, A_2\} \subseteq \mathcal{M}_D$.
Without lost of generality let us fix any two $T^n,\ T^{n-1} \in R_D$ such that $F^n,\ F^{n-1}$ have configurations A_1, A_2, respectively. Then condition (2) is true by Tab. 3 and condition (3) holds by Corollary 1. Thus, all assumption of Lemma 1 are fulfilled.

Case 2: $\mathrm{cr}_D(G) = 1$.

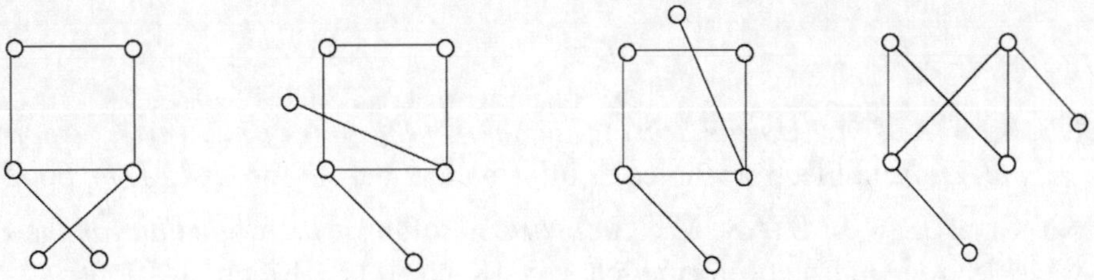

(a) (b) (c) (d)

Fig. 4. Four possible drawings of the graph G with one crossing among its edges.

Since $r \geq 1$, without lost of generality we assume $T^n \in R_D$. In all four possible drawing of the graph G it is possible to verify that $\mathrm{cr}_D(G \cup T^n, T^i) \geq 4$ for any subgraph T^i, $i = 1, \ldots, n-1$. Thus, by fixing of the graph F^n we obtain

$$\mathrm{cr}_D(G + D_n) = \mathrm{cr}_D(K_{6,n-1}) + \mathrm{cr}_D(K_{6,n-1}, G \cup T^n) + \mathrm{cr}_D(G \cup T^n) \geq$$

$$\geq 6\left\lfloor\frac{n-1}{2}\right\rfloor\left\lfloor\frac{n-2}{2}\right\rfloor + 4(n-1) + 1 \geq 6\left\lfloor\frac{n}{2}\right\rfloor\left\lfloor\frac{n-1}{2}\right\rfloor + 2\left\lfloor\frac{n}{2}\right\rfloor.$$

Case 3: $\mathrm{cr}_D(G) \geq 2$.

We are able to use the same idea as in the previous case for all possible drawing of the graph G with a possibility of an existence of a subgraph $T^i \in R_D$ in the considering drawing D.

This completes the proof of the main theorem. □

2.3 Corollaries

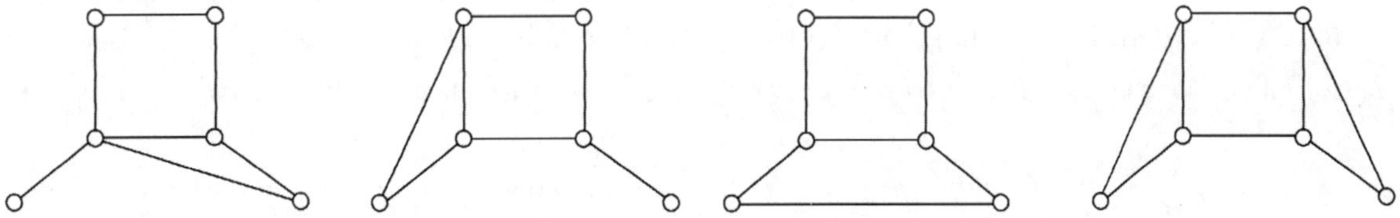

G_1 G_2 G_3 G_4

Fig. 5. Four graphs G_1, G_2, G_3, and G_4 by adding new edges to the graph G.

In Fig. 2 we are able to add some edges to the graph G without another crossings. So the drawing of the graphs $G_1 + D_n$, $G_2 + D_n$, $G_3 + D_n$, and $G_4 + D_n$ with $6\lfloor\frac{n}{2}\rfloor\lfloor\frac{n-1}{2}\rfloor + 2\lfloor\frac{n}{2}\rfloor$ crossings is obtained. Thus, the next results are obvious.

Collorary 2 $\mathrm{cr}(G_i + D_n) = 6\left\lfloor\frac{n}{2}\right\rfloor\left\lfloor\frac{n-1}{2}\right\rfloor + 2\left\lfloor\frac{n}{2}\right\rfloor$ *for* $n \geq 1$, *where* $i = 1, \ldots, 4$.

Remark that the crossing numbers of the graphs $G_3 + D_n$ and $G_4 + D_n$ were obtained in [8], [5] without using the vertex rotation.

3 Acknowledgments

The research was supported by the Slovak VEGA grant No. 1/0389/15. The research was also supported by the internal faculty research project no. FEI-2017-39.

References

[1] BEREŽNÝ, Š., STAŠ, M.: *On the crosing number of the join of five vertex graph G with the discrete graph D_n*, Acta Electrotechnica et Informatica, vol. 17, no. 3, (2017), pp. 27–32.

[2] BEREŽNÝ, Š., BUŠA, J., STAŠ, M.: *Software solution of the algorithm of the cyclic-order graph*, Acta Electrotechnica et Informatica, vol. 18, no. 1, (2018), pp. 3–10.

[3] HERNÁNDEZ-VÉLEZ, C., MEDINA, C., SALAZAR, G.: *The optimal drawing of $K_{5,n}$*, Electronic Journal of Combinatorics, vol. 21, no. 4, (2014), p. 29.

[4] KLEŠČ, M.: *The join of graphs and crossings numbers*, Electron. Notes Discrete Math., vol. 28, (2007), pp. 349–355.

[5] KLEŠČ, M.: *The crossing number of join of the special graph on six vertices with path and cycle*, Dicrete Math., vol. 310, (2010), pp. 1475–1481.

[6] KLEŠČ, M., SCHRÖTTER, Š.: *The crossing numbers of join products of paths with graphs of order four*, Discussiones Mathematicae Graph Theory, vol. 31, (2011), pp. 312–331.

[7] KLEŠČ, M., SCHRÖTTER, Š.: *The crossing numbers of join of paths and cycles with two graphs of order five*, Mathematical Modeling and Computational Science, vol. 7125, (2012), pp. 160–167.

[8] KLEŠČ, M., SCHRÖTTER, Š.: *On the crossing numbers of cartesian products of stars and graphs of order six*, Discussiones Mathematicae Graph Theory, vol. 33, (2013), pp. 583–597.

[9] KLEŠČ, M., KRAVECOVÁ, D., PETRILLOVÁ, J.: *The crossing number of join of special graphs*, Electrical Engineering and Informatics 2, (2011), pp. 522–527.

[10] KLEŠČ, M., VALO, M.: *Minimum crossings in join of graphs with paths and cycles*, Acta Electrotechnica et Informatica, vol. 12, (2012), pp. 32–37.

[11] STAŠ, M.: *On the crossing number of the join of the discrete graph with one graph of order five*, Mathematical Modelling and Geometry, vol. 5, no. 2, (2017), pp. 12–19.

[12] KLEITMAN, D. J.: *The crossing number of $K_{5,n}$*, J. Combinatorial Theory, vol. 9, (1970), pp. 315–323.

[13] WOODALL, D. R.: *Cyclic-order graphs and Zarankiewicz's crossings number conjucture*, J. Graph Theory, vol. 17, (1993), pp. 657–671.

Current address

Staš Michal, RNDr., PhD.
Department of Mathematics and Theoretical Informatics
Faculty of Electrical Engineering and Informatics
Technical University of Košice
Letná 9, 042 00 Košice, Slovak Republic
E-mail: michal.stas@tuke.sk

Slovak University of Technology in Bratislava
Faculty of Mechanical Engineering
17th Conference on Applied Mathematics **APLIMAT 2018**

Proceedings

TWO METHODS FOR ESTIMATING NUMBER OF INTERACTING VEHICLES

SZABOVÁ Zuzana (CZ), KRBÁLEK Milan (CZ)

Abstract. Knowledge of an interaction range in particle systems, especially in vehicular traffic could significantly contribute to modeling of traffic flow . Combination of simulation methods, analytical predictions of headway distribution, and correlation analysis led to several remarkable observations. We observe, that interaction range depends on both resistivity and type of repulsive potential. Moreover we introduce a novel method for detection of number of actively followed vehicles based on perturbation function.

Keywords: Headway distribution, interaction in vehicular traffic, correlation analysis

Mathematics subject classification: Primary 62H20; Secondary 62G07, 62P99

1 Introduction

Increasing traffic demand requires finding either numerical or theoretical models, capable to predict traffic scenarios such as traffic congestions. Recent studies and observations (e.g. [3], [5], [7], [8]) suggest that generally used premise on a short range of traffic interactions do not correspond to traffic reality. However, most traffic models [1], discrete as well as continuous presume that behavior and decision making of a driver is influenced by neighboring vehicles only. By means of correlation analysis we try to determine interaction range with restriction to fast lane and congested traffic phase, with intention to avoid overtaking and transitions between free and congested traffic phases. Moreover we aim to show that interaction range depends on quantities like traffic flow, traffic density or velocity and that correlation between traffic density and velocity plays a role in vehicular data. Further we suggest novel approach for detection of number of actively followed vehicles based on perturbation convolution function.

2 Data-sets and Processing

The used data-sets were provided by the Road and Motorway Directorate of the Czech Republic at the Expressway R1 in Prague, the Czech Republic. Data records contain single vehicle data recorded by double-loop detectors positioned under the expressway R1. For purposes of this research only fast-lane data have been taken into account. Following reasons lead us to this restriction. Firstly, very low proportion of long vehicles such as lorries or trucks and slow cars is present and more importantly, vehicles in fast lane cannot be overtaken, which could significantly influence traffic flow, especially in a transition phase between free flow and congested flow, when synchronization between lanes may occur. Measured microscopic quantities have been suited into a set of the form

$$S_j = \left\{(\tau_k^{(in)}, \tau_k^{(out)}, v_k) \in T^{(in)} \times T^{(out)} \times V \mid k = (j-1)m+1, (j-1)m+2, \ldots, jm\right\}, \quad (1)$$

where $j \in M = \lfloor \frac{N}{m} \rfloor$, $m = 50$ is the fixed sampling size and $N = \#\left\{\tau_k^{(in)} \mid k\right\}$. Sets $T^{(in)}$ and $T^{(out)}$ are defined as follows.

$$T^{(in)} = \left\{\tau_k^{(in)} \in \mathbb{R}_0^+ \mid k = 1, 2, \ldots, N\right\}$$

includes the chronologically-ordered times when the front bumper of k-th car has intersected the detector line. Respectively

$$T^{(out)} = \left\{\tau_k^{(out)} \in \mathbb{R}_0^+ \mid k = 1, 2, \ldots, N\right\}$$

includes the chronologically-ordered times when the rear bumper of k-th car has left the detector. Velocities of succeeding vehicles have been suited into a set

$$V = \left\{v_k \in \mathbb{R}_0^+ \mid k = 1, 2, \ldots, N\right\},$$

where v_k is a velocity of the k-th vehicle. For each sub-sample S_j, where $j = 1, 2, \ldots, m$, macroscopic quantity such as local flux (intensity)

$$Q_j = \frac{m}{\tau_{jm}^{(out)} - \tau_{(j-1)m+1}^{(in)}}, \quad (2)$$

representing number of vehicles passing the detector per fixed time interval (most commonly 1 hour) or average velocity

$$\bar{v}_j = \frac{1}{m} \sum_{k=(j-1)m+1}^{jm} v_k \quad (3)$$

can be calculated. These quantities can be further used to determine another macroscopic quantity, local vehicular density

$$\rho_j = \frac{Q_j}{\bar{v}_j}, \quad (4)$$

Which can be presented as number of vehicles passing the detector per fixed space unit (most commonly 1 km). Subsequently, time-clearances are scaled to sample-means equal to one and they are defined by[5]

$$t_k = \frac{m\left(\tau_k^{(in)} - \tau_k^{(out)}\right)}{\sum_{i=(\lceil k/m \rceil-1)m+1}^{\lceil k/m \rceil m} \tau_i^{(in)} - \sum_{i=(\lceil k/m \rceil-1)m+1}^{\lceil k/m \rceil m} \tau_i^{(out)}}. \quad (5)$$

We will profit from this re-scaling procedure when comparing empirical headway distributions to theoretical predictions having an estimated value equal to one.

At the very beggining we assume that in regions of congested traffic the interactions are more intense than in a free flow phase (which is meaningful premise), in which drivers are driving at their desired speed with desired time clearances. Earlier studies [10] show that both velocities correlations and time clearance correlations depend on traffic flow as well as on the traffic density. Moreover we would like to eliminate mixing of different traffic phases. Therefore we perform segmentation by local density ρ and flux Q (according to [4], [5]).

3 Segmentation of vehicular data by density and velocity

Study of the interaction range will be carried out separately for each flux-density region/window summarized in table 1 as in [11].

w	$\rho_{\min}$ [veh.km^{-1}]	$\rho_{\max}$ [veh.km^{-1}]	$Q_{\min}$ [veh.h^{-1}]	$Q_{\max}$ [veh.h^{-1}]	$\beta_w^{\log}$	β_w^{hyp}
1	25	30	500	2000	1.50	0.35
2	30	35	500	2500	2.55	0.95
3	35	40	500	3000	2.65	1.00
4	40	45	500	3000	3.15	1.30
5	45	50	500	3000	3.40	1.45
6	50	55	500	2500	3.65	1.60
7	55	60	500	2500	3.60	1.60
8	60	65	500	2500	4.50	1.90

Tab. 1. Basic information on the vehicular data used.

We also perform segmentation by velocity regions. Study of the interaction range will be carried out separately for each velocity region summarized in Tab. 2.

w	$\bar{v}_{j,\min}$ [km.h^{-1}]	$\bar{v}_{j,\max}$ [km.h^{-1}]	$\beta_w^{\log}$	β_w^{hyp}
1	20	30	2.90	1.45
2	30	35	3.50	1.35
3	35	40	3.75	1.50
4	40	45	3.25	1.35
5	45	50	3.00	1.35
6	50	55	3.80	1.25
7	55	60	2.90	1.15
8	60	70	2.45	0.95

Tab. 2. Basic information on the vehicular data used.

Basic insight into vehicular interactions can be obtained by means of correlation analysis. Two approaches have been considered. Pearson's correlation coefficient however disposes disadvantage that can obtain even negative values implying anti-correlation and more importantly zero value does not imply statistical independence. Brownian correlation coefficient [9] does not suffer from these shortcomings as has been proven in [9].

Definition 1 *Let X, Y denote random vectors with finite second moments. Then Brownian distance* *correlation $\mathscr{R}_{\mathscr{B}}(X, Y)$ is defined by*

$$\mathscr{R}_{\mathscr{B}}^2(X, Y) \begin{cases} \dfrac{\mathscr{V}_n^2(X, Y)}{\sqrt{\mathscr{V}_n^2(X)\mathscr{V}_n^2(Y)}}, & \mathscr{V}_n^2(X)\mathscr{V}_n^2(Y) > 0; \\ 0, & \mathscr{V}_n^2(X)\mathscr{V}_n^2(Y) = 0, \end{cases} \tag{6}$$

where $\mathscr{V}_n(X, Y)$ denotes distance covariance and $\mathscr{V}_n(X)$, $\mathscr{V}_n(Y)$ represent the associated variances.

However, this type of correlation coefficient also comes with deficiencies, such as strongly correlated variables are not causatively linked. Since, we will not be dealing with strongly correlated variables as was observed in [10], we will dismiss this disadvantage. From the definition of the brownian correlation coefficient it is clear that we will be working with vectors of random variables. Therefore we define

$$X_w := \{x_1^w, x_2^w, \ldots, x_\ell^w\} \tag{7}$$

and

$$X_w^{(n)} := \{x_{1+n}^w, x_{2+n}^w, \ldots, x_{\ell+n}^w \mid n \in \mathbb{N}\}, \tag{8}$$

Which are the sets of chronologically ordered individual clearances associated to the w-th flux-density window. Now we can proceed to evaluation of Brownian correlation coefficient $\mathscr{R}_{\mathscr{B}}(X_w, X_w^{(n)})$ with respect to number of separating vehicles. It follows from Fig. 1 that values of distance-correlation coefficient are mostly decreasing with n in all density regions as well as in all velocity regions.

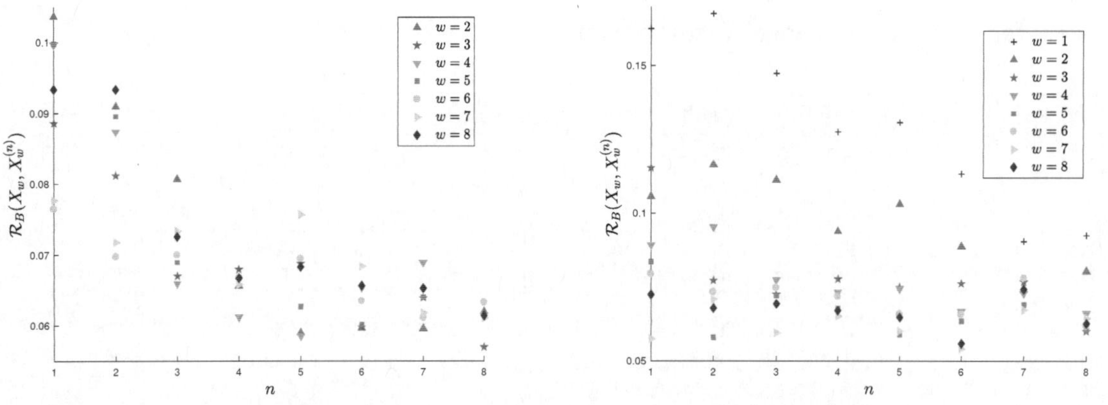

(a) (b)

Fig. 1. rownian distance-correlation of clearances in different density regions (see figure (a)) and velocity regions (see figure (b)).

Strongest correlation is detected for $n = 1$, means that clearances between neighboring vehicles are more strongly correlated compared to clearances measured between more distant vehicles (as in [3]). This is true especially in case of segmentation by density. Moreover, the lower the velocity, the hire the correlations are. As expected, when one finds himself in a strongly congested traffic, drivers are driving bumper to bumper, which generates strong correlation between clearances. On the contrary, when driving in a free phase, drivers tend to maintain desired gap to their predecessors, which corresponds to very little synchronization and therefore very little correlation between clearances. The isolated application of distance-correlations of clearances is as understandable ineffective for purposes of interaction range estimation. Therefore we focus our attention to the generalized local thermodynamic model allowing middle-ranged interactions. Main characteristics of the model are statistical resistivity β and thermodynamic potential. Short-ranged version of this model applying hyperbolic potential has been solved in [4]. One of the main results of the article is determination of the respective headway distribution, which reads

$$p(x) = \Theta(x) A \mathrm{e}^{-\frac{\beta}{x}} \mathrm{e}^{-Dx}, \tag{9}$$

where

$$D = \beta + \frac{3 - \sqrt{\beta}}{2}, \tag{10}$$

$$A = \left(2\sqrt{\frac{\beta}{D}} \mathscr{K}_1(2\sqrt{\beta D}) \right)^{-1}. \tag{11}$$

Here $\Theta(x)$ stands for Heaviside unit-step function and $\mathscr{K}_\nu(x)$ for Macdonald's function of the ν-th order. Analogously in article [4] has been determined a headway distribution for the thermodynamic model with short-ranged logarithmic potential, which reads

$$\eta(x) = \Theta(x) \frac{(\beta+1)^{\beta+1}}{\Gamma(\beta+1)} x^\beta \mathrm{e}^{-(\beta+1)x}. \tag{12}$$

Empirical values of resistivity β summarized in table 1 (columns 6 and 7) have been determined by means of MDE (minimum distance estimator)

$$\beta_w^h := \mathrm{argmin}_{\beta \in [0,\infty)} \int_0^\infty \left| H_w(x) - \Theta(x) A(\beta) \mathrm{e}^{-\frac{\beta}{x}} \mathrm{e}^{-D(\beta)x} \right| \mathrm{d}x,$$

$$\beta_w^{log} := \mathrm{argmin}_{\beta \in [0,\infty)} \int_0^\infty \left| H_w(x) - \Theta(x) \frac{(\beta+1)^{\beta+1}}{\Gamma(\beta+1)} x^\beta \mathrm{e}^{-(\beta+1)x} \right| \mathrm{d}x,$$

minimizing the statistical distance between theoretical prediction (9), (12) and histogram function $H_w(x)$, respectively. Steady states of the model with middle-ranged hyperbolic potential are not analytically calculated at the moment for any value of resistivity β. In contrast, steady states have been determined for model with middle-ranged logarithmic potential and specific values of resistivity β in [2] and [12]. However, numerical representation allows obtaining all the steady-distributions. Output of the model is vector of steady-state headways/clearances/distances (these terms are equivalent because particles are dimensionless) with average value equal to one

$$X = (x_1, x_2, \ldots, x_N),$$

where number of particles is equal to length of the circle $N = L$. Simultaneously, we define vector

$$X_n = (x_{n+1}, x_{n+2}, \ldots, x_{n+N}),$$

where n denotes number of particles lying between considered particles. For fixed interaction range and stochastic resistivity one can calculate Brownian distance-correlation coefficient $\mathscr{R}^n_{\mathscr{B}}(X, X_n)$. In Fig. 2 values of correlation coefficients $\mathscr{R}^n_{\mathscr{B}}(X, X_1)$ are plotted with respect to interaction range $k \in$

$\{1,2,3,4,5,6\}$ and resistivity β for both hyperbolic and logarithmic potential, respectively. Beside case $k=1$ yielding statistical independence among neighboring headways, one can detect values of distance-correlation coefficient increasing with the statistical resistivity. To estimate an interaction range, we plot empirical values of statistical resistivity β_w^h, β_w^{log} into these diagrams respectively and search for the closest correlation curve. To eliminate fluctuations, we use polynomial interpolation of correlation curves.

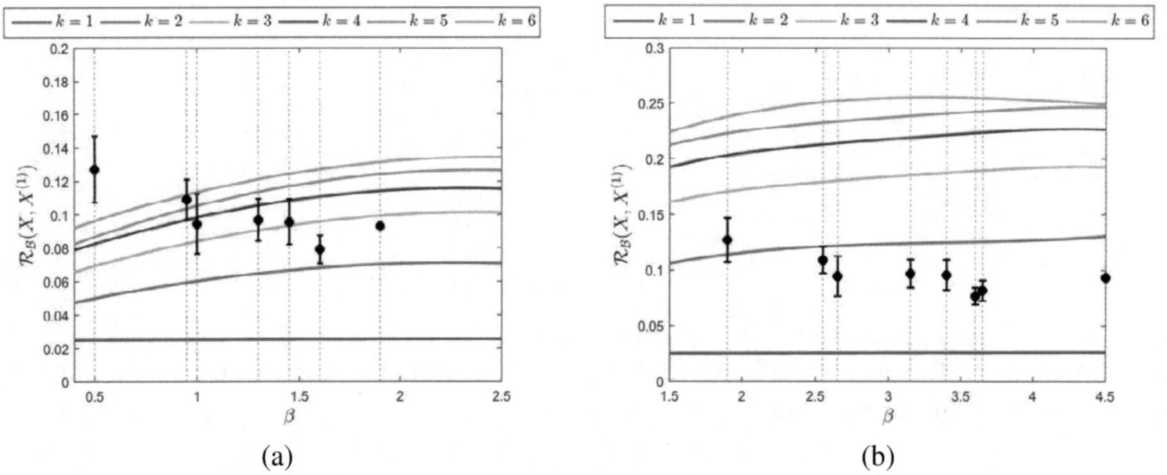

Fig. 2. Curves represent polynomial interpolations of distance-correlation coefficient. Black bullets represent empirical values of Brownian correlation coefficient obtained by MDE for hyperbolic potential (a) and logarithmic potential (b).

It is apparent that interaction range for both logarithmic and hyperbolic potential depends on traffic density. We can summarize that with increasing traffic density (in this case, with increasing statistical resistivity β) interaction range decreases, since correlation drops. In both cases, however, interaction is a middle-ranged in all density regions. Interaction is stronger when using hyperbolic potential. Eventhough we performed several restrictions (only fast lane data from congested phase were used omitting transition phases), we still can not guarantee homogeneity of velocities in each flux-density window. At this point it is convenient to study interactions with respect to velocities.

We use the same approach as for segmentation by velocities. Empirical values of resistivity β (summarized in Tab. 2) have been determined by means of MDE. Similarly to previous section we plot interpolated correlation curves and empirical values of stochastic resistivity in one diagram for both hyperbolic and logarithmic potential (see Fig. 3). In both cases with increasing resistivity, distance-correlation coefficients are increasing. It is also visible that all interactions are middle-ranged, as in the previous section.

4 Perturbation function

At this point it is essential to define multi-clearances of the order $\mu \in \mathbb{N}$ as follows

$$x_k \mid \mu = x_k + x_{k-1} + \cdots + x_{k-\mu}. \tag{13}$$

If the interactions in the system are short-ranged only, then random variables $x \mid \mu_1$ and $x \mid \mu_2$, where $\mu_1 \neq \mu_2$ are independent and the distribution of clearances among $\mu + 2$ particles is given by

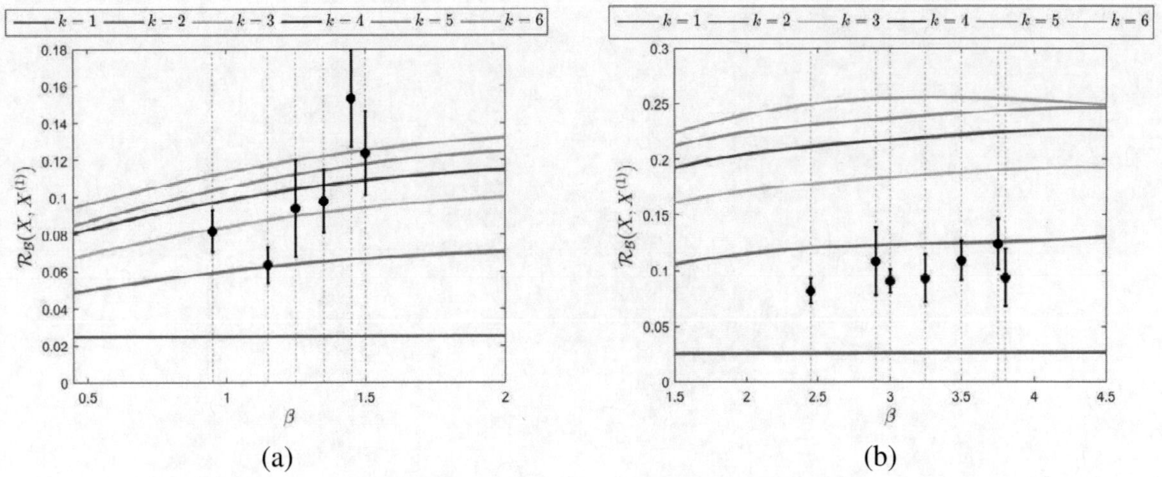

Fig. 3. Curves represent polynomial interpolations of distance-correlation coefficient. Black bullets represent empirical values of Brownian correlation coefficient obtained by MDE for hyperbolic potential (a) and logarithmic potential (b).

convolution formula

$$\rho(x \mid \mu) = \rho(x) \star \rho(x \mid \mu - 1) \equiv \int_{\mathbb{R}} \rho(s)\rho(x - s \mid \mu - 1)\mathrm{d}s. \tag{14}$$

Unfortunately, this formula holds true only for ensembles with short-ranged forces only. Therefore we introduce two-parametric family of probability densities defined in [6]

$$p(x) = \Theta(x) A x^{\alpha} \mathrm{e}^{-\frac{\beta}{x}} \mathrm{e}^{-Dx}, \tag{15}$$

where

$$D = \frac{\alpha}{\mu} + \frac{\beta}{\mu^2} + \frac{3 - \mathrm{e}^{-\sqrt{\frac{\beta}{\mu}}}}{2\mu}, \tag{16}$$

$$A = \left(2\sqrt{\frac{\beta}{D}} \mathscr{K}_{\alpha+1}(2\sqrt{\beta D}) \right)^{-1}. \tag{17}$$

Here $\Theta(x)$ stands for Heaviside unit-step function and $\mathscr{K}_\nu(x)$ for Macdonald's function of the ν-th order. To estimate the parameter α, β we use MDE method (Minimum Distance Estimator) given by formula

$$(\hat{\alpha}, \hat{\beta}) := \mathrm{argmin}_{(\alpha,\beta)\in C_\sigma} \left(\int_{\mathbb{R}} \mid H(x \mid \mu) - \hat{\rho}(x \mid \mu) \mid^2 \mathrm{d}x \right)^{1/2} \tag{18}$$

$$C_\sigma = \left\{ (\alpha, \beta) \in \Omega : \int_0^\infty x^2 \rho(x \mid (\alpha, \beta))\mathrm{d}x = \mu^2 + \sigma^2 \right\}, \tag{19}$$

where $H(x \mid \mu)$ is histogram function (distribution of clearances given by Metropolis-Hastings algorithm) and $\hat{\rho}(x \mid \mu)$ is MDE-estimation. In Fig. 4 we plot estimated values of parameters $\hat{\alpha}, \hat{\beta}$ with respect to degree of multi-clearance. We conclude, that with increasing degree of multi-clearance one obtains increasing values of parameters $\hat{\alpha}, \hat{\beta}$.

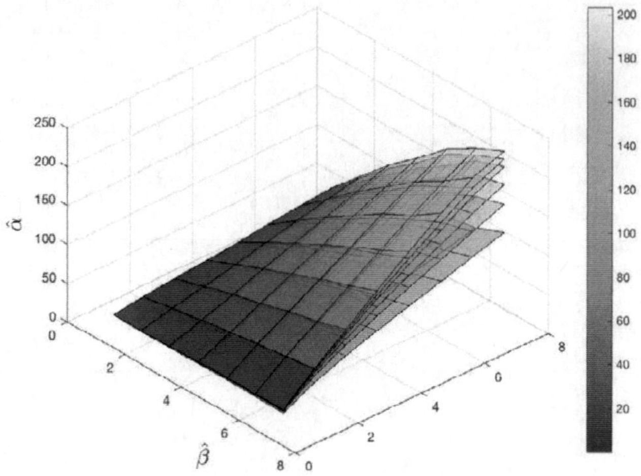

Fig. 4. Estimated values of parameters $\hat{\alpha}, \hat{\beta}$ with respect to degree of multi-clearance for different interactions, starting with short-ranged interaction (lowest plane) and increasing up to interaction of 7 vehicles (upper plane).

Further we would like to introduce a method for detection of interaction range. As we mentioned before, we can use a convolution rule for estimation of probability density $\rho(x \mid \mu), (\mu \neq 1)$ if random variables x and $x \mid \mu - 1$ are statisticaly independent. Therefore we introduce perturbation function

$$\psi(x \mid \mu) = \int_0^x (\hat{\rho}(y \mid \mu) - \hat{\rho}(y \mid 1) \star \hat{\rho}(y \mid \mu - 1))\mathrm{d}y, \tag{20}$$

which is used for testing statistical independence of different multi-clearances. Courses of perturbation functions for different values of stochastic resistivity β, interaction range and degree of multi-clearance are shown in Fig. 5.

It is desired to study how the perturbation function $\psi(x \mid \mu)$ deviates from zero value. For intentions of this quantification we use Kolmogorov distance defined by formula

$$G(\mu) = \sup_{x \in \mathbb{R}} \mid \psi(x \mid \mu) \mid . \tag{21}$$

Course of Kolmogorov distance with respect to interaction range, resistivity β and degree of multi clearance is shown on Fig. 6. Values of Kolmogorov distance are low for all interaction ranges and the associated neighboring planes are very close to each other. Kolmogorov distance planes have been created by means of simulation method (Metropolis-Hastings algorithm). Now question arises how precisely we are able to determine course of this planes. In Fig. 7(a) it is shown Kolmogorov distance plane with error bars for model with middle-ranged interaction (interaction of 5 vehicles) and in Fig. 7 (b) it is shown Kolmogorov distance plane with error bars for model with middle-ranged interaction (interaction of 5,6,7 vehicles, respectively). Since the upper error bar is intersecting Kolmogorov distance plane for interaction of 7 vehicles, at this point we would not be able to determine a range of interaction. Unambigously, this is true only for planes with interaction of more than 6 vehicles. For lower interaction the error bars are not intersecting neighboring distance planes.

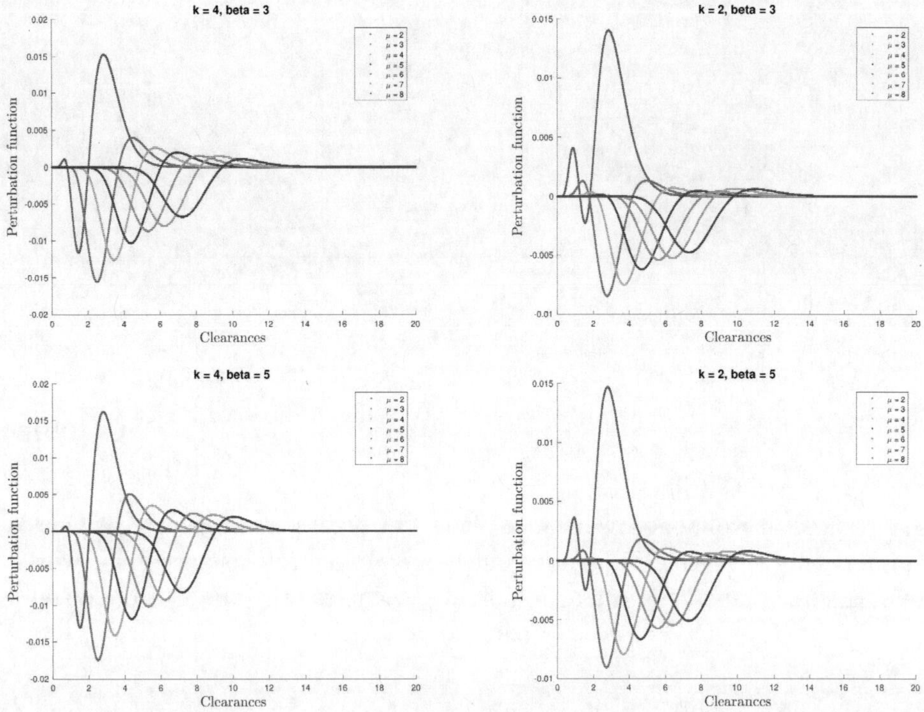

Fig. 5. Courses of perturbation functions for different values of stochastic resistivity β, interaction range and degree of multi-clearance.

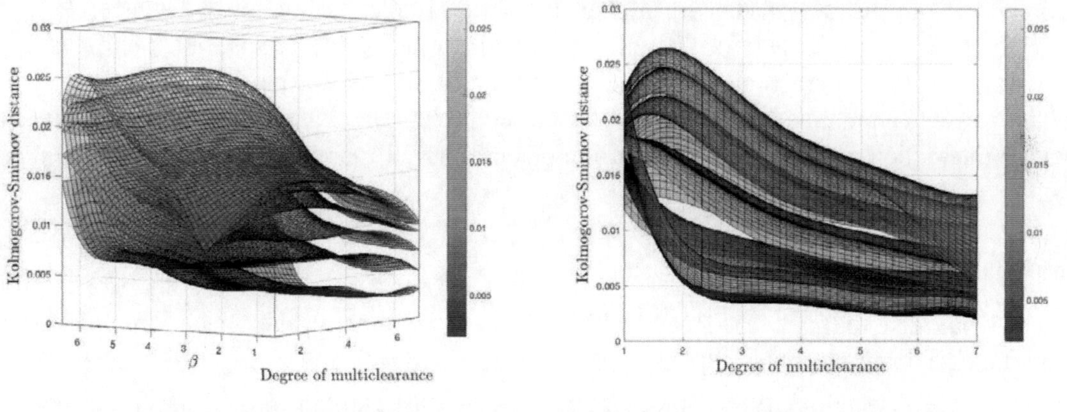

Fig. 6. Courses of Kolmogorov distance of perturbation function with respect to stochastic resistivity β, interaction range and degree of multi-clearance.

5 Conclusion

Main task of this paper was to estimate a number of actively followed vehicles by means of Brownian distance-correlation coefficients and numerical data obtained from simulations of one-dimensional particle gas with both middle-ranged hyperbolic and logarithmic potential. Segmentation of the empirical vehicular data by density and velocity has been performed. These separations lead to contradictory results. When segmentation by local density was used, we observed that with increasing stochastic resistivity, correlations were decreasing. On the contrary, segmentation by velocities lead to increasing correlations with increasing stochastic resistivity. This means, that either one of the seg-

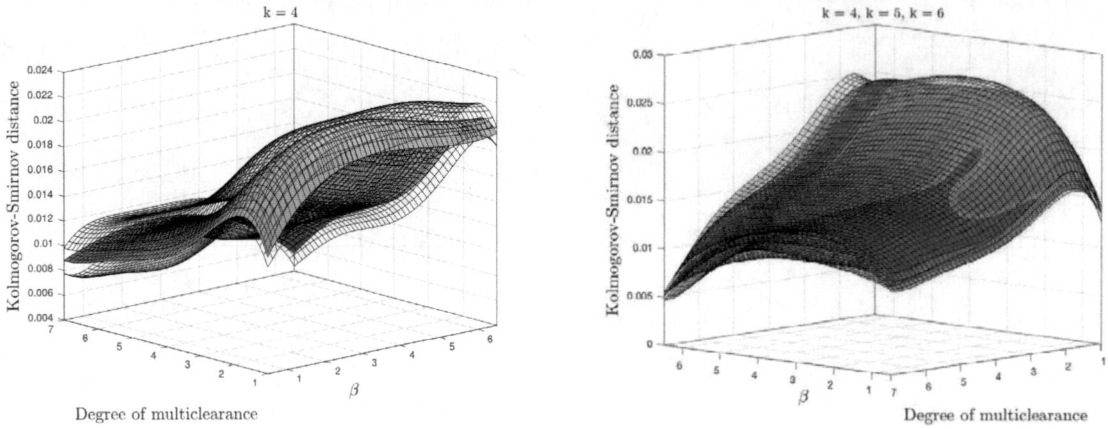

(a) (b)

Fig. 7. Course of Kolmogorov distance plane with error bars (transparent planes) for interaction of 5 vehicles (a) and course of Kolmogorov distance plane with error bars (red planes) for interaction of 5,6,7 vehicles respectively from down to up (b).

mentation methods is not appropriate or that a correlation between velocities and densities is present. Despite the contradictory results, we observe that in both cases the interactions are at least middle-ranged.

Method based on perturbation function needs to be tested and validated on empirical data. From simulations it is obvious so far, that for higher interaction range, Kolmogorov distance planes are too close to each other and therefore it will be difficult to determine precise interaction range (if it comes to high order interactions). On the other hand, it should not be difficult to determine at least character of interactions (short ranged, middle ranged), since Kolmogorov distance plane for short-range interaction is far from the rest of the planes derived for middle-ranged interactions (for $\mu > 1$).

Acknowledgements

The author would like to thank The Road and Motorway Directorate of the Czech Republic for providing traffic data analyzed in this paper.

Research presented in this work was supported by the grant 15-15049S provided by Czech Science Foundation (GA ČR). Partial support was provided by Czech Technical University in Prague within the project SGS15/214/OHK4/3T/14.

References

[1] D. HELBING. Traffic and related self-driven many-particle systems. *Rev. Mod. Phys.*, 73, 1067, 2001.

[2] E. B. BOGOMOLNY, U. GERLAND, C. SCHMIT. Short-range plasma model for intermediate spectral statistics. *Eur. Phys. J. B*, 19, 121, 2001.

[3] L. NEUBERT, L. SANTEN, A. SCHADSCHNEIDER, M. SCHRECKENBERG. Single-vehicle data of highway traffic: A statistical analysis. *Phys. Rev. E*, 60, 6480, 1999.

[4] M. KRBÁLEK. Equilibrium distributions in a thermodynamical traffic gas. *J. Phys. A: Math. Theor.*, 40, 5813, 2007.

[5] M. KRBÁLEK. Theoretical predictions for vehicular headways and their clusters. *J. Phys. A: Math. Theor.*, 46, 445101, 2013.

[6] M. KRBÁLEK, J. APELTAUER, T. APELTAUER, Z.SZABOVÁ. Three methods for estimating a range of vehicular interactions. *Physica A*, 491, 112-126, 2018.

[7] C. APPERT-ROLLAND. Experimental study of short-range interactions in vehicular traffic. *Phys. Rev. E*, 80, 036102, 2009.

[8] M. TREIBER, A.KESTING. *Traffic Flow Dynamics*. Berlin: Springer, 2013.

[9] G. J. SZÉKELY, M. L. RIZZO. Brownian distance covariance. *Annals of Applied Statistics*, 3/4, 1236, 2009.

[10] Z. SZABOVÁ. Correlation analysis in ensembles of empirical vehicular data. *Bachelor thesis*, 2016.

[11] Z. SZABOVÁ. Statistical detection of an interaction range in particle systems. *Proceedings of SPMS 2017*, Dobrichovice 2017.

[12] Z. SZABOVÁ. Statistical detection of an interaction range in particle systems. *Research task*, 2017.

Current address

Szabová Zuzana, Bc.
Department of mathematics
Faculty of Nuclear Sciences and Physical Engineering
Czech Technical University in Prague
Trojanova 13, Prague, Czech Republic
E-mail: szabov.zuzana@gmail.com

Krbálek Milan, doc. Mgr., PhD.
Department of mathematics
Faculty of Nuclear Sciences and Physical Engineering
Czech Technical University in Prague
Trojanova 13, Prague, Czech Republic
E-mail: milan.krbalek@fjfi.cvut.cz

Slovak University of Technology in Bratislava
Faculty of Mechanical Engineering
17th Conference on Applied Mathematics **APLIMAT 2018**

Proceedings

LINES ON COMPLETE INTERSECTION OF TWO QUADRICS IN $\mathbb{P}^4$

TEREŇOVÁ Zuzana (SK)

Abstract. In this paper we investigate a complete intersection of two quadrics in the projective 4-space over an algebraically closed field. This quartic surface contains one singular line. We determine all the lines on these surfaces and the arrangement of these lines.

Keywords: projective space, quadric, quartic, singular line, Plücker coordinates

Mathematics subject classification: Primary 14N10, 14Q10

1 Introduction

In the paper [2] there is the classification of all irreducible non-normal complete intersections of two quadrics that are not cones. Let $\mathbb{P}^r_K$ be the projective r-space over an algebraically closed field K of arbitrary characteristic. The authors of [2] proved the next theorem.

Theorem 1.1 *Let $X \subset \mathbb{P}^r_K$ be a complete intersection defined by two quadrics Q_1 and Q_2 in $S = K[x_0, \ldots, x_r]$. Suppose that X is irreducible, non-normal and not a cone.*
If char$K \neq 2$, then X is transformed into one of the following projectively non-equivalent cases:

(1) $r = 3$ *and* $Q_1 : x_0^2 + x_1x_2 = 0,\ Q_2 : x_3^2 + x_0x_2 = 0$

(2) $r = 3$ *and* $Q_1 : x_0^2 + x_1x_2 = 0,\ Q_2 : x_3^2 + x_0x_2 + x_0x_3 = 0$

(3) $r = 4$ *and* $Q_1 : x_0^2 + x_1x_2 = 0,\ Q_2 : x_3^2 + x_2x_4 = 0$

(4) $r = 4$ *and* $Q_1 : x_0^2 + x_1x_2 = 0,\ Q_2 : x_1x_3 + x_2x_4 = 0$

(5) $r = 4$ *and* $Q_1 : x_0^2 + x_1x_2 = 0,\ Q_2 : x_1^2 + x_0x_4 + x_2x_3 = 0$

(6) $r = 5$ *and* $Q_1 : x_0^2 + x_1x_2 = 0,\ Q_2 : x_0x_4 + x_1x_5 + x_2x_3 = 0.$

If char$K = 2$, then X is transformed into one of the cases $(1), \ldots, (6)$ and in addition:

(7) $r = 3$ *and* $Q_1 : x_0^2 + x_1x_2 = 0,\ Q_2 : x_1^2 + x_0x_1 + x_2x_3 = 0$

(8) $r = 4$ *and* $Q_1 : x_0^2 + x_1x_2 = 0,\ Q_2 : x_4^2 + x_0x_4 + x_2x_3 = 0$

(9) $r = 4$ *and* $Q_1 : x_0^2 + x_1x_2 + x_0x_2 = 0,\ Q_2 : x_1x_4 + x_2x_3 = 0.$

We were motivated by this results. In this paper we specialize in the cases (3), (4), (5) from Theorem 1.1. We work in the projective 4-space $\mathbb{P}^4_C$ over the complex numbers C. The intersection of two quadrics Q_1, Q_2 (hypersurfaces of 2nd degree) is a quartic surface $X = Q_1 \cap Q_2$ (surface of 4th degree).

Let $r = 4$, $X \subset \mathbb{P}^4_K$ be as in Theorem 1.1 and $\text{Sing}(X)$ is a set of singular points of $X = Q_1 \cap Q_2$. Then $\text{Sing}(X)$ is a line and the following is true (see [2]):

(3) If $Q_1 : x_0^2 + x_1x_2 = 0,\ Q_2 : x_3^2 + x_2x_4 = 0$ then $\text{Sing}(X) : x_0 = x_2 = x_3 = 0.$

(4) If $Q_1 : x_0^2 + x_1x_2 = 0,\ Q_2 : x_1x_3 + x_2x_4 = 0$ then $\text{Sing}(X) : x_0 = x_1 = x_2 = 0.$

(5) If $Q_1 : x_0^2 + x_1x_2 = 0,\ Q_2 : x_1^2 + x_0x_4 + x_2x_3 = 0$ then $\text{Sing}(X) : x_0 = x_1 = x_2 = 0.$

The 3rd author (P. Schenzel) of [2] asked if there are more lines on X than $\text{Sing}(X)$. By using two different methods we determined the lines on some cubic and quartic surfaces in $\mathbb{P}^3$ (see [3] and [5]). Here we want to apply one of these methods to add some properties of surfaces X given in the paper [2]. In this article we study the number of lines on X and the arrangement of these lines. The main result is in the following theorem.

Theorem 1.2 (Main Theorem) *Let* $X \subset \mathbb{P}^4_K$ *be as in Theorem 1.1. Then for* $X = Q_1 \cap Q_2$ *it holds*

a) *If* $Q_1 : x_0^2 + x_1x_2 = 0,\ Q_2 : x_3^2 + x_2x_4 = 0$ *then* X *contains the line* $\text{Sing}(X)$ *given by points* $(0, 1, 0, 0, 0), (0, 0, 0, 0, 1)$ *and no more lines.*

b) *If* $Q_1 : x_0^2 + x_1x_2 = 0,\ Q_2 : x_1x_3 + x_2x_4 = 0$ *then* X *contains infinitely many lines:*

- *the singular line* $\text{Sing}(X)$ *given by points* $(0, 0, 0, 1, 0), (0, 0, 0, 0, 1)$,
- *two lines given by points* $(0, 1, 0, 0, 0), (0, 0, 0, 0, 1)$ *and* $(0, 0, 1, 0, 0), (0, 0, 0, 1, 0)$ *which intersect the singular line,*
- *the family of lines* $m(u, v)$ *given by points* $(0, 0, 0, u^2, v^2), (-uv, -v^2, u^2, 0, 0),\ u, v \neq 0$. *Every line from this family intersects the singular line and every two lines* $m(u, v)$ *and* $m(-u, v)$ *given by* (u, v) *and* $(-u, v)$ *have a common point on the singular line. The lines of this family have no other common point.*

c) *If* $Q_1 : x_0^2 + x_1x_2 = 0,\ Q_2 : x_1^2 + x_0x_4 + x_2x_3 = 0$ *then* X *contains infinitely many lines:*

- *the singular line* $\text{Sing}(X)$ *given by points* $(0, 0, 0, 1, 0), (0, 0, 0, 0, 1)$,
- *the family of lines* $n(u, v)$ *given by points* $(0, 0, 0, u, v), (-uv^2, -u^2v, v^3, 0, u^3),\ v \neq 0$. *Every line from this family intersects the singular line and any two lines of this family are skew - they have no common point.*

We use the Plücker coordinates of the lines to prove this theorem (see also [1] and [4]).

2 Plücker coordinates of the lines

Definition 2.1 *Let the line $l \subset \mathbb{P}^4$ is given by points $(x), (y)$, with $(x) = (x_0, x_1, x_2, x_3, x_4)$, $(y) = (y_0, y_1, y_2, y_3, y_4)$. Then the Plücker coordinates of the line l are $(p_{01}, p_{02}, p_{03}, p_{04}, p_{12}, p_{13}, p_{14}, p_{23}, p_{24}, p_{34})$, where*

$$p_{ij} = x_i y_j - y_i x_j \ (0 \leq i < j \leq 4).$$

By means of the Plücker coordinates we map the lines of $\mathbb{P}^4$ to the points of $\mathbb{P}^9$. Every line $l \subset \mathbb{P}^4$ determines a point $(p_{ij}) \in \mathbb{P}^9$. Note that not every point $(p_{ij}) \in \mathbb{P}^9$ determines a line in $\mathbb{P}^4$.

Proposition 2.2 *The point $(p_{ij}) = (p_{01}, p_{02}, p_{03}, p_{04}, p_{12}, p_{13}, p_{14}, p_{23}, p_{24}, p_{34})$ of $\mathbb{P}^9$ corresponds to a line of $\mathbb{P}^4$ if it holds*

$$\begin{aligned}
p_{01}p_{23} - p_{02}p_{13} + p_{03}p_{12} &= 0 \\
p_{01}p_{24} - p_{02}p_{14} + p_{04}p_{12} &= 0 \\
p_{01}p_{34} - p_{03}p_{14} + p_{04}p_{13} &= 0 \\
p_{02}p_{34} - p_{03}p_{24} + p_{04}p_{23} &= 0 \\
p_{12}p_{34} - p_{13}p_{24} + p_{14}p_{23} &= 0.
\end{aligned} \tag{1}$$

Proof. Let $l = (x)(y)$, $(x) = (x_0, x_1, x_2, x_3, x_4)$, $(y) = (y_0, y_1, y_2, y_3, y_4)$ be a line of $\mathbb{P}^4$. For the matrix

$$\begin{pmatrix} x_0 & x_1 & x_2 & x_3 & x_4 \\ y_0 & y_1 & y_2 & y_3 & y_4 \\ x_0 & x_1 & x_2 & x_3 & x_4 \\ y_0 & y_1 & y_2 & y_3 & y_4 \end{pmatrix}$$

all the determinants 4×4 are equal to zero. These 5 determinants can be simplified to the equations (1). ∎

Remark 2.3 *The points $(p_{ij}) \in \mathbb{P}^9$, which coordinates satisfy conditions* (1), *are lying on the intersection of* 5 *hypesurfaces of* 2nd *degree in* $\mathbb{P}^9$.

We can determine the arrangement of two lines from the Plücker coordinates of these lines.

Proposition 2.4 *Let (p_{ij}), (q_{ij}) be the Plücker coordinates of two lines p, q of $\mathbb{P}^4$. The lines p, q are intersecting lines if and only if their Plücker coordinates satisfy the following conditions:*

$$\begin{aligned}
p_{01}q_{23} - p_{02}q_{13} + p_{03}q_{12} + p_{12}q_{03} - p_{13}q_{02} + p_{23}q_{01} &= 0 \\
p_{01}q_{24} - p_{02}q_{14} + p_{04}q_{12} + p_{12}q_{04} - p_{14}q_{02} + p_{24}q_{01} &= 0 \\
p_{01}q_{34} - p_{03}q_{14} + p_{04}q_{13} + p_{13}q_{04} - p_{14}q_{03} + p_{34}q_{01} &= 0 \\
p_{02}q_{34} - p_{03}q_{24} + p_{04}q_{23} + p_{23}q_{04} - p_{24}q_{03} + p_{34}q_{02} &= 0 \\
p_{12}q_{34} - p_{13}q_{24} + p_{14}q_{23} + p_{23}q_{14} - p_{24}q_{13} + p_{34}q_{12} &= 0.
\end{aligned} \tag{2}$$

Proof. Let $p = AB$, $q = CD$ be two lines with $A = (a_0, a_1, a_2, a_3, a_4)$, $B = (b_0, b_1, b_2, b_3, b_4)$, $C = (c_0, c_1, c_2, c_3, c_4)$, $D = (d_0, d_1, d_2, d_3, d_4)$ and M is 4×5 matrix

$$M = \begin{pmatrix} a_0 & a_1 & a_2 & a_3 & a_4 \\ b_0 & b_1 & b_2 & b_3 & b_4 \\ c_0 & c_1 & c_2 & c_3 & c_4 \\ d_0 & d_1 & d_2 & d_3 & d_4 \end{pmatrix}.$$

The lines $p,\ q$ are

- identical lines if $rank(M) = 2$
- intersecting lines if $rank(M) = 3$
- skew lines if $rank(M) = 4$.

The lines $p,\ q$ are intersecting lines if and only if $rank(M) = 3$, i. e. determinant of every 4×4 matrix is equal to zero. There are 5 determinants of 4×4 matrix. If we substitute $p_{ij} = a_i b_j - b_i a_j$ and $q_{ij} = c_i d_j - d_i c_j$ $(0 \leq i < j \leq 4)$ into these determinants we get the equations (2). ■

Now we are going to describe an algorithm for computing all the lines on a hypersurface. We use the Plücker coordinates of the lines.

Proposition 2.5 *Let $l \subset \mathbb{P}^4$ be a line with the Plücker coordinates (p_{ij}) and $Q(F) \subset \mathbb{P}^4$ be a hypersurface with the defining polynom F. The condition for a line l to lie on $Q(F)$ can be expressed by algebraic equations with p_{ij} and coefficients of the polynom F.*

Proof. We will follow [4], § 6, section 4. Let $(x),\ (y)$ be two different points of the line l. For an arbitrary point (u) of the line l we have

$$(u) = (x)f(y) - (y)f(x),$$

where f is a linear form with the coefficients $\alpha_0,\ \alpha_1,\ \alpha_2,\ \alpha_3,\ \alpha_4$, i. e. $f(x) = \sum_{i=0}^{4} \alpha_i x_i$ and $f(y) = \sum_{i=0}^{4} \alpha_i y_i$. For the coordinates of $(u) \in l$ we get

$$u_i = \sum_{j=0}^{4} \alpha_j p_{ij}, \quad \text{with } p_{ij} = x_i y_j - y_i x_j,$$

and therefore
$$\begin{pmatrix} u_0 \\ u_1 \\ u_2 \\ u_3 \\ u_4 \end{pmatrix} = \begin{pmatrix} 0 & p_{01} & p_{02} & p_{03} & p_{04} \\ -p_{01} & 0 & p_{12} & p_{13} & p_{14} \\ -p_{02} & -p_{12} & 0 & p_{23} & p_{24} \\ -p_{03} & -p_{13} & -p_{23} & 0 & p_{34} \\ -p_{04} & -p_{14} & -p_{24} & -p_{34} & 0 \end{pmatrix} \begin{pmatrix} \alpha_0 \\ \alpha_1 \\ \alpha_2 \\ \alpha_3 \\ \alpha_4 \end{pmatrix}.$$

By substituting this into the defining equation $F = 0$ of the hypersurface $Q(F)$ we get the equation $F(u_0, u_1, u_2, u_3, u_4) = F(\alpha_i) = 0$. The line l is on the hypersurface $Q(F)$ if and only if all the coefficients of monoms in α_i are zero. Thus we get a system of equations in p_{ij}. By solving this system of equations we get the Plücker coordinates of all lines on the hypersurface $Q(F)$. ■

Remark 2.6 *For a quadric hypersurface in $\mathbb{P}^4$ we get 15 equations of 2nd degree. To find all the lines on an intersection of two quadrics we have to solve a system of 30 equations of 2nd degree. Though it seems too difficult, one can be able in some situations to find the solutions "manually" since many of the coefficients equals to zero and the system becomes to be much simplier one (see e.g. Example 2 in [5] or the proof of the part a) of Theorem 1.2 in this paper). In case of more complicated systems we can use computer algebra software to solve this system (e.g. Singular or Reduce and Cali).*

3 Proof of the Main Theorem 1.2

Now we are going to prove Theorem 1.2, the main result of the paper. We apply the algorithm from Proposition 2.5 to find all the lines on the quartic surfaces $X = Q_1 \cap Q_2$. Then we determine the arrangement of these lines.

Proof. Let $X \subset \mathbb{P}^4_C$ be a complete intersection defined by two quadrics Q_1 and Q_2 in the projective 4-space $\mathbb{P}^4_C$ over the complex numbers C.

a) $Q_1 : x_0^2 + x_1x_2 = 0,\ Q_2 : x_3^2 + x_2x_4 = 0$
We substitute $u_i,\ i = 0, \ldots, 4$ from the proof of Proposition 2.5 into the defining equations of Q_1 and Q_2. For the line lying on the hypersurface it holds that all the coefficients of monoms in α_i are zero:

$$Q_1: \begin{array}{l} p_{01} = p_{02} = p_{12} = 0 \\ p_{03}^2 + p_{13}p_{23} = 0 \\ 2p_{03}p_{04} + p_{13}p_{24} + p_{14}p_{23} = 0 \\ p_{04}^2 + p_{14}p_{24} = 0 \end{array} \qquad Q_2: \begin{array}{l} p_{23} = p_{24} = p_{34} = 0 \\ p_{13}^2 + p_{12}p_{14} = 0 \\ 2p_{03}p_{13} + p_{02}p_{14} + p_{04}p_{12} = 0 \\ p_{03}^2 + p_{02}p_{04} = 0 \end{array}$$

From this conditions we can calculate the lines on $X = Q_1 \cap Q_2$. One can see that all p_{ij} vanish (equals to zero) except of p_{14}, so there is only one solution of this system of equations:

$$(0, 0, 0, 0, 0, 0, 1, 0, 0, 0).$$

These are the Plücker coordinates of the singular line $\mathrm{Sing}(X)$ on X. It is given as intersection of three hyperplanes $x_0 = x_2 = x_3 = 0$ or by two points $(0, 1, 0, 0, 0)$, $(0, 0, 0, 0, 1)$ in $\mathbb{P}^4$.

b) $Q_1 : x_0^2 + x_1x_2 = 0,\ Q_2 : x_1x_3 + x_2x_4 = 0$
The system of equations from Proposition 2.5 is

$$\begin{array}{l} p_{01} = p_{02} = p_{12} = 0 \\ p_{03}^2 + p_{13}p_{23} = 0 \\ 2p_{03}p_{04} + p_{13}p_{24} + p_{14}p_{23} = 0 \\ p_{04}^2 + p_{14}p_{24} = 0 \\ p_{03}p_{13} + p_{04}p_{23} = 0 \\ p_{03}p_{14} + p_{04}p_{24} = 0 \\ p_{13}^2 + p_{14}p_{23} = 0 \end{array} \qquad \begin{array}{l} p_{13}p_{14} + p_{14}p_{24} = 0 \\ p_{13}p_{23} + p_{23}p_{24} = 0 \\ p_{14}p_{23} + p_{24}^2 = 0 \\ p_{23}p_{34} = 0 \\ p_{13}p_{34} - p_{24}p_{34} = 0 \\ p_{14}p_{34} = 0 \end{array}$$

Solutions of this system give us the Plücker coordinates of all lines on X. Here are the results:

$$r : (0, 0, 0, 0, 0, 0, 0, 0, 0, 1),$$
$$m : (0, 0, u^3v, uv^3, 0, u^2v^2, v^4, -u^4, -u^2v^2, 0), \text{ with } (u, v) \neq (0, 0).$$

- The line r with Plücker coordinates $(0, 0, 0, 0, 0, 0, 0, 0, 0, 1)$ is a singular line $\mathrm{Sing}(X)$. It is given by equations $x_0 = x_1 = x_2 = 0$ or by two points $(0, 0, 0, 1, 0)$, $(0, 0, 0, 0, 1)$.
- The Plücker coordinates $(0, 0, u^3v, uv^3, 0, u^2v^2, v^4, -u^4, -u^2v^2, 0)$, with $(u, v) \neq (0, 0)$, determine lines $m(u, v)$ given by two points $(0, 0, 0, u^2, v^2)$, $(-uv, -v^2, u^2, 0, 0)$.

So there are infinitely many lines on $X = Q_1 \cap Q_2$.

Now we want to find the arrangement of these lines. It is clear that all of the lines $m(u, v)$ intersect the line r in the point $(0, 0, 0, u^2, v^2)$. The equations (2) are satisfied for the Plücker coordinates of the lines r and m.

Two lines m given by (u, v) and (u_1, v_1) intersect if and only if the equations (2) are satisfied. So we get the following equations

$$
\begin{array}{l}
-u^3vv_1^4 + uv^3u_1^2v_1^2 + u^2v^2u_1v_1^3 - v^4u_1^3v_1 = 0 \\
u^3vu_1^2v_1^2 - uv^3u_1^4 - u^4u_1v_1^3 + u^2v^2u_1^3v_1 = 0 \\
2u^2v^2u_1^2v_1^2 - v^4u_1^4 - u^4v_1^4 = 0
\end{array}
$$

and from this system we have the arrangement of these lines.

- For $u = 0$ we have the line with Plücker coordinates

$$(0, 0, 0, 0, 0, 0, 1, 0, 0, 0)$$

that does not intersect any line $m(u, v)$. This line is given by two points $(0, 1, 0, 0, 0)$, $(0, 0, 0, 0, 1)$.

- For $v = 0$ we have the line with Plücker coordinates

$$(0, 0, 0, 0, 0, 0, 0, 1, 0, 0)$$

that does not intersect any line $m(u, v)$. This line is given by two points $(0, 0, 1, 0, 0)$, $(0, 0, 0, 1, 0)$.

- For $u \neq 0,\ v \neq 0$ the line $m(u, v)$ intersects only the line $m(-u, v)$ in the point $(0, 0, 0, u^2, v^2)$ on the line $r = \mathrm{Sing}(X)$. These intersecting lines can be given by two points

$$
\begin{array}{l}
m(u, v) : \ (0, 0, 0, u^2, v^2),\ (-uv, -v^2, u^2, 0, 0) \\
m(-u, v) : \ (0, 0, 0, u^2, v^2),\ (uv, -v^2, u^2, 0, 0).
\end{array}
$$

We got the following result: For the points $A \in \mathrm{Sing}(X)$ it holds that there are two lines different from $\mathrm{Sing}(X)$ through every point A and lying on X except the points $(0, 0, 0, 1, 0)$ and $(0, 0, 0, 0, 1)$. For each of these two points there is exactly one line different from $\mathrm{Sing}(X)$ through them and lying on X (see Fig. 1).

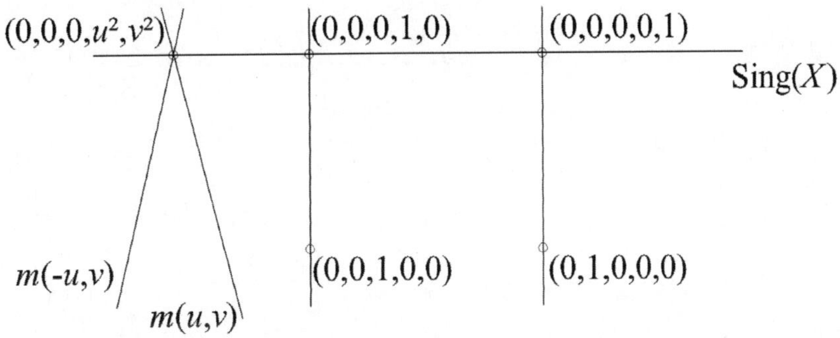

Fig. 1. The arrangement of lines on $X = Q_1 \cap Q_2$, $u, v \neq 0$.

Remark 3.1 *The lines which do not meet in the Figure are mutually skew.*

c) $Q_1 : x_0^2 + x_1x_2 = 0,\ Q_2 : x_1^2 + x_0x_4 + x_2x_3 = 0$

We get the Plücker coordinates of all the lines on $X = Q_1 \cap Q_2$ from the following system of equations (see Proposition 2.5)

$$
\begin{array}{ll}
p_{01} = p_{02} = p_{12} = 0 & p_{04}p_{14} + p_{13}p_{24} = 0 \\
p_{03}^2 + p_{13}p_{23} = 0 & p_{23}^2 + p_{03}p_{24} = 0 \\
2p_{03}p_{04} + p_{13}p_{24} + p_{14}p_{23} = 0 & p_{04}p_{24} + p_{23}p_{24} = 0 \\
p_{04}^2 + p_{14}p_{24} = 0 & p_{13}^2 - p_{03}p_{34} = 0 \\
p_{03}p_{04} + p_{03}p_{23} = 0 & 2p_{13}p_{14} - p_{04}p_{34} + p_{23}p_{34} = 0 \\
p_{04}^2 + p_{03}p_{24} = 0 & p_{14}^2 + p_{24}p_{34} = 0 \\
p_{03}p_{14} + p_{13}p_{23} = 0 &
\end{array}
$$

and the solution is

$$n : (0, 0, u^2v^2, uv^3, 0, u^3v, u^2v^2, -uv^3, -v^4, u^4), \text{ with } (u, v) \neq (0, 0).$$

These Plücker coordinates determine infinitely many lines given by two points

$$n(u, v) : \ (0, 0, 0, u, v), \ (-uv^2, -u^2v, v^3, 0, u^3).$$

For $v = 0$ we get the singular line which is given by equations $x_0 = x_1 = x_2 = 0$ or by two points $(0, 0, 0, 1, 0)$, $(0, 0, 0, 0, 1)$.

We get the arrangement of these lines from the equations (2) for two lines n given by (u, v) and (u_1, v_1):

$$
\begin{array}{l}
-2u^2v^2u_1^2v_1^2 + uv^3u_1^3v_1 + u^3vu_1v_1^3 = 0 \\
u^2v^2v_1^4 - 2uv^3u_1v_1^3 + v^4u_1^2v_1^2 = 0 \\
u^3vv_1^4 - u^2v^2u_1v_1^3 - uv^3u_1^2v_1^2 + v^4u_1^3v_1 = 0.
\end{array}
$$

From these conditions we have that the singular line $(0, 0, 0, 0, 0, 0, 0, 0, 0, 1)$ intersects every line $(0, 0, u^2v^2, uv^3, 0, u^3v, u^2v^2, -uv^3, -v^4, u^4)$ with $v \neq 0$. The point of intersection is $(0, 0, 0, u, v)$.

We got the following result: For the points $A \in \mathrm{Sing}(X)$ it holds that there is exactly one line different from $\mathrm{Sing}(X)$ through every point A and lying on X except the point $(0, 0, 0, 1, 0)$. For the point $(0, 0, 0, 1, 0)$ there is no line on X different from $\mathrm{Sing}(X)$ through this point (see Fig. 2).

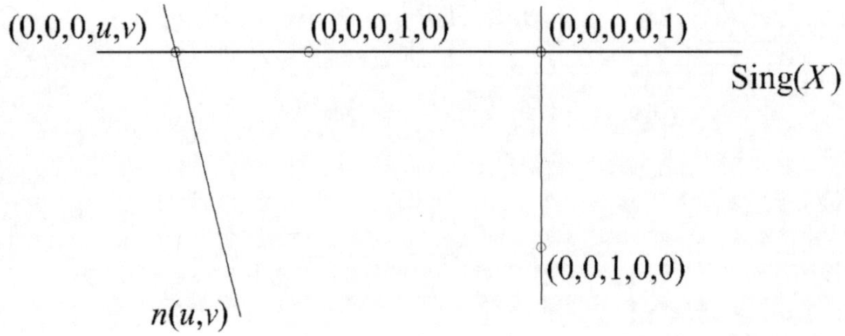

Fig. 2. The arrangement of lines on $X = Q_1 \cap Q_2$, $u, v \neq 0$.

■

4 Conclusion

We determined some properties of complete intersections of two quadrics in $\mathbb{P}^4$ over an algebraically closed field which can be taken as an addition to the results of the paper [2]. We applied the method which uses the Plücker coordinates of the lines. This method is universal but sometimes complicated calculation requires to use some algebra software.

References

[1] HODGE, W. V. D., PEDOE, D.: *Methods of Algebraic Geometry*. Cambridge University Press, 1952.

[2] LEE, W., PARK, E., SCHENZEL, P.: *On the classification of non-normal complete intersections of two quadrics*. Journal of Pure and Applied Algebra, Volume 216, Issue 5, 2012, pp. 1222-1234.

[3] SOLČAN, Š., TEREŇOVÁ, Z.: *On some quartic surfaces with triple lines*. In Beiträge zur Algebra und Geometrie / Contributions to Algebra and Geometry. Volume 52, Issue 2, 2011, pp. 285-296.

[4] ŠAFAREVIČ, I. R.: *O*snovy algebraičeskoj geometrii. Nauka, Moskva, 1972.

[5] TEREŇOVÁ, Z., SOLČAN, Š.: *Computing the lines on quartics*. In Aplimat 2007, Part I: 6th International Conference, Bratislava, 2007, pp. 251-260.

Current address

Tereňová Zuzana, RNDr., PhD.
Department of Mathematics and Descriptive Geometry
Faculty of Civil Engineering
Slovak University of Technology in Bratislava
Radlinského 11, 810 05 Bratislava, Slovak Republic
E-mail: zuzana.terenova@stuba.sk

Slovak University of Technology in Bratislava
Faculty of Mechanical Engineering
17th Conference on Applied Mathematics **APLIMAT 2018**

Proceedings

FORCED DUFFING EQUATION WITH A NON-STRICTLY MONOTONIC POTENTIAL

TOMICZEK Petr (CZ)

Abstract. This article is devoted to study the existence of a solution to the periodic nonlinear second order ordinary differential equation with damping

$$u''(x) + c\,u'(x) + g(x,u) = f(x)\,, \quad x \in [0,T]\,,$$
$$u(0) = u(T)\,,\ u'(0) = u'(T)\,,$$

where $c \in \mathbb{R}$, g is a Carathéodory function, $f \in L^1([0,T])$, a quotient $\frac{g(x,s)}{s}$ lies between 0 and $\frac{c^2}{4} + (\frac{\pi}{T})^2$ and a potential is a non-strictly monotonic function. The technique we use are variational method and critical point theorem.

Keywords: second order ODE, periodic problem, variational method, critical point

Mathematics subject classification: Primary 34G20; Secondary 35A15, 34K10

1 Introduction

The aim of this article is to provide new existence results for the nonlinear periodic boundary problem

$$u''(x) + c\,u'(x) + g(x,u) = f(x)\,, \quad x \in [0,T]\,, \qquad (1.1)$$
$$u(0) = u(T)\,,\ u'(0) = u'(T)\,,$$

where $c \in \mathbb{R}$, the nonlinearity $g\colon [0,T] \times \mathbb{R} \to \mathbb{R}$ is Carathéodory's function, $f \in L^1([0,T])$.

An equation of this form describes for example a spring–mass system with a damper in parallel; or the charge on the capacitor in a circuit containing resistance, capacitance, and inductance.

In papers [7], [14] authors used topological degree arguments and supposed that $\gamma(x) \le \liminf_{|s|\to\infty} \left|\frac{g(x,s)}{s}\right| \le \limsup_{|s|\to\infty} \left|\frac{g(x,s)}{s}\right| \le \Gamma(x)\,,\ \Gamma(x) \le \left(\frac{2\pi}{T}\right)^2$ with the strict inequality on a subset of $[0,T]$ of positive measure and $\gamma(x)$ satisfies $\int_0^T \gamma(x)\,dx \ge 0\,,\ \int_0^T \gamma^+(x)\,dx > 0$ where $\gamma^+(x) = \max_{x\in[0,T]} \{\gamma(x), 0\}\,.$

But we obtain the existence result to the equation (1.1) with the nonlinearity $g(x,u) = \arctan u$ (it follows that $\gamma(x) = 0$) if the right hand side f satisfies $-\frac{\pi T}{2} < \int_0^T f(x)\,dx < \frac{\pi T}{2}$.

Others have studied problem (1.1) with jumping nonlinearities [5], [2] using also topological method.

In this article, we choose another strategy of proof which rely essentially on a variational method (see also [3], [6], [11]). We will assume that the nonlinearity g satisfies $0 \le \liminf_{|s|\to\infty} \left|\frac{g(x,s)}{s}\right| < \limsup_{|s|\to\infty} \left|\frac{g(x,s)}{s}\right| < \frac{c^2}{4} + \left(\frac{\pi}{T}\right)^2$ and a potential of g is non-strictly monotonic. Precisely, let $u = u(a,x)$, $u : \mathbb{R} \times [0,T] \to \mathbb{R}, u(a,\cdot) \in C[0,T]$ for each $a \in \mathbb{R}$, $u(\cdot,x) \in C(\mathbb{R})$ for each $x \in [0,T]$ such that $\lim_{a\to\pm\infty} u(a,x) = \pm\infty$ uniformly on $[0,T]$. We denote

$$F(s) = \int_0^T \int_0^s [\,g(x,u(a,x)) - f(x)\,]\,da\,dx\,.$$

We suppose that for such $u = u(a,x)$ there exist constants $s_1 < s_2 < s_3 < s_4$ such that

$$F(s_1) \ge F(s_2) \quad \text{and} \quad F(s_3) \le F(s_4)\,.$$

We note this assumption is fulfilled if right hand side f satisfies orthogonal condition $\int_0^T f(x)\,dx = 0$ and g satisfies sign condition $g(x,s)s \ge 0$ (Fredholm alternative for a nonlinear equation). We generalize Landesman-Lazer type condition (see [2], [12]) for a resonance problems. We get solution to (1.1) also for strong resonant problem for g satisfying $\lim_{|s|\to\infty} sg(x,s) = 0$ and $\lim_{|s|\to\infty} F(x,s) = 0$, see [8], [9].

Similarly to [1] we firstly investigate the Dirichlet problem. Then, we apply this result for finding periodic solutions. In [1] author investigate problem $u''(x) + r(x)u'(x) + g(x,u) = f(x)$ under a Lipschitz condition on g. If we rewrite this condition with $r(x) = c$ we obtain $\frac{g(x,s)-g(x,t)}{s-t} \le c_T$ where $c_T < \left(\frac{\pi}{T}\right)^2$. We will assume that $\frac{g(x,s)-g(x,t)}{s-t} \le \frac{c^2}{4} + c_T$. Hence the nonlinearity g can also cross the eigenvalue $\left(\frac{2\pi}{T}\right)^2$ if $c^2 > 12\left(\frac{\pi}{T}\right)^2$.

We note that we can use our approach also to problem with impulses, see [3] and the existence and stability of periodic solutions to problem (1.1) with $f = 0$ is discussed in [13].

2 Preliminaries

Notation: We shall use the classical space $C^k(0,T)$ of functions whose k-th derivative is continuous and the space $L^p(0,T)$ of measurable real-valued functions whose p-th power of the absolute value is Lebesgue integrable. We denote H the Sobolev space of absolutely continuous functions $u : [0,T] \to \mathbb{R}$ such that $u' \in L^2(0,T)$, $u(0) = u(T) = 0$ endowed with the norm $\|u\| = \left(\int_0^T (u')^2\,dx\right)^{\frac{1}{2}}$.

By a solution to (1.1) we mean a function $u \in C^1(0,T)$ such that u' is absolutely continuous, u satisfies the boundary conditions and the equation (1.1) is satisfied a.e. on $(0,T)$.

Firstly we prove the existence of a solution to the Dirichlet problem

$$\begin{gathered} u''(x) + c\,u'(x) + g(x,u) = f(x)\,, \quad x \in [0,T]\,, \\ u(0) = u(T) = a\,, \end{gathered} \tag{2.1}$$

where $a \in \mathbb{R}$. Then, we apply this result for finding periodic solutions. To obtain an equation with potential we multiply (2.1) by the function $e^{\frac{c}{2}x}$. Then we put $w(x) = e^{\frac{c}{2}x}(u(x) - a)$ and get for w an equivalent problem

$$w''(x) - \frac{c^2}{4}\,w(x) + e^{\frac{c}{2}x}\,g(x, \frac{w}{e^{\frac{c}{2}x}} + a) = e^{\frac{c}{2}x}\,f(x)\,, \qquad (2.2)$$
$$w(0) = w(T) = 0\,.$$

We investigate (2.2) by using variational methods. More precisely, we find a critical point of the functional $J_a : H \to \mathbb{R}$, which is defined by

$$J_a(w) = \frac{1}{2}\int_0^T [(w')^2 + \frac{c^2}{4}w^2]\,dx - \int_0^T [e^{cx}G(x, \frac{w}{e^{\frac{c}{2}x}} + a) - e^{\frac{c}{2}x}fw]\,dx\,,$$

where

$$G(x,s) = \int_0^s g(x,t)\,dt\,.$$

We say that w_a is a critical point of J_a, if

$$\langle J_a'(w_a), z\rangle = 0 \quad \text{for all } \ z \in H.$$

We note

$$\begin{aligned} J_a(w+z) - J_a(w) = \ & \frac{1}{2}\int_0^T \left[2w'z' + (z')^2 + \frac{c^2}{4}(2wz + z^2)\right] dx \\ & - \int_0^T \left[e^{cx}\int_{\frac{w}{e^{\frac{c}{2}x}}+a}^{\frac{w+z}{e^{\frac{c}{2}x}}+a} g(x,t)\,dt - e^{\frac{c}{2}x}fz\right] dx\,, \end{aligned}$$

and by mean value theorem we get

$$\int_0^T [e^{cx}\int_{\frac{w}{e^{\frac{c}{2}x}}+a}^{\frac{w+z}{e^{\frac{c}{2}x}}+a} g(x,t)\,dt]\,dx = \int_0^T [e^{cx}g(x,\xi(x))\,\frac{z}{e^{\frac{c}{2}x}}]\,dx = \int_0^T [e^{\frac{c}{2}x}g(x,\xi(x))\,z]\,dx$$

where $\xi(x) \in (\frac{w}{e^{\frac{c}{2}x}} + a, \frac{w+z}{e^{\frac{c}{2}x}} + a)$.

Therefore every critical point $w \in H$ of the functional J_a satisfies

$$\int_0^T [w'z' + \frac{c^2}{4}\,wz]\,dx - \int_0^T [e^{\frac{c}{2}x}g(x, \frac{w}{e^{\frac{c}{2}x}} + a)z - e^{\frac{c}{2}x}fz]\,dx = 0 \quad \text{for all } \ z \in H, \qquad (2.3)$$

then w is also a weak solution to the Dirichlet problem (2.2) and vice versa. The usual regularity argument for ODE proves immediately (see Fučík [4]) that any weak solution to (2.2) is also a solution in the sense mentioned above.

We remark that for any function $w \in H$ holds

$$\int_0^T \left[(w')^2 - \left(\frac{\pi}{T}\right)^2 w^2\right] dx \ge 0\,. \qquad (2.4)$$

We will suppose that g satisfies the following growth restrictions.

There exist functions $a_+(x), a_-(x) \in L^1(0,\pi)$ and a constant $s_0 \in \mathbb{R}^+$ such that for a.e. $x \in (0,\pi)$

$$g(x,s) \le a_-(x) \quad \text{for} \quad s \le -s_0\,, \qquad g(x,s) \ge a_+(x) \quad \text{for} \quad s \ge s_0\,, \tag{2.5}$$

(hence $0 \le \liminf\limits_{|s|\to\infty} \frac{g(x,s)}{s}$),

there exists $c_T < \left(\frac{\pi}{T}\right)^2$ such that

$$\frac{g(x,s)-g(x,t)}{s-t} \le \frac{c^2}{4} + c_T \qquad \text{for } s,t \in \mathbb{R}\,,\ s \ne t\,,\ x \in [0,T]\,, \tag{2.6}$$

(hence $\limsup\limits_{|s|\to\infty} \frac{g(x,s)}{s} \le \frac{c^2}{4} + c_T$ uniformly on $[0,T]$) and there are $c_1 \in \mathbb{R}^+$, $q \in L^1(0,T)$ such that

$$|g(x,s)| \le c_1|s| + q(x) \qquad \text{for all } s \in \mathbb{R}\,,\ \text{for a. e. } x \in [0,T]\,. \tag{2.7}$$

Furthermore let $u = u(a,x)$, $u : \mathbb{R}\times[0,T] \to \mathbb{R}$, $u(a,\cdot) \in C[0,T]$ for each $a \in \mathbb{R}$, $u(\cdot,x) \in C(\mathbb{R})$ for each $x \in [0,T]$ such that $\lim\limits_{a\to\infty} u(a,x) = +\infty$ uniformly on $[0,T]$, $\lim\limits_{a\to-\infty} u(a,x) = -\infty$ uniformly on $[0,T]$. We denote

$$F(s) = \int_0^T\!\!\int_0^s [\,g(x,u(a,x)) - f(x)\,]\,da\,dx\,.$$

We suppose that for such $u = u(a,x)$ there exist constants $s_1 < s_2 < s_3 < s_4$ such that

$$F(s_1) \ge F(s_2) \quad \text{and} \quad F(s_3) \le F(s_4)\,. \tag{2.8}$$

In this section we introduce two lemmas which will be used in the proof of the main theorem.

Lemma 2.1 *(uniqueness) Let g satisfies (2.6) then the equation (2.2) has at most one solution.*

Proof. Let $w_1, w_2 \in H$ are two solutions to (2.2) then

$$\begin{aligned}\int_0^T \Big[(w_1-w_2)'z' + \frac{c^2}{4}(w_1-w_2)z\Big]\,dx \\ = \int_0^T \Big[e^{\frac{c}{2}x}\Big(g(x,\frac{w_1}{e^{\frac{c}{2}x}}+a) - g(x,\frac{w_2}{e^{\frac{c}{2}x}}+a)\Big)z\Big]\,dx\,,\end{aligned} \tag{2.9}$$

for all $z \in H$. We put $z = w_1 - w_2$ in (2.9) and using the assumption (2.6) we get

$$\int_0^T \Big((w_1-w_2)'\Big)^2\,dx \le \int_0^T \Big[c_T(w_1-w_2)^2\Big]\,dx\,. \tag{2.10}$$

From (2.10), (2.4) we conclude $w_1 = w_2$.

Lemma 2.2 *(continuity) Let $a_n \to a_0$ then a corresponding sequence (w_{a_n}) solutions to the equation (2.2) with $a = a_n$ contains subsequence $(w_{a_{n_k}})$ such that $w_{a_{n_k}} \to w_0$, $w_0 \in H$ and w_0 is a solution to (2.2) with $a = a_0$.*

Proof. The solution w_{a_n} to (2.2) satisfies

$$\int_0^T \left[w'_{a_n} z' + \frac{c^2}{4}\, w_{a_n} z\right] dx - \int_0^T \left[e^{\frac{c}{2}x}\, g(x, \frac{w_{a_n}}{e^{\frac{c}{2}x}} + a_n) z - e^{\frac{c}{2}x} f z\right] dx = 0\,, \tag{2.11}$$

for all $z \in H$.

We suppose that the sequence (w_{a_n}) is unbounded and we put $v_n = \frac{w_{a_n}}{\|w_{a_n}\|}$. Then there exists $v_0 \in H$ such that $v_n \rightharpoonup v_0$ in H and due to compact embedding H into $C([0,T])$ $v_n \to v_0$ in $C([0,T])$ (taking a subsequence if it is necessary). We divide (2.11) by $\|w_{a_n}\|$ and put $z = v_n$ then

$$\int_0^T \left[(v'_n)^2 + \frac{c^2}{4}\, v_n^2\right] dx - \int_0^T \left[\frac{e^{\frac{c}{2}x}\, g(x, \frac{w_{a_n}}{e^{\frac{c}{2}x}} + a_n) v_n}{\|w_{a_n}\|} - \frac{e^{\frac{c}{2}x} f v_n}{\|w_{a_n}\|}\right] dx = 0\,. \tag{2.12}$$

We use inequality $\int_0^T (v'_0)^2\, dx \le \liminf\limits_{n\to\infty} \int_0^T (v'_n)^2\, dx = 1$ (the weak sequential lower semi-continuity of the Hilbert norm) and pass to the limit in (2.12). According to (2.6), (2.7) we obtain

$$\int_0^T (v'_0)^2\, dx - \int_0^T \left[c_T (v_0)^2\right] dx \le 1 - \int_0^T \left[c_T (v_0)^2\right] dx \le 0\,,$$

a contradiction to the inequality (2.4).

Therefore the sequence (w_{a_n}) is bounded. Then there exists $w_0 \in H$ such that $w_{a_n} \rightharpoonup w_0$ in H, $w_{a_n} \to w_0$ in $L^2(0,T)$, $C([0,T])$ (taking a subsequence if it is necessary).

We put $n = m$ in (2.11) and subtract this equality from (2.11) (with n) we obtain

$$\lim_{\substack{n\to\infty\\ m\to\infty}} \left\{ \int_0^T \left[(w_{a_n} - w_{a_m})' z' + \frac{c^2}{4}\,(w_{a_n} - w_{a_m})\, z\right] dx \right. \tag{2.13}$$

$$\left. - \int_0^T \left[e^{\frac{c}{2}x} \left(g(x, \frac{w_{a_n}}{e^{\frac{c}{2}x}} + a_n) - g(x, \frac{w_{a_m}}{e^{\frac{c}{2}x}} + a_m)\right) z\right] dx \right\} = 0\,.$$

The convergency $w_{a_n} \to w_0$ in $C([0,T])$, (2.7) and $a_n \to a_0$ yield

$$\lim_{\substack{n\to\infty\\ m\to\infty}} \int_0^T \left[e^{\frac{c}{2}x}\Big(g(x, \frac{w_{a_n}}{e^{\frac{c}{2}x}} + a_n) - g(x, \frac{w_{a_m}}{e^{\frac{c}{2}x}} + a_m)\Big)\,(w_{a_n} - w_{a_m})\right] dx = 0\,. \tag{2.14}$$

We set $z = w_{a_n} - w_{a_m}$ in (2.13) then using (2.14) we get

$$\lim_{\substack{n\to\infty\\ m\to\infty}} \int_0^T \left[(w'_{a_n} - w'_{a_m})^2 + \frac{c^2}{4}\,(w_{a_n} - w_{a_m})^2\right] dx = 0\,. \tag{2.15}$$

Hence the strong convergence $w_{a_n} \to w_0$ in $L^2(0,T)$ and (2.15) imply the strong convergence $w_{a_n} \to w_0$ in H and we can pass to the limit in (2.11). We obtain

$$\int_0^T \left[w'_0 z' + \frac{c^2}{4}\, w_0 z\right] dx - \int_0^T \left[e^{\frac{c}{2}x}\, g(x, \frac{w_0}{e^{\frac{c}{2}x}} + a_0) z - e^{\frac{c}{2}x} f z\right] dx = 0\,, \tag{2.16}$$

for all $z \in H$. Hence w_0 is a critical point of J_{a_0} and a solution to (2.2) with $a = a_0$.

Remark 2.1 *We have proved that to each $a \in \mathbb{R}$ there exist function $u_a = \frac{w_a}{e^{\frac{c}{2}x}} + a$ such that the $A : \mathbb{R} \to H$, $A(a) = u_a$ is a continuous operator.*

3 Main result

Theorem 3.1 *Under the assumptions (2.5), (2.6), (2.7), (2.8), Problem (1.1) has at least one solution.*

Proof. We prove that J_a is a weakly coercive functional for each $a \in \mathbb{R}$ by a contradiction. Then there is a sequence $(w_n) \subset H$ such that $\|w_n\| \to \infty$ and a constant c_2 satisfying

$$\liminf_{n\to\infty} J_a(w_n) \le c_2 \,. \tag{3.1}$$

We put $v_n = \frac{w_n}{\|w_n\|}$ then there exists $v_0 \in H$ such that $v_n \rightharpoonup v_0$ in H, $v_n \to v_0$ in $C([0,T])$. We divide (3.1) by $\|w_n\|^2$ then

$$\begin{aligned} \frac{J_a(w_n)}{\|w_n\|^2} &= \frac{1}{2}\int_0^T \Big[(v_n')^2 + \frac{c^2}{4}v_n^2\Big]\,dx - \int_0^T \Big[\frac{e^{cx}G(x, \frac{w_n}{e^{\frac{c}{2}x}} + a)}{\|w_n\|^2} - \frac{e^{\frac{c}{2}x} f}{\|w_n\|} v_n\Big]\,dx \\ &\le \frac{c_2}{\|w_n\|^2} \,. \end{aligned} \tag{3.2}$$

Due to the assumptions (2.6), (2.7) we have

$$\lim_{n\to\infty} \int_0^T \Big[\frac{e^{cx}G(x, \frac{w_n}{e^{\frac{c}{2}x}} + a)}{\|w_n\|^2}\Big]\,dx \le \frac{1}{2}\int_0^T \Big[\Big(\frac{c^2}{4} + c_T\Big) v_0^2\Big]\,dx \,. \tag{3.3}$$

Using (3.3) and passing to the limit in (3.2) we get

$$\frac{1}{2}\Big(\int_0^T \Big[(v_0')^2 - c_T\,(v_0)^2\Big]\,dx\Big) \le \frac{1}{2}\Big(1 - \int_0^T c_T(v_0)^2\,dx\Big) \le 0 \,, \tag{3.4}$$

a contradiction to the inequality (2.4). Therefore J_a is a weakly coercive functional for each $a \in \mathbb{R}$.

By the standard arguments we can prove that J_a is a weakly sequentially lower semi-continuous functional on H. The weak sequential lower semi-continuity and the weak coercivity of the functional J_a imply (see Struwe [10]) the existence of a critical point w_a of the functional J_a. The usual regularity argument for ODE proves (see Fučík [4]) that w_a is also a solution to (2.2).

Now we prove the existence a classical solution to (1.1).

Let (a_n) be such sequence that $\lim_{n\to\infty} a_n = \infty$ and (w_{a_n}) be a corresponding sequence of the solutions to (2.2) with $a = a_n$. We denote $w_{a_n}^+(x) = \max\{w_{a_n}(x), 0\}$, $w_{a_n}^-(x) = \max\{-w_{a_n}(x), 0\}$ and multiply equation (2.2) by $w_{a_n}^-$. Then, integrating by parts, we have

$$\int_0^T \Big[({w_{a_n}^-}')^2 + \frac{c^2}{4}(w_{a_n}^-)^2\Big]\,dx = -\int_0^T \Big[e^{\frac{c}{2}x}\,(g(x, \frac{w_{a_n}}{e^{\frac{c}{2}x}} + a_n) - f)\,w_{a_n}^-\Big]\,dx \,. \tag{3.5}$$

By (2.6) for all $\varepsilon > 0$ there exists $r_0 > 0$ such that $g(x,s) \le (\frac{c^2}{4} + c_T + \varepsilon)s$ for all $s \ge r_0$ and $-g(x,s) \le -(\frac{c^2}{4} + c_T + \varepsilon)s$ for all $s \le -r_0$.

We can suppose that $r_0 \geq s_0$ and estimate by (2.5), (2.6), (2.7)

$$
\begin{aligned}
-\int\limits_{\frac{w_{a_n}}{e^{\frac{c}{2}x}}+a_n\leq -r_0} \Big[e^{\frac{c}{2}x}\,(g(x,\frac{w_{a_n}}{e^{\frac{c}{2}x}}+a_n))\,w_{a_n}^-\Big]\,dx &\overset{(2.6)}{\leq} \Big(\frac{c^2}{4}+c_T+\varepsilon\Big)\int\limits_{\frac{w_{a_n}}{e^{\frac{c}{2}x}}+a_n\leq -r_0}\Big[e^{\frac{c}{2}x}\Big(-\frac{w_{a_n}}{e^{\frac{c}{2}x}}-a_n\Big)w_{a_n}^-\Big]\,dx \\
&\leq \Big(\frac{c^2}{4}+c_T+\varepsilon\Big)\int\limits_0^T (w_{a_n}^-)^2\,dx\,,
\end{aligned}
$$

$$
\begin{aligned}
-\int\limits_{\left|\frac{w_{a_n}}{e^{\frac{c}{2}x}}+a_n\right|\leq r_0} \Big[e^{\frac{c}{2}x}\,(g(x,\frac{w_{a_n}}{e^{\frac{c}{2}x}}+a_n))\,w_{a_n}^-\Big]\,dx &\overset{(2.7)}{\leq} \int\limits_{\left|\frac{w_{a_n}}{e^{\frac{c}{2}x}}+a_n\right|\leq r_0}\Big[e^{\frac{c}{2}x}\Big(c_1\Big|\frac{w_{a_n}}{e^{\frac{c}{2}x}}+a_n\Big|+q(x)\Big)\,w_{a_n}^-\Big]\,dx \\
&\leq \int\limits_0^T [\,e^{\frac{c}{2}x}(c_1r_0+q(x))\,w_{a_n}^-\,]\,dx\,,
\end{aligned}
\tag{3.6}
$$

$$
\begin{aligned}
-\int\limits_{\frac{w_{a_n}}{e^{\frac{c}{2}x}}+a_n\geq r_0} \Big[e^{\frac{c}{2}x}\,(g(x,\frac{w_{a_n}}{e^{\frac{c}{2}x}}+a_n))\,w_{a_n}^-\Big]\,dx &\overset{(2.5)}{\leq} \int\limits_{\frac{w_{a_n}}{e^{\frac{c}{2}x}}+a_n\geq r_0}\Big[e^{\frac{c}{2}x}(-a_+(x))\,w_{a_n}^-\Big]\,dx \\
&\leq \int\limits_0^T \Big[e^{\frac{c}{2}x}|a_+(x)|\,w_{a_n}^-\Big]\,dx\,.
\end{aligned}
$$

Hence there exists $Q \in L^1(0,T)$ such that

$$
\begin{aligned}
-\int_0^T\Big[e^{\frac{c}{2}x}\,(g(x,\frac{w_{a_n}}{e^{\frac{c}{2}x}}+a_n)-f)\,w_{a_n}^-\Big]\,dx \leq \\
\Big(\frac{c^2}{4}+c_T+\varepsilon\Big)\int_0^T (w_{a_n}^-)^2\,dx + \int_0^T Q\,w_{a_n}^-\,dx\,.
\end{aligned}
\tag{3.7}
$$

We take ε such that $\hat{c}_T := c_T+\varepsilon < \left(\frac{\pi}{T}\right)^2$. Consequently, using (3.5), (3.7) we get

$$
\Big(1-\frac{\hat{c}_T}{(\frac{\pi}{T})^2}\Big)\int_0^T\Big[(w_{a_n}^-{}')^2\Big]\,dx \leq \int_0^T\Big[(w_{a_n}^-{}')^2-\hat{c}_T(w_{a_n}^-)^2\Big]\,dx \leq \int_0^T\Big[Q\,w_{a_n}^-\Big]\,dx\,. \tag{3.8}
$$

This yields that for $a_n \to \infty$ the sequence $(w_{a_n}^-)$ is bounded in $C([0,T])$ (due to compact embedding H into $C([0,T])$). Similarly we obtain that the sequence $(w_{a_n}^+)$ is bounded in $C([0,T])$ for $a_n \to -\infty$.

Now we denote $u_{a_n} = \dfrac{w_{a_n}}{e^{\frac{c}{2}x}} + a_n$ then u_{a_n} is a solution to

$$
\begin{aligned}
&u''(x)+c\,u'(x)+g(x,u)=f(x)\,, \quad x\in[0,T]\,,\\
&u(0)=u(T)=a_n\,.
\end{aligned}
\tag{3.9}
$$

Since (a_n) was an arbitrary sequence and (w_a^-) is bounded in $C([0,T])$ therefore $\lim\limits_{a\to\infty} u_a(x) = \infty$ uniformly on $[0,T]$. Similarly $\lim\limits_{a\to-\infty} u_a(x) = -\infty$ uniformly on $[0,T]$.

We denote

$$F(s) = \int_0^T \int_0^s [g(x, u_a(x)) - f(x)] \, da \, dx \, .$$

Using Lemma 2.2 we conclude $F \in C(\mathbb{R})$, $F'(s) = \int_0^T [g(x, u_s) - f] \, dx$. We get by (2.8) (where $u(a, x) = u_a(x)$) that there exist constants $s_1 < s_2 < s_3 < s_4$ such that $F(s_1) \geq F(s_2)$ and $F(s_3) \leq F(s_4)$.

Hence there exist a constant $\hat{s} \in \mathbb{R}$ and a solution $\hat{u}$ to (3.9) with $\hat{u}(0) = \hat{u}(T) = \hat{s}$ such that $F'(\hat{s}) = \int_0^T [g(x, \hat{u}) - f] \, dx = 0$. Integrating (3.9) over $[0, T]$ with $u = \hat{u}$ we obtain $\int_0^T \hat{u}'' \, dx = 0$. Hence $\hat{u}'(0) = \hat{u}'(T)$ and the function $\hat{u}$ is a solution to the periodic problem (1.1). The proof is completed.

Acknowledgements: This publication was supported by the project LO1506 of the Czech Ministry of Education, Youth and Sports.

References

[1] AMSTER P., Nonlinearities in a second order ODE USA-Chile Workshop on Nonlinear Analysis, Electronic Journal of Differential Equations, Conf. 06, 2001, pp. 13–21.

[2] DRÁBEK P., INVERNIZZI S., On the periodic BVP for the forced Duffing equation with jumping nonlinearity, Nonlinear Analysis 10 (1986), 643–650.

[3] DRÁBEK P., LANGEROVÁ M., On the second order periodic problem at resonance with impulses Journal of Mathematical Analysis and Applications, 428(2015) 1339—1353.

[4] FUČÍK S., Solvability of nonlinear equations and boundary value problems, D. Reidel Publishing Company, Holland 1980.

[5] HABETS P., Existence of periodic solutions of Duffing equations, Journal of Differential Equations 78 (1989), pp. 1=-32.

[6] MARIN M., On weak solutions in elasticity of dipolar bodies with voids, Journal of Computational and Applied Mathematics 82 (1997), pp. 291–297.

[7] MAWHIN J., WARD J.R., Nonuniform nonresonance condition at the two first eigenvalue for periodic solutions of forced Liénard and Duffing equations, Rocky Mountain Journal of Mathematics 12(1982), no.4, pp. 643–654.

[8] DA SILVA E. D., Quasilinear elliptic problems under strong resonance conditions, Nonlinear Analysis. 73 (2010), no. 8, 2451–2462

[9] DA SILVA E. D., Resonant elliptic problems under Cerami condition, arXiv:1205.2724.

[10] STRUWE M., Variational Methods, Springer, Berlin, (1996).

[11] TOMICZEK P., Forced Duffing equation with a resonance condition Advanced Nonlinear Studies 10 (2010), pp 573–580.

[12] TOMICZEK P., Periodic problem with a potential Landesman Lazer condition, Hindawi Publishing Corporation Boundary Value Problems (2010), Article ID 586971, 8 pages doi:10.1155/2010/586971.

[13] TORRES P.J., Existence and Stability of Periodic Solutions of a Duffing Equation by Using a New Maximum Principle, Mediterranean Journal of Mathematics, 1 (2004), 479-–486.

[14] WANG C., Multiplicity of periodic solutions for Duffing equations under nonuniform non-resonance condition, Proceedings of the American Mathematical Society, Vol. 126, No. 6, 1998, pp. 1725–1732.

Current address

Tomiczek Petr, RNDr., CSc.
European Centre of Excellence NTIS – New Technologies for Information Society
Department of Mathematics, Faculty of Applied Sciences, University of West Bohemia
Technická 8, 306 14 Plzeň, Czech Republic
E-mail: tomiczek@kma.zcu.cz

Slovak University of Technology in Bratislava
Faculty of Mechanical Engineering
17th Conference on Applied Mathematics **APLIMAT 2018**

Proceedings

TRAJAN'S COLUMN: MATH, ART AND CRYPTOGRAPHY

TORRE Matteo (IT)

Abstract. The paper discusses an original interpretation in cryptographic key of the continuous narrative of Trajan's Column historiated frieze, proposing a parallel between the column and the scytale, an ancient cryptographic technique.

Keywords: Roman Art, cryptography, scytale, Trajan's Column

Mathematics Subject Classification: Primary 01A20, 97A40, Secondary 97A80

1 Introduction

The history of civilization testifies to the interaction of Physics and Mathematics with Art and the influence that such disciplines have on it and on artists, for example, such as Poincaré and Penrose's theories, influenced the Escher's works on the organization of space-time, not to mention Dalí, Seurat, and many others. Feyerabend recognizes a common creative uniqueness between science and art [1] that is an effective way of science popularization. Even Ancient Rome was not insensible to the charm of art and understood its great communicative and universal character. In my opinion, one of the most significant artworks is the Trajan's Column because it is an artistic and, at the same time, scientific example of how mathematics can help art popularization to the people. Reading the art history books for high schools, it is very difficult to find the mathematical explanation of the model used for the decorative frieze (recalling the military campaigns of Trajan in Dacia), and it is even more difficult try to interpret, from a scientific point of view, the motivations that led the architect to choose that mathematical shape.

In this paper I propose a cryptographic interpretation of the mathematical model of the Trajan Column frieze. The idea born from the listening to a talk during Aplimat 2017 and the reading of the related paper [2] by Prof. Laura Tedeschini-Lalli and Prof. Alessandra Carlini. Some ideas on the mathematic interpretation of the Trajan's column frieze are taken directly from this paper [2]; while the cryptographic interpretation, supported by artistic and matematical motivations, is an original idea of mine. I am also grateful to Prof. Laura Tedeschini-Lalli and Porf. Alessandra Carlini to have been a source of inspiration, for having kindly accorded the use of some Trajan's Column images of their copyright and for their ability to popularize with simplicity and exactness the fascinating links between science and art.

2 Ancient cryptography

The word *cryptoghrapy* comes from ancient Greek *κρυπτοσ-γραφια* which means "hidden writing", a term that denotes the art, or rather, the science, to invent methods to hide messages for military, business, commercial or just for loving purpose (lover who writes a letter to his in lover). The message, i.e. the sequence of letters we want to transmit, is called *plaintext*, while the message that is really transmitted is called *chipertext*. The process of transforming the plaintext into chipertext is called *encryption* and is done by the sender; the reverse operation is called *decryption* and is made by the receiver who knows the encryption technique: this information is the encryption *key*. Summarize in a few lines the fascinating and complex history of cryptography is not so easy, so we'll just talk about the historical aspects that are related to the themes addressed in this paper, referring those interested to the most important sources [3], [14].

The oldest example of the use of cryptography was found in some ancient Egyptian hieroglyphics carved in some monuments during the Old Kingdom and dating back more than 4,500 years ago. These are *mutili hieroglyphs*, or signs created *ex novo*, or old signs coated with acronym to make mysterious a religious text and perhaps to incite reading through decryption, thereby winning the indifference of visitors and the sense of satiety produced by the monotony of repetitive formulas. These techniques can not be considered as serious examples of secret communications, but they are to be defined as attempts at mysterious, intriguing or extravagant scriptures made by literati of that time.

The earliest informations about cryptography come from ancient Greece and are on the *spartan scytale* or *lacedemonic scytale* or, more simply, scytale (from ancient greek *σκυτάλη*, stick). The Hellenistic Greek cryptographic systems do not have any connection to the Egyptians ones: Hellenistic cryptography is essentially military (as a part of the modern cryptography), while the Egyptian crypography privileges mystical, magical and religious purposes.

Scytale is one of the oldest known transposition cipher[1], the first, and probably the only, cryptographic tool that has had success with this method. The sender choses a stick, with a certain diameter, and wrapped around a leather ribbon[2], then wrote the message from one end of the stick to the other, parallel to the axis of the stick itself. Finally, the sender unrolled the leather ribbon and sent it to the receiver who, in order to correctly read the message, had to roll the ribbon onto another stick of the same diameter[3] (Fig. 1). It is not what we today would

[1]A transposing cipher is an encryption method in which the positions occupied by the plaintext units are changed according to a given pattern, so that the ciphertext constitutes a permutation of the plaintext. Mathematically speaking, a two-way matching function is used on character positions during the coding operation and an inverse function during decoding.

[2]As we shall see, there are differing opinions on the material of the ribbon.

[3]The cipher key is the diameter of the stick. Another important aspect on the safety of the method is in the ribbon: if was used a leather strip, it could easily be disguised as a belt.

call a "trusty algorithm" but, according to the Greek historians, it has been used successfully for many centuries.

From Plutarch evidence in *Parallel Lives* (*Lys*, XIX, 8-12) and Aulo Gellio in *Attic Nights* (XVII, IX)[4], we know that the scytale was used by the ephors, the five supreme magistrates of Sparta, to send secret messages to the generals involved in military expeditions, confirming also that (as Tucidide recalls[5]) confidentiality was an important component of the Sparta government. Plutarch also accurately describes the cryptographic method of scytale in *Lysander's life*, where he explains that with this term we mean both the parchment with the message and the stick used for his writing and deciphering[6]. Before writing the message, the ephors prepared a long and narrow parchment strip and wound it around a stick, which was exactly the same as length and diameter to another stick that the magistrates had previously provided to the receiver. After rolling parchment to the stick, be careful not to leave any space in which the wood was visible and at the same time avoiding overlapping several edges of the parchment itself, the ephors would write the message. Then, the ephors unrolled the strip from the stick and sent it to their emissary through a messenger.

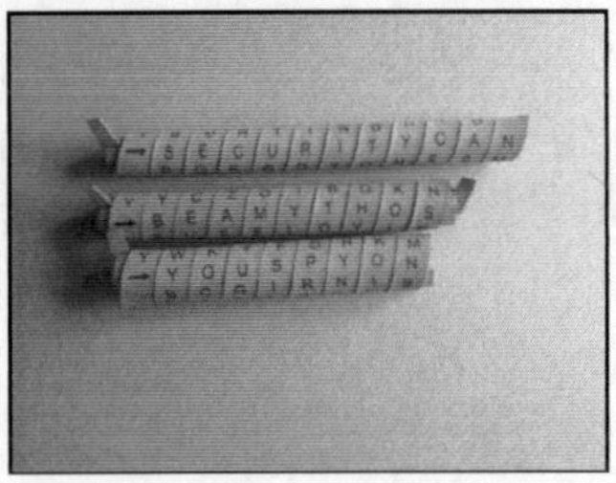

Fig. 1. Three different types ofscytales: the three sticks have different diameters, this corresponds to three different cryptographic keys.

The informations from Aulus Gellius in *Attic Nights* are very similar to those of Plutarch, but from this poem we can deduce an interesting information: the strip on which the encrypted message was written was of leather. Also A. Martin [4] considers unlikely that the ribbon was made of papyrus or parchment because this materials are too breakalble; a leather band instead was more easily transportable by the messenger (who could use it as a belt) and more resistant.

The cryptographic techniques evolved rapidly, and even the ancient Romans contributed to it. Back to this time is the Caesar chiper, the first one-alphabet substitution cipher. A substitute cipher operates in this way: substitution takes place letter by letter, scrolling text from start to finish, and each letter of the plaintext is replaced in the ciphertext by the letter that there is a certain number of positions later in the alphabet (Fig. 2). The Caesar chiper takes its name from Julius Caesar, who invented it and used it during the Gaul's war to protect his secret messages.

---

[4] *Gell.*, XVII, IX, 7: *Surculi duo erant teretes, oblonguli, pari crassamento eiusdem longitudinis, derasi atque ornati consimiliter.*

[5] During the 14th year of the Peloponnesian War (March 418 - February 417), the Argives and Spartans' army clashed in the Mantinea region. Thucydides confesses his inability to count the Spartan soldiers who took part in the war because the Spartans communicated via encrypted messages (*Thuc.*, V, LXVIII, 2).

[6] More precisely, Plutarch tells about a sending of a scytale to Lisandro with the ephors immediately called him back to Sparta to justify the looting and raids he had done in the Persian satrapia of Farnabazo II, where he was at that time.

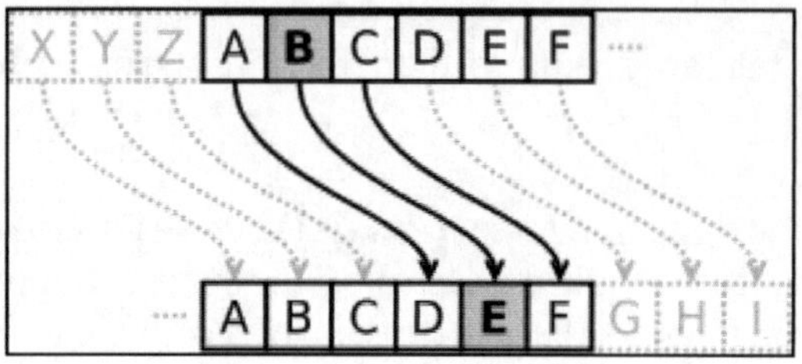

Fig. 2. mathematical diagram o f Caesar's cipher.

At that time this cryptographic technique was safe because common people were often unable to read a simple text and therefore not even a cipher one. From the discovery of alphabetic frequencies analysis, by the Arabic mathematician Al-Kindi in the XI century, all one-alphabetic substitution ciphers have become very easy to break. Over the centuries, cryptography evolved very quickly, but this brief summary of ancient cryptography is enough to the readers to understand the next paragraphs.

3 The Trajan's Column

The historical relief is the artistic type that most characterizes Roman art. It is a celebratory type used to emphasize important episodes (civil or military) of Roman history: winning campaigns, as in the Trajan and Marcus Aurelius columns, military triumphs, as in the Titus Arch, but also important events in civil life, such as the reliefs of the Trajan's Arch of Benevento, which celebrate the extension of the Appia Street to Brindisi. The celebration of military campaigns through artworks is not an exclusive of Roman art: it has precedents in Greek Hellenistic art, in which the victories of Alexander the Great are represented, through the myth and the real history, in the mosaic of Isso's battle. Typically of Roman art are, instead, some scenes such as the triumph cortege, the *adlucotio* (the speech) to the army, the sacrifice to the gods by the general, for which are fixed some canonical figurative formulas, which remain unchanged during the Roman art.

The Trajan's Column, with its long figured frieze, is among the monuments that best illustrate the strategies of political and military propaganda of the *princeps*. The Trajan's Column, built between 110 and 113 AD in the courtyard of the Ulpia Library (Fig. 3) to celebrate the two victorious campaigns of Trajan in Dacia, is a coclid column. This adjective is referred both to the spiral staircase that has internally (from Latin, *cochlis*), both for the helix resulting on their surface.

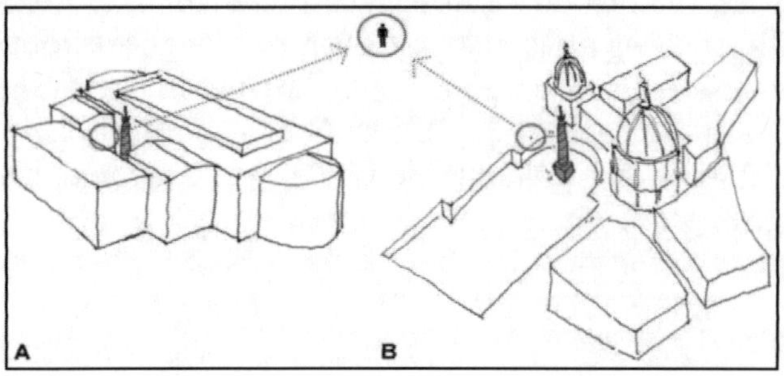

Fig. 3. on the left: onlooker at Trajan' s Column A) at the windows in the original forum complex; B) at street level in its urban context today. Image published in: A.Carlini, L.Tedeschini Lalli, Imperial roman cochlear columns and their historical narrative friezes, Proceedings Aplimat 2017, Slovak University of Technology Bratislava, Fig. N° 8 (© A. Carlini); on the right: an actual photo of Trajan's Column in Rome.

The column is a "centenary" one, that is 100 Roman feet high (29.78 meters, 39.86 meters, including the pedestal at the ground and the statue at the top) and consists of 17 colossal blocks in marble pario, each of which weighs about 40 tons and has a diameter of 3.83 meters. It has a shaft decorated with a spiral frieze, conceived as a superimposed ribbon covering it. This particular, undoubtedly mathematical nature, has created different interpretations of artistic and social nature; in the next paragraph we will also propose a cryptographic one.

In the top of the column, the frieze gives a look at the typical grooves of the Doric order. The order is, however, adapted because the column appears as a contamination between different models: a honorary column, with a statue on the top, and a votive column, adorned with reliefs in parallel bands. In addition, it associates a celebratory function of the Emperor to the funeral function, because within the base there is the cell destined to host the ashes of Trajan.

Thanks to its remarkable communication skills, the Trajan's Column has inspired the construction of other coclid columns such as the Marcus Aurelius Column, the Theodosius Column, the Arcadio's Column, the Justinian's Column, the Vendôme Column and the Alexander's Column.

The frieze: math and cryptography

The frieze of the Trajan's Column is made of bas-relief, has a length of about 200 meters and wraps 23 times around the shaft narrating the campaigns conducted by Trajan in Dacia.
The narration, realized in episodes, begins with the personification of the Danube (represented as an old man with a beard) and the passage of the Roman army on the bridge of boats (Fig. 4) and ends with the suicide of Decebalus, king of the Dacians, and the Roman conquest of the region.

In the 114 scenes, a continuous narration develops from bottom to top and places, facts and characters are described with meticulous care and the repetition of the protagonists marks the progress of the action. The rhythm with these scenes develop is so pressing that the critics agree that there is not a moment of repetitive weariness, of a repetition, of a void in the narrative context [5]. Traiano appears 59-60 times, always next to his soldiers or engaged in official acts and rituals.

Fig. 4. the passage of the Ro man army on the bridge of boats.

The frieze of the Trajan's Column can be considered as a "figurative ribbon" that wraps around the shaft of the column itself. This idea seems to be new and peculiar to Roman art.

Several hypothesis have been set forward to its genesis. We report those cited by Settis [9] and also proposed in [2]:

1. The ribbon is inspired by the *rotulo*, early form for a book, or *volumen* used in the ancient libraries and consisting of a strip of flexible material so that it could be rolled in various ways. In particular, some historians believe that the reference is to the *Commentarii*, Trajan's writings on the Dacia war, which have been lost.
2. The ribbon is inspired to the *tabulae triumphales*: pennants painted on cloths or wooden panels that were exposed during the triumphal cortege to explane to the people the places and the main episodes of the war. The ancient sources and some fragment of Roman painting make us guess that in these objects the story developed in a circular way and whit different registers.

In my opinion, the frieze of the Trajan's Column can be considered as the leather band that wraps around the wooden stick which, in this case, is represented by the column itself. For this interpretation I was ispired by the talk "*On imperial roman cochlear columns*" during the 16th Conference on Applied Mathematics [2].

The paper [2] describes in a significant and very clear way, how the narration of Trajan's military operations can be read from the bottom to the top following the continuous narration around the shaft of the column along the length of the frieze, but also following a vertical narration. In particular, the paper [2] suggests to follow three imaginary vertical line where there are aligned some significant moments: the Emperor's image, the Emperor's victory (the premonition of victory, the Winged Victory and the suicide of the Dacian king, sanctioning the defeat of the Dacians), the pontoon bridge for crossing by the troops.

For example, the very good sketch by Alessandra Carlini (Fig. 5) shows the vertical alignment of some episodes of which the lecture key is the Emperor's *adlocutio* to the army [12]. This alignment, avaiable to a standing onlooker in a particolar position, guarantees a quick interpretation of the ideological message of the frieze in accord, in my opinion, to the strategies of political and military propaganda of the *princeps*.

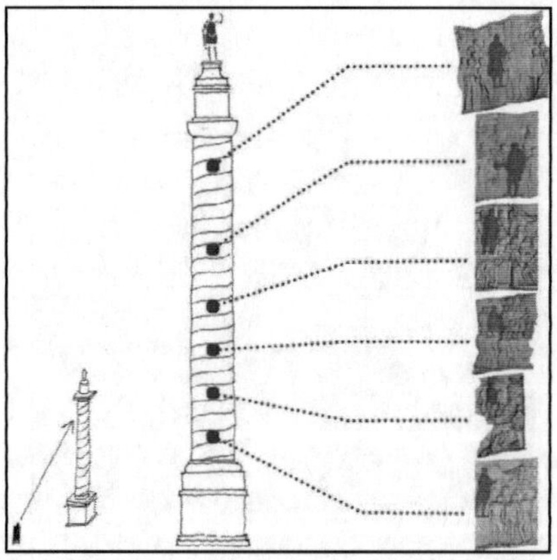

Fig. 5. Vertical alignment of E mperor's *adlocutio*, visually available to a standing onlooker.
Image published in:
A.Carlini, L.Tedeschini Lalli, *Imperial roman cochlear columns and their historical narrative friezes*, Proceedings Aplimat 2017, Slovak University of Technology Bratislava, Fig. n° 10 (© A. Carlini) .

This alternative reading of the frieze narration, a-temporal and by main facts, is in my opinion closely linked to the scytale as a tool to transmit secret information for several reasons:

1. The scytale and the Trajan's Column frieze have the same particular geometry: the scytale leather ribbon and the historiated frieze are helix and their matematically construction depending on the relation between its height and the slope of the helix, as we can read in [2] and also in [6] and [11] and the bibliography thereof.
2. Cryptography was known at Roman times and adding to it the mathematical characteristics of the frieze, it is possible to support the hypothesis that the author of the column has deliberately used the diameter of the column as a reading key (a cryptographic key) to decrypt the political propaganda message of the frieze.
3. According to Salomon Reinach [13] the bas-relief is divisible into 114 parts of equal width and, therefore, there are 5 scenes at every turn. Mathematically speaking, the vertical relation among the elements s_i of a sequence $S = \{s_1, s_2, \ldots, s_{114}\}$ of 114 scenes is almost insured in the actual Trajan column (5 scenes wrapped around 23 times) by a modular arithmetic with modulus 5. So, there are 5 different vertical alignments that are visible: for $i = 1 + 5 \cdot k$, that is $i \equiv 1 (mod\, 5)$, but also for $i \equiv 0, 2, 3, 4\ (mod\, 5)$. In my opinion, not all the alignements are interesting but only those connected to the political propaganda message (whose key image can be searched in [12]), like that of the Fig. 5, and this alignments are appreciable only for one standing onlooker.
4. Replacing letters (a finite set) of the scytale leather ribbon with scenes of the Trajan's Column frieze, it is not a problem and it is possible because there are some cryptographic examples [14] where in the scytale strip the encrypted message is not a writing letter by letter but several letters (Fig. 6) are written within the breadth of the strip. These examples show again and with considerable visual impact such how the link between scytale and Trajan's Column, through helix geometry, is not so forced.

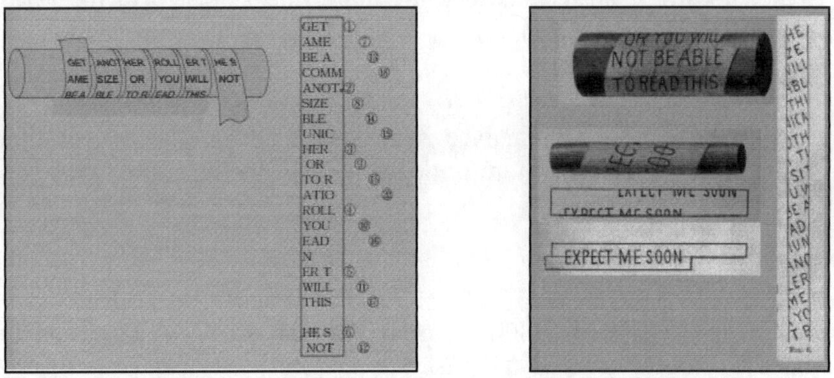

Fig. 6. Scytale as a transposition ciper by partial lines, not by letters.

Therefore becomes clear that the diameter of the column is inextricably linked to the meaning of the frieze continuous narration, but above all to the strategic position that certain scenes assume in relation to the position of the observer. Imagine to transfer the Trajan's Column frieze on the Marcus Aurelius Column (Fig. 7), which is also a centenary coclide column (that is 100 Roman feet high) but has a diameter ranging between 3.80 m and 3.65 m (which determines, therefore, a smaller number of turns of the frieze, 21 instead of 23): the frieze would lose its narrative meaning and the vertical alignment of the important scenes would no exist. It is the same situation that would happen to a spy who, intercepted the roll of skin with the encrypted message, tried to roll it on a stick with a different diameter from that used by the sender: the message would be meaningless and no interesting!

Fig. 7. Trajan's Column (on the left) and Column o f Marcus Aurelius (on the right). Image published in: A.Carlini, L.Tedeschini Lalli, *Imperial roman cochlear columns and their historical narrative friezes*, Proceedings Aplimat 2017, Slovak University of Technology Bratislava, Fig. n° 3 (© A. Carlini).

In my opinion, the motivations of the vertical alignment of the most significant scenes of the frieze can not only be linked to perspectives or artistic choices, but they must also be search in the symbolic sphere. In my opinion, for an artwork as a cochlid column, so markedly mathematical, the cryptography is an excellent link between helix geometry and art.

Conclusions and future developments

This cryptographic interpretation offers new possibilities to interpret not only the Trajan's Column, but also other cochlid columns and other monuments that contain a historiated frieze with continuous narration which I will deal with in a future paper. I want also to study in deep the analysis of the possible reading (and cryptographic) keys related to the various vertical alignments of the Trajan's Column, founding in [12], trying to verify if there are vertical alignments even for some other cochlid column.

Finally, another interesting aspect to analyze is use of scytale in other artistic monuments: I think this analysis should start with a more in-depth study of the history of the scytale.

References

[1] FEYERABEND P. THOMAS C., *Arte e Scienza*, Armando Editore, Roma, 2017.

[2] CARLINI A., TEDESCHINI-LALLI L., Imperial Roman cochlear columns and their historical narrative friezes, in *Proceedings of 16th Conference on Applied Mathematics*, APLIMAT 2017.

[3] BECKET B., *Introduction to Cryptology*, Blackwell Scientific Publications, 1988, ISBN 0-632-01836-4.

[4] MARTIN A., *Scytale*, DAGR IV, 2, 1918, p. 1162.

[5] BIANCHI BANDINELLI R., TORELLI M., *L'arte dell'antichità classica, Etruria-Roma*, Torino, Utet, 1976, ISBN 88-6008-053-3.

[6] CONVERSANO E., FRANCAVIGLIA M., LORENZI M.G., TEDESCHINI-LALLI L., Persistence of Form in Art and Architecture: Catenaries, Helicoids and Sinusoids, *APLIMAT J. of Appl. Math.*, 4 (2011).

[8] BIANCHI BANDINELLI R., *Il maestro delle imprese di Traiano*, Venezia, 2003.

[9] SETTIS S., MAFFEIS S., entry "Colonna", Enciclopedia dell'Arte Antica Treccani (1994).

[10] COLLARD B., La cryptographie dans l'Antiquité gréco-romaine. III. Le chiffrement par transposition, *Folia Electronica Classica*, Louvain-la-Neuve, N° 7, 2004.

[11] CONVERSANO E., FRANCAVIGLIA M., LORENZI M.G., TEDESCHINI-LALLI L., Geometric Forms that Persist in Art and Architecture, *Bridges 2011: Mathematics, Music, Art, Architecture, Culture*, pp. 463-466 (http://archive.bridgesmathart.org/2011/bridges2011-463.pdf).
[12] http://www.stoa.org/trajan/indices.html
[13] REINACH S., *La Colonne Trajane au Musée de Saint-Germain: Notice et explication*, Paris, 1886 (https://mediterranees.net/civilisation/armee_romaine/colonne_trajane/reinach2.html).
[14] HULME F. E., *Cryptography or, The history, principles, and practice of cipher-writing*, London, 1898.

Current address

Torre Matteo, high school teacher, free researcher
Liceo Scientifico Statale "L. B. Alberti"
Via Pontecurone, 17, 15142 Valenza (AL), Italy
E-mail: matteo.torre1984@gmail.com

Slovak University of Technology in Bratislava
Faculty of Mechanical Engineering
17th Conference on Applied Mathematics **APLIMAT 2018**

Proceedings

POSSIBLE COMPARISON BETWEEN TWO TIME SERIES

ŤOUPAL Tomáš (CZ), ŠEDIVÁ Blanka (CZ)

Abstract. This paper is focused on the possible methodologies for comparing two time series by estimating the probability that both time series will increase or decrease (of course in probability) in the same time.
It can be considered as the specific measure of dependence (more precisely concordance) between two random variables with non-parametric approach. Then the problem may arise with a statistical inference. The main idea of this approach is based on the transformation of observed data set into increasing and decreasing movement and then the Markov model (Markov chain) of transitions (increasing-increasing, increasing-decreasing, decreasing-increasing, decreasing-decreasing) is used with the removal of the assumption of independence.

Keywords: time series, Markov model, exchange rate, Czech crown, coefficient of concordance, dependence

Mathematics Subject Classification: Primary 62M05, 91B84; Secondary 62M10

1 Introduction

A frequent financial problem in the world is the assessment of how the financial assets denominated in some currency will behave in the currencies of that currency to the accounting currency (exchange rate loss or yield). This can be assessed on the base of knowledge of the set of past values and the assumption that there will be no significant change in the used probability model of the behaviour of such time series.
In general, it is a measure of the "power" of keeping or not keeping tendencies (monotony, growth, decrease) in two time series[1].

2 Basic terms

The presented text is motivated by the concept of positive and quadrant dependence (e.g. [1] and others). The bond strength between two time series can be measured in many ways. One of them can be preserving (or not preserving) the probability of monotone relationship. This is the

[1] Such theory can be done for more time series. There will be work only with pairs of time series.

probability that the values of one time series are increasing and the values of the second time series are also increasing and similarly for consistent decreases (analytically of course for discordance behaviour).
Next, here will be also work with time series whose values are continuous random variables (such a probability model, in many cases, does not distort reality). It means the following cases, which may occur:

Series A	increases	increases	decreases	decreases
Series B	increases	decreases	increases	decreases
Status code	1,1	1,0	0,1	0,0
Numeric code	3	2	1	0

Tab. 1. Possible time series states.

Of course, increase or decrease must be related to some, non-empty, time interval or to compare the value at one time point with the value at another time point. Here, time series are limited to time series with a discrete time axis and for comparison at two time points. Therefore, for some given τ the considered states are written in the following table.

Series A	$x_A(t) > x_A(t-\tau)$	$x_A(t) > x_A(t-\tau)$	$x_A(t) < x_A(t-\tau)$	$x_A(t) < x_A(t-\tau)$
Series B	$x_B(t) > x_B(t-\tau)$	$x_B(t) < x_B(t-\tau)$	$x_B(t) > x_B(t-\tau)$	$x_B(t) < x_B(t-\tau)$
Status code	1,1	1,0	0,1	0,0
Numeric code	3	2	1	0

Tab. 2. Possible comparison between two time series.

Thus, each realization of a pair of time series (for given τ and at each time t) is assigned its state. For the degree of consistency, the concordance rate, of these time series will then be considered the probability that there will be state $(1,1)$ or $(0,0)$ anytime. From this, it is obvious that there will be work with pairs of time series, for which this probability does not depend (functionally) on time t (but it may and it will be depend on time shift τ).

It is formally possible to write the relationship under the following assumptions:

$$\begin{aligned} concr(x_A, x_B; \tau) &= P([1,1]|_\tau \cup [0,0]|_\tau) = P([1,1]|_\tau) + P([0,0]|_\tau) \\ &\equiv P([3]|_\tau) + P([0]|_\tau). \end{aligned} \tag{2.1}$$

Obviously:

$$\begin{aligned} &concr(x_A, x_B; \tau) = concr(x_B, x_A; \tau), \\ &0 \le concr(x_A, x_B; \tau) \le 1, \\ &x_B(t) = h\big(x_A(t)\big) \Rightarrow concr(x_A, x_B; \tau) = 0,\ h(x) \textit{ is strictly decreasing}, \\ &concr(x, x_*; \tau) = 1, \textit{ where } x_* \textit{ is a copy of } x. \end{aligned} \tag{2.2}$$

In analogy with Kendal tau [2] a coefficient of concordance can be introduced as a difference of probability matching and probability discrepancy.

$$\begin{aligned} concc(x_A, x_B; \tau) &= P([1,1]|_\tau \cup [0,0]|_\tau) - P([1,0]|_\tau \cup [0,1]|_\tau) \\ &= 2concr(x_A, x_B; \tau) - 1. \end{aligned} \tag{2.3}$$

Obviously:

$$\begin{aligned}
&-1 \leq concc(x_A, x_B; \tau) \leq +1, \\
&concc(x_A, x_B; \tau) = 0, \textit{ if the decreases or increases in both series are independent,} \\
&concc(x_A, x_B; \tau) = -1, \textit{ if increases in one series implies decreases in second series} \\
&concc(x_A, x_B; \tau) = +1, \textit{ if increases in one series implies increases in second series.}
\end{aligned} \tag{2.4}$$

From theory's point of view, this introduction does not bring anything new. However, it is clear. If the coefficient is negative, the probability of discrepancy behaviour dominates the probability of the same. If the coefficient is positive, the probability of the same behaviour dominates over the probability of discrepancy. If it is not possible to make decision significantly from the behaviour of one series to the behaviour of the second series, it is zero or near zero.

3 Statistical inference

Under these assumptions, the state event are realized by a pair of time series in any state, at a given time (and at a given time shift), and equally distributed. However, the assumption of independence of observations will not satisfied for most real (pair) time series (this is needed if we want to use the approaches of random sample, i.i.d.). The pair of time series will contain a greater or lesser amount of memory (= probability dependency over time).
Therefore, we need to use some model of time dependence. This may be a Markov chain model (specifically for ordinary pairs of time series without other functional relationships, homogeneous and of course regular [2]). This can replace the assumption of the independence of observing states by assuming the independence of observed transitions from one state to another.

For this Markov chain model hold

$$p^T(t+1) = p^T(t)P, \tag{3.1}$$

where $p(t)$ is the probability vector of the occurrence of Markov chain in a given state, and P is the probability matrix of transition from state to state, obviously $dim\, p(t) = k$ and $dim\, P = k x k$, in our case $k = 4$.

The elements of the P - matrix can be (pointed) estimated by classical frequencies. So $P(i,j) = Probability\ (transition\ to\ state\ j\ from\ state\ i)$, i.e.:

$$\hat{P}(i,j) = \frac{n(i,j)}{n(i)}, \tag{3.2}$$

where $n(i,j)$ is the number of observed transitions from state i to state j and $n(i) = \sum_{l=1}^{k} n(i,l)$.

Against common tasks, here $\hat{P}(i,j)$ is not a proportion of random variable of the number of observations and a fixed range of random sample. This is the proportion of two random variables. This situation complicates the situation around this estimate. Nevertheless, such an estimate is a consistent[2] estimate of $P(i,j)$, which is sufficient for our purposes.

Another problem, in estimating, may be some specific properties of the matrix P. Here, from the declared use, is $P(i,j) > 0$. It means, that the probability of being in any state "in one step" can get into any other step or remain is nonzero. This would not be guaranteed by the above mentioned frequency approach (especially with a small number of observations = zero frequency problem). Because of this situation, it is used here a Bayesian approach assuming a uniform a priori distribution (e.g. [3]), where in this case is $k = 4$:

$$\hat{P}(i,j) = \frac{n(i,j)+1}{n(i)+k} = \frac{n(i,j)+1}{n(i)+4}; i,j = 0,1,2,3. \tag{3.3}$$

This is a consistent estimate again.

It is true (for regular strings) for any vector q of distribution of probability occurrences in the state of chain $\lim_{n\to\infty} q^T P^n = p^T$, where p is the single probability distribution vector in the state of chain with $p^T = p^T P$. It is only one stationary distribution vector $p_{stac} = p$.

Calculation in this way could be more complicated, and this is an important fact that stationary probabilities are the only ones for the regular chain. Therefore, another way of determining stationary probabilities (least squares method) is here possible:

$$p^T = p^T P \Rightarrow p^T(P - \boldsymbol{I}) = \boldsymbol{0}^T, \tag{3.4}$$

where $\boldsymbol{I}$ is an identical matrix of the same dimensions as P and $\mathbf{0}$ is a zero vector of the same size as p.

[2] Proof can be done with using the characteristics of multinomial distribution.

4 Obtained results

4.1 Table results

Exchanges rates of different currencies in one source currency

Data set is daily foreign exchange rates announced by the CNB for the period 3. 2. 2011 - 6. 4. 2017, www.cnb.cz[3] . $\tau = 5$ time shift is 5 quotation days.

CZK per EUR and CZK per USD

Estimation of transition matrix, From state (row), to state (column)

		0,0	0,1	1,0	1,1
		0	1	2	3
0,0	0	0,727	0,093	0,105	0,075
0,1	1	0,191	0,550	0,028	0,231
1,0	2	0,251	0,045	0,503	0,201
1,1	3	0,068	0,130	0,094	0,708

It is obvious that dominate the probability of staying in the state.

Stationary probablities		
p(0,0)=	*p(0)=*	***0,343***
p(0,1)=	*p(1)=*	***0,181***
p(1,0)=	*p(2)=*	***0,145***
p(1,1)=	*p(3)=*	***0,331***
Suma =		1,000
Concordance rate		**67,38%**
Concordance coeficient		**0,348**

With stationary probabilities, we can observe the dominance of probabilities of occurrence in concordant states. The probability of consistent behaviour is about 2/3.

Mean recurrence (stay) time		
t(0,0)=	*t(0)=*	3,67
t(0,1)=	*t(1)=*	2,22
t(1,0)=	*t(2)=*	2,01
t(1,1)=	*t(3)=*	3,43

Estimates of the mean times of stay in the state again prefer concordant states.

Matrix of mean transition times, From state (row), to state (column)

		0,0	0,1	1,0	1,1
		0	1	2	3
0,0	0	2,92	10,58	11,47	8,42
0,1	1	6,96	5,52	13,26	6,22
1,0	2	6,06	11,18	6,89	6,83
1,1	3	8,48	9,48	12,01	3,02

These values indicate that both time series tend more to conservative behaviour (in terms of mutual relationship).

If we make some qualitative statements about the long-term behaviour of the CZK against EUR and USD, there is no need to distinguish between these two target currencies.

Tab. 3. CZK per EUR and CZK per USD.

[3] The revaluation (or devaluation) here means appreciating the CZK against the currency in which the time series is.

CZK per CNY and CZK per JPY

Estimation of transition matrix, to state (column); From state (row)

From state (row)		0,0	0,1	1,0	1,1
		0	1	2	3
0,0	0	0,762	0,088	0,063	0,087
0,1	1	0,228	0,534	0,016	0,222
1,0	2	0,184	0,029	0,505	0,283
1,1	3	0,082	0,089	0,094	0,735

It is obvious that dominate the probability of staying in the state too.

Stationary probablities		
p(0,0)=	*p(0)=*	***0,361***
p(0,1)=	*p(1)=*	***0,147***
p(1,0)=	*p(2)=*	***0,122***
p(1,1)=	*p(3)=*	***0,370***
Suma =		1,000
Concordance rate		73,18%
Concordance coeficient		0,464

With stationary probabilities, we can observe the dominance of probabilities of occurrence in concordant states. The probability of consistent behaviour is about 73 %.

Mean recurrence (stay) time		
t(0,0)=	*t(0)=*	4,20
t(0,1)=	*t(1)=*	2,15
t(1,0)=	*t(2)=*	2,02
t(1,1)=	*t(3)=*	3,77

Estimates of the mean times of stay in the state again prefer concordant states, especially $t(0,0)$.

Matrix of mean transition times, to state (column); From state (row)

From state (row)		0,0	0,1	1,0	1,1
		0	1	2	3
0,0	0	2,77	12,41	15,31	7,89
0,1	1	6,46	6,82	16,29	6,19
1,0	2	7,25	13,76	8,23	5,31
1,1	3	8,52	12,49	13,96	2,70

Both time series tend to concordant states.

There are time series of foreign exchange currencies from partners from a geographically close environment, so both behave (in terms of changes) significantly similarly.

Tab. 4. CZK per CNY and CZK per JPY.

CZK per EUR and CZK per DKK

From state (row)		Estimation of transition matrix, to state (column)			
		0,0	0,1	1,0	1,1
		0	1	2	3
0,0	0	0,783	0,020	0,025	0,172
0,1	1	0,316	0,316	0,013	0,354
1,0	2	0,424	0,015	0,182	0,379
1,1	3	0,192	0,024	0,018	0,766

Significant dominance here have the probabilities of staying in concordant states.

Stationary probablities		
p(0,0)=	*p(0)=*	***0,493***
p(0,1)=	*p(1)=*	***0,031***
p(1,0)=	*p(2)=*	***0,026***
p(1,1)=	*p(3)=*	***0,450***
Suma =		1,000
Concordance rate		**94,38%**
Concordance coeficient		**0,888**

With stationary probabilities, we can observe the dominance of probabilities of occurrence in concordant states. The probability of consistent behaviour is very high. There is possible to say, that it is about 95 %.

Mean recurrence (stay) time		
t(0,0)=	*t(0)=*	4,61
t(0,1)=	*t(1)=*	1,46
t(1,0)=	*t(2)=*	1,22
t(1,1)=	*t(3)=*	4,27

Estimates of the mean times of stay in the state again prefer concordant states. Here is again significant dominance of concordant states.

From state (row)		Matrix of mean transition times, to state (column)			
		0,0	0,1	1,0	1,1
		0	1	2	3
0,0	0	2,03	46,54	46,12	5,46
0,1	1	4,11	32,63	47,15	4,06
1,0	2	3,60	46,66	39,06	4,13
1,1	3	4,97	46,03	46,94	2,22

Both time series tend to concordant states. By the biasing from the state, there is a much faster movement to a concordant rather than unconcordant state.

It is clear, that the CZK per DKK exchange rate is set in strong relationship to the CZK per EUR exchange rate (mediated through the market or directly by the method of determination).

Tab. 5. CZK per EUR and CZK per DKK.

CZK per EUR and CZK per PLN

Estimation of transition matrix, to state (column); From state (row)

		0,0	0,1	1,0	1,1
		0	1	2	3
0,0	0	0,676	0,128	0,140	0,056
0,1	1	0,165	0,611	0,034	0,190
1,0	2	0,208	0,044	0,580	0,168
1,1	3	0,072	0,137	0,143	0,648

Here, the probabilities of staying in position are greatly dominated.

Stationary probablities		
p(0,0)=	*p(0)=*	*0,305*
p(0,1)=	*p(1)=*	*0,219*
p(1,0)=	*p(2)=*	*0,210*
p(1,1)=	*p(3)=*	*0,266*
Suma =		1,000
Concordance rate		**57,16%**
Concordance coeficient		**0,143**

Probabilities of occurrence in individual states are practically uniformly distributed.

Mean recurrence (stay) time		
t(0,0)=	*t(0)=*	3,09
t(0,1)=	*t(1)=*	2,57
t(1,0)=	*t(2)=*	2,38
t(1,1)=	*t(3)=*	2,84

In terms of mean times of stay in the state, the individual states are similar.

Matrix of mean transition times, to state (column); From state (row)

		0,0	0,1	1,0	1,1
		0	1	2	3
0,0	0	3,27	9,17	8,79	9,27
0,1	1	7,15	4,58	10,58	7,18
1,0	2	6,42	10,51	4,76	7,73
1,1	3	8,24	8,98	8,75	3,76

Both time series tend to concordant states. Deviating from the state is a much faster movement to a concordant rather than unconcerned state.

Here are listed (in some ways) the opposite exchange rates. From the value of one exchange rate it is not possible to make "correct" conclusions about the second exchange rate (in terms of non-quantified changes, only increases and decreases).

Tab. 6. CZK per EUR and CZK per PLN.

4.2 Graphical results

The graphical results of the previous prepositions are shown in the following figures. The figures on the left side show clearly the relative frequencies, i.e. point estimates of stationary probabilities of individual states (status codes). Conversely, the figures on the right side the correlation between the considered exchange rates.

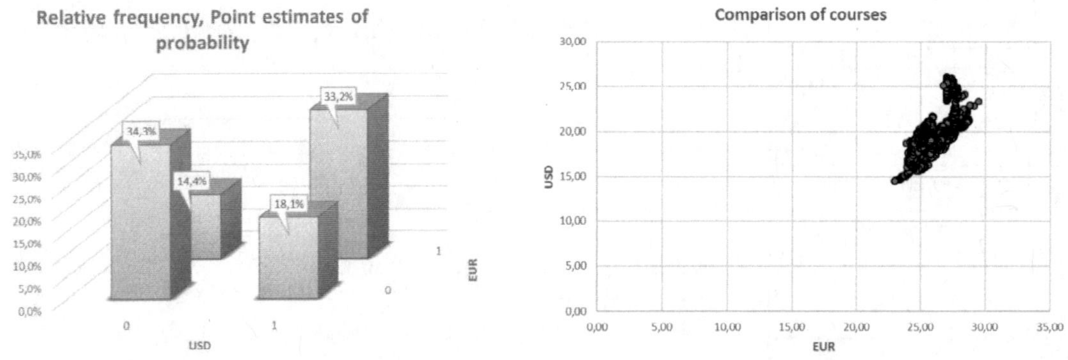

Fig. 1. Relative frequency and correlation among CZK, EUR and USD.

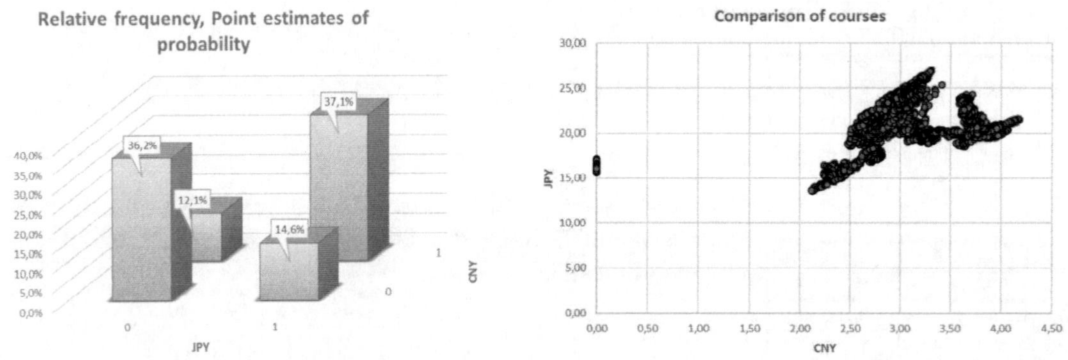

Fig. 2. Relative frequency and correlation among CZK, CNY and JPY.

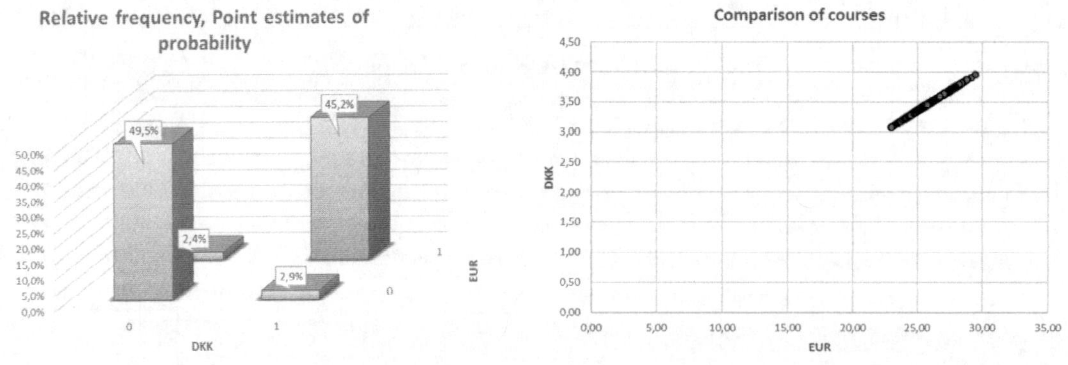

Fig. 3. Relative frequency and correlation among CZK, EUR and DKK.

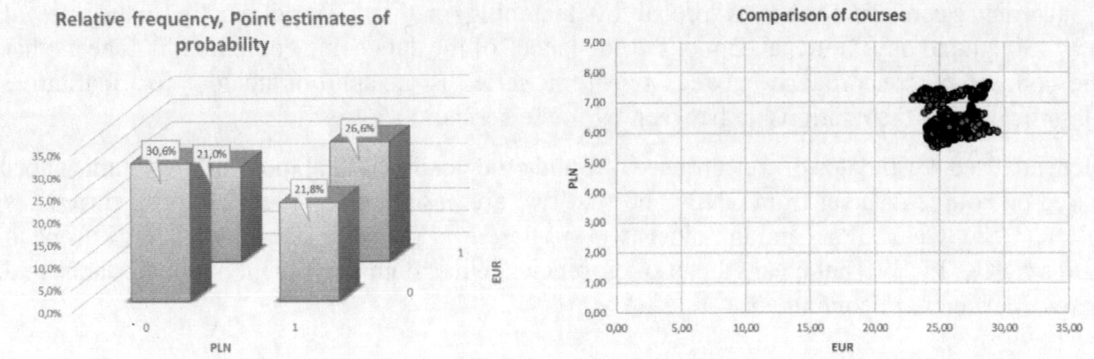

Fig. 4. Relative frequency and correlation among CZK, EUR and PLN.

Another problem in this approach may be the shift time τ. This is how the effect of the results of the shift time is affected. I.e. there is considered the question, how much time shift affect obtained results. The results are different for each observed pair of exchange rates. For better imagination, here is the first case i.e. CZK per EUR and CZK per USD.

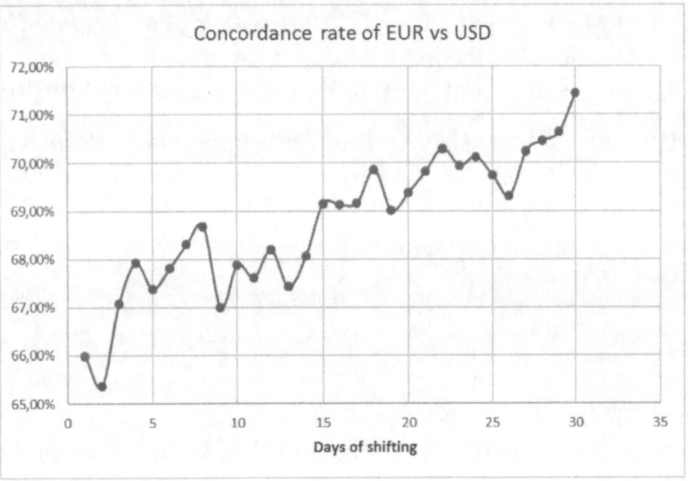

Fig. 5. Effect of time shift on concordance rates.

There is used a 5-day time shift for clarity on the base of observed simulations and for practical use. For the meaningfulness of these approaches, the maximum number of shifting days in simulations was limited to 30.

5 Conclusion

Computational modelling was done for daily published courses:

Source	Number of Exchange series	Source currency	From the period	to
ČNB	28	CZK	2. 1. 2007	6. 4. 2017

Tab. 7. Source data set.

Concordance can be described as "probability matching (qualitative)" practically (in the interval of 1 - 30 quoted days) independent of the "distance" of the date being compared. It appears that the concept of concordance between two time series is an appropriate tool for qualitative classification of the relationship between two time series.

Here are used 4 estimates of concordance rate and other statistics of all possibilities in this paper based on source data set from CNB. The first two are based on significant trading currencies (EUR, USD, CNY, JPY) and the other two are based on interesting results among European states (DKK, PLN). That means a pair of currencies with a high coefficient of concordance and a low coefficient of concordance.

Acknowledgement

This publication was supported by the project LO1506 of the Czech Ministry of Education, Youth and Sports.

References

[1] LEHMANN, E.L., *Some Concepts of Dependence*, 1966. [online] Ann. Math. Statist. Volume 37, Number p..1137-1153. Available at: <https://projecteuclid.org/euclid.aoms/1177699260> [Accessed 8 July 2017].

[2] GRINSTEAD, CH.M. and SNELL, J.L., *Introduction to Probability*, 2003 [online] Published by the American Mathematical Society. Available at: <https://math.dartmouth.edu/~prob/prob/prob.pdf> [Accessed 8 July 2017].

[3] WITTEN I. H. and BELL, T. C., *The Zero-Frequency Problem: Estimating the Probabilities of Novel Events in Adaptive Text Compression*, 1991. [online] IEEE TRANSACTIONS ON INFORMATION THEORY, VOL. 37, NO. 4. Available at: <http://ieeexplore.ieee.org/stamp/stamp.jsp?arnumber=87000> [Accessed 8 July 2017].

Current address

Ťoupal Tomáš, Ing., Ph.D.
European Centre of Excellence NTIS – New Technologies for Information Society
Faculty of Applied Sciences
University of West Bohemia
Univerzitní 8, 301 00 Plzeň, Czech Republic
Tel. number: +420 377 63 2674, e-mail: ttoupal@kma.zcu.cz

Šedivá Blanka, RNDr., Ph.D.
European Centre of Excellence NTIS – New Technologies for Information Society
Faculty of Applied Sciences
University of West Bohemia
Univerzitní 8, 301 00 Plzeň, Czech Republic
Tel. number: +420 377 63 2618, e-mail: sediva@kma.zcu.cz

Slovak University of Technology in Bratislava
Faculty of Mechanical Engineering
17th Conference on Applied Mathematics **APLIMAT 2018**

Proceedings

QUALITY COMPARISON BETWEEN HYBRID REGULARIZED EXPONENTIAL FORGETTING ALGORITHM WITH ALTERNATIVE COVARIANCE MATRIX AND SELECTED STANDARD LONG-RUN ON-LINE IDENTIFICATION METHODS OF INDUSTRIAL SYSTEMS

VACHÁLEK Ján (SK), MELICHER Markus (SK), VAŠEK Pavol (SK), ŠIŠMIŠOVÁ Dana (SK), VOLENSKÝ Tomáš (SK)

Abstract. This paper compares the quality of standardly used ON-LINE identification algorithms like directional forgetting and regularized exponential forgetting with the hybrid algorithm of exponential forgetting with an alternative covariance matrix. It points also to the behavior of selected algorithms in different deployment conditions and different run times. Particularly interesting is the long-run deployment, which correlates with use in industry. Based on the results obtained, decisions will be taken about the suitability of the hybrid regularized exponential forgetting algorithm with an alternative covariance matrix for selected industrial deployment scenarios.

Keywords: recursive algorithm, forgetting method, time varying parameters, long-run on-line identification, adaptive control

Mathematics Subject Classification: Primary 60A05, 08A72; Secondary 28E10

1 Introduction

The current character of real systems and their subsequent identification in practice clearly demonstrates the variability of their parameters [1,2,3]. This negative effect, in conjunction with adaptive control, prevents unlimited monitoring of the changes of parameters in time. The direct consequence of these negative effects is the impossibility of applying the standard recursive least squares method for estimating the linear regression model parameters. The reason is that simple, constant supply of new information causes, among other things, the system gain of the algorithm to converge to zero. This results in the destabilization of the numerical method for estimating the parameters, which is the "Blow up" problem [4].

The solution to this problem could be achieved by forgetting surplus information. The first techniques based on this principle include Exponential Forgetting (EF), Directional Forgetting (DF) (the basic idea was to reduce the concentration of the maximal reliability function or the „a posteriori" density function), or subsequent modifications based on different methods of regularization or stabilization of the covariance matrix. Regularized Exponential Forgetting (REF) or Stabilized Linear Forgetting (SLF) methods are examples of the latter.

Another possible solution to the problem of destabilization can be the use of new procedures using the preservation of „a priori" information about the system in the form of an alternative covariance matrix (ACM). The classic REF algorithm can be expanded with ACM to create the REFACM algorithm. Another developmental step is to combine the benefits of the REF and REFACM algorithms into a single block in the form of the Hybrid Regularized Exponential Forgetting Algorithm with Alternative Covariance Matrix (HREFACM). Quality aspects of our newly developed HREFACM algorithm is the subject of this article, which focuses on assessing its quality under various operating conditions.

2 Problem formulation

Let us consider a stochastic system, in which the observations are obtained at discrete time points k = 1,2 ....,. The directly controlled input u_k and the indirectly controlled output y_k (also may be multidimensional), can be arranged in the data pairs d_k = (uk, yk). The set of all observer data about the system up to time t is determined by $D_t = (d_1, d_2, ... d_t)$. The dependence of the new data pair (u_k, y_k) on the previous observations D_{k-1}, can be expressed by a conditional probability density function (p.d.f.) with the following structure:

$$p(y_k, u_k | D_{k-1}, \theta_k) = p(y_k | u_k, D_{k-1}, \theta_k) p(u_k | D_{k-1}) \tag{1}$$

The incomplete knowledge of the system is expressed by the vector of unknowns, which consists of time-varying parameters $\theta_k \in \theta$. Note that the input generator is described by the second part of this equation, and the first part characterizes the system.

The Bayesian approach:
If an unknown parameter, described by θ, is interpreted as a random variable, then the uncertainties θ, assuming the knowledge of the observed data D_t, is naturally expressed by the „a posteriori" conditional probability density function (p.d.f.) $p(\theta/D_t)$ conditionally dependent on D_t. This is generally determined by the Bayes theorem. If the input generator does not produce any new information compared to the information obtained from the observed data $p(u_k/D_{k-1}, \theta_k) = p(u_k/D_{k-1})$ k = 1,....t. The Bayes theorem is simplified [5]:

$$p(\theta | D_t) \propto p(\theta) \prod_{k=1}^{t} p(y_k | u_k, D_{k-1}, \theta) \tag{2}$$

An important special case in the first part of the equation (1) occurs when the output y_k depends on the previous data u_k, D_{k-1} over the known finite-dimensional vector function $\phi(u_k, D_{k-1}) = \phi_k$. This occurs in the following case:

$$p(\theta | D_t) \propto p(\theta) \prod_{k=1}^{t} p(y_k | u_k, D_{k-1}, \theta) \tag{3}$$

Based on the supposition that the „a priori" (p.d.f.) is normal $N(\widehat{\theta}_0, P_0)$ then it can be easily deduced that the „a posteriori" (p.d.f.) p(θ/D$_t$) produced by Bayes rule (2) is normal $N(\widehat{\theta}_t, P_t)$ with recursively modified statistics:

$$\widehat{\theta}_k = \widehat{\theta}_{k-1} + \frac{P_{k-1}\Phi_k}{\sigma^2 + \Phi'_k P_{k-1}\Phi_k}\widehat{e}_k, P_k^{-1} = P_{k-1}^{-1} + \frac{\Phi_k \Phi'_k}{\sigma^2} \tag{4}$$

where $\widehat{e}_u = y_k - \widehat{\theta}'_{k-1}\Phi_k$ expresses the prediction error. Now it can be seen that recursive relations (4) are identical to the famous recursive least square algorithm.

3 Universal recursive forgetting algorithm

Most methods designed for time-varying cases can be expressed as modifications of least squares recursive (LSR), typically with a method implementation of forgetting old data, to prevent gain of the algorithm to converge to zero. Each method can be expressed as a special case of the following universal forgetting algorithm [6]:

Measured data contribution:

$$\widehat{\theta}_{k|k} = \widehat{\theta}_{k|k-1} + \frac{P_{k|k-1}\Phi_k}{\sigma^2 + \Phi'_k P_{k|k-1}\Phi_k}\widehat{e}_k, P_{k|k}^{-1} = P_{k|k-1}^{-1} + \frac{\Phi_k \Phi'_k}{\sigma^2} \tag{5}$$

Time contribution:

$$\widehat{\theta}_{k+1|k} = \widehat{\theta}_{k|k}, P_{k+1|k} = F\{P_{k|k}\}(\geq P_{k|k}) \tag{6}$$

The contribution from the measured data is not affected by the parameters changed by the time. The time representation (6) represents the change of the parameter vector between two measurements. Time contribution equations can be derived analytically, if there is an explicit model of parameter changes. Such a case is extremely unsuitable, so an heuristic procedure must be used. It is usually assumed that θ is approximately constant in the time horizon N >> 0, and then an „ad hoc" method of forgetting is added, through the implementation of the forgetting operator F{·}.

The well-known method of exponential forgetting leads to this following contribution to the change of the covariance matrix:

$$P_{k+1|k} = \left(P_{k|k-1} - \frac{P_{k|k-1}\Phi_k \Phi'_k P_{k|k-1}}{1 + \Phi'_k P_{k|k-1}\Phi_k}\right)\frac{1}{\lambda} \tag{7}$$

The EF method was very widespread, achieving good results in a number of cases. However, if the data are not sufficiently informative for a long period of time (the regression vector sequence insufficiently builds up some parts of the parameter space), the corresponding matrix P eigenvalues will grow above all boundaries [7]. One way to remove a blow-up problem is to keep $P_{k+1/k}$ eigenvalues on a pre-specified interval [λ_0, λ_1] see [8]. The following time contribution is used:

$$P_{k+1|k} = \left(1 - \frac{\lambda_0}{\lambda_1}\right) P_{k|k} + \lambda_0 I \tag{8}$$

where ($P_{1/0} = \lambda_1 I$) and I is a „ones" matrix of this corresponding dimensions, $\lambda(\geq 0)$ and $\lambda(>0)$. Modification of the covariance matrix (8) was introduced to negate the convergence of the eigenvalues $P_{k+1/k}$ to zero. Generally, the $P_{k+1/k}$ matrix should be selected as:

$$P_{k|k} \leq P_{k|k+1} < \infty I \tag{9}$$

In this case, it is limited from the bottom and from the top. Then the exponential convergence of identification algorithms is guaranteed [8]. This can be compared with the time contribution for (EF) [9], in which old information are suppressed evenly in the parameter space. Kulhavý and Kárny [6] have designed an algorithm that forbids only part of the parameter space affected by the last data contribution. In this case, the time contribution is given as follows:

$$P_{k+1|k}^{-1} = P_{k|k}^{-1} - k_k \Phi_k \Phi'_k, k_k = (1-\lambda)\left[1 + \frac{1}{\Phi'_k P_{k|k-1} \Phi_k}\right] \tag{10}$$

For $\Phi'_k P_{k,k-1} \Phi_k > 0$. In general, it can be stated that in the recursive calculation of the covariance matrix there must be avoided the tending to lose tracking on one hand, and on the other hand the uncontrollable increase of the $P_{k+1/k}$ eigenvalues.

4 REF and SLF techniques

Assuming there is no explicit model of the process parameter changes the „a prioral" information can be quantified by introducing an alternative p.d.f. $p^*(\theta_{k+1}/D_k)$. The problem is then the construction of the p.d.f. $p(\theta_{k+1}/D_k)$ based on two hypotheses, which are described by the „a posteriori" p.d.f. $p(\theta_k/D_k)$ (this is the case without any parameter changes) and the alternative p.d.f. $p^*(\theta_{k+1}/D_k)$ (this is the case where the parameters change most significantly). For simplification we use $p_0(\theta)$, $p_1(\theta)$ and $p^*(\theta)$ for the „a posteriori", alternative and resulting p.d.f. as follows:

Kulhavý and Kraus [6] formed the task of selecting p^*, assuming the recognition of p_0 and p_1 as a Bayes theorem decision-making problem. In the following we recall the results of their solution. After adding the probabilities λ and $1-\lambda$ to the hypothesis p_0 and p_1. Considering p_0, p_1 belonging to the class of normal divisions N that can be expressed:

$$p_{\hat{\theta},P} = \frac{1}{\sqrt{2\Pi}} |P|^{-\frac{1}{2}} \exp\left(-\frac{1}{2} \left(\theta - \hat{\theta}\right)' P^{-1} \left(\theta - \hat{\theta}\right) \right) \tag{11}$$

the parameters $\hat{\theta}$ and P determine the mean value and covariance of the p.d.f. The results are shown below (considering $\hat{\theta}_1 = \hat{\theta}_0$)

EF:

$$\hat{\theta}^* = \hat{\theta}_0, P^{*-1} = \lambda P_0^{-1} + (1-\lambda) P_1^{-1} \tag{12}$$

LF:

$$\hat{\theta}^* = \hat{\theta}_0, P^* = \lambda P_0 + (1-\lambda) P_1 \tag{13}$$

Consider a model of a system with time-variable parameters θ_k (3).

Adding a standard LSM algorithm (5) with exponential or linear looping (12), (13) allows the changes of parameters to be monitored. The alternative mean value is set equal to a-posteriori $\hat{\theta}^{alt}_{k+1|k} = \hat{\theta}_{k|k}$ and the alternative covariance matrix is set equal to a-priori covariance $P^{alt}_{k+1|k} = P_{1,0} = Q$.

Using this selection we are able to apply a general recursive forgetting algorithm with the following form of the forgetting operator:

$$F\{P_{k|k}, Q\} = \left[\lambda P_{k|k}^{-1} + (1-\lambda) Q^{-1}\right]^{-1} \tag{14}$$

which creates a harmonious mean value for REF and:

$$F\{P_{k|k}, Q\} = \lambda P_{k|k} + (1-\lambda) Q \tag{15}$$

creating an algorithmic mean value for SLF. In both cases, the a-priori covariance matrix Q is not forgotten and is repeatedly taken into account in each step of the recursive identification of k.

5 Notes on Numerical Processing

The data contribution in the generalized recursive forgetting algorithm is given (5). Using a well- known lemma on inversion of matrix, considering the covariance matrix in L-D factorizations the form may be produced:

$$L'_{k|k} D_{k|k} L_{k|k} = L'_{k|k-1} D_{k|k-1} L_{k|k-1} - \frac{L'_{k|k-1} D_{k|k-1} L_{k|k-1} \Phi_k \Phi'_k L'_{k|k-1} D_{k|k-1} L_{k|k-1}}{\sigma^2 + \Phi'_k L'_{k|k-1} D_{k|k-1} L_{k|k-1} \Phi_k} \tag{16}$$

Instead of adjusting the covariance matrix as a whole, its expression can be modified in the factorized form to ensure its positive definition. The second element of the forgetting algorithm is its time post. The application of the regulation of the covariance matrix by an alternative matrix causes an increase in calculation requirements. In both cases (12), (13) a weighted sum of two fully ranked matrices P_0^{-1}, P_1^{-1} or P_0, P_1. This problem was solved by using a dyadic reduction, which is the core tool used for solving this type of tasks.

The dyadic reduction can be used to solve the weighted summation of two full matrices in factorized form. The equation (13) represents the time contribution to the change of the covariance matrix:

$$P = \lambda \sum_{i=1}^{\dim_{L0}} L'_{0,i} D_{o,i} L_{o,i} + (1-\lambda) \sum_{i=1}^{\dim_{L01}} L'_{1,i} D_{1,i} L_{1,i}, \dim_{L_0} = \dim_{L_1} \quad (17)$$

$L_{0,i}$ ($L_{1,i}$) are line vectors applied as the i-th line of the lower triangle of the triangular matrix L_0(L_1). The important factor is that the diagonal elements of both triangular matrices equal one, considering the dyadic reduction algorithm assumes that the element $f_0 = 1$ is rightly placed in the reducing line f. A closer look at algorithm performing a weighted sum of two covariance matrices can be found in the factorized expression [6].

6 Augmenting REF with ACM

The involved REF augmentation considers addition and keeping the initial information in the Alternative Covariance Matrix (ACM) form as shown in [8]. The augmentation is based on the modified Dyadic reduction algorithm. The main difference is that instead of adding a-priori covariance matrix Q, ACM is computed at each step. ACM stabilizes the evolution of matrix P(0) after the recursive update. This operation is necessary for the REF algorithms to be augmented by the stabilization component in the ACM form. The aforementioned stabilization component prevents the destabilization of the original algorithms in long running applications when slow time changes are to be expected in the observed parameters in relation to the sampling period. The modified REF algorithm augmented with ACM is to be called REFACM.

A hybrid algorithm combines the advantages of both previously presented algorithms. The application of the REF algorithm is used in short time simulations, while the advantages of the REFACM algorithm is utilized in long time simulations, emulating the long-run operation of time variant dynamic systems. For this purpose, we created a new block module in Matlab Simulink. When the quality of REFACM is dominating over REF, as shown in Fig. 1., the block module switches from the REF algorithm to the REFACM. The REF algorithm runs at the beginning and after the result samples of PE and IS are better than REF in several successive steps in a row the algorithm is switched to REFACM. The resulting hybrid algorithm has been named HREFACM. The choice between the REFACM and REF algorithm is based on the mixing ratio of integral sum of prediction error (PE) and the integral sum of the Euclidian norm of parameters (outputs) error (IS).

7 Simulational algorithm verification methodology

Model number one and model number two have a different approach to input excitation (input signal generator A and B). The models were created for the purpose of verification of the properties of the algorithms. All the algorithms were subjects of the same test with identical length using the two featured models. The test compared the quality of the algorithms by the observation of time variant parameters of a dynamic DF.

All the results were evaluated and analyzed using a table where algorithm quality was shown numerically through parameters IS and PE.

Description of models no. 1 and no. 2
Model number one is a second order model considering external disturbance v(t):

$$y_k = \sum_{i=1}^{2} a_i y_{k-i} + \sum_{i=0}^{2} b_i u_{k-i} + \sum_{i=0}^{2} d_i v_{k-i} + e_k, e_k \approx N(0,\sigma^2) \tag{18}$$

The values of constant parameters equal: a2 = -0.9, b0 = 0.5, b1 = -0.25, b2 = 0.1, d1 = 0.8, d2 = 0.2, σ = 0.1.

In the first half of the simulation the time variant parameter was set to a(1) = 0.98. This value was constant in the first half of the n simulation steps. Afterwards, at the time t = n / 2 the parameter value was changed to a(1) = − 0.98. The outside disturbance was simulated as a square signal periodically changing its value from +1 to −1 every hundred simulation steps. The main difficulty of the identification was the rarely occurring disturbances, which contained minimal information about the parameter d(i).

For the needs of the simulation, two input signal generators were assumed:

- Input signal generator A: discrete white noise generator
- Input signal generator B: the input signal has been generated using the following equation: $u_k^* = 0.8u_{k-1}^* + 0.2u_k$, where u*k is normally distributed white noise and u*k-1 is the previous input value. For model no. 2. only one change has been realized in comparison to model no. 1. This was carried out by altering the time variant parameter a(1,k) = 0.98 cos(2π k/250). In this case, two different input generators were considered as well:
  - Input signal generator A: discrete white noise generator
  - Input signal generator B: the input signal has been generated similarly to model no 1., where u(k) has been only chosen from the interval u(k) ~ (0.5, 1).

All the algorithms were subject to the same test of the identical length, using the featured model. All the results were graphically evaluated and analyzed in a table in which the algorithm quality was shown numerically through IS and PE. The model for verification was chosen from [6].

8 Verification – MATLAB Simulink

A universal user interface was created along with a set of S-Function libraries used for the verification of the simulation. The interface allows the user to select input data, simulated model and the observed algorithm. Output of the discussed simulations is a graphic representation of the observed parameters along with a data file containing the results for the analysis. Integral sum (IS) of the Euclidian norm of parameter error and prediction error PE has been shown, which is the amount exceeded by the interval ±3σ2. The simulation experiments will be marked by the character pair XY, where X is the number of the utilized model (no. 1 or no. 2) and Y represents the generator utilized (A, respective B).

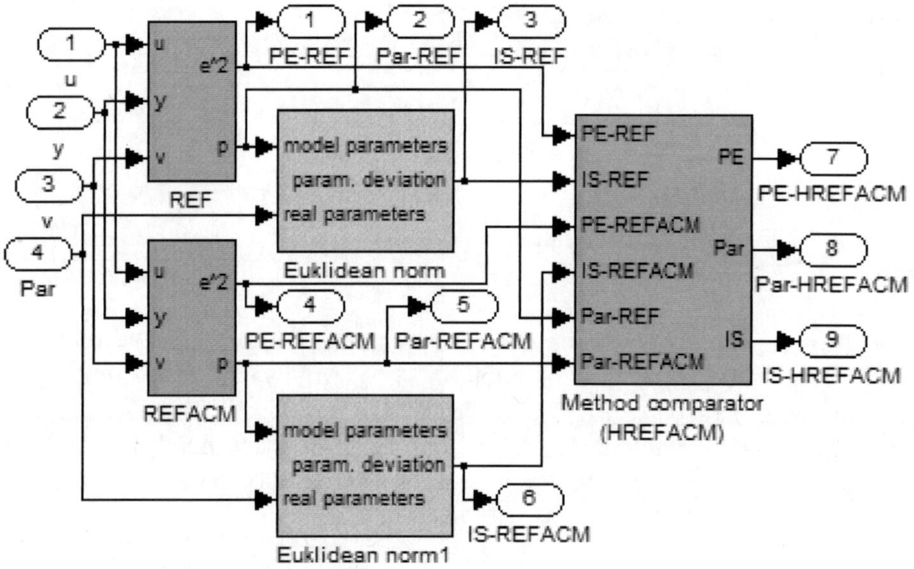

Fig. 1. HREFACM setup for benchmarks with algorithm in Matlab Simulink environment.

9 Evaluation of simulation results

The result section contains all the results in the form of a table. Tables (1) to (5) include detailed description and evaluation of the algorithm behavior during different simulations of duration n = 1 200, 6 000, 12 000, 60 000 and 120 000. The result confirms HREFACM algorithm quality in comparison to REF or REFACM. The results in Table (2); clearly state that at the simulation length n = 12 000 steps the properties of long lasting runs are already appearing., which in the case 1A achieved better results than REF. The data introduced in Table (5) confirm the HREFACM algorithm quality. The PF algorithm achieves excellent results also. It is clear, that using ACM as if a constraint has been enforced on parameter trending, which also implies the improvement of IS parameters in comparison to the results achieved by REF. The convergence of the REF covariance matrix is faster and finite in contrast to HREFACM, where the convergence is slower and the addition of excited ACM cannot be finite. The obtained simulation results and HREFACM algorithm behavior at 6 000 and 120 000 simulation steps show that as the running length increases the quality improves, in contrast to REF. The comparison is shown in Figure 1. The weighting factor λ used for the calculation of the quality of HREFACM algorithms equals λ=0.8.

1 200 steps		1A	1B	2A	2B
SF	IS	409,7	390,7	689,6	805,7
	PE	151	70	326	143
REF	IS	68,5	72,1	168,7	258,2
	PE	16	11	38	26
REFACM	IS	103,9	101,1	339,6	474,1
	PE	29	36	110	65
HREFACM	IS	68,5	72,1	168,7	258,2
	PE	16	11	38	26

Tab.1. IS and PE values of the observed algorithms during 1 200 simulation steps.

6 000 steps		1A	1B	2A	2B
SF	IS	356,6	957,3	3499,4	4016,0
	PE	130	131	1723	717
REF	IS	118,3	120,3	740,9	1336,7
	PE	18	12	174	103
REFACM	IS	127,4	177,8	963,8	1985,1
	PE	27	34	239	166
HREFACM	IS	127,1	176,4	961,7	1971,4
	PE	26	31	228	151

Tab.2. IS and PE values of the observed algorithms during 6 000 simulation steps.

12 000 steps		1A	1B	2A	2B
SF	IS	645,7	1604,6	6701,0	7474,5
	PE	253	214	3892	1488
REF	IS	168,9	253,9	1457,2	2525,3
	PE	11	12	303	124
REFACM	IS	162,9	296,3	1844,9	4507,3
	PE	19	24	523	317
HREFACM	IS	159,4	284,7	1784,9	4321,2
	PE	19	28	634	421

Tab.3. IS and PE values of the observed algorithms during 12 000 simulation steps.

60 000 steps		1A	1B	2A	2B
SF	IS	3734,8	8923,5	34241,6	33456,4
	PE	217	198	2002	745
REF	IS	512,5	2814,3	4403,2	2747,6
	PE	17	12	402	241
REFACM	IS	410,2	1241,2	3608,3	12607,4
	PE	21	31	654	439
HREFACM	IS	159,4	284,7	1364,7	4321,2
	PE	19	28	634	421

Tab.4. IS and PE values of the observed algorithms during 60 000 simulation steps.

120 000 steps		1A	1B	2A	2B
SF	IS	7234,8	17823,5	68601,8	76824,3
	PE	423	398	3892	1488
REF	IS	1035,4	3419,5	8976,2	33848,8
	PE	17	12	402	241
REFACM	IS	880,8	2914,3	1844,9	4507,3
	PE	51	29	654	439
HREFACM	IS	842,2	2914,3	1784,9	4321,2
	PE	46	26	634	421

Tab.5. IS and PE values of the observed algorithms during 120 000 simulation steps.

Conclusion

Based on our verified simulations using our universally developed test environment in MATLAB, we have come to the following conclusions. For short runtimes of the identification algorithm, classical methods like SF and REF are sufficient. The advantages of the tested HREFACM algorithm will only occur with long-term algorithms over 60,000 steps. Beyond that number of steps it is then advantageous to use it, because its hybrid properties are well adapted to the initial rise phase where it behaves as an REF algorithm. After that, in the long-term use it will again show its qualities when operating in the mode as the REFACM algorithm. Thus, the proposed hybrid algorithm combines the benefits of both algorithms and in the end is versatile and therefore suitable for industrial deployment.

Acknowledgment

The author(s) disclosed receipt of the following financial support for the research, authorship, and/or publication of this article: This work was supported by the Scientific Grant Agency VEGA of the Ministry of Education of Slovak Republic (grant number: 1/0317/17) and the Scientific Grant Agency KEGA (grant number: 027STU-4/2017).

References

[1] HULKÓ, G., BELAVÝ, C., ONDREJKOVIČ, K., BARTALSKÝ, L., BARTKO, M.: Control of technological and production processes as distributed parameter systems based on advanced numerical modeling. In Control Engineering Practice. Vol. 66, (2017), s. 23-38. ISSN 0967-0661.

[2] VASEK, V., KOLOMAZNIK, K., JANACOVA, D.: Tannery Wastes Processing Distributed Control System. In Proceedigs of 17th International Conference on CAD/CAM, Robotics and Factories of the Future. Durban : MRRG University of Natal, 2001, s. 894;900. ISBN 1-86840-432-3.

[3] VÍTEČKOVÁ, M., VÍTEČEK, A., BABIUCH, M.: Unified Approach to Analog and Digital Two-Degree-of-Freedom PI Controller Tuning for Integrating Plants with Time Delay. Acta Montanistica Slovaca, 2011, Vol. 1, pp. 89-94. ISNN 1335-1788

[4] PALENČÁR, R., SOPKULIAK, P., PALENČÁR, J., ĎURIŠ, S., SUROVIAK, E., HALAJ, M.: Application of Monte Carlo Method for evaluation of uncertainties of ITS-90 by Standard Platinum Resistance Thermometer. In Measurement Science Review. Vol. 17, no. 3 (2017), s. 108-116. ISSN 1335-8871.

[5] PETERKA, V.: Bayesian approach to system identification. Trends and Progress in System Identification., (P.Eykhoff, Ed.), Pergamon. Elmsford, Chap. 8:239-304, 1981.
[6] KULHAVÝ, R., KRAUS, F.J.: On Duality of regularized Exponencial and Linear Forgetting Automatica, vol. 32, 1996, No 10, pp. 1403-1415.
[7] PRALY, L.: Robustness of model reference adaptive control., 3th Yale WorkShop on application of adaptive system theory, 1993, Yale University, USA
[8] VACHÁLEK, J., BARTKO, M., BIZUB, J.: Long-run on-line identification with hybrid regularized exponential forgetting method. In Proceedings of the 15th International Carpathian Control Conference : ICCC 2014; Velké Karlovice, Czech Republic, May 28-30, 2014. [s.l.] : IEEE- Czechoslovakia Section of IEEE, 2014, CD-ROM, p. 635-640. ISBN 978-1-4799-3527-7.

Current address

Vachálek Ján, Doc. Ing., PhD.
Institute of Automation, Measurement and Applied Informatics,
Faculty of Mechanical Engineering,
Slovak University of Technology in Bratislava
Nám. slobody 17, 812 31 Bratislava, Slovak Republic
E-mail: jan.vachalek@stuba.sk

Melicher Markus, Ing.
Institute of Automation, Measurement and Applied Informatics,
Faculty of Mechanical Engineering,
Slovak University of Technology in Bratislava
Nám. slobody 17, 812 31 Bratislava, Slovak Republic
E-mail: markus.melicher@stuba.sk

Vašek Pavol, Ing.
Institute of Automation, Measurement and Applied Informatics,
Faculty of Mechanical Engineering,
Slovak University of Technology in Bratislava
Nám. slobody 17, 812 31 Bratislava, Slovak Republic
E-mail: pavol.vasek@stuba.sk

Šišmišová Dana, Ing., PhD.
Institute of Automation, Measurement and Applied Informatics,
Faculty of Mechanical Engineering,
Slovak University of Technology in Bratislava
Nám. slobody 17, 812 31 Bratislava, Slovak Republic
E-mail: dana.sismisova@stuba.sk

Volenský Tomáš, Ing., PhD.
Institute of Automation, Measurement and Applied Informatics,
Faculty of Mechanical Engineering,
Slovak University of Technology in Bratislava
Nám. slobody 17, 812 31 Bratislava, Slovak Republic
E-mail: tomas.volensky@stuba.sk

Slovak University of Technology in Bratislava
Faculty of Mechanical Engineering
17th Conference on Applied Mathematics **APLIMAT 2018**

Proceedings

PROTOTYPES IN PLANIMETRICS

VÁCLAVÍKOVÁ Zuzana (CZ)

Abstract. The paper is focused on the detection of a tendency of primary and secondary schools´ students to have a prototypical thought in planimetrics and consequences for mathematical thinking in geometric subjects of university students. The summary of the results which have been reached by questionnaires given to the selected schools are described (if the way used by students to sketch the specified units is similar to what they may have seen in the textbooks, eventually to results of their teachers, or if it depends on, for example, left-right orientation, sex, age or type of studied school). In addition, activities that are being prepared in this context for the students of our university study programme will be described.

Keywords: prototype, geometry, planimetrics

Mathematics Subject Classification: Primary 97-02, Secondary 97G60

1 Motivation

Let us think about this mathematical task: sketch a 5-gon ABCDE and a circumscribed circle (circumcircle) to it. Sketch a tangent line to the circle through point A.
If we do not think deeper about the task, we will automatically draw the sketch as Fig. 1 shows.

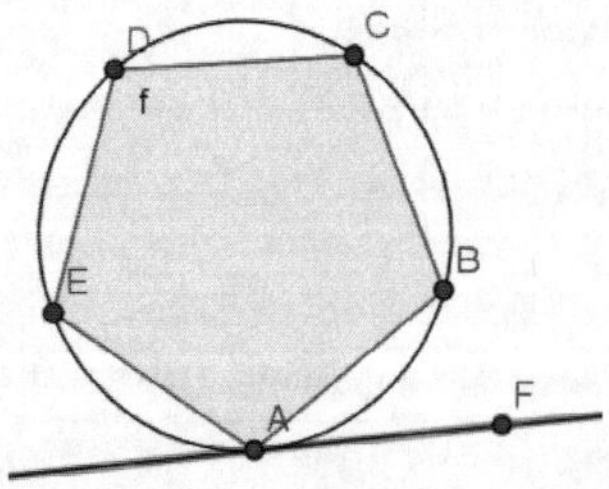

Fig. 1. 5-gon ABCDE and its circumcircle.

What is wrong? The solution is right but we have forgotten to think about the general situation. The solution should be: there is a case when it is possible to do it (the problem has a solution) but also a case when it is not possible (the problem has no solution). Because if we have a general 5-gon like Fig. 2 shows, there does not exist the circumcircle.

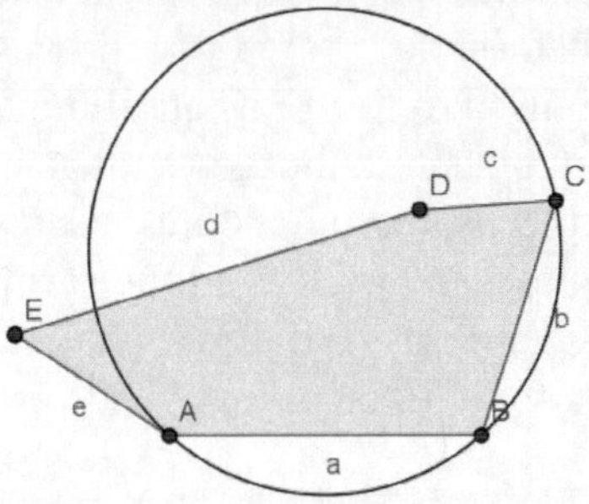

Fig. 2. General 5-gon ABCDE.

Why does it happen? The reason why we think this way is prototyping the objects. We do not use the general case for a 5-gon so often, more often we use a regular 5-gon that has a circumcircle as Fig. 1 shows. So when we hear a "5-gon", we imagine a regular one – as our "prototype" of a 5-gon.

In mathematics, like in other sciences, too strong tendency for prototyping or a bad choice of prototypes does not allow a general solution of tasks or it can significantly affect them [1], [2].

2 Prototype and theory of prototypes

A prototype, in general, is an early sample, model, or release of a product built to test a concept or process or to act as a thing to be replicated or learned from. It is an initial model of an object built to test a design. The word comes from a Greek word for "primitive form."

In context of learning, especially new terms or concepts, the prototype is the initial idea of a new, described object. If, after introducing a new concept, we begin to examine its properties and behaviour, we often follow this prototype and have a tendency to describe its properties and its behaviour.

The prototype theory is based on the notion that a given category is not defined by individual rules, but that there is some model or ideal example which characterises the given category – so-called prototype. However, a prototype may not necessarily be a real example which could be seen in real life. It can also be a mix of several other examples and such a model may appear more typical for the category than in any real creation. On the basis of similarity with the prototype, the individual decides whether or not the given object belongs to the category. Of course, the more typical the object is, the faster we can categorise it.

The prototype theory explains part of the process of knowing and classifying new concepts or objects. In our research we focused on a small area of geometry in which we tried to find out if the pupils really have prototypical ideas about some elementary geometric objects [3].

3 Research

We studied the prototyping of elementary terms in geometry between students of primary and secondary schools and potential factors of building their ideas about geometric objects - if the way used by students to sketch the specified objects is similar to what they may have seen in the textbooks, or to results of their teachers, or if it depends on, for example, left-right orientation, sex, age or type of studied school.
Testing of prototypes took place in the form of questionnaires. The basic principle in detecting prototypes is that the respondent cannot know that he/she is tested on prototypes, and in work with a given object, the prototype should be presented automatically, without deeper thinking. This is why the questionnaire was in the form of mathematical tasks where specific geometric objects were seemingly only marginal, and the text of the example concerned work or some calculation for a given object.

3.1 Preparation and creation of a questionnaire

Before the test itself, there was a discussion about the factors that could affect prototype building. The pupil's age and sex were selected, left-right orientation, assessment in mathematics, and an option for the students to inform us about a diagnosis they have (dysgraphia, dyslexia, etc.).
Together with each class, a similar questionnaire was filled in by a teacher, who also did not know that the main goal of questionnaire was testing the prototypes. It was assumed that the tendency to follow a particular prototype would be affected by the intensity of the prototype in the textbooks and by the prototype of the given object that the teacher has. The teachers also filled in the name of the textbooks which they work with in mathematical lessons.
Prior to creating the questionnaire, we also discussed whether we would require exact geometric constructions or sketches. The reason for choosing the sketches was that pupils are not influenced by laying of a ruler, so the sketches will show more about the prototype itself.
There were 8 tasks in the questionnaire from planimetrics. The tested objects were triangle, square, rectangle, right triangle, trapezoid, 5-gon, rhombus, and rhomboid.

3.2 Selected questions

As we have already mentioned, the questions were in the form of mathematical tasks so that the sketch of the prototyped object was automatic. Here are some of the questions from planimetrics:

- Sketch a triangle ABC and a circumcircle k. Sketch a tangent line to the circle through point A.
- Sketch a rhombus and write the relationship for calculating the perimeter and the area. Highlight everything you use to calculate it in your sketch.
- Sketch a right triangle, denote the sides and write the Pythagorean Theorem for it.

Every question had detailed criteria of evaluation for the prototype category of the sketch – for example the question about the triangle had four levels with respect to:

- type of the triangle (regular/equilateral triangle, isosceles triangle, acute triangle, obtuse triangle, right triangle, general, …),
- sides (lies on the side, stands at the vertice, ...),

- rotation (rotated left, right, no rotation),
- denoting the vertices (clockwise, counter-clockwise, no denoting).

Similarly, all the questions had different possible results and all questionnaires were processed manually because of the need for individual assessment of the sketches.

3.3 The results

3.3.1 Respondents

There were over 550 respondents from primary and secondary school aged 11-19 years, gender balanced, approximately 90% of students were right-oriented and 10% were left-oriented students, 2% of them wrote a diagnosis (ADHD, dysgraphia and dysortography, dyscalculia, dyslexia). The students were from primary school (144 respondents), multi-year gymnasium (104 respondents), four-year gymnasium (193 respondents) and secondary vocational school (121 respondents).

3.3.2 Selected results

Question 1: Sketch a triangle ABC and a circumcircle k. Sketch a tangent line to the circle through point A.

The question was focused on a general triangle. A triangle, as geometrical object that the pupil encounters since the youngest age, appears in the textbooks in many ways – as equilateral, isosceles, right one, convex and nonconvex, and variously rotated. We expected that pupils would have a tendency to sketch it in a general form. But the results were different. Approximately 85.71% of the respondents sketched an isosceles triangle (62.50% of the respondents sketched a regular/equilateral triangle, which is a special case of an isosceles triangle, and 23.21% of the respondents sketched an isosceles triangle that is not regular). Over 90% of the respondents sketched a triangle with the bottom side horizontally and over 86% of the respondents denoted the triangle counter-clockwise.

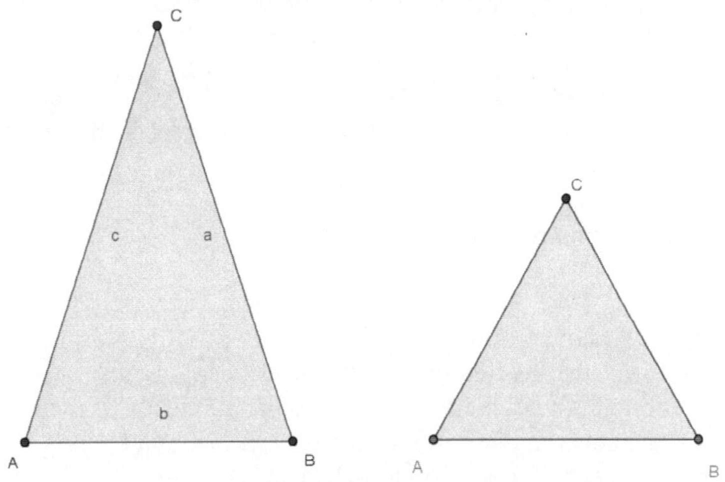

Fig. 3. Prototypes of a general triangle.

The main prototype of a general triangle is an isosceles triangle, lying on side AB parallel to the bottom edge of the paper, denoted counter-clockwise (Fig. 3 left). This case occurred in 80.54% of the questionnaires. What is more, in 57.14% of the questionnaires was sketched a regular triangle, lying on side AB parallel to the bottom edge of the paper, denoted counter-clockwise (Fig. 3 right).

Question 2: Sketch a 5-gon ABCDE and its circumcircle. Connect each two vertices to each other. How do we call the parts that have been created on the outside of the pentagon and are bounded by one side of the pentagon and a part of the circle?

A 5-gon is mentioned in the textbooks only marginally, most often in the form of a regular 5-gon. In this case, we supposed that almost all respondents would sketch a regular 5-gon, but it was sketched only in 49.84%. 18.61% of students drew the 5-gon like a "house" (see Fig. 4 right) and 13.41% drew a "diamond" (see Fig. 4 left). Three students drew a concave one. Of course, there does not exist a circumcircle, so they drew some circle through three or four points only or drew some closed curve that traversed all the vertices, but it was not a circle.

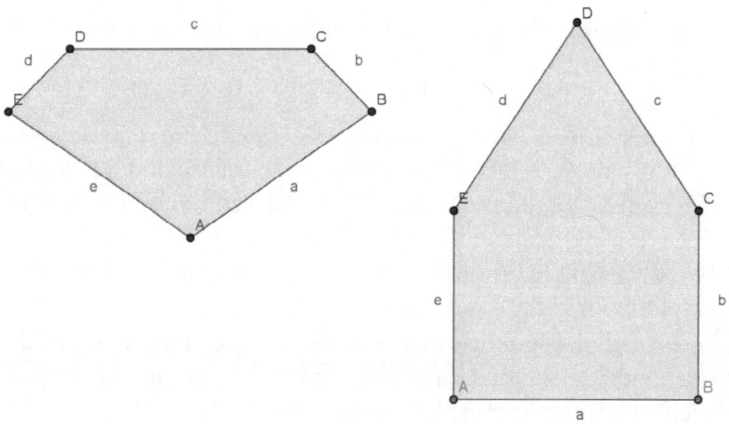

Fig. 4. Sketches of a general 5-gon.

Concerning squares and rectangles, prototypes were mostly sketched as an object with a horizontal bottom side (denoted as side AB) and counter-clockwise denoted. Much more general prototypes were in the case of not so often used objects like a trapezoid, 5-gon, rhombus, and rhomboid.

4 Conclusion of research and university students

Generally, we can state that the influence of gender, age, year, or school on the building prototypes in geometry was not confirmed.
It did not even confirm the assumption that the students with better assessment in mathematics would tend to draw geometrical objects more atypically or more generally.
There is no dependence comparing the results from the teachers' and pupils' questionnaires in the prototypes of basic objects like a triangle or square, and no dependence comparing the students' sketches with the textbooks [4].

Finally, we can proclaim that the more we know and use some kind of an object, the more generally we can think about it.
This finding should mean that university students, for whom the elementary geometric objects are well known, will have well-built prototypes and will be able to solve the tasks at a general level. But the reality is different.

5 Consequences for mathematical thinking of university students

We build up the prototypes very early, during the primary and secondary education, and that is the reason why university students have a tendency for similar prototypes in planimetrics.
At our university, we have two main study programmes – Applied Mathematics and Teacher of Mathematics. Both of them have at least three semesters of geometry and for all students of mathematics it is necessary to still keep in mind the general view of objects, because as we mentioned, the tendency for prototyping strongly affects our thinking. Especially a future teacher has to know good didactic tools how to teach this area so that his/her students will form the right prototypes. This skills and abilities need to be developed.
Textbooks for mathematical subjects for university assume that students already have a sufficient overview in the described area. But students from secondary schools are prepared to solve particular geometrical tasks that are often focused on a special case. The students are usually not prepared for discussion about generalisation of a problem with respect to other cases. Sometimes, it is simply enough to reformulate the task in such a way that will encourage discussion of various possible solutions with respect tothe object.

5.1 Mathematical tasks reflecting the prototyping

The easiest way to reformulate the tasks is using the parameters. Sometimes, it is more useful to reformulate the text in the task. Currently, we are working on finding examples of geometry tasks that support right forming of prototypes, or on converting some of the tasks to other forms to develop right prototyping. Especially in the subject Geometry Constructions, we are looking for this kind of tasks.
For example, a classical task on the construction of a circumcircle to a convex triangle can be rewritten to the following:
Draw any circle and its chord. Choose any other point on the circle and discuss the triangle made by these three points with regard to the convexity.
In this case, the student has to note that if the third point is chosen in a way that the centre of the circumcircle is outside the triangle, the triangle will be concave. If the third point is chosen in a way that the centre of the circumcircle is inside the triangle, the triangle will be convex, and if it is chosen in a way that the centre lies on the side of triangle, the triangle will be right triangle (and the circumcircle is a so-called Thales circle).
In parallel, with the students of teacher's programme in didactics subjects, we are working on collecting geometry tasks and examples for students of secondary school focused on developing the right prototypes. Future teachers discuss different examples from this point of view and learn the way how to generalise them by parameters or how to reformulate them so that the example are still understandable for secondary students but encourage discussion of various possible solutions.
At the end of our work, we hope that we will prepare a textbook with geometric examples that will reflect the prototype theory.

Acknowledgement

The paper summarizes the part of results of research that were made together with my diploma student Kristýna Šafářová and I would like to thank her for possibility to publish it.

References

[1] FERDIÁNOVÁ, V. a Poruba, J., Special 3D models for monge projection, APLIMAT 2016, 15th Conference on Applied Mathematics 2016, Proceedings, Bratislava, 2016, s. 349-361, ISBN 978-80-227-4531-4.

[2] FERDIÁNOVÁ, V., Using anaglyphs in descriptive geometry, Proceedings of the 15th European Conference on e-Learning ECEL, Praha, 2016, s. 194-200, ISBN 978-1-911218-17-3.

[3] STERNBERG, R. J., Kognitivní psychologie, Vyd. 2, Praha: Portál, 2009, ISBN 978-80-7367-638-4.

[4] ŠAFÁŘOVÁ, K., Prototypy v geometrii, diplomová práce, Ostravská univerzita, 2017.

Current address

Václavíková Zuzana, RNDr., Ph.D.
Department of Mathematics, Faculty of Science, University of Ostrava
30. dubna 22, 702 00 Ostrava, Czech Republic
E-mail: zuzana.vaclavikova@osu.cz

THE ROLE OF VISUALISATION IN SOLID GEOMETRY PROBLEM SOLVING

VÁGOVÁ Renáta (SK), KMEŤOVÁ Mária (SK)

Abstract. Visualising in mathematics is usually connected with drawing pictures or diagrams because these help learners to find a solution to the problem. The purpose of this paper is to provide qualitative research experiment on visualisation of hidden spatial objects during solid geometry problem solving. Subsequently, we offer a way how the teacher can easily demonstrate the problem via GeoGebra software and thus support the learner in problematic steps of problem solving. The usage of an interactive geometry software provides a great opportunity to demonstrate the correct image that leads to the solution.

Keywords: visualisation, mental image, mathematics problem solving, solid geometry, GeoGebra

Mathematics Subject Classification: Primary 97G40, Secondary 97D50.

1 Introduction

Students are very often afraid of solving mathematical problems. It can put a strain on some students and they can feel afraid of failing. This situation can result in loss of interest during the math lessons. However, on the other hand many teachers agree, that children like challenges. If the task is too easy, it is not challenging, so it is not interesting for them. Visualisation techniques help to solve this dilemma providing enough various tools to support mental imagery leading to the correct solution and make the task more attractive.

There are lots of research about visualisation playing an important role in the learning of mathematics. Nowadays, we can read several ideas about creating mental images and visualisation contributed by various researchers [1, 2, 3, 4, 5, 6]. However, in the early 20th century, the behaviorist revolution in psychological research led psychologists to deny the existence of mental images. From 1980s, constructivism strongly influenced the research in mathematics education. The importance of using mental imagery and visual processing in mathematics was increasingly recognised. The debate about real existence of an imprint of mental imagery in human brain took on a new form in the early 1990s, when neuroimaging became available. Recent empirical evidence [7] strongly supports the claim that humans can represent information in multiple ways, and that such representations can be used flexibly in working memory or during mental imagery.

2 Visual mental processing

The term visualisation has been used in various ways in the research literature. In [9], Presmeg gives an overview how it was used for approximately two decades. Thus visualisation is taken to include *"processes of constructing and transforming both visual mental imagery and all of the inscriptions of a spatial nature that may be implicated in doing mathematics"*. This characterisation is broad enough to include two aspects of spatial thinking, namely interpreting figural information and visual processing.

Presmeg [9] identified the next five types of imagery:

Concrete imagery - picture in the mind;
Kinaesthetic imagery – physical movement;
Dynamic imagery – the image itself is moved or transformed;
Memory images of formulae;
Pattern imagery - pure relationships stripped of concrete details.

If one visualises a problem it means understanding of the problem in terms of a visual mental representation. In accordance with this statement, the visualisation process includes visual imagery as an inseparable part of the method of solution. In [10], Kosslyn describes visualisation as a cognitive process involving visual imagery in which images are either generated, inspected, transformed or used for mathematical understanding. Thus he introduces four subsequent categories of this mental process [10]:

1. Image generation,
2. Image inspection,
3. Image transformation,
4. Image use.

In general, visualisation is a mental process which results in an identified product: visual image. These mental processes and images cannot be seen, because they are in the mind of learners, but they have to be understood through the study of the mind [2]. *"Visualisation offers a method of seeing the unseen and we are encouraged and should aspire to "see" not only what comes "within sight", but also what we are unable to see."* [3].

The findings cited above have very deep implications for teaching mathematics with understanding and reasoning. Visualisation helps teachers to make instructional decisions about how to teach, about the content and the nature of tasks. Furthermore, it aids teachers with the facilitation of lessons and with the ability to engage learners in realistic situations [2].

2.1 Learners and visualisation

Educators (teachers) teach their learners (students) the content of a subject, but the question is: "What is important and more effective for them?" They should teach them to think effectively about the subject and lead them to a deeper understanding of the base of mathematical problem, the base of mathematics. Understanding how learners think during problem solving is complex and it requires special training. Makina [2] argues that all mathematics teaching should be focused on "teaching for understanding", where the visualisation is the most important basis.

In the education of mathematics, the word problem solving has its significant place and it plays a prominent role in the present. Students often have a problem with understanding the text [1] and they can not identify the relation between the known and the unknown entities.

2.2 Purposes of visualisation

Visualisation plays an important role in problem solving. During this process, we create an image of the situation or we visualise a model which helps to represent the situation mathematically. Visualising in mathematics is regularly connected with drawing pictures or diagrams. These pictures help problem solvers find a solution to the problem. However, visualisation includes more functions, such as supporting the development of ideas, facilitating communication of results and understanding. In other words, it is not only about diagrams and pictures.

A particular emphasis is sometimes placed on using visualisation during problem solving. The problem solver identifies the key components and the relationships between them.

Piggott and Woodham [4] established these three purposes for visualising:

1. **Visualising to step into the problem** – is related to the idea of "getting started". The principal role of this step is to help problem solvers with understanding what the problem is about. The visualisation enables leaners to go deeper into the situation and it also supports their understanding.
2. **Visualising to model a situation** – is especially useful to solve physically unavailable situations. Piggott mentions for example the inside of a 3D object or considering a case involving a very large number.
3. **Visualising to plan ahead** – together with the second purpose go beyond the first purpose and it suggests that visualising occurs not only at the beginning. It occurs in the middle of problem solving, too. In this case asked questions are: "What will be the consequence if we do this?", and it is connected with "What would happen if ...?". This purpose introduces using visualising during the problem solving process to anticipate a solution of a problem.

3 Visualisation of solid geometry problems

Study of geometry supports not only the logical verification, but also the spatial orientation of learners, their spatial abilities and visual perception. However, at all levels of education, teaching of solid geometry remains almost unnoticed. It often happens that the geometric knowledge of the three-dimensional space that we observe at schools is limited to the knowledge of some basic solids (cube, sphere, polyhedron and cylinder) to learn some basic formulae for the calculation of areas and volumes [5].

Andrade and Montecino [6] analysed difficulties in understanding "the really spatial" relationships when teachers use iconic planar representations instead of real models during the lesson. This way of geometry teaching often deforms the building up a correct mental picture. From studying the history of teaching, we recognise using manipulative tangible objects. Wooden or metal structures of solids have been used for better understanding of solid geometry for a long time. Nowadays, with the rapid development of digital technologies, the integration of "3D virtual objects" into education has begun to overcome several obstacles in teaching solid geometry [5, 6]. For example, dynamic program GeoGebra supports creating anaglyphs [8] from given 3D pictures, so it allows us to see (using red-green glasses) spatial objects rotating really in space in front of a computer screen. Program GeoGebra turns out to be an excellent tool for experimentation and investigation in geometry. Work with GeoGebra in education allows us direct manipulation with virtual objects and to demonstrate different geometrical relationships immediately on the computer screen. We utilised this feature of the program also in our research experiment.

4 Qualitative research experiment on visualisation of hidden spatial objects

In our experiment we focus on visualisation of a hidden spatial object (a cube appearing in a geometric-algebraic problem) by an average learner aged 17 in the 3th grade at a grammar school. The idea of the given problem was taken from [2], but we completed it to a general geometric-algebraic problem. The goal of the experiment is to detect the learner`s strategy of visualisation during the problem solving. We aspire to follow detailed steps of mental process according to Kosslyn, mentioned above, and identify the types of imagery used from Presmeg`s classification (see above). At the very beginning of the creation of the mental image, we tried to catch, according to the learner's description, what he was thinking. The next steps of the mental image creation are depicted on the learner‘s paper notes. According to the notes we can determine the purposes of visualisation (see Piggott's classification above).

4.1 The problem of cube layering

We designed the following problem with tasks and questions divided into four levels: *Imagine that you have an unlimited number of small cubes (all the same size* $1\times1\times1$*) in different colours. Then, imagine you will build bigger cubes from these small cubes by adding layers (like in an onion or Russian matryoshka dolls) such that each layer has a different colour. Next, imagine the layers of small cubes and try to answer the following questions:*

1. *Imagine a cube the size of* $3\times3\times3$ *(each layer has a different colour).*
   a) *How many layers does the cube have?*
   b) *Sketch an outline how the small cubes of the outer layer touch the faces of the small cubes of the previous layers face. How many cubes of the outer layer have a face touching the inner layer faces?*
   c) *Sketch an outline how the small cubes of the outer layer touch the small cubes of the previous layer along the edges without touching any of the faces of the inner cubes. How many cubes of the outer layer touch the inner cubes along the edges?*
   d) *Sketch an outline how the small cubes of the outer layer touch the small cubes of the previous layer at the vertices without touching any of the edges of the inner cubes. How many cubes of the outer layer touch the inner cubes at the vertices?*
   e) *How many small cubes are there in total?*
   f) *Can you explain why?*
2. *Imagine a cube that has one more layer than the previous one (each layer has a different colour).*
   a) *How many small cubes did you add to the previous cube?*
   b) *Sketch an outline how the small cubes of the outer layer touch the faces of the small cubes of the previous layers face. How many cubes of the outer layer have a face touching the inner layer faces?*
   c) *Sketch an outline how the small cubes of the outer layer touch the small cubes of the previous layer along the edges without touching any of the faces of the inner cubes. How many cubes of the outer layer touch the inner cubes along the edges?*
   d) *Sketch an outline how the small cubes of the outer layer touch the small cubes of the previous layer at the vertices without touching any of the edges of the inner cubes. How many cubes of the outer layer touch the inner cubes at the vertices?*
   e) *How many small cubes are there in total?*
3. *Imagine a cube created from the cube of* $3\times3\times3$ *by adding next 5 layers. How many small cubes does the larger cube contain?*

4. *Imagine a cube created from the cube of* 3×3×3 *by adding next n layers. How many small cubes does the larger cube contain?*

The general solution of this problem is as follows:
The cube composed of k layers contains $(2k-1)^3$ small cubes.
(Remark: Adding one layer, we obtain a geometric proof of arithmetic recursive formula for odd numbers: $(2k-1)^3 + 6(2k-1)^2 + 12(2k-1) + 8 = (2k+1)^3$.)
Since the larger cube created from the cube 3×3×3 by adding next n layers has $n+2$ layers, it contains $(2(n+2) - 1)^3 = (2n+3)^3$ small cubes.

4.2 The learner's solution

At the beginning of solving the problem there weren't any models provided that could help a learner solve the problem. He had to "work" only with his imagination. The first substantial problem in the process of solution occurred immediately at the beginning of the problem solving.

4.2.1 Illustration of learner's solution of the task (a) at Level (1)

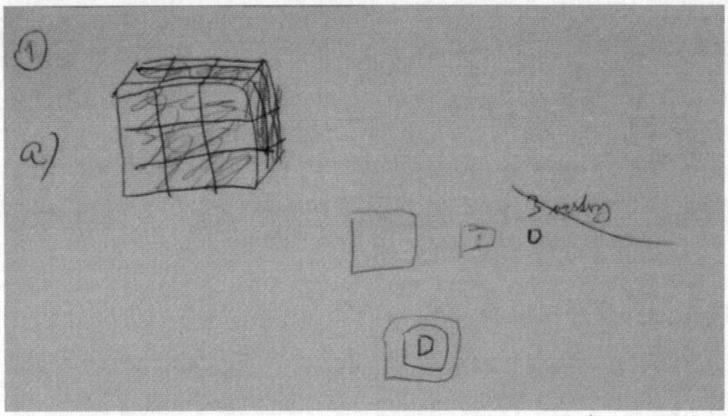

Fig. 1. Learner's solution.

The learner answered to the task (a) at Level (1) automatically, without hesitation that "The cube has 3 layers". When we asked him, if he is sure of his answer, he said, "Yes. Of course. Why not?"

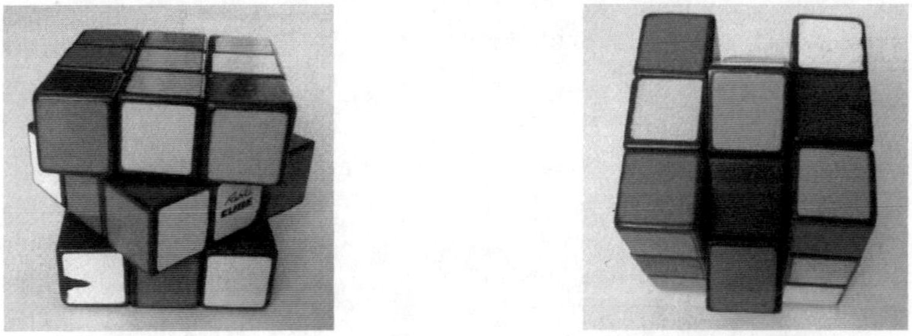

(a) (b)

Fig. 2. Learner's manipulation with the Rubik's cube.

This level of the problem is the most important because it reflects the solution of the rest of the problem. After the incorrect answer, we offered him a Rubik's cube as a model of a 3×3×3 cube.

When we gave the learner the Rubik's cube, he did not change his answer. However, the attitude of the learner changed when he started to manipulate with the floors of the Rubik's cube (see Fig. 2). He could see the problem from a different perspective. He found that the middle floor "is rotating only around one small cube". In fact, the middle floor of the real Rubik's cube does not have any small cube in the centre; nevertheless, the manipulation with the Rubik's cube helped the learner realise there could not be more layers inside.
Figure 1 represents the way the learner depicted his solution in the paper. It reflects his visual representation. The learner had problems to outline 3D objects, he prefered to sketch two-dimensional compensations, squares instead of cubes. This is the same problem as the one described by Andrade and Montecino in [6] - the learners can outline only the basic solids. When learners have to sketch a solid from different point of view, usually they do not know how to produce a correct image in their mind and how to engage their imagination into visualisation.

According to the Kosslyn's classification this level is integrated into the mental processes Image generation and Image inspection. The learner had to create the correct mental image for correct solving of the task (a) at Level (1) as well as all the other levels. In accordance with Presmeg's types of imagery, the learner creates Concrete imagery (picture in the mind) and applies Kinaesthetic imagery (physical movement) to solve this level. Finally, this problem is integrated into category Visualising to model a situation on the basis of the purposes of visualisation.

4.2.2 Illustration of learner's solution of the tasks (a), (b), (c) at Level (2)

The second problem appeared in the process of solution when the learner had to change their perspective to solve the problem. In the previous case, the learner had to look inside the cube but now he had to add the next layers to the basic cube (the size of 3×3×3). In Figure 3, the learner's solution at Level (2) is shown.

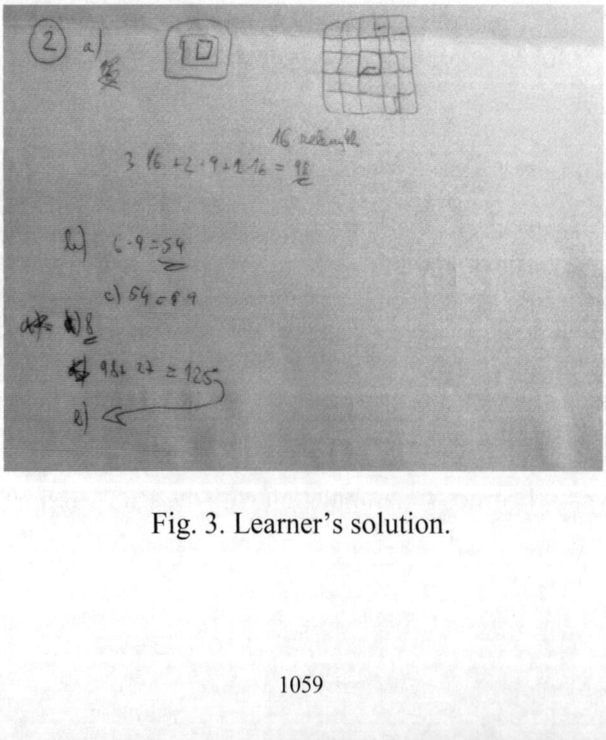

Fig. 3. Learner's solution.

In the learner's solution the basic arithmetic operations are dominant. The learner did not outline the 3D image of the cubes, he preferred to sketch only two-dimensional images. The learner sketched his mental image only of the task (a) at Level (2) and the rest of tasks was ignored. The learner's argument for it was the following, "It is too difficult for me. I do not know how to draw connected 3D objects. We did not learn it at school."
After that the learner grabbed the Rubik's cube automatically and started to manipulate with it. He tried to describe verbally the process of adding layers to the basic cube and he said, "The results are the same as in the first level!" The learner's incorrect answer reflects that he did not change his perspective. In Level (1) he manipulated with "the outer layer" and he had to imagine the inside of the cube. In Level (2) the Rubik's cube represented "the previous layer" and the learner had to add one more layer of small cubes a different colour. The most substantial problem occurred in tasks (b) and (c) at Level (2). The learner did not become aware of the number of the added cubes. There was not only one cube per face, but, actually, there were 9 cubes on every face. The same problem occurred in the task (c) at Level (2) because there were 3 cubes on every edge.

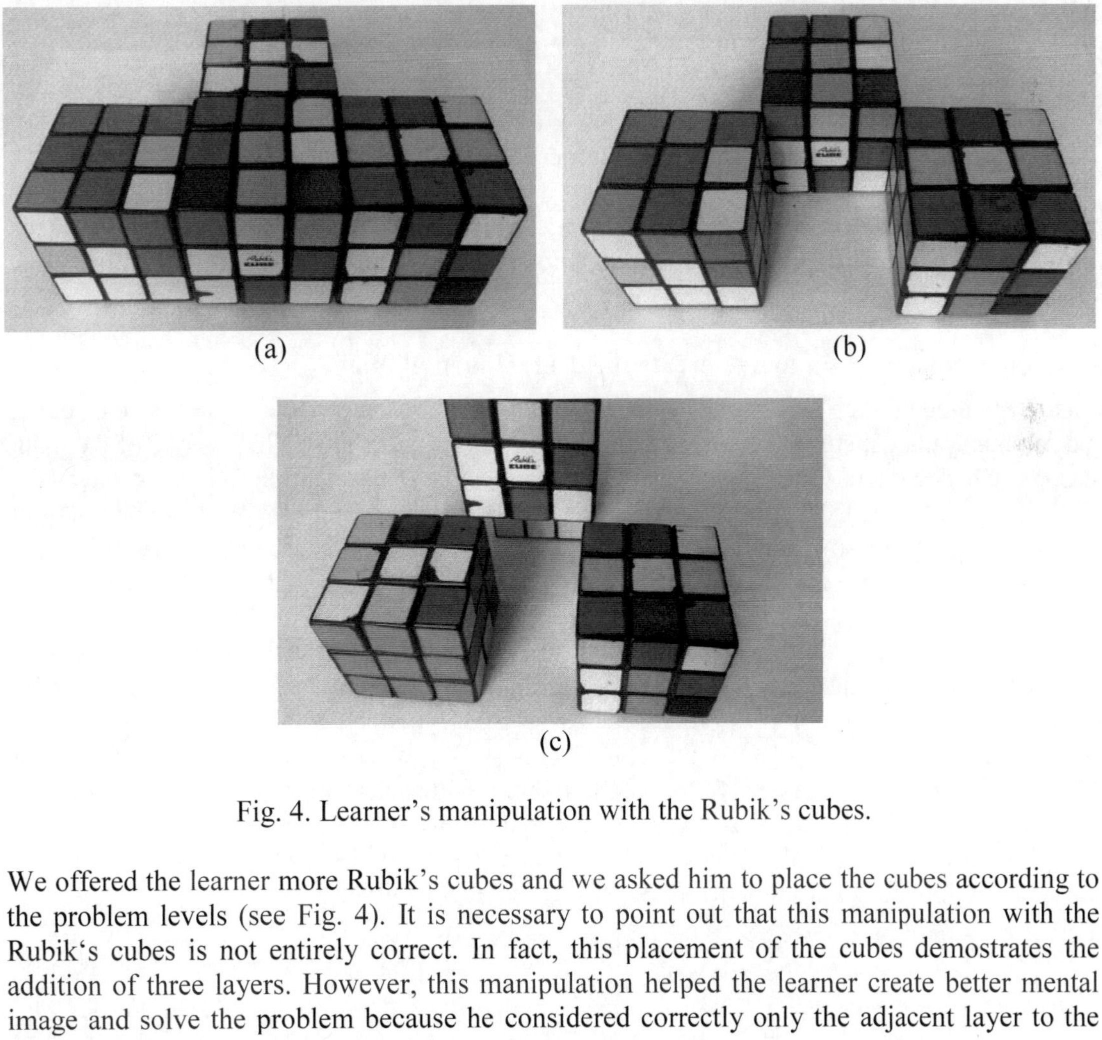

(a) (b)

(c)

Fig. 4. Learner's manipulation with the Rubik's cubes.

We offered the learner more Rubik's cubes and we asked him to place the cubes according to the problem levels (see Fig. 4). It is necessary to point out that this manipulation with the Rubik's cubes is not entirely correct. In fact, this placement of the cubes demostrates the addition of three layers. However, this manipulation helped the learner create better mental image and solve the problem because he considered correctly only the adjacent layer to the previous cube from the whole model.

According to the Kosslyn's classification these levels are integrated into the following mental processes: Image inspection, Image transformation and Image use. In accordance with Presmeg's types of imagery the learner creates Kinaesthetic imagery (physical movement) and Dynamic imagery (the image itself is moved or transformed) to solve these levels of the problem.

4.2.3 Illustration of learner's solution at Level (4)

The learner's solution of the task at Level (4) is not correct (see Fig. 5). The formula represents the number of small cubes of the cube the size of $(n+2)\times(n+2)\times(n+2)$. However, n has to represent the number of the added layers, not the number of the small cubes.

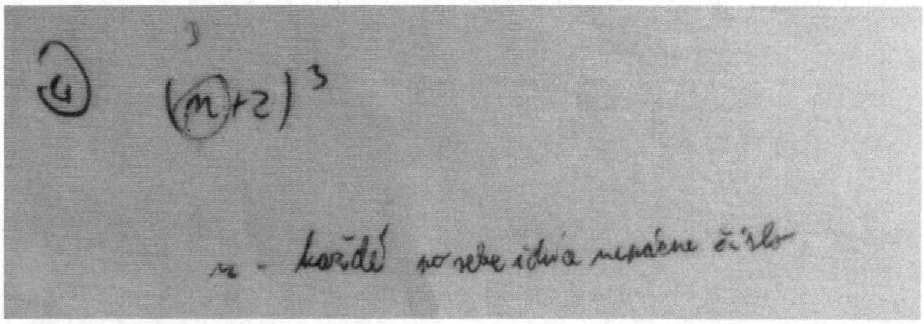

Fig. 5. Learner's solution.

In accordance with Presmeg's types of imagery, the learner creates Memory images of formulae and Pattern imagery (pure relationships stripped of concrete details) to solve these levels.

5 Using GeoGebra as a tool to create the correct mental image

Classic 3D models such as a cube, a sphere, etc. do not allow us to look at the centre of them. We can manipulate them, but we cannot break them into smaller parts. This is one of the main reasons why learners have incorrect mental images. When learners do not have the opportunity to "look to the base" of the objects, they will not know how to create a right mental image. This situation is difficult for learners, because they find themselves in a position where they are expected to imagine something that is "unimaginable" for them.

Modern information and communication technologies are powerful in representing information in different ways. This can be through text, pictures, tables and graphs or by helping visualisation of complex processes in sciences. Dynamic geometry programs allow us to manipulate objects, to change the colour or size, to rotate and to see the object from different viewpoints, etc. To solve the above-mentioned problem there is a great opportunity for using interactive geometry software. Subsequently, we offer a way how the teacher can easily demostrate the problem via GeoGebra software at Level (1) and Level (2). Figure 6 shows a possible demonstration of Level (1) via GeoGebra software.

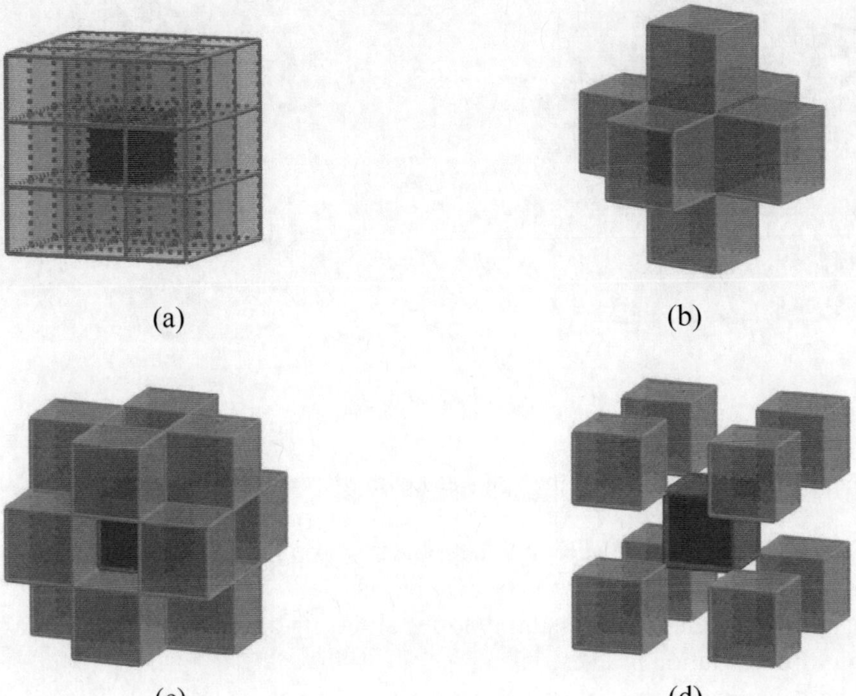

Fig. 6. The solution of the tasks at Level (1) via GeoGebra software.

Unlike Level (1), where the learner had to look inside the cube, the aim of Level (2) was to add a new layer of a different colour. The learner had to look at the cube from the outside. In other words, he had to change his perspective. Creation of new objects and manipulation with them in GeoGebra software is very simple for every learner. Thus the learners can change the perspective and they can see the object from every point of view. In this way the learners can "absorb" the correct image of the situation and create the correct mental visual image in their mind. The solution of the tasks at Level (2) via GeoGebra software (see Fig. 7 (a), (b), (c), (d)) was created by adding a new layer of purple colour small cubes to the cube in Level (1).

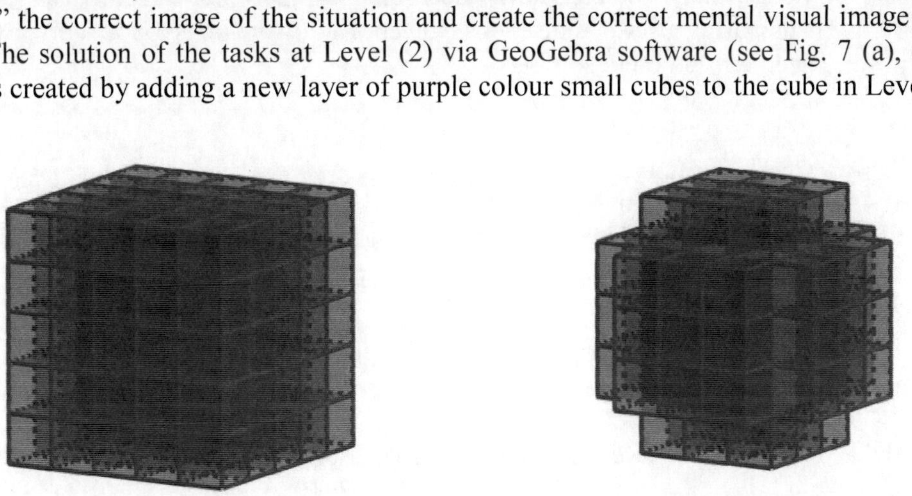

(a) (b)

(c)

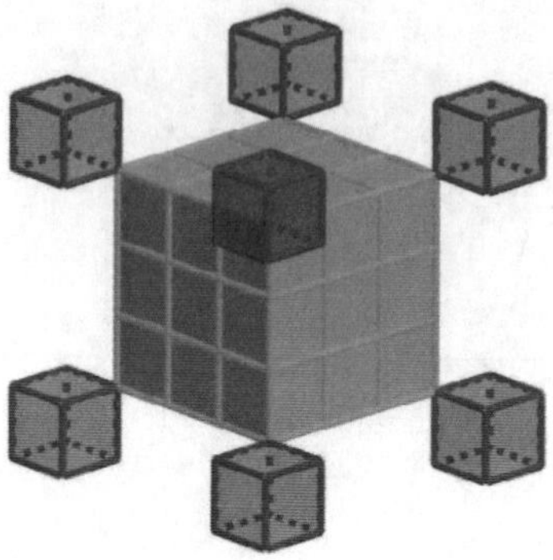

(d)

Fig. 7. The solution of the tasks at Level (2) via GeoGebra software.

Benefits of using GeoGebra software in this problem solving are the following:

- precise and ordered construction of 3D objects,
- transparency of surfaces (possibility to see all layers of the cube at the same time),
- colour differentiation of every layer,
- manipulation with cubes (possibility to change the perspective),
- demonstration of a correct image of the situation.

6 Conclusion

Qualitative research (especially case study [11]) is highly important in pedagogical sciences to see how individual students are thinking. In our experiment, we designed a solid geometry problem with hidden objects. We followed a learner`s solution step by step, focusing mainly on the mental visualisation process in this case study. We identified different categories of creating mental visual images at the learner according to appropriate theoretical literature. This helps us in preparing useful tools via GeoGebra software to support correct mental imagery in solid geometry problem solving.

Acknowledgement

The paper was supported by grant KEGA 016UKF – 4/2016 from Grant Agency of Ministry of Education of Slovak Republic.

References

[1] KASHEFI, H. et al., Visualisation in mathematics problem solving meta-analysis research. In: *International Conference on Social Sciences Research,* Malaysia, v.1, 2015, e-ISBN 978-967-0792-04-0.

[2] MAKINA, A. The role of visualisation in developing critical thinking in mathematics. In: *Perspectives in Education,* South Africa, v. 28, n. 1, 2010, p. 24-33, ISSN 0258 -2236.

[3] ARCAVI, A., The role of visual representations in the learning of mathematics. In: *Educational Studies in Mathematics*, Netherlands, v. 52, n. 3, 2003, pp. 215-241, e- ISSN 1573-0816.

[4] PIGGOTT, J., WOODHAM, L. Thinking through, and by, visualising. URL, https://nrich.maths.org/6447.

[5] DEL RÍO, L., PAULETICH, F., GeoGebra como herramienta para aprender y enseñar matemática en forma dinámica. URL, http://sedici.unlp.edu.ar/bitstream/handle/10915/59903/Documento_completo.pdf-PDFA2u.pdf?sequence=3.

[6] ANDRADE, M., MONTECINO, A. La problemática de la tridimensionalidad y su representación en el plano. In: *XIII Conferencia Interamericana de Educación Matemática;Recife,* Brasil, 2011. URL, https://www.researchgate.net/publication/283684064_La_problematica_de_la_tridimensionalidad_y_su_representacion_en_el_plano.

[7] PEARSON, J., KOSSLYN, S. L., The heterogeneity of mental representation: Ending the imagery debate. In: *Proceedings of the National Academy of Sciences of the United States of America*, USA, v. 112, n. 3, 2006, ISSN 0027-8424.

[8] KMEŤOVÁ, M., VITÉZOVÁ, Z., Anaglyfy a ich využitie vo vyučovaní geometrie, In: *G-Slovenský časopis pre geometriu a grafiku,* ročník 12, č. 23, 2015: s. 31-36, ISSN 1336-524X.

[9] PRESMEG, N., Research on visualization in learning and teaching mathematics. URL, http://www.kaputcenter.umassd.edu/.

[10] KOSSLYN, M.S., *Image and brain: The resolution of the imagery debate*. 1994, London, W.W. Norton.

[11] STARMAN, A. B., The case study as a type of qualitative research. *Journal of Contemporary Educational Studies*, v. 1, 2013, pp. 28–43.

Current addresses

Vágová Renáta, Mgr.
Department of Mathematics, Faculty of Natural Sciences,
Constantine the Philosopher University in Nitra
Tr. A. Hlinku 1, 949 74 Nitra, Slovak Republic
E-mail: renata.vagova@ukf.sk

Kmeťová Mária, doc. RNDr., PhD.
Department of Mathematics, Faculty of Natural Sciences,
Constantine the Philosopher University in Nitra
Tr. A. Hlinku 1, 949 74 Nitra, Slovak Republic
E-mail: mkmetova@ukf.sk

Slovak University of Technology in Bratislava
Faculty of Mechanical Engineering
17th Conference on Applied Mathematics **APLIMAT 2018**
Proceedings

COMPUTATIONAL OPTIMIZATION OF MATERIAL CHOICE FOR THERMAL CONTAINERS OF ADVANCED BUILDINGS

VALA Jiří (CZ)

Abstract. Optimization of performance of thermal containers installed in advanced buildings stimulates the multidisciplinary research, covering civil and mechanical engineering studies, material testing and physical, mathematical and computational analysis, up to the implementation of algorithms. This paper pays attention namely to the hot-wire identification of thermal characteristics of materials, with references to more general optimization tools.

Keywords: heat transfer, building design, inverse and optimization problems

Mathematics subject classification: Primary 35K05; Secondary 74A15, 35K10, 65K10

1 Introduction

Decreasing amount of energy sources, together with new achievement in the design of advanced materials, structures and technologies, stimulates the multidisciplinary research coupling civil and mechanical engineering studies, material testing and physical, mathematical and computational analysis. The European directive [2] forces the passive energy standard, whose proper formulation comes from [1], for both new and reconstructed buildings since 2020. One promising way for substantial reduction of energy requirements of buildings, namely those with controlled interior temperature, is connected with the design and construction of high-temperature thermal containers.

A representative example of such research activities seems to be the collaboration of the Brno University of Technology, Faculty of Civil Engineering (FCE BUT), with the Swedish specialists in fibre-optics-based heat production systems and with the Czech firm Alumistr a. s. Hrušovany (20 km southern from Brno), supported by the Technological Agency of the Czech Republic 2012–15. Apart from the successful construction and performance of a model house, supplied by the massive silicate thermal container, with no requirements to external energy sources, this project brings a strong motivation for further research in optimized choice of materials for thermal containers in the first step, followed by more general optimization of design of relevant structures and equipments. Since the technical details of the project solutions are not available for free access yet, the illustrative Fig. 1 shows just the first small model of the thermal container and documents one of the first series of high-temperature experiments with material specimens intended for the real container, namely for the evaluation of their basic thermal technical characteristics.

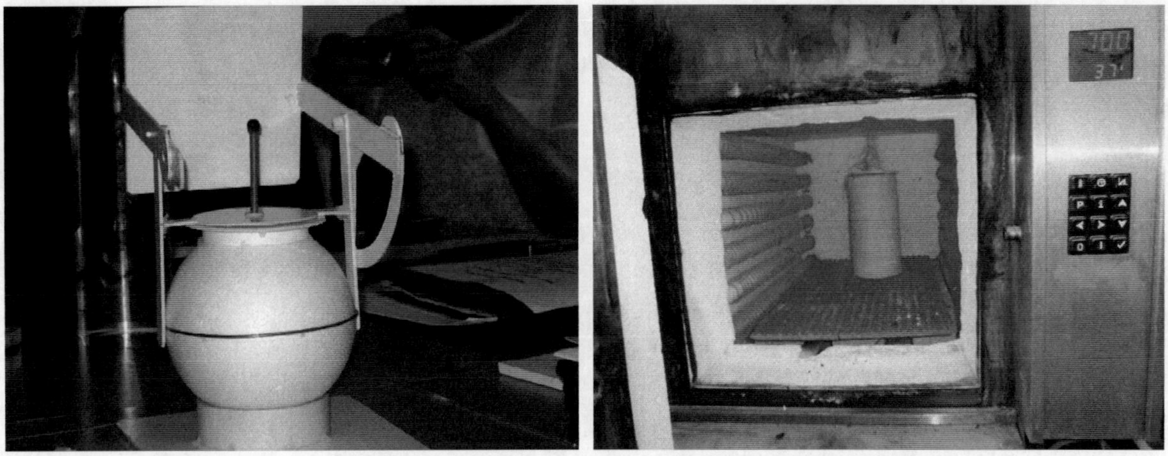

Fig. 1. A small model of thermal container (left photo). A cylindrical silicate specimen in the laboratory oven at the temperature 700 °C (right photo).

Much more consideration related to thermal collection and storage can be found in [4]. Practical equipments make use of i) sensible heat of high-temperature resistant materials by [5], or ii) materials involving phase-change components, appropriate for intended temperature levels, as parafine capsules, by [6]. In this paper we shall pay attention to i) with selected silicate materials, compatible with [7], namely to the computational evaluation of their thermal conductivity and diffusivity, as a substantial information for the reasonable design of the heat container; this will be supported by the hot-wire experimental setting, generalizing the approach of the European standard [8].

As a model problem for our physical, mathematical and computational analysis, we shall start with the study of of pure thermal conduction in an (at least macroscopically) isotropic medium without internal heat sources. However, the dependence of material characteristics on the temperature cannot be neglected, as usual in the case of temperature near to 20 °C, prescribed for building interiors – cf. [9]. This brings serious difficulties to the analysis of corresponding both direct and inverse problems, as well as to the relevant experiment planning.

2 A direct problem

Proper formulation of thermal processes starts with the conservation principles of selected scalar quantities, as for mass, (linear and angular) momentum and energy, from classical thermomechanics. Here, for simplicity, we shall work only with the conservation of thermal energy, with the absolute Kelvin temperature $\theta(x,t)$ as the reference variable. Here $x = (x_1, x_2, x_3)$ refers to the Cartesian coordinate system in the Euclidean space R^3, whereas t denotes the time from some interval $I = [0, T]$, assuming a finite positive T.

Let Ω be an open set (typically a domain) in R^3 with its boundary $\partial\Omega$ where the (formally outward) unit normal vector $\nu(x) = (\nu_1(x), \nu_2(x), \nu_3(x))$ can be introduced almost everywhere (which needs certain smoothness assumption – cf. [11], p. 16). We shall use upper dots instead of $\partial/\partial t$ for brevity, as well as Hamilton operators $\nabla = (\partial/\partial x_1, \partial/\partial x_2, \partial/\partial x_3)$ and central dots for scalar products in R^3. Consequently we can write e. g. Laplace operators briefly as $\Delta = \partial^2/\partial x_1^2 + \partial^2/\partial x_2^2 + \partial^2/\partial x_3^2 = \nabla\cdot\nabla$.

Following [10], p. 5, the classical formulation of energy conservation reads

$$\dot{\varepsilon} + \nabla \cdot q = 0 \qquad \text{on } \Omega \times I \tag{1}$$

where ε [J/m^3] represents the thermal energy (per unit volume) and $q = (q_1, q_2, q_3)$ the thermal flux [W/m^2]. Clearly (1) must be supplied by appropriate empirical constitutive equations coupling ε and q with θ. Here we shall suppose

$$\dot{\varepsilon} = \kappa\dot{\theta}\,, \qquad q = -\lambda\nabla\theta \qquad \text{on } \Omega \times I\,, \tag{2}$$

with two positive-valued material characteristics λ and κ: λ [W/(m.K)] is well-known as the thermal conductivity, κ [J/(m^3.K)] as the thermal capacity per unit volume; frequently the thermal capacity c [J/(kg.K)] is considered instead of κ, with the evident relation $\kappa = c\rho$ where ρ is the material density [kg/m^3] (evaluable from simple experiments in most cases). Let us notice that we are allowed to assume λ and κ independent of t (unlike x) directly, but their dependence on $\theta(x,t)$ may be substantial.

For simplicity, we shall consider only the Cauchy initial condition

$$\theta(.,0) = \theta_0 \qquad \text{on } \Omega \tag{3}$$

and the Neumann boundary one

$$q \cdot \nu = q_* \qquad \text{on } \partial\Omega \times I\,; \tag{4}$$

θ_0 must be prescribed on Ω and q_* on $\partial\Omega \times I$. Inserting (2) into (1), we have

$$\kappa\dot{\theta} - \nabla \cdot (\lambda\nabla\theta) = 0 \qquad \text{on } \Omega \times I\,, \tag{5}$$

and exploiting (2), too, similarly

$$-\lambda\nabla\theta \cdot \nu = q_* \qquad \text{on } \partial\Omega \times I\,. \tag{6}$$

Moreover, the thermal diffusivity α [m^2/s] is also introduced in many papers, alternatively to κ or c. The motivation is clear: for homogeneous materials, i. e. with constant λ and κ, taking $\alpha = \lambda/\kappa$, we can rewrite (5) to

$$\dot{\theta} - \alpha\Delta\theta = 0 \qquad \text{on } \Omega \times I\,. \tag{7}$$

However, to derive the analogy of (7) for inhomogeneous materials is not easy; this requires the application of the both enthalpy and Kirchhoff transforms by [11], pp. 68, 253, 291. Moreover, even for homogeneous materials, λ in (6) cannot be replaced by α (which does not allow the separate evaluation of α later in an inverse problem).

For numerical evaluation of θ, at least for all cases where no sufficiently simple forms of analytical solutions are available, the weak formulation of (5), (3) and (6) is useful. Using the standard notation of Lebesque, Sobolev, Bochner, etc. (abstract) function spaces, compatible with [11], pp. 10, 15, 23, and $(.,.)$ for scalar products in both $L^2(\Omega)$ and $L^2(\Omega)^3$ (Lebesgue intergals) and $\langle .,. \rangle$ for scalar products in $L^2(\partial\Omega)$ (Hausdorff integrals), we can write (5) and (6) in the integral form

$$(v, \kappa\dot{\theta}) + (\nabla v, \lambda\nabla\theta) = \langle v, q_* \rangle \qquad \text{on } I \tag{8}$$

for every test function $v \in W^{1,2}(\Omega)$; (3) remains to be consider separately. It is natural to assume $\theta_0 \in W^{1,2}(\Omega)$ and, at least for $\lambda, \kappa \in L^\infty(\Omega)$, moreover $q_* \in L^2(\partial\Omega)$. Consequently, for the linear parabolic system of evolution (8), using the convergence results for Rothe sequences by [11], p. 202, the existence and uniqueness of $\theta \in L^2(I, W^{1,2}(\Omega)$ with $\dot{\theta} \in L^2(I, L^2(\Omega)$ can be verified without difficulties. However, this is not true for λ and κ dependent of θ where additional growth conditions are necessary; for more details see [12], with numerous references to [11] (namely on the available results for quasilinear equations of evolution) again.

Let us remark that (5) and (6) can be derived from (8) using the Green-Ostrogradskiĭ theorem, at least in the distributional sense. However, this does not guarantee any regularity (which may be needed in applications), as discussed in [11], p. 215.

In particular, for constant λ and κ, various special (even classical) solutions of (5) can be found in the literature; they differs in the reasonable applicability of (6) for special geometrical configurations, often using numerical techniques again, as boundary elements on $\partial\Omega$ instead of finite elements on Ω. In this paper we shall utilize the following result of [13]:

$$\theta = \theta_0 + \frac{Q}{4\pi\lambda}\,\mathrm{Ei}\left(\frac{r^2}{4\alpha t}\right), \tag{9}$$

including the exponential integral

$$\mathrm{Ei}(\varsigma) = \int_{\varsigma}^{\infty} \frac{\exp(v)}{v}\,\mathrm{d}v = -\gamma - \ln\varsigma + \varsigma + \epsilon(\varsigma)\,, \qquad \epsilon(\varsigma) = \sum_{k=2}^{\infty} \frac{\varsigma^k}{k.k!} \tag{10}$$

where $\gamma \approx 0.577216$ is the Euler-Mascheroni constant (removable from practical computations), ς may be an arbitrary real number and $x_1 = r\cos\phi$, $x_2 = r\sin\phi$, $x_3 = z$ with $r \geq 0$, $0 \leq \phi < 2\pi$ and arbitrary real z, neglecting the dependence of θ on ϕ and z and assuming (as a special case of (3)) constant initial temperature θ_0 everywhere. Let us consider Ω covering R^3 except the axis x_3. Inserting (9) into (7), we can see (MAPLE, MATLAB toolbox *symbolic* or similar software support can be recommended) that (7), degenerated to

$$\frac{\dot{\theta}}{\alpha} = \theta'' + \frac{\theta'}{r} \qquad \text{on } \Omega \times I\,, \tag{11}$$

is satisfied automatically; primes here mean $\partial/\partial r$. Nevertheless, (4) must be replaced by the physically motivated limit formulation

$$\lim_{r\to 0} \frac{-\lambda\theta'}{q_*(r)} = 1\,, \qquad q_*(r) = \frac{Q}{2\pi r}\,, \tag{12}$$

assuming the thermal power Q [W/m], related to the unit length on the axis x_3, occurring everywhere at the position $r = 0$, starting from $t = 0$.

These considerations can be generalized in several directions. Especially for an infinitely long cylinder Ω with the missing axis x_3 and with a finite diameter a (thus $0 < r < a < \infty$) [14] demonstrates how the standard Laplace transform is able to convert (11) to the analysis of certain Bessel equation; consequently we have to work with the Bessel functions of the first and second kind instead of exponential integrals (the limit passage $a \to \infty$ comes back to (9) naturally). However, all relevant quasi-analytical computational formulae are much more complicated than (9), thus their exploitation in inverse problems (as the aim of this paper) is rather limited.

3 An inverse problem

As a model inverse problem, let us consider a direct problem introduced in the previous section with these modifications: i) no values of λ and κ are known a priori (except some their rough initial estimate), ii) in addition to (8) and (3) some values θ_* are prescribed on Γ, introduced as some surface (typically not included in $\partial\Omega$) in R^3. Considering some a priori weight functions $w \in L^\infty(\Gamma \times I)$ with values from [0,1] (related to preparation of real experiments), we can believe that the best available values of λ and κ correspond to the least-squares approach motivated minimum of

$$F(\lambda, \kappa) = \langle\langle \theta_{\lambda,\kappa} - \theta_*, w(\theta_{\lambda,\kappa} - \theta_*) \rangle\rangle \tag{13}$$

where $\langle\langle ., . \rangle\rangle$ refers to a scalar product in $L^2(\Gamma \times I)$ and $\theta_{\lambda,\kappa}$ is any solution of (8) and (3) for admissible λ and κ.

In particular, for constant values λ and κ we can introduce F in (13) as a (rather complicated) function of 2 real variables; in more general cases, namely those disturbing the linearity of (8), it must be considered as a functional defined on an appropriate infinite-dimensional function space. The pure linear case is studied in great details in [16] an in numerous references therein. For a priori given values of κ and sufficiently large Γ (with non-zero valuse of w) the dependence of λ on θ can be handled by [15] (although its approach has been suggested for capillary conduction, i. e. on the conservation of mass from the physical point of view, unlike the conservation of energy here).

Unfortunately, the quite general formulation (13) hides a lot of expectable difficulties: ill-possedness, numerical instability, need of artificial regularization, etc., as explained in [17], p. 255. Thus for practical identification of material characteristics the better choice is to pay attention to the careful organization of all experiments, to be able to make use of sufficiently simple computational formulae: in spherical coordinates for hot-ball measurements, in cylindrical coordinates for hot-wire measurements, like (9) here, in Cartesian coordinates for hot-plate measurements, etc. Another argument supporting this choice, important in this paper, is that an inexpensive experimental setting, covering all temperatures expectable in a thermal container, is not available, thus a series of measurements at several temperature levels, preserving nearly constant values of λ and κ, can represent a reasonable alternative.

Technical standards respect such arguments. By [8], coming from the above introduced cylindrical setting, we need:

a) a very long and thin heating wire of circular shape, whose thermal conductivity and capacity can be neglected,

b) a material specimen considerable as a very long rotational cylinder of large radius, whose axis lies just in the heating wire,

c) supplying to the specimen, starting from the zero time t, related to its unit length, to all directions, still the same thermal power Q, as needed by (9),

d) a material specimen prepared as homogeneous and isotropic for macroscopic calculations, with some effective values of scalar characteristics λ and α (and thus also κ),

e) all independent of θ in practice,

f) respecting (3) with constant θ_0.

Consequently (9) is applicable to corresponding direct calculations. For inverse ones, [8] works with the location of temperature sensor Γ degenerated to one point very close to the axis of rotation, i. e. lying just on the axis of the heating wire theoretically. This is impossible in practice, but needed only for the removal of all additive terms excepts the second one from the decomposition of Ei(.) in (10): if two temperatures by (9) in different times are subtracted, the first additive term vanishes and the third and all other terms are negligible, which gives

$$\theta(0,t) - \theta(0,t_\times) \approx \frac{Q}{4\pi\lambda}\ln\frac{t}{t_\times} \tag{14}$$

for any time $t \in I$ (except $t = 0$ naturally) and some comparative time $t_* \in I$, different from t. Since the values $\theta_*(t) - \theta_*(t_\times)$ for r close to zero are available, at least for a large number of time steps $t \in I$ and one appropriate $t_\times \in I$, (14) is applicable to the evaluation of λ using the linear regression analysis, interpretable as a special case of minimization of F from (13) with missing κ; for the repeatability and certain improvements of such evaluations see [18].

The principal drawback of the identification procedure of λ based on (14) is the ignorance of the influence of κ or α. Moreover, the temperature is usually recorded also later when no thermal power is active, which is not exploited. As a potential remedy, following [14], let us consider Γ containing two sensors at some locations $r = r_1$ and $r = r_2$ where $0 < r_1 < r_2 < \infty$; the temperature θ is recorded there for selected increasing positive values of time steps $t_1, \ldots, t_m, \ldots t_n$ where m and $n > m$ are some integer indices. We assume that the thermal power Q is active for $0 < t \leq t_m = \tau$, whereas the thermal power $-Q$ is active for $t \geq t_m$; thus the thermal power vanishes for $\tau < t \leq T$. The aim now is to evaluate of λ and α thanks to the above announced temperature values $\theta_{*1k}, \ldots, \theta_{*mk}, \ldots \theta_{*nk}$.

During the hot-wire experiment Q is usually constant from $t = 0$ up to $t = \tau$ where τ is some positive time. The temperature θ is recorded in some locations r_1 and r_2 for selected increasing positive values of time steps $t_1, \ldots, t_m, \ldots, t_n$ where m and $n > m$ are some integer indices: here we shall suppose that the thermal power Q is active for $0 \leq t \leq t_m = \tau$, whereas the thermal power $-Q$ is active for $t \geq t_m$; consequently the thermal power vanishes for $t > \tau$.

The main aim is to identify unknown values of λ and α thanks to the above announced temperature values $\theta_{*1k}, \ldots, \theta_{*mk}, \ldots, \theta_{*nk}$ where the last index $k \in \{1,2\}$ refers to the 1st or 2nd sensor. The corresponding computed values in the same time steps are denoted by $\theta_{1k}, \ldots, \theta_{mk}, \ldots, \theta_{nk}$. Considering (for simplicity here) the time step t_m as the reference one, consequently for any $j \in \{1, \ldots, m\}$ from (15) we obtain

$$\theta_j - \theta_m = \frac{Q}{4\pi\lambda}\ln\frac{t_j}{t_m} + \frac{Q}{4\pi\lambda}\cdot\frac{1}{4\alpha r_k^2}\left(\frac{1}{t_j} - \frac{1}{t_m}\right) \tag{15}$$
$$- \frac{Q}{4\pi\lambda}\left(\epsilon\left(\frac{1}{4\alpha r_k^2 t_j}\right) - \epsilon\left(\frac{1}{4\alpha r_k^2 t_m}\right)\right).$$

Repeating the same arguments to any $j \in \{m+1, \ldots, n\}$, we receive similarly

$$\theta_j - \theta_m = \frac{Q}{4\pi\lambda}\ln\frac{t_j}{t_j - t_m} + \frac{Q}{4\pi\lambda}\cdot\frac{1}{4\alpha r_k^2}\left(\frac{1}{t_j} - \frac{1}{t_j - t_m}\right) \tag{16}$$
$$- \frac{Q}{4\pi\lambda}\left(\epsilon\left(\frac{1}{4\alpha r_k^2 t_j}\right) - \epsilon\left(\frac{1}{4\alpha r_k^2 (t_j - t_m)}\right)\right).$$

Clearly the formulae (15) and (16) are not suitable to direct evaluation of λ and α because of the lack of unicity and of and the need of some relaxation of the effect of random (or even system) measurement errors. Let us utilize the least squares approach, motivated by (13), again, with non-zero weights w_j on discrete time steps t_j where some indices j are taken from $1, \ldots, m-1$ for (15) or from $m+1, \ldots, n$ for (16); such indices are intended as admissible ones. Thus the two-step algorithm is available:

1) considering some reasonable estimates of λ and α in all right-hand side additive terms of (15) and (16) except the first ones, multiplying (15) by $w_j \ln(t_j/t_m)$ and (16) by $w_j \ln(t_j/(t_j - t_m))$ formally and summing up over all admissible j, from the result we are able to compute just one improved value of λ,
2) in the analogous way, knowing the estimates of λ and α in all right-hand side additive terms of (15) and (16) except the second ones, multiplying (15) by $w_j(1/t_j - 1/t_m)$ and (16) by $w_j(1/t_j - 1/(t_j - t_m))$ formally and summing up over all admissible j, from the result we are able to compute just one improved value of α.

In the usual configuration with r_1 close to zero (for certain compatibility with [8]) and $r_1 \ll r_2$, the following iteration algorithm is suitable: from some estimates α and λ to improve λ from the first sensor measurements, then to improve α from the second sensor measurements and to repeat such cycle until the differences between iterated values are not negligible.

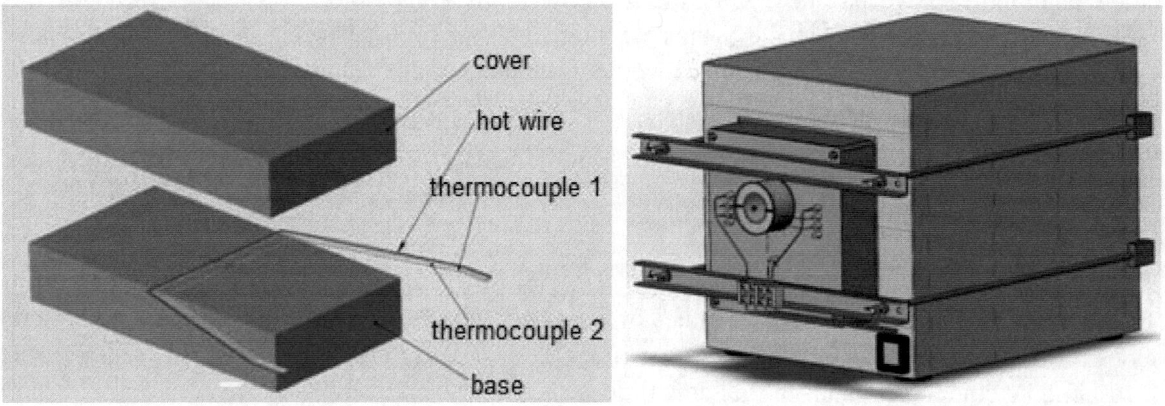

Fig. 2. Hot-wire experiment configuration (left scheme) and laboratory oven (right scheme).

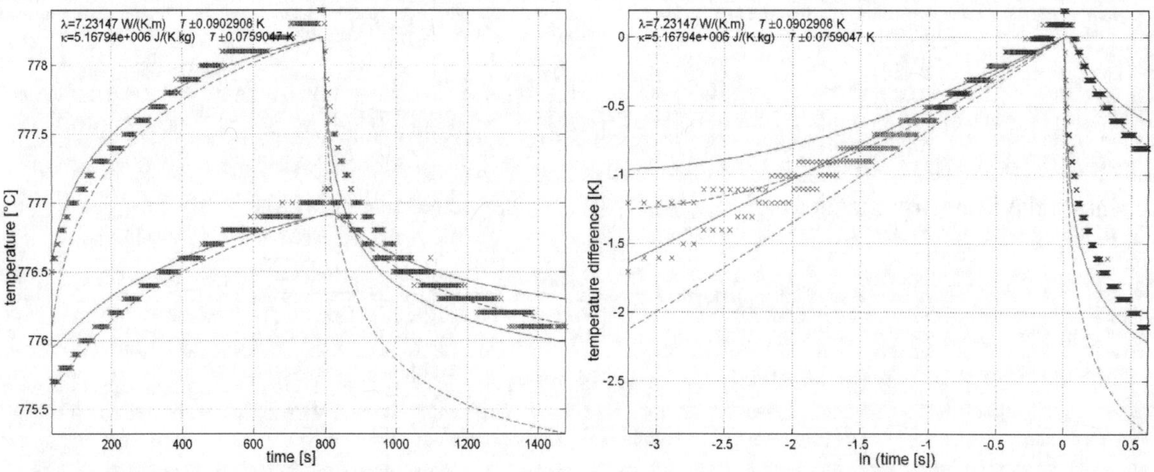

Fig. 3. Example of MATLAB-based identification procedure of λ and α for an experimental specimen at the reference temperature 786 °C: development of θ at 2 computational positions (left graph), also in the semilogarithmic scale (right graph).

4 Example of experimental and computational results

The practical configuration for high-temperature laboratory experiments is evident from Fig. 3; more technical details (including all oven components, their sizes, etc.) can be found in [7]. Unlike the heating kanthal (Fe, Cr, Al and Co alloy) wire, applied specimens have not (theoretically assumed) cylindrical shape, thus all experiments should work with T small enough, to suppress the effects of thermal transfer on their real exterior boundaries. The value of Q is not very substantial, but must be known a priori and stabilized carefully during every experiment.

The identification procedure for λ and κ (via α) has been written in MATLAB; no additional toolboxes are needed. Fig. 5 represents an illustrative example of its performance for one of a large number of tested specimens and temperature levels. The blue points correspond to the temperature records from both sensors with the sensibility 0.1 °C (thus they are seemingly piecewise constants). The dashed red curves correspond to the first rough estimate by (14); the linearity of the lower dashed curve in the semilogaritmic scale follows from (14) directly; all temperature differences are non-positive because of the reference temperature taken in the time τ (end of heating). The full red curves show the results obtained from the iteration procedure coming from (15) and (16).

Repeating such calculations several for a set of reference interior oven temperatures, here for (approximately) $\{20, 100, 200, 300, 400, 500, 600, 700, 800\}$ °C, we are able to obtain the temperature dependence of λ and κ for (8), consequently to evaluate its preference for the utilization in a real thermal container. Fig. 5 corresponds to the material chosen by [7] finally: both dependences of λ and κ on θ are nearly linear for temperature values between 300 and 800 °C. Higher temperatures than 800 °C cannot be recommended because of the real danger of unwanted structural changes both in the tested specimens and in particular components of experimental equipment, especially due to the applied heating wire.

5 More general optimization tools

Practical minimization of F by (13), with the aim of identification of λ and κ, including their dependence on θ, relies on substantial physical and geometrical simplification here and the consequent availability of some explicit computational formulae, based on (9), unlike the application of general (e. g. finite element) numerical techniques on (8) with (3). However, in the design of thermal containers, even in the choice of materials for their performance, this can be seen as a very particular optimization result only. Thus the development of a more general optimization tool, as robust, effective and inexpensive as possible, cannot be avoided. Most technical solutions declared as optimized (including [7], referenced in the previous section), apply a simple selection from a finite (rather low) number of variants only, in particular in such cases when both non-trivial experiments and computations are needed to receive separate values of rather complicated goal functions, generalizing F from (13). incorporating various additional conditions thanks to penalty terms.

Some authors avoid the analysis of differential or integral equations at all, referring to some soft-computing approaches, as genetic algorithms, neural networks, etc.; for much more information and references corresponding to various problems of thermal transfer see [19] and [20]. Most modern optimization algorithms usual some exact or approximate evaluations of gradients, needing many evaluations of F (quasi-Newton iterations, conjugate gradient techniques, etc.), as analyzed in [21] in all details. A promising (although not very quick) algorithm in such situation seems to be the nonlinear simplex Nelder-Mead one, suggested in [22] originally, modified by [23] slightly, with 5 gradient-free steps: i) reflection, ii) expansion, iii) outside contraction, iv) inside contraction, all with just one additional evaluation of F, and (as the last choice) v) shrink, with p additional evaluations of

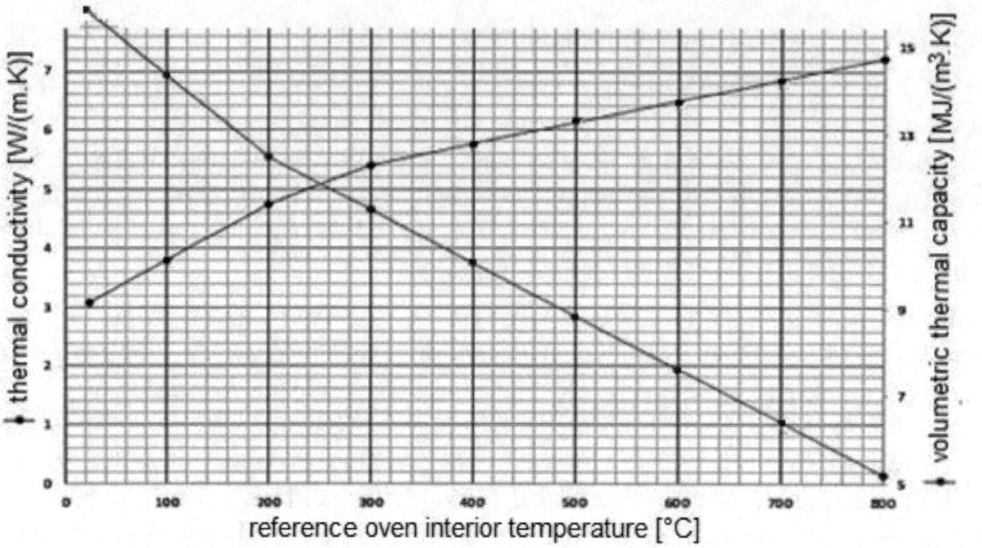

Fig. 4. Temperature dependence of λ and κ for the specimen with the following chemical composition: dominating 96.0 % MgO, further 2.2 % CaO, 0.5 % SiO_2, 0.5 % Fe_2O_3, 0.2 % Al_2O_3, etc. (0.6 % total); by the material data sheet of Slovmag a. s. Lubeník (Slovak Republic).

F where p refers to a number of optimized parameters. Unfortunately, the complete formal convergence proof of this algorithm is contained in [23] only for $p = 1$ (where every direction can be seen as a gradient one); the computer-supported 25-page proof for $p = 2$ in [24] does not allow any step ii) and cannot be extended to $p \in \{3, 4, \ldots\}$ in an intuitive way. Some artificial examples of stagnation or incorrect finishing of the algorithm can be overcome using certain adaptive parameters for steps i)–v), various conditional restarting, simplex regularization or reshaping strategies; most computations based on the MATLAB function *fminsearch* from the toolbox *optimization* and on similar software tools do not produce (at least evidently) wrong results. Consequently a frame-based method of [25] can be interpreted as a (rather complicated) always convergent variant of the Nelder-Mead algorithm. Moreover, the recent development (with still open questions) tries to combine some advantages of gradient and simplex algorithms, as sketched in [26].

Up to now, some partial results referenced in this section have been implemented to the design of controlled heating for passive, low-energy and similar buildings, depending on a rather low number p of optimized parameters, as documented by [27]. Their effective application to thermal containers needs deeper connection to advanced planning of numerous experiments and is still in development.

6 Conclusion

The paper documents the significance of development of the wide range of optimization methods, motivated by new development in civil engineering, compatible with increasing user comfort and sustainable development. The introduction to direct and inverse problems is accompanied by an instructive example of identification of material characteristics for thermal containers, as a part of research priorities at FCE BUT. However, partial results generate still other problems, stimulating the relevant research in the near future.

Acknowledgement. This research has been supported by the project LO1408 AdMaS UP (Advanced Materials, Structures and Technologies, Ministry of Education, Youth and Sports of the Czech Republic, National Sustainability Programme I).

References

[1] FEIST, W.: *Gestaltungsgrundlagen Passivhäuser*. Das Beispiel (Darmstadt), 1999. (In German.)

[2] Directive 2010/31/EU of the European Parliament and of the Council on the Energy Performance of Buildings. *Official Journal of the European Union* L 153/13, 2010.

[3] JAROŠOVÁ, P., and VALA, J.: New approaches to the thermal design of energy saving buildings. *Advanced Materials Research* 1126 (2015), 174–180.

[4] NAYAK, J. K., and SUKHATME, S. P.: *Principles of Thermal Collection and Storage*. Tata McGraw Hill, 2008.

[5] FERONE, C., COLANGELO, F., FRATTINI, D., ROVIELLO, G., CIOFFI, R., and DI MAGGIO, R.: Finite element method modeling of sensible heat thermal energy storage with innovative concretes and comparative analysis with literature benchmarks. *Energies* 7 (2014), 5291–5316.

[6] KUZNIK, F., DAVID, D., JOHANNES, K., and ROUX, J.-J.: A review on phase change materials integrated in building walls. *Renewable and Sustainable Energy Reviews 15* (2011), 379–391.

[7] ŠŤASTNÍK, S., VALA, J., and ŠOT, F.: Physical properties of selected silicates for a long-term heat container. *Advanced Materials Research*, submitted.

[8] EN ISO 8894-1: *Refractory materials – Determination of thermal conductivity – Part I: Hot-wire methods (cross-array and resistance thermometer)*. CEN, 2010.

[9] DOHNALOVÁ, L., TRHLÍKOVÁ, I., and ZMEŠKAL, O.: Temperature dependence of thermal parameters of building materials. *Thermophysics* 19 in Podkylava (Slovak Republic, 2013), Slovak Academy of Sciences, Bratislava, 2013, 194–207.

[10] BERMÚDEZ DE CASTRO, A.: *Continuum Termodynamics*. Birkhäuser, 2005.

[11] ROUBÍČEK, T.: *Nonlinear Partial Differential Equations with Applications*. Birkhäuser, 2005.

[12] VALA, J.: Computational approaches to some inverse problems from engineering practice. *Programs and Algorithms of Numerical Mathematics* 17 in Dolní Maxov (Czech Republic, 2014), Institute of Mathematics AS CR, Prague, 2015, 215–230.

[13] CARSLAW, H. C., and JAEGER, J. C.: *Conduction of Heat in Solids*. Oxford University, 1946.

[14] VALA, J., and JAROŠOVÁ, P.: Computational aspects of hot-wire identification of thermal conductivity and diffusivity under high temperature. *Thermophysics* 21 in Terchová (Slovak Republik, 2016), American Institute of Physics, 2016, 040029: 1–6.

[15] VALA, J., and JAROŠOVÁ, P.: Identification of the capillary conduction coefficient from experimental data. *Forum Statisticum Slovacum* 9 (2013), 250–255.

[16] ZABARAS, N.: Inverse problems in heat transfer, in: *Handbook on Numerical Heat Transfer* (Minkowycz, W. J., Sparrow, E. M., and Murthy, J. Y., eds.), Chap. 17. J. Wiley & Sons, 2004.

[17] ISAKOV, V.: *Inverse Problems for Partial Differential Equations*. Springer, 2006.

[18] BILEK, J., ATKINSON, J. K., and WAKEHAM, W. A.: Repeatabillity and refinement of transient hot-wire instrument for measuring the thermal conductivity of high-temperature melts. *International Journal of Thermophysics* 27 (2006), 1626–1637.

[19] COLACO, M. J., ORLANDE, H. R. B, and DUIKRAVICH, D. S.: Inverse and optimization problems in heat transfer. *Journal of the Brazilian Society of Mechanical Sciences and Engineering* 28 (2006), 1–24.

[20] PACHECO-VEGA, A.: *Soft Computing Applications in Thermal Energy Systems*. Springer, 2011.

[21] LUKŠAN, L.: *Numerické optimalizační metody*. Technical report 1152, Institute of Informatics AS CR, Prague, 2015. (In Czech.)

[22] NELDER, J. A., and MEAD, R.: Simplex method for function minimization, *Computer Journal* 7 (1965), 308–313.

[23] LAGARIAS, J. C., REEDS, J. A., WRIGTH, M. H., and WRIGTH, P. E.: Convergence properties of the Nelder-Mead simplex method in low dimensions. *SIAM Journal of Optimization* 9 (1998), 112–147.

[24] LAGARIAS, J. C., POONEN, B., and WRIGTH, M. H.: Convergence of the restricted Nelder-Mead algorithm in two dimensions. *SIAM Journal on Optimization* 22 (2012), 501–532.

[25] PRICE, C. J., COOPE, I. D., and BYATT, D.: A convergent variant of the Nelder-Mead algorithm. *Journal of Optimization Theory and Applications* 113 (2002), 5–19.

[26] PHAM, N., and WILAMOWSKI, B. M.: Improved Nelder-Mead's simplex method and applications. *Journal of Computing* 3 (2011), 55–63.

[27] VALA, J., and JAROŠOVÁ, P.: Computational optimization of energy consumption in the design of buildings. *International Conference on Numerical Analysis and Applied Mathematics (ICNAAM)* in Thessaloniki (2017), American Institute of Physics, 2018, in print.

Current address

Vala Jiří, prof. Ing., CSc.
Institute of Mathematics and Descriptive Geometry
Faculty of Civil Engineering
Brno University of Technology
Veveří 95, 602 00 Brno, Czech Republic
E-mail: vala.j@fce.vutbr.cz

Slovak University of Technology in Bratislava
Faculty of Mechanical Engineering
17th Conference on Applied Mathematics **APLIMAT 2018**

Proceedings

GEOGEBRA AND LOGARITHMIC SPIRAL IN EDUCATIONAL PROCESS

VALLO Dusan (SK), DURIS Viliam (SK)

Abstract. In this paper, we introduce a usage of dynamical geometry software Geogebra in teaching Mathematics and in teachers training education. For this purpose, we concerned some specific curves like spirals, especially logarithmic spiral. In software environment, we demonstrate some properties of this curve and its application in metal engineering. Some didactical remarks are included in the paper, too.

Keywords: curve, Geogebra, spirangle, spiral, logarithmic spiral, education

Mathematics subject classification: Primary 97D40; Secondary 97D40, 97D50, 97C20

The National Curriculum for Upper Secondary Education in Slovakia (ISCED 3A) classified the mathematical standards in Slovak high schools. One of the major topics in ISCED 3A is Geometry. The main role of education in Geometry is oriented to a development of geometric imagination of the students as one of the most important ability for their future profession. From this reason, various technologies and software products enter in the learning process. Pedagogical software like dynamical geometry software (DGS) has the most significant impact on its qualitative change. These software develop creative thinking, mathematical skills and lead to activity.
In this paper we introduce a usage of dynamical geometry software Geogebra in teaching Geometry. For this purpose, we pay an attention to algebraic curves - spirals, especially logarithmic spiral. In software environment we demonstrate some properties of this curve and its application in metal engineering. Didactical remarks are included in the paper, too.

1 Introduction

The use of ICT in teaching mathematics has its merit. Digital technologies offer a wide range of other, modern learning, knowledge transfer and skills. They are a driving force that gradually transforms the traditional school and its educational process. Various technologies and software products enter in the learning process, but pedagogical software like dynamical geometry software (DGS) has the most significant impact on its qualitative change. [1, 2, 3] They allow the development of teaching in the spirit of the principles of didactical constructivism. It means to teach mathematics and geometry with a constructivist approach, emphasizing the pupil's activity, in a creative environment, a good working atmosphere in the classroom, looking for different representations and detracting the formalism in the knowledge of students. [4]

In this paper, we will focus on usage of the dynamical geometry Geogebra in an investigation of properties of specific curves like spirals. In software environment, we demonstrate the attributes of the logarithmic spiral and we also suggest it as motivating factor in teaching. A few didactical remarks will be noted, too.

2 Spirals in generally

There exists many kinds of the spirals like easily constructed spirangles (Fig. 1) or continuous curves, whose constructions are more complicated. [5, 6] Some spirals are termed by the famous Mathemati-

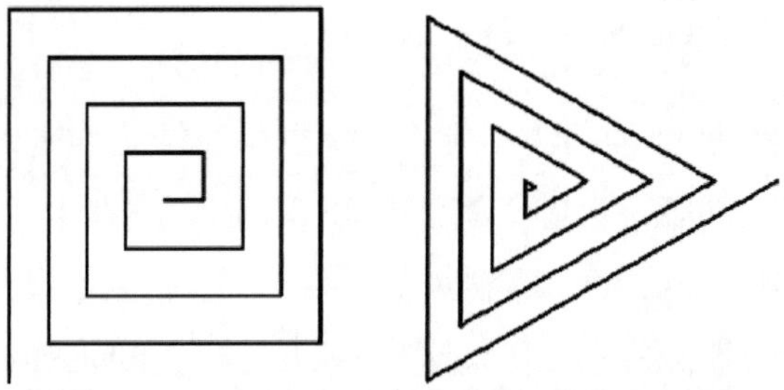

Fig. 1. Spirangles - figures related to spirals.

cians like Archimedes, Fermat, Galileo, etc.

Well-known are also spirals such as parabolic spiral, logarithmic spiral or sine spirals. [7, 8]. In (Fig. 2) are the curves constructed in Geogebra as geometrical loci of points. In general, spirals are planar curves generated by a continuous motion of a point on a line with prescribed velocity. The line rotates around a fixed center.

This free explanation indicates that the spirals can be well-defined by equations in polar coordinate system. In the following we present properties of the logarithmic spiral via software Geogebra.

3 Logarithmic spiral

This spiral is a curve whose polar equation is

$$\rho = a \cdot e^{b \cdot \varphi},$$

where ρ is radial distance from pole, φ is a counterclockwise polar angle from the positive semiaxis $x-$axis, $a, b \in R$, $a > 0$.

The point construction of the logarithmic spiral is in [7]. This synthetic approach can be replaced by the using software potential.

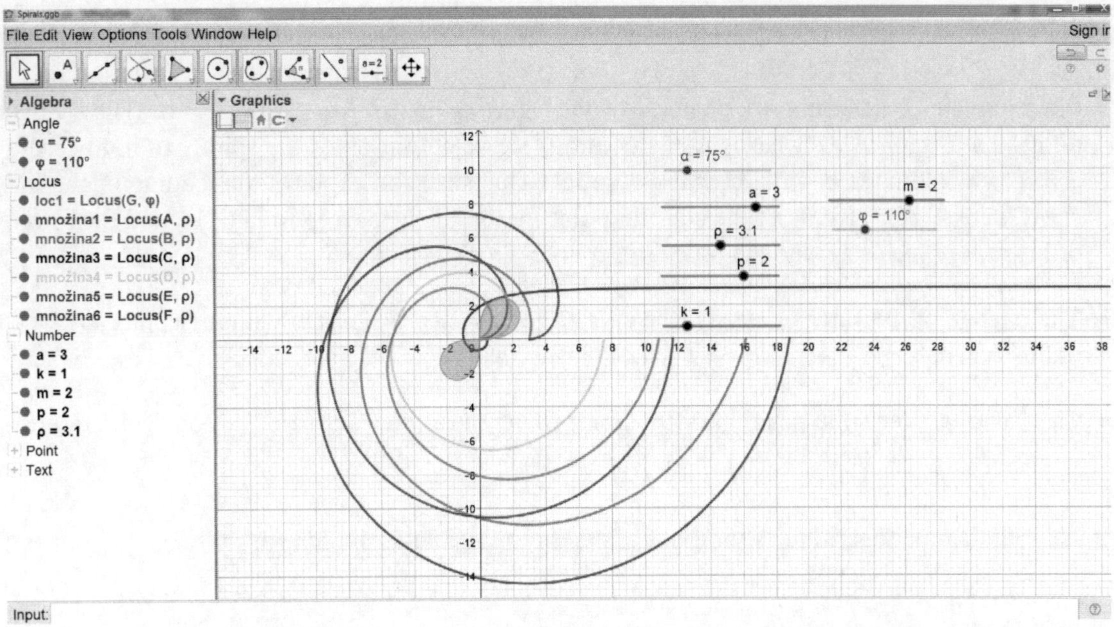

Fig. 2. A few spirals in Geogebra - Archimedes' (blue), Fermat's (green), Gallileo's (pink), Lemniscate of Bernoulli (orange; special case of sine spiral), logarithmic (red), hyperbolic (black), parabolic (brown).

In Geogebra graphic window we define the sliders a, b and $\varphi \in [0, 2\pi]$. In the next we re-define Cartesian coordinates of an arbitrary point M to polar coordinates and we re-write $M = (a \cdot e^{b \cdot \varphi}; \varphi)$. The point M lies on the logarithmic spiral. The tool *Trace on* allows us to visualize few its positions for φ. The tool *Locus* draws the implicit curve continuously (Fig. 3). The presented construction in

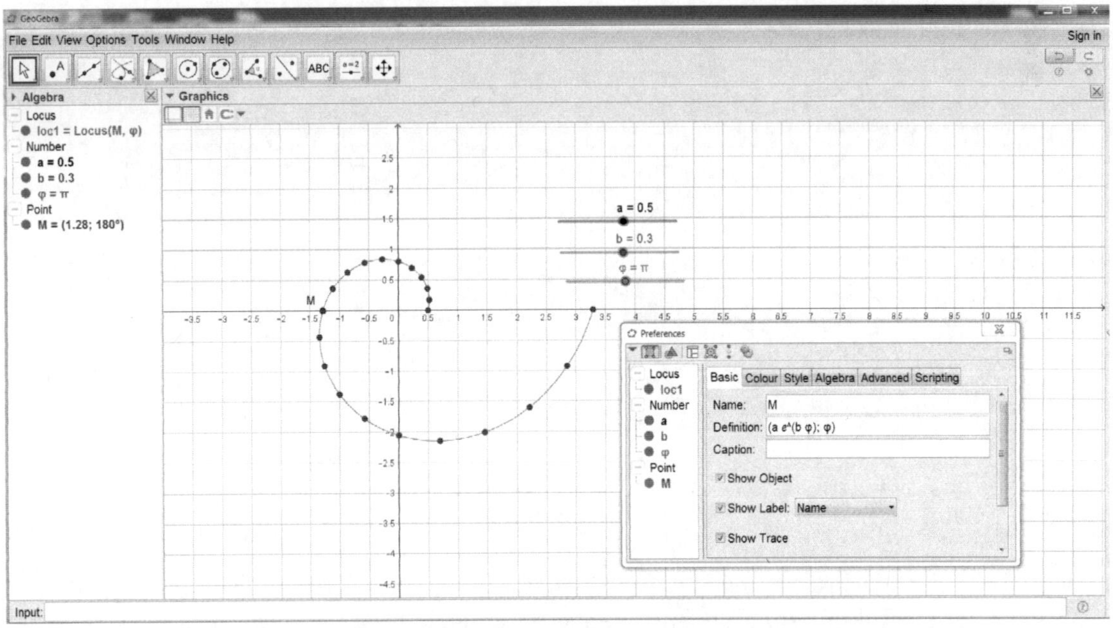

Fig. 3. Logarithmic spiral for $a = 0.5$, $b = 0.3$.

Geogebra software allows a user (student, teacher, ...) to change the parameters a, b, φ. It can be observed:

- the parameter a determines a position of the "starting" point $(a, 0)$ for $\varphi = 0$. The software allows us to set $a \leq 0$, what is in the contrary to the definition of the spiral. In both cases, if $a \neq 0$ we obtain the counter clockwise spiral. This situation explains the loading of the polar coordinate system. If $a < 0$, the starting point lies on the negative semiaxis o_x and the rotation of his radius vector around the pole O is counter clockwise, too.

- The change of the value b "opens" the spiral. For $b = 0$ it is a circle with the center O and the radius $r = a$.

- The value φ affects the position of the point M on the curve. It carries out one full circumvolution for $\varphi \in [0, 2\pi]$.

We must note that the logarithmic spiral has another remarkable attribute.
If we construct the tangent line t at the point M, we observe that the angle $\mu = \angle\left(\overrightarrow{OM}, t\right)$ is constant and it holds that

$$\tan \mu = \frac{1}{b}.$$

Remark. The tangent line at the point M is parallel to a hypotenuse RQ of triangle QRO, where $|OQ| = 1$, $|OR| = \frac{1}{b}$. The logical reasoning of this construction is in [7].

Logarithmic spiral has many applications in computer graphics like a modeling spirals in animal

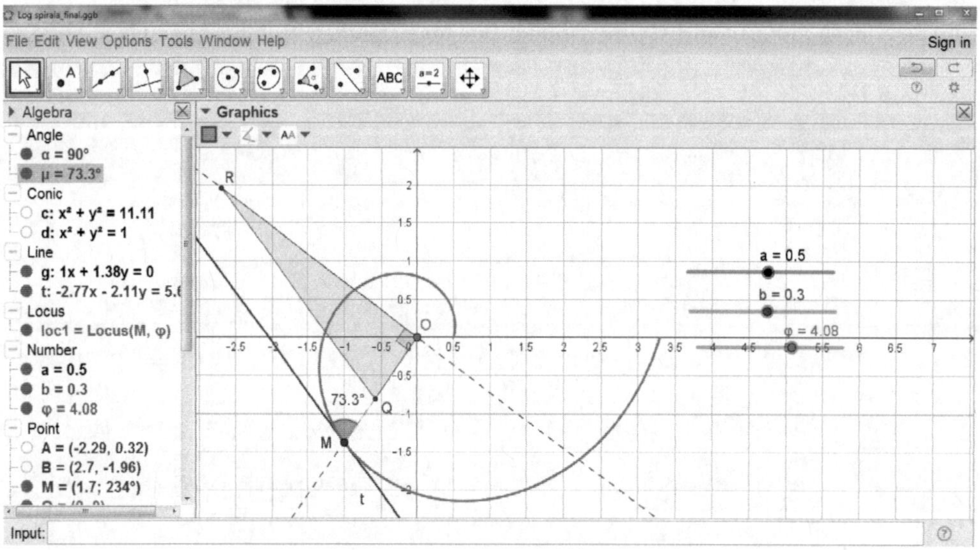

Fig. 4. Tangent line t at the point M of the logarithmic spiral for $a = 0.5$, $b = 0.3$.

structures [9] or in metal engineering, e.g. in the gear design of circular-saw blade (Fig. 5). [10, 11]
The attributes of logarithmic spiral mentioned above can be implemented in educational process as students' projects.
In Fig. 6 one can see a sample of design of the circular-saw blade. Its construction suits to the drawing of the logarithmic spiral like locus of points M for $\varphi \in \left[0, \frac{\pi}{4}\right]$. The n multiple repetitions of the arc

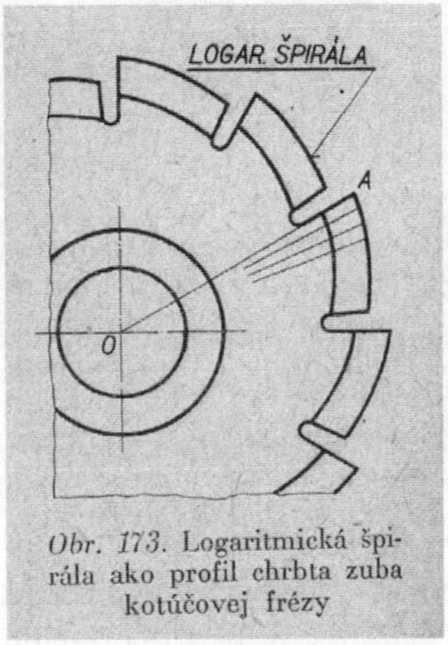

Fig. 5. Gear design of circular-saw blade. Source [10].

of spiral are constructed via the command *Sequence*in the form

$$Sequence\left(Rotate\left(M, \frac{2k\pi}{n}, O\right), k, 1, n, 1\right),$$

which produces the list of images of the point M.
The construction of the equidistant curve to the logarithmic spiral is based on the construction of the tangent line at M and the command *Sequence*, too.

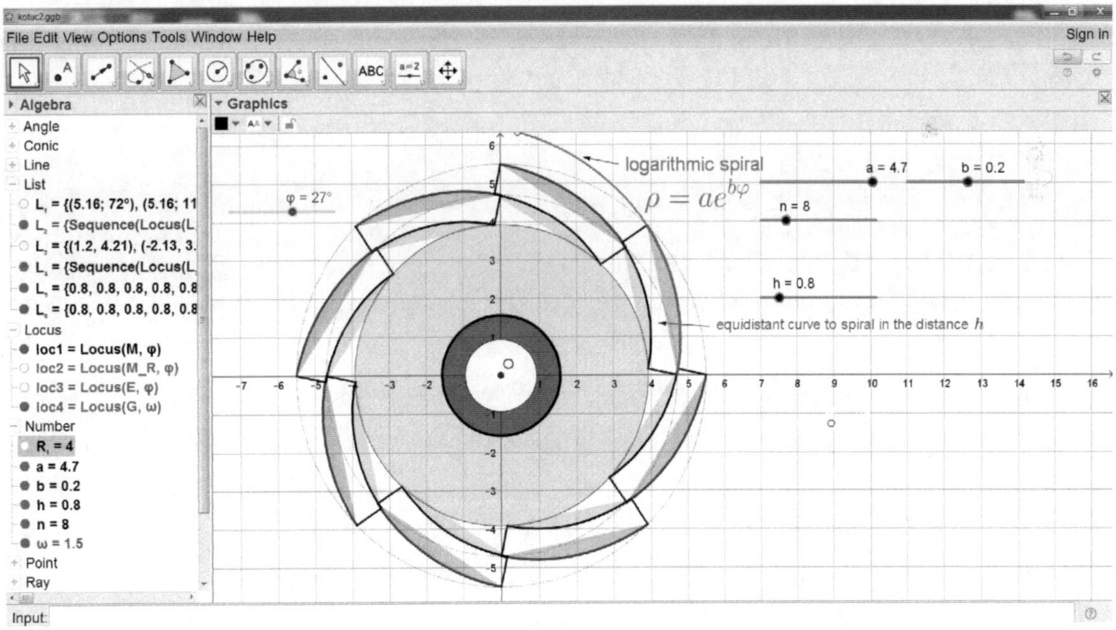

Fig. 6. Project of the gear of the circular-saw blade in GeoGebra.

4 Discusion

The National Curriculum for Upper Secondary Education in Slovakia (ISCED 3A) specifies the mathematical standards in Slovak high schools. One of the major topics in ISCED 3A is Geometry. The geometrical curriculum is focused on the teaching of backgrounds of planimetry and stereometry. The content of the curriculum was reduced in recent years, e.g. conic sections are no longer a part of the high school curriculum. From this perspective, it seems that the investigation of the properties of other algebraic curves, such as spirals, is counterproductive.

On the other hand, there is a psychological point of view. The theory of didactical constructivism states that the exploration and discovery have a great importance for students to motivate them to learn Mathematics. Especially, if it is related to an interesting topic with impact on a practice.

The main role of education in Geometry is oriented to a development of geometric imagination of the students as one of the most important abilities for their future profession as engineers, architects, structural engineers, etc. Learning supported with ICT motivates the pupils in many ways. Static models are replaced by dynamic ones, thus developing creative thinking, mathematical skills, leading students to act and develop the constructive thinking. Technology brings to students and their teacher's great opportunity to individualize learning, to eliminate formalism and to present examples, which do not belong to the standard topic in curriculum and therefore represent a challenge.

References

[1] LABORDE, J. M.: Dynamic geometry as a window on mathematical learning: empirical research on the use of Cabri-Geometry. *Proceedings of 2nd Mediterranean conference on mathematics education*, Nicosia, 2000, pp. 161 – 173

[2] JONES, K.: Learning Geometrical Concepts using Dynamic Geometry Software. *Mathematics Education Research: A catalyst for change*. Auckland, New Zealand: University of Auckland, 2001, pp. 50 - 58

[3] VELICHOVA, D.: Interactive maths with Geogebra. In. *International Journal of Emerging Technologies in Learning*, Vol. 6(2011), Special Issue 2, pp. 31-35

[4] HEJNY, M. KURINA, F.: *Child, school and mathematics: constructivist approach to teaching*, (Czech), Portal, Praha, 2001.

[5] URL: https://en.wikipedia.org/wiki/Spirangle.

[6] ZIATDINOV, R., MIURA, K. T.: On the Variety of Planar Spirals and Their Applications in Computer Aided Design. *European Researcher*, Vol. 27, No. 8-2, 2012: p. 1227–1232, E-ISSN 2224-0136.

[7] REKTORYS, K.: *Přehled užité matematiky* (Czech), SNTL, 1963.

[8] FRANK, L. et al.: *Matematika. Technický pruvodce 1* (Czech), SNTL Praha, 1973.

[9] HARARY, G., TAL, A.: The Natural 3D Spiral. *Eurographics*, Vol. 30, No. 2, 2011: p. 237-246, URL: http://onlinelibrary.wiley.com/doi/10.1111/j.1467-8659.2011.01855.x/pdf

[10] MAŠEK, K., ŠIMUNEK, P.: *Technické kreslenie II* (Slovak), SVTL Bratislava, 1963.

[11] GELLERT, W., KUSTNER, H., HELLWICH, M., KASTNER, H.: *Kleine Enzyklopedie Mathematik* (German), VEB Bibliographishes Institut Leipzig, 1974.

Acknowledgment

This paper was supported by project KEGA No. 016UKF-4/2016 MINE Slovak Republic titled *Implementation of Constructivist Approaches to Mathematics Teaching with a Focus on Active Acquisition of Knowledge by Pupils in the Context of Bilingual Education.*

Current address

Vallo Dusan, RNDr., PhD.
Department of Mathematics, Faculty of Natural Sciences
Constantine the Philosopher University in Nitra
Tr. A. Hlinku 1, 949 74 Nitra, Slovak Republic
Tel. number: +421 37 6408 690, E-mail:dvallo@ukf.sk

Duris Viliam, RNDr., PhD.
Department of Mathematics, Faculty of Natural Sciences,
Constantine the Philosopher University in Nitra
Tr. A. Hlinku 1, 949 74 Nitra, Slovak Republic
Tel. number: +421 37 6408 722, E-mail: vduris@ukf.sk

Slovak University of Technology in Bratislava
Faculty of Mechanical Engineering
17th Conference on Applied Mathematics **APLIMAT 2018**

Proceedings

CREATIVE USE OF MATHEMATICAL STRATEGIES IN PROOFS IN UNDERGRADUATE CALCULUS

VARGA Marek (SK), NAŠTICKÁ Zuzana (SK)

Abstract. Mathematical theories could be developed neither without mathematical proofs, nor without mathematical creativity. Undergraduate students should be, thus, shown multiple proofs of fundamental theorems of various mathematical branches, creatively applying various techniques and strategies. In this paper we present altogether nine proofs of Lagrange's Mean Value Theorem, employing various strategies which, in general, are useful in problem solving.

Keywords: mathematical proof, mathematical creativity, Lagrange's theorem, Rolle's theorem, Cauchy's theorem

Mathematics Subject Classification: 97 I 40.

1 Introduction

More than a half-century ago the well-known mathematician and mathematics educator George Pólya (1962) "defined mathematical knowledge as information and know-how" (in: Mann, 2006, p. 237), regarding the latter one "as the more important, defining it as the ability to solve problems requiring independence, judgement, originality, and creativity" (ibid). As for the creativity in mathematical thinking, it comprises "the ability to see new relationships between techniques and areas of application, and to make associations between possibly previously unrelated ideas" (Tammadge, 1979, in: Haylock, 1987, p. 60). Furthermore, mathematical creativity manifests itself in "the independent formulation of uncomplicated mathematical problems, finding ways and means of solving these problems, the invention of proofs and theorems, the independent deduction of formulas, and finding original methods of solving nonstandard problems" (Krutetskii, 1976, in: Haylock, 1987, p. 60).
In accordance with the above presented ideas developed by various mathematics education researchers, we suggest that undergraduate calculus lectures and seminars should pay more attention to proofs of mathematical theorems in multiple ways. In addition, encouraging the students to devise their own proofs using various techniques and strategies may improve not only their understanding of the theorems and theories, but also their problem solving skills and raise the level of their mathematical creativity.

2 Diversity in proofs of Lagrange's Mean Value Theorem

Differential calculus of one real variable plays an essential role in undergraduate calculus courses – its theoretical part allows for later generalization of the theory for multivariable functions, and its application part is widely used in solutions of problems outside mathematics. Many fundamental propositions in differential calculus bear the names of recognized mathematicians, such as Pierre de Fermat, Rolle, Jean Gaston Darboux, Joseph Louis Lagrange, Augustine Louis Cauchy, Guillaume Francois Antoine de l'Hospital, Brook Taylor. In this paper we focus on Lagrange's Theorem, more precisely Lagrange's Mean Value Theorem. After introducing its wording, we present several strategies of proving the validity of the theorem.

Theorem (*Lagrange*). Let a function f be continuous over a closed interval $\langle a;b\rangle$ and differentiable over an open interval $(a;b)$. Then there exists at least one number $\xi\in(a;b)$ such that

$$f'(\xi)=\frac{f(b)-f(a)}{b-a}.$$

Note. Geometrically speaking, the theorem says that if the function f satisfies all the assumptions, then there is a point $\xi\in(a;b)$ such that the tangent to the graph of the function f at the point $\Xi[\xi;f(\xi)]$ is parallel to the line $\overleftrightarrow{AB}$, where $A[a;f(a)]$, $B[b;f(b)]$ (see Fig. 1).

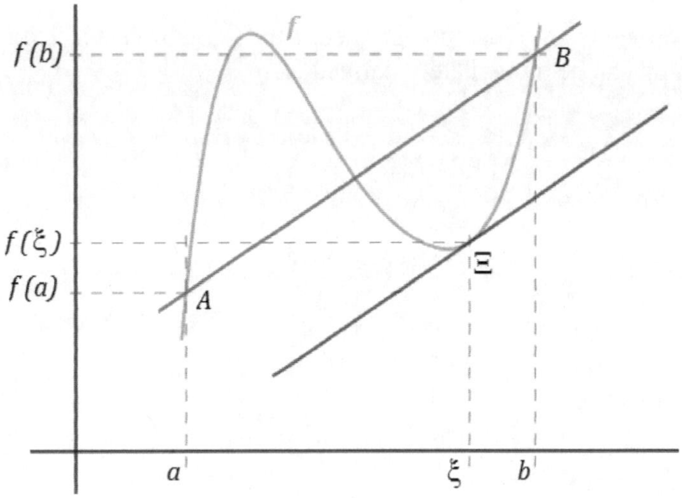

Fig. 1. Geometrical illustration of Lagrange's Mean Value Theorem..

In *Proofs 1-6*, other fundamental theorems of differential calculus and the strategy of using an auxiliary function are employed in order to prove Lagrange's Theorem. *Proof 7* shows an application of matrix determinant as a function of real variable. This proof might be of extraordinary value for university students, who, unfortunately, often misunderstand undergraduate algebra and calculus as very distant branches of mathematics. In *Proof 8* the strategy of 'special case' as an opposite of generalization is applied. *Proof 9* aims to facilitate students' understanding of Lagrange's Theorem and its obvious validity through a transition from abstract calculus to concrete interpretation in physics (or other sciences).

Proof 1.
Let us consider an auxiliary function defined as follows

$$\Phi(x) = f(x) - Kx, \text{ where } K \in \mathbf{R}.$$

The function Φ is continuous over a closed interval $\langle a;b\rangle$, and also differentiable over an open interval $(a;b)$. Taking *K* such that $\Phi(a) = \Phi(b)$ we obtain

$$K = \frac{f(b) - f(a)}{b - a}.$$

The function Φ now satisfies the assumptions of Rolle's theorem*. It follows that there must exist a $\xi \in (a;b)$ for which $\Phi'(\xi) = 0$. Since

$$\Phi'(\xi) = f'(\xi) - K = 0,$$

it is easily seen that $f'(\xi) = \dfrac{f(b) - f(a)}{b - a}$.

Proof 2.
A light tint of the previous proof can be obtained if the real number K is directly stated. Then, the auxiliary function can be immediately defined as

$$\Delta(x) = f(x) - \frac{f(b) - f(a)}{b - a}x,$$

and the rest of the proof is the same as shown above.

Proof 3.
Let us consider an auxiliary function defined as follows

$$\Gamma(x) = x\left[f(b) - f(a)\right] - (b - a)f(x).$$

This function is continuous over $\langle a;b\rangle$ and differentiable over $(a;b)$. It holds that

* *Rolle's Theorem:* Let a function F be continuous over $\langle a;b\rangle$, differentiable over $(a;b)$, and let $F(a) = F(b)$. Then there exists at least one number $\rho \in (a;b)$ such that $F'(\rho) = 0$.

$$\Gamma(a) = af(b) - bf(a) = \Gamma(b).$$

In other words, the function Γ satisfies the assumptions of Rolle's theorem, which implies that there exists a $\xi \in (a;b)$ such that $\Gamma'(\xi) = 0$. Obviously,

$$\Gamma'(\xi) = [f(b) - f(a)] - (b-a) f'(\xi) = 0,$$

which immediately implies that $f'(\xi) = \frac{f(b) - f(a)}{b-a}$.

Proof 4.
Indeed, the auxiliary function can assume various forms. Let us now consider

$$\Lambda(x) = f(x) - \frac{f(b) - f(a)}{b-a} x + \frac{af(b) - bf(a)}{b-a}.$$

Again, the function Λ is continuous over $\langle a;b \rangle$ and differentiable over $(a;b)$. In addition, it holds that $\Lambda(a) = 0 = \Lambda(b)$. Deriving from Rolle's theorem, there exists a $\xi \in (a;b)$ such that $\Lambda'(\xi) = 0$. Taking $x = \xi$ we get

$$\Lambda'(\xi) = f'(\xi) - \frac{f(b) - f(a)}{b-a},$$

and immediately $f'(\xi) = \frac{f(b) - f(a)}{b-a}$.

Proof 5.
Although an auxiliary function is applied in this proof again, let us refer back to the geometrical interpretation of Lagrange's Theorem (see Fig. 1). The line $\overleftrightarrow{AB}$ is described by the equation

$$y_{AB} = f(a) + \frac{f(b) - f(a)}{b-a}(x-a).$$

We will investigate now the difference $\Theta(x) = f(x) - y_{AB}$, i.e.

$$\Theta(x) = f(x) - f(a) - \frac{f(b) - f(a)}{b-a}(x-a).$$

The function Θ is continuous over $\langle a;b \rangle$ and differentiable over $(a;b)$. Evaluating Θ at a and b we see that $\Theta(a) = 0 = \Theta(b)$, implying that Θ satisfies all the three assumptions of

Rolle's Theorem. Hence, there is at least one $\xi \in (a;b)$ such that $\Theta'(\xi)=0$. Finally, we obtain

$$\Theta'(\xi)=f'(\xi)-\frac{f(b)-f(a)}{b-a}=0,$$

and consequently $f'(\xi)=\dfrac{f(b)-f(a)}{b-a}$.

Proof 6.
By a tiny modification of the function Θ from the previous proof we get a new way to prove Lagrange's Theorem. Suffice it to consider an auxiliary function

$$\Psi(x)=f(x)-\frac{f(b)-f(a)}{b-a}(x-a).$$

It can be easily justified that the function Ψ is also continuous over $\langle a;b\rangle$ and differentiable over $(a;b)$. In addition, it holds that $\Psi(a)=f(a)=\Psi(b)$. By Rolle's Theorem, there is $\xi \in (a;b)$ such that $\Psi'(x)=0$. Hence, we obtain $f'(\xi)=\dfrac{f(b)-f(a)}{b-a}$.

Proof 7.
Besides applying geometrical interpretations, algebraic concepts can also be wisely utilized. Let us define a determinant

$$\Omega(x)=\begin{vmatrix} f(x) & g(x) & h(x) \\ f(a) & g(a) & h(a) \\ f(b) & g(b) & h(b) \end{vmatrix}.$$

This function is also continuous over $\langle a;b\rangle$ and differentiable over $(a;b)$. Following the rules of calculations with determinants it is immediate that $\Omega(a)=0=\Omega(b)$. Thus, by Rolle's Theorem, there is a $\xi \in (a;b)$ such that $\Omega'(\xi)=0$. More precisely, we obtain

$$\begin{vmatrix} g(a) & g(b) \\ h(a) & h(b) \end{vmatrix} f'(\xi)-\begin{vmatrix} f(a) & f(b) \\ h(a) & h(b) \end{vmatrix} g'(\xi)+\begin{vmatrix} f(a) & f(b) \\ g(a) & g(b) \end{vmatrix} h'(\xi)=0.$$

Taking $g(x)=x$ and $h(x)=1$, the final equation assumes the form

$$\begin{vmatrix} a & b \\ 1 & 1 \end{vmatrix} f'(\xi)-\begin{vmatrix} f(a) & f(b) \\ 1 & 1 \end{vmatrix}=0, \text{ i.e. } f'(\xi)=\frac{f(b)-f(a)}{b-a}.$$

Proof 8.
The proof can be quite easily and quickly done by applying Cauchy's Theorem**. Lagrange's Theorem is, in fact, its special case. It is sufficient to take $g(x)=x$.

Proof 9.
Last but not least, if not as a mathematical proof, then surely as an intuitive consideration of validity of Lagrange's Theorem, let us employ one of its applications in elementary physics. Let $f=f(x)$ denote a function of point mass trajectory, the point changing its position from $A[a;f(a)]$ to $B[b;f(b)]$, where the continuous variable x denotes time, while $x\in\langle a;b\rangle$. Then, there must exist an instant ξ at which the instantaneous velocity, i.e. $f'(\xi)$, is equal to the average velocity, i.e. $\dfrac{f(b)-f(a)}{b-a}$.

Let us consider the motion described by the function $f=f(x)$ over the time interval $\langle a;b\rangle$. Roughly speaking, the instantaneous velocity of the point mass in motion cannot be smaller than the average velocity *over the whole interval in question*. Similarly, the instantaneous velocity of the point mass in motion cannot be greater than the average velocity at *all* instants of the considered time interval. Therefore, the instantaneous velocity of the point mass must inevitably assume values greater as well as smaller than the average velocity. Since the change of velocity over time is a continuous phenomenon, there must be an instant within the interval when the instantaneous velocity of the point mass assumes the value which is equal to the overall average velocity, i.e. there is a $\xi\in(a;b)$ such that $f'(\xi)=\dfrac{f(b)-f(a)}{b-a}$.

3 Conclusion

Every mathematical theory uses fundamental as well as less important theorems. Traditionally undergraduate calculus lectures aim to build students' knowledge of the theories gradually, systematically, including proofs of all theorems. We suggest this tradition be enriched by discussing with students multiple ways of proving the theorems. Diversity and creativity in techniques and strategies in mathematical proofs may satisfy students' needs, assuming that different students might prefer and better comprehend different approaches.

** *Cauchy's Theorem.* Let functions *F*, *G* be continuous over a closed interval $\langle a;b\rangle$, differentiable over an open interval $(a;b)$, and for all $x\in(a;b)$ let $g'(x)\neq 0$. Then there exists at least one $\gamma\in(a;b)$ such that $\dfrac{f(b)-f(a)}{g(b)-g(a)}=\dfrac{f'(\gamma)}{g'(x)}$. Incidentally, this theorem can be easily proved if we take $h(x)=1$ in *Proof 7*.

References

[1] BUDINSKÝ, B., CHARVÁT, J.: *Matematika I.* Praha: SNTL/Alfa, 1987. 504 s.

[2] FULIER, J., VRÁBEL, P.: *Diferenciálny počet. Nitra:* UKF, 1997. 300 s.

[3] GREBENČA, M. K., NOVOSELOV, S. I.: *Kurs matematičeskogo analiza 1.* Moskva: Učpedgiz, 1951. 543 s.

[4] HAYLOCK, D. W. A framework for assessing mathematical creativity in school children. In: *Educational Studies in Mathematics*. 1987, vol. 18, no. 1, p. 59-74. DOI: https://doi.org/10.1007/BF00367914

[5] IGNATJEVOVÁ, A. V.: *Kurs vyssej matematiky.* Moskva: Vysšaja škola, 1964. 684 s.

[6] KLUVÁNEK, I., MIŠÍK, L., ŠVEC, M.: *Matematika I.* Bratislava: SVTL, 1959. 760 s.

[7] KRUTETSKII, V. A. *The Psychology of Mathematical Abilities in Schoolchildren*, 1976. (J. Kilpatrick and I. Wirzup, eds.; J. Teller, trans.), University of Chicago Press, Chicago.

[8] KUDRJAVCEV, L. D. a kol.: *Kurs matematičeskogo analiza 1.* Moskva: Vysšaja škola, 1981. 687 s.

[9] MANN, E. L. Creativity: The Essence of Mathematics. In: *Journal for the Education of the Gifted.* 2006, Vol. 30, No. 2, p. 236-260. DOI: 10.4219/jeg-2006-264

[10] POLYA, G. *Mathematical Discovery: On Understanding, Learning and Teaching Problem Solving*. 1962, Vol. 1, Hoboken, NJ: John Wiley & Sons.

[11] TAMMADGE, A. Creativity: Presidential address to the Mathematical Association. In: *The Mathematical Gazette*. 1979, vol. 63, p. 145–163. DOI: 10.2307/3617885

[12] TRENCH, W. F.: *Introduction to real analysis.* NJ: Pearson, 2003. 583 s. ISBN 0-13-045786-8.

[13] URL, https://calculus.subwiki.org/wiki/Lagrange_mean_value_theorem.

Current address

Varga Marek, PaedDr., PhD.
Department of Mathematics, Faculty of Natural Sciences
Constantine the Philosopher University
Tr. A. Hlinku 1, 94974 Nitra, Slovak Republic
E-mail: mvarga@ukf.sk

Naštická Zuzana, Mgr.
Department of Mathematics, Faculty of Natural Sciences
Constantine the Philosopher University
Tr. A. Hlinku 1, 94974 Nitra, Slovak Republic
E-mail: zuzana.nasticka@ukf.sk

Slovak University of Technology in Bratislava
Faculty of Mechanical Engineering
17th Conference on Applied Mathematics **APLIMAT 2018**

Proceedings

SURFACE MEASURES REVISITED ON THE BOUNDARIES OF MATHEMATICS AND ART

VELICHOVÁ Daniela (SK)

Abstract. Paper brings some reflections on attempts to confront mathematical and artistic approaches to measurements of various aspects of creativity, imagination and our understanding of aesthetic values hidden in the produced artefacts in fine art and in science, namely mathematics. Phenomena that are carriers of these subtle values enable comparison of the concepts of measures and measurement strategies in maths and art. Short revision of the surface measure definitions and its application within the two environments is presented, together with continuous efforts of finding their common features and differences.

Keywords: mathematics and art, measure, truth, beauty, inner and outer measure

Mathematics Subject Classification: Primary 28A99; Secondary 91F99, 97M80.

1 Introduction

Science and art are two highly imaginative intellectual domains of human activity with no strict boundaries dividing the spheres of competencies for one or another. In both, such outstanding abilities of the human mind as abstraction, imagination, comparison, analysis and synthesis of pieces of knowledge and perceptions, generalisation, intuition, resourcefulness, aesthetic feelings or creativity are essential and obviously used and developed. The ability to convey a message on experienced and intentionally investigated events in the form of intellectually elaborated judgments and emotionally touched up pieces of art is unique to mankind, which enables us to express, exchange, share and reclaim the outcomes of mental activities of individuals, for the advantage of the whole community. Methods of these two distinct branches of systematic intellectual activity of mankind are different so as their working approaches differ. Nevertheless, the level of abstraction, timeless validity and recency of achieved results and articulated values of the conveyed messages of these abstract disciplines provide comparably important pieces of information on perception and understanding of our world, in which we are destined to live, work and create. The products of both of these seemingly antithetical domains embody the evidence of the level of development of the human society, since they reflect the society´s latest achievements in the never-ending efforts to find and perceive the truth, search for it, and understand it as the backbone of "knowing & feeling", undoubtedly present in harmony in both science and art.

My work has always tried to unite the true with the beautiful and when I had to choose one or the other, I usually chose the beautiful. Hermann Weyl

I am interested in mathematics only as a creative art. The mathematician's patterns, like the painter's or the poet's must be beautiful; the ideas, like the colours or the words must fit together in a harmonious way. Beauty is the first test: there is no permanent place in this world for ugly mathematics. G.H. Hardy

Mathematics, rightly viewed, possesses not only truth, but supreme beauty – a beauty cold and austere, like that of sculpture, without appeal to any part of our weaker nature, without the gorgeous trappings of painting or music, yet sublimely pure, and capable of a stern perfection such as only the greatest art can show. The true spiritual delight, the exaltation, the sense of being more than man, which is the touchstone of the highest excellence, is to be found in mathematics as surely as poetry. Bertrand Russell

The world around us is full of relationships, rhythms, correlations, patterns. And mathematics underlies all of these, and can be used to predict future outcomes. Our brains have evolved to survive in this world: to analyse the information it receives through our senses and spot patterns in the complexity around us. In fact, it's thought that the mathematical structure embedded in the rhythm and melody of music is what our brains latch on to, and that this is why we enjoy listening to it. It is perhaps not surprising then that there is a great deal of overlap between mathematics and the art that our brain finds so pleasing to look at.

Lewis Dartnell, [1]

Mathematical theorem is essentially an exactly formulated truth, which is a generalisation of a certain precisely recurring phenomena. It is a symbolic description of an observed behavioural feature of an object, presented in an abstract form, and simultaneously explaining also the circumstances under which the studied phenomena could appear. These circumstances are the assumptions of the proposition. Mathematical hypothesis is a symbolically coded message, often understandable only to experts, and subdued to strict laws of logics. Thus, the conveyed truth must be exactly verified by means of logical rules. It can be well comprehensible also to a layperson, when the main idea of a hypothesis is explained and illustrated, and its message can be visualised on a particular example, a geometric/graphical model preferably. Geometric interpretation helps to show the human face of mathematics. It serves as a platform for science and art matching, as a space for overlapping of the artistic and mathematical abstractions. However, their common intersection is not found only in the usage of geometric forms in fine arts. In much greater extent it can be found on the abstract level, in the search for a description of natural relations, principles and rules of our Universe.

An artist is also expressing and describing observed reality and truth by creating a masterpiece of any type – painting, musical piece, sculpture, film, or theatrical performance. Developing their own specific artistic approach she/he presents these observations in abstract form, symbolically, although "in her/his way". Reflecting subjective observations in mind and intimate emotional world, personalizing them, they are receiving a remarkable emotional mettle and generate thus the most remarkable power of the created pieces of art.

Fine art and mathematical abstract worlds are two parallels dealing with the understanding of our world and its laws, each from its own point of view and by its own means of expressions. Maths and Art represent two "on the edge" attitudes of describing the world's truth. They seem to be the most imaginative, but in spite of this common feature, also the most respectable achievements of human mind. Nonetheless, their naturally different character in attitude and work processes is rather unifying than separating them, as these two worlds are meeting on a higher abstract level, not clearly apparent at first sight to layperson. Mathematical proposition and fine art work are two different representations of the observed reality, two abstractions appearing seemingly on the opposite poles of human mind, two distinct illustrations of truth and effort to comprise it and present in an abstract way. These two representations of "world truths" are available to complement each other perfectly.

In the following section we give some examples on how the two disciplines differ and what are the limits of their convergence. Various illustrations of pieces of art will be presented in comparison to mathematical definitions and symbolic formulas representing the concept of measurement, as the background for the assessment of understanding, truthfulness and inner value.

2 Measures

In finding a measure of any kind of unknown, we get the feeling that we have reached the first level of understanding the matter itself.

Measuring is one of the rather natural human activities, when attempting to evaluate whatever we might come across, objects we are meeting, activities we are performing and their consequences and results, our relations and their manifestations, but firstly the space in which we are compelled to live or work. Finding a measure of unknown objects or feelings, we get the impression that we understand the matter. The concept of measure is a well defined phenomenon in mathematics, and measure theory represents a complex field of mathematical theory on a highly abstract level, with verified geometric models illustrating concept of different measures as length, area, or volume of geometric figures on particular examples and calculations. What exactly is a measure in mathematics?

Let X be a non-empty set and M a non-empty system of its subsets. M is called σ-algebra, if it is closed with respect to all complements and countable unions, i.e.

$$A \in M \Rightarrow X \backslash A \in M, \quad A_n \in M \Rightarrow \bigcup_{n=1}^{n} A_n \in M \tag{1}$$

If X is a topological space, then elements of σ-algebra B generated by system of all opened sets in X are called Borel sets.

Let M be a σ-algebra on a set X. Function $\mu : M \rightarrow [0,+\infty]$ is called (non-negative) measure, if it is σ-additive, i.e.

$$\mu\left(\bigcup_{n=1}^{n} A_n\right) = \sum_{n=1}^{\infty} \mu(A_n) \tag{2}$$

for any arbitrary system of mutually disjoint sets, while $\{A_n\}_{n=1}^{\infty} \subset M$ and $\mu(\varnothing) = 0$. Elements of M are called measurable sets, while triple (X, M, μ) is called space with a measure.

Let $A \subset \boldsymbol{E}_n$, and let $S = \{I_1 , ..., I_m\}$ be a system of closed intervals, in which no two different intervals have common interior points. Then we define $|A| = 0$, if set A does not contain interval, and if it contains some interval, then we define

$$|A|_* = \sup \textstyle\sum_{k=1}^{m} |I_k|, \quad \bigcup_{k=1}^{m} I_k \subset A, \quad |A|^* = \inf \textstyle\sum_{k=1}^{m} |I_k|, \quad A \subset \bigcup_{k=1}^{m} I_k \tag{3}$$

For any bounded set A the following relation is true: $|A|_* \leq |A|^*$.

Number $|A|_*$ is the inner Jordan measure of the set A, number $|A|^*$ is the outer Jordan measure of the set A. Bounded set A is measurable in the Jordan sense, if $|A|_* = |A|^* = |A|$.

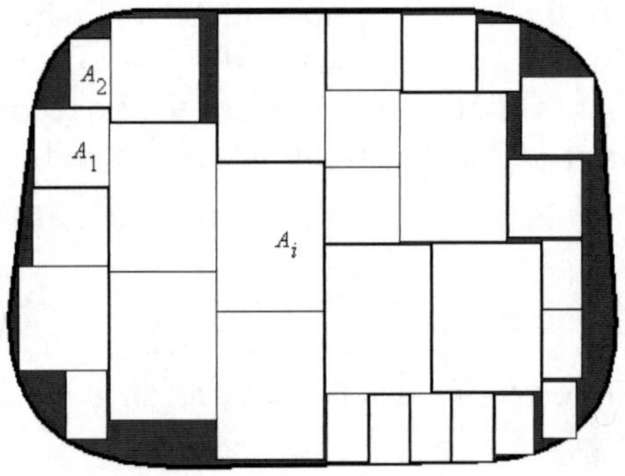

Fig. 1. Inner Jordan measure of set A - $|A|_*$.

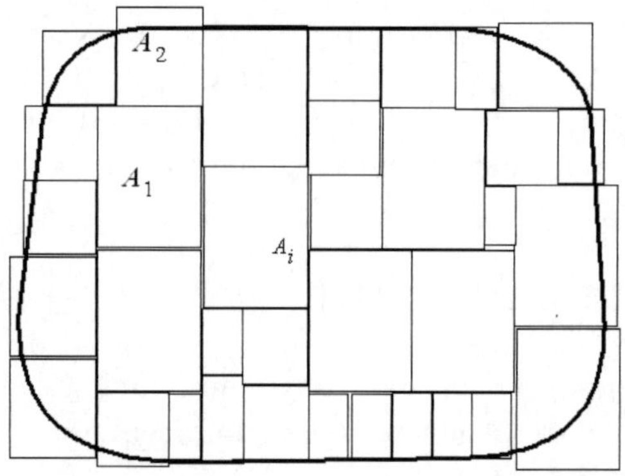

Fig. 2. Outer Jordan measure of set A - $|A|^*$.

Number $|A| \neq 0$ is called the Jordan measure of a set, it is represented as

$$|A| = \int_A \chi_A(x) dx = \int\int_A ... \int 1.dx_1 \, ... \, dx_n , \tag{4}$$

where $\chi_A(x)$ is a characteristic function such, that

$$\chi_A(x) = 1 \text{ for } x \in A, \ \chi_A(x) = 0 \text{ for } x \in E_n - A. \tag{5}$$

If (X, ρ) is a metric space and p is a positive number, then for any $A \subset X$ holds

$$\mu_p^*(A) = \sup_{\varepsilon > 0} \inf \{ \textstyle\sum_{n=1}^{\infty} (d(A_n))^p ; A = \bigcup_{n=1}^{\infty} A_n , d(A_n) < \varepsilon, n = 1, 2, ... \}. \tag{6}$$

Outer measure μ_p^* is called the Hausdorff p-dimensional measure.

Measure is a certain limit that can be reached in-between two different measurements based on similar, although not completely equal principles. Limit, in which inner and outer measures meet and overlap, thus finding an agreement reflected in the real value, must not necessarily exist in all cases.

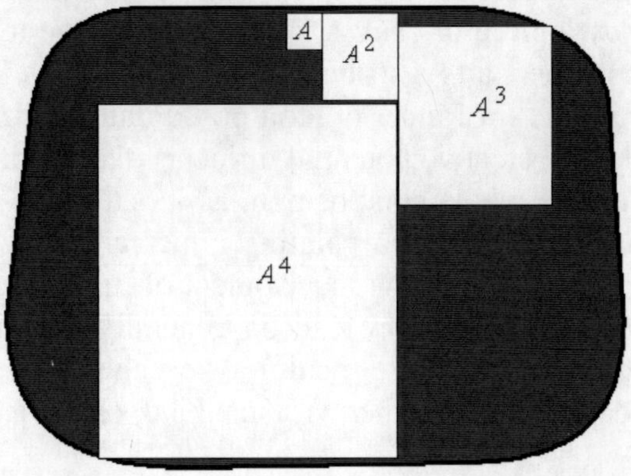

Fig. 3. Hausdorff *p*-dimensional measure of set *A*.

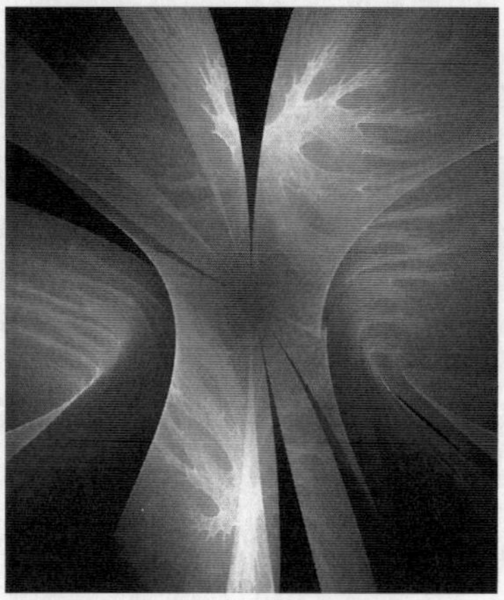

Fig. 4. Fractal.

Measure of an empty space is zero. The measurement units alone, objects filling the space, determine its non-zero measure. Consequently, these objects themselves, items representing our expectations, define the final measure value. The preconditions defined as assumptions influence a sort of the measure quality or type. Mathematics states these clearly by determining the initial conditions related to the specific desired type of measure, which are generally agreed and accepted, and serve as a certain guide for related measurements.

Let (X, M, μ) be a space with measure and let simultaneously X be a topological space. Measure μ is called

- Borel measure, if M contains all Borel sets, so $B \subset M$
- regular, if it is a Borel set and for all $A \in M$ holds
  - $\mu(A) = \inf\{\mu(U); A \subset U, U \text{ opened}\}$
  - $\mu(A) = \sup\{\mu(K); K \subset A, K \text{ compact}\}$
- (non-negative) Radon, if it is regular and for any compact $K \subset X$ holds $\mu(K) < \infty$
- complete, if for any $A \in M$ holds: if $\mu(A) = 0$ and $B \subset A$, then $B \in M$
- finite (or bounded), if $\mu(X) < \infty$
- σ-finite, if there exists $A_n \in M$ such, that $\mu(A_n) < \infty$ and $X = \bigcup_{n=1}^{\infty} A_n$.

Several other measures can be defined, such as arithmetic, Dirac, Carathéodory or Lebesgue-Stieltjes measures, we can speak about regular, complete, finite, σ-finite, arithmetic, and many other measures, which differ in the measuring approach and the measure quality for which they are applied, see [4], [5].

Do similar laws also exist in art, or is the imagination space of an artist completely unlimited? Are there certain obstacles, defined by authors themselves, imposed maybe intuitively due to cultural development of mankind, that are influencing creative work on the masterpiece, in the sense of its pre-defined inner desirable measure? Are the different measures reflected in different artistic styles that are approaching different aims and looking for different qualities?

Painters might apply a principle similar to mathematical definition of a measure, when filling the empty two-dimensional space represented by a piece of canvas, in order to change it into a painting. According to her/his own strategy she/he starts distributing separate objects in it, thus creating a unique composition reflecting her/his own inner criteria posed onto a surface, i.e. painting measure. Author´s emotional world and creative potential influence the quality of her/his measurements, it has a strong impact on the whole composition, as it determines the inner measure of the created piece of art. Outer measure of a painting is perceived by the observers. It comes out of the 2D space of the painting and it is independent of the viewer in certain sense. Although on individual level this outer measure can be generally perceived differently, it is defined and influenced by well formed rules of our cultural heritage applicable in painting, or sculpture and design, which are developing together with mankind´s knowledge, science and culture in a historical context.

It is very interesting and inspiring to discover, how a fine artist reacts to an abstract formalised mathematical proposition from the measure theory, coded in mathematical symbolism, and how a mathematician could understand his artistic illustration of the abstract mathematical imaginations in the fine art, shaping the intuitive vision of an artist. We still know only very little about general connections between different abilities of human brain to perceive pieces of art and to understand mathematical formulas. Is the former not just an illustrative emotional visualisation and a model of the latter one? Are they interfering and cooperating, or neglecting each other passing in parallels? How can we make these two abstract abilities to respect or influence each other, and thus utilise the synergetic effect of imaginative human abilities resulting from this cooperation? Do certain general rules exist, which human beings apply when measuring space by dividing it into smaller parts that actually inhabit it, and populating it make the space alive, being the carriers of interrelation with the exterior viewer? Are all these measurement processes intuitive, innate or natural, or are they only the side effects of our material obsession and desire to evaluate all our possessions?

Some of the questioned topics can be studied in the illustrations presenting artwork of the Slovak renowned female painters of the twentieth century.

Fig. 5. Ester Šimerová Martinčeková: Country.

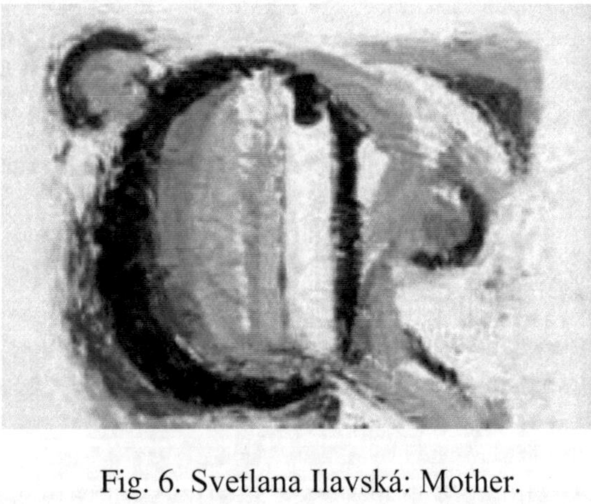

Fig. 6. Svetlana Ilavská: Mother.

The first presented painting in Fig. 5, "Country" by Ester Šimerová Martinčeková is an example of 2D picture with the 3D exterior measure that is visualising the 3D scene – snow-covered Slovak mountainous countryside. The chosen technique for the painting in shades of blue and grey and the simple division of the canvas area into several parts that determine the picture interior measure create in a synergic effect its 3D dimensional expression forming the outer measure of this oil painting on canvas.

The art-piece "Mother" by Svetlana Ilavská in Fig. 6 presents the unlimited power of a mother´s love. A mother, more dead than alive, carries the ill child on her back, doing her best to rescue him. The external dimension of this picture dwells in the depicted indescribable emotions - despair, belief and hope. The colour contrasts of mother and child silhouettes masterfully scratched by just few differently coloured domains in the two-dimensional space of canvas produce emotionally rich atmosphere and radiate this sensation out of the painting as its exterior dimension.

Fig. 7. Lýdia Jergušová-Vydarená: Antimass.

The third illustration in Fig. 7, “Antimass” by Lýdia Jergušová-Vydarená [4], is the artist´s visualisation of her own understanding of basic physical and mathematical concepts. The famous Einstein equation $E = mc^2$, representing physical law in mathematical symbolism, that matter can be turned into energy, and energy into matter, is presented in connection with the exploration of antimass, black matter and invisible energy concealed as the black holes of the Universe predicted by Dirac, and studied by Anderson. The artist´s imagination is expressed in a slightly adapted form of the equation with the doubled right side, thus expressing her belief in the existence of the antimass and its equivalence with the mass. Both are regarded merely as two different forms of energy that are available in the Universe. But the antimass is threatening and it makes us feel unsafe, as something unknown, mysterious, and perhaps dangerous. The sharp frame projecting from the mild background represents the antimass as the outer measure of this piece of art, which is earnestly and intrusively approaching the observer, making him/her feel insecurely. The inner measure in warm pleasant colours acts reassuringly and it calms our worry. Together they balance the story conveyed by the author in a common limit that defines the real value of the aesthetic measure of this piece of art.

The last examples in Fig. 8 represent work of young Slovak artist Žofia Dubová [2], who takes precisely into consideration the inner and the outer measure of her pieces of art. She is openly stressing the role of a visualised matter with the inner measure surrounded by the border area forming the frame, which itself is a part of the image as its outer measure. Thus, the observer is invited to perceive the artist´s message in the context of a framed object, realising both the inner and the outer measure, described explicitly by the author in the masterpiece. Experiencing both in the context of the openly articulated author´s message, gives the observer a chance to find the actual value measured as the limit balancing the inner and outer parts of the piece of art. This approach can be identified precisely with the main idea of an object measure as defined in mathematics, where both the inner and outer measures are given explicitly, e.g. in the Jordan measure definition.

Fig. 8. Žofia Dubová: Shadows (on the left), Over the mountain (on the right).

3 Conclusions

It is not easy to balance on the edge. Boundaries are very unsafe place, where you can never know where to turn, or which direction are you falling - inside or outside? This dilemma is very openly presented in the masterpieces of Mark Rothko, Fig. 9. In his later large canvases one or more rectangular fields of colour, which bear no relationship to geometry, appear floating in the indeterminate space over the background surface, yet they never stand out fully out from it. “Green on Maroon”, masterpiece belonging to the Museo Thyssen-Bornemisza in Madrid, is a colour field painting resembling more watercolour than oil, influenced by surrealists Carl Gustav Jung, Swiss psychiatrist and psychoanalyst who founded analytical psychology. Rothko leaves door open to people to perceive different emotions. Observers have chance to choose which part of the painting has to be regarded as the interior, and which one as the exterior. In “Lavender and Mulberry”, the lavender rectangular shape is arranged in the dark rectangle so that his colour is bleeding into the dark background. Or is it the other way round? Does the maroon background intrude into the green centre? Which of the depicted rectangulars is inside the figure, and which one is extended outside? Blurred edge on the boundary of the two colours gives observers a free choice for individual perception and determination of the outer and inner measures, where colour has a specific meaning. This was the main idea of Rothko´s paintings: to evoke moods of transcendental meditation about abstraction expressions, perception of different emotions, balancing on the edge of the interior and exterior measure.

Fig. 9. Mark Rothko: Lavender and Mulberry (left), Green on Maroon (right).

It is a pursuit of elegance that captures essence, and gives us a precise insight on relations.

Xah Lee

References

[1] DARTNELL, L., Maths and Art: The whistlestop tour. Available at: https://plus.maths.org/content/maths-and-art-whistlestop-tour
[2] DUBOVÁ, Ž., Online artist portfolio. Available at: https://www.works.io/zofia-dubova
[3] JERGUŠOVA-VYDARENÁ, L., Web. Available at: http://www.jergusova-vydarena.sk/
[4] VELICHOVÁ, D., Surface measures - on the boundary between mathematics and art. In *Journal of Mathematics & Design*, Volume 10. Number 1, 2010, The International Mathematics & Design Association, Buenos Aires 2011, Argentina, pp. 238-245.
[5] VELICHOVÁ, D., Surface Measures, In *Aplimat Journal of Applied Mathematics*, STU in Bratislava, Volume II, Number 1, Year 2009, ISSN 1337-6365, pp. 235 – 240.

Current address

Velichová Daniela, doc. RNDr., CSc.
Institute of Mathematics and Physics, FME, Slovak University of Technology in Bratislava
Nám. slobody 17, 812 31 Bratislava, Slovak Republic
E-mail: daniela.velichova@stuba.sk

THEATRE, LOVE AND GRAPHS

ZAWIŚLAK Stanisław (PL), KOPEĆ Jerzy (PL)

Abstract. In the paper, utilization of graphs in theatre-related issues is discussed. World-wide known plays are analysed: W. Shakespeare's "Measure for measure", M. Camoletti's "Boeing, Boeing" and Jarosław Murawski's "Humanka". The graphs represent relations among the heroes however they could be a part of scenography. Graph-based modelling allows for deeper understanding of attitudes and dependencies between dramatis personae, additionally unexpected visions of the plot are obtained.

Keywords: maggraph, clique, independent set, plot, deeper understanding

Mathematics Subject Classification: Primary 68R10, 05C90; Secondary 05C99

1 Introduction

The problems related to the topic "theatre and graph theory" are not recognized as classical ones. Sometimes, the search is even misleading because "operating theatre" is a place in a hospital. Moreover, theatre ventilating system or war theatre also could be found for these keywords. However, in our paper the word 'theatre' is considered as related to the art and it could be replaced by 'drama'. The phrase 'mathematics and theatre' is more promising, therefore several papers dedicated to graph application in theatrical matters can be listed, e.g.: [3-7]. The paper [3] gives a theoretical background for using graphs in modelling of drama related issues. It gives also a wide reference review and information how other theatre plays could be analysed via graph-based approach Works [4-6] of Italian scientist A.C. Sparavigna are extremely interested and fully fit to the discussed problems. The authors main interest is graph application in mechanics [11,14], however, special aspects were also propagated e.g. graph transformations utilization [10], design of technical artefacts (trusses) via evolutionary approach [12] as well as historical approach [9]. Moreover, writing plays' reviews is a hobby of the author [15]. The basic notions related to graph theory can be found in the classical book [8]. Graphs related to Shakespeare's plays can be seen in paper [3-6] and the webpage [2], which is dedicated to teaching of graph models of chosen plays at the University of Nebraska, Omaha, USA, whereas Greek tragedy in [17], respectively. Several graph-theatre assignment rules are given in this webpage, in [3] and in the present paper. Colour vertices and circle of different dimensions were used by Americans [2], whereas here different shapes of vertices as well as different line types are proposed.

The goal of the paper is to show how graph-based modelling is fruitful and adequate for theatre related issues. In general, the goal is also to propagate and promote graph theory learning and interest via showing unexpected but useful application. This direction of application was rather very narrow and relatively unknown. The new bust of interest is connected with the problem of a supported or an AI-supported story telling or scenarios creating for computer games [16] but also in general. So, it is recently intensively developed. Graphs turned into the useful algebraic structure to represent plot and maps of virtual world in the games. Also, these directions are under interest of the authors of the present work.

2 Short information on the considered theatre plays

In the present paper, three plays which are in the current repertoire of the theatre in Bielsko-Biała (Poland) – named "Teatr Polski" – are discussed. The first one is entitled "Humanka" – the word derived from 'human', whereas ending '-ka' (in Polish) is related to a female sex. To avoid placing quotation marks, the titles of the plays will be written in italics i.e. *Humanka*. The first drama author is a young Polish playwright Jarosław Murawski [1] who was graduated from the Warsaw University but also finished the course of scenario writing at The National Institute of Dramatic Art in Sydney in 2010. The play action takes place in next century when robots are fully identical to men. The hero is shy and a dweeb. So, he could not meet a faithful girl-friend – which is a dream of his mother. The model of 'humanka' bought by him is really intelligent, but mother is too clever to be fooled. Additionally, unexpectedly and baselessly the robot starts to be beloved in the hero. The love ends tragically. The second play which is discussed underneath is *Boeing, Boeing* by Marc Camoletti – famous French playwright. The play itself is known in tens of countries (over 50), being played for more than 50 years. At the moment it is played in Bielsko-Biala and in Trnava Theatre (in May 2017) under the title: *Tri letuški v Paríži* (Fig. 1).

Fig. 1. Bildings of theatres in Bielsko-Biala and Trnava where drama *Boeing, Boeing* was played. (Author of photos: Stan Zawiślak)

The plot is simple: Bernard – who in Bielsko is called Maks – has three 'love affairs' with three stewardesses: Janet from USA, Jola from Poland and Johana from Germany. It is rather 'affair' then 'love'. He has meetings with every one of them in the same time but not simultaneously, obviously. His school friend Paweł (Robert in the original version) comes. The matters start to be more complicated due to weather breakdown and consequently some delays of flights. Therefore three girls are coming to Maks simultaneously and due to Paweł help their meeting is initially avoided. They really need love, they are dreaming about gorgeous love. It seems, that they all are thinking seriously about requited love and marriage. It is well known American song where rhymes go: "Soldier, soldier! Will you marry me, with your musket, pipe and drum?" - but how Maks could marry three girls simultaneously? The solution to the problem is known by those who have seen this funny play.

The last considered masterpiece is the play *Measure for measure* by William Shakespeare. The plot is more complicated in this case. The drama was written in 1603. So, one of the heroes is prince. The procuress and prostitute occur, thereby so called (in Polish) 'paid love" exists in the world. However, in English language, there are rather phrases "to pay for sex" or "for providing sex". The action twists and turns. In general, the play heroes could be gathered in some groups of society: representatives of authorities, representatives of church as well as ordinary members of society, good and bad. The proposed graphs give an overview of this division or partitioning [13]. The drama ends in some marriages.

But what is love? According to the Bible "Love is patient, love is kind. It does not envy, it does not boast, it is not proud. It does not dishonour others, it is not self-seeking, it is not easily angered, it keeps no record of wrongs. Love does not delight in evil but rejoices with the truth. It always protects, always trusts, always hopes, always perseveres. Love never fails". Even for those who do not respect the Bible, these features of love could be easily accepted. So, true love – how it could be reach? In the analysed plays – according to the mentioned authors it ends in marriage in some phase of its passing along.

3 Short review of terms related to graph theory and new type of graphs

A graph G – G(V,E) – is a pair of sets (V,E), where nonempty set V denotes graph vertices and set E contains the graph edges. If E is empty, then there are not any edges, so the graph consists of remote vertices. The graphs could be simple and directed. Directed graph i.e. digraph has such edges that each edge has the beginning and the end. In graphical graphs' representations, these two graphs differs: edges of a simple graph are part of curves or parts of straight lines, whereas in digraphs edges are called arcs in which ends are marked by an arrow (more precisely arrowhead). A clique K_n or a full graph is such a graph which has all possible edges so if $|V(K_n)| = n$ than: $|E(K_n)| = m = 0.5 \cdot n \cdot (n-1)$.

The new notion is hereby introduced i.e. maggraph. The name is similar to 'digraph' which stands from '**di**rected graph' – new graph could be named **mag**ic graph or big (**mag**nitude) graph due to the fact that notions biggraph, maxgraph or exograph are already known, therefore: MG(MV, ME), is a pair of magvertices and magedges, so MV is set of vertices which are particular graphs themselves and ME is set of edges connecting these vertices e.g. $MV = \{G_1, G_2, G_3\}$; $ME = \{(G_1, G_2), (G_1, G_3), (G_2, G_3)\}$. We can consider also extended maggraph i.e. maggraph plus the union of all graphs being its vertices, then magedges and edges could be presented graphically, simultaneously.

An independent set of vertices is a set of the graph vertices which are not mutually connected i.e. the vertices are disconnected. Graphs are inseparably connected with marriage i.e. there is commonly known Hall's theorem on marriages [8]. According to Wikipedia: "The graph theoretic formulation (of this theorem) deals with a bipartite graph. It gives a necessary and sufficient condition for finding a matching that covers at least one side of the graph." The graphical forms of the graphs could be different, usually vertices are shown as dots, circles or other pictograms. The describtion of vertices are placed inside these circles or in the neighbourhood of them.

Graph is equivalent to /0,1/-matrix A and to relation R - considered as a subset of Cartesian product of set V i.e. subset of the structure V × V, $R \subset V \times V$. As it was mentioned, there are n vertices belonging to the set V. Therefore, the matrix A has n × n elements as well as there are n × n pairs belonging to the Cartesian product. An existing pair corresponds to existing adequate graph edge and the adequate element of adjacency matrix is equal to 1 (confirming existence of the edge). The described equivalence is a base for representing relationships between play heroes if they are represented via graph vertices. The basic notions connected with graphs are also: cycle, tree – graph connected without cycles, path, partitioning etc. Their applications will be discussed in different dramas.

4 Graphs related to the discussed plays

The graphs related to the aforementioned dramas will be given and analysed in the following subchapters.

a)

b)

Fig. 2. Newly delivered humanka Shi (a), dressed humanka (b).

It is worth to underline that these graphs differs essentially which exposes variable ideas to build plots, relations among characters and general message bring via a particular drama. Graph vertices are assigned to the characters, the edges represent relation among heroes: being colleagues, being members of family or being members of the same social group of the society.

4.1 *Humanka* – the love-less society

Like it was mentioned above, humanka is a human-like-robot. In Fig 2a, the newly unpacked humanka is shown, the switch button is placed on the neck. In Fig 2b, humanka is dressed. In Fig. 3a, Em's mother is presented, whereas a graph is visible in the background. The graphs are presented through the whole play on the rear side of the scene. It shows that future, novelty and technology according to scenographer is related to graphs which evolve in interesting manner. They are really attractive and important elements of dynamic scenography (delivered via an overhead beamer) which is rather rare in theatrical material-made artefacts.

4.1.1 Graphs as scenography

To our surprise, the graphs are a part of scenography in *Humanka*. During almost the whole play, on the rear wall of the small scene the graphs are shown via a special projector.

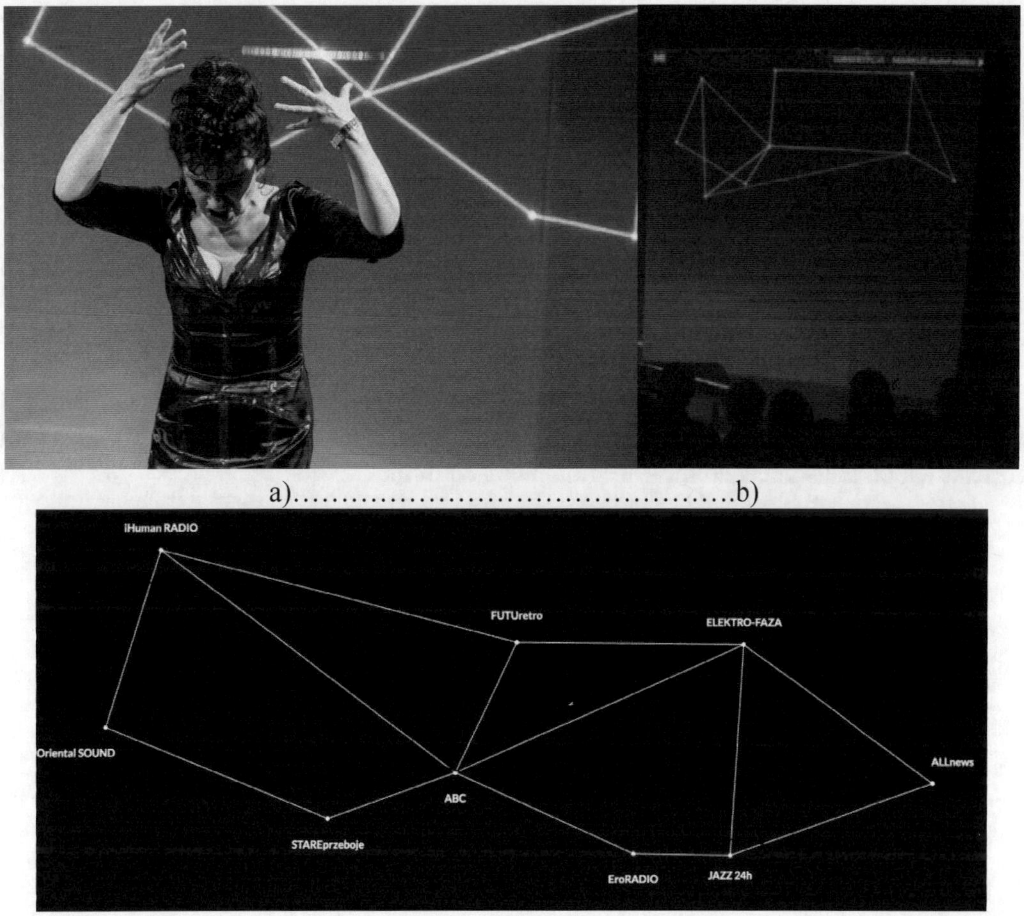

a)..b)

c)

Fig. 3. Fuorios Em's mather – graph in the background (a), graph as constant element of the secenery of *Humanka* – first version (b), second version (c).

The graphs are expose as a stable picture or they evolves being a network of local media or being a screen of a new videophone – device similar to our smartphones. The difference is in the dimensions of the screen. The graphs are presented in Fig. 3. It was really an inspiration to go further and utilize graphs for plays' analyses. The graph used as a videophone allows the hero to contact with some characters e.g. friend and father. Father (and others) even do not play as live actors, only movie - shown by means the videophone - informs the audience about the conversations of son and daddy.

4.1.2 Graphs representing the characters who are in some mutual connections

Two graphs representing the relations among drama heroes are presented in Fig. 4. In Fig. 4a, an initial situation is presented via a graph being a tree, especially a star. Father and mother are not connected because they are divorced. Em is in the centre of stage as well as in the middle of the action. The star like graph represents also a hub-like computer network where branches connect the hub with the so called slave computers. Humanka is really like a slave for him. However, he could be treated as an anti-star. His ex-girlfriend is prowling around but there is not any future for conciliation. She disappears eventually. Humanka Shi declares love to Em. He is furious, rejects love and resets the robot (clears her memory and feeling if they are stored separately)! Performance of the reset procedure is really tragically. Moreover it is shown in a touching way by the actress, and it is also tragic itself. Then, Em is selling the robot with cleared circuits and memory to his friend. It is an absurd and merciless act simultaneously. The audience realize that mother has a much much younger boyfriend. Such a pair is really popular in so called colour tabloids nowadays. Em's father has bought the same model of humanka as his son! Despite the fact that mother has a youth boyfriend, this information (passed to her by Em) makes her mad.

After this action the graph converts in a new one (Fig. 4b) where information about father's purchase and mother's boyfriend are taken into account. The final graph is also a tree. Em remains in the centre but he is even more lonely then previously. Father related vertex is shown as a rectangle because the father is known by theatre audience only from the video registration shown as a part of videophone connections. Some other actors appear only in the movies shown on the screen, so they are not enclosed in the graphs.

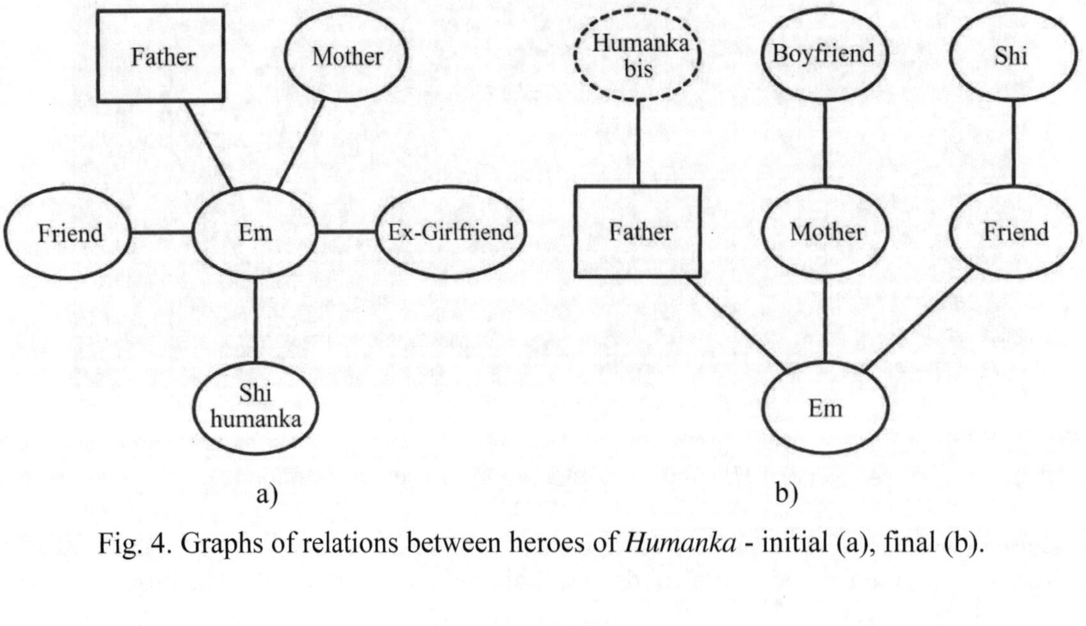

Fig. 4. Graphs of relations between heroes of *Humanka* - initial (a), final (b).

They are not taken into account in the considered graphs besides the father of Em which is also a subject of discussion the hero with mother. The vertex 'Humanka bis' is encircled via a dashed line to show that the character is only discussed by the remaining heroes. In both cases the graphs are symmetrical. Is only an incidence or it was deliberately planned by the author? The play is genially prepared by a theatre team. Actress Oriana Soyka was rewarded by the diploma of the Mayor of the town Bielsko-Biala at the beginning of 2017. Maybe, the spectators are forced to rethink their relationships with friends, family, spouse and beloved person. Such a world is suggested to be in a far far future, but it is even present nowadays?

4.2 *Boeing, Boeing* – desire for love

There are five heroes of the drama *Boeing, Boeing* at the beginning of the plot. The adequate graph (Fig. 5a) is given. We can distinguish two subgraphs related to the following two distinct vertex sets: V_1 = {Maks, Nadia} and V_2 = {Johana, Jola, Janet}. The subgraphs induced on these sets are clique K_2 and the independent set V_2. When the Maks' friend comes, the graph converts into similar one (Fig. 5b.).

There is a possibility to distinguish two subgraphs: K_3 and a subgraph containing three separate vertices, the same idea as previously. A possible meeting of three stewardesses could be a disaster for Maks but also for women themselves – realizing that they were deceived. In general, the stewardesses are the most intelligent women working in stressful situations, knowing foreign languages and having a good contact with plane passengers – so in case of Maks' fiancés they are obviously full of suspicions – the comedy consists in finding excuses for them. Sometimes such an excuse triggers new suspicion. Live in permanent lie is not too comfortable. The graph K_3 in known as a clique. The word 'clique' has a pejorative connotations in Polish language. In fact, our clique have to make tricks all the time. The idea of independent set is here also highlighted – the stewardesses do not know each other almost for the whole period of the drama.

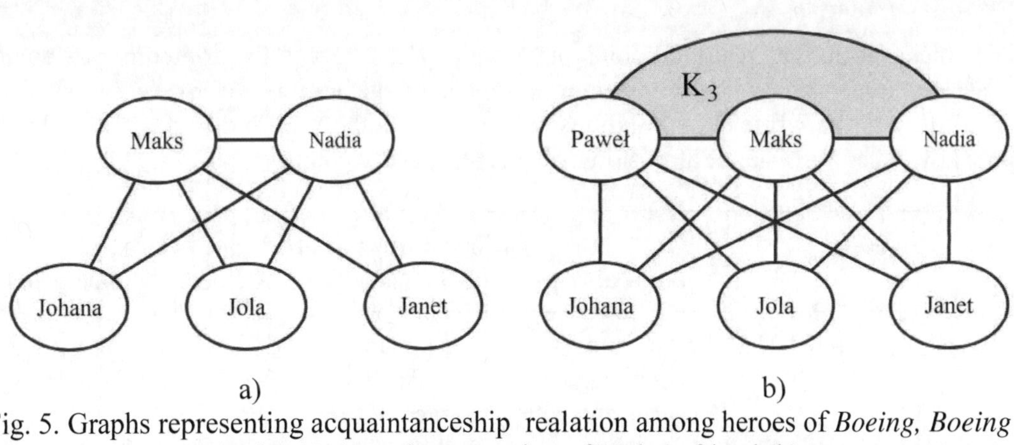

Fig. 5. Graphs representing acquaintanceship realation among heroes of *Boeing, Boeing* - initial (a), just after comming of Maks's friend (b).

Meetings of Paweł and Johana cause that their feelings evolve from interest, respect, finally into a deep love – just immediately, due to the idea of the drama author. It is even more really funny in Poland, because there are stereotypes that marriage of Pole and German is impossible. Nevertheless, the authors know some happy marriages of this type. Finally, the strongest feelings between Maks and Jola are announced. It seems for a moment that Janet remains without a pair. However, an unopened letter which had been circulating around, finally has been delivered to Janet. Its opening and reading brings an unexpected solution. Janet's American boyfriend declares love and proposes marriage. Janet is delighted and leaves Maks' home immediately, dreaming about the return journey to USA. Only Nadia remains alone. Happy end (Fig. 6a) is also for her, after a sound rise of her salary by Maks, she decided to remain as a home-keeper. In Fig, 6b, the maggraph for initial situation is presented. It represent ugly clique vs. innocent girls. Maggraph has two magvertices: MV = (G_1, G_2), G_1 is a smallest clique, G_2 is a disconnected graph. There is known a phrase "Nokia – connecting people", but here lack of connections means also that the girls are unknown one for another.

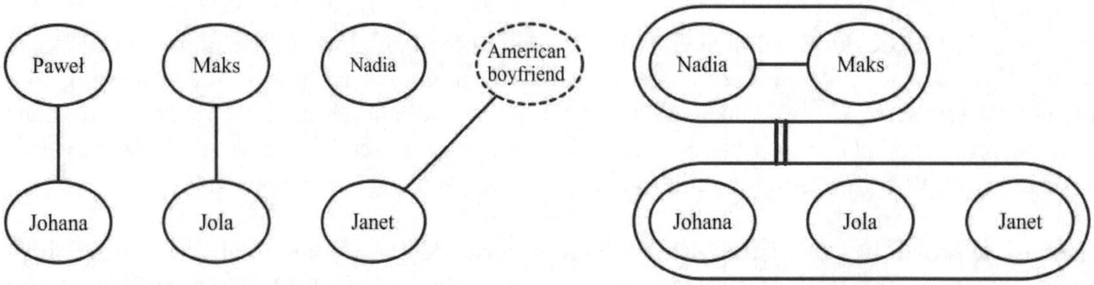

Fig. 6. Happy end of *Boeing, Boeing* (a), maggraph representing the play (b).

The comedy is played for over 50 years The story is really unbelievable. Nevertheless, it gives the spectators two hours of laugh and then time for rethinking. In case of graph representation – interesting is antithesis of clique and disconnected graph or independent set of vertices in original graph – depending on interpretation.

4.3 *Measure for Measure* – love and society

William Shakespeare was really a genius playwright. He observed the contemporary society with passion, understanding and admiration. In his drama entitled *Measure for measure*, published in 1623, he wrote about relationships between authorities and society. Introduction of a new law, faces a resistance almost in every case in every country.

Magedge	Despription of mutual attidudes between particular parts of the society according to Shakespeare
POWER → CHURCH	Power and church are separated and there is a lack of any tension between them. The Duke turns himself to a frair to observe an acction from coverture.
POWER → PEOPLE	Power would like to introduce new strict low, the Duke is afraid of the reaction of the people so gives temporary the power to Angelo to save his image as a good ruler.
CHURCH → POWER	Like it was previously mentioned, total separation. Rules of

	monastery are strict and they are related to internal matters of the monastery. Church is not the source or inspiration of new law however it is in agreement with the articles of faith. Church has to help the poor.
CHURCH → PEOPLE	Church does not force people. Monk being priest organizes the marriage ceremony. Isabella begs for merci for his brother and she would not resign of the church rules even when her brother will be beheaded.
PEOPLE → POWER	Part of society – criminals/offenders – are aginst new law. Their opinions is that new law is pointless because people will not obey it. The men are evil and they would like to have contacts with prostitutes. It was written around 400 years ago, but even nowadays the same opinions can be red.
PEOPLE → CHURCH	People respect the church.

Tab 1. Remarks on the meaning of magedges of maggraph shown in Fig. 7.

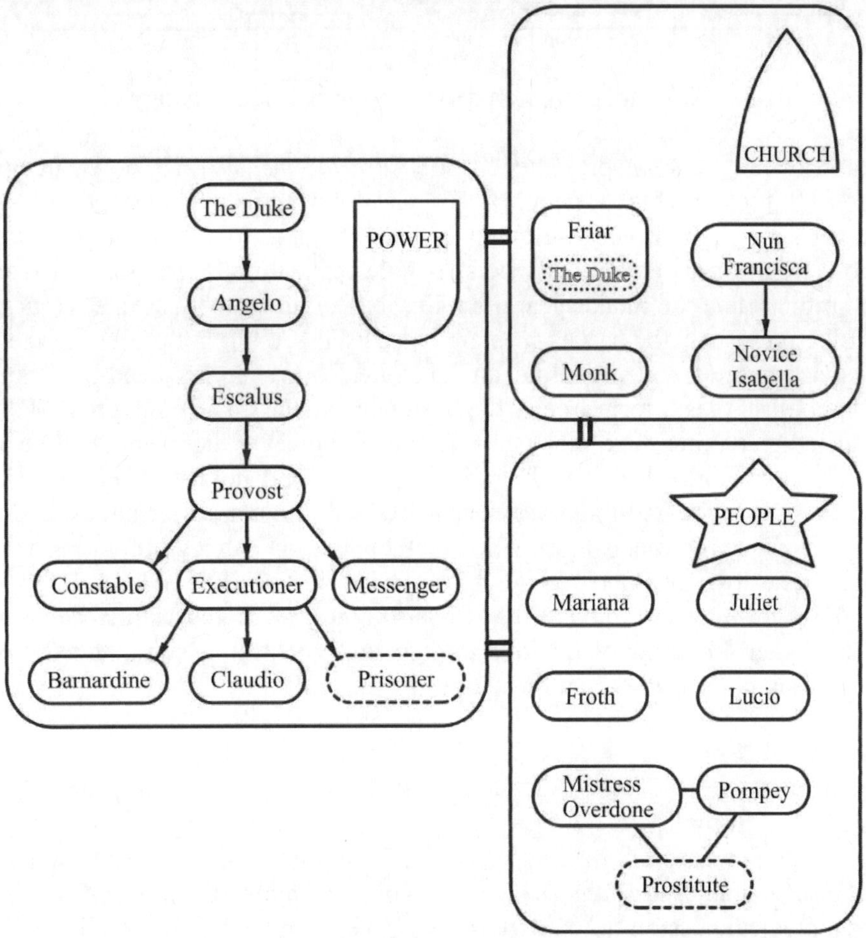

Fig. 7. First maggraph of *Measure for measure*.

In this case, new low is directed against promiscuity and for a **pure love** between men and women. The action is situated in Vienna, thence not in Britain where Anglicanism were introduced by the King Henry VIII. Those citizens who were surprised by new regulations have to be punished in as serious way as possible – even cruel one – to show determination of authorities in keeping the regulation. Claudio was sentenced for a dead punishment, at the beginning. The merciless prince would like to perform punishment. However, he hidden himself in other personality i.e. monk and observes the situation from the perspective of an ordinary citizen however belonging to a catholic church. Claudio's sister Isabell asks for forgiveness for her brother. She is a novice in convent, nevertheless she wears a nun uniform (habit). Usually, to be a nun, it is needed to swear an oath. So, even being a novice she looks like a regular nun.

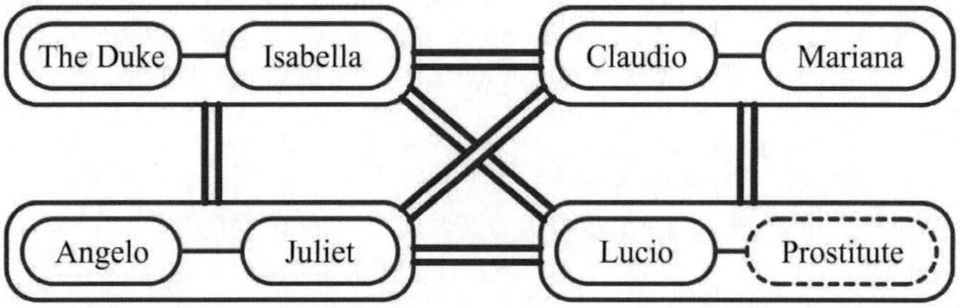

Fig. 8. Second maggraph of *Measure for measure* – happy end.

The maggraph of the situation presented in *Measure for measure* is shown in Fig. 7. The magvertices set = {power, church, people}. The set of magedges is described in Table 1. The MV_1 = 'power' is a digraph because a power i.e. the authorities have a clear structure of subordination, from top to bottom. In Bielsko-Biała realization, there are a few actors less than in an original text of Shakespeare. In this case, the graph has the vertical axis of symmetry. Like previously the heroes who are only discussed but not present in the scene are marked via dashed lines e.g. prostitute and prisoner (however his head was delivered to authorities to confirm the performance of the dead punishment). The magvertex MV_2 'church' is a disconnected graph: nuns and monks are separated, however the connectivity components 'nuns' is a directed subgraph due to hierarchy: prioress and novice. The magvertex MV_3 'people' is also a disconnected graph, society is divided into classes or honest and dishonest individuals. The clique represents those who break the law. The maggraph in Fig. 8 represents the marriages. It has two symmetry axes which symbolizes stability and harmony! The Duke forces the novice to leave the convent but it is also a gesture to reward her determination in rescue of her brother. Marriage of Lucio with pregnant prostitute is commanded or forced by the Duke. It is a sign of noble respect for women in general.

5 Conclusions

In the paper, three dramas were discussed. A real inspiration and a trigger to write a present paper was the scenography to *Humanka* where graphs are immanent part of decor, expressing new era, technological possibilities and world-wide web of connections. Additionally, several graphs were assigned to the theatre related matters for the discussed artistic works – e.g. representing the relationships among the heroes. All the considered plays are related to idea of love. Love is an essential idea of a human life therefore the playwrights feel correctly that this could attract the audience and makes that theatres are full of viewers. The graphs allow the spectators for deeper understanding of the plays' plots as well as give new insight in the plays via detection of variable types of graphs. What is the most important, a new type of graph was introduced giving a graphical view emphasising the special aspects and differences between plays and their heroes. Graphs (obviously in graphical manner but not only) show in one glimpse differences among dramas' plots e.g.: *Humanka's* tree shows main hero (in fact weak creature), the conversion of a tree onto a tree shows – in our opinion – some tragic aspect of selling a robot and remaining in the same situation. Contradistinction of a clique and an independent set is a clue of tension in *Boeing, Boeing*. Overcoming of this tension results in partitioning of graph i.e. splitting the group of heroes into marriages. The plot of *Measure for measure* is in Shakespeare style – full of trestles – but graph gives a clear vision of groups of heroes: power, church and people. Moreover internal division of these general entities into subgroups also helps to put in order the individual characters: nuns and monks, ordinary people and underclass (dregs). It is really amazing how fruitfully graphs fit to theatrical plot and characters' relationships. And even nomenclature like e.g. clique, independent set or star catch precisely the essences of particular problems. It seems that playwrights have to use graphs to think, rethink or overthink the plots and the figures in their arts.

Acknowledgement

The authors gratefully acknowledge the Theatre Polski in Bielsko-Biala, Poland – for the clips and photos from the play entitled "Humanka", especially Witold Mazurkiewicz – Managing Director and Małgorzata Mostek – Vice-Director.
The first author would like to express his cordial and kind gratitude to Actress Oriana Soyka for short nice meetings and exchange of opinions.

References

[1] URL, http://www.encyklopediateatru.pl/autorzy/6359/jaroslaw-murawski (open: 22 Nov. 2017)(in Polish).
[2] URL, https://www.teachengineering.org/activities/view/uno_graphtheory_lesson01_activity2, (open: 22 Nov. 2017).
[3] MORETTI F., Network Theory, Plot Analysis. *New Left Review* , 68 (2011): 80-102.
[4] SPARAVIGNA A. C., Using Time Series and Graphs in the Analysis of Dante's Divine Comedy. *International Journal of Sciences*, 3 (12), 2014, pp. 33-40.
[5] SPARAVIGNA A. C., On Social Networks in Plays and Novels. *International Journal of Sciences*, 2(10), 2013, pp. 20-25.
[6] SPARAVIGNA A. C., MARAZZATO R., Graph visualization software for networks of characters in plays. *International Journal of Sciences*, 3(2), 2014, pp. 69-79.
[7] SOLOMON, M., The theatre of mathematics and the mathematics of the theatre. In *Proceedings of BRIDGES – Mathematical Connections in Art, Music and Science*. South-western College, 1999. pp. 293-294.
[8] WILSON R. J., *Introduction to graph theory*. The Open University/Pearson, 5-th edition, 2010, (Polish translations were published several times, last – in 2017).
[9] WOJNAROWSKI J., ZAWIŚLAK S., Kazimierz Kuratowski—Biography and Genesis of the Theorem on Planar Graphs. In: Zawiślak S., Rysiński J. (Eds), *Graph-Based Modelling in Engineering. Mechanisms and Machine Science*, Springer, Cham, 2017.

[10] ZAWIŚLAK S., et. al., Some applications of graph transformations in modelling of mechanical systems, *7-th International Workshop on Graph Transformation and Visual Modelling Techniques* (GT-VMT Budapest, Hungary), 2008, pp. 332-345.
[11] ZAWIŚLAK S., RYSIŃSKI J. (Editors): *Graph-based modelling in engineering*. Springer, Cham, 2017.
[12] ZAWIŚLAK S., MATUSIAK D., Multi-objective evolutionary design of trusses using graph-based method for stress calculation, *Transactions of the Universities of Košice*, 3, 2009, pp.181-184.
[13] ZAWIŚLAK S., FREJ G., An influence of parameters of the evolutionary algorithm applied for the graph k-partitioning problem. *Studia Informatica*, 24(4), 2003, pp. 165-187.
[14] ZAWIŚLAK S., *The graph-based methodology as an artificial intelligence aid for mechanical engineering design*, Habilitation-thesis, Publishing House of the University of Bielsko-Biala, Bielsko-Biała, Poland, 2010, p. 289.
[15] ZAWIŚLAK S., Wesele, wesele (in Polish), review of the Anton Chekhov play in a student newspaper, *Margines*, ATH, Bielsko-Biala, 2017.
[16] YOUNG R.M., et. al., An architecture for integrating plan-based behavior generation with interactive game environments, *Journal of Game Development*, 1, 2004.
[17] RYDBERG-COX J., Social networks and the language of Greek tragedy. *Journal of the Chicago Colloquium on Digital Humanities and Computer Science*. 1(3), 2011.

Appendix: Names of Actresses and Actors

It is difficult to list all the creators of art like e.g. at the end of every movie where the list of all people involved in production are usually listed. In our case, we give the names of Actresses and Actors (in alphabetical order) to pay them the due respect:
(a) *Humanka:* Michał Czaderna, Piotr Gajos, Barbara Guzińska, Sławomir Miska, Adam Myrczek, Daria Polasik-Bułka, Anisa Raik, Grzegorz Sikora, Oriana Soyka; Zofia Schwinke, Mateusz Wojtasiński;
(b) *Boeing, Boeing*: Anita Jancia, Agnieszka Rose, Rafał Sawicki, Oriana Soyka, Wiktoria Węgrzyn, Mateusz Wojtasiński;
(c) *Measure for measure*: Michał Czaderna, Kazimierz Czapla, Piotr Gajos, Łukasz Kaczmarek, Jagoda Krzywicka, Maciej Kulig, Tomasz Lorek, Grzegorz Margas, Sławomir Miska, Adam Myrczek, Daria Polasik-Bułka, Bogdan Słomiński, Grzegorz Sikora, Oriana Soika, Maria Suprun, Wiktoria Węgrzyn, Mateusz Wojtasiński.

Current addresses

Zawiślak Stanisław, dr hab. inż., University Professor
Chair of Informatics and Automation, Faculty of Mechanical Engineering and Computer Science, University of Bielsko-Biala, Willowa 2, 43-309 Bielsko-Biała, Poland
E-mail: szawislak@ath.bielsko.pl

Kopeć Jerzy, dr inż.
Chair of Machine Building, Faculty of Mechanical Engineering and Computer Science, University of Bielsko-Biala, Willowa 2, 43-309 Bielsko-Biała, Poland
E-mail: jkopec@ath.bielsko.pl

Slovak University of Technology in Bratislava
Faculty of Mechanical Engineering
17th Conference on Applied Mathematics **APLIMAT 2018**

Proceedings

GENERAL THEORY OF RETAIL GRAVITATION FIELD

ZEITHAMER Tomáš R. (CZ), POSPÍŠIL Jiří (CZ)

Abstract. The work examines space weather phenomena that affect the transfer of mass, momentum, jerk, energy and charge of heliospheric magnetized plasma into the geosphere and its sub-spheres. In addition to phenomena involving transfers between individual parts of the geosphere, space weather also affects biological and technological systems found both in the heliosphere and below the Kármán line, as well as biological and technological systems on the ground. The economic implications of the transfer mechanisms for space business are also discussed.

Keywords: cognitive informatics, physical economics, space economics, space geodesy, space weather

Mathematics Subject Classification: Primary 83C99, 91F10; Secondary 92B99.

1 Introduction

The current age of the Sun is estimated on 4.6 billion Earth´s years (Bonanno, Schlattl & Paternò, 2002), while its lifetime in the main sequence of the Hertzsprung-Russell diagram is estimated at 10 billion Earth´s years (Adams, Bodenheimer & Laughlin, 2005). In a time interval of five billion Earth´s years it will be necessary for people to find a new home on another body of the Solar System or outside the Solar System. The Apollo Research Program has created the conditions for launching intensive preparatory work for the colonization of the Moon and other bodies of the Solar System.

Analysis of satellite data and remote sensing of the Moon have provided indications of the presence of water on lunar surface (e.g. in lunar pyroclastic deposits) (Milliken & Shuai Li, 2017), (Pieters et al., 2009). The lunar crust is composed primarily of oxygen, silicon, magnesium, iron, calcium, and aluminium, but important minor and trace elements such as titanium, uranium, thorium, potassium, and hydrogen are present as well. Based on geophysical techniques, the crust is estimated to be on average about 50 km thick (Wieczorek et al., 2006). It means for example from the chemical point of view that the combination of chemical elements mentioned above will allow the production of fuel directly on the Moon.

2 Informal background of finitary relations theory

In this paper a relational concept is expressed as the set. An n–ary finitary relation over the sets S_1, S_2, …, S_n is a subset P of their Cartesian product, i.e. $P \subseteq S_1 \times S_2 \times \ldots \times S_n$. Finitary relations are classified according to the number of sets in the Cartesian product, i.e. $\rho\, x_1$ denotes a unary relation, where $x_1 \in S_1$; $\rho x_1 x_2$ or $x_1 \rho x_2$ denote a binary relation, where $x_1 \in S_1,$ $x_2 \in S_2$; $\rho x_1 x_2 x_3$ denotes a ternary relation, where $x_1 \in S_1,$ $x_2 \in S_2$, $x_3 \in S_3$; $\rho x_1 x_2 x_3 x_4$ denotes quaternary relation, where $x_1 \in S_1, x_2 \in S_2$, $x_3 \in S_3$, $x_4 \in S_4$. Finitary relations with more than four terms are usually referred to as n–ary (e.g. 7–ary relation). It means that a n–ary finitary relation is simply a set of n–tuples (Fraissé, 2000).

The principle of correspondence that for two different scientific fields may also be called the binary principle of correspondence is a simple representation that assigns quantifiable variable from one scientific field to a quantifiable variable of a second scientific field based on the same qualitative importance of mechanisms of transfer of mass, momentum, jerk, energy and charge describing the development of systems states in two different scientific fields. Finding a principle of correspondence requires a high degree of understanding of the two different scientific fields. If a principle of correspondence is derived between n scientific fields, then this principle of correspondence is called the n–ary principle of correspondence.

3 Results

3.1 Binary principle of correspondence in space weather

The Lausanne School of Economics and the Cambridge School of Economics were among the first schools of economics to systematically describe and analyze economic processes using methods and models of non-relativistic theoretical physics, especially those of mechanics and thermodynamics (Zeithamer 2012, a), (Zeithamer 2012, b). Both of these schools of economics laid the foundations for a new field of science that incorporates the current state of knowledge of basic research in the areas of physics, economics, sociology, neurophysiology, biophysics and cognitive informatics and is known as physical economics. The term physical economics is used in the book title "Econophysics and Physical Economics", authored by Peter Richmond, Jürgen Mimkes and Stefan Hutzler (Richmond, Mimkes & Hutzler, 2013).

In heliosphere the Sun generates space weather phenomena that physically involve the transfer of mass, momentum, jerk, energy and charge in heliospheric magnetized plasma that interacts with the geosphere and its sub-spheres. Quantifying future states of magnetized heliospheric plasma sub-systems in space weather is analogous to developing methods for meteorological forecasts (Siscoe, 2007). One approach to quantifying the size of storm systems in terrestrial meteorology is to classify them into three basic categories in descending order based on the extent of the Earth's surface that is affected: extratropic cyclones, hurricanes and tornadoes. Storms in cosmic weather can also be classified in descending order into three basic sizes: M-region storm, coronal mass ejection, super auroral electrojet. The correspondence principle for forecasting space weather and terrestrial weather can be formulated as follows: extratropical cyclone ↔ M-region storm; hurricane ↔ coronal mass ejection; tornado ↔ super auroral electrojet (Siscoe, 2007).

3.2 Binary principle of correspondence in price theory

A comparison of the analytical structure of economics and the analytical structure of non-relativistic theoretical mechanics has led to the following conclusion: path s travelled by a body over time t corresponds with price n of the commodity at time t, i.e. $n(t) \leftrightarrow s(t)$ for $t \in \langle 0, +\infty)$. There is a fundamental difference between the real functions of real variable t "commodity price" $n(t)$ and "the path travelled by the body" $s(t)$. The path travelled by the body is a non-decreasing real function of time, but the commodity price may over certain time intervals increase or decrease as a function of time. The first derivative of path s according to time t, i.e. $\frac{ds}{dt}(t)$, is the path travelled by the body over a unit of time, called the instantaneous magnitude of velocity of the body. A change in commodity price over a unit of time is the first derivative of commodity price n over time t, i.e. $\frac{dn}{dt}(t)$. For the time being in this work, a change in commodity price over a unit of time is called the "tempo of commodity price change" or "price tempo". The change in commodity price over a unit of time may be positive, zero or negative, while a change in the path per unit of time is nonnegative (positive or zero). The second derivative of the path of the body over time means the first derivative of the velocity of the body is a change in the velocity of the body over a unit of time, called the acceleration of the body. The second derivative of commodity price over time means the first derivative of price tempo over time and in this work is called "acceleration of commodity price", or price acceleration, or "velocity of price tempo". The third derivative of the path of the body over time, i.e. the first derivative of the acceleration of the body over time is the instantaneous magnitude of jerk of the body (body jerk). The third derivative of the commodity price over time means the first derivative of the price acceleration, or price jerk (Zeithamer, 2015 a). The economic significance of price jerk is the change of price acceleration over a unit of time. The fourth derivative of the path of the body over time is, from a physical perspective, the change in magnitude of jerk of the body over a unit of time (i.e. also the speed with which body jerk changes). There is no established Czech term for this, but in electrical engineering the fourth derivative of a signal is often called the signal vibration. The fourth derivative of commodity price over time, $\frac{d^4n}{dt^4}(t)$, that expresses the change in commodity price jerk over a unit of time is the tempo of price jerk, or price vibration (quake) or commodity pricequake. The fifth derivative of the path of the body over time means the physical change in speed of body jerk over a unit of time, or the acceleration of the body jerk. The fifth derivative of commodity price over time is the economic change in tempo of commodity price jerk over a unit of time, or price crackle.

In this section of the work it is necessary to mention the following notes:

1) In physics the quantities of velocity, acceleration and jerk are vector quantities. For example this pertains to railway engineering, construction of machinery, civil engineering and aviation engineering.

2) This work uses the scalar meaning of price. An article devoted to the vector meaning of price is being prepared for publication.

3.3 Axiomatic approach to a stationary field of retail gravitation

This work assumes that preferences exhibit the characteristics given by the following axioms: I.1. Axiom of reflexivity; I.2. Axiom of completeness; I.3. Axiom of transitivity; I.4. Axiom of continuity; I.5. Axiom of convexity; II.1. Axiom of non-satiation (Never Get Enough); III.1. Axiom of continuity of the retail gravitation field.

This work uses an ECEF Cartesian frame of reference $\sum_3$, the origin of which is located at the gravitational center of the Earth with axes fixed to the geoid (Earth-Centered, Earth-Fixed frame of reference). For the description of retail gravitation this coordinate system is a sufficiently accurate approximation of the inertial coordinate system (see references (Burša & Kostelecký, 1999), (Leick, 2004)).

Let A be a city on the geoid with population P_a. Because retail flows are realized in three-dimensional space and time, we define the stationary potential of retail gravitation of city A by the relation

$$\varphi(x, y, z) = \varphi(D_a) = \alpha(A, E_A)\frac{N_a}{D_a}, \tag{1}$$

Where D_a is the distance of the observation point from city A measured along the geoid. If the distance measured along the geoid is such that the geoid curvature can be neglected, then the distance of the observation point from city A is given by the relation $D_a = \sqrt{(x - x_A)^2 + (y - y_A)^2 + (z - z_A)^2}$, where x, y, z are the coordinates of the observation point and x_A, y_A and z_A are the coordinates of city A; $\vec{D}_a = (x - x_A, y - y_A, z - z_A)$, i.e. $\vec{D}_a = (x - x_A)\vec{i} + (y - y_A)\vec{j} + (z - z_A)\vec{k}$ is the position vector of the observation point, from which retail trade is attracted along the geoid in the direction of city A, or to mass point A; $\vec{i}, \vec{j}$ and $\vec{k}$ are orthogonal unit vectors. Proportionality constant $\alpha(A, E_A)$ is expressed in units $[\alpha] = c.u.\ m^2\ pers.^{-2}$, where $c.u.$ = currency unit, m^2 = square meter, $pers.$= person; A is mass point with coordinates x_A, y_A and z_A, i.e. mass point $A = [x_A, y_A, z_A]$ depicts city A; E_A– is an economic facility of city A; N_a is the number of long-term economically active adult inhabitants in city A at time t. Vector field $\vec{K}_a$ of the intensity of retail gravitation is determined by the negative gradient of potential of retail gravitation φ, i.e.

$$\vec{K}_a = \left(-\frac{\partial\varphi}{\partial x}(D_a), -\frac{\partial\varphi}{\partial y}(D_a), -\frac{\partial\varphi}{\partial z}(D_a)\right). \tag{2}$$

This means that the magnitude K_a of the intensity vector of retail gravitation is given by the relation $K_a = |\vec{K}_a| = \alpha(A, E_A)\,\frac{N_a}{D_a^2}$. For the intensity vector of retail gravitation the following relations apply $div\,\vec{K}_a = 0$, $rot\,\vec{K}_a = \vec{0}$, and the potential of retail gravitation is obtained by solving the Laplace homogeneous partial differential equation $\Delta\varphi = 0$. Any vector field of retail gravitation $\vec{M}_a$ can be considered the superposition of two fields, one of which is determined by scalar potential while the second is determined by vector potential $\vec{A}$. Let us designate the field of retail gravitation determined by scalar potential φ as $\vec{K}_a$ and the field of retail gravitation determined by vector potential $\vec{A}$ as $\vec{L}_a$, then $\vec{M}_a = \vec{K}_a + \vec{L}_a$, it means that the vector of retail gravitation at the observation point is expressed by the relation

$$\vec{M}_a = -\, grad\, \varphi + rot\, \vec{A}. \quad (3)$$

The first summand in relation (3) determines the amount of retail trade that can be attracted to city A by a single inhabitant who is at the observation point. The second summand in relation (3) determines the amount of retail trade that can be relocated from city A to the observation point per single inhabitant found at the observation point. The presented theory of a stationary field of retail gravitation implicitly contains axiom III.1., which concerns the continuity of the field of retail gravitation, i.e. the scalar and vector stationary potentials are continuous functions of spatial coordinates and time.

Conclusion

Basic Czech research into the principle of correspondence between physics and economics began in the early 1970s. Pioneers of this basic research in the Czech school of economics included prof. Dr. Ing. Pavel Hrubý and his co-worker Ing. Jaromír Kálal. With periodic breaks of varying length, the research have continued to the present day and is steadily incorporating current findings obtained in the field of Sun-Earth relations. The experimental work has resulting in numerous studies, the most important of which are cited in this work; study authors are prof. Dr. Ing. Pavel Hrubý, Ing. Jaromír Kálal, doc. Ing. František Drozen, CSc. (VŠE in Prague), prof. Ing. Jiří Pospíšil, CSc. (ČVUT in Prague), Ing. Tomáš Zeithamer, Ph.D. (VŠE in Prague) (Hrubý & Kálal, 1974), (Drozen, 2008), (Pospíšil, 2013), (Zeithamer, 1986, 1988, 1990, 2012 a, 2012 b, 2013, 2014 a, 2014 b, 2015 a, 2015 b, 2016 a, 2016 b), (Zeithamer & Pospíšil, 2015, 2016 a, 2016 b).The following articles have also been fully utilized in this article: (Zeithamer & Pospíšil, 2017 a, b, c, d).

The transfer of mass, momentum, jerk, energy and charge in Sun-Earth relations is the fundamental process that affects the majority of communication technologies we encounter on dynamically changing markets. Therefore, it is essential for the scientific community to understand the axioms of market dynamics in its basic research, and based on this understanding play an active and independent role in the changing business environment.

Acknowledgements

The authors are grateful to Mrs. Pavla Jará, Mrs. Věra Hrabovská, Mrs. Hana Košťáková, Ing. Jiří Henzl, Mrs. Radka Hůjová, Mgr. Edita Jindráková, Mgr. Luděk Škrabal and the National Library of Technology for their great effort and excellent work, which was indispensable in the completion of a large portion of this work. Authors thank for technical assistance to Managing Director Mr. Jiří Břichnáč and to the firm Tomados Ltd. without their contribution would not be possible to finish the article. This paper is dedicated to Mrs. Věra Ruml Zeithamer, Mr. Josef Ruml Zeithamer, Mrs. Anna Ruml, Mr. František Ruml and Ing. Milan Pospíšil.

References

[1] ADAMS, F. C., BODENHEIMER, P. & LAUGHLIN, G. (2005). M dwarfs: planet formation and long term evolution. *Astron. Nachr.*, *AN 326*(10), 913–919.

[2] BONANNO, A., SCHLATTL, H. & PATERNÒ, L. (2008). The age of the Sun and relativistic corrections in the EOS. *Astronomy and Astrophysics*, *390*(3), 1115–1118.

[3] BURŠA, M. & KOSTELECKÝ, J. (1999). *Space Geodesy and Space Geodynamics*. Prague: Czech Republic, Ministry of Defense of the Czech Republic.

[4] DROZEN, F. (2008). Modelling of price dynamics and appreciation. *Ekonomický časopis* (*Journal of Economics*), *56*(1), 1033–1044.

[5] FRAISSÉ, R (2000). *Theory of Relations*. Amsterdam: North-Holland.

[6] HRUBÝ, P. & KÁLAL, J. (1974). *Methods of Economic Time*. 1st ed., Prague: Czech Republic, Institute of the Czech Committee for Scientific Management. (in Czech)

[7] LEICK, A. (2004). *GPS Satellite Surveying*. New York: Wiley.

[8] MILIKEN, R. E. & SHUAI, L. (2009). Remote detection of widespread indigenous water in lunar pyroclastic deposits. *Nature Geoscience*, *56*, 561–565.

[9] PIETERS, C. M. et al. (2009). Character and Spatial Distribution of OH/H2O on the Surface of the Moon Seen by M3 on Chandrayaan-1, *Science*, *326*, (5952), 568–572.

[10] POSPÍŠIL, J. (2013). Possible Uses of Newton´s laws of Motion in Commodity Price Theory and the Training of Expert Appraisers at Universities. *Procedia – Social and Behavioral Sciences*, *106*, 2071–2079.

[11] RICHMOND, P., MIMKES, J. & HUTZLER, S. (2013). *Econophysics and Physical Economics*. Oxford: UK, Oxford University Press.

[12] SISCOE, G. (2007). Space Weather Forecasting Historically Viewed through the Lens of Meteorology. In: V. Bothmer, I. A. Daglis (Eds.) *Space Weather – Physics and Effects*, Berlin: Springer.

[13] WIECZOREK, M. et al. (2006). The Constitution and Structure of the Lunar Interior *Reviews in Mineralogy and Geochemistry*, *60*, 221–364.

[14] ZEITHAMER,T. (1986). Úloha fyziky vztahů Slunce – Země ve světové ekonomii. In: V. Mlejnek (Ed.) *Člověk ve svém pozemském a kosmickém prostředí*, Úpice: Czech Republic, Observatory of Úpice. (in Czech)

[15] ZEITHAMER, T. (1988). Dynamics of Synergetic Systems in Solar – Terrestrial Relationships. In: *Proc. of the Symp. on Human Biometeorology*, Štrbské pleso: Slovak Republic, pp.452–456.

[16] ZEITHAMER, T. (1990). Structural Stability of the Earth Magnetosphere. In: *Proc. Int. Cong. on Geo – Cosmic Relations*, 19 – 22 April 1989, Amsterdam. Wageningen: The Netherlands, PUDOC, pp. 321–326.

[17] ZEITHAMER, T. R. (2012 a). Economic Phenomena from the Viewpoint of the Mechanics of Materials. *Procedia – Social and Behavioral Sciences*, *55*, 547–553.

[18] ZEITHAMER, T. R. (2012 b). Analytical Theory of Monotone Commodity State Development with Inflexion. *Procedia – Social and Behavioral Sciences*, *55*, 445–450.

[19] ZEITHAMER, T. R. (2013). Possible Uses of Deterministic Equations of Motion in Commodity Price Theory and for Training Appraisers. *Procedia – Social and Behavioral Sciences*, *106*, 2063–2070.

[20] ZEITHAMER, T. R. (2014 a). Methodology of Theoretical Physics in Economics: The Principle of Correspondence between Economic Variables and Kinematic Variables of Nonrelativistic Mechanics. In: *Proceedings from IX. International Conference on*

Applied Business Research, ICABR 2014, 06. 10. 2014 – 10. 10. 2014, Talca [online]. Brno: Czech Republic, Mendel University in Brno, pp. 1138–1146.

[21] ZEITHAMER, T. R. (2014 b). The Correspondence Principle in Theoretical Regional Economics. In: *Sborník příspěvků z mezinárodní vědecké konference Region v rozvoji společnosti 2014*. 23. 10. 2014, Brno [online]. Brno: Czech Republic, Mendelova univerzita v Brně, pp. 1016–1023.

[22] ZEITHAMER, T. R. (2015 a). Methodology of Theoretical Physics in Economics: Examining Price Jerk. *Procedia – Social and Behavioral Sciences, 176*, 933–939.

[23] ZEITHAMER, T. R. (2015 b). Methodology of Theoretical Physics in Economics: Vector Theory of Retail Gravitation Law. In: *Proceedings of The International Scientific Conference INPROFORUM 2015*, 05. 11. 2015 – 06. 11. 2015, České Budějovice [online]. České Budějovice: Czech Republic, University of South Bohemia, pp. 199–203.

[24] ZEITHAMER, T. R. (2016 a). Methodology of Theoretical Physics in Economics: Non-relativistic Physics as rediscovered approach to economics. In: *1st International Conference on Lifelong Learning and Leadership for All*, 29. 10. 2015 – 31. 10. 2015, Olomouc [online]. Sakarya: Turkey, Sakarya University, pp. 160–167.

[25] ZEITHAMER, T. R. (2016 b). Theoretical Background for the Globalization of Retail Gravitation. In: T. Klieštik (Ed.) *Proceedings of the 16th International Scientific Conference on Globalization and Its Socio-Economic Consequences*, 05 – 06 October 2016, Rajecké Teplice. Žilina: Slovak Republic, University of Žilina, pp. 2482–2489.

[26] ZEITHAMER, T. R. & POSPÍŠIL, J. (2015). Newton's Laws of Motion in Commodity Price Theory. In: *Proceedings of the 2nd International Conference on Finance and Economics 2015 – (ICFE 2015)*, 04. 06. 2015 – 06. 06. 2015, Ho Chi Minh City, [online]. Prague: Czech Republic, Melandrium, pp. 265–274.

[27] ZEITHAMER, T. R. & POSPÍŠIL, J. (2016 a). Newton´s Laws of Motion in Economics. In: *The 10th International Conference on Applied Business Research* (*ICABR 2015*), 14. 09. 2015 – 18. 09. 2015, Madrid [online]. Brno: Czech Republic, Mendel University in Brno, pp. 1165–1178.

[28] ZEITHAMER, T. R. & POSPÍŠIL, J. (2016 b). Theoretical Background for the Law of Retail Gravitation. In: *1st InternationalConference on Lifelong Learning and Leadership for All*, 29. 10. 2015 – 31. 10. 2015, Olomouc, [online]. Sakarya: Turkey, Sakarya University, pp. 160–167.

[29] ZEITHAMER, T. R. & POSPÍŠIL, J. (2017 a). Axiomatic Theory of a Retail Gravitation Field. In: D. Szarková, P. Letavaj, D. Richtáriková, M. Prašilová (Eds.) *Proceedings of the 16th Conference on Applied Mathematics, APLIMAT 2017*, 31. 01. 2017 – 02. 02. 2017, Bratislava, [online]. Bratislava: Slovak Republic, Spektrum Publishing House of the Slovak University of Technology in Bratislava (STU), pp. 1772–1777.

[30] ZEITHAMER, T. R. & POSPÍŠIL, J. (2017 b). Space Weather in the Context of Education in Cyber-Security. In: O. Titrek (Ed.) *Proceedings of the 3rd International Conference on Lifelong Learning and Leadership for All* (*ICLEL 2017*), 12. 09. 2017 – 14. 09. 2017, Porto, [online]. Sakarya: Turkey, Sakarya University, pp. 170–177 (will be published).

[31] ZEITHAMER, T. R. & POSPÍŠIL, J. (2017 c). Stationary Fields of Retail Gravitation in Mechanisms of Globalization Relating to the Sun - Earth Relations. In: T. Klieštik (Ed.) *Proceedings of the 17th International Scientific Conference: Globalization and Its*

Socio-Economic Consequences, 04. 10. 2017 – 05. 10. 2017, Rajecké Teplice, Slovak Republic. Žilina: Slovak Republic, University of Žilina, pp. 3053–3060.

[32] ZEITHAMER, T. R. & POSPÍŠIL, J. (2017 d). The Main Effects of Space Weather on the Economy. In: *Proceedings of the 11th International Scientific Conference INPROFORUM 2017*, 09. 11. 2017 - 10. 11. 2017, České Budějovice. České Budějovice: Czech Republic, Faculty of Economics, University of South Bohemia in České Budějovice, pp. 217–225 (will be published).

Current address

Zeithamer Tomáš R., Ing., Ph.D.
Department of Mathematics, Faculty of Informatics and Statistics
University of Economics
Ekonomická 957, 140 00 Prague, Czech Republic
IIPESP, Vimperk Square 316, 148 00 Prague, Czech Republic
E-mail: zeithamer.tomas@gmail.com

Pospíšil Jiří, prof. Ing., CSc.
Department of Special Geodesy, Faculty of Civil Engineering
Czech Technical University in Prague
Thákurova 7, 166 29 Prague 6, Czech Republic
E-mail: pospisil@fsv.cvut.cz

AUTHOR INDEX

AUTHOR INDEX